Cellular, Molecular, Physiological, and Behavioral Aspects of Spinal Cord Injury

Cellular, Molecular, Physiological, and Behavioral Aspects of Spinal Cord Injury

The Neuroscience of Spinal Cord Injury

Edited by

Rajkumar Rajendram

Department of Medicine, King Abdulaziz Medical City, King Abdulaziz International Medical Research Center, Ministry of National Guard - Health Affairs, Riyadh, Saudi Arabia

College of Medicine, King Saud bin Abdulaziz University of Health Sciences, Riyadh, Saudi Arabia

Victor R. Preedy

King's College London, London, United Kingdom

Colin R. Martin

Institute for Health and Wellbeing, University of Suffolk, United Kingdom

ELSEVIER

ACADEMIC PRESS

An imprint of Elsevier

Academic Press is an imprint of Elsevier
125 London Wall, London EC2Y 5AS, United Kingdom
525 B Street, Suite 1650, San Diego, CA 92101, United States
50 Hampshire Street, 5th Floor, Cambridge, MA 02139, United States
The Boulevard, Langford Lane, Kidlington, Oxford OX5 1GB, United Kingdom

ISBN 978-0-12-822427-4

SET ISBN 978-0-323-99198-8

For information on all Academic Press publications
visit our website at https://www.elsevier.com/books-and-journals

Publisher: Nikki Levy
Acquisitions Editor: Natalie Farra
Editorial Project Manager: Timothy Bennett
Production Project Manager: Paul Prasad Chandramohan
Cover Designer: Mark Rogers

Typeset by STRAIVE, India

Dedication

I dedicate this book to my wonderful daughter, Dr. Caragh Brien,
of whom I am so incredibly proud.
Colin R. Martin

Contents

Contributors

Numbers in parentheses indicate the pages on which the authors' contributions begin.

Cigdem Acioglu (153), Reynolds Family Spine Laboratory, Department of Neurosurgery, New Jersey Medical School-Rutgers, The State University of New Jersey, Newark, NJ, United States

Kiran Aftab (11), Section of Neurosurgery, Department of Surgery, The Aga Khan University, Karachi, Pakistan

Safwan Al Omari (425), Department of Neurosurgery, American University of Beirut Medical Center, Beirut, Lebanon

Beatriz Souza de Albuquerque Cacique New York (35), Department of Physiotherapy, Federal University of Rio Grande do Norte, Lagoa Nova, Brazil

Safwan Alomari (133), Department of Neurosurgery, American University of Beirut Medical Center, Beirut, Lebanon

Nadia Al-Sammarraie (263), Department of Pathology, Microbiology, and Immunology, University of South Carolina School of Medicine, Columbia, SC, United States

Herb Ames (509), Department of Spinal Cord Injury, Michael E. DeBakey VA Medical Center; Department of Physical Medicine and Rehabilitation, Baylor College of Medicine, Houston, TX, United States

Ali Amine (133), Department of Neurosurgery, American University of Beirut Medical Center, Beirut, Lebanon

Jesús Amo-Aparicio (105), Department of Medicine, University of Colorado Denver, Aurora, CO, United States

Malcolm Anderson (459), School of Nursing and Health, Avondale University, Cooranbong, NSW, Australia

Amanda Vitória Lacerda de Araújo (35), Department of Neurology, University of São Paulo, São Paulo, Brazil

Angel Arevalo-Martin (239), Laboratory of Neuroinflammation, Hospital Nacional de Paraplejicos, SESCAM, Toledo, Spain

Namrah Aziz (11), Medical College, Aga Khan University Hospital, Karachi, Pakistan

Arcangelo Barbonetti (377), Andrology Unit, Department of Life, Health and Environmental Sciences, University of L'Aquila, L'Aquila, Italy

Valéria Ribeiro Nogueira Barbosa (35), Department of Physiotherapy, State University of Paraiba R. Baraúnas, University, Paraíba, Brazil

M.G. Barderas (143), Department of Vascular Physiopathology, Hospital Nacional de Parapléjicos (HNP), SESCAM, Toledo, Spain

John C. Bradley (509), Department of Psychiatry, Harvard Medical School, Boston; Department of Mental Health, VA Boston Healthcare System, Brockton; Department of Psychiatry, Boston University, Boston, MA, United States

Hélène Bras (249), Faculté de Médecine de la Timone, Institut de Neurosciences de la Timone, UMR7298 CNRS and AMU, Marseille, France

James Bresnahan (J.) (339), Department of Rehabilitation Medicine, Thomas Jefferson University Hospital, Philadelphia, PA, United States

Frédéric Brocard (249), Faculté de Médecine de la Timone, Institut de Neurosciences de la Timone, UMR7298 CNRS and AMU, Marseille, France

Shadi Bsat (133, 425), Department of Neurosurgery, American University of Beirut Medical Center, Beirut, Lebanon

Maggi A. Budd (509), Department of Spinal Cord Injury/Disorders, Boston VA Healthcare System; Department of Psychiatry, Harvard Medical School, Boston, MA; Department of Bioethics, Harvard Medical School, Boston, United States

Candice Care-Unger (459), Royal Rehab, Ryde, NSW, Australia

Chiara Castellini (377), Andrology Unit, Department of Life, Health and Environmental Sciences, University of L'Aquila, L'Aquila, Italy

Humberto A. Cerrel Bazo (51), Ospedale Riabilitativo di Alta Specializzazione (ORAS)-ULSS 2 TV, Motta di Livenza, TV, Italy; Department of

Neuroscience—Physical Medicine and Rehabilitation, University of Padova, Medical School, Padova, Italy

Hani Chanbour (133), Faculty of Medicine, Lebanese University, Beirut, Lebanon

Jonathon Chon Teng Chio (119), Institute of Medical Science, Faculty of Medicine, University of Toronto; Division of Genetics and Development, University Health Network, Toronto, ON, Canada

N. Corbacho-Alonso (143), Department of Vascular Physiopathology, Hospital Nacional de Parapléjicos (HNP), SESCAM, Toledo, Spain

Settimio D'Andrea (377), Andrology Unit, Department of Life, Health and Environmental Sciences, University of L'Aquila, L'Aquila, Italy

Maria M. D'souza (327), Division of NMR, Institute of Nuclear Medicine and Allied Sciences, Delhi, India

Fernando da Silva Fiorin (165, 447), Graduate Program in Neuroengineering, Edmond and Lily Safra International Institute of Neuroscience, Santos Dumont Institute, Macaíba, RN, Brazil

Robin M. Daly (389, 405), Institute for Physical Activity and Nutrition (IPAN), School of Exercise and Nutrition Sciences, Faculty of Health, Deakin University, Burwood, VIC, Australia

Katherine J. Desneves (389, 405), Institute for Physical Activity and Nutrition (IPAN), School of Exercise and Nutrition Sciences, Faculty of Health, Deakin University, Burwood; Nutrition and Dietetics Department, Division of Allied Health, Austin Health, Heidelberg, VIC, Australia

Nejada Dingu (249), Faculté de Médecine de la Timone, Institut de Neurosciences de la Timone, UMR7298 CNRS and AMU, Marseille, France

Caroline Cunha do Espírito Santo (165, 447), Neuromotor System Laboratory, Spinal Cord Injury Rehabilitation Research Group, NuLeMe, Santa Catarina State University, Florianópolis, SC, Brazil

Débora Araújo do Nascimento (35), Department of Physiotherapy, State University of Paraiba R. Baraúnas, University, Paraíba, Brazil

Mirelly dos Santos Abilio (35), Health Sciences Center, Federal University of Pernambuco, Recife, Brazil

Anusha Dravid (215), School of Pharmacy, Faculty of Medical and Health Sciences, University of Auckland, Auckland, New Zealand

Jeanne Maria Dsouza (327), KMC Manipal Medical College, Madhav Nagar, Manipal, Karnataka, India

W. Brent Edwards (315), Human Performance Lab and McCaig Institute for Bone and Joint Health, University of Calgary, Calgary, AB, Canada

Mohamad Nabih El Houshiemy (133, 425), Department of Neurosurgery, American University of Beirut Medical Center, Beirut, Lebanon

Stella Elkabes (153), Reynolds Family Spine Laboratory, Department of Neurosurgery, New Jersey Medical School-Rutgers, The State University of New Jersey, Newark, NJ, United States

Syed Ather Enam (11), Section of Neurosurgery, Department of Surgery, The Aga Khan University, Karachi, Pakistan

Jana Fedorova (187), Department of Neurodegeneration, Plasticity and Repair, Institute of Neurobiology, Biomedical Research Center of the Slovak Academy of Sciences, Kosice, Slovakia

Michael G. Fehlings (119), Division of Neurosurgery, Toronto Western Hospital, University Health Network, Toronto, ON, Canada

Felice Francavilla (377), Andrology Unit, Department of Life, Health and Environmental Sciences, University of L'Aquila, L'Aquila, Italy

Sandro Francavilla (377), Andrology Unit, Department of Life, Health and Environmental Sciences, University of L'Aquila, L'Aquila, Italy

Louna Ftouni (425), Faculty of Medicine, Beirut Arab University, Beirut, Lebanon

Daniel Garcia-Ovejero (239), Laboratory of Neuroinflammation, Hospital Nacional de Paraplejicos, SESCAM, Toledo, Spain

Stefan Golaszewski (301), Department of Neurology, Christian Doppler Klinik, Salzburg, Austria

Jaime Gosálvez (93), Genetics Unit, Department of Biology, Universidad Autónoma de Madrid, Madrid, Spain

Lukas Grassner (239), Department of Neurosurgery, Medical University Innsbruck, Innsbruck; Institute of Molecular Regenerative Medicine; Spinal Cord Injury and Tissue Regeneration Center Salzburg (SCI-TReCS), Paracelsus Medical University; ParaMove, SCI Research Unit, BG Trauma Center Murnau, Murnau, Germany, and Paracelsus Medical University, Salzburg, Austria; Spinal Cord Injury Center, Trauma Center Murnau, Murnau, Germany

Jason H. Gumbel (289), Department of Anatomical Sciences and Neurobiology, University of Louisville, Louisville, KY, United States

Ifaz T. Haider (315), Human Performance Lab and McCaig Institute for Bone and Joint Health, University of Calgary, Calgary, AB, Canada

Adham Halaoui (133, 425), Department of Neurosurgery, American University of Beirut Medical Center, Beirut, Lebanon

Inbo Han (229), Department of Neurosurgery, CHA University School of Medicine, CHA Bundang Medical Center, Seongnam-si, Gyeonggi-do, Republic of Korea

Robert F. Heary (153), Department of Neurological Surgery, Hackensack Meridian School of Medicine, Nutley, NJ, United States

James Hong (119), Division of Genetics and Development, University Health Network, Toronto, ON, Canada

Philip J. Horner (51), Department of Neurosurgery, Center for Neuroregeneration, Houston Methodist Research Institute, Houston, TX, United States

Ning Huang (199), Synaptic Function Section, National Institute of Neurological Disorders and Stroke (NINDS), National Institutes of Health (NIH), Bethesda, MD, United States

Charles H. Hubscher (289), Department of Anatomical Sciences and Neurobiology; Kentucky Spinal Cord Injury Research Center, University of Louisville, Louisville, KY, United States

Joji Inamasu (3), Department of Neurosurgery, Saiseikai Utsunomiya Hospital, Utsunomiya, Japan

Hiroyuki Ishiguro (177), Department of Orthopaedic Surgery, National Hospital Organization Osaka National Hospital, Osaka, Japan

Justin Y. Jeon (23), Department of Sport Industry Studies & Department of Medicine, Cancer Prevention Center, Shinchon Severance Hospital, Yonsei University, Seoul, Korea

Stephen D. Johnston (93), School of Agriculture and Food Sciences, The University of Queensland, Gatton, QLD, Australia

Kate Jones (459), Institute for Ethics and Society, University of Notre Dame Australia, Sydney, NSW, Australia

Ignacio Jure (275), Laboratorio de Bioquímica Neuroendócrina, Instituto de Biología y Medicina Experimental, CONICET, Buenos Aires, Argentina

Takashi Kaito (177), Department of Orthopaedic Surgery, Osaka University Graduate School of Medicine, Osaka, Japan

Prabhjot Kaur (327), Division of NMR, Institute of Nuclear Medicine and Allied Sciences, Delhi, India

Sarah Kawtharani (133, 425), Department of Neurosurgery, American University of Beirut Medical Center, Beirut, Lebanon

Erika Kellerova (187), Department of Neurodegeneration, Plasticity and Repair, Institute of Neurobiology, Biomedical Research Center of the Slovak Academy of Sciences, Kosice, Slovakia

Kristina A. Kigerl (435), Center for Brain and Spinal Cord Repair; Belford Center for Spinal Cord Injury; Department of Neuroscience, The Ohio State University College of Medicine, Columbus, OH, United States

Dong-il Kim (23), Division of Health and Kinesiology, Incheon National University, Incheon, Korea

Nicole Kiss (389, 405), Institute for Physical Activity and Nutrition (IPAN), School of Exercise and Nutrition Sciences, Faculty of Health, Deakin University, Burwood, VIC, Australia

Firas Kobeissy (133, 425), Department of Biochemistry and Molecular Genetics, Faculty of Medicine, American University of Beirut, Beirut, Lebanon

Radha Korupolu (351), Physical Medicine and Rehabilitation, University of Texas Health Science Centre; TIRR Memorial Hermann Hospital, Houston, TX, United States

Hemant Kumar (229), Department of Pharmacology and Toxicology, National Institute of Pharmaceutical Education and Research (NIPER)-Ahmedabad, Gandhinagar, Gujarat, India

Pawan Kumar (327), Division of NMR, Institute of Nuclear Medicine and Allied Sciences, Delhi, India

Florencia Labombarda (275), Laboratorio de Bioquímica Neuroendócrina, Instituto de Biología y Medicina Experimental, CONICET, Buenos Aires, Argentina

Asma Akbar Ladak (11), Medical College, Aga Khan University Hospital, Karachi, Pakistan

Jeongmin Lee (23), Department of Sport Industry Studies, Yonsei University, Seoul, Korea

Lun Li (153), Reynolds Family Spine Laboratory, Department of Neurosurgery, New Jersey Medical School-Rutgers, The State University of New Jersey, Newark, NJ, United States

Roger Liu (H.) (339), Department of Rehabilitation Medicine, Thomas Jefferson University Hospital, Philadelphia, PA, United States

Eldon Loh (497), Lawson Health Research Institute, Parkwood Institute Main Bldg., Department of Physical Medicine and Rehabilitation; St. Joseph's Health Care London, Department of Physical Medicine and Rehabilitation, London, ON, Canada

Swati Mehta (497), Lawson Health Research Institute, Parkwood Institute Main Bldg., Department of Physical Medicine and Rehabilitation; St. Joseph's Health Care London, Parkwood Institute Main Bldg., Department of Psychology; St. Joseph's Health Care London, Department of Physical Medicine and Rehabilitation, London, ON, Canada

Farah Mneimneh (425), Department of Engineering and Architecture, American University of Beirut, Beirut, Lebanon

Eduardo Molina-Holgado (239), Laboratory of Neuroinflammation, Hospital Nacional de Paraplejicos, SESCAM, Toledo, Spain

Robert W. Motl (471), Department of Physical Therapy, School of Health Professions, University of Alabama at Birmingham, Birmingham, AL, United States

L. Mourino-Alvarez (143), Department of Vascular Physiopathology, Hospital Nacional de Parapléjicos (HNP), SESCAM, Toledo, Spain

Charbel Moussalem (133, 425), Department of Neurosurgery, American University of Beirut Medical Center, Beirut, Lebanon

Fatima Mubarak (11), Department of Radiology, The Aga Khan University, Karachi, Pakistan

Batool Mujtaba (11), Section of Neurosurgery, Department of Surgery, The Aga Khan University, Karachi, Pakistan

Raffaele Nardone (301), Department of Neurology, Hospital of Merano (SABES-ASDAA), Merano-Meran, Italy; Department of Neurology, Christian Doppler Klinik, Salzburg, Austria

Simon J. O'Carroll (215), Department of Anatomy and Medical Imaging, Faculty of Medical and Health Sciences, School of Medical Sciences, University of Auckland, Auckland, New Zealand

Ibrahim Omeis (133, 425), Department of Neurosurgery, American University of Beirut Medical Center, Beirut, Lebanon; Department of Neurosurgery, Baylor College of Medicine, Houston, TX, United States

Steve Orenczuk (497), St. Joseph's Health Care London, Parkwood Institute Main Bldg., Department of Psychology, London, ON, Canada

Patricia Orme (339), Department of Rehabilitation Medicine, Thomas Jefferson University Hospital, Philadelphia, PA, United States

İlknur Özkan (483), İnternal Medicine Nursing Department, Kumluca Faculty of Health Sciences, Akdeniz University, Kumluca-Antalya, Turkey

Jaroslav Pavel (187), Department of Neurodegeneration, Plasticity and Repair, Institute of Neurobiology, Biomedical Research Center of the Slovak Academy of Sciences, Kosice, Slovakia

Clara Penas (105), Department of Cell Biology, Physiology and Immunology, Facultat de Medicina, Institut of Neurosciences, Centro de Investigación Biomédica en Red sobre Enfermedades Neurodegenerativas (CIBERNED), Universitat Autònoma de Barcelona, Bellaterra, Spain

James Pendleton (339), Department of Rehabilitation Medicine, Thomas Jefferson University Hospital, Philadelphia, PA, United States

Phillip G. Popovich (435), Center for Brain and Spinal Cord Repair; Belford Center for Spinal Cord Injury; Department of Neuroscience, The Ohio State University College of Medicine, Columbus, OH, United States

Julie Pryor (459), Royal Rehab, Ryde, NSW, Australia

Swapan K. Ray (263), Department of Pathology, Microbiology, and Immunology, University of South Carolina School of Medicine, Columbia, SC, United States

Marcelo Riberto (65), Departamento de Ortopedia e Anestesia—Faculdade de Medicina de Ribeirão Preto, Universidade de São Paulo—Ribeirão Preto, Ribeirão Preto/SP, CEP, Brazil

Elizabeth J. Richardson (413), Department of Behavioral & Social Sciences, University of Montevallo, Montevallo, AL, United States

Emma V. Richardson (471), School of Sport and Exercise Science, University of Worcester, Worcester, United Kingdom

Luiz Fernando Freire Royes (165, 447), Exercise Biochemistry Laboratory, Center of Physical Education and Sports, Federal University of Santa Maria, Santa Maria, RS, Brazil

Rana Sarieddine (425), Department of Neurosurgery, American University of Beirut Medical Center, Beirut, Lebanon

T. Sastre-Oliva (143), Department of Vascular Physiopathology, Hospital Nacional de Parapléjicos (HNP), SESCAM, Toledo, Spain

Lumy Adams Sawaki (351), Department of Physical Medicine and Rehabilitation, University of Kentucky, Lexington, KY, United States

Dimitry G. Sayenko (51), Department of Neurosurgery, Center for Neuroregeneration, Houston Methodist Research Institute, Houston, TX, United States

Thomas J. Schnitzer (315), Department of Physical Medicine and Rehabilitation, Feinberg School of Medicine, Northwestern University Feinberg School of Medicine, Chicago, IL, United States

Benjamin Scoblionko (R.) (339), Department of Rehabilitation Medicine, Thomas Jefferson University Hospital, Philadelphia, PA, United States

Zu-Hang Sheng (199), Synaptic Function Section, National Institute of Neurological Disorders and Stroke

(NINDS), National Institutes of Health (NIH), Bethesda, MD, United States

Stephanie L. Silveira (471), Department of Physical Therapy, School of Health Professions, University of Alabama at Birmingham, Birmingham, AL, United States

Narina Simonian (315), Department of Physical Medicine and Rehabilitation, Feinberg School of Medicine; Northwestern University Clinical and Translational Sciences Institute, Northwestern University Feinberg School of Medicine, Chicago, IL, United States

Grahame Simpson (459), John Walsh Centre for Rehabilitation Research, Sydney School of Medicine, University of Sydney, NSW, Australia

Thomas Sinkjær (365), Department of Health Science and Technology, Aalborg University, Aalborg, Denmark

Daniel Rubio de Souza (65), Divisão de Medicina Física, Instituto de Ortopedia e Traumatologia da Faculdade de, Medicina da Universidade de São Paulo, R. Ovídio Pires de Campos, Cerqueira Cesar, São Paulo/SP, CEP, Brazil

Argyrios Stampas (351), Physical Medicine and Rehabilitation, University of Texas Health Science Centre; TIRR Memorial Hermann Hospital, Houston, TX, United States

Darren Svirskis (215), School of Pharmacy, Faculty of Medical and Health Sciences, University of Auckland, Auckland, New Zealand

Giuliano Taccola (51), Department of Neurosurgery, Center for Neuroregeneration, Houston Methodist Research Institute, Houston, TX, United States; Neuroscience Department, International School for Advanced Studies (SISSA), Bonomea, Trieste, Italy

Sudha S. Tallavajhula (351), TIRR Memorial Hermann Hospital; Department of Neurology, University of Texas Health Science Centre, Houston, TX, United States

Seçil Taylan (483), Surgical Nursing Department, Kumluca Faculty of Health Sciences, Akdeniz University, Kumluca-Antalya, Turkey

Robert Teasell (497), Lawson Health Research Institute, Parkwood Institute Main Bldg; St. Joseph's Health Care London, Department of Physical Medicine and Rehabilitation, London, ON, Canada

Aiko K. Thompson (365), Department of Health Sciences and Research, College of Health Professions, Medical University of South Carolina, Charleston, SC, United States

Amirali Toossi (119), Division of Genetics and Development, University Health Network, Toronto, ON, Canada

Eugen Trinka (301), Department of Neurology, Christian Doppler Klinik, Salzburg, Austria

Zina Trost (413), Department of Physical Medicine and Rehabilitation, Virginia Commonwealth University, Richmond, VA, United States

Eduardo Vargas-Baquero (93), Sexual and Fertility Unit, Hospital Nacional de Parapléjicos de Toledo, Toledo, Spain

Leigh C. Ward (389, 405), School of Chemistry and Molecular Biosciences, The University of Queensland, St Lucia, QLD, Australia

Ligia Jia Lin Wu (65), Divisão de Medicina Física, Instituto de Ortopedia e Traumatologia da Faculdade de, Medicina da Universidade de São Paulo, R. Ovídio Pires de Campos, Cerqueira Cesar, São Paulo/SP, CEP, Brazil

Ariane Zamarioli (81), Department of Orthopaedics and Anesthesiology, Ribeirao Preto Medical School, University of Sao Paulo, Ribeirao Preto, Brazil

Mohammad-Masoud Zavvarian (119), Institute of Medical Science, Faculty of Medicine, University of Toronto; Division of Genetics and Development, University Health Network, Toronto, ON, Canada

Preface

Spinal injury affects about 10 million people annually worldwide. Many of these injuries are preventable and occur due to falls, violence, and road traffic accidents. However, spinal cord damage may also result from nontraumatic injury, for example, due to toxins, infection, cancer, intervertebral disc damage, and vascular disease.

Spinal injuries may cause significant, lifelong disabilities and thereby impact the family unit. Symptoms are varied and include paresthesia, spasticity, and loss of motor control. In extreme cases, affected individuals are chair- or bedbound for life. These debilitating injuries may also be accompanied by severe pain. Loss of bladder and bowel control increases dependency and further reduces quality of life.

Diagnosis is generally based on symptoms and various imaging techniques in the clinical context of a traumatic injury to the spinal cord. Various syndromes are associated with spinal cord injury (e.g., central, anterior, and posterior cord syndromes). However, various pathological processes are involved in the initiation of primary and secondary damage to the spinal cord.

Physical trauma may be accompanied by multiorgan cellular and biochemical injury. These injuries include neurodegeneration, free radical damage, changes in gene expression, and physiological changes such as loss of the regulation of blood pressure. So the manifestation of symptoms and signs from any given injury may be highly variable.

The patient's functional outcome predominantly depends on the level of spinal column injury (i.e., cervical, thoracic, lumbar, or sacral), the location of the damage within the spinal cord (e.g., central, anterior, and posterior), and the severity. This functional outcome may be modified to some extent by medical therapies and rehabilitation.

To fully comprehend and positively influence the trajectory of patients' outcomes, it is necessary to understand the fundamental principles of the conditions that arise as a result of spinal cord injury. Presently, the availability of much of this information is sporadic, in different scientific domains, and designed for different scientific specialities. Spinal cord injuries are diverse, so a multidisciplinary approach is needed. This is addressed in the two-volume set *The Neuroscience of Spinal Cord Injury* comprising the 2 books:

Diagnosis and Treatment of Spinal Cord Injury
Cellular, Molecular, Physiological, and Behavioral Aspects of Spinal Cord Injury

This book, *Cellular, Molecular, Physiological, and Behavioral Aspects of Spinal Cord Injury*, comprises the following four sections:

- *Setting the scene and introductory chapters*
- *Cellular and molecular aspects of spinal injury*
- *Physiological and metabolic effects*
- *Behavioral and psychological effects*

Each chapter has the following sections:

- *Abstract (published online)*
- *Key facts*
- *Mini-dictionary of terms*
- *Applications to other areas of neuroscience*
- *Summary points*

The sections *Key facts*, *Mini-dictionary of terms*, and *Summary points* enable the reader to cross the transintellectual and transdisciplinary divides. The section *Applications to other areas of neuroscience* pertains to the translational aspects of the chapter and the applicability of the information.

The Neuroscience of Spinal Cord Injury is designed for research and teaching purposes. It is suitable for neurologists, surgeons, trauma specialists, psychologists, health scientists, public health workers, doctors, pharmacologists, and research scientists. It is valuable as a personal reference book and also for academic libraries that cover the domains of trauma and neurology. Contributions are from leading national and international experts including those from world-renowned institutions. It is suitable for undergraduate and postgraduate students as well as lecturers and academic professors.

Rajkumar Rajendram, Victor R. Preedy, and Colin R. Martin (Editors)

Section A

Setting the scene and introductory chapters

Chapter 1

Causes of spinal injury: Motor vehicle accidents and beyond

Joji Inamasu

Department of Neurosurgery, Saiseikai Utsunomiya Hospital, Utsunomiya, Japan

List of abbreviations

AACN Advanced Automatic Collision Notification
MVA motor vehicle accident

Introduction

As has been known widely, motor vehicle accident (MVA) is one of the most common causes of spinal injury, together with accidental fall and sports injury (Blackmore, Goswami, & Chancey, 2012) (Fig. 1). Take the cervical spine injury for example: it has recently been reported that approximately 869,000 traffic crash-related cervical spine injuries were seen in hospitals in the United States annually, including 841,000 sprain/strain injuries, 2800 disc injuries, 23,500 spine fractures, 2800 spinal cord injuries, and 1500 facet joint dislocations (Freeman & Leith, 2020). There are numerous occupant-specific, vehicle-specific, and environmental factors that may influence the location and severity of MVA-related spinal injuries (Table 1).

While the circumstances of MVAs are quite diverse, the number of subjects with a MVA-associated spinal injury was greatest among automobile drivers, followed by pedestrians hit by cars, car passengers, and motorcycle riders (Wang et al., 2016). As a consequence, male aged 20–60 are the population most likely to sustain MVA-related spinal injury. Historically, spinal injuries sustained by automobile drivers have been studied most extensively because they are relatively easy to simulate or reproduce in laboratory settings, and may be preventable or at least minimizable with the use of protective devices such as seatbelt and airbag. Numerous basic research studies to simulate the spinal injury have been conducted for this purpose (Xu et al., 2018). For example, dummies which are equipped with various sensors and transducers have been used in crash tests for decades (Fig. 2), providing useful information to automobile manufacturers and researchers. More recently, computational models of the human spine (Wang, Feng, and Hu, 2018) have been so developed that they can reliably simulate and estimate quantity of stress sustained by each vertebral column and ligament (Fig. 3).

In a large retrospective study conducted in the Wisconsin State, as many as 12.5% of drivers who were admitted after experiencing a severe MVA had been found to have a spinal injury (Wang, Pintar, Yoganandan, & Maiman, 2009). The seat position within an automobile may not influence the likelihood of spinal injury: there was no statistical difference in the frequency of spinal injury between drivers and passengers in that study. Spinal injury patterns substantially differ between the four-wheeled vehicles and two-wheeled vehicles: in other words, it may reflect differing mechanisms of injury between the automobile occupants who are mostly restrained and motorcycle riders who are mostly unrestrained. While the thoracic spine is most frequently injured in motorcycle riders, the cervical spine is most frequently injured in automobile occupants (Robertson, Branfoot, Barlow, & Giannoudis, 2002). In addition, the motorcycle riders are more likely to sustain multiple spinal injuries compared to automobile drivers. Interestingly, spinal injuries sustained by bicycle riders are most likely to develop at the lower cervical spine (Broe, Kelly, Groarke, Synnott, & Morris, 2018). Unlike automobile occupants, the characteristics of spinal injury in pedestrians have been reported much less often. A study from the California State described that the incidence of cervical spine injury was 2.1% among 8401 pedestrian injuries caused by automobiles, and the incidence increased with age and severity of head trauma (Yanar et al., 2007).

Cellular, Molecular, Physiological, and Behavioral Aspects of Spinal Cord Injury. https://doi.org/10.1016/B978-0-12-822427-4.00001-0

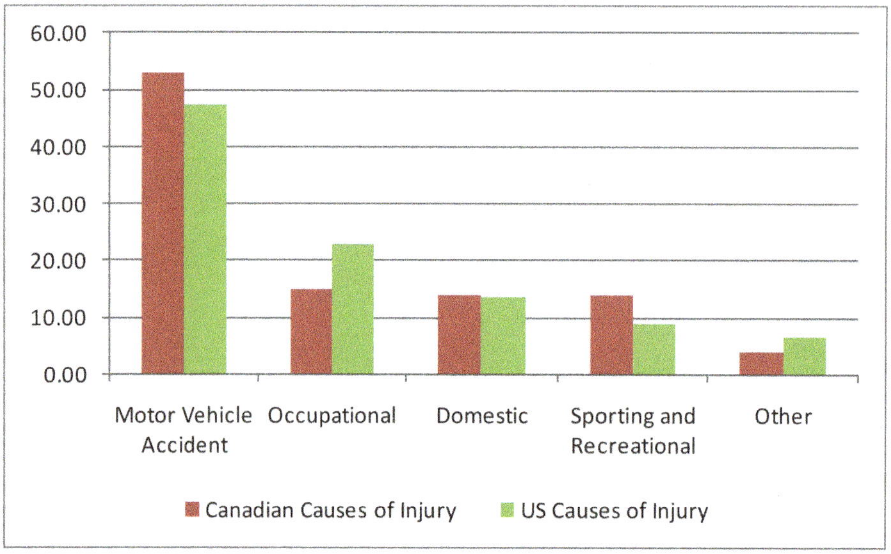

FIG. 1 Activities causing cervical spine injuries. Causes of spine injury in Canada (*brown*) and the United States (*green*) in percentages. *From: Blackmore, M.E., Goswami, T., & Chancey C. (2012). Cervical spinal injuries and risk assessment. In T. Goswani (Ed),* Injury and skeletal biomechanics *(*Chapter 7*). Reprint with permission. ©2012 IntechOpen Ltd.*

TABLE 1 Factors that may influence location and severity of spinal injury in car passengers.

- Type of vehicle (weight, shape, year-model)
- Type of collision (frontal, lateral, rear-end, rollover)
- Speed and degree of deceleration of the vehicle
- Type and condition of road
- Weather
- Use of seatbelt by car passengers
- Use of medication or drugs that may influence the mental status of a driver
- Stature, weight, age, and sex of car passengers
- Position (seat) of car passengers

Difference in the type of four-wheeled vehicle may not substantially influence the frequency or severity of spine injury. There was no significant difference between the frequency of cervical spine injury among drivers of sedan, light truck, and sports utility vehicle (Stein et al., 2011). More recently, a study from Japan found that the frequency of spinal injury did not differ significantly between drivers of sedan and those of K-cars (Inamasu, Kujirai, Izawa, Kase, & Shinozaki, 2019).

Spinal injury in automobile drivers

While spinal column mechanics and stability are typically concerned with compression forces, the cervical column is also exposed to a variety of other loading modes such as flexion/extension, lateral bending, axial rotation, tension, and shear during physiologic and traumatic loading scenarios (Stemper, Pintar, & Rao, 2011). In anatomical perspective, the lower cervical spine and the thoracolumbar junction are the most common locations for spine fractures among automobile drivers who experience a severe MVA. However, the frequency of spinal cord injury is much higher in the former group: while 33% of drivers with cervical spine fractures sustained a spinal cord injury, only 18% of drivers with thoracolumbar spine injury sustained a spinal cord injury (Smith, Siegel, & Siddiqi, 2005).

Automobile crashes may be quadrichotomized into frontal, lateral, rear-end, and rollover crash depending on the direction of the impact to the vehicle. Among them, frontal crash is one of the most frequently encountered automobile

Model name	Hybrid III	THOR-M	SID	SID-IIs
Figure			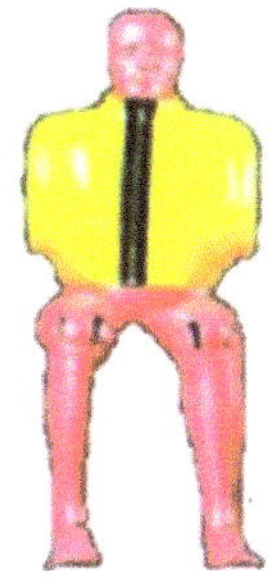	
Application	Frontal impact	Frontal impact	Side impact	Side impact
Model name	BioSID	EuroSID-II	WorldSID	BioRID
Figure				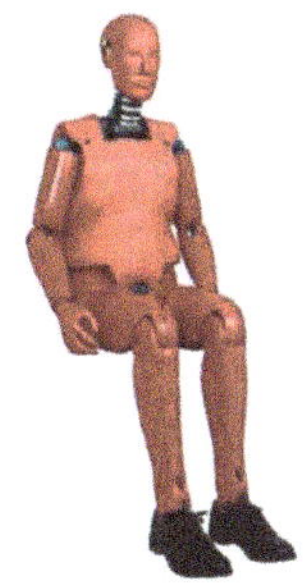
Application	Side impact	Side impact	Side impact	Rear impact

FIG. 2 Various crash test dummies. Various crash test dummies used by automobile manufacturers are presented. *From: Xu, T., Sheng, X., Zhang, T., Liu, H., Liang, X., & Ding, A. (2018). Development and validation of dummies and human models used in crash test.* Applied Bionics and Biomechanics, *3832850, Table 3 Reprint with permission. ©2018 Hindawi Publishing Co.*

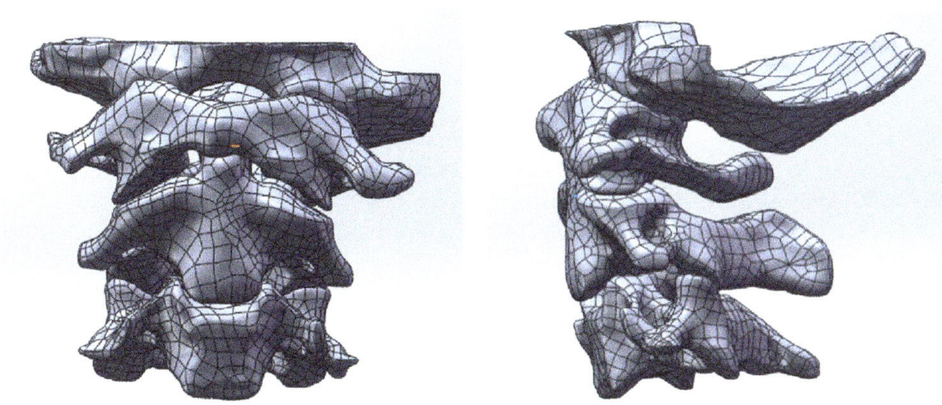

FIG. 3 3-D model of the upper cervical spine. An example of computational 3-D model of the upper cervical spine for evaluation of mechanical stress is presented. *From: Wang, X.D., Feng, M.S., Hu, Y.C. (2018). Establishment and finite element analysis of a three-dimensional dynamic model of upper cervical spine instability.* Orthopaedic Surgery, *11, 500–509, Fig. 4. Reprint with permission ©2019 John Wiley & Sons.*

crashes (Parenteau & Viano, 2014). In the frontal crash, impact transmitted to the spine of a restrained driver is mostly in the sagittal direction. During a frontal crash, the vehicle and its seat bases abruptly decelerate upon impact. Inertia of the torso causes the trunk to accelerate forward (Svensson et al., 2000). Head inertia and cervical spine flexibility cause relative forward acceleration and motion between the head and torso. The relative acceleration and motion may in turn cause the cervical spine injury. In a biomechanical study during simulated frontal crash, the greatest flexion motion beyond the physiologic limit was observed at C7-T1, followed by C5–6 (Ivancic, 2016). The finding was compatible to studies using finite element analysis (Panzer, Fice, & Cronin, 2011). Cervical spine injuries after frontal crashes are more likely to be of flexion-extension type. The finite element analysis study also found that the annulus fibrosus of the cervical spine was the anatomical component which was primarily injured after the frontal impact (Panzer et al., 2011).

Regarding the thoracolumbar spine injury, it has been reported that compression and/or flexion injuries frequently occurred in frontal crashes due to seat pan and vertical loading (Pintar, Yoganandan, Maiman, Scarboro, & Rudd, 2012). The thoracolumbar spine was most commonly fractured at either the T12 or L1 level. Majorly, burst-type fractures occurred predominantly at T12, L1, or L5; wedge fractures were most common at L1. Most of automobile occupants in that study had been restrained with seatbelt. Interestingly, there was an increasing trend in incidence rate of thoracolumbar fractures in frontal impact crashes as a function of vehicle model year from 1986 to 2008 reason of which remained unknown (Pintar et al., 2012). Our group also made an investigation on thoracolumbar fractures caused during frontal crashes, and found that the incidence of neurological deficit in the unrestrained automobile occupants with a thoracolumbar fracture was approximately six times higher compared with restrained automobile occupants with a thoracolumbar fracture (Inamasu & Guiot, 2007).

During a lateral crash, the struck vehicle and its seat bases accelerate laterally. Inertia of the occupant's torso causes it to displace laterally toward the point of impact. Prior studies have demonstrated that occupants of lateral crashes are at higher risk of more severe injuries than those of frontal crashes because the greater crash-related energy is transferred to the occupant in the former group (Inamasu, Nakaya, Kujirai, Mayanagi, & Nakatsukasa, 2018). Additionally, seatbelt is less effective in preventing injuries due to lateral loading. Cervical spine motions beyond physiologic limits were observed in the lateral bending at C1–2 and C3–4 through C7-T1 (Ivancic, 2016). Multiplanar soft tissue injuries were also observed at C3–4 to C7-T1 due to simulated lateral crashes (Ivancic, 2016).

Most researchers agree that the risk of spinal injury is greatest in rollover collision. It has recently been reported that among the four categories of automobile crash, the rate of spinal cord injury was highest in rollover crash, and was lowest in frontal crash (Parenteau & Viano, 2014). In another study from Baltimore, the risk of cervical spine injury was 5.3 times higher among drivers who experienced a rollover crash compared with those who experienced a frontal crash (Stein et al., 2011). Roof deformation and protrusion has been thought responsible for this disproportionally high rate of cervical spine injury after rollover crashes (Bidez, Cochran Jr., King, & Burke, 2007). Our group investigated the difference in the injury pattern between restrained and unrestrained drivers who sustained a thoracolumbar junction fracture after rollover crash (Inamasu & Guiot, 2009). Neurologic deficits were present exclusively in the unrestrained group, and rotational thoracolumbar spine injury was present exclusively in the unrestrained group.

Rear-end collision is unique that severity of the spinal injury is usually modest and fractures occur less often compared with other types of collision. This is mainly because the speed of the vehicles involved is usually lower in rear-end collisions. However, there are many automobile occupants who suffer from whiplash injury as a sequela of rear-end crash. The whiplash injury mostly consists of damages to the soft tissue such as muscles, ligaments, discs, and joint capsules, which persists for months to years. Occasionally, elderly automobile occupants with pre-existing cervical canal stenosis such as ossification of the posterior longitudinal ligament may sustain a cord injury without fracture after low-speed rear-end collision (Hussain, Abu-Khumra, Alnajjar, & Abdo, 2016).

Role of seatbelt and airbag in protecting spine

There is little doubt about the importance of seatbelt (three-point seatbelt) and airbag in prevention of the spinal injury for automobile occupants. Nevertheless, spine injuries do occur in drivers wearing a seatbelt and for whom an airbag has deployed. In a recent US study, 89 of the 1067 (8%) drivers with MVA-related cervical spine injuries had been restrained and an airbag deployed (Wang et al., 2009). The combination of seatbelt and airbag is considered to be most effective. There have been several studies which evaluated efficacy of seatbelt and/or airbag in the reduction of spinal injury (Claytor, MacLennan, McGwin Jr., Rue 3rd, & Kirkpatrick, 2004; Reed et al., 2006; Wang et al., 2009; Williams et al., 2008). Those results as relative risk reduction of seatbelt and/or airbag are summarized in Table 2. While outcome parameters were somewhat different among studies, the relative risk reduction of cervical spine injury in belted, airbag-deployed drivers compared with non-restrained drivers was unequivocal, which ranged from 0.19 to 0.67. While the incidence of thoracic spine injury was also significantly reduced in belted, airbag-deployed drivers, there was no significant difference in the incidence of lumbosacral spine injury between the two groups (Wang et al., 2009). Ejection of automobile occupants from their vehicle is problematic particularly in rollover crashes. The efficacy of seatbelt is quite obvious in this setting. In restrained automobile occupants, partial ejection and complete ejection occurred in 2.4% and 0.03%. On the other hand, 3.8% and 20% of unrestrained automobile occupants sustained partial and complete ejection, respectively (Funk et al., 2012).

TABLE 2 Relative risk reduction of seatbelt and/or airbag against spine injury.

Author, year	Parameter	Unrestrained (Ref)	Seatbelt only	Airbag only	Seatbelt and airbag
Claytor et al. (2004)	Cervical spine injury	1	0.40	1.02	0.19
Reed et al. (2006)	Spinal cord injury	1	0.57	1.14	0.25
Williams et al. (2008)	Cervical spine injury	1	0.57	0.49	0.41
Wang et al. (2009)	Spine injury	1	1.17	1.05	0.67

It has been known for many years that airbag alone may not afford adequate protection to automobile occupants. A study from Pittsburg demonstrated that airbag-deployed but unrestrained drivers were 6.7 times more likely to suffer from the cervical cord injury compared with those using both protective devices (Donaldson 3rd., Hanks, Nassr, Vogt, & Lee, 2008). The risk of thoracic organ injury may also be increased in that situation (Khouzam, Al-Mawed, Farah, & Mizeracki, 2014). Airbag may potentially be deleterious to children and adult drivers with short stature (Blacksin, 1993). We recently reported that population might particularly be at elevated risk for hyper-extension injury of the upper cervical spine after severe impact, even though they were properly restrained by seatbelt, because their jaws are forcefully lifted upward owing to a rapidly expanding airbag (Inamasu & Kato, 2018). The mechanism of such airbag-induced spine injury and representative case is illustrated in Fig. 4. Many traffic safety guidelines strongly advise that children <12 years of age should not be seated in the front seat for fear of this type of injury. Catastrophic atlanto-occipital dislocations were more common than any other type of dislocation for 8- to 12-year-olds, even though they were seated in rear-seats (Mallory, Stammen, & Zhu, 2019).

Historically, Chance fracture, flexion-extension injuries of the thoracolumbar spine that extend to involve all three (anterior, middle, posterior) spinal columns, have been associated with the use of lap (two-point) seatbelt (Fig. 5). Since use of the two-point lap belt in automobile has been outlawed in many countries, Chance fracture has rarely been encountered in emergency departments nowadays. Interestingly, several cases of Chance fracture have been reported after crash landing of commercial flights, in which two-point seatbelt is equipped with passenger seat (Lee et al., 2015).

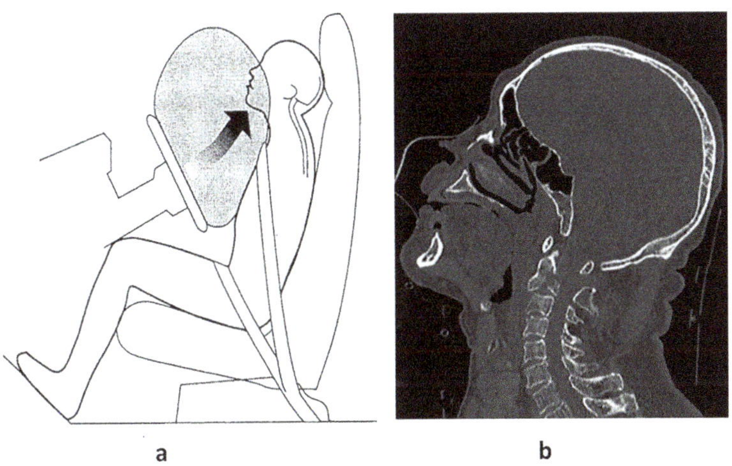

FIG. 4 Airbag-induced cervical spine injury. (A): An illustration showing mechanism of airbag-induced hyperextension injury of the cervical spine. (B): A sagittal CT scan of the spine showing an atlanto-axial dislocation caused by excessive inflation of an airbag. *Panel (A) From: Blacksin MF (1993). Patterns of fracture after air bag deployment.* Journal of Trauma, 35, *840–843, Fig. 4 Reprint with permission ©1993 Wolters Kluver.*

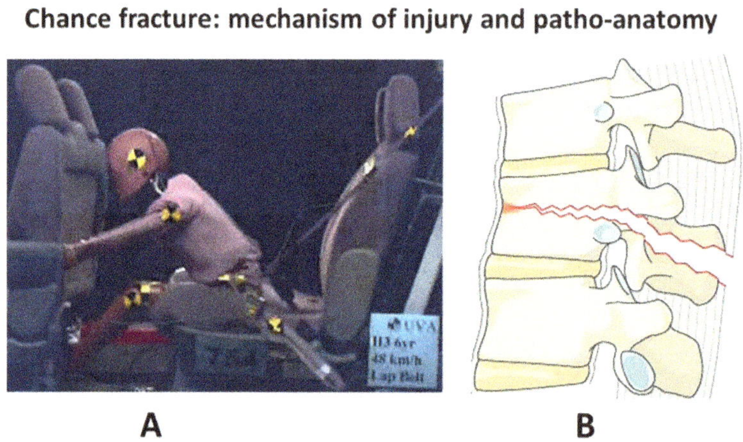

FIG. 5 Chance fracture. (A): A frontal crash test on a dummy wearing a two point seatbelt to simulate Chance fracture. (B): A schema of the Chance fracture of the lumbar spine illustrating extent of bony and soft tissue damage.

Prevention of spinal injury among motorcycle riders

Unlike automobile drivers, there have been no universally accepted protective devices for motorcyclists. Several researchers suggest that hard-shell back protective device may be effective: in a study from Italy, the percentage of spinal injuries among those who used a high-level safety device (hard-shell back protective device and/or airbags) was reduced by approximately 50% compared to those who used only protective clothing (Giustini et al., 2014). Because of lack of studies, however, it remains unclear to what degree those protective devices are effective. Helmets, which are intended to protect the head of the motorcycle rides from impact, may have protective effect against cervical spine injury. According to a recent study from South Korea, the use of helmet was associated with approximately 50% decrease in the frequency of cervical spine injury among motorcyclists (Park et al., 2019).

Application to other areas of neuroscience

In the automobile industry, biomechanical engineers have been working hard to create safer cars which minimize damage inflicted to automobile occupants. Thanks to a recent rapid development in information technology, more and more new vehicles are equipped with Advanced Automatic Collision Notification (AACN), which is a system on a motor vehicle that notifies a public safety answering point, either directly or through a third party, that the vehicle has had a crash. The AACN system enables earlier notification of an MVA and provides an injury prediction that can help dispatchers and first responders make better decisions about how and where to transport the injured passengers, thus getting the patients to definitive care sooner (Lee, Wu, Kang, & Craig, 2017). Considering that the spinal injury is both medical and surgical emergency for which prompt therapeutic intervention is required, we expect that the outcomes of patients with MVA-related spinal injury be improved with further expansion and refinement of the AACN system. Moreover, the concept of active safety is increasingly being introduced to avoid or minimize the impact of a crash. The active safety research today focuses primarily on sensor-based systems, such as advanced driver-assistance systems including adaptive cruise control and collision-warning/avoidance/mitigation systems (Li et al., 2019). It is expected that those sophisticated systems will dramatically reduce the incidence of MVA-related spine injury in the near future.

Mini-dictionary of terms

Spinal injury: in this chapter, the word is used to include spine (vertebral) injury, spinal cord injury, and soft tissue injury.

Frontal crash: it is a crash where the front end of a vehicle is impacted. It can further be classifiable to offset and non-offset collisions.

Lateral (side) crash: it is a crash where the side of one or more vehicles is impacted, and may be classifiable to near-side crash and far-side crash. It frequently occurs at intersections.

Rollover crash: is a crash in which a vehicle tips over onto its side or roof.

Rear-end crash: it is a crash that occurs when a vehicle crashes into the one in front of it.

Seatbelt: it is a vehicle safety device designed to secure the car occupants against harmful movement that may result during a crash. In an automobile setting, it usually refers to three-point seatbelt.

Key facts of spinal injury simulation

- Simulation of spinal injury consists of crash test dummies and computational models.
- Historically, crash test dummies have played an important role in simulation of the cervical spine injury.
- Crash test dummies may be classifiable into three categories depending on the direction of impact (frontal, lateral, and rear-end).
- Crash test dummies are designed based on American and European population, and damage prediction may be limited in other regions.
- The computational models began with finite element models of the body parts in the 1970s, evolving into more sophisticated full human body models.

Summary points

- While the circumstances of motor vehicle accidents (MVAs) are quite diverse, the number of MVA-associated spinal injury was greatest among automobile drivers.
- The frequency and severity of spinal injury is influenced by crash patterns, and rollover crash is associated with highest frequency of spinal injury.
- The combination of seatbelt and airbag is considered to be most efficacious in preventing or minimizing the spinal injury of automobile occupants.
- Various crash test dummies and computational models have been developed to simulate the spinal injury sustained by automobile occupants.
- Advanced Automatic Collision Notification (AACN), which is a system on a motor vehicle that notifies a public safety answering point that the vehicle has had a crash, is expected to improve the outcomes of patients with MVA-related spinal injury.

References

Bidez, M. W., Cochran, J. E., Jr., King, D., & Burke, D. S., 3rd. (2007). Occupant dynamics in rollover crashes: Influence of roof deformation and seat belt performance on probable spinal column injury. *Annals of Biomedical Engineering*, *35*, 1973–1988.

Blackmore, M. E., Goswami, T., & Chancey, C. (2012). Cervical spinal injuries and risk assessment. In T. Goswani (Ed.), *Injury and skeletal biomechanics (chapter 7)*. London: IntechOpen Ltd.

Blacksin, M. F. (1993). Patterns of fracture after air bag deployment. *Journal of Trauma*, *35*, 840–843.

Broe, M. P., Kelly, J. C., Groarke, P. J., Synnott, K., & Morris, S. (2018). Cycling and spinal trauma: A worrying trend in referrals to a national spine Centre. *The Surgeon*, *16*, 202–206.

Claytor, B., MacLennan, P. A., McGwin, G., Jr., Rue, L. W., 3rd, & Kirkpatrick, J. S. (2004). Cervical spine injury and restraint system use in motor vehicle collisions. *Spine*, *29*, 386–389.

Donaldson, W. F., 3rd., Hanks, S. E., Nassr, A., Vogt, M. T., & Lee, J. Y. (2008). Cervical spine injuries associated with the incorrect use of airbags in motor vehicle collisions. *Spine (Phila Pa 1976)*, *15*, 631–634.

Freeman, M. D., & Leith, W. M. (2020). Estimating the number of traffic crash-related cervical spine injuries in the United States; an analysis and comparison of national crash and hospital data. *Accident Analysis & Prevention*, *142*, 105571.

Funk, J. R., Cormier, J. M., Bain, C. E., Wirth, J. L., Bonugli, E. B., & Watson, R. A. (2012). Factors affecting ejection risk in rollover crashes. *Annals of Advances in Automotive Medicine*, *56*, 203–211.

Giustini, M., Cedri, S., Tallon, M., Roazzi, P., Formisano, R., & Pitidis, A. (2014). Use of back protector device on motorcycles and mopeds in Italy. *International Journal of Epidemiology*, *43*, 1921–1928.

Hussain, K., Abu-Khumra, S. K. A., Alnajjar, F. J. K., & Abdo, M. M. (2016). Triple trouble: A case of traumatic cervical spinal cord injury in a patient with ossification of posterior longitudinal ligament and disc prolapse. *Turkish Journal of Emergency Medicine*, *16*, 129–131.

Inamasu, J., & Guiot, B. H. (2007). Thoracolumbar junction injuries after motor vehicle collision: Are there differences in restrained and non-restrained front seat occupants? *Journal of Neurosurgery. Spine*, *7*, 311–314.

Inamasu, J., & Guiot, B. H. (2009). Thoracolumbar junction injuries after rollover crashes: Difference between belted and unbelted front seat occupants. *European Spine Journal*, *18*, 1464–1468.

Inamasu, J., & Kato, M. (2018). Airbag deployment and cervical spine injury in restrained drivers following motor vehicle collisions. *Neuroradiology*, *60*, 1307–1313.

Inamasu, J., Kujirai, D., Izawa, Y., Kase, K., & Shinozaki, H. (2019). Difference in severity and distribution of bodily injuries following collision between drivers of K-cars and standard vehicles. *Injury*, *50*, 205–209.

Inamasu, J., Nakaya, M., Kujirai, D., Mayanagi, K., & Nakatsukasa, M. (2018). Frequency and characteristics of traumatic brain injury in restrained drivers involved in road traffic accidents. *Acta Neurochirurgica (Wien)*, *160*, 1921–1929.

Ivancic, P. C. (2016). Mechanisms and mitigation of head and spinal injuries due to motor vehicle crashes. *The Journal of Orthopaedic and Sports Physical Therapy*, *46*, 826–833.

Khouzam, R. N., Al-Mawed, S., Farah, V., & Mizeracki, A. (2014). Next-generation airbags and the possibility of negative outcomes due to thoracic injury. *Canadian Journal of Cardiology*, *30*, 396–404.

Lee, H. J., Moon, B. J., Pennant, W. A., Shin, D. A., Kim, K. N., Yoon, D. H., et al. (2015). Three cases of spine fractures after an airplane crash. *Korean Journal of Neurotrauma*, *11*, 195–200.

Lee, E., Wu, J., Kang, T., & Craig, M. (2017). Estimate of mortality reduction with implementation of advanced automatic collision notification. *Traffic Injury Prevention*, *29*(Sup. 1), S24–S30.

Li, F., Liu, N., Li, H., Zhang, B., Tian, S., Tan, M., et al. (2019). A review of neck injury and protection in vehicle accidents. *Transportation Safety and Environment*, *1*, 89–105.

Mallory, A., Stammen, J., & Zhu, M. (2019). Cervical and thoracic spine injury in pediatric motor vehicle crash passengers. *Traffic Injury Prevention*, *20*, 84–92.

Panzer, M. B., Fice, J. B., & Cronin, D. S. (2011). Cervical spine response in frontal crash. *Medical Engeneering & Physics*, *33*, 1147–1159.

Parenteau, C. S., & Viano, D. C. (2014). Spinal fracture-dislocations and spinal cord injuries in motor vehicle crashes. *Traffic Injury Prevention*, *15*, 694–700.

Park, G. J., Shin, J., Kim, S. C., Na, D. S., Lee, H. J., Kim, H., et al. (2019). Protective effect of helmet use on cervical injury in motorcycle crashes: A case-control study. *Injury*, *50*, 657–662.

Pintar, F. A., Yoganandan, N., Maiman, D. J., Scarboro, M., & Rudd, R. W. (2012). Thoracolumbar spine fractures in frontal impact crashes. *Annals of Advances in Automotive Medicine*, *56*, 277–283.

Reed, M. A., Naftel, R. P., Carter, S., MacLennan, P. A., McGwin, G., Jr., & Rue, L. W., 3rd. (2006). Motor vehicle restraint system use and risk of spine injury. *Traffic Injury Prevention*, *7*, 256–263.

Robertson, A., Branfoot, T., Barlow, I. F., & Giannoudis, P. V. (2002). Spinal injury patterns resulting from car and motorcycle accidents. *Spine (Phila Pa 1976)*, *27*, 2825–2830.

Smith, J. A., Siegel, J. H., & Siddiqi, S. Q. (2005). Spine and spinal cord injury in motor vehicle crashes: A function of change in velocity and energy dissipation on impact with respect to the direction of crash. *Journal of Trauma*, *59*, 117–131.

Stein, D. M., Kufera, J. A., Ho, S. M., Ryb, G. E., Dischinger, P. C., O'Connor, J. V., et al. (2011). Occupant and crash characteristics for case occupants with cervical spine injuries sustained in motor vehicle collisions. *Journal of Trauma*, *70*, 299–309.

Stemper, B. D., Pintar, F. A., & Rao, R. D. (2011). The influence of morphology on cervical injury characteristics. *Spine (Phila Pa 1976)*, *36*(25 Suppl), S180–S186.

Svensson, M. Y., Boström, O., Davidsson, J., Hansson, H. A., Håland, Y., Lövsund, P., et al. (2000). Neck injuries in car collisions: A review covering a possible injury mechanism and the development of a new rear-impact dummy. *Accident Analysis & Prevention*, *32*, 167–175.

Wang, X. D., Feng, M. S., & Hu, Y. C. (2018). Establishment and finite element analysis of a three-dimensional dynamic model of upper cervical spine instability. *Orthopaedic Surgery*, *11*, 500–509.

Wang, H., Liu, X., Zhao, Y., Ou, L., Zhou, Y., Li, C., et al. (2016). Incidence and pattern of traumatic spinal fractures and associated spinal cord injury resulting from motor vehicle collisions in China over 11 years: An observational study. *Medicine (Baltimore)*, *95*, e5220.

Wang, M. C., Pintar, F., Yoganandan, N., & Maiman, D. J. (2009). The continued burden of spine fractures after motor vehicle crashes. *Journal of Neurosurgery. Spine*, *10*, 86–92.

Williams, R. F., Fabian, T. C., Fischer, P. E., Zarzaur, B. L., Magnotti, L. J., & Croce, M. A. (2008). Impact of airbags on a level I trauma center: Injury patterns, infectious morbidity, and hospital costs. *Journal of the American College of Surgeons*, *206*, 962–968.

Xu, T., Sheng, X., Zhang, T., Liu, H., Liang, X., & Ding, A. (2018). Development and validation of dummies and human models Used in crash test. *Applied Bionics and Biomechanics*, 3832850.

Yanar, H., Demetriades, D., Hadjizacharia, P., Nomoto, S., Salim, A., Inaba, K., et al. (2007). Pedestrians injured by automobiles: Risk factors for cervical spine injuries. *Journal of the American College of Surgeons*, *205*, 794–799.

Chapter 2

Magnetic resonance imaging (MRI) findings in spinal cord injury during acute and chronic phases

Kiran Aftab[a], Namrah Aziz[b], Batool Mujtaba[a], Asma Akbar Ladak[b], Fatima Mubarak[c], and Syed Ather Enam[a]

[a]*Section of Neurosurgery, Department of Surgery, The Aga Khan University, Karachi, Pakistan,* [b]*Medical College, Aga Khan University Hospital, Karachi, Pakistan,* [c]*Department of Radiology, The Aga Khan University, Karachi, Pakistan*

List of abbreviations

DTI diffusion tensor imaging
DWI diffusion weighted imaging
FA fractional anisotropy
fMRI functional magnetic resonance imaging
GRE gradient echo images
MPM multiparametric mapping
SCI spinal cord injury
SPAM subacute progressive ascending myelopathy
STIR short-tau inversion recovery
T1W T1-weighted
T2W T2-weighted

Introduction

MRI is considered the gold standard in imaging soft tissue structures in spinal trauma (Bozzo, Marcoux, Radhakrishna, Pelletier, & Goulet, 2011). It plays an especially important role in cases of spinal cord injury without radiological abnormality (SCIWORA) where first-line imaging, i.e., CT and radiography fail to show features of injury despite neurological deficits (Shen et al., 2007) In cases where both MRI and CT are required, it can help as an initial complex imaging in guiding a restricted CT scan for assessment of bony structures, preventing excessive radiation dose in spinal cord injury (SCI) patients who are mostly young (Chandra et al., 2012).

In the acute phase of injury, MRI plays a crucial role in diagnostic workup, surgical planning, and prognostication (Bozzo et al., 2011). It allows better visualization of the cord as well as the discoligamentous complex and guides the clinician in evaluating the site, severity, extent, and mechanism of the injury (Chandra et al., 2012; Freund et al., 2019; Kumar & Hayashi, 2016). Features including cord hematoma, severity of maximal cord compression and length of cord edema on MRI correlate significantly with functional outcomes (Chandra et al., 2012). MRI in the subacute and chronic phases of injury guides in the assessment of recovery and neuro-rehabilitation (Chandra et al., 2012; Freund et al., 2019). In this chapter, we elaborate various MRI sequences applicable to SCI management, findings observed in the acute, subacute and chronic phases of SCI, and their correlation with clinical features.

MRI sequences used in SCI

With the advancements in imaging technology, a multitude of sequences of MRI have been developed which range from conventional to novel and quantitative MRI techniques for the evaluation of varied aspects of SCI (Table 1).

Cellular, Molecular, Physiological, and Behavioral Aspects of Spinal Cord Injury. https://doi.org/10.1016/B978-0-12-822427-4.00002-2

TABLE 1 Applications of MRI sequences in phases of SCI.

Phase of SCI	MRI sequences	Applications
Acute	Sagittal T1W, sagittal and axial T2W, sagittal and axial T2W GRE (injury site), sagittal STIR sequences (whole spine)	diagnostic workup, surgical planning, prognostication
Subacute	Sagittal T1W, sagittal and axial T2W (injury site)	evaluation of postoperative complications or ascending neurological deficits
Chronic	T1W, T2W, DTI, Magnetization transfer, Multiparametric mapping, task-based fMRI	evaluation of late complications and recovery, clinical trials on neurorehabilitation.

Conventional MRI

Conventional MRI sequences used for evaluating SCI include T1-weighted (T1W) and T2-weighted images (T2W) to observe the primary pathological changes at the injury epicentre (Freund et al., 2019). These sequences elucidate the level of the injury and the length of intramedullary or extramedullary damage seen as edema and hemorrhage, the degree of spinal cord compression, extent of disk herniation, and the ligamentous damage (Freund et al., 2019). Axial T2-weighted is used for classifying acute cord injury in the Brain and Spinal Injury Center (BASIC) score (Freund et al., 2019). Lesions such as disk herniation and spinal cord compression are also evident on axial T2 sequences and guide the decision for surgical decompression (Bozzo et al., 2011; Freund et al., 2019). These sequences also have a role in follow up of SCI patients to see the changes evolving in the subacute phase of injury such as posttraumatic cyst (Freund et al., 2019). Moreover, the sagittal T2-weighted MRI shows midsagittal tissue bridges which predict sensorimotor recovery and can be used to stratify patients in clinical trials (Freund et al., 2019; Vallotton et al., 2019).

Use of MRI may be limited by the high cost incurred by many sequences and issues of logistics in trauma (Bozzo et al., 2011). Besides, image quality may be compromised by susceptibility artefacts due to metallic clips and dental implants, cerebrospinal fluid (CSF) flow artefact, Gibbs artefact, and chemical shift (Chandra et al., 2012; Kumar & Hayashi, 2016).

Novel and quantitative MRI sequences

Recently developed sequences like gradient echo images (GRE) offer better visualization of hemorrhage (Bozzo et al., 2011; Chandra et al., 2012), while short-tau inversion recovery (STIR), a fat suppressed version of T2-weighted, and gradient recall acquisition steady state (GRASS) can be useful for ligamentous injuries (Bozzo et al., 2011). Despite their utility, these sequences are yet to be included in injury classification scales. Functional MRI, based on blood flow, is being used in research to study the affected sensorimotor networks and plasticity occurring in the brain and spinal cord (Freund et al., 2019).

Quantitative MRI techniques have a role in demonstrating secondary microstructural changes reflecting neurodegeneration and neuroplasticity occurring throughout the neuraxis. Latest software-based image processing enables the calculation of various parameters at the voxel level (Freund et al., 2019). Diffusion weighted imaging (DWI), which utilizes information from the Brownian motion of water molecules, is sensitive to changes in axons and myelin. DWI can detect white matter tract changes as hyperintense lesions in acute SCI when conventional MRI fails to detect any abnormality (Shen et al., 2007). Diffusion Tensor Imaging (DTI) uses tensor models to depict these changes quantitatively (Zaninovich et al., 2019). Values of DTI parameters depict microstructural and degenerative changes that correlate with the severity of SCI and prognostic scores (Cheran et al., 2011; Ellingson, Ulmer, Kurpad, & Schmit, 2008). However, DWI is less established in the cord region due to small tissue volume, CSF flow and motion artefacts (Chandra et al., 2012). Moreover, DTI metrics are seen to be inconsistent across different SCI studies (Zaninovich et al., 2019). Machine learning and improved imaging methods are needed to improve DTI results (Zaninovich et al., 2019). Multiparameter mapping is another quantitative technique that combines magnetization transfer and relaxation mapping to indicate demyelination and iron deposition in the cervical cord and the brain (Freund et al., 2019).

Combined with conventional sequences, these techniques enable quantification of the degeneration and repair processes at various endpoints. Thus, they are useful biomarkers to predict the long-term outcomes as well as in clinical trials on spinal cord regeneration (Seif, Wheeler-Kingshott, Cohen-Adad, Flanders, & Freund, 2019).

Protocols for MRI in SCI

It is recommended to perform the imaging within 24–72 h of the trauma (Bozzo et al., 2011; Freund et al., 2019) Specifically, the sagittal T2W MRI should be done in the acute phase of SCI for prognostication of neurological outcome and clinical decision-making (Bozzo et al., 2011). The length of edema on sagittal T2W MRI increases proportionally with the delay in time of imaging in the acute phase (Leypold, Flanders, & Burns, 2008). After the acute phase, MRI sensitivity to detect ligamentous injury is also decreased (Kumar & Hayashi, 2016).

Recommended MRI protocol for acute spinal injury includes sagittal fast spin echo T1W, sagittal and axial T2W, sagittal and axial T2W gradient recalled echo (GRE) sequences of the primary injury site, and sagittal whole spine short tau inversion recovery (STIR) sequences for non-contiguous remote vertebral injuries (Chandra et al., 2012; Kumar & Hayashi, 2016). In the subacute phase, only targeted T1W and T2W images of the injury site are required for evaluation of post-operative complications or ascending neurological deficits (Chandra et al., 2012).

MRI findings in acute SCI

MRI reveals important features of acute SCI such as edema, hematoma, and compression as well as damage to extramedullary structures. These features are important when considering the acute management and deciding between surgical and nonsurgical options (Bozzo et al., 2011; Chandra et al., 2012). In obtundated patients with neurological deficit, MRI may be useful in picking up cervical injuries missed on CT scan and lead to surgical interventions (Bozzo et al., 2011). These findings are also integral to prognostication of SCI (Kulkarni, Bondurant, Rose, & Narayana, 1988).

Extra-medullary findings

Extra-medullary lesions are located within the dural sac but lie outside the spinal cord. MRI is important for evaluating the level and extent of extra-medullary lesions and spinal stability to guide clinical decision-making (Freund et al., 2019).

Ligamentous injury

Although CT displays most imaging findings of instability such as widening of the facet joints, it is unable to detect ligamentous injuries (Kumar & Hayashi, 2016). MRI, specially STIR, proves to be a more sensitive imaging modality for direct visualization and evaluation of soft tissue structures including subtle ligamentous injuries (Izzo et al., 2019; Kumar & Hayashi, 2016). Normal spinal ligaments are visualized as continuous bands of low signal intensity. Tears in ligaments can be either partial or complete; the former appears as hyperintense areas on STIR images due to injury-related edema and hemorrhage with variable intact fibers, while the latter differs in that there is a complete lack of intact fibers and hence appears discontinuous (Kumar & Hayashi, 2016).

Disc damage

Damage to intervertebral discs can be classified as an internal disc injury or disc herniation. Damage to the internal structure of the disc substance leading to edema or hemorrhage appears as an irregular narrowing or widening of a disk space with an abnormal high-intensity signal within the disc on T2W or STIR images (Dundamadappa & Cauley, 2012; Provenzale, 2007). MRIs are well suited for evaluating traumatic disc herniation due to a good signal contrast between the vertebral body, disc material, and cerebrospinal fluid. Additionally, sagittal plane imaging aids in determining if a fragment of the disc has been detached which is useful in surgical planning (Provenzale, 2007).

Other soft tissues

MR imaging also allows for visualization of prevertebral soft tissue injuries, which appears abnormally thickened due to associated edema and hemorrhage and is seen as a high intensity signal on T2-weighted or STIR images (Provenzale, 2007). Muscle strains are accompanied by edema which also appear as hyperintense signals on STIR images, whereas muscle hemorrhages present with a heterogeneous intensity signal due to a combination of hemorrhage and edema (Kumar & Hayashi, 2016).

Vascular injury

Typically, a normal artery demonstrates flow void on long TE sequences on an axial MRI. Hence, most vascular injuries will present as a disruption or loss of normal flow void on T2W images (Dundamadappa & Cauley, 2012; Kumar &

Hayashi, 2016). In the presence of an arterial thrombus or retarded flow, axial T2W images will demonstrate signal hyperintensity. Axial fat-suppressed T1W or T2W imaging along with MR angiography is better suited in cases of intramural hematoma associated with dissection for better luminal narrowing visualization and the hemorrhage is seen as a hyper intense signal on T1W or T2W images (Dundamadappa & Cauley, 2012).

Extramedullary hematoma

Spinal extradural hematomas are considered an emergency when they begin to compress the spinal cord. MRI is the preferred diagnostic tool compared to CT due to superior tissue contrast between the hematoma and adjacent soft tissues and structures (Izzo et al., 2019). Epidural hematomas will typically appear as an isointense to hyperintense signal on T1W images and hyperintense on T2W images (Fig. 1). The hematoma may also demonstrate heterogeneity on T2W imaging depending on its age (Ghasemi, Haddadi, & Shad, 2015; Sklar, Post, & Falcone, 1999). In addition, sagittal MRI allows visualization of the extent of the hematoma along the entire cranio-caudal axis (Izzo et al., 2019). Similarly, subdural hematoma and subarachnoid hemorrhage are also seen as varying signal intensities (Kumar & Hayashi, 2016; Sklar et al., 1999).

Fractures

Although CT is considered the imaging modality for diagnosing fractures, it is challenging to distinguish acute from chronic fractures. MRI can do so due to the presence of accompanying edema and hemorrhage in acute fractures. Bone marrow of acute fractures appears as hypointense signals on T1W images and hyperintense on T2W and STIR images, indicating marrow edema (Provenzale, 2007). In chronic fractures, the fatty marrow will produce an increased signal on both T1W and T2W images (Kumar & Hayashi, 2016). MR imaging can also distinguish various pathological fractures. Osteoporotic fractures show hypointense signal bands on T1W and T2W images and spared normal fatty marrow signal intensity along with a retropulsed posterior bone fragment. Metastatic compression fractures are characterized by an abnormal signal intensity involving the pedicles or posterior elements, convexity of the posterior vertebral wall and presence of epidural and paraspinal masses (Jung, Jee, McCauley, Ha, & Choi, 2003).

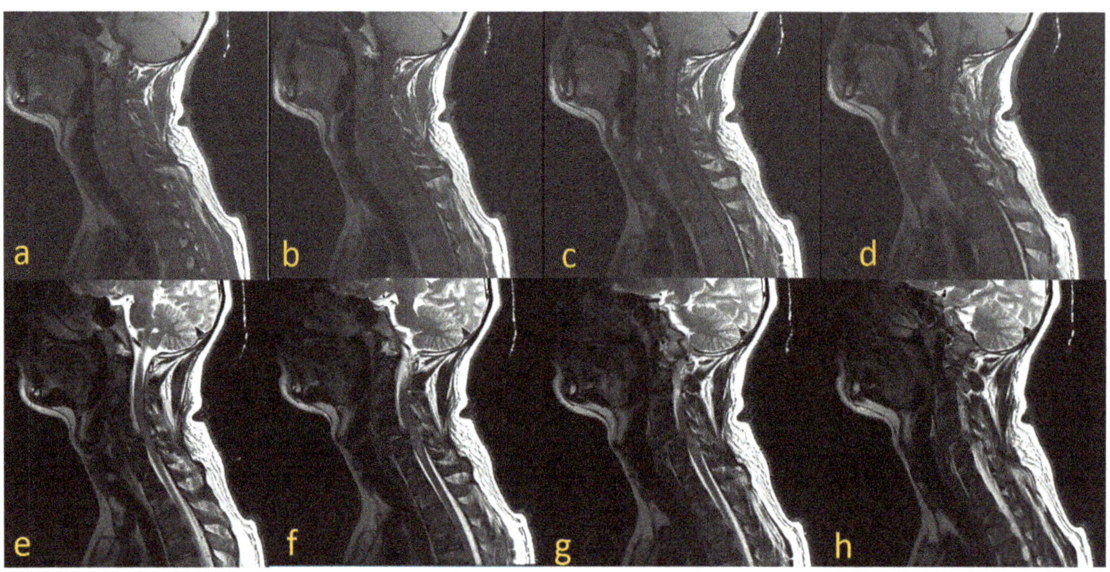

FIG. 1 Epidural hematoma and C5–C6 Spondylolisthesis with cord edema. (A)–(D) Sagittal T1W and (E)–(H) Sagittal T2W cervical spine show swollen C5–C6 intervertebral disc with heterogenous T2 hyperintense signals which is bulging posteriorly resulting in canal narrowing and mild cord compression. On T2W, heterogenous hyperintensity is observed at C5–C7 cord representing acute cord edema. Diffuse abnormal T2 hyperintense signals identified in epidural space from C2 to C4 level represent epidural hematoma.

Intramedullary findings

In the acute phase, cord may be affected with varying severity ranging from mild concussion to edema, hemorrhage, and even complete transection. MRI provides an excellent evidence base to determine the patient's severity level with a good radio–pathological correlation (Chandra et al., 2012).

Cord concussion

Spinal cord concussion is a temporary disturbance of the cord function present with or without vertebral damage, mostly associated with contact sports and whiplash injuries (Nesnídal, Stulík, & Barna, 2012). It often presents along with pre-existing vertebral abnormalities resulting in areas of hypermobility or narrowing of spinal cord with motor weakness and sensory impairment that resolves between two days to two months (Chandra et al., 2012). Cord concussion may show no signal change in either T1W or T2W images throughout the clinical course and has a very favorable outcome (Kumar & Hayashi, 2016; Nesnídal et al., 2012).

Cord compression

Spinal cord compression is an emergency which develops when the vertebral body exerts pressure on the spinal cord following traumatic impact or a pathological fracture. Timely diagnosis of cord compression and surgical decompression is required to resolve the neurological deficit (Alizadeh, Dyck, & Karimi-Abdolrezaee, 2019). The degree of cord compression and maximum canal stenosis at the site of injury are both linked to degree of neurological deficit and functional outcome (Miyanji, Furlan, Aarabi, Arnold, & Fehlings, 2007). On MRI, displaced or fractured vertebral bodies are seen compressing the cord which shows no signal change on T2W. However, a high signal within the soft tissues on sagittal STIR images is seen (Chandra et al., 2012).

Cord edema

Cord edema is the swelling of the spinal cord after spinal cord injury (Cho, Hachem, & Fehlings, 2017). Cord edema is not static; it occurs in the acute phase or often 2–3 days after the injury and may resolve after 14 days of injury (Cho et al., 2017; Leypold et al., 2008). Spinal cord edema spreads from the initial site of injury as one vertebral segment per 30 hours in the first three days post-SCI (Chandra et al., 2012; Cho et al., 2017). This can be detected as a high T2 signal and a normal or a low T1 signal inside the cord on MRI (Fig. 1) (Chandra et al., 2012). STIR images are particularly sensitive in detecting edema because of more uniform fat suppression (Kumar & Hayashi, 2016). The length of swelling is associated with the neurological presentation and outcome of the patient. Patients with transient edema have a favorable neurological outcome whereas worse baseline American Spinal Injury Association (ASIA) motor scores and neurological recovery are seen with persistent edema (Chandra et al., 2012).

Hemorrhagic contusion

Spinal cord injuries that occur as high-energy blunt trauma cause contusion injuries with crushed cord tissue and bleeding (Ju, Wang, Wang, & Zhao, 2014). On MRI, it is seen as a central intramedullary focus of low signal in the cord that is surrounded by a thick rim of hyperintense signals on T2W image (Fig. 2). Hemorrhage is best seen on T2W GRE images. However, T1W images can show subtle increase in signal denoting acute hemorrhage (Kulkarni et al., 1988). Presence of hemorrhage is associated with complete cord injury and worse ASIA motor score, and indicates poor prognosis (Chandra et al., 2012; Kumar & Hayashi, 2016). The extent of hemorrhage correlates with the outcome, a small hemorrhage of less than 4 mm has a better clinical and radiological improvement chances on follow-up imaging (Chandra et al., 2012).

Cord hematoma

Spinal cord hematoma is a severe and rare condition of the cord that often leads to permanent neurological deficit if not treated promptly (Kreppel, Antoniadis, & Seeling, 2003). On MRI, it shows up as a large (>4 mm) focus of intramedullary low signal that is surrounded by a thin rim of high signal, best seen on T2W GRE but also visible on T2W imaging (Chandra et al., 2012).

Cord ischemia

Trauma is one of the reasons for spinal cord ischemia with a poor prognosis. On MRI, axial sequences are useful to differentiate it from other causes (Vargas et al., 2015). Findings on MRI show up as hyperintense signal on T2W with cord

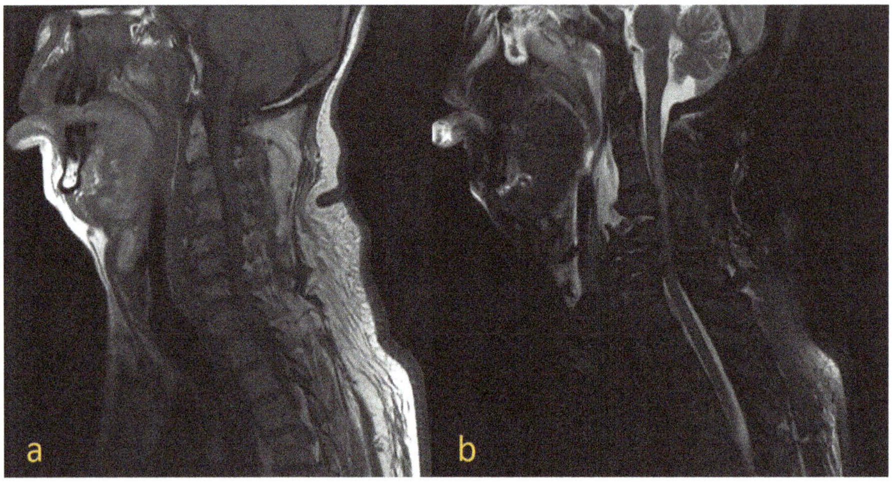

FIG. 2 Cord contusion. (A) Sagittal T1W and (B) Sagittal T2W cervical spine show prevertebral space collection extending from C2 to C6 vertebra. Compression fracture of C5 vertebra is seen with grade I retrolisthesis of C6 over C7 vertebra resulting in moderate spinal canal stenosis and severe bilateral radicular compression. On T2W, hyperintensity in the cord extending from C2–C3 intervertebral disc level to C6–C7 intervertebral disc level represents cord contusion.

enlargement (Chandra et al., 2012). Also, where traditional MRI techniques may fail to show any abnormalities, DWI shows promise in showing reduced apparent diffusion coefficient value in transverse plane specifying restricted diffusion (Chandra et al., 2012; Nogueira et al., 2012). Diffusion restriction correlates with unfavorable outcomes (Nedeltchev et al., 2004).

Cord transection

Spinal cord transection is the most severe and rare SCI, seen with unstable spinal fractures with associated spondylolisthesis or penetration injury with not much apparent bony injury (Baliyan et al., 2016). The former can be due to blast injury like gunshot or due to a direct tear (Chandra et al., 2012). It results in irreversible neurological impairment (Baliyan et al., 2016). On MRI, it appears as a complete disruption of the cord with high-signal CSF seen between the severed ends of the cord on T2W images (Fig. 3) (Mostafa, 2019).

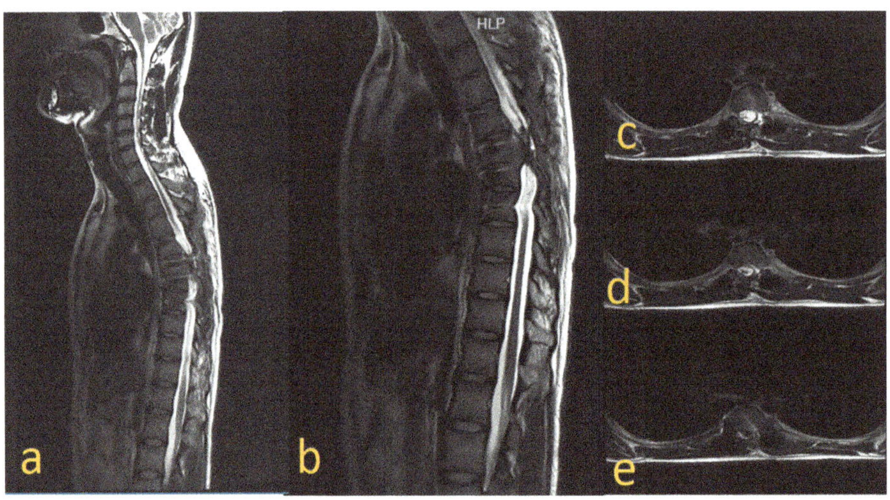

FIG. 3 Complete disruption of the dorsal cord. (A) Sagittal T2W whole spine (B) Sagittal T2W dorsal spine (C)–(E) Axial T2W dorsal spine show long segment of complete disruption of the cord extending from T5 to T9 vertebral levels. Multilevel fixation screws with residual kyphosis can be seen.

MRI findings in subacute SCI

Following the acute phase of injury, spinal cord lesions such as edema and hemorrhage evolve as well as begin to resolve as the cord is infiltrated by macrophages and blood–brain barrier starts to repair (Hausmann, 2003). In the first month post-injury, posttraumatic cyst begins to form, and small tissue bridges develop around the cyst (Huber, Lachappelle, Sutter, Curt, & Freund, 2017).

Subacute progressive ascending myelopathy

Subacute progressive ascending myelopathy (SPAM) is a rare SCI complication with an incidence of 1.8–6% (Planner, Pretorius, Graham, & Meagher, 2008). SPAM presents as delayed ascending neurological deterioration traumatic within 30 days of SCI and is often associated with early surgical interventions and infections (Planner et al., 2008). Although the exact cause is not well understood, it may develop due to mechanical insult, altered CSF flow, vascular injury or immune-mediated mechanisms (Zhang & Wang, 2017). Most patients show some recovery; however, it is associated with a 10% mortality rate (Zhang & Wang, 2017).

MRI shows typical progression of edema more than four segments cephalad the original injury in a tapered fashion, sometimes extending up till medulla oblongata (Zhang & Wang, 2017). Edema often subsides, but some edema remains above the initial level (Chandra et al., 2012). It shows characteristic sparing of the cord periphery while the hyperintense signal abnormality remains central on T2W image. SPAM shows heterogeneous intramedullary signal on T1W image along with an expanded spinal cord (Farooque, Kandwal, & Gupta, 2014; Zhang & Wang, 2017).

Early syrinx

Syrinx is a longitudinal fluid filled structure, as opposed to a small and localized posttraumatic cyst, formed in 3% of patients after a period of months to years post SCI. It may span many segments of the cord leading to progressive myelopathy (Ahuja et al., 2017). Syrinx has tapered cephalad and caudal extents which expand beyond the injured section and gives off isointense signals to CSF on sagittal T2W MRI (Chandra et al., 2012). In patients with Chiari 1 malformation, MRI shows compression of retro-cerebellar CSF spaces differentiating it from early syrinx due to SCI (Klekamp, 2018).

MRI findings in chronic SCI

The chronic phase of SCI spans months to years after onset of injury and it involves neuroplastic and degenerative changes in the spinal cord and brain. Despite conventional MRI's unprecedented importance in acute SCI management, it provides little information about the ongoing microstructural neurodegeneration and neuroplasticity in the chronic phase of injury (Freund et al., 2016). In contrast, quantitative MRI techniques like magnetization transfer, magnetic resonance relaxation mapping, and diffusion imaging have shown to play a vital role in detection of spinal cord atrophy, demyelination and iron deposition, painting a picture of progressive neurodegeneration (Freund et al., 2019). Tensor-based morphometry and multiparametric mapping (MPM) have been used as quantitative methods to measure volumetric changes and microstructural demyelination post SCI. This further aids clinicians in prognosticating the outcome of patients and exploiting the sensitivity and specificity of MRI findings as biomarkers in clinical trials (Freund et al., 2019).

Neurodegeneration and demyelination of the spinal cord post SCI

Features in chronic SCI seen on conventional MRI include myelomalacia, atrophy, and formation of cystic cavities and tethering of cord. Myelomalacia is seen on MRI as a poorly defined area of signal intensity between CSF and normal cord (Fig. 4) (Burks et al., 2019). Burks et al. observed changes in chronic (>1 year) SCI patients on sagittal T2W images and found myelomalacia as the most frequent finding, followed by cysts, septation, and cord tethering. Another study using T2W MRI showed a transition of acute changes (edema and hemorrhage) to subacute intramedullary lesion expansion in the span of a year. After resolution of the acute changes, a posttraumatic cyst appeared after a month of the injury and small tissue bridges were detected around the cyst. The quantification and location of these bridges on T2W MRI scans can serve as a prognostic tool as they provide valuable information on tract-specific electrophysiology and long-term functional recovery (Huber et al., 2017).

In a prospective longitudinal study, signs of remote spinal cord atrophy at the C2/C3 level were observed on T1W images over one year. Compared to controls, patients had a significantly greater rate of change in spinal cord area (decrease

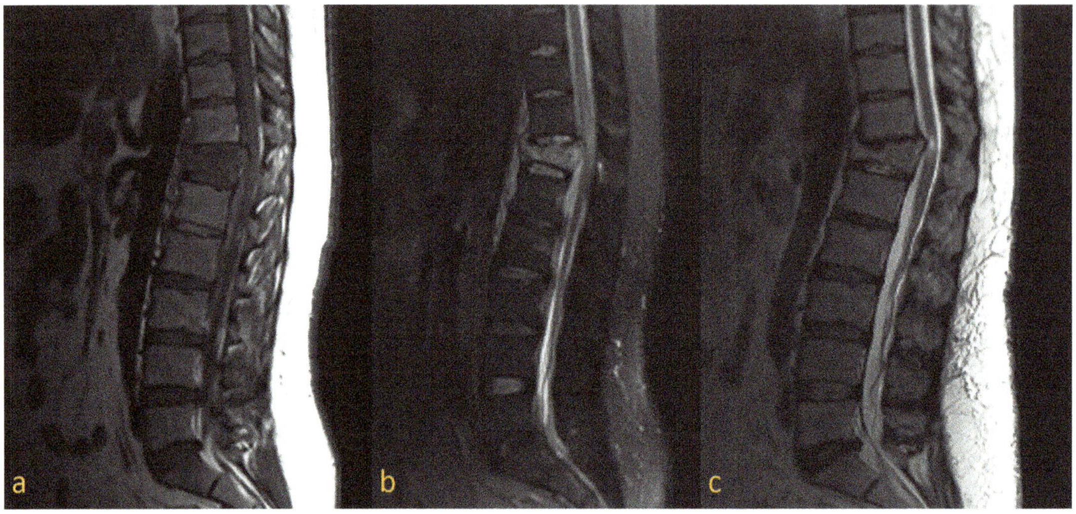

FIG. 4 D12 fracture secondary to road traffic accident. (A) Sagittal T1W Lumber spine and (B) Sagittal T2W Lumber spine show compression fracture of D12 vertebra. This is associated with retropulsed bony fragment, which is causing significant pressure over the anterior theca at this level. T2-hypointense signals at this level indicate cord edema. (C) Sagittal T2W Lumber spine of the same patient after 8 years show hyperintense T2 signals suggesting focal myelomalacia at D12 level.

of 0·46 mm^2 per month) (Freund et al., 2013). Huber et al. showed evidence of remote neurodegeneration in dorsal horns, ventral horns, and white matter at cervical spinal cord in patients with chronic spinal cord injury on T2W images. Atrophy of dorsal and ventral horns correlated clinically with sensory and motor outcomes, respectively (Huber et al., 2018). It is still unclear whether the rate of spinal cord atrophy is linked to level of the lesion or severity of the injury (Freund et al., 2019). MPM studies demonstrated that spinal cord atrophy was also paralleled with an increase in iron content and decreased myelin content (Ziegler et al., 2018). Interestingly, histological analysis also showed that the iron accumulated in phagocytic macrophages 28–35 days after spinal cord injury (Freund et al., 2019).

Altered DTI metrics have been observed at injury level as well as remotely, specifically in the corticospinal tract and dorsal columns corresponding to axonal degeneration in acute and chronic SCI (David, Freund, & Mohammadi, 2017). Similar DTI patterns have been noted in in the lumbar enlargement indicating trans-synaptic degeneration of motor neuron pools deprived of input (Koskinen et al., 2014). Ellingson et al. showed low fractional anisotropy (FA) values at injury level, these values were directly proportional to ASIA classification. While FA values were low throughout the cord compared to controls, a sharp drop in FA at the injury site implies utility of DTI to localize the lesion (Ellingson et al., 2008). A decreased fractional anisotropy and an increased ADC are associated with neuropathological changes like axonal loss, demyelination, gliosis, astrocytic scarring and an increase in extracellular matrix (Pierpaoli et al., 2001). Moreover, fMRI studies of the cord show preservation of task-related spinal activities above and below the site of injury (Freund et al., 2019).

Neurodegeneration of the brain post-SCI

Studies have shown a link between brain atrophy and SCI most likely due to retrograde degeneration and demyelination of tracts (Freund et al., 2019). Brain atrophy after spinal cord injury is particularly prominent across the cranial projections of the corticospinal tracts, primary motor cortex, insula, anterior cingulate gyrus, and thalamus (Fig. 5) (Freund et al., 2013; Grabher et al., 2015; Ziegler et al., 2018). Parallel analysis using MPM related these volume changes to demyelination of axons and their cell bodies (Freund et al., 2013). Volumetric changes in structural T1W MRI have been linked with clinical outcomes and can be used to monitor treatment effects. Significant volume reductions in the brainstem have been associated with poorer prognosis for lower limb motor function recovery; whereas lower limb training in patients with SCI showed increases in the brain volume owing to reorganization processes (Villiger et al., 2015).

DTI studies in chronic SCI patients also show altered metrics in the white matter of the brain reflecting compromised structural connectivity (Zaninovich et al., 2019). Statistically significant differences in DTI parameters are observed between chronic SCI injury patients and healthy controls in areas such as posterior area of the centrum semiovale, corona radiata, cerebral peduncles, and internal capsule (Koskinen et al., 2014; Zaninovich et al., 2019). Functional reorganization of cortical sensorimotor networks in chronic SCI can be investigated using fMRI techniques in order to understand the process of plasticity and recovery in these patients (Freund et al., 2019).

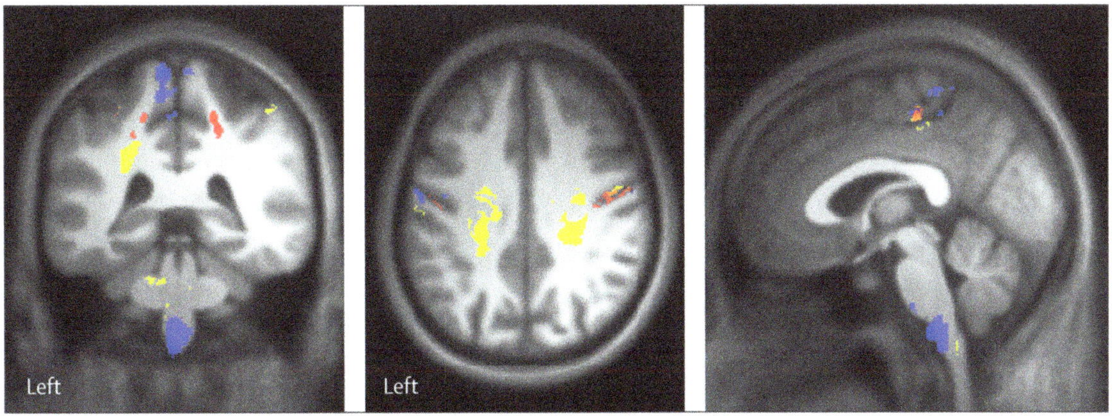

FIG. 5 Changes at 12 months shown by voxel-wise analysis of microstructure and volume. Overlay of statistical parametric maps (uncorrected $P < 0{\cdot}001$, shown for descriptive purposes, masked by the union of the cranial corticospinal tract and the bilateral sensorimotor cortex) showing regions of reduced volume of grey and white matter (*blue*) in patients compared with controls at 12-month follow-up. The reductions in longitudinal relaxation rate R1 (*yellow*) and magnetization transfer MT (*red*) in patients compared with controls at 12-month follow-up suggest changes in myelination. *Courtesy Freund, P., Weiskopf, N., Ashburner, J., Wolf, K., Sutter, R., Altmann, D. R., et al. (2013). MRI investigation of the sensorimotor cortex and the corticospinal tract after acute spinal cord injury: A prospective longitudinal study.* The Lancet Neurology, 12 *(9), 873–881; under Creative Commons.*

Thus, advanced MRI techniques show huge potential in allowing clinicians to plan and monitor treatment protocols by getting valuable microstructural information.

Conclusion

MRI plays a pivotal role in determining the underlying cause and extent of neurological deficits in the acute phase of SCI for a prompt diagnosis and prognostication. Conventional MRI sequences elaborate findings including extramedullary damage, edema, hemorrhage, and compression of the cord that are present at the time of injury as well as syrinx, SPAM, atrophy, and myelomalacia which develop in the subacute and chronic phases of injury. Advanced quantitative MRI techniques show microstructural changes that occur in the long term and reflect degeneration and plasticity, offering potential as biomarkers to monitor treatment and evaluate outcomes in clinical trials. There is a need to develop methods to overcome the high cost associated with MRI. Currently, advanced techniques like DTI, MPM, and fMRI are used in research only. Enhanced signal-to-noise ratio and machine learning approaches for consistent results will enable translation of these techniques from clinical trials to the bedside management of SCI.

Applications to other areas of neuroscience

MRI supersedes all other imaging modalities when it comes to investigating the structure of spinal cord and associated soft structures such as nerve roots and ligaments. Examples of spinal disorders with a central role for MRI include nontraumatic spinal cord emergencies, inflammatory disorders and spinal cord neoplasms. Nontraumatic spinal cord emergencies requiring MRI studies include extramedullary compressive causes such as epidural abscess and intramedullary etiologies such as transverse myelitis and spinal cord infarct. MRI findings can give clues regarding the diagnosis and guide on the level of compression for treatment planning (Flanagan & Pittock, 2017). Inflammatory conditions such as multiple sclerosis and neuromyelitis optica (NMO) show hyperintense signals in the cord on T2W MRI (Ciccarelli et al., 2019). Atrophy and myelomalacia are also seen in NMO (Ciccarelli et al., 2019). Contrary to SCI, gadolinium contrast enhancement is an important feature in these conditions (Bozzo et al., 2011; Ciccarelli et al., 2019). In spinal tumors, MRI findings have accuracy in determining the site, i.e., extradural, intradural, extramedullary, or intramedullary, as well as tissue characterization aided with contrast enhancement (Yadav et al., 2016). In COVID—19-associated acute necrotizing myelitis, T2W MRI shows hyperintense lesion involving medulla and cervical cord with patchy enhancing lesions (Sotoca & Rodríguez-Álvarez, 2020). Quantitative MRI has useful information on degeneration and comprises an important part of work-up and research for other degenerative disorders. For example, MPM of spinal cord in Amyotrophic Lateral Sclerosis enabled quantification of cord atrophy which can predict disease progression (El Mendili et al., 2014). Similarly, in chronic traumatic brain injury (TBI), quantification of grey matter volumes using T1W scans and correlating these with neuropsychological testing helps in understanding the neural correlates of cognitive impairment experienced in chronic TBI (Levine et al., 2013).

Mini-dictionary of terms

Conventional MRI is standard MRI imaging which includes T1W and T2W images.
Quantitative MRI is an imaging technique which obtains maps of meaningful physical or chemical variables that can be measured in physical units and compared between tissue regions and among subjects like magnetic resonance relaxation times, magnetization transfer, and diffusion of water molecules.
Cord concussion is a type of mild spinal cord injury which manifests as transient paraplegia, variable degrees of sensory impairment and motor weakness which typically resolve within 72 h.
Cord compression is a condition that occurs due to external pressure on the spinal cord often due to a malignant growth or disc herniation. This often leads to pain, numbness and weakness symptoms below the level of the compression.
Hemorrhagic contusion of the spinal cord is a discrete focus of hemorrhage within the substance of the spinal cord often secondary to a crush injury. This most commonly is present within the central grey matter of the spinal cord at the point of impact.
Cord transection is a transverse cut within the spinal cord due to significant traumatic injury. This leads to permanent loss of central control of motor, sensory, and autonomic functions below the level of injury.
Sub-acute progressive ascending myelopathy is a sequela of spinal cord injury starting within the first few weeks of the injury. There is ascending neurological deterioration till at least four vertebral levels above the level of injury.
Syrinx is a fluid filled central cavity within the spinal cord.

Key facts of SPAM

- Subacute progressive ascending myelopathy (SPAM) is a rare complication arising within 3 weeks of spinal cord injury, presenting with ascending neurological deficit.
- It was first described by Frankel in 1969 with an incidence of 1% in 808 admissions with SCI.
- It mostly affects young and middle-aged patients with a ~5.1 male to female ratio.
- MRI is the gold standard imaging and shows cord changes at least four vertebral levels above the initial site of insult.
- It has been postulated to arise from venous thrombosis, arterial occlusion, fibrocartilage embolism, CSF flow disruption, mechanical insult, infection, and immune-mediated response.
- Risk factors include complete SCI, initial injury site at thoracolumbar junction, asymptomatic low blood pressure and early postoperative orthostatic mobilization.
- It is associated with a mortality of 9%. Recovery is generally present, but this improvement is up to a vertebral level higher than original site of injury.
- MRI shows resolution of changes in recovered patients with a small focal region of myelomalacia seen at the level of initial insult and vertebral level above it.

Summary points

- MRI is the modality of choice for imaging soft tissue in SCI.
- Conventional T1W and T2W MRI along with newer GRE and STIR sequences reveal important characteristics associated with intramedullary as well as extramedullary spinal cord injuries.
- In the acute phase, these include edema, hemorrhage, hematoma, and compression. While in the subacute phase, edema is seen to resolve with formation of posttraumatic cysts and syrinx.
- Novel quantitative MRI techniques including volumetry, DTI and magnetization transfer demonstrate microstructural changes in the chronic phase of SCI.
- These changes indicate neurodegeneration, providing important information regarding spinal cord atrophy, demyelination, and iron deposition. Thus, these advanced techniques have a potential role in neurorehabilitation and clinical trials.

References

Ahuja, C. S., Wilson, J. R., Nori, S., Kotter, M. R., Druschel, C., Curt, A., et al. (2017). Traumatic spinal cord injury. *Nature Reviews. Disease Primers*, *3*(1), 1–21.

Alizadeh, A., Dyck, S. M., & Karimi-Abdolrezaee, S. (2019). Traumatic spinal cord injury: An overview of pathophysiology, models and acute injury mechanisms. *Frontiers in Neurology*, *10*, 282.

Baliyan, V., Shylendran, S., Ajay, K. Y., Kumar, A., Gamanagatti, S., & Sinha, S. (2016). Unusual cord transection in a patient with traumatic spondylolisthesis. *Asian Journal of Neurosurgery*, *11*(1), 72.

Bozzo, A., Marcoux, J., Radhakrishna, M., Pelletier, J., & Goulet, B. (2011). The role of magnetic resonance imaging in the management of acute spinal cord injury. *Journal of Neurotrauma*, *28*(8), 1401–1411.

Burks, J. D., Gant, K. L., Guest, J. D., Jamshidi, A. G., Cox, E. M., Anderson, K. D., et al. (2019). Imaging characteristics of chronic spinal cord injury identified during screening for a cell transplantation clinical trial. *Neurosurgical Focus*, *46*(3), E8.

Chandra, J., Sheerin, F., De Heredia, L. L., Meagher, T., King, D., Belci, M., et al. (2012). MRI in acute and subacute post-traumatic spinal cord injury: Pictorial review. *Spinal Cord*, *50*(1), 2–7.

Cheran, S., Shanmuganathan, K., Zhuo, J., Mirvis, S. E., Aarabi, B., Alexander, M. T., et al. (2011). Correlation of MR diffusion tensor imaging parameters with ASIA motor scores in hemorrhagic and nonhemorrhagic acute spinal cord injury. *Journal of Neurotrauma*, *28*(9), 1881–1892.

Cho, N., Hachem, L. D., & Fehlings, M. G. (2017). Spinal cord edema after spinal cord injury: From pathogenesis to management. In *Brain edema* (pp. 261–275). Elsevier.

Ciccarelli, O., Cohen, J. A., Reingold, S. C., Weinshenker, B. G., Amato, M. P., Banwell, B., et al. (2019). Spinal cord involvement in multiple sclerosis and neuromyelitis optica spectrum disorders. *The Lancet Neurology*, *18*(2), 185–197.

David, G., Freund, P., & Mohammadi, S. (2017). The efficiency of retrospective artifact correction methods in improving the statistical power of between-group differences in spinal cord DTI. *NeuroImage*, *158*, 296–307.

Dundamadappa, S. K., & Cauley, K. A. (2012). MR imaging of acute cervical spinal ligamentous and soft tissue trauma. *Emergency Radiology*, *19*(4), 277–286.

El Mendili, M.-M., Cohen-Adad, J., Pelegrini-Issac, M., Rossignol, S., Morizot-Koutlidis, R., Marchand-Pauvert, V., et al. (2014). Multi-parametric spinal cord MRI as potential progression marker in amyotrophic lateral sclerosis. *PLoS One*, *9*(4), e95516.

Ellingson, B., Ulmer, J., Kurpad, S., & Schmit, B. (2008). Diffusion tensor MR imaging in chronic spinal cord injury. *American Journal of Neuroradiology*, *29*(10), 1976–1982.

Farooque, K., Kandwal, P., & Gupta, A. (2014). Subacute post-traumatic ascending myelopathy (SPAM): Two cases of SPAM following surgical treatment of thoracolumbar fractures. *Neurology India*, *62*(2), 192.

Flanagan, E., & Pittock, S. J. (2017). Diagnosis and management of spinal cord emergencies. In *Vol. 140. Handbook of clinical neurology* (pp. 319–335). Elsevier.

Freund, P., Friston, K., Thompson, A. J., Stephan, K. E., Ashburner, J., Bach, D. R., et al. (2016). Embodied neurology: An integrative framework for neurological disorders. *Brain*, *139*(6), 1855.

Freund, P., Seif, M., Weiskopf, N., Friston, K., Fehlings, M. G., Thompson, A. J., et al. (2019). MRI in traumatic spinal cord injury: From clinical assessment to neuroimaging biomarkers. *The Lancet Neurology*, *18*(12), 1123–1135.

Freund, P., Weiskopf, N., Ashburner, J., Wolf, K., Sutter, R., Altmann, D. R., et al. (2013). MRI investigation of the sensorimotor cortex and the corticospinal tract after acute spinal cord injury: A prospective longitudinal study. *The Lancet Neurology*, *12*(9), 873–881.

Ghasemi, A., Haddadi, K., & Shad, A. A. (2015). Comparison of diagnostic accuracy of MRI with and without contrast in diagnosis of traumatic spinal cord injuries. *Medicine*, *94*(43), e1942.

Grabher, P., Callaghan, M. F., Ashburner, J., Weiskopf, N., Thompson, A. J., Curt, A., et al. (2015). Tracking sensory system atrophy and outcome prediction in spinal cord injury. *Annals of Neurology*, *78*(5), 751–761.

Hausmann, O. B. (2003). Post-traumatic inflammation following spinal cord injury. *Spinal Cord*, *41*(7), 369–378.

Huber, E., David, G., Thompson, A. J., Weiskopf, N., Mohammadi, S., & Freund, P. (2018). Dorsal and ventral horn atrophy is associated with clinical outcome after spinal cord injury. *Neurology*, *90*(17), e1510–e1522.

Huber, E., Lachappelle, P., Sutter, R., Curt, A., & Freund, P. (2017). Are midsagittal tissue bridges predictive of outcome after cervical spinal cord injury? *Annals of Neurology*, *81*(5), 740–748.

Izzo, R., Popolizio, T., Balzano, R. F., Pennelli, A. M., Simeone, A., & Muto, M. (2019). Imaging of cervical spine traumas. *European Journal of Radiology*, *117*, 75–88.

Ju, G., Wang, J., Wang, Y., & Zhao, X. (2014). Spinal cord contusion. *Neural Regeneration Research*, *9*(8), 789.

Jung, H. S., Jee, W. H., McCauley, T. R., Ha, K. Y., & Choi, K. H. (2003). Discrimination of metastatic from acute osteoporotic compression spinal fractures with MR imaging. *Radiographics*, *23*(1), 179–187.

Klekamp, J. (2018). How should syringomyelia be defined and diagnosed? *World Neurosurgery*, *111*, e729–e745.

Koskinen, E., Hakulinen, U., Brander, A., Luoto, T., Ylinen, A., & Öhman, J. (2014). Clinical correlates of cerebral diffusion tensor imaging findings in chronic traumatic spinal cord injury. *Spinal Cord*, *52*(3), 202–208.

Kreppel, D., Antoniadis, G., & Seeling, W. (2003). Spinal hematoma: A literature survey with meta-analysis of 613 patients. *Neurosurgical Review*, *26*(1), 1–49.

Kulkarni, M. V., Bondurant, F. J., Rose, S., & Narayana, P. A. (1988). 1.5 tesla magnetic resonance imaging of acute spinal trauma. *Radiographics*, *8*(6), 1059–1082.

Kumar, Y., & Hayashi, D. (2016). Role of magnetic resonance imaging in acute spinal trauma: A pictorial review. *BMC Musculoskeletal Disorders*, *17*(1), 310.

Levine, B., Kovacevic, N., Nica, E. I., Schwartz, M. L., Gao, F., & Black, S. E. (2013). Quantified MRI and cognition in TBI with diffuse and focal damage. *NeuroImage: Clinical*, *2*, 534–541.

Leypold, B., Flanders, A., & Burns, A. (2008). The early evolution of spinal cord lesions on MR imaging following traumatic spinal cord injury. *American Journal of Neuroradiology*, *29*(5), 1012–1016.

Miyanji, F., Furlan, J. C., Aarabi, B., Arnold, P. M., & Fehlings, M. G. (2007). Acute cervical traumatic spinal cord injury: MR imaging findings correlated with neurologic outcome—Prospective study with 100 consecutive patients. *Radiology*, *243*(3), 820–827. https://doi.org/10.1148/radiol.2433060583.

Mostafa, M. A. (2019). Traumatic cervical spinal cord transection. *BJR Case Reports*, *5*(1), 20180043.

Nedeltchev, K., Loher, T. J., Stepper, F., Arnold, M., Schroth, G., Mattle, H. P., et al. (2004). Long-term outcome of acute spinal cord ischemia syndrome. *Stroke*, *35*(2), 560–565.

Nesnídal, P., Stulík, J., & Barna, M. (2012). Spinal cord concussion: A retrospective study of twenty-four patients. *Acta Chirurgiae Orthopaedicae et Traumatologiae Cechoslovaca*, *79*(2), 150–155.

Nogueira, R. G., Ferreira, R., Grant, P. E., Maier, S. E., Koroshetz, W. J., Gonzalez, R. G., et al. (2012). Restricted diffusion in spinal cord infarction demonstrated by magnetic resonance line scan diffusion imaging. *Stroke*, *43*(2), 532–535.

Pierpaoli, C., Barnett, A., Pajevic, S., Chen, R., Penix, L., Virta, A., et al. (2001). Water diffusion changes in Wallerian degeneration and their dependence on white matter architecture. *NeuroImage*, *13*(6), 1174–1185.

Planner, A., Pretorius, P., Graham, A., & Meagher, T. (2008). Subacute progressive ascending myelopathy following spinal cord injury: MRI appearances and clinical presentation. *Spinal Cord*, *46*(2), 140–144.

Provenzale, J. (2007). MR imaging of spinal trauma. *Emergency Radiology*, *13*(6), 289–297.

Seif, M., Wheeler-Kingshott, C. A. G., Cohen-Adad, J., Flanders, A. E., & Freund, P. (2019). Guidelines for the conduct of clinical trials in spinal cord injury: Neuroimaging biomarkers. *Spinal Cord*, *57*(9), 717–728.

Shen, H., Tang, Y., Huang, L., Yang, R., Wu, Y., Wang, P., et al. (2007). Applications of diffusion-weighted MRI in thoracic spinal cord injury without radiographic abnormality. *International Orthopaedics*, *31*(3), 375–383.

Sklar, E. M. L., Post, J. M. D., & Falcone, S. (1999). MRI of acute spinal epidural hematomas. *Journal of Computer Assisted Tomography*, *23*(2), 238–243.

Sotoca, J., & Rodríguez-Álvarez, Y. (2020). COVID-19-associated acute necrotizing myelitis. *Neurology® Neuroimmunology & Neuroinflammation*, *7*(5), e803.

Vallotton, K., Huber, E., Sutter, R., Curt, A., Hupp, M., & Freund, P. (2019). Width and neurophysiologic properties of tissue bridges predict recovery after cervical injury. *Neurology*, *92*(24), e2793–e2802.

Vargas, M., Gariani, J., Sztajzel, R., Barnaure-Nachbar, I., Delattre, B., Lovblad, K., et al. (2015). Spinal cord ischemia: Practical imaging tips, pearls, and pitfalls. *American Journal of Neuroradiology*, *36*(5), 825–830.

Villiger, M., Grabher, P., Hepp-Reymond, M.-C., Kiper, D., Curt, A., Bolliger, M., et al. (2015). Relationship between structural brainstem and brain plasticity and lower-limb training in spinal cord injury: A longitudinal pilot study. *Frontiers in Human Neuroscience*, *9*, 254.

Yadav, P., Sharma, A., Sikarwar, J., Goyal, S., Soni, V. K., & Chandravanshi, M. (2016). Magnetic resonance imaging in spinal tumours and its histopathological correlation. *Journal of Evolution of Medical and Dental Sciences*, *5*(104), 7616–7622.

Zaninovich, O. A., Avila, M. J., Kay, M., Becker, J. L., Hurlbert, R. J., & Martirosyan, N. L. (2019). The role of diffusion tensor imaging in the diagnosis, prognosis, and assessment of recovery and treatment of spinal cord injury: A systematic review. *Neurosurgical Focus*, *46*(3), E7.

Zhang, J., & Wang, G. (2017). Subacute posttraumatic ascending myelopathy: A literature review. *Spinal Cord*, *55*(7), 644–650.

Ziegler, G., Grabher, P., Thompson, A., Altmann, D., Hupp, M., Ashburner, J., et al. (2018). Progressive neurodegeneration following spinal cord injury: Implications for clinical trials. *Neurology*, *90*(14), e1257–e1266.

Chapter 3

Exercise programs and spinal cord injury (SCI): Linking the clinical, physiological, and psychological consequences of SCI

Jeongmin Lee[a], Dong-il Kim[b], and Justin Y. Jeon[c]

[a]*Department of Sport Industry Studies, Yonsei University, Seoul, Korea,* [b]*Division of Health and Kinesiology, Incheon National University, Incheon, Korea,* [c]*Department of Sport Industry Studies & Department of Medicine, Cancer Prevention Center, Shinchon Severance Hospital, Yonsei University, Seoul, Korea*

List of abbreviations

ANS	autonomic nervous system
CVD	cardiovascular disease
FES	functional electrical stimulation
FM	fat mass
HDL-C	high-density lipoprotein cholesterol
LB	lower body
LDL-C	low-density lipoprotein cholesterol
PA	physical activity
QoL	quality of life
SCI	spinal cord injury
TG	triglycerides
UB	upper body
WB	whole body

Prevalence and consequences of spinal cord injury

A spinal cord injury (SCI) is a major life event leading to serious physical disability and secondary clinical problems (Kang et al., 2017). Worldwide, the incidences of SCI are rising with an annual estimate of 10.4–83 cases per million worldwide (Jazayeri, Beygi, Shokraneh, Hagen, & Rahimi-Movaghar, 2015). SCI management requires significant healthcare and can place a substantial financial burden through rehabilitation services, expensive personal assistance, lost productivity due to disability, and social isolation. Given the effects of SCI, there is a pressing need to identify effective methods to manage these injuries and reduce the extent of future disability.

Clinical, physiological, and psychological consequences of SCI (Fig. 1)

Primary consequences of SCI: Paralysis and ANS

Paralysis of the voluntary musculature is the most obvious consequence of SCI (Maynard, Karunas, & Waring, 1990). Damage to the descending motor tracts, anterior horn cells, or nerve roots leads to an impaired capacity to contract the skeletal muscles at or below the level of the lesion (Assinck, Duncan, Hilton, Plemel, & Tetzlaff, 2017). This paralysis results in a loss of control over the trunk and extremities. Because muscles below injury levels are denervated, these paralyzed muscles undergo rapid changes, including the loss of muscle mass as well as biochemical and morphological changes. These biochemical and morphological changes include the loss of capillarization and changes in the muscle fiber

Cellular, Molecular, Physiological, and Behavioral Aspects of Spinal Cord Injury. https://doi.org/10.1016/B978-0-12-822427-4.00003-4

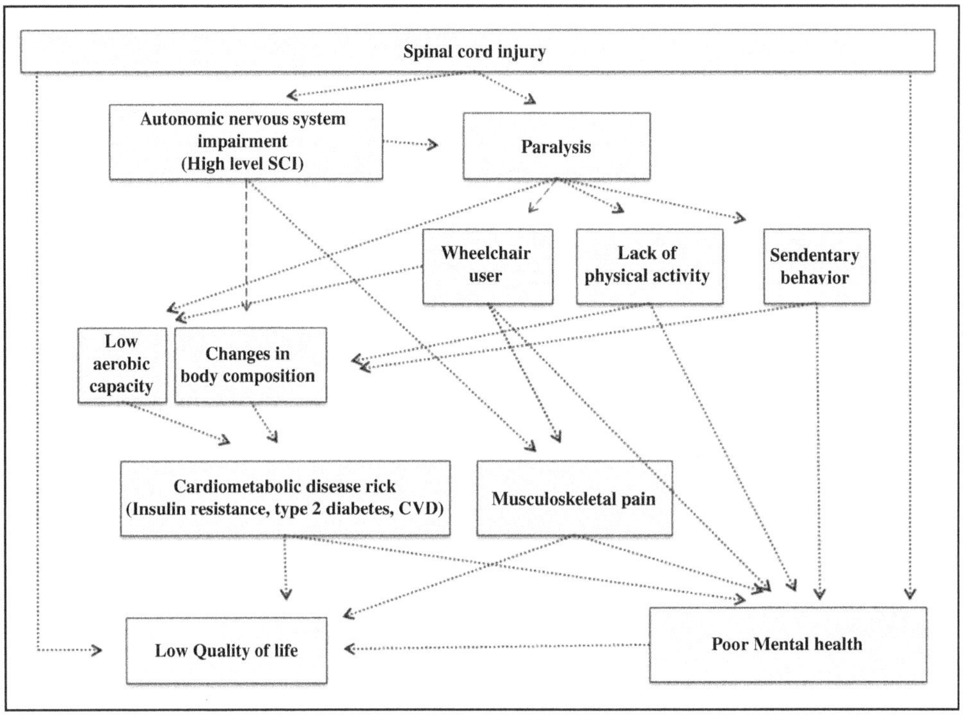

FIG. 1 Overview of the pathophysiology and health consequences after SCI. SCI is an acquired disorder that leads to dramatic changes in a person's lifestyle and requires continuous observation and treatment. Due to paralysis and ANS dysfunction, people with SCI have lower physical activity and higher sedentary behavior as well as increased risk of chronic diseases compared to the able-bodied. Also, people with spinal cord injuries use their shoulders excessively because they use wheelchairs for transportation. For this reason, most experience shoulder joint problems such as rotator cuff tears and subacromial impingement syndrome. Besides an increased risk of developing physical health consequences, people with SCI also have mental health problems such as fatigue, anxiety, and depression.

type from oxidative to glycolytic (i.e., Type 1 to IIb) (Burnham et al., 1997). Interestingly, both reductions in muscle loss and changes in muscle characteristics are associated with poor cardiometabolic profiles, most importantly insulin resistance (Gater, Farkas, & Tiozzo, 2021).

In addition to devastating paralysis, SCI also results in significant dysfunctions of the ANS (Wang et al., 2000). The ANS regulates many functions, in particular cardiovascular functions such as coronary blood flow, cardiac contractility, heart rate, and peripheral vasomotor responses (Garstang & Miller-Smith, 2007). In addition, ANS controls blood flow to the skeletal muscles, kidneys, splanchnic circulation, and skin. Impaired ANS regulation caused by SCI leads to many of the clinical issues seen in people with SCI. SCI above T1 causes a complete disruption of the sympathetic pathways and results in a variety of problems, including bradycardia, neurogenic pulmonary edema, arrhythmias, and hypotension (Sweis & Biller, 2017). SCI above T6 causes an altered splanchnic outflow, which causes hypotension and altered vascular regulation (Garstang & Miller-Smith, 2007). Even people with injuries below T6 have changes in cardiovascular response as a consequence of the altered ANS regulation (Garstang & Miller-Smith, 2007). Thus, the role of autonomic dysfunction in people with SCI is crucially important when designing an exercise program for them because many aspects of the altered physiology seen in these individuals are directly caused by ANS dysregulation.

Secondary consequences of SCI

Physical inactivity and sedentary behavior

Due to the loss of motor, sensory, and ANS below the lesion level, people with SCI are often restricted in their activity of daily living. Most people with SCI are considered to be at high risk for an inactive lifestyle (Fernhall, Heffernan, Jae, & Hedrick, 2008). One prospective study on the course of everyday physical activity (PA) levels with 40 people with SCI reported that the PA levels increased during inpatient rehabilitation, but shortly after discharge from the rehabilitation center, the level decreased significantly (van den Berg-Emons et al., 2008). Furthermore, Van den Berg-Emons et al. reported that the PA levels of people with SCI were only 40% of the PA levels of able-bodied peers and were even lower than those of people with other chronic diseases (van den Berg-Emons, Bussmann, & Stam, 2010). Besides insufficient PA, sedentary time is much greater, which poses a greater threat to the health of people with SCI. One cross-sectional study with 695 people with SCI (male = 531, female = 164) reported that an estimated 50% of people with SCI live completely sedentary lives (Ginis et al., 2010). These findings may especially apply to people with SCI who are often wheelchair bound and spend large amounts of time sitting compared to the general population.

Body composition and poor metabolic profile

The clinical consequences of SCI, paired with insufficient PA and sedentary behavior, often result in a reduction in muscle mass and an increase in FM. Compared to the general population, people with SCI show increased levels of intramuscular fat at 6 weeks post-injury, and this continues to increase over the next 3 months (Gorgey & Dudley, 2007). A loss of skeletal muscle in people with SCI could be the result of paralysis, but it could also be secondary from a lack of PA and exercise. Because skeletal muscle mass is the most important determinant of the resting metabolic rate, a reduction in muscle mass increases the risk of obesity (Zurlo, Larson, Bogardus, & Ravussin, 1990). Spungen et al. demonstrated that people with SCI have an average of 5 kg more FM and 50% more total body fat while also being 13.1% fatter per unit of body mass index (BMI) compared to able-bodied controls matched by age, height, and ethnicity (Spungen et al., 2003). Knowing that an intact SNS is required for the anti-obesity hormone leptin, people with SCI may have a greater risk of obesity and obesity-associated cardiometabolic diseases (Jeon et al., 2003). Indeed, the deterioration of body composition after SCI such as an increase in FM and a loss of lean mass have been associated with a higher risk of type 2 diabetes (Bauman & Spungen, 2001). Furthermore, many individuals with SCI have increased levels of triglycerides (TG), low-density lipoprotein cholesterol (LDL-C), and decreased high-density lipoprotein cholesterol (HDL-C) (Bauman et al., 1992; Bauman & Spungen, 2001). Bauman et al. reported that more than 37% of individuals with SCI have less than 35 mg/dL of HDL-C. Also, 18% of this population has an LDL-C level greater than 160 mg/dL (Bauman et al., 1992). This may explain why people with SCI are at a higher risk of developing type 2 diabetes and cardiovascular disease (CVD) than the general population.

Poor musculoskeletal conditions

As a consequence of lower extremities weakness or paralysis after SCI, the increased use of the upper extremities has been associated with high incidences of musculoskeletal pain in the shoulder, elbow, and wrist with a prevalence ranging from 5% to 78% among people with SCI (Capoor & Stein, 2005). After SCI, they are forced to bear weight on the upper extremity joints for mobility and transferring, pushing up, and hand-propelling the wheelchair (Riek, Ludewig, & Nawoczenski, 2008). Previous studies have reported that musculoskeletal pain causes poorer psychological functioning and has a negative impact on the quality of life (QoL) among people with SCI (Ullrich et al., 2013). Therefore, it is important to set priorities and design effective programs to manage musculoskeletal pain and improve the QoL in people with SCI.

Poor mental health and low QoL

While muscle paralysis may be the most obvious consequence of SCI, people with SCI are susceptible to several functional complications such as sexual dysfunction, spontaneous urine loss, incontinence, and pain, which lead to considerable changes in the QoL (Ahuja et al., 2017). Furthermore, people with SCI have significantly elevated risks of chronic fatigue, anxiety, and depression (Le & Dorstyn, 2016). A recent meta-analysis indicated that the prevalence of depression after SCI is substantially greater than that in the general medical population (Williams & Murray, 2015). Consequently, people with SCI have a higher prevalence of psychological morbidity and suicide (Kennedy & Garmon-Jones, 2017). Therefore, seeking ways to improve the psychological well-being in people with SCI is of importance. In this regard, participation in PA and exercise might be an effective strategy to improve the psychological well-being in people with SCI.

Exercise and SCI (Fig. 2)

As mentioned previously, SCI may result in various health consequences. Some of these health consequences are primary consequences of paralysis (ANS, muscle fiber type, muscle mass loss in the paralyzed muscle) and some are secondary consequences of SCI, such as a lack of PA and exercise (FM increase, cardiometabolic profiles) (Harvey & Graves, 2011). Likewise, some variables related to the psychological well-being change due to physical impairment, but some are related to a lack of PA and exercise. Thus, it is logical to think that exercise and PA participation would reverse some secondary consequences of SCI. Some exercises such as functional electrical stimulation (FES)-assisted exercise, either FES-assisted contraction of the paralyzed muscle alone or combined with voluntary muscle contraction of the unparalyzed

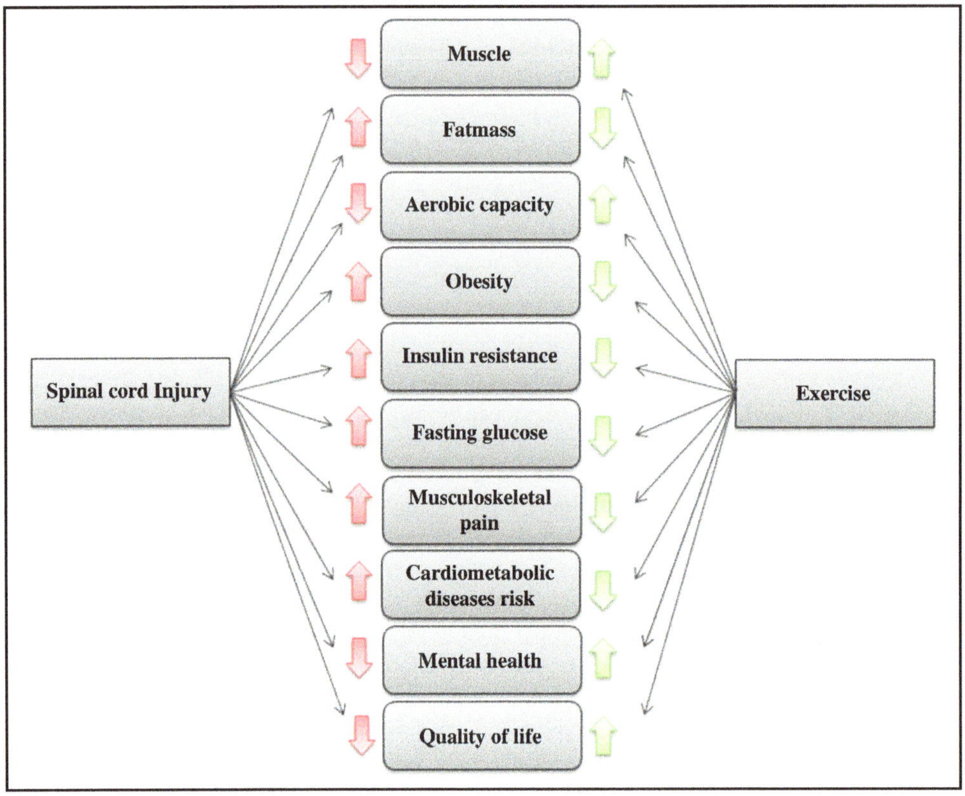

FIG. 2 Mechanism of exercise-associated reduction in spinal cord injury. Increasing PA and regular participation in exercise in people with SCI have a positive physiological effect such as improving insulin resistance. In particular, participation in aerobic and/or resistance exercise in people with SCI is reported to prevent cardiovascular complications and improve cardiovascular function. Moreover, people with SCI have many shoulder joint and muscle problems due to excessive shoulder use. Therefore, these problems can be prevented in advance by doing shoulder strengthening exercises such as shoulder extension and elbow flexion exercises. Additionally, increasing PA and exercise in people with spinal cord injury is critical to improving their psychological well-being and life satisfaction.

muscle, may reverse both the direct and indirect consequences of SCI. Further, other exercises such as wheeling and participation in sport activities generally increase PA levels, which could reverse the secondary consequences of paralysis. In some cases, specialized therapeutic exercises could be applied to reduce pain or improve joint condition, such as the elbow and shoulder joints of people with SCI. There is substantial evidence that increased PA participation is associated with increased aerobic capacity and muscle strength as well as a smaller risk of CVD in people with SCI (Nooijen et al., 2012). Moreover, people with SCI who participate in more PA have lower levels of depression and anxiety as well as higher levels of social support compared to those who don't participate in PA as much (Kim, Lee, Park, & Jeon, 2020). In this next section, types of exercises and their potential health benefits on the primary and secondary consequences of SCI will be discussed. Previous studies that investigated the effect of exercise programs in people with SCI are summarized in Table 1.

Exercise programs for people with SCI

According to the current literature, we can identify four different types of exercises in which people with SCI can participate: (1) upper-body (UB) aerobic exercises (arm ergometer, wheeling), (2) UB resistance exercises (free weights, weight machines, and thera-bands), (3) FES-assisted lower-body (LB) exercises (FES cycling), and (4) whole-body (WB) exercises (hybrid FES cycling and arm ergometer as well as FES rowing). In addition to the exercise type, the exercise intensity is also important when developing an exercise program for optimal health benefits. In that regard, when the lesion level is high (i.e., above C6–7), it is difficult to perform a high-intensity exercise. In this case, FES-assisted cycling or FES rowing may enable people to increase the exercise intensity to improve cardiopulmonary and cardiometabolic health. Likewise, people with shoulder problems such as shoulder impingement syndrome due to scapular abduction may benefit from thera-band and rowing exercises, which may strengthen the scapular adductors and external rotators. Likewise, the exercise program should be designed according to each individual's physical condition.

TABLE 1 Previous studies of effect and types of exercise programs in people with SCI.

No	Author (year)	Subject	Injury level	Exercise type	Duration	Results (significant)
1	Fornusek, Davis, and Russold (2013)	Eight individuals with chronic SCI	C7 ~ T11	Isokinetic FES cycle ergometer	6 weeks, 3 days/ week	– Muscle hypertrophy ↑ – Muscle strength ↑
2	Giangregorio et al. (2012)	34 Individuals with SCI	C2 ~ T12/ASIA: C or D	FES assisted walking	16 weeks, 2 days/ week	– Body fat ↓ – Muscle strength ↑
3	Jeon et al. (2010)	Six male participants with paraplegia	T4–5 and T10	FES rowing	12 weeks, 3–4 days/ week	– Peak oxygen consumption ↑ – Leptin level ↓ – Glucose levels ↓
5	Johnston, Smith, Mulcahey, Betz, and Lauer (2009)	30 Children with SCI	C4 ~ T11/ASIA: A, B, or C	FES cycling	6 months, 3 days/ week	– Oxygen uptake ($\dot{V}O_2$) ↑ – Cholesterol levels ↓
6	Griffin et al. (2009)	18 Subjects with SCI	C4, T3	FES cycling	10 weeks, 2–3 days/ week	– Muscle strength ↑ – Thigh sensory scores ↑ – IL-6, TNF-a, CRP ↓
7	Johnston, Smith, Betz, and Lauer (2008)	Four subjects with SCI	C7 ~ T6/ASIA: A	FES cycling passive motion cycling	6 months, 3 days/ week	– Left femoral neck ↑ – Distal femur ↑ – Resting heart rate ↓
8	Frotzler et al. (2008)	17 Subjects with spinal cord injury	T3 ~ T12	FES cycling	12 months, 5 days/ weekly	– BMD ↑ – Muscle CSA ↑ – CSA fat ↓
9	Zbogar et al. (2008)	Four subjects with SCI	C4, C5, T4, T7	FES leg cycle ergometer	3 days/ weeks Total: 36 times	– Small artery compliance ↑
10	Castello, Louis, Cheng, Armento, and Santos (2012)	Six subjects with SCI	C3, C3, C4, C5, T4	FES cycling	9 months, 3 days/ week	– Pediatric quality of life – Inventory ↑ – Bone mineral density (BMD) ↑
11	Janssen and Pringle (2008)	12 With chronic SCI	C4 ~ T11	Electrical stimulation induced leg cycle ergometer	6 weeks, 3 days/ week	– Cardiorespiratory responses, stroke volume, peak values for power output ↑
12	Latimer, Ginis, Hicks, and McCartney (2004)	21 Individuals with SCI	C3 ~ T7	Stretching, aerobic, arm ergometer exercise, resistance exercise	9 months, 2 days/ week	– Pain ↓ – Stress ↓ – Depression ↓
13	Silva et al. (1998)	12 People with SCI and 12 able-bodied controls	T1 ~ T12	Arm crank ergometer training	6 weeks, 3 days/ week	– Forced vital capacity ↑ – Ventilatory muscle endurance test ↑

Continued

TABLE 1 Previous studies of effect and types of exercise programs in people with SCI—cont'd

No	Author (year)	Subject	Injury level	Exercise type	Duration	Results (significant)
14	Liusuwan, Widman, Abresch, Styne, and McDonald (2007)	20 Adolescents (aged 11–18) with spinal cord dysfunction	Non	Aerobic (arm ergometer), strengthening exercises	16 weeks, 3 days/ week	– Shoulder extension strength ↑ – Maximum power output ↑ – Whole body lean tissue ↑ – Work efficiency ↑ – Resting oxygen uptake ↑
15	Mahoney et al. (2005)	Five men with SCI	C5 ~ T9	Knee extensions by electric stimulation unit	12 weeks, 2 days/ week	– Skeletal muscle cross-sectional area (CSA) in the right and left quadriceps femoris ↑
16	Giangregorio et al. (2005)	Five individuals with traumatic SCI	C3, C5, C6, C6, C8	Body weight supported treadmill	6–8 months, 2 days/ week Total: 48 sessions	– Muscle CSAs ↑ – Lumbar spine BMD ↑
17	Giangregorio et al. (2006)	14 Patients with SCI	C4 ~ T12	Body weight supported treadmill training	12 months, 3 days/ week	– Whole-body lean mass ↑ – Muscle CSAs ↑ – Whole-body bone density ↑
18	Wirz et al. (2005)	20 Chronic patients with SCI	ASIA: C ($n = 9$) and D ($n = 11$)	Locomotor training (using robotic-assisted, body weight supported treadmill training)	8 weeks, 3–5 days/ week	– Velocity, endurance, and performance of functional tasks ↑
19	Field-Fote and Tepavac (2002)	14 Subjects with incomplete SCI	C4 ~ T7	Combining body weight support, electrical stimulation, and treadmill training	12 weeks, 3 days/ week	– Over ground and treadmill walking speed ↑
20	Ditor et al. (2003)	Seven subjects with SCI	C5 ~ T12/ASIA: A ~ D	Supervised exercise training (arm ergometer, stretching, resistance exercise)	3 months, 2 days/ week	– Exercise adherence ↓ – PQOL ↓ – Pain ↓ – Stress ↓
21	Hicks et al. (2005)	14 Individuals with incomplete SCI	C4 ~ T12	Body-weight-supported treadmill training	12 months, 3 days/ week	– Treadmill walking ability ↑ – Treadmill walking speed ↑ – Treadmill walking distance ↑ – Capacity to walk over ground ↑ – Physical function ↑
22	Hicks et al. (2003)	34 Subjects with traumatic SCI	C4 ~ L/ASIA: A ~ D	Supervised exercise (arm ergometer, stretching, resistance exercise)	2 days/ week	– Submaximal arm ergometer power output ↑ – Upper body muscle strength ↑ – Pain, stress and depression ↓

UB aerobic exercise includes the use of an arm ergometer and a wheelchair, whereas UB resistance exercises include free weights, weight machines, and thera-bands. LB exercise includes cycles that use electrical stimulation, such as FES and WB exercises that include a combination of those exercises and a rowing machine.

UB aerobic and strength exercise

UB exercises have been widely studied as an effective form of aerobic and strength exercise for individuals with SCI. Aerobic exercise programs such as the arm ergometer and hand bike are effective in improving body composition and cardiopulmonary fitness as well as reducing insulin resistance in people with SCI (Kim et al., 2019). In addition to the well-established benefits of UB aerobic exercise as an effective means to improve fitness in people with SCI, arm ergometers also offer an exercise modality that is accessible and does not require overly specialized expertise or infrastructure. Recently, a randomized controlled trial study with an exercise program on indoor hand bikes equipped with an information technology system for people with SCI showed significantly improved cardiopulmonary fitness and UB strength as well as reduced BMI, waist circumference, insulin levels, and HOMA-IR levels (Kim, Lee, Lee, Kim, & Jeon, 2015). Moreover, aerobic exercise using a manual wheelchair, in particular backward wheeling exercises, have been shown to effectively strengthen the scapular retractor muscles (Olenik, Laskin, Burnham, Wheeler, & Steadward, 1995).

In terms of UB resistance exercises, such as free weights, weight machines, and thera-band exercise programs has shown to improve the cardiorespiratory fitness, muscle strength and the tolerance of shoulder pain in people with SCI (Jacobs, Nash, & Rusinowski, 2001). As the upper extremities are used for most daily activities, shoulder problems may lead to difficulties in propelling wheelchairs and transferring the body (i.e., a wheelchair to a bed). Thus, strengthening the shoulder muscles may also reduce shoulder pain (i.e., rotator cuff impingement syndrome and general shoulder pain). Recently, Kim et al. reported that 6 weeks of UB resistance exercise improved UB muscle strength, as measured by shoulder flexion, extension, abduction, and adduction, as well as elbow flexion (Kim et al., 2019). Kim et al. further reported that UB resistance exercise significantly increased the HDL-C levels, UB muscle strength, and VO_2 peaks of the participants (Kim et al., 2019).

Despite this benefit of UB exercises, potential issues related to UB overuse injuries, such as shoulder joint ligament tears, muscle strains, tendon sprains, or stress fractures, may occur (Dyson-Hudson, Sisto, Bond, Emmons, & Kirshblum, 2007). These concerns are based on repetitive upper limb activities such as wheelchair propulsion and other activities of daily living such as transfers and vertical lifts involves pushing actions (shoulder flexion and horizontal adduction) (Van Drongelen et al., 2005). Therefore, backward wheeling, rowing exercises, and thera-band exercises that require shoulder extensions, external rotations, and scapula retractions would be beneficial exercises for people with SCI (Kim et al., 2019).

FES-assisted LB and/or WB exercise

Using FES would reverse some primary consequences of SCI such as changes in muscle mass, skeletal muscle capillarization, and fiber type (Collins, 2007). Currently, FES is the most common technique used to generate muscle contractions to maintain muscle quality and produce purposeful movement after SCI (Ibitoye, Hamzaid, Hasnan, Abdul Wahab, & Davis, 2016). For LB exercises, FES cycling has been shown to be effective in increasing cardiopulmonary fitness and the protein expression of glucose transporters in the paralyzed muscle while decreasing insulin resistance. It is also beneficial to dyslipidemia via muscle fiber type changes, capillarization, and even the protein expression of glucose transporter 1 and 4 in the vastus lateralis muscle (Jeon et al., 2002). Especially, FES-assisted cycling has been tested most extensively on acute and chronic physiological responses (Castello et al., 2012; Frotzler et al., 2008; Griffin et al., 2009; Jeon et al., 2002; Johnston et al., 2008; Johnston et al., 2009; Zbogar et al., 2008).

In terms of WB exercises, a combination of FES cycling and arm ergometer exercises (Verellen et al., 2001; Verellen, Vanlandewijck, Andrews, & Wheeler, 2007) as well as FES rowing exercises (Kim, Park, Lee, & Jeon, 2014) has been reported to be effective for improving the cardiovascular system and imparting positive effects to people with SCI. Hooker et al. previously demonstrated that hybrid exercise, which uses both upper and LB muscles, is a more effective exercise modality to improve cardiovascular fitness than the arm ergometer and FES cycle alone (Hooker et al., 1992). Specifically, rowing exercises require a strong physical condition for fast muscle contraction because the mobilization of anaerobic energy sources is higher than in other exercises (Wagenmakers, Coakley, & Edwards, 1990). The FES rowing exercise has been reported to be effective at improving cardiovascular fitness, UB strength, and body composition of people with SCI (Jeon et al., 2010; Kim et al., 2014). Furthermore, backward wheeling, and isolated and strengthening the scapular retractor muscle during the FES rowing exercise was effective and would improve the scapular stability which is important in prevention and treatment of shoulder problems in people with SCI (Kim et al., 2014; Olenik et al., 1995). Individuals with higher levels of SCI may benefit the most from FES-assisted exercise; however, FES rowing, cycling, and the arm ergometer have limitations due to the high cost of equipment and lack of availability.

Tailored exercise for people with SCI

Although exercise have been shown to confer beneficial physiological and mental effects to people with SCI, the levels of SCI can influence the individual's ability to perform exercise limiting type, intensity, and duration of exercise. Furthermore, to maximize the health benefits of exercise, the type, duration, and intensity should be prescribed according to participant comorbidities (obesity, type 2 diabetes, CVD) as well as shoulder and elbow joint conditions. However, most studies that demonstrated the beneficial effect of exercise on the cardiometabolic profile and joint problems in individuals with SCI applied the same exercise protocols regardless of participants' level of injuries, degree of adiposity, and comorbidities. Therefore, it is important to develop personalized exercise programs for people with SCI by considering the multiple factors of their exercise ability and the potential benefits they could reasonably obtain from such exercise programs. In this regard, Kim et al. attempted to personalize exercise programs for people with SCI. They personalized an exercise program for SCI based on four factors: level of SCI (A1–7), comorbidities (B1–6), purpose of exercise (C1–4), and joint conditions (D1–3). Therefore, there could be 504 different exercise programs based on their levels of injury, comorbidities, purpose of exercise, and joint condition (Table 2). This exercise program consisted of 25 min of warm up, 30 min of exercise, and 5 min of cool down. A total of 17 people with SCI participated in this pilot randomized controlled trial. These participants' fasting insulin levels, surrogate measures of insulin resistance (HOMA-IR), and FM were significantly decreased. Furthermore, UB muscle strength, measured by elbow flexion/extension and shoulder adduction/abduction/flexion/extension, were significantly improved in all measurements. Lastly, the psychological well-being improved significantly among those in the exercise group; however, these results have not been published in a peer-review journal yet (Kim, 2014). Although this study was a pilot randomized controlled trial, studies like this should be expanded further with a larger sample size and the addition of FES (if available).

Conclusion and future perspective

In this chapter, the effects of different exercises on health outcomes in people with SCI were summarized and discussed. In general, exercise and PA are beneficial on their cardiometabolic, musculoskeletal, and psychosocial health. Yet, due to their paralysis, people with SCI have less choices on the mode of exercise. They also have to be more careful not to injure their joints because UB joints are essential for their mobility and daily living. In the future, more personalized exercise program should be applied according to their levels of injury, comorbidities, and purpose of exercise participation.

Applications to other areas of neuroscience

The primary consequences of SCI are paralysis and autonomic nervous system dysfunction while the secondary consequences including physical inactivity, changes in body composition and metabolic profiles, poor musculoskeletal problems, and poor mental health. Various exercise programs such as UB aerobic and resistance exercise and LB and

TABLE 2 Development of tailored exercise programs for people with different levels of SCI.

	Option 1	Option 2	Option 3	Option 4
Screening	Level of injury	Chronic disease	Exercise purpose	Joint stability
Contents	Component of resistance exercise	Exercise frequency and intensity level	Ratio of aerobic, resistance, and circuit training	Rehabilitation exercise
Category	A type	B type	C type	D type
	A1: C4–6 A2: C7–8 A3: T1–6 A4: T7–L2 A5: L3 A6: L4–5 A7: S1–2	B1: Metabolic syndrome B2: Hypertension B3: Diabetes B4: Osteoporosis B5: Obesity B6: None	C1: Functional fitness C2: Weight loss C3: Muscle strength C4: Cardiovascular fitness	D1: Shoulder D2: Elbow D3: Wrist

To maximize the health benefits of exercise for people with SCI, the exercise should be tailored to the presence of chronic diseases (metabolic syndrome, hypertension, type 2 diabetes, osteoporosis, and obesity) and the existence of joint problems while the level of SCI should also be considered.

WB were discussed. Regular exercise and increased PA in people with SCI seem beneficial for overall fitness and psychological well-being, and can even help those with high spinal cord lesions. Exercise is beneficial in all neurological disorders or conditions whether they are progressive (i.e., Parkinson's disease) or non-progressive (i.e., traumatic brain injury and cerebral palsy). Evidence exists that exercise slows down the progression of neurologic disorders while improving physical function and the QoL of people with different neurological conditions. However, no evidence-based personalized exercise program has yet been developed for people with other neurological disorders. Because the level of injury can influence the patient's ability to perform particular exercises, exercise programs for people with neurological disorders vary according to their types and levels of injury. Furthermore, the level of adiposity, joint problems, and the presence of chronic diseases such as type 2 diabetes, hypertension, and CVD may also influence the type, intensity, duration, and frequency of exercise in which people with SCI can participate. Just as exercise should be personalized for people with SCI, exercise should be personalized for different neurological disorders to maximize benefits while minimizing side effects.

Mini-dictionary of terms

- **Spinal cord injury (SCI):** Damage to the spinal cord resulting from trauma, disease, or degeneration. Symptoms of SCI depend on the severity of injury and its location on the spinal cord. Symptoms may include partial or complete loss of sensory function or motor control of the upper body, lower body, and/or the whole-body.
- **Exercise:** Physical activity that is planned, structured, and repetitive for the purpose of conditioning the body. Exercise consists of aerobic, resistance, and flexibility.
- **Body mass index:** Person's weight in kilograms divided by the square of height in meters. It's an easy screening method for weight category (underweight, healthy weight, overweight, and obese).
- **Type 2 diabetes mellitus:** The most common type of diabetes, which is a disease that occurs when your blood glucose is too high.
- **Cardiovascular disease:** Any disease of the heart and its associated blood vessels, most commonly coronary heart disease, stroke, and peripheral vascular disease.
- **Muscle strength:** Defined as the amount of force you can put out or the amount of weight you can lift.
- **Functional electrical stimulation:** Treatment that involves the application of a mild electrical stimulus to a muscle to produce functional movement after paralysis.

Key facts on exercises and people with SCI

- The clinical consequences of SCI, paired with a reduction in physical activity and increased sedentary time, often result in a deterioration of body composition and metabolic profile.
- Significant changes in both body composition and metabolic profiles after SCI can result in significant health consequences and increased risk of chronic disease.
- Exercise has been shown to improve muscle strength, aerobic capacity, and psychological well-being in people with SCI.
- Wheelchair propulsion as well as transfers cause and increase musculoskeletal problems, such as shoulder, elbow, and wrist pain in people with SCI.
- Musculoskeletal problems in people with SCI may be partially a result of overuse; however, certain exercises and routines can help reduce pain while strengthening and stretching the muscles used for joint function.
- Personalized exercise programs for people with a SCI should be developed by considering multiple factors capable of influencing their exercise ability, and the potential benefits they could reasonably obtain from such exercise programs are important.

Summary points

- This chapter focuses on the effects of exercise and various exercise programs for people with SCI.
- SCI is a catastrophic injury with potential devastating impacts, including far-reaching physical and psychological consequences.
- Lack of exercise and physical inactivity are the primary causes of chronic diseases such as cardiovascular disease, type 2 diabetes, hypertension, and metabolic syndrome in people with SCI.

- Evidence suggests that different types of upper body exercise and lower body and whole body exercise with functional electrical stimulation programs is effective in the physical and psychological well-being of people with SCI.
- However, the levels of SCI can influence the individual's ability in regard to the type, intensity, and duration of exercise.
- Therefore, personalized exercise program should be applied according to their levels of injury, comorbidities, and purpose of exercise participation.

References

Ahuja, C. S., Wilson, J. R., Nori, S., Kotter, M. R., Druschel, C., Curt, A., & Fehlings, M. G. (2017). Traumatic spinal cord injury. *Nature Reviews Disease Primers*, *3*(1), 1–21.

Assinck, P., Duncan, G. J., Hilton, B. J., Plemel, J. R., & Tetzlaff, W. (2017). Cell transplantation therapy for spinal cord injury. *Nature Neuroscience*, *20* (5), 637–647.

Bauman, W., Spungen, A., Zhong, Y.-G., Rothstein, J., Petry, C., & Gordon, S. (1992). Depressed serum high density lipoprotein cholesterol levels in veterans with spinal cord injury. *Spinal Cord*, *30*(10), 697–703.

Bauman, W. A., & Spungen, A. M. (2001). Carbohydrate and lipid metabolism in chronic spinal cord injury. *Journal of Spinal Cord Medicine*, *24*(4), 266–277. https://doi.org/10.1080/10790268.2001.11753584.

Burnham, R., Martin, T., Stein, R., Bell, G., MacLean, I., & Steadward, R. (1997). Skeletal muscle fibre type transformation following spinal cord injury. *Spinal Cord*, *35*(2), 86–91. https://doi.org/10.1038/sj.sc.3100364.

Capoor, J., & Stein, A. B. (2005). Aging with spinal cord injury. *Physical Medicine and Rehabilitation Clinics*, *16*(1), 129–161.

Castello, F., Louis, B., Cheng, J., Armento, M., & Santos, A. M. (2012). The use of functional electrical stimulation cycles in children and adolescents with spinal cord dysfunction: A pilot study. *Journal of Pediatric Rehabilitation Medicine*, *5*(4), 261–273. https://doi.org/10.3233/PRM-2012-00221.

Collins, D. F. (2007). Central contributions to contractions evoked by tetanic neuromuscular electrical stimulation. *Exercise and Sport Sciences Reviews*, *35*(3), 102–109.

Ditor, D. S., Latimer, A. E., Ginis, K. A. M., Arbour, K. P., McCartney, N., & Hicks, A. L. (2003). Maintenance of exercise participation in individuals with spinal cord injury: Effects on quality of life, stress and pain. *Spinal Cord*, *41*(8), 446–450. https://doi.org/10.1038/sj.sc.3101487.

Dyson-Hudson, T. A., Sisto, S. A., Bond, Q., Emmons, R., & Kirshblum, S. C. (2007). Arm crank ergometry and shoulder pain in persons with spinal cord injury. *Archives of Physical Medicine and Rehabilitation*, *88*(12), 1727–1729.

Fernhall, B., Heffernan, K., Jae, S. Y., & Hedrick, B. (2008). Health implications of physical activity in individuals with spinal cord injury: A literature review. *Journal of Health and Human Services Administration*, *30*(4), 468–502.

Field-Fote, E. C., & Tepavac, D. (2002). Improved intralimb coordination in people with incomplete spinal cord injury following training with body weight support and electrical stimulation. *Physical Therapy*, *82*(7), 707–715.

Fornusek, C., Davis, G. M., & Russold, M. F. (2013). Pilot study of the effect of low-cadence functional electrical stimulation cycling after spinal cord injury on thigh girth and strength. *Archives of Physical Medicine and Rehabilitation*, *94*(5), 990–993. https://doi.org/10.1016/j.apmr.2012.10.010.

Frotzler, A., Coupaud, S., Perret, C., Kakebeeke, T. H., Hunt, K. J., Donaldson, N. D. N., & Eser, P. (2008). High-volume FES-cycling partially reverses bone loss in people with chronic spinal cord injury. *Bone*, *43*(1), 169–176. https://doi.org/10.1016/j.bone.2008.03.004.

Garstang, S. V., & Miller-Smith, S. A. (2007). Autonomic nervous system dysfunction after spinal cord injury. *Physical Medicine and Rehabilitation Clinics of North America*, *18*(2), 275–296.

Gater, D. R., Jr., Farkas, G. J., & Tiozzo, E. (2021). Pathophysiology of neurogenic obesity after spinal cord injury. *Topics in Spinal Cord Injury Rehabilitation*, *27*(1), 1–10.

Giangregorio, L., Craven, C., Richards, K., Kapadia, N., Hitzig, S. L., Masani, K., & Popovic, M. R. (2012). A randomized trial of functional electrical stimulation for walking in incomplete spinal cord injury: Effects on body composition. *Journal of Spinal Cord Medicine*, *35*(5), 351–360. https://doi.org/10.1179/2045772312Y.0000000041.

Giangregorio, L. M., Hicks, A. L., Webber, C. E., Phillips, S. M., Craven, B. C., Bugaresti, J. M., & McCartney, N. (2005). Body weight supported treadmill training in acute spinal cord injury: Impact on muscle and bone. *Spinal Cord*, *43*(11), 649–657. https://doi.org/10.1038/sj.sc.3101774.

Giangregorio, L. M., Webber, C. E., Phillips, S. M., Hicks, A. L., Craven, B. C., Bugaresti, J. M., & McCartney, N. (2006). Can body weight supported treadmill training increase bone mass and reverse muscle atrophy in individuals with chronic incomplete spinal cord injury? *Applied Physiology, Nutrition, and Metabolism*, *31*(3), 283–291. https://doi.org/10.1139/h05-036.

Ginis, K. A. M., Latimer, A. E., Arbour-Nicitopoulos, K. P., Buchholz, A. C., Bray, S. R., Craven, B. C., … Potter, P. J. (2010). Leisure time physical activity in a population-based sample of people with spinal cord injury part I: Demographic and injury-related correlates. *Archives of Physical Medicine and Rehabilitation*, *91*(5), 722–728.

Gorgey, A., & Dudley, G. (2007). Skeletal muscle atrophy and increased intramuscular fat after incomplete spinal cord injury. *Spinal Cord*, *45*(4), 304–309.

Griffin, L., Decker, M. J., Hwang, J. Y., Wang, B., Kitchen, K., Ding, Z., & Ivy, J. L. (2009). Functional electrical stimulation cycling improves body composition, metabolic and neural factors in persons with spinal cord injury. *Journal of Electromyography and Kinesiology*, *19*(4), 614–622. https://doi.org/10.1016/j.jelekin.2008.03.002.

Harvey, L., & Graves, D. (2011). International standards for the neurological classification of spinal cord injury. *Journal of Physiotherapy*, *57*(2), 129.

Hicks, A., Martin, K., Ditor, D., Latimer, A., Craven, C., Bugaresti, J., & McCartney, N. (2003). Long-term exercise training in persons with spinal cord injury: Effects on strength, arm ergometry performance and psychological well-being. *Spinal Cord*, *41*(1), 34–43.

Hicks, A. L., Adams, M. M., Martin Ginis, K., Giangregorio, L., Latimer, A., Phillips, S. M., & McCartney, N. (2005). Long-term body-weight-supported treadmill training and subsequent follow-up in persons with chronic SCI: Effects on functional walking ability and measures of subjective well-being. *Spinal Cord*, *43*(5), 291–298. https://doi.org/10.1038/sj.sc.3101710.

Hooker, S. P., Figoni, S. F., Rodgers, M. M., Glaser, R. M., Mathews, T., Suryaprasad, A., & Gupta, S. (1992). Metabolic and hemodynamic responses to concurrent voluntary arm crank and electrical stimulation leg cycle exercise in quadriplegics. *Journal of Rehabilitation Research and Development*, *29*(3), 1–11.

Ibitoye, M. O., Hamzaid, N. A., Hasnan, N., Abdul Wahab, A. K., & Davis, G. M. (2016). Strategies for rapid muscle fatigue reduction during FES exercise in individuals with spinal cord injury: A systematic review. *PLoS One*, *11*(2), e0149024.

Jeon, J. Y., Hettinga, D., Steadward, R. D., Wheeler, G. D., Bell, G., & Harber, V. (2010). Reduced plasma glucose and leptin after 12 weeks of functional electrical stimulation–rowing exercise training in spinal cord injury patients. *Archives of Physical Medicine and Rehabilitation*, *91*(12), 1957–1959.

Jeon, J. Y., Steadward, R. D., Wheeler, G. D., Bell, G., McCargar, L., & Harber, V. (2003). Intact sympathetic nervous system is required for leptin effects on resting metabolic rate in people with spinal cord injury. *Journal of Clinical Endocrinology and Metabolism*, *88*(1), 402–407. https://doi.org/10.1210/jc.2002-020939.

Jacobs, P. L., Nash, M. S., & Rusinowski, J. W. (2001). Circuit training provides cardiorespiratory and strength benefits in persons with paraplegia. *Medicine & Science in Sports & Exercise*, *33*(5), 711–717. https://doi.org/10.1097/00005768-200105000-00005.

Janssen, T. W. J., & Pringle, D. D. (2008). Effects of modified electrical stimulation-induced leg cycle ergometer training for individuals with spinal cord injury. *Journal of Rehabilitation Research and Development*, *45*(6), 819–830. https://doi.org/10.1682/jrrd.2007.09.0153.

Jazayeri, S. B., Beygi, S., Shokraneh, F., Hagen, E. M., & Rahimi-Movaghar, V. (2015). Incidence of traumatic spinal cord injury worldwide: A systematic review. *European Spine Journal*, *24*(5), 905–918. https://doi.org/10.1007/s00586-014-3424-6.

Jeon, J., Weiss, C., Steadward, R., Ryan, E., Burnham, R., Bell, G., … Wheeler, G. (2002). Improved glucose tolerance and insulin sensitivity after electrical stimulation-assisted cycling in people with spinal cord injury. *Spinal Cord*, *40*(3), 110–117.

Johnston, T. E., Smith, B. T., Betz, R. R., & Lauer, R. T. (2008). Exercise testing using upper extremity ergometry in pediatric spinal cord injury. *Pediatric Physical Therapy*, *20*(2), 146–151. https://doi.org/10.1097/PEP.0b013e318171faa1.

Johnston, T. E., Smith, B. T., Mulcahey, M. J., Betz, R. R., & Lauer, R. T. (2009). A randomized controlled trial on the effects of cycling with and without electrical stimulation on cardiorespiratory and vascular health in children with spinal cord injury. *Archives of Physical Medicine and Rehabilitation*, *90*(8), 1379–1388. https://doi.org/10.1016/j.apmr.2009.02.018.

Kang, Y., Ding, H., Zhou, H., Wei, Z., Liu, L., Pan, D., & Feng, S. (2017). Epidemiology of worldwide spinal cord injury: A literature review. *Journal of Neurorestoratology*, *6*, 1–9.

Kennedy, P., & Garmon-Jones, L. (2017). Self-harm and suicide before and after spinal cord injury: A systematic review. *Spinal Cord*, *55*(1), 2–7.

Kim, D.-I. (2014). *(The) effect of exercise on fasting insulin and fitness levels in people with spinal cord injury*. Doctoral dissertation Seoul: Graduate School, Yonsei University. Retrieved from http://www.riss.kr/link?id=T13599229.

Kim, D.-I., Lee, H., Lee, B.-S., Kim, J., & Jeon, J. Y. (2015). Effects of a 6-week indoor hand-bike exercise program on health and fitness levels in people with spinal cord injury: A randomized controlled trial study. *Archives of Physical Medicine and Rehabilitation*, *96*(11), 2033–2040. e2031.

Kim, D. I., Park, D., Lee, B., & Jeon, J. (2014). A six-week motor-driven functional electronic stimulation rowing program improves muscle strength and body composition in people with spinal cord injury: A pilot study. *Spinal Cord*, *52*(8), 621–624.

Kim, D.-I., Taylor, J. A., Tan, C. O., Park, H., Kim, J. Y., Park, S.-Y., … Jeon, J. Y. (2019). A pilot randomized controlled trial of 6-week combined exercise program on fasting insulin and fitness levels in individuals with spinal cord injury. *European Spine Journal*, *28*(5), 1082–1091.

Kim, D., Il, Lee, J., Park, H., & Jeon, J. Y. (2020). The relationship between physical activity levels and mental health in individuals with spinal cord injury in South Korea. *International Journal of Environmental Research and Public Health*, *17*(12). https://doi.org/10.3390/ijerph17124423.

Latimer, A. E., Ginis, K. A. M., Hicks, A. L., & McCartney, N. (2004). An examination of the mechanisms of exercise-induced change in psychological well-being among people with spinal cord injury. *Journal of Rehabilitation Research and Development*, *41*(5), 643–652. https://doi.org/10.1682/jrrd.2003.04.0043.

Le, J., & Dorstyn, D. (2016). Anxiety prevalence following spinal cord injury: A meta-analysis. *Spinal Cord*, *54*(8), 570–578.

Liusuwan, R. A., Widman, L. M., Abresch, R. Ted., Styne, D. M., & McDonald, C. M. (2007). Body composition and resting energy expenditure in patients aged 11 to 21 years with spinal cord dysfunction compared to controls: Comparisons and relationships among the groups. *Journal of Spinal Cord Medicine*, *30*(Suppl 1), S105–S111. https://doi.org/10.1080/10790268.2007.11754613.

Mahoney, E. T., Bickel, C. S., Elder, C., Black, C., Slade, J. M., Apple, D., Jr., & Dudley, G. A. (2005). Changes in skeletal muscle size and glucose tolerance with electrically stimulated resistance training in subjects with chronic spinal cord injury. *Archives of Physical Medicine and Rehabilitation*, *86*(7), 1502–1504. https://doi.org/10.1016/j.apmr.2004.12.021.

Maynard, F., Karunas, R., & Waring, W., 3rd. (1990). Epidemiology of spasticity following traumatic spinal cord injury. *Archives of Physical Medicine and Rehabilitation*, *71*(8), 566–569.

Nooijen, C. F., de Groot, S., Postma, K., Bergen, M. P., Stam, H. J., Bussmann, J., & Van den Berg-Emons, R. (2012). A more active lifestyle in persons with a recent spinal cord injury benefits physical fitness and health. *Spinal Cord*, *50*(4), 320–323.

Olenik, L., Laskin, J., Burnham, R., Wheeler, G., & Steadward, R. (1995). Efficacy of rowing, backward wheeling and isolated scapular retractor exercise as remedial strength activities for wheelchair users: Application of electromyography. *Spinal Cord*, *33*(3), 148–152.

Riek, L., Ludewig, P., & Nawoczenski, D. (2008). Comparative shoulder kinematics during free standing, standing depression lifts and daily functional activities in persons with paraplegia: Considerations for shoulder health. *Spinal Cord*, *46*(5), 335–343.

Silva, A. C., Neder, J. A., Chiurciu, M. V., Pasqualin, D. C., da Silva, R. C., Fernandez, A. C., … Tufik, S. (1998). Effect of aerobic training on ventilatory muscle endurance of spinal cord injured men. *Spinal Cord*, *36*(4), 240–245. https://doi.org/10.1038/sj.sc.3100575.

Spungen, A. M., Adkins, R. H., Stewart, C. A., Wang, J., Pierson, R. N., Jr., Waters, R. L., & Bauman, W. A. (2003). Factors influencing body composition in persons with spinal cord injury: A cross-sectional study. *Journal of Applied Physiology (1985)*, *95*(6), 2398–2407. https://doi.org/10.1152/japplphysiol.00729.2002.

Sweis, R., & Biller, J. (2017). Systemic complications of spinal cord injury. *Current Neurology and Neuroscience Reports*, *17*(1), 1–8.

Ullrich, P. M., Lincoln, R. K., Tackett, M. J., Miskevics, S., Smith, B. M., & Weaver, F. M. (2013). Pain, depression, and health care utilization over time after spinal cord injury. *Rehabilitation Psychology*, *58*(2), 158.

van den Berg-Emons, R. J., Bussmann, J. B., Haisma, J. A., Sluis, T. A., van der Woude, L. H., Bergen, M. P., & Stam, H. J. (2008). A prospective study on physical activity levels after spinal cord injury during inpatient rehabilitation and the year after discharge. *Archives of Physical Medicine and Rehabilitation*, *89*(11), 2094–2101.

van den Berg-Emons, R. J., Bussmann, J. B., & Stam, H. J. (2010). Accelerometry-based activity spectrum in persons with chronic physical conditions. *Archives of Physical Medicine and Rehabilitation*, *91*(12), 1856–1861.

Van Drongelen, S., Van der Woude, L. H., Janssen, T. W., Angenot, E. L., Chadwick, E. K., & Veeger, D. H. (2005). Mechanical load on the upper extremity during wheelchair activities. *Archives of Physical Medicine and Rehabilitation*, *86*(6), 1214–1220.

Verellen, J., Kerby, B., Olijnyk, B., Saab, S., Smith, G., Jeon, J., … Wheeler, G. (2001). A comparison of the peak physical work capacity during arm ergometry, Fes cycling and two hybrid exercise conditions in spinal cord injured. In *Paper presented at the 7th Vienna international workshop on functional electrical stimulation*.

Verellen, J., Vanlandewijck, Y., Andrews, B., & Wheeler, G. D. (2007). Cardiorespiratory responses during arm ergometry, functional electrical stimulation cycling, and two hybrid exercise conditions in spinal cord injured. *Disability and Rehabilitation: Assistive Technology*, *2*(2), 127–132.

Wagenmakers, A., Coakley, J., & Edwards, R. (1990). Metabolism of branched-chain amino acids and ammonia during exercise: Clues from McArdle's disease. *International Journal of Sports Medicine*, *11*(S2), S101–S113.

Wang, Y.-H., Huang, T.-S., Lin, J.-L., Hwang, J.-J., Chan, H.-L., Lai, J.-S., & Tseng, Y.-Z. (2000). Decreased autonomic nervous system activity as assessed by heart rate variability in patients with chronic tetraplegia. *Archives of Physical Medicine and Rehabilitation*, *81*(9), 1181–1184.

Williams, R., & Murray, A. (2015). Prevalence of depression after spinal cord injury: A meta-analysis. *Archives of Physical Medicine and Rehabilitation*, *96*(1), 133–140.

Wirz, M., Zemon, D. H., Rupp, R., Scheel, A., Colombo, G., Dietz, V., & Hornby, T. G. (2005). Effectiveness of automated locomotor training in patients with chronic incomplete spinal cord injury: A multicenter trial. *Archives of Physical Medicine and Rehabilitation*, *86*(4), 672–680. https://doi.org/10.1016/j.apmr.2004.08.004.

Zbogar, D., Eng, J. J., Krassioukov, A. V., Scott, J. M., Esch, B. T. A., & Warburton, D. E. R. (2008). The effects of functional electrical stimulation leg cycle ergometry training on arterial compliance in individuals with spinal cord injury. *Spinal Cord*, *46*(11), 722–726. https://doi.org/10.1038/sc.2008.34.

Zurlo, F., Larson, K., Bogardus, C., & Ravussin, E. (1990). Skeletal muscle metabolism is a major determinant of resting energy expenditure. *Journal of Clinical Investigation*, *86*(5), 1423–1427.

Chapter 4

Use of anodal transcranial direct current stimulation: Features, facets, and applications to incomplete spinal cord injury

Amanda Vitória Lacerda de Araújo[a], Mirelly dos Santos Abilio[d], Débora Araújo do Nascimento[b], Beatriz Souza de Albuquerque Cacique New York[c], and Valéria Ribeiro Nogueira Barbosa[b]

[a]*Department of Neurology, University of São Paulo, São Paulo, Brazil,* [b]*Department of Physiotherapy, State University of Paraiba R. Baraúnas, University, Paraiba, Brazil,* [c]*Department of Physiotherapy, Federal University of Rio Grande do Norte, Lagoa Nova, Brazil,* [d]*Health Sciences Center, Federal University of Pernambuco, Recife, Brazil*

List of abbreviations

AEs	adverse effects
ANS	autonomic nervous system
ASIA	American Spinal Injury Association Scale
a-tDCS	anodal transcranial direct current stimulation
BDNF	Brain-derived neurotrophic factor
CE	cortical excitability
cSCI	complete spinal cord injury
DC	direct current
iSCI	incomplete spinal cord injury
M1	primary motor cortex
mA	milliampere
mA/cm^2	milliampere by square squares
MEP	motor evoked potential
MF	motor function
NMDA	*N*-methyl-D-aspartate
NP	neuropathic pain
SCI	spinal cord injury
tDCS	transcranial Direct Current Stimulation

Introduction

Discovery of the direct current therapeutic effect and technological progress

It is estimated that in the first century before Christ the first Direct Current (DC) generator was created, an invention attributed to the ancient Persian civilization probably used for therapeutic purposes. After, in 1800, Alessandro Volta created an electric battery called "voltaic battery" and it is making possible the transcranial application. DC was applied to a farmer who suffered from major depression and the symptoms were reduced after a few weeks (Sarmiento, San-Juan, & Prasath, 2016). Thus, the use of transcranial DC opened the door for experiments on psychiatric and neurological conditions. Fig. 1 shows the progress of the technologies for DC over time.

In 1883, methods and dosages to enhance the benefits of DC application through experimental research were established. Initially, the technological limitation and the protocols variability prevented researchers from understanding the DC effects (Erb, 1883). Only in 1964, experimental researchers about the difference between two types of transcranial

Cellular, Molecular, Physiological, and Behavioral Aspects of Spinal Cord Injury. https://doi.org/10.1016/B978-0-12-822427-4.00004-6

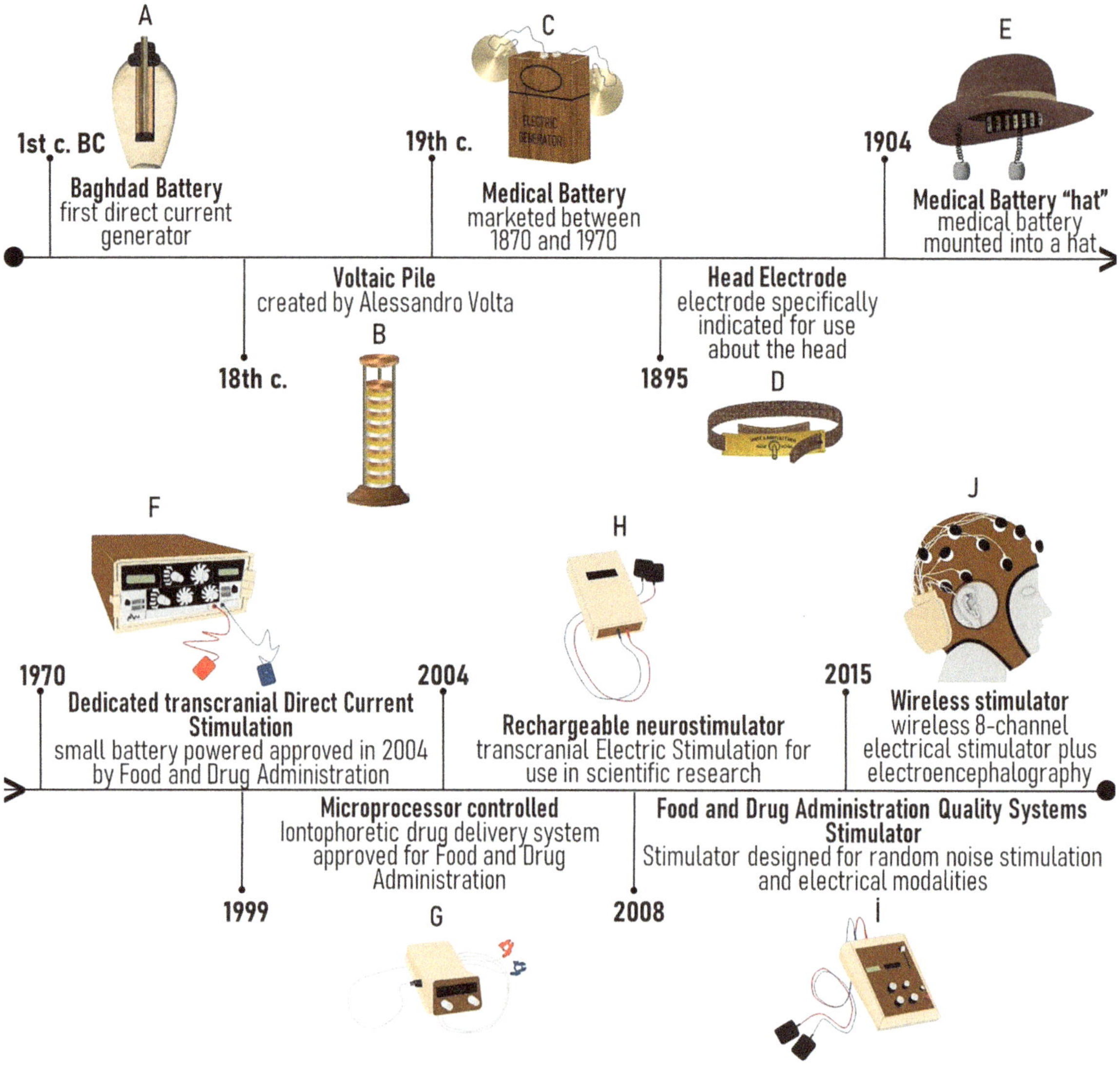

FIG. 1 Timeline of technological progress for direct current over time. Timeline shows the direct current devices in different forms from 1st BC until the year 2015 based on the report of the first studies using each. Unpublished. tDCS, transcranial Direct Current Stimulation.

DC, anodal and cathodal, were started in some clinical outcomes. The initial evidence provided knowledge of the transcranial DC effects on cortical excitability (CE) as a specific modulator of polarity from the neurons (Lippold & Redfearn, 1964). Later, in 2000, the term transcranial Direct Current Stimulation (tDCS) was cited for the first time by Nitsche and Paulus (2000) in an experiment that showed a significant increase in motor CE during anodal stimulation (a-tDCS) and a significant decrease during cathodal stimulation.

Based on the current evidence, tDCS can be considered a noninvasive, painless, and safe neurological stimulation. The tDCS uses a low-intensity DC between 1 and 4 mA distributed to the skull and brain through electrodes on specific cortical areas (Renga, 2020). The current flow depends on the electrode site. In a-tDCS, the anode is on a target brain area, and the cathode on a referential area. Thus, the current flow occurs from the anode to the cathode with a depolarization threshold reduction in the neuronal membrane. Instead, when the cathode is targeted on a brain area the current flow is the inverse, and the stimulation causes hyperpolarization in the neurons (Renga, 2020).

Neurophysiological basis of the a-tDCS

The neurophysiological basis of the a-tDCS can be divided into primary or secondary effects. Primary effects have been related to modulation of sodium and calcium-dependent voltage channel activity in the neuronal membrane and, therefore,

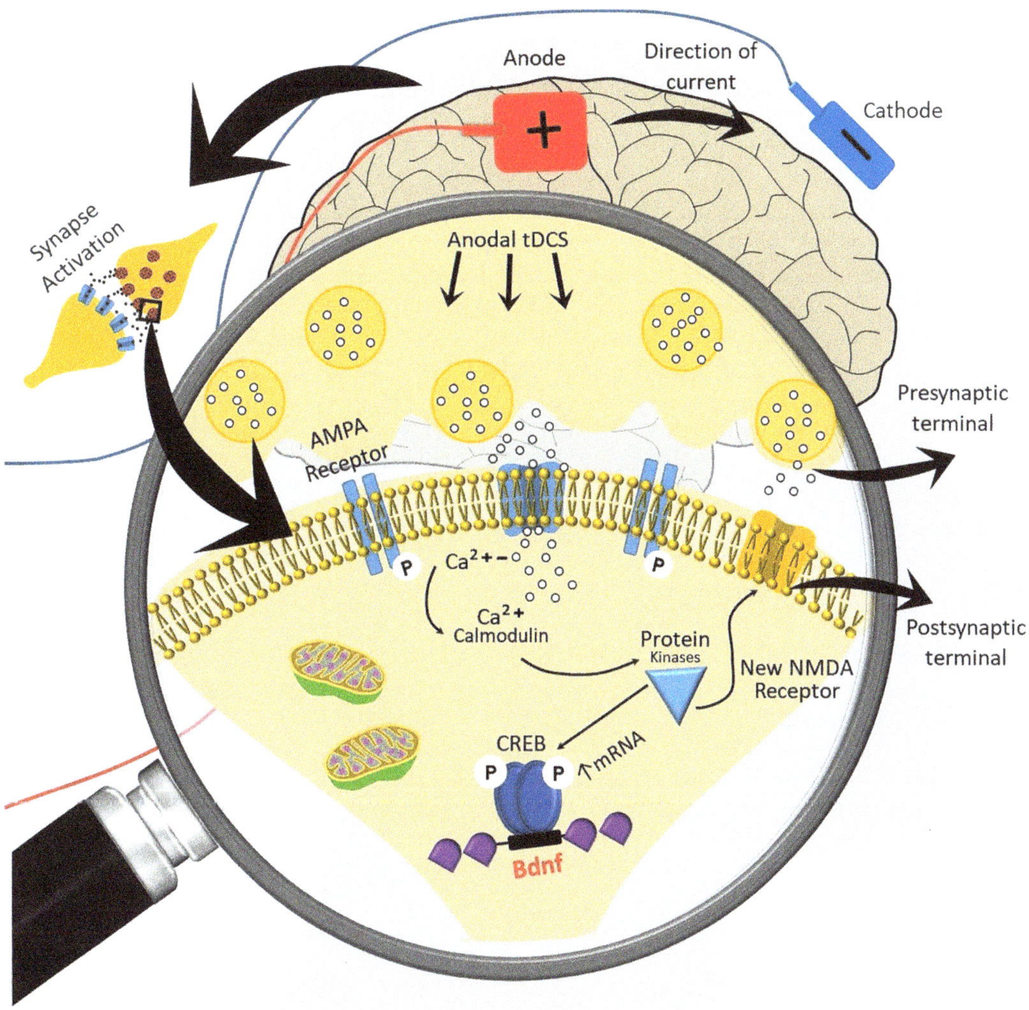

FIG. 2 Neurophysiological basis of a-tDCS primary effects. a-tDCS current flow modulates dependent voltage channel causing depolarization of the presynaptic membrane. The neurotransmitters release activates the NMDA receptors in the postsynaptic membrane and increases the influx of calcium, activating protein kinases signalization to a BDNF gene transcription. New NMDA receptors and, therefore, neuroplasticity like a long-term potentiation occurs. Unpublished. *tDCS*, transcranial Direct Current Stimulation; *NMDA*, *N*-methyl-D-aspartate; *AMPA*, alpha-amino-3-hydroxy-methyl-5-4-isoxazol-propionic; *Ca*, Calcium; *P*, Potassium; *CREB*, cAMP response element-binding protein, an intracellular protein related to expression of genes; *mRNA*, messenger RiboNucleic Acid; *Bndf*, Brain-derived neurotrophic factor.

with an influx of sodium (Horvath, Vogrin, Carter, Cook, & Forte, 2016; Jamil et al., 2020). These effects are closely linked to the potential in reducing the depolarization threshold of neurons because axons demonstrate susceptibility to changes in membrane potential (Jamil et al., 2020; Lefaucheur et al., 2017). When the DC induced by the a-tDCS reaches the neurons a modification in the transmembrane conduction, axonal transport and the structural conformation of plasmalemma may occur. Thus, a-tDCS facilitates neuronal depolarization, the activity of neurons, and, therefore, enhances CE (Horvath et al., 2016; Jamil et al., 2020). Fig. 2 illustrates the primary effects of a-tDCS in the neuronal membrane.

a-tDCS primary effects at the neuronal membrane induce the secondary effects (Lefaucheur et al., 2017) linked to neuroplastic changes in the neuronal circuit. The neuroplasticity induced by a-tDCS appears to be related to a repetitive activation of the *N*-methyl-D-aspartate (NMDA) receptors, which induces an influx of calcium in the postsynaptic neurons. As a consequence, a cascade of neurophysiological changes may be activating a signalization to a Brain-derived neurotrophic factor (BDNF) gene transcription to new NMDA receptors (Jamil et al., 2020; Lefaucheur et al., 2017; Podda et al., 2016). Finally, dependent activity neuroplasticity like a long-term potentiation in the glutamatergic neurons at the postsynaptic membrane may occur. Besides that, a-tDCS has been demonstrated to reduce the gabaergic transmission by a blockage of the receptors of gamma-aminobutyric acid, conducting a reinforcement of the glutamatérgic synapses. Therefore, a-tDCS interferes with the functional connectivity of cortical and subcortical networks.

Previous studies observed a-tDCS effects in the glial, endothelial, and lymphatic cells also. Effects on endothelial cells were related to increased cerebral blood flow. Moreover, glial cells respond to a-tDCS improving the regulation of the neurons. These effects were pointed to as predictors of long-term effects stimulation due to the enhancement of substrates for neuronal functioning (Jamil et al., 2020).

In the iSCI, the a-tDCS potential to neuroplasticity by modulating the CE and motor descending pathways such as the corticospinal tract have an important place in rehabilitation. Previous studies showed the a-tDCS effect on the CE underlying the stimulation site, corticospinal pathways, and medullary neurons (Bunday & Perez, 2012; Lefaucheur et al., 2017). For example, the stimulation on the hand or leg motor area in the primary motor cortex (M1) appears to increase the CE and fire neurons from the descending motor circuit to the medullary neurons, which will generate muscle responses (Bunday & Perez, 2012; Cortes et al., 2017; Lefaucheur et al., 2017; Potter-Baker et al., 2018; Yamaguchi et al., 2016; Yozbatiran et al., 2016). Thus, the knowledge of the a-tDCS neuroplasticity effects is important to formulate an ideal intervention to promote effective rehabilitation. The neuronal circuit activated by a-tDCS on M1 is shown in Fig. 3.

Importance of the a-tDCS after iSCI

Impairments in motor and sensory functions after iSCI generates a significant reduction in the functionality and quality of life. Besides, morbidities, psychological stress, and the changes imposed by the injury impair the reintegration of affected individuals into activities of daily living and work (Khorasanizadeh et al., 2019). Therefore, effective interventions and adequate management should be a priority to the rehabilitation after iSCI.

Epidemiological data help to understand the profile of iSCI affected individuals and appropriate rehabilitation. The International Spinal Cord Society points to a global incidence rate of 23 cases per million, resulting in approximately 179 thousand new cases annually. Overall, men have been more affected than women with odds ratios ranging from 2.4:1 to 4:1. Incomplete injuries were more frequent—ranging between 47.8% and 80.5%. A high number of men over the age of 40 and working-age individuals have been victims in developed and developing countries, respectively. Findings motivate researchers for effective rehabilitation. Thus, the a-tDCS has been pointed out as a potential intervention to reach better functionality and quality of life.

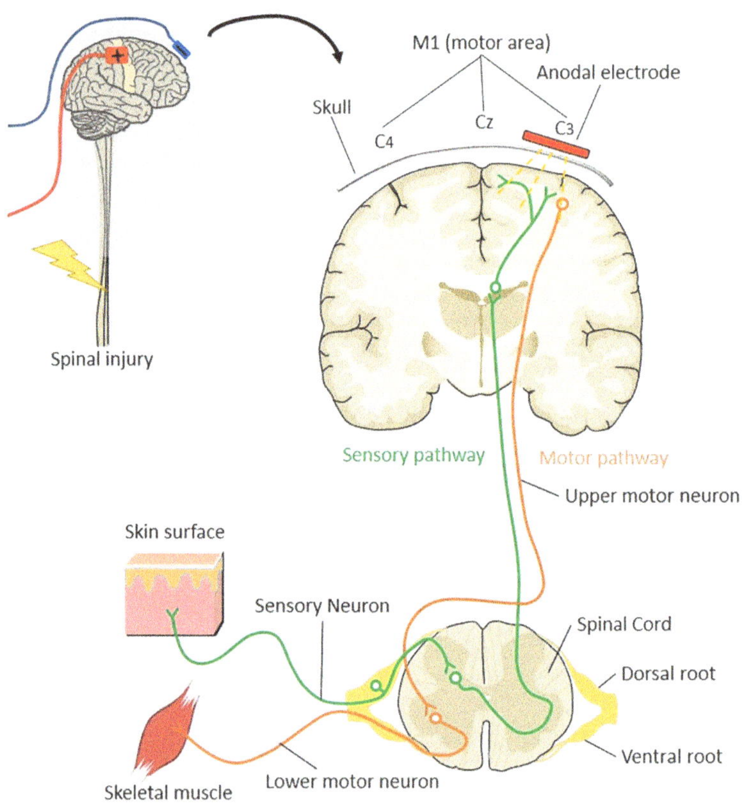

FIG. 3 Neuronal circuit activated by a-tDCS on M1 and electrode site. a-tDCS at M1 increases the CE and fire neurons from the descending motor circuit to the medullary neurons, generating muscle and sensory responses. Unpublished. *iSCI*: incomplete spinal cord injury; *a-tDCS*: anodal transcranial direct current stimulation; *CE*: cortical excitability.

A-tDCS outcomes after iSCI

CE and neuroplasticity-related outcomes

The CE was found as an outcome related to a-tDCS effects in protocols to improve motor function (MF). Two previous studies reported increases in the CE after one or five sessions at 1 or 2 mA during 20/60 min on the M1 (Murray et al., 2015; Potter-Baker et al., 2018) (see Table 1). Murray et al. showed modulation of CE through the motor evoked potential (MEP) measurements in the extensor carpi radialis immediately after a-tDCS on M1, but not maintained as a long-term effect.

TABLE 1 a-tDCS protocols parameters applied to iSCI.

Study (year)	Outcomes	Anodal electrode site	Cathodal electrode site	Electrode size	Intensity	Current density	Session duration	Sessions (n)
Soler et al. (2010)	Neuropathic Pain	C3 or C4	Contralateral supraorbital area	35 cm^2	2 mA	0.0571 mA/cm^2	20 min	10
Kumru et al. (2013)	Neuropathic Pain	C3 or C4	Contralateral supraorbital area	35 cm^2	2 mA	0.0571 mA/cm^2	20 min	10
Yoon et al. (2014)	Neuropathic Pain	C3	Contralateral supraorbital area	–	2 mA	–	20 min	20
Ngernyam et al. (2015)	Neuropathic Pain	C3 or C4	Contralateral shoulder	35 cm^2	2 mA	0.0571 mA/cm^2	20 min	1
Li, Stampas, Frontera, and Davis, (2018); Li, Fan, Yang, He, and Li, (2018)	Neuropathic Pain	C3	Contralateral supraorbital area	35 cm^2	2 mA	0.0571 mA/cm^2	20 min	1
Soler, Moriña, Kumru, Vidal, and Navarro (2020)	Neuropathic Pain	C3 or C4	Contralateral supraorbital area	25 cm^2	2 mA	0.08 mA/cm^2	20 min	10
Thibaut, Carvalho, Morse, Zafonte, and Fregni (2017)	Neuropathic Pain/Quality of life	M1	Contralateral supraorbital area	35 cm^2	2 mA	0.0571 mA/cm^2	20 min	5–10
Salmon et al. (2014)	Motor Function	M1	Contralateral supraorbital area	25 cm^2	2 mA	0.08 mA/cm^2	20 min	12
Murray et al. (2015)	Cortical excitability	C3	Contralateral supraorbital area	3.14 cm^2	1 mA	0.31-or 0.63 mA/cm^2	20 min	1
Kumru, Murillo, Benito-Penalva, Tormos, and Vidal (2016)	Motor Function/ Functionality	M1	Contralateral supraorbital area	35 cm^2	2 mA	0.0571 mA/cm^2	20 min	20

Continued

TABLE 1 a-tDCS protocols parameters applied to iSCI—cont'd

Study (year)	Outcomes	Anodal electrode site	Cathodal electrode site	Electrode size	Intensity	Current density	Session duration	Sessions (n)
Raithatha et al. (2016)	Motor Function/ Functionality	M1	Contralateral supraorbital area	25 cm^2	2 mA	0.08 mA/cm^2	20 min	36
Yamaguchi et al. (2016)	Motor Function/ Functionality	M1	Ipsilateral supraorbital area	35 cm^2	1 mA	0.028 mA/cm^2	20 min	1
Yozbatiran et al. (2016)	Motor Function/ Functionality	C3 or C4	Contralateral supraorbital area	35 cm^2	2 mA	0.0571 mA/cm^2	20 min	10
Yozbatiran et al. (2017)	Motor Function/ Functionality	C3 or C4	Contralateral supraorbital area	35 cm^2	2 mA	0.0571 mA/cm^2	20 min	10
Cortes et al. (2017)	Motor Function/ Functionality	C3 or C4	Contralateral supraorbital area	3 cm^2	1 mA	0.6 or 0.3 mA/cm^2	20 min	1
Potter-Baker et al. (2017)	Motor Function/ Functionality	C3 or C4	Contralateral supraorbital area	35 cm^2	2 mA	0.0571 mA/cm^2	30 min	10
da Silva et al. (2017)	Autonomic modulation	Cz	Occipital protuberance area	35 cm^2	2 mA	0.055 mA/cm^2	12 min	1

Table shows a-tDCS parameters applied in each study based on the main outcomes. Unpublished.
a-tDCS, transcranial direct current stimulation; *iSCI*, incomplete spinal cord injury; *C3 or C4*, hand motor area; *M1*, primary motor cortex; *Cz*, leg motor area; *cm^2*, square centimeters; *mA*, milliampere; – not available; *min*, minutes; *n*, number of sessions.

Increases in the CE has been associated with a boost of the neuroplasticity on the spinal cord, which facilitates the movement of weak muscles below the level of injury (Carmel, Berrol, Brus-Ramer, & Martin, 2010) and contribute to more lasting improvements in motor function after a-tDCS (Potter-Baker et al., 2018). Potter-Baker et al. demonstrated the increased MEPs after stimulation on M1 at a site related to the most affected and weakest muscle below the level of the injury. Improvements in MF on chronic iSCI, which were maintained for 3 months, appears to be related to neuroplasticity in the volume of the motor map related to the weak muscles and CE enhancements. The results were stronger in this study due to the greater number of the sessions as well as the a-tDCS uses paired with massed practice training.

Motor function and functionality outcomes

The crucial role of neuroplasticity of the residual cortical and spinal pathways to reach a recovery of the MF has been reported (Fawcett et al., 2007). Stimuli to neuroplasticity occurs by the modulation of CE and reinforcement of the synapses (Martin, 2016). Thus, a-tDCS have been shown as a potential intervention that might be able to promote neuroplasticity, MF rehabilitation, and functionality in iSCI individuals.

The current evidence shows some studies reporting positive a-tDCS effects on MF or functionality after iSCI. A systematic review and meta-analysis based on the six randomized clinical trials showed a significant improvement in functionality, but not muscle strength (De Araújo et al., 2020) after a-tDCS at 2 mA from 20 to 30 min with the anode on the leg or hand motor area. Overall, studies applied to the a-tDCS paired with conventional therapies, which appears to reinforce the effects (see Table 1).

Some studies showed a significant improvement on the hand and leg MF immediately after isolated a-tDCS at 2 mA or when the stimulation was paired with specific upper or lower extremities training. Improvements in muscle strength and functionality were observed immediately and lasted for up to 3 months after the intervention (Cortes et al., 2017; Raithatha et al., 2016; Yozbatiran et al., 2017). Under these results, the study of Yamaguchi et al. demonstrated a significantly increased number of ankle movements immediately after the a-tDCS at 1 mA on the leg motor area. On the other hand, Kumru et al. despite reporting a significant improvement in the muscle strength of lower extremities, the results did not significantly differ from the group of participants who received the a-tDCS in the sham mode. Thus, the results of the previous studies suggest a small moderate short-term effect of a-tDCS following iSCI (de Araújo et al., 2020), which needs to be confirmed by future randomized clinical trials.

Studies reported positive effects of the a-tDCS on the MF or functionality based on finds that stimulation could modulate excitability on the cortical motor area and corticospinal pathways, which might cause spinal plasticity and, therefore, improvements in hand or leg functions on iSCI (Cortes et al., 2017; Potter-Baker et al., 2018; Yamaguchi et al., 2016; Yozbatiran et al., 2016). Overall, most studies reported positive results of the a-tDCS in people with chronic iSCI, which needs to reinforce the neuroplasticity to continued gains in the MF (de Araújo et al., 2020).

Chronic neuropathic pain (NP)

Chronic NP has been frequently related to iSCI because of the injuries on the nervous system affecting the somatosensory system (Treede et al., 2008). In the current evidence has been reported a-tDCS effects in iSCI individuals chronic NP. The a-tDCS was paired with breathing-controlled electrical stimulation, and results pointed to a reduction of pain (Li, 2013; Li, Stampas, Frontera, & Davis, 2018). The pain reduction was related to the greater number of sessions, which induced lasting neuroplasticity and synaptic strengthening. Moreover, it appears that the higher the pain levels, the greater the reduction in pain after a-tDCS in iSCI individuals (Thibaut et al., 2017) (see Table 1). Despite the positive effects pointed out in this study, few researchers are aiming to investigate the a-tDCS effects of NP reduction after iSCI. Previous studies were formed by a mixed sample with iSCI and complete SCI (cSCI). For example, Yoon et al. (2014) showed a significant NP reduction, but not clinic reduction in the pain level after 20 sessions at 2 mA a-tDCS for 20 min. Similarly, Ngernyam et al. (2015) reported significant NP reduction in iSCI and cSCI individuals immediately after one a-tDCS session.

A meta-analysis evidenced the positive effects of a-tDCS in the NP in SCI but only 32.5% of the sample had iSCI. Results indicated a reduction of NP and highlighted a trend to greater positive effects in individuals with pain duration less than 5 years. Moreover, some studies pointed out that the a-tDCS effects appear more evident in the long term (Mehta, Mcintyre, Guy, Teasell, & Loh, 2015). Thus, the pain was smaller when measured after 1 week from the stimulation. Curiously, a meta-analysis recently published presented controversial results and reported no positive effect of the a-tDCS in NP after iSCI (Yu, Qiu, Li, Zhong, & Li, 2020).

Some studies have applied the a-tDCS at 2 mA paired with visual illusion in which individuals were in front of a mirror with projected legs simulating walking (Soler et al., 2010). Findings suggested the positive effects to reduce the perception of heat pain in rostral dermatomes at the level of injury (Kumru et al., 2013). However, the a-tDCS results in the NP must be observed with caution because are heterogeneous, based on nonstandardized protocols, and in a mixed sample (i.e., iSCI and cSCI).

Other a-tDCS outcomes

Some additional outcomes have been studied in the current researches with a-tDCS on iSCI individuals. Thibaut et al. (2017) applied five to 10 a-tDCS sessions at 2 mA to evaluate the quality of life, without significant results. Although the quality of life is a clinically relevant outcome impacted by the iSCI impairments, few studies aimed to observe this outcome. We highlight the need for future studies to provide evidence of a-tDCS effects on the quality of life.

Moreover, the autonomic nervous system (ANS) impairments due to immobility postinjury and physical inactivity (Serra-Añó et al., 2015) have been aimed by a few studies with a-tDCS after iSCI. Findings pointed to the positive a-tDCS effects on ANS. A heart rate variability modulation regardless of gender, type of injury, and time of injury was suggested and related to an increased CE and reduction on ANS dysfunctions (Da Silva et al., 2017). However, these results cannot be generalized for the entire iSCI population because they were carried out with a mixed sample.

Although in recent years the a-tDCS has been increasingly studied, there is still a shortage of studies, especially in some specific outcomes. To facilitate monitoring the development of studies with a-tDCS after iSCI, Table 2 summarizes some clinical trials in progress registered at clinicaltrials.gov or not yet published, which aims to assess motor and/or functional effects, reinforcing that this has been the main focus of research.

Characteristics of the iSCI individuals under a-tDCS protocols

The a-tDCS protocols for the rehabilitation should consider the profile of iSCI individuals (de Araújo et al., 2020). Studies presented samples with men, cervical or thoracic traumatic injuries, B and C injuries levels by the American Spinal Injury

TABLE 2 Clinical trials registered at clinicaltrials.gov in the year 2020.

Study Title	Register Number	Status	Sponsor	Experimental Intervention
Association Between tDCS and Lokomat Training in Patients with Incomplete Spinal Cord Injury	NCT02562001	Not recruiting	University of Sao Paulo General Hospital	tDCS during 20 min, before robotic device to gait training.
Combined Robotic Training and tDCS in Chronic SCI	NCT03555838	Completed (in 2019)	Burke Medical Research Institute	tDCS (2 mA) on M1 of the more affected arm, before robotic training.
Improving Spinal Cord Injury Rehabilitation Interventions by Retraining the Brain with Stimulation	NCT01539109	Completed (in 2020), a pilot study published	The Cleveland Clinic	tDCS (2 mA) during rehabilitation exercises of weak upper limbs.
Improving Hand Recovery with Neuromodulation in Tetraplegia	NCT03954496	Recruiting	University of Kentucky	tDCS during 20 min (2,5 mA) followed by intensive motor therapy of the more affected UE.
Improving SCI Rehabilitation Interventions by Retraining the Brain	NCT03892746	Recruiting	The Cleveland Clinic	tDCS (2 mA) to the motor cortex of the weaker upper limb during task-oriented practice.
Conditioning Neural Circuits to Improve UE Function	NCT02611375	Completed (in 2020)	Shepherd Center	tDCS (2 mA) over the hand area of the M1 while completing a functional task practice.

Data were based on the ongoing protocols published in the clinicaltrials.gov but not published in a scientific journal yet. Unpublished.
tDCS, transcranial direct current stimulation; *SCI*, spinal cord injury; *UE*, upper extremity; *M1*, primary motor cortex; *mA*, milliampere.

Association Scale (ASIA) as B and C, and chronic stage, predominantly. Table 3 summarizes the characteristics of the iSCI sample from the current evidence. It is noticed that the profile of iSCI individuals was similar to indicated by the epidemiological global data. Overall, previous studies excluded individuals with drug instability, progressive neurodegenerative disorders, concomitant neurological conditions, inability to cooperate, individuals with absolute contraindication to a-tDCS (i.e., history of epilepsy, metallic implants in the head, pacemaker, and pregnant women) (Li, Stampas, et al., 2018; Murray et al., 2015; Potter-Baker et al., 2018).

TABLE 3 Profile of the iSCI individuals in the a-tDCS protocols and paired therapies.

		iSCI characteristics				
Study (year)	Outcomes	Sample with iSCI (iSCI plus cSCI)	Mean age	Injury level	Postinjury stage	a-tDCS paired therapies
Soler et al. (2010)[a]	Neuropathic Pain	8 (39)	45	Cervical and thoracic	Chronic	Visual illusion
Kumru et al. (2013)[a]	Neuropathic Pain	30 (52)	49.4	Cervical and thoracic	Chronic	Visual illusion
Yoon et al. (2014)[a]	Neuropathic Pain	6 (16)	44.1	Cervical and thoracic	Chronic	None
Ngernyam et al. (2015)[a]	Neuropathic Pain	11 (20)	44.5	Cervical and thoracic	Chronic	None
Li, Stampas, Frontera, and Davis, (2018); Li, Fan, Yang, He, and Li, (2018)	Neuropathic Pain	12	43.4	Cervical and thoracic	Chronic	Breathing-controlled electrical stimulation

TABLE 3 Profile of the iSCI individuals in the a-tDCS protocols and paired therapies—cont'd

Study (year)	Outcomes	Sample with iSCI (iSCI plus cSCI)	Mean age	iSCI characteristics		a-tDCS paired therapies
				Injury level	Postinjury stage	
Soler et al. (2020)[a]	Neuropathic Pain	70 (130)	49	Cervical and thoracic	Chronic	Visual illusion
Thibaut et al. (2017)	Neuropathic Pain/ Quality of life	33	51.1	–	Chronic	None
Salmon et al. (2014)	Motor Function	2	46	Cervical	Chronic	Task-oriented UE training
Murray et al. (2015)	Cortical excitability	9	40.7	Cervical	Chronic	None
Kumru et al. (2016)	Motor Function/ Functionality	24	51.2	Cervical and thoracic	Acute	Gait training
Raithatha et al. (2016)	Motor Function/ Functionality	15	47.5	Cervical, thoracic, and lumbar	Chronic	Gait training
Yamaguchi et al. (2016)[a]	Motor Function/ Functionality	11 (21)	51.8	Cervical and thoracic	Chronic	Patterned Electrical Stimulation
Yozbatiran et al. (2016)	Motor Function/ Functionality	9	52.7	Cervical	Chronic	Robotic-Assisted Arm Training
Yozbatiran et al. (2017)	Motor Function/ Functionality	4	49.7	Cervical	Chronic	Robotic-Assisted Arm Training
Cortes et al. (2017)	Motor Function/ Functionality	11	44.9	Cervical	Chronic	None
Potter-Baker (2017, 2018)	Motor Function/ Functionality	8	53.5	Cervical	Chronic	Massed practice training
da Silva et al. (2017)	Autonomic modulation	18	32.9	Thoracic	Chronic	None

Data of age and injury level were based on the mean and percentage, respectively. Unpublished.
iSCI, incomplete spinal cord injury; *cSCI*, complete spinal cord injury; –, not available; *a-tDCS*, anodal transcranial direct current stimulation; *UE*, upper extremities.
[a]*Mixed sample (cSCI plus iSCI).*

Parameters of the a-tDCS after iSCI

Overall, the skin impedance level, neuropathological status, concomitant medications or therapies, and anatomical characteristics can be factors related to the resultant a-tDCS effect (Fregni et al., 2020). Besides that, the a-tDCS parameters such as intensity, current density, stimuli duration, number of sessions, targeted brain area, electrode size, and the electrode site should be planned to improve the effectiveness of the stimulation for a specific neurological condition.

The intensity of a-tDCS reaches a range of 1–2 mA. Although the relationship between a-tDCS parameters and mechanisms associated is not fully understood and not necessarily linear, there is evidence that the magnitude of excitability promoted by the a-tDCS is intensity-dependent (Cortes et al., 2017; Murray et al., 2015). Despite some protocols with 4 mA have been applied in human studies, none of these were applied after iSCI.

The electrode type used in a-tDCS presented a rubber composition involved in sponges soaked in saline. Overall, the electrode size ranges from 25 to 35 cm^2. Previous studies pointed to a relationship between the electrode size and the current density. The ratio between electrode size and current density (mA/cm^2) should be carefully observed to guarantee the correct quantity of direct current reaching the neuronal circuits (Elsner, Kugler, Pohl, & Mehrholz, 2019; Fregni et al., 2020). The most frequent densities to the a-tDCS in iSCI range from 0.05–0.63 mA/cm^2 and studies showed that high densities were related to feelings of discomfort (Fregni et al., 2020).

Regarding the stimuli duration, some studies reported the negative effect of the prolonged a-tDCS in the CE, therefore, the maximum stimuli duration should be between 10 and 20 min when applied for more than one session or 20–40 min in the case of a single session (Truong & Bikson, 2018). None of the protocols applied in iSCI reached a duration above 20 min. Also, a-tDCS sessions were observed ranging from one to 36 sessions in the current protocols (de Araújo et al., 2020; Mehta et al., 2015). However, there is not a consensus about the more effective number of sessions.

The electrode site was pointed in a range of studies as a variable responsible to change the current flow. The brain target has been defined by the International 10/20 Electroencephalogram System to find the brain target area (Rich & Gillick, 2019). In the 10/20 system, the target brain areas are found from the location of four individual anatomical landmarks: (a) nasion (lower depression between the forehead and the nose); (b) inion (the lowest point of the skull at the back of the head) and (c) preauricular points of each ear (Rich & Gillick, 2019). In general, the brain areas most stimulated after iSCI were located on the M1 (Fig. 3).

In the a-tDCS, the anode should be placed in a target brain area, while the cathode is placed over the supraorbital region (i.e., bicephalic montage) or on the trapezius muscle (i.e., monocephalic montage) (Elsner et al., 2019; Truong & Bikson, 2018). Bicephalic montages have been used in most a-tDCS studies in iSCI. The electrode site should be maintained through a safety electrode fixation because small deviations in the position can lead to significant changes in the current density (Fregni et al., 2020). The a-tDCS parameters applied to iSCI individuals based on the current evidence are summarized in Table 1.

Short and long-term effects of the a-tDCS

a-tDCS short-term effects have been referred to as immediate effects or in an early period, whereas the long-term effects corresponded to a prolonged effect after stimulation. One recent systematic review aiming to investigate the a-tDCS effects on MF and functionality on iSCI reported a lack of studies with assessments of long-term effects (de Araújo et al., 2020). Similarly, a systematic review of the a-tDCS effects on NP found a lack of adequate follow-up to observe the long-term effects (Mehta et al., 2015). Thus, the current findings of a-tDCS effects after iSCI have been restricted to last up to 3 months.

Overall, the a-tDCS effects appear to be reinforced when paired with other therapies. A systematic review found that prolonged effects were associated with a-tDCS paired with conventional therapies (de Araújo et al., 2020). Besides that, improvements on proximal arm movement and a higher amount of use to upper extremities with a lasting effect of at least 2 months were observed after a-tDCS paired with Robotic-Assisted Arm Training (Yozbatiran et al., 2016, 2017). The paired a-tDCS and Patterned Electrical Stimulation induced a significantly higher number of ankle movements 20 min after the interventions (Yamaguchi et al., 2016). Whereas a-tDCS with massed practice shows a trend in improving the strength of muscles last up to 3 months (Potter-Baker et al., 2018). Similarly, a-tDCS paired with lower extremity training, a significant improvement was evident for both lower extremities and gait velocity for at least 1 month (Raithatha et al., 2016). On the other hand, Kumru et al. (2016) performed a similar protocol but did not obtain positive results in the same outcomes (see Table 3).

A-tDCS safety and adverse effects (AEs)

AEs can be classified as the level of severity in a range of grades. Grade 1 corresponds to AEs without the need for medical intervention. In grade 2, AEs need noninvasive medical intervention. On the other hand, AEs into grades 3 and 4 are defined as serious. In these occurs the need for hospitalization or prolonged hospitalization, or life risk, respectively. AEs with a higher grade are associated with an event that results in death (Antal et al., 2017).

No serious AEs (i.e., 3, 4, or higher grade) were reported in the current evidence about the a-tDCS after iSCI. The AEs most observed were mild with grade 1 such as tingling, burning sensation, discomfort, headache, and itching (Kumru et al., 2016; Lefaucheur et al., 2017; Murray et al., 2015; Ngernyam et al., 2015). Some studies reported mild headache, redness, fatigue, dizziness, mild skin irritation, and somnolence also (Fregni et al., 2020; Lefaucheur et al., 2017; Yozbatiran et al., 2016). The AEs were related to electrode positioning, preexisting skin conditions, high impedance, long durations, and repeated sessions, or indiscriminate home use.

Thus, additional care during a-tDCS needs to be taken. First, to avoid skin irritation, by contact of metal or conductive rubber with the skin, the thickness of the sponge must be controlled (Truong & Bikson, 2018). Second, to prevent increased impedance on the scalp the patient should be advised to avoid using skin and/or hair products on stimulation days (Fregni et al., 2020). Moreover, the correct skin preparation, use of a suitable electrode, nonapplication on sensitive skin regions, and instructions for the patient not to perform skin abrasion and to report any discomfort quickly can prevent the appearance of AEs (Antal et al., 2017; Lefaucheur et al., 2017). A study reported the interference of AEs in the adherence rate to the a-tDCS protocol (Carvalho et al., 2018). Thus, the care to avoid the occurrence of AEs is important for better adherence to treatment as well as the correct identification.

Clinical practice based on the a-tDCS

The current evidence highlights a-tDCS as a promising therapy to induce a change in the nervous system excitability and, consequently, neuroplasticity. In addition to the initial evidence of positive a-tDCS effects, their use has been reported by the ease of administration, possibility of self-application, and relatively low price. A meta-analysis suggests the significant a-tDCS effect on MF on iSCI individuals, but the limited number of studies included in the study makes it necessary to carry out future studies (de Araújo et al., 2020). Similarly, a recommendation of a level C of evidence to a-tDCS effects in the NP reduction was made (Lefaucheur et al., 2017). Despite being identified as a potential therapy, there are still some questions unanswered about the a-tDCS effects after iSCI. Currently, there are no bases to support or refute the use of stimulation in daily clinical practice.

Some aspects highlighted regarding the paths for future studies were: evaluation of the neurophysiological mechanisms of a-tDCS through neuroimaging; a consensus of the protocol parameters related to a better response in the iSCI; definition of the time postinjury, lesion level, and type of injury affecting the a-tDCS results; designing of studies to evaluate the long-term effects; greater sample size; a combination of a-tDCS with standard therapies (Cortes et al., 2017; de Araújo et al., 2020; Jamil et al., 2020); AEs, cost-effectiveness, adherence level and factors contributing to the retention of the individuals in the protocols (Carvalho et al., 2018; Kim, 2011). Similar to the a-tDCS, other interventions experimentally have shown initial results of effects after iSCI (See Table 4).

Applications to other areas of neuroscience

The anodal transcranial direct current stimulation has been applied in a range of neurological conditions. In addition to the previously reported outcomes, some studies have shown the effects of the stimulation on depression. Prevalence of depression after incomplete spinal cord injury has been high and, therefore, it is important to discuss the effects in these cases (Williams & Murray, 2015). Borrione, Moffa, Martin, Loo, and Brunoni (2018) showed improvements in the depressive symptoms after anodal transcranial direct current stimulation and probable effectiveness when the anode is positioned in the left dorsolateral prefrontal cortex.

Considering neurological conditions with similar involvement to incomplete spinal cord injury, the postpolio syndrome appears to respond positively to anodal transcranial direct current stimulation with a reduction in muscle fatigue and improvements in quality of life (Matsushima, Hachisuka, Itoh, Sugimoto, & Saeki, 2019). Moreover, in the stroke, previous studies have shown a moderate increase in cortical excitability, improvements in the motor function, mobility, and muscle strength promoting effective short-term motor performance poststimulation, but not in reducing upper extremity spasticity, walking speed, and muscle endurance (Elsner et al., 2019; Li, Fan, Yang, He, & Li, 2018). The stimulation to Parkinson's disease also over M1 resulted in immediate motor gains, better walking performance, and decreased time to perform movements of the upper limbs with a sustained effect for 3 months (Broeder et al., 2015).

Dictionary of terms

American Spinal Injury Association Scale: Standardized scale for the classification of spinal cord injury by assessing motor skills and sensitivity.

Autonomic nervous system: System controlling vegetative functions such as breathing, blood circulation, temperature control, and digestion.

Cortical excitability: Changes in the cortical neuronal activation by the reduction of the depolarization threshold.

Corticospinal tract: Tract with nerve fibers from the central nervous system to the spinal cord to controls muscles.

TABLE 4 Experimental interventions similar to a-tDCS used after iSCI.

Experimental modality	Characteristics of the intervention	Outcomes
Transcutaneous Spinal Direct Current Stimulation (tSCS)	Electrodes are placed over the spinal process and other electrodes are placed abdominally. The devices work with biphasic pulses. Abdominal electrodes work as an anode in the first and as a cathode in the second phase	Voluntary locomotor activity, motor or sensory function, and cortical excitability
Transcranial Magnetic Stimulation (TMS)	Electrical transcranial stimuli are delivered to induce changes in brain activity through single or repetitive pulses varying between <1 Hz and 20 Hz. A coil is frequently placed over a targeted brain area	Spasticity, motor, and sensory function
Theta-burst stimulation (iTBS)	The magnetic stimuli are applied through a circular coil over a brain area. Three stimuli at 50 Hz repeated at 200 ms intervals for 2 s were applied	Spasticity and motor function
Spinal Associative Stimulation (Peripheral nerve stimulation plus TMS)	TMS applied through a coil to the brain area paired with electrical stimulation delivered with surface bipolar electrodes at single 200 μs rectangular pulses	Spinal excitability
Deep Brain Stimulation	The invasive technique is applied by a microelectrode surgically. The microelectrode is implanted and connected to a pulse generator—generally located in the clavicle, similarly a pacemaker—to induce current to the tissue	Neuropathic Pain
Pelvic neuromodulation	The invasive technique in which electrodes are inserted into the third sacral foramina and are stimulated by an external pulse generator. The pulse generator can be implanted in the patient's buttock	Pelvic function and global health

Data were based on the studies with positive effects in individuals with iSCI published until 2020. Unpublished.
a-tDCS, transcranial direct current stimulation; *iSCI,* incomplete spinal cord injury.

Direct current: A current with a flow of electrons in a single direction through the presence of a potential difference, from the negative pole to the positive pole.
Functionality: Ability of the motor function to respond correctly and to generate movements to assist the activities of daily living.
Long-term potentiation: Persistent increasements in synaptic strength, making it more effective and lasting.
Motor Evoked Potential. Electrical response evoked in a muscle or motor nerve by electrical stimulation.
Motor function. Ability to learn or to demonstrate the skillful, maintenance, modification, and control of voluntary postures and movement patterns.
Neuroplasticity. Ability to the adaptation to the nervous system, as a result of repetitive activations.
Neuropathic Pain. Pain that occurs along the course or distribution of a peripheral nerve or sensitive nerves.

Key facts of anodal transcranial direct current stimulation

- Direct current was created before Christ. However, it was the creation of a voltaic battery in 1800, that made viable transcranial stimulation.
- The brain area to be stimulated must be chosen based on the International 10/20 Electroencephalogram System, which allows the stimulation of specific areas related to the function that is to be improved.
- When applied on the motor primary cortex, anodal transcranial direct current stimulation activates cortical neurons, descending motor circuits, and spinal neurons, which makes possible the use as a potential intervention to improve rehabilitation.
- Long-term potentiation induced by the anodal transcranial direct current has been identified as one of the most important mechanisms of neuroplasticity.
- Simultaneous application of anodal transcranial direct current stimulation and standard therapies appears to produce more intense and longer effects.

Summary points

- Motor and sensory function impairments after an incomplete spinal cord injury cause a reduction in the functionality and quality of life, which should be targeted by effective rehabilitation.
- Transcranial direct current stimulation uses a low-intensity direct current distributed to the brain through two electrodes, anode, and cathode.
- Current flow depends on the electrode site, in which anodal stimulation occurs from the anode to the cathode causing a depolarization threshold reduction in the neuronal membrane.
- Neurophysiological effects of the anodal transcranial direct current stimulation are related to the modulation of neuronal excitability and long-term effects linked to neuroplasticity.
- Anodal transcranial direct current stimulation protocol parameters must be planned considering the desired neurophysiological effects. Thus, frequency, intensity, electrode size, and position must be specific for a targeted outcome.
- Effects on the cortical excitability, neuroplasticity, motor function, or functionality, and chronic neuropathic pain appear to be a potential effective rehabilitation to reduce impairments after incomplete spinal cord injury.
- Future studies should observe the short-, medium-, and long-term effects potentially related to the anodal transcranial direct current stimulation to define the therapy value to clinical practice, cost-effectiveness, and adherence level.

References

Antal, A., Alekseichuk, I., Bikson, M., Brockmöller, J., Brunoni, A. R., Chen, R., … Paulus, W. (2017). Low intensity transcranial electric stimulation: Safety, ethical, legal regulatory, and application guidelines. *Clinical Neurophysiology*, *128*(9), 1774–1809. https://doi.org/10.1016/j.clinph.2017.06.001.

Borrione, L., Moffa, A. H., Martin, D., Loo, C. K., & Brunoni, A. R. (2018). Transcranial direct current stimulation in the acute depressive episode: A systematic review of current knowledge. *The Journal of ECT*, *34*(3), 153–163. https://doi.org/10.1097/YCT.0000000000000512.

Broeder, S., Nackaerts, E., Heremans, E., Vervoort, G., Meesen, R., Verheyden, G., & Nieuwboer, A. (2015). Transcranial direct current stimulation in Parkinson's disease: Neurophysiological mechanisms and behavioral effects. *Neuroscience and Biobehavioral Reviews*, *57*, 105–117. https://doi.org/10.1016/j.neubiorev.2015.08.010.

Bunday, K. L., & Perez, M. A. (2012). Motor recovery after spinal cord injury is enhanced by strengthening corticospinal synaptic transmission. *Current Biology*, *22*(24), 2355–2361. https://doi.org/10.1016/j.cub.2012.10.046.

Carmel, J. B., Berrol, L. J., Brus-Ramer, M., & Martin, J. H. (2010). Chronic electrical stimulation of the intact corticospinal system after unilateral injury restores skilled locomotor control and promotes spinal axon outgrowth. *Journal of Neuroscience*, *30*(32), 10918–10926. https://doi.org/10.1523/JNEUROSCI.1435-10.2010.

Carvalho, S., Leite, J., Jones, F., Morse, L. R., Zafonte, R., & Fregni, F. (2018). Study adherence in a tDCS longitudinal clinical trial with people with spinal cord injury. *Spinal Cord*, *56*(5), 502–508. https://doi.org/10.1038/s41393-017-0023-5.

Cortes, M., Medeiros, A. H., Gandhi, A., Lee, P., Krebs, H. I., Thickbroom, G., & Edwards, D. (2017). Improved grasp function with transcranial direct current stimulation in chronic spinal cord injury. *NeuroRehabilitation*, *41*(1), 51–59. https://doi.org/10.3233/NRE-171456.

Da Silva, F. T. G., Browne, R. A. V., Pinto, C. B., Saleh Velez, F. G., Do Egito, E. S. T., Do Rêgo, J. T. P., … Fregni, F. (2017). Transcranial direct current stimulation in individuals with spinal cord injury: Assessment of autonomic nervous system activity. *Restorative Neurology and Neuroscience*, *35*(2), 159–169. https://doi.org/10.3233/RNN-160685.

De Araújo, A. V. L., Ribeiro, F. P. G., Massetti, T., Potter-Baker, K. A., Cortes, M., Plow, E. B., … de Mello Monteiro, C. B. (2020). Effectiveness of anodal transcranial direct current stimulation to improve muscle strength and motor functionality after incomplete spinal cord injury: A systematic review and meta-analysis. *Spinal Cord*, *58*(6), 635–646. https://doi.org/10.1038/s41393-020-0438-2.

Elsner, B., Kugler, J., Pohl, M., & Mehrholz, J. (2019). Transcranial direct current stimulation (tDCS) for improving aphasia in adults with aphasia after stroke. *Cochrane Database of Systematic Reviews*, *2019*(5). https://doi.org/10.1002/14651858.CD009760.pub4.

Erb, W. H. (1883). *Handbook of electro-therapeutics. Vol. 27.* Retrieved from https://books.google.com.br/books/about/Handbook_of_Electro_Therapeutics.html?id=vf1DswEACAAJ&redir_esc=y.

Fawcett, J. W., Curt, A., Steeves, J. D., Coleman, W. P., Tuszynski, M. H., Lammertse, D., … Short, D. (2007). Guidelines for the conduct of clinical trials for spinal cord injury as developed by the ICCP panel: Spontaneous recovery after spinal cord injury and statistical power needed for therapeutic clinical trials. *Spinal Cord*, *45*(3), 190–205. https://doi.org/10.1038/sj.sc.3102007.

Fregni, F., El-Hagrassy, M. M., Pacheco-Barrios, K., Carvalho, S., Leite, J., Simis, M., … Brunoni, A. R. (2020). Evidence-based guidelines and secondary meta-analysis for the use of transcranial direct current stimulation (tDCS) in neurological and psychiatric disorders. *The International Journal of Neuropsychopharmacology*. https://doi.org/10.1093/ijnp/pyaa051.

Horvath, J. C., Vogrin, S. J., Carter, O., Cook, M. J., & Forte, J. D. (2016). Effects of a common transcranial direct current stimulation (tDCS) protocol on motor evoked potentials was found to be highly variable within individuals over 9 testing sessions. *Experimental Brain Research*, *234*(9), 2629–2642. https://doi.org/10.1007/s00221-016-4667-8.

Jamil, A., Batsikadze, G., Kuo, H., Meesen, R. L. J., Dechent, P., Paulus, W., & Nitsche, M. A. (2020). Current intensity- and polarity-specific online and aftereffects of transcranial direct current stimulation: An fMRI study. *Human Brain Mapping*, *41*(6), 1644–1666. https://doi.org/10.1002/hbm.24901.

Khorasanizadeh, M. H., Yousefifard, M., Eskian, M., Lu, Y., Chalangari, M., Harrop, J. S., … Rahimi-Movaghar, V. (2019). Neurological recovery following traumatic spinal cord injury: A systematic review and meta-analysis. *Journal of Neurosurgery: Spine*, *30*(5), 683–699. https://doi.org/10.3171/2018.10.SPINE18802.

Kim, Y. (2011). Missing data handling in chronic pain trials. *Journal of Biopharmaceutical Statistics*, *21*(2), 311–325. https://doi.org/10.1080/10543406.2011.550112.

Kumru, H., Murillo, N., Benito-Penalva, J., Tormos, J. M., & Vidal, J. (2016). Transcranial direct current stimulation is not effective in the motor strength and gait recovery following motor incomplete spinal cord injury during Lokomat® gait training. *Neuroscience Letters*, *620*, 143–147. https://doi.org/10.1016/j.neulet.2016.03.056.

Kumru, H., Soler, D., Vidal, J., Navarro, X., Tormos, J. M., Pascual-Leone, A., & Valls-Sole, J. (2013). The effects of transcranial direct current stimulation with visual illusion in neuropathic pain due to spinal cord injury: An evoked potentials and quantitative thermal testing study. *European Journal of Pain*, *17*(1), 55–66. https://doi.org/10.1002/j.1532-2149.2012.00167.x.

Lefaucheur, J. P., Antal, A., Ayache, S. S., Benninger, D. H., Brunelin, J., Cogiamanian, F., … Paulus, W. (2017). Evidence-based guidelines on the therapeutic use of transcranial direct current stimulation (tDCS). *Clinical Neurophysiology*, *128*(1), 56–92. https://doi.org/10.1016/j.clinph.2016.10.087.

Li, S. (2013). Breathing-controlled Electrical Stimulation (BreEStim) for management of neuropathic pain and spasticity. *Journal of Visualized Experiments*, *71*. https://doi.org/10.3791/50077.

Li, Y., Fan, J., Yang, J., He, C., & Li, S. (2018). Effects of transcranial direct current stimulation on walking ability after stroke: A systematic review and meta-analysis. *Restorative Neurology and Neuroscience*, *36*(1), 59–71. https://doi.org/10.3233/RNN-170770.

Li, S., Stampas, A., Frontera, J., & Davis, M. (2018). Combined transcranial direct current stimulation and breathing-controlled electrical stimulation for management of neuropathic pain after spinal cord injury. *Journal of Rehabilitation Medicine*, *50*(9), 814–820. https://doi.org/10.2340/16501977-2379.

Lippold, O. C., & Redfearn, J. W. (1964). Mental changes resulting from the passage of small direct currents through the human brain. *The British Journal of Psychiatry: The Journal of Mental Science*, *110*, 768–772. https://doi.org/10.1192/bjp.110.469.768.

Martin, J. H. (2016). Harnessing neural activity to promote repair of the damaged corticospinal system after spinal cord injury. *Neural Regeneration Research*, *11*(9), 1389–1391. https://doi.org/10.4103/1673-5374.191199.

Matsushima, Y., Hachisuka, A., Itoh, H., Sugimoto, K., & Saeki, S. (2019). Safety and feasibility of transcranial direct current stimulation for patients with post-polio syndrome. *Brain Stimulation*, *12*(2), 431. https://doi.org/10.1016/j.brs.2018.12.396.

Mehta, S., Mcintyre, A., Guy, S., Teasell, R. W., & Loh, E. (2015). Effectiveness of transcranial direct current stimulation for the management of neuropathic pain after spinal cord injury: A meta-analysis. *Spinal Cord*, *53*, 780–785. https://doi.org/10.1038/sc.2015.118.

Murray, L. M., Edwards, D. J., Ruffini, G., Labar, D., Stampas, A., Pascual-Leone, A., & Cortes, M. (2015). Intensity dependent effects of transcranial direct current stimulation on corticospinal excitability in chronic spinal cord injury. *Archives of Physical Medicine and Rehabilitation*, *96*(4), S114–S121. https://doi.org/10.1016/j.apmr.2014.11.004.

Ngernyam, N., Jensen, M. P., Arayawichanon, P., Auvichayapat, N., Tiamkao, S., Janjarasjitt, S., … Auvichayapat, P. (2015). The effects of transcranial direct current stimulation in patients with neuropathic pain from spinal cord injury. *Clinical Neurophysiology*, *126*(2), 382–390. https://doi.org/10.1016/j.clinph.2014.05.034.

Nitsche, M. A., & Paulus, W. (2000). Excitability changes induced in the human motor cortex by weak transcranial direct current stimulation. *Journal of Physiology*, *527*(3), 633–639. https://doi.org/10.1111/j.1469-7793.2000.t01-1-00633.x.

Podda, M. V., Cocco, S., Mastrodonato, A., Fusco, S., Leone, L., Barbati, S. A., … Grassi, C. (2016). Anodal transcranial direct current stimulation boosts synaptic plasticity and memory in mice via epigenetic regulation of Bdnf expression. *Scientific Reports*, *6*(1), 1–19. https://doi.org/10.1038/srep22180.

Potter-Baker, K. A., et al. (2017). Transcranial direct current stimulation (tDCS) paired with massed practice training to promote adaptive plasticity and motor recovery in chronic incomplete tetraplegia: A pilot study. *Journal of Spinal Cord Medicine*, *41*(5), 503–517. https://doi.org/10.1080/10790268.2017.1361562.

Potter-Baker, K. A., et al. (2018). Variability of motor evoked potentials in stroke explained by corticospinal pathway integrity. *Brain Stimulation*, *11*(4).

Raithatha, R., Carrico, C., Powell, E. S., Westgate, P. M., Chelette, K. C., Lee, K., … Sawaki, L. (2016). Non-invasive brain stimulation and robot-assisted gait training after incomplete spinal cord injury: A randomized pilot study. *NeuroRehabilitation*, *38*(1), 15–25. https://doi.org/10.3233/NRE-151291.

Renga, V. (2020). Electricity, neurology, and noninvasive brain stimulation: Looking back, looking ahead. *Neurology Research International*. https://doi.org/10.1155/2020/5260820.

Rich, T. L., & Gillick, B. T. (2019). Electrode placement in transcranial direct current stimulation—How reliable is the determination of C3/C4? *Brain Sciences*, *9*(3). https://doi.org/10.3390/brainsci9030069.

Salmon, E., Carrico, C., Nichols, L., Reddy, L., Salles, S., & Sawaki, L. (2014). Transcranial direct current stimulation to enhance motor function in spinal cord injury: Pilot data. In *Vol. 2014. 2014 IEEE 16th international conference on E-health networking, applications and services* (pp. 1–6). Healthcom. https://doi.org/10.1109/HealthCom.2014.7001804.

Sarmiento, C. I., San-Juan, D., & Prasath, V. B. S. (2016). Letter to the Editor: Brief history of transcranial direct current stimulation (tDCS): From electric fishes to microcontrollers. *Psychological Medicine*, *46*, 3259–3261. https://doi.org/10.1017/S0033291716001926.

Serra-Añó, P., Montesinos, L. L., Morales, J., López-Bueno, L., Gomis, M., García-Massó, X., & González, L. M. (2015). Heart rate variability in individuals with thoracic spinal cord injury. *Spinal Cord*, *53*(1), 59–63. https://doi.org/10.1038/sc.2014.207.

Soler, M. D., Kumru, H., Pelayo, R., Vidal, J., Tormos, J. M., Fregni, F., … Pascual-Leone, A. (2010). Effectiveness of transcranial direct current stimulation and visual illusion on neuropathic pain in spinal cord injury. *Brain*, *133*(9), 2565–2577. https://doi.org/10.1093/brain/awq184.

Soler, D., Moriña, D., Kumru, H., Vidal, J., & Navarro, X. (2020). Transcranial direct current stimulation and visual illusion effect according to sensory phenotypes in patients with spinal cord injury and neuropathic pain. *The Journal of Pain*. https://doi.org/10.1016/j.jpain.2020.06.004.

Thibaut, A., Carvalho, S., Morse, L. R., Zafonte, R., & Fregni, F. (2017). Delayed pain decrease following M1 tDCS in spinal cord injury: A randomized controlled clinical trial. *Neuroscience Letters*, *658*, 19–26. https://doi.org/10.1016/j.neulet.2017.08.024.

Treede, R. D., Jensen, T. S., Campbell, J. N., Cruccu, G., Dostrovsky, J. O., Griffin, J. W., ... Serra, J. (2008). Neuropathic pain: Redefinition and a grading system for clinical and research purposes. *Neurology*, *70*(18), 1630–1635. https://doi.org/10.1212/01.wnl.0000282763.29778.59.

Truong, D. Q., & Bikson, M. (2018). Physics of transcranial direct current stimulation devices and their history. *The Journal of ECT*, *34*(3), 137–143. https://doi.org/10.1097/YCT.0000000000000531.

Williams, R., & Murray, A. (2015). Prevalence of depression after spinal cord injury: A meta-analysis. *Archives of Physical Medicine and Rehabilitation*, *96*(1), 133–140. https://doi.org/10.1016/j.apmr.2014.08.016.

Yamaguchi, T., Fujiwara, T., Tsai, Y. A., Tang, S. C., Kawakami, M., Mizuno, K., ... Liu, M. (2016). The effects of anodal transcranial direct current stimulation and patterned electrical stimulation on spinal inhibitory interneurons and motor function in patients with spinal cord injury. *Experimental Brain Research*, *234*(6), 1469–1478. https://doi.org/10.1007/s00221-016-4561-4.

Yoon, E. J., Kim, Y. K., Kim, H. R., Kim, S. E., Lee, Y., & Shin, H. I. (2014). Transcranial direct current stimulation to lessen neuropathic pain after spinal cord injury: A mechanistic PET study. *Neurorehabilitation and Neural Repair*, *28*(3), 250–259. https://doi.org/10.1177/1545968313507632.

Yozbatiran, N., Keser, Z., Davis, M., Stampas, A., O'Malley, M. K., Cooper-Hay, C., ... Francisco, G. E. (2016). Transcranial direct current stimulation (tDCS) of the primary motor cortex and robot-assisted arm training in chronic incomplete cervical spinal cord injury: A proof of concept sham-randomized clinical study. *NeuroRehabilitation*, *39*(3), 401–411. https://doi.org/10.3233/NRE-161371.

Yozbatiran, N., Keser, Z., Hasan, K., Stampas, A., Korupolu, R., Kim, S., ... Francisco, G. E. (2017). White matter changes in corticospinal tract associated with improvement in arm and hand functions in incomplete cervical spinal cord injury: Pilot case series. *Spinal Cord Series And Cases*, *3*(1), 1–5. https://doi.org/10.1038/scsandc.2017.28.

Yu, B., Qiu, H., Li, J., Zhong, C., & Li, J. (2020). Noninvasive brain stimulation does not improve neuropathic pain in individuals with spinal cord injury: Evidence from a meta-analysis of 11 randomized controlled trials. *American Journal of Physical Medicine and Rehabilitation*, *99*(9), 811–820. https://doi.org/10.1097/PHM.0000000000001421.

Further reading

Kashiwagi, F. T., El Dib, R., Gomaa, H., Gawish, N., Suzumura, E. A., Da Silva, T. R., ... Bazan, R. (2018). Noninvasive brain stimulations for unilateral spatial neglect after stroke: A systematic review and meta-analysis of randomized and nonrandomized controlled trials. *Neural Plasticity*, *2018*. https://doi.org/10.1155/2018/1638763.

Chapter 5

Neuromodulation and restoration of motor responses after severe spinal cord injury

Dimitry G. Sayenko[a], Humberto A. Cerrel Bazo[b,c], Philip J. Horner[a], and Giuliano Taccola[a,d]

[a]Department of Neurosurgery, Center for Neuroregeneration, Houston Methodist Research Institute, Houston, TX, United States, [b]Ospedale Riabilitativo di Alta Specializzazione (ORAS)-ULSS 2 TV, Motta di Livenza, TV, Italy, [c]Department of Neuroscience—Physical Medicine and Rehabilitation, University of Padova, Medical School, Padova, Italy, [d]Neuroscience Department, International School for Advanced Studies (SISSA), Bonomea, Trieste, Italy

List of abbreviations

CNS	central nervous system
CPGs	central pattern generators
ILS	infralesional segment
IM	injured metamere
MRI	magnetic resonance imaging
NMES	neuromuscular electrical stimulation
SLSs	supralesional segments
TSS	transcutaneous spinal stimulation

Neuromodulation exploits intrinsic information processing

The term neuromodulation refers to any intervention that affects the strength of the information transfer within a neuronal circuit. Neuromodulation is also defined as a physiological processing of synaptic input that is sustained by endogenous chemicals, named neuromodulators, which activate metabotropic receptors, that in turn determine a cascade of molecules within the cell. Chemical neuromodulators target the membrane, the cytoplasm or the nucleus, and can, in turn, affect network excitability through both gene expression and plastic synaptic events (Marder, O'Leary, & Shruti, 2014). The importance of chemical neuromodulators in regulating neuronal transmission is largely exploited for pharmacotherapy, as confirmed by the wide range of pharmacological agents available to activate metabotropic receptors, ranking among the most represented classes of drugs available.

Similar modulatory actions can also be achieved through several physical forces supplied to the neuronal networks, in the form of mechanical, magnetic, and electrical pulses.

In particular, the intensity of electrical pulses can be strong enough to directly trigger an action potential in excitable cells, thus generating a stereotyped response (also called reflex) to each delivered input. In this case, we refer to the term of **electrical stimulation**. For instance, electrical stimulation of the limb surface is used in clinics to depolarize motor end-plates and elicit passive contractions for functional purposes (recently reviewed by Günter, Delbeke, & Ortiz-Catalan, 2019) and even to directly activate paralyzed muscles after denervation (recently reviewed by Kern & Carraro, 2020). Repetitive electrical stimulation has been applied to the spinal cord of spinalized animals to evoke the patterned sequence of activation of multiple muscles, enabling the automatic execution of rhythmic and coordinated motor tasks. Each task, e.g., stepping, corresponds to a defined rhythmic motor program that is driven by the activity of a remarkable ensemble of spinal neurons, named **central pattern generators** (CPGs, Kiehn, 2006). In particular, locomotor CPGs are responsible for the expression of rhythmic electrical discharges, alternating among pools of motoneurons, even in the absence of any input from either the supraspinal centres or sensory feedback from the periphery (Fig. 1).

Cellular, Molecular, Physiological, and Behavioral Aspects of Spinal Cord Injury. https://doi.org/10.1016/B978-0-12-822427-4.00005-8

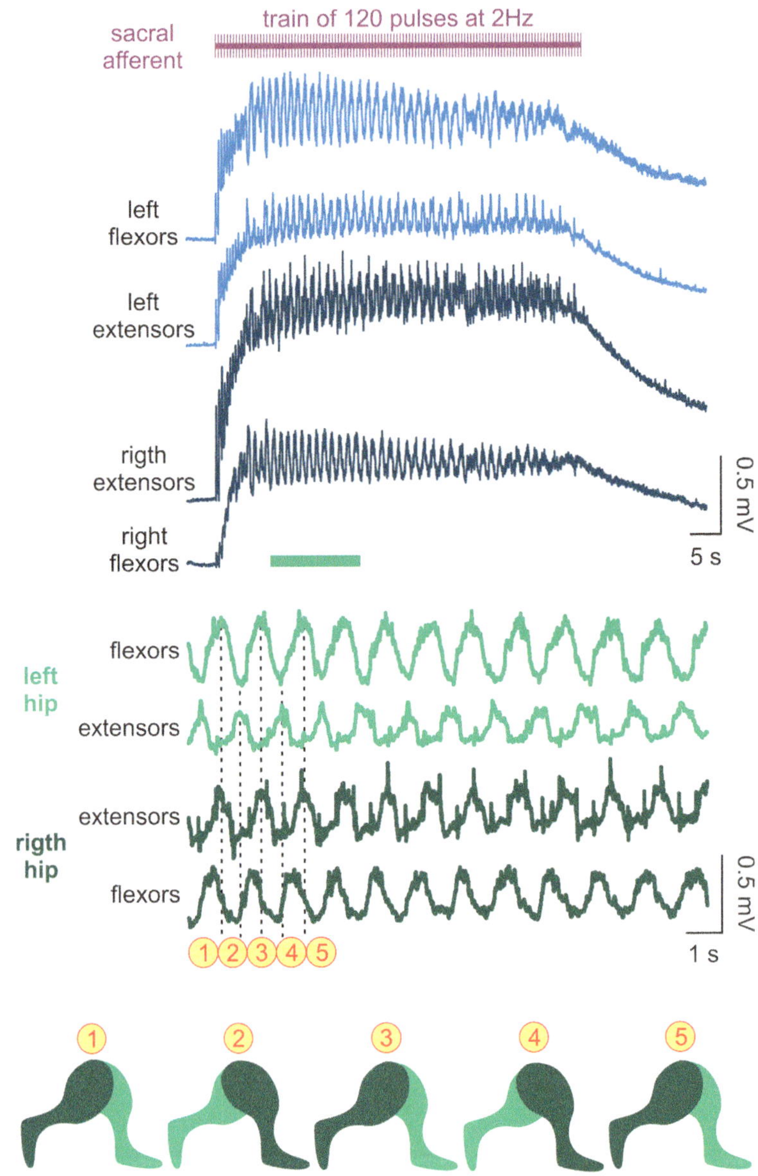

FIG. 1 Spinal central pattern generators for locomotion rhythmically drive the coordinated pattern of activation of motor pools during the locomotor program. A convincing proof of the existence of central pattern generators for locomotion comes from the isolated spinal cord of a neonatal rat, where an episode of rhythmic patterns is elicited in response to a nonpatterned stereotyped train of electrical pulses (120 stimuli at 2 Hz) applied to a single sacral afferent. The electrical activity of hip flexor and extensor motor pools is extracellularly derived from bilateral lumbar (L)2 and L5 ventral roots, respectively. A distinct feature of this locomotor-like episode is the appearance of a cumulative depolarization with superimposed rhythmic discharges that alternate between the two sides of the cord and between flexor- and extensor-related motor pools. The schematic representation at the bottom of the page visualizes how electrical oscillations derived from the motor pools of the neonatal cord in vitro correspond to the different phases of real gait (unpublished observations by GT).

Locomotor CPGs have finally been experimentally established in humans and electrical stimulation of the spinal cord generates episodes of coordinated EMG signals in lower limbs. However, this motor pattern alone was insufficient to enable persons with SCI to deambulate overground. Although electrical stimulation failed to recover automatic functional stepping in paraplegics, it has been shown to facilitate limited volitional motor control of paralyzed limbs, such as an increased volitional control of limb movement in the absence of load (Angeli et al., 2018; Gill et al., 2018, Harkema et al., 2011) and thereby has been proposed as a new rehabilitative practice termed **electrical neuromodulation**.

As opposed to electrical stimulation, electrical neuromodulation does not directly activate motoneurons. Rather, electrical neuromodulation modifies the probability for cells in spinal circuits to reach the threshold for triggering an action potential in response to any upcoming synaptic input. Thus, the spinal cord is more prone to integrate all afferent and propriospinal input with any spared descending commands.

For decades, spinal neuromodulation has been used to limit pain transmission, exploiting physiological gates that regulate pulse conduction along spinal *ascending* tracts (Shealy, Mortimer, & Reswick, 1967). It is noteworthy that a similar technique and the same implanted hardware at the dorsal spinal cord surface are used to generate the two opposite outcomes of pain inhibition vs motor recovery. In particular, the inhibition of pain tracts or, alternatively, the facilitation of motor pathways, are obtained by varying the paradigms (mostly frequency) of trains of electrical pulses, without however significantly modifying the location of epidural electrodes on the spinal cord. Indeed, while high frequency is delivered for alleviating pain (above 100 Hz), 20–40 Hz have been traditionally supplied to facilitate the excitability of motor circuitry and improve volitional motor control. It could thus be inferred that pain and locomotion are both controlled by a group of cells with distinct membrane properties, which show a different responsivity to repetitive pulses delivered at different frequencies (Michelson, Eles, Vazquez, Ludwig, & Kozai, 2019).

Furthermore, when similar paradigms of stimulation are applied to adjacent sites of the spinal cord, multiple outcomes can be obtained. For example, identical electrical paradigms applied with similar electrode locations can modulate seemingly disparate modalities including muscular spasticity (Hofstoetter et al., 2019), bladder functions (Sysoev et al., 2020), standing posture (Sayenko et al., 2019), and volitional stepping activity (Gill et al., 2018). Likely, spinal networks responsible for these tasks are not strictly compartmentalized in the spinal cord, but are in fact largely overlapping. This interconnection can even represent a hurdle in the sequenced restoration of multiple functional deficits after a SCI represented by the challenge of conferring selectivity to electrical neuromodulation to evoke only distinct outcomes (e.g., locomotion or upright standing without micturition). On the other hand, this overlap in the function and structure of spinal networks can also be beneficial. Indeed, once properly targeted, the overlap can lead to a single therapeutic intervention, which can thus yield multiple clinical and functional outcomes and restore both somatosensory and autonomic functions.

A residual functional potential remains after SCI

The changes occurring following an SCI include the structural and functional reorganization of the CNS (Dimitrijevic, McKay, & Sherwood, 1997). These changes result from altered post-SCI motor and sensory control. In this new setting, the supralesional segments (SLSs) of the spinal cord sustain voluntary motor control, sensation and autonomic functions. However, sensory, motor, metabolic and autonomic system deprivation does occur at the injured metamere (IM) where the first impact occurred, as well as in the neighboring spinal cord or infralesional segment (ILS). Muscles corresponding to the IM may become denervated and atrophic. IM and ILS descending and ascending axonal tracts may be interrupted, although the circumferential white matter is often spared. In addition, at the IM, surviving local terminals derived from a different set of inputs can spontaneously form new synapses (Raisman, 1969). Although long debated, it is now clear that this type of synapse and circuit reorganization occurs spontaneously after all forms of CNS injury, including SCI, and that it can be associated with either adaptive or maladaptive functional changes. For instance, spontaneous synapse turnover and circuit reorganization can lead to maladaptive consequences, such as **spasticity** (Roy & Edgerton, 2012), **autonomic dysreflexia** (Hou & Rabchevsky, 2014), or **neuropathic pain** (Kramer et al., 2017; Tan & Waxman, 2012). In other cases, spontaneous circuit reorganization can be adaptive, supporting the restoration of function after incomplete SCI (i.e., **Brown Sequard Syndrome**), in spite of the permanent loss of descending supraspinal connections on the injured side (Ballermann & Fouad, 2006; Bareyre et al., 2004; Courtine et al., 2008; Filli, Zorner, Weinmann, & Schwab, 2011; Little & Halar, 1985; Rosenzweig et al., 2010). On the other hand, even at SLS, SCI leads to the death of cells, including neurons, oligodendrocytes, astrocytes, and precursor cells (Horky, Galimi, Gage, &, Horner, 2006) that can be compensated by various forms of cortical, brainstem, and spinal plasticity contributing to functional recovery (Raineteau & Schwab, 2001; Weidner, Ner, Salimi, & Tuszynski, 2001).

Central nervous system reorganization after SCI

The central nervous system of individuals with SCI is susceptible to substantial reorganization as cortical, subcortical, and much of the local spinal cord circuitry remain largely intact and partially interconnected through unlesioned nerve fibres (Kakulas, 1984; Raineteau & Schwab, 2001; Sherwood, Dimitrijevic, & Barry McKay, 1992). Although any of various adaptive reorganizations can contribute to the recovery of motor control, the role of cortical, as well as corticospinal reconfigurations seem to be crucial and potentially the most impactful among the mechanisms associated with the regaining of voluntary function.

Continuous cortical reorganization occurs during sensorimotor learning (Holtmaat & Svoboda, 2009) and is also commonly observed in humans after SCI, although it becomes particularly extreme after injuries that lead to massive deafferentation (Endo, Spenger, Tominaga, Brene, & Olson, 2007; Ghosh et al., 2010; Jain, Catania, & Kaas, 1997; Kokotilo, Eng, & Curt, 2009; Moxon, Oliviero, Aguilar, & Foffani, 2014). In particular, the decrease in the integrity of the corticospinal tract and in the volume of cortical gray matter is directly correlated to **spinal cord atrophy** (Freund et al., 2012; Freund, Curt, Friston, & Thompson, 2013), suggesting that trauma-induced spinal degenerative processes spread all the way to the brain. Human magnetic resonance imaging (MRI) studies confirm that SCI can cause progressive reduction in gray matter volume, not only in the sensorimotor cortex, but also in brain regions not directly connected to spinal circuitry (Nicotra, Critchley, Mathias, & Dolan, 2005; Wrigley et al., 2008). Mapping studies with transcranial magnetic stimulation (TMS) and functional MRI reveal an enlargement of cortical sensorimotor areas that represent muscles preserved above the level of injury in individuals with quadriplegia (Kokotilo et al., 2009), as well as an enhanced excitability of motor pathways targeting muscles rostral to the level of a spinal lesion in persons with paraplegia (Topka, Cohen, Cole, & Hallett, 1991). In nonhuman primates, it has been shown that after complete cervical dorsal column transection, neurons in the deafferented contralateral primary somatosensory cortex become initially unresponsive to stimulation of the hand followed by innervation of cortical representation of the face (Jain, Qi, Collins, & Kaas, 2008). This is an example of how the cortical reorganization after a dorsal column transection is not limited to the primary somatosensory cortex, but also extends to the secondary somatosensory cortex and to the parietal ventral area (Kokotilo et al., 2009). From a translational viewpoint, the complexity of these interactions highlights the importance to better understand the anatomical level and degree of transection of spared pathways with the goal of designing approaches that minimize maladaptive plasticity and maximize functional recovery after SCI. Despite the heterogeneity in residual motor functions associated with human SCIs, little has been done so far to investigate the function undertaken by each specific neural structures after injury. It seems imperative to adopt training approaches aimed at reversing maladaptive **cortical remapping** when defining training programs for the recovery of voluntary motor control. The fact that the regaining of voluntary motor control can occur within a few sessions, and prior to repetitive training sessions (Angeli, Edgerton, Gerasimenko, & Harkema, 2014), indicates that the cortex is highly plastic and prepared for a "good" reorganization even months or years after an SCI.

Mechanisms of spinal neuromodulation

It is widely assumed that the mechanisms of spinal neuromodulation consist of reactivating dormant connections by mimicking afferent (mainly proprioceptive) input through the electrical activation of Aβ-fibres that are located in dorsal roots and represent low-threshold afferents characterized by a large diameter, high myelination and ability for rapid signal conduction (Holsheimer, 2002). Starting from this broad definition, some clarifications must be introduced to gain a better understanding of neuromodulation.

A first clarification must be made around the idea of *dormant* connections. Indeed, the activation of dormant connections would require the presence of **silent synapses**, which have not yet been reported in the spinal cord after injury. Indeed, silent synapses have been clearly identified so far in the hippocampus (Voronin & Cherubini, 2004), as a functional reserve of synapses that, at rest, are not engaged in information transfer but can be converted in functional ones, for example, in response to plastic events driven by a lesion (Zhang et al., 2018).

Second, although Aβ-fibres are inevitably activated by spinal neuromodulation due to their low threshold of activation, a selective stimulation of Aβ-fibbers in dorsal roots is likewise improbable. Indeed, the amount of current operatively applied to recover volitional control recruits many more types of dorsal root fibbers and passes across several spinal segments (Swiontek et al., 1976) that are populated by a myriad of excitable targets. As a result, spinal neuromodulation likely generates higher neuronal excitability, faster pulse conduction through axons, and an increased synaptic release with profound changes in the chemical composition of the extracellular fluid.

In spinal networks, the mechanism of neuromodulation can be mainly ascribed to the facilitation of the presynaptic input (Eccles, Kostyuk, & Schmidt, 1962) from peripheral afferents and from the cortical tracts spared by the lesion. Indeed, due to neuromodulation, network synapses depolarize and approach the threshold for activating calcium channels, in turn increasing the probability for any presynaptic input to activate neurotransmitter release onto postsynaptic targets. Moreover, for neuromodulation to be effective in allowing functional recovery after SCI, the presynaptic action requires that some functional synapses are spared by the injury and can be recruited in the presence of neuromodulation by the volitional attempt to move paralyzed limbs (Taccola, Sayenko, Gad, Gerasimenko, & Edgerton, 2018). Notably, strong electrical pulses can directly depolarize the resting membrane potential of postsynaptic targets (i.e., motoneurons and their axons), without recruiting any interneuronal circuitry, and thus lead to massive muscle activation. These stereotyped muscle responses rely on motoneuronal integrity, even in case of a large loss of premotoneuronal interneurons.

Fibre regeneration and sprouting of new synapses are both able to circumvent the interruption of descending tracts at the site of the lesion, and have been reported as the main mechanisms supporting functional recovery after SCI (Courtine et al., 2008). Indeed, functional anatomical changes are a consequence of activity-dependent plasticity triggered by repetitive sessions of neuromodulation, especially when associated with physical rehabilitation (Jack, Hurd, Martin, & Fouad, 2020). However, clinical benefits of neuromodulation have been reported as early as after one session (Gill et al., 2018), a time span theoretically too short to be accounted for by significant fibre remodeling but potentially indicating an acute functional reconfiguration of spared pathways including physiological or molecular changes to synaptic efficacy.

For the above-mentioned reasons, the present authors find it more rigorous to define electrical neuromodulation for the restoration of volitional motor control after a severe injury as a physical perturbation of many excitable structures in the spinal cord. As a result, neuromodulation profoundly affects the performance and organization of spared elements in injured networks by bringing them to a new homeostasis that, even if not necessarily physiological, is functionally beneficial for the expression of certain motor functions.

It is however questionable whether neuromodulation could generate a novel physiological state in spinal networks. For example, it is still to be demonstrated whether neuromodulation could also improve motor performances in neurologically intact subjects, thereby indicating potential to exploit the supraphysiological functionality of spinal networks.

Nevertheless, the broad activation of multiple neural structures during electrical neuromodulation might indeed alter the balance of injured networks by activating multiple plastic and compensatory events that, in sum, result in practical benefits for accomplishing the impaired motor tasks. A similar phenomenon, called **metaplasticity,** has been described for the conditioning of muscle reflexes elicited by repetitive electrical stimulation of peripheral nerves (Wolpaw, 2018).

Similar to the scrambling of pain fibre signals, also neuromodulation, at least when used acutely, likely "scrambles" the nonfunctional and aberrant motor patterns resulting from the injury, instead of adding new and previously missing sources of motor signals (Marineo, 2019).

The ideal candidate for the restoration of volitional motor responses through neuromodulation

Following the above interpretation, neuromodulation can be advantageous as long as at least some neural connections remain across the lesion. In this regard, Dimitrijevic et al. (Sherwood et al., 1992) have demonstrated that some individuals with a complete SCI are still capable of subclinical motor control below the level of injury and may even regain mastery of controlled movement when the spinal cord is appropriately stimulated. Neurophysiological evidence of subclinical brain influence on spinal cord function below the lesion has been found in "so-called" clinically **discomplete SCIs** (Sherwood et al., 1992). People with a discomplete SCI syndrome may have similar motor patterns in terms of brain control and/or the presence of subclinical motor activity as persons with incomplete SCIs (Tang, Tuel, McKay, & Dimitrijevic, 1994). A predictive neurophysiological assessment for the functional outcome of individuals with SCI is thus essential before defining any treatment to promote recovery (Tansey, 2012).

However, current diagnostic tools do not explicitly define the completeness of a SCI resulting in a rationale to test neuromodulation on the vast majority of persons with clinically diagnosed complete lesions in addition to those that maintain some residual anatomical connection.

A more sensitive surface EMG recording technology (Campanini, Disselhorst-Klug, Rymer, & Merletti, 2020), together with more detailed and advanced tools of spinal imaging (Kaushal, Shabani, Budde, & Kurpad, 2019), would clarify what extent of spared pathways is required to selectively address the benefits of neuromodulating therapies. Thus, the ideal candidate for the restoration of volitional motor responses using neuromodulation after a severe SCI, should be a subject with an incomplete and/or discomplete SCI, with minimal musculoskeletal atrophy below the level of injury, minimal joint limitations, and good trunk stability, as well as good cardiovascular reflex responses, cognitive and emotional stability, and strong motivation.

Neuromuscular electrical stimulation for the recovery of independent stepping

Neuromuscular electrical stimulation (NMES) has been intensively investigated in the 1980s (Cybulski, Penn, & Jaeger, 1984) with the intent to bypass the lesioned area by electrically stimulating certain paralyzed peripheral nerves, to regain some "artificial" motor control, recondition paralyzed limbs to muscle fatigue, reactivate the paralyzed muscle pump, and condition the cardiopulmonary and circulatory systems through muscle stimulation in nondenervated subjects with SCI. The Parastep walking NMES system, designed in Dr. Graupe's lab in Chicago, was the first system created for independent indoor and outdoor walking for persons with complete and nearly complete SCI at thoracic level (T1 to T12), having no motor function nor sensation below the lesion.

At the same time, allowing individuals with a complete SCI to stand and/or walk by means of NMES is a puzzle building procedure. Indeed, it allows, and increases over time, proper body alignment and functional range of motion, muscle strength, physiological cardio-pulmonary responses, and proper integration of the NMES activity with the body as a whole. The training methodology with Parastep was not univocal, so initially every rehabilitation centre came up with its own procedure to train Parastep users in walking. Cost and convenience were obviously an important factor. Hence, training was attempted for a short period of time, but the overall walking performance was not good enough to sustain its use for functional walking for the vast majority of individuals with SCI.

Parastep users were trained to stand and walk against gravity only after participating in a well-structured task-oriented rehabilitation program. The program consisted in a multistep methodology (Fig. 2) aimed at: (1) strengthening upper body muscles and conditioning the cardio-pulmonary and circulatory system; (2) maintaining lower extremities within a functional range of motion for standing and walking; (3) reversing muscle atrophy, stopping muscle degeneration and increasing muscle force and endurance both through NMES and associated training that includes repetitive (and increasing) weightlifting, kinematic exercise equipment and NMES cycling equipment; (4) learning to walk through a task-oriented-approach; (5) repetitive standing/walking exercise initially on a flat surface, followed by training on a treadmill; and (6) only then, progressing toward walking on diverse terrain surfaces (slope or obstacles, including stairs) and to building up confidence

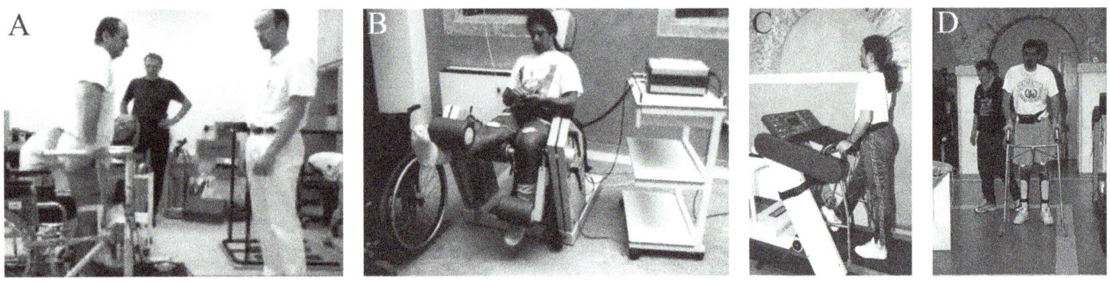

E

Subjects	Height (m)	Weight (Kg)	SPWS (m/min)	**Work rate (watt)**
Parastep® (n = 5)	1.70 ± 0.04	63 ± 6.3	18.40 ± 0.89	**311.8 ± 62**
Controls (n = 5)	1.79 ± 0.09	75.2 ± 14	16.30 ± 2.30	**303.0 ± 51**

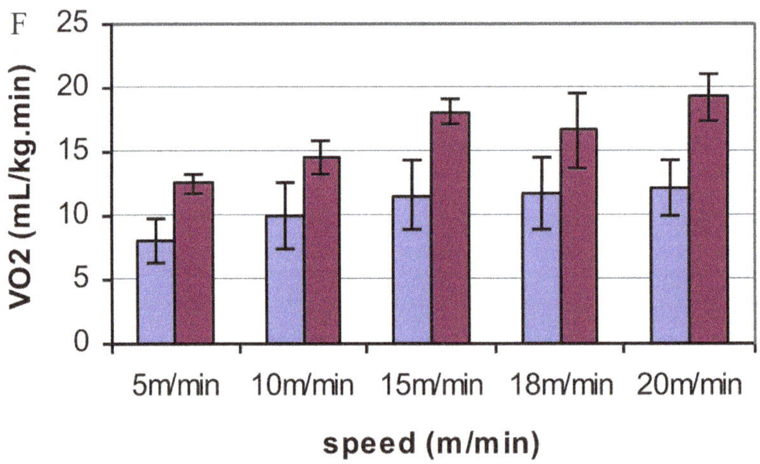

FIG. 2 Neuromuscular electrical stimulation (NMES) in persons with complete spinal cord injuries (SCIs). Training methodology with Parastep for the recovery of independent stepping in persons with complete SCIs at thoracic-level. The protocol progresses along several sessions, starting from repetitive standing exercises to improve trunk control and body re-alignment (A), strengthening leg muscles with NMES + cycling (B), walking initially over a controlled environment (C) and only then, progressing toward autonomously walking overground, showing the integration of electrically induced motor limb movements. (D) Characteristics of 10 research subjects (5 SCI + NMES; 5 able body controls) showing similar biomechanics and self-preferred walking speeds (SPWS). Although complete paralysis, the organism's work rate expressed during walking with NMES is equal to a physiological one. (E) Benefits of gait with NMES over traditional reciprocal gait orthosis on the cardio-pulmonary and circulatory systems is reported as volume of oxygen (VO_2) consumption per mL/Kg at different walking speeds (F). Ventilatory responses expressed during NMES trials ($n = 5$, *dark purple*) were higher than those pertaining to the group using traditional reciprocal gait orthosis ($n = 5$, *light purple*). Differences were statistically significant at $P \leq 0.01$ (unpublished observations by HACB).

in walking in the absence of direct feedback sensation from the NMES stimulated paralyzed limbs (Carmick, 1993; Graupe & Bazo, 2015; Waters, 1984). The methodology facilitated the physiological trunk posture and hip extension, properly integrating body systems above and below the level of injury to load or unload, to maximize the goal-oriented task of standing, balancing, and walking. The sequencing and timing of the movement thus became automatic, shifting from direct visual control to a more internalized form of control.

Rationale for combining NMES with other neuromodulation modalities

In individuals with a discomplete SCI, the NMES training methodology promoted sensorimotor integration and regaining of voluntary motor control for intentional cycling and/or walking. In individuals with a complete SCI, NMES promoted secure step, trunk control, and body re-alignment, showing the integration of electrically induced motor limb movements with SLS postural and voluntary motor control activity. It can be concluded that NMES can facilitate spinal integration of sensory afferent feedback coming from SLS and ILS to improve walking performance after SCI.

In summary, NMES can activate paralyzed muscles, although its use may present several disadvantages: (1) many of the key muscles lie deep in the legs and trunk and are not accessible with surface electrodes; (2) even with the percutaneous implantation system, multiple channels are required to stimulate all muscles; (3) peripheral neuromuscular stimulation inevitably results in significant fatigue because this stimulation paradigm recruits the fastest and most fatigable fibres first; and (4) complex motor tasks, such as standing and stepping, may require coordination by an external computer-controlled closed-loop system, when the body's position and velocity are continually monitored, and the level of stimulation is adjusted accordingly on a moment-to-moment basis (Ho et al., 2014). Therefore, while noninvasive NMES is preferred for therapeutic purposes, individual responses to this technique are generally not strong enough for individuals with SCI to regain sufficient motor functioning. Indeed, while NMES can reactivate paralyzed muscles, the effects may be insufficient to induce full body weight bearing during self-assisted stepping or standing. At the same time, a combinatory approach aimed at integrating, for instance, NMES and other neuromodulation modalities, may present an appealing and effective method to engage neural circuitries and regain robust self-assisted and weight-bearing activities in individuals with SCI. In particular, **transcutaneous spinal stimulation** (TSS), another noninvasive electrical stimulation technique, has recently drawn the attention of clinicians and researchers to activate and enable multiple paralyzed muscle groups at once (Sayenko et al., 2019). However, it has also been shown that individual responses to TSS can vary greatly, depending, for instance, on the strength, excitability or asymmetry of lower limb muscles due to SCI. For some participants, this can result in an insufficient bilateral muscle activation during activity. Therefore, the potential of combining various modalities to promote motor functions are worth being explored in future studies (Fig. 3).

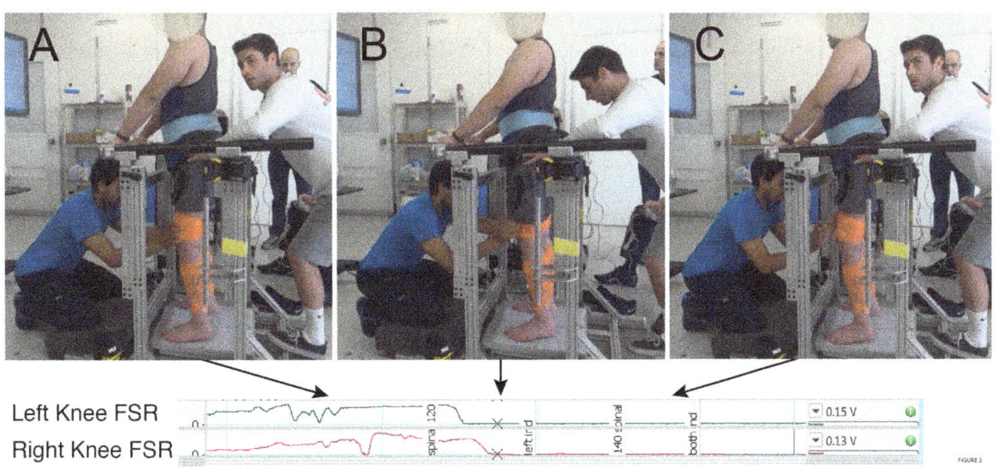

FIG. 3 An example of a combinatorial application of neuromuscular electrical stimulation (NMES) and transcutaneous spinal stimulation (TSS) in individual with complete spinal cord injury. (A) In a subject with chronic (4 years postinjury) thoracic (T2) SCI, only TSS was applied (L1, 15 Hz, and 120 mA). (B) TSS was combined with NMES on the plantarflexors (30 Hz and 70 mA). (C) TSS intensity was increased to 140 mA, with NMES on the plantarflexors (30 Hz and 70 mA). The traces at the bottom indicate the signals from two force sensitive resistance sensors on both knees. When TSS is combined with NMES, the load of the force sensors got reduced as well as the therapist's hands are detached from the knee shown on (B) and (C), indicating that manual assistance to the knees is no longer needed (unpublished observations by DGS).

Regaining of motor function after severe spinal cord injury

It has been implied that both daily training using epidural stimulation with stand training, and home-based voluntary training with **epidural stimulation** resulted in improvements, characterized by higher forces and lower stimulation voltages required to reach the thresholds that enable voluntary motor responses. Most recent studies of epidural spinal stimulation have revealed that neuromodulation of lumbosacral sensorimotor networks in combination with intensive training enables individuals with a chronic motor complete spinal cord injury (SCI) to perform self-assisted stepping utilizing paralyzed muscles (Angeli et al., 2018; Gill et al., 2018; Wagner et al., 2018).

Although these results demonstrate that spinal circuitries below a paralyzing injury have a functional potential that far exceeds what has ever been thought possible, two critical questions arise: (1) Is a repetitive activity-based training necessary to recover the voluntary motor control? and (2) Is there any specific training approach that can promote recovery of function-specific voluntary motor control?

Multimodal rehabilitation

It seems critical that approaches to regain voluntary motor control must engage the broadest variety of motor programs and neural networks, and not be limited to a single movement or function, to provide sufficient challenge of descending spinal tracts to activate spared supra- and intraspinal network. From this viewpoint, it is not surprising that somewhat limited clinical improvements can be seen following **conventional locomotor training**, primarily due to the limited spinal and supraspinal circuitry engagement during the intervention. Therefore, although positive physiological effects of locomotor training are indisputable (Dietz, Colombo, Jensen, & Baumgartner, 1995), this rehabilitative approach appears to target only rhythmogenic properties of the intrinsic spinal network, and may not promote, if applied alone, the volitional control of movement.

Postural verticalization is another conventional therapeutic modality that has been extensively adopted for decades in the rehabilitation of persons with a spinal lesion, though without being linked to any neurological recovery. It has been suggested that the standing posture has a number of therapeutic and functional benefits aimed at overcoming physiological problems, such as bladder infections, spasticity, and blood pressure homeostasis (Harkema, Ferreira, van den Brand, & Krassioukov, 2008). Surprisingly, minimally assisted verticalization of a subject with compromised sensorimotor control of the trunk or lower limbs may represent a comparable or even more dynamic experience (Sayenko et al., 2019) than locomotor training. Indeed, a sudden postural perturbation can trigger multiple sensorimotor signals and commands to recover equilibrium, which can result in a cascade of compelling ascending and descending inputs with a higher modulation frequency, compared to inputs evoked by stereotyped stepping on a treadmill, with passive weight bearing (Fig. 4). Based on these considerations, it can be proposed that stand training must engage all components of the postural sensorimotor network from supraspinal centres to spinal circuits (Sayenko et al., 2019). Stand training with visual feedback and voluntary

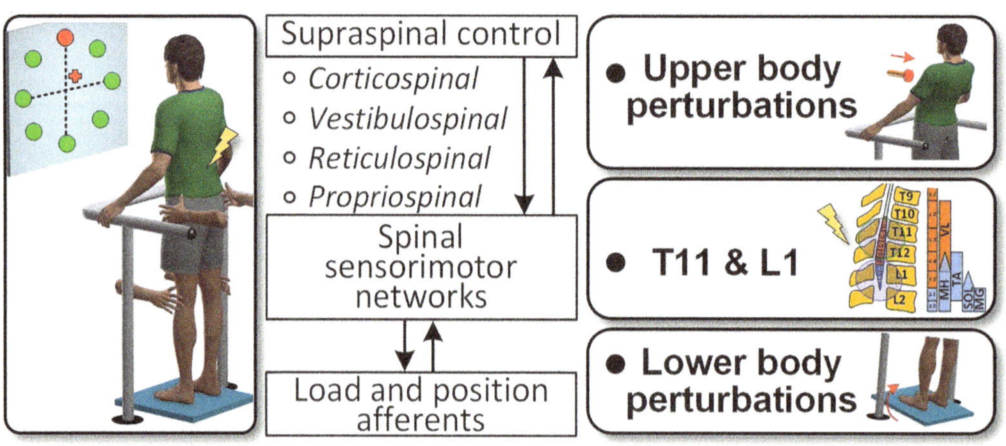

FIGURE 4

FIG. 4 Postural control and stand training. Main elements of stand training include spinal stimulation-induced minimally assisted standing, voluntary control of body position using visual feedback, and postural perturbations applied to upper and lower body. Multimodal approach activates sub-functional longitudinal fibres of the spinal cord and promotes the emerging responsiveness of spinal networks to descending commands and sensory inputs via corticospinal, vestibulospinal, reticulospinal, and propriospinal pathways (unpublished observations by DGS).

control of body movements, as well as postural perturbations, can provide a diffused background of descending and ascending somatosensory inputs that reach the spinal cord and can facilitate the activation of the corticospinal, vestibulospinal, reticulospinal, and propriospinal networks, all critical for postural regulation and motor control (Sayenko et al., 2019).

In this regard, the data by Gill et al. (2018) demonstrate that **multimodal rehabilitation**, which consists of the dynamic approach of stand-and-step training during each session, is highly efficient to enable both independent standing and stepping. Moreover, the authors observed that although task-specific epidural stimulation parameters enabled independent standing, when parameters were adjusted to enable stepping, standing was still possible. When the subject changed his intention from standing to stepping, he was able to independently step with both legs on a treadmill. Subsequent to stepping, spinal stimulation parameters were adjusted to enable optimal standing performance, and as a result of the dynamic training of spinal sensorimotor networks during multimodal rehabilitation, independent standing was achieved (Fig. 5).

To summarize, it is now clear that the majority of studies utilizing epidural spinal stimulation for motor recovery after motor complete SCI have been successful so far in generating an almost *immediate* regaining of muscle-specific control

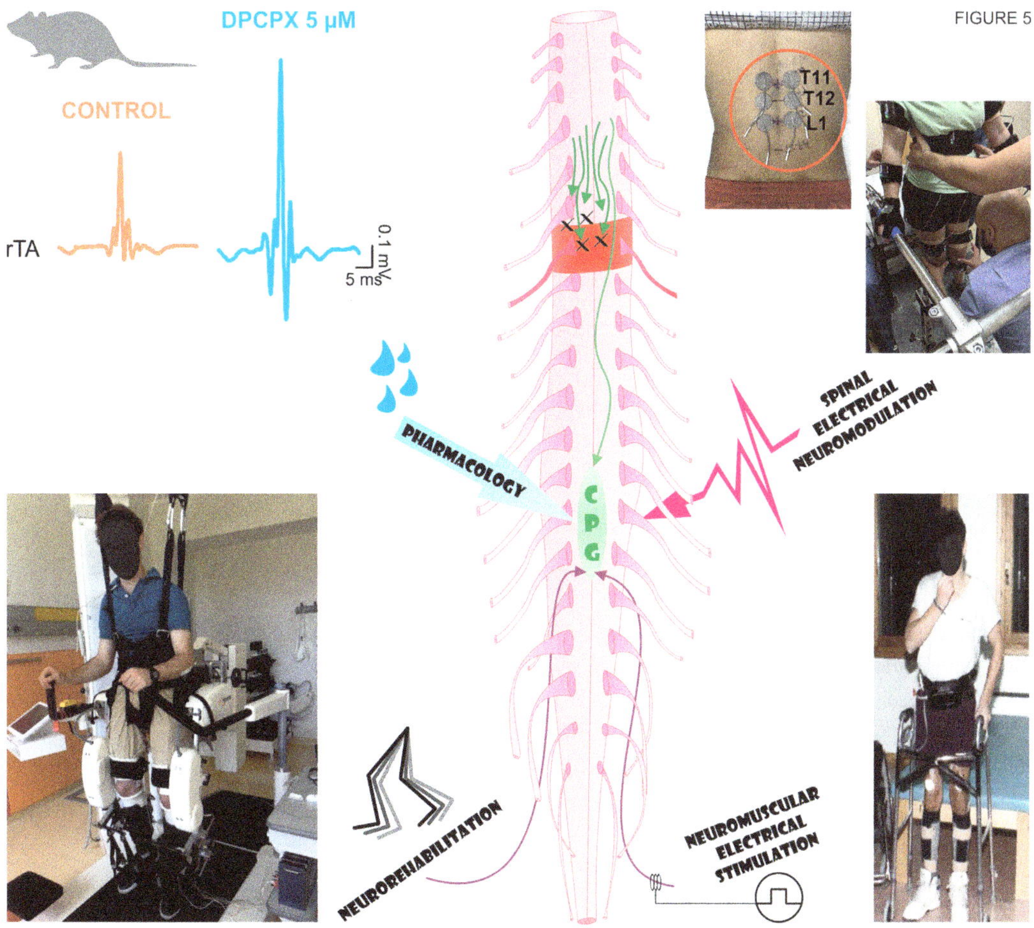

FIG. 5 Multilevel approach to the treatment of persons with spinal cord injury. The cartoon summarizes the different protocols that are experimentally or clinically used after spinal cord injury, including neuromodulation. A discomplete injury to the cord is depicted as a red segment, blocking (*black* "Xs") the majority of descending volitional commands (*green* downward arrows). A weak spared input passes the block (*pale green* downwards arrow) but is unable to activate the Central Pattern Generator (CPG) for locomotion located in upper lumbar segments. Sensory input from the periphery reaches the CPG from the sacrocaudal cord through ascending projections (*purple* upward arrows). CPG excitability might reach the threshold for motor control recovery by synergizing different approaches, which are schematized in clockwise order, as follows. Direct spinal cord neuromodulation is applied using transcutaneous electrodes placed on the back at the level of the thoracolumbar vertebrae (T11 to L1) to elicit standing posture in persons with chronic spinal cord injury. Neuromuscular electrical stimulation of peripheral afferents can be applied through surface electrode on the skin of lower limbs for the "artificial" control of standing and stepping. Robotic exoskeletons driving repetitive the passive motion of legs over a sliding treadmill generate afferent input that in turn modulate the CPG. Finally, in animal experiments, the direct application of neurochemicals (i.e., DPCPX, the selective pharmacological antagonist of adenosinergic A1 receptors; Taccola et al., 2020) modulates electrically-evoked motor responses from hindlimb muscles (i.e., *TA*, tibialis anterior; (unpublished materials by DGS, HACB, and GT).

below the lesion in the presence of stimulation. However, successful performance of more complex motor tasks, such as standing and stepping, most likely requires intensive training to promote integration of the *vast variety* of neural networks above, across, and below the level of neurological lesion.

Applications to other areas of neuroscience

Understanding of the neuromodulation mechanisms expands our knowledge on the basic rules pertaining to the recruitment of networks of neurons. In particular, the insights about the nonlinearity of neuronal circuits, according to which a quite stereotyped electrical input above a distinct threshold shapes a patterned motor output, could be exploited to design brain machine interfaces and robotics.

In this chapter, we reviewed the effects of electrical neuromodulation in restoring motor responses after spinal cord injury. Electrical spinal stimulation through transcutaneous or epidural electrodes is also an emerging practice in the treatment of several neuromotor disorders. Spinal stimulation for the recovery of motor control can been explored for lesions of upper motoneurons, such as stroke, cerebral palsy, Parkinson's disease and multiple sclerosis.

Noteworthy, spinal neuromodulation for the restoration of motor responses after severe spinal cord injury shares multiple features with epidural electrical stimulation for alleviating central pain, electrical stimulation of the peripheral nerve, and deep brain stimulation. Therefore, advancements in any of these closely related fields can potentially contribute to improving treatment of the other distinct neurological conditions, as well.

Although the neurophysiological processes at the base of electrical neuromodulation still need to be fully identified, advances in hardware design, rehabilitation training and pharmacological therapies might suggest new avenues on how electrical field interacts with a network of excitable neural targets.

Mini-dictionary of terms

Autonomic dysreflexia: Increased systolic blood pressure greater than 20–30 mmHg.

Brown-Sequard Syndrome: A partial damage to the cord, affecting half of a spinal segment and causing lateral sensory-motor deficits.

Central pattern generators: Networks of neurons that, due to their intrinsic membrane properties or their distinct synaptic wiring among the elements of the circuit, are able to express a rhythmic functional output even in the absence of any central or peripheral input.

Conventional locomotor training: Manually assisted stepping on a treadmill or overground with or without body weight support.

Cortical remapping: Changes in brain activation after spinal cord injury in terms of intensity, volume, and somatotopic localization, as well as preservation of activation during attempted and/or imagined movements.

Discomplete lesion: A clinically complete absence of motor activity after spinal cord injury, with or without any partial preservation of sensory functions, showing a subclinical brain influence on spinal cord functions below lesion.

Epidural stimulation: Electrical stimulation of the central nervous system through electrodes placed over the outmost meninges, which are called dura mater.

Metaplasticity: Homeostatic compensatory plastic events consequent to a rise in the excitability induced by the activation of previously nonresponsive injured networks.

Multimodal rehabilitation: Dynamic training of spinal stimulation-enabled functions in multiple environments, including stimulation-enabled stepping and standing.

Neuromuscular electrical stimulation: Electrical stimulation of motor endplates to induce muscle contraction for functional purposes.

Neuropathic pain: Chronic pain syndrome originated from the central nervous system.

Spasticity: A motor disorder with a velocity-dependent increase in tonic stretch reflexes and increased muscle activity during passive stretch, with clinical representation of hyperreflexia, spasms, and clonus.

Spinal cord atrophy: A reduction in the cross-sectional integrity of the cord.

Spinal metamere: A horizontal segment of the spinal cord that includes a pair of bilateral dorsal roots and a pair of bilateral ventral roots.

Transcutaneous spinal stimulation: A noninvasive electrical stimulation technique delivering pulses to the spinal cord through surface electrodes applied to the skin over the backbone.

Key facts of neuromodulation and restoration of motor responses after severe spinal cord injury

- Excitable cells use digital electrical input to transmit information.
- Rhythmic motor activities are automatically generated by distinct circuits of spinal neurons.
- Commands for the volitional control of motor tasks are generated by the brain.
- The spinal cord is not a simple tube of fibres, but contains networks of neurons that express smart decision-making properties.
- Sensory input from the periphery modulates the motor behavior mainly through unconscious mechanisms.
- The broad stimulation of spinal segments below the level of injury elicits the recovery of motor skills that require the sequenced activation of multiple motor pools.
- The rhythmic afferent feedback generated by physical training supports the reactivation of spinal networks through direct electrical stimulation.

Summary points

- A spinal cord injury determines a maladaptive reorganization of the motor cortex, which can be reverted by training programs aimed at regain voluntary motor control.
- Some individuals with a complete spinal cord injury are still capable of subclinical motor control below the level of injury and may regain motor control when the spinal cord is appropriately stimulated.
- Electric charges applied to the central nervous system affect the excitability of neuronal networks.
- Neuromodulation, as a direct physical perturbation of all excitable structures in the spinal cord, reconfigures spared elements in injured networks to restore the control of certain volitional motor functions after injury.
- The electrical stimulation of certain peripheral nerves facilitates spinal integration of sensory afferent feedback and allows some "artificial" motor control of paralyzed limbs after spinal cord injury.
- A multimodal intensive rehabilitation, which combines complex motor tasks with the electrical modulation of both spinal cord and peripheral nerves, restores motor control by exploiting the integration of multiple neural networks above, across, and below the level of a neurological lesion.

References

Angeli, C. A., Boakye, M., Morton, R. A., Vogt, J., Benton, K., Chen, Y., et al. (2018). Recovery of over-ground walking after chronic motor complete spinal cord injury. *The New England Journal of Medicine*, *379*, 1244–1250.

Angeli, C. A., Edgerton, V. R., Gerasimenko, Y. P., & Harkema, S. J. (2014). Altering spinal cord excitability enables voluntary movements after chronic complete paralysis in humans. *Brain*, *137*(Pt 5), 1394–1409. https://doi.org/10.1093/brain/awu038.

Ballermann, M., & Fouad, K. (2006). Spontaneous locomotor recovery in spinal cord injured rats is accompanied by anatomical plasticity of reticulospinal fibers. *The European Journal of Neuroscience*, *23*, 1988–1996.

Bareyre, F. M., Kerschensteiner, M., Raineteau, O., Mettenleiter, T. C., Weinmann, O., & Schwab, M. E. (2004). The injured spinal cord spontaneously forms a new intraspinal circuit in adult rats. *Nature Neuroscience*, *7*, 269–277.

Campanini, I., Disselhorst-Klug, C., Rymer, W. Z., & Merletti, R. (2020). Surface EMG in clinical assessment and neurorehabilitation: Barriers limiting its use. *Frontiers in Neurology*, *11*, 934.

Carmick, J. (1993). Clinical use of neuromuscular electrical stimulation for children with cerebral palsy. Part 1: Lower extremity. *Physical Therapy*, *73*, 514–522.

Courtine, G., Song, B., Roy, R. R., Zhong, H., Herrmann, J. E., Ao, Y., et al. (2008). Recovery of supraspinal control of stepping via indirect propriospinal relay connections after spinal cord injury. *Nature Medicine*, *14*, 69–74.

Cybulski, G. R., Penn, R. D., & Jaeger, R. J. (1984). Lower extremity functional neuromuscular stimulation in cases of spinal cord injury. *Neurosurgery*, *15*, 132–146.

Dietz, V., Colombo, G., Jensen, L., & Baumgartner, L. (1995). Locomotor capacity of spinal cord in paraplegic patients. *Annals of Neurology*, *37*, 574–582.

Dimitrijevic, M. R., McKay, W. B., & Sherwood, A. M. (1997). Motor control physiology below spinal cord injury: Residual volitional control of motor units in paretic and paralyzed muscles. *Advances in Neurology*, *72*, 335–345.

Eccles, J. C., Kostyuk, P. G., & Schmidt, R. F. (1962). The effect of electric polarization of the spinal cord on central afferent fibres and on their excitatory synaptic action. *The Journal of Physiology*, *162*, 138–150.

Endo, T., Spenger, C., Tominaga, T., Brene, S., & Olson, L. (2007). Cortical sensory map rearrangement after spinal cord injury: fMRI responses linked to Nogo signalling. *Brain*, *130*, 2951–2961.

Filli, L., Zorner, B., Weinmann, O., & Schwab, M. E. (2011). Motor deficits and recovery in rats with unilateral spinal cord hemisection mimic the Brown-Sequard syndrome. *Brain*, *134*, 2261–2273.

Freund, P., Curt, A., Friston, K., & Thompson, A. (2013). Tracking changes following spinal cord injury: Insights from neuroimaging. *The Neuroscientist*, *19*, 116–128.

Freund, P., Wheeler-Kingshott, C. A., Nagy, Z., Gorgoraptis, N., Weiskopf, N., Friston, K., et al. (2012). Axonal integrity predicts cortical reorganisation following cervical injury. *Journal of Neurology, Neurosurgery, and Psychiatry*, *83*, 629–637.

Ghosh, A., Haiss, F., Sydekum, E., Schneider, R., Gullo, M., Wyss, M. T., et al. (2010). Rewiring of hindlimb corticospinal neurons after spinal cord injury. *Nature Neuroscience*, *13*, 97–104.

Gill, M. L., Grahn, P. J., Calvert, J. S., Linde, M. B., Lavrov, I. A., Strommen, J. A., et al. (2018). Neuromodulation of lumbosacral spinal networks enables independent stepping after complete paraplegia. *Nature Medicine*, *24*, 1677–1682.

Graupe, D., & Bazo, H. A. C. (2015). Thoracic level complete paraplegia–walking performance, training and medical benefits with the PARASTEP FES system. *International Journal of Physical Medicine & Rehabilitation*, *3*, 5.

Günter, C., Delbeke, J., & Ortiz-Catalan, M. (2019). Safety of long-term electrical peripheral nerve stimulation: Review of the state of the art. *Journal of Neuroengineering and Rehabilitation*, *16*, 13.

Harkema, S., Gerasimenko, Y., Hodes, J., Burdick, J., Angeli, C., Chen, Y., et al. (2011). Effect of epidural stimulation of the lumbosacral spinal cord on voluntary movement, standing, and assisted stepping after motor complete paraplegia: A case study. *Lancet*, *377*, 1938–1947.

Harkema, S. J., Ferreira, C. K., van den Brand, R. J., & Krassioukov, A. V. (2008). Improvements in orthostatic instability with stand locomotor training in individuals with spinal cord injury. *Journal of Neurotrauma*, *25*(12), 1467–1475. https://doi.org/10.1089/neu.2008.0572.

Ho, C. H., Triolo, R. J., Elias, A. L., Kilgore, K. L., DiMarco, A. F., Bogie, K., et al. (2014). Functional electrical stimulation and spinal cord injury. *Physical Medicine and Rehabilitation Clinics of North America*, *25*, 631–654.

Hofstoetter, U. S., Freundl, B., Danner, S. M., Krenn, M. J., Mayr, W., Binder, H., et al. (2019). Transcutaneous spinal cord stimulation induces temporary attenuation of spasticity in individuals with spinal cord injury. *Journal of Neurotrauma*, *37*, 481–493.

Holsheimer, J. (2002). Which neuronal elements are activated directly by spinal cord stimulation. *Neuromodulation*, *5*, 25–31.

Holtmaat, A., & Svoboda, K. (2009). Experience-dependent structural synaptic plasticity in the mammalian brain. *Nature Reviews Neuroscience*, *10*, 647–658.

Horky, L. L., Galimi, F., Gage, F. H., & Horner, P. J. (2006). Fate of endogenous stem/progenitor cells following spinal cord injury. *The Journal of Comparative Neurology*, *498*, 525–538.

Hou, S., & Rabchevsky, A. G. (2014). Autonomic consequences of spinal cord injury. *Comprehensive Physiology*, *4*, 1419–1453.

Jack, A. S., Hurd, C., Martin, J., & Fouad, K. (2020). Electrical stimulation as a tool to promote plasticity of the injured spinal cord. *Journal of Neurotrauma*, *37*, 1933–1953.

Jain, N., Catania, K. C., & Kaas, J. H. (1997). Deactivation and reactivation of somatosensory cortex after dorsal spinal cord injury. *Nature*, *386*, 495–498.

Jain, N., Qi, H.-X., Collins, C. E., & Kaas, J. H. (2008). Large-scale reorganization in the somatosensory cortex and thalamus after sensory loss in macaque monkeys. *Journal of Neuroscience*, *28*, 11042–11060.

Kakulas, B. A. (1984). Pathology of spinal injuries. *Central Nervous System Trauma*, *1*, 117–129.

Kaushal, M., Shabani, S., Budde, M., & Kurpad, S. (2019). Diffusion tensor imaging in acute spinal cord injury: A review of animal and human studies. *Journal of Neurotrauma*, *36*, 2279–2286.

Kern, H., & Carraro, U. (2020). Home-based functional electrical stimulation of human permanent Denervated muscles: A narrative review on diagnostics, managements, results and byproducts revisited. *Diagnostics*, *10*, 529.

Kiehn, O. (2006). Locomotor circuits in the mammalian spinal cord. *Annual Review of Neuroscience*, *29*, 279–306.

Kokotilo, K. J., Eng, J. J., & Curt, A. (2009). Reorganization and preservation of motor control of the brain in spinal cord injury: A systematic review. *Journal of Neurotrauma*, *26*, 2113–2126.

Kramer, J. L., Minhas, N. K., Jutzeler, C. R., Erskine, E. L., Liu, L. J., & Ramer, M. S. (2017). Neuropathic pain following traumatic spinal cord injury: Models, measurement, and mechanisms. *Journal of Neuroscience Research*, *95*, 1295–1306.

Little, J. W., & Halar, E. (1985). Temporal course of motor recovery after Brown-Sequard spinal cord injuries. *Paraplegia*, *23*, 39–46.

Marder, E., O'Leary, T., & Shruti, S. (2014). Neuromodulation of circuits with variable parameters: Single neurons and small circuits reveal principles of state-dependent and robust neuromodulation. *Annual Review of Neuroscience*, *37*, 329–346.

Marineo, G. (2019). Inside the scrambler therapy, a noninvasive treatment of chronic neuropathic and cancer pain: From the gate control theory to the active principle of information. *Integrative Cancer Therapies*, *18*, 1–17.

Michelson, N. J., Eles, J. R., Vazquez, A. L., Ludwig, K. A., & Kozai, T. D. Y. (2019). Calcium activation of cortical neurons by continuous electrical stimulation: Frequency dependence, temporal fidelity, and activation density. *Journal of Neuroscience Research*, *97*, 620–638.

Moxon, K. A., Oliviero, A., Aguilar, J., & Foffani, G. (2014). Cortical reorganization after spinal cord injury: Always for good? *Neuroscience*, *283*, 78–94.

Nicotra, A., Critchley, H. D., Mathias, C. J., & Dolan, R. J. (2005). Emotional and autonomic consequences of spinal cord injury explored using functional brain imaging. *Brain*, *129*, 718–728.

Raineteau, O., & Schwab, M. E. (2001). Plasticity of motor systems after incomplete spinal cord injury. *Nature Reviews. Neuroscience*, *2*, 263–273.

Raisman, G. (1969). Neuronal plasticity in the septal nuclei of the adult rat. *Brain Research*, *14*, 25–48.

Rosenzweig, E. S., Courtine, G., Jindrich, D. L., Brock, J. H., Ferguson, A. R., Strand, S. C., et al. (2010). Extensive spontaneous plasticity of corticospinal projections after primate spinal cord injury. *Nature Neuroscience*, *13*, 1505–1510.

Roy, R. R., & Edgerton, V. R. (2012). Neurobiological perspective of spasticity as occurs after a spinal cord injury. *Experimental Neurology*, *35*, 116–122.

Sayenko, D. G., Rath, M., Ferguson, A. R., Burdick, J. W., Havton, L. A., Edgerton, V. R., et al. (2019). Self-assisted standing enabled by non-invasive spinal stimulation after spinal cord injury. *Journal of Neurotrauma*, *36*, 1435–1450.

Shealy, C. N., Mortimer, J. T., & Reswick, J. B. (1967). Electrical inhibition of pain by stimulation of the dorsal columns: Preliminary clinical report. *Anesthesia and Analgesia*, *46*, 489–491.

Sherwood, A. M., Dimitrijevic, M. R., & Barry McKay, W. (1992). Evidence of subclinical brain influence in clinically complete spinal cord injury: Discomplete SCI. *Journal of the Neurological Sciences*, *110*, 90–98.

Swiontek, T. J., Sances, A., Jr., Larson, S. J., Ackmann, J. J., Cusick, J. F., Meyer, G. A., et al. (1976). Spinal cord implant studies. *IEEE Transactions on Biomedical Engineering*, *23*, 307–312.

Sysoev, Y., Bazhenovam, E., Lyakhovetskii, V., Kovalev, G., Shkorbatova, P., Islamova, R., et al. (2020). Site-specific neuromodulation of detrusor and external urethral sphincter by epidural spinal cord stimulation. *Frontiers in Systems Neuroscience*, *14*, 47.

Taccola, G., Salazar, B. H., Apicella, R., Hogan, M. K., Horner, P. J., & Sayenko, D. (2020). Selective antagonism of A1 adenosinergic receptors strengthens the neuromodulation of the sensorimotor network during epidural spinal stimulation. *Frontiers in Systems Neuroscience*, *14*, 44.

Taccola, G., Sayenko, D., Gad, P., Gerasimenko, Y., & Edgerton, V. R. (2018). And yet it moves: Recovery of volitional control after spinal cord injury. *Progress in Neurobiology*, *160*, 64–81.

Tan, A. M., & Waxman, S. G. (2012). Spinal cord injury, dendritic spine remodeling, and spinal memory mecha- nisms. *Experimental Neurology*, *235*, 142–151.

Tang, S. F., Tuel, S. M., McKay, W. B., & Dimitrijevic, M. P. (1994). Correlation of motor control in the supine position and assistive device used for ambulation in chronic incomplete spinal cord-injured persons. *American Journal of Physical Medicine & Rehabilitation*, *73*, 268–274.

Tansey, K. E. (2012). Profiling motor control in spinal cord injury: Moving towards individualized therapy and evidence-based care progression. *The Journal of Spinal Cord Medicine*, *35*, 305–309.

Topka, H., Cohen, L. G., Cole, R. A., & Hallett, M. (1991). Reorganization of corticospinal pathways following spinal cord injury. *Neurology*, *41*, 1276–1283.

Voronin, L. L., & Cherubini, E. (2004). 'Deaf, mute and whispering' silent synapses: Their role in synaptic plasticity. *The Journal of Physiology*, *557*, 3–12.

Wagner, F. B., Mignardot, J. B., Le Goff-Mignardot, C. G., Demesmaeker, R., Komi, S., Capogrosso, M., et al. (2018). Targeted neurotechnology restores walking in humans with spinal cord injury. *Nature*, *563*, 65–71.

Waters, R. L. (1984). The enigma of "carry-over". *International Rehabilitation Medicine*, *6*, 9–12.

Weidner, N., Ner, A., Salimi, N., & Tuszynski, M. H. (2001). Spontaneous corticospinal axonal plasticity and functional recovery after adult central nervous system injury. *Proceedings. National Academy of Sciences. United States of America*, *98*, 3513–3518.

Wolpaw, J. R. (2018). The negotiated equilibrium model of spinal cord function. *The Journal of Physiology*, *596*, 3469–3491.

Wrigley, P. J., Gustin, S. M., Macey, P. M., Nash, P. G., Gandevia, S. C., Macefield, V. G., et al. (2008). Anatomical changes in human motor cortex and motor pathways following complete thoracic spinal cord injury. *Cerebral Cortex*, *19*, 224–232.

Zhang, Q., Li, S., Wong, H. C., He, X. J., Beirl, A., Petralia, R. S., et al. (2018). Synaptically silent sensory hair cells in zebrafish are recruited after damage. *Nature Communications*, *9*, 1388.

Further reading

Delattre, V., Denison, T., Lambert, H., Minassian, K., Bloch, J., & Courtine, G. (2018). Targeted neurotechnology restores walking in humans with spinal cord injury. *Nature*, *563*, 65–71.

Hachmann, J. T., Grahn, P. J., Calvert, J. S., Drubach, D. I., Lee, K. H., & Lavrov, I. A. (2017). Electrical neuromodulation of the respiratory system after spinal cord injury. *Mayo Clinic Proceedings*, *92*, 1401–1414.

McKay, W. B., Lim, H. K., Priebe, M. M., Stokic, D. S., & Sherwood, A. M. (2004). Clinical neurophysiological assessment of residual motor control in post-spinal cord injury paralysis. *Neurorehabilitation and Neural Repair*, *18*, 144–153.

Chapter 6

Rehabilitation and wheelchair users after spinal cord injury: An overview

Marcelo Riberto[a], Ligia Jia Lin Wu[b], and Daniel Rubio de Souza[b]

[a]*Departamento de Ortopedia e Anestesia—Faculdade de Medicina de Ribeirão Preto, Universidade de São Paulo—Ribeirão Preto, Ribeirão Preto/SP, CEP, Brazil,* [b]*Divisão de Medicina Física, Instituto de Ortopedia e Traumatologia da Faculdade de, Medicina da Universidade de São Paulo, R. Ovídio Pires de Campos, Cerqueira Cesar, São Paulo/SP, CEP, Brazil*

List of abbreviations

AD	autonomic dysreflexia
HO	heterotopic ossification
NB	neurogenic bladder
PU	pressure ulcers
SCI	spinal cord injury

Introduction

Spinal cord injury (SCI) is one of the most complex clinical conditions that occur in humans. It involves motor, sensory, and autonomic changes with functional, emotional, economic, and social implications. This is a costly condition for the individual, his family and society. In Canada, it has been estimated to cost 1.47 to 3.03 million dollars per individual over a lifetime (Kang et al., 2018).

Numerous actions regarding clinical treatment and prevention of complications should be taken as soon as possible to facilitate clinical management and reduce morbidity and mortality (Riberto, Pinto, Sakamoto, & Battistella, 2005). These patients should be transferred to a specialized rehabilitation unit as soon as they are clinically stable (Bickenbach et al., 2013; Kirshblum & BrooksIn, 2010) where they must be assisted by a specialized interdisciplinary team to achieve desired therapeutic functional and clinical objectives (Nas, Yazmalar, Şah, et al., 2015).

Regarding mobility, robotic devices can offer highly repetitive training, coupled interactive strategies to enhance engagement, thus enhancing neuroplastic reorganization. Alternatively, powered exoskeletons are already commercially available and have proven their use in social activities and exercises.

Epidemiology

Incidence and prevalence of SCI have increased over the expansion of human activity. The epidemiological profile varies according to the region of the world and the most common etiologies are traumatic, with falls and traffic accidents prevailing, followed by sports accidents and violence. The incidence varies between 13 and 163.4/million and may be higher in some developing countries. There are few reports on the prevalence, but it is estimated to be between 490 and 526/million. Traumatic etiology predominates among younger men with ages varying between 14 and 68 years, but those figures are narrower in developing countries (between 29 and 46 years). The most common anatomical location is the cervical spine, therefore the neurological deficit is tetraplegia. There is a predominance of the most severe neurological lesions (AIS A and B) (Kang et al., 2018).

Other causes of SCI include infections, vascular injuries, genetic, and oncologic etiologies. This is a different group of patients, which include varied ages and certainly much more women. Nontraumatic SCI usually progresses slowly and causes incomplete harm to the spinal cord, which implies in varied clinic-functional pictures and prognosis (Almeida, Coelho, & Riberto, 2016). Mortality among patients with traumatic SCI was reported 28.8 times higher than in the Brazilian

Cellular, Molecular, Physiological, and Behavioral Aspects of Spinal Cord Injury. https://doi.org/10.1016/B978-0-12-822427-4.00006-X

population without SCI and infectious causes were responsible for 55.3% of deaths (Leite, Souza, Imamura, & Battistella, 2019).

Classification and prognosis

The American Spinal Cord Injury Association (ASIA) recommends use of ASIA Impairment Scale (AIS) and a standardization of the physical examination of muscle strength and cutaneous sensibility to pain and touch to define motor and sensitive levels, respectively. This neurological examination should be repeated periodically to assess the evolution of the level and severity of the injury. In general, most patients will experience some degree of improvement after SCI, especially in the first six months, and may continue up to 2 years (Kirshblum, Burns, Biering-Sorensen, et al., 2011).

The initial classification after the end of spinal shock can provide the first prognostic information. The rate of neurological recovery is extremely low in patients with complete and incomplete sensory injuries, and some degree of gait recovery occurs in 0.9–8.3% of AIS A patients and can reach 67.2–97.3% in AIS D. Younger patients have better prognosis of neurological recovery but most patients will have permanent sequelae and will lose the ability to walk (van Middendorp et al., 2011).

The recovery of voiding control is also related to the motor recovery of lower limbs (Pavese et al., 2016). The prediction of manual functional recovery is even more uncertain. It is assumed that the earlier and more intense the motor and sensory recovery, the greater the chance of manual function recovery (Velstra et al., 2014).

Objectives in spinal cord rehabilitation

These objectives must be shared among all rehabilitation team members, who must establish a plan, considering the needs and patient perspectives:

- Education patients/families about SCI and its consequences for functioning.
- Emotional support for the patient and family.
- Promote autonomy and independence in activities of daily living and mobility, considering the functional capacity and adaptation of individual needs.
- Pain and spasticity management.
- Rehabilitation of neurogenic bladder and bowel.
- Sexual rehabilitation.
- Prevention of musculoskeletal deformities.
- Prescription of orthoses and other assistive technology.
- Support for social and family reorganization.

Robotic rehabilitation of movement

Mobility after SCI is the most expected aim of recovery and is believed to be responsible for the highest improvement in quality of life. Individuals with tetraplegia affirm hand function as their priority of recovery to regain their lives back, while those with paraplegia, rank walking as their preferences.

When SCI motor level is above C8, complete or not, hand function is extremely limited, thus emphasis in rehabilitation relies on accessing one's environmental factors, which includes technology and much support from caretakers for many daily activities. Subsequently, rehabilitation for the upper limbs stresses on strengthening any intact motor capacities to the arm and hand as well as bringing awareness to areas with impaired sensibility e proprioception, to prevent secondary skin and musculoskeletal injuries.

Several different robotic devices are currently used for neurorehabilitation of the upper extremities following SCI. The most striking aspect of robotic therapy is the possibility to deliver a high volume of high-quality movement repetitions, which promote functional recovery and may potentially facilitate adaptive plasticity, reducing therapist burden and cost of care (Mekki, Delgado, Fry, Putrino, & Huang, 2018). The most studied robotic devices for upper limb rehabilitation in the market are Armeo, InMotion ARM/WRIST, The Hand of Hope, ReoGo, and the Arm Light Exoskeleton (ALEx) (Fig. 1). They share the use of different support systems or exoskeletons to facilitate movement by the subject. These devices can initiate, drive, or finalize the strength efforts in the motor tasks. Alternatively, the hand and forearm may be strapped into a molded device and use electromyography sensors on the forearm to control the hand for several different tasks. Also, they engage the user with a visual feedback component and interactive games that can be used to improve the

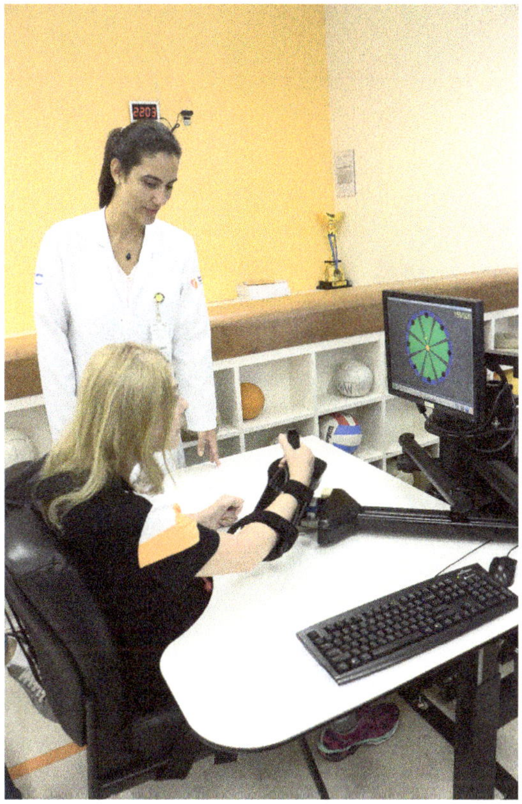

FIG. 1 Robotic rehabilitation upper limbs.

quality of therapeutic sessions. The Armeo Spring has been shown to be a reliable clinical tool to assess movement in the workplace (Rudhe, Albisser, Starkey, Curt, & Bolliger, 2012) and substitute other assessment tools (Zariffa et al., 2012), which added a novel use of robots as an extremely sensible measurement of functional achievements in SCI. Most of the published articles reporting their use so far are case reports or series with reports of motor improvements in chronic incomplete SCI subjects (Lu, Tong, Shin, Stampas, & Zhou, 2017; Siedziewski, Schaaf, & Mount, 2012).

In incomplete paraplegia, orthoses augment walking function, although they require a significant amount of strength and significantly increase energy expenditure (Mekki et al., 2018).

Moving around for AIS A or B paraplegic subjects, with motor levels above L3, has traditionally been limited to the wheelchair. While wheelchairs provide a modified level of independence and amplified access to the environment, wheelchair users continue to face difficulties in access and mobility. Usual rehabilitation with orthoses is highly demanding in physical aspects either on users, but also on therapists and caregivers (Fig. 2).

Body weight supported treadmill training may reduce this problem, but is not portable and therefore not recommended for community ambulation. In such scenarios, robotic gait rehabilitation can increase the amount of practice in a controlled system and may add an interactive engaging environment (Carpino, Pezzola, Urbano, & Guglielmelli, 2018) (Fig. 3).

Wearable exoskeletons are motorized or robotic devices that initiate or augment movement and may improve mobility and independence in nonambulatory people, thus possibly reducing secondary health conditions related to sedentariness. They facilitate recovery by delivering fully weight-bearing, repetitive, symmetric locomotor training efficiently and in high volume (Arazpour, Bani, Hutchins, & Jones, 2013).

Wearable lower limb exoskeletons for gait rehabilitation are still in their early stages of development and randomized control trials are needed to demonstrate their clinical efficacy. These robotic devices are used either for therapeutic purposes with repetitive, functional movement or as a mobility technology beyond wheelchairs and other walking aids. A recent report described the context in which one brand of exoskeletons have been used by 14 chronic complete SCI after an eight-week period of training. Subjects reported good satisfaction with the device in terms of comfort, adjustment, dimensions, durability and services, and less satisfaction with weight, effectiveness, and ease of use. The exoskeleton

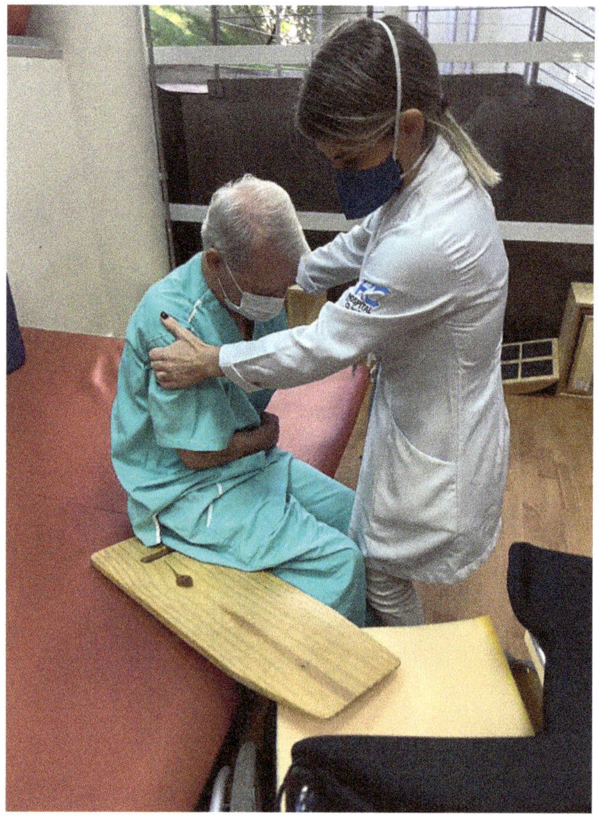

FIG. 2 Transfer training in SCI.

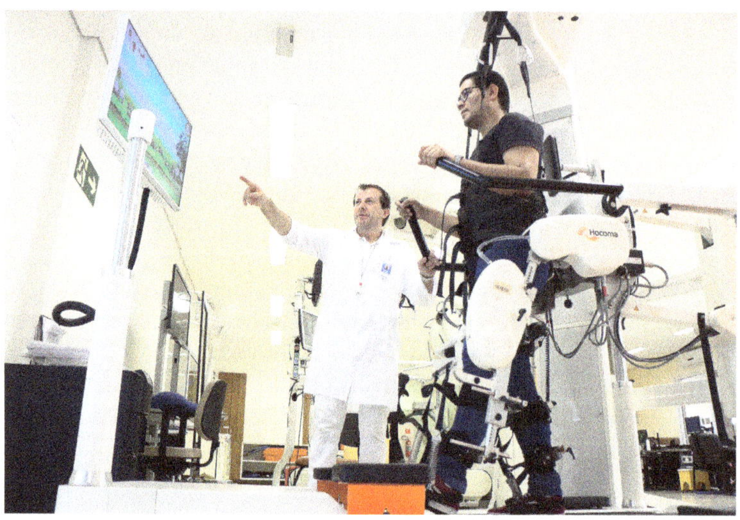

FIG. 3 Robotic rehabilitation lower limbs.

was used primarily for exercises and social events, and 48% of times in outdoor environments (van Dijsseldonk, van Nes, Geurts, & Keijsers, 2020).

Bone fractures and falls are possible adverse effects of the use of powered exoskeletons in SCI. Bone fractures have been associated with decreased body mass density as well as the unfamiliar forces placed on the paralyzed limbs or misalignment and excessive torques of the exoskeleton during gait, coupled with a lack of pain response (Miller, Zimmermann, & Herbert, 2016).

Main clinical aspects for rehabilitation

Pain

Pain is a highly disabling and extremely frequent complication that can occur at any moment after SCI, interfering with rehabilitation and quality of life (Burke, Fullen, Stokes, et al., 2017).

Musculoskeletal pain is perceived in regions with preserved sensitivity. In the acute phase, it affects about 50% of patients and is related to the muscle, joint, tendon, and bone structures involved in spinal trauma, surgical manipulations, and posture. After 6 months, it affects 40–58% of patients with SCI (Siddal & Loeser, 2001). Myofascial pain, tendinitis and bursitis are frequent and generally related to poor positioning in bed, poor sitting posture, repetitive strain injury or degenerative-musculoskeletal processes. Subjects in wheelchairs and those with walking aids may overuse their upper extremities when moving around or during transfers. Although the muscles in their shoulders become stronger with time, this is a source of musculoskeletal injuries and pain (Freitas, Santana, Manoel, Serenza, & Riberto, 2019) (Fig. 4).

The diagnosis of musculoskeletal nociceptive pain is based on the anamnesis, physical exam, and complementary exams. Treatment consists of removing triggering factors, prescribing analgesics, opioids, antiinflammatories, and muscle relaxants in association with analgesic measures, such as acupuncture, blocking myofascial trigger points, massage, and physical means. Therapeutic exercises (physiotherapy and occupational therapy) are also recommended for muscle and postural rebalancing (Siddal & Loeser, 2001).

Neuropathic pain challenges the management of 50% SCI patients in the acute phase and rises to as much as 83% in the chronic phase and affects mostly the lower limbs, being characterized as burning, shock or pressure, present in patients (Burke et al., 2017). When it occurs above the level of the SCI, it is usually related to the root and peripheral nerve compression or with complex regional pain syndrome. Allodynia, which is the sensation of pain at any sensory stimulus such as touch, heat, or cold, is seen in 14% of acutely injured patients and tends to decrease in the first 6 months, being more frequent in incomplete injuries. A band-like pattern of neuropathic pain at the neurologic level is reported early after SCI in half of the patients. It is characterized by burning pain, shock, or pins-and-needles, which can be confused with visceral cramps when it is in the abdominal region. It can be associated with passive or active movement (Siddal & Loeser, 2001).

Drug treatment must consider the classification of pain and chronicity. Musculoskeletal nociceptive pain may respond to the use of analgesics, opioids, antiinflammatories, and muscle relaxants. Neuropathic pain should be addressed with the use of antidepressants (tricyclic antidepressants and duloxetine may be good options) and anticonvulsants (gabapentin and pregabalin are first-line options) (Mehta, McIntyre, Janzen, et al., 2016). Drug treatment of neuropathic pain offers several options, but with results it is not always satisfactory, justifying the combination of drugs. Their side effects should be considered as well as drug interactions since patients with SCI often receive multiple medications. Transcranial electrical stimulation in the motor cortex has been studied and appears to show satisfactory effects, with improvement in neurogenic pain (Fregni, Boggio, Lima, et al., 2006).

Selected cases, refractory to these previous interventions can be subjected to electrical stimulation of the spinal cord, with implantation of electrodes in the dorsal column, but pain improvement noticed in the initial phase loses effectiveness over time. Other surgical procedures include intrathecal pumps of morphine, lidocaine, and baclofen (Mehta et al., 2016), as

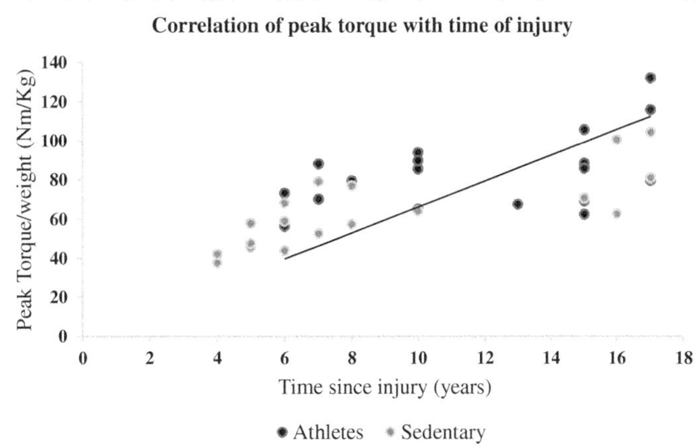

FIG. 4 Correlation between time (years) since injury and peak torque/weight (Nm/kg) of the internal rotators of the dominant shoulder of athletes and sedentary SCI subjects at 60°/s.

well as ablative surgeries—rhizotomy, cordotomy, and cordectomy with variable results. Complications of the most radical surgeries comprise loss of residual sensory, motor functions, urinary control, or sexual function (Mehta et al., 2016).

Spasticity

Spasticity is a motor dysfunction characterized by an involuntary, speed-dependent increase in muscle tone, as an exaggerated response to myotendinous stretching (exacerbation of the myotendinous reflex). It results from injuries to the pyramidal pathway leading to loss of inhibitory control of the medullary interneurons and modulation of cortical descending tracts. Spasticity is only observed in lesions above the medullary cone, after the phase of spinal shock (Elovic, Eisenberg, & Jasey Jr, 2010). The earlier the onset of spasticity, the greater its intensity. It is estimated that up to 80% of patients will have hypertonia and involuntary movements 1 year after the injury and 37% will need medications (Adams & Hicks, 2005).

Clinically, it is manifested by an increase in muscle resistance to sudden stretching (jackknife rigidity), clonus (rapid and repetitive involuntary muscle contractions), muscle automatisms (involuntary contractions) and deep tendon hyperreflexia. Muscle spasms can occur after any irritating nociceptive stimulus below the injury level, the most common are sudden movements, bladder repletion, intestinal constipation, urinary tract infection, pressure sores, ingrown nails, tight clothing, and inadequate positioning, among others (Elovic et al., 2010).

Spasticity can interfere in the control of movements of patients with partial motor control, worsening their mobility, namely in transfers, bed and chair positioning, standing and moving around. Muscle spasms and hypertonia hinder adequate positioning and activities such as clothing and bathing. In the long run, it can cause viscoelastic changes in the muscle leading to fibrosis, atrophy, and shortening and can cause deformities. However, mild to moderate spasticity can be useful as a facilitating factor for standing, sitting, handgrip, and postural changes (Elovic et al., 2010).

In the management of spasticity, the decision on the type of treatment to be performed depends on how much it interferes with functioning and quality of life (Taricco, Adone, Pagliacci, & Telaro, 2000). Treatment includes noninvasive therapeutic options, such as physiotherapy and oral medications, or more invasive techniques such as neurochemical blocks and orthopedic and neurosurgical procedures (Carvajal, Silva, Maeda, & Riberto, 2019; Elbasiouny, Moroz, Bakr, et al., 2010).

Drugs used in the systemic treatment may not guarantee the ideal control of spasticity and can cause systemic dose-dependent side effects like drowsiness and mental confusion. Most used drugs to treat spasticity are baclofen, benzodiazepines, and tizanidine. Anticonvulsants such as gabapentin can also be used, because of its gabaergic effect, as much as medications such as carbamazepine (which acts more by decreasing the frequency of spasms) (Taricco et al., 2000).

Regional treatment of spasticity can be performed with chemical neurolysis with phenol (5%) or alcohol (40–100%), which causes denaturation of the myelin sheath, leading to selective relaxation of target muscles. Muscle relaxation is immediate and can last up to 5 months. The procedure can be repeated when necessary, without restricting the interval between applications at the maximum dose of 15 mL with good results (Carvajal et al., 2019) (Fig. 5).

Botulinum toxin prevents the release of acetylcholine at the neuromuscular junction and successfully reduces spasticity. It is injected intramuscularly, in the motor plate, causing partial neurolysis. The effects are observed within 24–72 h. and can be temporary from 3 to 6 months, according to the range of complimentary treatment with psychical therapy or orthoses (Carvajal et al., 2019). When spasticity is refractory to drug and physical treatments, surgical procedures such as tenotomy, implantation of intrathecal baclofen or morphine infusion pumps, rhizotomies, cordotomies and cordectomies may be indicated (Elovic et al., 2010).

Heterotopic ossification

Heterotopic ossification (HO) is the pathological formation of ectopic bone in soft tissues. It occurs mainly in the first 12 months after the injury (with a peak up to months), always below the SCI level and close to the hips, knees, elbows and shoulders (Brady, Schultz, McDonald, & O'Brien, 2018). Clinical manifestations include low fever, regional oedema, hyperemia, pain in incomplete SCI, reduced range of motion and increased spasticity (Castro & Greve, 2003). Although physiopathology is not fully understood, an inflammatory modulation has been suggested to interfere on connective progenitor cells leading to differentiation into osteoblasts. Four to six weeks later there is calcification of the extracellular matrix produced by the newly differentiated cell (Ranganathan et al., 2015). Bone reabsorption is not noticed, and HO can only be removed surgically. The most striking impact of HO over functioning refers to the occurrence of deformities, which can further limit gait or adequate positioning in bed or wheelchair (Kirshblum & BrooksIn, 2010).

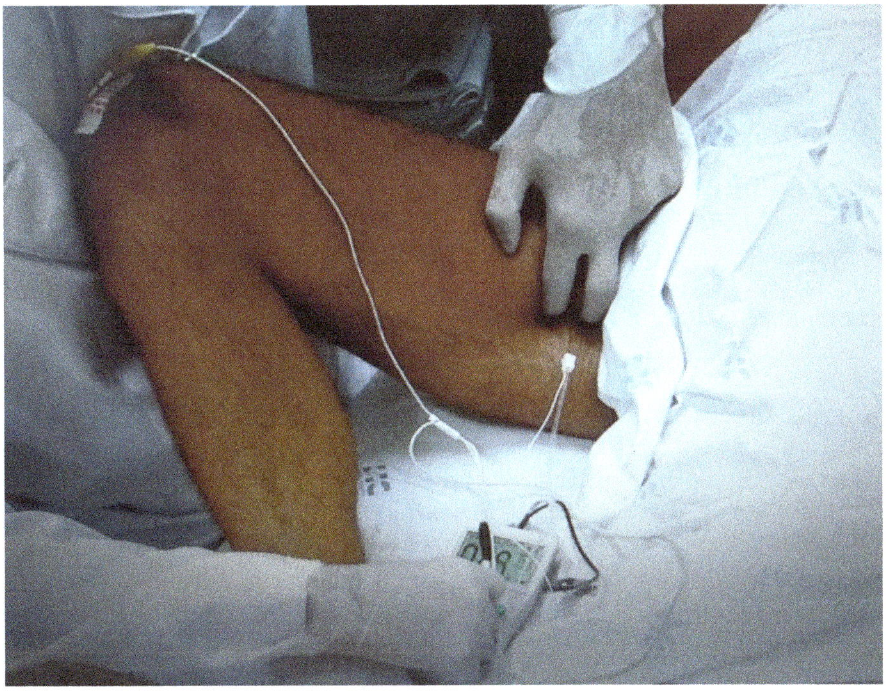

FIG. 5 Phenolic block of the anterior branch of the obturator nerve aiming to reduce adductor spasticity.

Early diagnosis is based on the clinical suspicion in a SCI subject with reduced range of motion and signal of limb oedema and hyperemia. Dosage of alkaline phosphatase, serum phosphorus, creatinophosfokinase (CPK) are sensitive, but nonspecific findings. Until a radiographic image is present after 6 weeks, bone scans are the only diagnostic resource available. Prophylactic use of nonsteroidal antiinflammatory agents or a single dose of 8 Grays radiation have been suggested and are routinely performed in some services (Popovic, Agarwal, Zhang, et al., 2014; Teasell, Mehta, Aubut, Ashe, et al., 2010) (Fig. 6).

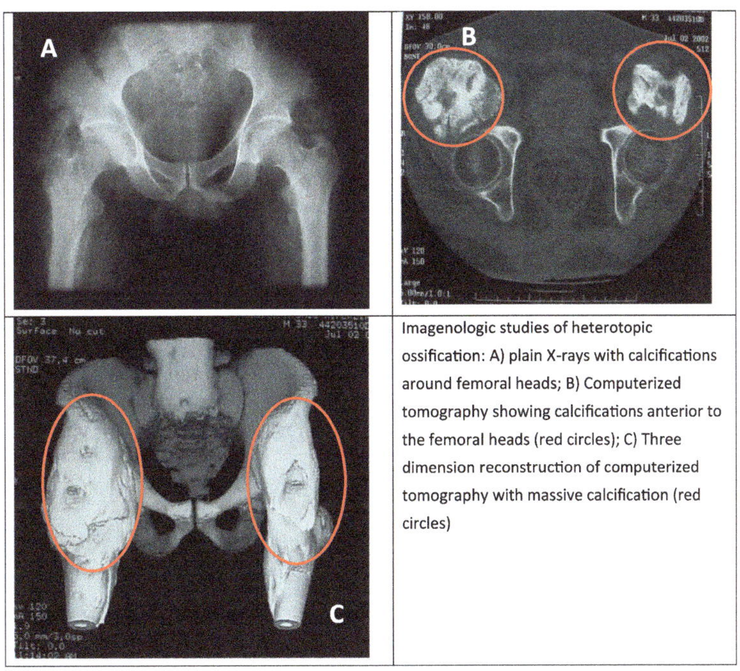

FIG. 6 Heterotopic ossification.

Autonomic dysreflexia

Autonomic dysreflexia (AD) is the sudden onset of arterial hypertension (increase of 20–40 mmHg in baseline blood pressure in adults or 15–20 mmHg in pediatric clinics) which occurs in patients with either complete or incomplete SCI above T6 associated with phenomena of cutaneous vasodilation in the head and neck (above the level of injury), characterized by facial flushing, nasal congestion and increased sweating, excruciating headaches, piloerection below the level of the lesion and, eventually, bradycardia (Cowan, Lakra, & Desai, 2020).

Early recognition is mandatory because of the risk of lethal complications such as acute myocardial infarction, pulmonary edema, hypertensive encephalopathy, seizures, and intracranial hemorrhage. Triggers of AD include nociceptive stimuli below the level of injury like the distention of hollow viscera, inflammatory and infectious processes, cutaneous injuries, or inadequate positioning (Eldahan & Rabchevisky, 2018). Chronic maintenance of repetitive episodes of AD are related with cardiovascular disease in people with SCI (Cragg, Noonan, Krassioukov, & Borisoff, 2013; Krassioukov, Warburton, Teasell, et al., 2009; West, Popok, Crawford, & Krassioukov, 2015).

Treatment of AD episodes starts with the correction of the triggering factors. Positioning of the patient to stimulate orthostatic hypotension, loosening of clothes and orthoses, as well as emptying the bladder and bowel are the recommended initial maneuvers. Other painful sources should be sought for and treated. If blood pressure remains elevated, with systolic values above 150 mmHg and/or diastolic values above 120 mmHg, antihypertensive pharmacological treatment with rapid-onset drugs—nitrates, channel blockers calcium (nifedipine) and alpha-blockers (hydralazine)—can be initiated, followed by the investigation of the triggering factors. If the hypertensive crisis is sustained and the triggering factor cannot be removed, the patient must be hospitalized and monitored until blood pressure normalizes (Cowan et al., 2020).

Neurogenic bladder

Urination is a process controlled by a complex neural network that involves multiple segments of the central nervous system (cerebral cortex, brainstem, spinal cord), peripheral nervous system and autonomic nervous system. In the SCI, ascending or descending tracts and the autonomic nervous system are compromised, altering the entire voiding cycle (Lisenmeyer, Bodner, Creasey, et al., 2006). Most common consequences of lesions in the upper motor neuron (i.e., above the medullary cone) are:

- Detrusor hyperreflexia: when the detrusor musculature contracts with minimal volumes of bladder repletion. These contractions can lead to hypertrophy of the detrusor musculature, formation of bladder diverticula and reduce bladder compliance.
- Voiding residue: the sphincter remains closed and opens momentarily at the expense of high intravesical pressure. Therefore, there is an accumulation of urine which predisposes the subject to infections and urinary stones.
- Vesical sphincter dyssynergy: under physiological conditions, the detrusor contracts as the urinary sphincter relaxes. In this pathologic condition, the detrusor contracts independently from sphincter opening and it is usual that paradoxical hypertonia of the sphincter is noticed (Lucas, 2019). Increased intravesical pressure that can lead to vesicoureteral reflux, elevation of pressure in the ureters and renal pelvis, causing dilation of the ureters and loss of renal function.

Periodic urodynamic study and urinary tract images are recommended after the acute phase of the SCI (Kreydin, Welk, Chung, Clemens, et al., 2018). Treatment of detrusor hyperreflexia requires tricyclic antidepressants, antimuscarinic drugs, or beta 3 agonists which differ among themselves in the frequency of side effects like blurry vision, impaired cognition, dry mouth, constipation, and tolerability (Milligan, Goetz, & Kennelly, 2020). The direct injection of botulinum toxin to the detrusor has been shown to significantly improve the bladder capacity without adverse effects, reducing the frequency of intermittent catheterization. Alpha-blockers reduce tension in sphincter and bladder neck muscles and can be combined in the treatment, especially in subjects with AD (Abreu-Mendes, Cruz, & Martins-Silva, 2020).

To guarantee an adequate bladder emptying, without post-voiding residue and low intravesical pressure, the most appropriate method is clean intermittent catheterization by the patient or a caregiver soon after the patient is clinically stable (Romo, Smith, Cox, et al., 2018). Complications comprise recurrent urinary tract infection and urethral trauma. For male patients with the condom catheter can use reflex voiding or bladder expression to push urine to the collector system, but recent systematic reviews (Groen, Pannek, Castro Diaz, et al., 2016; Kreydin et al., 2018) recommended clean intermittent catheterization as the primary method for bladder management, once theses maneuvers are associated with elevated intravesical pressure, vesicoureteral reflux, and upper tract disease (Wyndaele, Madersbacher, & Kovindha, 2001).

In lower motor neuron injuries, the biggest problem will be stress urinary incontinence, which occurs even with minimal movements. The detrusor musculature will be atonic and there will be great bladder capacity. Sometimes emptying maneuvers (Crede, Valsalva or a combination of these) will be insufficient for the complete emptying of the bladder but emptying by clean intermittent catheterization is also stimulated (Romo et al., 2018).

Neurogenic bowel dysfunction

Neurogenic bowel dysfunction results from autonomic nervous system injury and lack of central nervous system control. These disorders are highly prevalent in SCI and involve constipation (prevalence ranging between 56% and 80%), fecal incontinence (range 42%–75%), hemorrhoids (36%) and bloating (31%) (Krassioukov, Eng, Claxton, et al., 2010). Typically, constipation will occur in injuries above T11/T12 region, whereas incontinence follows lesions below that level (Rodriguez, 2016). Because traumatic SCI usually occurs at the cervical or thoracic levels, the spinal defecation center, located in the sacral spinal cord, is intact (Callaghan, Furness, & Pustovit, 2018).

In lesions above the sacral segments (medullary cone), the connections between the spinal cord and the bowel are maintained, the external sphincter and pelvic floor are hypertonic, causing stool retention. Since nerve connections between the spinal cord and the colon are preserved, bowel movement can be triggered by reflexes, so an irritating suppository or digital stimulation are therapeutic strategies to stimulate defecation (Adriaansen, van Asbeck, & van Kuppevelt, 2015).

Lesions below the sacral segments or cauda equina are characterized by loss of spinal cord mediation of the peristalsis, the propulsion of the fecal bolus is sluggish, there is intestinal areflexia, decreased bowel transit and sphincter tone, leading constipation associated with fecal incontinence. Digital stool extraction is quite common (Krassioukov et al., 2010).

Guidelines on emptying maneuvers include diet recommendations (Bernardi et al., 2020), choosing the best position for defecation (sitting in a hygienic chair), digital rectal stimulation or suppository which triggers the defecation reflex, relaxes the external anal sphincter, and can facilitate digital stool extraction. Abdominal massage and Valsalva maneuver are also indicated. Drug treatment is based on the use of laxatives, preferably mass forming (natural or synthetic fibers). Osmotic laxatives (lactulose) and emollients (mineral oil) are also prescribed when intestinal emptying is unsatisfactory with previous methods. The use of stimulants (bisacodyl) or irritants (senna) is avoided because chronic use can cause damage to the myenteric plexus (Krassioukov et al., 2010). More invasive interventions include trans-anal irrigation, antegrade colonic irrigation, nerve stimulation of anterior sacral roots and, finally, a colostomy, which provides good results and greater independence, even having good acceptance by patients (Hultling, 2020).

Pressure ulcers

Pressure ulcers (PU) are common complications with enormous impact on morbidity and mortality after SCI. They can be life-threatening and are a major economic burden, causing prolonged hospital stay, and compromised functional prognosis. They often occur over bony prominence in patients with altered sensitivity who remain in the same position for a long time (Atkinson & Cullum, 2018). Apparently, skin healing response to pressure differs in subjects with SCI as much as skin temperature, moistness, presence of feces and urine which interfere in the pathogenesis. A recent systematic review could not identify sufficient evidence for optimal bed or seated positions, or pressure relief maneuvers to prevent PUs in SCI persons (Groah, Schladen, Pineda, & Hsieh, 2015), although there is consensus that avoiding the 90° lateral position reduces PU risk over the trochanters. For wheelchair users, pressures are redistributed from the sitting area during recline and tilt, although reclining increases the risk of shear forces on the skin. Thus, technology with of cushions and wheelchairs is decisive in the prevention of PUs (He & Shi, 2020).

Facing an established PU, the treatment should focus on patient education to avoid any body support on the injured area, followed by local care to warrant a clean wound environment. Nutritional support is mandatory because of increased catabolic metabolism, risk of anemia and hypoalbuminemia (Bettex, Philandrianos, Jaloux, Bertrand, & Casanova, 2019). The control of automatisms, spasticity and prevention of deformities are key aspects to reduce pressure and shear forces (Atkinson & Cullum, 2018). The occurrence of infections on PUs is not rare, though swab cultures are often inconclusive and multi-bacterial (Dana & Bauman, 2015). If there is bony exposure, the recommendation is to collect bony material for culture and perform magnetic resonance studies to clarify the diagnosis of osteomyelitis and plan the surgical approach (Bettex et al., 2019) (Fig. 7).

Applications to other areas of neuroscience

Despite the enormous efforts of science, there is still no effective measure to cure SCI. The possibility of finding ways to provide neurological recovery depends on further clarifying the neurophysiological mechanisms related to SCI and recovery after trauma. We can say the same about the treatment of its main complications such as neuropathic pain,

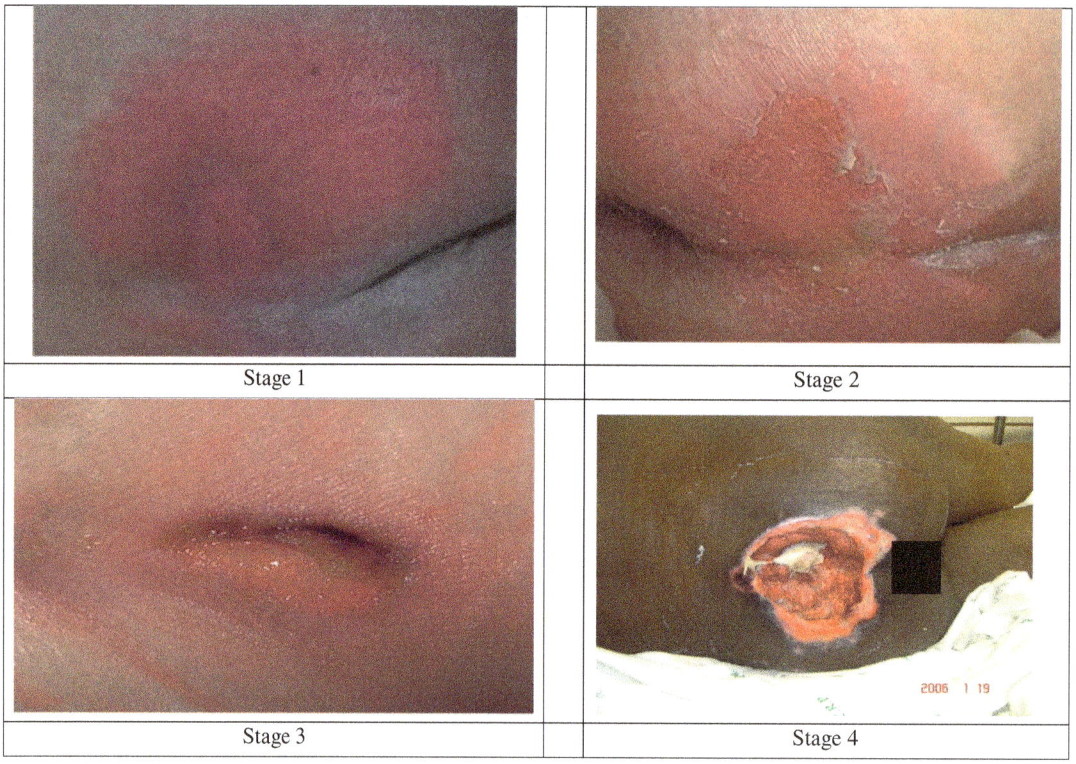

FIG. 7 Classification of pressure-induced skin injuries according to severity and deepness of the tissues involved.

spasticity, autonomic changes, and sphincter dysfunctions. The best management and reduction of their negative functional impacts also depends on a greater understanding of the complex mechanisms involved, knowing that structural and functional transformations of the neural axis and autonomic nervous system are involved. An improvement in the understanding of cellular and neurophysiological events may bring more effective treatments and greater possibilities of rehabilitation for patients with SCI.

Once structural and physiologic recovery are still not possible, rehabilitation is the main resource to attend the needs of these individuals. As much as in other central nervous system injuries, strategies to improve neuroplasticity based on the repetition are recommenced. This is particularly true in robotic therapy and use of powered exoskeletons for mobility or upper limb use. Future expectations rely on the crescent investigation of brain–machine interface. Also, alternative means of accomplishing the daily tasks should be texted and reinforced by a multi-professional rehabilitation team, considering the growing contribution of environmental facilitators and assistive technology to enhance social participation.

Mini-dictionary of terms

Activities of daily Living: term used to collectively describe fundamental skills that are required to independently care for oneself such as eating, bathing, and mobility.

Disability: The condition of being unable to perform an action because of an impairment in a specific environment and personal context.

Orthosis: an artificial or mechanical aid, such as a brace, to support or assist movement of a weak or injured part of the body.

Robotic devices: Powered instruments designed to help the subject to complete motor tasks or achieve better performance.

Urodynamic study: Clinical physiological study of the responses of the detrusor muscle to the accumulation of fluid in the bladder.

Key facts of clinical classification of SCI and mobility

- SCI is an extremely complex clinical condition that involves motor, sensory, and autonomic changes.
- The traumatic etiology responds for at least half of the cases, which affect mainly young adults and generate many costs to the individual, family, and society.
- The American Spinal Injury Association (ASIA) classification should be used to warrant standardization in diagnosis and comparison of treatments in different countries.
- Complete injuries are defined by the absence of sensitivity and motor skills in the last sacral roots and are less likely to have neurological recovery.
- Incomplete injuries manifest some motor or sensory preservation in the sacral levels and usually have better functional prognosis.
- Rehabilitation programs aim for improvement of functioning, prevention of secondary impairments, health promotion and education for patients and caregivers.
- The use of robotic equipment and powered exoskeletons are useful strategies in SCI because of a high volume of high-quality movement repetitions and reduced therapist burden.
- Access to orthoses and assistive technology allows people with spinal cord injuries to have greater independence and social participation.

Key facts on clinical complications of SCI

- Pain after SCI is frequent and may require centrally acting drugs or neurosurgical interventions in the most refractory cases.
- Spasticity treatments include physiotherapy, medications, neurolytic blocks, and the use of orthoses.
- The absence of sensitivity and movement in this population determines a high risk of pressure ulcers.
- Heterotopic ossification is the growth of an ectopic bone usually close to large joints which limits range of motion and hinders proper positioning and transfers and mobility.
- SCI usually results in loss of voluntary control of urination thus requiring bladder emptying through a catheter to prevent urinary loss and social restriction.
- Appropriate management of the neurogenic bladder prevents complications such as urinary stones or renal failure.
- Impairment of bowel and fecal sphincter control heightens the risk of constipation and eventual fecal incontinence.

Summary points

- Rehabilitation of SCI persons requires multi-professional skilled teams.
- Management of rehabilitation of SCI persons should use multi-dimensional assessment tools.
- Robotic devices promote high amounts of repetition and engage patients in motor tasks in interactive environments.
- Powered exoskeletons are already commercially available and can enhance exercise practice, social participation although there is still room for substantial improvement.
- Secondary complications of SCI may affect multiple systems and require the clinician to be aware of preventive and therapeutic interventions.

References

Abreu-Mendes, P., Cruz, F., & Martins-Silva, C. (2020). Treatment of neurogenic lower urinary tract symptoms: Main contributions from 2018 and 2019. *Current Opinion in Urology*, *30*(4), 486–490.

Adams, M. M., & Hicks, A. L. (2005). Spasticity after spinal cord injury. *Spinal Cord*, *43*, 577–586.

Adriaansen, J. J., van Asbeck, F. W., & van Kuppevelt, D. (2015). Outcomes of neurogenic bowel management in individuals living with a spinal cord injury for at least 10 years. *Archives of Physical Medicine and Rehabilitation*, *96*(5), 905–912.

Almeida, C., Coelho, J. C., & Riberto, M. (2016). Applicability, validation, and reproducibility of the spinal cord Independence measure version III (SCIM III) in patients with non-traumatic spinal cord lesions. *Disability and Rehabilitation*, *38*(22), 2229–2234.

Arazpour, M., Bani, M. A., Hutchins, S. W., & Jones, R. K. (2013). The physiological cost index of walking with mechanical and powered gait orthosis in patients with spinal cord injury. *Spinal Cord*, *51*(5), 356–359.

Atkinson, R. A., & Cullum, N. A. (2018). Interventions for pressure ulcers: A summary of evidence for prevention and treatment. *Spinal Cord*, *56*(3), 186–198.

Bernardi, M., Fedullo, A. L., Bernardi, E., Munzi, D., Peluso, I., Myers, J., et al. (2020). Diet in neurogenic bowel management: A viewpoint on spinal cord injury. *World Journal of Gastroenterology*, *26*(20), 2479–2497.

Bettex, Q., Philandrianos, C., Jaloux, C., Bertrand, B., & Casanova, D. (2019). La chirurgie secondaire des escarres chez le blessé médullaire [Surgical treatment of recurrent pressure ulcers in spinal cord injured patients]. *Annales de Chirurgie Plastique et Esthétique*, *64*(5–6), 674–684.

Bickenbach, J., Officer, A., Shakespeare, T., von Groote, P., World Health Organization, et al. (2013). *International perspectives on spinal cord injury: Summary*. World Health Organization. Available at https://apps.who.int/iris/handle/10665/94192.

Brady, R. D., Schultz, S. R., McDonald, S. J., & O'Brien, T. J. (2018). Neurological heterotopic ossification: Current understanding and future direction. *Bone*, *109*, 35–42.

Burke, D., Fullen, B. M., Stokes, D., et al. (2017). Neuropathic pain prevalence following spinal cord injury: A systematic review and meta-analysis. *European Journal of Pain*, *21*(1), 29–44.

Callaghan, B., Furness, J. B., & Pustovit, R. V. (2018). Neural pathways for colorectal control, relevance to spinal cord injury and treatment: A narrative review. *Spinal Cord*, *56*, 199–205.

Carpino, G., Pezzola, A., Urbano, M., & Guglielmelli, E. (2018). Assessing effectiveness and costs in robot-mediated lower limbs rehabilitation: A meta-analysis and state of the art. *Journal of Healthcare Engineering*, *2018*, 12.

Carvajal, D. V. T., Silva, N. O., Maeda, L. S., & Riberto, M. (2019). Phenol as a neurolytic agent in the treatment of focal and multifocal spasticity. *Acta Fisiatrica*, *26*(1), 1.

Castro, A. W., & Greve, J. M. D. (2003). Heterotopic ossification in patients with spinal cord injuries: Association with HLA-system antigens. *Acta Ortopedica Brasileira.*, *11*(2), 102–108.

Cowan, H., Lakra, C., & Desai, M. (2020). Autonomic dysreflexia in spinal cord injury. *BMJ*, *371*, m3596.

Cragg, J. J., Noonan, V. K., Krassioukov, A., & Borisoff, J. (2013). Cardiovascular disease and spinal cord injury: Results from a national population health survey. *Neurology*, *81*, 723–728.

Dana, A. N., & Bauman, W. A. (2015). Bacteriology of pressure ulcers in individuals with spinal cord injury: What we know and what we should know. *The Journal of Spinal Cord Medicine*, *38*(2), 147–160.

Elbasiouny, S. M., Moroz, D., Bakr, M. M., et al. (2010). Management of spasticity after spinal cord injury: Current techniques and future directions. *Neurorehabilitation and Neural Repair*, *24*(1), 23–33.

Eldahan, K. C., & Rabchevisky, A. G. (2018). Autonomic Dysreflexia after spinal cord injury: Systemic pathophysiology and methods of management. *Autonomic Neuroscience*, *209*(1), 59–70.

Elovic, E. P., Eisenberg, M. E., & Jasey, N. N., Jr. (2010). Spasticity and muscle Overactivity as components of the upper motor neuron syndrome. In *DeLisa's physical-medicine-rehabilitation: Principles and practice* (pp. 1319–1344). Philadelphia: Lippincott Williams & Wilkins.

Fregni, F., Boggio, P. S., Lima, M. C., et al. (2006). A sham-controlled, phase II trial of transcranial direct current stimulation for the treatment of central pain in traumatic spinal cord injury. *Pain*, *122*, 197–209.

Freitas, P. S., Santana, T. S., Manoel, L. S., Serenza, F. S., & Riberto, M. (2019). A comparison of isokinetic rotator cuff performance in wheelchair basketball athletes vs. non-athletes with spinal cord injury. *The Journal of Spinal Cord Medicine*, *16*, 1–6.

Groah, S. L., Schladen, M., Pineda, C. G., & Hsieh, C. H. (2015). Prevention of pressure ulcers among people with spinal cord injury: A systematic review. *PM & R: The Journal of Injury, Function, and Rehabilitation*, *7*(6), 613–636.

Groen, J., Pannek, J., Castro Diaz, D., et al. (2016). Summary of European Association of Urology (EAU) guidelines on neuro-urology. *European Urology*, *69*(2), 324–333.

He, C., & Shi, P. (2020). Interface pressure reduction effects of wheelchair cushions in individuals with spinal cord injury: A rapid review. *Disability and Rehabilitation*, *23*, 1–8.

Hultling, C. (2020). Neurogenic bowel management using trans anal irrigation by persons with spinal cord injury. *Physical Medicine and Rehabilitation Clinics of North America*, *31*, 305–318.

Kang, Y., Ding, H., Zhou, H., Wei, Z., Liu, L., Pan, L., et al. (2018). Epidemiology of worldwide spinal cord injury: A literature review. *Journal of Neurorestoratology.*, *6*, 1–9.

Kirshblum, S., & BrooksIn, M. (2010). Rehabilitation of spinal cord injury. In J. DeLisa (Ed.), *DeLisa's physical-medicine-rehabilitation: Principles and practice* (pp. 665–716). Philadelphia: Lippincott Williams & Wilkins.

Kirshblum, S. C., Burns, S. P., Biering-Sorensen, F., et al. (2011). International standards of neurological classification of spinal cord injury (revised 2001). *The Journal of Spinal Cord Medicine.*, *34*(6), 535–546.

Krassioukov, A., Eng, J., Claxton, G., et al. (2010). Neurogenic bowel management after spinal cord injury: A systematic review of the evidence. *Spinal Cord*, *48*(10), 718–733.

Krassioukov, A., Warburton, D. E. R., Teasell, R., et al. (2009). The SCIRE research team. A systematic review of the Management of Autonomic Dysreflexia Following Spinal Cord Injury. *Archives of Physical Medicine and Rehabilitation*, *90*(4), 682–695.

Kreydin, E., Welk, B., Chung, D., Clemens, Q., et al. (2018). Surveillance and management of urologic complications after spinal cord injury. *World Journal of Urology*, *36*(10), 1545–1553.

Leite, V. F., Souza, D. R., Imamura, M., & Battistella, L. R. (2019). Post-discharge mortality in patients with traumatic spinal cord injury in a Brazilian hospital: A retrospective cohort. *Spinal Cord*, *57*(2), 134–140.

Lisenmeyer, T. A., Bodner, D. R., Creasey, H. G., et al. (2006). The consortium for spinal cord medicine—Bladder Management for Adults with spinal cord injury: A clinical practice guideline for health-care providers. *The Journal of Spinal Cord Medicine*, *29*(5), 527–573.

Lu, Z., Tong, K. Y., Shin, H., Stampas, A., & Zhou, P. (2017). Robotic hand-assisted training for spinal cord injury driven by myoelectric pattern recognition: A case report. *American Journal of Physical Medicine & Rehabilitation*, *96*(10 Suppl 1), S146–s9.

Lucas, E. (2019). Medical Management of Neurogenic Bladder for children and adults: A review. *Topics in Spinal Cord Injury Rehabilitation*, *25*(3), 195–204.

Mehta, S., McIntyre, A., Janzen, S., et al. (2016). Systematic review of pharmacologic treatments of pain after spinal cord injury: An update. *Archives of Physical Medicine and Rehabilitation*, *97*(8), 1381–1391.

Mekki, M., Delgado, A. D., Fry, A., Putrino, D., & Huang, V. (2018). Robotic rehabilitation and spinal cord injury: A narrative review. *Neurotherapeutics*, *15*, 604–617.

Miller, L. E., Zimmermann, A. K., & Herbert, W. G. (2016). Clinical effectiveness and safety of powered exoskeleton-assisted walking in patients with spinal cord injury: Systematic review with meta-analysis. *Medical Devices (Auckland, N.Z.)*, *22*(9), 455–466.

Milligan, J., Goetz, L. L., & Kennelly, M. J. (2020). A primary care Provider's guide to management of neurogenic lower urinary tract dysfunction and urinary tract infection after spinal cord injury. *Topics in Spinal Cord Injury Rehabilitation*, *26*(2), 108–115.

Nas, K., Yazmalar, L., Şah, V., et al. (2015). Rehabilitation of spinal cord injuries. *World Journal of Orthopedics*, *6*(1), 8–16.

Pavese, C., Schneider, M. P., Schubert, M., Curt, A., Scivoletto, G., Finazzi-Agro, E., et al. (2016). Prediction of bladder outcomes after traumatic spinal cord injury: A longitudinal cohort study. *PLoS Medicine*, *13*(6), 1.

Popovic, M., Agarwal, A., Zhang, L., et al. (2014). Radiotherapy for prophylaxis of heterotopic ossification: A systematic review and meta-analysis of published data. *Radiotherapy and Oncology*, *113*(1), 10–17.

Ranganathan, K., Loder, S., Agarwal, S., Wong, V. W., Forsberg, J., Davis, T. A., et al. (2015). Heterotopic ossification: Basic-science principles and clinical correlates. *The Journal of Bone and Joint Surgery. American Volume*, *97*(13), 1101–1111.

Riberto, M., Pinto, P. P. N., Sakamoto, H., & Battistella, L. R. (2005). Functional independence of spinal cord injured patients. *Acta Fisiátrica*, *12*(2), 61–66.

Rodriguez, G. M. (2016). Bowel function after spinal cord injury. *Archives of Physical Medicine and Rehabilitation*, *97*, 339–340.

Romo, P. G. B., Smith, C. P., Cox, A., et al. (2018). Non-surgical urologic management of neurogenic bladder after spinal cord injury. *World Journal of Urology*, *36*(10), 1555–1568.

Rudhe, C., Albisser, U., Starkey, M. L., Curt, A., & Bolliger, M. (2012). Reliability of movement workspace measurements in a passive arm orthosis used in spinal cord injury rehabilitation. *Journal of Neuroengineering and Rehabilitation*, *9*, 37.

Siddal, P. J., & Loeser, J. D. (2001). Pain following spinal cord injury. *Spinal Cord*, *39*, 63–73.

Siedziewski, L., Schaaf, R. C., & Mount, J. (2012). Use of robotics in spinal cord injury: A case report. *The American Journal of Occupational Therapy*, *66*(1), 51–58.

Taricco, M., Adone, R., Pagliacci, C., & Telaro, E. (2000). Pharmacological interventions for spasticity following spinal cord injury. *Cochrane Database of Systematic Reviews*, *2*.

Teasell, R. W., Mehta, S., Aubut, J. L., Ashe, M. C., et al. (2010). A systematic review of the therapeutic interventions for heterotopic ossification after spinal cord injury. *Spinal Cord*, *48*, 512–521.

van Dijsseldonk, R. B., van Nes, I. J. W., Geurts, A. C. H., & Keijsers, N. L. W. (2020). Exoskeleton home and community use in people with complete spinal cord injury. *Scientific Reports*, *10*(1), 15600.

van Middendorp, J. J., Hasman, A. J. F., Danders, A. R. T., Pouw, N. H., Ditunno, J. F., Curt, A., et al. (2011). A clinical prediction rule for ambulation outcomes after traumatic spinal cord injury: A longitudinal cohort study. *The Lancet*, *377*, 1004–1010.

Velstra, I. M., Bolliger, M., Tanadini, L. G., Baumberger, M., Abel, R., Rietman, J. S., et al. (2014). Prediction e stratification of upper limb function and sef-care in acute cervical spinal cord injury with the graded redefined assessment of strength, sensibility and prehension (GRASSP). *Neurorehabilitation and Neural Repair*, *28*(7), 632–642.

West, C. R., Popok, D., Crawford, M. A., & Krassioukov, A. V. (2015). Characterizing the temporal development of cardiovascular dysfunction in response to spinal cord injury. *Journal of Neurotrauma*, *32*(12), 922–930.

Wyndaele, J. J., Madersbacher, H., & Kovindha, A. (2001). Conservative treatment of the neuropathic bladder in spinal cord injured patients. *Spinal Cord*, *39*, 294–300.

Zariffa, J., Kapadia, N., Kramer, J. L., Taylor, P., Alizadeh-Meghrazi, M., Zivanovic, V., et al. (2012). Relationship between clinical assessments of function and measurements from an upper-limb robotic rehabilitation device in cervical spinal cord injury. *IEEE transactions on Neural Systems and Rehabilitation Engineering: A Publication of the IEEE Engineering in Medicine and Biology Society*, *20*(3), 341–350.

Section B

Cellular and molecular aspects of spinal injury

Chapter 7

Gene expression and bone loss following spinal cord injury

Ariane Zamarioli
Department of Orthopaedics and Anesthesiology, Ribeirao Preto Medical School, University of Sao Paulo, Ribeirao Preto, Brazil

List of abbreviations

BMD	bone mineral density
BV/TV	fractional bone volume
CCL2	C—C motif chemokine ligand 2
CDC20	cell division cycle 20
CDK1	cyclin-dependent kinase 1
CCNA2	cyclin A2
CCNB1	cyclin B1
COL1a2	collagen type 1, alpha 2
Conn.D	connectivity density
CSA	cross-sectional area
CSF2	colony-stimulating factor 2
CTX	C-terminal telopeptide
DNA	deoxyribonucleic acid
DKK1	Dickkopf-1
EGR1	early growth response protein 1
FGF2	fibroblast growth factor
FOS	fos proto-oncogene
GFAP	glial fibrillary acidic protein
IL-β	interleukin 1 beta
IL6	interleukin 6
ITGAM	integrin subunit alpha M
JUN	jun proto-oncogene
lncRNA	long non-coding RNA
LRP-5	low-density lipoprotein receptor-related protein 5
LRP-6	low-density lipoprotein receptor-related protein 6
MBOAT4	membrane-bound O-acyltransferase domain containing 4
miRNA	micro RNA
mRNA	messenger RNA
MYC	MYC proto-oncogene
OPG	osteoprotegerin
OSX	osterixPOLE DNA polymerase epsilon, catalytic subunit
RANKL	receptor activator of NF-kB ligand
RNA	ribonucleic acid
RUNX2	runt-related transcription factor 2
SCI	spinal cord injury
sFRPs	secreted frizzled-related proteins
SMI	structure model index
STAT3	signal transducer and activator of transcription 3
Tb.Sp	trabecular separation
Tb.N	trabecular number

Cellular, Molecular, Physiological, and Behavioral Aspects of Spinal Cord Injury. https://doi.org/10.1016/B978-0-12-822427-4.00007-1

TGFBR2	transforming growth factor β receptor 2
TNFα	tumor necrosis factor alpha
TOP2A	DNA topoisomerase II alpha
TRAP	tartrate-resistant acid phosphatase TNFtumor necrosis factor
Wnt/β-catenin	canonical Wnt or beta (β)-catenin pathway
VEGF	vascular endothelial growth factor
VEGFR2	vascular endothelial growth factor receptor 2

Introduction

If a spinal cord is injured, several molecular and cellular processes are impaired, resulting in systemic changes that affect several organs and tissues, and leads to a complex disease that requires multi-professional care treatment. Although the motor and sensory deficits are the most evident and life quality impacting changes due to spinal cord injury (SCI), autonomic and cardiovascular abnormalities occur and lead to several secondary complications (Eldahan & Rabchevsky, 2018). Aside from the immobility due to SCI, these individuals have a disconnection between autonomic circuits and supraspinal control, leading to significant liability in blood pressure and vascular injuries (Popa et al., 2010). Therefore, ranging from extreme hypotension during episodes of orthostatic hypotension to severe hypertension during episodes of autonomic dysreflexia (Cragg, Noonan, Krassioukov, & Borisoff, 2013; Karlsson, 1999). Of note, cardiovascular and respiratory complications secondary to SCI are among the most common underlying and contributing causes of death in this population (Myers, Lee, & Kiratli, 2007). In the past, renal failure was a leading cause of death in SCI individuals. However, proper care of the urological tract has decreased mortality rate, but require early, intense, and continuous care to avoid deterioration of renal function over time (Nseyo & Santiago-Lastra, 2017; Panicker, Fowler, & Kessler, 2015; Samson & Cardenas, 2007; Welk et al., 2018). Other complications include hepatic dysfunction (Goodus & McTigue, 2020), gastrointestinal and metabolic disorder (Kigerl, Zane, Adams, Sullivan, & Popovich, 2020), thermoregulation (Price & Trbovich, 2018), and pressure ulcer formation (Baron et al., 2018), not to mention the mental and emotional impact, as well as the economic burdens to individuals, families, and societies (Ramer, Ramer, & Bradbury, 2014). This chapter aims to gain a better understanding of the different mechanisms responsible for systemic changes following a spinal cord injury and to specifically underly the relation between gene expression and SCI-induced bone loss.

Incidence and pathophysiological mechanisms of spinal cord injury

Approximately half a million people damage their spinal cord every year worldwide (Holmes, 2017), being approximately 18,000 new SCI cases per year in the United States, excluding those whose death occurred immediately post-injury (National Spinal Cord Injury Statistical Center, 2020). Aside from the non-traumatic SCI, a recent study evidenced the global incidence of traumatic spinal injury as 10.5 cases per 100,000 persons, resulting in an estimated 768,473 new cases every year. Road traffic accidents, followed by falls, were the most common mechanisms of injury. Furthermore, SCI incidence has been shown to be higher in low- and middle-income countries (8.72 per 100,000 persons) compared with high-income countries (13.69 per 100,000 persons) (Kumar et al., 2018). Accordingly, previous studies have shown that the prevalence of SCI is estimated as much as 71 new cases per million in Brazil (de Campos et al., 2008) and 318 in Iran, annually (Soleyman-Jahi et al., 2018).

SCI is a highly debilitating disease and is increasingly being recognized as a significant global health priority. Yet, the mechanisms underlying SCI have not been fully elucidated, and effective therapies for SCI are lacking (Wang et al., 2019). What is known is that its pathophysiological mechanism has two phases. The primary damage is the mechanical injury itself, which occurs in response to the force directly impacted to the spinal cord during the mechanical injury, which leads to disruption in axons, blood vessels, and cell membranes (Venkatesh, Ghosh, Mullick, Manivasagam, & Sen, 2019). The secondary damage phase results from biochemical processes following the primary damage, and it is characterized by vascular dysfunction, edema, ischemia, excitotoxicity, electrolyte shifts, free radical production, inflammation, and cell death, representing an important therapeutic target at preventing injury progression (Rowland, Hawryluk, Kwon, & Fehlings, 2008). In the initial stage of the secondary phase, activation of microglia, T-cells, and astrocytes and upregulation of proinflammatory cytokines like the tumor necrosis factor alpha (TNFα) and interleukin 1 beta (IL-β) occur. As a result, it leads to disrupted endothelial cells, increased permeability of the blood–brain barrier during the inflammation, and hemorrhage within the gray matter, thereby resulting in necrotic cell death or ischemia (Pineau & Lacroix, 2007). Another stage of the secondary phase refers to the Ca–dependent, glutamate-associated cell death and the formation of free radicals and nitric oxide, leading to impaired proteins, nucleic acids, lipids, and extracellular matrix proteins such as glycosaminoglycans. Consequently, resulting in neuronal cell death and function loss (Bao & Liu, 2002), as well as an overproduction of excitatory neurotransmitters such as glutamate and aspartate, which causes apoptosis of glial cells and neurons (Li & Stys, 2000; Park, Velumian, & Fehlings, 2004). Regeneration of the spinal cord does not occur after the injury due to intrinsic inhibitory factors expressed on central myelin and the extracellular matrix of the post-traumatic gliotic scar (Rowland et al., 2008).

Altered gene expression due to spinal cord injury

The expression of different genes has been identified in several diseases after the development of microarray and high-throughput sequence technologies (Venkatesh et al., 2019). Non-coding ribonucleic acid (RNA) genes, such as ribosomal RNAs, transfer RNAs, small nuclear RNAs, micro RNAs (miRNAs), and long non-coding RNAs (lncRNAs), have been identified in human and rodent genomes. MiRNAs are known to be differentially expressed after an SCI and may be a therapeutic target for the diagnosis, treatment, and prognosis of these lesions. Previous authors found that the serum *miR-9*, *miR-219*, and *miR-384-5p* levels may be promising biomarkers for predicting the severity of SCI (Hachisuka et al., 2014). LncRNAs constitute the largest portion of mammalian non-coding RNAs, with a critical role in the central nervous system, and that have risen as important regulators of several disease processes (Iyer et al., 2015). Specifically, previous authors reported that seven lncRNAs were found to be differentially expressed after an SCI, in which the *RGD1559747* (named as *lncSCIR1*) exhibited a significant downregulation at 1, 4, and 7 days following the injury (Wang et al., 2015). Recently, Zhou et al. have identified the expression of 772 lncRNA and 992 messenger RNA (mRNA) 2 h post-SCI, in which the interleukin 6 (IL6), membrane-bound o-acyltransferase domain containing 4 (MBOAT4), fos proto-oncogene (FOS), tumor necrosis factor (TNF), jun proto-oncogene (JUN), signal transducer and activator of transcription 3 (STAT3), colony-stimulating factor 2 (CSF2), MYC proto-oncogene (MYC), C—C motif chemokine ligand 2 (CCL2), and fibroblast growth factor 2 (FGF2) may be important targets in the immediate phase of SCI (Zhou et al., 2018). Changes in lncRNAs expression were also evidenced at the acute stages of SCI in a mouse model of spinal cord contusion. The authors detected few changes in the lncRNA expression levels at day 1 post-SCI, but the differential changes in LncRNA expression peaked 1 week after SCI and subsequently declined until 3 weeks post-SCI. The authors reported that differentially expressed mRNAs were involved in transport, cell adhesion, ion transport, and metabolic processes (Ding, Song, & Liu, 2016). In a very recent study, Shi et al. identified 3193 differentially expressed lncRNAs and 4308 differentially expressed mRNAs in an SCI rat model 2 days post-injury. The authors also detected 10 core genes, including interleukin 6 (IL6); deoxyribonucleic acid (DNA) topoisomerase II alpha (TOP2A); cyclin-dependent kinase 1 (CDK1); DNA polymerase epsilon, catalytic subunit (POLE); cyclin B1 (CCNB1); tumor necrosis factor (TNF); cyclin A2 (CCNA2); cell division cycle 20 (CDC20); integrin subunit alpha M (ITGAM); and MYC proto-oncogene (MYC), which may provide novel insights into the molecular mechanism of the early acute phase of SCI (Shi et al., 2019). In a recent and original study, Duran et al. detected critical pathways and networks that exhibit sustained alterations at the sub-chronic and chronic stages post-SCI. The authors identified 137 differentially expressed lncRNAs 1-month post-SCI, 239 in the third month, and 179 in the sixth month following injury, in which the molecules with the greatest number of interactions include transforming growth factor β receptor 2 (TGFBR2), glial fibrillary acidic protein (GFAP), signal transducer and activator of transcription 3 (STAT3), and early growth response protein 1 (EGR1), among others involved in astrogliosis and fibrosis. TGFBR2 exhibited a seven- to ninefold increase in gene expression from 1 month to 6 months post-SCI and had the most connections in the network, suggesting its essential roles in gliosis. The authors also identified several lncRNAs that may potentially be involved in the progression of SCI. These results indicate a high level of transcriptional abnormalities, which persisted in the sub-chronic and chronic stages, and the altered genes were related to several pathways such as immune and inflammatory responses (Duran et al., 2017).

Bone remodeling and osteometabolic dysfunction due to SCI

Bone remodeling corresponds to an ingenious mechanism, in which the bone contributes to the circulatory levels of calcium, and in parallel, ensures bone renewal. In ideal conditions, adult individuals restore bone resorption by an adequate amount of bone formation, guaranteeing bone strength maintenance (Zamarioli, de Andrade Staut, & Volpon, 2020). Any imbalance in bone turnover leads to osteometabolic dysfunction, and several mechanisms are involved according to each disease. Disuse osteoporosis is characterized by bone loss due to inactivity, and it is prevalent in several neurological disorders, including SCI, as well as in long-term bed rest and during microgravity exposure experienced by astronauts.

The osteocytes are bone cells embedded into the mineralized matrix, with the capability to sense mechanical stimuli and react to these loads by transducing these signals into biological responses, leading to the secretion of several factors, thus balancing bone turnover (Uda, Azab, Sun, Shi, & Pajevic, 2017). In able-bodied individuals, bone responds to its loading demand, in which higher mechanical load results in bone apposition and lower load results in bone resorption. SCI impairs motor function, thus resulting in immediate and permanent decrease of bone loading experienced by the lower extremities (Troy & Morse, 2015). The mechanisms underlying SCI-induced bone loss are complex and differ from other types of osteoporosis. Mechanical disuse in response to motor dysfunction, combined with post-injury neurogenic, vascular, and hormonal impairments, results in an imbalance of the natural resorption and formation of bone (Haider, Lobos, Simonian, Schnitzer, & Edwards, 2018).

Molecular mechanisms of SCI-induced bone loss

Although the mechanisms leading to SCI-induced bone loss are yet to be fully elucidated, the mechanical unloading may be a causative factor related to SCI-induced bone loss, mediated by osteocyte. Therefore, unloading conditions have not only been associated with osteocyte and osteoblast apoptosis but also with overactivity in bone resorption, providing a cellular mechanism for the suppressed bone formation and rapid bone loss (Gifre et al., 2015). The association between rapid bone loss and increased bone resorption rate, including increased alkaline phosphatase, urinary calcium, hydroxyproline, zinc, and constituents of bone matrix has been reported in the literature for a long time (Garland et al., 1992). Accordingly, a recent study with a rodent model of complete transection of the spinal cord evidenced an overactivity of bone resorption, parallel with a down expression of osteoblastic-related genes at different stages of cell maturation (Butezloff et al., 2019). Similarly, an uncoupled bone turnover was identified in SCI individuals as early as 2 days post-injury, with a predominance of resorption over formation, in which levels of serum C-terminal telopeptide (CTX) are increased by approximately 1.5-fold, in the absence of a coordinated change in serum osteocalcin (Peng et al., 2020). Bone resorption is stimulated by the release of receptor activator of NF-kB ligand (RANKL) by osteoblast lineage cells, thus stimulating differentiation and activity of osteoclasts. Osteocytes represent a major source of RANKL in unloaded bones, reflecting its fundamental role in mechanotransduction. The canonical Wnt or beta (β)-catenin pathway (Wnt/β-catenin) pathway regulates bone homeostasis by adjusting osteoblast function according to mechanical loading/unloading. Previous studies have reported upregulation of sclerostin, a Wnt inhibitor, in people with SCI (L. R. Morse et al., 2013). Consequently, leading to a decrease in Wnt/β-catenin signaling in osteoblasts, RANKL upregulation, osteoprotegerin (OPG) downregulation (Wijenayaka et al., 2011), therefore increasing bone resorption activity. Besides sclerostin, Dickkopf-1 (DKK1) and frizzled-related proteins (sFRPs) are Wnt inhibitors related to bone metabolism that exhibit important roles in the pathogenesis of disuse bone loss. Wnt binds to a co-receptor complex involving low-density lipoprotein receptor-related protein 5 or 6 (LRP-5 or LRP-6), present on osteoblasts, resulting in translocation of β-catenin, which activates gene transcription and promotes osteoblast proliferation, differentiation, and function, thereby leading to new bone formation. Sclerostin and DKK1 bind to LRP5/6 and suppress Wnt signaling, whereas sFRPs bind Wnts to reduce their biological activity (Peng et al., 2020).

Bone loss following SCI is certainly unique in its magnitude and rapidity, which may be explained by the concomitant effects of the absence of mechanical loading to the vascular changes, muscle atrophy, neural, and endocrine components. Appropriate vascular supply is essential for bone health, in which several pro-angiogenic factors, including vascular endothelial growth factor (VEGF) and its receptor VEGFR2, are crucial in the cross talk between bone cells and endothelial cells (Veeriah, Paone, Chatterjee, Teti, & Capulli, 2019). SCI-induced muscle atrophy also impairs bone homeostasis where the locoregional elastic deformation of the bone in response to muscle contraction is absent, thereby not leading to bone stimulation by means of mechanotransduction (Colaianni et al., 2017). Accordingly, a recent investigation in rats with muscle atrophy identified increased vascular porosity, with a reduction in osteocyte lacunar density, which resulted in increased intracortical porosity (Gatti et al., 2019), highly correlated to bone fragility (Iori et al., 2019). Bone–fat interactions are also involved in ongoing bone loss in chronic SCI. Of note, adiponectin, an adipokine produced by visceral fat, is negatively related to bone mass and strength in individuals with SCI, likely due to direct actions of adiponectin on bone cell receptors in regulating osteoblastogenesis and osteoclastogenesis (Doherty et al., 2014).

Acute and chronic stages of bone loss following SCI

SCI is known to cause rapid and intense osteoporosis, which is most severe below the level of injury (Morse et al., 2019) and may lead to a 5- to 23-fold increase in fracture risk. The mechanisms leading to SCI-induced bone loss involve two phases. A rapid and acute bone loss occurs in the first phase and reaches a plateau approximately 2 years post-injury. The second phase is characterized by a chronic and slower ongoing bone loss, which may continue for decades following SCI (Troy & Morse, 2015). It is also well recognized and reported in the literature that lower bone mass and skeletal microarchitecture deterioration are consequences of SCI. Table 1 summarizes changes in bone quality in animals and patients with SCI, previously reported in the literature. As of note, bone loss starts acutely post-lesion and continues at a rate of nearly 4% per month in areas where the trabecular bone is dominant and 2% per month in regions mostly occupied by cortical bone (Wilmet, Ismail, Heilporn, Welraeds, & Bergmann, 1995). Recently, Peng et al. detected bone mass loss and microarchitectural deterioration of trabecular bone occurring as early as 2 days post-SCI, becoming more evident on

TABLE 1 Previous scientific evidence for bone loss following spinal cord injury.

Authors, year	Clinical or experimental design and time post-SCI	Assessed outcomes evidencing bone loss
Haider et al. (2018)	Computed tomography measures were obtained from 101 individuals, who experienced an SCI 30 days to 50 years prior to participation.	Bone loss showed an exponential decay over time until steady-state levels (3.6 years post-SCI). SCI also induced rapid and profound reductions in bone stiffness at the knee.
Gifre et al. (2015)	42 patients with a recent (<6 months) and complete SCI. Dickkopf-1 (DKK1), bone turnover marker C-terminal telopeptide (CTX), and bone mineral density (BMD) were assessed at baseline, at 6 and 12 months.	SCI patients exhibited a marked increase in CTX and DKK1 throughout the study, concomitant with important decrease in BMD. Patients with higher DKK1 values showed higher sublesional BMD loss.
Garland et al. (1992)	BMD of the knee was measured in 55 subjects. 10 were controls (non-SCI); 25 acutely injured (114 days post-SCI), with 12 reexamined 16 months after injury; and 20 chronic (>5 years post-SCI).	SCI-induced bone loss occurs throughout the entire skeleton, except the skull. Regression techniques demonstrated early, rapid, and linear decline of bone below the pelvis.
Butezloff et al. (2019)	Gene expression, bone quality and quantity were assessed in 36 rats, which sustained an SCI for 24 days prior to analysis.	SCI downregulated osteoblastic-related gene expression, associated with a twofold increase in osteoclasts and overexpression of receptor activator of NF-kB ligand (RANKL), resulting in lower bone mass, impaired microarchitecture, and weaker bones.
Peng et al. (2020)	20–24 rats were subjected to complete transection of the spinal cord. Blood, hindlimb bone samples, and bone marrows were collected at 2 and 7 days after SCI.	SCI-induced bone loss occurs as early as 2 days after injury and the skeletal defects become more evident at 7 days, with increased CTX levels.
Doherty et al. (2014)	The association between BMD and circulating adipokines (adiponectin and leptin) was investigated in 149 men with chronic SCI.	Adiponectin is negatively related to bone mass in individuals with SCI, likely due to direct actions of adiponectin on bone cell receptors in regulating osteoblastogenesis and osteoclastogenesis.
Wilmet et al. (1995)	One-year longitudinal study of bone mineral measurements and soft tissue composition in supra- and infra-lesional areas of 31 patients with SCI.	The authors observed a rapid decrease of bone mass in the paralyzed areas, (−4%/month during the first year in areas rich in trabecular bone and −2%/month in areas containing mainly compact bone).
Eser et al. (2004)	89 individuals between 2 months and 50 years post-SCI (24 tetraplegics and 65 paraplegics) were measured by peripheral quantitative computed tomography to quantify bone loss following SCI.	In the femur and tibia, bone mass, total and trabecular bone mineral density of the epiphyses, as well as bone mass and cortical cross-sectional area of the diaphyses showed an exponential decrease with time after injury in the SCI subjects. The decreasing bone parameters reached new steady states after 3–8 years, depending on the parameter.
Zamarioli et al. (2013)	Bone quality and quantity were assessed in 31 rats, which sustained an SCI for 33 days prior to analysis.	The authors observed lower mass, increased osteoclastic activity, and weaker bone integrity due to SCI.
Rodriguez et al. (2021)	The incidence of musculoskeletal morbidities was compared between adults with (9081) and without SCIs (1,474,232). Incidence estimates of common musculoskeletal morbidities (e.g., osteoporosis, sarcopenia, osteoarthritis, fractures, etc.) were compared at 5 years of enrollment.	Adults living with traumatic SCIs had a higher incidence of musculoskeletal morbidities (82% vs. 47%) as compared to adults without SCI, and differences were to a clinically meaningful extent.

Previous experimental and clinical studies reported bone changes in animals and individuals with spinal cord injury.

day 7 after injury. Specifically, the authors identified a rapid reduction of sublesional cancellous bone mineral density (BMD), a tridimensional structural disruption, including lower bone volume (BV/TV), increased trabecular separation (Tb.Sp), and higher structure model index (SMI) for rod-like geometry, as well as lower trabecular number (Tb.N) and connectivity density (Conn.D), which resulted in compromised structural integrity and reduced mechanical strength (Peng et al., 2020). Bone losses are more significant around the knee (distal femur and proximal tibia) and previous studies have documented reduction in bone mineral content (BMC) of up to 50% within the first 5 years post-SCI, reflecting important clinical implications as SCI individuals are more than twice as likely to fracture a sublesional bone at some point of their life compared to able-bodied controls (Haider et al., 2018). The chronic stage of bone loss following SCI (more than 2 years after injury) is poorly defined as ongoing bone loss is not usually assessed until a fracture occurs. Chronic SCI is associated with significant decreases in bone mass, microstructure, bone strength, and muscle mass in the lower limbs. As of note, previous investigations have identified reductions in femoral neck BMD of 27%–40%, knee BMD of 37%–70%, and tibia cortical diaphysis BMC of 7%–25%. Trabecular microarchitectural parameters such as BV/TV and Tb.N at the distal femur and proximal tibia are decreased by 20%–27% compared to able-bodied healthy individuals (Troy & Morse, 2015). Mean cortical thickness decreased at a loss rate of 0.23 mm/year in the femur and 0.28 mm/year in the tibia. The reduction of cortical cross-sectional area (CSA), concomitant with stable total CSA suggests mostly endocortical resorption of the femoral and tibial shaft (Eser et al., 2004), which leads to severe mechanical integrity impairment.

Changes in bone quality and quantity following SCI

Previous outcomes regarding bone loss following SCI have been published in the literature by the author's research group (Butezloff et al., 2019; Zamarioli et al., 2013, 2014; Zamarioli et al., 2020). Here, we illustrate bone changes in response to complete transection of the spinal cord at the vertebral T10 level in rats after 30 days of injury in comparison with age-matched, sham-operated rats. Fig. 1 represents the uncoupled turnover in SCI rats in which bone resorption exceeds formation, compared to a balanced bone turnover in sham-operated rats. The increased osteoclastic activity is demonstrated histologically by tartrate-resistant acid phosphatase (TRAP) staining that specifically identifies osteoclasts cells, in red (Fig. 2). As a result of the increased resorption activity over bone formation, Fig. 3 shows bone microarchitecture deterioration in rats sustaining a complete SCI compared to the shams. Accordingly, Fig. 4 evidences that such deterioration occurs both in trabecular and cortical bone. These tridimensional figures are outcomes obtained by microcomputed tomography. Assessments performed by means of scanning electron microscopy also detected a time-dependent behavior of SCI-induced bone loss, demonstrated in Fig. 5. In this image, a substantial bone loss may be seen in paraplegic rats at both 30 and 60 days after injury. However, bone loss is more intense on day 60 than 30 post-SCI. The rapid and intense bone loss following SCI will ultimately lead to secondary complications, including bone fracture. Recently, it has been demonstrated changes in bone healing in response to complete transection of the spinal cord, which may be related to a down expression

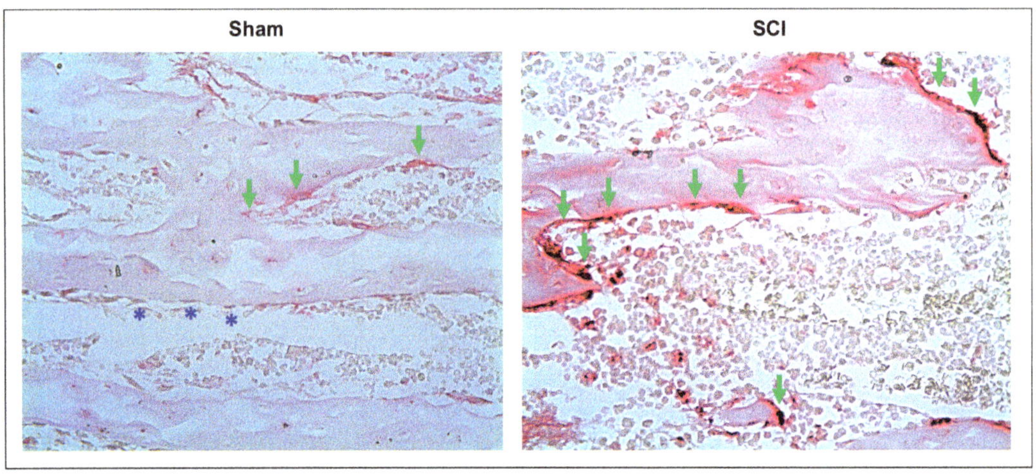

FIG. 1 Uncoupled bone in rats with spinal cord injury (SCI). A balanced turnover is observed in the sham images by a proportional activity between bone formation and resorption. Conversely, SCI images show an uncoupled bone turnover in which resorption exceeds formation, thereby leading to bone loss. Bone lining cells are represented by asterisks and indicate bone formation. The increased osteoclastic activity is demonstrated histologically by TRAP staining that specifically identifies osteoclast cells, in red (arrowheads), responsible for bone resorption. Magnification of 400×. Unpublished and original data.

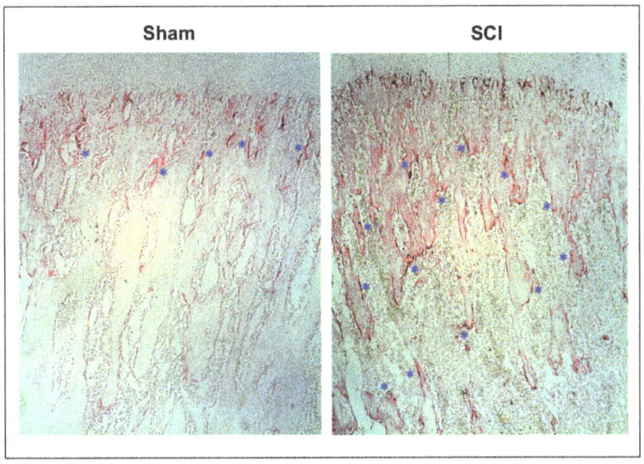

FIG. 2 Increased bone resorption activity in rats with spinal cord injury (SCI). TRAP histological slides specifically stain osteoclasts cells that may be identified as red cells (asterisks) and are responsible for bone resorption. The SCI group exhibited significantly more TRAP-positive cells per area of trabecular bone at the proximal tibial metaphysis than controls. Magnification of 100×. Unpublished and original data.

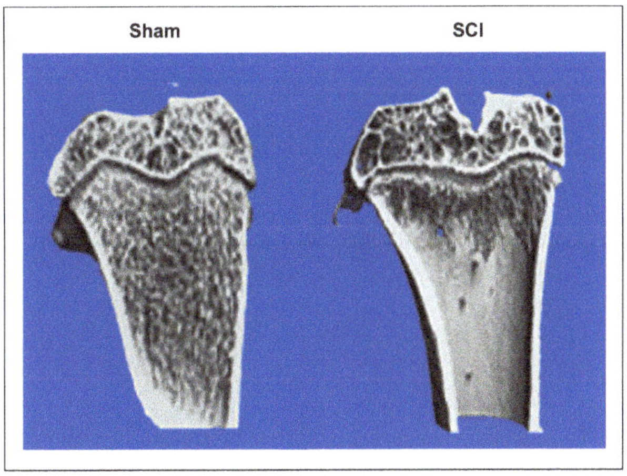

FIG. 3 Bone microarchitecture deterioration post-spinal cord injury. Fig. 3 shows tridimensional microarchitecture deterioration assessed by microcomputed tomography in response to increased resorption activity over bone formation. Unpublished and original data.

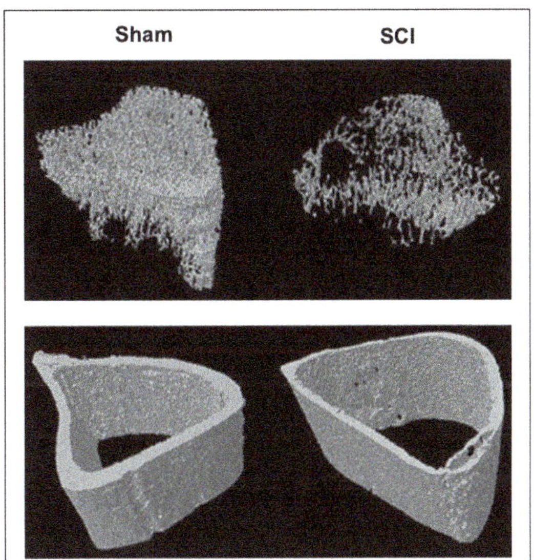

FIG. 4 Trabecular and cortical microstructure impairment due to paraplegia. Bone microstructure deterioration occurs both in trabecular and cortical bone following complete transection of the spinal cord, thus compromising its mechanical integrity. Unpublished and original data.

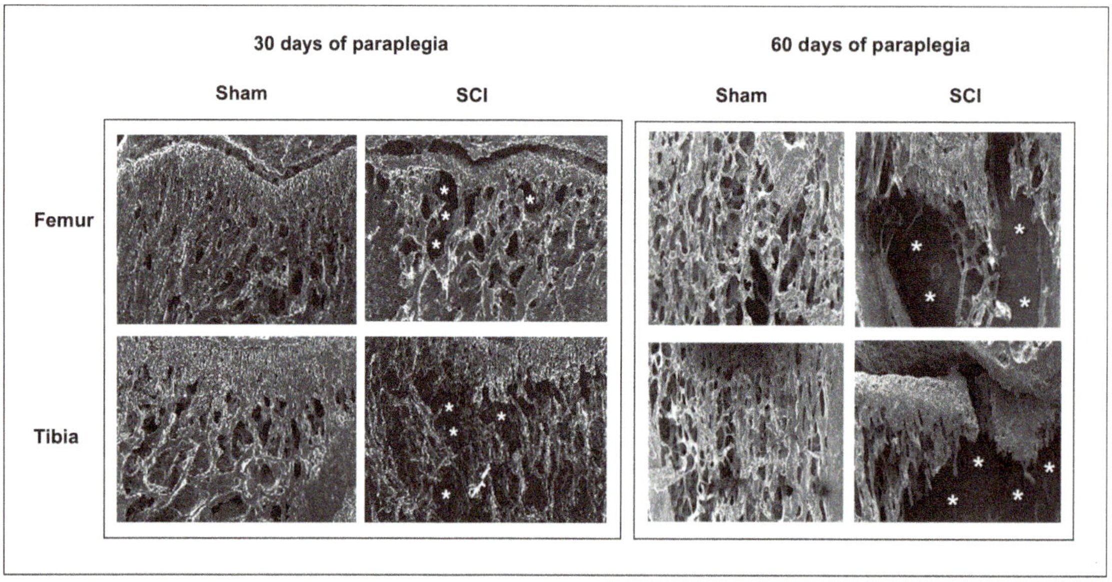

FIG. 5 Spinal cord injury (SCI)-induced bone loss is time-dependent on the injury. Scanning electron microscopy images detected a time-dependent behavior of SCI-induced bone loss. Although a substantial bone loss (asterisks) may be seen in paraplegic rats at both 30 and 60 days after injury, the loss is more intense on day 60 post-injury. Unpublished and original data.

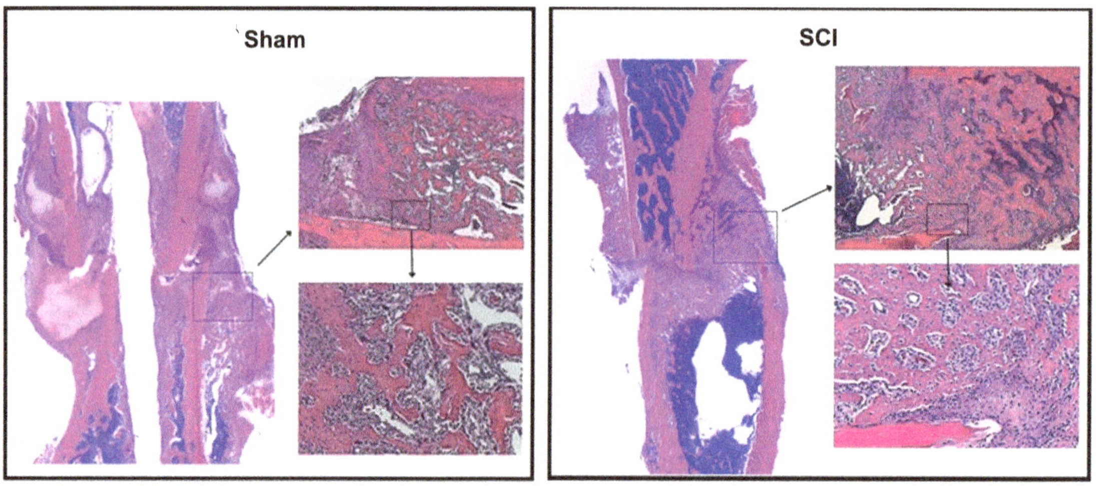

FIG. 6 Bone fragility in response to uncoupled turnover post-spinal cord injury (SCI). HE-stained histological images of the fracture evidencing a larger presence of woven bone and trabeculae than the sham rats, concomitant with lower cartilaginous tissue. Although bone healing may be accelerated in SCI rats, the newly formed callus is weaker, has thinner trabeculae, and has lower density. Magnifications of ×12.5, ×50, and ×200. Unpublished and original data.

in osteoblastic-related genes, including collagen type 1, alpha 2 (Col1a2), runt-related transcription factor 2 (Runx2), and osterix (Osx) (Butezloff et al., 2019). Butezloff et al. (2019) showed bone callus formation of lower quality in the SCI rats than the shams (Fig. 6), characterized by a lower density tissue, with thinner trabeculae and weaker at sustaining mechanical load until failure.

Secondary complication due to SCI-induced bone loss

Although mortality rates are declining for cancer, heart disease, stroke, arterial diseases, pulmonary embolus, urinary diseases, digestive diseases, and suicide among SCI individuals, these gains are being offset by increasing mortality rates for musculoskeletal disorders, accidents, endocrine, metabolic and nutritional conditions (National Spinal Cord Injury Statistical Center, 2020). In a very recent study, Rodriguez et al. reported that adults with SCI have a significantly higher risk and incidence of musculoskeletal morbidities (e.g., osteoporosis, bone fractures, sarcopenia, etc.) in comparison with non-SCI adults (Rodriguez et al., 2021). Recently, the incidence of bone fracture has been reported as high as 7.4% per year (Troy & Morse, 2015), and has been estimated that more than 50% of SCI individuals will experience an osteoporotic fracture at some point in their life post-injury. The rates of fractures range from 1.2 to 3.4 per 100 patients per year (Morse et al., 2019). Both the distal femur and the proximal tibia seem to fracture at about the same rate. These fractures frequently occur from relatively minimal trauma, which would not cause any fracture in able-bodied individuals, and are known as low-energy/impact fractures. Causes of low-impact fractures include routine events such as rolling in bed, transferring from a wheelchair, being turned in bed, incorrect transfer maneuvers, as well as during rehabilitation techniques (e.g., passive stretching exercises). Another common cause of fracture is when the foot falls from the wheelchair, hits an obstacle, and the wheelchair conductor forces a torsional loading at the knee, causing a spiral fracture. The rate of secondary complications related to bone fracture after SCI is high, which impose devastating consequences for patient morbidity and mortality, including prolonged hospitalization, delayed union or non-union, cellulitis, skin breakdown with pressure ulcer formation, increased pain and spasticity, lower extremity amputation, and premature death (Haider et al., 2018; Troy & Morse, 2015). Therefore, any effort to maintain skeletal integrity and decrease secondary complications due to bone loss is imperative following SCI to fully benefit from future advances in paralysis cure research and robotic exoskeletons, brain–computer interfaces, and other evolving technologies (Morse et al., 2019).

Conclusions

Although the pathophysiology of SCI-induced bone loss remains inconclusive, studies are rapidly expanding with hopes of elucidating mechanisms and pathways responsible for this disorder and further therapies. SCI has a unique feature to cause a rapid and intense uncoupled bone turnover, with an increase in resorption activity synergistically with suppression in bone formation, resulting in a significant bone loss following injury. SCI-induced bone loss is linked to a high incidence of low-energy fracture, which may invariably result in healing complications, prolonged hospitalizations, increased morbidity and mortality, and a high cost for the medical system. Here, we discussed the mechanisms leading to SCI-induced bone loss, as well as the phenotypic consequences observed in the bone tissue.

Applications to other areas of neuroscience

In this chapter, we have reviewed the mechanisms and pathways responsible for bone loss due to spinal cord injury (SCI). Bone loss due to disuse also occurs in other conditions such as long-term bed rest (Rolvien et al., 2020), and microgravity exposure during spaceflight missions (Stavnichuk, Mikolajewicz, Corlett, Morris, & Komarova, 2020). However, the interconnection between neurological injury and the lack of mechanical loading (disuse osteoporosis) increases the speed and amount of bone loss, which may also be seen in other neurological conditions (e.g., cerebral palsy, stroke) (Lee, Varghese, Brooks, & Turpen, 2020; Shin, Yoon, Chung, Rhee, & Cho, 2017). Bone loss in neurologic disorders is dependent on the time from injury and extent of sequela. Accordingly, Shi et al. have shown that patients with ambulatory cerebral palsy have higher bone quality than patients with non-ambulatory neurological disorder. Likewise, SCI individuals with complete motor paralysis have lower bone mass than patients who can walk with orthosis and sustain load into their limbs (weight-bearing activity) (Karimi, Esrafilian, Esrafilian, Sadigh, & Amiri, 2013). Bone loss mainly occurs in the first year post-SCI, and fractures are more common around the knee. Due to the low bone quality in SCI individuals, a fracture does not require high force, and routine activities may lead to bone failure. Health professionals caring for individuals with neurological paralysis must be alert for fragility fractures and consider referral for surgical evaluation and management.

Mini-dictionary of terms

- SCI-induced bone loss: a loss in bone quality due to spinal cord injury.
- Low-energy fracture: fractures that occur from relatively minimal trauma, which would not cause any fracture in able-bodied individuals.
- Osteoporotic fracture: a fracture in an osteoporotic bone.
- Osteocytes: bone cells embedded in the bone matrix; derived from the osteoblast cell lineage; and the principal sensors for mechanical loading of bone.
- Mechanotransduction: is how the bone cells sense physical forces and translate them into biochemical and biological responses.

- Osteoblasts: bone cells primarily responsible for bone formation.
- Osteoclasts: bone cells primarily responsible for bone resorption.
- Bone turnover: is the process of resorption followed by replacement with new bone that occurs throughout a person's life. An imbalance in bone turnover where resorption exceeds formation and results in bone loss.
- Secondary disease: is a disease that is a sequela or complication of a primary disease. For example, osteoporosis is a secondary disease of the spinal cord injury (primary disease).

Key facts of gene expression, bone loss, and spinal cord injury

- Spinal cord injury (SCI) is a highly disabling condition, which commonly results in significant long-term disability as it deprives many patients of their arm and leg function, bowel, bladder, and sexual function.
- Bone loss is a secondary disease of SCI, which may severely decrease life quality and increase morbidity and mortality rate.
- Fractures may occur during daily routines, such as transferring from bed or wheelchair, rolling or being turned in bed, fall, incorrect transfer maneuvers, as well as during medical care or rehabilitation treatment.
- More than 50% of SCI individuals will sustain a bone fracture in a sublesional bone at some point of their life post-injury.
- Bone healing in SCI individuals commonly leads to several complications, including delayed union, prolonged hospitalization, and pressure ulcer formation.

Summary points

- Spinal cord injury (SCI) is a highly debilitating disease and is increasingly being recognized as an important global health priority.
- After the development of microarray and high-throughput sequence technologies, the expression of different genes has been identified in several diseases, including SCI.
- Non-coding and long non-coding ribonucleic acids (RNAs) have been identified to be differentially expressed after SCI and may be a therapeutic target for the diagnosis, treatment, and prognosis of these lesions.
- The mechanisms underlying SCI have not yet been fully elucidated, and effective therapies for SCI are lacking.
- Two phases characterize the damages following the injury.
- The primary damage is the mechanical injury itself, and the secondary damage phase results from biochemical processes following the primary damage.
- Bone loss following SCI is certainly unique in its magnitude and rapidity, which may be explained by the concomitant effects of the absence of mechanical loading to the vascular changes, muscle atrophy, neural, and endocrine components.
- Uncoupled bone turnover was identified in SCI individuals and animals, with a predominance of resorption over formation.
- The highly uncoupled bone turnover due to SCI, associated with muscle atrophy and vascular disorder, potentially increases the risk for low-impact fractures in the sublesional bones in these individuals.
- Maintaining bone integrity and preventing fractures are crucial for individuals with SCI in order to fully benefit from future advances in paralysis cure research and robotic exoskeletons, brain–computer interfaces, and other evolving technologies.

References

Bao, F., & Liu, D. (2002). Peroxynitrite generated in the rat spinal cord induces neuron death and neurological deficits. *Neuroscience*. https://doi.org/10.1016/S0306-4522(02)00506-7.

Baron, J. S., Sullivan, K. J., Swaine, J. M., Aspinall, A., Jaglal, S., Presseau, J., … Grimshaw, J. M. (2018). Self-management interventions for skin care in people with a spinal cord injury: Part 2—A systematic review of use of theory and quality of intervention reporting. *Spinal Cord*. https://doi.org/10.1038/s41393-018-0136-5.

Butezloff, M. M., Volpon, J. B., Ximenez, J. P. B., Astolpho, K., Correlo, V. M., Reis, R. L., … Zamarioli, A. (2019). Gene expression changes are associated with severe bone loss and deficient fracture callus formation in rats with complete spinal cord injury. *Spinal Cord*. https://doi.org/10.1038/s41393-019-0377-y.

Colaianni, G., Mongelli, T., Cuscito, C., Pignataro, P., Lippo, L., Spiro, G., … Grano, M. (2017). Irisin prevents and restores bone loss and muscle atrophy in hind-limb suspended mice. *Scientific Reports*. https://doi.org/10.1038/s41598-017-02557-8.

Cragg, J. J., Noonan, V. K., Krassioukov, A., & Borisoff, J. (2013). Cardiovascular disease and spinal cord injury results from a national population health survey. *Neurology*. https://doi.org/10.1212/WNL.0b013e3182a1aa68.

de Campos, M. F., Ribeiro, A. T., Listik, S., de Pereira, C. A. B., de Andrade Sobrinho, J., & Rapoport, A. (2008). Epidemiology of spine injuries. *Revista do Colégio Brasileiro de Cirurgiões*. https://doi.org/10.1590/s0100-69912008000200005.

Ding, Y., Song, Z., & Liu, J. (2016). Aberrant LncRNA expression profile in a contusion spinal cord injury mouse model. *BioMed Research International*. https://doi.org/10.1155/2016/9249401.

Doherty, A. L., Battaglino, R. A., Donovan, J., Gagnon, D., Lazzari, A. A., Garshick, E., … Morse, L. R. (2014). Adiponectin is a candidate biomarker of lower extremity bone density in men with chronic spinal cord injury. *Journal of Bone and Mineral Research*. https://doi.org/10.1002/jbmr.2020.

Duran, R. C. D., Yan, H., Zheng, Y., Huang, X., Grill, R., Kim, D. H., … Wu, J. Q. (2017). The systematic analysis of coding and long non-coding RNAs in the sub-chronic and chronic stages of spinal cord injury. *Scientific Reports*. https://doi.org/10.1038/srep41008.

Eldahan, K. C., & Rabchevsky, A. G. (2018). Autonomic dysreflexia after spinal cord injury: Systemic pathophysiology and methods of management. *Autonomic Neuroscience: Basic & Clinical*. https://doi.org/10.1016/j.autneu.2017.05.002.

Eser, P., Frotzler, A., Zehnder, Y., Wick, L., Knecht, H., Denoth, J., & Schiessl, H. (2004). Relationship between the duration of paralysis and bone structure: A pQCT study of spinal cord injured individuals. *Bone*. https://doi.org/10.1016/j.bone.2004.01.001.

Garland, D. E., Stewart, C. A., Adkins, R. H., Hu, S. S., Rosen, C., Liotta, F. J., & Weinstein, D. A. (1992). Osteoporosis after spinal cord injury. *Journal of Orthopaedic Research*. https://doi.org/10.1002/jor.1100100309.

Gatti, V., Ghobryal, B., Gelbs, M. J., Gerber, M. B., Doty, S. B., Cardoso, L., & Fritton, S. P. (2019). Botox-induced muscle paralysis alters intracortical porosity and osteocyte lacunar density in skeletally mature rats. *Journal of Orthopaedic Research*. https://doi.org/10.1002/jor.24276.

Gifre, L., Vidal, J., Carrasco, J. L., Filella, X., Ruiz-Gaspà, S., Muxi, A., … Peris, P. (2015). Effect of recent spinal cord injury on wnt signaling antagonists (Sclerostin and Dkk-1) and their relationship with bone loss. A 12-month prospective study. *Journal of Bone and Mineral Research*. https://doi.org/10.1002/jbmr.2423.

Goodus, M. T., & McTigue, D. M. (2020). Hepatic dysfunction after spinal cord injury: A vicious cycle of central and peripheral pathology? *Experimental Neurology*. https://doi.org/10.1016/j.expneurol.2019.113160.

Hachisuka, S., Kamei, N., Ujigo, S., Miyaki, S., Yasunaga, Y., & Ochi, M. (2014). Circulating microRNAs as biomarkers for evaluating the severity of acute spinal cord injury. *Spinal Cord*. https://doi.org/10.1038/sc.2014.86.

Haider, I. T., Lobos, S. M., Simonian, N., Schnitzer, T. J., & Edwards, W. B. (2018). Bone fragility after spinal cord injury: Reductions in stiffness and bone mineral at the distal femur and proximal tibia as a function of time. *Osteoporosis International*. https://doi.org/10.1007/s00198-018-4733-0.

Holmes, D. (2017). Spinal-cord injury: Spurring regrowth. *Nature*, *552*(7684), S49. https://doi.org/10.1038/d41586-017-07550-9.

Iori, G., Schneider, J., Reisinger, A., Heyer, F., Peralta, L., Wyers, C., … Raum, K. (2019). Large cortical bone pores in the tibia are associated with proximal femur strength. *PLoS One*. https://doi.org/10.1371/journal.pone.0215405.

Iyer, M. K., Niknafs, Y. S., Malik, R., Singhal, U., Sahu, A., Hosono, Y., … Chinnaiyan, A. M. (2015). The landscape of long noncoding RNAs in the human transcriptome. *Nature Genetics*. https://doi.org/10.1038/ng.3192.

Karimi, M. T., Esrafilian, O., Esrafilian, A., Sadigh, M. J., & Amiri, P. (2013). Determination of the influence of walking with orthosis on bone osteoporosis in paraplegic subjects based on the loads transmitted through the body. *Clinical Biomechanics*. https://doi.org/10.1016/j.clinbiomech.2013.01.001.

Karlsson, A. K. (1999). Autonomic dysreflexia. *Spinal Cord*. https://doi.org/10.1038/sj.sc.3100867.

Kigerl, K. A., Zane, K., Adams, K., Sullivan, M. B., & Popovich, P. G. (2020). The spinal cord-gut-immune axis as a master regulator of health and neurological function after spinal cord injury. *Experimental Neurology*. https://doi.org/10.1016/j.expneurol.2019.113085.

Kumar, R., Lim, J., Mekary, R. A., Rattani, A., Dewan, M. C., Sharif, S. Y., … Park, K. B. (2018). Traumatic spinal injury: Global epidemiology and worldwide volume. *World Neurosurgery*. https://doi.org/10.1016/j.wneu.2018.02.033.

Lee, J., Varghese, J., Brooks, R., & Turpen, B. J. (2020). A primary care provider's guide to accessibility after spinal cord injury. *Topics in Spinal Cord Injury Rehabilitation*. https://doi.org/10.46292/SCI2602-79.

Li, S., & Stys, P. K. (2000). Mechanisms of ionotropic glutamate receptor-mediated excitotoxicity in isolated spinal cord white matter. *Journal of Neuroscience*. https://doi.org/10.1523/jneurosci.20-03-01190.2000.

Morse, L. R., Biering-Soerensen, F., Carbone, L. D., Cervinka, T., Cirnigliaro, C. M., Johnston, T. E., … Craven, B. C. (2019). Bone mineral density testing in spinal cord injury: 2019 ISCD official position. *Journal of Clinical Densitometry*. https://doi.org/10.1016/j.jocd.2019.07.012.

Morse, L. R., Sudhakar, S., Lazzari, A. A., Tun, C., Garshick, E., Zafonte, R., & Battaglino, R. A. (2013). Sclerostin: A candidate biomarker of SCI-induced osteoporosis. *Osteoporosis International*. https://doi.org/10.1007/s00198-012-2072-0.

Myers, J., Lee, M., & Kiratli, J. (2007). Cardiovascular disease in spinal cord injury: An overview of prevalence, risk, evaluation, and management. *American Journal of Physical Medicine and Rehabilitation*. https://doi.org/10.1097/PHM.0b013e31802f0247.

National Spinal Cord Injury Statistical Center. (2020). *Facts and figures at a glance*. Birmingham, AL: University of Alabama at Birmingham, National Spinal Cord Injury Statistical Center.

Nseyo, U., & Santiago-Lastra, Y. (2017). Long-term complications of the neurogenic bladder. *Urologic Clinics of North America*. https://doi.org/10.1016/j.ucl.2017.04.003.

Panicker, J. N., Fowler, C. J., & Kessler, T. M. (2015). Lower urinary tract dysfunction in the neurological patient: Clinical assessment and management. *The Lancet Neurology*. https://doi.org/10.1016/S1474-4422(15)00070-8.

Park, E., Velumian, A. A., & Fehlings, M. G. (2004). The role of excitotoxicity in secondary mechanisms of spinal cord injury: A review with an emphasis on the implications for white matter degeneration. *Journal of Neurotrauma*. https://doi.org/10.1089/0897715041269641.

Peng, Y., Zhao, W., Hu, Y., Li, F., Guo, X. E., Wang, D., … Qin, W. (2020). Rapid bone loss occurs as early as 2 days after complete spinal cord transection in young adult rats. *Spinal Cord*. https://doi.org/10.1038/s41393-019-0371-4.

Pineau, I., & Lacroix, S. (2007). Proinflammatory cytokine synthesis in the injured mouse spinal cord: Multiphasic expression pattern and identification of the cell types involved. *Journal of Comparative Neurology*. https://doi.org/10.1002/cne.21149.

Popa, C., Popa, F., Grigorean, V. T., Onose, G., Sandu, A. M., Popescu, M., … Sinescu, C. (2010). Vascular dysfunctions following spinal cord injury. *Journal of Medicine and Life*, *3*(3), 275–285.

Price, M. J., & Trbovich, M. (2018). Thermoregulation following spinal cord injury. In *Handbook of clinical neurology*. https://doi.org/10.1016/B978-0-444-64074-1.00050-1.

Ramer, L. M., Ramer, M. S., & Bradbury, E. J. (2014). Restoring function after spinal cord injury: Towards clinical translation of experimental strategies. *The Lancet Neurology*. https://doi.org/10.1016/S1474-4422(14)70144-9.

Rodriguez, G., Berri, M., Lin, P., Kamdar, N., Mahmoudi, E., & Peterson, M. D. (2021). Musculoskeletal morbidity following spinal cord injury: A longitudinal cohort study of privately-insured beneficiaries. *Bone*. https://doi.org/10.1016/j.bone.2020.115700.

Rolvien, T., Milovanovic, P., Schmidt, F. N., von Kroge, S., Wölfel, E. M., Krause, M., … Busse, B. (2020). Long-term immobilization in elderly females causes a specific pattern of cortical bone and osteocyte deterioration different from postmenopausal osteoporosis. *Journal of Bone and Mineral Research*. https://doi.org/10.1002/jbmr.3970.

Rowland, J. W., Hawryluk, G. W. J., Kwon, B., & Fehlings, M. G. (2008). Current status of acute spinal cord injury pathophysiology and emerging therapies: Promise on the horizon. *Neurosurgical Focus*. https://doi.org/10.3171/FOC.2008.25.11.E2.

Samson, G., & Cardenas, D. D. (2007). Neurogenic bladder in spinal cord injury. *Physical Medicine and Rehabilitation Clinics of North America*. https://doi.org/10.1016/j.pmr.2007.03.005.

Shi, Z., Ning, G., Zhang, B., Yuan, S., Zhou, H., Pan, B., … Feng, S. (2019). Signatures of altered long noncoding RNAs and messenger RNAs expression in the early acute phase of spinal cord injury. *Journal of Cellular Physiology*. https://doi.org/10.1002/jcp.27560.

Shin, Y. K., Yoon, Y. K., Chung, K. B., Rhee, Y., & Cho, S. R. (2017). Patients with non-ambulatory cerebral palsy have higher sclerostin levels and lower bone mineral density than patients with ambulatory cerebral palsy. *Bone*. https://doi.org/10.1016/j.bone.2017.07.015.

Soleyman-Jahi, S., Yousefian, A., Maheronnaghsh, R., Shokraneh, F., Zadegan, S. A., Soltani, A., … Rahimi-Movaghar, V. (2018). Evidence-based prevention and treatment of osteoporosis after spinal cord injury: A systematic review. *European Spine Journal*. https://doi.org/10.1007/s00586-017-5114-7.

Stavnichuk, M., Mikolajewicz, N., Corlett, T., Morris, M., & Komarova, S. V. (2020). A systematic review and meta-analysis of bone loss in space travelers. *NPJ Microgravity*. https://doi.org/10.1038/s41526-020-0103-2.

Troy, K. L., & Morse, L. R. (2015). Measurement of bone: Diagnosis of SCI-induced osteoporosis and fracture risk prediction. *Topics in Spinal Cord Injury Rehabilitation*. https://doi.org/10.1310/sci2104-267.

Uda, Y., Azab, E., Sun, N., Shi, C., & Pajevic, P. D. (2017). Osteocyte mechanobiology. *Current Osteoporosis Reports*. https://doi.org/10.1007/s11914-017-0373-0.

Veeriah, V., Paone, R., Chatterjee, S., Teti, A., & Capulli, M. (2019). Osteoblasts regulate angiogenesis in response to mechanical unloading. *Calcified Tissue International*. https://doi.org/10.1007/s00223-018-0496-z.

Venkatesh, K., Ghosh, S. K., Mullick, M., Manivasagam, G., & Sen, D. (2019). Spinal cord injury: Pathophysiology, treatment strategies, associated challenges, and future implications. *Cell and Tissue Research*. https://doi.org/10.1007/s00441-019-03039-1.

Wang, J., Hu, B., Cao, F., Sun, S., Zhang, Y., & Zhu, Q. (2015). Down regulation of lncSCIR1 after spinal cord contusion injury in rat. *Brain Research*. https://doi.org/10.1016/j.brainres.2015.07.052.

Wang, F., Liu, J., Wang, X., Chen, J., Kong, Q., Ye, B., & Li, Z. (2019). The emerging role of lncRNAs in spinal cord injury. *BioMed Research International*. https://doi.org/10.1155/2019/3467121.

Welk, B., Schneider, M. P., Thavaseelan, J., Traini, L. R., Curt, A., & Kessler, T. M. (2018). Early urological care of patients with spinal cord injury. *World Journal of Urology*. https://doi.org/10.1007/s00345-018-2367-7.

Wijenayaka, A. R., Kogawa, M., Lim, H. P., Bonewald, L. F., Findlay, D. M., & Atkins, G. J. (2011). Sclerostin stimulates osteocyte support of osteoclast activity by a RANKL-dependent pathway. *PLoS One*. https://doi.org/10.1371/journal.pone.0025900.

Wilmet, E., Ismail, A. A., Heilporn, A., Welraeds, D., & Bergmann, P. (1995). Longitudinal study of the bone mineral content and of soft tissue composition after spinal cord section. *Paraplegia*. https://doi.org/10.1038/sc.1995.141.

Zamarioli, A., Battaglino, R. A., Morse, L. R., Sudhakar, S., Maranho, D. A. C., Okubo, R., … Shimano, A. C. (2013). Standing frame and electrical stimulation therapies partially preserve bone strength in a rodent model of acute spinal cord injury. *American Journal of Physical Medicine and Rehabilitation*, *92*(5). https://doi.org/10.1097/PHM.0b013e318287697c.

Zamarioli, A., de Andrade Staut, C., & Volpon, J. B. (2020). Review of secondary causes of osteoporotic fractures due to diabetes and spinal cord injury. *Current Osteoporosis Reports*. https://doi.org/10.1007/s11914-020-00571-w.

Zamarioli, A., Maranho, D. A., Butezloff, M. M., Moura, P. A., Volpon, J. B., & Shimano, A. C. (2014). Anatomic changes in the macroscopic morphology and microarchitecture of denervated long bone tissue after spinal cord injury in rats. *BioMed Research International*, *2014*. https://doi.org/10.1155/2014/853159.

Zhou, H., Shi, Z., Kang, Y., Wang, Y., Lu, L., Pan, B., … Feng, S. (2018). Investigation of candidate long noncoding RNAs and messenger RNAs in the immediate phase of spinal cord injury based on gene expression profiles. *Gene*. https://doi.org/10.1016/j.gene.2018.03.074.

Chapter 8

Sperm DNA fragmentation and its relevance to men with spinal cord injury

Jaime Gosálvez[a], Eduardo Vargas-Baquero[b], and Stephen D. Johnston[c]

[a]*Genetics Unit, Department of Biology, Universidad Autónoma de Madrid, Madrid, Spain,* [b]*Sexual and Fertility Unit, Hospital Nacional de Parapléjicos de Toledo, Toledo, Spain,* [c]*School of Agriculture and Food Sciences, The University of Queensland, Gatton, QLD, Australia*

List of abbreviations

ICSI Intracytoplasmic sperm injection
PVS Penile vibratory stimulation
ROS Reactive oxygen species
SDF Sperm DNA fragmentation
SCI Spinal cord injury

Introduction

DNA integrity and stability are crucial for normal embryonic development leading to a successful pregnancy. However, as DNA is an active and dynamic molecule, it is constantly under pressure from physicochemical and biological forces that can induce changes or mutations to its native base-pair information. Of the adverse changes affecting the DNA molecule, one of the most profound and severe, is DNA fragmentation. DNA fragmentation is the separation or breaking of a continuous DNA strand into pieces such that genetic information and the control of effective replication and translation are compromised. To overcome this problem, somatic cells in eukaryotic organisms use mechanisms of DNA repair but these mechanisms are silent in mature spermatozoa (Gonzalez-Marin, Gosálvez, & Roy, 2012). Consequently, any breakage of the sperm DNA molecule is susceptible of being incorporated into the zygote following fertilization, unless it is repaired by the oocyte. If the damage is too severe, then this may result in potential errors during embryo development leading to failed pregnancy (García-Rodríguez, Gosálvez, Agarwal, Roy, & Johnston, 2019). In general, while all ejaculates possess a proportion of spermatozoa with damaged DNA, the higher the percentage of sperm with DNA fragmentation, the greater the probability of infertility (Deng et al., 2019). Human males presenting with a high proportion of sperm DNA fragmentation (SDF) are regarded as having impaired fertility once their SDF exceeds 30% (Esteves et al., 2020), whereas individuals considered fertile or presenting with acceptable levels of SDF typically present in the order of $\leq$20% (Gosálvez, Fernández, et al., 2015). Although the precise relationship between SDF and pregnancy outcome is still debatable, the extremely high incidence of sperm DNA damage reported in spinal cord injury (SCI) males (Vargas-Baquero et al., 2020) makes this condition, its causes and consequences, highly relevant to the SCI community. Consequently, understanding SDF in males with SCI is vital not only for establishing a clinical prognosis but also for developing effective therapies to overcome infertility. Given that the impact of SCI on male fertility is likely to be a complex synergistic combination of neural and lifestyle factors, it will also be important to compare how men with SCI differ from fertile and infertile males without SCI in terms of SDF and what elements of their medical condition might predispose them to high levels of SDF. This review will present the current status of information regarding SDF to males with SCI by analyzing the limited studies on this phenomenon and comparing these observations to other infertility etiologies associated with high levels of SDF. Additionally, we will define the primary mechanisms of SDF, discuss the possible etiologies of SDF that appear unique to SCI males, and finally propose clinical strategies designed to reduce SDF in SCI males to improve the probability of a successful pregnancy.

Cellular, Molecular, Physiological, and Behavioral Aspects of Spinal Cord Injury. https://doi.org/10.1016/B978-0-12-822427-4.00008-3

How does SDF in men with SCI compare to other causes of infertility?

SDF in men with SCI has to date been limited to six studies, three using the sperm chromatin dispersion (SCD) test (Bartolomé-Nebreda et al., 2020; Qiu et al., 2012; Vargas-Baquero et al., 2020), one by means of the sperm chromatin structure assay (Brackett et al., 2008) and two using the terminal deoxynucleotidyl transferase dUTP nick end labeling (TUNEL) assay procedure (Restelli et al., 2007; Talebi, Khalili, Vahidi, Ghasemzadeh, & Tabibnejad, 2013). Despite the methodology, the conclusion of all studies was that SDF in men with SCI is higher than in men without SCI. The next question then is how different are the levels of SDF in men with SCI compared to other clinical situations in men without SCI? To analyze this, we have selected data published by Esteves et al. (2015) in males without SCI (Fig. 1); to avoid differences based on the methodology, all cases were assessed for SDF using the SCD. In Fig. 1, we have included data from men without SCI that have experienced at least two failed assisted reproduction cycles, that have been diagnosed with leukocytospermia, *Chlamydia trachomatis* infection, testicular cancer, and varicocele and compared these to a control group (sperm donors) and men with SCI, previously published by Vargas-Baquero et al. (2020). Fig. 1 reveals that males without SCI but with other pathologies had SDF values approximately 2.5 × that of sperm donors but in the case of men with SCI their SDF values were 5 × higher than the sperm donors. Although these high SDF values position men with SCI as the most affected clinical group with SDF pathology, it should be noted that in almost every ejaculate of males with SCI, there is at least a small proportion of spermatozoa that were not affected by DNA damage, that theoretically, could be recovered to establish a pregnancy.

Molecular mechanisms of SDF

There are three fundamental theories thought to explain the presence of SDF in the ejaculate (Gosálvez, López-Fernández, Fernández, Esteves, & Johnston, 2015). The first hypothesis proposes that DNA breakage is associated with improper replacement of histone with protamine during spermiogenesis. Precise histone–protamine transition allows greater

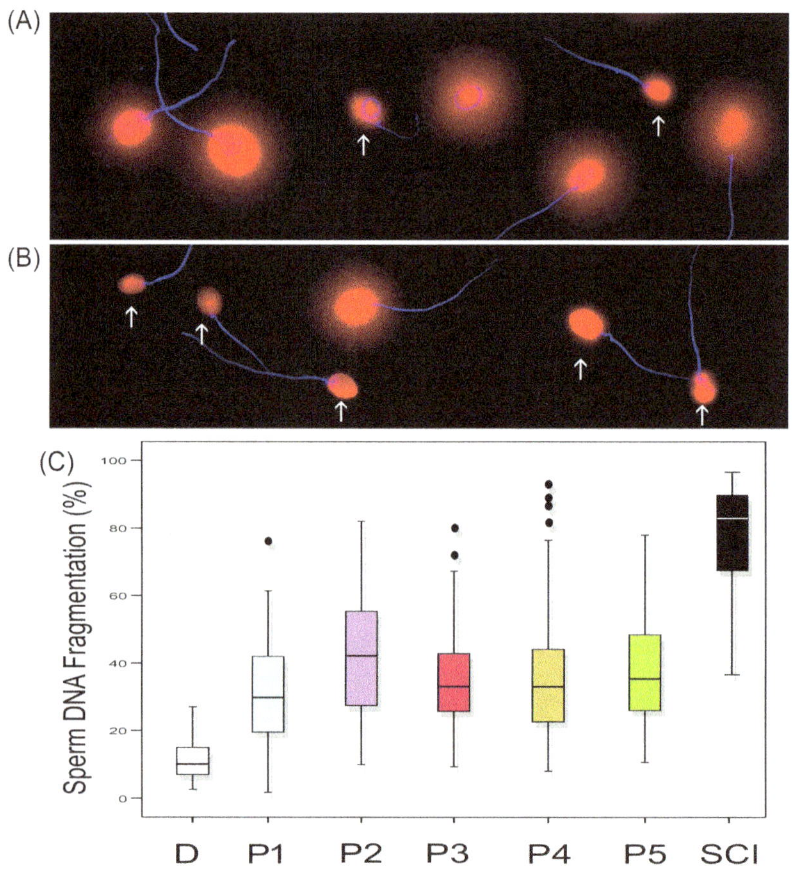

FIG. 1 Sperm DNA fragmentation as visualized with the Sperm Chromatin Dispersion test after dual fluorescence staining. (A) Normozoospermic non-SCI individual showing a high proportion of sperm presenting large haloes of chromatin dispersion (non-fragmented DNA). (B) SCI patients presenting absence of haloes (fragmented DNA: arrow). (C) Box and whisker plots in a series of males without SCI affected by infertility (colored). (P1: *blue*) failure to establish a pregnancy after two cycles of assisted reproduction, (P2: *pink*) leukocytospermia, (P3: *red*) *Chlamydia* infection, (P4: *yellow*) testicular cancer, and (P5: *green*) varicocele to be compared with sperm donors (D: *white*) and males with spinal cord injury (SCI: *black*).

compaction of the DNA molecule than that which occurs in the somatic cell line and facilitates protection of the DNA molecule to exist *ex soma*, for when it will experience external stressors associated with sperm transport in the female. García-Peiró et al. (2011) have shown misbalance in the ratio protamine-1/protamine-2 in human spermatozoa is associated with loss of DNA longevity, confirming that precise protamination is important for chromatin stability. Using animal models with a spinal contusion, Huang, Li, Wang, Wang, and Ottenweller (2003) showed a decrease in mRNA transcripts for spermatid-specific protamine-1 and transition protein-2. If this same phenomenon is also occurring in men with SCI, then it might explain the altered levels of chromatin condensation and low levels of chromatin stability in these individuals (Talebi et al., 2013). The second hypothesis suggests that SDF is a consequence of excessive oxidative stress acting at different levels of the male reproductive tract (Aitken, 2020); this hypothesis, will be analyzed in more detail later in the chapter. The third hypothesis highlights apoptotic events that occur during sperm maturation within the epididymis and are similar in mechanism to those described in somatic cells. Essentially, once caspase or annexin-V is detected at the sperm membrane, double-strand DNA breakage results, controlled by specific DNases (Gorczyca, Traganos, Jesionowska, & Darzynkiewicz, 1993). Some studies have referenced this mechanism as being operative in males with SCI (Restelli et al., 2007; Talebi et al., 2013) but have only used toluidine blue and chromomycin A or TUNEL to make this conclusion; specific protocols for apoptotic marker visualization or other molecular approaches have yet to be performed on the ejaculates of men with SCI. While all hypotheses that explain SDF in males without SCI are equally attributable to men with SCI, there are also likely to be additional effects associated with SCI that are potentially synergistically compounding SDF.

Etiologies of SDF in men with SCI

Hormone alterations

The relationship between altered hormone production and SDF is controversial, even in males without SCI. Some males without SCI show an increase in SDF when both extremely low and high levels of FSH and LH are diagnosed, with SDF being negatively correlated with testosterone concentration (Wdowiak, Raczkiewicz, Stasiak, & Bojar, 2014). In other cases, a direct role of FSH level has been implicated to affect SDF such that patients administered with recombinant FSH (150 IU) show a significant reduction in SDF and an improvement in gonadal function (Colacurci et al., 2018). Some of this variation may also be related to the fact that gonadotrophins associated with spermatogenesis are under homeostatic control such that both too high and too low levels of FSH and LH could have equally adverse effects on sperm production. Although Naderi and Safarinejad (2003) have reported lower reproductive hormones associated with hypothalamus-pituitary-gonadal function in men with SCI, any strong relationship between SDF and variable levels of hormones in SCI patients has yet to be established (Brackett, Nash, & Lynne, 1996), such that a direct role on SDF would be speculative. Odum, Sonksen, and Biering-Sorensen (1995) have also shown that somatostatin, which is important for the normal regulation of the endocrine system, is significantly reduced in the seminal plasma of males with SCI above thoracic vertebrate T6.

Elevated scrotal temperature

Men with scrotal temperatures above the normal range have elevated levels of sub or infertility associated with an increase in the proportion of abnormal and immature spermatozoa (Mieusset et al., 1987). Studies in mice have shown a direct link between heat stress and altered DNA quality, chromatin packaging and embryonic loss (e.g., Banks, King, Irvine, & Saunders, 2005); this damage may occur at the level of the testis (spermatogenesis) and/or the epididymis (sperm maturation). Hyperthermia can also result in decreased DNA synthesis and degradation of mRNAs and proteins necessary for cell survival (Durairajanayagam, Agarwal, & Ong, 2015). Increased SDF may also be associated with heat stress linked to gene expression changes, increased reactive oxygen species (ROS), and alterations in sperm DNA repair mechanisms during spermatogenesis (Durairajanayagam et al., 2015). Given that men with SCI spend many hours in a wheelchair, one might expect that testicular hyperthermia could represent a cause of poor sperm quality (Linsenmeyer & Perkash, 1991), potentially including elevated levels of SDF. While some studies have demonstrated men with SCI in wheelchairs have elevated scrotal temperature compared to men without SCI (Wang, Huang, & Lien, 1992), only one study found an inverse correlation of scrotal temperature with sperm motility and a further study found no difference (Brackett, Lynne, Weizman, Bloch, & Padron, 1994). In addition, men with SCI, but not bound to wheelchairs, still had poor semen quality, suggesting that there is likely to be some other underlying aspect of SCI contributing to lower sperm motility

(Brackett, 2012). As sperm DNA damage in mice can be induced with as little as 30 min exposure at 42 °C (Banks et al., 2005), further studies of men with SCI, examining SDF, might reveal more subtle differences in DNA damage not apparent in other semen parameters.

Ejaculation frequency

It has been suggested that anejaculation in men with SCI may lead to sperm accumulation and deterioration in the cauda epididymis (Vargas-Baquero et al., 2020) but the evidence is controversial. While studies report an improvement in sperm motility in men with SCI after frequent ejaculation (Hamid, Patki, Bywater, Shah, & Craggs, 2006), others have failed to demonstrate improvement (Das et al., 2006). Although there are currently no specific studies with respect to the effect of anejaculation in men with SCI on SDF, Gosálvez, Gonzalez-Martınez, López-Fernández, Fernández, and Sanchez-Martın (2011) have demonstrated a decrease in the level of SDF in ejaculates of normozoospermic men with a corresponding improvement in other semen characteristics and the rate of pregnancy (Sánchez-Martín, Sánchez-Martín, González-Martínez, & Gosálvez, 2013). A similar strategy for improving the DNA quality of the semen sample might be to use the first ejaculated sub-fraction; sperm quality in humans recovered from the first ejaculated fraction, compared to the second fraction show improved measures of sperm concentration, motility, and SDF (de la Torre, Sánchez-Martín, Gosálvez, & Crespo, 2016; Hebles, Dorado, Gallardo, González-Martínez, & Sánchez-Martín, 2015). Prolonged anejaculation may also induce anti-sperm antibodies but the role of these molecules on the fertility of men with SCI is still unresolved (Dimitriadis et al., 2010).

Sperm chromatin maturity

While Hou, Chen, and Jeyendran (1995) was unable to show any effect of SCI on sperm nuclear maturity, Engh, Clausen, Purvis, and Stien (1993) revealed a high degree of abnormal chromatin condensation and reduced binding of a fluorescent acrosomal marker in men with SCI. Investigation of the molecular biology of the rat testis associated with SCI has shown a reduction in the Sertoli cell protein transferrin-mRNA but no change in androgen binding protein-mRNA, or of the mRNA transcripts of spermatogenic cell-specific hemiferrin or spermatid transition protein-2 and protamine-1 and a decline in mRNA transcripts for spermatid-specific protamine-1 and transition protein-2 (Huang et al., 1997). Huang et al. (2003) also noted a precocious expression of CREM (cAMP-responsive element modulator) in early spermatocytes and an absence of it in young spermatids during the acute phase after SCI, which Dimitriadis et al. (2010) have suggested could disrupt spermiogenesis resulting in spermatozoa with abnormal morphology and/or function, including SDF.

Reactive oxygen species

The ejaculates of men with SCI appear to have increased levels of ROS, reduced levels of superoxide dismutase, and increased concentrations of cytokines (de Lamirande & Gagnon, 1993). ROS disequilibrium in the ejaculate is thought to have a significant effect on sperm membranes because in contrast to somatic cells, sperm cells are composed of a high percentage (50%) of fatty acids which are highly susceptible to lipid peroxidation (Henkel, 2011). While the plasma membrane is likely to be the first component of the sperm cell to be adversely affected by oxidative stress, this pathology likely facilitates subsequent ROS damage to the DNA molecule. Such a scenario is also consistent with the relatively high levels of necrozoospermia observed in the semen of SCI patients (Mallidis et al., 2000), where damage to the sperm membrane serves as a precursor to SDF. However, as the intensity of REDOX unbalance, as well as the capacity of the chromatin to be affected by this stressor, appears to be highly dependent on sperm protein composition (Gosálvez, López-Fernández, Fernández, Gouraud, & Holt, 2011), elevated SDF is also likely to be a consequence of the cooperative participation of ROS-related injury to the sperm membrane and incomplete sperm chromatin protamination (García-Peiró et al., 2011).

Leukocytospermia and ROS-related damage

In men with SCI, it is common to find an elevated leukocyte concentration in the semen samples (Jung et al., 2016) even without accompanying urogenital tract infection (UTI). Dimitriadis et al. (2010) have estimated that recurrent UTIs and subsequent leukocytospermia may affect up to 60–70% of men with SCI. Leucocytes can produce abnormally high levels of ROS generating an unbalance between the pro-oxidant and anti-oxidant capacity of the sperm environment (Henkel, 2011). However, although it is well established that ROS is directly involved in cell DNA damage (Poetsch, 2020), the adverse effect of leukocytospermia in men with SCI requires further investigation. For example, it has been reported that not all leucocytes present in the ejaculate of men with SCI show a high production of ROS (Vargas-Baquero et al., 2020). Given the high incidence of SDF observed in the ejaculates of men with SCI, it would be interesting to examine the dynamic relationship between loss of sperm membrane quality and the increase of SDF during in vitro incubation. In men with SCI that have a high level of leukocytospermia, we might predict DNA damage to be accelerated associated with a loss of sperm membrane integrity (necrozoospermia).

Sperm concentration

Sperm concentration tends to be high in men with SCI but it is unlikely to have any benefit to overall semen quality; on the contrary, it may be detrimental to the other semen parameters. For example, a positive correlation between sperm concentration and SDF in the ejaculates of males with SCI has been reported by Vargas-Baquero et al. (2020). This effect has also been reported in human and animal species without spinal injury (López-Fernández, Johnston, Fernández, Wilson, & Gosálvez, 2010; Tvrdá, Arroyo, Ďuračka, López-Fernández, & Gosálvez, 2019). It is difficult to know whether this adverse effect is attributable to the accumulation of toxic metabolic products, active free enzymes such as those found in the acrosome, ROS initially present at the ejaculate, and/or additional ROS generated after sperm plasma membrane degradation.

Seminal vesicular secretions

El-Demiry and James (1988) and Harrison, Barratt, Robinson, Kessopoulou, and Cooke (1991) have referred to shielding components in the seminal fluids that protect the spermatozoa from the effects of leukocytospermia. As the ejaculates of men with SCI are characterized by high numbers of leukocytes compared to men without SCI, it has been proposed that dysfunction of compounds released from the seminal vesicles may interfere with the normal inhibition of leukocyte activation and accumulation (Dimitriadis et al., 2010). This would explain, in part at least, why some leukocytes do not show a large production of ROS (Vargas-Baquero et al., 2020). It is also possible this dysfunction subsequently interrupts the normal secretion of superoxide dismutase and catalase by the seminal vesicles leading to an increase in the adverse effects of ROS and thereby contributing to reduced motility, damaged sperm plasma membranes, and ultimately elevated SDF.

Immune function and cytokine production

Basu et al. (2002) have shown that semen samples of men with SCI have increased numbers of T-cells that co-express human leucocyte antigen HLA-DR and interleukin-2 receptor alpha chain (CD25). Basu, Aballa, Ferrell, Lynne, and Brackett (2004) have also shown that men with SCI typically have seminal plasma with elevated levels of the cytokines, with those with Th1 effector function (IL1β, IL12, IFNγ, TNFα, TNF) being predominant over those with Th2 effector function (IL4, IL6, IL10, TGFβ1) when compared to non-SCI healthy controls; these observations suggest a cell-mediated, rather than a humoral immunological basis for infertility (Dimitriadis et al., 2010). There is also evidence to suggest that the adverse effects of inflammation on sperm motility can be reduced (neutralized) with the use of monoclonal antibodies to cytokines in SCI men (Brackett, Cohen, Ibrahim, Aballa, & Lynne, 2007) but no such data currently exists for their effect on SDF. da Silva et al. (2013) have explored proteomic differences in the seminal plasma protein composition of men with and without SCI and revealed high levels of proteins in SCI patients that appeared to be related to a decrease in sperm motility. They also noted that proteins that initiate the cytokine cascade related to proinflammatory processes were enhanced in men with SCI, as well as elevated concentrations of the cytokines IL1β, IL6, IL12, IFNc, and TNFα but lower concentrations of IL4 and TGFβ1. An apoptosis-associated speck-like protein containing a caspase activation and recruitment domain (ASC) has been detected and found to be elevated in the seminal plasma of men with SCI (Zhang et al., 2013); this same study revealed caspase-1 to be amplified in men with SCI and that both proteins were correlated with increased expression of the proinflammatory cytokines IL1β and IL18. Elevated levels of ASC were also found in the semen of men with SCI almost free of leukocytes. Both ASC and caspase-1 expression were also found in the sperm cell, as was a negative correlation of these proteins with motility (Zhang et al., 2013) but its effect on sperm DNA has not yet been examined. Hirsch, Sedor, Callahan, and Staas (1992) have also reported elevated sperm antibodies in men with SCI.

Other seminal plasma biochemical compounds

Platelet-activating factor acetylhydrolase (PAFah) is derived from the prostate and seminal vesicles and has higher activity in the seminal plasma of men with SCI (Hirsch, Jeyendran, Sedor, Rosecrans, & Staas, 1991), and is negatively correlated with sperm motility (Dimitriadis et al., 2010). PAFah is known to catalyze hydrolysis of esterified PAF to give the biologically active lyso-PAF which acts as an autocrine modulator of sperm motility and enhances both sperm capacitation and the acrosome reaction (Dimitriadis et al., 2010). While there are no current studies linking high PAFah in the semen of men with SCI and SDF, premature capacitation, and acrosome reaction could lead to detrimental plasma membrane changes that pre-dispose the DNA to early ROS degradation.

Based on rat studies, Huang et al. (1997) have shown a temporal impairment in autoregulation of the androgen receptor mRNA shortly after induction of SCI and an accompanying continuing elevation of testosterone repressed prostate message 2 m-RNA (TRPM-2); the gene product of TRPM-2 has a role in the control of apoptosis. Huang et al. (1997) found that elevation of TRPM-2 is associated with a decrease in prostate epithelial cell height and prostate weight in SCI rats. Men with SCI typically present with significantly smaller prostates and produce less prostatic fluid than fertile males without SCI (Hvarness, Jakobsen, & Biering-Sorensen, 2007; Pannek, Berges, Cubick, Meindl, & Senge, 2003). This reduction in prostate size appears to be related to secretory dysfunction and corresponds with lower concentrations of prostate-specific antigen (PSA) in the semen, but higher or equivalent levels systemically (Alexandrino, Rodrigues, & Matsuo, 2004; Lynne et al., 1999). Hirsch et al. (1991) have noted lower levels of the energy substrates fructose and albumin and of the enzymes glutamic oxaloacetic transaminase and alkaline phosphatase in the seminal plasma of men with SCI, along with a higher concentration of chloride. Alexandrino, Rodrigues, Matsuo, Gregório, and Santilli (2011) have reported lower levels of Zinc in the ejaculates of men with SCI. Dimitriadis et al. (2010) have suggested that decreased seminal PSA in men with SCI could result in reduced concentrations of citrate, zinc, and cholesterol in the seminal plasma; molecules which are thought to be important for motility. The relationship between seminal PSA and SDF is yet to be investigated.

The presence of cell-free DNA (cfDNA) has recently been described in the ejaculates of men with SCI (Bartolomé-Nebreda et al., 2020). Using standard gel electrophoresis, a cfDNA band of ~150 bp was observed in all SCI patients analyzed (Fig. 2); this band was occasionally accompanied by another band of ~90 bp but both these bands were not observed in normozoospermic donors. Despite these findings, no correlation was observed between the intensity of the two DNA bands and the level of SDF in males with SCI. Neither was there any correlation with the occurrence of DNase activity present in the seminal plasma?

Improving sperm DNA quality in men with SCI

Success rates for the recovery of spermatozoa from men with SCI have been summarized (See Table 1 in the review of Dimitriadis et al., 2010). Spermatozoa have been recovered nonsurgically using pharmaceuticals, by prostatic massage, but the most common approach for men with SCI is by penile vibratory stimulation (PVS) or electroejaculation. Spermatozoa may also be recovered surgically from the testis (testicular sperm extraction—TESE), epididymis, vas deferens, or

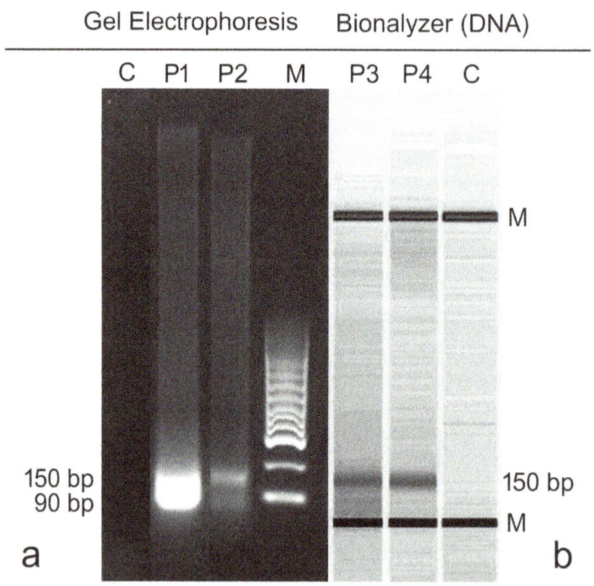

FIG. 2 (A) Gel electrophoresis showing cfDNA in the seminal plasma of two men with SCI (P1/P2) and the presence of two bands (150 bp and 90 bp) of cfDNA and its absence in normozoospermic a non-SCI individual (C). (B) High-resolution automated electrophoresis of cf-DNA from the seminal plasma of two SCI patients (P3/P4). M: DNA markers. In the middle 100 bp DNA ladder. C: control subjects.

TABLE 1 Summary of the differences in seminal plasma chemistry of men with and without SCI.

Seminal Plasma Chemistry	Relative difference in men with SCI	Physiological process impacted	References
Albumin	Lower	Energy substrate	Hirsch et al. (1991)
Chloride	Higher	Multiple functions	Hirsch et al. (1991)
DNase activity	Unchanged	Unknown	Bartolomé-Nebreda et al. (2020)
DNA—Free circulating	Higher	Unknown	Bartolomé-Nebreda et al. (2020)
Fructose	Lower	Energy substrate	Hirsch et al. (1991)
Glutamic oxaloacetic transaminase and alkaline phosphatase	Lower	Energetic metabolisms	Hirsch et al. (1991)
Hormones—Luteinising hormone, Follicle stimulating hormone, Testosterone, Prolactin	Lower	Hypothalamus-pituitary-gonad axis dysfunction	Naderi and Safarinejad (2003)
Interleukin-1b Interleukin-6, interleukin-12, and Tumor necrosis factor-A	Higher	Inflammatory cytokines	Basu et al. (2004) da Silva et al. (2013)
Interleukin-4 and Transforming growth factor-b	Lower	Inflammation	Basu et al. (2004) da Silva et al. (2013)
Interferon gamma	Higher	Reduce sperm motility	Basu et al. (2004) da Silva et al. (2013)
Immunoglobulin G immunoglobulin A	Higher	Anti-sperm antibodies	Hirsch et al. (1992)
Platelet activating factor acetyl hydrolase	Higher	Sperm motility	Hirsch et al. (1991)
Prostatic serum levels	Equivalent	Prostate and seminal vesicle function	Hirsch et al. (1991)
Prostate-specific antigen	Lower	Prostate and seminal vesicle function	Lynne et al. (1999) Alexandrino et al. (2004)
Somatostatin	Lower	Undefined role	Odum et al. (1995)
T-lymphocytes	Higher	Oxidative stress	Basu et al. (2002)
Zinc	Lower	Multiple functions	Alexandrino et al. (2011)

implanted sperm reservoirs. The advantage of TESE is that it can be applied regardless of the level of the spinal cord lesion, patient age, time-period post-injury, hormonal concentration, and semen analysis (Elliott et al., 2000).

The most efficient approach in obtaining a pregnancy in SCI patients is that of Intracytoplasmic Sperm Injection (ICSI), as this negates the need for sperm motility, which is a requirement of IVF, and only requires the selection of a single sperm cell. Under these conditions, a sperm cell free of fragmented DNA is essential. Therefore, there needs to be a reduction in the proportion of spermatozoa with SDF so the likelihood of accidentally selecting a spermatozoon with damaged DNA is reduced. Consequently, we propose a strategy that draws together the combined use of preejaculation and postejaculation procedures to increase the proportion of spermatozoa without DNA damage in the semen of men with SCI for ICSI sperm selection (See Fig. 3).

Prior to semen collection, we suggest the use of recurrent ejaculation and/or use of the first ejaculated fraction to reduce the level of SDF in the ejaculate. Immediately after ejaculation, supportive sperm media (containing anti-oxidant compounds) should be added to reduce the high sperm concentration and the effect of oxidative stress, especially, if there is accompanying leukocytospermia. The semen sample can then be treated to remove immature sperm cells using Density-Gradient Centrifugation (DGC) to further reduce SDF (Gosálvez, Gonzalez-Martınez, et al., 2011). If the resulting number and relative motility of the reconstituted sample are suitable then it could be prepared for intrauterine insemination (IUI) or in vitro fertilization (IVF); if the sample is not suitable then ICSI would be employed.

There are a number of assisted reproductive strategies that can be used in association with the ICSI procedure to further improve the probability of selecting a sperm without fragmented DNA; these include magnetic-activated cell sorting (MACS) to remove apoptotic cells, Physiological-ICSI (PICSI) which differentially selects mature sperm based on hyaluronic acid-binding and intracytoplasmic morphologically selected sperm injection (IMSI) which selects sperm based on their morphological appearance; MACS and PICSI have both been developed to reduce the chance of selecting sperm with damaged DNA. If all this fails, then it is possible to harvest testicular sperm and use these cells in combination with ICSI. When used in men without SCI it has been shown to dramatically reduce the level of SDF (Esteves, Sanchez-Martín, Sanchez-Martín, Schneider, & Gosálvez, 2015); the technique has already resulted in successful pregnancies for men with SCI (Kanto et al., 2009).

Conclusion

This review has illustrated that men with SCI typically present with levels of SDF substantially higher than men without SCI, including those with male factor infertility and that the ultimate causes of this phenomenon are difficult to disentangle from the primary causes of high SDF in men without SCI. The etiologies of SDF in men with SCI are likely to be the result of synergistic pathologies so that the specifics of what causes the higher levels of SDF will require further investigation and are probably best suited to controlled studies involving animal models. Nevertheless, the take-home message of this review is that despite high SDF, the majority of men with SCI contain sufficient spermatozoa with an orthodox DNA molecule, that if clinically selected, could be used for successful assisted reproduction.

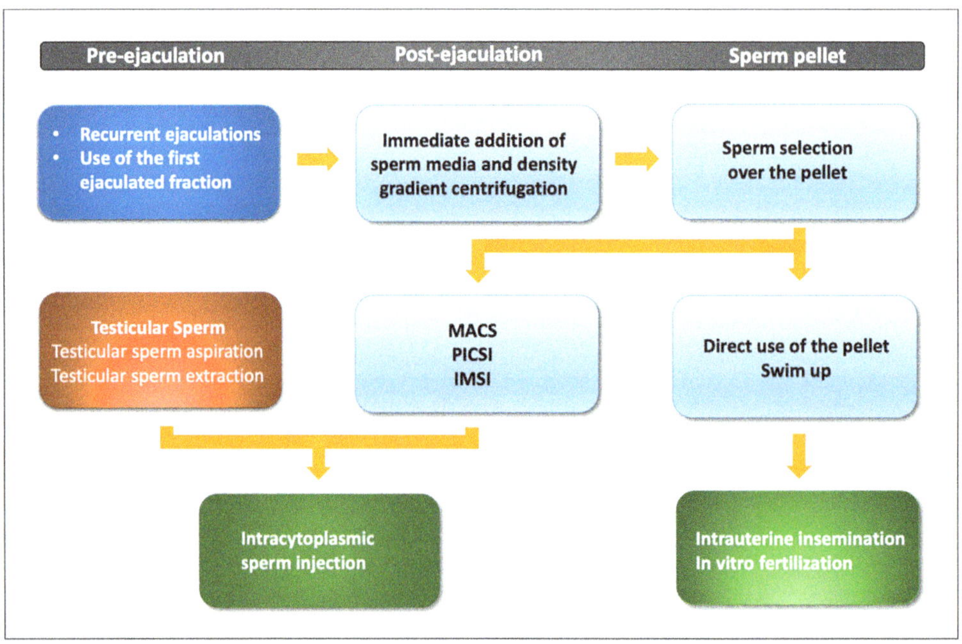

FIG. 3 Recommended protocol for an effective sperm DNA fragmentation reduction in SCI men. *TESE*, testicular sperm extraction; *TESA*, testicular sperm aspiration; *DCG*, density gradient centrifugation; *IVF*, in vitro fertilization; *ICSI*, intracytoplasmic sperm injection; *IUI*, intrauterine insemination; *MACS*, magnetic activated cell sorting; *PICSI*, physiological ICSI; *IMSI*, intracytoplasmic morphologically selected sperm injection.

Applications to other areas of neuroscience

As we have discovered in this chapter, the etiology of sperm DNA damage in the ejaculates of men with SCI is likely to be multifactorial and difficult to attribute to a specific neurophysiological pathology. The loss of sexual function and realization of the long-term impact of SCI can have a devastating impact on the male partner's psychological well-being (increased susceptibility to depression) and self-worth, not to mention the additional stress that the consequences of the injury may have on the female partner and the couple's desire to have a family. Nevertheless, most men with SCI have a proportion of spermatozoa in their ejaculate suitable for artificial reproductive technology procedures such as intracytoplasmic sperm injection that can be applied to obtain a successful pregnancy. Such fulfilment in establishing a family through assisted reproductive technology despite their injury is likely to be a powerful motivator in their recovery and rehabilitation process that should not be underestimated. In addition, some of the causes attributed to poor semen quality in men with SCI do not appear to be accounted for by their physiological dysfunction, suggesting that there are other subtle neural explanations and interactions (e.g., parasympathetic system) that we yet do not understand. In this respect studies of the effects of SCI on male fertility may also provide to be a useful research model for insights into SDF in men without SCI or the effects of oxidative damage on cellular function. SCI is typically accompanied by motor, vegetative, and sensitive dysfunction associated with the production of reactive oxygen species that can be extremely aggressive to surrounding tissues. We propose that sperm cells and sperm DNA fragmentation may serve as a useful bio-indicator for studying oxidative damage to better understand the pathophysiology of free radicals and for developing treatment protocols that reduce tissue necrosis, inflammation, infection, and apoptosis.

Mini-dictionary of terms

Apoptosis: a form of programmed cell death
Caspase: a family of protease enzymes essential to programmed cell death
Electroejaculation: method for collecting a semen sample
Histone: highly basic proteins that pack and order DNA into structural units called nucleosomes—in sperm replaced by protamine during spermiogenesis.
Leukocytospermia: abnormally high numbers of white cells in the semen
Necrozoospermia: ejaculate contains mainly dead spermatozoa
Normozoospermic: ejaculate which contains normal characteristics of seminogram
Oxidative stress: an imbalance between free radicals and anti-oxidants
Penile vibratory stimulation: method for collecting semen samples
Protamine: DNA binding proteins that facilitate the packaging of sperm chromatin
SCD assay: Sperm chromatin dispersal assay—simple rapid individual sperm assay to assess fragmented DNA in spermatozoa—based on the movement of fragmented DNA in a microgel
SCSA: Sperm chromatin structure assay—sperm assay based on flow cytometry to assay fragmented DNA in spermatozoa
TUNEL assay—Allows for the visualization and quantification of apoptotic cells
Varicocele—enlargement of the veins within pampiniform plexus

Key facts of sperm DNA fragmentation

- Sperm DNA Fragmentation (SDF) is a result of broken or fragmented DNA in the sperm nucleus without the capacity to be repaired.
- SDF has been shown to reduce reproductive outcome.
- SDF is high in ejaculates of men with spinal cord injury (SCI).
- The cause of elevated SDF in men with SCI is likely to be multifactorial.
- The high levels of SDF in men with SCI present problems in achieving pregnancy.
- Sperm selection reproductive technology can be used to increase the proportion of sperm with nonfragmented DNA in order to obtain a successful pregnancy.

Summary points

- Sperm DNA Fragmentation (SDF) occurs when the DNA molecule of the sperm cell is broken or fragmented in the form of single or double-stranded DNA breakages.

- SDF may arise in the testis during spermiogenesis, during sperm maturation and storage within the epididymis, iatrogenically during processing of the ejaculate for assisted fertilization or within the female reproductive tract.
- If not corrected by the oocyte following syngamy, SDF has been shown to result in embryonic failure and reduced pregnancy rate.
- Men with SCI have ejaculates that typically contain extremely high levels of SDF, up to 5 × higher than those of fertile men without SCI and 2 × higher than men without SCI but with evidence of male factor infertility.
- Previous reviews have identified etiologies in men with SCI that are thought to be associated with poor semen quality and which include alternations in the hormonal milieu of the testis, scrotal hyperthermia, anejaculation, differences in molecular biology of the testis, leukocytospermia, oxidative damage, abnormal seminal vesicular secretions, and abnormal seminal plasma chemistry.
- The majority of studies on the ejaculate of men with SCI have focused on classical sperm parameters (sperm motility and necrozoospermia) with little attention given to SDF; consequently, we explored the possible interaction with the etiologies indicated above and how these might influence the expression of SDF.
- We speculate that SDF is closely related to sperm plasma membrane damage linked to ROS formation leading to a secondary attack on sperm chromatin.
- The most efficient solution for overcoming infertility in men with SCI is likely to be through the use of intracytoplasmic sperm injection that requires the selection of a single spermatozoon without DNA damage.

References

Aitken, R. J. (2020). Impact of oxidative stress on male and female germ cells: Implications for fertility. *Reproduction*, *159*, 189–201.

Alexandrino, A. P., Rodrigues, M. A., & Matsuo, T. (2004). Evaluation of serum and seminal levels of prostate specific antigen in men with spinal cord injury. *Journal of Urology*, *171*, 2230–2232.

Alexandrino, A. P., Rodrigues, M. A. F., Matsuo, T., Gregório, E. P., & Santilli, J. C. (2011). Evaluation of seminal zinc levels by atomic absorption in men with spinal cord injury. *Spinal Cord*, *49*, 435–438.

Banks, S., King, S. A., Irvine, D. S., & Saunders, P. T. (2005). Impact of a mild scrotal heat stress on DNA integrity in murine spermatozoa. *Reproduction*, *129*, 505–514.

Bartolomé-Nebreda, J., Vargas-Baquero, E., López-Fernández, C., Fernández, J. L., Johnston, S. D., & Gosálvez, J. (2020). Free circulating DNA and DNase activity in the ejaculates of men with spinal cord injury. *Spinal Cord*. https://doi.org/10.1038/s41393-020-0518-3.

Basu, S., Aballa, T. C., Ferrell, S. M., Lynne, C. M., & Brackett, N. L. (2004). Inflammatory cytokine concentrations are elevated in seminal plasma of men with spinal cord injuries. *Journal of Andrology*, *25*, 250–254.

Basu, S., Lynne, C. M., Ruiz, P., Aballa, T. C., Ferrell, S. M., & Brackett, N. L. (2002). Cytofluorographic identification of activated T-cell subpopulations in the semen of men with spinal cord injuries. *Journal of Andrology*, *23*, 551–556.

Brackett, N. L. (2012). Infertility in men with spinal cord injury: Research and treatment. *Scientifica*, *12*. https://doi.org/10.6064/2012/578257, ID578257.

Brackett, N. L., Cohen, D. R., Ibrahim, E., Aballa, T. C., & Lynne, C. M. (2007). Neutralization of cytokine activity at the receptor level improves sperm motility in men with spinal cord injuries. *Journal of Andrology*, *28*, 717–721.

Brackett, N. L., Ibrahim, E., Grotas, J. A., Teodoro, C., Aballa, T. C., & Lynne, C. M. (2008). Higher sperm DNA damage in semen from men with spinal cord injuries compared with controls. *Journal of Andrology*, *29*, 93–99.

Brackett, N. L., Lynne, C. M., Weizman, M. S., Bloch, W. E., & Padron, O. F. (1994). Scrotal and oral temperatures are not related to semen quality or serum gonadotropin levels in spinal cord injured men. *Journal of Andrology*, *15*, 614–619.

Brackett, N. L., Nash, M. S., & Lynne, C. M. (1996). Male fertility following spinal cord injury: Facts and fiction. *Physical Therapy*, *76*, 1221–1231.

Colacurci, N., De Leo, V., Ruvolo, G., Piomboni, P., Caprio, F., Pivonello, R., et al. (2018). Recombinant FSH improves sperm DNA damage in male infertility: A phase II clinical trial. *Frontiers in Endocrinology*, *9*, 383.

Das, S., Dodd, S., Soni, B. M., Sharma, S. D., Gazvani, R., & Lewis-Jones, D. I. (2006). Does repeated electro-ejaculation improve sperm quality in spinal cord injured men? *Spinal Cord*, *44*, 753–756.

El-Demiry, M., & James, K. (1988). Lymphocyte subsets and macrophages in the male genital tract in health and disease. A monoclonal antibody-based study. *European Urology*, *14*, 226–235.

Deng, C., Li, T., Xie, Y., Guo, Y., Yang, Q., Liang, X., et al. (2019). Sperm DNA fragmentation index influences assisted reproductive technology outcome: A systematic review and meta-analysis combined with a retrospective cohort study. *Andrologia*, *51*, e13263.

Dimitriadis, F., Karakitsios, K., Tsounapi, P., Tsambalas, S., Loutradis, D., Kanakas, N., et al. (2010). Erectile function and male reproduction in men with spinal cord injury: A review. *Andrologia*, *42*, 139–165.

Durairajanayagam, D., Agarwal, A., & Ong, C. (2015). Causes, effects and molecular mechanisms of testicular heat stress. *Reproductive Biomedicine Online*, *30*, 14–27.

Elliott, S. P., Orejuela, F., Hirsch, I. H., Lipshultz, L. I., Lamb, D. J., & Kim, E. D. (2000). Testis biopsy findings in the spinal cord injured patient. *Journal of Urology*, *163*, 792–795.

Engh, E., Clausen, O. P., Purvis, K., & Stien, R. (1993). Sperm quality assessed by flow cytometry and accessory sex gland function in spinal cord injured men after repeated vibration-induced ejaculation. *Paraplegia*, *31*, 3–12.

Esteves, S. C., Cho, C. L., Majzoub, A., Zini, A., Coward, R. M., Evenson, D. P., et al. (2020). Clinical practice guidelines for sperm DNA fragmentation testing. *Andrologia*. https://doi.org/10.1111/and.13874.

Esteves, S. C., Gosálvez, J., López-Fernández, C., Núñez-Calonge, R., Caballero, P., Agarwal, A., et al. (2015). Diagnostic accuracy of sperm DNA degradation index (DDSi) as a potential noninvasive biomarker to identify men with varicocele-associated infertility. *International Urology and Nephrology*, *47*, 1471–1477.

Esteves, S. C., Sanchez-Martín, F., Sanchez-Martín, P. T., Schneider, D. T., & Gosálvez, J. (2015). Comparison of reproductive outcome in oligozoospermic men with high sperm DNA fragmentation undergoing intracytoplasmic sperm injection with ejaculated and testicular sperm. *Fertility and Sterility*, *104*, 1398–1405.

García-Peiró, A., Martínez-Heredia, J., Oliver-Bonet, M., Abad, C., Amengual, M. J., Navarro, J., et al. (2011). Protamine P1/P2 ratio correlates with dynamic aspects of DNA fragmentation in human sperm. *Fertility and Sterility*, *95*, 105–109.

García-Rodríguez, A., Gosálvez, J., Agarwal, A., Roy, R., & Johnston, S. (2019). DNA damage and repair in human reproductive cells. *International Journal of Molecular Sciences*, *20*, 31.

Gonzalez-Marin, C., Gosálvez, J., & Roy, R. (2012). Types, causes, detection and repair of DNA fragmentation in animal and human sperm cells. *International Journal of Molecular Sciences*, *13*, 14026–14052.

Gorczyca, W., Traganos, F., Jesionowska, H., & Darzynkiewicz, Z. (1993). Presence of DNA strand breaks and increased sensitivity of DNA in situ to denaturation in abnormal human sperm cells: Analogy to apoptosis of somatic cells. *Experimental Cell Research*, *207*, 202–205.

Gosálvez, J., Fernández, J. L., Yaniz, J., de la Casa, M., López-Fernández, C., & Johnston, S. D. (2015). A comparison of sperm DNA damage in the neat ejaculate of sperm donors and males presenting for their initial seminogram. *Austin Journal of Reproductive Medicine and Infertility*, *2*, 1014.

Gosálvez, J., Gonzalez-Martınez, M., López-Fernández, C., Fernández, J. L., & Sanchez-Martın, P. (2011). Shorter abstinence decreases sperm deoxyribonucleic acid fragmentation in ejaculate. *Fertility and Sterility*, *96*, 1083–1086.

Gosálvez, J., López-Fernández, C., Fernández, J. L., Gouraud, A., & Holt, W. V. (2011). Relationships between the dynamics of iatrogenic DNA damage and genomic design in mammalian spermatozoa from eleven species. *Molecular Reproduction and Development*, *78*, 951–961.

Gosálvez, J., López-Fernández, C., Fernández, J. L., Esteves, S. C., & Johnston, S. D. (2015). Unpacking the mysteries of sperm DNA fragmentation: Ten frequently asked questions. *Journal of Reproductive Biotechnology and Fertility*, *4*, 1–16.

Hamid, R., Patki, P., Bywater, H., Shah, P. J. R., & Craggs, M. D. (2006). Effects of repeated ejaculations on semen characteristics following spinal cord injury. *Spinal Cord*, *44*, 369–373.

Harrison, P. E., Barratt, C. L., Robinson, A. J., Kessopoulou, E., & Cooke, I. D. (1991). Detection of white blood cell populations in the ejaculates of fertile men. *Journal of Reproductive Immunology*, *19*, 95–98.

Hebles, M., Dorado, M., Gallardo, M., González-Martínez, M., & Sánchez-Martín, P. (2015). Seminal quality in the first fraction of ejaculate. *Systems Biology in Reproductive Medicine*, *61*, 113–116.

Henkel, R. R. (2011). Leukocytes and oxidative stress: Dilemma for sperm function and male fertility. *Asian Journal of Andrology*, *13*, 43–52.

Hirsch, I. H., Jeyendran, R. D., Sedor, J., Rosecrans, R. R., & Staas, W. E. (1991). Biochemical analysis of electroejaculates in spinal cord injured men: Comparison to normal ejaculates. *Journal of Urology*, *145*, 73–76.

Hirsch, I. H., Sedor, J., Callahan, H. J., & Staas, W. E. (1992). Antisperm antibodies in seminal plasma of spinal cord-injured men. *Urology*, *39*, 243–247.

Hou, J. W., Chen, D., & Jeyendran, R. S. (1995). Sperm nuclear maturity in spinal cord-injured men: Evaluation by acidic aniline blue stain. *Archives of Physical Medicine and Rehabilitation*, *76*, 444–445.

Huang, H. F., Li, M. T., Linsenmeyer, T. A., Ottenweller, J. E., Pogach, L. M., & Irwin, R. J. (1997). The effects of spinal cord injury on the status of messenger ribonucleic acid for TRPM 2 and androgen receptor in the prostate of the rat. *Journal of Andrology*, *18*, 250–256.

Huang, H. F., Li, M. T., Wang, S., Wang, G., & Ottenweller, J. E. (2003). Spinal cord contusion impairs sperm motility in the rat without disrupting spermatogenesis. *Journal of Andrology*, *24*, 371–380.

Hvarness, H., Jakobsen, H., & Biering-Sorensen, F. (2007). Men with spinal cord injury have a smaller prostate than men without. *Scandinavian Journal of Urology and Nephrology*, *41*, 120–123.

Jung, J. H., Kim, M. H., Kim, J., Baik, S. K., Koh, S. B., Park, H. J., et al. (2016). Treatment of leukocytospermia in male infertility: A systematic review. *World Journal of Mens Health*, *34*, 165–172.

Kanto, S., Uto, H., Toya, M., Ohnuma, T., Arai, Y., & Kyono, K. (2009). Fresh testicular sperm retrieved from men with spinal cord injury retains equal fecundity to that from men with obstructive azoospermia via intracytoplasmic sperm injection. *Fertility and Sterility*, *92*, 1333–1336.

de la Torre, J., Sánchez-Martín, P., Gosálvez, J., & Crespo, F. (2016). Equivalent seminal characteristics in human and stallion at first and second ejaculated fractions. *Andrologia*. https://doi.org/10.1111/and.12708.

de Lamirande, E., & Gagnon, C. (1993). A positive role for the superoxide anion in triggering hyperactivation and capacitation of human spermatozoa. *International Journal of Andrology*, *16*, 21–25.

Linsenmeyer, T. A., & Perkash, I. (1991). Infertility in men with spinal cord injury. *Archives of Physical Medicine and Rehabilitation*, *72*, 747–754.

López-Fernández, C., Johnston, S. D., Fernández, J. L., Wilson, R. J., & Gosálvez, J. (2010). Fragmentation dynamics of frozen-thawed ram sperm DNA is modulated by sperm concentration. *Theriogenology*, *74*, 1362–1370.

Lynne, C. M., Aballa, T. C., Wang, T. J., Rittenhouse, H. G., Ferrell, S. M., & Brackett, N. L. (1999). Serum and seminal plasma prostate specific antigen (PSA) levels are different in young spinal cord injured men compared to normal controls. *Journal of Urology*, *162*, 89–91.

Mallidis, C., Lim, T. C., Hill, S. T., Skinner, D. J., Brown, D. J., Johnston, W. I. H., et al. (2000). Necrospermia and chronic spinal cord injury. *Fertility and Sterility*, *74*, 221–227.

Mieusset, R., Bujan, L., Mondinat, C., Mansat, A., Pontonnier, F., & Grandjean, H. (1987). Association of scrotal hyperthermia with impaired spermatogenesis in infertile men. *Fertility and Sterility*, *48*, 1006–1011.

Naderi, A. R., & Safarinejad, M. R. (2003). Endocrine profiles and semen quality in spinal cord injured men. *Clinical Endocrinology*, *58*, 177–184.

Odum, L., Sonksen, J., & Biering-Sorensen, F. (1995). Seminal somatostatin in men with spinal cord injury. *Paraplegia*, *33*, 374–376.

Pannek, J., Berges, R. R., Cubick, G., Meindl, R., & Senge, T. (2003). Prostate size and PSA serum levels in male patients with spinal cord injury. *Urology*, *62*, 845–848.

Poetsch, A. R. (2020). The genomics of oxidative DNA damage, repair, and resulting mutagenesis. *Computational and Structural Biotechnology Journal*, *18*, 207–219.

Qiu, Y., Wang, L. G., Zhang, L. H., Li, J., Zhang, A. D., & Zhang, M. H. (2012). Sperm chromosomal aneuploidy and DNA integrity of infertile men with anejaculation. *Journal of Assisted Reproduction and Genetics*, *29*, 185–194.

Restelli, A. E., Borrelli, M., Jr., Spaine, D. M., Bertolla, R. P., Ortiz, V., & Cedenho, A. P. (2007). Sperm nuclear apoptotic DNA fragmentation in men with spinal cord injury. *Fertility and Sterility*, *88*(Suppl. 1), S374.

Sánchez-Martín, P., Sánchez-Martín, F., González-Martínez, M., & Gosálvez, J. (2013). Increased pregnancy after reduced male abstinence. *Systems Biology in Reproductive Medicine*, *59*, 256–260.

da Silva, B. F., Souza, G. H., Lo Turco, E. G., Del Giudice, P. T., Soler, T. B., Spaine, D. M., et al. (2013). Differential seminal plasma proteome according to semen retrieval in men with spinal cord injury. *Fertility and Sterility*, *100*, 959–969.

Talebi, A. R., Khalili, M. A., Vahidi, S., Ghasemzadeh, J., & Tabibnejad, N. (2013). Sperm chromatin condensation, DNA integrity, and apoptosis in men with spinal cord injury. *Journal of Spinal Cord Medicine*, *36*, 140–146.

Tvrdá, E., Arroyo, F., Ďuračka, M., López-Fernández, C., & Gosálvez, J. (2019). Dynamic assessment of human sperm DNA damage II: The effect of sperm concentration adjustment during processing. *Journal of Assisted Reproduction and Genetics*, *36*, 799–807.

Vargas-Baquero, E., Johnston, S., Sánchez-Ramos, A., Arévalo-Martin, A., Wilson, R., & Gosálvez, J. (2020). The incidence and etiology of sperm DNA fragmentation in the ejaculates of males with spinal cord injuries. *Spinal Cord*, *58*, 803–810.

Wang, Y. H., Huang, T. S., & Lien, I. N. (1992). Hormone changes in men with spinal cord injuries. *American Journal of Physical Medicine and Rehabilitation*, *71*, 328–332.

Wdowiak, A., Raczkiewicz, D., Stasiak, M., & Bojar, I. (2014). Levels of FSH, LH and testosterone, and sperm DNA fragmentation. *Neuro Endocrinology Letters*, *35*, 73–79.

Zhang, X., Ibrahim, E., de Rivero-Vaccari, J. P., Lotocki, G., Aballa, T. C., Dietrich, W. D., et al. (2013). Involvement of the inflammasome in abnormal semen quality of men with spinal cord injury. *Fertility and Sterility*, *99*, 118–124.

Chapter 9

Beneficial and detrimental effects of cytokines after spinal cord injury

Jesús Amo-Aparicio[a] and Clara Penas[b]

[a]*Department of Medicine, University of Colorado Denver, Aurora, CO, United States,* [b]*Department of Cell Biology, Physiology and Immunology, Facultat de Medicina, Institut of Neurosciences, Centro de Investigación Biomédica en Red sobre Enfermedades Neurodegenerativas (CIBERNED), Universitat Autònoma de Barcelona, Bellaterra, Spain*

Abbreviations

CCL	chemokine (C-C motif) ligand
CSPGs	chondroitin sulfate proteoglycans
CXCL	chemokine (C-X-C motif) ligand
HDAC	histone deacetylase
IFN	interferon
IL	interleukin
MAG	myelin-associated glycoprotein
MCAO	middle cerebral artery occlusion
MCP	monocyte chemoattractant protein
M-CSF	macrophage colony-stimulating factor
MIP	macrophage inflammatory protein
miRNA	microRNA
MMPs	matrix metalloproteinases
MS	multiple sclerosis
NO	nitric oxide
NOS	nitric oxide synthase
OMgp	oligodendrocyte myelin glycoprotein
SCI	spinal cord injury
TGF	transforming growth factor
TNF	tumor necrosis factor
vMIPII	broad-spectrum chemokine receptor antagonist

Introduction

Cytokines are signaling proteins involved in intercellular communication, and mediate the regulation of a large variety of immune and nonimmune cells. These molecules have been mainly associated with an immune action, although they play pleiotropic effects such as proliferation, differentiation, growth, survival, and cell death. Thus, it is not surprising that cytokines play crucial roles in the pathophysiology after spinal cord injury (SCI). After SCI, there is an activation of resident glial cells, such as microglia and astrocytes, and infiltration of blood immune cells, such as macrophages, granulocytes, and lymphocytes. These cells produce sequential segregation of huge amounts of cytokines which stimulate diverse detrimental as well as beneficial processes after SCI.

After the primary impact on the spinal cord, within few hours after the lesion, there is an initial spike of proinflammatory cytokines secretion which promotes and exacerbates inflammation. This initial release of pro-inflammatory cytokines produces an increase of vascular permeability, cellular infiltration from peripheral blood, and enhances proliferation and hypertrophy of astrocytes. Then, activated astrocytes generate the glial scar which acts as a barrier for axonal regeneration. All these processes contribute to the secondary injury, where neuronal loss and functional outcomes are represented. Few hours after this initial spike, there is also a small expression of antiinflammatory cytokines which regulates the extent of

Cellular, Molecular, Physiological, and Behavioral Aspects of Spinal Cord Injury. https://doi.org/10.1016/B978-0-12-822427-4.00009-5

peripheral cell infiltration, inhibits proinflammatory cytokine production, and provides trophic support to neurons allowing axonal growth. Therefore, to some extent, these antiinflammatory cytokines produce neuroprotection and functional recovery. However, the net imbalance toward antiinflammatory cytokine production drives to an extension of the secondary lesion and persisted inflammation which remains chronically impairing regeneration and functional recovery after SCI.

The main goal of this present book chapter is to describe the role of cytokines within the pathogenesis of SCI as well as to describe the main research findings that envisage them as therapeutic targets. Through the years, there has been important research that aims to alter the expression or activity of cytokines after neural trauma to modulate neurodegeneration, promote repair, and improve functional outcomes after SCI.

Cytokine expression after SCI

After the injury, cytokines are released by multiple different cells which proliferate or infiltrate into the spinal cord (Fig. 1A). Local glial cells (microglia and astrocytes) are the first to be activated after damage. In normal conditions, microglia display numerous cytoplasmic processes on their surface that extend and retract to monitor central nervous system (CNS) homeostasis. Minutes to hours after damage detection, microglial cells proliferate and their processes establish a dense network surrounding the lesion site. Macrophage colony-stimulating factor (M-CSF) is the main cytokine involved in the proliferation of microglia that peak at 7 days after injury (Francos-Quijorna et al., 2017). Meanwhile, the release of ATP from astrocytes promotes the rapid extension of microglia toward the injury site (Davalos et al., 2005). As they transition to the next step of phagocytosis, microglia retract their processes and adopt an amoeboid morphology in response to high levels of nitric oxide (NO) (Dibaj et al., 2010).

Then, neutrophils infiltrate from the peripheral blood followed by monocytes, which differentiate into macrophages. Neutrophils reach a maximum at 24 h, whereas macrophages and astrocytes have a delayed maximum peak number within the first week after injury. Chemokine (C—C motif) ligand (CCL)-2 is an important cytokine for the recruitment of these leukocytes since mice lacking CCL2 receptors showed a reduction in the infiltration of these cells after CNS damage (Conductier, Blondeau, Guyon, Nahon, & Rovere, 2010).

Lymphocytes are the last cells to be infiltrated reaching a plateau of infiltration by 3 days postinjury (Donnelly & Popovich, 2008). Studies describe that all these local and blood-infiltrated cells remain within the spinal cord parenchyma chronically after injury promoting a persistent and abnormal inflammation.

Regarding the analysis of released cytokines within the spinal cord parenchyma after traumatic injury, the majority of studies available have been performed in murine models of SCI (Fig. 1B). During the first hours after injury, there is an enhanced expression of proinflammatory cytokines, such as interleukin (IL)-6, chemokine (C-X-C motif) ligand (CXCL)-1, CXCL2, CCL-2, and CXCL10 (Francos-Quijorna, Amo-Aparicio, Martinez-Muriana, & Lopez-Vales, 2016; Stammers, Liu, & Kwon, 2012) (Fig. 1B). These cytokines increase between 100 and 800-fold, reaching their maximum expression by 6 h after the injury. Other proinflammatory cytokines are also highly expressed between 5 and 25-fold, such as CCL1,

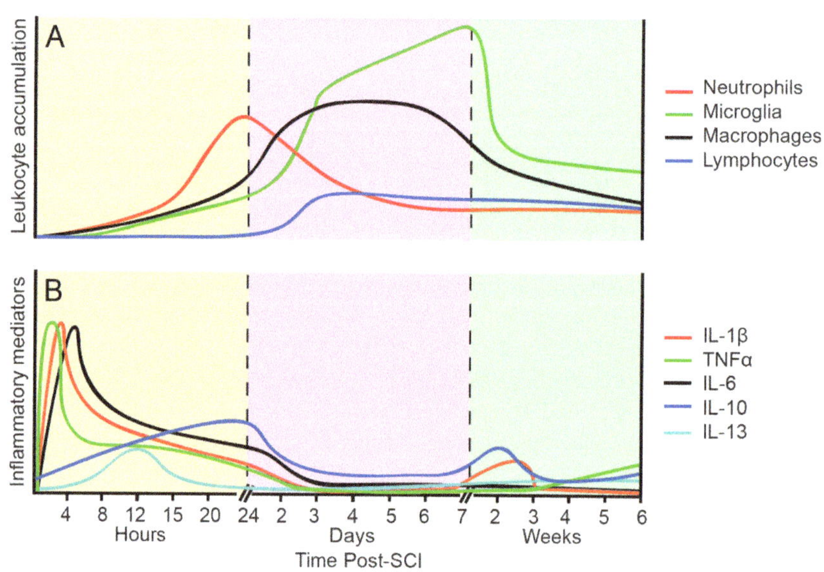

FIG. 1 Dynamics of the inflammatory response in the mouse spinal cord after injury. Immune cells (A) and inflammatory mediators (B) from hours to weeks after SCI. Values on the vertical axis represent relative changes and are not to scale. *Adapted from Donnelly, D. J., & Popovich, P. G. (2008). Inflammation and its role in neuroprotection, axonal regeneration and functional recovery after spinal cord injury.* Experimental Neurology, 209 *(2), 378–388. doi:10.1016/j.expneurol.2007.06.009 with data from Francos-Quijorna, I., Santos-Nogueira, E., Gronert, K., Sullivan, A. B., Kopp, M. A., Brommer, B., … Lopez-Vales, R. (2017). Maresin 1 promotes inflammatory resolution, neuroprotection, and functional neurological recovery after spinal cord injury.* The Journal of Neuroscience, 37*(48), 11731–11743. https://doi.org/10.1523/JNEUROSCI.1395-17.2017.*

CCL3, CCL4, CCL5 IL-1α, IL-9, IL-15, macrophage inflammatory protein (MIP)-1α, and tumor necrosis factor (TNF)-α with a peak between 6 and 24 h after injury. Contrasting, the number of antiinflammatory cytokines released is minimal. IL-13 and IL-10 increase by 5- to –15-fold between 12 and 24 h after lesion, respectively (Francos-Quijorna et al., 2016). Other antiinflammatory cytokines such as IL-4 do not alter their level of expression or even decrease. Therefore, a severe imbalance toward proinflammatory cytokine expression is described, which has a critical impact on the pathophysiology of SCI.

Several studies in humans correlate cytokine levels on the cerebrospinal fluid at 24 h after SCI with the level of the American Spinal Injury Association Impairment Scale (AIS) grade that patients reached. In addition, cytokine levels on human cerebrospinal fluid were also significantly different between those who improved an AIS grade over 6 months and those who did not improve (Dalkilic et al., 2018; Kwon et al., 2017). Therefore, the authors were able to predict the AIS grade progression through cerebrospinal fluid analyses of cytokines. These results emphasize the need to find therapeutic acute interventions to decrease the expression of inflammatory cytokines after lesion.

Cytokines promoting cell death and neurodegeneration after SCI

The initial trauma of the spinal cord produces massive necrosis of neurons and glial cells. Then, apoptosis plays an important role in the subsequent sustained loss of cells, which spreads tissue damage after the initial injury (Crowe, Bresnahan, Shuman, Masters, & Beattie, 1997). Several cytokines have been related to the activation of this secondary wave of cell death, mediated by apoptosis, and thus are related to the neurodegenerative events occurring after SCI.

Microglia, astrocytes, neurons, and infiltrating macrophages produce a rapid increase of TNFα expression in the lesion site, within 1 h after SCI, which remains significantly elevated up to 3 days posttrauma (Fig. 1B) (Harrington, Messier, Levine, Szmydynger-Chodobska, & Chodobski, 2005). The biological effects of TNFα are mediated by two receptors, the TNFR1 and TNFR2. Both receptors can mediate cell survival and cell death. However, after SCI, the balance is tipped toward cell death (Fig. 2). In fact, the initial peak of TNFα is followed by a significant increase on the expression of both receptors in neurons and oligodendrocytes at 3 and 6 h after the insult, which coincides with a high increment of apoptotic cells within the spinal cord parenchyma (Inukai et al., 2009). The accumulation of TNFα mediates apoptotic cell death of neurons and oligodendrocytes, in part, by inducing NO synthase (NOS) expression, and thus, NO production after injury (Lee et al., 2000).

Another important family of cytokines playing a key role in neurodegeneration is the IL-1 family. From this extensive family, IL-1α and IL-1β have been found to be important after SCI, exerting their pro-inflammatory actions by binding to the IL-1 type I receptor (IL-1R1). These cytokines are produced by resident spinal cells such as microglia, astrocytes, and neurons and promote tissue infiltration of neutrophils as well as astrogliosis. IL-1α protein expression in the injured spinal cord peaks at 4 h and it is mostly localized in microglia in the site of injury (Bastien et al., 2015). Its importance releases on studies that have shown that deletion of IL-1α gene protected against oligodendrocyte cell death and functional disabilities, but only at early stages after damage (Bastien et al., 2015). In contrast, IL-1β is upregulated later than IL-1α after SCI, peaking at 24 h and it is produced mainly by infiltrated neutrophils and macrophages (Bastien et al., 2015). IL-1β induces apoptosis, in part, by p38 MAPK induction and subsequent activation of caspase-3 (Wang, Kong, Qi, Ye, & Song, 2005).

Also, IL-17 mediates the recruitment and activation of mainly neutrophils, and stimulates chemokine and proinflammatory cytokine release, such as IL-1 and IL-6. In fact, a study with a knockout mice for IL-17 demonstrated that these injured mice have reduced cell infiltration in the injured spinal cord accompanied by reduced lesion size and enhanced functional performance (Hill, Kim, Gorrie, & Moalem-Taylor, 2011).

Finally, chemokines are a family of small cytokines which are able to induce chemotaxis in nearby responsive cells. Their importance after SCI is denoted by a study where they found that infusion of the broad-spectrum chemokine receptor antagonist (vMIPII) reduced monocyte infiltration into the spinal cord (Ghirnikar, Lee, & Eng, 2001). This reduction on infiltration was accompanied by a decrease on neuronal death, axonal degeneration, and demyelination within the injured spinal cord at 2 and 4 weeks after trauma.

Other studies also reflect the important role of specific chemokines, such as CXCL10, which has an important role in triggering T-lymphocyte and macrophage infiltration after trauma. CXCL10 expression peaks at 6 h after the trauma and remains elevated at 14 days after injury (Gonzalez, Glaser, Liu, Lane, & Keirstead, 2003). Inhibition of CXCL10 signaling by using specific neutralizing antibody results in a great reduction of lymphocyte and macrophage infiltration, tissue loss, and locomotor deficit (Gonzalez et al., 2003).

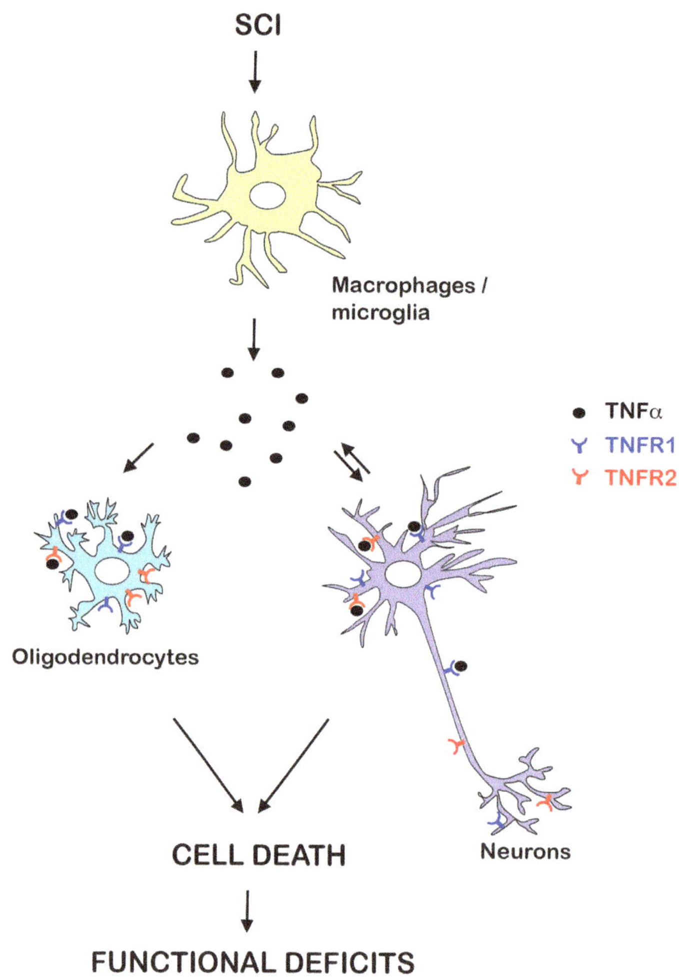

FIG. 2 TNFα effects after SCI. After the primary impact, resident cells such as microglia, neurons, and infiltrated macrophages secrete TNFα. Then TNFR1 and TNFR2 are upregulated in oligodendrocytes and neurons, promoting cell death after TNFα binding.

Cytokines in inflammation and glial scar formation

Macrophage and microglia polarization

Cytokines have an important role on macrophage and microglia functional state. Macrophages and microglia participate in the removal of cellular debris and the reestablishment of tissue homeostasis. However, there are many discrepancies about the contribution of these cells after injury. Several studies showed that the abolition of macrophages through different approaches could reduce the damage after SCI (Letellier et al., 2010; Lopez-Vales et al., 2005). Contrary, there are other lines of evidence showing that macrophages and microglia are protective and pro-regenerative (Yong, Rawji, Ghorbani, Xue, & Yong, 2019). These opposite actions of macrophages and microglia reflex a dual function that can be explained by their remarkable plasticity.

The term "polarization" refers to the plasticity of macrophages and microglia to acquire different and even opposite phenotypes (Fig. 3), and in which cytokines actively participate. Traditionally, two distinct polarization states have been described. Stimulation of macrophages and microglia with the prototypical Th1 cytokine Interferon (IFN)-γ or LPS induces M1 (or *classic*) polarization in which cells secrete pro-inflammatory cytokines (IL-1β, TNFα, and IL-6), cytotoxic mediators (iNOS and NO), and extracellular matrix-degrading enzymes (or matrix metalloproteinases, MMPs). On their surface, they express phagocytic receptors (CD16/32) and costimulatory molecules (CD86) (David & Kroner, 2011). They are therefore often referred to as pro-inflammatory cells. In contrast, stimulation with the Th2 cytokines, IL-4 or IL-13, induces the M2 (or *alternative*) polarization. This M2 phenotype is characterized by the upregulation of antiinflammatory cytokines (IL-10, transforming growth factor (TGF)-β), wound healing enzymes (Arg1), expression of clearance receptors (CD204 and CD206), and defective NF-κB activation (David & Kroner, 2011; Martinez, Helming, & Gordon, 2009). There are

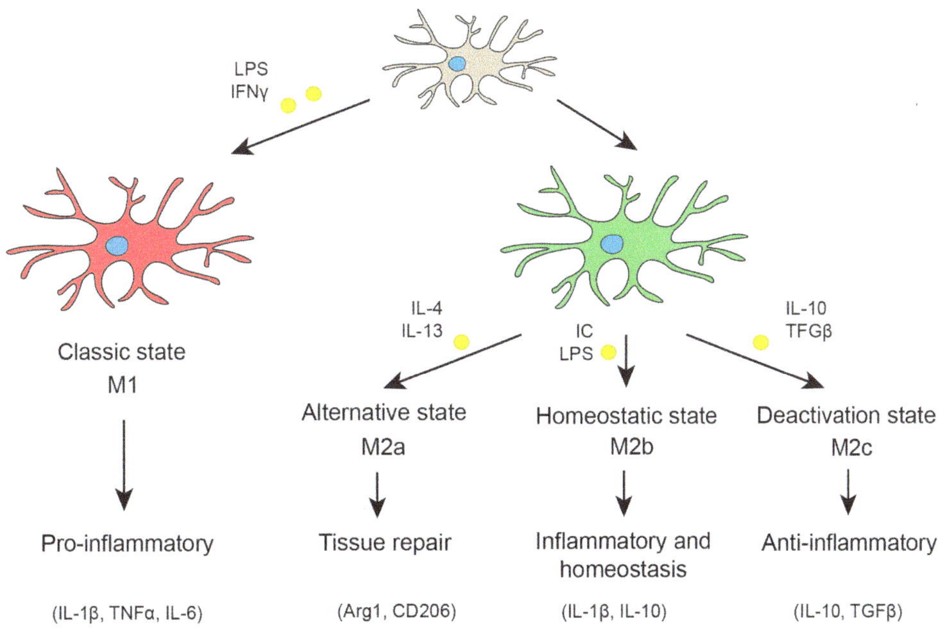

FIG. 3 Polarization of macrophages and microglia toward M1 and M2 phenotypes. M2 phenotype can be divided into M2a, M2b, and M2c profiles. Stimuli that induce polarization, general roles of these states, and produced factors are indicated.

therefore referred to as antiinflammatory cells. M2 polarization is more complex than M1 and, for some authors, it should be broken down into M2a, M2b, and M2c based on the specific profile of secreted cytokines (David, 2015).

M1 macrophages and microglia fight bacterial infections and remove cellular debris. Traditionally, they have been related to the detrimental effects of the inflammatory response since they secrete cytotoxic factors that could damage healthy tissue and contributes to secondary damage. M2 macrophages and microglia promote tissue healing and therefore they are related to the beneficial effects of the inflammatory response (David & Kroner, 2011). It is necessary to have in mind that the M1/M2 classification of macrophages and microglia is an experimental construct that was originally described using in vitro studies (Martinez et al., 2009). In vivo, cells are influenced by multiple additional factors leading to a wide spectrum of intermediate phenotypes where cells can express opposite markers at the same time (Morganti, Riparip, & Rosi, 2016).

Glial scar formation

Besides clearing up cellular debris and promote tissue healing, macrophages, and microglia also establish a burden of massive deposition of molecules called a glial scar. The glial scar is an evolving structure with different cells arriving and participating at different stages. At the first stages, the glial scar is formed by macrophages and microglia. However, the final structure is formed mainly by astrocytes (Fawcett & Asher, 1999). Although this glial structure delimitates the injury site avoiding its expansion, it also produces an inhibitory environment where axons fail to regenerate. Characterization of CNS myelin leads to the identification of several components that exert a potent inhibitory action on neurite outgrowth. These molecules include myelin-associated glycoprotein (MAG), oligodendrocyte myelin glycoprotein (OMgp), and chondroitin sulfate proteoglycans (CSPGs) (Fawcett & Asher, 1999). Astrocytes can express receptors and respond to a large variety of cytokines and growth factors involved in the proliferation and formation of the glial scar. From all the cytokines that participate in this process, TGF-β and IL-6 are the most important (Hamby & Sofroniew, 2010).

TGF-β is rapidly upregulated after CNS injury and promotes the activation and process extension of astrocytes, producing astrogliosis. TFG-β modulates the inflammatory response and induces the glial scar formation by stimulating the deposition of new matrix proteins (collagen, fibronectin, and proteoglycan) (Kohta, Kohmura, & Yamashita, 2009; Susarla et al., 2011). In agreement with these functions, intrathecal administration of neutralizing antibody to TGF-β into spinal cord injured rats reduced glial scar formation and enhanced axons growth (Kohta et al., 2009).

IL-6 is also involved in the formation of the glial scar. IL-6 shows an acute increase within the 12 h after injury (Francos-Quijorna et al., 2016), and induces astrogliosis through STAT3 signaling (Sofroniew, 2009). In vitro studies showed that IL-6 acts on neural stem cells to induce their differentiation into astrocytes (Bonni et al., 1997). Moreover, different authors showed that astrogliosis can be targeted through excessive activation or blockade of IL-6/IL-6R pathways (Brunello et al., 2000; Okada et al., 2004).

Cytokines in neuroprotection and repair

During the first 2 weeks after SCI, macrophages and microglia display a predominant pro-inflammatory phenotype (Francos-Quijorna et al., 2016; Kroner et al., 2014). This unfavorable polarization can be explained by the low levels of antiinflammatory cytokines observed in the spinal epicenter. Aimed at modulating this injury environment, different cytokine-based strategies have been proposed.

Enhancement of antiinflammatory cytokine levels

Interleukin 4 (IL-4) is undetected in the spinal cord during the first 28 days after injury (Francos-Quijorna et al., 2016). Since this cytokine induces the M2 polarization of macrophages and microglia in vitro (Fig. 3) (David & Kroner, 2011; Martinez et al., 2009), promoting its levels after an injury has been used as a therapeutic approach. Single administration of recombinant IL-4 into the spinal cord at 48 h after injury resulted in the antiinflammatory polarization of macrophages and microglia based on the increased levels of Arg1 and CD206 (Francos-Quijorna et al., 2016) (Fig. 4A). At the histological level, recombinant IL-4 promoted tissue protection by increasing myelin sparing and preservation of central axonal pathways (Fig. 4B). All these changes translated into improved functional recovery after injury (Fig. 4C). Importantly, these results were not obtained with the acute administration of recombinant IL-4 within the first minutes after injury evidencing the unfavorable environment during the first days after damage (Francos-Quijorna et al., 2016).

IL-37 is a cytokine of the interleukin 1 family with antiinflammatory properties. Unlike other members of the IL-1 family, the open reading frame for the human *IL37* gene has not been found in the mouse genome yet (Taylor, Renshaw, Garka, Smith, & Sims, 2002). Besides that, transgenic mice expressing the human form of *IL37* gene exhibited reduced tissue damage and functional deficits after SCI in comparison with wild-type mice (Coll-Miro et al., 2016).

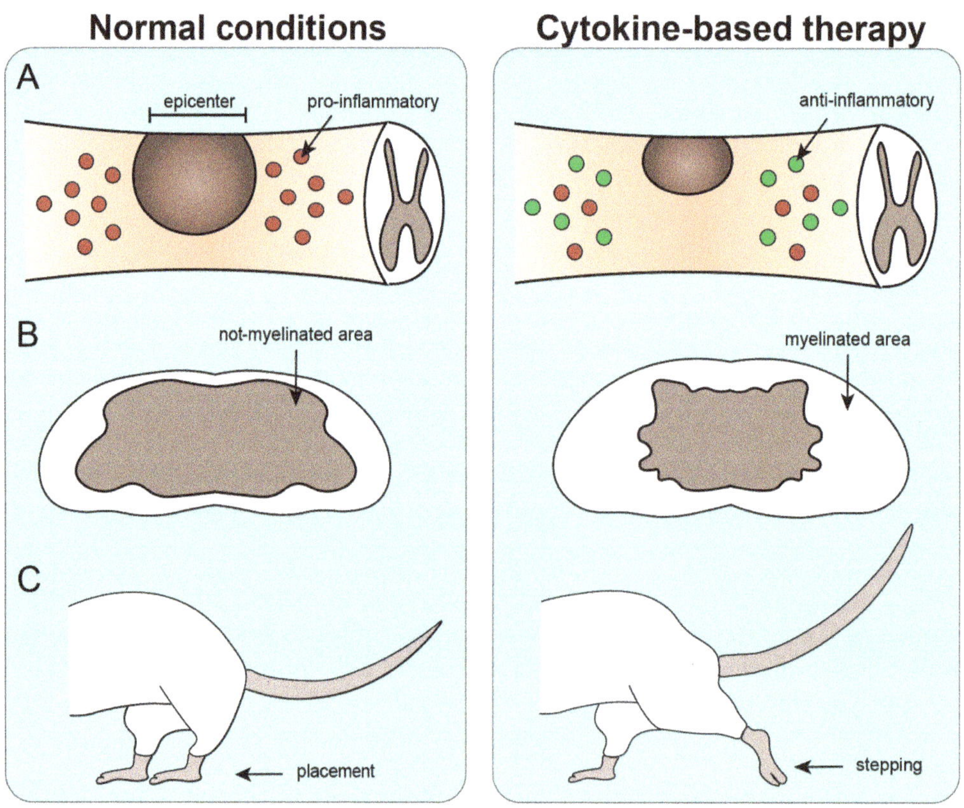

FIG. 4 Results of some cytokine-based therapies after SCI in animal models. Changes in the inflammatory environment (A), lesion volume (B), and locomotor recovery (C) due to the administration of antiinflammatory cytokines or blockade of proinflammatory cytokines.

This beneficial effect was associated with the moderated concentration of inflammatory cytokines and chemokines in the injury environment. Changes in axonal regeneration were not observed. Similar to transgenic expression, single administration of recombinant IL-37 protein in the spinal cord immediately after injury also resulted in functional recovery (Coll-Miro et al., 2016). Since recombinant IL-37 protein acts only through the extracellular pathway of this cytokine, it seems that this route is the main responsible for the beneficial effects of IL-37. Moreover, the abolition of an extracellular pathway in IL-37 transgenic mice resulted into the lack of protection after injury (Amo-Aparicio et al., 2020).

Inhibition of proinflammatory cytokine pathways

Not only promotion of antiinflammatory cytokines can be used to modulate the inflammatory environment after injury, strategies based on the inhibition of proinflammatory cytokines have also been studied. As one of the main proinflammatory cytokines after SCI, IL-1β has been targeted to promote protection. *Il1b* gene-deficient mice significantly improved locomotor recovery after SCI in comparison with wild-type mice (Boato et al., 2013). Histological analysis also revealed that the lack of *Il1b* gene produced smaller lesion size, reduced lesion width, and decreased glial scar. Moreover, the number of corticospinal tract fibers increased significantly in caudal regions (Boato et al., 2013). Despite these results, other publications using *Il1b* knockout mice showed no changes in the lesion volume or locomotor outcomes although proinflammatory macrophages and neutrophils in the spinal cord were reduced (Bastien et al., 2015).

Knocking out the *Tnf* gene in mice resulted in the amelioration of the proinflammatory environment after SCI (Kroner et al., 2014). Seven days after contusion injury, *Tnf* null mice showed a shift in macrophage polarization toward antiinflammatory phenotype based on the increase of Arg1 and CD206, and the reduction of CD16/32 markers. These changes resulted in improved tissue preservation and locomotor outcomes (Kroner et al., 2014). Similar results were observed with the intraperitoneal administration of TNF antagonist etanercept in spinal cord injured rats (Genovese et al., 2006).

Similarly, blockade of IL-6/IL-6R engagement through MR16-1, an antimouse IL-6R monoclonal antibody, resulted in protective effects. Single intraperitoneal dose of MR16-1 immediately after SCI increased the number of CD206-positive macrophages in the injury site (Guerrero et al., 2012). Moreover, the levels of IFNγ and TNFα in the lesioned area were reduced, whereas the levels of IL-4 and IL-13 were increased. All these changes correlated with the promotion of spared myelin, enhancement of the functional outcome (Guerrero et al., 2012), and reduction of astrogliosis (Okada et al., 2004). However, in contrast to expected, complete blockade of IL-6 through knockout mice showed detrimental effects on the regeneration after spinal cord injury suggesting a time-dependent dual effect (Cafferty et al., 2004).

Epigenetic regulation of cytokine expression

Epigenetics refers to inherited and reversible alterations to chromatin that alter gene expression without changing the DNA sequence. These mechanisms are regulated by two major epigenetic modifications: chemical modifications of DNA (DNA methylation) and covalent modification of histones associated with DNA (histone modification). From these, histone modifications, in particular acetylations, have been postulated to modulate cytokine expression after SCI. In addition, a third system included in epigenetic alterations is the expression of noncoding RNA (ncRNA)-associated with gene silencing.

Histone acetylation

Histone deacetylases (HDACs) belong to a family of enzymes that remove acetyl groups from lysine residues located on the amino-terminal tails of histone proteins, producing chromatin compaction, and thus, reduced gene expression. HDACs pan-inhibitors such as VPA and TSA promote antiinflammatory effects, reducing the expression of antiinflammatory cytokines from stimulated microglia/macrophages (Chen et al., 2018; Halili et al., 2010) and producing neuroprotection (Penas et al., 2011). Similarly, the class I HDAC inhibitor CI-994 tends to decrease pro-inflammatory cytokine production, inhibits the accumulation of neutrophils, reduces neuronal loss, and enhances functional recovery from SCI (Zhang, Fujita, Matsuzaki, & Yamashita, 2018). However, class IIa inhibition produces opposite effects, aggravating tissue damage, and impairing the functional recovery post-SCI (Qi & Wang, 2018).

There have been also some studies about epigenetic enzymes regarding these acetylated proteins. Bromodomain and extra-terminal domain (BET) proteins bind to acetylated lysine residues in both histone and nonhistone proteins, recruit transcriptional complexes to regulate gene expression. In particular, it has been observed that BET inhibition reduces proinflammatory cytokine expression and enhances antiinflammatory cytokine production after SCI (Rudman et al., 2018; Sanchez-Ventura, Amo-Aparicio, Navarro, & Penas, 2019) (Fig. 5). As a result, BET inhibition reduces inflammation, promotes neuroprotection, and enhances functional outcomes after SCI (Sanchez-Ventura et al., 2019). BET family

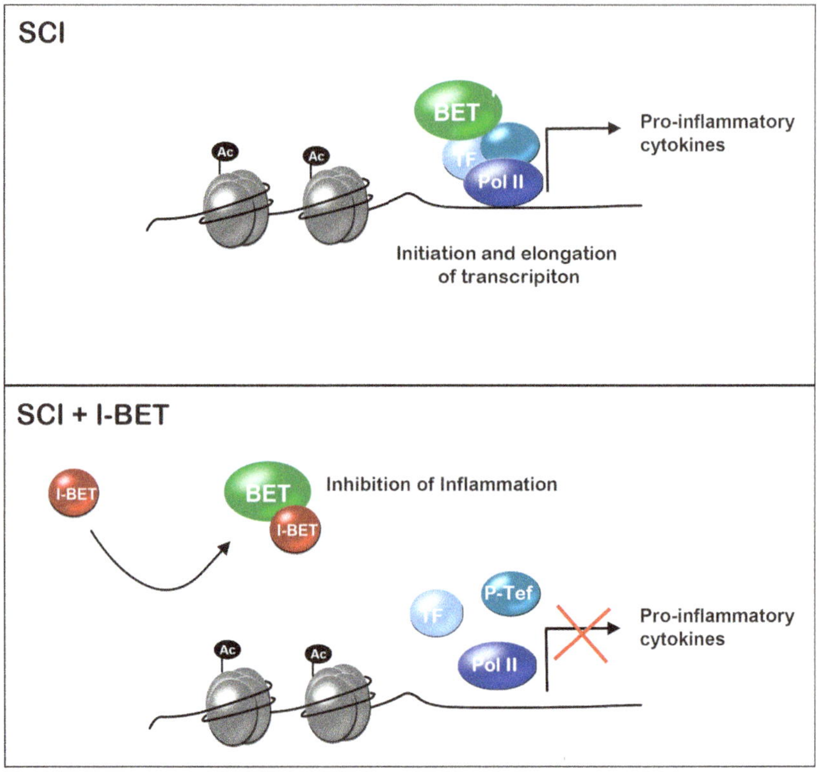

FIG. 5 Inhibition of BET proteins reduces inflammation after SCI. (Up) After the impact to the spinal cord, BET proteins recruit transcription factors (TF, P-Tef) and the RNA polymerase II (Pol II) to the chromatin, promoting the expression of proinflammatory cytokines. (Down) Treatment with a BET inhibitor (I-BET), displace BET proteins from the chromatin, inhibiting proinflammatory cytokine expression.

includes BRD2, BRD3, BRD4, and BRDT, and among them, BRD4 seems to be the main responsible for pro-inflammatory gene expression (Wang et al., 2019). Specific BRD4 knockdown suppresses inflammatory response via negatively modulating of NF-κB and MAPK signaling pathways (Wang et al., 2019).

Noncoding RNAs

Noncoding RNAs are diverse types of RNA molecules that are not translated into proteins. From these, microRNAs (miRNAs) have been deeply studied. miRNAs are small RNAs of about 22 nucleotides, which function in RNA silencing and posttranscriptional regulation of gene expression. These miRNAs function via complementary base-pairing sequences with mRNA molecules, which become silenced. After SCI, 138 miRNAs were found to be significantly expressed at 6 h postinjury, among which 87 miRNAs were upregulated and 51 miRNAs were downregulated (He et al., 2017). These altered miRNAs have opposite effects on inflammatory processes (Fig. 6). Literature shows that some decreased miRNAs

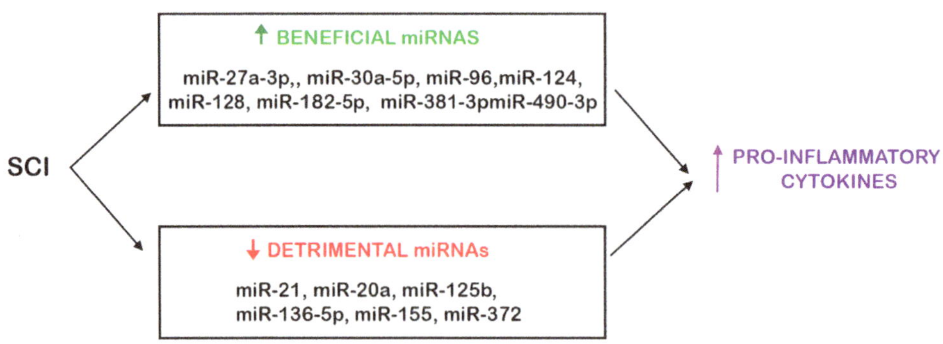

FIG. 6 MiRNAs affect the expression of cytokines after SCI. After trauma, there are several miRNAs that become upregulated and reduce the expression of proinflammatory cytokines, having a beneficial effect after SCI. Other miRNAs are downregulated, leading the expression of proinflammatory cytokines to become enhanced, and therefore, have a detrimental effect after SCI.

after SCI downregulate NF-κB pathway. Thus, its overexpression produces decreased cytokine expression and promotes reduced apoptosis, as well as enhanced functional recovery after SCI (Chen, Luo, Cao, Cheng, & He, 2018; Deng et al., 2018; He et al., 2017). Others, such as miR-21, which is increased after SCI, targets the PI3K/Akt/mTOR pathway and induces the expression of cytokines such as TGF-β, TNF-α, and IL-1β, promoting astrocytic and microglial immune response. Therefore, miR-21 has detrimental effects after SCI (Liu et al., 2018; Xie et al., 2018).

Applications to other areas of neuroscience

In addition, from spinal cord injury, some of the clinical approaches mentioned above have been demonstrated to be applicable to other neurological diseases.

Recent results showed that IL-37 exerts therapeutic effects in a mouse model of multiple sclerosis (MS). Similar to SCI, transgenic expression of IL-37 or recombinant protein administration protected against neurological deficits and myelin loss after MS. These changes were attributed to an attenuation in the inflammatory response on the spinal cord and lymph nodes. Authors also found that levels of IL-37 were ineffectively induced in the brain lesions of MS patients demonstrating the promising potential of this cytokine as a therapeutic approach (Sánchez-Fernández et al., 2020).

Regarding amyotrophic lateral sclerosis, using the mice model $SOD1^{G93A}$, a IL-4 lentiviral injection promoted microglia cell proliferation and modulated the inflammatory response by the reduction of pro-inflammatory cytokines and the increase on antiinflammatory cytokines (Rossi et al., 2018). All these changes translated into the slowed progression of motor impairments and delayed disease onset. Besides that, life span of $SOD1^{G93A}$ mice was not altered (Rossi et al., 2018).

For ischemic stroke, blockade of IL-1 β was shown to be protective in more than 25 publications (Banwell, Sena, & Macleod, 2009; McCann, Cramond, Macleod, & Sena, 2016). Rats and mice undergoing permanent or transient middle cerebral artery occlusion (MCAO) were treated with recombinant IL-1Ra (anakinra) by central or peripheral administration. Overall, anakinra reduced infarct volume by 38.2% (Banwell et al., 2009). Anakinra has also been studied in phase II clinical trials with ischemic stroke patients. Results showed a reduction of the inflammatory markers in plasma associated with worse outcomes after stroke (Smith et al., 2018). However, further studies are required to fully elucidate the efficacy and possible interaction of IL-1Ra with tissue plasminogen activator (Smith et al., 2018).

Finally, cytokines have been also found to have beneficial effects promoting axonal regeneration after nerve injury. Although peripheral axons are able to regenerate, peripheral regeneration is limited, especially in humans, due to the large distances that axons have to range. For instance, it has been demonstrated that the impairment of the inflammation via blockade of Toll-like receptor signaling impairs axonal outgrowth (Boivin et al., 2007) and that inhibition of TNF-α affects the survival and growth of sensory neurons (Fernyhough et al., 2005; Saleh et al., 2011). Importantly, antiinflammatory cytokines, secreted in the distal injured stump, have also a clear role in enhancing the capacity of proximal neurons to cross the critical zone. For example, it is known that animals lacking IL-10 regenerate worse than wild-type animals (Siqueira Mietto et al., 2015) and that the administration of IL-4 and IL-10 enhances axonal regeneration (Atkins et al., 2007; Vidal, Lemmens, Dooley, & Hendrix, 2013).

Mini-dictionary of terms

- Chemotaxis: Migration in response to stimuli. In the context of SCI, immune cells migrate toward the injury site in response to chemokines released by damaged tissue.
- Neurodegeneration: Progressive atrophy and loss of function of neurons.
- Neuroprotection: Preservation of neural structure and/or function.
- Secondary damage: Loss of neural tissue that takes weeks, and even months, after the initial trauma and expands the degenerative process of the lesion site through healthy neighboring areas.
- Polarization: Plasticity through macrophages and microglia acquire different, and even opposite, phenotypes. Traditionally, two opposite phenotypes have been described named M1 and M2.
- Proinflammatory cytokine: Cytokine that promotes the initiation or propagation of the inflammatory response
- Antiinflammatory cytokine: Cytokine that promotes the reduction of the inflammatory response

Key facts

Key facts of glial scar

- It is an evolving structure, cellularly composed by macrophages and microglia at the beginning and by astrocytes at late-stage
- Remains chronically after SCI
- Its extracellular matrix contains MAG, OMgp, and CSPGs
- Its major beneficial effect is to limit the propagation of tissue degeneration processes
- Its major detrimental effect is the inhibition of axonal regeneration

Key facts of inflammation after SCI

- Remains chronically after SCI
- It is imbalanced toward a proinflammatory environment
- Produces cytotoxic factors that damage healthy tissue
- Contributes to secondary damage
- Increases the lesion size and disabilities produced by the original trauma

Key facts on the major effects of cytokines

- Promotes chemotaxis of immune cells to the injury site
- Induces glial reactivity
- Promotes cell death and tissue damage
- Enhances vascular permeability allowing cell infiltration from blood
- Induces cell proliferation

Summary points

- Resident glial cells and blood-borne immune cells produce a high quantity of cytokines acutely after SCI.
- The major effects of cytokines after SCI are detrimental, including the promotion of cell infiltration from the blood, cell death, and glial scar formation.
- Cytokines have also beneficial effects including neuroprotection and axonal regeneration.
- Giving the importance of cytokines after SCI, they are being used as therapeutic targets after SCI.
- There are currently several lines of investigation studying the epigenetic mechanisms that regulate cytokine expression after SCI.

Acknowledgments

This work was supported by grant PID2019-108496RA-I00 from Ministerio de Ciencia e Innovación and Agencia Estatal de Investigación of Spain, CIBERNED (CB06/05/1105) and TERCEL (RD16/0011/0014) funds from the Instituto de Salud Carlos III of Spain, co-funded by European Union (ERDF/ESF, "Investing in your future").

References

Amo-Aparicio, J., Sanchez-Fernandez, A., Li, S., Eisenmesser, E. Z., Garlanda, C., Dinarello, C. A., et al. (2020). Extracellular and nuclear roles of IL-37 after spinal cord injury. *Brain, Behavior, and Immunity*, *91*, 194–201 (in press).

Atkins, S., Loescher, A. R., Boissonade, F. M., Smith, K. G., Occleston, N., O'Kane, S., et al. (2007). Interleukin-10 reduces scarring and enhances regeneration at a site of sciatic nerve repair. *Journal of the Peripheral Nervous System*, *12*(4), 269–276. https://doi.org/10.1111/j.1529-8027.2007.00148.x.

Banwell, V., Sena, E. S., & Macleod, M. R. (2009). Systematic review and stratified meta-analysis of the efficacy of interleukin-1 receptor antagonist in animal models of stroke. *Journal of Stroke and Cerebrovascular Diseases*, *18*(4), 269–276. https://doi.org/10.1016/j.jstrokecerebrovasdis.2008.11.009.

Bastien, D., Bellver Landete, V., Lessard, M., Vallieres, N., Champagne, M., Takashima, A., et al. (2015). IL-1alpha gene deletion protects oligodendrocytes after spinal cord injury through upregulation of the survival factor Tox3. *The Journal of Neuroscience*, *35*(30), 10715–10730. https://doi.org/10.1523/JNEUROSCI.0498-15.2015.

Boato, F., Rosenberger, K., Nelissen, S., Geboes, L., Peters, E. M., Nitsch, R., et al. (2013). Absence of IL-1beta positively affects neurological outcome, lesion development and axonal plasticity after spinal cord injury. *Journal of Neuroinflammation*, *10*, 6. https://doi.org/10.1186/1742-2094-10-6.

Boivin, A., Pineau, I., Barrette, B., Filali, M., Vallieres, N., Rivest, S., et al. (2007). Toll-like receptor signaling is critical for Wallerian degeneration and functional recovery after peripheral nerve injury. *The Journal of Neuroscience*, *27*(46), 12565–12576. https://doi.org/10.1523/JNEUROSCI.3027-07.2007.

Bonni, A., Sun, Y., Nadal-Vicens, M., Bhatt, A., Frank, D. A., Rozovsky, I., et al. (1997). Regulation of gliogenesis in the central nervous system by the JAK-STAT signaling pathway. *Science*, *278*(5337), 477–483. https://doi.org/10.1126/science.278.5337.477.

Brunello, A. G., Weissenberger, J., Kappeler, A., Vallan, C., Peters, M., Rose-John, S., et al. (2000). Astrocytic alterations in interleukin-6/soluble interleukin-6 receptor alpha double-transgenic mice. *The American Journal of Pathology*, *157*(5), 1485–1493. https://doi.org/10.1016/s0002-9440(10)64787-6.

Cafferty, W. B., Gardiner, N. J., Das, P., Qiu, J., McMahon, S. B., & Thompson, S. W. (2004). Conditioning injury-induced spinal axon regeneration fails in interleukin-6 knock-out mice. *The Journal of Neuroscience*, *24*(18), 4432–4443. https://doi.org/10.1523/JNEUROSCI.2245-02.2004.

Chen, W. C., Luo, J., Cao, X. Q., Cheng, X. G., & He, D. W. (2018). Overexpression of miR-381-3p promotes the recovery of spinal cord injury. *European Review for Medical and Pharmacological Sciences*, *22*(17), 5429–5437. https://doi.org/10.26355/eurrev_201809_15802.

Chen, S., Ye, J., Chen, X., Shi, J., Wu, W., Lin, W., et al. (2018). Valproic acid attenuates traumatic spinal cord injury-induced inflammation via STAT1 and NF-kappaB pathway dependent of HDAC3. *Journal of Neuroinflammation*, *15*(1), 150. https://doi.org/10.1186/s12974-018-1193-6.

Coll-Miro, M., Francos-Quijorna, I., Santos-Nogueira, E., Torres-Espin, A., Bufler, P., Dinarello, C. A., et al. (2016). Beneficial effects of IL-37 after spinal cord injury in mice. *Proceedings of the National Academy of Sciences of the United States of America*, *113*(5), 1411–1416. https://doi.org/10.1073/pnas.1523212113.

Conductier, G., Blondeau, N., Guyon, A., Nahon, J. L., & Rovere, C. (2010). The role of monocyte chemoattractant protein MCP1/CCL2 in neuroinflammatory diseases. *Journal of Neuroimmunology*, *224*(1–2), 93–100. https://doi.org/10.1016/j.jneuroim.2010.05.010.

Crowe, M. J., Bresnahan, J. C., Shuman, S. L., Masters, J. N., & Beattie, M. S. (1997). Apoptosis and delayed degeneration after spinal cord injury in rats and monkeys. *Nature Medicine*, *3*(1), 73–76. https://doi.org/10.1038/nm0197-73.

Dalkilic, T., Fallah, N., Noonan, V. K., Salimi Elizei, S., Dong, K., Belanger, L., et al. (2018). Predicting injury severity and neurological recovery after acute cervical spinal cord injury: A comparison of cerebrospinal fluid and magnetic resonance imaging biomarkers. *Journal of Neurotrauma*, *35*(3), 435–445. https://doi.org/10.1089/neu.2017.5357.

Davalos, D., Grutzendler, J., Yang, G., Kim, J. V., Zuo, Y., Jung, S., et al. (2005). ATP mediates rapid microglial response to local brain injury in vivo. *Nature Neuroscience*, *8*(6), 752–758. https://doi.org/10.1038/nn1472.

David, S. (2015). *Neuroinflammation: New insights into beneficial and detrimental functions*. John Wiley & Sons. ISBN: 1118732774, 9781118732779.

David, S., & Kroner, A. (2011). Repertoire of microglial and macrophage responses after spinal cord injury. *Nature Reviews. Neuroscience*, *12*(7), 388–399. https://doi.org/10.1038/nrn3053.

Deng, G., Gao, Y., Cen, Z., He, J., Cao, B., Zeng, G., et al. (2018). miR-136-5p regulates the inflammatory response by targeting the IKKbeta/NF-kappaB/A20 pathway after spinal cord injury. *Cellular Physiology and Biochemistry*, *50*(2), 512–524. https://doi.org/10.1159/000494165.

Dibaj, P., Nadrigny, F., Steffens, H., Scheller, A., Hirrlinger, J., Schomburg, E. D., et al. (2010). NO mediates microglial response to acute spinal cord injury under ATP control in vivo. *Glia*, *58*(9), 1133–1144. https://doi.org/10.1002/glia.20993.

Donnelly, D. J., & Popovich, P. G. (2008). Inflammation and its role in neuroprotection, axonal regeneration and functional recovery after spinal cord injury. *Experimental Neurology*, *209*(2), 378–388. https://doi.org/10.1016/j.expneurol.2007.06.009.

Fawcett, J. W., & Asher, R. A. (1999). The glial scar and central nervous system repair. *Brain Research Bulletin*, *49*(6), 377–391. https://doi.org/10.1016/s0361-9230(99)00072-6.

Fernyhough, P., Smith, D. R., Schapansky, J., Van Der Ploeg, R., Gardiner, N. J., Tweed, C. W., et al. (2005). Activation of nuclear factor-kappaB via endogenous tumor necrosis factor alpha regulates survival of axotomized adult sensory neurons. *The Journal of Neuroscience*, *25*(7), 1682–1690. https://doi.org/10.1523/JNEUROSCI.3127-04.2005.

Francos-Quijorna, I., Amo-Aparicio, J., Martinez-Muriana, A., & Lopez-Vales, R. (2016). IL-4 drives microglia and macrophages toward a phenotype conducive for tissue repair and functional recovery after spinal cord injury. *Glia*, *64*(12), 2079–2092. https://doi.org/10.1002/glia.23041.

Francos-Quijorna, I., Santos-Nogueira, E., Gronert, K., Sullivan, A. B., Kopp, M. A., Brommer, B., et al. (2017). Maresin 1 promotes inflammatory resolution, neuroprotection, and functional neurological recovery after spinal cord injury. *The Journal of Neuroscience*, *37*(48), 11731–11743. https://doi.org/10.1523/JNEUROSCI.1395-17.2017.

Genovese, T., Mazzon, E., Crisafulli, C., Di Paola, R., Muia, C., Bramanti, P., et al. (2006). Immunomodulatory effects of etanercept in an experimental model of spinal cord injury. *The Journal of Pharmacology and Experimental Therapeutics*, *316*(3), 1006–1016. https://doi.org/10.1124/jpet.105.097188.

Ghirnikar, R. S., Lee, Y. L., & Eng, L. F. (2001). Chemokine antagonist infusion promotes axonal sparing after spinal cord contusion injury in rat. *Journal of Neuroscience Research*, *64*(6), 582–589. https://doi.org/10.1002/jnr.1110.

Gonzalez, R., Glaser, J., Liu, M. T., Lane, T. E., & Keirstead, H. S. (2003). Reducing inflammation decreases secondary degeneration and functional deficit after spinal cord injury. *Experimental Neurology*, *184*(1), 456–463. https://doi.org/10.1016/s0014-4886(03)00257-7.

Guerrero, A. R., Uchida, K., Nakajima, H., Watanabe, S., Nakamura, M., Johnson, W. E., et al. (2012). Blockade of interleukin-6 signaling inhibits the classic pathway and promotes an alternative pathway of macrophage activation after spinal cord injury in mice. *Journal of Neuroinflammation*, *9*, 40. https://doi.org/10.1186/1742-2094-9-40.

Halili, M. A., Andrews, M. R., Labzin, L. I., Schroder, K., Matthias, G., Cao, C., et al. (2010). Differential effects of selective HDAC inhibitors on macrophage inflammatory responses to the Toll-like receptor 4 agonist LPS. *Journal of Leukocyte Biology*, *87*(6), 1103–1114. https://doi.org/10.1189/jlb.0509363.

Hamby, M. E., & Sofroniew, M. V. (2010). Reactive astrocytes as therapeutic targets for CNS disorders. *Neurotherapeutics*, *7*(4), 494–506. https://doi.org/10.1016/j.nurt.2010.07.003.

Harrington, J. F., Messier, A. A., Levine, A., Szmydynger-Chodobska, J., & Chodobski, A. (2005). Shedding of tumor necrosis factor type 1 receptor after experimental spinal cord injury. *Journal of Neurotrauma*, *22*(8), 919–928. https://doi.org/10.1089/neu.2005.22.919.

He, J., Zhao, J., Peng, X., Shi, X., Zong, S., & Zeng, G. (2017). Molecular mechanism of MiR-136-5p targeting NF-kappaB/A20 in the IL-17-mediated inflammatory response after spinal cord injury. *Cellular Physiology and Biochemistry*, *44*(3), 1224–1241. https://doi.org/10.1159/000485452.

Hill, F., Kim, C. F., Gorrie, C. A., & Moalem-Taylor, G. (2011). Interleukin-17 deficiency improves locomotor recovery and tissue sparing after spinal cord contusion injury in mice. *Neuroscience Letters*, *487*(3), 363–367. https://doi.org/10.1016/j.neulet.2010.10.057.

Inukai, T., Uchida, K., Nakajima, H., Yayama, T., Kobayashi, S., Mwaka, E. S., et al. (2009). Tumor necrosis factor-alpha and its receptors contribute to apoptosis of oligodendrocytes in the spinal cord of spinal hyperostotic mouse (twy/twy) sustaining chronic mechanical compression. *Spine (Phila Pa 1976)*, *34*(26), 2848–2857. https://doi.org/10.1097/BRS.0b013e3181b0d078.

Kohta, M., Kohmura, E., & Yamashita, T. (2009). Inhibition of TGF-beta1 promotes functional recovery after spinal cord injury. *Neuroscience Research*, *65*(4), 393–401. https://doi.org/10.1016/j.neures.2009.08.017.

Kroner, A., Greenhalgh, A. D., Zarruk, J. G., Passos Dos Santos, R., Gaestel, M., & David, S. (2014). TNF and increased intracellular iron alter macrophage polarization to a detrimental M1 phenotype in the injured spinal cord. *Neuron*, *83*(5), 1098–1116. https://doi.org/10.1016/j.neuron.2014.07.027.

Kwon, B. K., Streijger, F., Fallah, N., Noonan, V. K., Belanger, L. M., Ritchie, L., et al. (2017). Cerebrospinal fluid biomarkers to stratify injury severity and predict outcome in human traumatic spinal cord injury. *Journal of Neurotrauma*, *34*(3), 567–580. https://doi.org/10.1089/neu.2016.4435.

Lee, Y. B., Yune, T. Y., Baik, S. Y., Shin, Y. H., Du, S., Rhim, H., et al. (2000). Role of tumor necrosis factor-alpha in neuronal and glial apoptosis after spinal cord injury. *Experimental Neurology*, *166*(1), 190–195. https://doi.org/10.1006/exnr.2000.7494.

Letellier, E., Kumar, S., Sancho-Martinez, I., Krauth, S., Funke-Kaiser, A., Laudenklos, S., et al. (2010). CD95-ligand on peripheral myeloid cells activates Syk kinase to trigger their recruitment to the inflammatory site. *Immunity*, *32*(2), 240–252. https://doi.org/10.1016/j.immuni.2010.01.011.

Liu, R., Wang, W., Wang, S., Xie, W., Li, H., & Ning, B. (2018). microRNA-21 regulates astrocytic reaction post-acute phase of spinal cord injury through modulating TGF-beta signaling. *Aging (Albany NY)*, *10*(6), 1474–1488. https://doi.org/10.18632/aging.101484.

Lopez-Vales, R., Garcia-Alias, G., Fores, J., Udina, E., Gold, B. G., Navarro, X., et al. (2005). FK 506 reduces tissue damage and prevents functional deficit after spinal cord injury in the rat. *Journal of Neuroscience Research*, *81*(6), 827–836. https://doi.org/10.1002/jnr.20605.

Martinez, F. O., Helming, L., & Gordon, S. (2009). Alternative activation of macrophages: An immunologic functional perspective. *Annual Review of Immunology*, *27*, 451–483. https://doi.org/10.1146/annurev.immunol.021908.132532.

McCann, S. K., Cramond, F., Macleod, M. R., & Sena, E. S. (2016). Systematic review and meta-analysis of the efficacy of Interleukin-1 receptor antagonist in animal models of stroke: An update. *Translational Stroke Research*, *7*(5), 395–406. https://doi.org/10.1007/s12975-016-0489-z.

Morganti, J. M., Riparip, L. K., & Rosi, S. (2016). Call off the dog(ma): M1/M2 polarization is concurrent following traumatic brain injury. *PLoS One*, *11*(1). https://doi.org/10.1371/journal.pone.0148001, e0148001.

Okada, S., Nakamura, M., Mikami, Y., Shimazaki, T., Mihara, M., Ohsugi, Y., et al. (2004). Blockade of interleukin-6 receptor suppresses reactive astrogliosis and ameliorates functional recovery in experimental spinal cord injury. *Journal of Neuroscience Research*, *76*(2), 265–276. https://doi.org/10.1002/jnr.20044.

Penas, C., Verdu, E., Asensio-Pinilla, E., Guzman-Lenis, M. S., Herrando-Grabulosa, M., Navarro, X., et al. (2011). Valproate reduces CHOP levels and preserves oligodendrocytes and axons after spinal cord injury. *Neuroscience*, *178*, 33–44. https://doi.org/10.1016/j.neuroscience.2011.01.012.

Qi, X., & Wang, P. (2018). Class IIa HDACs inhibitor TMP269 promotes M1 polarization of macrophages after spinal cord injury. *Journal of Cellular Biochemistry*, *119*(4), 3081–3090. https://doi.org/10.1002/jcb.26446.

Rossi, C., Cusimano, M., Zambito, M., Finardi, A., Capotondo, A., Garcia-Manteiga, J. M., et al. (2018). Interleukin 4 modulates microglia homeostasis and attenuates the early slowly progressive phase of amyotrophic lateral sclerosis. *Cell Death & Disease*, *9*(2), 250. https://doi.org/10.1038/s41419-018-0288-4.

Rudman, M. D., Choi, J. S., Lee, H. E., Tan, S. K., Ayad, N. G., & Lee, J. K. (2018). Bromodomain and extraterminal domain-containing protein inhibition attenuates acute inflammation after spinal cord injury. *Experimental Neurology*, *309*, 181–192. https://doi.org/10.1016/j.expneurol.2018.08.005.

Saleh, A., Smith, D. R., Balakrishnan, S., Dunn, L., Martens, C., Tweed, C. W., et al. (2011). Tumor necrosis factor-alpha elevates neurite outgrowth through an NF-kappaB-dependent pathway in cultured adult sensory neurons: Diminished expression in diabetes may contribute to sensory neuropathy. *Brain Research*, *1423*, 87–95. https://doi.org/10.1016/j.brainres.2011.09.029.

Sánchez-Fernández, A., Zandee, S., Amo-Aparicio, J., Charabati, M., Prat, A., Garlanda, C., et al. (2020). IL-37 exerts therapeutic effects in experimental autoimmune encephalomyelitis through the receptor complex IL-1R5/IL-1R8. *Theranostic*, *11*, 1–13 (in press).

Sanchez-Ventura, J., Amo-Aparicio, J., Navarro, X., & Penas, C. (2019). BET protein inhibition regulates cytokine production and promotes neuroprotection after spinal cord injury. *Journal of Neuroinflammation*, *16*(1), 124. https://doi.org/10.1186/s12974-019-1511-7.

Siqueira Mietto, B., Kroner, A., Girolami, E. I., Santos-Nogueira, E., Zhang, J., & David, S. (2015). Role of IL-10 in resolution of inflammation and functional recovery after peripheral nerve injury. *The Journal of Neuroscience*, *35*(50), 16431–16442. https://doi.org/10.1523/JNEUROSCI.2119-15.2015.

Smith, C. J., Hulme, S., Vail, A., Heal, C., Parry-Jones, A. R., Scarth, S., et al. (2018). SCIL-STROKE (subcutaneous Interleukin-1 receptor antagonist in ischemic stroke): A randomized controlled phase 2 trial. *Stroke*, *49*(5), 1210–1216. https://doi.org/10.1161/STROKEAHA.118.020750.

Sofroniew, M. V. (2009). Molecular dissection of reactive astrogliosis and glial scar formation. *Trends in Neurosciences*, *32*(12), 638–647. https://doi.org/10.1016/j.tins.2009.08.002.

Stammers, A. T., Liu, J., & Kwon, B. K. (2012). Expression of inflammatory cytokines following acute spinal cord injury in a rodent model. *Journal of Neuroscience Research*, *90*(4), 782–790. https://doi.org/10.1002/jnr.22820.

Susarla, B. T., Laing, E. D., Yu, P., Katagiri, Y., Geller, H. M., & Symes, A. J. (2011). Smad proteins differentially regulate transforming growth factor-beta-mediated induction of chondroitin sulfate proteoglycans. *Journal of Neurochemistry*, *119*(4), 868–878. https://doi.org/10.1111/j.1471-4159.2011.07470.x.

Taylor, S. L., Renshaw, B. R., Garka, K. E., Smith, D. E., & Sims, J. E. (2002). Genomic organization of the interleukin-1 locus. *Genomics*, *79*(5), 726–733. https://doi.org/10.1006/geno.2002.6752.

Vidal, P. M., Lemmens, E., Dooley, D., & Hendrix, S. (2013). The role of "anti-inflammatory" cytokines in axon regeneration. *Cytokine & Growth Factor Reviews*, *24*(1), 1–12. https://doi.org/10.1016/j.cytogfr.2012.08.008.

Wang, J., Chen, J., Jin, H., Lin, D., Chen, Y., Chen, X., et al. (2019). BRD4 inhibition attenuates inflammatory response in microglia and facilitates recovery after spinal cord injury in rats. *Journal of Cellular and Molecular Medicine*, *23*(5), 3214–3223. https://doi.org/10.1111/jcmm.14196.

Wang, X. J., Kong, K. M., Qi, W. L., Ye, W. L., & Song, P. S. (2005). Interleukin-1 beta induction of neuron apoptosis depends on p38 mitogen-activated protein kinase activity after spinal cord injury. *Acta Pharmacologica Sinica*, *26*(8), 934–942. https://doi.org/10.1111/j.1745-7254.2005.00152.x.

Xie, W., Yang, S. Y., Zhang, Q., Zhou, Y., Wang, Y., Liu, R., et al. (2018). Knockdown of MicroRNA-21 promotes neurological recovery after acute spinal cord injury. *Neurochemical Research*, *43*(8), 1641–1649. https://doi.org/10.1007/s11064-018-2580-1.

Yong, H. Y. F., Rawji, K. S., Ghorbani, S., Xue, M., & Yong, V. W. (2019). The benefits of neuroinflammation for the repair of the injured central nervous system. *Cellular & Molecular Immunology*, *16*(6), 540–546. https://doi.org/10.1038/s41423-019-0223-3.

Zhang, S., Fujita, Y., Matsuzaki, R., & Yamashita, T. (2018). Class I histone deacetylase (HDAC) inhibitor CI-994 promotes functional recovery following spinal cord injury. *Cell Death & Disease*, *9*(5), 460. https://doi.org/10.1038/s41419-018-0543-8.

Chapter 10

Neurovascular pathology following traumatic spinal cord injury

Mohammad-Masoud Zavvarian[a,b], James Hong[b], Jonathon Chon Teng Chio[a,b], Amirali Toossi[b], and Michael G. Fehlings[c]

[a]*Institute of Medical Science, Faculty of Medicine, University of Toronto, Toronto, ON, Canada,* [b]*Division of Genetics and Development, University Health Network, Toronto, ON, Canada,* [c]*Division of Neurosurgery, Toronto Western Hospital, University Health Network, Toronto, ON, Canada*

Abbreviations

ASA	anterior spinal artery
ASIA	American Spinal Injury Association
BBB	blood–brain barrier
BSCB	blood–spinal cord barrier
CNS	central nervous system
CSPG	chondroitin sulfate proteoglycans
EC	endothelial cells
IgG	immunoglobulin G
IVIG	intravenous immunoglobulin G
MPSS	methylprednisolone sodium succinate
MSC	mesenchymal stem cell
NG2	nerve/glial antigen 2
NVU	neurovascular unit
SCI	spinal cord injury
SCI-IDS	injury-induced immune depression syndrome
TJ	tight junction
ZO	zona occludens

Introduction

Traumatic spinal cord injury (SCI) damages the local neuronal, glial, and vascular cells, which impairs the transmission and facilitation of autonomic and sensorimotor signals between the brain and peripheral organs (Zavvarian, Hong, & Fehlings, 2020). While the initial mechanical damage to the neuronal and glial cells directly impacts neurotransmission, vascular damage manifests in a prolonged pathology influencing the postinjury response and recovery (Tator & Fehlings, 1991). In the noninjured spinal cord, the spinal microvascular network maintains a steady supply of oxygen and nutrients to the neurons and adjacent glial cells, while excreting waste products. It also prevents the infiltration of harmful circulatory molecules and immune cells into the nervous system. The blood–spinal cord barrier (BSCB), and by extension the blood–brain barrier (BBB), form a tight seal between the central nervous system (CNS) and the bloodstream to regulate the passage of oxygen, nutrients, and waste products across the CNS microvasculature (Winkler et al., 2013). Recent studies have explored the interactions between vascular, glial, and neuronal cells, characterized as the neurovascular unit (NVU) model, to describe the BSCB function in both healthy and injured spinal cords (Xu et al., 2018).

The NVU damage following traumatic SCI stems from the direct disruption of vessels due to shear injury, leading to hemorrhage and vasospasm (Ahuja et al., 2017). This vascular disruption will subsequently result in the loss of autoregulation of spinal cord blood flow, which is exacerbated by damage to the descending sympathetic input into cardiac control centers in the upper thoracic cord. These mechanisms exacerbate ischemia. Furthermore, sympathetic nervous system impairment due to the loss of supraspinal input at the T6 level or higher injuries results in hypotension, autonomic dysreflexia, and cardiac arrhythmias (Sharif & Hou, 2017). Furthermore, traumatic SCI increases the risk of pulmonary

Cellular, Molecular, Physiological, and Behavioral Aspects of Spinal Cord Injury. https://doi.org/10.1016/B978-0-12-822427-4.00010-1

embolism and deep venous thrombosis due to loss of mobility. Addressing these vascular-related complications is paramount for the mitigation of further spinal cord damage (Furlan & Fehlings, 2008).

Despite the high prevalence of SCI—ranging between 250 and 906 cases per million—our understanding of SCI-induced vascular damage and treatment options available to patients remains limited (Singh, Tetreault, Kalsi-Ryan, Nouri, & Michael, 2014). Although the first line of defense is the prevention of traumatic incidents, enhancing therapeutics aimed at restoring BSCB function would have a drastic impact on the patient's quality of life. Accordingly, this chapter will first provide an overview of the role of vascular damage following traumatic SCI and then explore therapeutic targets with the potential for clinical translation.

Secondary pathogenesis

The traumatic compression and contusion of the spinal cord cause immediate damage and long-term consequences, referred to as primary and secondary injury respectively (Fig. 1). Immediately following injury, the neuronal, glial, and vascular cells present at the injury site are damaged and result in the production of toxic cellular debris. Ischemia, hemorrhage, vasospasm, and edema are important consequences of vascular disruption in the injured spinal cord (Figley, Khosravi, Legasto, Tseng, & Fehlings, 2014). These lead to oxidative damage and the infiltration of reactive circulatory cells and molecules that induce the secondary injury pathogenesis. While the infiltrating immune cells are necessary to clear the cellular debris following injury, their continued presence is an important mediator of secondary SCI damage (Schwab, Zhang, Kopp, Brommer, & Popovich, 2014). Over the course of secondary injury, the injured spinal cord forms three distinct histological sections (Fig. 2), evident in both preclinical studies and postmortem analyses (O'Shea, Burda, & Sofroniew, 2017). The injury epicenter contains fibrotic scar and fluid-filled cavitations formed by cellular debris clearance. Immediately surrounding the lesion core, there is the astrocytic scar border which is formed from the proliferation of new astrocytes intermingled with extracellular molecules in order to encapsulate the immune cells in the lesion core. Expanding away from the astrocytic scar border is the spared neural tissue, which undergoes synaptic plasticity and axonal sprouting (Courtine & Sofroniew, 2019).

The secondary responses following ischemic-reperfusion damage due to traumatic SCI are classified into acute (48 h), subacute (2–14 days), intermediate (14–56 days), and chronic phases (more than 56 days) (Ahuja, Wilson, et al., 2017). In the acute phase of injury, the compromised microvascular network limits the blood supply at the lesion core, leading to the

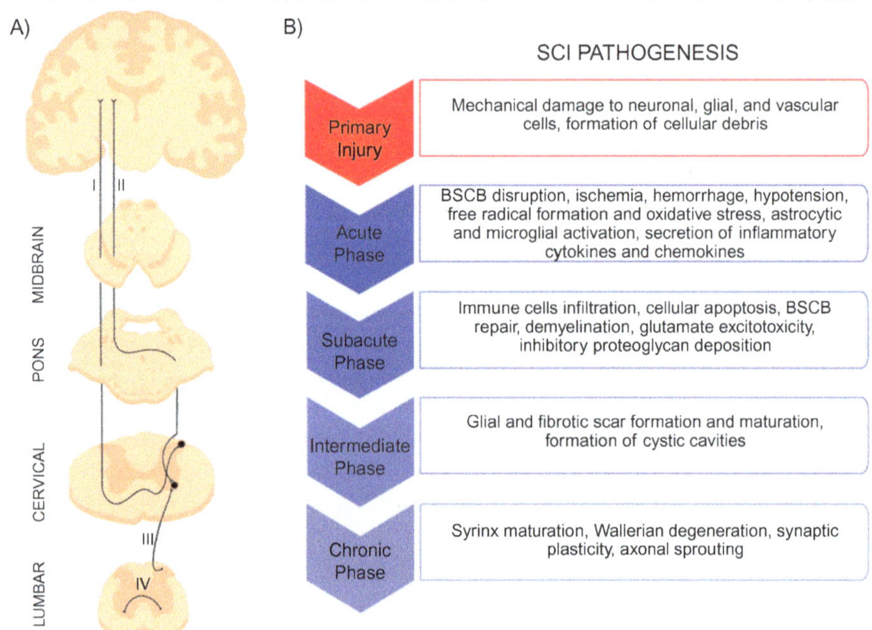

FIG. 1 Traumatic SCI hinders the ability of the spinal neural circuits to transmit and process autonomic and sensorimotor signals. (A) Representative schematic of spinal neural circuits; I: spinothalamic tract, II: corticospinal tract, III: propriospinal neurons, IV: commissural interneuron. (B) The secondary SCI pathogenesis is divided into four phases. The acute encompasses the first 48 h after injury. The sub-acute phase is a period between 48 h to 2 weeks postinjury. The intermediate phase refers to 2 weeks to 2 months after injury, and the chronic phase refers to greater than 2 months post-SCI.

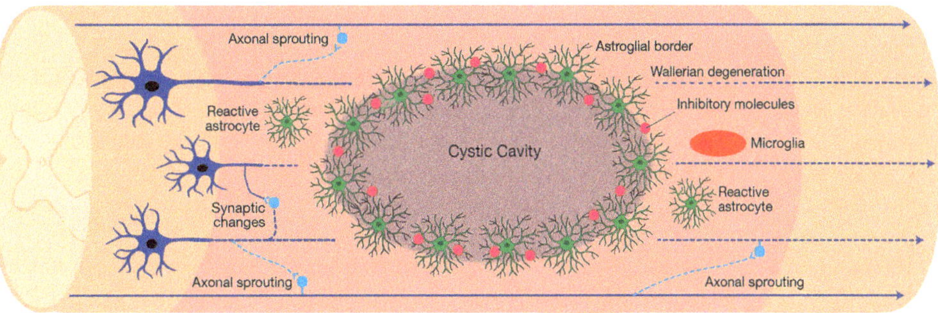

FIG. 2 The injured spinal cord consists of three distinct histological compartments, including a nonneuronal lesion core, a surrounding astroglial scar border, and pre-lesional tissue with spared but reactive neural tissue. Each compartment consists of unique cellular composition and exhibits distinct pathophysiology following traumatic SCI. *Pictures edited from Zavvarian, M.-M., Hong, J., & Fehlings, M. G. (2020). The functional role of spinal interneurons following traumatic spinal cord injury.* Frontiers in Cellular Neuroscience, 14, *127. doi:10.3389/fncel.2020.00127; Zavvarian, M.-M., Toossi, A., Khazaei, M., Hong, J., & Fehlings, M. (2020). Novel innovations in cell and gene therapies for spinal cord injury.* F1000Research, 9, *279. https://doi.org/10.12688/f1000research.21989.1.*

induction of hypoxia, ischemic injury, and hemorrhage (Mautes, Weinzierl, Donovan, & Noble, 2000). Moreover, reactive circulatory cells and molecules infiltrate the neural tissue, further complicating the injury pathology. In parallel, the produced cellular debris increases the proapoptotic signaling and triggers the activation of local microglia and astrocytes to clear the introduced debris and recruit additional blood-borne immune cells (Ahuja, Wilson, et al., 2017). In the subacute phase, the BSCB integrity is reestablished and the scar tissue begins to form in order to encapsulate the infiltrated immune cells. Ultimately, during the intermediate and chronic phase, the extended clearance of debris by microglia and macrophages leads to the development of the cystic cavity and continued axonal degradation. In many SCI cases, the formation of a longitudinal fluid-filled cavity occurs across many segments of the spinal cord (syringomyelia) and results in progressive myelopathy. The cause of syringomyelia occurrence is unknown, but it is affected by the gradual accumulation of intraparenchymal cerebrospinal fluid (Fig. 1B) (Siddiqui, Khazaei, & Fehlings, 2015).

SCI is a highly heterogeneous condition and the cellular response to trauma varies based on the severity and level of injury (Fig. 3) (Liu et al., 2019). The damage to the spinal cord can either be complete or incomplete (Raineteau & Schwab, 2001). The American Spinal Injury Association (ASIA) impairment scale uses a four-letter scheme to characterize the injury severity. Grade A refers to the complete loss of sensorimotor function below the site of injury. Grade B indicates

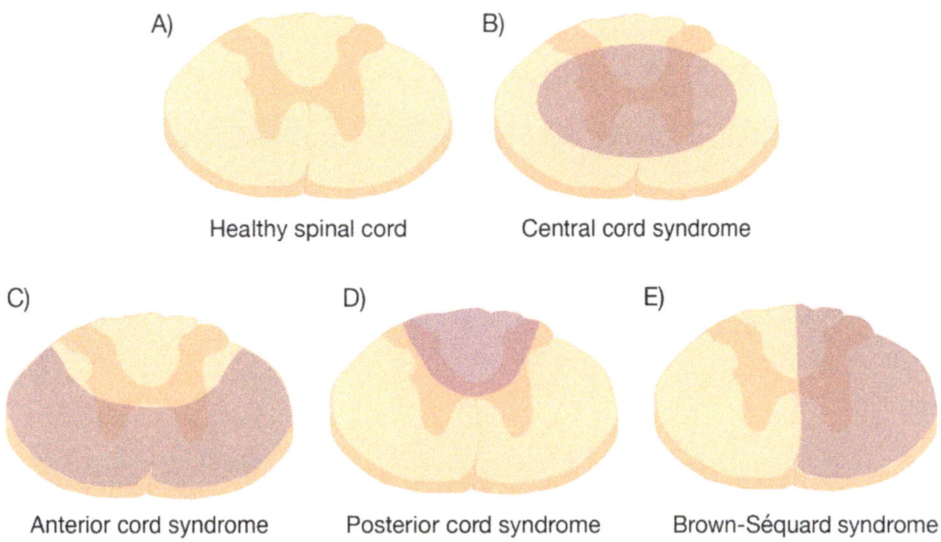

FIG. 3 SCI patients exhibit differential patterns of functional due to damage to specific spinal cord tracts while sparing other tracts. (A) A healthy spinal cord that consists of ascending and descending tracks. (B) Patients with central cord syndrome present with disproportionately greater upper limb motor impairment as compared to the lower limbs. (C) Patients with anterior cord syndrome suffer from complete motor paralysis and nociception loss due to damage to the corticospinal tract and spinothalamic tract, respectively. However, these patients continue to possess proprioception. (D) Posterior cord syndrome results in proprioception loss but preservation of sensorimotor function. (E) Brown-Séquard syndrome characterized by nociception loss contralateral to the lesion site and ipsilateral sensorimotor loss.

loss of motor function while sensation is still present. Grade C and D are indicative of preserved motor movement but differ from each other in the extent of mobility (Ahuja, Wilson, et al., 2017). Second, there are level-specific variations to SCI pathogenesis and the spinal cord varies in the degree of vascularization and myelination across different levels (Wilcox et al., 2017). While the most common level of injury is at the cervical level, the majority of preclinical studies have utilized a thoracic injury model. This introduces a knowledge gap in our understanding of cervical-specific vascular responses following SCI. This is a growing area of research as the majority of deaths following SCI are due to breathing dysfunction, which results from upper cervical injuries.

Hemodynamic response

The spinal cord is a highly vascularized organ, which requires an extensive network of blood vessels to maintain its ability to transmit and integrate signals between the brain and peripheral organs (Fig. 4). Traumatic damage to the spinal cord hinders this blood supply and results in areas of low perfusion (Mautes et al., 2000). The hemodynamic alterations to the injured spinal cord vary depending on the site of injury. This is due to the differences in the blood supply pattern across the spinal cord in order to account for necessary blood demands for each region. For instance, the anterior spinal artery (ASA) is largest in the cervical and lumbar regions, to correspond with the increased requirement for supplying the enlarged ganglions at these levels (Fig. 4). Additionally, the central sulcal arteries are both larger and increased in number in the cervical and lumbar regions, relative to the thoracic region. Furthermore, the significant narrowing of the spinal cord in the mid-thoracic region corresponds with the narrowing of its major longitudinal arteries and the reduced perfusion requirements from T4 to approximately the T9 level. The relatively reduced arterial supply renders this area more susceptible to ischemic damage upon vascular insult (Dommisse, 1974; Turnbull, 1971). However, the collateral circulation in any given area may be as relevant to the local blood supply as the number of arteries (Crosby & Lauer, 1962).

The hemodynamic disruption of the spinal cord vasculature results from the flattening of the spinal cord anteroposteriorly and it's widening laterally, as a direct consequence of a traumatic insult. This results in the lengthwise compression and shortening of perforant branches of the pial plexus running directly in the anteroposterior direction, which supplies the anterior and posterior columns. In contrast, the lateral perforant branches supplying the lateral columns are stretched lengthwise and flattened, decreasing their diameter. Hence, while the shortened anterior and posterior perforant branches do not alter blood supply, the flattened lateral perforant branches decrease the blood supply. The same effect is observed in the central arterial system, in which the lateral segments are flattened, and the anteroposterior segment in the anterior medial fissure is less dramatically affected (Turnbull, 1971). The varying hemodynamics among regions of the cord result

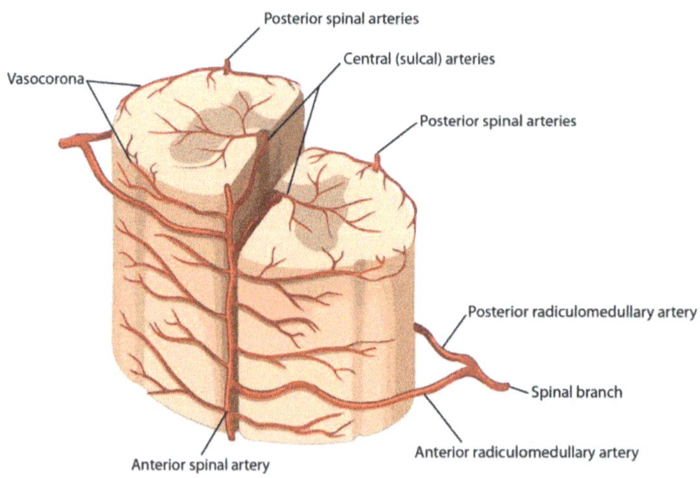

FIG. 4 An overview of the major spinal arteries. The vertebral arteries are the main source of blood to the spinal cord. However, the following figure shows the arteries that branch from the vertebral arteries to directly supply the spinal cord itself. It consists of one anterior spinal artery, the central sulcal arteries, two posterior spinal arteries, the anterior and posterior radiculomedullary arteries, and the arterial vasocorona—the anastomose between the spinal arteries. *Pictures edited from Santillan, A., Nacarino, V., Greenberg, E., Riina, H. A., Gobin, Y. P., & Patsalides, A. (2012). Vascular anatomy of the spinal cord.* Journal of NeuroInterventional Surgery, *4(1), 67–74. https://doi.org/10.1136/neurintsurg-2011-010018.*

in several watershed regions with relatively low perfusion, such as the upper thoracic cord between the cervical and lumbar limb enlargements, the anterolateral surface between the coronary perimedullary arches, and the intramedullary gray–white junction between the centripetal and centrifugal arterial systems (Bosmia, Hogan, Loukas, Tubbs, & Cohen-Gadol, 2015). Previous studies demonstrate that the reduction in perfusion due to compression is associated with vessel distance from the dura mater and suggest that posterior compression may have a greater likelihood of producing ischemic injury when compared to anterior compression (Alshareef et al., 2014).

Neurovascular unit (NVU)

The NVU is an important tri-partite structure consisting of astrocytes, pericytes, and endothelial cells, which is involved in the coupling of local perfusion and neuronal activity (Cauli, 2004; Chio et al., 2019; Gordon, Mulligan, & Macvicar, 2007). It regulates the rate of blood flow to match the metabolic demands of active neurons (Fig. 5). The NVU's neural, glial, and vascular components work together to mediate the spinal blood flow and vascular permeability (McConnell, Kersch, Woltjer, & Neuwelt, 2017). The neurons and astrocytes signal to pericytes and endothelial cells (and vice versa) to modify the vascular tone, which affects the delivery of blood, oxygen, and glucose to the brain and spinal cord (Attwell et al., 2010; McConnell et al., 2017).

As blood flows from superficial to deeper regions in the CNS, the capillaries lose their smooth muscle cell and pia mater coverage between the endothelial cells and astrocytes (McConnell et al., 2017). Instead, they gain pericyte coverage between endothelial cells and astrocytes; the three cell constituents of the NVU. In close proximity to each other, neurons can release glutamate at the synapses, which can bind to metabotropic glutamate receptors on astrocytes. This induces a rise in intracellular calcium levels in astrocytes, which causes the release of vasoactive molecules such as prostaglandin, nitric oxide, serotonin, or neuropeptide Y) (McConnell et al., 2017). These compounds also induce changes in pericyte contractility (Itoh & Suzuki, 2012). Tight junction (TJ) proteins, such as occludin, zona occludens (ZO), and claudin protein families hold the NVU's cellular components together and allow them to communicate with each other (Kaplan, Chow, & Gu, 2020). In addition, these TJ proteins help form the BSCB, which is a regulated structure that partially separates the spinal cord from peripheral blood flow (Kaplan et al., 2020; Stamatovic, Johnson, Keep, & Andjelkovic, 2016). TJ proteins are associated with, but not entirely responsible for, the BSCB's integrity, as the knockout of occludin (Saitou et al., 1998, 2000), ZO (Katsuno et al., 2008) and claudin (Castro Dias, Coisne, Baden, et al., 2019; Castro Dias, Coisne, Lazarevic, et al., 2019) does not significantly increase permeability. Rather, the BSCB in these genetic models seems intact based on electron microscopy (Kaplan et al., 2020).

Pericytes are largely responsible for TJ protein expression (Thomsen, Burkhart, & Moos, 2015; Underly et al., 2017). While aforementioned studies have identified that the loss of TJ proteins does not result in significant increases in BSCB permeability (Castro Dias, Coisne, Baden, et al., 2019; Castro Dias, Coisne, Lazarevic, et al., 2019; Katsuno et al., 2008; Saitou et al., 1998, 2000; Umeda et al., 2006; Underly et al., 2017), pericytes are important members of the NVU for maintaining BSCB integrity and controlling blood flow in CNS tissue. After ischemia, the loss of BSCB integrity commonly originates in regions formerly covered by pericytes, as they die or migrate away from ischemic sites in a matrix

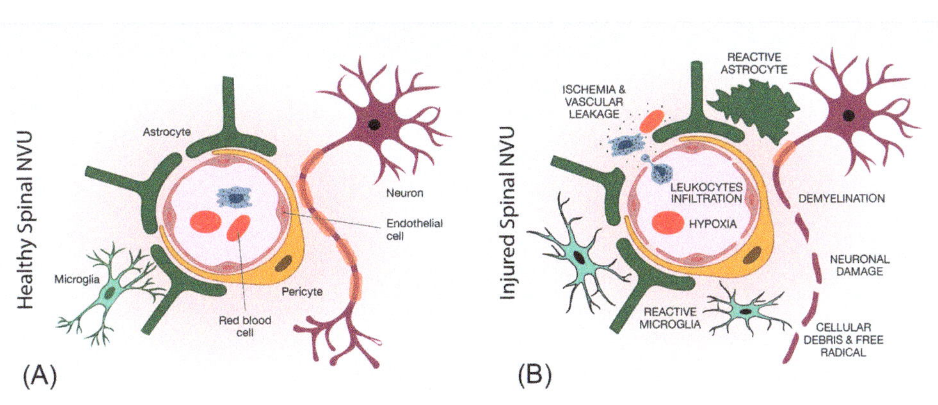

FIG. 5 Traumatic SCI impairs neurotransmission and NVU function. (A) The structure of a healthy spinal NVU comprised of endothelial cells, pericytes, astrocytes, and the surrounding cells. (B) The structure of an injured spinal NVU resulting in secondary SCI-induced pathologies. *Pictures edited from "biorender.com."*

metalloproteinase-dependent manner (Underly et al., 2017). Overall, regulation of BBB and BSCB integrity by the NVU is an active signaling process between NVU cells (Kaplan et al., 2020). The cell-specific responses that result in the loss of BSCB integrity are discussed below.

Endothelial cell (EC) response

ECs line up the arteries and capillaries, and are tightly associated with pericytes (Kaplan et al., 2020) to maintain homeostatic interactions with pericytes through numerous pathways (Park et al., 2017). ECs in the CNS have the highest pericyte coverage and markedly less fluid-phase transcytosis (Andreone et al., 2017; Ben-Zvi et al., 2014; Chow & Gu, 2017; McConnell et al., 2017), which are unique features that distinguish brain endothelium from peripheral tissues and allow ECs to adapt to the previously described unique requirements of the CNS (Li et al., 2011). As stated previously, ECs help to form the tri-partite BSCB, which regulates the entry and exit of molecules to and from the CNS. ECs have multiple signaling pathways that are integral to the proper formation and maintenance of the BSCB, such as retinoic acid (Mizee et al., 2013; Pollock, Xie, Bell, & Anand-Apte, 2018), N-cadherin-Notch (Li et al., 2011), Sonic Hedgehog-Sonic Hedgehog receptors (Alvarez et al., 2011), transforming growth factor-β (Vladimir, Dan, Jonathan, Rotem, & Adiel, 2019), and WNT-Frizzled receptors (Daneman et al., 2009; Kaplan et al., 2020). Outcomes stemming from a deficiency in these pathways are implicated in greater vascular permeability under pathological conditions.

Expressed on other cells, WNT ligands bind to Frizzled receptors on ECs and enable translocation of β-catenin to the nucleus (Kaplan et al., 2020). This allows transcription of genes associated with a less permeable BSCB. A rodent model of multiple sclerosis demonstrated that sub-optimal WNT-Frizzled receptor interactions, as mediated by secretion of WNT inhibitory factor 1 (Wif1), disrupts barrier integrity and triggers CNS inflammation (Niu et al., 2019). This results in ECs secreting less TJ proteins. Attempts to upregulate the expression of WNT and β-catenin in non-CNS tissues result in a greater presence of BBB-related genes and decreased barrier permeability, but still inferior to the integrity of native CNS tissues (Benz et al., 2019; Lengfeld et al., 2017; Wang et al., 2019; Wang et al., 2020). TGF-β is an anti-inflammatory cytokine that is upregulated after ischemia induced by SCI and other CNS injuries (Benton, Maddie, Dincman, Hagg, & Whittemore, 2009). SMADs are signal transducers for intracellular pathways mediated by TGF-β (Derynck & Zhang, 2003). TGF-β can cause ECs to be proliferative or quiescent (Kaplan et al., 2020), with proliferative and quiescent ECs associated with a leaky or non-porous BBB and BSCB, respectively. Binding between TGF-β and ALK1 causes phosphorylation of SMAD1 as well as SMAD5 and leads to proliferative ECs. On the contrary, interactions between TGF-β and ALK5 result in SMAD2 as well as SMAD3 phosphorylation and lead to quiescent ECs (ten Dijke & Arthur, 2007; van Meeteren & ten Dijke, 2012). ALK1 and ALK5 signals antagonize each other (Finnson, Parker, ten Dijke, Thorikay, & Philip, 2008). Intrathecal administration of TGF-β increases the protein expression of ZO-1 and occludin after nerve injury (Echeverry, Shi, Rivest, & Zhang, 2011). Overall, aberrant TGF-β signaling leads to vascular pathologies (Walshe et al., 2009).

The Notch-N-cadherin pathway is associated with pathologies due to aberrant TGF-β and WNT signaling (Li et al., 2011). The Notch pathway is important for angiogenic sprouting, vascular remodeling, and arteriovenous differentiation. N-cadherin mediates cell–cell adhesion through the formation of adherens junctions (Miyamoto, Sakane, & Hashimoto, 2015). Influence of the crosstalk between TGF-β and Notch pathways on vascular integrity was demonstrated when ECs deficient in SMAD4 had lower expression of Notch receptors and N-cadherin. Even though TJ proteins were detected in SMAD4 deficient ECs, the ECs adopted a proliferative state, which is a phenotype associated with a porous BBB. This substantiates findings that TJ proteins are associated with, but do not cause, a porous barrier. Furthermore, EC expression of N-cadherin occurs when ECs and pericytes are in contact with each other; with a lack of N-cadherin resulting in defective pericyte adhesion to the vasculature. Overall, these results indicate how a deficiency in SMAD4, TGF-β, and Notch pathways disrupts EC-pericyte interactions and leads to barrier disruptions.

Pericytic response

The role of pericytes in both the healthy and injured NVU was largely unknown and, until recently, was thought to only provide structural support to the vascular integrity. However, this view is being updated, particularly with the advent and wide-spread use of single-cells RNA-sequencing that enables the identification, classification, and expressional analysis of pericytes in the spinal cord (Dias et al., 2018). These studies demonstrate that not only are pericytes involved in secondary SCI pathogenesis, but they are also one of the key regulators of the acute secondary response.

Pericytes are composed of several subtypes, which differ in their functionality, expressional profile, distribution, and lineage. Common molecular and cellular markers to identify spinal pericytes include platelet-derived growth factor receptor B and nerve/glial antigen 2 (NG2) proteoglycan (Picoli et al., 2019). Pericyte Type A is involved in the formation of fibrotic scar, which is found in the lesion core. These cells represent nearly 10% of all pericytes in the injured spinal cord. Fibrotic scar formation is important to regain tissue integrity after SCI. However, deposition of extracellular matrix proteins in the fibrotic scar limits regeneration and a moderate reduction of fibrotic scar-producing pericytes can increase axonal sprouting (Dias et al., 2018). In addition, pericytes regulate hemodynamics and capillary tone in the injured rat spinal cord through vasoconstriction (Picoli et al., 2019). Pericytes are also critical for regulating inflammation through lymphocyte activation, adhesion molecule expression, chemokine secretion, toxic cellular byproduct clearance, and phagocytosis (Picoli et al., 2019).

Astrocytic response

Astrocytes are among the early responders to the injury and play a diverse set of roles from immune modulation and scar formation to neuronal support. In the lesion core, astrocytes, along with microglia, initiate the early local response to the traumatic damage by releasing cytokines and chemokines to further advance the immune response to the injury (O'Shea et al., 2017). In the subacute phase of secondary SCI, hypertrophic astrocytes proliferate and intermingle with inhibitory molecules, which forms a barrier-like structure (Bradbury & Burnside, 2019).

This astrocytic barrier structure consists of several other cell types, including reactive oligodendrocyte precursor cells. These reactive glial cells release inhibitory molecules, such as tenascin and chondroitinase sulfate proteoglycans (CSPGs). CSPGs can be in the form of brevican, phosphacan, neurocan, versican, and neural/glial antigen 2 (NG2) proteoglycans. These inhibitory molecules are secreted from the reactive glia and activate the Rho–ROCK signaling pathway in neurons, which attenuates the repair and regeneration process (Zavvarian, Toossi, Khazaei, Hong, & Fehlings, 2020). Hence, the astrocytic scar was initially known as a barrier to recovery and regeneration. As such, the enzymatic degradation of inhibitory molecules in the astrocytic scar, such as CSPGs, resulted in significantly improved axonal sprouting and regeneration (Burnside et al., 2018). However, recent studies indicate that the astrocytic scar plays a diverse role in the injured spinal cord. Hence, it is difficult to identify it under a binary classification of either "good" or "bad" (Bradbury & Burnside, 2019). Namely, the genetic knockout of STAT3 to inhibit astrocytic scar formation results in reduced axonal growth and regeneration (Anderson et al., 2016).

Level-specific differences in the vascular architecture of the spinal cord

As stated previously, the vascular response to traumatic SCI varies according to the level of injury. The cervical region has a larger amount of gray as well as white matter and requires a greater blood supply compared to the thoracic region. The increased blood supply may correspond with greater vascular disruption in the cervical region following SCI. This is supported by greater cavitation and lesional area in cervical injuries when compared to thoracic injuries. Additionally, injury at the C5 level causes neuronal death and ventral horn atrophy over three segments (C6–C8), while injury at C6 impacts two segments (C8-T1) and C7 caused injury over one segmental level (T2) (Wilcox et al., 2017). These results are consistent with the higher vascular demands from elevated grey matter content in the limb enlargements, which would be disproportionately affected by secondary injury stemming from vascular disruption. Notably, due to a decreased amount of pericytes, the BSCB is inherently more permeable in the grey matter of the cervical region as compared to the thoracic region (Winkler, Sengillo, Bell, Wang, & Zlokovic, 2012).

Muradov et al. have shown that the extent of axonal and oligodendrocyte loss is associated with focal microvascular damage (Muradov, Ewan, & Hagg, 2013). Preclinical investigations indicate that the grey matter is more vulnerable to blood flow disturbances following SCI as compared to white matter, and typically contains relatively smaller vessels (Figley et al., 2014). The white matter adjacent to hemorrhagic grey matter is also especially prone to ischemic damage (Tator & Fehlings, 1991). Overall, the varying arterial and gray matter density throughout the cord has implications for primary and secondary injury after SCI. Regions with reduced vessel density, associated with regions of lesser gray matter, may display reduced focal hemorrhage and its associated secondary effects, including focal ischemic compromise. However, these regions may also be more susceptible to widespread ischemic compromise arising directly from the primary injury. In contrast, areas with increased arterial and microvascular density may display increased focal hemorrhage and secondary injury, while being less susceptible to widespread ischemic compromise from the primary injury.

The increased mobility, smaller vertebrae, and reduced stabilizing musculature of the cervical spine may make it more susceptible to injury (Fehlings, Ulndreaj, & Badner, 2017; Sekhon & Fehlings, 2001). The cervical region is also more susceptible to contusion injury due to the relatively smaller subdural space (Stokes & Jakeman, 2002). A recent study notes that the limited distance between components of the lower thoracic and lumbar spine may cause especially extended tissue damage in this region (Liu et al., 2019). However, Vaccaro et al. report that a spinal canal with a larger transverse diameter was predictive of neurologic deficit after burst fractures, and thus that larger canals are not necessarily protective of the spinal cord (Vaccaro et al., 2001).

The level of injury also influences the systemic inflammatory response and, by extension, the local inflammatory environment. Notably, the expression of both proinflammatory and antiinflammatory compounds may be influenced by the level of injury. Hong, Chang, Liu, Wang, and Fehlings (2019) showed reductions in systemic circulatory levels of proinflammatory molecules with lesions in the cervical region when compared to other regions. SCI above the mid-thoracic level disrupts connections to sympathetic preganglionic neurons below that level, which control secondary lymphoid tissues and systemic immune function. While this dysfunction may lead to spinal cord injury-induced immune depression syndrome (SCI-IDS), it may also result in reduced chronic autoimmunity (Brommer et al., 2016; Fehlings, Ulndreaj, & Badner, 2017). Ibbara et al. demonstrated that the level of injury influences the functions of T cells, which are implicated in the severe infections accompanying SCI-IDS. The type of SCI (complete vs incomplete) may also be an important factor in the immunologic depression observed after SCI (Ibarra, Jiménez, Cortes, & Correa, 2007). As recently demonstrated by Hong et al., contusion injuries may increase splenic function (Hong et al., 2019). This contrasts with previous studies that used transection injuries and resulted in decreased peripheral immune function.

Therapeutic approach

Early intervention is crucial for improved recovery after traumatic SCI. Our multicenter investigation of surgical decompression following traumatic SCI demonstrates that the first 24–36 h after injury constitutes a critical time window to achieve optimal neurological recovery with decompressive surgery. Furthermore, the vascular complications following traumatic SCI require immediate attention to mitigate further damage (Badhiwala et al., 2021). The current guidelines recommend first monitoring the cardiac and hemodynamic parameters following injury, inotropic support to maintain a minimum mean arterial blood pressure of 85 mmHg, detection and treatment of neurogenic shock and cardiac arrhythmias, and immediate treatment of autonomic dysreflexia (Furlan & Fehlings, 2008). The current standard of care also involves the administration of methylprednisolone sodium succinate (MPSS) acutely to mitigate inflammation at the site of injury, improve spinal cord blood flow (SCBF), and enhance functional recovery (Fehlings et al., 2017).

Pharmacological interventions

Several pharmacological agents have been tested in both preclinical and clinical settings to attenuate secondary pathogenesis following SCI-induced vascular pathologies. Notable drug candidates include riluzole, which is a sodium channel blocker currently used in amyotrophic lateral sclerosis (ALS) (Fehlings et al., 2020). Riluzole was developed as an approach to address the consequences of ischemic damage. Preclinical studies examining the impact of riluzole following traumatic SCI reveal its ability to enhance tissue preservation and recovery (Nagoshi, Nakashima, & Fehlings, 2015). The current Phase2/3 clinical trial (ClinicalTrials.gov identifier: NCT01597518) will further examine riluzole as a promising pharmacological candidate for SCI patients.

Acute infusion of mesenchymal stromal cells to attenuate vascular disruption

Mesenchymal stem cells (MSCs) are an intensely studied population of multipotent mesodermal progenitors, which are relatively easy to isolate and display homing properties to the site of injury (Vawda et al., 2019). They differentiate quickly into chondrocytes, adipocytes, myocytes, and osteoblasts (Ahuja et al., 2017; Cofano et al., 2019). MSCs can be derived from various sources including bone-marrow (BM-MSC), umbilical cord (UC-MSC), and adipose-derived MSCs (AD-MSC). MSC treatments display both antiinflammatory properties, notably through macrophage phenotype modulation, and provide trophic support for neuroprotection and regeneration (Nakajima et al., 2012; Spejo et al., 2018). Interestingly, in both human and animal models, they also decrease peripheral inflammatory cell infiltration through protection of the BSCB (Ahuja, Nori, et al., 2017). Various factors have been reported to contribute to this function, notably IL-10, TSG-6, and TIMP-6 (Badner et al., 2016; Li et al., 2018; Watanabe et al., 2013). Recently, MSCs have been reported to reduce intraparenchymal hemorrhage and increase systemic levels of IL-10 in a spleen-dependent manner, which was further supported by altered splenic cytokine levels after MSC infusion (Badner et al., 2018). El-keir et al. report that in a phase I/II controlled single-blind trial (ID: NCT00816803) where chronic cervical and thoracic SCI patients were administered autologous adherent BM-BMCs, smaller lesions, and greater functional improvements were seen in thoracic SCI in comparison to cervical injuries (El-Kheir et al., 2014). There are several clinical trials currently assessing MSCs.

Notably, a phase II/II trial for BM-MSCs in chronic SCI (NCT01676441) and a phase I trial for AD-MSCs (NCT03308565). UC-MSCs are also being tested for use in sub-acute and chronic injury in phase I/II trials (NCT03521323, NCT03505034).

Acute IgG infusion for immunomodulation and rescue of BSCB disruption

Intravenous immunoglobulin G (IVIG) consists of pooled serum immunoglobulin G (IgG) from healthy donors. Despite a relatively uncertain mechanism of action, it is currently approved by the FDA for the treatment of autoimmune and immunodeficiency disorders, namely primary humoral deficiency idiopathic, thrombocytopenic purpura, and chronic inflammatory demyelinating polyneuropathy (Chio et al., 2019; Fehlings, Ulndreaj, & Badner, 2017). Brennan et al. report that IVIG decreased lesion enlargement and axonal degeneration after contusion SCI in a mouse model. This study also suggests that IVIG may achieve these effects through attenuation of the complement response and a decrease in the phagocytic activity of macrophages (Brennan et al., 2016). Nguyen et al. reported that in a rat model of cervical compression injury, IVIG administration reduced pro-inflammatory signaling (IL-1b and IL-6). This may have had neuroprotective effects via the BSCB, as demonstrated by the reduction in neutrophil infiltration and reductions in the level of BSCB-damaging metalloproteinase-9 (MMP-9) (Nguyen et al., 2012). This is further supported by reports of neurovascular protection through increases in tight junction protein expression and significant decreases in MMP-9. The improved BSCB integrity was associated with increased serum levels of inflammatory cytokines (IL-8, MIP-1α, CCL-2/MCP-1, and IL-5). IgG is also colocalized with vascular cell adhesion molecule 1 (VCAM-1) without reducing its expression. VCAM-1 is used by neutrophils in extravasation (Chio et al., 2019). Notably, IgG co-localized with astrocytes and pericytes, which suggests a role in modulating immune cell infiltration through the BSCB. Administration of 2 g/kg resulted in increased tissue preservation, blood flow, and behavioral recovery; effects comparable with MPSS treatment.

Conclusions

This chapter described a brief overview of the cellular and extracellular elements of the neurovascular unit, discussed their physiological functions, and reviewed their functional implications in the context of SCI. Traumatic SCI presents a complicated and heterogeneous condition, and this chapter emphasized the segmental anatomy of the neurovasculature of the spinal cord and discussed their implications on the region-specific pathophysiology of traumatic SCI. These are important considerations for the management of neurovascular pathology following traumatic SCI. Recent discoveries in vascular-related therapeutic strategies have greatly improved our ability to enhance protection against damage to the spinal neurovasculature and BSCB, and ultimately improve functional recovery.

Application to other areas of neuroscience

The secondary response following traumatic SCI is a complex process and involves a variety of cell types in the spinal cord. While the discussions provided in this chapter were primarily presented in the context of traumatic SCI, similar mechanisms are at play across the CNS. The endothelial, astrocytic, and pericytic responses demonstrated are active in other CNS-related neurovascular injuries. Therefore, the discussed vascular-mediated treatment strategies may be explored and considered for other neurological injuries and diseases, such as traumatic brain injuries, amyotrophic lateral sclerosis (ALS), multiple sclerosis (MS), and stroke. Many of the therapeutics discussed in this chapter are clinically approved for conditions other than SCI, which ease the clinical translation and repurposing for multiple conditions. Furthermore, the elucidation of vascular-related SCI pathogenesis will improve our understanding of the cellular structure and function of the spinal cord. This can have a significant impact for improving our understanding of non-traumatic SCI disorders, as these present the most common form of SCI in the clinic.

Mini-dictionary of terms

Autonomic dysreflexia: Sudden episodic elevation in blood pressure accompanied by bradycardia due to loss of supraspinal input into sympathetic preganglionic neurons.

Neurovascular unit: A model that characterizes the interaction between endothelial, astrocytic, pericytic, and other vascular cells to maintain the necessary oxygen and nutrient supply to the nervous system, while protecting it from foreign and toxic molecules.

Inotropic support: Altering the force of the heart's contraction to stabilize circulation.

Fibrotic scar: The deposition of stromal-derived fibroblasts and extracellular matrix molecules at the SCI epicenter.
Astrocytic scar: The accumulation of hypertrophic astrocytes intermingled with inhibitory molecules which surround the SCI lesion core.

Key facts

Key facts of the American Spinal Injury Association (ASIA) Impairment Scale

- ASIA impairment scale categorizes the SCI severity in the clinical setting ranging from Grade A to Grade E.
- Grade A indicates the complete loss of autonomic and sensorimotor functions distal to the site of injury.
- Grade B: Loss of motor function but not sensory function distal to the injury site. No motor function is present more than three levels below the neurological level, on either side of the body.
- Grade C: Preserved motor function distal to the neurological level of injury with more than half of the key muscles having a grade of <3 on the ASIA motor score.
- Grade D: Preserved motor function distal to the neurological level of injury with more than half of the key muscles having a grade of ≥ 3 on the ASIA motor score.
- Grade E: Neurologically intact sensorimotor function in all segments in patients with previous deficits due a suspected SCI.

Key facts of the spinal neurovascular unit (NVU)

- The neurovascular unit (NVU) is comprised of endothelial cells, pericytes, and astrocytes, which work alongside neurons to regulate spinal blood flow and vascular permeability.
- Occludin, zona occludens, and claudin are important tight junction (TJ) proteins forming the blood–spinal cord barrier (BSCB) and keeping NVU cells together.
- Molecular signaling cascades involving WNT, TGF-β, SMAD4, and Notch molecules regulate BSCB function and integrity.
- The full extent of pericytic and astrocytic responses to SCI continues to be under investigation. In addition to structural support, these cells play critical roles in scar formation.

Summary points

1. Aside from primary mechanical damage to neuronal, glial, and vascular cells, traumatic spinal cord injury impairs the local neuronal, glial, and vascular cells, which results in a cascade of secondary pathologies.
2. Vascular damage is a key mediator of this secondary response and exacerbates injury-induced tissue and functional loss, as it disrupts the homeostatic exchange of oxygen and essential nutrients.
3. Current clinically approved neuroprotective treatments that involve acute administration of MPSS have limited effectiveness to treat SCI patients.
4. The endothelial, pericytic, and astrocytic responses to injury continues to be under investigation, as they present promising therapeutic targets for SCI patients.
5. Recent advances in mesenchymal stromal cell administration or acute infusion of IgG demonstrate great efficacy to mitigate SCI-induced vascular pathologies and improve recovery.

References

Ahuja, C. S., Nori, S., Tetreault, L., Wilson, J., Kwon, B., Harrop, J., et al. (2017). Traumatic spinal cord injury—Repair and regeneration. *Clinical Neurosurgery*, *80*(3), S22–S90. https://doi.org/10.1093/neuros/nyw080.

Ahuja, C. S., Wilson, J. R., Nori, S., Kotter, M. R. N., Druschel, C., Curt, A., et al. (2017). Traumatic spinal cord injury. *Nature Reviews. Disease Primers*, *3* (1), 17018. https://doi.org/10.1038/nrdp.2017.18.

Alshareef, M., Krishna, V., Ferdous, J., Alshareef, A., Kindy, M., Kolachalama, V. B., et al. (2014). Effect of spinal cord compression on local vascular blood flow and perfusion capacity. *PLoS One*, *9*(9). https://doi.org/10.1371/journal.pone.0108820, e108820.

Alvarez, J. I., Dodelet-Devillers, A., Kebir, H., Ifergan, I., Fabre, P. J., Terouz, S., et al. (2011). The hedgehog pathway promotes blood-brain barrier integrity and CNS immune quiescence. *Science*, *334*(6063), 1727–1731. https://doi.org/10.1126/science.1206936.

Anderson, M. A., Burda, J. E., Ren, Y., Ao, Y., O'Shea, T. M., Kawaguchi, R., et al. (2016). Astrocyte scar formation aids central nervous system axon regeneration. *Nature*, *532*(7598), 195–200. https://doi.org/10.1038/nature17623.

Andreone, B. J., Chow, B. W., Tata, A., Lacoste, B., Ben-Zvi, A., Bullock, K., et al. (2017). Blood-brain barrier permeability is regulated by lipid transport-dependent suppression of caveolae-mediated transcytosis. *Neuron*, *94*(3), 581–594.e5. https://doi.org/10.1016/j.neuron.2017.03.043.

Attwell, D., Buchan, A. M., Charpak, S., Lauritzen, M., MacVicar, B. A., & Newman, E. A. (2010). Glial and neuronal control of brain blood flow. *Nature*, *468*(7321), 232–243. https://doi.org/10.1038/nature09613.

Badhiwala, J. H., Wilson, J. R., Witiw, C. D., Harrop, J. S., Vaccaro, A. R., Aarabi, B., et al. (2021). The influence of timing of surgical decompression for acute spinal cord injury: A pooled analysis of individual patient data. *The Lancet Neurology*, *20*(2), 117–126. https://doi.org/10.1016/S1474-4422(20)30406-3.

Badner, A., Hacker, J., Hong, J., Mikhail, M., Vawda, R., & Fehlings, M. G. (2018). Splenic involvement in umbilical cord matrix-derived mesenchymal stromal cell-mediated effects following traumatic spinal cord injury. *Journal of Neuroinflammation*, *15*(1). https://doi.org/10.1186/s12974-018-1243-0.

Badner, A., Vawda, R., Laliberte, A., Hong, J., Mikhail, M., Jose, A., et al. (2016). Early intravenous delivery of human brain stromal cells modulates systemic inflammation and leads to vasoprotection in traumatic spinal cord injury. *Stem Cells Translational Medicine*, *5*(8), 991–1003. https://doi.org/10.5966/sctm.2015-0295.

Benton, R. L., Maddie, M. A., Dincman, T. A., Hagg, T., & Whittemore, S. R. (2009). Transcriptional activation of endothelial cells by TGFβ coincides with acute microvascular plasticity following focal spinal cord ischaemia/reperfusion injury. *ASN Neuro*, *1*(3). https://doi.org/10.1042/AN20090008, AN20090008.

Benz, F., Wichitnaowarat, V., Lehmann, M., Germano, R. F., Mihova, D., Macas, J., et al. (2019). Low wnt/β-catenin signaling determines leaky vessels in the subfornical organ and affects water homeostasis in mice. *eLife*, *8*. https://doi.org/10.7554/eLife.43818, e43818.

Ben-Zvi, A., Lacoste, B., Kur, E., Andreone, B. J., Mayshar, Y., Yan, H., et al. (2014). Mfsd2a is critical for the formation and function of the blood–brain barrier. *Nature*, *509*(7501), 507–511. https://doi.org/10.1038/nature13324.

Bosmia, A. N., Hogan, E., Loukas, M., Tubbs, R. S., & Cohen-Gadol, A. A. (2015). Blood supply to the human spinal cord: Part I. Anatomy and hemodynamics. *Clinical Anatomy*, *28*(1), 52–64. https://doi.org/10.1002/ca.22281.

Bradbury, E. J., & Burnside, E. R. (2019). Moving beyond the glial scar for spinal cord repair. *Nature Communications*, *10*(1), 3879. https://doi.org/10.1038/s41467-019-11707-7.

Brennan, F. H., Kurniawan, N. D., Vukovic, J., Bartlett, P. F., Käsermann, F., Arumugam, T. V., et al. (2016). IVIg attenuates complement and improves spinal cord injury outcomes in mice. *Annals of Clinical Translational Neurology*, *3*(7), 495–511. https://doi.org/10.1002/acn3.318.

Brommer, B., Engel, O., Kopp, M. A., Watzlawick, R., Müller, S., Prüss, H., et al. (2016). Spinal cord injury-induced immune deficiency syndrome enhances infection susceptibility dependent on lesion level. *Brain*, *139*(3), 692–707. https://doi.org/10.1093/brain/awv375.

Burnside, E. R., De Winter, F., Didangelos, A., James, N. D., Andreica, E.-C., Layard-Horsfall, H., et al. (2018). Immune-evasive gene switch enables regulated delivery of chondroitinase after spinal cord injury. *Brain*, *141*(8), 2362–2381. https://doi.org/10.1093/brain/awy158.

Castro Dias, M., Coisne, C., Baden, P., Enzmann, G., Garrett, L., Becker, L., et al. (2019). Claudin-12 is not required for blood–brain barrier tight junction function. *Fluids and Barriers of the CNS*, *16*(1), 30. https://doi.org/10.1186/s12987-019-0150-9.

Castro Dias, M., Coisne, C., Lazarevic, I., Baden, P., Hata, M., Iwamoto, N., et al. (2019). Claudin-3-deficient C57BL/6J mice display intact brain barriers. *Scientific Reports*, *9*(1), 203. https://doi.org/10.1038/s41598-018-36731-3.

Cauli, B. (2004). Cortical GABA interneurons in neurovascular coupling: Relays for subcortical vasoactive pathways. *Journal of Neuroscience*, *24*(41), 8940–8949. https://doi.org/10.1523/JNEUROSCI.3065-04.2004.

Chio, J. C. T., Wang, J., Badner, A., Hong, J., Surendran, V., & Fehlings, M. G. (2019). The effects of human immunoglobulin G on enhancing tissue protection and neurobehavioral recovery after traumatic cervical spinal cord injury are mediated through the neurovascular unit. *Journal of Neuroinflammation*, *16*(1), 141. https://doi.org/10.1186/s12974-019-1518-0.

Chow, B. W., & Gu, C. (2017). Gradual suppression of transcytosis governs functional blood-retinal barrier formation. *Neuron*, *93*(6), 1325–1333.e3. https://doi.org/10.1016/j.neuron.2017.02.043.

Cofano, F., Boido, M., Monticelli, M., Zenga, F., Ducati, A., Vercelli, A., et al. (2019). Mesenchymal stem cells for spinal cord injury: Current options, limitations, and future of cell therapy. *International Journal of Molecular Sciences*, *20*(11). https://doi.org/10.3390/ijms20112698.

Courtine, G., & Sofroniew, M. V. (2019). Spinal cord repair: Advances in biology and technology. *Nature Medicine*, *25*(6), 898–908. https://doi.org/10.1038/s41591-019-0475-6.

Crosby, H., & Lauer. (1962). *Correlative Anatomy of the Nervous System*.

Daneman, R., Agalliu, D., Zhou, L., Kuhnert, F., Kuo, C. J., & Barres, B. A. (2009). Wnt/-catenin signaling is required for CNS, but not non-CNS, angiogenesis. *Proceedings of the National Academy of Sciences of the United States of America*, *106*(2), 641–646. https://doi.org/10.1073/pnas.0805165106.

Derynck, R., & Zhang, Y. E. (2003). Smad-dependent and Smad-independent pathways in TGF-b family signalling. *Nature*, *425*, 8.

Dias, D. O., Kim, H., Holl, D., Werne Solnestam, B., Lundeberg, J., Carlén, M., et al. (2018). Reducing pericyte-derived scarring promotes recovery after spinal cord injury. *Cell*, *173*(1), 153–165.e22. https://doi.org/10.1016/j.cell.2018.02.004.

Dommisse, G. F. (1974). The blood supply of the spinal cord. A critical vascular zone in spinal surgery. *Journal of Bone and Joint Surgery—Series B*, *56*(2), 225–235.

Echeverry, S., Shi, X. Q., Rivest, S., & Zhang, J. (2011). Peripheral nerve injury alters blood-spinal cord barrier functional and molecular integrity through a selective inflammatory pathway. *Journal of Neuroscience*, *31*(30), 10819–10828. https://doi.org/10.1523/JNEUROSCI.1642-11.2011.

El-Kheir, W. A., Gabr, H., Awad, M. R., Ghannam, O., Barakat, Y., Farghali, H. A. M. A., et al. (2014). Autologous bone marrow-derived cell therapy combined with physical therapy induces functional improvement in chronic spinal cord injury patients. *Cell Transplantation*, *23*(6), 729–745. https://doi.org/10.3727/096368913X664540.

Fehlings, M. G., Badhiwala, J. H., Ahn, H., Farhadi, H. F., Shaffrey, C. I., Nassr, A., et al. (2020). Safety and efficacy of riluzole in patients undergoing decompressive surgery for degenerative cervical myelopathy (CSM-protect): A multicentre, double-blind, placebo-controlled, randomised, phase 3 trial. *The Lancet Neurology*. https://doi.org/10.1016/S1474-4422(20)30407-5. S1474442220304075.

Fehlings, M. G., Ulndreaj, A., & Badner, A. (2017). Promising neuroprotective strategies for traumatic spinal cord injury with a focus on the differential effects among anatomical levels of injury. In *Vol. 6. F1000Research* Faculty of 1000 Ltd. https://doi.org/10.12688/f1000research.11633.1.

Fehlings, M. G., Wilson, J. R., Harrop, J. S., Kwon, B. K., Tetreault, L. A., Arnold, P. M., et al. (2017). Efficacy and safety of methylprednisolone sodium succinate in acute spinal cord injury: A systematic review. *Global Spine Journal*, *7*(3_suppl), 116S–137S. https://doi.org/10.1177/2192568217706366.

Figley, S. A., Khosravi, R., Legasto, J. M., Tseng, Y.-F. F., & Fehlings, M. G. (2014). Characterization of vascular disruption and blood-spinal cord barrier permeability following traumatic spinal cord injury. *Journal of Neurotrauma*, *31*(6), 541–552. https://doi.org/10.1089/neu.2013.3034.

Finnson, K. W., Parker, W. L., ten Dijke, P., Thorikay, M., & Philip, A. (2008). ALK1 opposes ALK5/Smad3 signaling and expression of extracellular matrix components in human chondrocytes. *Journal of Bone and Mineral Research*, *23*(6), 896–906. https://doi.org/10.1359/jbmr.080209.

Furlan, J. C., & Fehlings, M. G. (2008). Cardiovascular complications after acute spinal cord injury: Pathophysiology, diagnosis, and management. *Neurosurgical Focus*, *25*(5). https://doi.org/10.3171/FOC.2008.25.11.E13, E13.

Gordon, G. R. J., Mulligan, S. J., & Macvicar, B. A. (2007). Astrocyte control of the cerebrovasculature. *Glia*, *55*, 1214–1221.

Hong, J., Chang, A., Liu, Y., Wang, J., & Fehlings, M. G. (2019). Incomplete spinal cord injury reverses the level-dependence of spinal cord injury immune deficiency syndrome. *International Journal of Molecular Sciences*, *20*(15), 3762. https://doi.org/10.3390/ijms20153762.

Ibarra, A., Jiménez, A., Cortes, C., & Correa, D. (2007). Influence of the intensity, level and phase of spinal cord injury on the proliferation of T cells and T-cell-dependent antibody reactions in rats. *Spinal Cord*, *45*(5), 380–386. https://doi.org/10.1038/sj.sc.3101972.

Itoh, Y., & Suzuki, N. (2012). Control of brain capillary blood flow. *Journal of Cerebral Blood Flow & Metabolism*, *32*(7), 1167–1176. https://doi.org/10.1038/jcbfm.2012.5.

Kaplan, L., Chow, B. W., & Gu, C. (2020). Neuronal regulation of the blood–brain barrier and neurovascular coupling. *Nature Reviews Neuroscience*, *21*(8), 416–432. https://doi.org/10.1038/s41583-020-0322-2.

Katsuno, T., Umeda, K., Matsui, T., Hata, M., Tamura, A., Itoh, M., et al. (2008). Deficiency of zonula occludens-1 causes embryonic lethal phenotype associated with defected yolk sac angiogenesis and apoptosis of embryonic cells. *Molecular Biology of the Cell*, *19*(6), 2465–2475. https://doi.org/10.1091/mbc.e07-12-1215.

Lengfeld, J. E., Lutz, S. E., Smith, J. R., Diaconu, C., Scott, C., Kofman, S. B., et al. (2017). Endothelial Wnt/β-catenin signaling reduces immune cell infiltration in multiple sclerosis. *Proceedings of the National Academy of Sciences of the United States of America*, *114*(7), E1168–E1177. https://doi.org/10.1073/pnas.1609905114.

Li, F., Lan, Y., Wang, Y., Wang, J., Yang, G., Meng, F., et al. (2011). Endothelial Smad4 maintains cerebrovascular integrity by activating N-cadherin through cooperation with notch. *Developmental Cell*, *20*(3), 291–302. https://doi.org/10.1016/j.devcel.2011.01.011.

Li, R., Liu, W., Yin, J., Chen, Y., Guo, S., Fan, H., et al. (2018). TSG-6 attenuates inflammation-induced brain injury via modulation of microglial polarization in SAH rats through the SOCS3/STAT3 pathway. *Journal of Neuroinflammation*, *15*(1). https://doi.org/10.1186/s12974-018-1279-1.

Liu, S.-J., Wang, Q., Tang, H.-H., Bai, J.-Z., Wang, F.-Y., Lv, Z., et al. (2019). Heterogeneity among traumatic spinal cord injuries at the thoracolumbar junction: Helping select patients for clinical trials. *Spinal Cord*, *57*(11), 972–978. https://doi.org/10.1038/s41393-019-0317-x.

Mautes, A. E., Weinzierl, M. R., Donovan, F., & Noble, L. J. (2000). Vascular events after spinal cord injury: Contribution to secondary pathogenesis. *Physical Therapy*, *80*(7), 673–687. https://doi.org/10.1093/ptj/80.7.673.

McConnell, H. L., Kersch, C. N., Woltjer, R. L., & Neuwelt, E. A. (2017). The Translational significance of the neurovascular unit. *Journal of Biological Chemistry*, *292*(3), 762–770. https://doi.org/10.1074/jbc.R116.760215.

Miyamoto, Y., Sakane, F., & Hashimoto, K. (2015). N-cadherin-based adherens junction regulates the maintenance, proliferation, and differentiation of neural progenitor cells during development. *Cell Adhesion & Migration*, *9*(3), 183–192. https://doi.org/10.1080/19336918.2015.1005466.

Mizee, M. R., Wooldrik, D., Lakeman, K. A. M., van het Hof, B., Drexhage, J. A. R., Geerts, D., et al. (2013). Retinoic acid induces blood-brain barrier development. *Journal of Neuroscience*, *33*(4), 1660–1671. https://doi.org/10.1523/JNEUROSCI.1338-12.2013.

Muradov, J. M., Ewan, E. E., & Hagg, T. (2013). Dorsal column sensory axons degenerate due to impaired microvascular perfusion after spinal cord injury in rats. *Experimental Neurology*, *249*, 59–73. https://doi.org/10.1016/j.expneurol.2013.08.009.

Nagoshi, N., Nakashima, H., & Fehlings, M. (2015). Riluzole as a neuroprotective drug for spinal cord injury: From bench to bedside. *Molecules*, *20*(5), 7775–7789. https://doi.org/10.3390/molecules20057775.

Nakajima, H., Uchida, K., Guerrero, A. R., Watanabe, S., Sugita, D., Takeura, N., et al. (2012). Transplantation of mesenchymal stem cells promotes an alternative pathway of macrophage activation and functional recovery after spinal cord injury. *Journal of Neurotrauma*, *29*, 1614–1625. https://doi.org/10.1089/neu.2011.2109.

Nguyen, D. H., Cho, N., Satkunendrarajah, K., Austin, J. W., Wang, J., & Fehlings, M. G. (2012). Immunoglobulin G (IgG) attenuates neuroinflammation and improves neurobehavioral recovery after cervical spinal cord injury. *Journal of Neuroinflammation*, *9*. https://doi.org/10.1186/1742-2094-9-224.

Niu, J., Tsai, H.-H., Hoi, K. K., Huang, N., Yu, G., Kim, K., et al. (2019). Aberrant oligodendroglial–vascular interactions disrupt the blood–brain barrier, triggering CNS inflammation. *Nature Neuroscience*, *22*(5), 709–718. https://doi.org/10.1038/s41593-019-0369-4.

O'Shea, T. M., Burda, J. E., & Sofroniew, M. V. (2017). Cell biology of spinal cord injury and repair. *Journal of Clinical Investigation*, *127*(9), 3259–3270. https://doi.org/10.1172/JCI90608.

Park, D. Y., Lee, J., Kim, J., Kim, K., Hong, S., Han, S., et al. (2017). Plastic roles of pericytes in the blood–retinal barrier. *Nature Communications*, *8*(1), 15296. https://doi.org/10.1038/ncomms15296.

Picoli, C. C., Coimbra-Campos, L. M. C., Guerra, D. A. P., Silva, W. N., Prazeres, P. H. D. M., Costa, A. C., et al. (2019). Pericytes act as key players in spinal cord injury. *The American Journal of Pathology*, *189*(7), 1327–1337. https://doi.org/10.1016/j.ajpath.2019.03.008.

Pollock, L. M., Xie, J., Bell, B. A., & Anand-Apte, B. (2018). Retinoic acid signaling is essential for maintenance of the blood-retinal barrier. *The FASEB Journal*, *32*(10), 5674–5684. https://doi.org/10.1096/fj.201701469R.

Raineteau, O., & Schwab, M. E. (2001). Plasticity of motor systems after incomplete spinal cord injury. *Nature Reviews Neuroscience*, *2*(4), 263–273. https://doi.org/10.1038/35067570.

Saitou, M., Fujimoto, K., Doi, Y., Itoh, M., Fujimoto, T., Furuse, M., et al. (1998). Occludin-deficient embryonic stem cells can differentiate into polarized epithelial cells bearing tight junctions. *Journal of Cell Biology*, *141*(2), 397–408. https://doi.org/10.1083/jcb.141.2.397.

Saitou, M., Furuse, M., Sasaki, H., Schulzke, J.-D., Fromm, M., Takano, H., et al. (2000). Complex phenotype of mice lacking occludin, a component of tight junction strands. *Molecular Biology of the Cell*, *11*(12), 4131–4142. https://doi.org/10.1091/mbc.11.12.4131.

Schwab, J. M., Zhang, Y., Kopp, M. A., Brommer, B., & Popovich, P. G. (2014). The paradox of chronic neuroinflammation, systemic immune suppression, autoimmunity after traumatic chronic spinal cord injury. *Experimental Neurology*, *258*, 121–129. https://doi.org/10.1016/j.expneurol.2014.04.023.

Sekhon, L. H. S., & Fehlings, M. G. (2001). Epidemiology, demographics, and pathophysiology of acute spinal cord injury. *Spine*, *26*(24 Suppl), S2–12. https://doi.org/10.1097/00007632-200112151-00002.

Sharif, H., & Hou, S. (2017). Autonomic dysreflexia: A cardiovascular disorder following spinal cord injury. *Neural Regeneration Research*, *12*(9), 1390. https://doi.org/10.4103/1673-5374.215241.

Siddiqui, A. M., Khazaei, M., & Fehlings, M. G. (2015). Translating mechanisms of neuroprotection, regeneration, and repair to treatment of spinal cord injury. In *Vol. 218. Progress in brain research* (pp. 15–54). Elsevier. https://doi.org/10.1016/bs.pbr.2014.12.007.

Singh, A., Tetreault, L., Kalsi-Ryan, S., Nouri, A., & Michael, F. (2014). Global prevalence and incidence of traumatic spinal cord injury. *Clinical Epidemiology*, *309*. https://doi.org/10.2147/CLEP.S68889.

Spejo, A. B., Chiarotto, G. B., Ferreira, A. D. F., Gomes, D. A., Ferreira, R. S., Barraviera, B., et al. (2018). Neuroprotection and immunomodulation following intraspinal axotomy of motoneurons by treatment with adult mesenchymal stem cells. *Journal of Neuroinflammation*, *15*(1), 230. https://doi.org/10.1186/s12974-018-1268-4.

Stamatovic, S. M., Johnson, A. M., Keep, R. F., & Andjelkovic, A. V. (2016). Junctional proteins of the blood-brain barrier: New insights into function and dysfunction. *Tissue Barriers*, *4*(1). https://doi.org/10.1080/21688370.2016.1154641, e1154641.

Stokes, B., & Jakeman, L. (2002). Experimental modelling of human spinal cord injury: A model that crosses the species barrier and mimics the spectrum of human cytopathology. *Spinal Cord*, *40*(3), 101–109. https://doi.org/10.1038/sj.sc.3101254.

Tator, C. H., & Fehlings, M. G. (1991). Review of the secondary injury theory of acute spinal cord trauma with emphasis on vascular mechanisms. *Journal of Neurosurgery*, *75*(1), 15–26. https://doi.org/10.3171/jns.1991.75.1.0015.

ten Dijke, P., & Arthur, H. M. (2007). Extracellular control of TGFβ signalling in vascular development and disease. *Nature Reviews Molecular Cell Biology*, *8*(11), 857–869. https://doi.org/10.1038/nrm2262.

Thomsen, L. B., Burkhart, A., & Moos, T. (2015). A triple culture model of the blood-brain barrier using porcine brain endothelial cells, astrocytes and pericytes. *PLoS One*, *10*(8). https://doi.org/10.1371/journal.pone.0134765, e0134765.

Turnbull, I. M. (1971). Microvasculature of the human spinal cord. *Journal of Neurosurgery*, *35*(2), 141–147. https://doi.org/10.3171/jns.1971.35.2.0141.

Umeda, K., Ikenouchi, J., Katahira-Tayama, S., Furuse, K., Sasaki, H., Nakayama, M., et al. (2006). ZO-1 and ZO-2 independently determine where claudins are polymerized in tight-junction strand formation. *Cell*, *126*(4), 741–754. https://doi.org/10.1016/j.cell.2006.06.043.

Underly, R. G., Levy, M., Hartmann, D. A., Grant, R. I., Watson, A. N., & Shih, A. Y. (2017). Pericytes as inducers of rapid, matrix metalloproteinase-9-dependent capillary damage during ischemia. *The Journal of Neuroscience*, *37*(1), 129–140. https://doi.org/10.1523/JNEUROSCI.2891-16.2016.

Vaccaro, A. R., Nachwalter, R. S., Klein, G. R., Sewards, J. M., Albert, T. J., & Garfin, S. R. (2001). The significance of thoracolumbar spinal canal size in spinal cord injury patients. *Spine*, *26*(4), 371–376. https://doi.org/10.1097/00007632-200102150-00013.

van Meeteren, L. A., & ten Dijke, P. (2012). Regulation of endothelial cell plasticity by TGF-β. *Cell and Tissue Research*, *347*(1), 177–186. https://doi.org/10.1007/s00441-011-1222-6.

Vawda, R., Badner, A., Hong, J., Mikhail, M., Lakhani, A., Dragas, R., et al. (2019). Early intravenous infusion of mesenchymal stromal cells exerts a tissue source age-dependent beneficial effect on neurovascular integrity and neurobehavioral recovery after traumatic cervical spinal cord injury. *Stem Cells Translational Medicine*, *8*(7), 639–649. https://doi.org/10.1002/sctm.18-0192.

Vladimir, S., Aaron, F., Dan, M., Jonathan, O., Rotem, S.-A., & Adiel, C. (2019). Blood-brain barrier dysfunction in aging induces hyperactivation of TGF signaling and chronic yet reversible neural dysfunction. *Science Translational Medicine*, *11*, 15.

Walshe, T. E., Saint-Geniez, M., Maharaj, A. S. R., Sekiyama, E., Maldonado, A. E., & D'Amore, P. A. (2009). TGF-β is required for vascular barrier function, endothelial survival and homeostasis of the adult microvasculature. *PLoS One*, *4*(4). https://doi.org/10.1371/journal.pone.0005149, e5149.

Wang, Z., Liu, C.-H., Huang, S., Fu, Z., Tomita, Y., Britton, W. R., et al. (2020). Wnt signaling activates MFSD2A to suppress vascular endothelial transcytosis and maintain blood-retinal barrier. *Science Advances*, *6*(35). https://doi.org/10.1126/sciadv.aba7457, eaba7457.

Wang, Y., Sabbagh, M. F., Gu, X., Rattner, A., Williams, J., & Nathans, J. (2019). Beta-catenin signaling regulates barrier-specific gene expression in circumventricular organ and ocular vasculatures. *eLife*, *8*. https://doi.org/10.7554/eLife.43257, e43257.

Watanabe, J., Shetty, A. K., Hattiangady, B., Kim, D. K., Foraker, J. E., Nishida, H., et al. (2013). Administration of TSG-6 improves memory after traumatic brain injury in mice. *Neurobiology of Disease*, *59*, 86–99. https://doi.org/10.1016/j.nbd.2013.06.017.

Wilcox, J. T., Satkunendrarajah, K., Nasirzadeh, Y., Laliberte, A. M., Lip, A., Cadotte, D. W., et al. (2017). Generating level-dependent models of cervical and thoracic spinal cord injury: Exploring the interplay of neuroanatomy, physiology, and function. *Neurobiology of Disease*, *105*, 194–212. https://doi.org/10.1016/j.nbd.2017.05.009.

Winkler, E. A., Sengillo, J. D., Bell, R. D., Wang, J., & Zlokovic, B. V. (2012). Blood-spinal cord barrier pericyte reductions contribute to increased capillary permeability. *Journal of Cerebral Blood Flow and Metabolism*, *32*(10), 1841–1852. https://doi.org/10.1038/jcbfm.2012.113.

Winkler, E. A., Sengillo, J. D., Sullivan, J. S., Henkel, J. S., Appel, S. H., & Zlokovic, B. V. (2013). Blood–spinal cord barrier breakdown and pericyte reductions in amyotrophic lateral sclerosis. *Acta Neuropathologica*, *125*(1), 111–120. https://doi.org/10.1007/s00401-012-1039-8.

Xu, J., Long, H., Chen, W., Cheng, X., Yu, H., Huang, Y., et al. (2018). Ultrastructural features of neurovascular units in a rat model of chronic compressive spinal cord injury. *Frontiers in Neuroanatomy*, *11*, 136. https://doi.org/10.3389/fnana.2017.00136.

Zavvarian, M.-M., Hong, J., & Fehlings, M. G. (2020). The functional role of spinal interneurons following traumatic spinal cord injury. *Frontiers in Cellular Neuroscience*, *14*, 127. https://doi.org/10.3389/fncel.2020.00127.

Zavvarian, M.-M., Toossi, A., Khazaei, M., Hong, J., & Fehlings, M. (2020). Novel innovations in cell and gene therapies for spinal cord injury. *F1000Research*, *9*, 279. https://doi.org/10.12688/f1000research.21989.1.

Chapter 11

Protein degradome in spinal cord injury

Shadi Bsat[a], Hani Chanbour[b], Ali Amine[a], Charbel Moussalem[a], Mohamad Nabih El Houshiemy[a], Sarah Kawtharani[a], Adham Halaoui[a], Firas Kobeissy[c], Safwan Alomari[a], and Ibrahim Omeis[a,d]

[a]*Department of Neurosurgery, American University of Beirut Medical Center, Beirut, Lebanon,* [b]*Faculty of Medicine, Lebanese University, Beirut, Lebanon,* [c]*Department of Biochemistry and Molecular Genetics, Faculty of Medicine, American University of Beirut, Beirut, Lebanon,* [d]*Department of Neurosurgery, Baylor College of Medicine, Houston, TX, United States*

Abbreviations

BDPs breakdown products
CSPGs chondroitin sulfate proteoglycans
GFAP glial fibrillary acidic protein
HA hyaluronan
I-NaP increased persistent sodium current
MMPs matrix metalloproteinases
Nav voltage-gated sodium
NF200 neurofilament
SBDPs αII-spectrin breakdown product
SCI spinal cord injury
UPS ubiquitin-proteasome system
USP4 ubiquitin-specific protease 4

Introduction

Degradomics is a subdiscipline of proteomics, it refers to the study of proteases, the enzymes responsible for the hydrolysis of peptide bonds, and the study of breakdown products (BDPs) ensuing from the targeted substrates (Overall & Dean, 2006) (Fig. 1). Among the techniques that were used to decipher the degradomic profiling in neurotrauma, mass spectrometry has proved to be the most efficacious in conveying the precise measurements of BDPs, proteases, and substrates (Lopez-Otin & Overall, 2002). Several disease states produce an array of BDPs resulting from an uncontrolled protease-substrate repertoire, that could be further used in the diagnosis, assessment, treatment, and follow-up (Savickas & Keller, 2017).

Spinal Cord Injury (SCI) can be divided into a primary injury, where the shear forces of the direct trauma disrupt the axonal relays, causing disability (McKinley, Santos, Meade, & Brooke, 2007), and a secondary injury, where neurotoxicity plays a key role (Tator & Fehlings, 1991). The latter can exacerbate the former injury and slows the recovery, which will lead to a downward spiral of cellular and molecular activation (Tator & Fehlings, 1991), alterations of proteolytic enzymes, upregulation of cytotoxic neurotransmitters (Veeravalli, Dasari, & Rao, 2012), and impairment of blood–CNS barrier (Nasser et al., 2016).

The mechanism of spinal cord injury is complex and interrelated. This chapter sheds light on the degradomics implicated in SCI, it uncovers hidden pathophysiological processes after the identification and characterization of BDPs, resulting from the overactivation of proteases. It describes signature substrates and their BDPs involved in SCI, including ion channels, cytoskeletal proteins, cellular junction proteins, and others. Different enzymes responsible for the proteolysis will be discussed, such as caspases, calpains, cathepsins, and Matrix Metalloproteinases (MMPs). Taken together, it can improve our understanding of the secondary injury, and guide new therapeutic approaches.

Protease-substrate repertoires

Proteolysis of specific substrates during SCI was extensively studied in the past few years (Fig. 2). Circulating cytokines after SCI activate the proteolytic cascade in the spinal cord (Supinski, Wang, & Callahan, 2009), this was evident when

Cellular, Molecular, Physiological, and Behavioral Aspects of Spinal Cord Injury. https://doi.org/10.1016/B978-0-12-822427-4.00011-3

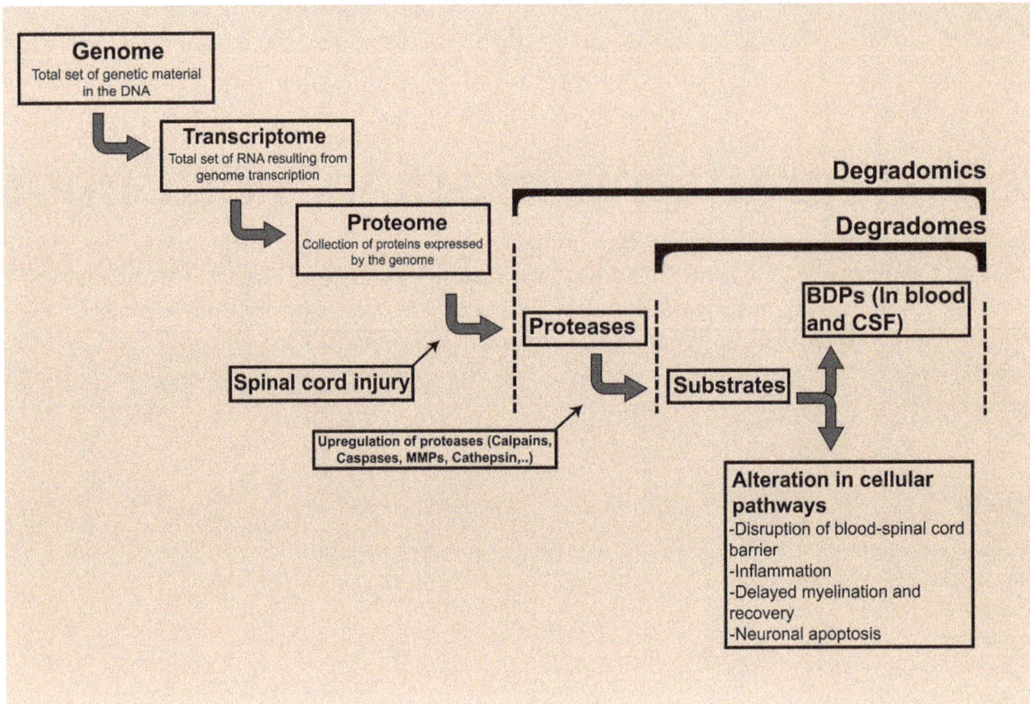

FIG. 1 Showing protease-substrate repertoires during cellular injury after spinal cord injury. *aSAH*, aneurysmal subarachnoid hemorrhage; *BDPs*, breakdown products; *MMPs*, matrix metalloproteinase.

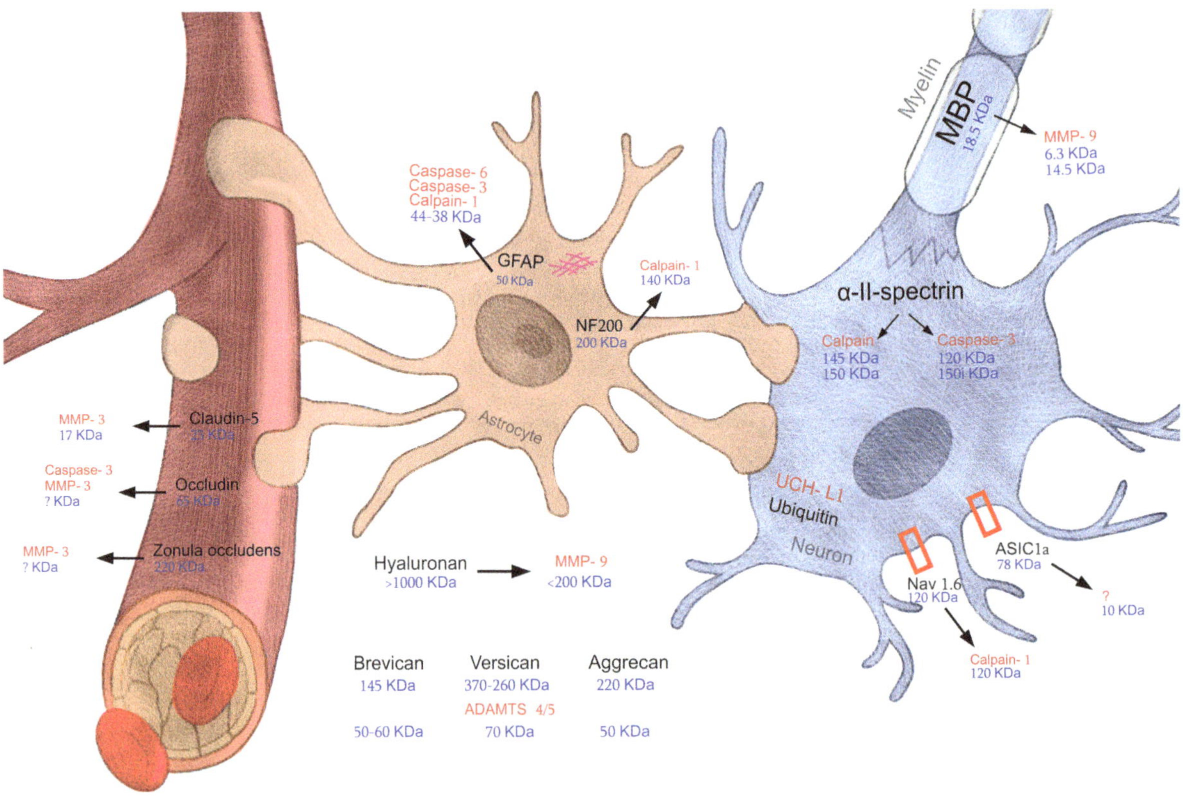

FIG. 2 Showing the location of different substrates, their proteases, and the resulting BDPs implicated in spinal cord injury. *ASIC1a*, mAcid-sensing ion channel 1a; *GFAP*, glial fibrillary acidic protein; *KDa*, kilodalton; MMP, matrix metalloproteinase; UCH-L1, ubiquitin C terminal hydrolase 1.

caspase-3, a cysteine aspartic protease, was found elevated at 1-day post-SCI, and calpain-3, a nonlysosomal cysteine protease, was elevated at day 7 post-SCI, as the extent of injury progressed (Banik et al., 1998; Gill et al., 2014).

Cathepsin-B and cathepsin-D are aspartyl proteases that were linked to SCI (Ellis, Earnhardt, Hayes, Wang, & Anderson, 2004; Hashimoto et al., 2005). 37-kDa cathepsin-B proenzyme was increased at 48, 72, and 168 h, in SCI post-contusion in rats (Ellis et al., 2004), and 31-kDa cathepsin-D were detected at 3 and 7 days in microglia post-SCI (Hashimoto et al., 2005). This contributed to the increased phagocytosis and lysosomal activation during the spinal cord insult (Moon et al., 2008). The endoplasmic reticulum injury was also studied as one of the reasons for apoptosis following injury, where caspase-12 and caspase-3 are expressed to accelerate the process (Liu et al., 2015).

Matrix Metalloproteinases (MMPs) are proteases that have an important vital function, including wound healing, neuronal growth and plasticity, and other physiological functions. Cytokines and growth factors are considered the main regulators of MMPs (Nagase & Woessner, 1999). An increase in MMP-1 and MMP-2 levels was found in macrophages of traumatized human spinal cords at the site of injury (Buss et al., 2007). Interestingly, MMP-2 had a major contribution to the extent of glial scarring (Veeravalli et al., 2009). In a study involving MMP-9 null mice vs wild-type mice with SCI, MMP-9 null mice were associated with a more stable blood–spinal cord barrier, decreased neutrophils at the site of injury, and a better locomotor activity (Noble, Donovan, Igarashi, Goussev, & Werb, 2002). A similar effect on locomotor activity was also seen in MMP-12 null mice compared with wild-type controls, as well as a higher motor strength days after SCI (Wells et al., 2003). MMPs are major contributors to the secondary damage after SCI, and a crucial cause of disability, where cell apoptosis exacerbates the underlying neuronal damage. Furthermore, when rats with SCI were injected intrathecally with SB-3CT (MMP-2/MMP-9 inhibitor), apoptotic cell death was inhibited, thus reducing secondary damage (Yu, Kamada, Niizuma, Endo, & Chan, 2008). This can guide further human clinical trials, to prevent MMPs' remodeling effect of the spinal cord, and prevent superimposed injuries.

Membrane type 1-MMP (MT1-MMP) and Glutamine Synthetase (GS) share an intimate correlation during SCI (Zou et al., 2011). This study noted a decreased GS levels acutely post-SCI. It is degraded by MT1-MMP, which amplifies glial scar formation and increases neurotoxicity, given that GS is highly expressed in astrocytes and plays a paramount protective effect by eliminating cytotoxic agents such as glutamate and ammonia (Zou et al., 2011).

Other proteases had a positive role in reducing injury and improving functional recovery, this is the case of few serine proteases like tissue-type Plasminogen Activator (tPA) and Urokinase PA (uPA). They reduce immune cell activation, soothe the destructive process, and promote recovery following the traumatic injury (Seeds et al., 2011). In contrast, neuropsin (kallikrein-like serine protease) promotes oligodendrocyte death, demyelination, and axonal degeneration after spinal cord injury (Terayama et al., 2007).

The Ubiquitin-Proteasome System (UPS) includes the proteasome and ubiquitin ligases, hydrolases, and ubiquitin itself (Gong, Radulovic, Figueiredo-Pereira, & Cardozo, 2016), which organizes many aspects of cellular behavior and function (Gong et al., 2016). It is found to be upregulated during acute SCI. The phosphorylation of Ubiquitin Carboxyl-terminal hydrolase L1 (UCH-L1) and conjugated ubiquitin is increased, which in turn, regulates neurogenesis and promotes neuronal progenitor cells (Sakurai et al., 2006). On the other hand, Ubiquitin-specific protease 4 (USP4) belongs to the deubiquitinase family which inhibits the NF-κB pathway. During SCI, USP4 is downregulated, which promotes the NF-κB pathway and increases microglial proliferation and amplifies the inflammatory process (Jiang et al., 2017).

Recently, UPS activation post-SCI was found to be the origin of the decline of heart function in terms of contractility and cardiomyocyte size weeks after the initial injury to the spinal cord, and preclinical trials targeting the UPS activation yielded promising results, using phosphodiesterase-4 inhibitors and dibutyryl-cAMP (Myeku, Wang, & Figueiredo-Pereira, 2012; Squair et al., 2018).

Extensive data analysis showed increased Presenilin (PS1) expression, the catalytic subunit of gamma-secretase, on day 1 after SCI (Kobayashi et al., 2010). PS1 is responsible for the formation of amyloid plaque due to its action on amyloid precursor protein involved in synaptogenesis and iron transport, and it is highly implicated in Alzheimer's disease (Murphy & LeVine, 2010). Theoretically, the development of gamma-secretase inhibitors would probably be of great interest to decrease secondary injury caused by B-amyloid plaque deposition.

Other protease inhibitors showed promising results in rats, but none has reached clinical or preclinical studies. Carfilzomib is a proteasome inhibitor that interacts with PSMB10, improved SCI injury in rats (Sharma et al., 2015). As previously stated, caspase-3 mediates apoptosis of neurons and glial cells after SCI which exacerbates the initial insult (Gill et al., 2014), few protease inhibitors such as nafamostat mesilate and Mesencephalic astrocyte-derived Neurotrophic Factor (MANF) reduced caspase-3-mediated apoptosis and showed improved functional recovery after SCI in rats (Gao et al., 2018). Due to the similarity of molecular changes after SCI between rats and humans, it is crucial to transfer these findings into human clinical trials to minimize the damage of proteolytic reactions and reduce neurotoxicity.

Degradomes in SCI

Degradomes constitute the substrates which the proteases act upon and the breakdown products generated from these proteolytic reactions. These degradomes modulate the cellular behavior in a number of ways, they carry important structural and functional cellular roles, including cytoskeleton, signaling pathways, enzymes, and mitotic or apoptotic properties (Rawlings, Barrett, & Finn, 2016).

Cytoskeletal proteins

Calpains are responsible for the neurochemical cascade that ends with glutamate neurotoxicity and degradation of the cytoskeleton (Veeravalli et al., 2012). αII-spectrin breakdown product (SBDP150) was found to increase at 3 and 6 h post-SCI in the thoracic spine in mice after high-pressure air blast injury, due to the action of calpains on αII-spectrin (280 kDa) (del Mar et al., 2015). In addition, 145- and 150-kDa SBDPs generated from the activity of calpain-1 were also found in rats with spinal cord contusion at T-10 level (Xiong, Rabchevsky, & Hall, 2007). Recently, SBDP120 was also detected at day 7 postspinal cord trauma (Yang et al., 2018).

Neurofilament (NF200) is an important 200,000 molecular weight protein that stabilizes and maintains neuronal structure (Banik, Matzelle, Gantt-Wilford, Osborne, & Hogan, 1997). Calpain-1 elevated levels at 2 h post-SCI cleave NF200 generating 140,000 fragments (Banik et al., 1997). In ovariectomized rats, Gonadotropin-releasing hormone treatment improves locomotor activity, 160- and 200-kDa neurofilament protein expression, and urinary function after spinal cord injury (Calderon-Vallejo & Quintanar, 2012).

Glial Fibrillary Acidic Protein (GFAP) is an intermediate filament located in astrocytes and Schwann cells, it has a molecular weight of 50 kDa. It is acutely cleaved during SCI into 38–44-kDa fragments by calpain-1 (Abou-El-Hassan et al., 2017). GFAP is found elevated during SCI, mainly due to protease-activated receptor 2 (PAR2) (Radulovic et al., 2015). PAR2 is a suppressor of myelin, its inhibition by lipopeptide inhibitors reduced ERK1/2 signaling, and the upcoming neurotoxicity (Yoon et al., 2013).

Other cytoskeletal proteins fluctuated after 12 h of SCI, there's an increase of α-tubulin and decrease of both dynamin 1 and the novel fascin, but they return to normal after 24 h of the primary injury (Zhu et al., 2013).

The clinical effect of the cytoskeletal protein degradation by the upregulated proteases is referred to the structural importance of these proteins in maintaining the stability of neuronal and glial cells. The inhibition of this proteolysis can lead to a decrease in neuronal loss and subsequent injury. A summary of cytoskeletal proteins is found in Table 1.

TABLE 1 Cytoskeletal proteins in SCI.

Precursor	Weight (kDa)	Location	Role	Proteolytic enzyme	Proteolytic product
αII-spectrin	280	Cytoplasm	Cytoskeletal protein, maintains cellular integrity, and stabilizes axons	Calpain-1	150-kDa SBDP
				Calpain-2	150-kDa SBDP 145-kDa SBDP
				Calpain-3	150i-kDa SBDP 120-kDa SBDP
NF200	200	Cytoplasm	Intermediate filaments, neuronal, and axonal structure	Calpain-1 cathepsin	140 kDa
GFAP	50	Cytoplasm	Intermediate filament, glial cell cytoskeleton	Calpain-1 Caspase-3 Caspase-6	38–44 kDa

GFAP, glial fibrillary acidic protein; *NF200*, neurofilament.

Extracellular matrix

After the initial SCI, there's a blood–spinal cord barrier damage that paves the way for a massive neuroinflammatory response. Extracellular matrix deposition is deemed one of the most important factors that regulate the entry of inflammatory cells and amplify the immune response (Deng et al., 2014). It is regulated by releasing and presenting peptides to increase cellular response and produce further tissue damage (Deng et al., 2014).

MMPs, as previously stated, are found in higher levels during SCI (Buss et al., 2007). They play an important role in Degrading Hyaluronan (HA), which produces sulfated Proteoglycans (PG) and tenascin fragments (Gaudet & Popovich, 2014). Further studies have shown a significant recovery of SCI in rats after adding HA, this has led to a decrease in macrophage density and reactive gliosis, with improved myelination and nerve regeneration. It also prevented the further deposition of proteoglycans with the ensuing inflammatory reactions (Park et al., 2009).

Axonal growth following SCI is inhibited mostly by Chondroitin Sulfate Proteoglycans (CSPGs), which increase perineuronal scarring. The proposed mechanism is narrowed down to the axonal tips being wedged and embedded within the inhibitory matrix (Nazari-Robati, Golestani, & Asadikaram, 2016). Aggrecan (220 kDa), versican (260–370 kDa), and brevican (145 kDa) are all considered parts of CSPGs (Nazari-Robati et al., 2016). Many successful attempts were used to inhibit CSPGs and showed promising results that can be further translated into the clinical routine in the management of SCI, aiming to improve neuroplasticity and nerve regeneration.

Myelin Basic Protein (MBP) is considered a key factor for proper neuro-regeneration and the restoration of normal nerve function. Adenovirus-mediated delivery of intact MBP improved neurological function when compared with inactive MBP that is cleaved after SCI (Lutz et al., 2016).

Extracellular matrix changes during SCI provide an early insight into the vast molecular pathways that intervene with neuro-regeneration. It is also a target for several therapeutic agents that can modulate the axonal myelination and decrease further neuronal damage. A summary of the extracellular matrix is found in Table 2.

Cell junction proteins

MMPs emerged to play a very important role in blood–spinal cord barrier breakdown, through their action on junction proteins. Specifically, 48-kDa MMP-3 peaks at 1-day postspinal cord injury, causing a decrease of zonula occludens-1 (220 kDa) at 4–8 h, occludin (65 kDa) at day 1, and claudin-5 (23 kDa) at 4–8 h (Lee, Choi, Ahn, Ju, & Yune, 2014). Hence, higher levels of junction proteins were detected with MMP-3 inhibition (Lee et al., 2014). Similarly, another study showed a protective role of metformin on the blood–spinal cord barrier by shutting down the breakdown of junctional proteins, by decreasing AMP-activated protein kinase (AMPK)-dependent MMP-9, ICAM-1, and neutrophil infiltration (Zhang et al., 2017). Likewise, mithramycin (an anticancer drug) and anti-Ly6G inhibited MMP-9 expression and thus decreased tight junction proteins degradation and blood–spinal cord barrier instability (Lee, Choi, Park, Ju, & Yune, 2018; Lee, Rosen, Weinstein, van Rooijen, & Noble-Haeusslein, 2011). In another study, MMP-8 inhibitor reduced significantly the

TABLE 2 Extracellular matrix in SCI.

Precursor	Weight (kDa)	Location	Role	Proteolytic enzyme	Proteolytic product
Hyaluronan	>1000	ECM	ECM, maintains CNS structure	MMP-9	Sulfated proteoglycan < 200 kDa Tenascin fragments < 200 kDa
Aggrecan	220	ECM	ECM-Chondrocyte function, part of CSPGs	ADAMTS4/5	50 kDa
Versican	260–370	ECM	Cell adhesion and migration, part of CSPGs		70 kDa
Brevican	145	ECM	Neuronal growth Part of CSPGs		50–60 kDa
Myelin basic protein	18.5	Myelin membrane	Myelination of nervous tissue	MMP-9	6.3, 7.0, 7.3, 8.3, 10.2, 14.5 kDa

proteolysis of occluding and zonula occludens-1 protein (Kumar et al., 2018). In brief, MMPs play a crucial role in destabilizing junction proteins, allowing the entry of inflammatory cells through the breakage of the blood–spinal cord barrier, further decapacitating the healing process and exacerbating the injury. A summary of cell junction proteins is found in Table 3.

Ion channels

In the past few years, ion channelopathies have taken a huge role in explaining many of the cellular injuries following ischemic insults. Neuron transmembrane potential and ion gradients are highly influenced by the alteration of ion channels, resulting in pericellular blebbing and plasma membrane damage (Liu et al., 2012). Proteolytic enzyme activation by calcium influx in SCI is one of the proposed mechanisms of neurotoxic injuries (Oliveira et al., 2014). Administration of calcium channel inhibitors had major neuroprotective properties in rat models. For example, Omega-conotoxin, attenuated neuronal death, by inhibiting voltage-dependent calcium channels, hence decreasing caspase-3 activation (Oliveira et al., 2014). Similarly, inhibition of sodium–calcium exchanger enhanced recovery and protected spectrin proteins from degradation by calpains, using bepridil and KB-R7943 (Li, Jiang, & Stys, 2000).

mAcid-sensing ion channel 1a (ASIC1a) allows sodium and calcium influx. It is upregulated in acute hypoxia in SCI, causing mitochondrial dysfunction and apoptosis, therefore amplifying the secondary injury (Yermolaieva, Leonard, Schnizler, Abboud, & Welsh, 2004). Its inhibition using spider-venom peptide PcTx1 leads to a better functional outcome (Koehn et al., 2016).

Another serious disability following SCI is spasticity. Molecular studies showed that calpain-mediated proteolysis of 250-kDa voltage-gated sodium (Nav) 1.6 channels increased persistent sodium current (I-NaP), the origin of increased spasticity (Brocard et al., 2016). Inhibition of calpain-mediated proteolysis of Nav was achieved using MDL28170, which downregulated I-NaP and relieved spasticity (Plantier & Brocard, 2017).

The pathophysiology of neuropathic pain following SCI was considerably studied, revealing an electric instability of the neuronal membrane, and a disruption of the structural and functional integrity of myelin (Dubin & Patapoutian, 2010). This activates the nociceptive circuitry in the small unmyelinated C-nociceptive and thinly myelinated Aδ afferents. This mechanism is mediated by MMPs' action on MBP (Dubin & Patapoutian, 2010). The severity of demyelination was reduced by tamoxifen, a selective estrogen receptor modulator, which also had a positive effect on upregulating aquaporin 4 (AQP4) water channels, decreasing edema, and improving recovery. It emphasized the positive effect of estrogen as a mediator of neuroprotection (Guptarak et al., 2014). A summary of ion channels is found in Table 4.

Clinical perspectives

Since the development of sophisticated methods and analytical tools to discover the molecular pattern of various medical conditions, new standards of characterizing patients suffering from SCI have emerged. Treatment can be personalized depending on each case based on the clinical symptomatology and severity, enabling specific targeted treatment that aims to reduce the secondary injury following SCI. Discoveries in animal models can be translated to humans via clinical trials and molecular studies that acknowledge the molecular and cellular behavior of neurons following the primary injury. Protein's identification is now available due to the advancement in proteomic studies, including quantitative high-resolution two-dimensional (2D-PAGE-isoelectric focusing and SDSPAGE) protein separation technique, matrix-assisted laser

TABLE 3 Cellular junction proteins in SCI.

Precursor	Weight (kDa)	Location	Role	Proteolytic enzyme	Proteolytic product
Zonula occludens-1	220 65 23	Cell membrane	Close paracellular pathway	MMP-3	?
Occluding	65	Cell membrane	Close paracellular pathway	MMP-3	?
Claudin-5	23	Cell membrane	Close paracellular pathway	MMP-3	17 kDa

TABLE 4 Ion channels in SCI.

Precursor	Weight (kDa)	Location	Role	Proteolytic enzyme	Proteolytic product
ASIC1a	78	Cell membrane	Sodium and calcium influx	?	10 kDa
Nav1.6	250	Cell membrane	Voltage-gated channel	Calpain-1	120 kDa

ASIC1a, mAcid-sensing ion channel 1a; *Nav*, voltage-gated sodium.

desorption/ionization (MALDI) TOF mass spectrometry, and nano ESI-MS/MS (Abou-El-Hassan et al., 2020). The protease-substrate repertoire was evident using a two-hybrid screening assay and high-throughput protein microarrays, which was considered a step further toward a better understanding of secondary injury (Abou-El-Hassan et al., 2020).

Application to other neuroscience areas

Degradomics' impact on CNS injury was also studied in traumatic brain injury (Abou-El-Hassan et al., 2017). The activation of proteases following the calcium influx and mitochondrial dysfunction leads to disruption of key components of cellular recovery, such as cytoskeleton integrity, synaptic transmissions, and neuronal survival (Zhou, Xu, Liao, Bi, & Baudry, 2009). In addition, brain ischemia and the subsequent neurotoxicity were also linked to calpain activation (Sharma et al., 2015).

Lately, emerging biomarkers involving aneurysmal subarachnoid hemorrhage were studied (Bsat et al., 2020, 2021). The early classification of patients according to the complications and severity of the aneurysmal rupture can guide an early and aggressive treatment to prevent further ischemic damage. Cerebral vasospasm is considered the main complication of aneurysmal subarachnoid hemorrhage, and the proposed mechanism is also related to calcium influx and a proteolytic cascade that ends with apoptosis (Abou-El-Hassan et al., 2017). Breakdown products of many cellular components can then be detected in blood, following proteolysis and apoptosis. The BDPs poured in blood and CSF reflect the severity of the damage and can predict the outcome. As an example, higher levels of SBDP150, SBDP145, and SBDP120 levels were significantly linked to higher WFNS (World federation of neurological surgeons) grades of severity and Glasgow coma scale (Papa et al., 2018).

Degradomics remains a new material for neurosurgeons, but it opens a variety of molecular patterns that can be the target of newer treatment after SCI. Further studies need to assess the efficacy of treating secondary injury in humans, but promising results are already coming to light.

Mini-dictionary of terms

Proteomics: The study of proteins, the vital parts of the living organisms, on a system-wide scale.
Degradomics: A subdiscipline in biology involving the analysis of proteases and their targeted substrates, their inhibitors, and the breakdown product.
Breakdown products: Peptides resulting from the action of proteases on their targeted substrates.
Channelopathy: The dysregulation of ion channel subunits, or the disturbance of proteins that regulate them.
Ubiquitin-proteasome system: Eukaryotic system concerned in the degradation and turnover of intracellular proteins through the action of proteases. Proteins destined for proteolysis are marked and linked by ubiquitin.

Key facts of spinal cord degradomics

- Degradomics constitutes a subdiscipline of proteomics.
- It is the study of proteases and their target substrates, that generate breakdown products.
- Breakdown products can be detected in serum or CSF of patients with SCI.
- Proteases play a key role in exacerbating spinal cord injury.
- Substrate's degradation influences neuronal and axonal metabolism, leading to demyelination and apoptosis.
- Targeted treatment can be implemented, to decrease the molecular cascade and inhibit neuronal apoptosis.

Summary points

- This chapter focuses on the secondary injury following spinal cord injury, induced by degradomics.
- Proteases such as caspase-3, calpains, cathepsins, and matrix metalloproteinases (MMPs) are found elevated after spinal cord injury.
- Ubiquitin-proteasome system (UPS) is considered protective, improving myelination, and nerve regeneration.
- Calpain degrades cytoskeletal proteins such as αII-spectrin, glial fibrillary acidic protein (GFAP), and neurofilament.
- Hyaluronan, a key component of extracellular matrix, increase myelination. and decrease inflammation. It is cleaved by MMPs after SCI.
- Chondroitin sulfate proteoglycans (CSPGs) decrease axonal regeneration and increase perineuronal scarring. Its inhibition potentiates neuroplasticity.
- Ion channelopathies increase neurotoxicity after spinal cord injury. Increased persistent sodium current (I-NaP) after calpains induced proteolysis of voltage-gated sodium (Nav) causes spasticity.
- Neuropathic pain after the axonal injury is exacerbated by myelin disruption through the action of MMPs on myelin basic protein (MBP), in addition to the electrical instability.

References

Abou-El-Hassan, H., Bsat, S., Sukhon, F., Assaf, E. J., Mondello, S., Kobeissy, F., et al. (2020). Protein degradome of spinal cord injury: Biomarkers and potential therapeutic targets. *Molecular Neurobiology*, *57*(6), 2702–2726. https://doi.org/10.1007/s12035-020-01916-3.

Abou-El-Hassan, H., Sukhon, F., Assaf, E. J., Bahmad, H., Abou Abbass, H., Jourdi, H., et al. (2017). Degradomics in neurotrauma: Profiling traumatic brain injury. *Methods in Molecular Biology*, *1598*, 65–99. https://doi.org/10.1007/978-1-4939-6952-4_4.

Banik, N. L., Matzelle, D. C., Gantt-Wilford, G., Osborne, A., & Hogan, E. L. (1997). Increased calpain content and progressive degradation of neurofilament protein in spinal cord injury. *Brain Research*, *752*(1–2), 301–306. https://doi.org/10.1016/S0006-8993(96)01488-6.

Banik, N. L., Shields, D. C., Ray, S., Davis, B., Matzelle, D., Wilford, G., et al. (1998). Role of calpain in spinal cord injury: Effects of calpain and free radical inhibitors(a). *Annals of the New York Academy of Sciences*, *844*(1), 131–137. https://doi.org/10.1111/j.1749-6632.1998.tb08228.x.

Brocard, C., Plantier, V., Boulenguez, P., Liabeuf, S., Bouhadfane, M., Viallat-Lieutaud, A., et al. (2016). Cleavage of Na(+) channels by calpain increases persistent Na(+) current and promotes spasticity after spinal cord injury. *Nature Medicine*, *22*(4), 404–411. https://doi.org/10.1038/nm.4061.

Bsat, S., Chanbour, H., Alomari, S., Moussalem, C., Houshiemy, M. N. E., & Sunna, T. (2020). Can we improve the prediction of complications and outcome in aneurysmal subarachnoid hemorrhage? The clinical implications of serum proteomics. *Journal of Integrative Neuroscience*, *19*(4), 711–717. https://doi.org/10.31083/j.jin.2020.04.370.

Bsat, S., Chanbour, H., Bsat, A., Alomari, S., Moussalem, C., Houshiemy, M. N. E., et al. (2021). Clinical utility of degradomics as predictors of complications and clinical outcome in aneurysmal subarachnoid hemorrhage. *Journal of Integrative Neuroscience*, *20*(2), 489–497. https://doi.org/10.31083/j.jin2002052.

Buss, A., Pech, K., Kakulas, B. A., Martin, D., Schoenen, J., Noth, J., et al. (2007). Matrix metalloproteinases and their inhibitors in human traumatic spinal cord injury. *BMC Neurology*, *7*, 17. https://doi.org/10.1186/1471-2377-7-17.

Calderon-Vallejo, D., & Quintanar, J. L. (2012). Gonadotropin-releasing hormone treatment improves locomotor activity, urinary function and neurofilament protein expression after spinal cord injury in ovariectomized rats. *Neuroscience Letters*, *515*(2), 187–190. https://doi.org/10.1016/j.neulet.2012.03.052.

del Mar, N., von Buttlar, X., Yu, A. S., Guley, N. H., Reiner, A., & Honig, M. G. (2015). A novel closed-body model of spinal cord injury caused by high-pressure air blasts produces extensive axonal injury and motor impairments. *Experimental Neurology*, *271*, 53–71. https://doi.org/10.1016/j.expneurol.2015.04.023.

Deng, W. P., Yang, C. C., Yang, L. Y., Chen, C. W., Chen, W. H., Yang, C. B., et al. (2014). Extracellular matrix-regulated neural differentiation of human multipotent marrow progenitor cells enhances functional recovery after spinal cord injury. *The Spine Journal*, *14*(10), 2488–2499. https://doi.org/10.1016/j.spinee.2014.04.024.

Dubin, A. E., & Patapoutian, A. (2010). Nociceptors: The sensors of the pain pathway. *The Journal of Clinical Investigation*, *120*(11), 3760–3772. https://doi.org/10.1172/JCI42843.

Ellis, R. C., Earnhardt, J. N., Hayes, R. L., Wang, K. K., & Anderson, D. K. (2004). Cathepsin B mRNA and protein expression following contusion spinal cord injury in rats. *Journal of Neurochemistry*, *88*(3), 689–697. https://doi.org/10.1046/j.1471-4159.2003.02197.x.

Gao, L., Xu, W., Fan, S., Li, T., Zhao, T., Ying, G., et al. (2018). MANF attenuates neuronal apoptosis and promotes behavioral recovery via Akt/MDM-2/p53 pathway after traumatic spinal cord injury in rats. *BioFactors*. https://doi.org/10.1002/biof.1433.

Gaudet, A. D., & Popovich, P. G. (2014). Extracellular matrix regulation of inflammation in the healthy and injured spinal cord. *Experimental Neurology*, *258*, 24–34. https://doi.org/10.1016/j.expneurol.2013.11.020.

Gill, L. C., Ross, H. H., Lee, K. Z., Gonzalez-Rothi, E. J., Dougherty, B. J., Judge, A. R., et al. (2014). Rapid diaphragm atrophy following cervical spinal cord hemisection. *Respiratory Physiology & Neurobiology*, *192*, 66–73. https://doi.org/10.1016/j.resp.2013.12.006.

Gong, B., Radulovic, M., Figueiredo-Pereira, M. E., & Cardozo, C. (2016). The ubiquitin-proteasome system: Potential therapeutic targets for Alzheimer's disease and spinal cord injury. *Frontiers in Molecular Neuroscience*, *9*, 4. https://doi.org/10.3389/fnmol.2016.00004.

Guptarak, J., Wiktorowicz, J. E., Sadygov, R. G., Zivadinovic, D., Paulucci-Holthauzen, A. A., Vergara, L., et al. (2014). The cancer drug tamoxifen: A potential therapeutic treatment for spinal cord injury. *Journal of Neurotrauma*, *31*(3), 268–283. https://doi.org/10.1089/neu.2013.3108.

Hashimoto, M., Koda, M., Ino, H., Yoshinaga, K., Murata, A., Yamazaki, M., et al. (2005). Gene expression profiling of cathepsin D, metallothioneins-1 and -2, osteopontin, and tenascin C in a mouse spinal cord injury model by cDNA microarray analysis. *Acta Neuropathologica*, *109*(2), 165–180. https://doi.org/10.1007/s00401-004-0926-z.

Jiang, X., Yu, M., Ou, Y., Cao, Y., Yao, Y., Cai, P., et al. (2017). Downregulation of USP4 promotes activation of microglia and subsequent neuronal inflammation in rat spinal cord after injury. *Neurochemical Research*, *42*(11), 3245–3253. https://doi.org/10.1007/s11064-017-2361-2.

Kobayashi, S., Sasaki, T., Katayama, T., Hasegawa, T., Nagano, A., & Sato, K. (2010). Temporal-spatial expression of presenilin 1 and the production of amyloid-beta after acute spinal cord injury in adult rat. *Neurochemistry International*, *56*(3), 387–393. https://doi.org/10.1016/j.neuint.2009.11.005.

Koehn, L. M., Noor, N. M., Dong, Q., Er, S. Y., Rash, L. D., King, G. F., et al. (2016). Selective inhibition of ASIC1a confers functional and morphological neuroprotection following traumatic spinal cord injury. *F1000Research*, *5*, 1822. https://doi.org/10.12688/f1000research.9094.2.

Kumar, H., Jo, M. J., Choi, H., Muttigi, M. S., Shon, S., Kim, B. J., et al. (2018). Matrix metalloproteinase-8 inhibition prevents disruption of blood-spinal cord barrier and attenuates inflammation in rat model of spinal cord injury. *Molecular Neurobiology*, *55*(3), 2577–2590. https://doi.org/10.1007/s12035-017-0509-3.

Lee, J. Y., Choi, H. Y., Ahn, H. J., Ju, B. G., & Yune, T. Y. (2014). Matrix metalloproteinase-3 promotes early blood-spinal cord barrier disruption and hemorrhage and impairs long-term neurological recovery after spinal cord injury. *The American Journal of Pathology*, *184*(11), 2985–3000. https://doi.org/10.1016/j.ajpath.2014.07.016.

Lee, J. Y., Choi, H. Y., Park, C. S., Ju, B. G., & Yune, T. Y. (2018). Mithramycin A improves functional recovery by inhibiting BSCB disruption and hemorrhage after spinal cord injury. *Journal of Neurotrauma*, *35*(3), 508–520. https://doi.org/10.1089/neu.2017.5235.

Lee, S. M., Rosen, S., Weinstein, P., van Rooijen, N., & Noble-Haeusslein, L. J. (2011). Prevention of both neutrophil and monocyte recruitment promotes recovery after spinal cord injury. *Journal of Neurotrauma*, *28*(9), 1893–1907. https://doi.org/10.1089/neu.2011.1860.

Li, S., Jiang, Q., & Stys, P. K. (2000). Important role of reverse Na(+)Ca(2+) exchange in spinal cord white matter injury at physiological temperature. *Journal of Neurophysiology*, *84*(2), 1116–1119. https://doi.org/10.1152/jn.2000.84.2.1116.

Liu, S., Sarkar, C., Dinizo, M., Faden, A. I., Koh, E. Y., Lipinski, M. M., et al. (2015). Disrupted autophagy after spinal cord injury is associated with ER stress and neuronal cell death. *Cell Death & Disease*, *6*. https://doi.org/10.1038/cddis.2014.527, e1582.

Liu, H., Shiryaev, S. A., Chernov, A. V., Kim, Y., Shubayev, I., Remacle, A. G., et al. (2012). Immunodominant fragments of myelin basic protein initiate T cell-dependent pain. *Journal of Neuroinflammation*, *9*, 119. https://doi.org/10.1186/1742-2094-9-119.

Lopez-Otin, C., & Overall, C. M. (2002). Protease degradomics: A new challenge for proteomics. *Nature Reviews. Molecular Cell Biology*, *3*(7), 509–519. https://doi.org/10.1038/nrm858.

Lutz, D., Kataria, H., Kleene, R., Loers, G., Chaudhary, H., Guseva, D., et al. (2016). Myelin basic protein cleaves cell adhesion molecule L1 and improves regeneration after injury. *Molecular Neurobiology*, *53*(5), 3360–3376. https://doi.org/10.1007/s12035-015-9277-0.

McKinley, W., Santos, K., Meade, M., & Brooke, K. (2007). Incidence and outcomes of spinal cord injury clinical syndromes. *The Journal of Spinal Cord Medicine*, *30*(3), 215–224. https://doi.org/10.1080/10790268.2007.11753929.

Moon, C., Lee, T. K., Kim, H., Ahn, M., Lee, Y., Kim, M. D., et al. (2008). Immunohistochemical study of cathepsin D in the spinal cords of rats with clip compression injury. *The Journal of Veterinary Medical Science*, *70*(9), 937–941. https://doi.org/10.1292/jvms.70.937.

Murphy, M. P., & LeVine, H., 3rd. (2010). Alzheimer's disease and the amyloid-beta peptide. *Journal of Alzheimer's Disease*, *19*(1), 311–323. https://doi.org/10.3233/JAD-2010-1221.

Myeku, N., Wang, H., & Figueiredo-Pereira, M. E. (2012). cAMP stimulates the ubiquitin/proteasome pathway in rat spinal cord neurons. *Neuroscience Letters*, *527*(2), 126–131. https://doi.org/10.1016/j.neulet.2012.08.051.

Nagase, H., & Woessner, J. F., Jr. (1999). Matrix metalloproteinases. *The Journal of Biological Chemistry*, *274*(31), 21491–21494. https://doi.org/10.1074/jbc.274.31.21491.

Nasser, M., Bejjani, F., Raad, M., Abou-El-Hassan, H., Mantash, S., Nokkari, A., et al. (2016). Traumatic brain injury and blood-brain barrier cross-talk. *CNS & Neurological Disorders Drug Targets*, *15*(9), 1030–1044.

Nazari-Robati, M., Golestani, A., & Asadikaram, G. (2016). Improvement of proteolytic and oxidative stability of chondroitinase ABC I by cosolvents. *International Journal of Biological Macromolecules*, *91*, 812–817. https://doi.org/10.1016/j.ijbiomac.2016.06.030.

Noble, L. J., Donovan, F., Igarashi, T., Goussev, S., & Werb, Z. (2002). Matrix metalloproteinases limit functional recovery after spinal cord injury by modulation of early vascular events. *The Journal of Neuroscience*, *22*(17), 7526–7535. https://doi.org/10.1523/jneurosci.22-17-07526.2002.

Oliveira, K. M., Lavor, M. S., Silva, C. M., Fukushima, F. B., Rosado, I. R., Silva, J. F., et al. (2014). Omegaconotoxin MVIIC attenuates neuronal apoptosis in vitro and improves significant recovery after spinal cord injury in vivo in rats. *International Journal of Clinical and Experimental Pathology*, *7*(7), 3524–3536.

Overall, C. M., & Dean, R. A. (2006). Degradomics: Systems biology of the protease web. Pleiotropic roles of MMPs in cancer. *Cancer Metastasis Reviews*, *25*(1), 69–75. https://doi.org/10.1007/s10555-006-7890-0.

Papa, L., Rosenthal, K., Silvestri, F., Axley, J. C., Kelly, J. M., & Lewis, S. B. (2018). Evaluation of alpha-II-spectrin breakdown products as potential biomarkers for early recognition and severity of aneurysmal subarachnoid hemorrhage. *Scientific Reports*, *8*(1). https://doi.org/10.1038/s41598-018-31631-y.

Park, J., Lim, E., Back, S., Na, H., Park, Y., & Sun, K. (2009). Nerve regeneration following spinal cord injury using matrix metalloproteinase-sensitive, hyaluronic acid-based biomimetic hydrogel scaffold containing brain-derived neurotrophic factor. *Journal of Biomedical Materials Research Part A*, *9999A*. https://doi.org/10.1002/jbm.a.32519.

Plantier, V., & Brocard, F. (2017). Calpain as a new therapeutic target for treating spasticity after a spinal cord injury. *Medical Science*, *33*(6–7), 629–636. https://doi.org/10.1051/medsci/20173306020.

Radulovic, M., Yoon, H., Wu, J., Mustafa, K., Fehlings, M. G., & Scarisbrick, I. A. (2015). Genetic targeting of protease activated receptor 2 reduces inflammatory astrogliosis and improves recovery of function after spinal cord injury. *Neurobiology of Disease*, *83*, 75–89. https://doi.org/10.1016/j.nbd.2015.08.021.

Rawlings, N. D., Barrett, A. J., & Finn, R. (2016). Twenty years of the MEROPS database of proteolytic enzymes, their substrates and inhibitors. *Nucleic Acids Research*, *44*, D343–D350. https://doi.org/10.1093/nar/gkv1118.

Sakurai, M., Ayukawa, K., Setsuie, R., Nishikawa, K., Hara, Y., Ohashi, H., et al. (2006). Ubiquitin C-terminal hydrolase L1 regulates the morphology of neural progenitor cells and modulates their differentiation. *Journal of Cell Science*, *119*(Pt. 1), 162–171. https://doi.org/10.1242/jcs.02716.

Savickas, S., & Keller, U. A. D. (2017). Targeted degradomics in protein terminomics and protease substrate discovery. *Biological Chemistry*. https://doi.org/10.1515/hsz-2017-0187.

Seeds, N., Mikesell, S., Vest, R., Bugge, T., Schaller, K., & Minor, K. (2011). Plasminogen activator promotes recovery following spinal cord injury. *Cellular and Molecular Neurobiology*, *31*(6), 961–967. https://doi.org/10.1007/s10571-011-9701-6.

Sharma, H. S., Muresanu, D. F., Lafuente, J. V., Sjoquist, P. O., Patnaik, R., & Sharma, A. (2015). Nanoparticles exacerbate both ubiquitin and heat shock protein expressions in spinal cord injury: Neuroprotective effects of the proteasome inhibitor carfilzomib and the antioxidant compound H-290/51. *Molecular Neurobiology*, *52*(2), 882–898. https://doi.org/10.1007/s12035-015-9297-9.

Squair, J. W., DeVeau, K. M., Harman, K. A., Poormasjedi-Meibod, M. S., Hayes, B., Liu, J., et al. (2018). Spinal cord injury causes systolic dysfunction and cardiomyocyte atrophy. *Journal of Neurotrauma*, *35*(3), 424–434. https://doi.org/10.1089/neu.2017.4984.

Supinski, G. S., Wang, W., & Callahan, L. A. (2009). Caspase and calpain activation both contribute to sepsis-induced diaphragmatic weakness. *Journal of Applied Physiology*, *107*(5), 1389–1396. https://doi.org/10.1152/japplphysiol.00341.2009.

Tator, C. H., & Fehlings, M. G. (1991). Review of the secondary injury theory of acute spinal cord trauma with emphasis on vascular mechanisms. *Journal of Neurosurgery*, *75*(1), 15–26. https://doi.org/10.3171/jns.1991.75.1.0015.

Terayama, R., Bando, Y., Murakami, K., Kato, K., Kishibe, M., & Yoshida, S. (2007). Neuropsin promotes oligodendrocyte death, demyelination and axonal degeneration after spinal cord injury. *Neuroscience*, *148*(1), 175–187. https://doi.org/10.1016/j.neuroscience.2007.05.037.

Veeravalli, K. K., Dasari, V. R., & Rao, J. S. (2012). Regulation of proteases after spinal cord injury. *Journal of Neurotrauma*, *29*(13), 2251–2262. https://doi.org/10.1089/neu.2012.2460.

Veeravalli, K. K., Dasari, V. R., Tsung, A. J., Dinh, D. H., Gujrati, M., Fassett, D., et al. (2009). Human umbilical cord blood stem cells upregulate matrix metalloproteinase-2 in rats after spinal cord injury. *Neurobiology of Disease*, *36*(1), 200–212. https://doi.org/10.1016/j.nbd.2009.07.012.

Wells, J. E., Rice, T. K., Nuttall, R. K., Edwards, D. R., Zekki, H., Rivest, S., et al. (2003). An adverse role for matrix metalloproteinase 12 after spinal cord injury in mice. *The Journal of Neuroscience*, *23*(31), 10107–10115. https://doi.org/10.1523/jneurosci.23-31-10107.2003.

Xiong, Y., Rabchevsky, A. G., & Hall, E. D. (2007). Role of peroxynitrite in secondary oxidative damage after spinal cord injury. *Journal of Neurochemistry*, *100*(3), 639–649.

Yang, Z., Bramlett, H. M., Moghieb, A., Yu, D., Wang, P., Lin, F., et al. (2018). Temporal profile and severity correlation of a panel of rat spinal cord injury protein biomarkers. *Molecular Neurobiology*, *55*(3), 2174–2184. https://doi.org/10.1007/s12035-017-0424-7.

Yermolaieva, O., Leonard, A. S., Schnizler, M. K., Abboud, F. M., & Welsh, M. J. (2004). Extracellular acidosis increases neuronal cell calcium by activating acid-sensing ion channel 1a. *Proceedings of the National Academy of Sciences of the United States of America*, *101*(17), 6752–6757. https://doi.org/10.1073/pnas.0308636100.

Yoon, H., Radulovic, M., Wu, J., Blaber, S. I., Blaber, M., Fehlings, M. G., et al. (2013). Kallikrein 6 signals through PAR1 and PAR2 to promote neuron injury and exacerbate glutamate neurotoxicity. *Journal of Neurochemistry*, *127*(2), 283–298. https://doi.org/10.1111/jnc.12293.

Yu, F., Kamada, H., Niizuma, K., Endo, H., & Chan, P. H. (2008). Induction of MMP-9 expression and endothelial injury by oxidative stress after spinal cord injury. *Journal of Neurotrauma*, *25*(3), 184–195. https://doi.org/10.1089/neu.2007.0438.

Zhang, D., Tang, Q., Zheng, G., Wang, C., Zhou, Y., Wu, Y., et al. (2017). Metformin ameliorates BSCB disruption by inhibiting neutrophil infiltration and MMP-9 expression but not direct TJ proteins expression regulation. *Journal of Cellular and Molecular Medicine*, *21*(12), 3322–3336. https://doi.org/10.1111/jcmm.13235.

Zhou, M., Xu, W., Liao, G., Bi, X., & Baudry, M. (2009). Neuroprotection against neonatal hypoxia/ischemia-induced cerebral cell death by prevention of calpain-mediated mGluR1alpha truncation. *Experimental Neurology*, *218*, 75–82. https://doi.org/10.1016/j.expneurol.2009.04.006.

Zhu, B., Li, Y., Li, M., Yang, X., Qiu, B., Gao, Q., et al. (2013). Dynamic proteome analysis of spinal cord injury after ischemia reperfusion in rabbits by two-dimensional difference gel electrophoresis. *Spinal Cord*, *51*(8), 610–615. https://doi.org/10.1038/sc.2013.24.

Zou, J., Wang, Y. X., Mu, H. J., Xiang, J., Wu, W., Zhang, B., et al. (2011). Down-regulation of glutamine synthetase enhances migration of rat astrocytes after in vitro injury. *Neurochemistry International*, *58*(3), 404–413. https://doi.org/10.1016/j.neuint.2010.12.018.

Chapter 12

Proteomics of pressure ulcers in spinal cord injury

L. Mourino-Alvarez, N. Corbacho-Alonso, T. Sastre-Oliva, and M.G. Barderas

Department of Vascular Physiopathology, Hospital Nacional de Parapléjicos (HNP), SESCAM, Toledo, Spain

List of abbreviations

IMS	imaging mass spectrometry
LC-MS/MS	liquid chromatography tandem mass spectrometry
MS	mass spectrometry
PTMs	post-translational modifications
PBS	phosphate buffered saline
PUs	pressure ulcers
SCI	spinal cord injury
SRM	selected reaction monitoring
TMT	tandem mass tags
WB	Western blot

Background

Spinal cord injury (SCI) is the result of trauma at any level of the spinal cord causing temporary or permanent damage, with a significant effect on the patient's physical and psychosocial well-being (Van Middendorp, Goss, Urquhart, et al., 2011). SCI is related to considerable healthcare costs, morbi-mortality, especially when it reaches advanced stages. The most common medical classification to characterize SCI is the International Standards for Neurological Classification of Spinal Cord Injury (ISNCSCI) to classify subjects using the ASIA Impairment Scale (AIS) and the neurological level of SCI. ISNCSCI categorizes injuries by severity into complete (patient without sacral preservation; defined as ASIA A), and incomplete (patients with sensorimotor sacral preservation) injuries, which is further subdivided into different degrees of involvement (B, C, D, and E) (Schuld et al., 2016). Currently, there is no treatment available for SCI once it is established, and rehabilitation is the only therapy for functional improvement. The life-changing consequences, along with the high social cost and the lack of treatment options, make the study of SCI and its adjacent pathologies an important scientific and medical objective (Alizadeh, Dyck, & Karimi-Abdolrezaee, 2019).

Even with advances in healthcare, pressure ulcers (PUs) remain a more common complication in patients with SCI. PU formation is a complex process, and its incidence significantly grows with time post-injury (Charlifue, Jha, & Lammertse, 2010; Chen, DeVivo, & Jackson, 2005). Pressure ulcers are defined as lesions on any skin surface that result from localized shear and/or compression for a prolonged period over bony prominences at certain anatomic locations (Black, Baharestani, Cuddigan, et al., 2007). This occurrence leads to ischemia of overlying soft tissues, which can finally result in necrosis (Bogie, Nuseibeh, & Bader, 1995). Frequently, severe PUs do not respond to traditional wound therapy and surgery is required to avoid further tissue damage (Bhattacharya & Mishra, 2015). These limitations in the therapeutic strategies used for PUs highlight the urgent need for new treatments for this grave public health problem.

While enormous efforts have been expended to better understand the main risk factors for PUs and to improve prevention, the course of these lesions hampers an accurate and individualized evaluation (Martin-Rojas et al., 2020). Accordingly, new tools are desirable to further our knowledge on the cellular/molecular subjacent mechanisms of PU development (Baldan-Martin et al., 2020).

Cellular, Molecular, Physiological, and Behavioral Aspects of Spinal Cord Injury. https://doi.org/10.1016/B978-0-12-822427-4.00012-5

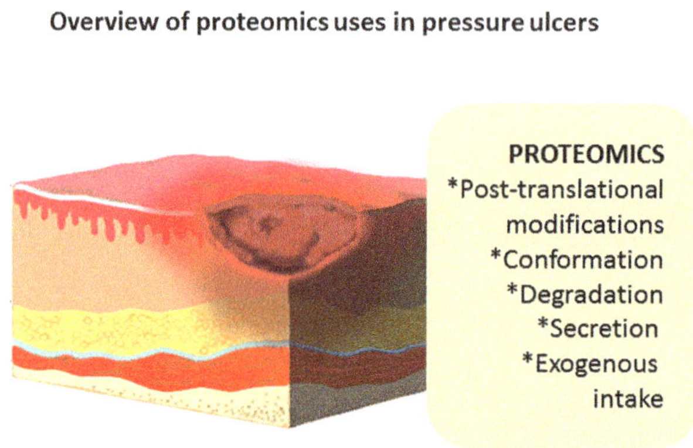

FIG. 1 Overview of proteomics uses: Proteomics gives an insight into the estate of a system, as affected by stimulus. In the case of pressure ulcers in spinal cord injury patients, researchers hope to observe the effect of these pressure ulcers on the types and quantities of proteins.

Proteomics technologies offer considerable chances to identify significant changes in protein abundance (Pandey et al., 2015) during the progression of PU formation in an unbiased manner. Improvements in proteomics technology allow the molecular determinants of complex samples like tissue to be analyzed using mass spectrometry (MS) (Marshall & Williams, 2002). The use of these methods can permit the identification of important biological processes that are altered in PU and may enable the discovery of new therapeutic options to improve clinical management of these patients (Fig. 1).

Patient's management

PUs are complex and chronic wounds in patients with SCI and no gold standard has yet been established for their prevention and treatment. PUs are difficult to prevent and manage and can lead to a decline in the overall well-being of patients in SCI (Dwivedi et al., 2017) (Fig. 2).

Chronic wounds place the person with SCI at high risk of infections, sepsis, and death. Skin health and disruption is individual and multifactorial, thus prevention requires personalized education direct to patient predilections and objectives (Rosin, Tabibi, Trimbath, & Henzel, 2020). Assessment requires a correct description of wound-type/PUs stage, location, size, wound bed, wound margin, epithelialization, exudate, and peri-wound condition (Canadian Agency for Drugs and Technologies in Health, 2013). PUs should be staged using the National Pressure Injury Advisory Panel (NPIAP) staging system. Successful treatment requires optimal wound bed preparation, pressure off-loading, and access to surgical specialists if needed (Medical Advisory Secretariat. Pressure ulcer prevention, 2009). People with spinal cord injury, due

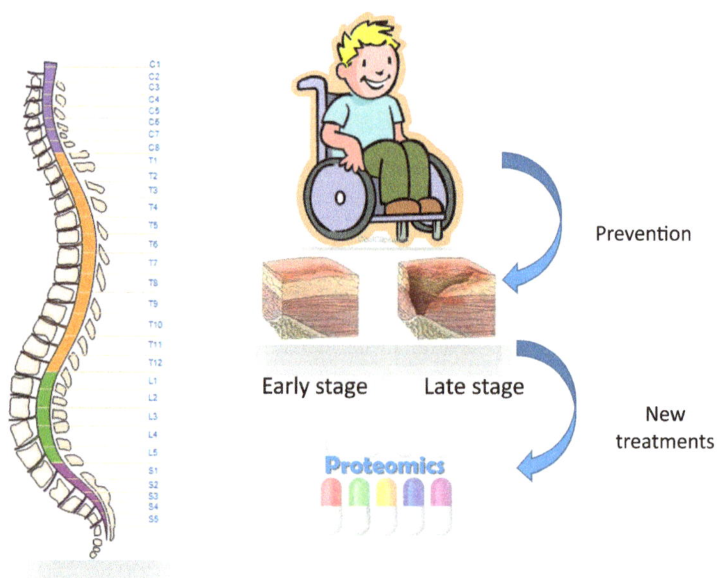

FIG. 2 Pressure ulcers are a more common complication in patients with SCI: The use of proteomics can permit the identification of important biological processes that are altered in PU and may enable the discovery of new therapeutic options to improve clinical management of these patients.

to impaired mobility, are most vulnerable to PUs. Furthermore, and taking into account the Canadian Agency for Drugs and Technologies in Health guides, conditions such as poor nutrition, poor sensation, urinary and fecal incontinence, and poor overall physical and mental health are predisposing factors to PU formation.

Prevalence of PUs is considered as an indicator of quality for long-term care facilities, and develop of PUs in hospitalized patients is often considered an unnecessary complication defining failure of inpatient management. Normally, higher prevalence is reported for the elderly, the acutely ill, and those who have sustained SCI (O'Meara, Cullum, Majid, & Sheldon, 2000).

Exits several paths that may be employed in PU wound care: (i) reduction or elimination of underlying contributing conditions such as modifying support surfaces and providing nutritional support; (ii) provision of local wound care, including but not limited to wound dressing and topical applications to promote healing; (iii) surgical repair of the ulcer, where appropriate (Jaul, 2010).

The main objective of therapy in most cases consists of complete healing with the restoration of functional integrity of skin to the highest extent possible. Sometimes the goal of therapy may be palliative for certain patients such as the terminally ill, focusing on reducing discomfort, and/or deterioration of the PU (Evans, Andrews, Chutka, Fleming, & Garness, 1995). In any case, dressing a specific PU wound care could help protect PU contamination and improve the quality of life of these kinds of patients. A wide variety of dressings are available; including many with various combinations of properties such as wound bed preparation (debridement), antimicrobial activity, and moisture control (Lachenbruch, Ribble, Emmons, & Van Gilder, 2016).

Clinical proteomics

Clinical proteomics is based on the application of proteomics technologies to diagnosis, prognosis, treatment, and course of human diseases (Dominguez, Lopes, & Torres, 2007). It is predicted that the 20,300 protein-coding genes enclosed in the human genome are responsible for more than 1 million different protein isoforms derived by DNA recombination, alternative splicing, and diverse post-translational modifications (PTMs) that vary across time and location, and are impacted by different kinds of perturbations (Chambers, Lawrie, Cash, & Murray, 2000; Hooff et al., 2011; Legrain et al., 2011). These isoforms and PTMs are detectable only by studying the proteins directly and can be indicative of specific protein functions.

The proteomics strategy is somehow "blind" at the discovery phase. This means that no key target is pre-selected, all the protein found in the discovery phase are investigated as a whole (Fig. 3). In this way, not only particular pathways or

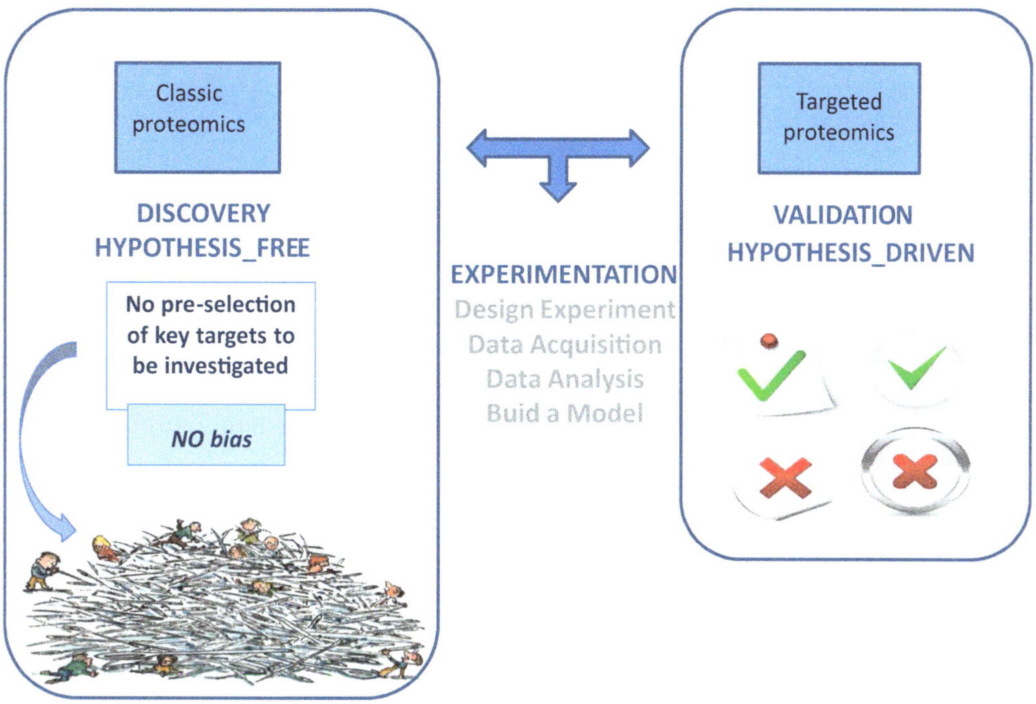

FIG. 3 The proteomics strategy: In the discovery phase, no potential marker, no key target is pre-selected, but all the protein sets are investigated as a whole. In the validation phase, the discovery phase makes all sense, confirming previous results in wider cohorts and setting valid conclusions for the clinical practice.

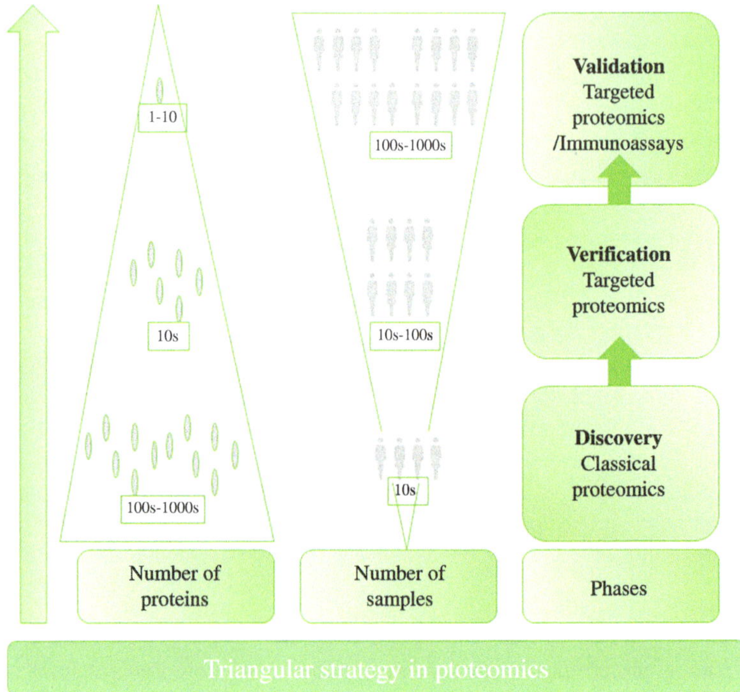

FIG. 4 The inverted triangle strategy: The inverted triangle strategy is a very visual concept to easily understand the phases that make up a clinical proteomics study. In the discovery phase, we start with an "n" of very small subjects, dozens, and we will have a very high number of identified proteins. In the verification phase and with an independent cohort of patients, the "n" will be increased to hundreds. In this case, dozens of proteins could be verified. Finally, in the validation phase, the "n" will be of several miles of subjects, and we could validate 1 to 10 proteins as pathology markers.

responding molecules commonly measured in routine biochemical patient's analysis are being investigated but also those whose relationship with the pathophysiological processes taking place are still unknown.

The action for identifying new decisive biomarkers of disease is based on three steps: (1) discovery, where the differential expression of specific proteins between states is defined; (2) verification, where hundreds of patients with environmental, genetic, and biological differences are analyzed; and (3) validation, where the discovery phase makes all the sense and it is then when candidate makers are further investigated, confirming previous results in wider cohorts and setting valid conclusions for the clinical practice (Paulovich et al., 2008. Rifai, Gillette, & Carr, 2006) (Fig. 4).

Proteomics in pressure ulcers

Since the growth development of mass spectrometry (MS) during the past 20 years, proteomics has become a powerful tool in biomedical research (Lindsey, Mayr, Gomes, Delles, et al., 2015). Proteomics has enabled the measurement of multiple properties of thousands of proteins, including their abundance, subcellular localization, post-translational modifications, and interactions (Wilhelm et al., 2014). While traditional molecular biology methods detect a limited number of proteins based on signaling or metabolic pathways, proteomics has become a systematic approach to qualitative and quantitative localization of entire proteomes in large-scale research (Yates, Ruse, & Nakorchevsky, 2009). In the area of traumatology, proteomics for comparing protein expression profiles between normal and disease states has seldom been applied to obtain unique protein-expression profiles of PUs, let alone deep tissues. Furthermore, recent advances in proteomic approaches have allowed the study of proteins to understand pathophysiological mechanisms in SCI disease and PUs (Albayar et al., 2019; Baldan-Martin et al., 2020; Liu, Cui, Hu, & Zhang, 2020; Martin-Rojas et al., 2020). Nevertheless, the search for biomarkers for clinical implementation by high-throughput proteomics requires multiple stages, from shotgun to targeted approaches, and nowadays, there is a lack of clinically useful protein biomarkers due to sample optimization and technological limitation not only in PUs but also in other diseases. Despite this, proteome analysis is currently helping to better understand and define the pathophysiology of PUs, contributing to the clinical needs of treatment of PUs in SCI patients (Baldan-Martin et al., 2020).

A recent investigation by Baldan-Martin et al. (2020) performed tissue proteomics from PUs obtained from SCI patients. They identified a total of 4504 proteins, of which 76 were differentially expressed in cases from controls using TMT-based multiplexed isobaric labeling followed by LC-MS/MS (Fig. 5). According to the molecular function, it was remarkable that a substantial number of proteins were implicated in enzyme regulator activity, including peptidase regulator or lipase inhibitor activity categories.

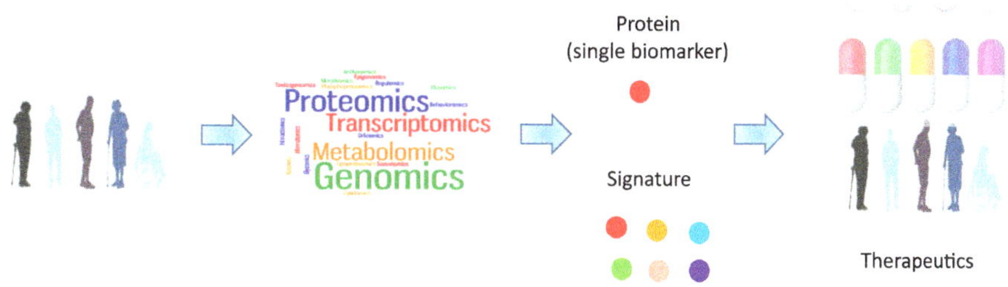

FIG. 5 Precision medicine is fundamentally changing the way therapies are being developed. Personalized medicine is directed toward personalized treatment. Patients are different and so their treatment. Personalized medicine aims to give the right treatment to the right patients. For example, make subgroups of patients to try to identify those who are more likely to benefit from a specific treatment, identify those who may suffer side effects, etc.

Following the analysis of tissue samples, another study performed in the muscle of patients with PUs (Liu et al., 2018) identified 2558 proteins, 520 of which were identified with differential expression between samples from patients with deep PU and normal muscle tissues by iTRAQ LC-MS/MS, particularly, JAK2 was downregulated. These results demonstrated that JAK2 may play an important protective role in muscular ischemia and reperfusion injury during DTI development by inhibition of autophagy and apoptosis through the AKT and ERK1/2 pathways. Furthermore, Liu et al. (2020), using ITRAQ LC-MS/MS, found a number of proteins that were differentially expressed in PU muscle samples compared with the normal and identified unique proteins expression patterns between these two groups. Importantly, cathepsin B and D were identified, which indicated that they might be key drug targets for PUs.

Another proteomics technique, imaging mass spectrometry (IMS), was employed for the analysis of frozen skin biopsies to investigate the differences between stage IV pressure ulcers that remain stalled, stagnant, and unhealed versus those exhibiting clinical and histological signs of improvement. Taverna, Pollins, Sindona, Caprioli, and Nanney (2015) revealed a rich diversity of proteins that are dynamically modulated, and they selectively highlight a family of calcium-binding proteins (S-100 molecules) including calcyclin (S100-A6), calgranulins A (S100-A8) and B (S100-A9), and calgizzarin (S100-A11). IMS allowed defining three discrete regions of interest: the wound bed, adjacent dermis, and hypertrophic epidermis. The discovery-based approach with IMS augments current knowledge of the molecular signatures within PUs while providing a rationale for a focused examination of the role of calcium modulators within the context of impaired wound healing.

All in all, integration of biomarker information coming from several proteomic strategies is still a challenge. Nonetheless, once we get an integrative approach we will be able to develop a "multimarker" approach which may achieve higher sensitivity and specificity to therapy which, probably, would be more appropriate than a single marker for complex conditions such as Pus (Table 1).

The importance of precision medicine

The objectives of clinical practice, both in PUs from patients with SCI and in other diseases, have always been addressed to the patients, to aid from pain, to treat the disease, and to help prevent illness from occurring. Normally, reductionism in medicine presupposes that patients with common signs and symptoms share the same disease pathophenotype and will respond similarly to medical interventions tested in groups of similar subjects (Leopold & Loscalzo, 2018).

Contrarily, precision medicine is dedicated to selecting the proper treatment for each patient according to their personal pathophenotype. Consequently, precision medicine combines standard clinical evaluation, including synthesis of data, imaging, or laboratory tests, with different -omics approaches such as genomics, transcriptomics, proteomics, or metabolomics, for deep phenotyping (Antman & Loscalzo, 2016; Lee et al., 2012) (Fig. 6). Defining the phenotype before disease onset may help risk stratification, which are procedures used for risk assessment that regulate individuals based on the probability of developing a specific outcome. Identifying distinct risk phenotypes would allow the screening in patients with high risk, or avoid gratuitous treatments with corresponding side effects. Precision medicine, in the treatment of patients, aims to select the best therapy for each individual, considering that different groups of patients may respond differently to the same drug in terms of dose, length, or side effects (Garmaroudi et al., 2016; Manrai et al., 2016).

It is hoped that the application of precision medicine in PUs generally, but in SCI patients, principally, will improve patient management by promoting adequate prevention and treatment options, which should reduce the number of surgeries as well as the economic costs (Fuster, 2014; Mirnezami, Nicholson, & Darzi, 2012). Indeed, PUs in SCI patients is a perfect candidate for precision medicine.

TABLE 1 Compilation of studies reported so far in the field of proteomics in pressure ulcers.

Most relevant proteins (validated)	Abbreviation	Methodology	Reference
ANTITHROMBIN III	ANT3	TMT-LC/MS-MS	Baldan-Martin et al. (2020)
ALPHA-A-ANTITRYPSIN	A1AT	TMT-LC/MS-MS	Baldan-Martin et al. (2020)
KININOGEN-1	KNG1	TMT-LC/MS-MS	Baldan-Martin et al. (2020)
ALPHA-2-MACROGLOBULIN	A2MG	TMT-LC/MS-MS	Baldan-Martin et al. (2020)
FIBRONECTIN	FINC	TMT-LC/MS-MS	Baldan-Martin et al. (2020)
APOLIPROTEIN AI	APOA1	TMT-LC/MS-MS	Baldan-Martin et al. (2020)
COLLAGEN ALPHA-1 (XII) CHAIN	COCA	TMT-LC/MS-MS	Baldan-Martin et al. (2020)
HAPTOGLOBIN	HPT	TMT-LC/MS-MS	Baldan-Martin et al. (2020)
APOLIPROTEIN B-100	APOB	TMT-LC/MS-MS	Baldan-Martin et al. (2020)
COMPLEMENT FACTOR B	CFAB	TMT-LC/MS-MS	Baldan-Martin et al. (2020)
CATHEPSIN B	CTSB	iTRAQ LC/MS-MS	Liu et al. (2020)
CATHEPSIN D	CTSD	iTRAQ LC/MS-MS	Liu et al. (2020)
JANUS KINASE 2	JAK2	iTRAQ LC/MS-MS	Liu et al. (2018)
CALGRANULINS A	S100A8	IMAGING-MS	Taverna et al. (2015)
CALGRANULINS B	S100A9	IMAGING-MS	Taverna et al. (2015)
CALCICYN	S100A6	IMAGING-MS	Taverna et al. (2015)
CALGIZZARIN	S100A11	IMAGING-MS	Taverna et al. (2015)

Limitations of proteomics

Despite the progressively realized power of proteomics, several problems exist for the above-mentioned methods. In MS, the cost of equipment is a determinant, and separation and ionization techniques may affect findings. To overcome this, proteomics can be used to discover novel proteins, and orthogonal techniques such as SRM or immunoassays are used to quantitate larger sample sets even in platforms capable of absolute quantification (Jiang, Zhang, Yu, et al., 2014). Although all platforms provide insights into relative protein regulation, absolute quantification is only possible in MS and, in a proximity extension assay, these assays require the use of labeled standards that may not be feasible in untargeted experiments (Jarchum, 2016; Pan, Aebersold, Chen, et al., 2009; Van de Merbel, 2019).

It is also important to consider the sample source when designing experiments. Although peripheral blood sera samples are readily collected, the blood proteome may be influenced by a large number of systemic issues that may or may not be related to the system of interest. Proteomics conducted on tissue samples, although more difficult to obtain, may better reflect the proteome of the system of interest.

Conclusions

The study of PUs through proteomic approach, and the subsequent integration of data, can provide molecular-level information of the molecular pathways that are more active in that tissue. The characterization of physiological proteins in this tissue and the creation of a complete database with the results from the descriptive studies may serve as a reference material for further studies. This makes it easier the search for new therapeutic options by supplementation, which may enhance the physiological response of wound healing.

Application to other areas of neuroscience

- Great efforts have been made in recent years to understand the mechanisms leading to the development of different neurological diseases.
- Proteomics techniques are widely applied in most areas of biology and medicine.

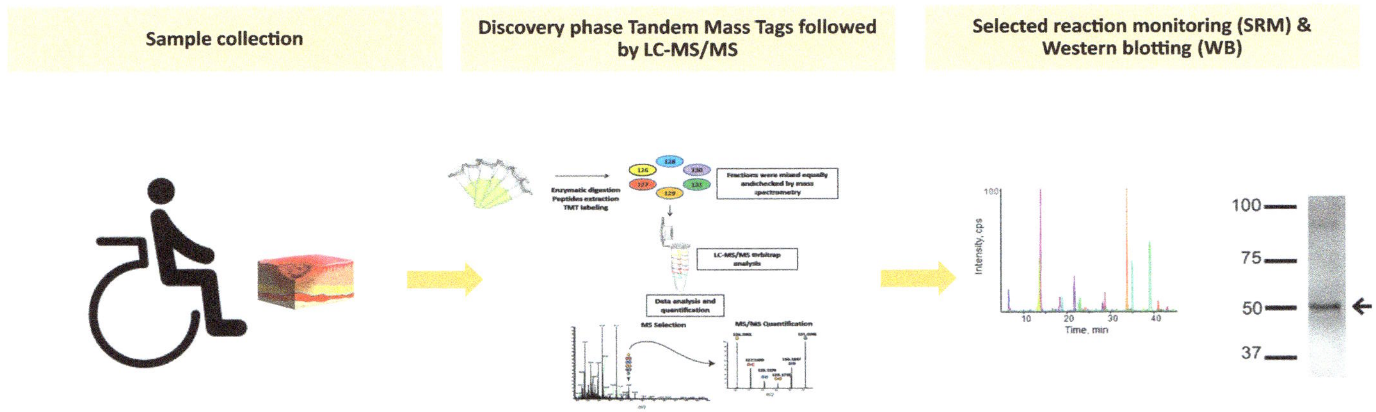

FIG. 6 Schematic representation of proteomics analysis in PUs of SCI patients. (A) Samples collection. Tissue samples were collected from subjects with SCI who were scheduled for PU surgery. (B) Discovery phase using Tandem Mass Tags (TMT) labeling followed by liquid chromatography tandem mass spectrometry (LC-MS/MS). TMT are isobaric chemical tags that provide multiplexing capabilities for relative quantitative proteomics analysis. This ability to perform concurrent MS analysis of multiple samples increases throughput and enables relative quantitation of up to 16 different samples derived from cells, tissues, or biological fluids. The tags consist of an MS/MS reporter group, a spacer arm, and an amine-reactive group. The amine-reactive group binds to the N-terminus of a peptide or to a lysine residue during labeling. In relative quantitation experiments, different isobaric tags are used to label different systemic conditions. Once labeled, all samples are mixed and analyzed in a single liquid chromatography-mass spectrometry (LC-MS) experiment. Because the isobaric tags possess the same chemical properties, all peptides from different TMT-labeled samples co-elute during LC separation. Once the peptides enter the mass spectrometer, they are detected simultaneously as a single and indistinguishable precursor ion peak. (C) Verification phase employing two orthogonal techniques, selected reaction monitoring (SRM) and western blotting (WB). SRM is a method used in tandem mass spectrometry in which an ion of a particular mass is selected in the first stage of a tandem mass spectrometer and an ion product of a fragmentation reaction of the precursor ion is selected in the second mass spectrometer stage for detection. Using a WB, researchers are able to identify specific proteins from a complex mixture of proteins extracted from cells. The technique uses three elements to accomplish this task: (1) separation by size, (2) transfer to a solid support, and (3) marking target protein using proper primary and secondary antibodies to visualize.

- The powerful "omic" technologies (mass spectrometry (MS)-based proteomics, focusing on the analysis of whole proteomes, protein-based interactions, and post-translational modifications) have provided an important amount of results in different diseases including neurological diseases.
- Complexity of the nervous system at the molecular level remains one of the central challenges in neuroscience.
- Proteomic tools offer a new platform for studies of complex biological functions involving large numbers and networks of proteins. Intracellular networks of proteins perform key functions in neurons and glia.
- Proteomics has applications in neurology and neurosciences, for example, studying neurotoxicology, neuroendocrinology, neuropsychiatric, neurodegenerative disorders, neurometabolism among others.
- Proteomics studies the determination of specific proteomic aspects of individual brain areas and body fluids in neurodegeneration too.
- The identification of new protein patterns involved in the pathogenesis of this kind of disease may lead to better treatment and new therapeutic interventions.
- In this innovative chapter, we explained how proteomics could have utility in therapy by supplementation, which may enhance the physiological response of neurological diseases.
- Although numerous challenges still exist, proteomics approach will increase the understanding of pathological mechanisms involved in different areas of neurosciences.

Key facts of proteomics in PUs of spinal cord patients

- Proteomics technologies offer considerable chances to identify significant changes in protein abundance during the progression of PU formation in an unbiased manner.
- Proteomics technologies allow the molecular determinants of complex samples like tissue obtained of PUs to be analyzed using mass spectrometry.
- Proteomics methods can permit the identification of important biological processes that are altered in PU.
- Proteomics may enable the discovery of new therapeutic options to improve clinical management of these kinds of patients.
- Proteomics helps to identify distinct risk phenotypes which would allow the screening in patients with high risk or to avoid gratuitous treatments with corresponding side effects.

Mini-dictionary of terms

- Spinal cord injury (SCI) is the result of trauma at any level of the spinal cord causing temporary or permanent damage, with a significant effect on the patient's physical and psychosocial well-being.
- International Standards for Neurological Classification of Spinal Cord Injury is the most common medical classification to characterize SCI.
- Pressure ulcers are defined as lesions on any skin surface that result from localized shear and/or compression for a prolonged period over bony prominences at certain anatomic locations.
- Proteomics is the large-scale study of proteomes. Proteomics can provide significant biological information for many biological problems.
- Proteome is a set of proteins produced in an organism, system, or biological context. The proteome is not constant; it differs from cell to cell and changes over time.
- MS proteomic methods provide important information by identifying many proteins and detecting post-translational modifications.
- Liquid chromatography-mass spectrometry/mass spectrometry (LC-MS/MS) is an analytical chemistry technique that combines the physical separation capabilities of liquid chromatography (or HPLC) with the mass analysis capabilities of mass spectrometry (MS).
- Selected reaction monitoring (SRM) is a method used in tandem mass spectrometry in which an ion of a particular mass is selected in the first stage of a tandem mass spectrometer and an ion product of a fragmentation reaction of the precursor ion is selected in the second mass spectrometer stage for detection.
- iTRAQ LC-MS/MS isobaric tags for relative and absolute quantitation (iTRAQ) is an isobaric labeling method used in quantitative proteomics by tandem mass spectrometry to determine the amount of proteins from different sources in a single experiment (Gafken & Lampe, 2006; Ross et al., 2004; Zieske, 2006). It uses stable isotope-labeled molecules that can be covalent bonded to the N-terminus and side chain amines of proteins.
- TMT-based, multiplexed isobaric labeling followed by LC-MS/MS. The Thermo Scientific Tandem Mass Tag Reagents are designed to enable identification and quantitation of proteins in different samples using tandem mass

spectrometry (MS). All mass tagging reagents within a set have the same nominal mass (i.e., are isobaric) and a chemical structure composed of an amine-reactive NHS ester group, a spacer arm (mass normalizer), and a mass reporter.

- Precision medicine is an emerging approach for disease treatment and prevention that takes into account individual variability in genes, environment, and lifestyle for each person.

Summary points

- Proteomic study of tissue, cells, and/or biological fluids will transform the diagnostic methods currently used in medicine.
- Advancements in proteomics allow researchers to measure a multitude of proteins simultaneously with excellent sensitivity and selectivity to detect low abundance proteins.
- This chapter focuses on the identification of important biological processes that are altered in pressure ulcers.
- These identifications of proteins may enable the discovery of new therapeutic options to improve clinical management of patients with spinal cord injury.
- A deeper understanding of the pathophysiology of disease would allow us to develop a precision medicine based on each patient, optimizing the effectiveness of pharmacological treatments.

Acknowledgments and sources of funding

This work was supported by grants from the Instituto de Salud Carlos III (PI14/01917, PI18/00995). Sociedad Española de Cardiología para la Investigación Básica 2017. Grant PRB3 (IPT17/0019—ISCIII-SGEFI/ERDF).

References

Albayar, A. A., Roche, A., Swiatkowski, P., Antar, S., Ouda, N., Emara, E., et al. (2019). Biomarkers in spinal cord injury: Prognostic insights and future potentials. *Frontiers in Neurology*, *10*, 27. Jan 29.

Alizadeh, A., Dyck, S. M., & Karimi-Abdolrezaee, S. (2019). Traumatic spinal cord injury: An overview of pathophysiology, models and acute injury mechanisms. *Frontiers in Neurology*, *10*, 282.

Antman, E. M., & Loscalzo, J. (2016). Precision medicine in cardiology. *Nature Reviews. Cardiology*, *13*(10), 591–602.

Baldan-Martin, M., Martin-Rojas, T., Corbacho-Alonso, N., Lopez, J. A., Sastre-Oliva, T., Gil-Dones, F., et al. (2020). Comprehensive proteomic profiling of pressure ulcers in patients with spinal cord injury identifies a specific protein pattern of pathology. *Advances in Wound Care (New Rochelle)*, *9*(5), 277–294. May 1.

Bhattacharya, S., & Mishra, R. K. (2015). Pressure ulcers: Current understanding and newer modalities of treatment. *Indian Journal of Plastic Surgery*, *48* (1), 4–16.

Black, J., Baharestani, M. M., Cuddigan, J., et al. (2007). National Pressure Ulcer Advisory Panel's updated pressure ulcer staging system. *Advances in Skin & Wound Care*, *20*(5), 269–274.

Bogie, K. M., Nuseibeh, I., & Bader, D. L. (1995). Early progressive changes in tissue viability in the seated spinal cord injured subject. *Paraplegia*, *33*, 141.

Canadian Agency for Drugs and Technologies in Health. (2013). *Dressing materials for the treatment of pressure ulcers in patients in long-term care facilities: A review of the comparative clinical effectiveness and guidelines*. Ottawa, ON: Canadian Agency for Drugs and Technologies in Health. [Internet]. Nov 18.

Chambers, G., Lawrie, L., Cash, P., & Murray, G. I. (2000). Proteomics: A new approach to the study of disease. *The Journal of Pathology*, *192*(3), 280–288. Nov.

Charlifue, S., Jha, A., & Lammertse, D. (2010). Aging with spinal cord injury. *Physical Medicine and Rehabilitation Clinics of North America*, *21*(2), 383–402.

Chen, Y., DeVivo, M. J., & Jackson, A. B. (2005). Pressure ulcer prevalence in people with spinal cord injury: Age-period-duration effects. *Archives of Physical Medicine and Rehabilitation*, *86*(6), 1208–1213.

Dominguez, D. C., Lopes, R., & Torres, M. L. (2007). Proteomics: Clinical applications. *Clinical Laboratory Science*, *20*(4), 245–248.

Dwivedi, M. K., Bhagat, A. K., Srivastava, R. N., Jain, A., Baghel, K., & Raj, S. (2017). Expression of MMP-8 inpressure injuries in spinal cord injury patients managed by negative pressure wound therapyor conventional wound care: A randomized controlled trial. *Journal of Wound, Ostomy, and Continence Nursing*, *44*, 343–349.

Evans, J. M., Andrews, K. L., Chutka, D. S., Fleming, K. C., & Garness, S. L. (1995). Pressure ulcers: Prevention and management. *Mayo Clinic Proceedings*, *70*(8), 789–799. Aug.

Fuster, V. (2014). A first dilemma in cardiovascular medicine: Adherence versus personalized therapy. *Journal of the American College of Cardiology*, *64* (10), 1059–1060.

Gafken, P. R., & Lampe, P. D. (2006). Methodologies for characterizing phosphoproteins by mass spectrometry. *Cell Communication & Adhesion*, *13*(5–6), 249–262.

Garmaroudi, F. S., et al. (2016). Systems pharmacology and rational polypharmacy: Nitric oxide-cyclic GMP signaling pathway as an illustrative example and derivation of the general case. *PLoS Computational Biology*, *12*(3), e1004822.

Hooff, G. P., van Huizen, N. A., Meesters, R. J., Zijlstra, E. E., Abdelraheem, M., Abdelraheem, W., et al. (2011). Analytical investigations of toxic p-phenylenediamine (PPD) levels in clinical urine samples with special focus on MALDI-MS/MS. *PLoS One*, *6*(8), e22191.

Jarchum, I. (2016). Proteomics: Proteome quantification compared. *Nature Methods*, *13*, 972.

Jaul, E. (2010). Assessment and management of pressure ulcers in the elderly: Current strategies. *Drugs & Aging*, *27*(4), 311–325. Apr 1.

Jiang, H., Zhang, L., Yu, Y., et al. (2014). A pilot study of angiogenin in heart failure with preserved ejection fraction: A novel potential biomarker for diagnosis and prognosis? *Journal of Cellular and Molecular Medicine*, *18*, 2189–2197.

Lachenbruch, C., Ribble, D., Emmons, K., & Van Gilder, C. (2016). Pressure ulcer risk in the incontinent patient: Analysis of incontinence and hospital-acquired pressure ulcers from the International Pressure Ulcer Prevalence™ Survey. *Journal of Wound, Ostomy, and Continence Nursing*, *43*(3), 235–241. May-Jun.

Lee, M.-S., et al. (2012). Personalized medicine in cardiovascular diseases. *Korean Circulation Journal*, *42*(9), 583–585.

Legrain, P., et al. (2011). The human proteome project: Current state and future direction. *Molecular & Cellular Proteomics*, *10*(7), M111.009993.

Leopold, J. A., & Loscalzo, J. (2018). Emerging role of precision medicine in cardiovascular disease. *Circulation Research*, *122*(9), 1302–1315.

Lindsey, M. L., Mayr, M., Gomes, A. V., Delles, C., et al. (2015). Transformative impact of proteomics on cardiovascular health and disease: A scientific statement from the American Heart Association. *Circulation*, *132*(9), 852–872.

Liu, Z., Cui, X., Hu, Y., & Zhang, P. (2020). Quantitative iTRAQ LC-MS/MS reveals muscular proteome profiles of deep pressure ulcers. *Bioscience Reports*, *40*(6). Jun 26. BSR20200563.

Liu, Z., Ren, L., Cui, X., Guo, L., Jiang, B., Zhou, J., et al. (2018). Muscular proteomic profiling of deep pressure ulcers reveals myoprotective role of JAK2 in ischemia and reperfusion injury. *American Journal of Translational Research*, *10*(11), 3413–3429. Nov 15.

Manrai, A. K., et al. (2016). Genetic misdiagnoses and the potential for health disparities. *The New England Journal of Medicine*, *375*(7), 655–665.

Marshall, T., & Williams, K. M. (2002). Proteomics and its impact upon biomedical science. *British Journal of Biomedical Science*, *59*(1), 47–64.

Martin-Rojas, T., Sastre-Oliva, T., Esclarín-Ruz, A., Gil-Dones, F., Mourino-Alvarez, L., Corbacho-Alonso, N., et al. (2020). Effects of growth hormone treatment and rehabilitation in incomplete chronic traumatic spinal cord injury: Insight from proteome analysis. *Journal of Personalized Medicine*, *10*(4), 183.

Medical Advisory Secretariat. (2009). Pressure ulcer prevention: An evidence-based analysis. *Ontario Health Technology Assessment Series*, *9*(2), 1–104.

Mirnezami, R., Nicholson, J., & Darzi, A. (2012). Preparing for precision medicine. *The New England Journal of Medicine*, *366*(6), 489–491.

O'Meara, S., Cullum, N., Majid, M., & Sheldon, T. (2000). Systematic reviews of wound care management: Antimicrobial agents for chronic wounds; diabetic foot ulceration. *Health Technology Assessment*, *4*(21), 1–237.

Pan, S., Aebersold, R., Chen, R., et al. (2009). Mass spectrometry based targeted protein quantification: Methods and applications. *Journal of Proteome Research*, *8*, 787–797.

Pandey, A., Garg, S., Khunger, M., Darden, D., Ayers, C., Kumbhani, D. J., et al. (2015). Dose-response relationship between physical activity and risk of heart failure: A meta-analysis. *Circulation*, *132*(19), 1786–1794. 10.

Paulovich, A. G., et al. (2008). The interface between biomarker discovery and clinical validation: The tar pit of the protein biomarker pipeline. *Proteomics. Clinical Applications*, *2*(10 – 11), 1386–1402.

Rifai, N., Gillette, M. A., & Carr, S. A. (2006). Protein biomarker discovery and validation: The long and uncertain path to clinical utility. *Nature Biotechnology*, *24*(8), 971–983.

Rosin, N. R., Tabibi, R. S., Trimbath, J. D., & Henzel, M. K. (2020). A primary care provider's guide to prevention and management of pressure injury and skin breakdown in people with spinal cord injury. *Topics in Spinal Cord Injury Rehabilitation*, *26*(3), 177–185. Summer.

Ross, P. L., Huang, Y. N., Marchese, J. N., Williamson, B., Parker, K., Hattan, S., et al. (2004). Multiplexed protein quantitation in *Saccharomyces cerevisiae* using amine-reactive isobaric tagging reagents. *Molecular & Cellular Proteomics*, *3*(12), 1154–1169.

Schuld, C., Franz, S., Brüggemann, K., Heutehaus, L., Weidner, N., Kirshblum, S. C., et al. (2016). International standards for neurological classification of spinal cord injury: Impact of the revised worksheet (revision 02/13) on classification performance. *The Journal of Spinal Cord Medicine*, *39*, 504–512.

Taverna, D., Pollins, A. C., Sindona, G., Caprioli, R. M., & Nanney, L. B. (2015). Imaging mass spectrometry for assessing cutaneous wound healing: Analysis of pressure ulcers. *Journal of Proteome Research*, *14*(2), 986–996. Feb 6.

Van de Merbel, N. C. (2019). Protein quantification by LC–MS: A decade of progress through the pages of bioanalysis. *Bioanalysis*, *11*, 629–644.

Van Middendorp, J. J., Goss, B., Urquhart, S., et al. (2011). Diagnosis and prognosis of traumatic spinal cord injury. *Global Spine Journal*, *1*(1), 1–8.

Wilhelm, M., Schlegl, J., Hahne, H., Gholami, A. M., Lieberenz, M., Savitski, M. M., et al. (2014). Mass-spectrometry-based draft of the human proteome. *Nature*, *509*, 582–587.

Yates, J. R., Ruse, C. I., & Nakorchevsky, A. (2009). Proteomics by mass spectrometry: Approaches, advances, and applications. *Annual Review of Biomedical Engineering*, *11*, 49–79.

Zieske, L. R. (2006). A perspective on the use of iTRAQ reagent technology for protein complex and profiling studies. *Journal of Experimental Botany*, *57*(7), 1501–1508.

Chapter 13

Innate immune responses of glia and inflammatory cells in spinal cord injury

Lun Li[a,*], Cigdem Acioglu[a,*], Robert F. Heary[b], and Stella Elkabes[a]

[a]*Reynolds Family Spine Laboratory, Department of Neurosurgery, New Jersey Medical School-Rutgers, The State University of New Jersey, Newark, NJ, United States,* [b]*Department of Neurological Surgery, Hackensack Meridian School of Medicine, Nutley, NJ, United States*

Abbreviations

AQP4	aquaporin-4
ASC	apoptosis-associated speck-like protein containing a CARD
ATP	adenosine triphosphate
bFGF	basic fibroblast growth factor
CARD	caspase activation and recruitment domain
CLR	C-type lectin receptor
CNS	central nervous system
CpG-ODN	cytosine-phosphate-guanine oligodeoxynucleotides
CSPG	chondroitin sulfate proteoglycan
DAMP	danger-associated molecular pattern
ECM	extracellular matrix
HA	hyaluronan
HMGB1	high-mobility group box 1
HSP	heat shock protein
IL	interleukin
Mincle	macrophage-inducible C-type lectin
NLR	NOD-like receptor
NLRP	NLR with pyrin domain
NLRC	NLR with CARD
NOD	nucleotide-binding oligomerization domain
ODN	oligodeoxynucleotides
OPC	oligodendrocyte progenitor cell
P2X7	P2X purinoreceptor 7
PAMP	pathogen-associated molecular pattern
Panx1	pannexin 1
PRR	pattern recognition receptor
RLR	retinoic acid-inducible gene 1 (RIG-1)-like receptor
ROS	reactive oxygen species
SC	spinal cord
SCI	spinal cord injury
TGF-β	transforming growth factor-beta
TIR	toll/interleukin-1 receptor
TLR	toll-like receptor
TnC	tenascin-C
TnR	tenascin-R

* Equal contribution.

Cellular, Molecular, Physiological, and Behavioral Aspects of Spinal Cord Injury. https://doi.org/10.1016/B978-0-12-822427-4.00013-7

Introduction

The central nervous system (CNS) was historically considered an immune-privileged organ. However, this concept has changed as compelling evidence demonstrates that CNS infections and injuries as well as neurodegenerative diseases provoke innate immune responses in the brain and spinal cord (SC). Signals derived from infectious agents, pathogenic proteins in neurodegenerative diseases, and danger signals released by stressed and damaged cells following neural injuries activate innate immunity in the CNS via engagement of pattern recognition receptors (PRRs) (Russo & McGavern, 2015). Both neurons and glia express PRRs. Stimulation of PRR signaling in resident CNS cells leads to neuroinflammation, with beneficial and detrimental outcomes that are context-, CNS region-, cell type-, PRR-, and ligand-dependent. Microglia and astrocytes are the principal innate immune cells of the CNS. However, PRRs in neurons and oligodendrocytes mediate non-immune functions and contribute to the overall functional outcomes.

During traumatic spinal cord injury (SCI), a cascade of deleterious mechanisms follows the initial mechanical damage and inflicts secondary injury to the adjacent spared tissue. SCI elicits microglial activation and astrocyte reactivity at the lesion site and in regions remote from the injury epicenter (Pallottie et al., 2018). Activated/reactive glia play both protective and detrimental roles that are context-dependent. Glial activation encompasses morphological, molecular and functional changes, increased proliferation and migration to the affected site, and altered gene expression (Sofroniew, 2015). Stimulation of PRR signaling in activated microglia and reactive astrocytes leads to cytokine and chemokine release, which initiates neuroinflammation and promotes chemotaxis of immune cells to the injury site. The breaching of the blood-spinal cord barrier and the release of interleukin 1β (IL-1β) by activated microglia and reactive astrocytes enhances vascular permeability and the extravasation of immune cells (Fahey & Doyle, 2019). Reactive astrocytes foster the formation of a glial scar around the lesion (Burda & Sofroniew, 2014). The glial scar limits secondary tissue damage by confining toxic molecules released by damaged cells to the lesion site, and by restricting the infiltration of immune cells. The glial scar is also considered as an impediment to axonal regeneration. The heterogeneity and phenotypic plasticity of astrocytes and microglia, as well as the crosstalk between these glial subtypes influence neuroinflammation and neural dysfunction (Liddelow, Marsh, & Stevens, 2020).

Monocytes and neutrophils are the first cells that infiltrate the lesion site followed by macrophages and T-lymphocytes (Stirling & Yong, 2008). Neutrophils have also been implicated in secondary tissue loss in SCI (Anwar, Al Shehabi, & Eid, 2016). However, following SCI, depletion of neutrophils impairs the wound-healing response of astrocytes, reduces white matter sparing and axonal preservation, and impairs functional recovery. This suggests that neutrophils could help recovery and repair. Neutrophils and macrophages express PRRs (Thomas & Schroder, 2013). It is highly probable that activation of PRRs on these cells contributes to the neuroinflammatory response in SCI.

Toll-like receptors (TLRs), nucleotide-binding oligomerization domain (NOD)-like receptors (NLRs), retinoic acid-inducible gene 1 (RIG-1)-like receptors (RLRs), and the C-type lectin receptors (CLRs) are among the PRR families (Takeuchi & Akira, 2010). This chapter will provide a brief overview of the major PRRs expressed in the CNS. As TLRs are the first identified and best-characterized PRRs, their role in SCI will be discussed.

Overview of PRRs

PRRs are highly conserved among species and recognize molecules derived from pathogens, collectively called pathogen-associated molecular patterns (PAMPs). PRRs are also activated by endogenous ligands released by damaged or dead cells, namely, danger-associated molecular patterns (DAMPs) (Amarante-Mendes et al., 2018). The PAMPs that activate PRRs, include bacterial lipopolysaccharides, lipoproteins and lipopeptides, flagellin, single-or double-stranded RNA, unmethylated CpG DNA, and zymosan. DAMPs include high-mobility group box 1 (HMGB1), adenosine triphosphate (ATP), heat shock proteins (HSPs), mitochondrial DNA, histones, interleukin 1α (IL-1α), IL-33, amyloid β, uric acid, crystallin, and extracellular matrix (ECM) components such as hyaluronic acid, fibronectin, and fibrinogen.

TLRs

TLRs are the mammalian homologs of Toll receptors, which were first identified in *Drosophila melanogaster*. To date, 10 functional TLRs (TLR1–10) have been identified in humans and 12 TLRs (TLR1–9 and TLR11–13) in the mouse. TLRs 1, 2, 4–6, 11–12 are localized at the cell surface whereas TLRs 3, 7–9, and 13 are intracellular receptors found in endosomal compartments and endoplasmic reticulum (Fig. 1; (Heiman, Pallottie, Heary, & Elkabes, 2014). The initiation of TLR signaling necessitates the formation of either homodimers or heterodimers. TLR2 forms heterodimers with TLR1 or TLR6.

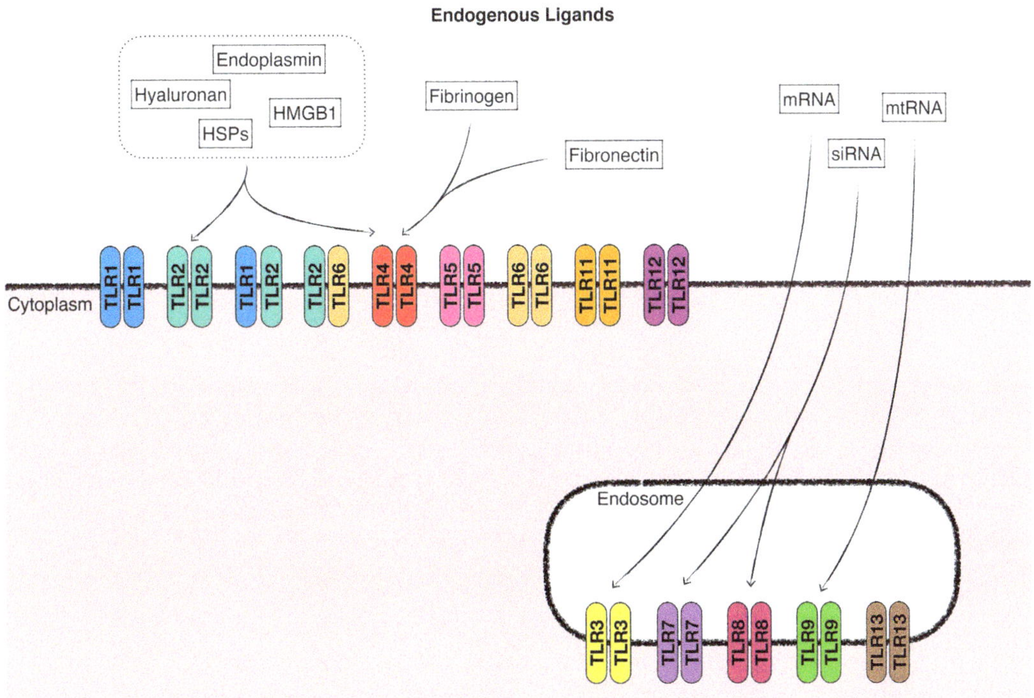

FIG. 1 Cell surface and intracellular Toll-like receptors (TLRs) in the mouse. Toll-like receptors form homodimers or heterodimers. The heterodimers are exemplified by TLR2/TLR1 and TLR2/TLR6. Humans express only TLR1–10. Mice do not express functional TLR10. Examples of endogenous ligands are provided. HSPs, heat shock proteins; HMGB1, high-mobility group box 1.

The type of heterodimer formed influences the response elicited by specific stimuli. Some TLRs also utilize co-receptors. CD14 is one of the best-known co-receptors and it interacts with TLR4.

All TLRs, except TLR3, utilize the MyD88-signaling pathway. TLR3 signals through the Toll/interleukin-1 receptor (TIR)-domain-containing adapter-inducing interferon-β (TRIF)-dependent signaling pathway. TLR4 utilizes both the MyD88-dependent and TRIF-dependent signaling pathways. Activation of TLR signaling leads to the transcription of inflammatory cytokines and type I interferons (Heiman et al., 2014). Synergy and counteractions between different TLRs have been reported (Rosenberger et al., 2014).

NLRs, inflammasome complexes, and TLR-inflammasome cooperation

NOD-like receptors are cytoplasmic PRRs that detect a wide range of PAMPs and DAMPs. Members of the NLR family have a C-terminal leucine-rich repeat (LRR) domain, a central nucleotide-binding oligomerization domain (NACHT), and a variable interaction domain at the amino terminus. The NLR family is divided into various sub-families distinguishable by their N-terminal effector domains. Two NLR families, NLR with pyrin domain (NLRP) and NLR with caspase activation and recruitment domain (CARD) (NLRC), will be further discussed, as they form inflammasomes (Walsh, Muruve, & Power, 2014).

Inflammasomes are cytoplasmic multi-protein complexes that regulate inflammation. The complex contains a cytoplasmic PRR, pro-caspase-1, and an adaptor protein such as apoptosis-associated speck-like protein containing a CARD (ASC), which mediates the interaction between the PRR and pro-caspase-1. In some inflammasome complexes, the PRR and pro-caspase-1 do not require an adaptor protein and interact directly. The assembly of the inflammasome complex leads to the auto-proteolytic cleavage of pro-caspase-1, which results in caspase-1 activation. Active caspase-1 cleaves pro-IL-1β and pro-IL-18 to generate mature IL-1β and IL-18. Upon release, IL-1β initiates neuroinflammation and neuronal death, whereas caspase-1 induces pyroptosis (Walsh et al., 2014). Although inflammasome signaling has been predominantly studied in microglia, neurons, astrocytes, and oligodendrocytes also express the components of inflammasome complexes. NLRP1, NLRP3, and NLRC4 are primarily found in microglia, whereas NLRP2 is mostly localized in astrocytes. Neurons express NLRP1 (Mamik & Power, 2017). NLRP1 forms an inflammasome by interacting directly with pro-caspase-1, and its expression is increased following SCI (de Rivero Vaccari, Lotocki, Marcillo, Dietrich, & Keane, 2008). The NLRP2 inflammasome has been detected in human primary astrocyte cultures and is activated by ATP. The effects of ATP are mediated by the P2X purinoreceptor 7 (P2X7) and the membrane channel pannexin 1 (Panx1). Inhibition of Panx1 and P2X7 attenuates ATP-dependent NLRP2 inflammasome activation. Studies suggest that astrocytic NLRP2 inflammasome is an important component of the inflammatory response following CNS injuries (Minkiewicz, de Rivero Vaccari, & Keane, 2013). The NLRP3 inflammasome is the most widely studied and best-characterized inflammasome. Activation of the NLRP3 inflammasome requires a two-step process. A first signal, e.g., a TLR ligand, primes the cell and enhances pro-IL-1β and NLRP3 expression. Subsequently, a second signal triggers the assembly and activation of the NLRP3 inflammasome complex. Increased K^+ efflux, ATP, and release of reactive oxygen species (ROS) by damaged mitochondria are among the triggers that activate NLRP3 inflammasome (Voet, Srinivasan, Lamkanfi, & van Loo, 2019; Walsh et al., 2014). The cooperation between TLRs and NLRP3 inflammasome is essential for IL-1β production and the initiation of inflammation (Fig. 2). Although the role of the NLRP3 inflammasome in neurodegenerative diseases and CNS infections is well established, its contribution to SCI remains inadequately investigated. NLRP3 expression is increased in neurons, microglia, and astrocytes following SCI. Inhibition of the NLRP3 inflammasome attenuates neuroinflammation, microglial activation, motor neuron death, and tissue damage, and improves recovery of locomotor function in mice with SCI (Jiang, Li, He, Zhou, & Zhu, 2017). NLRC4 has also been implicated in sterile inflammation associated with CNS injury and diseases (Voet et al., 2019). Although DAMPs that activate NLRC4 have not been fully identified, lysophosphatidylcholine, a major component of the low-density protein, induces NLRP4 inflammasome activation in microglia and astrocytes leading to sterile inflammation (Freeman et al., 2017).

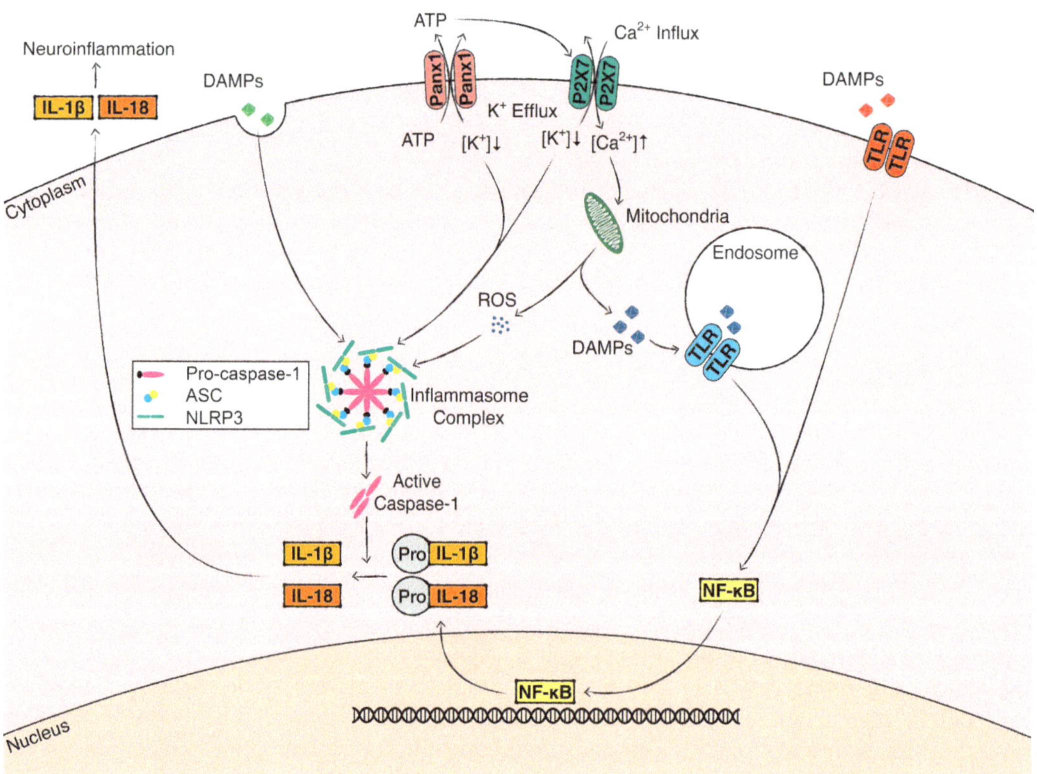

FIG. 2 Cooperation between NLRP3 inflammasome and TLRs in macrophages. Activation of TLR signaling promotes transcription of pro-IL-1β and pro-IL-18. Extracellular ATP, released by damaged cells or through Panx1 channels, activates P2X7 purinergic receptors, and enhances K^+ efflux. The drop in intracellular K^+, ROS released by damaged mitochondria, and other DAMPs initiate the assembly of the inflammasome complex. This is followed by cleavage of pro-caspase-1 and activation of caspase-1. Active caspase-1 cleaves pro-IL-1β and pro-IL-18. Mature IL-1β and IL-18 are released and promote neuroinflammation. ATP, adenosine triphosphate; ASC, apoptosis-associated speck-like protein containing a CARD; DAMPs, danger-associated molecular patterns; NLRP3, nucleotide-binding oligomerization domain-like receptor with pyrin domain; IL, interleukin; P2X7, P2X purinoreceptor 7; Panx1, pannexin 1; ROS, reactive oxygen species.

CLRs and RLRs

C-type lectin receptors are a family of transmembrane and soluble receptors that bind carbohydrates. They trigger multiple intracellular signaling pathways leading to nuclear factor kappa-light-chain-enhancer of activated B cells (NF-κB)-mediated type I interferon and cytokine production. Dectin-1 is a CLR subtype. Stimulation of dectin-1 in the intact rodent SC promotes macrophage activation, demyelination, and axonal injury. Accordingly, axonal dieback is lower in dectin-1-knockout mice sustaining a SCI compared to injured wild type controls. These negative effects of dectin-1 are alleviated by simultaneous activation of TLR2 (Gensel et al., 2015), suggesting crosstalk between TLR2 and dectin-1. Another CLR, macrophage-inducible C-type lectin (Mincle), is expressed in brain microglia and neurons, and SC neurons (Yang, Dai, Chen, & Cui, 2020). However, in Mincle-deficient mice sustaining a SCI, locomotor recovery and lesion size are not different than in injured wild-type controls (Arumugam et al., 2017). Nevertheless, further studies are needed to determine whether ligand-induced activation of Mincle impacts the outcomes of SCI. Even though synergy between TLRs and Mincle has been reported (van Haren et al., 2016), it is still unknown whether such synergistic interactions occur in SCI.

In the CNS, both neurons and glia express RLRs, which are cytoplasmic receptors that recognize RNA viruses and self-RNA. SCI and stretch injury of cultured astrocytes activate RLR signaling, whereas ligand-induced stimulation of RLRs upregulates markers of reactive astrocytes (de Rivero Vaccari et al., 2012). These studies indicate that RLRs modulate the innate immune response of astrocytes.

It is possible that multiple PRRs are simultaneously activated in SCI. The interactions between different TLRs, the crosstalk between TLRs and other PRRs (Pashenkov, Murugina, Budikhina, & Pinegin, 2019), and the cooperation between innate immune receptor signaling pathways (Vajjhala, Ve, Bentham, Stacey, & Kobe, 2017) likely modulate the inflammatory reaction, glial reactivity, and cell death in SCI.

Role of TLRs in SCI

Even though all TLRs are expressed in the SC, TLR2, TLR4, and TLR9 are the best studied in pre-clinical rodent models of traumatic SCI (Li, Acioglu, Heary, & Elkabes, 2020). The expression of cell surface TLRs and the intracellular TLR7 is increased following SCI (Kigerl et al., 2007). In a mouse model of mid-thoracic SC contusion injury, TLR9 transcript levels progressively increase within 7-days post-injury (dpi) and remain high even at 35 dpi (Fig. 3).

TLR2 in SCI

Initial studies on the roles of TLR2 in SCI were performed on whole body TLR2 knockout mice sustaining a SCI. These investigations revealed that the recovery of gait and motor coordination was impaired in TLR2 knockout mice as compared to wild-type controls (Kigerl et al., 2007). Axonal dieback was also more pronounced in TLR2 knockout mice (Stirling et al., 2014). These findings suggested that TLR2 plays beneficial roles in SCI. In fact, in an ex vivo laser-induced SCI model, stimulation of TLR2 by use of the agonist Pam2CSK4 mitigated axonal dieback and secondary degeneration of axons by promoting the microglial neuroprotective properties (Stirling et al., 2014). The protective role of TLR2 was

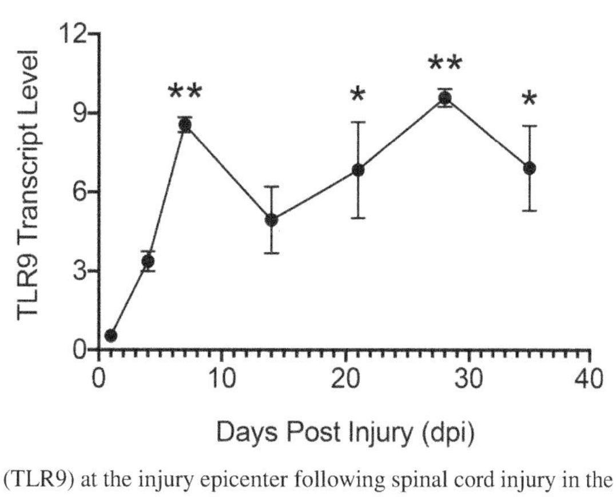

FIG. 3 Expression of Toll-like receptor 9 (TLR9) at the injury epicenter following spinal cord injury in the mouse. TLR9 transcript levels significantly increase by 7 days post-injury (dpi) and remain high until 35 dpi. Significantly different from 1 dpi by one-way ANOVA followed by Tukey's post hoc test, $*P < 0.05$, $**P < 0.01$. Values represent fold increase compared to the housekeeping gene and are presented as mean ± S.E.M. $n = 3$–4 (unpublished data).

further corroborated by the use of another TLR2-specific agonist, Pam3CSK4, in a similar SCI model (Stivers et al., 2017). In a rat model of SC crush injury, intraspinal injection of Pam2CSK4 increased macrophage activation, reduced axonal dieback, and decreased lesion size (Gensel et al., 2015). Since earlier investigations showed that activated macrophages promote axonal dieback (Evans et al., 2014), the simultaneous occurrence of enhanced macrophage reactivity and reduced axonal dieback raised the possibility that the TLR2 agonist attenuates axonal dieback by altering macrophage effector function.

TLR4 in SCI

Early investigations on the contribution of TLR4 to SCI were performed on TLR4-mutant mice that sustained a moderate SC contusion injury. Recovery of locomotor coordination was inferior in the TLR4-mutant mice compared to injured wild-type controls, suggesting that expression of TLR4 supports functional recovery (Kigerl et al., 2007). Increased demyelination and oligodendrocyte loss, reduced oligodendrogenesis, delayed myelin debris clearance, enhanced astrogliosis, and altered regulation of cytokine and growth factor expression were also observed in TLR4-mutant mice following SCI (Church, Kigerl, Lerch, Popovich, & McTigue, 2016; Kigerl et al., 2007). Although intraspinal injection of TLR4 agonists into the intact SC resulted in acute oligodendrocyte loss, this was attributed to the toxicity of activated microglia/macrophages at the injection site. This initial oligodendrocyte loss was followed by proliferation of oligodendrocyte progenitor cells (OPCs) and oligodendrogenesis (Church et al., 2016; Schonberg, Popovich, & McTigue, 2007), supporting the idea that TLR4 signaling can induce repair mechanisms. Studies have also shown that TLR4 positive astrocytes are found in the vicinity of the injury epicenter but not at distant locations or in the intact controls (Nishimura et al., 2013), indicating that astrocytes upregulate TLR4 expression in response to local changes elicited by SCI. It is possible that the endogenous TLR4 ligand HMGB1, which is dramatically increased at the lesion site (Kigerl, Lai, Wallace, Yang, & Popovich, 2018), is one of the triggers that increase astroglial TLR4 expression, as suggested by in vitro investigations (Famakin et al., 2020). In astrocyte cultures, which were maintained under conditions that simulate hypoxic injury, activation of TLR4 by HMGB1 resulted in the upregulation of the water channel Aquaporin-4 (AQP4) and astrocyte swelling. These findings suggest that TLR4 modulates a key mechanism underlying CNS edema (Sun et al., 2017). TLR4 and MyD88 have also been implicated in the necroptosis of reactive astrocytes located around the lesion core following SCI. The expression of TLR4 and MyD88 increased in reactive astrocytes undergoing necroptosis in vivo, and in a model of astrocyte necroptosis in vitro. Inhibition of MyD88-dependent signaling attenuated necroptosis in cultured astrocytes (Fan et al., 2016). The aforementioned findings, taken together, support the notion that TLR4 signaling influences various mechanisms following SCI. Some of these mechanisms foster repair whereas others contribute to secondary injury. However, the net effect of TLR4 signaling in SCI appears to be beneficial.

TLR9 in SCI

In contrast to TLR2 and TLR4, stimulation of TLR9 by the agonist ODN 1826 increases the number of inflammatory cells and the expression of the pro-inflammatory cytokine TNF-α at the injury epicenter following a severe SC contusion injury (David et al., 2013). The agonist also exacerbates bladder dysfunction (unpublished results). Therefore, it was hypothesized that inhibition of TLR9 signaling could positively impact the outcomes of SCI by preventing the activation of TLR9 by endogenous ligands. Indeed, intrathecal administration of a TLR9 antagonist, ODN 2088, to mice sustaining a SCI, decreased inflammatory cell number and the expression of pro-inflammatory mediators at the injury epicenter, reduced proliferation of reactive astrocytes at the glial scar, preserved injured proximal axons, increased the percentage of the anti-inflammatory M2 macrophages, ameliorated tissue sparing, and decreased the levels of chondroitin sulfate proteoglycans (CSPGs), ECM proteins which have been implicated in the inhibition of axonal regeneration (David et al., 2013; David et al., 2014; Li, Ni, Eugenin, Heary, & Elkabes, 2019; Li, Ni, Heary, & Elkabes, 2020). At the functional level, intrathecal treatment with the TLR9 antagonist alleviated urinary retention and pain hypersensitivity but did not improve locomotor function (David et al., 2013, 2014).

In summary, the studies on TLR2, TLR4, and TLR9 indicate that TLR signaling can lead to both beneficial or deleterious effects following SCI depending on the TLR involved and the cell-type engaged (Fig. 4). Some of these effects could be the result of overall changes in the inflammatory milieu, whereas others could be due to alterations in glial and neuronal function via direct activation of TLR signaling in these cells.

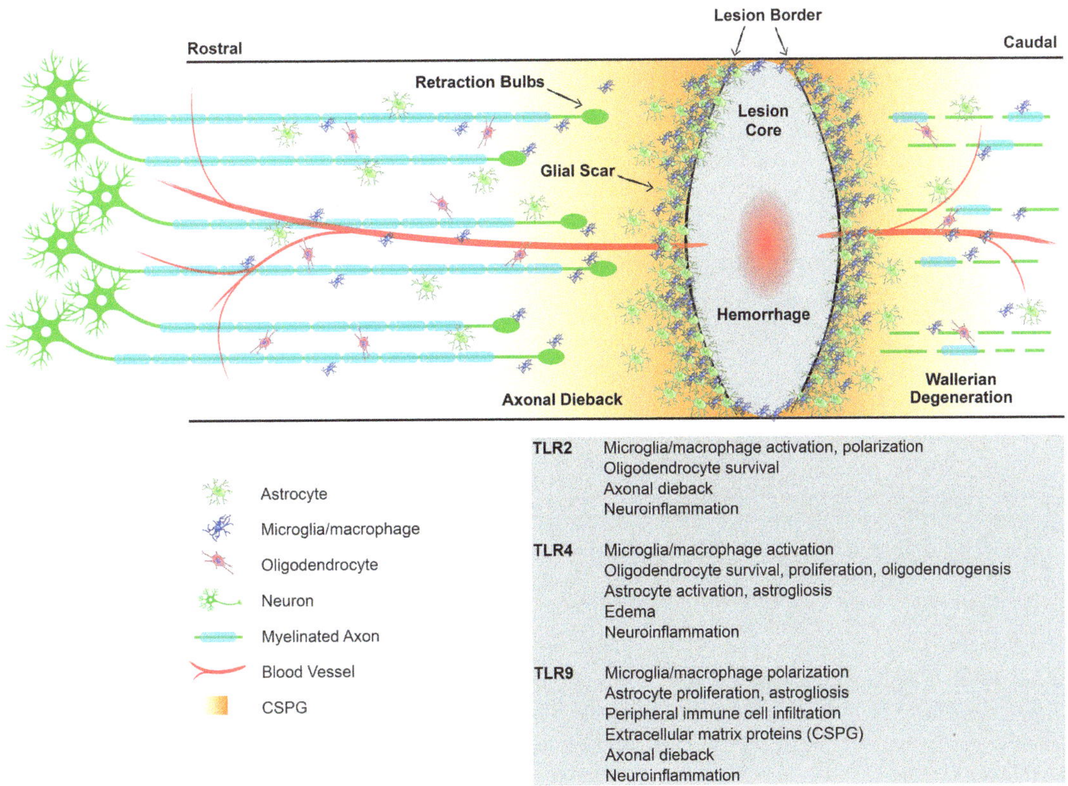

FIG. 4 Role of Toll-like receptors (TLRs) in secondary injury mechanisms following spinal cord injury. Injured proximal axons retract from the injury site (axonal dieback). A glial scar forms around the lesion core. Chondroitin sulfate proteoglycans (CSPGs) increase at the lesion site. The mechanisms influenced by TLR2, TLR4, and TLR9 are listed.

Non-immune functions of TLRs in neurons: Relevance to SCI

Mouse neurons express TLR2–4 and TLR7–9 (Acioglu et al., 2016; Carty & Bowie, 2011; Liu et al., 2013; Ma et al., 2006). This raises the possibility that TLRs exert non-immune functions in the CNS. In fact, neuronal TLR signaling modulates neurite outgrowth and neuronal survival, in vitro, processes that have relevance to SCI as well as CNS development. TLR3 and TLR8 are highly expressed in growth cones. Agonist-elicited activation of TLR3 and TLR8 causes growth cone collapse and irreversible inhibition of neurite outgrowth, followed by neuronal apoptosis in vitro (Cameron et al., 2007; Liu et al., 2013; Ma et al., 2006). It is plausible that inhibition of neuronal TLR3 and TLR8 signaling facilitates axonal regrowth or sprouting following SCI, a possibility that has not been investigated. Interestingly, neuronal TLR9 antagonism protects against excitotoxic death, in vitro, through the modulation of the endoplasmic reticulum stress response (Acioglu et al., 2016). Moreover, TLR2- and TLR4-deficiency renders neurons more resistant to death provoked by energy deprivation, suggesting that TLR2 and TLR4 increase the vulnerability of neurons to insults. Even though TLR2 and TLR4 agonists play an overall positive role in SCI by fostering microglia/macrophage-mediated neuroprotection, it is conceivable that they also induce neuronal death through direct effects on these cells (Fig. 5). These reparative versus neurodegenerative effects of TLR2 and TLR4 signaling could occur in parallel. This possibility has not been addressed and warrants further investigations.

Modulation of ECM proteins by TLRs: Relevance to SCI

The ECM regulates cell proliferation, survival, migration, and differentiation (Bonnans, Chou, & Werb, 2014). In the adult CNS, the ECM is primarily composed of hyaluronan (HA), sulfated proteoglycans, which include CSPGs, and Tenascin-R (TnR) (Haggerty, Marlow, & Oudega, 2017). However, after SCI, the molecular composition of the ECM is altered. Hyaluronan is degraded, CSPG expression is increased, and de novo synthesis of Tenascin-C (TnC) is induced (Yang et al., 2020). These changes influence both injury and repair mechanisms. Reactive astrocytes are an abundant source of CSPGs. In CNS pathologies, CSPGs modulate inflammation by acting on immune cells and by binding cytokines and chemokines

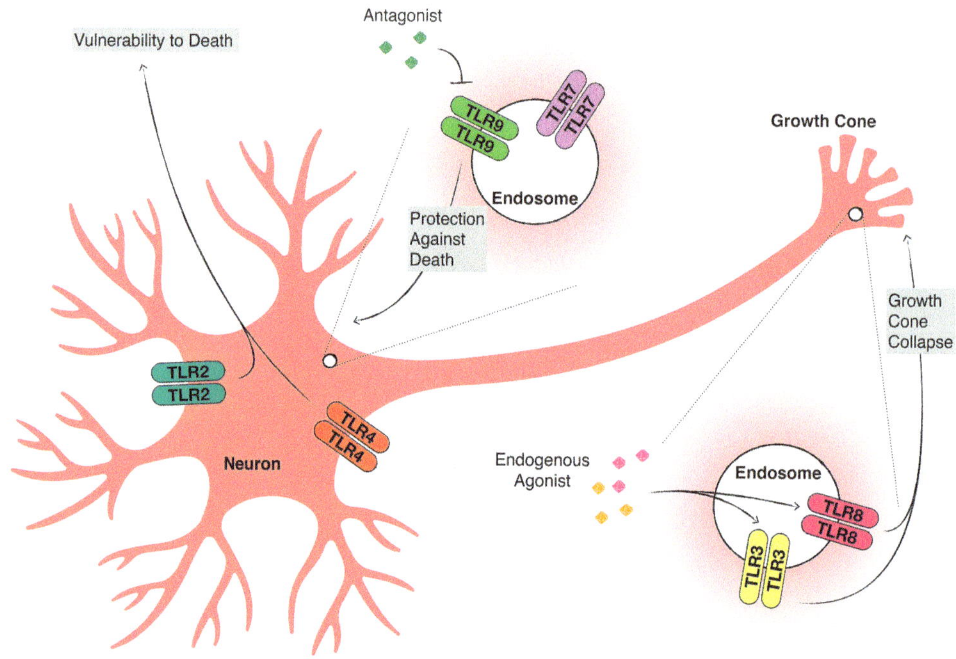

FIG. 5 Non-immune functions of neuronal Toll-like receptors (TLRs). Activation of neuronal TLR2 and TLR4 signaling has been associated with neuronal death whereas inhibition of TLR9 increases the resistance of spinal cord neurons to excitotoxic death. In addition, TLR3 accumulates in growth cones and causes growth cone collapse, whereas activation of TLR8 inhibits neurite outgrowth, in vitro.

(Stephenson & Yong, 2018). CSPGs inhibit axonal regeneration by excessive adhesion and immobilization of regenerating axons (Yang et al., 2020). Elimination of CSPGs at the glial scar promotes axonal regeneration and functional recovery following SCI (Bradbury et al., 2002). As indicated above, the TLR9 antagonist reduces CSPG levels at the glial scar following SCI (Li et al., 2019).

Tenascin-C is expressed during CNS development but is downregulated in adulthood (Wiese, Karus, & Faissner, 2012). However, after SCI, there is de novo synthesis of TnC around the lesion site (Zhang, Winterbottom, Schachner, Lieberman, & Anderson, 1997). TnC-deficient mice exhibit poor axonal regrowth and sprouting, and impaired locomotor recovery following SCI, suggesting that TnC facilitates axonal regeneration (Chen et al., 2010). Astrocyte TnC expression is increased by TGF-β and bFGF (Smith & Hale, 1997). Both growth factors are robustly upregulated at the injury epicenter following SCI (Smith & Strunz, 2005). Interestingly, TLR9 antagonism in astrocytes regulates the TGF-β and bFGF-elicited increase in TnC expression in an opposite manner (Fig. 6). The antagonist suppresses TGF-β-elicited increase

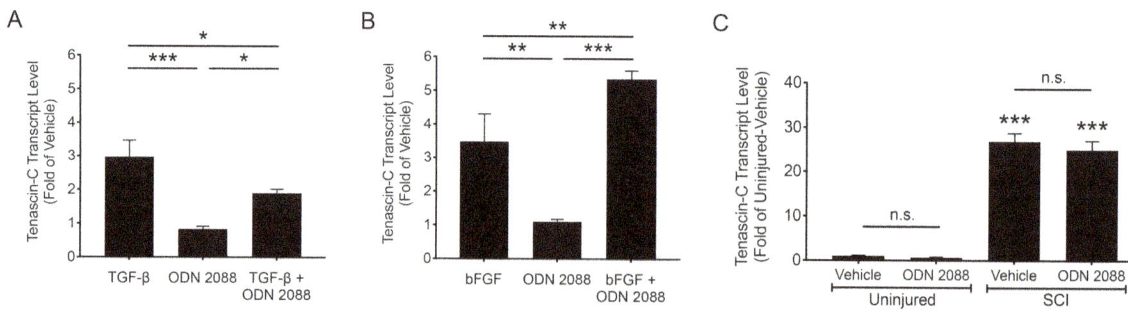

FIG. 6 Effects of a TLR9 antagonist on Tenascin (TnC) expression in spinal cord astrocytes, in vitro, and at the lesion site following spinal cord injury (SCI). (A) TnC transcript levels are increased in Transforming growth factor-beta (TGF-β)-treated spinal cord astrocytes, in vitro. The TLR9 antagonist, ODN 2088, inhibits the TGF-β-elicited increase. (B) TnC transcript levels are increased when astrocytes are treated with basic fibroblast growth factor (bFGF). TLR9 antagonism enhances the increase elicited by bFGF. (C) TnC levels are significantly increased at the lesion site following SCI. Intrathecal ODN 2088 treatment does not alter the increase in TnC expression. It is likely, that in vivo, ODN 2088 exerts opposite effects on TnC expression, depending on the stimuli that induce an increase, as indicated by the studies shown in A and B. The net outcome could be unaltered TnC expression at the injury site. The graphs in A and B show the mean of three independent experiments ± S.E.M. The graph in C shows a representative experiment with $n = 4$/group. Significantly different by one-way ANOVA followed by Tukey's post-hoc test $^{*}P < 0.05$, $^{**}P < 0.01$, $^{***}P < 0.001$. n.s. = not significant. The figure shows previously unpublished data.

in TnC transcript levels, whereas it enhances the bFGF-elicited increase in TnC expression, in vitro (Fig. 6A and B). Therefore, it is not surprising that ODN 2088 does not alter SCI-induced increase in TnC transcript levels in mice, since this could be the net outcome of such opposite effects (Fig. 6C).

Applications to other areas of neuroscience

TLRs have been implicated in a broad range of CNS pathologies (Li, Acioglu, et al., 2020) including injury, epilepsy, Alzheimer's disease, Parkinson's disease, multiple sclerosis, and neuropathic pain (Deerhake, Biswas, Barclay, & Shinohara, 2019; Gesuete, Kohama, & Stenzel-Poore, 2014; Gross et al., 2017; Kouli, Horne, & Williams-Gray, 2019; Momtazmanesh, Perry, & Rezaei, 2020; Shi, Hua, Kong, Stein, & Hua, 2019). In the majority of these pathological conditions, TLRs have been investigated in the context of neuroinflammation and glial activation. Although in each of these disorders, the affected brain or spinal cord regions and the neuronal sub-populations are distinct, and the stimuli that activate TLRs are different, there could be commonalities in the way that TLRs drive the innate immune reaction. Therefore, deciphering the cellular and molecular mechanisms underlying TLR-dependent innate immunity in SCI could provide insights into events that lead to sterile inflammation in neurodegenerative diseases and other pathologies.

TLRs also modulate neural progenitor cell proliferation, cell differentiation, and neurite outgrowth (Okun et al., 2012). These processes are important for CNS development. Therefore, aberrant TLR function could play a role in developmental disorders of the CNS. The impairment of sensory-motor function in TLR9 knockout mice (Khariv et al., 2013) and the alterations in learning and memory in TLR4 deficient mice (Okun et al., 2012) highlight the importance of TLRs in development.

Conclusions

The studies described above, collectively show that TLRs modulate the functional and histopathological outcomes of SCI. However, further investigations are needed to elucidate the complex cellular and molecular mechanisms underlying the effects of TLRs in SCI. This is an important, yet challenging question to address in vivo, given the fact that neurons and all glial subtypes express TLRs. The net outcome of activating or inhibiting a TLR in the injured SC is likely to reflect the combination of responses mounted by various cell types. In addition, the crosstalk among different TLRs and the interactions between TLRs and other PRRs further compound this complexity. The simultaneous use of in vitro and in vivo approaches, and the generation of mice with cell-type-specific TLR deletion could facilitate the delineation of distinct molecular and cellular mechanisms. As stated before, studies on TLRs in SCI have been primarily focused on TLR2, TLR4, and TLR9. Further investigations are needed to interrogate the role of other TLRs in SCI.

There is also a need for advancing the understanding of the non-immune role of TLRs in CNS cells. The modulation of neuronal viability and function through the direct effects of TLR agonists or antagonists on neurons (Acioglu et al., 2016; Korgaonkar et al., 2020; Li et al., 2015) and the regulation of astrocyte (Li et al., 2019; Li, Ni, et al., 2020) and oligodendrocyte function by direct activation of TLR signaling warrant further investigations.

It is worth noting that the development of TLR modulators has been an active research field primarily for the treatment of cancers and infectious diseases (Anwar, Shah, Kim, & Choi, 2019). The availability of such TLR modulators could also be useful as therapeutic agents in SCI.

Mini-dictionary of terms

- **Axonal dieback:** The progressive retraction of injured proximal axons from the lesion site.
- **Astrogliosis:** The changes observed in astrocytes in response to injury, which include alterations in gene expression, proliferation, and morphology.
- **Chemotaxis:** Migration of cells in response to extracellular signals.
- **Dystrophic axons:** Axons that show pathological changes such as swellings in disease and injury.
- **Retraction bulbs:** The oval or round swellings that form at the end of injured proximal axons.
- **Microglia and macrophage polarization:** The acquisition of pro-inflammatory and detrimental (M1 phenotype) versus anti-inflammatory and beneficial (M2 phenotype) properties under the influence of various stimuli.
- **Necroptosis:** Programmed necrotic cell death induced by inflammation.
- **Proximal axons:** Injured axons which remain connected to the cell body.

- **Pyroptosis:** A form of inflammatory programmed cell death.
- **Wallerian degeneration:** The mechanism by which injured distal axons, disconnected from the cell body, degenerate following SCI.

Key facts of innate immunity in central nervous system injury

- Injury provokes innate immune responses in the CNS.
- PRRs are the principal mediators of innate immunity and are activated by self-ligands released by damaged or stress cells following injury.
- Cells intrinsic to the spinal cord and immune cells that infiltrate the injury site express functional PRRs and release inflammatory effectors in response to activation of these receptors.
- TLRs are the best-characterized family of PRRs.
- Reactive astrocytes and activated microglia are key players in TLR-mediated innate immunity in the injured CNS.

Summary points

- TLRs modulate the innate immune and non-immune functions of glia and neurons following SCI.
- Stimulation of TLRs leads to enhanced production of type I interferons, cytokines, and chemokines, which initiate neuroinflammation.
- The crosstalk between TLRs and other PRRs, and the interactions between different TLRs are important determinants of neuroinflammation.
- TLR2, TLR4, and TLR9 are the best-studied TLRs in SCI.
- TLR2 and TLR4 are associated with improvement of function, protection, and repair in SCI.
- In contrast, TLR9 antagonism protects injured axons, alters astrocyte function, polarizes macrophages into a beneficial phenotype, and improves the functional outcomes of SCI.
- Elucidating the contribution of TLRs to SCI could provide insights into their roles in pathogenic mechanisms associated with neurodegenerative diseases, brain injuries, and potentially developmental disorders of the CNS.

Acknowledgments

The work in the authors' laboratory was supported by New Jersey Commission on Spinal Cord Research Grants CSCR12IRG007 and CSCR17IRG007 and the Reynolds Family Spine Laboratory funds.

References

Acioglu, C., Mirabelli, E., Baykal, A. T., Ni, L., Ratnayake, A., Heary, R. F., et al. (2016). Toll like receptor 9 antagonism modulates spinal cord neuronal function and survival: Direct versus astrocyte-mediated mechanisms. *Brain, Behavior, and Immunity*, *56*, 310–324.

Amarante-Mendes, G. P., Adjemian, S., Branco, L. M., Zanetti, L. C., Weinlich, R., & Bortoluci, K. R. (2018). Pattern recognition receptors and the host cell death molecular machinery. *Frontiers in Immunology*, *9*, 2379.

Anwar, M. A., Al Shehabi, T. S., & Eid, A. H. (2016). Inflammogenesis of secondary spinal cord injury. *Frontiers in Cellular Neuroscience*, *10*, 98.

Anwar, M. A., Shah, M., Kim, J., & Choi, S. (2019). Recent clinical trends in Toll-like receptor targeting therapeutics. *Medicinal Research Reviews*, *39*, 1053–1090.

Arumugam, T. V., Manzanero, S., Furtado, M., Biggins, P. J., Hsieh, Y. H., Gelderblom, M., et al. (2017). An atypical role for the myeloid receptor Mincle in central nervous system injury. *Journal of Cerebral Blood Flow and Metabolism*, *37*, 2098–2111.

Bonnans, C., Chou, J., & Werb, Z. (2014). Remodelling the extracellular matrix in development and disease. *Nature Reviews. Molecular Cell Biology*, *15*, 786–801.

Bradbury, E. J., Moon, L. D., Popat, R. J., King, V. R., Bennett, G. S., Patel, P. N., et al. (2002). Chondroitinase ABC promotes functional recovery after spinal cord injury. *Nature*, *416*, 636–640.

Burda, J. E., & Sofroniew, M. V. (2014). Reactive gliosis and the multicellular response to CNS damage and disease. *Neuron*, *81*, 229–248.

Cameron, J. S., Alexopoulou, L., Sloane, J. A., DiBernardo, A. B., Ma, Y., Kosaras, B., et al. (2007). Toll-like receptor 3 is a potent negative regulator of axonal growth in mammals. *The Journal of Neuroscience*, *27*, 13033–13041.

Carty, M., & Bowie, A. G. (2011). Evaluating the role of Toll-like receptors in diseases of the central nervous system. *Biochemical Pharmacology*, *81*, 825–837.

Chen, J., Joon Lee, H., Jakovcevski, I., Shah, R., Bhagat, N., Loers, G., et al. (2010). The extracellular matrix glycoprotein tenascin-C is beneficial for spinal cord regeneration. *Molecular Therapy*, *18*, 1769–1777.

Church, J. S., Kigerl, K. A., Lerch, J. K., Popovich, P. G., & McTigue, D. M. (2016). TLR4 deficiency impairs oligodendrocyte formation in the injured spinal cord. *The Journal of Neuroscience*, *36*, 6352–6364.

David, B. T., Ratnayake, A., Amarante, M. A., Reddy, N. P., Dong, W., Sampath, S., et al. (2013). A Toll-like receptor 9 antagonist reduces pain hypersensitivity and the inflammatory response in spinal cord injury. *Neurobiology of Disease*, *54*, 194–205.

David, B. T., Sampath, S., Dong, W., Heiman, A., Rella, C. E., Elkabes, S., et al. (2014). A Toll-like receptor 9 antagonist improves bladder function and white matter sparing in spinal cord injury. *Journal of Neurotrauma*, *31*, 1800–1806.

Deerhake, M. E., Biswas, D. D., Barclay, W. E., & Shinohara, M. L. (2019). Pattern recognition receptors in multiple sclerosis and its animal models. *Frontiers in Immunology*, *10*, 2644.

Evans, T. A., Barkauskas, D. S., Myers, J. T., Hare, E. G., You, J. Q., Ransohoff, R. M., et al. (2014). High-resolution intravital imaging reveals that blood-derived macrophages but not resident microglia facilitate secondary axonal dieback in traumatic spinal cord injury. *Experimental Neurology*, *254*, 109–120.

Fahey, E., & Doyle, S. L. (2019). IL-1 family cytokine regulation of vascular permeability and angiogenesis. *Frontiers in Immunology*, *10*, 1426.

Famakin, B. M., Tsymbalyuk, O., Tsymbalyuk, N., Ivanova, S., Woo, S. K., Kwon, M. S., et al. (2020). HMGB1 is a potential mediator of astrocytic TLR4 signaling activation following acute and chronic focal cerebral ischemia. *Neurology Research International*, *2020*.

Fan, H., Zhang, K., Shan, L., Kuang, F., Chen, K., Zhu, K., et al. (2016). Reactive astrocytes undergo M1 microglia/macrohpages-induced necroptosis in spinal cord injury. *Molecular Neurodegeneration*, *11*.

Freeman, L., Guo, H., David, C. N., Brickey, W. J., Jha, S., & Ting, J. P. (2017). NLR members NLRC4 and NLRP3 mediate sterile inflammasome activation in microglia and astrocytes. *The Journal of Experimental Medicine*, *214*, 1351–1370.

Gensel, J. C., Wang, Y., Guan, Z., Beckwith, K. A., Braun, K. J., Wei, P., et al. (2015). Toll-like receptors and dectin-1, a c-type lectin receptor, trigger divergent functions in CNS macrophages. *The Journal of Neuroscience*, *35*, 9966–9976.

Gesuete, R., Kohama, S. G., & Stenzel-Poore, M. P. (2014). Toll-like receptors and ischemic brain injury. *Journal of Neuropathology and Experimental Neurology*, *73*, 378–386.

Gross, A., Benninger, F., Madar, R., Illouz, T., Griffioen, K., Steiner, I., et al. (2017). Toll-like receptor 3 deficiency decreases epileptogenesis in a pilocarpine model of SE-induced epilepsy in mice. *Epilepsia*, *58*, 586–596.

Haggerty, A. E., Marlow, M. M., & Oudega, M. (2017). Extracellular matrix components as therapeutics for spinal cord injury. *Neuroscience Letters*, *652*, 50–55.

van Haren, S. D., Dowling, D. J., Foppen, W., Christensen, D., Andersen, P., Reed, S. G., et al. (2016). Age-specific adjuvant synergy: Dual TLR7/8 and mincle activation of human newborn dendritic cells enables Th1 polarization. *Journal of Immunology*, *197*, 4413–4424.

Heiman, A., Pallottie, A., Heary, R. F., & Elkabes, S. (2014). Toll-like receptors in central nervous system injury and disease: A focus on the spinal cord. *Brain, Behavior, and Immunity*, *42*, 232–245.

Jiang, W., Li, M., He, F., Zhou, S., & Zhu, L. (2017). Targeting the NLRP3 inflammasome to attenuate spinal cord injury in mice. *Journal of Neuroinflammation*, *14*, 207.

Khariv, V., Pang, K., Servatius, R. J., David, B. T., Goodus, M. T., Beck, K. D., et al. (2013). Toll-like receptor 9 deficiency impacts sensory and motor behaviors. *Brain, Behavior, and Immunity*, *32*, 164–172.

Kigerl, K. A., Lai, W., Rivest, S., Hart, R. P., Satoskar, A. R., & Popovich, P. G. (2007). Toll-like receptor (TLR)-2 and TLR-4 regulate inflammation, gliosis, and myelin sparing after spinal cord injury. *Journal of Neurochemistry*, *102*, 37–50.

Kigerl, K. A., Lai, W., Wallace, L. M., Yang, H., & Popovich, P. G. (2018). High mobility group box-1 (HMGB1) is increased in injured mouse spinal cord and can elicit neurotoxic inflammation. *Brain, Behavior, and Immunity*, *72*, 22–33.

Korgaonkar, A. A., Li, Y., Sekhar, D., Subramanian, D., Guevarra, J., Swietek, B., et al. (2020). Toll-like receptor 4 signaling in neurons enhances calcium-permeable alpha-amino-3-hydroxy-5-methyl-4-isoxazolepropionic acid receptor currents and drives post-traumatic epileptogenesis. *Annals of Neurology*, *87*, 497–515.

Kouli, A., Horne, C. B., & Williams-Gray, C. H. (2019). Toll-like receptors and their therapeutic potential in Parkinson's disease and α-synucleinopathies. *Brain, Behavior, and Immunity*, *81*, 41–51.

Li, Y., Korgaonkar, A. A., Swietek, B., Wang, J., Elgammal, F. S., Elkabes, S., et al. (2015). Toll-like receptor 4 enhancement of non-NMDA synaptic currents increases dentate excitability after brain injury. *Neurobiology of Disease*, *74*, 240–253.

Li, L., Acioglu, C., Heary, R. F., & Elkabes, S. (2020). Role of astroglial Toll-like receptors (TLRs) in central nervous system infections, injury and neurodegenerative diseases. *Brain, Behavior, and Immunity*, *91*, 740–755.

Li, L., Ni, L., Eugenin, E. A., Heary, R. F., & Elkabes, S. (2019). Toll-like receptor 9 antagonism modulates astrocyte function and preserves proximal axons following spinal cord injury. *Brain, Behavior, and Immunity*, *80*, 328–343.

Li, L., Ni, L., Heary, R. F., & Elkabes, S. (2020). Astroglial TLR9 antagonism promotes chemotaxis and alternative activation of macrophages via modulation of astrocyte-derived signals: Implications for spinal cord injury. *Journal of Neuroinflammation*, *17*, 73.

Liddelow, S. A., Marsh, S. E., & Stevens, B. (2020). Microglia and astrocytes in disease: Dynamic duo or partners in crime? *Trends in Immunology*, *41*, 820–835.

Liu, H. Y., Hong, Y. F., Huang, C. M., Chen, C. Y., Huang, T. N., & Hsueh, Y. P. (2013). TLR7 negatively regulates dendrite outgrowth through the Myd88-c-Fos-IL-6 pathway. *The Journal of Neuroscience*, *33*, 11479–11493.

Ma, Y., Li, J., Chiu, I., Wang, Y., Sloane, J. A., Lu, J., et al. (2006). Toll-like receptor 8 functions as a negative regulator of neurite outgrowth and inducer of neuronal apoptosis. *The Journal of Cell Biology*, *175*, 209–215.

Mamik, M. K., & Power, C. (2017). Inflammasomes in neurological diseases: Emerging pathogenic and therapeutic concepts. *Brain*, *140*, 2273–2285.

Minkiewicz, J., de Rivero Vaccari, J. P., & Keane, R. W. (2013). Human astrocytes express a novel NLRP2 inflammasome. *Glia*, *61*, 1113–1121.

Momtazmanesh, S., Perry, G., & Rezaei, N. (2020). Toll-like receptors in Alzheimer's disease. *Journal of Neuroimmunology*, *348*, 577362.

Nishimura, Y., Natsume, A., Ito, M., Hara, M., Motomura, K., Fukuyama, R., et al. (2013). Interferon-β delivery via human neural stem cell abates glial scar formation in spinal cord injury. *Cell Transplantation*, *22*, 2187–2201.

Okun, E., Barak, B., Saada-Madar, R., Rothman, S. M., Griffioen, K. J., Roberts, N., et al. (2012). Evidence for a developmental role for TLR4 in learning and memory. *PLoS One*, *7*, e47522.

Pallottie, A., Ratnayake, A., Ni, L., Acioglu, C., Li, L., Mirabelli, E., et al. (2018). A Toll-like receptor 9 antagonist restores below-level glial glutamate transporter expression in the dorsal horn following spinal cord injury. *Scientific Reports*, *8*, 8723.

Pashenkov, M. V., Murugina, N. E., Budikhina, A. S., & Pinegin, B. V. (2019). Synergistic interactions between NOD receptors and TLRs: Mechanisms and clinical implications. *Journal of Leukocyte Biology*, *105*, 669–680.

de Rivero Vaccari, J. P., Lotocki, G., Marcillo, A. E., Dietrich, W. D., & Keane, R. W. (2008). A molecular platform in neurons regulates inflammation after spinal cord injury. *The Journal of Neuroscience*, *28*, 3404–3414.

de Rivero Vaccari, J. P., Minkiewicz, J., Wang, X., De Rivero Vaccari, J. C., German, R., Marcillo, A. E., et al. (2012). Astrogliosis involves activation of retinoic acid-inducible gene-like signaling in the innate immune response after spinal cord injury. *Glia*, *60*, 414–421.

Rosenberger, K., Derkow, K., Dembny, P., Kruger, C., Schott, E., & Lehnardt, S. (2014). The impact of single and pairwise Toll-like receptor activation on neuroinflammation and neurodegeneration. *Journal of Neuroinflammation*, *11*, 166.

Russo, M. V., & McGavern, D. B. (2015). Immune surveillance of the CNS following infection and injury. *Trends in Immunology*, *36*, 637–650.

Schonberg, D. L., Popovich, P. G., & McTigue, D. M. (2007). Oligodendrocyte generation is differentially influenced by Toll-like receptor (TLR) 2 and TLR4-mediated intraspinal macrophage activation. *Journal of Neuropathology and Experimental Neurology*, *66*, 1124–1135.

Shi, H., Hua, X., Kong, D., Stein, D., & Hua, F. (2019). Role of Toll-like receptor mediated signaling in traumatic brain injury. *Neuropharmacology*, *145*, 259–267.

Smith, G. M., & Hale, J. H. (1997). Macrophage/microglia regulation of astrocytic tenascin: Synergistic action of transforming growth factor-β and basic fibroblast growth factor. *The Journal of Neuroscience*, *17*, 9624–9633.

Smith, G. M., & Strunz, C. (2005). Growth factor and cytokine regulation of chondroitin sulfate proteoglycans by astrocytes. *Glia*, *52*, 209–218.

Sofroniew, M. V. (2015). Astrogliosis. *Cold Spring Harbor Perspectives in Biology*, *7*, a020420.

Stephenson, E. L., & Yong, V. W. (2018). Pro-inflammatory roles of chondroitin sulfate proteoglycans in disorders of the central nervous system. *Matrix Biology*, *71–72*, 432–442.

Stirling, D. P., Cummins, K., Mishra, M., Teo, W., Yong, V. W., & Stys, P. (2014). Toll-like receptor 2-mediated alternative activation of microglia is protective after spinal cord injury. *Brain*, *137*, 707–723.

Stirling, D. P., & Yong, V. W. (2008). Dynamics of the inflammatory response after murine spinal cord injury revealed by flow cytometry. *Journal of Neuroscience Research*, *86*, 1944–1958.

Stivers, N. S., Pelisch, N., Orem, B. C., Williams, J., Nally, J. M., & Stirling, D. P. (2017). The Toll-like receptor 2 agonist Pam3CSK4 is neuroprotective after spinal cord injury. *Experimental Neurology*, *294*, 1–11.

Sun, L., Li, M., Ma, X., Feng, H., Song, J., Lv, C., et al. (2017). Inhibition of HMGB1 reduces rat spinal cord astrocytic swelling and AQP4 expression after oxygen-glucose deprivation and reoxygenation via TLR4 and NF-κB signaling in an IL-6-dependent manner. *Journal of Neuroinflammation*, *14*, 231.

Takeuchi, O., & Akira, S. (2010). Pattern recognition receptors and inflammation. *Cell*, *140*, 805–820.

Thomas, C. J., & Schroder, K. (2013). Pattern recognition receptor function in neutrophils. *Trends in Immunology*, *34*, 317–328.

Vajjhala, P. R., Ve, T., Bentham, A., Stacey, K. J., & Kobe, B. (2017). The molecular mechanisms of signaling by cooperative assembly formation in innate immunity pathways. *Molecular Immunology*, *86*, 23–37.

Voet, S., Srinivasan, S., Lamkanfi, M., & van Loo, G. (2019). Inflammasomes in neuroinflammatory and neurodegenerative diseases. *EMBO Molecular Medicine*, *11*, e10248.

Walsh, J. G., Muruve, D. A., & Power, C. (2014). Inflammasomes in the CNS. *Nature Reviews. Neuroscience*, *15*, 84–97.

Wiese, S., Karus, M., & Faissner, A. (2012). Astrocytes as a source for extracellular matrix molecules and cytokines. *Frontiers in Pharmacology*, *3*, 120.

Yang, T., Dai, Y., Chen, G., & Cui, S. (2020). Dissecting the dual role of the glial scar and scar-forming astrocytes in spinal cord injury. *Frontiers in Cellular Neuroscience*, *14*, 78.

Zhang, Y., Winterbottom, J. K., Schachner, M., Lieberman, A. R., & Anderson, P. N. (1997). Tenascin-C expression and axonal sprouting following injury to the spinal dorsal columns in the adult rat. *Journal of Neuroscience Research*, *49*, 433–450.

Chapter 14

The role of oxidative stress in spinal cord injury animal models: A focus on nuclear factor erythroid-2 related factor 2

Fernando da Silva Fiorin[a], Caroline Cunha do Espírito Santo[b], and Luiz Fernando Freire Royes[c]

[a]*Graduate Program in Neuroengineering, Edmond and Lily Safra International Institute of Neuroscience, Santos Dumont Institute, Macaíba, RN, Brazil,*
[b]*Neuromotor System Laboratory, Spinal Cord Injury Rehabilitation Research Group, NuLeMe, Santa Catarina State University, Florianópolis, SC, Brazil,*
[c]*Exercise Biochemistry Laboratory, Center of Physical Education and Sports, Federal University of Santa Maria, Santa Maria, RS, Brazil*

Abbreviations

3-NT	3-nitrotyrosine
4-HNE	4-hidroxinonenal
ARE	antioxidant response element
ATP	adenosine triphosphate
Bax	Bcl-2-like protein 4
BBB	Basso, Beattie, and Bresnahan
Bcl-2	anti-apoptotic B-cell lymphoma
BDNF	brain-derived neurotrophic factor
BMS	Basso mouse scale
CAT	catalase
CK2	casein kinase 2
DNA	deoxyribonucleic acid
EAAT1	excitatory amino acids transport
EAAT2	excitatory amino acids transport
ERK	extracellular signal-regulated kinase
GPx	peroxidase glutathione
GSH	reduced glutathione
GSK-3	glycogen synthase kinase
GSSG	oxidized glutathione
HO-1	hemeoxygenase 1
IL-6	interleukin-6
JNK	c-jun N-terminal kinase
KEAP1	Kelch-like ECH-associated protein 1
MAF	musculo aponeurotic fibrosarcoma
MDA	malondialdehyde
mRNA	messenger RNA
NADH	nicotinamine adenine dinucleotide
NF-κB	nuclear factor kappa B
NLRP3	Nod-like receptor pyrin domain-containing 3 complex
NMDA	N-metil-D-aspartate
NQO1	NA(*D*)PH quinone oxidoreductase 1
Nrf2	nuclear factor erythroid-2 related factor 2
P38 MAPK	p38 mitogen-activated protein kinase
PI3K:phosphoinositide 3-kinase	
PKC	protein kinase C

Cellular, Molecular, Physiological, and Behavioral Aspects of Spinal Cord Injury. https://doi.org/10.1016/B978-0-12-822427-4.00014-9

RNS	reactive nitrogen species
ROS	reactive oxidative species
SCI	spinal cord injury
SOD	superoxide dismutase
TBARS	thiobarbituric acid reactive substances
TLR-4	toll-like receptor 4
TNF-α	tumoral necrosis factor-α

Introduction

Paraplegia and tetraplegia are major consequences induced by spinal cord trauma and affect many young adults (Jia et al., 2012). Movement loss, sensory deficits, and autonomic dysfunction are common aggravating factors after spinal cord injury (SCI), and numerous pre-clinical studies have been performed to better understand the mechanisms underlying the lesion in order to improve rehabilitation programs. Tissue lesions can occur by compression, contusion, or laceration of the spinal cord, leading to transitory tissue disruption associated with primary cellular mechanisms, including acute neuronal loss by necrosis immediately after trauma (Beattie, Farooqui, & Bresnahan, 2000).

Primary lesions are followed by molecular and cellular cascades in the damage site, such as glutamatergic excitotoxicity, ischemia/reperfusion, inflammatory processes, edema, and oxidative stress (Table 1) that are classified in the secondary mechanism of the lesion and can cause late cell apoptosis (Beattie et al., 2000). Indeed, the molecular and cellular alterations that accompany initial tissue disruption can hinder recovery and are directly involved with social problems and debilities in the injured individuals' quality of life.

Regarding SCI pathophysiology, it is important to note that the spared nerve fibers are essential in the functional recovery process. However, the number of fibers remaining depends on the volume of axonal bundles lost at the time of injury and Wallerian degeneration in response to secondary mechanisms. Free radical and reactive oxygen species and reactive nitrogen species (ROS/RNS) production are part of normal metabolisms in the physiological systems because, under normal conditions, these compounds are kept under control by cellular defense mechanisms. Moreover, oxidative molecules are necessary for several biochemical reactions in physiological systems during homeostasis (Droge, 2002). On the other hand, the excessive increase of oxidant molecules is part of many secondary mechanisms that accompany SCI (from minutes to weeks) and may be associated with tissue degeneration and inhibited recovery (Woźniak, Woźniak, Mila-Kierzenkowska, & Kasprzak, 2016). This process can begin in the spinal cord by excessive calcium (Ca^{2+}) influx through *N*-methyl-D-aspartate (NMDA) receptors mediated by glutamate neurotransmitters, damaging the mitochondrial matrix and increasing free radicals and ROS (Ma et al., 2020; Springer et al., 1997).

Excessive oxidant molecules and mitochondrial damage cause cell apoptosis by cytochrome C release and neuronal loss. Although most of these cell characteristics occur after compression, contusion, or laceration of the spinal cord, further knowledge on the effects of oxidants and antioxidant molecules after SCI is still necessary. Therefore, the present chapter describes the main oxidants released in the medullary lesion and antioxidant interventions in pre-clinical research that should be useful for future therapeutic treatments.

TABLE 1 Secondary damage after spinal cord injury.

Spinal cord injury Secondary damage
Excitotoxicity
Ischemia/reperfusion
Oxidative stress
Edema
Inflammation

Molecular and cellular cascades in the damage site.
Source: Authors.

Oxidative stress characteristics and mechanisms

Oxidative stress is characterized by the oxidation of biomolecules that may lead to the loss of biological functions in healthy tissues (Halliwell & Whiteman, 2004). This process occurs when the production of free radicals or ROS/RNS is higher than the antioxidant defenses present in the physiological system. Free radicals are atoms or molecules that, when formed, have one or more unpaired electrons in their valence shell (Halliwell, 1989), while ROS and RNS are equally reactive compounds but do not have unpaired electrons (Droge, 2002). Even though the electrons are unpaired, ROS and RNS are highly reactive and seek stability during their brief existence, reacting with other molecules and damaging cell membranes, proteins, and deoxyribonucleic acid (DNA) of cells (Halliwell, 2012).

Oxidative phosphorylation is the principal ROS generator, in which the first free radical formed is superoxide ($O_2^{-\bullet}$), mainly in complex I (NADH dehydrogenase) and complex III (ubiquinone cytochrome C oxidase) of the respiratory chain, where Flavin's mononucleotide coenzyme and ubiquinol reduce molecular oxygen (O_2) and form the free radical (Navarro & Boveris, 2007). Reactive oxygen species such as hydrogen peroxide (H_2O_2) and other free radicals, including hydroxyl radical ($^-OH^\bullet$), which is the most reactive radical, can be formed from $O_2^{-\bullet}$. In addition, H_2O_2 can react with metals such as copper (Cu^{+1}) and iron (Fe^{+2}) in the so-called "Fenton reaction," forming $^-OH^\bullet$ (Jia et al., 2012). This dangerous molecule can act as an electrophile or nucleophile and attack organic molecules by hydrogen abstraction or coupling in double bonds and aromatic rings (hydroxylation), even in substituted positions, causing reactions such as deamination (Fukushima, Tatsumi, & Morimoto, 2000) and decarboxylation (Chen, Ma, He, & Zhao, 2002). Additionally, H_2O_2 can also react with $O_2^{-\bullet}$ itself in the so-called "Haber-Weiss reaction," forming $^-OH^\bullet$(Jia et al., 2012). Moreover, $O_2^{-\bullet}$ can react with nitric oxide (NO) formed by nitric oxide synthase (NOS) and form peroxynitrite ($ONOO^-$), an RNS that damages membranes, proteins, and cellular DNA (Lau & Tymianski, 2010).

The excessive formation of these ROS is controlled by an antioxidant defense system composed of enzymatic antioxidants, such as superoxide dismutase (SOD), catalase (CAT), and glutathione peroxidase (GPx). All antioxidant defenses are involved with a system of auxiliary molecules. There are also non-enzymatic antioxidants [vitamins A, C, D, and E and reduced glutathione (GSH)] (Halliwell & Gutteridge, 1995). Among enzymatic antioxidants, SOD is found in two isoforms (SOD1 or CuZinSOD)and uses copper and zinc as cofactors, being located mainly in the cell cytoplasm, although it is also present in peroxisomes, lysosomes, and the mitochondrial intermembrane space. Other isoforms such as SOD2 or MnSODuse manganese as a cofactor and are located in the mitochondrial matrix (Jia et al., 2012). These two isoforms catalyze the $O_2^{-\bullet}$ dismutation reaction to form H_2O_2. Following these reactions, the enzymes CAT and GPx catalyze the degradation reaction of H_2O_2 into H_2O and O_2. The GPx enzyme is found in two isoforms: one is present in mitochondria and cytosol and uses selenium as a cofactor, while the other is only located in the cytosol and does not use selenium as a cofactor. For the GPx reaction, the reduced glutathione substrate (GSH) is required for the enzyme, forming oxidized glutathione (GSSG) and the reaction product of the hydroperoxide. To restore the GSH molecule, the GPx enzyme works coupled with glutathione reductase (GR), which is an enzyme that reduces the GSSG molecule to GSH using nicotinamide adenine dinucleotide phosphate (NADPH) (Maiorino, Gregolin, & Ursini, 1990). Thus, the antioxidant system can reduce the action of oxidant molecules (Fig. 1).

However, excitotoxicity is directly related to the excessive free radical formation (ROS and RNS) caused by mitochondrial damage (Navarro & Boveris, 2007) and calcium-dependent enzyme activation, such as phospholipase A2 (Miller et al., 2010) and xanthine oxidase (Allen et al., 2012), which hinder the blockage of the antioxidant system. Among the damage resulting from the exacerbated production of free radicals, lipid peroxidation also stands out (Allen et al., 2012). Lipid peroxidation produces aldehydes, hydrocarbon gases, and various chemical residues, including malondialdehyde (MDA), conjugated dienes, and 4-hydroxynonenal (4-HNE) (Hotz, Hoet, & Lauwerys, 1987). In this sense, the cell membranes are composed of lipids, and research has shown that soon after the SCI model (1–24 h), there is an increase in lipid peroxidation that inhibits glutamate uptake, causing excitotoxicity in injured spinal cords (Springer et al., 1997).

Nonetheless, it is important to note that the overstimulation of glutamatergic receptors also represents the main mediator of oxidative stress in several neurological disease models (Kumar, Kalonia, & Kumar, 2011). In this scenario, reducing antioxidant enzymes in the brain, such as SOD, CAT, and GPx, contributes to developing various neurodegenerative diseases through free radicals in pathological conditions. Therefore, experimental findings have described the protective role of antioxidants in animal models of traumatic injury, strokes, Huntington's, Parkinson's, and Alzheimer's diseases (Halliwell, 1989).

Different isoforms of glutamate transporters show sensitivity to biological oxidants and may share one or more vulnerable sites to oxidants. Oxidants also inhibit the transport activity of the principal excitatory amino acid transports (EAAT1 and EAAT2) by directly affecting glutamate transport proteins. A common primary target for $O_2^{-\bullet}$, H_2O_2, $ONOO^-$, and the by-product $^-OH^\bullet$ is the oxidation of sulfhydryl cysteine groups (Radi, Beckman, Bush, & Freeman,

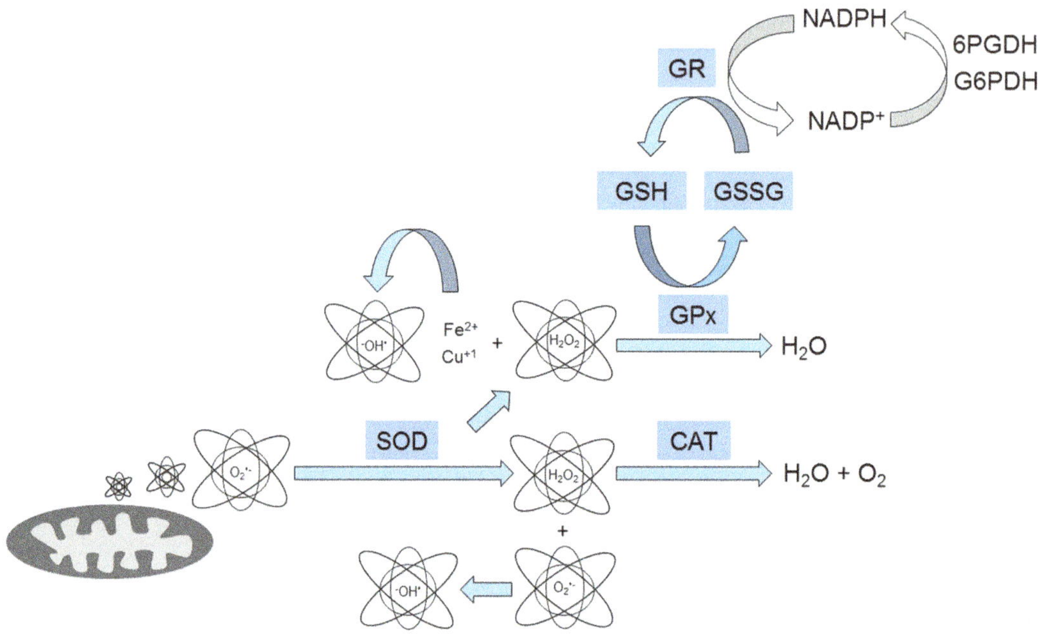

FIG. 1 Oxidant and antioxidant actions in cellular homeostasis. Oxidative phosphorylation generates the free radical $O_2^{-\bullet}$; SOD form H_2O_2 from $O_2^{-\bullet}$; H_2O_2 can react with $O_2^{-\bullet}$ and generate $^{-}OH^{\bullet}$ (Haber-Weiss reaction) or react with Cu^{+1} or Fe^{+2} ions to form $^{-}OH^{\bullet}$ (Fenton reaction); CAT and GPx degrade H_2O_2 in H_2O and O_2; GPx uses GSH as a substrate for the enzyme, forming GSSG; GR reduces GSSG to GSH using NADPH, forming $NADP^{+}$; NADPH is restored from G6PDH, forming 6GPDH. $O_2^{-\bullet}$: superoxide radical; GPx: peroxidase glutathione; H_2O_2: hydrogen peroxide; $^{-}OH^{\bullet}$: hydroxyl radical; SOD: superoxide dismutase; GSH: reduced glutathione; GSSG: oxidized glutathione; GR: reductase glutathione; NADPH: reduced nicotinamide adenine dinucleotide phosphate; $NADP^{+}$: oxidized nicotinamide adenine dinucleotide phosphate; G6PDH: Glycose 6 phosphate dehydrogenase; 6GPDH: 6-phosphogluconate dehydrogenase; H_2O: water; O_2: oxygen. *(Source: Authors.)*

1991). In addition, by decomposing it in neutral pH, $ONOO^{-}$ generates nitronium ion and nitrogen dioxide, which promote nitrosylation and nitration of tyrosine residues and alter the phosphorylation processes in these residues (Radi et al., 1991). However, the physiological system may protect the cells in some situations in which homeostasis is disrupted. This protection system occurs via endogenous cascades of transcription of defense molecules, which can be exogenously stimulated.

The implications of oxidative damage in SCI pathophysiology

One of the leading causes of motor deficits after SCI is the irreversible loss of motor neurons during primary damage. However, sensorimotor recovery is also impaired by fiber loss associated with late degeneration processes. Regarding SCI pathophysiology, white matter injury occurs due to primary and secondary damage, and it is important to note that therapeutic intervention and attenuation in such late secondary phases is still possible to reduce the white matter loss (Hayta & Elden, 2018). In this sense, from a translational perspective for motor recovery, which is a primary need for SCI patients, in rodents, at least 10% of white matter must be spared to improve the motor system after SCI (Kloos, Fisher, Detloff, Hassenzahl, & Basso, 2005).

Furthermore, oxidative and antioxidative balance marked by reactive substances such as thiobarbituric acid reactive substances (TBARS), conjugated dienes concentrations, and enzymatic antioxidant system (SOD, CAT, and GPx) is deregulated in the blood of individuals with cervical SCI (Woźniak et al., 2016). Reactive oxygen species-induced protein oxidation immediately after injury is a critical secondary SCI pathophysiology marked by protein carbonyl formation and 3-nitrotyrosine (3-NT) in spinal cord tissues (Visavadiya, Patel, VanRooyen, Sullivan, & Rabchevsky, 2016). Severe SCI increases $O_2^{-\bullet}$, H_2O_2, and NO levels in isolated mitochondria and decreases GSH content with continuous production of NO levels in cell parameters (Visavadiya et al., 2016). Consequently, NO can react with $O_2^{-\bullet}$ to form $ONOO^{-}$, causing nitration and the oxidation and nitrosylation of several mitochondrial compounds that cause drastic consequences in proteins, lipids, and nucleic acid formation (Szabó, Ischiropoulos, & Radi, 2007).

Several biological degradation products are released during oxidative stress and bring characteristics that allow establishing which intracellular tissue organelles were affected. In this scenario, lipid peroxidation and protein carbonyls are relevant biochemical markers and indications of oxidative damage in lipids and proteins, respectively (Luo, Borgens, & Shi, 2004; Visavadiya et al., 2016). Although the oxidative process is more intense in damage epicenter in experimental models, adjacent segments can also suffer degradation in plasmatic membrane lipids. For instance, experimental data by Luo et al. (2004) revealed that a rat thoracic compression model with SCI in the T10–T11 segments presented spread lipid peroxidation in rostral and caudal segments T6–T7, T8–T9, T12–T13, and L1–L2, respectively, in addition to the epicenter (T10–T11).

Membrane degradation can deregulate Na^+/K^+ homeostasis in axoplasm and mitochondria (LoPachin et al., 1999), leading to higher Ca^{2+} influx inside cells and mitochondrial matrix that consequently cause over-production of ROS (Luo et al., 2004). Additionally, mitochondrial swelling is involved with ROS production and lipid peroxidation after experimental SCI in rats (Luo et al., 2004). Besides inducing higher production of free radicals, the mitochondria are also targeted by oxidant molecules, leading to loss of function in transport electrons and reduced ATP synthesis. The constant opening of the mitochondrial membrane permeability transition pore and transport of cytochrome C to cell cytosol induce caspase activation with consequent cell apoptosis (Ma et al., 2020). In this sense, the intrasynaptosomal Ca^{2+} influx soon after SCI increases in the lesion area, which is associated with reduced synaptosomal function (Luo et al., 2004). Moreover, Ca^{2+} and H_2O_2 cause permeability transition and mitochondrial swelling in isolated mitochondria, similar to what happened in synaptosomes of injured tissues in a compression SCI model (Luo et al., 2004). Thus, oxidative stress can cause degeneration of remaining neurons and oligodendrocytes (Wallerian degeneration), leading to impairment in ascending and descending (sensorimotor) information of axons of the injured spinal cord and proximal segments (Beattie et al., 2000).

This way, tissue protection in secondary phases of SCI is vital for maintaining cell survival of sensory and motor tracts and simultaneously provide better conditions for rehabilitating injured individuals. Hence, the transcription factor Nrf2 should be a target to reduce the continuous damage caused by oxidative stress in tracts of the spinal cord and contribute to patient recovery.

Protection mechanisms by endogenous antioxidants via Nrf2-ARE system

During evolution, cells developed defense systems against harmful endogenous processes and exogenous substances. Several transcription factors are involved in increasing cell defenses. A complex system of phase I, II, and III genes involves antioxidant enzyme activation. The main regulator of phase II genes and some phase III genes of antioxidant enzymes is the nuclear factor erythroid-2-related factor 2 (Nrf2) (Itoh et al., 1997). The principle of the system is to keep the Nrf2 protein low under normal cellular conditions, with the possibility of rapid induction in the event of increased cellular oxidation. Furthermore, Nrf2 is a member of the family of transcription factors that includes Nrf1, Nrf3, p45 NF-E2, Bach1, and Bach2 (Motohashi, O'Connor, Katsuoka, Engel, & Yamamoto, 2002). Nrf2 is composed of six functional domains known as Nrf2-ECH homologies (Neh) and classified as Neh1–6, with each Neh domain having its own designated function (Baird & Dinkova-Kostova, 2011). Under quiescent conditions, Nrf2 is maintained by a cytosolic repressor, Keap1 (Kelch-like ECH-associated protein 1), a cytoskeletal protein that anchors and suppresses transcription activity (Itoh et al., 1999; Tong et al., 2007). Keap1 promotes rapid proteasomal degradation of Nrf2 through ubiquitination and acts as an oxidative and electrophilic stress sensor (Itoh et al., 1999).

Research has suggested that changes in the structure of Keap1 lead to the dissociation of the Nrf2-Keap1 complex (Motohashi & Yamamoto, 2004), although a specific modification in the Keap1 site may also alter ubiquitin ligase E3 function and consequently reduce Nrf2 degradation (Tong et al., 2007). Additionally, a pathway for ubiquitination and degradation of Nrf2 independent of Keap1 is also proposed (Rojo et al., 2012). In this model, GSK-3 phosphorylates serine residues in the Neh 6 domain in Nrf2 (Rada et al., 2012), and phosphorylated Neh 6 can be recognized by a ubiquitin ligase E3. The dissociation of the Nrf2-Keap1 complex in the cytoplasm is a widely accepted mechanism for Nrf2 activation. In summary, the oxidation of cysteine residues modifies the conformation of Keap1, thus initiating Nrf2 activation (Tong et al., 2007). In this way, oxidative products improve Nrf2 stability and increase phase II gene expression (McMahon, Thomas, Itoh, Yamamoto, & Hayes, 2006). The phosphorylation in the serine/threonine residues in Nrf2 may be a distinct mechanism that Nrf2 dissociates from Keap1 (Surh, Kundu, & Na, 2008). Protein kinases including protein kinase C (PKC), c-jun N-terminal kinase (JNK), extracellular signal-regulated kinase (ERK), casein kinase 2 (CK2), p38 mitogen-activated protein kinase (p38 MAPK), and phosphoinositide 3-kinase (PI3K) appear to modulate the import and export of Nrf2 from the nucleus (Jain & Jaiswal, 2007; Surh et al., 2008). Therefore, the oxidation of cysteine residues that are sensitive to the redox state in Keap1 constitutes the molecular basis for Nrf2 activation (Tong et al., 2007).

The induction of phase II antioxidant defense enzymes by Nrf2 is regulated by the antioxidant response element (ARE) or electrophile response element (EpRE) (Itoh et al., 1997; Venugopal & Jaiswal, 1998). The Nrf2-ARE bond can initiate the transcription of hundreds of cytoprotective genes, including SOD, CAT (Cho et al., 2011), and a series of phase II antioxidants that include the glutathione defense system, NA(*D*)PH quinone oxidoreductase 1 (NQO1), and heme oxygenase-1 (HO-1) (Zhang et al., 2012).

Notably, Nrf2 is a transcription factor able to induce the expression of other types of protective proteins in the CNS. Among them are the brain-derived neurotrophic factor (BDNF) (Sakata et al., 2012) and anti-apoptotic B-cell lymphoma (Bcl-2) (Niture & Jaiswal, 2012), which are extremely important proteins for cell maintenance and survival.

Hence, Nrf2 in the brain can act as one of the most decisive defenses against oxidative stress by modulating microglia (Rojo et al., 2010), protecting astrocytes (Vargas & Johnson, 2009) and neurons (Lee, Shih, Murphy, & Johnson, 2003) from toxic insults, and regulating the expression of inflammatory markers (Innamorato et al., 2008) and antioxidant enzymes (Shah et al., 2007; Yan et al., 2009). In this context, it is clear that Nrf2 plays a protective role in neurodegenerative disorders, including Parkinson's (Cuadrado, Moreno-Murciano, & Pedraza-Chaverri, 2009), Alzheimer's (Kanninen et al., 2008), and Huntington's diseases (Stack et al., 2010). Moreover, Nrf2 is also a regulator in secondary mechanisms of lesions in traumatic brain injury (Yan et al., 2009) and SCI (Ma et al., 2020).

Involvement of Nrf2 in blockage of oxidative stress post-SCI induced by exogenous stimuli

Many studies have attempted to reduce neurodegenerative processes and improve functionality in rat SCI models, including treatments that decrease oxidative stress and inflammation via Nrf2 signaling pathway activation (Table 2). It is important to note that damage in the CNS, such as traumatic brain injury, can elevate p-Nrf2/Nrf2 ratio and Nrf2 levels in a "compensation system" due to higher ROS production after injury (da Silva Fiorin et al., 2016; Yan et al., 2009). However, Nrf2 activation is unable to prevent prolonged pathophysiologic cascades after the damage that leads to behavioral deficits and brain cell loss (da Silva Fiorin et al., 2016). Nevertheless, increased Nrf2 in spinal cord damage is accompanied by increased Bax apoptotic protein expression, decreased Bcl-2 anti-apoptotic protein expression, and elevated apoptosis rates compared to non-injured tissues (Liu et al., 2020; Lv et al., 2019). After the contusion SCI model, the increased Nrf2 levels were unable to reduce inflammatory markers, such as interleukin-6 (IL-6), tumoral necrosis factor-α (TNF-α), and lesion-induced ROS production, in addition to there being no protection in the antioxidant system due to the lower SOD expression levels (Liu et al., 2020). On the other hand, when physical stimuli or pharmacological treatments increased p-Nrf2/Nrf2 ratio or Nrf2 levels in injured animals, the antioxidant system was able to induce protection of the injured nervous tissues (da Silva Fiorin et al., 2016; Liu et al., 2020; Lv et al., 2019) (Fig. 2).

In this sense, Ma et al. (2020) reported significant motor recovery during 28 days of rosmarinic acid (RA) administration after compression of the spinal cord in rats. Interestingly, an integration between oxidative and inflammatory mechanisms leads to degenerative processes through apoptosis (Ma et al., 2020). Furthermore, the authors also proposed that high nuclear factor kappa B (NF-κB) activation is induced by toll-like receptor 4 (TLR4) activation (Ma et al., 2020). The consequent inflammatory process leads to apoptosis caused by mitochondrial damage and caspase 3, 8, and 9 activation that are blocked due to musculoaponeurotic fibrosarcoma (MAF), HO-1, and NQO1 dependents of Nrf2 (Ma et al., 2020). In fact, crosstalk was observed between NF-κB and Nrf2 after white matter injury caused by hypoxia with reduced NF-κB signaling pathway via Nrf2-induced antioxidant response (Daverey & Agrawal, 2020).

The NFκB-TLR4 pathway is up-regulated after SCI, and this signaling pathway appears to be modulated by phosphorylation in p38 (p-p38) and JNK (p-JNK) proteins (Guan, Chen, Zhong, Liu, & Chen, 2020). Hence, higher phosphorylation was observed in the p-IKKα/β, p-Ikβα, and nuclear p65 subunits after experimental SCI showing active NF-κB signaling pathway and inducing transcription of pro-inflammatory mediators. On the other hand, antioxidant treatment up-regulated Nrf2 and decreased the NF-κB signaling pathway with increased p65 subunit expression in the cell cytoplasm, indicating lower activation and less nuclear transcription (Ma et al., 2020). Furthermore, p-NF-κB decreased after the ketogenic diet in SCI rats associated with less synthesis and release of pro-inflammatory mediators and increased Nrf2 levels (Lu et al., 2018). Corroborating such data, the compression SCI model has already been reported to cause pro-inflammatory cytokine release for 28 days after injury with spread systemic inflammation (do Espírito Santo et al., 2019).

Although many mechanisms of Nrf2-activation by antioxidant and anti-inflammatory compounds are unclear, many of those molecules can induce protective effects via Nrf2 at the cellular level, including reduced apoptosis, lower loss of ventral motor neurons at rostral, caudal, and epicenter segments, and more preserved tissue (Zhang et al., 2019). Therefore, it is important to note that after the severe SCI model, the Nrf2/HO-1 pathway activation in neurons and astrocytes promotes functional recovery through Basso, Beattie, and Bresnahan (BBB) locomotor rating scale (Lin et al., 2019). On the other hand, Li et al. (2019) reported that the motor effects did not improve, despite the coenzyme Q10 suppressing oxidative

TABLE 2 Effects of various antioxidants composed in biochemical mechanisms in SCI animal models.

Author	SCI model	Animal species	Proposed treatment	Biochemical outcomes
Wang et al., 2020	Compression (T9 vertebra)	Adult female Sprague-Dawley rats	Metformin	Increased Nrf2; Improved mitochondrial function; Decreased apoptosis
Ma et al., 2020	Compression (T9 vertebra)	Adult female Sprague-Dawley rats	Rosmarinic acid	Increased Nrf2 and HO-1; Inhibited NF-kB-p65 translocation to the nucleus; Reduced apoptosis
Liu et al., 2020	Contusion (T10 vertebra)	Adult male and female Sprague-Dawley rats	Imatinib	Increased Nrf2, HO-1, and SOD; Decreased IL-6, TNF-α, and ROS; Reduced apoptosis
Li et al., 2020	Contusion (T9 vertebra)	Adult C57BL/6 J mice	Zinc	Increased Nrf2 and HO-1; Decreased NLRP3 (inflammation-associated protein) and caspase-1
Mao et al., 2020	Weight drop (T8 vertebra)	Adult male Sprague-Dawley rats	Eriodictyol	Increased Nrf2 and HO-1; Decreased p-NF-kB/NF-kB
Dai, Tang, Zhao, Dai, & Huang, 2020	Compression (unspecified vertebra)	Adult female C57BL/6 J mice	Electroacupuncture	Increased Nrf2, HO-1, and NQO1
Zhang et al., 2019	Contusion (T9 vertebra)	Adult female Sprague-Dawley rats	Sinomenine	Increased Nrf2; Decreased inflammation and oxidative stress
Lv et al., 2019	Weight drop (T8 vertebra)	Adult male Sprague-Dawley rats	Polydatin	Increased Nrf2; Decreased oxidative stress; Reduced apoptosis
Lin et al., 2019	Compression (between T10 and T12 vertebrae)	Adult male Sprague-Dawley rats	2-(-2-Benzofuranyl)-2-imidazoline	Increased Nrf2, HO-1, SOD, GPx, and Bcl-2; Decreased Bax and caspase-3; Reduced apoptosis
Li et al., 2019	Weight drop (between T9 and T10 vertebra)	Adult male Sprague-Dawley rats	Coenzyme Q10	Increase Nrf2, NQO-1, SOD, GSH, and Bcl-2; Decreased Bax and caspase-3; Reduced apoptosis
Lu et al., 2018	Hemi-contusion (C7 vertebra)	Adult male Sprague-Dawley rats	Ketogenic diet	Increased Nrf2 and SOD; Decreased p-NF-κB and pro-inflammatory markers

T: thoracic; p: phosphorylation; p65: phosphorylation on the 65 subunit; NLRP3: Nod-like receptor pyrin domain-containing 3 complex; SOD: superoxide dismutase; GSH: reduced glutathione; Nrf2: nuclear factor erythroid-2 related factor 2; HO-1: hemeoxygenase 1; NQO1: NA(*D*)PH quinone oxidoreductase 1; GPx: peroxidase glutathione; NF-κB: nuclear factor kappa B; Bcl-2: anti-apoptotic B-cell lymphoma; Bax: Bcl-2-like protein 4; ROS: reactive oxygen species; IL-6: interleukin-6; TNF-α: tumor necrosis factor alpha.
Source: Authors.

stress and apoptosis after SCI in rats. However, it is important to note the differences in the SCI model that may explain some discrepancies (Table 2).

Using the Basso mouse scale (BMS) after experimental SCI, Li et al. (2020) demonstrated that zinc improved motor function in injured animals during 28 days of evaluation. After the injury, the levels of both Nrf2/HO-1 and Nod-like receptor pyrin domain-containing 3 complex (NLRP3) increase, although after treatment, only Nrf2/HO-1 continues to

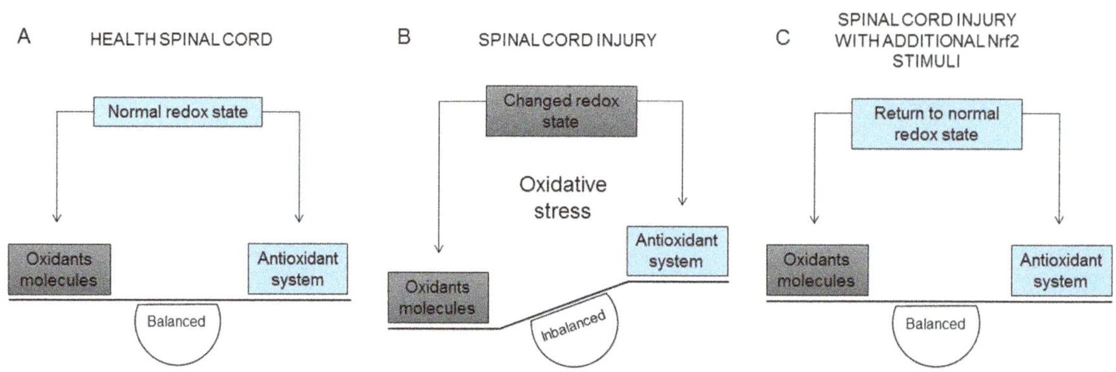

FIG. 2 Balanced and imbalanced redox state before and after SCI. (A) the oxidant and antioxidant molecules are in balance in the healthy spinal cord. (B) SCI increases oxidant molecules and causes a redox state imbalance, leading to oxidative stress. (C) Additional stimuli increase the antioxidant system via Nrf2 and balance redox state after SCI. SCI: spinal cord injury; Nrf2: nuclear factor erythroid-2-related factor 2. *(Source: Authors.)*

increase while NLRP3 decreases (Li et al., 2020). Intervention soon after SCI appears to be promising and beneficial for reorganizing and protecting the cellular microenvironment. Hence, the elevated Nrf2 mRNA expression levels presented in rats treated with antioxidants after spinal cord damage corroborates increased Nrf2 protein levels (Liu et al., 2020). Moreover, decreased pro-inflammatory cytokine release, antioxidant improvement, ROS reduction, and apoptosis protection via down-regulation of Bax, and up-regulation of Bcl-2 proteins were also observed (Liu et al., 2020). After SCI in animals, electroacupuncture stimulates Nrf2 levels, reduces pathological scores evaluated by histomorphology near the epicenter of damage, and induces protection in axons 28 days after injury (Dai, Tang, Zhao, Dai, & Huang, 2020). Reduced oxidative and inflammatory processes improved the BMS scale during the evaluation periods and electroacupuncture treatment (Dai, Tang, Zhao, Dai, & Huang, 2020).

In summary, when the up-regulation of Nrf2 is stimulated, this transcription factor can act against oxidative and inflammatory crosstalk at the cellular level after injury of the spinal cord characterized by several markers of oxidative damage, antioxidant system, and inflammatory pathway. The blockage of the oxidant and inflammatory signaling pathway improves neuronal and glial cell protection, reducing the loss of the remaining ascending and descending fibers. This protection occurs at the epicenter of damage and at the rostral and caudal segments and may lead to motor recovery in some SCI rat models (Fig. 3).

Concluding remarks

Regarding the complexity of SCI pathophysiology, many points still need to be clarified. Oxidative stress is only one of them, and the protection in targets of this secondary complication may accompany technological development to improve therapeutic treatments. Moreover, oxidative damage is a debilitating factor to the cellular environment, although it can be an "alert system" for tissue prevention or damage compensation. Furthermore, it is not yet clear if oxidative stress is caused or a consequence of tissue degeneration; therefore, additional studies are required.

In view of this, further comprehensive scientific research on cellular mechanisms, signaling pathways, and treatments are crucial and may contribute to the maintenance of cells in injured tissue. Despite advances in this research area, oxidative stress is still not a promising treatment in clinical studies due to difficult motor recovery, even with improved therapies. Nevertheless, the role of oxidative stress in injured tissues may hamper the evolution of the injured individual, and more knowledge on oxidative processes may help improve the effectiveness of current technological therapies, such as neuromodulation and brain-machine interface for the sensorimotor recovery of SCI patients.

Application to others areas of neuroscience

This chapter focuses on describing the effects of oxidative stress after SCI. Oxidative stress comprises a complex process of reactive molecules implicated in cell degeneration. Here, we also associated oxidative and inflammatory events after SCI. However, it is accepted that oxidant processes and inflammation are involved in the pathogenesis of diseases and tissue injury of the central nervous system, leading to cell loss. For example, in Parkinson's (Cuadrado et al., 2009), Alzheimer's (Kanninen et al., 2008), and Huntington's diseases (Stack et al., 2010), the reduction of oxidative damage by antioxidants can reduce brain tissue damage. Brain areas that are significantly involved in normal processes are debilitated by oxidative

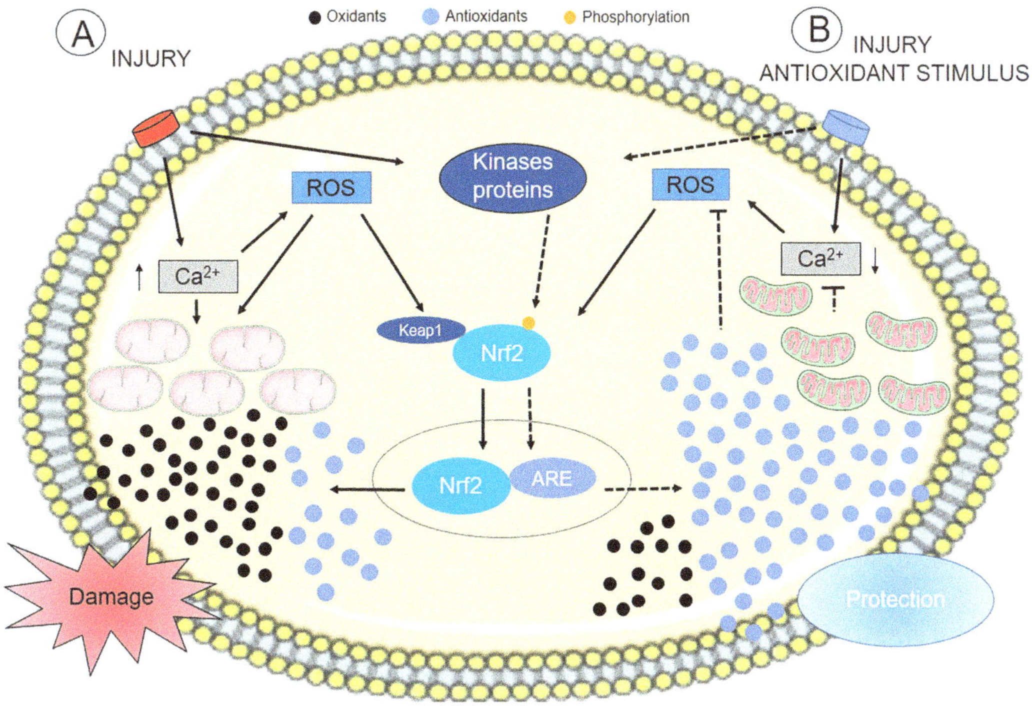

FIG. 3 Schematic representation of the redox signaling pathway after SCI and antioxidant treatment. (A) Injury causes an elevation in intracellular Ca^{2+} levels that induce ROS production through Ca^{2+}-dependent enzymes and mitochondrial damage. Oxidation of cysteine residues modifies the conformation of Keap1 and, together with phosphorylation, initiates the activation of Nrf2. The Nrf2-ARE system is activated but is not effective due to excessive ROS synthesis that causes increased Ca^{2+} influx and cellular and mitochondrial damage. (B) Antioxidant stimuli soon after the injury potentializeNrf2-ARE system activation and increase antioxidant molecule expression against oxidative molecules, decreasing intracellular Ca^{2+} influx and maintaining the tissue and mitochondrial protection. ROS: reactive oxygen species; Keap1: Kelch-like ECH-associated protein 1; Nrf2: nuclear factor erythroid-2-related factor 2; ARE: antioxidant response element; Ca^{2+}: calcium; Continuous arrows: dangerous; Dotted arrows: protective. *(Source: Authors.)*

stress in neurodegenerative disorders. Additionally, reduced dopaminergic pathways in motor areas in Parkinson's disease (Cuadrado et al., 2009) and beta amyloid accumulation in essential memory areas in Alzheimer's disease (Kanninen et al., 2008)also occur due to oxidative stress. Moreover, endogenous and exogenous injuries such as stroke (Halliwell, 1989) and traumatic brain injury (Yan et al., 2009) lead to cascades of signaling events by excessive production of oxidant molecules with consequent functional deficits. The progressive damage induced by oxidative stress appears to play a critical role in axonal degeneration and cell apoptosis after several central nervous system disorders.

Mini-dictionary of terms

Apoptosis: the phenomenon of programmed cell death.
Axoplasm: a cytoplasmic portion of neural axons.
Compression SCI model: compression of the spinal cord using a neurovascular clip, forceps, or balloon.
Contusion SCI model: an impact on the spinal cord using an impactor apparatus.
Excitotoxicity: overstimulation of glutamatergic receptors causing higher calcium influx inside the cells.
Mitochondrial swelling: a marker of mitochondrial damage.
Oxidants: reactive molecules that seek stability during their brief existence through electron attraction.
Oxidative stress: when oxidant molecules are higher than antioxidants, causing cell damage.
Redox state: relative changes in oxidation and reduction state of molecules that are not well defined or quantified.
Wallerian degeneration: a process of axon degeneration associated with the disruption of the myelin sheath.

Key facts of oxidative stress

Oxidative stress causes cell damage by oxidization in proteins, lipids, and nucleic acids after SCI.
In several neurodegenerative disorders and tissue damage, there is oxidative stress-induced cell damage.

Oxidative stress activates the cascade of events that leads to cell apoptosis.
Wallerian degeneration is caused by secondary damage after tissue injury, including oxidative stress.
Nuclear factor erythroid-2-related factor 2 stimulation can increase endogenous antioxidant defenses that may protect cells from oxidative stress damage.

Summary points

This chapter focuses on degeneration mechanisms and protection of nervous fibers after SCI.
Oxidative damage is involved in nervous fiber degeneration after SCI.
The transcription factor nuclear factor erythroid-2 related factor 2 can protect ascending and descending fibers from degeneration by inducing AREs.
Nuclear factor erythroid-2-related factor 2 induces the activation of protective molecules, such as anti-apoptotic B-cell lymphoma and BDNF.
Inflammatory processes reduce with nuclear factor erythroid-2-related factor 2 activation through crosstalk between antioxidant and anti-inflammatory systems.

References

Allen, M., Zou, F., Chai, H. S., Younkin, C. S., Miles, R., Nair, A. A., et al. (2012). Glutathione S-transferase omega genes in Alzheimer and Parkinson disease risk, age-at-diagnosis and brain gene expression: an association study with mechanistic implications. *Molecular Neurodegeneration*, *7*(1), 1–12.

Baird, L., & Dinkova-Kostova, A. T. (2011). The cytoprotective role of the Keap1-Nrf2 pathway. *Archives of Toxicology*, *85*(4), 241–272.

Beattie, M. S., Farooqui, A. A., & Bresnahan, J. C. (2000). Review of current evidence for apoptosis after spinal cord injury. *Journal of Neurotrauma*, *17* (10), 915–925.

Chen, F., Ma, W., He, J., & Zhao, J. (2002). Fenton degradation of malachite green catalyzed by aromatic additives. *The Journal of Physical Chemistry A*, *106*(41), 9485–9490.

Cho, B. O., Ryu, H. W., Jin, C. H., Choi, D. S., Kang, S. Y., Kim, D. S., … Jeong, I. Y. (2011). Blackberry extract attenuates oxidative stress through up-regulation of Nrf2-dependent antioxidant enzymes in carbon tetrachloride-treated rats. *Journal of Agricultural and Food Chemistry*, *59*(21), 11442–11448.

Cuadrado, A., Moreno-Murciano, P., & Pedraza-Chaverri, J. (2009). The transcription factor Nrf2 as a new therapeutic target in Parkinson's disease. *Expert Opinion on Therapeutic Targets*, *13*(3), 319–329.

da Silva Fiorin, F., de Oliveira Ferreira, A. P., Ribeiro, L. R., Silva, L. F., de Castro, M. R., da Silva, L. R., … Franco, J. L. (2016). The impact of previous physical training on redox signaling after traumatic brain injury in rats: A behavioral and neurochemical approach. *Journal of Neurotrauma*, *33*(14), 1317–1330.

Dai, N., Tang, C., Zhao, H., Dai, P., & Huang, S. (2020). *Effect of electroacupuncture on spinal cord injury in mice: Inhibition of inflammatory response and oxidative stress via ApoE and Nrf2* (pp. 1–25). Research Square. https://doi.org/10.21203/rs.3.rs-22504/v1.

Daverey, A., & Agrawal, S. K. (2020). Curcumin protects against white matter injury through NF-κB and Nrf2 cross talk. *Journal of Neurotrauma*, *37*(10), 1255–1265.

do Espírito Santo, C. C., da Silva Fiorin, F., Ilha, J., Duarte, M. M. M. F., Duarte, T., & Santos, A. R. S. (2019). Spinal cord injury by clip-compression induces anxiety and depression-like behaviours in female rats: The role of the inflammatory response. *Brain, Behavior, and Immunity*, *78*, 91–104.

Droge, W. (2002). Free radicals in the physiological control of cell function. *Physiological Reviews*, *82*(1), 47–95.

Fukushima, M., Tatsumi, K., & Morimoto, K. (2000). The fate of aniline after a photo-Fenton reaction in an aqueous system containing iron (III), humic acid, and hydrogen peroxide. *Environmental Science & Technology*, *34*(10), 2006–2013.

Guan, B., Chen, R., Zhong, M., Liu, N., & Chen, Q. (2020). Protective effect of oxymatrine against acute spinal cord injury in rats via modulating oxidative stress, inflammation and apoptosis. *Metabolic Brain Disease*, *35*(1), 149–157.

Halliwell, B. (1989). Oxidants and the central nervous system: Some fundamental questions. Is oxidant damage relevant to Parkinson's disease, Alzheimer's disease, traumatic injury or stroke? *Acta Neurologica Scandinavica*, *80*, 23–33.

Halliwell, B. (2012). Free radicals and antioxidants: Updating a personal view. *Nutrition Reviews*, *70*(5), 257–265.

Halliwell, B., & Gutteridge, J. M. (1995). The definition and measurement of antioxidants in biological systems. *Free Radical Biology & Medicine*, *18* (1), 125.

Halliwell, B., & Whiteman, M. (2004). Measuring reactive species and oxidative damage in vivo and in cell culture: How should you do it and what do the results mean? *British Journal of Pharmacology*, *142*(2), 231–255.

Hayta, E., & Elden, H. (2018). Acute spinal cord injury: A review of pathophysiology and potential of non-steroidal anti-inflammatory drugs for pharmacological intervention. *Journal of Chemical Neuroanatomy*, *87*, 25–31.

Hotz, P., Hoet, P., & Lauwerys, R. (1987). Lipid peroxidation in human pathology: Evaluation of data in literature. *Pathologie Biologie*, *35*(7), 1067–1073.

Innamorato, N. G., Rojo, A. I., García-Yagüe, Á. J., Yamamoto, M., De Ceballos, M. L., & Cuadrado, A. (2008). The transcription factor Nrf2 is a therapeutic target against brain inflammation. *The Journal of Immunology*, *181*(1), 680–689.

Itoh, K., Chiba, T., Takahashi, S., Ishii, T., Igarashi, K., Katoh, Y., … Yamamoto, M. (1997). An Nrf2/small Maf heterodimer mediates the induction of phase II detoxifying enzyme genes through antioxidant response elements. *Biochemical and Biophysical Research Communications*, *236*(2), 313–322.

Itoh, K., Wakabayashi, N., Katoh, Y., Ishii, T., Igarashi, K., Engel, J. D., & Yamamoto, M. (1999). Keap1 represses nuclear activation of antioxidant responsive elements by Nrf2 through binding to the amino-terminal Neh2 domain. *Genes & Development*, *13*(1), 76–86.

Jain, A. K., & Jaiswal, A. K. (2007). GSK-3β acts upstream of Fyn kinase in regulation of nuclear export and degradation of NF-E2 related factor 2. *Journal of Biological Chemistry*, *282*(22), 16502–16510.

Jia, Z., Zhu, H., Li, J., Wang, X., Misra, H., & Li, Y. (2012). Oxidative stress in spinal cord injury and antioxidant-based intervention. *Spinal Cord*, *50*(4), 264–274.

Kanninen, K., Malm, T. M., Jyrkkänen, H. K., Goldsteins, G., Keksa-Goldsteine, V., Tanila, H., … Koistinaho, J. (2008). Nuclear factor erythroid 2-related factor 2 protects against beta amyloid. *Molecular and Cellular Neuroscience*, *39*(3), 302–313.

Kloos, A. D., Fisher, L. C., Detloff, M. R., Hassenzahl, D. L., & Basso, D. M. (2005). Stepwise motor and all-or-none sensory recovery is associated with nonlinear sparing after incremental spinal cord injury in rats. *Experimental Neurology*, *191*(2), 251–265.

Kumar, P., Kalonia, H., & Kumar, A. (2011). Expression of concern: Role of LOX/COX pathways in 3-nitropropionic acid-induced Huntington's disease-like symptoms in rats: Protective effect of licofelone. *British Journal of Pharmacology*, *164*(2b), 644–654.

Lau, A., & Tymianski, M. (2010). Glutamate receptors, neurotoxicity and neurodegeneration. *Pflügers Archiv: European Journal of Physiology*, *460*(2), 525–542.

Lee, J. M., Shih, A. Y., Murphy, T. H., & Johnson, J. A. (2003). NF-E2-related factor-2 mediates neuroprotection against mitochondrial complex I inhibitors and increased concentrations of intracellular calcium in primary cortical neurons. *Journal of Biological Chemistry*, *278*(39), 37948–37956.

Li, D., et al. (2020). Zinc promotes functional recovery after spinal cord injury by activating Nrf2/HO-1 defense pathway and inhibiting inflammation of NLRP3 in nerve cells. *Life Sciences*, *245*, 117351.

Li, X., et al. (2019). Coenzyme Q10 suppresses oxidative stress and apoptosis via activating the Nrf-2/NQO-1 and NF-κB signaling pathway after spinal cord injury in rats. *American Journal of Translational Research*, *11*(10), 6544.

Lin, X., et al. (2019). Treatment with 2-BFI attenuated spinal cord injury by inhibiting oxidative stress and neuronal apoptosis via the Nrf2 signaling pathway. *Frontiers in Cellular Neuroscience*, *13*, 567.

Liu, L., et al. (2020). Imatinib inhibits oxidative stress response in spinal cord injury rats by activating Nrf2/HO-1 signaling pathway. *Experimental and Therapeutic Medicine*, *19*(1), 597–602.

LoPachin, R. M., Gaughan, C. L., Lehning, E. J., Kaneko, Y., Kelly, T. M., & Blight, A. (1999). Experimental spinal cord injury: Spatiotemporal characterization of elemental concentrations and water contents in axons and neuroglia. *Journal of Neurophysiology*, *82*(5), 2143–2153.

Lu, Y., Yang, Y. Y., Zhou, M. W., Liu, N., Xing, H. Y., Liu, X. X., & Li, F. (2018). Ketogenic diet attenuates oxidative stress and inflammation after spinal cord injury by activating Nrf2 and suppressing the NF-κB signaling pathways. *Neuroscience Letters*, *683*, 13–18.

Luo, J., Borgens, R., & Shi, R. (2004). Polyethylene glycol improves function and reduces oxidative stress in synaptosomal preparations following spinal cord injury. *Journal of Neurotrauma*, *21*(8), 994–1007.

Lv, R., et al. (2019). Polydatin attenuates spinal cord injury in rats by inhibiting oxidative stress and microglia apoptosis via Nrf2/HO-1 pathway. *Life Sciences*, *217*, 119–127.

Ma, Z., et al. (2020). Rosmarinic acid exerts a neuroprotective effect on spinal cord injury by suppressing oxidative stress and inflammation via modulating the Nrf2/HO-1 and TLR4/NF-κB pathways. *Toxicology and Applied Pharmacology*, 115014.

Maiorino, M., Gregolin, C., & Ursini, F. (1990). [47] phospholipid hydroperoxide glutathione peroxidase. *Methods in Enzymology*, *186*, 448–457.

Mao, X., et al. (2020). Eriodictyol attenuates spinal cord injury by activating Nrf2/HO-1 pathway and inhibiting NF-κB pathway. *Tropical Journal of Pharmaceutical Research*, *19*(8).

McMahon, M., Thomas, N., Itoh, K., Yamamoto, M., & Hayes, J. D. (2006). Dimerization of substrate adaptors can facilitate cullin-mediated ubiquitylation of proteins by a "tethering" mechanism a two-site interaction model for the Nrf2-Keap1 complex. *Journal of Biological Chemistry*, *281*(34), 24756–24768.

Miller, J. P., Holcomb, J., Al-Ramahi, I., De Haro, M., Gafni, J., Zhang, N., … Botas, J. (2010). Matrix metalloproteinases are modifiers of huntingtin proteolysis and toxicity in Huntington's disease. *Neuron*, *67*(2), 199–212.

Motohashi, H., O'Connor, T., Katsuoka, F., Engel, J. D., & Yamamoto, M. (2002). Integration and diversity of the regulatory network composed of Maf and CNC families of transcription factors. *Gene*, *294*(1–2), 1–12.

Motohashi, H., & Yamamoto, M. (2004). Nrf2-Keap1 defines a physiologically important stress response mechanism. *Trends in Molecular Medicine*, *10* (11), 549–557.

Navarro, A., & Boveris, A. (2007). Brain mitochondrial dysfunction in aging: Conditions that improve survival, neurological performance and mitochondrial function. *Frontiers in Bioscience*, *12*, 1154–1163.

Niture, S. K., & Jaiswal, A. K. (2012). Nrf2 protein up-regulates antiapoptotic protein Bcl-2 and prevents cellular apoptosis. *Journal of Biological Chemistry*, *287*(13), 9873–9886.

Rada, P., Rojo, A. I., Evrard-Todeschi, N., Innamorato, N. G., Cotte, A., Jaworski, T., … Hayes, J. D. (2012). Structural and functional characterization of Nrf2 degradation by the glycogen synthase kinase 3/β-TrCP axis. *Molecular and Cellular Biology*, *32*(17), 3486–3499.

Radi, R., Beckman, J. S., Bush, K. M., & Freeman, B. A. (1991). Peroxynitrite-induced membrane lipid peroxidation: The cytotoxic potential of superoxide and nitric oxide. *Archives of Biochemistry and Biophysics*, *288*(2), 481–487.

Rojo, A. I., Innamorato, N. G., Martín-Moreno, A. M., De Ceballos, M. L., Yamamoto, M., & Cuadrado, A. (2010). Nrf2 regulates microglial dynamics and neuroinflammation in experimental Parkinson's disease. *Glia*, *58*(5), 588–598.

Rojo, A. I., Medina-Campos, O. N., Rada, P., Zúñiga-Toalá, A., López-Gazcón, A., Espada, S., … Cuadrado, A. (2012). Signaling pathways activated by the phytochemical nordihydroguaiaretic acid contribute to a Keap1-independent regulation of Nrf2 stability: Role of glycogen synthase kinase-3. *Free Radical Biology and Medicine*, *52*(2), 473–487.

Sakata, H., Niizuma, K., Yoshioka, H., Kim, G. S., Jung, J. E., Katsu, M., … Chan, P. H. (2012). Minocycline-preconditioned neural stem cells enhance neuroprotection after ischemic stroke in rats. *Journal of Neuroscience*, *32*(10), 3462–3473.

Shah, Z. A., Li, R. C., Thimmulappa, R. K., Kensler, T. W., Yamamoto, M., Biswal, S., & Doré, S. (2007). Role of reactive oxygen species in modulation of Nrf2 following ischemic reperfusion injury. *Neuroscience*, *147*(1), 53–59.

Springer, J. E., Azbill, R. D., Mark, R. J., Begley, J. G., Waeg, G., & Mattson, M. P. (1997). 4-Hydroxynonenal, a lipid peroxidation product, rapidly accumulates following traumatic spinal cord injury and inhibits glutamate uptake. *Journal of Neurochemistry*, *68*(6), 2469–2476.

Stack, C., Ho, D., Wille, E., Calingasan, N. Y., Williams, C., Liby, K., … Beal, M. F. (2010). Triterpenoids CDDO-ethyl amide and CDDO-trifluoroethyl amide improve the behavioral phenotype and brain pathology in a transgenic mouse model of Huntington's disease. *Free Radical Biology and Medicine*, *49*(2), 147–158.

Surh, Y. J., Kundu, J. K., & Na, H. K. (2008). Nrf2 as a master redox switch in turning on the cellular signaling involved in the induction of cytoprotective genes by some chemopreventive phytochemicals. *Planta*, *74*(13), 1526–1539.

Szabó, C., Ischiropoulos, H., & Radi, R. (2007). Peroxynitrite: Biochemistry, pathophysiology and development of therapeutics. *Nature Reviews Drug Discovery*, *6*(8), 662–680.

Tong, K. I., Padmanabhan, B., Kobayashi, A., Shang, C., Hirotsu, Y., Yokoyama, S., & Yamamoto, M. (2007). Different electrostatic potentials define ETGE and DLG motifs as hinge and latch in oxidative stress response. *Molecular and Cellular Biology*, *27*(21), 7511–7521.

Vargas, M. R., & Johnson, J. A. (2009). The Nrf2-ARE cytoprotective pathway in astrocytes. *Expert Reviews in Molecular Medicine*, *11*.

Venugopal, R., & Jaiswal, A. K. (1998). Nrf2 and Nrf1 in association with Jun proteins regulate antioxidant response element-mediated expression and coordinated induction of genes encoding detoxifying enzymes. *Oncogene*, *17*(24), 3145–3156.

Visavadiya, N. P., Patel, S. P., VanRooyen, J. L., Sullivan, P. G., & Rabchevsky, A. G. (2016). Cellular and subcellular oxidative stress parameters following severe spinal cord injury. *Redox Biology*, *8*, 59–67.

Wang, H., et al. (2020). Metformin promotes axon regeneration after spinal cord injury through inhibiting oxidative stress and stabilizing microtubule. *Oxidative Medicine and Cellular Longevity*, *2020*.

Woźniak, B., Woźniak, A., Mila-Kierzenkowska, C., & Kasprzak, H. A. (2016). Correlation of oxidative and antioxidative processes in the blood of patients with cervical spinal cord injury. *Oxidative Medicine and Cellular Longevity*, *2016*.

Yan, W., Wang, H. D., Feng, X. M., Ding, Y. S., Jin, W., & Tang, K. (2009). The expression of NF-E2-related factor 2 in the rat brain after traumatic brain injury. *Journal of Trauma and Acute Care Surgery*, *66*(5), 1431–1435.

Zhang, F., Wang, S., Zhang, M., Weng, Z., Li, P., Gan, Y., … Sporn, M. B. (2012). Pharmacological induction of heme oxygenase-1 by a triterpenoid protects neurons against ischemic injury. *Stroke*, *43*(5), 1390–1397.

Zhang, L., et al. (2019). Sinomenine attenuates traumatic spinal cord injury by suppressing oxidative stress and inflammation via Nrf2 pathway. *Neurochemical Research*, *44*(4), 763–775.

Chapter 15

Novel agent ONO-2506 suppresses astrocytic activation and attenuates post-spinal cord injury pain

Hiroyuki Ishiguro[a] and Takashi Kaito[b]

[a]Department of Orthopaedic Surgery, National Hospital Organization Osaka National Hospital, Osaka, Japan, [b]Department of Orthopaedic Surgery, Osaka University Graduate School of Medicine, Osaka, Japan

Abbreviations

CNS	central nervous system
CSPGs	chondroitin sulfate proteoglycans
EAAT	excitatory amino acid transporter
GABA	γ-aminobutyric acid
GAD	glutamic acid decarboxylase
GFAP	glial fibrillary acidic protein
GLAST	glutamate aspartate transporter
GLT	glutamate transporter
IL	interleukin
iNOS	inducible nitric oxide synthase
pMCAO	permanent middle cerebral artery occlusion
SCI	spinal cord injury
TNF	tumor necrosis factor

Introduction

Neuropathic pain occurs in 53% of traumatic spinal cord injury (SCI) patients (Burke, Fullen, Stokes, & Lennon, 2017). Post-SCI neuropathic pain is characterized by burning, stabbing, or electric-like pain and often persists chronically (Siddall, McClelland, Rutkowski, & Cousins, 2003). This severe pain markedly affects patients' activities of daily living (ADL) and quality of life (QOL) and can be a cause of depression and even suicide (Cairns, Adkins, & Scott, 1996; Harris, Barraclough, Grundy, Bamford, & Inskip, 1996).

Post-SCI neuropathic pain includes peripherally evoked pain such as hyperalgesia (exaggerated pain evoked by noxious stimuli), allodynia (pain evoked by non-noxious stimuli), and spontaneous pain. Pain is classified into three regions: above-level pain, at-level pain, and below-level pain (Siddall, Taylor, & Cousins, 1997). Despite a variety of pharmacological, surgical, and behavioral therapeutic strategies, it remains challenging to treat post-SCI neuropathic pain effectively (Hatch, Cushing, Carlson, & Chang, 2018; Hulsebosch, Hains, Crown, & Carlton, 2009).

In SCI, an immune reaction represented by astrocytic activation, characterized by the aberrant proliferation and phenotypic change of astrocytes, occurs after mechanical damage. Several studies have demonstrated that astrocytic activation plays important role in post-SCI neuropathic pain (Kramer et al., 2017; Putatunda, Hala, Chin, & Lepore, 2014), and inhibition of this activation has attracted attention as a new therapeutic target for post-SCI neuropathic pain (Nakagawa & Kaneko, 2010).

Cellular, Molecular, Physiological, and Behavioral Aspects of Spinal Cord Injury. https://doi.org/10.1016/B978-0-12-822427-4.00015-0

Secondary injury of the spinal cord and astrocytic activation

Primary and secondary injury of the spinal cord

Traumatic spinal injury results in various patterns of mechanical damage to the spinal cord tissue, including compression, contusion, and laceration. After these primary injuries, the invasion of immune cells through the damaged blood-spinal cord barrier occurs. These cells induce secondary injury of the spinal cord by causing an immune reaction (inflammation). Neutrophils, macrophages, T-lymphocytes, and microglia are reported to invade the injured spinal cord and release pro-inflammatory cytokines such as tumor necrosis factor (TNF)-α and interleukin (IL)-1β, oxidative and proteolytic enzymes, nitric oxide (NO), and free radicals. This secondary injury expands the injured area through the necrosis and apoptosis of neurons and glial cells around the focal region and results in further neurological deterioration. In addition, a fluid-filled cystic cavity that is formed after clearance of the injured area by microglia can be a barrier for axonal regrowth (Ahuja et al., 2017; Fleming et al., 2006; Macaya & Spector, 2012). Currently, the administration of high-dose methylprednisolone sodium succinate is the only available treatment for secondary injury although its use is not recommended because of the limited effectiveness and risk of complications (Hurlbert, 2000) (Table 1).

Astrocytes and their activation

Astrocytes comprise the largest population of glial cells and play a critical role in the secondary injury of the spinal cord. Astrocytes have various important functions, such as maintaining the physical structure of the central nervous system (CNS), formation of the blood-spinal cord barrier, maintaining the extracellular homeostasis of the ionic environment, the release of various neurotrophic factors and cytokines, and uptake of neurotransmitters. Inflammation around the injured site can activate astrocytes and change their phenotype to reactive astrocytes. Reactive astrocytes upregulate glial fibrillary acidic protein (GFAP), key intermediate filaments as astrocyte markers, and exhibit morphologic characteristics such as expanded cytoplasm and extended processes. They produce various pro-inflammatory cytokines (such as IL-1β, IL-6, and TNF-α) to induce further inflammation and neurotoxic agents (such as cyclooxygenase [COX]-2, inducible nitric oxide synthase [iNOS], S100B, and glutamate), which induce neuronal hyperexcitability and apoptosis (Gaudet & Fonken, 2018).

Reactive astrocytes and glial scarring

Reactive astrocytes migrate to the injured site, accumulate around the cystic cavities, and form high-density scar tissue, called glial scars (Fig. 1). Glial scars create physical barriers for axonal regeneration and inhibit the recovery of motor and sensory functions (Silver & Miller, 2004). During scar formation, the reactive astrocytes upregulate the production of chondroitin sulfate proteoglycans (CSPGs; aggrecan, brevican, neurocan, versican, phosphacan, and NG2) (Yiu & He, 2006). CSPGs are extracellular matrix molecules composed of a protein core with glycosaminoglycan (GAG) side chains, which inhibit axonal regrowth. Treatment using chondroitinase ABC (ChABC), an enzyme that removes GAG chains from CSPGs, improved the axonal regeneration and recovery of sensorimotor function in a rat SCI model (Bradbury et al., 2002), and the administration of ChABC is thought to be a potential treatment option for SCI by modulating glial scar formation. The significance of glial scar formation is controversial, however, because recent reports have suggested that

TABLE 1 **Pharmacological treatments for post-SCI neuropathic pain.**

Clinically available drugs	
Pregabalin, Gabapentin, Lamotrigine	Antiepileptics
Amitriptyline	Tricyclics
Tramadol	Opioids
Duloxetine	Serotonin noradrenaline reuptake inhibitors (SNRI)
Potential therapeutic agents	
ONO-2506 (Ishiguro et al., 2019)	Inhibitor of astrocytic activation
Propentofylline (Gwak et al., 2008)	Inhibitor of microglial and astrocytic activation

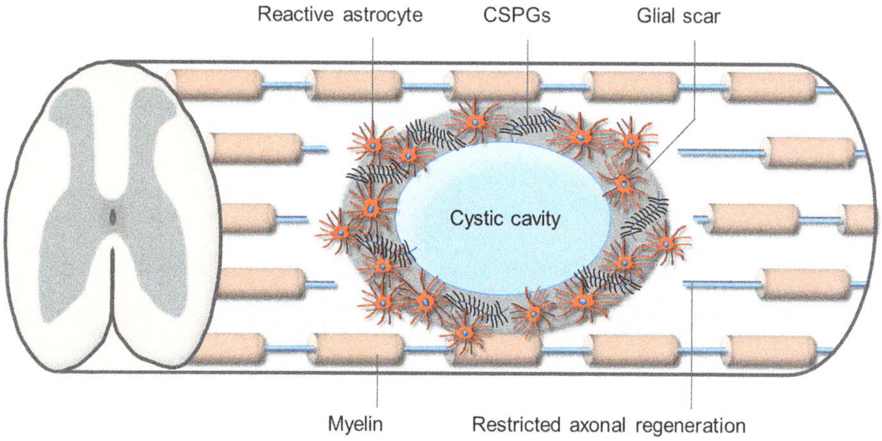

FIG. 1 Glial scar and cystic cavity at the injured site of the spinal cord. Reactive astrocytes form a glial scar with deposition of chondroitin sulfate proteoglycans (CSPGs) around a cystic cavity at the injured site of the spinal cord. The scar tissue creates physical barriers for axonal regeneration.

glial scars might have beneficial functions in neural protection, inhibiting the migration of immune cells and localizing tissue degeneration (Anderson et al., 2016; Faulkner et al., 2004).

Astrocytic activation and post-SCI neuropathic pain

Recent evidence indicates that astrocytic activation plays a major role in post-SCI neuropathic pain. The reduction of reactive astrocytes by the viral overexpression of IL-10 attenuated hindlimb hyperalgesia and allodynia in a rat SCI model (Lau et al., 2012). Below-level allodynia after SCI was alleviated in connexin-43 (a gap junction protein predominantly expressed by astrocytes) knockout mice (Chen et al., 2012). Several studies have reported that reactive astrocytes are observed in the below-level dorsal horn as well as at the regional level after SCI. These findings indicate that astrocytic activation at each level of the spinal cord may alter synaptic activity at the dorsal horn (Carlton et al., 2009; Gwak, Kang, Unabia, & Hulsebosch, 2012). The absence of reactive astrocytes at the below-level dorsal horn after SCI was also reported, however (Ishiguro et al., 2019).

Mechanisms of post-SCI neuropathic pain by astrocytic activation

The detailed mechanisms of post-SCI neuropathic pain associated with astrocytic activation remain largely unknown. Possible causes of the pain include changes in extracellular glutamate and extracellular γ-aminobutyric acid (GABA).

Increase in extracellular glutamate

Glutamate is a neurotransmitter that is released from the presynaptic terminal of primary afferent neurons to secondary neurons. Increased extracellular glutamate concentration in the CNS not only has a harmful influence on synaptic transport but also has the potential to induce neuronal death due to excessive activation of glutamate receptors, which is termed excitotoxicity (Choi, 1988). The glutamate release to extracellular fluid is mediated by excitatory amino acid transporters (EAATs). Among the variety of EAATs, EAAT1 (glutamate aspartate transporter; GLAST) and EAAT2 (glutamate transporter-1; GLT-1), which are selectively expressed by astrocytes, are the main pathways of glutamate uptake (Zhou & Danbolt, 2014) (Fig. 2).

Increased extracellular glutamate concentration at the spinal dorsal horn leads to excitation of the spinothalamic tract neurons (Gwak & Hulsebosch, 2011). The inhibition of glutamate uptake through the intrathecal injection of glutamate transporter blockers DL-threo-beta-benzyloxyaspartate to an intact spinal cord induced post-SCI pain-related behaviors (Liaw et al., 2005).

Post-SCI astrocytic activation causes an elevation in extracellular glutamate concentration in the spinal cord. Thermal hyperalgesia in a rat cervical contusion SCI model demonstrated that the activation of astrocytes was accompanied by the reduced expression of GLT-1 at the dorsal horn (Putatunda et al., 2014; Watson, Hala, Putatunda, Sannie, & Lepore, 2014). In contrast, the viral overexpression of GLT-1 in astrocytes attenuated thermal hyperalgesia in this model (Falnikar, Hala, Poulsen, & Lepore, 2016). These findings suggest that neuropathic pain is caused by the dysregulation of extracellular

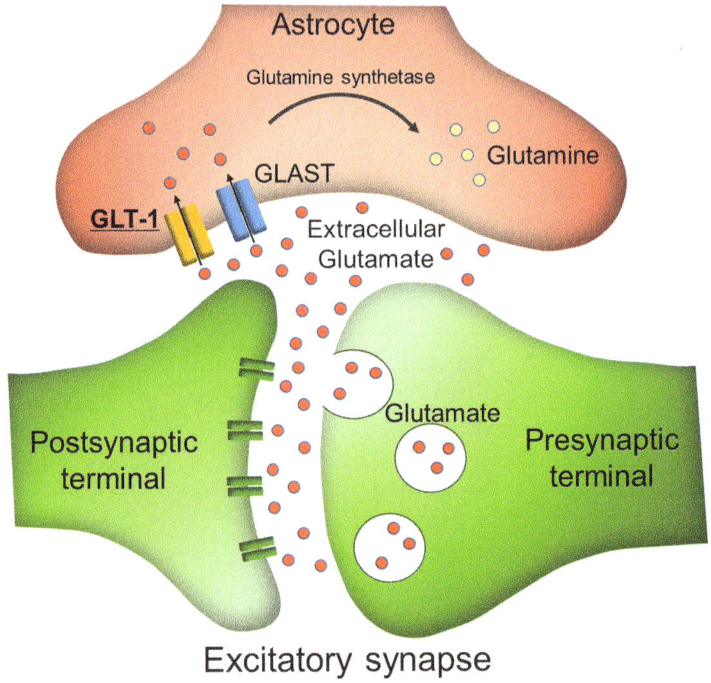

FIG. 2 The uptake of extracellular glutamate by astrocyte glutamate transporters. The glutamate released into an extracellular fluid is mainly mediated by glutamate aspartate transporter (GLAST) and glutamate transporter-1 (GLT-1), which are expressed in astrocytes. Astrocytic activation causes the reduction of GLT-1 expression at the dorsal horn and induces neuropathic pain through the elevation of extracellular glutamate concentration in the spinal cord. GLAST: glutamate aspartate transporter; GLT: glutamate transporter.

glutamate uptake and the subsequent increase of extracellular glutamate concentration. The enhanced release of glutamate from the reactive astrocytes also contributes to the increase in the concentration (Chao, Hu, & Peterson, 1995).

Decrease in extracellular GABA

GABA is a neurotransmitter that plays a crucial role in inhibitory synaptic transmission in the CNS. GABA is produced from glutamate through decarboxylation by glutamic acid decarboxylase (GAD), which is expressed by GABAergic inhibitory neurons. GABA released from the presynaptic terminal to the synaptic cleft acts via ionotropic $GABA_A$ receptors or G-protein-coupled $GABA_B$ receptors. Astrocytes modulate extracellular GABA concentration through the uptake of released GABA via GABA transporter (GAT) (Fig. 3). Gwak et al. reported that astrocytic activation was accompanied by reduced expression of GAD_{65} at the dorsal horn in a rat thoracic SCI model with below-level allodynia. They also demonstrated that the intrathecal administration of propentophylline, a glial modulator that inhibits phosphodiesterase and adenosine uptake, prevented the decrease of GAD_{65} expression at the lumbar dorsal horn and attenuated the astrocytic activation and mechanical allodynia following SCI. These findings suggest that astrocytic activation induces the decreases of GABA production via the reduction of GAD_{65} expression, suppresses the GABAergic inhibitory tone, and induces the sensitization of secondary afferent neurons (Gwak, Crown, Unabia, & Hulsebosch, 2008; Gwak & Hulsebosch, 2011).

Attenuation of post-SCI neuropathic pain by ONO-2506

S100B

The S100 proteins are a family of 8–14 kDa EF-hand Ca^{2+}-binding proteins that have more than 20 subfamilies (Heizmann, Fritz, & Schäfer, 2002). S100B, a homodimer of the S100β subunits, is a member of the S100 protein family that is produced by astrocytes. Reactive astrocytes in the injured CNS due to ischemia or trauma enhance S100B production and secretion, resulting in elevation of the blood and tissue concentration of S100B. This protein is widely used as a reliable clinical biomarker that reflects the volume of cerebral infarction (Missler, Wiesmann, Friedrich, & Kaps, 1997; Thelin, Nelson, & Bellander, 2017). The correlation between serum and cerebrospinal fluid levels of S100B and the severity of SCI has also been reported (Cao et al., 2008). An animal study using S100B overexpressing transgenic mice showed the exacerbation of the infarct volume and neuronal deficit after cerebral ischemia (Mori et al., 2008). In contrast, the neuronal dysfunction due to cerebral infarction was attenuated in S100B knockout mice (Wainwright et al., 2004). Thus, S100B is considered to play an important role in central nerve injury.

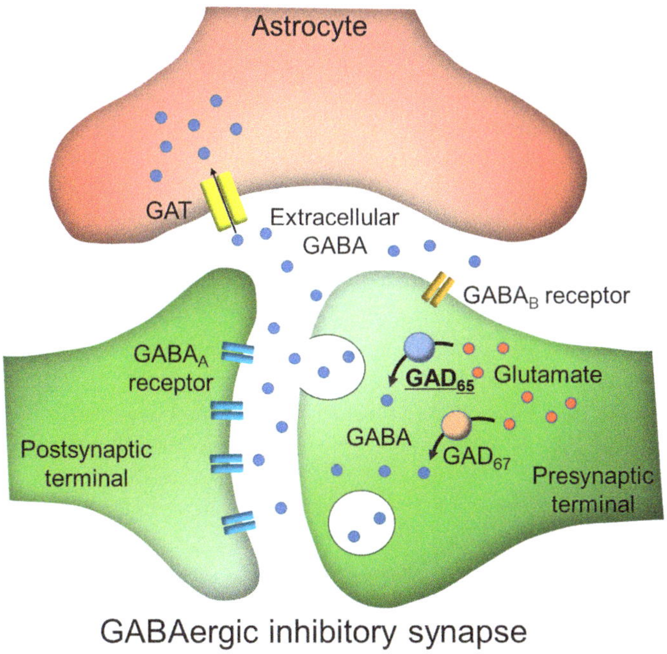

FIG. 3 The GABAergic system in the inhibitory synapse. γ-aminobutyric acid (GABA) is produced from glutamate through decarboxylation by glutamic acid decarboxylase (GAD), which is expressed by GABAergic inhibitory neurons. Astrocytic activation reduces the expression of GAD_{65} at the dorsal horn, which decreases GABA production and induces the sensitization of secondary afferent neurons.

S100B is known to exert both positive and negative effects according to the concentration. Under physiological conditions of low concentration (nanomolar level), S100B exhibit neurotrophic effects, such as the prevention of neuron death, promotion of synapse formation, enhancement of neurite extension, and modulation of long-term neuronal synaptic plasticity (Iwasaki, Shiojima, & Kinoshita, 1997; Kligman & Marshak, 1985; Nishiyama, Knopfel, Endo, & Itohara, 2002). In contrast, a high concentration (micromolar level) of S100B exhibits neurotoxic and pro-inflammatory effects, including the induction of glutamate-induced excitotoxicity and NO-dependent neuronal death (Hu, Ferreira, & Van Eldik, 1997). In addition, S100B promotes astrocytic activation through the receptor for advanced glycation endproducts (RAGE) in astrocytes. This mechanism configures an amplification loop of astrocytic activation and enhances the production of iNOS and pro-inflammatory factors, such as COX-2, IL-1β, IL-6, and TNF-α (Michetti et al., 2019; Villarreal et al., 2014) (Fig. 4).

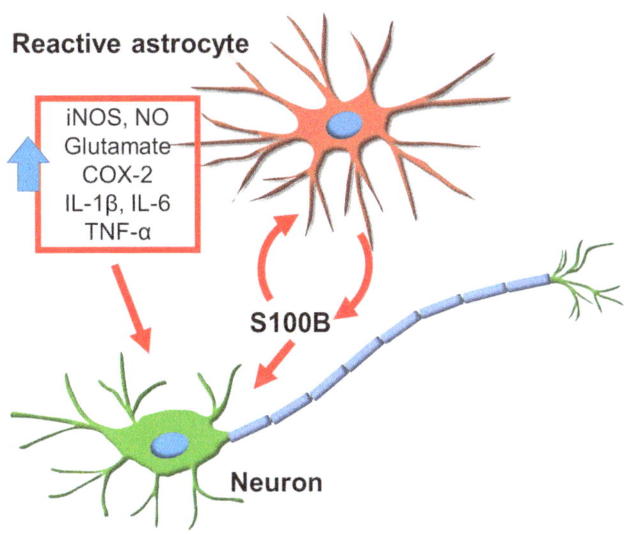

FIG. 4 The effects of S100B secreted by reactive astrocytes. S100B secreted by reactive astrocytes exhibits neurotoxic effects. In addition, S100B promotes astrocytic activation and enhances the production of iNOS, glutamate, and pro-inflammatory factors, such as cyclooxygenase (COX)-2, interleukin (IL)-1β, IL-6, and tumor necrosis factor (TNF)-α, which induce neuronal hyperexcitability and apoptosis.

ONO-2506

ONO-2506 (arundic acid, [R]-[−]-2-npropyloctanoic acid) is an agent that inhibits spontaneous S100B expression in cultured astrocytes (Fig. 5). ONO-2506 selectively inhibits S100B production and suppresses astrocytic activation. In this respect, ONO-2506 differs from other antiastrocytic agents such as pentamidine, which blocks the interaction between S100B and p53 (Cirillo et al., 2015), and propentofylline, which inhibits adenosine reuptake (Tawfik et al., 2008). No significant side effects of ONO-2506 on other functions of the central nervous, cardiopulmonary, renal, or digestive systems were observed in animal experiments (Asano et al., 2005).

Action mechanisms of ONO-2506

ONO-2506 suppresses the expression of iNOS and COX-2 in astrocytes by inhibiting the excessive production of S100B in vitro. In a co-culture system of astrocytes and neurons, ONO-2506 attenuated NO-induced neuronal death through the suppression of iNOS expression in astrocytes (Asano et al., 2005). Furthermore, ONO-2506 improves glutamate uptake through the enhancement of GLAST and GLT-1 expression in astrocytes and suppresses glutamate excitotoxicity (Karki et al., 2018; Mori et al., 2004).

The positive effects of ONO-2506 on cerebral ischemia have been confirmed in several studies. In cerebral ischemia, astrocytic activation is known to affect the expansion of the infarct region and the subsequent neurological deterioration. In a rat permanent middle cerebral artery occlusion (pMCAO) model, the administration of ONO-2506 significantly suppressed the delayed expansion of the infarct region. In this model, the increase of S100B at the peri-infarct area and in the cerebrospinal fluid was significantly suppressed by the administration of ONO-2506. Moreover, better neurological recovery within 24 h following the administration of ONO-2506 was observed (Tateishi et al., 2002). In another animal study using a rat transient middle cerebral artery occlusion model, ONO-2506 prevented the reduction of the mRNA expression of glial glutamate transporters (GLT-1 and GLAST) and promoted neurological recovery (Mori et al., 2004). This effect was also confirmed in a monkey model of pMCAO (Asano et al., 2005).

As for the effects of ONO-2506 on SCI, Hanada et al. reported that the administration of ONO-2506 suppressed the astrocytic expression of S100 protein at the injured site and improved motor function after SCI (Hanada et al., 2014).

Effects of ONO-2506 on post-SCI neuropathic pain

Recently, we investigated the effects of ONO-2506 on post-SCI neuropathic pain in a rat model of SCI (Ishiguro et al., 2019). The daily administration of ONO-2506 or saline was continued for 1 week following SCI, and the below-level mechanical and thermal allodynia were evaluated. At post-SCI week 6, the injured site and the dorsal horn at L4/5 were evaluated by fluorescent immunohistochemical analysis using S100B and GFAP antibodies.

The below-level mechanical and thermal allodynia were significantly attenuated by the administration of ONO-2506 throughout the entire 6 weeks after SCI. Fluorescent immunohistochemical assessment at the injured site revealed a significant reduction in the cross-sectional area of the cysts, and a high fluorescence intensity area of S100B and GFAP were confirmed in the ONO-2506 group (Fig. 6). The morphological changes, including cytoplasm expansion and cell process extension in GFAP-positive astrocytes in the saline group, were markedly reduced in the ONO-2506 group. These findings indicate that ONO-2506 attenuates post-SCI neuropathic pain by suppressing secondary injury at the injured site, which occurs within 1 week after SCI. In contrast, reactive astrocytes were not observed at the L4/5 dorsal horn in either group, which suggests that incomplete SCI does not induce below-level astrocytic activation at the dorsal horn. Therefore, the changes at the injured site are the key components in post-SCI neuropathic pain rather than the changes at the lumbar dorsal horn. In addition, a strong correlation was found between the severity of allodynia and the area of the S100B high-intensity lesion at the injured site. These findings suggest that ONO-2506, an inhibitor of S100B production in astrocytes, may be a promising therapeutic agent for post-SCI neuropathic pain.

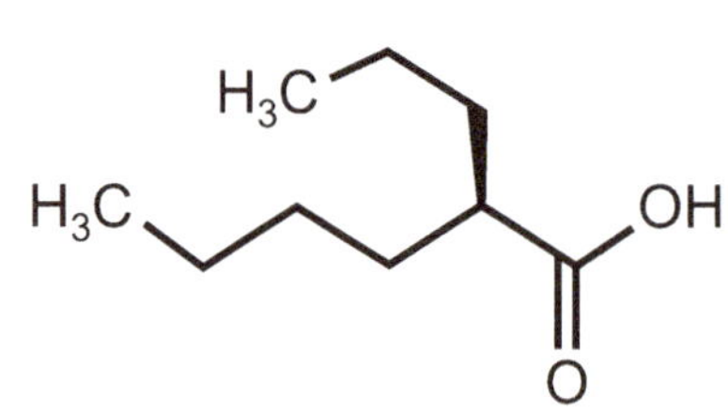

FIG. 5 Chemical structure of ONO-2506.

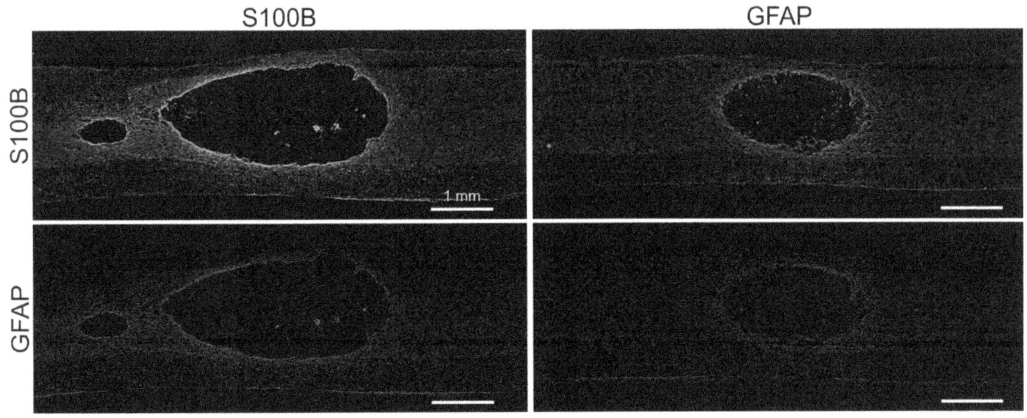

FIG. 6 Immunohistochemical images at the injured site. Fluorescent immunostaining for S100B (red) and glial fibrillary acidic protein (GFAP; green) in the sagittal section of the spinal cord at the injured level. A significant reduction in the cross-sectional area of the cysts and a high fluorescence intensity area of S100B and GFAP were confirmed in the ONO-2506 group. GFAP: glial fibrillary acidic protein.

As mentioned above, astrocytic activation at the dorsal horn has been regarded as the main cause of post-SCI neuropathic pain in previous studies. However, in our study, astrocytic activation was not detected at the below-level dorsal horn. Although the relatively low contusion energy applied to the spinal cord may contribute to this difference, this finding offers a new perspective. First, the reduction of glial scar formation around the lesion by ONO-2506 could have led to the alleviation of neuropathic pain. Since glial scarring prevents neuronal regeneration, this effect may have promoted the regeneration of secondary afferent nociceptive neurons or descending inhibitory neurons. Second, the increase in S100B at the injured site may be critically involved in the development of sensory nervous system disorder after SCI. Further studies are required to identify the detailed underlying mechanisms of astrocytic activation and S100B on post-SCI neuropathic pain.

Post-SCI neuropathic pain due to SCI is a serious disorder that significantly affects patients' ADL and QOL. To develop an effective treatment for post-SCI pain relief, a more detailed understanding of the pathology is essential. As described in this chapter, it is clear that astrocytic activation plays a crucial role in the development of post-SCI neuropathic pain. Although the detailed mechanisms of astrocytic activation on neuropathic pain remain unclear, ONO-2506 may be a promising candidate for the attenuation of post-SCI neuropathic pain by the suppression of astrocytic activation.

Applications to other areas of neuroscience

In this chapter, we have reviewed the mechanisms of post-SCI neuropathic pain, focusing on astrocytic activation associated with secondary injury of the spinal cord. Astrocytic activation causes hyperexcitability of afferent neurons through alteration of the environment of glutamate and gamma-aminobutyric acid (GABA) in the spinal cord, resulting in neuropathic pain such as allodynia and hyperalgesia. Administration of ONO-2506, an agent that suppresses astrocytic activation by inhibition of S100B production, maybe a novel treatment option for post-SCI neuropathic pain.

Enhancement of S100B production in astrocytes is observed not only in cerebral ischemia (Tateishi et al., 2002) and traumatic brain injury (Kleindienst et al., 2005; Sandhir, Onyszchuk, & Berman, 2008), but also in epilepsy (Yang & Sun, 2008); neurodegenerative disorders such as Alzheimer's disease (Doraiswamy & Xiong, 2006; Mori et al., 2006), Parkinson's disease (Kato, Kurosaki, Oki, & Araki, 2004; Sathe et al., 2012), and amyotrophic lateral sclerosis (Serrano et al., 2017; Traynor et al., 2006); and psychological disorders such as schizophrenia (Katsel et al., 2011; Steiner et al., 2014), major or minor depressive disorder, and bipolar disorder (Gos et al., 2013). Further studies using agents that modulate astrocytic S100B production, including ONO-2506, are expected to provide effective treatment options for a variety of neurological and psychological disorders.

Mini-dictionary of terms

- **Allodynia:** Pain evoked by non-noxious stimuli.
- **Astrocytic activation:** A reaction with morphological and functional changes of astrocytes.

- **Excitotoxicity:** A pathological process triggered by the overactivation of receptors for glutamate that induces neuronal death.
- **Gamma-aminobutyric acid (GABA):** A neurotransmitter that plays a crucial role in inhibitory synaptic transmission in the CNS.
- **Glial fibrillary acidic protein (GFAP):** The key intermediate filament as an astrocyte marker. Reactive astrocytes upregulate this protein.
- **Glial scar:** High-density scar tissue formed by reactive astrocytes that creates physical barriers to axonal regeneration.
- **Glutamate:** An excitatory neurotransmitter in the CNS. An elevated concentration of extracellular glutamate is toxic to neurons.
- **Hyperalgesia:** Exaggerated pain evoked by noxious stimuli.
- **Secondary injury:** An expansion of the injured region due to an immune reaction after mechanical damage to the spinal cord.
- **S100B:** A member of the S100 protein family that is produced by astrocytes. Reactive astrocytes in the injured CNS enhance S100B production.

Key facts of ONO-2506

- ONO-2506 is a novel agent that suppresses astrocytic activation by inhibiting S100B production from astrocytes.
- ONO-2506 suppresses nitric oxide-induced neuronal death through the reduction of the inducible nitric oxide synthase expression of astrocytes.
- ONO-2506 ameliorates the reduction of glutamate transporters in reactive astrocytes.
- Studies on cerebral ischemia suggest that ONO-2506 significantly suppresses the delayed expansion of the infarct region due to astrocytic activation.
- The attenuation of post-SCI neuropathic pain by ONO-2506 is demonstrated in animal models of SCI.

Summary points

- Post-SCI neuropathic pain is an important problem that markedly affects patients' ADL and QOL.
- Studies suggest that astrocytic activation plays important role in post-SCI neuropathic pain.
- Changes in the extracellular concentration of glutamate and γ-aminobutyric acid (GABA) have been reported to be involved in astrocytic activation associated post-SCI neuropathic pain.
- ONO-2506 is a novel agent that suppresses astrocytic activation by inhibiting the S100B production from astrocytes.
- Recently, ONO-2506 was reported to attenuate post-SCI neuropathic pain in a rat model of SCI.
- ONO-2506 is a promising candidate for attenuating post-SCI neuropathic pain.

CSPGs: chondroitin sulfate proteoglycans.

GABA: γ-aminobutyric acid, GAD: glutamic acid decarboxylase, GAT: GABA transporter.

iNOS: inducible nitric oxide synthase, NO: nitric oxide, COX: cyclooxygenase, IL: interleukin, TNF: tumor necrosis factor.

References

Ahuja, C. S., Wilson, J. R., Nori, S., Kotter, M. R. N., Druschel, C., Curt, A., et al. (2017). Traumatic spinal cord injury. *Nature Reviews. Disease Primers*, *3*, 17018.

Anderson, M. A., Burda, J. E., Ren, Y., Ao, Y., O'Shea, T. M., Kawaguchi, R., et al. (2016). Astrocyte scar formation aids central nervous system axon regeneration. *Nature*, *532*(7598), 195–200.

Asano, T., Mori, T., Shimoda, T., Shinagawa, R., Satoh, S., Yada, N., et al. (2005). Arundic acid (ONO-2506) ameliorates delayed ischemic brain damage by preventing astrocytic overproduction of S100B. *Current Drug Targets. CNS and Neurological Disorders*, *4*(2), 127–142.

Bradbury, E. J., Moon, L. D., Popat, R. J., King, V. R., Bennett, G. S., Patel, P. N., et al. (2002). Chondroitinase ABC promotes functional recovery after spinal cord injury. *Nature*, *416*(6881), 636–640.

Burke, D., Fullen, B. M., Stokes, D., & Lennon, O. (2017). Neuropathic pain prevalence following spinal cord injury: A systematic review and meta-analysis. *European Journal of Pain*, *21*(1), 29–44.

Cairns, D. M., Adkins, R. H., & Scott, M. D. (1996). Pain and depression in acute traumatic spinal cord injury: Origins of chronic problematic pain? *Archives of Physical Medicine and Rehabilitation*, *77*(4), 329–335.

Cao, F., Yang, X. F., Liu, W. G., Hu, W. W., Li, G., Zheng, X. J., et al. (2008). Elevation of neuron-specific enolase and S-100beta protein level in experimental acute spinal cord injury. *Journal of Clinical Neuroscience*, *15*(5), 541–544.

Carlton, S. M., Du, J., Tan, H. Y., Nesic, O., Hargett, G. L., Bopp, A. C., et al. (2009). Peripheral and central sensitization in remote spinal cord regions contribute to central neuropathic pain after spinal cord injury. *Pain*, *147*(1–3), 265–276.

Chao, C. C., Hu, S., & Peterson, P. K. (1995). Glia, cytokines, and neurotoxicity. *Critical Reviews in Neurobiology*, *9*(2–3), 189–205.

Chen, M. J., Kress, B., Han, X., Moll, K., Peng, W., Ji, R. R., et al. (2012). Astrocytic CX43 hemichannels and gap junctions play a crucial role in development of chronic neuropathic pain following spinal cord injury. *Glia*, *60*(11), 1660–1670.

Choi, D. W. (1988). Glutamate neurotoxicity and diseases of the nervous system. *Neuron*, *1*(8), 623–634.

Cirillo, C., Capoccia, E., Iuvone, T., Cuomo, R., Sarnelli, G., Steardo, L., et al. (2015). S100B inhibitor pentamidine attenuates reactive gliosis and reduces neuronal loss in a mouse model of Alzheimer's disease. *BioMed Research International*, *2015*, 508342.

Doraiswamy, P. M., & Xiong, G. L. (2006). Pharmacological strategies for the prevention of Alzheimer's disease. *Expert Opinion on Pharmacotherapy*, *7* (1), 1–10.

Falnikar, A., Hala, T. J., Poulsen, D. J., & Lepore, A. C. (2016). GLT1 overexpression reverses established neuropathic pain-related behavior and attenuates chronic dorsal horn neuron activation following cervical spinal cord injury. *Glia*, *64*(3), 396–406.

Faulkner, J. R., Herrmann, J. E., Woo, M. J., Tansey, K. E., Doan, N. B., & Sofroniew, M. V. (2004). Reactive astrocytes protect tissue and preserve function after spinal cord injury. *The Journal of Neuroscience*, *24*(9), 2143–2155.

Fleming, J. C., Norenberg, M. D., Ramsay, D. A., Dekaban, G. A., Marcillo, A. E., Saenz, A. D., et al. (2006). The cellular inflammatory response in human spinal cords after injury. *Brain*, *129*(Pt. 12), 3249–3269.

Gaudet, A. D., & Fonken, L. K. (2018). Glial cells shape pathology and repair after spinal cord injury. *Neurotherapeutics*, *15*(3), 554–577.

Gos, T., Schroeter, M. L., Lessel, W., Bernstein, H. G., Dobrowolny, H., Schiltz, K., et al. (2013). S100B-immunopositive astrocytes and oligodendrocytes in the hippocampus are differentially afflicted in unipolar and bipolar depression: A postmortem study. *Journal of Psychiatric Research*, *47*(11), 1694–1699.

Gwak, Y. S., Crown, E. D., Unabia, G. C., & Hulsebosch, C. E. (2008). Propentofylline attenuates allodynia, glial activation and modulates GABAergic tone after spinal cord injury in the rat. *Pain*, *138*(2), 410–422.

Gwak, Y. S., & Hulsebosch, C. E. (2011). GABA and central neuropathic pain following spinal cord injury. *Neuropharmacology*, *60*(5), 799–808.

Gwak, Y. S., Kang, J., Unabia, G. C., & Hulsebosch, C. E. (2012). Spatial and temporal activation of spinal glial cells: Role of gliopathy in central neuropathic pain following spinal cord injury in rats. *Experimental Neurology*, *234*(2), 362–372.

Hanada, M., Shinjo, R., Miyagi, M., Yasuda, T., Tsutsumi, K., Sugiura, Y., et al. (2014). Arundic acid (ONO-2506) inhibits secondary injury and improves motor function in rats with spinal cord injury. *Journal of the Neurological Sciences*, *337*(1–2), 186–192.

Harris, E. C., Barraclough, B. M., Grundy, D. J., Bamford, E. S., & Inskip, H. M. (1996). Attempted suicide and completed suicide in traumatic spinal cord injury. Case reports. *Spinal Cord*, *34*(12), 752–753.

Hatch, M. N., Cushing, T. R., Carlson, G. D., & Chang, E. Y. (2018). Neuropathic pain and SCI: Identification and treatment strategies in the 21st century. *Journal of the Neurological Sciences*, *384*, 75–83.

Heizmann, C. W., Fritz, G., & Schäfer, B. W. (2002). S100 proteins: Structure, functions and pathology. *Frontiers in Bioscience*, *7*, d1356–d1368.

Hu, J., Ferreira, A., & Van Eldik, L. J. (1997). S100beta induces neuronal cell death through nitric oxide release from astrocytes. *Journal of Neurochemistry*, *69*(6), 2294–2301.

Hulsebosch, C. E., Hains, B. C., Crown, E. D., & Carlton, S. M. (2009). Mechanisms of chronic central neuropathic pain after spinal cord injury. *Brain Research Reviews*, *60*(1), 202–213.

Hurlbert, R. J. (2000). Methylprednisolone for acute spinal cord injury: An inappropriate standard of care. *Journal of Neurosurgery*, *93*(1 Suppl), 1–7.

Ishiguro, H., Kaito, T., Hashimoto, K., Kushioka, J., Okada, R., Tsukazaki, H., et al. (2019). Administration of ONO-2506 suppresses neuropathic pain after spinal cord injury by inhibition of astrocytic activation. *The Spine Journal*, *19*(8), 1434–1442.

Iwasaki, Y., Shiojima, T., & Kinoshita, M. (1997). S100 beta prevents the death of motor neurons in newborn rats after sciatic nerve section. *Journal of the Neurological Sciences*, *151*(1), 7–12.

Karki, P., Hong, P., Johnson, J., Jr., Pajarillo, E., Son, D. S., Aschner, M., et al. (2018). Arundic acid increases expression and function of astrocytic glutamate transporter EAAT1 via the ERK, Akt, and NF-κB pathways. *Molecular Neurobiology*, *55*(6), 5031–5046.

Kato, H., Kurosaki, R., Oki, C., & Araki, T. (2004). Arundic acid, an astrocyte-modulating agent, protects dopaminergic neurons against MPTP neurotoxicity in mice. *Brain Research*, *1030*(1), 66–73.

Katsel, P., Byne, W., Roussos, P., Tan, W., Siever, L., & Haroutunian, V. (2011). Astrocyte and glutamate markers in the superficial, deep, and white matter layers of the anterior cingulate gyrus in schizophrenia. *Neuropsychopharmacology*, *36*(6), 1171–1177.

Kleindienst, A., McGinn, M. J., Harvey, H. B., Colello, R. J., Hamm, R. J., & Bullock, M. R. (2005). Enhanced hippocampal neurogenesis by intraventricular S100B infusion is associated with improved cognitive recovery after traumatic brain injury. *Journal of Neurotrauma*, *22*(6), 645–655.

Kligman, D., & Marshak, D. R. (1985). Purification and characterization of a neurite extension factor from bovine brain. *Proceedings of the National Academy of Sciences of the United States of America*, *82*(20), 7136–7139.

Kramer, J. L., Minhas, N. K., Jutzeler, C. R., Erskine, E. L., Liu, L. J., & Ramer, M. S. (2017). Neuropathic pain following traumatic spinal cord injury: Models, measurement, and mechanisms. *Journal of Neuroscience Research*, *95*(6), 1295–1306.

Lau, D., Harte, S. E., Morrow, T. J., Wang, S., Mata, M., & Fink, D. J. (2012). Herpes simplex virus vector-mediated expression of interleukin-10 reduces below-level central neuropathic pain after spinal cord injury. *Neurorehabilitation and Neural Repair*, *26*(7), 889–897.

Liaw, W. J., Stephens, R. L., Jr., Binns, B. C., Chu, Y., Sepkuty, J. P., Johns, R. A., et al. (2005). Spinal glutamate uptake is critical for maintaining normal sensory transmission in rat spinal cord. *Pain*, *115*(1–2), 60–70.

Macaya, D., & Spector, M. (2012). Injectable hydrogel materials for spinal cord regeneration: A review. *Biomedical Materials*, *7*(1), 012001.

Michetti, F., D'Ambrosi, N., Toesca, A., Puglisi, M. A., Serrano, A., Marchese, E., et al. (2019). The S100B story: From biomarker to active factor in neural injury. *Journal of Neurochemistry*, *148*(2), 168–187.

Missler, U., Wiesmann, M., Friedrich, C., & Kaps, M. (1997). S-100 protein and neuron-specific enolase concentrations in blood as indicators of infarction volume and prognosis in acute ischemic stroke. *Stroke*, *28*(10), 1956–1960.

Mori, T., Tan, J., Arendash, G. W., Koyama, N., Nojima, Y., & Town, T. (2008). Overexpression of human S100B exacerbates brain damage and periinfarct gliosis after permanent focal ischemia. *Stroke*, *39*(7), 2114–2121.

Mori, T., Tateishi, N., Kagamiishi, Y., Shimoda, T., Satoh, S., Ono, S., et al. (2004). Attenuation of a delayed increase in the extracellular glutamate level in the peri-infarct area following focal cerebral ischemia by a novel agent ONO-2506. *Neurochemistry International*, *45*(2–3), 381–387.

Mori, T., Town, T., Tan, J., Yada, N., Horikoshi, Y., Yamamoto, J., et al. (2006). Arundic acid ameliorates cerebral amyloidosis and gliosis in Alzheimer transgenic mice. *The Journal of Pharmacology and Experimental Therapeutics*, *318*(2), 571–578.

Nakagawa, T., & Kaneko, S. (2010). Spinal astrocytes as therapeutic targets for pathological pain. *Journal of Pharmacological Sciences*, *114*(4), 347–353.

Nishiyama, H., Knopfel, T., Endo, S., & Itohara, S. (2002). Glial protein S100B modulates long-term neuronal synaptic plasticity. *Proceedings of the National Academy of Sciences of the United States of America*, *99*(6), 4037–4042.

Putatunda, R., Hala, T. J., Chin, J., & Lepore, A. C. (2014). Chronic at-level thermal hyperalgesia following rat cervical contusion spinal cord injury is accompanied by neuronal and astrocyte activation and loss of the astrocyte glutamate transporter, GLT1, in superficial dorsal horn. *Brain Research*, *1581*, 64–79.

Sandhir, R., Onyszchuk, G., & Berman, N. E. (2008). Exacerbated glial response in the aged mouse hippocampus following controlled cortical impact injury. *Experimental Neurology*, *213*(2), 372–380.

Sathe, K., Maetzler, W., Lang, J. D., Mounsey, R. B., Fleckenstein, C., Martin, H. L., et al. (2012). S100B is increased in Parkinson's disease and ablation protects against MPTP-induced toxicity through the RAGE and TNF-α pathway. *Brain*, *135*(Pt. 11), 3336–3347.

Serrano, A., Donno, C., Giannetti, S., Perić, M., Andjus, P., D'Ambrosi, N., et al. (2017). The astrocytic S100B protein with its receptor RAGE is aberrantly expressed in SOD1(G93A) models, and its inhibition decreases the expression of proinflammatory genes. *Mediators of Inflammation*, *2017*, 1626204.

Siddall, P. J., McClelland, J. M., Rutkowski, S. B., & Cousins, M. J. (2003). A longitudinal study of the prevalence and characteristics of pain in the first 5 years following spinal cord injury. *Pain*, *103*(3), 249–257.

Siddall, P. J., Taylor, D. A., & Cousins, M. J. (1997). Classification of pain following spinal cord injury. *Spinal Cord*, *35*(2), 69–75.

Silver, J., & Miller, J. H. (2004). Regeneration beyond the glial scar. *Nature Reviews. Neuroscience*, *5*(2), 146–156.

Steiner, J., Schmitt, A., Schroeter, M. L., Bogerts, B., Falkai, P., Turck, C. W., et al. (2014). S100B is downregulated in the nuclear proteome of schizophrenia corpus callosum. *European Archives of Psychiatry and Clinical Neuroscience*, *264*(4), 311–316.

Tateishi, N., Mori, T., Kagamiishi, Y., Satoh, S., Katsube, N., Morikawa, E., et al. (2002). Astrocytic activation and delayed infarct expansion after permanent focal ischemia in rats. Part II: Suppression of astrocytic activation by a novel agent (R)-(−)-2-propyloctanoic acid (ONO-2506) leads to mitigation of delayed infarct expansion and early improvement of neurologic deficits. *Journal of Cerebral Blood Flow and Metabolism*, *22*(6), 723–734.

Tawfik, V. L., Regan, M. R., Haenggeli, C., Lacroix-Fralish, M. L., Nutile-McMenemy, N., Perez, N., et al. (2008). Propentofylline-induced astrocyte modulation leads to alterations in glial glutamate promoter activation following spinal nerve transection. *Neuroscience*, *152*(4), 1086–1092.

Thelin, E. P., Nelson, D. W., & Bellander, B. M. (2017). A review of the clinical utility of serum S100B protein levels in the assessment of traumatic brain injury. *Acta Neurochirurgica*, *159*(2), 209–225.

Traynor, B. J., Bruijn, L., Conwit, R., Beal, F., O'Neill, G., Fagan, S. C., et al. (2006). Neuroprotective agents for clinical trials in ALS: A systematic assessment. *Neurology*, *67*(1), 20–27.

Villarreal, A., Seoane, R., Gonzalez Torres, A., Rosciszewski, G., Angelo, M. F., Rossi, A., et al. (2014). S100B protein activates a RAGE-dependent autocrine loop in astrocytes: Implications for its role in the propagation of reactive gliosis. *Journal of Neurochemistry*, *131*(2), 190–205.

Wainwright, M. S., Craft, J. M., Griffin, W. S., Marks, A., Pineda, J., Padgett, K. R., et al. (2004). Increased susceptibility of S100B transgenic mice to perinatal hypoxia-ischemia. *Annals of Neurology*, *56*(1), 61–67.

Watson, J. L., Hala, T. J., Putatunda, R., Sannie, D., & Lepore, A. C. (2014). Persistent at-level thermal hyperalgesia and tactile allodynia accompany chronic neuronal and astrocyte activation in superficial dorsal horn following mouse cervical contusion spinal cord injury. *PLoS One*, *9*(9), e109099.

Yang, L., & Sun, R. (2008). ONO-2506 may both prevent and treat epilepsy. *Medical Hypotheses*, *71*(3), 465–466.

Yiu, G., & He, Z. (2006). Glial inhibition of CNS axon regeneration. *Nature Reviews. Neuroscience*, *7*(8), 617–627.

Zhou, Y., & Danbolt, N. C. (2014). Glutamate as a neurotransmitter in the healthy brain. *Journal of Neural Transmission (Vienna)*, *121*(8), 799–817.

Chapter 16

Neural tissue loss after spinal cord injury

Jaroslav Pavel, Jana Fedorova, and Erika Kellerova
Department of Neurodegeneration, Plasticity and Repair, Institute of Neurobiology, Biomedical Research Center of the Slovak Academy of Sciences, Kosice, Slovakia

List of abbreviations

ASA	anterior spinal artery
BSCB	blood–spinal cord barrier
CSF	cerebrospinal fluid
FLA	fluorescein-labeled albumin
ISP	intraspinal pressure
MAP	mean arterial pressure
PSAs	posterior spinal arteries
SCBF	spinal cord blood flow
SCI	spinal cord injury

Introduction

Even if the spinal cord is naturally well protected by the spine, cerebrospinal fluid (CSF) buffering the dural sac movements, and epidural fat unevenly distributed along the spinal canal, spinal trauma can induce a sudden or slow and gradual pressure on the spinal cord resulting in the interruption of information that transfer between brain and periphery. The severity and duration of compression determine the manifestation of clinical symptoms. The progress of histopathological changes arising from the initial insult (primary injury) and following secondary injury can be classified into a distinct, but in certain degree overlapping periods: immediate phase (initial 1–2 h), acute phase (early acute 2–48 h, subacute 48 h to 14 days), intermediate phase (14 days to 6 months), and chronic injury phase (>6 months) (Rowland, Hawryluk, Kwon, & Fehlings, 2008). Following pathological events of acute traumatic injury occurring during early and/or late phases significantly contribute to the progressive degeneration and subsequently to the neural tissue loss of the spinal cord.

Dysfunction of vascular system

The spinal cord is equipped with rich, highly variable, and extensively anastomotic vascular network providing continuous delivery of oxygen and nutrients that are necessary for the maintenance of high metabolic demands. Therefore, systemic and local dysfunction of vascular system, resulting in post-traumatic ischemia and subsequent spinal cord infarction, represents a crucial factor affecting the extent of secondary injury.

The reduced blood supply of spinal cord

Since the spinal cord is vulnerable to suboptimal perfusion, it is therefore equipped with pressure autoregulatory capacity, also called *vascular autoregulation*. The synergy between sympathetic and parasympathetic nervous systems plays an important role in the control of cardiovascular system. While parasympathetic outflow is preserved, the direct mechanical insult to the spinal cord can cause the disruption of efferent pathways from the cardiovascular center in the brainstem to the spinal sympathetic centers resulting in failure of short-term blood pressure regulation. It is dependent on injury level, with more severe in cervical and high thoracic SCI. Sudden loss of autonomic tone due to the SCI can typically lead to *neurogenic shock*. Its typical symptoms are bradycardia and hypotension with decreased peripheral resistance and depressed cardiac output (Kiss & Tator, 1993). Persisted hypotension as a manifestation of impairment to the neuraxial blood flow autoregulation markedly increases the risk of secondary ischemia. Experimental study of autoregulation of spinal cord

Cellular, Molecular, Physiological, and Behavioral Aspects of Spinal Cord Injury. https://doi.org/10.1016/B978-0-12-822427-4.00016-2

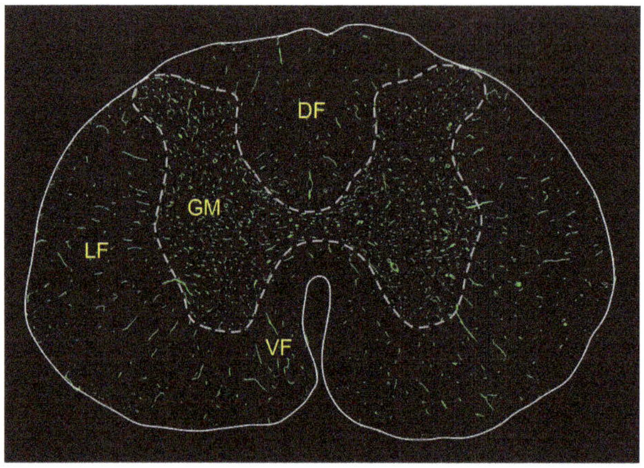

FIG. 1 Vascular density in the spinal cord. Vascular network is much denser in the gray matter (GM) than in the white one. Although blood flow in the dorsal (DF), lateral (LF), and ventral funiculus (VF) of the white matter is relatively similar, blood flow in the gray matter regions is approximately 2–5 times higher.

blood flow (SCBF) in rats at the lesion site demonstrated that vascular autoregulatory mechanisms are maintained after mild SCI, but were lost in the severe SCI (Guha, Tator, & Rochon, 1989).

Measurement of SCBF during experimental SCI revealed significant post-traumatic ischemia at the injury site and adjacent segments, and the blood flow reduction correlated with the severity of the injury (Koyanagi, Tator, & Lea, 1993). Interestingly, utilizing 3D model demonstrated that anterior SCI results in the least reduction of SCBF, while posterior SCI contributes to the greatest reduction of SCBF (Alshareef et al., 2014). Detailed study of blood flow following SCI in rats revealed a severe reduction of SCBF 15 min after injury that lasted for at least 24 h in both white and gray matter, but predominantly in the gray one (Fig. 1) (Rivlin & Tator, 1978). Similarly, pathological changes in vascular endothelium can lead to progressive deterioration of SCBF. Ultrastructural studies of injured spinal cord revealed the integrity loss of endothelial tight junctions in capillaries and larger vessels with smooth muscle cells (Goodman, Bingham Jr, & Hunt, 1976). Other factors associated with spinal vasculature disruption, such as tissue swelling and hemorrhages, can substantially support the development of secondary ischemia. Besides, since venous blood return usually promoted by the skeletal muscle pumping effect is disrupted due to motor deficit, the venous stasis can be one of the factors contributing to the risk of deep vein thrombosis.

Recently, real-time monitoring of microvascular SCBF using laser speckle contrast imaging in human patients after the SCI determined three distinct patterns of SCBF: (1) *the necrosis-penumbra SCBF pattern* characterized by a central region of very low SCBF (necrosis, defined as <25% of maximum flux), intermediate SCBF above and below (penumbra, defined as 25%–75% of maximum flux), and normal SCBF more distally; (2) *the hyperperfusion SCBF pattern* characterized by very high SCBF throughout the injury site, and was only seen in cervical SCI; and (3) *the patchy-perfusion SCBF pattern* characterized by irregular regions of ischemic and hyperemic SCBF (Gallagher, Hogg, Zoumprouli, Papadopoulos, & Saadoun, 2019). Interestingly, the intervention to elevate mean arterial pressure (MAP) by 20 mmHg, in order to normalize perfusion of injured spinal cord, increased overall SCBF at injury site only in some patients, and blood flow increased in some regions but decreased in the others (phenomenon termed as *blood pressure-induced local steal*). Moreover, some parts of the injured spinal cord were well perfused in systole, but inadequately perfused in diastole (known as *diastolic ischemia*). These findings suggested that intervention to increase MAP may induce detrimental responses in post-traumatic SCBF, at least, in some cases.

Intraparenchymal hemorrhages

The instant traumatic event causes mechanical damage to spinal parenchyma at the lesion center, including the direct disruption of local vascular integrity. The rupture of vessel results in *primary hemorrhage*, a leakage of blood that pressurizes on surrounding spinal parenchyma. The size of hemorrhage is maximal at the lesion epicenter, and is directly proportional to the severity of initial insult (Mautes, Weinzierl, Donovan, & Noble, 2000). Intraparenchymal hemorrhage occupies predominantly the gray matter and extends radially into the adjacent white matter in a variable proportion (Losey, Young, Krimholtz, Bordet, & Anthony, 2014). The extent of hemorrhage within injured spinal cord is commonly associated with the formation of intraparenchymal cavities. Resulting from progressive failure of structural capillary integrity, there is an

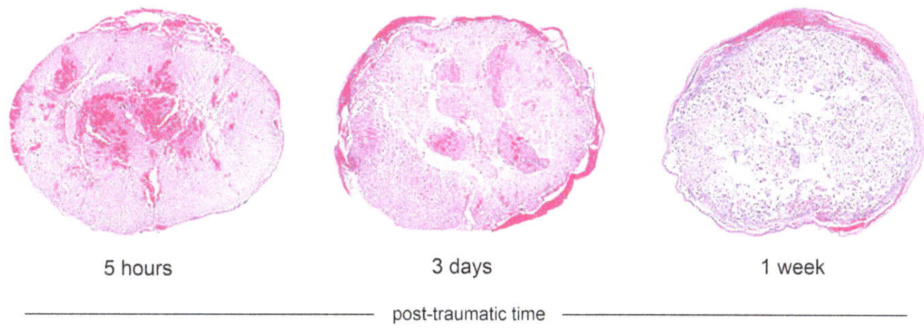

FIG. 2 Petechial hemorrhages. Coronal sections demonstrating numerous discrete petechial hemorrhages observed early after SCI that disappeared over time. *(Reproduced from Leonard, A. V., Thornton, E., & Vink, R. (2015). The relative contribution of edema and hemorrhage to raised intrathecal pressure after traumatic spinal cord injury.* Journal of Neurotrauma, 32*(6), 397–402.)*

early observed formation of small *petechial hemorrhages* (Fig. 2) in the spinal cord tissue surrounding the site of primary lesion, which within time increase in number and coalesce into a hemorrhagic lesion (Gerzanich et al., 2009). A marked reduction of microcirculation in both gray and white matter, especially in hemorrhagic regions and in adjacent zones, was observed (Tator & Fehlings, 1991). Since the predominance of hemorrhage, gray matter was considered to be the origin of the spinal tissue damage, which slowly spreads to the surrounding white matter. However, it was suggested that initial trauma also disrupts the functional ability of white matter and the injury is not simply a result of spreading from gray to white matter (Martirosyan et al., 2011).

Previous microangiographic studies showed that the vascular damage in human SCI occurs primarily in the intramedullary vascular system (Tator & Koyanagi, 1997). Contrary, traumatic occlusion or thrombosis of major superficial arteries after trauma, including the anterior spinal artery (ASA) and posterior spinal arteries (PSAs), is extremely rare (Jellinger, 1976). A physical modeling of spinal cord contusion revealed that compression force on spinal cord produces longitudinal stress being most intense in the center (Blight & Decrescito, 1986). Thus, it explains why the superficial vessels of the spinal cord are relatively spared from initial insults, whereas microvasculature in the central gray matter is subjected to stretching and shear stress. Additionally, the spinal cord is anisotropic and the elasticity of white and gray matter is different (Popovich, Lemeshow, Gensel, & Tovar, 2012).

It seems that central (sulcal) arteries are the most frequently affected by traumatic insult, especially in acute compression-induced SCI. Since they represent the main blood supply of the spinal cord, their disruption and occlusion subsequently interrupt the blood supply to the capillary beds and cause extensive spinal cord ischemia (Fig. 3). Regular narrowed sulcal arteries at the injury site and sporadic mild narrowing of the ASA were observed by scanning electron microscopy technique after the application of aneurysm clip compression (Koyanagi et al., 1993). Fibrinoid necrosis of sulcal arteries at the lesion site immediately and aneurysmal dilatation and rupture of intramedullary arteries in the lateral spinal cord columns at 4 and 8 h were described after experimental traumatic insult (Balentine, 1978). The sulcal arteries supplying area closely correspond with the distribution of both hemorrhagic necroses in the acute phase and the central myelomalacia and cavitation in later phases (Tator & Koyanagi, 1997). The results published by Tator and Koyanagi (1997) showed that the gray matter at the lesion site is more prone to be hemorrhagic, whereas non-hemorrhagic degenerative changes, including decreased staining, periaxonal and axonal swelling, and myelin disruption on most of the white matter are distributed around these hemorrhagic gray matter and extended rostrally and caudally from the lesion site, especially in the posterior white columns. Thus, the surrounding white matter lesions can be partially explained by disrupted sulcal centrifugal arterial system in gray matter that leads to a subsequent interruption of blood flow to the inner half of the white matter.

The initial trauma affects arterial as well as venous system. Electron microscopy studies showed the distention of muscular venules in the gray matter within 5 min after injury (Dohrmann, Wagner Jr., & Bucy, 1971). Small hemorrhages with extravasation of erythrocytes into the perivascular spaces of post-capillary and muscular venules are observed between 15 and 30 min post-traumatic period. In addition, it was suggested that the disruption of venous drainage by either direct mechanical or secondary to arterial or capillary occlusion can markedly contribute to the non-hemorrhagic lesions of the white matter.

The formation of intraparenchymal hemorrhage in lesioned spinal cord can substantially contribute to the following secondary injury at least by (1) direct compression of surrounding tissue resulting in reduced blood flow (secondary ischemia) and (2) toxicity of extravascular blood components such as elevated thrombin (Hua et al., 2006), erythrocyte lysis, and iron toxicity (Garton, Keep, Hua, & Xi, 2016).

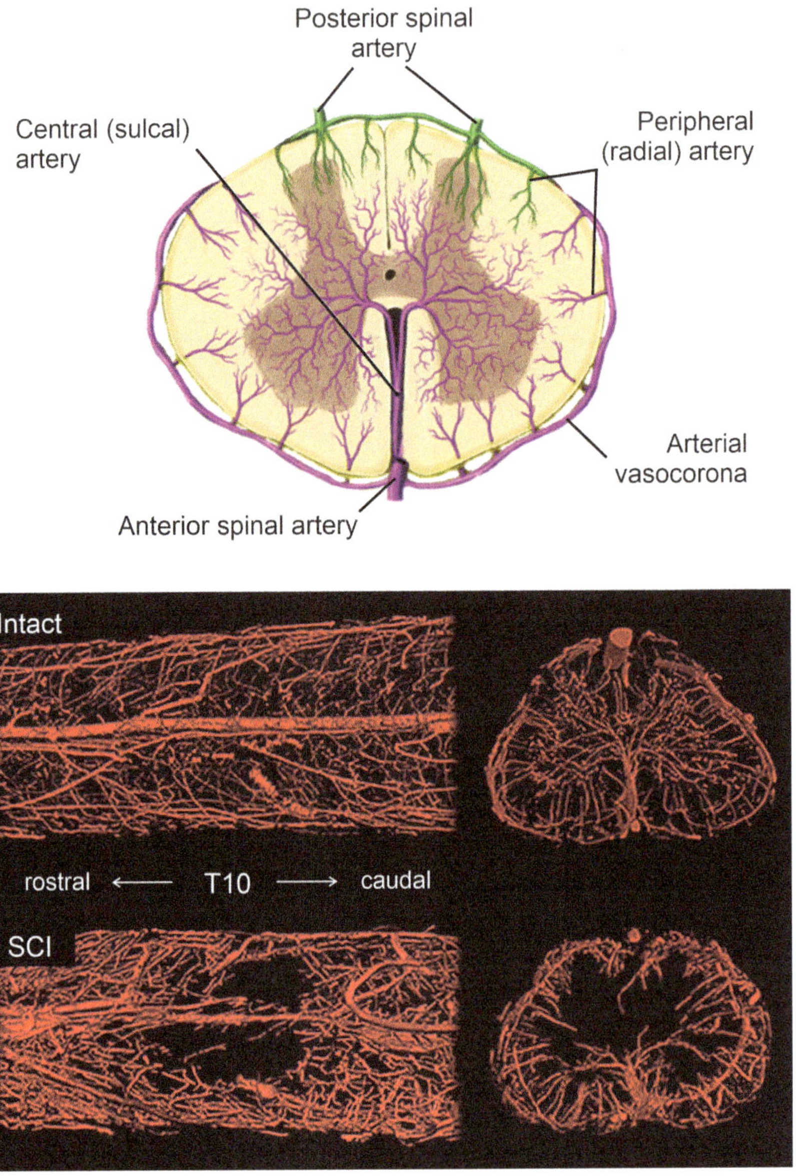

FIG. 3 Intramedullary arterial system after the SCI. *Upper:* The spinal cord parenchyma is instantly supplied by two types of intramedullary feeding arteries: the central (sulcal) arteries arising from the ASA and the penetrating peripheral (radial) arteries arising from the vasocorona, which are formed by branches of the PSAs. In humans, the central arteries centrifugally supply the major part of the spinal cord, which consists of all anterior gray matter, the anterior half of the posterior gray matter, approximately the inner half of the anterior and lateral white matter columns, and the anterior half of the posterior white columns. The peripheral arteries centripetally supply the outer part of the ventral and lateral columns, and dorsal columns in white matter, and the dorsal part of the dorsal gray matter. *Lower*: 3D morphology (longitudinal and transverse views) of the rat spinal cord vasculature of intact and injured spinal cord 1 day after trauma. Mechanical impact was applied dorsally at the T10 segmental level. It can be noted on the transverse section that avascular area predominantly corresponds with the area supplied by central (sulcal) artery. *Upper: Modified from Martirosyan, N. L., Feuerstein, J. S., Theodore, N., Cavalcanti, D.D., Spetzler, R. F., & Preul, M. C. (2011). Blood supply and vascular reactivity of the spinal cord under normal and pathological conditions.* Journal of Neurosurgery: Spine, 15*(3), 238–251. Lower: Reproduced from Cao, Y., Wu, T., Yuan, Z., Li, D., Ni, S., Hu, J. et al. (2015). Three-dimensional imaging of microvasculature in the rat spinal cord following injury.* Scientific Reports, 5, *12643.*

Spinal cord edema

Direct vascular damage by initial trauma as well as ischemia-induced changes in capillary permeability combined with inflammatory response can result in edema causing spinal cord tissue swelling. Post-traumatic edema is clinically considered as a serious complication markedly contributing to the secondary injury. The precise pathology of edema formation is still not very well understood. Spinal cord edema develops during early acute phase and generally resolves within 14 days when the initiation of astroglial scar and blood–spinal cord barrier (BSCB) repair occurs (Rowland et al., 2008). It seems that long-lasting, post-traumatic edema is independent of ischemia, because although experimental studies performed on primates showed that spinal cord edema persisted for 15 days, ischemic conditions were prominent only up to 18 h following injury (Locke, Yashon, Feldman, & Hunt, 1971; Yashon, Bingham Jr, Faddoul, & Hunt, 1973). The degree of edema formation and its longitudinal extension is directly related to the severity of initial trauma (Wagner Jr & Stewart, 1981). In the model of permanent focal ischemia, magnetic resonance imaging (MRI) technique demonstrated the presence of edema first in peri-infarct regions that are still perfused, suggesting that tissue swelling requires active blood flow providing a source of water, ions, and blood (Simard, Kent, Chen, Tarasov, & Gerzanich, 2007). Driving force for the movement of edema fluid into the unperfused tissue is the concentration gradient for moving constituents, such as Na^+, Cl^-, and water.

Traumatic spinal cord injury results in edema formation ultrastructurally showing both vasogenic and cytotoxic features (Goodman et al., 1976). V*asogenic edema* characterized by increased vascular permeability associated with BSCB breakdown is primarily due to direct traumatic impact and secondary due to release of endogenous vasoactive agents such as bradykinin, pro-inflammatory cytokines/chemokines, histamines, and prostaglandins (Abbott, 2000). In case that vascular leakage resulting from partial disconnection of endothelial cells, the basement membrane serves as a filter allowing passive diffusion of water, but retains serum proteins. White matter seems to be more vulnerable to vasogenic edema development than gray one (Kimelberg, 1995).

On contrary, the *cytotoxic edema*, otherwise known as cellular edema, is characterized by intracellular accumulation of Na^+ (primary driver), Cl^- and water (secondary participants), mainly due to post-traumatic depletion of ATP that leads to failure of the Na^+-K^+ ATPase (Simard et al., 2007). Cytotoxic edema itself does not result in spinal cord swelling, but the extravasation of fluid is evoked by the disruption of osmotic pressure gradient resulting from depletion of the extracellular Na^+, Cl^-, and water (Michinaga & Koyama, 2015). Its formation accelerates intravascular Na^+ outflow inducing extravasation of fluid without BSCB disruption, and it leads to extracellular fluid accumulation that is known as *ionic edema*.

The formation and recovery of vasogenic and cytotoxic edema are different in time. Even if both events are equally common during post-traumatic period, cytotoxic edema is considered as more prominent during the early phases (Hudak et al., 2014). Cellular swelling begins early after induction of focal ischemia particularly around capillaries (Liang, Bhatta, Gerzanich, & Simard, 2007). The most likely mediators of glial cell swelling are acidosis and elevated extracellular K^+ (Sykova, Svoboda, Polak, & Chvatal, 1994). Swelling of astrocytes is considered as much more prominent than neuronal ones, since they are involved in the clearance of K^+ and glutamate from extracellular space which promotes water inflow (Liang et al., 2007). When compensatory mechanisms in the plasma membrane fail, swollen cell undergoes the *oncosis*, a necrotic cell death mediated by ATP depletion and cell swelling. In clinical practice, the presence of post-traumatic hematoma and/or edema provides clinicians information regarding the severity and extent of injury and helps predict functional recovery (Chandra et al., 2012). Intradural MRI findings supported the concept that an edematous spinal cord has more potential in functional recovery than hemorrhagic (Tarawneh, D'Aquino, Hilis, Eisa, & Quraishi, 2020).

Raised intraspinal pressure

Since the spinal cord is located in rigid spinal canal and relative inelasticity of dura mater, the abnormal increase of spinal cord tissue volume due to post-traumatic hemorrhages and edema elevates the *intraspinal pressure* (ISP). Under normal conditions, a small increase in intraspinal volume does not greatly increase ISP, because of the elasticity of spinal cord, blood, and CSF. However, when capacity of compensatory mechanisms is exceeded, further increase in volume of spinal tissue due to post-traumatic hemorrhage and edema elevates ISP and causes a secondary compression of neurovascular structures resulting in a profound reduction of SCBF and subsequent ischemic tissue damage. Real-time monitoring of ISP after SCI in rats revealed dynamic changes (Dong et al., 2016). The ISP fluctuated over time and peaked at 1 and 48 h after mild and moderate SCI, but remained at high level during severe injury. It was demonstrated its positive correlation with the injury degree. Elevated intraspinal pressure after the SCI is initially driven by hemorrhage, however, latter is driven by edema (Leonard, Thornton, & Vink, 2015).

Breakdown of BSCB

BSCB provides a stable microenvironment that is necessary for the physiological function of the nervous system. Structurally, the BSCB is composed of tight junction-connected endothelial cell layer and its basement membrane that are surrounded by astrocytes foot processes and pericytes. Around small penetrating vessels are interstitial fluid-filled cavities, known as *perivascular spaces* or Virchow–Robin spaces (Fig. 4), which function as spinal drainage system. The preferential presence of spinal cord edema in white matter can be partially explained by a lower vascular density, and thus smaller capacity of fluid outflow via perivascular spaces occurring in the spinal white matter as compared with gray one. Using weight-drop SCI model in cats, it was detected a centrifugal spreading of extravasated fluorescein-labeled albumin (FLA) from gray to white matter (Wagner Jr & Stewart, 1981). It well corresponds with the disruption of central (sulcal) arteries that centrifugally supply the major part of the spinal cord. Extravasated FLA extended rostro-caudally from lesioned site along white matter columns, but the greatest longitudinal extent was usually seen in the lateral columns. It is likely associated with the development of tissue pressure gradients between epicenter and rostral and caudal areas.

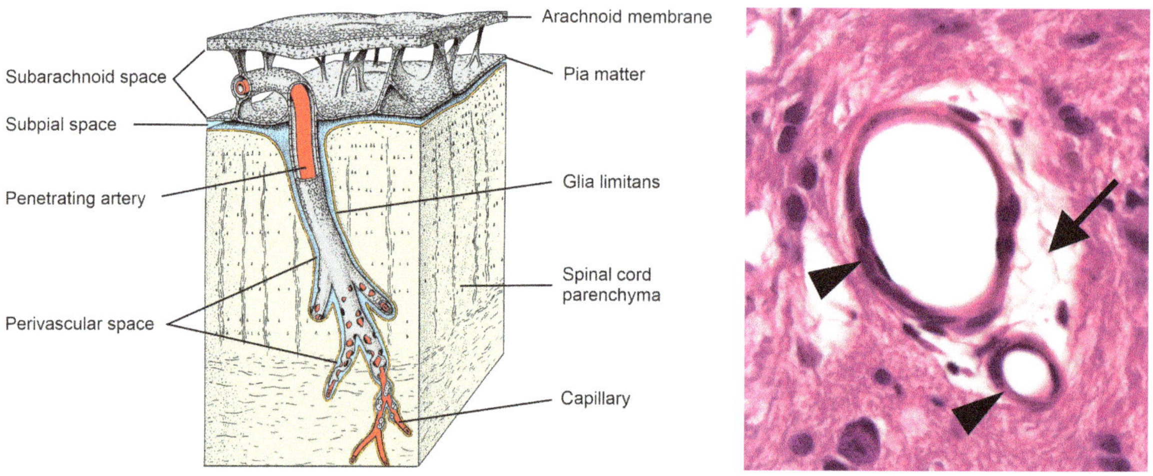

FIG. 4 Perivascular spaces and fluid exchange. *Left:* Anatomical illustration demonstrating a pivotal role of the perivascular (Virchow–Robin) spaces in the fluid exchange between the spinal cord tissue (interstitial fluid) and surrounding subarachnoid spaces (CSF). *Right:* Photomicrograph of coronal section from ischemia-injured spinal cord in rabbit stained with hematoxylin-eosin showing vessels (arrowheads) with surrounding enlarged perivascular space (arrow). *Left: Modified from Zhang, E. T., Inman, C. B., & Weller, R. O. (1990). Interrelationships of the pia mater and the perivascular (Virchow-Robin) spaces in the human cerebrum.* Journal of Anatomy, 170, *111–123.*

It was reported that the SCI induces a biphasic opening of the BSCB (Whetstone, Hsu, Eisenberg, Werb, & Noble-Haeusslein, 2003). The first abnormal leaking occurs within the first several hours after the injury, whereas a second one is evident within 3–7 days, a time period corresponding to marked post-traumatic revascularization. The authors expect that first robust leakage reflects direct damage to blood vessels and activated inflammatory processes can contribute to the more delayed second phase of barrier breakdown. It is assumed that the prolonged vascular permeability observed after the SCI is an inherent part of normal wound-healing response and reflects ongoing angiogenesis.

The inflammatory response and elimination of necrotic debris

The clearance of cellular debris and extravasated blood is indispensably required to maintain the homeostasis of surrounding spared neuronal and glial cells. Comprehensive post-mortem analysis of human spinal cord after traumatic injury provided valuable knowledge of cell types participating in the *post-traumatic inflammation* (Fleming et al., 2006). It showed a correlation between necrotic areas and the presence of inflammatory cells. Surprisingly, they often extended into more intact tissue adjacent to the lesion epicenter suggesting their role in the expansion of the lesion size. First immune cells coming very early from the ruptured blood vessels into lesioned tissue where phagocyte and clear tissue debris are *neutrophils*. In these human studies, extravascular neutrophils were found in the areas of hemorrhage, necrosis, and tissue fragmentation. It looks that these immune cells can initiate the secondary injury, because their maximal prevalence occurring 1–3 days corresponded with abrupt tissue damage. Neutrophil infiltration of lesion site ceased after 10 days. Perivascular neutrophils are an important source of inflammatory matrix metalloproteinase-9 (Rosell et al., 2008), the enzyme specifically degrading components of basal lamina of the BSCB. Therefore, it is likely that they contribute to the early secondary tissue damage and hemorrhagic injury.

Predominant inflammatory cells at and beyond 5–10 days are *activated microglia* and *macrophages* considered as another part of phagocytic population (Popovich et al., 2002). Phagocytic macrophages were detected in necrotic areas and along the border of cystic cavities, whereas activated microglia surrounded necrotic area and lesion site. The phagocytic activity of macrophages decreases within weeks after human SCI. The primary role of macrophages is to remove tissue and cellular debris, which enable spinal cord repair. It was demonstrated that decreased number of macrophages accumulating at the lesion site of SCI resulted in reduced cavitation and myelin damage (Popovich et al., 1999). Besides phagocytic function, macrophages also play an important role in wound healing. Macrophages can be classified into two main groups: M1 contributing to the tissue inflammation and damage and M2 contributing to wound healing and tissue remodeling. Predominant macrophages/microglial cells in the lesioned spinal cord are the M1 subtype, and the M2 subtype occurs transiently and in small amounts (Kigerl et al., 2009). Recent study investigating the phagocytic response of infiltrating macrophages and resident microglia and their viability at lesioned spinal cord concluded that (1) activated microglia contact and engulf degenerating axons early after SCI, before the infiltration of peripheral macrophages; (2) infiltrating macrophages are more susceptible than microglia to phagocytosis of myelin-induced apoptosis after the SCI; and

(3) microglial proliferation is about sevenfold greater than peripheral macrophages (Greenhalgh & David, 2014). It has been shown that infiltrating macrophages after injury become resident cells of CNS (Simard et al., 2007). Experimental SCI studies in rats comparing gray and white matter showed that macrophages and activated microglia were predominantly found in degenerating white matter tracts (Popovich, Wei, & Stokes, 1997), demonstrating a more extensive inflammatory response in the white matter regions and probable involvement of these cells in the initial phases of Wallerian degeneration. Majority of gray matter loss after the SCI occurs within hours and is mostly complete by 24 h, however, loss of white matter extends over several days post-injury (Ek et al., 2010). Despite progress in the last decades, a better understanding of the role of resident and peripheral immune cells in beneficial and detrimental processes of post-traumatic inflammatory response is still necessary.

Development of cysts/cavities and syrinx

An initial mechanical insult followed by secondary degenerative processes can initiate a serious chronic complication of SCI associated with delayed neurological deterioration known as *post-traumatic syringomyelia* (PTS). It is characterized by the presence of fluid-filled cysts (called syrinxes) that can expand over time, resulting in additional compression. Its time of onset after trauma ranges from months to decades, and it is developed in up to one-third of SCI patients (Berliner et al., 2020). Typical histological features of PTS are a cavity with a thick gliotic wall and enlarged perivascular spaces. MRI scanning of many SCI patients demonstrates fluid-filled cysts or microcysts that do not represent PTS, but rather myelomalacia, "softening" of the injured spinal cord (Shields, Zhang, & Shields, 2012). Retrospective MRI analysis of the compressive myelomalacia in humans suggested that early stage is most frequently characterized by the presence of cord edema, intermediate stage by varying degrees of cystic necrosis of the central gray matter following prolonged cord edema, and central cystic degeneration of the spinal cord and syrinx formation was typical for late stage (Ramanauskas, Wilner, Metes, Lazo, & Kelly, 1989). Moreover, it was concluded that early stage may be reversible (depends on severity of initial injury), but intermediate and late stages are progressive and most likely irreversible. Despite progress in research, there is still no explanation why in some SCI patients is PTS developed and in others not. Even if its pathogenesis is still not completely understood, the post-traumatic formation of spinal cord cysts can be divided into two following steps: (1) initial cavity formation and (2) cyst extension. The changes resulting from primary and secondary injuries, such as hemorrhagic necrosis, ischemic damage, and formation and coalescence of microcysts, are factors that predispose the syrinx formation.

The presence of small cavities at the injury level is very common (Fig. 5), but not all of them enlarge to form syrinxes. The syrinxes developed in spinal parenchyma after the SCI are usually extracanalicular (separated from the central canal)

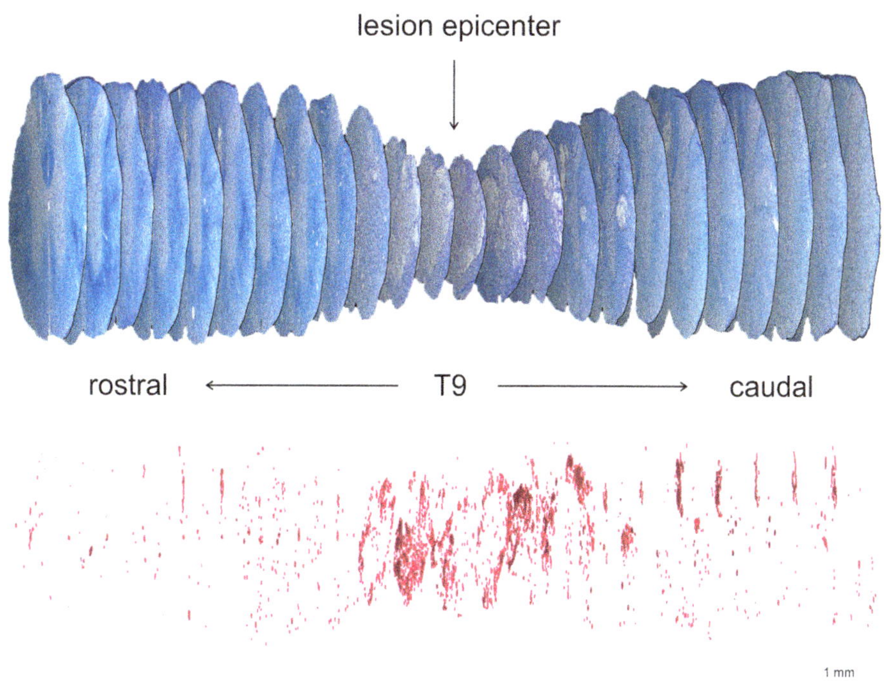

FIG. 5 Post-traumatic spinal cord degeneration and formation of cavities. 3D reconstructions demonstrating the longitudinal development of secondary tissue damage (*upper image*) and corresponding cavity formation (*lower image*) 28 days after severe spinal cord trauma at T9 segmental level in rats.

and are typically found in spinal watershed zones (Milhorat, 2000). Despite many theories, it seems that severe inflammatory response, arachnoiditis, and the impairment of CSF flow are the most probable factors in the development of PTS. According to current prevailing theory, the increased pulse pressure of CSF in the subarachnoid space above tethering of the spinal cord at the lesion site can force CSF through perivascular spaces toward syrinx (Greitz, 2006). The detailed experimental study of rat PTS model revealed perivascular microcavities or entirely enlarged spaces around venules, arterioles, and capillaries (Berliner et al., 2020). Spinal cord tissue surrounding cavities was characterized by integrity loss and enlarged extracellular spaces in the gray and white matter, consistent with severe parenchymal edema. In addition to that, the impairment of BSCB can increase the extracellular fluid and provide a potential source for fluid flow into post-traumatic syrinxes. It is supported by previous experimental study that suggests the impaired BSCB in the spinal cord tissue adjacent to a cyst (Josephson, Greitz, Klason, Olson, & Spenger, 2001).

The histological analysis of human cadavers with tubular cavitations of spinal cord revealed that most of the primary parenchymal cavitations were detected in central and dorsolateral gray matter, the watershed zone of the spinal cord (Milhorat, Capocelli Jr, Anzil, Kotzen, & Milhorat, 1995). Extracanalicular syrinx consisted mostly of irregularly shaped tubular cavities lined by densely packed glial or fibroglial tissue, with the presence of hemosiderin-laden macrophages and microglia in their wall. They were also associated with various degrees of focal necrosis and Wallerian degeneration. Cavitations in spinal gray matter displayed central chromatolysis, neuronophagia, and gliosis. Demyelinization in the white matter columns was accompanied by gliosis and the infiltration of foamy fat-laden macrophages. Cavitations present in the anterior horn of gray matter were associated with denervation atrophy in the diaphragm and segmental muscles. Atrophic cavitations do not propagate, and they can be related to degenerative changes leading to the formation of microcysts, intramedullary clefts, and localized dilations of the central spinal canal (Milhorat, 2000).

The post-traumatic spinal cord shrinkage

Predominantly damaging processes of progressive secondary injury involving a wide range of pathological events and initiated by mechanical insult lead to spinal cord tissue destruction. Therefore, the major target of SCI research is to develop effective and clinically acceptable therapeutic intervention that could reverse the destructive processes or reduce the final damage of spinal cord tissue. One of the common experimental parameters characterizing the severity and extent of SCI as well as determining the effectiveness of therapeutic approach in correlation with functional motor deficits is the amount of lost or spared neural tissue. However, removal of tissue debris can cause an extensive reduction of neural tissue at lesioned site that can be seen as spinal cord shrinkage, especially during later phases of SCI (Fig. 6). Therefore, the assessment of spared or lost tissue by direct measurement of total cross-sectional area without knowing pre-injury area in lesioned region

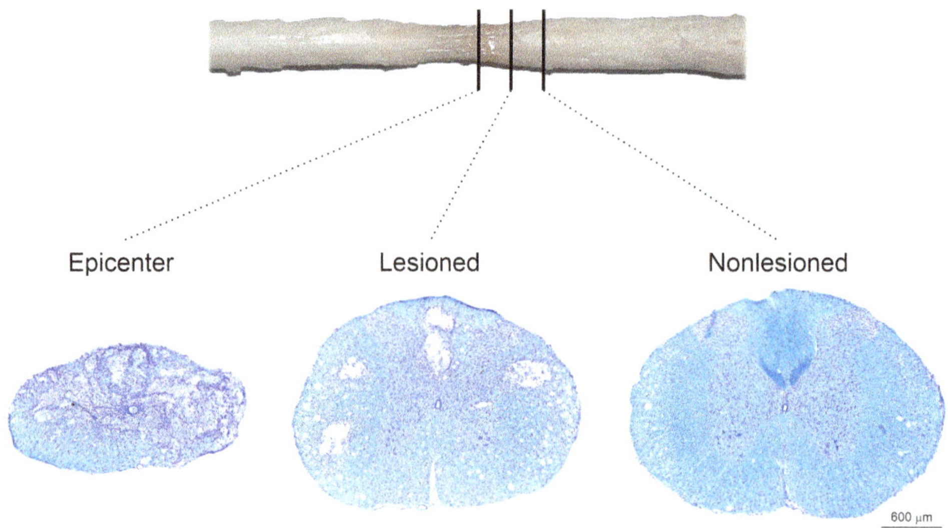

FIG. 6 The neural tissue loss after SCI. Demonstration of spinal cord tissue loss after severe posterior compression in rat detected 28 days post-injury (*upper image*). Progress of damage can be observed as longitudinal tissue shrinkage peaking at the lesion epicenter. Transverse sections stained with the Luxol fast blue and Cresyl Violet from three differently injured regions demonstrate various degrees of tissue damage resulting in spinal cord tissue loss (*lower images*). Numerous cysts and cavities can be noted predominantly in the white matter columns of lesioned region.

may result in an underestimation of tissue loss or an overestimation of tissue sparing. Recently, it was suggested that the new quantification approach overcomes the issue of post-traumatic spinal cord shrinkage and is relatively fast, accurate, and optimally reproducible (Fedorova & Pavel, 2019).

Applications to other areas of neuroscience

In this chapter, we have briefly reviewed the major pathological changes that substantially can contribute to post-traumatic neural tissue loss of the spinal cord. Some non-traumatic SCI related to metabolic, vascular, and inflammatory causes can share same neuropathological aspects, such as ischemia, hemorrhages, and edema. However, it is necessary to be aware of several important facts.

Ischemic injuries to the spinal cord are very uncommon since the spinal cord is equipped with a longitudinally continuous, highly variable, and extensively collateral vascular network. The incidence of spinal cord infarction usually results from ischemia originating in feeding extravertebral arteries. Since a loss of local vascular integrity, intervention to increase blood pressure to prevent persisted hypotension applied generally in human SCI patients is controversial. Intramedullary spinal cord hemorrhages are most commonly found in association with spinal cord trauma. Non-traumatic intramedullary spinal cord hemorrhages are rare and are often caused by spinal vascular malformations. According to clinical experience, spinal cord swelling indicates a severe spinal cord damage. If swelling is not apparent, neurological improvement can be expected even if severe deficits are initially apparent. Elimination of causal factors of increased intraspinal pressure has beneficial effects. In addition, post-traumatic spinal cord shrinkage should be taken into account during the assessment of lesion volumes in experimental studies.

Mini-dictionary of terms

Vascular autoregulation. The intrinsic ability of the spinal cord through arteries (specifically arterioles) to maintain a constant blood flow despite changes in perfusion pressure.

Neurogenic shock. A life-threatening condition associated with autonomic dysregulation resulting in inadequate spinal tissue perfusion.

Petechial hemorrhage. Capillary bleeding into spinal cord parenchyma manifested as tiny red spots.

Vasogenic edema. A pathophysiological process characterized by extravasation and extracellular accumulation of fluid with high protein content in spinal cord parenchyma caused by the BSCB breakdown.

Cytotoxic edema. A pathophysiological process that involves cellular swelling due to the movement of osmotically active molecules from the extracellular to intracellular space.

Ionic edema. A form of edema associated with cytotoxic edema, and represents the transcapillary movement of sodium ions and water into the extracellular space of spinal cord parenchyma.

Watershed zone. Spinal cord region vulnerable in case of vascular obstruction and do not receive a direct blood supply, but is supplied by overlapping vascular fields coming from neighboring arteries.

Key facts of post-traumatic loss of spinal cord tissue

The dysfunction of blood supply after traumatic SCI represents a crucial factor affecting the extent of secondary injury and subsequent tissue loss.

Vascular damage in human SCI occurs primarily in the intramedullary vascular system involving central (sulcal) arteries.

Primary hemorrhages are maximal at lesion center, whereas petechial hemorrhages predominantly surround the site of primary lesion.

Although post-traumatic edema shows both vasogenic and cytotoxic features, cytotoxic edema is considered as more prominent during the early phases.

Hemorrhages are responsible for raised intraspinal pressure initially, latter is driven by edema.

Both hemorrhages and edema cause additional compression of trauma-injured spinal cord and with inflammatory response result in more profound tissue damage.

First immune cells infiltrating lesioned tissue are neutrophils, whereas phagocytic microglia and macrophages are predominant later.

Majority of gray matter loss occurs within hours after the SCI, but loss of white matter extends over several days.

Summary points

- Final damage of spinal cord is highly variable and depends on aspects of traumatic insult as well as on local metabolic demands and morphological arrangement, including specific neuronal groups and local vascular characteristics.
- Major post-traumatic events contributing to neural tissue loss involve a systemic and local dysfunction of vascular system (including neurogenic shock, hemorrhage, impairment of blood flow autoregulation), edema formation, elevation of intraspinal pressure, BSCB breakdown, the inflammatory response associated with the elimination of necrotic debris, and the development of cysts/cavities and syrinx.
- Adequate vascular perfusion and regulated inflammatory response are essential factors for survival of reversibly injured spinal cord tissue.
- Hemorrhage early and edema later after SCI are major factors that are responsible for raised ISP. Therefore, its clinical monitoring is needed to ensure adequate vascular perfusion in post-traumatic period.
- The underestimation of lesion volume due to progressive shrinkage of the spinal cord represents a risk of consistent and reproducible quantification.

Acknowledgment

The authors would like to thank Dr. Ivo Vanicky for providing stained histological sections. Supported by APVV-18-0163 (J.P.).

References

Abbott, N. J. (2000). Inflammatory mediators and modulation of blood-brain barrier permeability. *Cellular and Molecular Neurobiology*, *20*(2), 131–147.

Alshareef, M., Krishna, V., Ferdous, J., Alshareef, A., Kindy, M., Kolachalama, V. B., et al. (2014). Effect of spinal cord compression on local vascular blood flow and perfusion capacity. *PLoS One*, *9*(9), e108820.

Balentine, J. D. (1978). Pathology of experimental spinal cord trauma. I. The necrotic lesion as a function of vascular injury. *Laboratory Investigation*, *39* (3), 236–253.

Berliner, J., Hemley, S., Najafi, E., Bilston, L., Stoodley, M., & Lam, M. (2020). Abnormalities in spinal cord ultrastructure in a rat model of post-traumatic syringomyelia. *Fluids and Barriers of the CNS*, *17*(1), 11.

Blight, A. R., & Decrescito, V. (1986). Morphometric analysis of experimental spinal cord injury in the cat: The relation of injury intensity to survival of myelinated axons. *Neuroscience*, *19*(1), 321–341.

Chandra, J., Sheerin, F., Lopez de Heredia, L., Meagher, T., King, D., Belci, M., et al. (2012). MRI in acute and subacute post-traumatic spinal cord injury: Pictorial review. *Spinal Cord*, *50*(1), 2–7.

Dohrmann, G. J., Wagner, F. C., Jr., & Bucy, P. C. (1971). The microvasculature in transitory traumatic paraplegia. An electron microscopic study in the monkey. *Journal of Neurosurgery*, *35*(3), 263–271.

Dong, X., Yang, D., Li, J., Liu, C., Yang, M., Du, L., et al. (2016). Intramedullary pressure changes in rats after spinal cord injury. *Spinal Cord*, *54*(11), 947–950.

Ek, C. J., Habgood, M. D., Callaway, J. K., Dennis, R., Dziegielewska, K. M., Johansson, P. A., et al. (2010). Spatio-temporal progression of grey and white matter damage following contusion injury in rat spinal cord. *PLoS One*, *5*(8), e12021.

Fedorova, J., & Pavel, J. (2019). An accurate method for histological determination of neural tissue loss/sparing after compression-induced spinal cord injury with optimal reproducibility. *Journal of Neurotrauma*, *36*(18), 2665–2675.

Fleming, J. C., Norenberg, M. D., Ramsay, D. A., Dekaban, G. A., Marcillo, A. E., Saenz, A. D., et al. (2006). The cellular inflammatory response in human spinal cords after injury. *Brain*, *129*(Pt 12), 3249–3269.

Gallagher, M. J., Hogg, F., Zoumprouli, A., Papadopoulos, M. C., & Saadoun, S. (2019). Spinal cord blood flow in patients with acute spinal cord injuries. *Journal of Neurotrauma*, *36*(6), 919–929.

Garton, T., Keep, R. F., Hua, Y., & Xi, G. (2016). Brain iron overload following intracranial haemorrhage. *Stroke and Vascular Neurology*, *1*(4), 172–184.

Gerzanich, V., Woo, S. K., Vennekens, R., Tsymbalyuk, O., Ivanova, S., Ivanov, A., et al. (2009). De novo expression of Trpm4 initiates secondary hemorrhage in spinal cord injury. *Nature Medicine*, *15*(2), 185–191.

Goodman, J. H., Bingham, W. G., Jr., & Hunt, W. E. (1976). Ultrastructural blood-brain barrier alterations and edema formation in acute spinal cord trauma. *Journal of Neurosurgery*, *44*(4), 418–424.

Greenhalgh, A. D., & David, S. (2014). Differences in the phagocytic response of microglia and peripheral macrophages after spinal cord injury and its effects on cell death. *The Journal of Neuroscience*, *34*(18), 6316–6322.

Greitz, D. (2006). Unraveling the riddle of syringomyelia. *Neurosurgical Review*, *29*(4), 251–263.

Guha, A., Tator, C. H., & Rochon, J. (1989). Spinal cord blood flow and systemic blood pressure after experimental spinal cord injury in rats. *Stroke*, *20*(3), 372–377.

Hua, Y., Wu, J., Keep, R. F., Nakamura, T., Hoff, J. T., & Xi, G. (2006). Tumor necrosis factor-alpha increases in the brain after intracerebral hemorrhage and thrombin stimulation. *Neurosurgery*, *58*(3), 542–550.

Hudak, A. M., Peng, L., Marquez de la Plata, C., Thottakara, J., Moore, C., Harper, C., et al. (2014). Cytotoxic and vasogenic cerebral oedema in traumatic brain injury: Assessment with FLAIR and DWI imaging. *Brain Injury*, *28*(12), 1602–1609.

Jellinger, K. (1976). Neuropathology of cord injuries. In P. J. Vinken, & G. W. Bruyn (Eds.), *Handbook of clinical neurology Vol. 25, injuries of the spine and spinal cord part 1* (pp. 43–121). Amsterdam: North-Holland Publishing Co.

Josephson, A., Greitz, D., Klason, T., Olson, L., & Spenger, C. (2001). A spinal thecal sac constriction model supports the theory that induced pressure gradients in the cord cause edema and cyst formation. *Neurosurgery*, *48*(3), 636–646.

Kigerl, K. A., Gensel, J. C., Ankeny, D. P., Alexander, J. K., Donnelly, D. J., & Popovich, P. G. (2009). Identification of two distinct macrophage subsets with divergent effects causing either neurotoxicity or regeneration in the injured mouse spinal cord. *The Journal of Neuroscience*, *29*(43), 13435–13444.

Kimelberg, H. K. (1995). Current concepts of brain edema. Review of laboratory investigations. *Journal of Neurosurgery*, *83*(6), 1051–1059.

Kiss, Z. H. T., & Tator, C. H. (1993). Neurogenic shock. In E. R. Geller (Ed.), *Shock and resuscitation* (pp. 421–440). New York: McGraw-Hill Education.

Koyanagi, I., Tator, C. H., & Lea, P. J. (1993). Three-dimensional analysis of the vascular system in the rat spinal cord with scanning electron microscopy of vascular corrosion casts. Part 2: Acute spinal cord injury. *Neurosurgery*, *33*(2), 285–292.

Leonard, A. V., Thornton, E., & Vink, R. (2015). The relative contribution of edema and hemorrhage to raised intrathecal pressure after traumatic spinal cord injury. *Journal of Neurotrauma*, *32*(6), 397–402.

Liang, D., Bhatta, S., Gerzanich, V., & Simard, J. M. (2007). Cytotoxic edema: Mechanisms of pathological cell swelling. *Neurosurgical Focus*, *22* (5), E2.

Locke, G. E., Yashon, D., Feldman, R. A., & Hunt, W. E. (1971). Ischemia in primate spinal cord injury. *Journal of Neurosurgery*, *34*(5), 614–617.

Losey, P., Young, C., Krimholtz, E., Bordet, R., & Anthony, D. C. (2014). The role of hemorrhage following spinal-cord injury. *Brain Research*, *1569*, 9–18.

Martirosyan, N. L., Feuerstein, J. S., Theodore, N., Cavalcanti, D. D., Spetzler, R. F., & Preul, M. C. (2011). Blood supply and vascular reactivity of the spinal cord under normal and pathological conditions. *Journal of Neurosurgery: Spine*, *15*(3), 238–251.

Mautes, A. E., Weinzierl, M. R., Donovan, F., & Noble, L. J. (2000). Vascular events after spinal cord injury: Contribution to secondary pathogenesis. *Physical Therapy*, *80*(7), 673–687.

Michinaga, S., & Koyama, Y. (2015). Pathogenesis of brain edema and investigation into anti-edema drugs. *International Journal of Molecular Sciences*, *16*(5), 9949–9975.

Milhorat, T. H. (2000). Classification of syringomyelia. *Neurosurgical Focus*, *8*(3), E1.

Milhorat, T. H., Capocelli, A. L., Jr., Anzil, A. P., Kotzen, R. M., & Milhorat, R. H. (1995). Pathological basis of spinal cord cavitation in syringomyelia: Analysis of 105 autopsy cases. *Journal of Neurosurgery*, *82*(5), 802–812.

Popovich, P. G., Guan, Z., McGaughy, V., Fisher, L., Hickey, W. F., & Basso, D. M. (2002). The neuropathological and behavioral consequences of intraspinal microglial/macrophage activation. *Journal of Neuropathology and Experimental Neurology*, *61*(7), 623–633.

Popovich, P. G., Guan, Z., Wei, P., Huitinga, I., van Rooijen, N., & Stokes, B. T. (1999). Depletion of hematogenous macrophages promotes partial hindlimb recovery and neuroanatomical repair after experimental spinal cord injury. *Experimental Neurology*, *158*(2), 351–365.

Popovich, P. G., Lemeshow, S., Gensel, J. C., & Tovar, C. A. (2012). Independent evaluation of the effects of glibenclamide on reducing progressive hemorrhagic necrosis after cervical spinal cord injury. *Experimental Neurology*, *233*(2), 615–622.

Popovich, P. G., Wei, P., & Stokes, B. T. (1997). Cellular inflammatory response after spinal cord injury in Sprague-Dawley and Lewis rats. *The Journal of Comparative Neurology*, *377*(3), 443–464.

Ramanauskas, W. L., Wilner, H. I., Metes, J. J., Lazo, A., & Kelly, J. K. (1989). MR imaging of compressive myelomalacia. *Journal of Computer Assisted Tomography*, *13*(3), 399–404.

Rivlin, A. S., & Tator, C. H. (1978). Regional spinal cord blood flow in rats after severe cord trauma. *Journal of Neurosurgery*, *49*(6), 844–853.

Rosell, A., Cuadrado, E., Ortega-Aznar, A., Hernández-Guillamon, M., Lo, E. H., & Montaner, J. (2008). MMP-9-positive neutrophil infiltration is associated to blood-brain barrier breakdown and basal lamina type IV collagen degradation during hemorrhagic transformation after human ischemic stroke. *Stroke*, *39*(4), 1121–1126.

Rowland, J. W., Hawryluk, G. W., Kwon, B., & Fehlings, M. G. (2008). Current status of acute spinal cord injury pathophysiology and emerging therapies: Promise on the horizon. *Neurosurgical Focus*, *25*(5), E2.

Shields, C. B., Zhang, Y. P., & Shields, L. B. (2012). Post-traumatic syringomyelia: CSF hydrodynamic changes following spinal cord injury are the driving force in the development of PTSM. *Handbook of Clinical Neurology*, *109*, 355–367.

Simard, J. M., Kent, T. A., Chen, M., Tarasov, K. V., & Gerzanich, V. (2007). Brain oedema in focal ischaemia: Molecular pathophysiology and theoretical implications. *The Lancet Neurology*, *6*(3), 258–268.

Sykova, E., Svoboda, J., Polak, J., & Chvatal, A. (1994). Extracellular volume fraction and diffusion characteristics during progressive ischemia and terminal anoxia in the spinal cord of the rat. *Journal of Cerebral Blood Flow and Metabolism*, *14*(2), 301–311.

Tarawneh, A. M., D'Aquino, D., Hilis, A., Eisa, A., & Quraishi, N. A. (2020). Can MRI findings predict the outcome of cervical spinal cord injury? A systematic review. *European Spine Journal*, *29*(10), 2457–2464.

Tator, C. H., & Fehlings, M. G. (1991). Review of the secondary injury theory of acute spinal cord trauma with emphasis on vascular mechanisms. *Journal of Neurosurgery*, *75*(1), 15–26.

Tator, C. H., & Koyanagi, I. (1997). Vascular mechanisms in the pathophysiology of human spinal cord injury. *Journal of Neurosurgery*, *86*(3), 483–492.

Wagner, F. C., Jr., & Stewart, W. B. (1981). Effect of trauma dose on spinal cord edema. *Journal of Neurosurgery*, *54*(6), 802–806.

Whetstone, W. D., Hsu, J. Y., Eisenberg, M., Werb, Z., & Noble-Haeusslein, L. J. (2003). Blood-spinal cord barrier after spinal cord injury: Relation to revascularization and wound healing. *Journal of Neuroscience Research*, *74*(2), 227–239.

Yashon, D., Bingham, W. G., Jr., Faddoul, E. M., & Hunt, W. E. (1973). Edema of the spinal cord following experimental impact trauma. *Journal of Neurosurgery*, *38*(6), 693–697.

Chapter 17

Remodeling mitochondrial transport and cellular energetics in axonal regeneration and spinal cord injury

Ning Huang and Zu-Hang Sheng

Synaptic Function Section, National Institute of Neurological Disorders and Stroke (NINDS), National Institutes of Health (NIH), Bethesda, MD, United States

List of abbreviations

ATP	adenosine triphosphate
CNS	central nervous system
CST	corticospinal tract
DH	dorsal hemisection
EMG	electromyography
FRET	Försterresonance energy transfer
MT	microtubule
OMM	outer mitochondrial membrane
ROS	reactive oxygen species
SCI	spinal cord injury
SNPH	syntaphilin
TMRE	tetramethylrhodamine ethyl ester

Introduction

While young neurons possess the capacity for robust axon growth and synaptogenesis, mature neurons in the central nervous system (CNS) typically fail to survive and regenerate after injury, leading to permanent neurological impairment. Many studies have suggested that mature CNS neurons have lost their regrowth capacity due to an intrinsic decline in permissive conditions for regeneration (Bradbury & McMahon, 2006; Case & Tessier-Lavigne, 2005; Harel & Strittmatter, 2006). Neuronal regeneration is a highly energy-demanding process. To succeed in regeneration following an injury, mature neuronal axons must reseal injured terminals, reform growth cones, reorganize the cytoskeleton network, synthesize and transport raw building materials, assemble axon components, and make new synaptic connections (He & Jin, 2016). All of these cellular events require a high level of energy consumption. However, injury is an acute insult that damages local mitochondria leading to an energy crisis, thus raising a fundamental question as to whether injury-induced energetic deficits contribute to the intrinsic restriction that accounts for regeneration failure in the mature CNS.

Mitochondria are the main cellular bioenergetic organelles that convert glucose and pyruvate into adenosine triphosphate (ATP) through the electron transport chain and oxidative phosphorylation (Mattson, Gleichmann, & Cheng, 2008). In the brain, ~93% of ATP is produced in mitochondria, while glycolysis generates ~7% of total ATP (Harris, Jolivet, & Attwell, 2012). Thus, a constant energy supply by mitochondria in the brain is essential to power neuronal growth, function, and regeneration. Mitochondrial biogenesis and turnover primarily take place in the cell body of neurons. Due to the polarized structures and high-energy demand in axons, ATP diffusion capacity through long axons is rather limited to maintain energy homeostasis (Sun, Qiao, Pan, Chen, & Sheng, 2013). Given the far distance between injured axons and their cell body, neurons face exceptional challenges in maintaining energy supply in injured axons and growth cones. Injury-damaged mitochondria not only fail to produce ATP but also release toxic reactive oxygen species (ROS) to induce neuronal degeneration. It is essential for neurons to adopt specialized mechanisms that efficiently deliver the appropriate

Cellular, Molecular, Physiological, and Behavioral Aspects of Spinal Cord Injury. https://doi.org/10.1016/B978-0-12-822427-4.00017-4

number of healthy mitochondria into regenerative tips and remove damaged mitochondria to meet the enhanced energy consumption for ensuring neuronal survival and axonal regeneration where energy is in high demand (Sheng, 2017).

Although mitochondria transplantation and enhanced mitochondrial biogenesis have surfaced as possible avenues for therapeutic development for CNS injuries including spinal cord injury (SCI), genetic and cellular remodeling of enhanced delivery of healthy mitochondria into injured axons has emerged as one of the robust mechanisms for boosting local energy supply and thus facilitating axon regeneration in various CNS injury models (Cartoni et al., 2016; Han et al., 2020; Han, Baig, & Hammarlund, 2016; Zhou et al., 2016). In this chapter, we first briefly introduce the microtubule (MT)-based trafficking and anchoring mechanisms that control mitochondrial delivery and distribution into distal axons. Next, we provide an in-depth discussion of recent emerging findings on how genetic and cellular regulation of axonal mitochondrial transport promotes the energy-demanding axon regeneration and functional recovery in various injury models. In the final part of this chapter, we provide perspective views on the potential therapeutic strategies by remodeling mitochondrial transport and/or by elevating cellular energetic metabolism for SCI. For additional insights into mitochondrial trafficking and neuronal energy maintenance, we refer readers to other reviews (Devine & Kittler, 2018; Misgeld & Schwarz, 2017; Sheng & Cai, 2012).

Complex motility patterns of axonal mitochondria

To monitor mitochondrial transport in live neurons, expression of mitochondria-tagged proteins (e.g., DsRed-Mito) or loading mitochondrial dyes (e.g., Mitotracker Green) have routinely been applied to label mitochondria for time-lapse imaging. Axonal mitochondria display complex motility patterns: they move bi-directionally over long distances, pause and change direction frequently, and can be re-mobilized and re-distributed in axons in response to changes in metabolic needs and growth status (Fig. 1). Given frequent pauses and changes in direction, the mean velocities of axonal mitochondrial motility are highly variable ranging from ~0.2 to 2.0 μm per second (Sheng, 2014). We and others consistently observed a significant difference in the total motility of axonal mitochondria between young and mature neurons. Approximately 30%–45% of axonal mitochondria are mobile and the remaining two-thirds are stationary in young developing neurons. These motility profiles decline and become relatively stable after neurons mature. These complex motility patterns indicate a unique mechanistic interplay between anterograde and retrograde motors in driving mitochondria bi-directional transport along with anchoring proteins in positioning mitochondria in distal axons or at synapses.

Molecular motors driving mitochondrial bi-directional transport in axons

Long-distance mitochondrial transport along MT-tracks is driven by ATP-hydrolyzing molecular motors. MTs within axons display uniform polarity: their plus-ends are directed to distal axons and their minus-ends toward the soma. Thus,

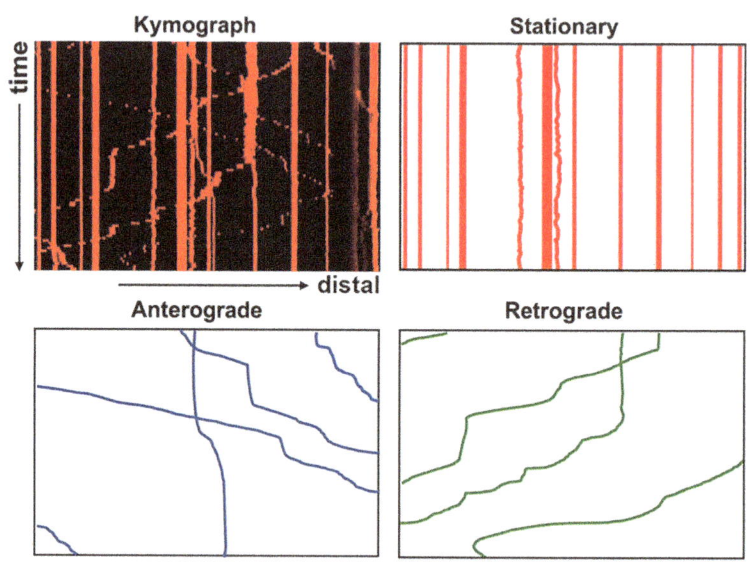

FIG. 1 Complex motility patterns of axonal mitochondria. Representative kymograph showing bi-directional motility patterns of axonal mitochondria. In the kymograph, vertical lines (upper panels) represent stationary mitochondria; oblique lines or curves to the right (lower panels, *blue*) indicate anterograde transport toward distal terminals, whereas those to the left (lower panels, *green*) denote retrograde transport toward the cell body.

kinesin motors drive anterograde transport from the soma toward distal axons, while dynein motors mediate retrograde transport back toward the soma. Kinesin-1 family members, also known as KIF5A, KIF5B, and KIF5C, are the primary motors driving neuronal mitochondrial transport (Pilling, Horiuchi, Lively, & Saxton, 2006; Tanaka et al., 1998). Kinesin-1 motors are a hetero tetramerized complex composed of two heavy chains (KHC) and two light chains (KLC). The KHC contains a motor domain at the N-terminal region that has ATPase activity and is directly engaged with MTs, whereas the C-terminal domain of KHC binds to KLC or directly associates with cargoes (Hirokawa, Niwa, & Tanaka, 2010). Cytoplasmic dynein motors contain two motor-containing dynein heavy chains (DHC) and several other chains including dynein intermediate chains (DIC), light chains (DLC), and dynactin that mainly regulate motility or associate with their cargoes. In *Drosophila*, dynein motors are recruited on mitochondria and mediate retrograde trafficking along axons (Pilling et al., 2006).

Mitochondrial motor adaptors and receptors

Mitochondria recruit molecular motors through their specific adaptors and/or receptors that associate with the outer mitochondrial membrane (OMM). Mitochondrial adaptor proteins help to connect kinesin-1 motors and mitochondria, thus regulating mitochondrial transport. In *Drosophila*, Milton is the adaptor that bridges KIF5 motors with Miro, an OMM receptor (Glater, Megeath, Stowers, & Schwarz, 2006). *Milton* mutation impairs anterograde transport of mitochondria without affecting other axonal cargoes. Extracellular glucose alters axonal mitochondrial transport through GlcNAcylation modification of Milton (Pekkurnaz, Trinidad, Wang, Kong, & Schwarz, 2014). Two mammalian Milton orthologs, Trak1 and Trak2, regulate axonal and dendritic mitochondrial trafficking and outgrowth (van Spronsen et al., 2013). Trak1 facilitates bi-directional mitochondrial trafficking along axons through interacting with both kinesin-1 and dynein motors, while Trak2 mediates dendritic mitochondrial transport via preferentially binding to dynein motors.

Miro is a Rho-GTPase with two GTPase domains and two Ca^{2+}-binding EF-hand motifs, which allows mitochondrial transport to be regulated by Ca^{2+} signaling and synaptic activity (Fransson, Ruusala, & Aspenstrom, 2006). In *Drosophila*, *Miro* mutation alters anterograde mitochondrial transport (Guo et al., 2005). In mammals, there are two Miro isoforms: Miro1 and Miro2. By using a*miro1* deletion mouse model, Lopez-Domenech et al. revealed that Miro1, but not Miro2, mediates mitochondrial trafficking (Lopez-Domenech et al., 2016). Recent studies revealed that Miro1 expression is facilitated by two Parkinson's disease (PD)-related kinases PINK1 and LRRK2 (Hsieh et al., 2016). Interestingly, a new splicing variant of Miro1 was found to localize to peroxisomes and mediate long-distance peroxisome trafficking together with Trak2 (Okumoto et al., 2018). Miro1 and Miro2 have also been suggested to play a role in the peroxisomal fission process, supporting the crosstalk between peroxisomes and mitochondria (Covill-Cooke et al., 2020). Together, KIF5 (motor), Milton/Trak (adaptor), and Miro (receptor) comprise the mitochondrial transport machinery that drives anterograde transport.

Syntabulin is an attractive motor-adaptor that targets the OMM via its carboxyl-terminal mitochondria-targeting domain and interacts with and thus recruits KIF5 motors to mitochondria (Cai, Gerwin, & Sheng, 2005). In hippocampal neurons, knocking down syntabulin with siRNA or interfering with syntabulin-KIF5 interaction impairs mitochondrial transport from the soma into axons. Several other KIF5 adaptors, such as FEZ1 and RanBP2, have also been reported to mediate neuronal mitochondrial transport (Cho et al., 2007; Fujita et al., 2007). The existence of multiple pairs of motor-adaptor complexes may suggest diverse mechanisms in coordinating mitochondrial trafficking and positioning in response to various physiological conditions or axonal injury.

Compared to the well-characterized kinesin-adaptor transport complexes, dynein-adaptor complexes are less well-defined. Several studies suggest that Miromay also associates with dynein motors for mediating retrograde transport. While the loss of *dmiro* in *Drosophila* neurons impairs bi-directional mitochondrial transport (Guo et al., 2005), overexpressing dMiro enhances axonal mitochondrial transport in both directions (Russo et al., 2009). Consistent with findings in *Drosophila*, mammalian Trak1/2 have the binding capacity for both KIF5B and dynein/dynactin (van Spronsen et al., 2013). Based on these findings, an attractive model was proposed: bi-directional mitochondrial transport is driven by opposing motors kinesin-1 and dynein likely through the same adaptor and receptor complexes (Fig. 2).

Declined axonal mitochondrial transport in mature neurons

During early neuronal development, axon outgrowth, branching, and synaptogenesis require bulk synthesis and delivery of cellular proteins and organelles into growing tips. These developmental events are highly energy demanding, and thus require robust mitochondrial transport from the soma into distal axons (Morris & Hollenbeck, 1993; Ruthel & Hollenbeck, 2003). In young developing neurons, we and others consistently found that $\sim$30%–45% of axonal

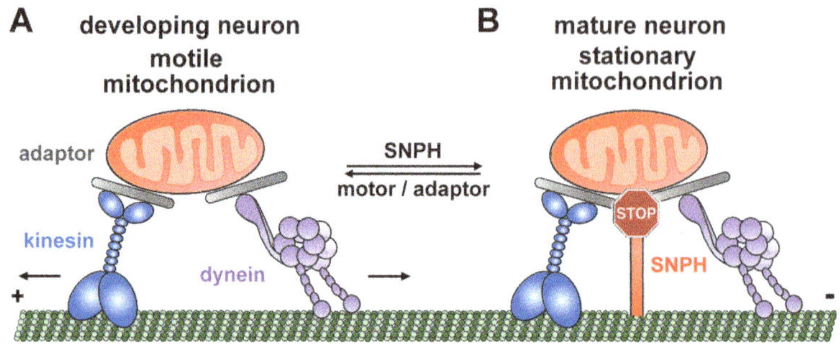

FIG. 2 Mechanistic interplay between mitochondrial trafficking and anchoring proteins. Schematic models showing motor-adaptor complexes and anchoring protein SNPH in regulating axonal mitochondrial motility. (A) Long-distance axonal mitochondrial transport is driven by MT-based motors: the plus-end directed kinesin and the minus-end directed dynein. Axonal MTs are uniformly arranged so that their plus-ends are directed distally. (B) SNPH acts as a "static anchor" that arrests axonal mitochondrial transport by anchoring them to MTs. The balance of motile versus stationary mitochondria pools depends on the relative action of the motor/adaptor and SNPH to axonal mitochondria. Progressive elevation of SNPH expression with neuron maturation contributes to declined motility of axonal mitochondria in mature neurons and adult brains.

mitochondria are mobile and the remaining two-thirds are stationary. When neurons fully mature, the majority of axonal mitochondria enter a steady phase and remain stationary (Zhou et al., 2016). Given that synaptic transmission and action potential propagation are two major functional features in mature neurons, axonal mitochondria, instead of maintaining a uniform distribution, display increased density at synapses and nodes of Ranvier, two active sites of energy metabolism. An increasing majority of axonal mitochondria remain stationary in matured axonal branches or axonal terminals, which were consistently observed in rodent nervous systems in vitro and in vivo from several independent studies (Lewis Jr., Turi, Kwon, Losonczy, & Polleux, 2016; Smit-Rigter et al., 2016; Takihara et al., 2015; Zhou et al., 2016). These studies suggest that axonal mitochondrial motility undergoes a progressive regulation from developing neurons (or brains) to mature neurons (or adult brains). This declined axonal mitochondrial motility raises an interesting and fundamental question as to whether mitochondrial trafficking or its anchoring mechanisms are progressively regulated with neuronal maturation and age.

SNPH immobilizes axonal mitochondria in mature neurons and adult brains

Mitochondrial anchoring mechanisms are essential for arresting mobile mitochondria within axons. This anchoring mechanism was recently revealed by our series of studies demonstrating that syntaphilin (SNPH) acts as a "static anchor" for axonal mitochondria (Chen & Sheng, 2013; Kang et al., 2008; Zhou et al., 2016). SNPH selectively associates with axonal mitochondria through its OMM-insertion domain and axon-sorting sequence and anchors mitochondria on MTs through its MT-binding domain. Deleting *snph* robustly enhances axonal mitochondrial transport (by increasing the total motility up to ~78%) in mature neurons in vitro and adult brains in vivo; overexpressing SNPH in mouse cortical and hippocampal neurons abolishes axonal mitochondrial motility. Recent optogenetic analysis further confirmed that the balance between mobile vs stationary pools of axonal mitochondria is controlled by selective recruitment of the KIF5 motor vs SNPH onto axonal mitochondria, respectively (van Bergeijk, Adrian, Hoogenraad, & Kapitein, 2015) (Fig. 2).

SNPH-mediated anchoring is the key mechanism contributing to this maturation-associated decline of axonal mitochondrial motility. Our recent parallel study of axonal mitochondrial motility and SNPH expression in mouse cortical neurons demonstrated that axonal mitochondria display progressively reduced motility from young developing neurons at DIV7 ($47.60 \pm 4.60\%$) to well-matured neurons at DIV18 ($22.88 \pm 3.09\%$, $p < 0.001$) (Fig. 3A). In contrast, SNPH expression is hardly detectable in neurons before DIV3, and in mouse embryonic brains, becomes readily detectable after DIV9 in culture and gradually reaches its peak at DIV22 and in adult mouse brains (Fig. 3B) (Zhou et al., 2016). This progressively elevated expression is specific for SNPH; as other mitochondrial trafficking motor-adaptor proteins including KIF5, DIC, Miro1/2, and Trak2 show consistent levels of expression throughout neuronal development and maturation. The elevated SNPH expression pattern was also confirmed in rat brains: it is undetectable at embryonic stages, gradually increases during brain maturation, and peaks in adulthood (Das, Boczan, Gerwin, Zald, & Sheng, 2003). Such progressively elevated SNPH expression with neuron and brain maturation supports the notion that SNPH-mediated anchoring contributes to declined motility of axonal mitochondria in mature neurons and adult brains. Therefore, motile mitochondria in developing neurons are recruited to the stationary pool during neuronal maturation, thus maintaining energy metabolism to sustain key neuronal activities by positioning mitochondria at synapses and nodes of Ranvier. Deleting *snph* in mice disrupts this anchoring mechanism, thus robustly increasing axonal mitochondria in motile pools: 78% of axonal mitochondria are motile in cultured hippocampal neurons at DIV14, and 71% of axonal mitochondria are motile in axonal bundles of sciatic nerves in 2-month-old adult mice (Chen & Sheng, 2013; Kang et al., 2008; Zhou et al., 2016). Given

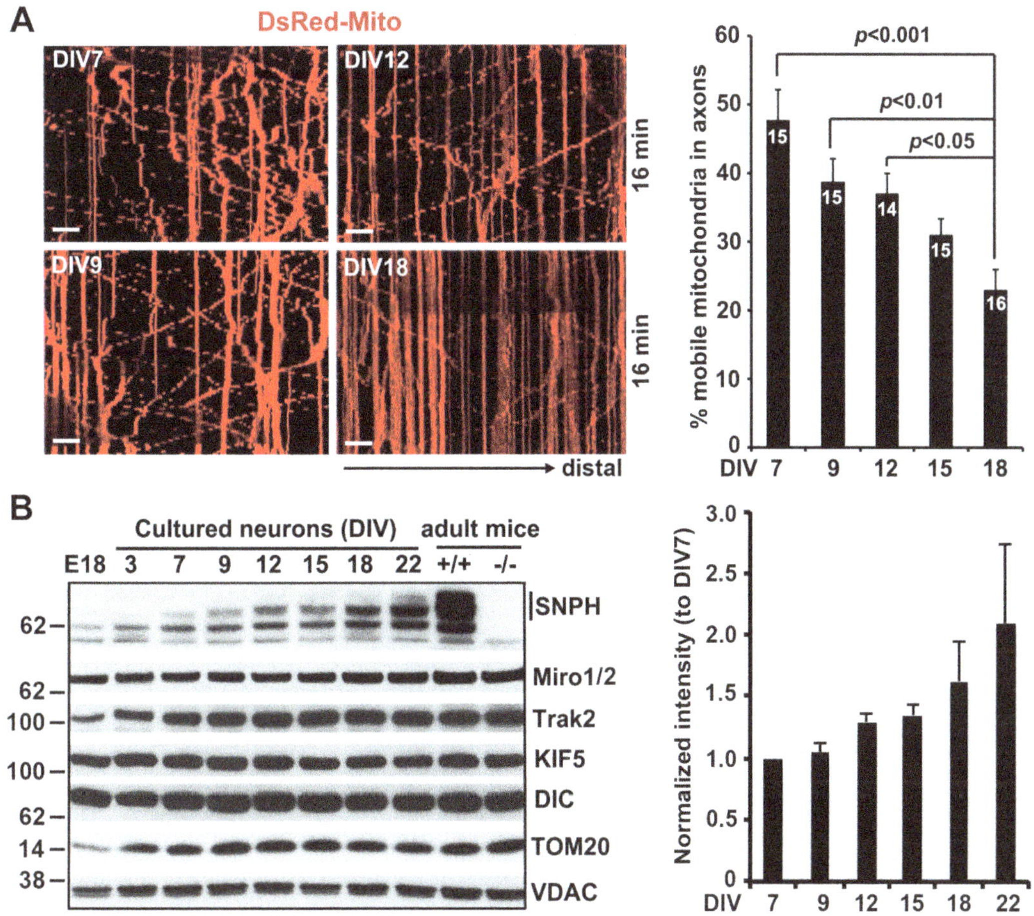

FIG. 3 SNPH contributes to declined motility of axonal mitochondria in mature neurons. (A) Kymographs showing a progressive decline of axonal mitochondrial motility with neuron maturation. Cortical neurons were transfected with DsRed-Mito, followed by time-lapse images at DIV7, 9, 12, or 18. Data were analyzed from the total number of neurons indicated within bars and expressed as mean ± SEM. Bars, 10 μm. (B) Progressive increase in SNPH expression with neuron maturation. Cortical neurons isolated from E18 mouse brains were cultured for 3, 7, 9, 12, 15, 18, and 22 days. Equal amounts of lysates were sequentially immunoblotted with various antibodies. Brain lysates from E18 WT, adult WT, and *snph* KO mouse brain homogenates were used as controls. The intensity of SNPH bands was quantified from three repeats, calibrated to TOM20 levels, and then normalized to SNPH expression at DIV7. *(Adapted with permission from Zhou, B., Yu, P., Lin, M. Y., Sun, T., Chen, Y., & Sheng, Z. H. (2016). Facilitation of axon regeneration by enhancing mitochondrial transport and rescuing energy deficits.* Journal of Cell Biology, 214*(1), 103–119. https://doi.org/10.1083/jcb.201605101.)*

the fact that SNPH plays a major role in anchoring the majority of axonal mitochondria in mature neurons and adult mouse brains, *snph* knockout (KO or $snph^{-/-}$) mice serve as an ideal genetic model for investigations into whether genetically enhanced transport of axonal mitochondria ensures matured neurons and adult brains regain regenerative capacity by reversing the injury-induced energy crisis.

Deleting SNPH anchoring boosts axon regeneration in vitro and in vivo

While developing neurons possess robust axon growth and dynamic mitochondrial transport, mature neurons in the CNS display a declined mitochondrial motility and typically fail to regrow after injury. A widely accepted concept is that the limited regrowth capacity of the mature CNS is due to an intrinsic decline of permissive conditions. Thus, it is critical to elucidate these intrinsic mechanisms accounting for declined regrowth capacity in mature neurons. As we discussed above, neural regeneration is a highly energy-demanding process starting from resealing axonal terminals, reorganizing the cytoskeleton, synthesizing and delivering building materials into new growth cones, and finally re-connecting synapses and restoring functions (He & Jin, 2016). Given that (1) the majority of axonal mitochondria in the mature CNS remain stationary (Lewis Jr. et al., 2016; Smit-Rigter et al., 2016; Takihara et al., 2015; Zhou et al., 2016), and (2) brain injury is an acute insult that damages local mitochondria leading to an energy crisis (Cavallucci et al., 2014; O'Donnell, Vargas, &

Sagasti, 2013), proper delivery of healthy mitochondria into injured axons ensures an adequate supply of ATP. Two fundamental questions emerge: (1) do energetic deficits, caused by impaired ATP production combined with high ATP consumption, contribute to one of the intrinsic restrictions that account for regeneration failure in mature CNS neurons? If this is the case, then (2) does reversing declined mitochondrial transport enable mature neurons to regain regrowth capacity following in vitro and in vivo injury?

Our recent study addressed these two issues by live imaging mitochondrial transport, mitochondrial membrane potential, and local ATP levels within injured axons of mature cortical neurons cultured in a microfluidic chamber system (Zhou et al., 2016). The study demonstrated that axon injury is an acute stress signal that rapidly depolarizes mitochondria, leading to a local energy crisis in the vicinity of the injured site. We further applied *snph*$^{-/-}$ mice as a genetic model to investigate whether enhanced mitochondrial transport in mature neurons regains their regenerative capacity after injury. Surprisingly, robustly enhancing mitochondrial transport in mature *snph*$^{-/-}$ axons helps remove damaged mitochondria from injured axons and replenish with healthy ones, thus effectively reversing local energy deficits. Thus, maintaining ATP supply in injured axons via recruiting healthy mitochondria is critical to meeting enhanced energy requirements during regeneration. Significantly, such enhanced mitochondrial trafficking not only facilitates axonal regeneration in mature cortical neurons after axotomy in vitro (Fig. 4A and B), but also boosts regeneration of dorsal root ganglion (DRG) neurons in vivo after crush injury of sciatic nerves in adult mice (Fig. 4E). Alternatively, manipulating an enhanced axonal mitochondrial motility in wild-type neurons by overexpressing motor receptor Miro1, instead of deleting anchoring protein SNPH, also promotes axon regeneration. As a negative control, our study further confirmed that re-expressing SNPH arrests almost all mitochondria and thus abolishes axon regeneration (Fig. 4C and D). This study provides mechanistic insights into how elevated SNPH expression in the mature CNS neurons and an injury-induced energy crisis contribute to regeneration failure.

Enhanced mitochondrial transport facilitates axon regeneration

SNPH and Miro1 become attractive candidate targets for investigations into the roles of mitochondrial trafficking and anchoring and thus recovery of energy metabolism in the regeneration capacity of the adult CNS. The enhanced axonal regeneration found in *snph*$^{-/-}$ mice was supported by three recent in vitro and in vivo studies demonstrating that enhanced mitochondrial transport and recruitment play critical roles in nerve regeneration in both worm and mouse models. By studying an enhanced regeneration mouse model with SOCS3 and PTEN double dKO, Cartoni et al. reported that a mitochondria-targeted protein named Armcx1 promotes mitochondrial transport in mouse retinal ganglion cells (RGCs) likely through its interaction with mitochondrial motor receptor Miro1 (Cartoni et al., 2016). Interestingly, Armcx1 expression is elevated in RGCs with high regenerative capacity. Overexpressing Armcx1 enhances axonal mitochondrial transport in RGCs and supports the survival and regeneration of axotomized neurons. Conversely, depleting Armcx1 in this dKO model decreases its regeneration capacity. This study highlighted a regeneration-relevant mitochondrial protein that enhances axonal mitochondrial transport in response to injury, thus adding an additional line of evidence showing a critical role of mitochondrial transport in nerve regeneration. A second study by Kalinski et al. revealed an unexpected role of histone deacetylase 6 (HDAC6) that is thought to deacetylate α-tubulin. HDAC6-mediated deacetylation of Miro1 decreases axonal mitochondrial transport and thus inhibits DRG neuron axon growth (Kalinski et al., 2019). Thus, this study provided a new line of evidence demonstrating that proper mitochondrial trafficking is crucial for maintaining neural regeneration capacity. In a third study using *C. elegans* as a model organism, Han et al. revealed that axotomy in GABA motor neurons induces axonal energy stress that triggers an enhanced mitochondrial density within injured axons, thus meeting the increased ATP demand to support axon survival and regeneration. Interestingly, they found the dual leucine zipper kinase 1 (DLK-1) as a candidate signaling pathway that is able to recruit mitochondria to the injury site (Han et al., 2016).

The notion that enhanced mitochondrial transport benefits nerve regeneration after injury is also supported by several earlier in vivo studies in different nervous systems. By examining the proximal segments of transected intercostal nerves, Misgeld et al. found that injury induces mitochondrial anterograde transport and re-distribution at the growth cone (Misgeld, Kerschensteiner, Bareyre, Burgess, & Lichtman, 2007). In the *C. elegans ric-7* mutant where mitochondrial transport is impaired, degeneration of injured axons is inhibited by delivering mitochondria into axons (Rawson et al., 2014). The axon-protective Wld^{S} protein reduces the decline of axonal ATP levels by partially rescuing glycolysis and mitochondrial respiration after axonal injury (Godzik & Coleman, 2015). It was also reported that Wld^{S}-induced mitochondrial flux protects axons from Wallerian degeneration after injury in both *Drosophila* and mouse models (Avery et al., 2012).

In addition, several regeneration-associated signaling pathways have been reported to crosstalk with the mechanisms that drive mitochondrial transport. In zebrafish, pharmacologically activating cyclic adenosine monophosphate (cAMP)

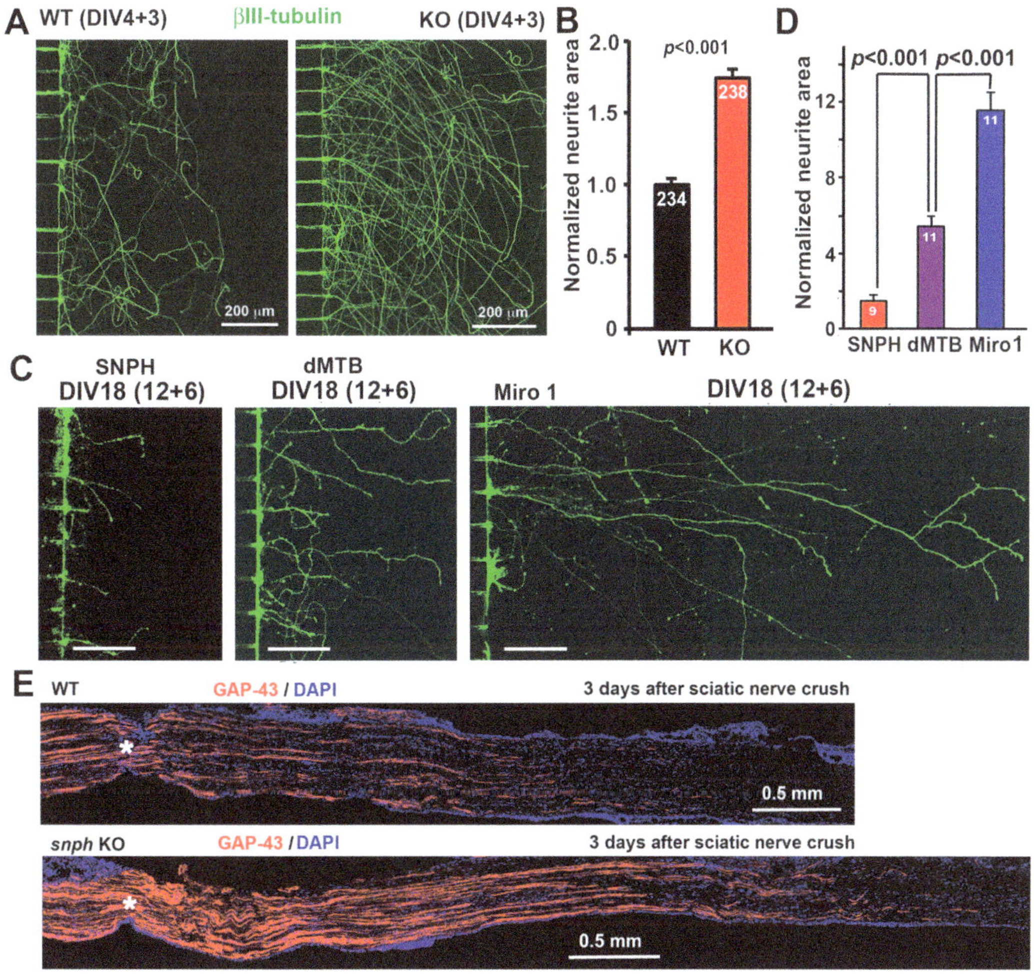

FIG. 4 Enhancing mitochondrial transport boosts axon regrowth capacity after injury in vitro and in vivo. (A, B) Enhancing axonal mitochondrial transport in *snph* KO cortical neurons facilitates axon regrowth after axotomy. Neurons cultured in microfluidic chambers were axotomized at DIV4 and allowed to regrow for 3 days. Axons were stained with βIII-tubulin at DIV7. (C, D) Mature neurons regain regrowth capacity by enhancing axonal mitochondrial transport. Neurons expressing SNPH, SNPH mutant dMTB, or Miro1 were grown in microfluidic chambers for 12 days before axotomy. Regeneration was evaluated 6 days after axotomy (DIV18). Note that abolishing mitochondrial transport by overexpressing SNPH shows failed axon regrowth, whereas enhancing mitochondrial transport by overexpressing Miro1 boosts axon regrowth after injury. (E) SNPH KO mice display enhanced regenerative capacity of sciatic nerves after crush injury in vivo. Adult WT and *snph* KO mice were subjected to a sciatic nerve crush and sacrificed at 3 days post-injury. Regenerating axons were visualized by expressing GAP-43 (*red*) on sciatic nerve longitudinal sections (*crush site). Total number of microgroove channels (B) or chambers (D) are indicated within bars. Scale bars: 200 μm (A), 100 μm (C), 500 μm (E). *(Adapted with permission from Zhou, B., Yu, P., Lin, M. Y., Sun, T., Chen, Y., & Sheng, Z. H. (2016). Facilitation of axon regeneration by enhancing mitochondrial transport and rescuing energy deficits.* Journal of Cell Biology, 214*(1), 103–119. https://doi.org/10.1083/jcb.201605101.)*

enhances axonal mitochondrial trafficking and axon regrowth in Mauthner cells (Xu, Chen, Hu, Huang, & Hu, 2017). Furthermore, the cAMP-PKA-Kinesin1 axis was reported to regulate mitochondrial flux in an aging *Drosophila* model (Vagnoni & Bullock, 2018). A second PI3K-AKT axis was found to negatively regulate PTEN, thus playing an opposite role in mitochondrial motility and axon regeneration (Huang et al., 2019; Park et al., 2008). Although mechanisms as to how these signaling pathways facilitate mitochondrial transport have not been fully elucidated, these studies support an emerging concept: activating an intrinsic "regeneration program" requires the recovery of energy supply by reprogramming mitochondrial transport and recruitment.

Remodeling mitochondrial transport promotes regeneration after SCI

CNS injury and particularly SCI are often characterized by the initial traumatic injury, followed by a secondary injury cascade that could last progressively over weeks (Ahuja et al., 2017). While the primary injury ruptures the axon plasma membrane and elevates intracellular calcium leading to acute mitochondrial damage (Bradke, Fawcett, & Spira, 2012), widespread secondary injury may also cause progressive mitochondrial dysfunction due to chronic inflammation, excitotoxicity, and oxidative response. Although restoring local energy supply by enhancing mitochondrial transport has proven to be arobust strategy in promoting axon survival and regeneration in cultured neurons and the peripheral nerve system (Zhou et al., 2016), it is rather challenging to reverse the axonal energy crisis and support regeneration and synaptic re-connection in injured long-projection corticospinal tract (CST) neurons, and thus restore motor functions after SCI. By collaborating with Xiao-Ming Xu's group at the University of Indiana, we recently established that SCI-induced mitochondrial dysfunction and the local energy crisis are intrinsic restrictions linked to regeneration failure. Using three different murine CNS injury models, we provided in vivo evidence that reversing energy deficits by enhancing mitochondrial transport using *snph*$^{-/-}$ micerepresents a promising therapeutic direction to stimulate axonal regeneration and functional recovery after SCI (Han et al., 2020).

First, Han et al. (2020) used a 5th cervical (C5) dorsal hemisection (DH) injury model to study CST regeneration. While WT mice showed little axonal regrowth beyond the lesion, *snph*$^{-/-}$ mice exhibited robust axon regeneration past the lesion, extending into the caudal spinal cord (Fig. 5). SCI impairs synaptic connections between CST axons and their postsynaptic targets, and thus disrupts the cortico-motoneuronal circuit. By combing light and electron microscopy imaging along with electromyography (EMG) recording, the study demonstrated that these regenerated CSTs in *snph*$^{-/-}$ mice are able to establish cortico-spinal motor connections and conduct synaptic transmission caudal to the injury (Fig. 6). One of the most promising findings was that *snph*$^{-/-}$ mice display notable forelimb dexterous improvement in the single-pellet retrieval test after the C5 SCI. Second, Han et al. (2020) utilized a unilateral pyramidotomy model in which CST axons were severed unilaterally, leading to complete loss of CST innervation on the contralesional spinal cord. WT mice display no change in the termination pattern of CST collaterals after pyramidotomy; most axons remain ipsilateral to the lesion at cervical, thoracic, and lumbar spinal cord levels. In contrast, after pyramidotomy, *snph*$^{-/-}$ mice showed enhanced CST sprouting across the midline and extend into the denervated side at cervical, thoracic, and lumbar spinal cord levels. This striking difference indicates that enhancing axonal mitochondrial transport by deleting *snph* in mice facilitates robust CST sprouting to the denervated contralesional hemicord after a unilateral pyramidotomy. To explore whether increased regrowth could be applied to other CNS pathways, the study applied the third injury model of a thoracic 8 (T8) spinal complete transection to investigate regeneration of both CST and monoaminergic pathways. CST axons dramatically died back from the rostral lesion border in all WT mice. Notably, such typical retraction did not occur in *snph*$^{-/-}$ mice. Instead, CST axon terminals

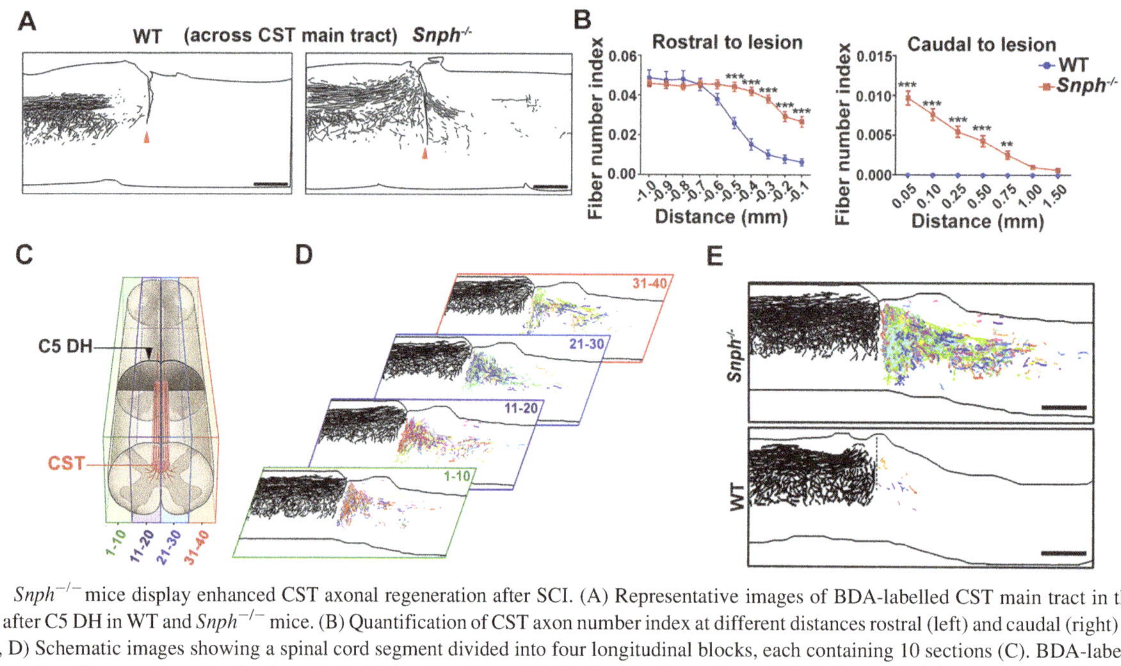

FIG. 5 *Snph*$^{-/-}$ mice display enhanced CST axonal regeneration after SCI. (A) Representative images of BDA-labelled CST main tract in the dorsal column after C5 DH in WT and *Snph*$^{-/-}$ mice. (B) Quantification of CST axon number index at different distances rostral (left) and caudal (right) to the C5 DH. (C, D) Schematic images showing a spinal cord segment divided into four longitudinal blocks, each containing 10 sections (C). BDA-labelled CST axons in each section were traced and color-coded with Imaris software (D). Each of the four color-coded blocks on the right was generated by overlaying 10 consecutive sections. Each color represents a trace of BDA-labeled axons caudal to the lesion in a single section of one block. (E) Reconstruction of a stack of all 40 color-coded images showing CST regeneration across and beyond the lesion in WT and *Snph*$^{-/-}$ mice at 8 weeks post SCI. *(Adapted with permission from Han, Q., Xie, Y., Ordaz, J. D., Huh, A. J., Huang, N., Wu, W., Liu, N., Chamberlain, K. A., Sheng, Z. H., & Xu, X. M. (2020). Restoring cellular energetics promotes axonal regeneration and functional recovery after spinal cord injury.* Cell Metabolism, 31*(3), 623–641 e628. https://doi.org/10.1016/j.cmet.2020.02.002.)*

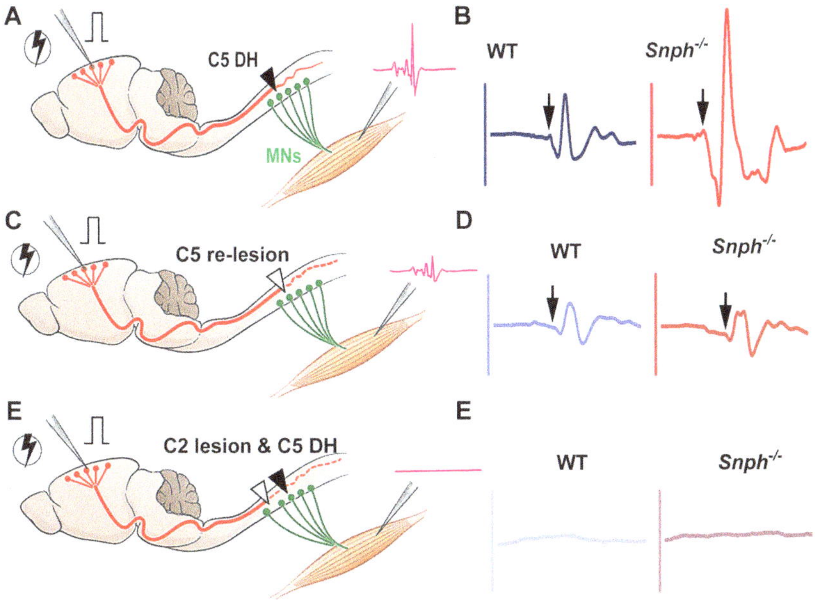

FIG. 6 *Snph*$^{-/-}$ mice display enhanced corticomotoneuron connection after SCI. Schematic model and sample traces of electromyography (EMG) in WT and *Snph*$^{-/-}$ mice recorded from forelimb biceps in response to motor cortex single-pulse stimulation after the C5 DH (A, B). The increased EMG responses in *Snph*$^{-/-}$ mice were abolished after the C5 re-lesion (C, D). EMG activity in both WT and *Snph*$^{-/-}$ mice was completely abolished after a second C2 lesion rostral to the C5 DH in mice (E, F). Scale: 400 μm. *(Adapted with permission from Han, Q., Xie, Y., Ordaz, J. D., Huh, A. J., Huang, N., Wu, W., Liu, N., Chamberlain, K. A., Sheng, Z. H., & Xu, X. M. (2020). Restoring cellular energetics promotes axonal regeneration and functional recovery after spinal cord injury.* Cell Metabolism, 31*(3), 623–641 e628. https://doi.org/10.1016/j.cmet.2020.02.002.)*

were sustained at the rostral lesion border. In addition, we provided in vivo evidence that enhancing mitochondrial transport helps deliver healthy mitochondria from the motor cortex into the injured and regenerating CST axons, thus recovering local axonal mitochondrial integrity. Altogether, restored CNS regeneration in vivo in the three injury models (Han et al., 2020), combined with our findings in in vitro cortical neuron injury and in vivo DRG neurons after crush injury of sciatic nerves (Zhou et al., 2016), supports an emerging concept: enhancing mitochondrial transport via deleting mitochondrial anchoring protein SNPH enables a robust regenerative response to injury in both the PNS and CNS in adult mice.

Boosting energetic metabolism promotes regeneration after SCI

Instead of enhancing mitochondrial transport, boosting energy metabolism in injury neurons represents an alternative promising therapeutic direction. To further test the "injury-induced energy crisis" model, we elevated energy metabolism using creatine, an FDA-approved blood-brain-barrier permeable bioenergetic compound that regenerates ATP from ADP independent of the mitochondrial transport chain (Tarnopolsky & Beal, 2001). While saline-treated WT mice showed no CST axon regeneration after C5 DH, creatine-treated WT mice exhibited significant regeneration of CSTs past the lesion site. However, most regenerated CST axons in creatine-treated WT mice stopped at 0.5 mm caudal to the lesion. In contrast, regenerated axons in *snph*$^{-/-}$ mice extended further up to 1.5 mm in the caudal spinal cord. One explanation is that creatine may induce a rather limited recovery of energy metabolism because of the majority of damaged mitochondria stay within injured axons. Thus, enhanced delivery of healthy mitochondria into injured axons by deleting *snph* likely plays a more persistent and predominant role in recovering axonal energy supply. This hypothesis was supported by the combination of *snph*$^{-/-}$ mice with creatine treatment, which offers even greater effects to boost axon regeneration and functional improvement after SCI (Han et al., 2020).

These regeneration studies in *snph*$^{-/-}$ mice (Han et al., 2020; Zhou et al., 2016) reveal for the first time an "injury-induced energy crisis" model and provide new mechanistic insights as to how enhancing mitochondrial transport enables the mature CNS to regain regenerative capacity after both PNS and CNS injury. However, such genetic manipulation is rather artificial and not suitable for translational development and clinical application in treating SCI. Thus, it is urgent to develop a therapeutic strategy that specifically targets energetic signaling pathways that could either enhance mitochondrial transport and/or elevate energetic metabolism in local axons in response to the brain and spinal injury. Several signaling pathways, including LKB1-Nuak1, cAMP-PKA, and PTEN, have been reported to recruit axonal mitochondria by enhancing their motility in response to various physiological and pathological conditions (Cartoni et al., 2016; Courchet et al., 2013; Vagnoni & Bullock, 2018). A recent high-content screen on mitochondrial motility revealed that pharmacological inhibition of Tripeptidyl peptidase 1 (TPP1) or Aurora Kinase B (AurKB) enhances axonal mitochondrial motility in rat hippocampal neurons and iPSC-derived human cortical neurons (Shlevkov et al., 2019), which provides therapeutic potential in mobilizing axonal mitochondria after SCI.

Creatine is one of the therapeutic compounds that have been applied in rodent SCI models and SCI patients (Tarnopolsky & Beal, 2001). In vertebrates, creatine is converted into phosphocreatine for rapid ATP generation through creatine kinase (CK) independent of the mitochondrial transport chain, and thus CK activity is critical to maintaining creatine concentration and energy demand in tissues (Wyss & Kaddurah-Daouk, 2000). Oral creatine supplementation elevates the phosphocreatine level in the rat nervous system. After SCI, creatine supplementation significantly reduced the scar tissue surrounding the lesion and promoted locomotor capacity (Hausmann, Fouad, Wallimann, & Schwab, 2002). In patients with complete cervical-level SCI, creatine supplementation was reported to enhance their exercise capacity (Jacobs, Mahoney, Cohn, Sheradsky, & Green, 2002), although it remains unknown whether this is the effect of creatine-enhanced regeneration and/or muscle strength. Han et al. (2020) demonstrated that creatine treatment helps astrocytes bridge the lesion gap and promotes CST axon regeneration passing through the lesion in the C5 DH SCI model. However, creatine-induced CST axon regeneration in WT mice is rather limited after SCI when compared to enhanced mitochondrial transport in *snph*$^{-/-}$ mice. Thus, enhanced delivery of healthy mitochondria into injured axons would play a more persistent and predominant role in recovering energy supply than systemic administration of creatine.

Conclusions and new challenges

Mitochondria are the main cellular powerhouses that produce ATP essential for neuronal survival and regeneration. Since injury triggers acute mitochondrial dysfunction, the "injury-induced energy crisis" model has emerged as a new concept of intrinsic restriction that accounts for regeneration failure in SCI and other CNS injuries. Recent studies demonstrate that enhancing mitochondrial transport and replacing damaged mitochondria could reverse these energy deficits and promote axon regeneration and functional recovery, thus providing an emerging therapeutic target for treating SCI. While recent preclinical SCI studies have focused on cell-to-cell mitochondrial transfer and/or enhancement of mitochondrial biogenesis (Simmons, Scholpa, & Schnellmann, 2020), future investigations into the energy-sensing regulation of axonal mitochondrial trafficking and anchoring is a new and promising frontier in neurobiology and neurology (Sheng, 2017). These self-regulation pathways within injured neurons and local axons seem more effective and robust to reverse the energy crisis following acute injury. However, there are many fundamental questions that remain to be addressed. Specifically, while the majority of axonal mitochondria are in the stationary status in mature neurons and adult or aging brains, how could mitochondrial trafficking and anchoring mechanisms be reprogrammed by energetic repairing signals in order to sense the local energetic crisis and regrowth status after SCI and brain injury? Revealing and reprogramming such intrinsic "energy repairing programs" is an emerging frontier for therapeutic investigations. In addition, it remains largely unknown whether bioenergetic compounds could be efficiently delivered into injured axons to elevate local mitochondrial energetic metabolism. Given our findings that creatine-induced axon regeneration after SCI is rather limited when compared to enhanced mitochondrial transport in *snph*$^{-/-}$ mice (Han et al., 2020), future development of more effective bioenergetic compounds would be beneficial to quickly reverse the energy crisis following CNS injuries. Technically, a more sensitive ATP sensor suitable for in vivo real-time tracking of energy levels in the CNS needs to be developed. The current available ATP reporters or biosensors such as Perceval HR, which monitors intracellular ATP/ADP ratios (Tantama, Martinez-Francois, Mongeon, & Yellen, 2013), and GO-ATeam 2, a Förster Resonance Energy Transfer (FRET)-based fluorescent probe to measure cytosolic ATP levels (Nakano, Imamura, Nagai, & Noji, 2011), have been widely applied to examine intracellular ATP levels in live neurons. However, these ATP imaging tools have limited capacity for directly assessing in vivo energy deficits, thus preventing the evaluation of therapeutic strategies in rescuing the energy crisis after SCI.

Applications to other areas of neuroscience

In this chapter, we focused on providing updates on recent studies into how enhanced axonal mitochondrial transport promotes the highly energy-demanding axon regeneration and functional recovery in in vitro and in vivo injury models, and provide prospective views on the potential therapeutic strategy of remodeling mitochondrial transport and elevating cellular energetic metabolism for SCI. In fact, acute mitochondrial damage and altered mitochondrial transport has also been observed in other CNS injury models including traumatic brain injury (Simmons et al., 2020) and ischemia (Yang, Mukda, & Chen, 2018). Preclinical trials that restore axonal mitochondrial integrity and/or local energy homeostasis through exogenous mitochondrial transplantation or pharmacological treatment have surfaced as mitochondrial therapeutic strategies against these CNS injuries. Mitochondrial transplantation is a cellular surgical approach that replaces damaged mitochondria through in situ injections of exogenously isolated healthy ones. The concept of mitochondrial transplantation comes from the fact that after CNS ischemia-reperfusion injury, astrocyte mitochondria are released into the extracellular space, and transferred into injured neurons to support oxidative phosphorylation. Recent studies suggest that such mitochondrial transplantation could increase total ATP levels, oxygen uptake, and expression of key mitochondrial proteins in injected tissues without inducing an inflammatory response. However, mitochondrial transplantation in SCI models is quite challenging given the very limited efficiency in transferring mitochondria into the injured spinal cord axons. Thus, pharmacological activation of the signaling pathways that elevate local mitochondrial biogenesis within injury sites is an attractive direction for future preclinical studies against CNS injury. Several studies have reported that increased mitochondrial biogenesis could slow injury progression and promote functional recovery. Furthermore, pharmacological compounds that enhance mitochondrial biogenesis have already been approved by the FDA for the treatment in rodent SCI models. Thus, these compounds benefit therapeutic development against traumatic brain injury, ischemia, and other CNS injuries. In addition, chronic mitochondrial dysfunction and impaired transport have also been implicated in major neurodegenerative diseases and neurological disorders. Aging-associated accumulation of chronically damaged mitochondria not only produces energy less efficiently but also releases harmful reactive oxygen species. The removal of defective mitochondria from axons and synapses by enhancing their transport constitutes a critical step of mitochondrial quality control (Lin et al., 2017). Therefore, revealing and reprogramming "energy repairing programs" in SCI also benefits the therapeutic development of aging-associated neurodegeneration and neurological disorders.

Mini-dictionary of terms

Adenosine triphosphate (ATP): An organic compound that provides energy to support cellular processes. In the brain, ~93% of ATP is supplied by mitochondria, and the remaining ~7% of ATP is generated through glycolysis.

Mitochondria: The main cellular powerhouses that convert glucose and pyruvate into cellular energy ATP through the electron transport chain and oxidative phosphorylation, and also buffers cytosol calcium.

Axonal mitochondrial motility: Axonal mitochondria move bi-directionally along axonal microtubules; they can pause and change direction frequently and re-mobilize and re-distribute in axons. The mean velocities of axonal mitochondrial motility are highly variable ranging from ~0.2 to 2.0 μm per second. In developing neurons, ~30%–45% of axonal mitochondria are mobile and the remaining two-thirds are stationary. These motility profiles decline and become relatively stable after neurons mature.

Mitochondrial adaptors: A class of mitochondria-associated proteins that connect mitochondria with their transport motor proteins.

Kinesin-1: A class of the plus-end directed and microtubule-based motor proteins that hydrolyze ATP to power anterograde axonal transport of membrane cargoes from the soma toward distal axons in neurons.

Cytosolic dynein: A family of the minus-end directed and microtubule-based motor proteins that hydrolyze ATP to power retrograde axonal transport of membrane cargoes from distal axons back toward the soma in neurons.

Microtubules: A class of cytoskeleton that is formed by polymerized tubulin. Axonal mitochondrial trafficking and anchoring machineries are mainly based on axonal MT tracks.

Mitochondrial membrane potential: An essential functional component of mitochondria that maintains ATP generation and storage capacity during oxidative phosphorylation and is widely used as a marker of mitochondrial integrity.

ROS: Reactive oxygen species are produced by mitochondria during oxidative phosphorylation. High levels of mitochondrial ROS induce cellular stress and apoptosis as well as neuronal degeneration.

Local energy homeostasis: An energetic status that maintains cytosolic ATP to fully power various cellular processes in local areas.

Energy deficits: An insufficient ATP supply status when mitochondria are damaged and/or during increased energy consumption such as after axonal injury and during regeneration.

Creatine: An FDA-approved blood–brain barrier permeable bioenergetic compound that is converted into phosphocreatine for rapid ATP generation from ADP through creatine kinase (CK) independent of the mitochondrial transport chain.

SNPH-mediated axonal mitochondrial anchoring: Syntaphilin (SNPH) acts as a 'static anchor' that selectively associates with axonal mitochondria through its OMM-insertion domain and axon-sorting sequence, and anchors mitochondria on MTs through its MT-binding domain.

Mitochondrial transplantation: A cellular surgical approach that replaces damaged mitochondria through in situ injections of exogenously isolated healthy ones.

Key facts of SNPH-mediated decline of axonal mitochondrial transport in mature neurons

- Axonal mitochondrial motility progressively declines from developing neurons (brains) to mature neurons or adult brains.
- In developing neurons, 30%–45% of axonal mitochondria are mobile and the remaining two-thirds are stationary; when neurons mature, axonal mitochondria largely remain stationary.
- Mitochondrial anchoring mechanisms are essential for arresting mobile axonal mitochondria.
- SNPH acts as a "static anchor" for axonal mitochondria and immobilizes axonal mitochondria.
- Progressive elevation of SNPH expression with neuron maturation contributes to declined motility of axonal mitochondria in mature neurons and adult brains.

Key facts of the energy crisis that accounts for regeneration failure

- Neural regeneration is a highly energy-demanding process, while brain injury is an acute insult that damages local mitochondria.
- An insufficient ATP supply when mitochondria are damaged and increased energy consumption during regeneration collectively contribute to the energy crisis in local injured axons.
- Enhancing mitochondrial transport in *snph*$^{-/-}$ axons helps remove damaged mitochondria from injured axons and replenish with healthy ones, thus reversing local energy deficits.
- Reversing the energy crisis by deleting *snph* enables mature neurons to regain regrowth capacity following in vitro and in vivo injury.
- Enhancing mitochondrial transport and replacing damaged mitochondria reverse the energy deficits and promote axon regeneration and functional recovery after SCI.

Key facts of boosting local energy supply in injured axons

- In addition to enhancing mitochondrial transport, boosting energy metabolism in injured neurons represents an alternative therapeutic direction for stimulating regeneration after SCI.
- Boosting energy metabolism using the bioenergetic compound creatine modestly facilitates CST axon regeneration after SCI.
- The combination of enhanced mitochondrial transport in *snph*$^{-/-}$ mice with creatine treatment offers even greater effects at boosting axon regeneration and functional improvement after SCI.
- Reprogramming intrinsic "energy repairing programs" that enhance mitochondrial transport in response to the brain and spinal injury is an emerging frontier for therapeutic investigations.
- Pharmacological activation of the signaling pathways that elevate mitochondrial biogenesis within injured axons is an attractive direction for future preclinical studies against CNS injury.

Summary points

- The mature CNS typically fails to regenerate after injury leading to permanent neurological impairment.
- Regeneration requires high-energy consumption in the form of ATP that is mainly produced in mitochondria.
- Injury is an acute insult that damages local mitochondria leading to an energy crisis.
- The "injury-induced energy crisis" concept has emerged as one of the intrinsic restrictions that accounts for regeneration failure in SCI and other CNS injuries.
- Axonal mitochondrial transport progressively declines from developing neurons to mature neurons or adult brains.
- SNPH acts as a "static anchor" contributing to declined motility of axonal mitochondria in mature neurons and adult brains.
- Enhancing mitochondrial transport in *snph*$^{-/-}$ axons helps remove damaged mitochondria and replenish them with healthy ones, thus reversing the local energy crisis.
- Reversing this energy crisis enables mature neurons to regain regrowth capacity and promote axon regeneration and functional recovery after SCI.

- Boosting energy metabolism using the bioenergetic compound creatine facilitates CST axon regeneration after SCI.
- Remodeling the delivery of healthy mitochondria into injured axons has emerged as one of the robust mechanisms for boosting axon regeneration after SCI.

Acknowledgment

This work was supported by the Intramural Research Program of NINDS, NIH ZIA NS003029 and ZIA NS002946 (Z-H. Sheng).

References

Ahuja, C. S., Wilson, J. R., Nori, S., Kotter, M. R. N., Druschel, C., Curt, A., et al. (2017). Traumatic spinal cord injury [Primer]. *Nature Reviews. Disease Primers*, *3*, 17018.

Avery, M. A., Rooney, T. M., Pandya, J. D., Wishart, T. M., Gillingwater, T. H., Geddes, J. W., et al. (2012). WldS prevents axon degeneration through increased mitochondrial flux and enhanced mitochondrial Ca2+ buffering. *Current Biology*, *22*(7), 596–600.

Bradbury, E. J., & McMahon, S. B. (2006). Spinal cord repair strategies: Why do they work? *Nature Reviews. Neuroscience*, *7*(8), 644–653.

Bradke, F., Fawcett, J. W., & Spira, M. E. (2012). Assembly of a new growth cone after axotomy: The precursor to axon regeneration [review article]. *Nature Reviews. Neuroscience*, *13*(3), 183–193.

Cai, Q., Gerwin, C., & Sheng, Z. H. (2005). Syntabulin-mediated anterograde transport of mitochondria along neuronal processes. *The Journal of Cell Biology*, *170*(6), 959–969.

Cartoni, R., Norsworthy, M. W., Bei, F., Wang, C., Li, S., Zhang, Y., et al. (2016). The mammalian-specific protein Armcx1 regulates mitochondrial transport during axon regeneration. *Neuron*, *92*(6), 1294–1307.

Case, L. C., & Tessier-Lavigne, M. (2005). Regeneration of the adult central nervous system. *Current Biology*, *15*(18), R749–R753.

Cavallucci, V., Bisicchia, E., Cencioni, M. T., Ferri, A., Latini, L., Nobili, A., et al. (2014). Acute focal brain damage alters mitochondrial dynamics and autophagy in axotomized neurons. *Cell Death & Disease*, *5*, e1545.

Chen, Y., & Sheng, Z. H. (2013). Kinesin-1-syntaphilin coupling mediates activity-dependent regulation of axonal mitochondrial transport. *The Journal of Cell Biology*, *202*(2), 351–364.

Cho, K. I., Cai, Y., Yi, H., Yeh, A., Aslanukov, A., & Ferreira, P. A. (2007). Association of the kinesin-binding domain of RanBP2 to KIF5B and KIF5C determines mitochondria localization and function. *Traffic*, *8*(12), 1722–1735.

Courchet, J., Lewis, T. L., Jr., Lee, S., Courchet, V., Liou, D. Y., Aizawa, S., et al. (2013). Terminal axon branching is regulated by the LKB1-NUAK1 kinase pathway via presynaptic mitochondrial capture. *Cell*, *153*(7), 1510–1525.

Covill-Cooke, C., Toncheva, V. S., Drew, J., Birsa, N., Lopez-Domenech, G., & Kittler, J. T. (2020). Peroxisomal fission is modulated by the mitochondrial Rho-GTPases, Miro1 and Miro2. *EMBO Reports*, *21*(2), e49865.

Das, S., Boczan, J., Gerwin, C., Zald, P. B., & Sheng, Z. H. (2003). Regional and developmental regulation of syntaphilin expression in the brain: A candidate molecular element of synaptic functional differentiation. *Brain Research. Molecular Brain Research*, *116*(1–2), 38–49.

Devine, M. J., & Kittler, J. T. (2018). Mitochondria at the neuronal presynapse in health and disease. *Nature Reviews. Neuroscience*, *19*(2), 63–80.

Fransson, S., Ruusala, A., & Aspenstrom, P. (2006). The atypical Rho GTPases Miro-1 and Miro-2 have essential roles in mitochondrial trafficking. *Biochemical and Biophysical Research Communications*, *344*(2), 500–510.

Fujita, T., Maturana, A. D., Ikuta, J., Hamada, J., Walchli, S., Suzuki, T., et al. (2007). Axonal guidance protein FEZ1 associates with tubulin and kinesin motor protein to transport mitochondria in neurites of NGF-stimulated PC12 cells. *Biochemical and Biophysical Research Communications*, *361*(3), 605–610.

Glater, E. E., Megeath, L. J., Stowers, R. S., & Schwarz, T. L. (2006). Axonal transport of mitochondria requires milton to recruit kinesin heavy chain and is light chain independent. *The Journal of Cell Biology*, *173*(4), 545–557.

Godzik, K., & Coleman, M. P. (2015). The axon-protective WLD(S) protein partially rescues mitochondrial respiration and glycolysis after axonal injury. *Journal of Molecular Neuroscience*, *55*(4), 865–871.

Guo, X., Macleod, G. T., Wellington, A., Hu, F., Panchumarthi, S., Schoenfield, M., et al. (2005). The GTPase dMiro is required for axonal transport of mitochondria to Drosophila synapses. *Neuron*, *47*(3), 379–393.

Han, S. M., Baig, H. S., & Hammarlund, M. (2016). Mitochondria localize to injured axons to support regeneration. *Neuron*, *92*(6), 1308–1323.

Han, Q., Xie, Y., Ordaz, J. D., Huh, A. J., Huang, N., Wu, W., et al. (2020). Restoring cellular energetics promotes axonal regeneration and functional recovery after spinal cord injury. *Cell Metabolism*, *31*(3), 623–641. e628.

Harel, N. Y., & Strittmatter, S. M. (2006). Can regenerating axons recapitulate developmental guidance during recovery from spinal cord injury? *Nature Reviews. Neuroscience*, *7*(8), 603–616.

Harris, J. J., Jolivet, R., & Attwell, D. (2012). Synaptic energy use and supply. *Neuron*, *75*(5), 762–777.

Hausmann, O. N., Fouad, K., Wallimann, T., & Schwab, M. E. (2002). Protective effects of oral creatine supplementation on spinal cord injury in rats. *Spinal Cord*, *40*(9), 449–456.

He, Z., & Jin, Y. (2016). Intrinsic control of axon regeneration. *Neuron*, *90*(3), 437–451.

Hirokawa, N., Niwa, S., & Tanaka, Y. (2010). Molecular motors in neurons: Transport mechanisms and roles in brain function, development, and disease. *Neuron*, *68*(4), 610–638.

Hsieh, C. H., Shaltouki, A., Gonzalez, A. E., Bettencourt da Cruz, A., Burbulla, L. F., St Lawrence, E., et al. (2016). Functional impairment in miro degradation and mitophagy is a shared feature in familial and sporadic Parkinson's disease. *Cell Stem Cell*, *19*(6), 709–724.

Huang, H., Miao, L., Yang, L., Liang, F., Wang, Q., Zhuang, P., et al. (2019). AKT-dependent and -independent pathways mediate PTEN deletion-induced CNS axon regeneration. *Cell Death & Disease*, *10*(3), 203.

Jacobs, P. L., Mahoney, E. T., Cohn, K. A., Sheradsky, L. F., & Green, B. A. (2002). Oral creatine supplementation enhances upper extremity work capacity in persons with cervical-level spinal cord injury. *Archives of Physical Medicine and Rehabilitation*, *83*(1), 19–23.

Kalinski, A. L., Kar, A. N., Craver, J., Tosolini, A. P., Sleigh, J. N., Lee, S. J., et al. (2019). Deacetylation of Miro1 by HDAC6 blocks mitochondrial transport and mediates axon growth inhibition. *The Journal of Cell Biology*, *218*(6), 1871–1890.

Kang, J. S., Tian, J. H., Pan, P. Y., Zald, P., Li, C., Deng, C., et al. (2008). Docking of axonal mitochondria by syntaphilin controls their mobility and affects short-term facilitation. *Cell*, *132*(1), 137–148.

Lewis, T. L., Jr., Turi, G. F., Kwon, S. K., Losonczy, A., & Polleux, F. (2016). Progressive decrease of mitochondrial motility during maturation of cortical axons in vitro and in vivo. *Current Biology*, *26*(19), 2602–2608.

Lin, M. Y., Cheng, X. T., Tammineni, P., Xie, Y., Zhou, B., Cai, Q., et al. (2017). Releasing syntaphilin removes stressed mitochondria from axons independent of mitophagy under pathophysiological conditions. *Neuron*, *94*, 595–610.

Lopez-Domenech, G., Higgs, N. F., Vaccaro, V., Ros, H., Arancibia-Carcamo, I. L., MacAskill, A. F., et al. (2016). Loss of dendritic complexity precedes neurodegeneration in a mouse model with disrupted mitochondrial distribution in mature dendrites. *Cell Reports*, *17*(2), 317–327.

Mattson, M. P., Gleichmann, M., & Cheng, A. (2008). Mitochondria in neuroplasticity and neurological disorders. *Neuron*, *60*(5), 748–766.

Misgeld, T., Kerschensteiner, M., Bareyre, F. M., Burgess, R. W., & Lichtman, J. W. (2007). Imaging axonal transport of mitochondria in vivo. *Nature Methods*, *4*(7), 559–561.

Misgeld, T., & Schwarz, T. L. (2017). Mitostasis in neurons: Maintaining mitochondria in an extended cellular architecture. *Neuron*, *96*(3), 651–666.

Morris, R. L., & Hollenbeck, P. J. (1993). The regulation of bidirectional mitochondrial transport is coordinated with axonal outgrowth. *Journal of Cell Science*, *104*(Pt 3), 917–927.

Nakano, M., Imamura, H., Nagai, T., & Noji, H. (2011). Ca(2)(+) regulation of mitochondrial ATP synthesis visualized at the single cell level. *ACS Chemical Biology*, *6*(7), 709–715.

O'Donnell, K. C., Vargas, M. E., & Sagasti, A. (2013). WldS and PGC-1alpha regulate mitochondrial transport and oxidation state after axonal injury. *The Journal of Neuroscience*, *33*(37), 14778–14790.

Okumoto, K., Ono, T., Toyama, R., Shimomura, A., Nagata, A., & Fujiki, Y. (2018). New splicing variants of mitochondrial Rho GTPase-1 (Miro1) transport peroxisomes. *The Journal of Cell Biology*, *217*(2), 619–633.

Park, K. K., Liu, K., Hu, Y., Smith, P. D., Wang, C., Cai, B., et al. (2008). Promoting axon regeneration in the adult CNS by modulation of the PTEN/mTOR pathway. *Science*, *322*(5903), 963–966.

Pekkurnaz, G., Trinidad, J. C., Wang, X., Kong, D., & Schwarz, T. L. (2014). Glucose regulates mitochondrial motility via Milton modification by O-GlcNAc transferase. *Cell*, *158*(1), 54–68.

Pilling, A. D., Horiuchi, D., Lively, C. M., & Saxton, W. M. (2006). Kinesin-1 and Dynein are the primary motors for fast transport of mitochondria in Drosophila motor axons. *Molecular Biology of the Cell*, *17*(4), 2057–2068.

Rawson, R. L., Yam, L., Weimer, R. M., Bend, E. G., Hartwieg, E., Horvitz, H. R., et al. (2014). Axons degenerate in the absence of mitochondria in *C. elegans*. *Current Biology*, *24*(7), 760–765.

Russo, G. J., Louie, K., Wellington, A., Macleod, G. T., Hu, F., Panchumarthi, S., et al. (2009). Drosophila Miro is required for both anterograde and retrograde axonal mitochondrial transport. *The Journal of Neuroscience*, *29*(17), 5443–5455.

Ruthel, G., & Hollenbeck, P. J. (2003). Response of mitochondrial traffic to axon determination and differential branch growth. *The Journal of Neuroscience*, *23*(24), 8618–8624.

Sheng, Z. H. (2014). Mitochondrial trafficking and anchoring in neurons: New insight and implications. *The Journal of Cell Biology*, *204*(7), 1087–1098.

Sheng, Z. H. (2017). The interplay of axonal energy homeostasis and mitochondrial trafficking and anchoring. *Trends in Cell Biology*, *27*(6), 403–416.

Sheng, Z. H., & Cai, Q. (2012). Mitochondrial transport in neurons: Impact on synaptic homeostasis and neurodegeneration. *Nature Reviews. Neuroscience*, *13*(2), 77–93.

Shlevkov, E., Basu, H., Bray, M. A., Sun, Z., Wei, W., Apaydin, K., et al. (2019). A high-content screen identifies TPP1 and Aurora B as regulators of axonal mitochondrial transport. *Cell Reports*, *28*(12), 3224–3237.e3225.

Simmons, E. C., Scholpa, N. E., & Schnellmann, R. G. (2020). Mitochondrial biogenesis as a therapeutic target for traumatic and neurodegenerative CNS diseases. *Experimental Neurology*, *329*, 113309.

Smit-Rigter, L., Rajendran, R., Silva, C. A., Spierenburg, L., Groeneweg, F., Ruimschotel, E. M., et al. (2016). Mitochondrial dynamics in visual cortex are limited in vivo and not affected by axonal structural plasticity. *Current Biology*, *26*(19), 2609–2616.

Sun, T., Qiao, H., Pan, P. Y., Chen, Y., & Sheng, Z. H. (2013). Motile axonal mitochondria contribute to the variability of presynaptic strength. *Cell Reports*, *4*(3), 413–419.

Takihara, Y., Inatani, M., Eto, K., Inoue, T., Kreymerman, A., Miyake, S., et al. (2015). In vivo imaging of axonal transport of mitochondria in the diseased and aged mammalian CNS. *Proceedings of the National Academy of Sciences of the United States of America*, *112*(33), 10515–10520.

Tanaka, Y., Kanai, Y., Okada, Y., Nonaka, S., Takeda, S., Harada, A., et al. (1998). Targeted disruption of mouse conventional kinesin heavy chain, kif5B, results in abnormal perinuclear clustering of mitochondria. *Cell*, *93*(7), 1147–1158.

Tantama, M., Martinez-Francois, J. R., Mongeon, R., & Yellen, G. (2013). Imaging energy status in live cells with a fluorescent biosensor of the intracellular ATP-to-ADP ratio. *Nature Communications*, *4*, 2550.

Tarnopolsky, M. A., & Beal, M. F. (2001). Potential for creatine and other therapies targeting cellular energy dysfunction in neurological disorders. *Annals of Neurology*, *49*(5), 561–574.

Vagnoni, A., & Bullock, S. L. (2018). A cAMP/PKA/Kinesin-1 axis promotes the axonal transport of mitochondria in aging drosophila neurons. *Current Biology*, *28*(8), 1265–1272.e1264.

van Bergeijk, P., Adrian, M., Hoogenraad, C. C., & Kapitein, L. C. (2015). Optogenetic control of organelle transport and positioning. *Nature*, *518*(7537), 111–114.

van Spronsen, M., Mikhaylova, M., Lipka, J., Schlager, M. A., van den Heuvel, D. J., Kuijpers, M., et al. (2013). TRAK/Milton motor-adaptor proteins steer mitochondrial trafficking to axons and dendrites. *Neuron*, *77*(3), 485–502.

Wyss, M., & Kaddurah-Daouk, R. (2000). Creatine and creatinine metabolism. *Physiological Reviews*, *80*(3), 1107–1213.

Xu, Y., Chen, M., Hu, B., Huang, R., & Hu, B. (2017). In vivo imaging of mitochondrial transport in single-axon regeneration of zebrafish mauthner cells. *Frontiers in Cellular Neuroscience*, *11*, 4.

Yang, J. L., Mukda, S., & Chen, S. D. (2018). Diverse roles of mitochondria in ischemic stroke. *Redox Biology*, *16*, 263–275.

Zhou, B., Yu, P., Lin, M. Y., Sun, T., Chen, Y., & Sheng, Z. H. (2016). Facilitation of axon regeneration by enhancing mitochondrial transport and rescuing energy deficits. *The Journal of Cell Biology*, *214*(1), 103–119.

Chapter 18

Neurotrophins and their role in axonal outgrowth following spinal cord injury

Anusha Dravid[a], Simon J. O'Carroll[b], and Darren Svirskis[a]

[a]*School of Pharmacy, Faculty of Medical and Health Sciences, University of Auckland, Auckland, New Zealand,* [b]*Department of Anatomy and Medical Imaging, Faculty of Medical and Health Sciences, School of Medical Sciences, University of Auckland, Auckland, New Zealand*

List of abbreviations

BDNF	brain-derived neurotrophic factor
BSCB	blood-spinal-cord barrier
CST	corticospinal tract
CNS	central nervous system
GC	growth cone
k_d	receptor-neurotrophin dissociation constant
NGF	nerve growth factor
NT-3	neurotrophin-3
NT-4/5	neurotrophin-4/5
SCI	spinal cord injury
Trk	tropomyosin-regulated kinase

Introduction

Throughout neuronal development, neurotrophic expression is essential for the survival of growing axonal projections. Following the discovery of the prototypic neurotrophic factor, nerve growth factor (NGF) (Levi-Montalcini & Hamburger, 1951), researchers have identified a family of structurally related molecules termed "neurotrophins" including NGF, brain-derived neurotrophic factor (BDNF), neurotrophin-3 (NT-3) and neurotrophin-4/5 (NT-4/5). A repertoire of experimental studies have since established these biomolecules as key modulators of neuronal differentiation (Berkemeier et al., 1991), growth (Boyd & Gordon, 2003) plasticity (Brock et al., 2010), and axonal guidance (Gundersen & Barrett, 1979). However, contrary to embryonic development, neurotrophin production is drastically reduced within the adult spinal cord (Widenfalk, Lundströmer, Jubran, Brené, & Olson, 2001). Amongst other factors, this deficiency in trophic support contributes to the absence of spontaneous axonal regeneration following traumatic spinal cord injury (SCI) (Fawcett, 2020). However, despite limited endogenous production, adult neurons still express neurotrophin receptors and retain their capacity to respond. Hence, exogenous stimulation has been explored as a potential avenue to promote regeneration after an SCI (Widenfalk et al., 2001). As receptor expression differs across ascending and descending spinal populations, each neurotrophin has been shown to vary in effect on distinct spinal tracts (Keefe, Sheikh, & Smith, 2017).

The initial traumatic insult associated with SCI, precedes a secondary injury cascade, cystic cavitation, and the formation of a growth-inhibitory lesion known as a "glial scar" (Fitch & Silver, 2008). Supporting spared axons to bridge this injury zone and innervate existing neuronal targets is a challenge to restoring function after injury (McCall, Weidner, & Blesch, 2012). Although propriospinal relay connections have been attributed to some spontaneous functional recovery (Courtine et al., 2008), re-establishment of original synaptic connections requires spared axons to grow beyond the injury site and re-enter the host spinal cord towards denervated targets (McCall et al., 2012). In anatomically incomplete injuries, this path can either be through or around the lesion. Although localized neurotrophin administration has been

Cellular, Molecular, Physiological, and Behavioral Aspects of Spinal Cord Injury. https://doi.org/10.1016/B978-0-12-822427-4.00018-6

shown to support the extension of injured axons through the lesion, it is now understood that delivery in the form of a spatiotemporal concentration gradient is a key strategy for promoting axons to exit the lesion zone and grow towards pre-injury targets (Lu, Yang, Jones, Filbin, & Tuszynski, 2004).

This chapter discusses the role of neurotrophins in stimulating axonal outgrowth following SCI, with a focus on the application of neurotrophin concentration gradients. We detail the structure and function of the neurotrophins with respect to the spinal cord, followed by an explanation of growth cone navigation in response to neurotrophin gradients. Finally, we outline current progress and future considerations towards clinically appropriate neurotrophin delivery in *in vivo* settings.

Structure and function of the neurotrophins

The neurotrophins are an endogenously produced group of polypeptides and the most investigated growth factors in SCI studies. This family consists of NGF, BDNF, NT-3, and NT-4/5, all sharing near 50% homology in their primary structure (Hauburger, Kliemannel, Madsen, Rudolph, & Schwarz, 2007). The neurotrophins are synthesized on the rough endoplasmic reticulum as a "pre-pro-neurotrophin" precursor form (Al-Qudah & Al-Dwairi, 2016). Subsequent translocation to the Golgi apparatus and cleavage of the "pre-region" produces the ~30 kDa pro-neurotrophin precursor which is known to exert apoptotic effects (Chao & Bothwell, 2002; Kowiański et al., 2018). Further proteolytic cleavage of the pro-neurotrophin occurs intracellularly and extracellularly by plasmin and furin molecules to generate the final mature neurotrophin – a 13 kDa non-covalently linked homodimer molecule (Bibel & Barde, 2000) which is responsible for neurotrophic activity.

Neurotrophin function is mediated by the transmembrane tropomyosin-regulated kinase (Trk) and p75 neurotrophin receptors ($p75^{NTR}$). The low-affinity $p75^{NTR}$, a member of the tumor necrosis superfamily group, non-selectively binds to all the neurotrophins in the mature and uncleaved pro-neurotrophin form (Meeker & Williams, 2015). On the other hand, each isoform of the Trk family, consisting of TrkA, TrkB, and TrkC, exhibits selective, high-affinity binding for a specified mature neurotrophin (Fig. 1). Namely, NGF binds TrkA; BDNF and NT-4/5 both bind TrkB; while NT-3 binds primarily to TrkC, and a lesser degree with TrkA and TrkB (Chao & Bothwell, 2002). Neurotrophin binding results in autophosphorylation of intracellular tyrosine residues of the Trk receptor. This, in turn, initiates key downstream signaling cascades such as the phosphatidylinositol 3-kinase/protein kinase B (PI3K-Akt), mitogen-activated protein kinase (MAPK), extracellular signal-regulated kinases (ERK), and phospholipase C-γ (PLCγ) pathways (Meeker & Williams, 2015) which are responsible for neurite outgrowth, cell survival, cell differentiation, and synaptic plasticity. In contrast, $p75^{NTR}$ receptor binding results in cellular apoptosis through Jun N-terminal kinase (JNK) and p53 activation (Dechant & Barde, 2002).

Receptor distribution of the Trk isoforms varies between the different neuronal subpopulations within the spinal cord. Hence, each neurotrophin differs in the extent to which it affects distinct spinal tracts (Dravid, Parittotokkaporn, Aqrawe, O'Carroll, & Svirskis, 2020; Keefe et al., 2017), as illustrated in Fig. 2. Knowledge about Trk receptor expression in the injured spinal cord is still developing. Liebl et al. have previously investigated Trk expression in a contusion rat model, observing reduced levels within the first week of trauma at and near the injury site, but not within the proximal and distal spinal cord (Liebl, Huang, Young, & Parada, 2001). Similarly, King et al. have also reported a reduction of receptor expression at the lesion site of a hemisection rat model, while expression in the undamaged regions of the cord remained

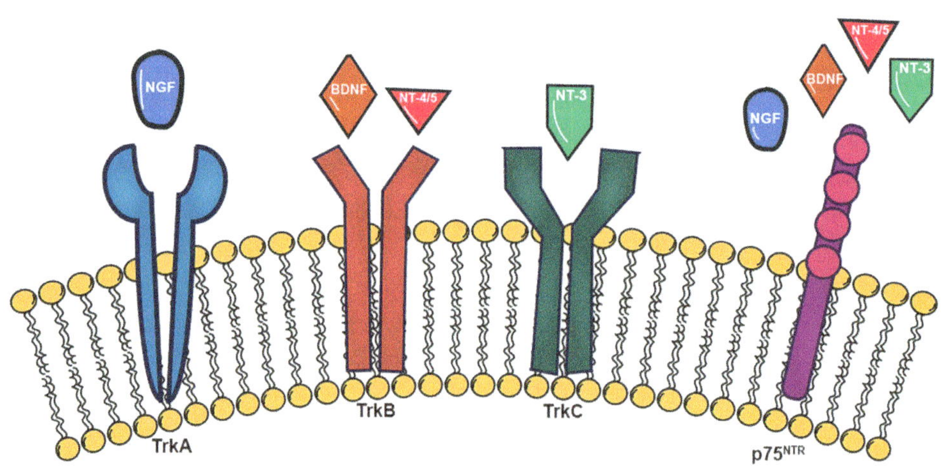

FIG. 1 Schematic illustration of neurotrophin-Trk receptor preferential binding interactions. Each neurotrophin binds with high affinity to a specific Trk receptor isoform. The $p75^{NTR}$ binds all neurotrophins with low affinity. NGF, nerve growth factor; BDNF, brain-derived neurotrophic factor; NT-3, neurotrophin-3; NT-4/5, neurotrophin-4/5.

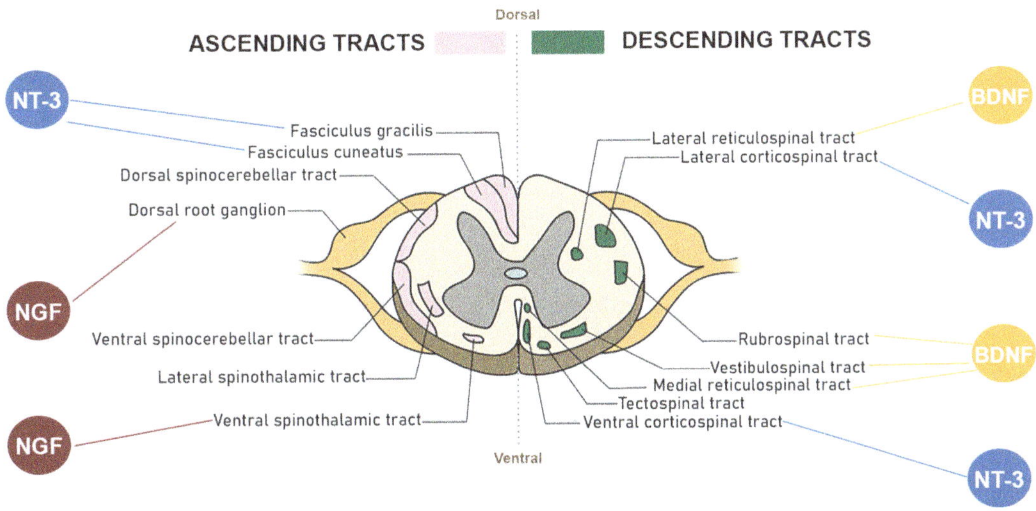

FIG. 2 Spinal cord segment highlighting neurotrophin sensitivities. NGF, nerve growth factor; BDNF, brain-derived neurotrophic factor; NT-3, neurotrophin-3; NT-4/5, neurotrophin-4/5. *(Adapted with permission from Dravid, A., Parittotokkaporn, S., Aqrawe, Z., O'Carroll, S. J., & Svirskis, D. (2020). Determining neurotrophin gradients in vitro to direct axonal outgrowth following spinal cord injury.* ACS Chemical Neuroscience, *11(2), 121–132. doi:10.1021/acschemneuro.9b00565. Copyright (2020) American Chemical Society.)*

normal (King, Bradbury, McMahon, & Priestley, 2000). These observations support the application of a neurotrophin gradient, with the highest concentrations delivered near uninjured regions of the cord where Trk receptors are expressed after injury, as a strategy to encourage growth towards pre-injury targets.

Nerve growth factor

The findings of a soluble, diffusible factor enhancing the outgrowth of adjacent sympathetic and sensory neuronal populations led to the discovery and isolation of NGF, a pivotal moment in modern neurobiology (Levi-Montalcini & Hamburger, 1951). As the oldest member of the neurotrophin family, numerous studies have aimed to harness its neurotrophic effects for therapeutic outcomes. NGF is a robust promoter of sensory neuron survival and growth following injury (Oudega & Hagg, 1996, 1999; Tuszynski et al., 1994), with the TrkA receptor expressed widely across the ventral and lateral spinothalamic neuron population groups. Additionally, growth of small diameter CGRP$^+$ sensory and TH$^+$ cerulospinal neurons after an injury has been observed in response to NGF (Grill, Blesch, & Tuszynski, 1997).

Brain-derived neurotrophic factor

The discovery of NGF initiated a pursuit for other nerve growth factors, achieving success decades later in 1982 with the identification of BDNF. Demonstrating an ability to evoke neuroprotective and growth responses in a diversity of cell populations, including those unaffected by NGF, the neurotrophin BDNF has become one of the most investigated growth factor therapies in SCI models (Weishaupt, Blesch, & Fouad, 2012). The BDNF receptor, TrkB, is widely distributed in spinal populations involved in producing motor function (Kishino, Ishige, Tatsuno, Nakayama, & Noguchi, 1997). Specifically, the 5-HT$^+$ raphaespinal, reticulospinal, rubrospinal, vestibulospinal, and TH$^+$ cerulospinal have all exhibited axonal growth in response to BDNF stimulation (Hollis 2nd & Tuszynski, 2011; Koda et al., 2004; Lu, Jones, & Tuszynski, 2005).

Neurotrophin-3

NT-3 has also been widely investigated in experimental SCI, as the only factor to demonstrate significant growth-promoting effects in the corticospinal tract (CST) (Blits, Dijkhuizen, Boer, & Verhaagen, 2000) (Tuszynski et al., 2003). Unlike the other neurotrophins, NT-3 has a more versatile binding capacity, demonstrating low-affinity binding to the TrkA and TrkB isoforms alongside its preferential binding with TrkC. NT-3 has also been shown to promote regeneration of dorsal column proprioceptive sensory axons and the 5-HT$^+$ raphaespinal neuron populations (Taylor, Jones, Tuszynski, & Blesch, 2006).

Neurotrophin-4/5

As with BDNF, the neurotrophin NT-4/5 also signals through the TrkB receptor, affecting the same spinal populations based on receptor expression. Despite binding to the same receptor, some studies have alluded to the possibility of axonal growth differing in response to BDNF and NT-4/5 due to the presence of TrkB splice variants or variations in downstream signaling cascades (Blesch, Yang, Weidner, Hoang, & Otero, 2004). Thorough investigations of this factor in experimental SCI models are still wanting, with current studies suggesting roles in promoting sensory and motor neuron outgrowth (Kobayashi et al., 1997) (Schmalbruch & Rosenthal, 1995) (Stucky, Shin, & Lewin, 2002).

Neurotrophic gradients for guiding regenerating axons

In 1892 neuroscientist Ramón y Cajal proposed his "neurotropic hypothesis", detailing the ability of the growth cone (GC) to be guided by diffusible chemotactic gradients secreted by target cells. A body of *in vitro* gradient-based assays have validated this theory, demonstrating how axons alter their trajectory when presented with a chemotactic stimulus. The GC is a dynamic sensory feature located at the terminal end of elongating axons. We now know these actin-rich structures to be responsible for (i) detecting spatial gradients of chemical stimuli in the extracellular environment and (ii) navigating axonal trajectory towards the direction of the chemoattractive stimulus.

Amongst well-established roles in axonal outgrowth and survival, the neurotrophins have been shown to act as axonal guidance cues for developing and regenerating neurons. Early investigations with chick dorsal root ganglion (DRG) first elicited the chemotactic effects of NGF in vitro, with the axons reorienting their growth towards the direction of the NGF source (Gundersen & Barrett, 1979). Similarly, BDNF and NT-3 demonstrated the ability to influence GC-mediated turning of *Xenopus* spinal neurons (Ming, Lohof, & Zheng, 1997). Following Trk receptor activation at the GC, the downstream signaling cascades involved in survival and outgrowth have also been thought to contribute to axonal guidance. In fact, Ming et al. have previously ascribed the effect of NGF on *Xenopus* spinal neuron turning to the co-activation of the PI3K-Akt and PLCγ signaling pathways at the GC (Ming et al., 1999).

Delivery of the neurotrophins in the form of a concentration gradient (Fig. 3) has increasingly presented itself as an attractive approach to encourage axonal bridging of the injury lesion (McCall et al., 2012). For severed axons to regenerate

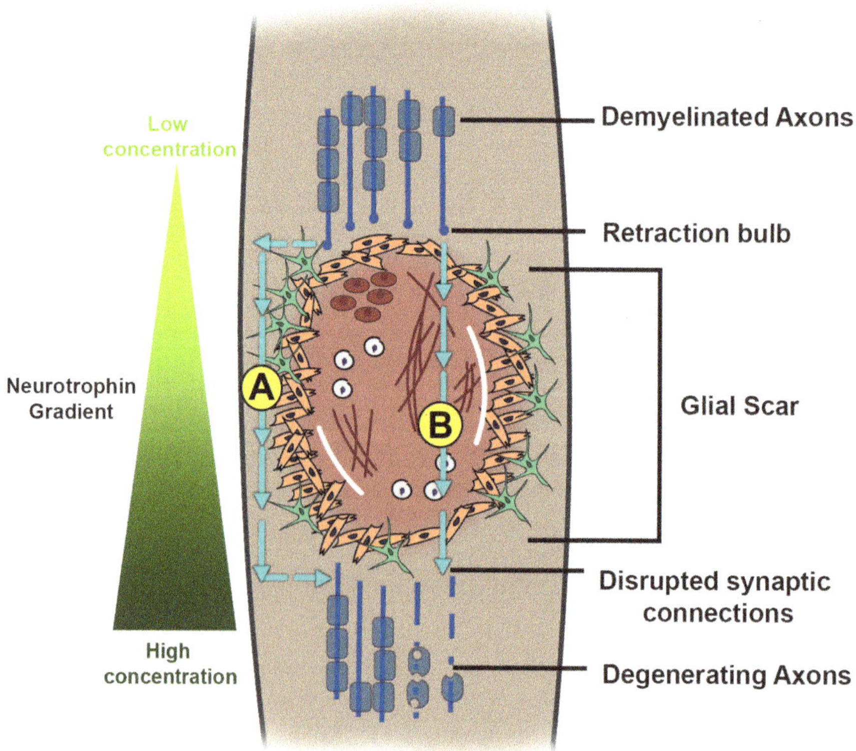

FIG. 3 Schematic illustration depicting the effect of a neurotrophin gradient on axonal guidance of descending projections through glial scar. A neurotrophic gradient may facilitate the growth of severed axons either (A) around or (B) through the lesion towards denervated targets.

TABLE 1 Receptor-neurotrophin dissociation constants for the three key neurotrophins investigated in experimental SCI studies.

Receptor-neurotrophin	k_d	References
TrkA + NGF	2.3×10^{-11} M	Rodríguez-Tébar, Dechant, Götz, and Barde (1992), Sutter, Riopelle, Harris-Warrick, and Shooter (1979)
TrkB + BDNF	1.7×10^{-11} M	Rodriguez-Tébar, Dechant, and Barde (1990), Sakuragi, Tominaga-Yoshino, and Ogura (2013)
TrkC + NT-3	1.8×10^{-11} M	Rodríguez-Tébar et al. (1992)

NGF, nerve growth factor; *BDNF*, brain-derived neurotrophic factor; *NT-3*, neurotrophin-3; *NT-4/5*, neurotrophin-4/5.
Adapted with permission from Dravid, A., Parittotokkaporn, S., Aqrawe, Z., O'Carroll, S. J., & Svirskis, D. (2020). Determining neurotrophin gradients in vitro to direct axonal outgrowth following spinal cord injury. *ACS Chemical Neuroscience, 11*(2), 121–132. https://doi.org/10.1021/acschemneuro.9b00565.

and re-connect with denervated pre-injury targets, the neurotrophins must be delivered in a dynamic pattern. Initially, this would require a sufficient concentration delivered within the lesion, to promote growth within this inhibitory region (Jones, Oudega, Bunge, & Tuszynski, 2001). Following this, the neurotrophins should then be applied in the form of a concentration gradient, with the highest concentrations expressed beyond the injury site (Jones et al., 2001). This gradient may be able to guide the growing axons over a length of several millimeters (Giger, Hollis 2nd, & Tuszynski, 2010).

For a GC to respond to a concentration gradient, it must be able to detect a significant location-dependent concentration difference of the ligand across its width. The width of a GC has commonly been assumed as 10 μm–20 μm in existing gradient-based axonal guidance assays. Further, the absolute concentration (i.e., the amount of neurotrophin per unit of volume) must be within a suitable range with respect to the receptor–neurotrophin dissociation constant (k_d) (Table 1). A concentration higher than the k_d may lead to receptor saturation across both sides of the GC, impeding its directional sensing (Goodhill & Baier, 1998). However, a concentration too low may not overcome noise created by thermal fluctuations in ligand number, nor will a sufficient number of receptors be activated to initiate downstream signaling pathways (Goodhill & Baier, 1998). Within the current literature, two models have been reported to describe the mechanism of GC detection (Goodhill, 1998; Goodhill & Baier, 1998). The first model proposes that the GC responds to the absolute change across its width (ΔC), while the second model suggests that the fractional or relative change ($\Delta C/C$) is detected by the GC. In the latter model, "C" refers to the concentration at the GC midpoint. The fractional change can also be represented as a percentage, known as the steepness. While *in vivo* quantification of such gradient parameters are complicated, *in vitro* studies have focused on determining optimal gradient steepness values to promote axonal turning over biased outgrowth.

Preclinical models of neurotrophin delivery to promote axonal outgrowth and regeneration following SCI

The effect of various neurotrophin delivery approaches (Fig. 4) has been widely investigated using in vivo SCI models. To date, few studies have aimed to achieve a gradient release profile for guiding axonal outgrowth beyond the lesion. Delivery of trophic support to the lesion site has been explored in animal models using direct protein injections, osmotic mini-pumps, viral-vector mediated gene therapy, and biomaterial-based approaches. Selected studies and key outcomes using these approaches have been summarised in Table 2.

Acute localized injections and continuous infusions

Traditional approaches of neurotrophin delivery involved acute localized injections or continuous osmotic pump infusions to administer the neurotrophins as recombinant proteins and bypass the blood–spinal cord barrier (BSCB). These injections have previously been administered via intravenous, intrathecal, and intraparenchymal routes. Schnell et al. investigated the effects of NGF, BDNF, and NT-3 injected locally on the CST (Schnell et al., 1994) observing sprouting of the projections in response to NT-3, but not BDNF. Similarly, Bradbury et al. used a mini-osmotic pump to deliver BDNF and NT-3, noting a growth response of the lesioned dorsal column axons only to NT-3 (Bradbury et al., 1999). The invasive nature of this

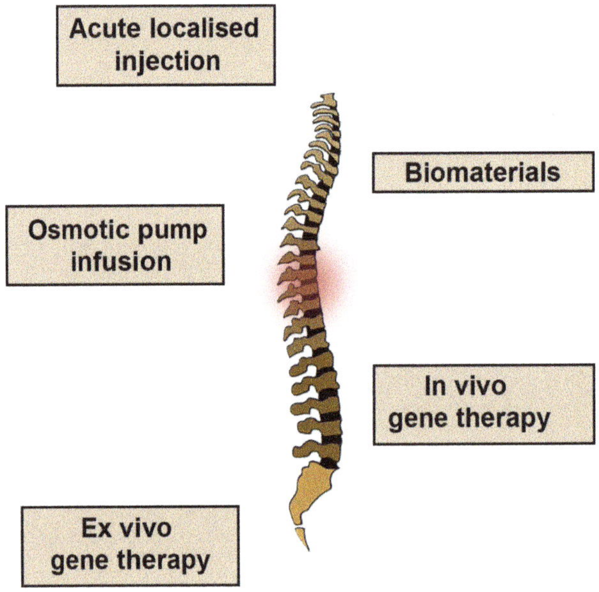

FIG. 4 Current approaches for neurotrophin delivery to the injured spinal cord. Each approach has been investigated in animal models for delivering the neurotrophins to the spinal cord.

TABLE 2 Overview of selected studies investigating neurotrophin delivery in preclinical models.

Neurotrophin	Delivery mode	Animal	Injury	Key results	Reference
NGF BDNF NT-3	Direct protein injection	Rat (P2) Lewis	Bilateral transection of the dorsal halves of the lower thoracic spinal cord	BDNF had no effect; NT-3 increased regenerative sprouting of transected CST axons although the distance of growth was limited; NT-3 combined with antibody against myelin-associated inhibitory protein improved elongation distance	Schnell, Schneider, Kolbeck, Barde, and Schwab (1994)
BDNF NT-3	Mini-osmotic pump	Rat (Adult) Wistar	T6 dorsal column crush	BDNF had no growth-promoting effects on dorsal column axons;NT-3 significantly promoted growth of dorsal column axons after injury with fibers extending into and beyond the epicenter of the injury	Bradbury et al. (1999)
NGF	Mini-osmotic pump	Rat (Adult) Sprague Dawley	T10 contusion injury	NGF delivered directly at injury site. Zone of injury reduced in response to NGF, however, this was not statistically significant	Lee, Green, Dietrich, and Yezierski (1999)
NGF	Ex-vivo gene therapy (Fibroblasts)	Rat (Adult) Fischer 344	T7 dorsal hemi section lesion	By 2 weeks, significant penetration of CGRP+ sensory neurites was observed in response to NGF secretion; only 46.2% of ChAT+ motor neurons penetrated the NGF-secreting graft	Tuszynski et al. (1996)

TABLE 2 Overview of selected studies investigating neurotrophin delivery in preclinical models—cont'd

Neurotrophin	Delivery mode	Animal	Injury	Key results	Reference
BDNF	Ex-vivo gene therapy (bone marrow stromal cells)	Rat (Adult) Fischer 344	Dorsal column bilateral transection	Axonal penetration into BDNF grafts was 2.8 greater than the control at 3 months post-surgery, however, was not greater than the growth observed in BDNF grafts at 1 month; axons in the graft were TH+ coerulospinal, 5-HT+ raphaespinal, CGRP+ sensory and ChAT motor; despite axonal growth into the grafts, no functional recovery was observed	Lu et al. (2005)
BDNF	In vivo gene therapy (AAV vector)	Rat (Adult) Wistar	T8 complete transection	Partial regrowth of transected descending rubrospinal axons and some functional recovery observed in response to BDNF; Highest Basso, Beattie, and Bresnahan (BBB) score in BDNF treated groups was 7, compared to 1 in control	Koda et al. (2004)
NT-3	In vivo gene therapy (LV vector)	Rat (Adult) Fischer 344	C1 dorsal partial laminectomy	LV vector used in combination with mesenchymal stem cell grafts; LV-NT-3 injected into the nucleus gracilis promoted regeneration and synaptic formation	Alto et al. (2009)
NT-3	Biomaterial (Collagen)	Rat (Adult) Wistar	T9 dorsal laminectomy	NT-3 containing collagen fibers promoted the directional growth of CST fibers (22 ± 6%) compared to the control (7 ± 2%). No fibers seen caudal to the lesion. NT-3 treated animals exhibited functional recovery in the gridwalk test, although this was not due to the regeneration of CST axons	Houweling, Lankhorst, Gispen, Bär, and Joosten (1998)
BDNF	Biomaterial (Agarose)	Rat (Adult) Fischer 344	C4 dorsal column aspiration lesion	Hydrogel scaffolds support the linear growth of injured axons, compared to disorganized growth observed in non-scaffold controls; BDNF increases the density of axonal penetration; fibrin-filled scaffolds with BDNF-MSCs had a greater density (1.5 ×) of axons compared to Matrigel	Stokols et al. (2006)

The method of neurotrophin delivery, animal and injury model used and key outcomes have been summarised in the columns. *NGF*, nerve growth factor; *BDNF*, brain-derived neurotrophic factor; *NT-3*, neurotrophin-3; *NT-4/5*, neurotrophin-4/5; *CST*, corticospinal tract.

delivery approach in conjunction with limited protein stability and the need for repetitive injections has limited its progression as a potential intervention. Hence, the development of other delivery strategies has now superseded direct protein injections and infusions.

Ex-vivo gene therapy

Using this approach, cells are first genetically modified in vitro to express a neurotrophin and subsequently transplanted within the injury lesion site. Fibroblasts, Schwann cells, bone marrow stromal cells, and neural stem cells have previously been genetically modified to express neurotrophins in SCI models. Grill et al. reported an increase in corticospinal growth in response to NT-3 secreting fibroblasts (Grill, Murai, Blesch, Gage, & Tuszynski, 1997). Neurite outgrowth of sensory and noradrenergic populations has been observed in response to NGF-secreting fibroblasts grafted into a hemisection lesion (Tuszynski et al., 1996). Schwann cell grafts secreting BDNF improved the regenerative response of reticulospinal and raphaespinal neurons (Menei, Montero-Menei, Whittemore, Bunge, & Bunge, 1998). While this approach can provide a supportive environment for axonal extension into the graft, axons seldom exit the lesion and long-distance regeneration is unlikely to be observed.

In vivo gene therapy

The *in vivo* gene therapy approach involves the use of viral vectors to directly transduce host cell populations in vivo in the region of interest to express a specific gene. The adenovirus (AV), lentivirus (LV), adeno-associated virus (AAV), and herpes simplex virus (HSV) are commonly used to express neurotrophins in SCI models. Zhang et al. observed that by administering AV containing NT-3 cDNA into the ventral horn of the spinal cord, the transgenes could be strongly expressed in glial cells with dorsal root axonal regeneration observed (Zhang, Dijkhuizen, Anderson, Lieberman, & Verhaagen, 1998). Koda et al. used AAV gene transfer of BDNF in a rat transection model, observing a regenerative response in descending rubrospinal axons and recovery of hindlimb function (Koda et al., 2004). Chemotropic guidance of regenerating axons was investigated using LV expressing NT-3 in C1 lesioned rats (Alto et al., 2009). Notably, re-innervation of brainstem targets was reported in response to NT-3 delivery.

Biomaterial-based approaches

Biomaterials are materials engineered to interface with biological tissue. They are attractive candidates for neurotrophin delivery due to their biocompatibility and tuneable properties which permit the fabrication of sustained drug delivery systems. However, ensuring the stability of the neurotrophin within a polymeric drug delivery system during fabrication and application is important for maintaining bioactivity. NT-3 released from fibrin scaffolds 2 weeks following injury was shown to enhance neural fiber sprouting (Johnson, Parker, & Sakiyama-Elbert, 2009). Similarly, Nguyen et al. sustained NT-3 release from collagen scaffolds integrated with poly (ε-caprolactone-*co*-ethyl ethylene phosphate) (PCLEEP) nanofibers (Nguyen et al., 2017), observing regeneration 1 week after injury.

Challenges and considerations for neurotrophin delivery and clinical translation

Development of a clinically translatable neurotrophin delivery system has been a problematic pursuit, demonstrating limited efficacy in human clinical trials. Current preclinical investigations have been conducted using rodent and mice animal models, which possess several anatomical and functional discrepancies compared to nonhuman primates and other large animals (Nardone et al., 2017). Hence, optimal safe and effective dosages determined on cellular and animal models may not accurately translate to humans. Although no animal model can completely recapitulate what is expected in humans, further investigations of motor and sensory outcomes in primates would hold a great benefit in determining the safety and efficacy of the neurotrophins and delivery systems in vivo for subsequent human translation (Hollis 2nd & Tuszynski, 2011).

As the cellular response to an SCI occurs over extended time periods (i.e., months) an ideal platform would enable localized and sustained delivery of therapeutic concentrations to produce the desired biological effects. Overcoming the challenges associated with delivering these factors into the CNS is still a work in progress. The neurotrophins are large molecules, cationic in nature, and poor permeability across the BSCB (Weissmiller & Wu, 2012). This issue is further complicated by a short biological half-life, poor pharmacokinetic properties, enzymatic degradation, and off-target effects. NGF is known to regulate nociceptive pathways, with acute pain and hyperalgesia reported as a common and significant adverse effect in clinical trials and rodent models (Eriksdotter Jönhagen et al., 1998). Similarly, systemic administration of BDNF has been linked to chronic pain (Boudes & Menigoz, 2009) and weight gain (Lin et al., 2008). Systemic routes typically require high doses to be administered to attain therapeutic levels at the target region. Additionally, suboptimal neurotrophin dose and release kinetics can lead to aberrant axonal growth (Madduri & Gander, 2012). Hence, a focus on localized and smart drug delivery systems may be one approach to mitigate such off-target effects and improve permeability across the BSCB. Any delivery system designed to release neurotrophins over an extended time period would be required to maintain the stability of the bioactive payload. As previously mentioned, the spatial availability of these growth factors in the form of a gradient is central to directing axonal growth towards pre-injury targets. In addition, the size and position of the gradient must be flexible to change over time (i.e. temporal availability). The requirement of such precise release patterns necessitates the development of smart drug delivery platforms.

Finally, the pathophysiological progression of an SCI involves several distinct stages (i.e. acute, sub-acute, and chronic) each with distinct features that can affect the efficacy of therapeutic interventions. Administration within the acute stage may have benefits for minimizing axonal retraction, neuronal atrophy, and supporting neuronal survival (Awad, Carmody, & Steinmetz, 2015). However, this may not be the most pragmatic approach in a clinical setting, where there is a time delay between the occurrence of the initial injury, stabilization, and treatment initiation (Lee, Ryu, & Vig, 2019). As such, there would be significant value in investigating the efficacy of these factors delivered at the chronic stage of injury.

Concluding remarks

Exogenous application of the neurotrophins demonstrates promise in promoting axonal regeneration, outgrowth, and survival within the growth-inhibitory injured CNS environment. While the complexity of an SCI will undoubtedly require a combinatorial therapeutic approach to target the various facets of injury, the provision of trophic support could help encourage neuronal survival and re-innervation of natural post-synaptic targets when delivered as a concentration gradient. Additionally, knowledge of receptor expression and distribution on impacted cellular populations can inform which neurotrophins should be applied for delivery. Improvements are still needed in the delivery of these growth factors and further studies conducted using nonhuman primate animal models may be more informative about the human response, increasing the chances of successful translation for human therapy in the future.

Applications to other areas of neuroscience

In this chapter, we have explored the application of neurotrophins on axonal outgrowth in spinal neurons. We describe the structure and function of the neurotrophins and detail their role as chemotactic and regenerative factors for application following spinal cord injury. Further, we outline current progress and challenges associated with neurotrophin delivery to the spinal cord.

As the neurotrophins are essential to neuronal growth and survival, their therapeutic potential is applicable to other neurotraumatic and neurodegenerative conditions where the neuronal function is perturbed (Nagahara & Tuszynski, 2011). Such conditions include Stroke (Ploughman et al., 2009), Traumatic Brain Injury (Marshall et al., 2017), cochlear neuron survival in deafness (Ramekers, Versnel, Grolman, & Klis, 2012), Alzheimer's Disease (Mitra, Behbahani, & Eriksdotter, 2019), Amyotrophic Lateral Sclerosis (Henriques, Pitzer, & Schneider, 2010), Parkinson's Disease (Siegel & Chauhan, 2000), Huntington's disease (Alberch, Pérez-Navarro, & Canals, 2004) and Peripheral Neuropathy (Apfel et al., 1998). As with SCI, one of the factors affecting the progression of neurotrophin-based therapies for these other conditions has been the challenges associated with drug delivery to the CNS. Doses and delivery durations may also vary between SCI and neurodegenerative diseases, with the latter potentially requiring more prolonged periods of neurotrophin delivery based on disease pathology. An understanding of neurotrophin structure, function, and challenges for therapeutic translation can help inform the development of novel drug delivery systems to treat various neuronal diseases.

Mini-dictionary of terms

Absolute change ($\Delta C/x$): Defined as the change in concentration over the average width of a growth cone (typically $x = 10\ \mu m$).

Fractional change ($\Delta C/C$): Defined as the change in concentration (ΔC) over the concentration at the midpoint of the growth cone (C).

Gradient: Occurs when the concentration of a factor is higher in one region compared to another. The gradient is the gradual difference in concentration between the two regions.

Growth Cone: A "fan-shaped" actin structure present at the tip of elongating axons, responsible for navigating axonal growth.

Neurotrophic stimulation: Neurotrophin-induced activation of the Trk receptor and downstream signaling pathways which increase neurite outgrowth and axonal regeneration.

Steepness: The slope of the gradient. It can be described using the absolute change or fractional change model.

Key facts of neurotrophins in spinal cord injury

- Neurotrophins are chemotrophic growth factors that are essential to the regeneration, growth, and survival of neurons after a spinal cord injury.
- Discovery of the first neurotrophin, nerve growth factor, was a pivotal moment in neurobiology—receiving the 1986 Nobel Prize in Physiology or Medicine.

- Each neurotrophin targets a different population of spinal neurons based on Trk receptor expression.
- Ramón y Cajal proposed his "neurotropic hypothesis" in 1892, describing how the axon growth cone can be navigated towards a certain target.
- Neurotrophic gradients delivered to the injured spinal cord could help to regenerate axons grow beyond the site of the lesion and re-connect with host targets.

Summary points

- To re-establish original functional connectivity regenerating axons should extend beyond the injury lesion and re-connect with pre-injury targets.
- The neurotrophins are chemoattractive growth factors with a demonstrated role in axonal outgrowth and survival following spinal cord injury.
- Delivery of the neurotrophins in the form of a concentration gradient could be used to direct axonal outgrowth.
- *In vivo* neurotrophin delivery has been explored using intrathecal injections, osmotic pumps, *ex vivo* gene therapy, and biomaterial-based approaches. However, limited studies have achieved spatiotemporal gradient delivery.
- Sustained, localized delivery of the neurotrophins must be achieved to mitigate off-target adverse effects.

References

Alberch, J., Pérez-Navarro, E., & Canals, J. M. (2004). Neurotrophic factors in Huntington's disease. *Progress in Brain Research*, *146*, 195–229. https://doi.org/10.1016/s0079-6123(03)46014-7.

Al-Qudah, M. A., & Al-Dwairi, A. (2016). Mechanisms and regulation of neurotrophin synthesis and secretion. *Neurosciences (Riyadh, Saudi Arabia)*, *21*(4), 306–313. https://doi.org/10.17712/nsj.2016.4.20160080.

Alto, L. T., Havton, L. A., Conner, J. M., Hollis Ii, E. R., Blesch, A., & Tuszynski, M. H. (2009). Chemotropic guidance facilitates axonal regeneration and synapse formation after spinal cord injury. *Nature Neuroscience*, *12*(9), 1106–1113. https://doi.org/10.1038/nn.2365.

Apfel, S. C., Kessler, J. A., Adornato, B. T., Litchy, W. J., Sanders, C., & Rask, C. A. (1998). Recombinant human nerve growth factor in the treatment of diabetic polyneuropathy. NGF Study Group. *Neurology*, *51*(3), 695–702. https://doi.org/10.1212/wnl.51.3.695.

Awad, B. I., Carmody, M. A., & Steinmetz, M. P. (2015). Potential role of growth factors in the management of spinal cord injury. *World Neurosurgery*, *83*(1), 120–131. https://doi.org/10.1016/j.wneu.2013.01.042.

Berkemeier, L. R., Winslow, J. W., Kaplan, D. R., Nikolics, K., Goeddel, D. V., & Rosenthal, A. (1991). Neurotrophin-5: A novel neurotrophic factor that activates trk and trkB. *Neuron*, *7*(5), 857–866. https://doi.org/10.1016/0896-6273(91)90287-A.

Bibel, M., & Barde, Y. A. (2000). Neurotrophins: Key regulators of cell fate and cell shape in the vertebrate nervous system. *Genes & Development*, *14*(23), 2919–2937. https://doi.org/10.1101/gad.841400.

Blesch, A., Yang, H., Weidner, N., Hoang, A., & Otero, D. (2004). Axonal responses to cellularly delivered NT-4/5 after spinal cord injury. *Molecular and Cellular Neuroscience*, *27*(2), 190–201. https://doi.org/10.1016/j.mcn.2004.06.007.

Blits, B., Dijkhuizen, P. A., Boer, G. J., & Verhaagen, J. (2000). Intercostal nerve implants transduced with an adenoviral vector encoding neurotrophin-3 promote regrowth of injured rat corticospinal tract fibers and improve Hindlimb function. *Experimental Neurology*, *164*(1), 25–37. https://doi.org/10.1006/exnr.2000.7413.

Boudes, M., & Menigoz, A. (2009). Non-neuronal BDNF, a key player in development of central sensitization and neuropathic pain. *The Journal of Physiology*, *587*(Pt 10), 2111–2112. https://doi.org/10.1113/jphysiol.2009.172130.

Boyd, J. G., & Gordon, T. (2003). Neurotrophic factors and their receptors in axonal regeneration and functional recovery after peripheral nerve injury. *Molecular Neurobiology*, *27*(3), 277–324. https://doi.org/10.1385/mn:27:3:277.

Bradbury, E. J., Khemani, S., Von, R., King, Priestley, J. V., & McMahon, S. B. (1999). NT-3 promotes growth of lesioned adult rat sensory axons ascending in the dorsal columns of the spinal cord. *European Journal of Neuroscience*, *11*(11), 3873–3883. https://doi.org/10.1046/j.1460-9568.1999.00809.x.

Brock, J. H., Rosenzweig, E. S., Blesch, A., Moseanko, R., Havton, L. A., Edgerton, V. R., et al. (2010). Local and remote growth factor effects after primate spinal cord injury. *Journal of Neuroscience*, *30*(29), 9728–9737. https://doi.org/10.1523/JNEUROSCI.1924-10.2010.

Chao, M. V., & Bothwell, M. (2002). Neurotrophins: To cleave or not to cleave. *Neuron*, *33*(1), 9–12. https://doi.org/10.1016/S0896-6273(01)00573-6.

Courtine, G., Song, B., Roy, R. R., Zhong, H., Herrmann, J. E., Ao, Y., et al. (2008). Recovery of supraspinal control of stepping via indirect propriospinal relay connections after spinal cord injury. *Nature Medicine*, *14*(1), 69–74. https://doi.org/10.1038/nm1682.

Dechant, G., & Barde, Y.-A. (2002). The neurotrophin receptor p75NTR: Novel functions and implications for diseases of the nervous system. *Nature Neuroscience*, *5*(11), 1131–1136. https://doi.org/10.1038/nn1102-1131.

Dravid, A., Parittotokkaporn, S., Aqrawe, Z., O'Carroll, S. J., & Svirskis, D. (2020). Determining neurotrophin gradients in vitro to direct axonal outgrowth following spinal cord injury. *ACS Chemical Neuroscience*, *11*(2), 121–132. https://doi.org/10.1021/acschemneuro.9b00565.

Eriksdotter Jönhagen, M., Nordberg, A., Amberla, K., Bäckman, L., Ebendal, T., Meyerson, B., et al. (1998). Intracerebroventricular infusion of nerve growth factor in three patients with Alzheimer's disease. *Dementia and Geriatric Cognitive Disorders*, *9*(5), 246–257. https://doi.org/10.1159/000017069.

Fawcett, J. W. (2020). The struggle to make CNS axons regenerate: Why has it been so difficult? *Neurochemical Research, 45*(1), 144–158. https://doi.org/10.1007/s11064-019-02844-y.

Fitch, M. T., & Silver, J. (2008). CNS injury, glial scars, and inflammation: Inhibitory extracellular matrices and regeneration failure. *Experimental Neurology, 209*(2), 294–301. https://doi.org/10.1016/j.expneurol.2007.05.014.

Giger, R. J., Hollis, E. R., 2nd, & Tuszynski, M. H. (2010). Guidance molecules in axon regeneration. *Cold Spring Harbor Perspectives in Biology, 2*(7), a001867. https://doi.org/10.1101/cshperspect.a001867.

Goodhill, G. J. (1998). Mathematical guidance for axons. *Trends in Neurosciences, 21*(6), 226–231. https://doi.org/10.1016/s0166-2236(97)01203-4.

Goodhill, G. J., & Baier, H. (1998). Axon guidance: Stretching gradients to the limit. *Neural Computation, 10*(3), 521–527. https://doi.org/10.1162/089976698300017638.

Grill, R. J., Blesch, A., & Tuszynski, M. H. (1997). Robust growth of chronically injured spinal cord axons induced by grafts of genetically modified NGF-secreting cells. *Experimental Neurology, 148*(2), 444–452. https://doi.org/10.1006/exnr.1997.6704.

Grill, R., Murai, K., Blesch, A., Gage, F. H., & Tuszynski, M. H. (1997). Cellular delivery of neurotrophin-3 promotes corticospinal axonal growth and partial functional recovery after spinal cord injury. *Journal of Neuroscience, 17*(14), 5560–5572. https://doi.org/10.1523/JNEUROSCI.17-14-05560.1997.

Gundersen, R., & Barrett, J. (1979). Neuronal chemotaxis: Chick dorsal-root axons turn toward high concentrations of nerve growth factor. *Science (New York, N.Y.), 206*(4422), 1079–1080. https://doi.org/10.1126/science.493992.

Hauburger, A., Kliemannel, M., Madsen, P., Rudolph, R., & Schwarz, E. (2007). Oxidative folding of nerve growth factor can be mediated by the propeptide of neurotrophin-3. *FEBS Letters, 581*(22), 4159–4164. https://doi.org/10.1016/j.febslet.2007.07.063.

Henriques, A., Pitzer, C., & Schneider, A. (2010). Neurotrophic growth factors for the treatment of amyotrophic lateral sclerosis: Where do we stand? *Frontiers in Neuroscience, 4*, 32. https://doi.org/10.3389/fnins.2010.00032.

Hollis, E. R., 2nd, & Tuszynski, M. H. (2011). Neurotrophins: Potential therapeutic tools for the treatment of spinal cord injury. *Neurotherapeutics : The Journal of the American Society for Experimental NeuroTherapeutics, 8*(4), 694–703. https://doi.org/10.1007/s13311-011-0074-9.

Houweling, D. A., Lankhorst, A. J., Gispen, W. H., Bär, P. R., & Joosten, E. A. (1998). Collagen containing neurotrophin-3 (NT-3) attracts regrowing injured corticospinal axons in the adult rat spinal cord and promotes partial functional recovery. *Experimental Neurology, 153*(1), 49–59. https://doi.org/10.1006/exnr.1998.6867.

Johnson, P. J., Parker, S. R., & Sakiyama-Elbert, S. E. (2009). Controlled release of neurotrophin-3 from fibrin-based tissue engineering scaffolds enhances neural fiber sprouting following subacute spinal cord injury. *Biotechnology and Bioengineering, 104*(6), 1207–1214. https://doi.org/10.1002/bit.22476.

Jones, L. L., Oudega, M., Bunge, M. B., & Tuszynski, M. H. (2001). Neurotrophic factors, cellular bridges and gene therapy for spinal cord injury. *The Journal of Physiology, 533*(Pt 1), 83–89. https://doi.org/10.1111/j.1469-7793.2001.0083b.x.

Keefe, K. M., Sheikh, I. S., & Smith, G. M. (2017). Targeting neurotrophins to specific populations of neurons: NGF, BDNF, and NT-3 and their relevance for treatment of spinal cord injury. *International Journal of Molecular Sciences, 18*(3). https://doi.org/10.3390/ijms18030548.

King, V. R., Bradbury, E. J., McMahon, S. B., & Priestley, J. V. (2000). Changes in truncated trkB and p75 receptor expression in the rat spinal cord following spinal cord hemisection and spinal cord hemisection plus neurotrophin treatment. *Experimental Neurology, 165*(2), 327–341. https://doi.org/10.1006/exnr.2000.7480.

Kishino, A., Ishige, Y., Tatsuno, T., Nakayama, C., & Noguchi, H. (1997). BDNF prevents and reverses adult rat motor neuron degeneration and induces axonal outgrowth. *Experimental Neurology, 144*(2), 273–286. https://doi.org/10.1006/exnr.1996.6367.

Kobayashi, N. R., Fan, D. P., Giehl, K. M., Bedard, A. M., Wiegand, S. J., & Tetzlaff, W. (1997). BDNF and NT-4/5 prevent atrophy of rat rubrospinal neurons after cervical axotomy, stimulate GAP-43 and Talpha1-tubulin mRNA expression, and promote axonal regeneration. *Journal of Neuroscience, 17*(24), 9583–9595. https://doi.org/10.1523/JNEUROSCI.17-24-09583.1997.

Koda, M., Hashimoto, M., Murakami, M., Yoshinaga, K., Ikeda, O., Yamazaki, M., et al. (2004). Adenovirus vector-mediated in vivo gene transfer of brain-derived neurotrophic factor (BDNF) promotes rubrospinal axonal regeneration and functional recovery after complete transection of the adult rat spinal cord. *Journal of Neurotrauma, 21*(3), 329–337. https://doi.org/10.1089/089771504322972112.

Kowiański, P., Lietzau, G., Czuba, E., Waśkow, M., Steliga, A., & Moryś, J. (2018). BDNF: A key factor with multipotent impact on brain signaling and synaptic plasticity. *Cellular and Molecular Neurobiology, 38*(3), 579–593. https://doi.org/10.1007/s10571-017-0510-4.

Lee, T. T., Green, B. A., Dietrich, W. D., & Yezierski, R. P. (1999). Neuroprotective effects of basic fibroblast growth factor following spinal cord contusion injury in the rat. *Journal of Neurotrauma, 16*(5), 347–356. https://doi.org/10.1089/neu.1999.16.347.

Lee, H. J., Ryu, J. S., & Vig, P. J. (2019). Current strategies for therapeutic drug delivery after traumatic CNS injury. *Therapeutic Delivery, 10*(4), 251–263. https://doi.org/10.4155/tde-2019-0006.

Levi-Montalcini, R., & Hamburger, V. (1951). Selective growth stimulating effects of mouse sarcoma on the sensory and sympathetic nervous system of the chick embryo. *Journal of Experimental Zoology, 116*(2), 321–361. https://doi.org/10.1002/jez.1401160206.

Liebl, D. J., Huang, W., Young, W., & Parada, L. F. (2001). Regulation of Trk receptors following contusion of the rat spinal cord. *Experimental Neurology, 167*(1), 15–26. https://doi.org/10.1006/exnr.2000.7548.

Lin, J. C., Tsao, D., Barras, P., Bastarrachea, R. A., Boyd, B., Chou, J., et al. (2008). Appetite enhancement and weight gain by peripheral administration of TrkB agonists in non-human primates. *PLoS One, 3*(4), e1900. https://doi.org/10.1371/journal.pone.0001900.

Lu, P., Jones, L. L., & Tuszynski, M. H. (2005). BDNF-expressing marrow stromal cells support extensive axonal growth at sites of spinal cord injury. *Experimental Neurology, 191*(2), 344–360. https://doi.org/10.1016/j.expneurol.2004.09.018.

Lu, P., Yang, H., Jones, L. L., Filbin, M. T., & Tuszynski, M. H. (2004). Combinatorial therapy with neurotrophins and cAMP promotes axonal regeneration beyond sites of spinal cord injury. *Journal of Neuroscience, 24*(28), 6402–6409. https://doi.org/10.1523/JNEUROSCI.1492-04.2004.

Madduri, S., & Gander, B. (2012). Growth factor delivery systems and repair strategies for damaged peripheral nerves. *Journal of Controlled Release*, *161*(2), 274–282. https://doi.org/10.1016/j.jconrel.2011.11.036.

Marshall, J., Szmydynger-Chodobska, J., Rioult-Pedotti, M. S., Lau, K., Chin, A. T., Kotla, S. K. R., et al. (2017). TrkB-enhancer facilitates functional recovery after traumatic brain injury. *Scientific Reports*, *7*(1), 10995. https://doi.org/10.1038/s41598-017-11316-8.

McCall, J., Weidner, N., & Blesch, A. (2012). Neurotrophic factors in combinatorial approaches for spinal cord regeneration. *Cell and Tissue Research*, *349*(1), 27–37. https://doi.org/10.1007/s00441-012-1388-6.

Meeker, R. B., & Williams, K. S. (2015). The p75 neurotrophin receptor: At the crossroad of neural repair and death. *Neural Regeneration Research*, *10*(5), 721–725. https://doi.org/10.4103/1673-5374.156967.

Menei, P., Montero-Menei, C., Whittemore, S. R., Bunge, R. P., & Bunge, M. B. (1998). Schwann cells genetically modified to secrete human BDNF promote enhanced axonal regrowth across transected adult rat spinal cord. *European Journal of Neuroscience*, *10*(2), 607–621. https://doi.org/10.1046/j.1460-9568.1998.00071.x.

Ming, G.-L., Lohof, A. M., & Zheng, J. Q. (1997). Acute morphogenic and chemotropic effects of neurotrophins on cultured embryonic Xenopus spinal neurons. *Journal of Neuroscience*, *17*(20), 7860–7871. https://doi.org/10.1523/JNEUROSCI.17-20-07860.1997.

Ming, G.-L., Song, H.-J., Berninger, B., Inagaki, N., Tessier-Lavigne, M., & Poo, M.-M. (1999). Phospholipase C-γ and phosphoinositide 3-kinase mediate cytoplasmic signaling in nerve growth cone guidance. *Neuron*, *23*(1), 139–148. https://doi.org/10.1016/S0896-6273(00)80760-6.

Mitra, S., Behbahani, H., & Eriksdotter, M. (2019). Innovative therapy for Alzheimer's disease-with focus on biodelivery of NGF. *Frontiers in Neuroscience*, *13*(38). https://doi.org/10.3389/fnins.2019.00038.

Nagahara, A. H., & Tuszynski, M. H. (2011). Potential therapeutic uses of BDNF in neurological and psychiatric disorders. *Nature Reviews Drug Discovery*, *10*(3), 209–219. https://doi.org/10.1038/nrd3366.

Nardone, R., Florea, C., Höller, Y., Brigo, F., Versace, V., Lochner, P., et al. (2017). Rodent, large animal and non-human primate models of spinal cord injury. *Zoology (Jena, Germany)*, *123*, 101–114. https://doi.org/10.1016/j.zool.2017.06.004.

Nguyen, L. H., Gao, M., Lin, J., Wu, W., Wang, J., & Chew, S. Y. (2017). Three-dimensional aligned nanofibers-hydrogel scaffold for controlled non-viral drug/gene delivery to direct axon regeneration in spinal cord injury treatment. *Scientific Reports*, *7*(1), 42212. https://doi.org/10.1038/srep42212.

Oudega, M., & Hagg, T. (1996). Nerve growth factor promotes regeneration of sensory axons into adult rat spinal cord. *Experimental Neurology*, *140*(2), 218–229. https://doi.org/10.1006/exnr.1996.0131.

Oudega, M., & Hagg, T. (1999). Neurotrophins promote regeneration of sensory axons in the adult rat spinal cord. *Brain Research*, *818*(2), 431–438. https://doi.org/10.1016/S0006-8993(98)01314-6.

Ploughman, M., Windle, V., MacLellan, C. L., White, N., Doré, J. J., & Corbett, D. (2009). Brain-derived neurotrophic factor contributes to recovery of skilled reaching after focal ischemia in rats. *Stroke*, *40*(4), 1490–1495. https://doi.org/10.1161/STROKEAHA.108.531806.

Ramekers, D., Versnel, H., Grolman, W., & Klis, S. F. L. (2012). Neurotrophins and their role in the cochlea. *Hearing Research*, *288*(1), 19–33. https://doi.org/10.1016/j.heares.2012.03.002.

Rodriguez-Tébar, A., Dechant, G., & Barde, Y. A. (1990). Binding of brain-derived neurotrophic factor to the nerve growth factor receptor. *Neuron*, *4*(4), 487–492. https://doi.org/10.1016/0896-6273(90)90107-q.

Rodríguez-Tébar, A., Dechant, G., Götz, R., & Barde, Y. A. (1992). Binding of neurotrophin-3 to its neuronal receptors and interactions with nerve growth factor and brain-derived neurotrophic factor. *EMBO Journal*, *11*(3), 917–922.

Sakuragi, S., Tominaga-Yoshino, K., & Ogura, A. (2013). Involvement of TrkB- and p75(NTR)-signaling pathways in two contrasting forms of long-lasting synaptic plasticity. *Scientific Reports*, *3*, 3185. https://doi.org/10.1038/srep03185.

Schmalbruch, H., & Rosenthal, A. (1995). Neurotrophin- 4/5 postpones the death of injured spinal motoneurones in newborn rats. *Brain Research*, *700*(1), 254–260. https://doi.org/10.1016/0006-8993(95)00982-V.

Schnell, L., Schneider, R., Kolbeck, R., Barde, Y.-A., & Schwab, M. E. (1994). Neurotrophin-3 enhances sprouting of corticospinal tract during development and after adult spinal cord lesion. *Nature*, *367*(6459), 170–173. https://doi.org/10.1038/367170a0.

Siegel, G. J., & Chauhan, N. B. (2000). Neurotrophic factors in Alzheimer's and Parkinson's disease brain. *Brain Research Reviews*, *33*(2), 199–227. https://doi.org/10.1016/S0165-0173(00)00030-8.

Stokols, S., Sakamoto, J., Breckon, C., Holt, T., Weiss, J., & Tuszynski, M. H. (2006). Templated agarose scaffolds support linear axonal regeneration. *Tissue Engineering*, *12*(10), 2777–2787. https://doi.org/10.1089/ten.2006.12.2777.

Stucky, C. L., Shin, J. B., & Lewin, G. R. (2002). Neurotrophin-4: A survival factor for adult sensory neurons. *Current Biology*, *12*(16), 1401–1404. https://doi.org/10.1016/s0960-9822(02)01072-2.

Sutter, A., Riopelle, R. J., Harris-Warrick, R. M., & Shooter, E. M. (1979). Nerve growth factor receptors. Characterization of two distinct classes of binding sites on chick embryo sensory ganglia cells. *Journal of Biological Chemistry*, *254*(13), 5972–5982.

Taylor, L., Jones, L., Tuszynski, M. H., & Blesch, A. (2006). Neurotrophin-3 gradients established by lentiviral gene delivery promote short-distance axonal bridging beyond cellular grafts in the injured spinal cord. *Journal of Neuroscience*, *26*(38), 9713–9721. https://doi.org/10.1523/jneurosci.0734-06.2006.

Tuszynski, M. H., Gabriel, K., Gage, F. H., Suhr, S., Meyer, S., & Rosetti, A. (1996). Nerve growth factor delivery by gene transfer induces differential outgrowth of sensory, motor, and noradrenergic neurites after adult spinal cord injury. *Experimental Neurology*, *137*(1), 157–173. https://doi.org/10.1006/exnr.1996.0016.

Tuszynski, M. H., Grill, R., Jones, L. L., Brant, A., Blesch, A., Löw, K., et al. (2003). NT-3 gene delivery elicits growth of chronically injured corticospinal axons and modestly improves functional deficits after chronic scar resection. *Experimental Neurology*, *181*(1), 47–56. https://doi.org/10.1016/S0014-4886(02)00055-9.

Tuszynski, M. H., Peterson, D. A., Ray, J., Baird, A., Nakahara, Y., & Gages, F. H. (1994). Fibroblasts genetically modified to produce nerve growth factor induce robust neuritic ingrowth after grafting to the spinal cord. *Experimental Neurology*, *126*(1), 1–14. https://doi.org/10.1006/exnr.1994.1037.

Weishaupt, N., Blesch, A., & Fouad, K. (2012). BDNF: The career of a multifaceted neurotrophin in spinal cord injury. *Experimental Neurology*, *238*(2), 254–264. https://doi.org/10.1016/j.expneurol.2012.09.001.

Weissmiller, A. M., & Wu, C. (2012). Current advances in using neurotrophic factors to treat neurodegenerative disorders. *Translational Neurodegeneration*, *1*(1), 14. https://doi.org/10.1186/2047-9158-1-14.

Widenfalk, J., Lundströmer, K., Jubran, M., Brené, S., & Olson, L. (2001). Neurotrophic factors and receptors in the immature and adult spinal cord after mechanical injury or kainic acid. *Journal of Neuroscience*, *21*(10), 3457–3475. https://doi.org/10.1523/JNEUROSCI.21-10-03457.2001.

Zhang, Y., Dijkhuizen, P. A., Anderson, P. N., Lieberman, A. R., & Verhaagen, J. (1998). NT-3 delivered by an adenoviral vector induces injured dorsal root axons to regenerate into the spinal cord of adult rats. *Journal of Neuroscience Research*, *54*(4), 554–562.

Chapter 19

The neuroscience of transient receptor potential vanilloid type 4 (TRPV4) and spinal cord injury

Hemant Kumar[a] and Inbo Han[b]

[a]*Department of Pharmacology and Toxicology, National Institute of Pharmaceutical Education and Research (NIPER)-Ahmedabad, Gandhinagar, Gujarat, India,* [b]*Department of Neurosurgery, CHA University School of Medicine, CHA Bundang Medical Center, Seongnam-si, Gyeonggi-do, Republic of Korea*

List of abbreviations

BSCB	blood–spinal cord barrier
CNS	central nervous system
CVOs	circumventricular organs
DRGs	dorsal root ganglia
ECs	endothelial cells
ECM	extracellular matrix
IL-1β	interleukin-1 beta
IP3R	inositol triphosphate receptors
mPOA	medial preoptic area
NVC	neurovascular coupling
PNS	peripheral nervous system
PVN	paraventricular nucleus
RVD	regulatory volume decrease.
SCI	spinal cord injury
TG	trigeminal ganglia
TJ	tight Junctions
TNF-α	tumor necrosis factor alpha
TRPV4	transient receptor potential vanilloid type 4
OVLT	organum vasculosum lamina terminalis
ZO-1	zona occludens

Introduction

Transient receptor potential vanilloid type 4 (TRPV4)

Transient receptor potential (TRP) cation channels form a superfamily of inimitable sensory proteins that play a substantial role in varied homeostatic functions. TRP channels are well established for their involvement in sensory transduction (Damann, Voets, & Nilius, 2008; Montell, Birnbaumer, & Flockerzi, 2002; Talavera, Nilius, & Voets, 2008), and are broadly divided into six subfamilies in mammals: TRPA (ankyrin), TRPC (canonical), TRPML (mucolipin), TRPM (melastatin), TRPP (polycystin), and TRPV (vanilloid) (Damann et al., 2008). TRPV4 channels belong to the TRPV class and are mainly expressed in mammalian tissues, including the PNS and the CNS (Montell et al., 2002). TRPV4 ion channels have six transmembrane alpha-helices (S1-S6) stretching across the lipid bilayer. Both NH_2 and COOH terminals are localized in the cytoplasm. The NH_2 terminus is a proline-rich region and plays a crucial role in mediating the effects of pacsin-3, a synaptic vesicular membrane trafficking protein that inhibits the basal expression of TRPV4. The COOH terminus region of TRPV4 maintains channel protein folding, maturation, and trafficking (Fig. 1). They are

Cellular, Molecular, Physiological, and Behavioral Aspects of Spinal Cord Injury. https://doi.org/10.1016/B978-0-12-822427-4.00019-8

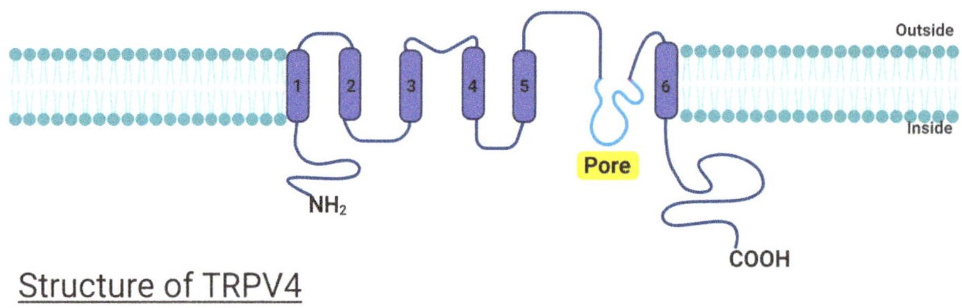

FIG. 1 TRPV4 Ion channels have six transmembrane alpha-helices (S1–S6) spread out across the lipid bilayer. Interestingly, both NH_2 and COOH terminals are contained in the cytoplasm. The proline-rich region (NH_2 terminus) plays a fundamental role in mediating the effects of pacsin-3, a synaptic vesicular membrane trafficking protein that inhibits the basal expression of TRPV4. COOH terminus region of TRPV4 maintains channel protein folding, maturation, and trafficking. The COOH region interacts with MAP-7. The pore that allows the cations is located between S5 and S6 domains.

permeable to many cations, with relatively high permeability to Ca^{2+} (5–10 times higher than the permeability to Na^+; $P_{Ca}/P_{Na} = 6$) (Kumar, Lee, Kim, Zeng, & Han, 2018; Plant & Strotmann, 2007; Watanabe et al., 2003). TRPV4 is nonspecific and is primarily recognized as a calcium-permeable non-selective cation channel, but its non-selective nature allows it to play a role as a polymodal cation channel permeable to other ions, notably Na^+. A comprehensive range of stimuli activates TRPV4, and it serves as an osmosensor and thermosensor (activated by innocuous warm temperatures in the range 27–35°C). It can also be activated directly or sensitized indirectly via intracellular signaling pathways (Kumar, Lee, et al., 2018; Liedtke et al., 2000; Plant & Strotmann, 2007; Strotmann, Harteneck, Nunnenmacher, Schultz, & Plant, 2000; Vennekens, Owsianik, & Nilius, 2008; Vriens et al., 2004; Wegierski, Lewandrowski, Muller, Sickmann, & Walz, 2009).

The expression of TRPV4 in the DRG cell bodies and the sensory nerve terminals projecting to laminae I and II of the spinal cord in the peripheral nociceptive neurons with cell bodies located in the TGs and DRGs allowing it to act as a sensor for noxious mechanical and osmotic stimuli (Alessandri-Haber et al., 2003). Interestingly, TRPV4 is transported distally toward peripheral nerve endings and act as an osmosensor, mechanosensor, and thermosensor. TRPV4-mediated Ca^{2+} oscillations in the endfeet arbitrate NVC (the connection between neurons and their vascular supply), suggesting that TRPV4 is crucial in controlling the brain's vascular tone and local perfusion (Dunn, Hill-Eubanks, Liedtke, & Nelson, 2013). Synaptic transmission, a process by which a neuron communicates to target cells across the synapse. Interestingly, astrocytes play an essential role in the regulation of synaptic transmission. These cell–cell communications in astroglia are finely tuned by intracellular Ca^{2+} responses elicited by extracellular signaling molecules and via activation of ion channels including TRPV4. TRPV4 is majorly found to be expressed in astrocytic membranes at the interface between the brain and extracerebral fluid spaces (Benfenati et al., 2007). It is obvious to predict that TRPV4 is a vital protein in controlling the astroglial osmosensation and plays a central role in brain-volume homeostasis (Benfenati et al., 2007). Not all astrocytes in the CNS express TRPV4; one of the studies reported that roughly 30% of astrocytes express TRPV4, signifying astrocytic subtypes' subsistence (Shibasaki, Ikenaka, Tamalu, Tominaga, & Ishizaki, 2014). Interestingly, astrocytes with TRPV4 expression ($TRPV4^+$ astrocytes) are exclusively accountable for instigating the excitatory gliotransmitter release that further augments the synaptic transmission, forms a core assembly of excitatory glia in the brain to augment the neuronal excitation efficiently (Shibasaki et al., 2014). Therefore, activation of TRPV4 plays an imperative role in the interaction between astrocytes and thereby improves neural activity (Shibasaki & Tominaga, 2007).

Spinal cord injury

Spinal cord injury (SCI) pathology is multifaceted, involving primary and secondary injury. The primary trauma causes tissue disruption, leading to secondary injury progression. The SCI pathology's significant events involve infiltration of neutrophils, the release of cytokines and chemokines leading to inflammation, followed by the death of ECs, neurons, and glial activation, glial, fibrotic scar formation, ECM remodeling, and cavitation (Kumar et al., 2020). SCI pathology also involves the loss of microcirculation, loss of structural organization, disruption of BSCB, endothelial cell, and vascular remodeling (Kumar, Ropper, Lee, & Han, 2017; Tator & Koyanagi, 1997). Vascular damage following SCI augments secondary damage, and vascular protection or maintaining vascular integrity through various ways has favorable effects (Kumar et al., 2017). Endothelial cells participate in all facets of vascular homeostasis; they play a wide variety of crucial roles in controlling vascular functions, including thrombosis, inflammation, or vascular wall remodeling. The death of endothelial cells disengages the vascular network, and ischemia results in apoptosis and cell death of CNS cells because they cannot survive without an adequate blood supply.

Inflammation is entrenched in SCI pathology and directly contributes to fibrosis scarring in part via altering numerous pathways. The continued presence of activated glial cells points toward a pro-inflammatory milieu concurrent with augmented neurotoxicity and weakens wound healing. The pathology of SCI is multifarious and diverse mechanisms might contribute to axonal conduction defects. Ion channels are an essential target that directly or indirectly contributes to sensory and motor functions. The role of several ion channels in regulating the pathology of SCI is established, and targeting TRP(s) seems fascinating to attenuate the SCI pathology (Gerzanich et al., 2009; Kumar et al., 2020; Shimizu et al., 2018).

TRPV4 and spinal cord injury

SCI is primarily caused by trauma, with the maximum incidence due to accidents or falls. There is a mechanical impact following the injury. It tends to change/excite the mechanosensor that can sense mechanical pressure and eventually increase the expression of ion channels and/or proteins, which can finally aggravate pathology. TRPV4 is an established mechanosensor, osmosensor, and thermosensor and plays a diverse role (Table 1). TRPV4 expression was increased during the inflammatory/acute phase and was correlated with injury severity at the epicenter after SCI (Fig. 2). After SCI, TRPV4 activation disrupted the EC organization, and the ECs were protected from SCI-induced damage in the TRPV4 absence (Kumar et al., 2020).

Similarly, TRPV4 deletion substantially impacted SCI-induced inflammatory cytokines and chemokine, BSCB integrity, the extent of scarring, and spontaneous locomotor recovery. Markedly, chemical inhibition with specific antagonists or deficiency of TRPV4 is propitious for endothelial factors, such as neurotrophins, angiopoietin, and BDNF, after SCI (Kumar et al., 2020). Notably, ECs detachment, which continues throughout the acute SCI phase (Koyanagi, Tator, & Lea, 1993), contributes to neuronal and glial cells' death (Casella, Marcillo, Bunge, & Wood, 2002). The SCI-induced expression of TGF-β/CD-68 was mitigated in TRPV4 KO mice signifying that increased TRPV4 levels contribute to the pathology and lessened SCI recovery (Kumar et al., 2020).

Interestingly, the injection of GSK, a TRPV4 agonist, brings about EC damage and scarring. Accordingly, the TRPV4 $^{-/-}$ mice showed reduced reactive gliosis, decrease hyperalgesia toward heat, and quicker functional recovery after SCI than WT mice (Kumar et al., 2020). Activation of TRPV4 showed a differential effect in peripheral nerves, and CNS facilitated neurotrophic factor caused neuritogenesis in peripheral nerves (Jang et al., 2012), and induced neuronal apoptosis in the CNS (Jie et al., 2015). After SCI, axon regeneration necessitates synchronized neuronal homeostasis changes besides local cytoskeletal modifications necessary to promote growth cone formation. The neuron allows for proper mitochondrial trafficking, activation of transcription factors regulating pro-regenerative programs, and epigenetic modifications. Reduced scarring is seen in TRPV4 KO mice compared with that in WT mice after enhanced neuronal markers expression accompanied SCI. However, these increases were attenuated in TRPV4 KO mice, portentous that the enhanced TRPV4 expression in WT animals augmented the scarring (glial and fibrotic) process (Kumar et al., 2020).

After SCI, it leads to ischemia, which is well established to trigger several known and unknown pathways. Neutrophils arrive at the site of injury, and they start the release of chemokines and cytokines, which finally leads to disruption of ECs, TJs, and eventually affects BSCB integrity. TRPV4 is known to be expressed at astrocyte endfeet TRPV4 expression and is increased during the acute/inflammatory phase, which augments SCI's pathology by either contributing to the release of Ca^{2+} augmenting the pro-inflammatory cytokines or chemokines finally damage the ECs integrity (Fig. 3).

TRPV4, a non-selective cation channel expressed by astrocytes and neurons, is a critical factor of endothelial disruption and pathology after SCI. However, several mechanisms might be responsible for which these effects are observed in SCI's experimental model. (i) The upsurge in TRPV4 expression is induced by mechanical stimulation due to cell swelling, as TRPV4 also serves as a volume-sensitive mechanosensor (Nilius, Prenen, Wissenbach, Bödding, & Droogmans, 2001). TRPV4 is involved in mechanotransduction, and it functions as a mechanosensitive channel, and the traumatic injury eventually will cause mechanical stretch, thereby increasing the TRPV4 expression after SCI.

TABLE 1 Role of TRPV4 in different cells.

Astrocytes	Neurons	Microglia	Endothelial cells
• Control cell volume • Amplify NVC • Regulate Ca^{2+} release	• Regulates excitability • Apoptosis	• Increase M1 phenotypes • Promotes inflammation	• Activation damages ECs • Regulates signals controlling barrier property

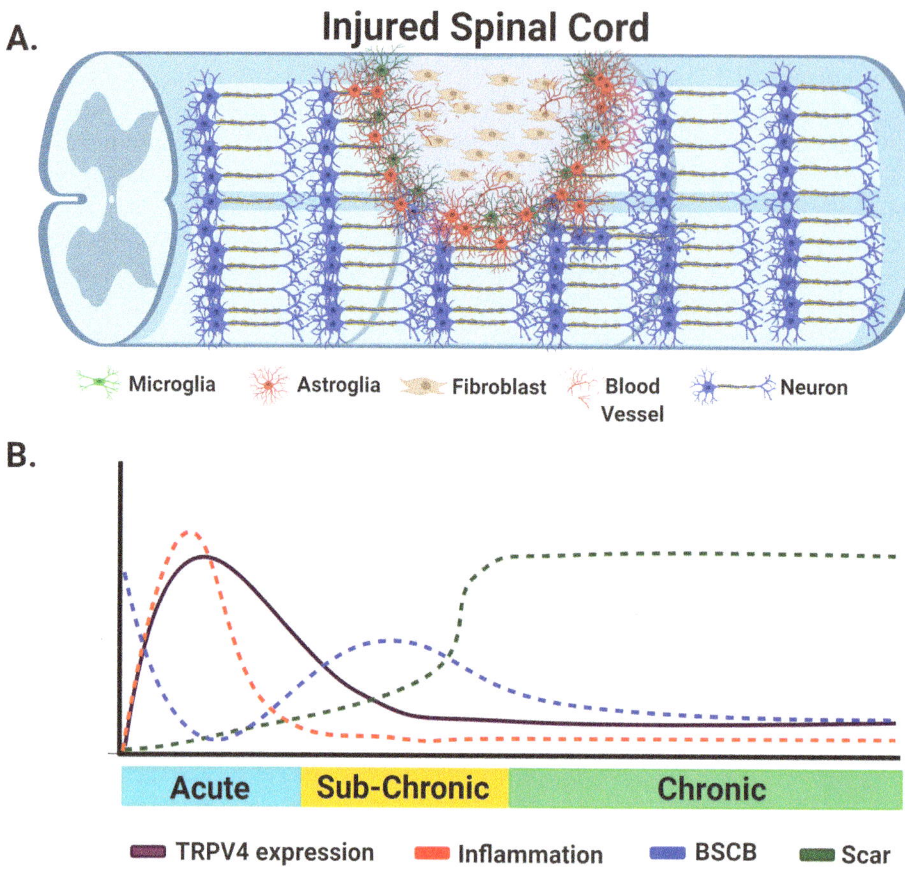

FIG. 2 TRPV4 expression in spinal cord injury pathology. TRPV4 expression is increased during the inflammatory phase. Both neuronal and non-neuronal cells express functional TRPV4, which is a response to a variety of stimuli. SCI leads to infiltration of neutrophils at the epicenter, followed by an increase in cytokines and chemokines. TRPV4 was reported to increase during the inflammatory phase, increasing secondary damage after spinal cord injury.

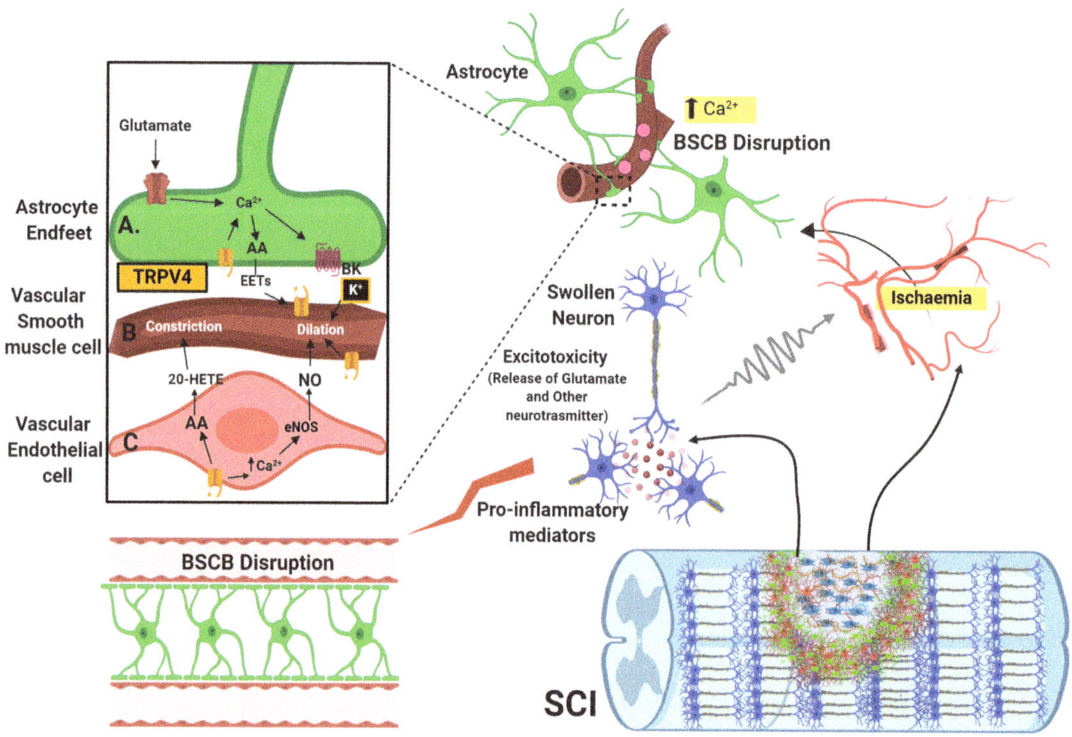

FIG. 3 Role of TRPV4 in the pathology of spinal cord injury (SCI). TRPV4 expression is upregulated after SCI. TRPV4 is expressed at the astrocyte end foot and endothelial cells. Increased expression of TRPV4 is linked with inflammation, endothelial damage, and blood–spinal cord barrier damage. SCI leads to ischemia, the release of pro-inflammatory mediators, and excitotoxicity, directly linked to endothelial damage and BSCB disruption. TRPV4 plays a crucial role in increasing calcium levels and promoting inflammation.

Reports indicate that an increase in TRPV4 activity disintegrated the blood-cerebrospinal fluid barrier (Narita et al., 2015). TRPV4 expression followed injury-dependent expression after mild, moderate, and severe SCI. Thus, a recent study indicates that escalating TRPV4 expression might be correlated with predicting CNS injury severity (Kumar et al., 2020).

The TJ proteins expression (occludin, ZO-1, and claudins) is reported to be downregulated after CNS injury, including SCI (Kumar et al., 2018), suggesting disturbances in the BSCB integrity. Pacsin-3, a synaptic vesicular membrane trafficking protein that impedes TRPV4 basal expression (D'hoedt et al., 2008), decreased after SCI [17], portentous that the SCI-induced upregulation of TRPV4 has functional relevance. Vascular ECs express TRPV4 in rodents (Watanabe et al., 2002) and humans (Sullivan, Francis, Pitts, Taylor, & Earley, 2012). Interestingly, TRPV4 presence has been confirmed in vascular smooth muscle cells (Senadheera et al., 2013) and in the astrocytic endfeet processes that wrap around blood vessels in the CNS (Benfenati et al., 2011). Activation of TRPV4 in the astrocytic endfeet contributes to NVC via regulating the Ca^{2+} entry (Dunn, Bonev, & Nelson, 2011). Neurovascular units include neurons, pericytes, and glia and are closely connected and functionally assembled with ECs in CNS. (Iadecola, 2004). Injection of specific TRPV4 agonists directly into the spinal cord and TRPV4 pharmacological activation and reduced the expression of TJ markers (ZO-1, ZO-2, and claudin-1). The TRPV4 agonist injection increases fibrosis and scarring, and it is enhanced in the presence of injury. The probable reason is in vivo application of GSK1016790A, a TRPV4 agonist, leading to the damage to ECs and basal lamina deposition, decreasing angiogenesis, and promoting cystic cavity formation, thereby enhancing EC damage after SCI. Interestingly, TRPV4 inhibition via RN-1734, a specific antagonist, attenuated inflammatory cascade, reduced free $[Ca^{2+}]_i$ levels, and preserved the BSCB (Kumar et al., 2020).

TJs proteins (Zona occludens, Claudins, and Occludins) are crucial for maintaining the integrity and preventing leakiness and disruption of these proteins that lead to BSCB instability and augmented permeability. TJs proteins' selective leakiness promotes the infiltration of inflammatory cytokines and chemokines, which are significant factors for secondary injury after SCI. Ion channels play a crucial role in maintaining the ECs integrity after SCI. After SCI, the secondary damage has been attributed to the augmented expression of sulfonylurea receptor 1-regulated $NC_{Ca\text{-}ATP}$ channels, such as Trpm4, by ECs (Gerzanich et al., 2009). Additionally, ECs express the TRPV4 receptor, which has also been associated with EC dysfunction and death during SCI-induced secondary damage (Kumar et al., 2020). TRPV4 KO mice ECs were better protected after SCI-induced damage, suggestive that SCI-induced TRPV4 activation might directly affect EC survival. A conceivable explanation is that TRPV4-mediated entry of Ca^{2+} into ECs regulates NO production and responds to inflammatory reaction/signals by changing the barrier properties (Tiruppathi et al., 2002) (Fig. 3). The absence of TRPV4 lessened SCI-induced inflammatory cascade, including the synthesis of chemokines and cytokines, in addition to enhanced EC integrity, reduced permeability, and reduced apoptosis (Kumar et al., 2020).

Spinal cord microvessels not only support neural parenchyma by bestowing adequate blood supply and oxygenation, but the CNS endothelium continuously secretes neurotrophins such as brain-derived neurotrophic factor (BDNF) and angiopoietins that are well established to support regenerating spinal cord tissue. The ECs permeability depends on several proteins that can directly or indirectly modulate the vascular permeability/inflammation. Angiopoietin-1 (ANG-1) is involved in maintaining vascular permeability, strengthening the endothelial associated molecules, and regulating the interendothelial adhesion. ANG-1 expression was decreased in the experimental model of SCI, and the reduction in ANG-1 was mitigated in TRPV4 KO mice (Kumar et al., 2020). The survival of EC depends on the continued presence and support of neurotrophins such as BDNF and NT-3, in addition to levels of NG-2, SMA, and vWF. Furthermore, TRPV4 KO mice preserved neurotrophins' levels and endorsed EC survival (Kumar et al., 2020).

TRPV4 role in non-injured conditions

CNS tissue responds to the different stressors by modulating the function of the protein or specific ion channels. Osmotic homeostasis also termed osmoregulation, is a vital physiological process, and the CNS serves as a central site for osmoregulation. The major areas controlling osmotic homeostasis are the OVLT, CVOs, mPOA of the hypothalamus, and the PVN. TRPV4 is osmotically activated and is articulated in the CVOs in the mammalian CNS (Liedtke & Friedman, 2003). The expression of TRPV4 is abundant in the astrocytic membranes unraveling the brain and extracellular liquid spaces, and therefore likely plays a critical role in brain-volume homeostasis (Benfenati et al., 2007). A crucial study observed the abnormalities in osmoregulation in TRPV4 null (*TRPV4*$^{-}$/$^{-}$) mice suggesting that TRPV4 is essential for the typical response to osmotic pressure changes and serves as an osmosensor in the CNS (Liedtke & Friedman, 2003). TRPV4 expression is also reported in other neurosensory structures, including Merkel cells, inner-ear hair cells, and sensory neurons. *Trpv4*$^{-/-}$ mice showed diminished nociceptive responses to pressure (Liedtke & Friedman, 2003; Suzuki, Mizuno, Kodaira, & Imai, 2003). Knockdown or deletion of TRPV4 lessens nociceptive responses to hypotonic and mildly hypertonic stimuli (Alessandri-Haber et al., 2003) and in SCI's experimental model (Kumar et al., 2020). Thus, it's evident that TRPV4 functions to sense both osmotic and mechanical stimuli.

RVD is another process directly controlled by TRPV4 expression. RVD plays a crucial role in maintaining brain-volume homeostasis at the cellular and organ levels and prevents detrimental swelling in response to hypo-osmotic stress. Astrocytes-containing TRPV4/AQP4 complex plays a central role in maintaining RVD in the brain (Benfenati et al., 2011; Simard & Nedergaard, 2004).

Neurovascular coupling (NVC) is critical for neuronal survival and homeostasis. Astrocytes are considered an essential player in coordinating NVC (Koehler, Roman, & Harder, 2009). They perform this function in coordination with specific proteins/channels. Every astrocyte has no less than one process with endfeet surrounding a blood vessel (Simard, Arcuino, Takano, Liu, & Nedergaard, 2003). Interestingly, in contrast to astrocytes, neuronal processes are rarely in direct contact with intraparenchymal blood vessels; instead, they are indirectly connected. This anatomical association supports the hypothesis that astrocytes play a pivotal role in cerebral circulation's dynamic regulation (Paulson & Newman, 1987). TRPV4 channel presence has been well-known in astrocytes, and TRPV4 activation within the astrocytic endfeet contributes to NVC via Ca^{2+} entry (Dunn et al., 2011; Kumar et al., 2020).

Conversely, incubation of brain slices with HC067047, a selective TRPV4 antagonist, attenuates the Ca^{2+} oscillations (Dunn et al., 2011). These results suggest that TRPV4 selectively expressed at perivascular astrocytic endfeet regulates local cerebral blood flow and regulates the endfeet $[Ca^{2+}]_i$ to optimize NVC (Dunn et al., 2011; Dunn, Baylie, & Nelson, 2012). TRPV4 channels hasten Ca^{2+}-induced Ca^{2+} release from IP_3R in astrocytic endfeet and strengthen the NVC response (Dunn et al., 2013). As soon as the IP_3R-dependent Ca^{2+} wave reaches the astrocytic endfeet, the increase in endfeet $[Ca^{2+}]_i$ triggers Ca^{2+}-responsive pathways to liberate the vasoactive substances that commence the adjoining arteriole to dilate (Filosa et al., 2006; Straub & Nelson, 2007). The endfeet $[Ca^{2+}]_i$ is thus coupled to the vascular diameter and local perfusion, but there is a duality to this relationship: a medium increase in endfeet $[Ca^{2+}]_i$ dilates parenchymal arterioles, whereas high endfeet $[Ca^{2+}]_i$ constrict them (Girouard et al., 2010). This bidirectional vascular response suggests that endfoot intracellular Ca^{2+} concentration must be finely regulated to ensure adequate perfusion and maintain cerebral homeostasis. Similar to IP3Rs, TRPV4 channel activity is sensitive to modest $[Ca^{2+}]_i$ elevations and inhibited by high $[Ca^{2+}]_i$ (Strotmann, Schultz, & Plant, 2003).

Applications to other areas of neuroscience

TRPV4 expression exhibits age-related spinal cord changes and also shows alterations in cortical pyramidal neurons, the thalamus, and the cerebellum's basal nuclei during aging (Lee & Choe, 2014). Failure of neuronal TRPV4 expression in these brain regions may underlie some TRPV4-related pathological conditions. $TRPV4^{R269C}$, a TRPV4 mutant, expression in vivo causes neuronal dysfunction and axonal degeneration by increasing the intracellular Ca^{2+} through a CAMKII. Interestingly, TRPV4 antagonists may reverse the damage suggesting that inhibition of TRPV4 signaling may ameliorate TRPV4-linked neurodegenerative diseases (Woolums et al., 2020).

TRPV4 activates neuropeptide release from afferent nerves and stimulates neurogenic inflammation (Vergnolle et al., 2010). Interestingly, TRPV4 is also reported to be triggered by changes in osmolarity, causing increased release of pro-inflammatory cytokines (IL-1β and IL-6) in intervertebral discs (Walter et al., 2016). GSK2193874, a TRPV4 antagonist, attenuated the LPS-induced surge in pro-inflammatory cytokines (TNF-α, IL-1β, and IL-6), signifying that blocking TRPV4 markedly decreases the production of cytokines in the development of sepsis (Dalsgaard, Sonkusare, Teuscher, Poynter, & Nelson, 2016). The role of TRPV4 in other areas in neuroscience is summarized in Table 2.

TABLE 2 TRPV4 roles/applications to other areas of neuroscience.

Disease	Role	Reference
Stroke	TRPV4 channels contribute to calcium influx into astrocytes and neurons and subsequent extracellular glutamate accumulation during PIDs	Rakers, Schmid, and Petzold (2017)
	Trpv4$^{-/-}$ mice showed reduced ischemia-induced lesion volume, water content, and Evans blue leakage and prevented TJs' loss compared to WT mice	Tanaka et al. (2020)
AD	TRPV4 expression in the brain of control mice, early-stage, and late-stage AD model mice significantly increased in the cerebral cortex, hippocampal formation, striatum, and thalamus of AD model mice compared with control mice. Interestingly, amyloid-β is reported to activate TRPV4 expression	Bai and Lipski (2014), Lee and Choe (2016)

TABLE 2 TRPV4 roles/applications to other areas of neuroscience—cont'd

Disease	Role	Reference
TBI	TBI causes augmented TRPV4 expression and then induced severe brain edema and neuronal damage by activating the MAPK cascade and Akt-related signaling pathway	Lu, Huang, Tsai, and Yang (2017)
MS	TRPV4 expression was increased in the mouse model of MS, and specific inhibition using a TRPV4 antagonist alleviated demyelination. Interestingly, TRPV4-antagonist inhibited glial activation and attenuated the production of TNF-α and IL-1β, suggesting that TRPV4 plays a role in CNS demyelination	Liu et al. (2018)
Leukodystrophy	Megalencephalic leukoencephalopathy with subcortical cysts (MLC), a rare leukodystrophy characterized by macrocephaly, subcortical fluid cysts, and myelin vacuolation. MLC1 gene is highly expressed in astrocytes, and it cooperates with the TRPV4	Lanciotti et al. (2012)
Migraine	TRPV4 activation within the meninges generates afferent nociceptive signaling that contributes to migraine headaches	Wei, Edelmayer, Yan, and Dussor (2011)
Hydrocephalus (an excessive accumulation of CSF)	TRPV4 antagonists inhibit the development of ventriculomegaly in a rodent model of communicating hydrocephalus, the Wpk rat (Tmem67^{-}/$^{-}$)	Hochstetler et al. (2020)
Epilepsy	Activation of TRPV4 induces neuronal injury, enhances inflammatory response, and promotes the pro-inflammatory cytokines. TRPV4 activation with specific agonists causes microglia and astrocyte activation. It increases NLPR3 inflammasome and pro-inflammatory cytokines, whereas selective TRPV4 inhibition increases the number of surviving cells and attenuated the activated microglia and astrocyte	Wang et al. (2019)
	The frequencies of epileptic electroencephalogram in WT mice were significantly larger than those in Trpv4$^{-/-}$ mice in the experimental model of epilepsy. These results suggest that activation of TRPV4 contributes to the disease progression of epilepsy	(Shibasaki et al. (2020)
Amyotrophic lateral sclerosis	TRPV4 expression was evident in the cerebral cortex, hippocampal formation, thalamus, cerebellum, and spinal cord of symptomatic SOD1^{G93A} transgenic mice, an animal model of amyotrophic lateral sclerosis	Lee, Joo, Choe, and Cha (2012)

Mini-dictionary of terms

Spinal cord injury: A injury or damage to the spinal cord caused either by trauma or non-trauma. It can cause permanent or life-long disability.

TRPV4: A non-selective cationic channel that tunes the functions of diverse tissues.

Gliosis: Also known as glial scar in which reactive changes in the glial cells occur in response to damage to the CNS including SCI.

Fibrosis: Also known as fibrotic scarring, is a pathological process where connective tissue like fibroblasts replaces normal parenchymal leading to substantial tissue remodeling and the development of permanent scar. Fibrotic scar forms the core of the glial scar after SCI.

Blood–spinal cord barrier: Analogous to BBB, BSCB is formed by specialized small blood vessels that surround the spinal cord.

Endothelial cells: These are single cells layers that surround blood vessels and control the blood flow and nutritional exchange.

Tight junctions: These are the cells that seal the paracellular pathway. Thereby tightly regulates the transport and prevents leakage.

Neurovascular coupling: the relationship between local neural activity and resultant changes in cerebral blood flow.

Regulatory volume decrease: Cells can endure the changes in cell volume; specific proteins' expression tightly controls the mechanism to resist these changes. These proteins return cell volume to baseline following an osmolarity-induced swelling response. This volume recovery is termed RVD.

TRPV4 null mice: Transgenic mouse in which TRPV4 gene has been disrupted or deleted and there is no functional gene product. These are also known as knock-out mice.

Key facts about SCI

- SCI is a multifacet pathology mostly caused by traumatic and sometimes non-traumatic events.
- SCI is often associated with traumatic injury primarily caused by trauma/accidents, is synchronized with mechanical injury, shear stress, and osmolarity changes.
- The SCI involved a primary phase where inflammatory components predominate; whereas, the secondary phase involves scarring and secondary damages.
- During inflammatory phase, there is a release of chemokines, cytokines, and ECs damage which can cause BSCB disruption.
- During secondary phase, there is scarring where the core is made up of fibrotic tissues, and periphery is mostly formed by glial cells.

Key facts about TRPV4

- TRPV4 is a non-selective cationic channel whose activation is typically linked with the influx of Ca^{2+}. However, it also allows for the transport of other cations.
- TRPV4 acts as a thermosensor, mechanosensory, and osmosensor and is expressed in neuronal as well as non-neuronal cells. It controls physiological function, and TRPV4 activation is linked with pathological changes.
- TRPV4 expression is linked with CNS injuries pathology, including SCI, and linked with the injury's severity.
- Knockdown of TRPV4 lessens scarring (glial and fibrotic), and SCI-injured animals recovered faster than animals that had an expression of TRPV4.
- Targeting TRPV4 will be an excellent strategy to prevent endothelial damage, cell damage, and attenuate inflammation.

Summary points

- TRPV4, a non-selective cation channel, is commonly expressed in both excitable and non-excitable cells. After SCI, TRPV4 activation is correlated with the injury's severity at the epicenter at the time points that were reported to be crucial for repair/treatment. TRPV4 activation damages ECs that form the BSCB and thus contribute to scarring (glial and fibrotic).
- Crucially, TRPV4 inhibition/knockdown vetoed SCI-induced cell damage, inflammation, and scarring. Thus, the manipulation of TRPV4 signaling might lead to new therapeutic strategies or combinatorial therapies to protect ECs and augment repair after SCI. Subsequently, the maneuvering of TRPV4 signaling and downstream pathways might lead to new therapeutic approaches after CNS injuries, including SCI.

References

Alessandri-Haber, N., Yeh, J. J., Boyd, A. E., Parada, C. A., Chen, X., Reichling, D. B., et al. (2003). Hypotonicity induces TRPV4-mediated nociception in rat. *Neuron*, *39*(3), 497–511.

Bai, J.-Z., & Lipski, J. (2014). Involvement of TRPV4 channels in Aβ40-induced hippocampal cell death and astrocytic Ca2+ signalling. *Neurotoxicology*, *41*, 64–72.

Benfenati, V., Amiry-Moghaddam, M., Caprini, M., Mylonakou, M. N., Rapisarda, C., Ottersen, O. P., et al. (2007). Expression and functional characterization of transient receptor potential vanilloid-related channel 4 (TRPV4) in rat cortical astrocytes. *Neuroscience*, *148*(4), 876–892. https://doi.org/10.1016/j.neuroscience.2007.06.039.

Benfenati, V., Caprini, M., Dovizio, M., Mylonakou, M. N., Ferroni, S., Ottersen, O. P., et al. (2011). An aquaporin-4/transient receptor potential vanilloid 4 (AQP4/TRPV4) complex is essential for cell-volume control in astrocytes. *Proceedings of the National Academy of Sciences*, *108*(6), 2563–2568.

Casella, G. T., Marcillo, A., Bunge, M. B., & Wood, P. M. (2002). New vascular tissue rapidly replaces neural parenchyma and vessels destroyed by a contusion injury to the rat spinal cord. *Experimental Neurology*, *173*(1), 63–76.

Dalsgaard, T., Sonkusare, S. K., Teuscher, C., Poynter, M. E., & Nelson, M. T. (2016). Pharmacological inhibitors of TRPV4 channels reduce cytokine production, restore endothelial function and increase survival in septic mice. *Scientific Reports*, *6*, 33841. https://doi.org/10.1038/srep33841.

Damann, N., Voets, T., & Nilius, B. (2008). TRPs in our senses. *Current Biology*, *18*(18), R880–R889.

D'hoedt, D., Owsianik, G., Prenen, J., Cuajungco, M. P., Grimm, C., Heller, S., et al. (2008). Stimulus-specific modulation of the cation channel TRPV4 by PACSIN 3. *Journal of Biological Chemistry*, *283*(10), 6272–6280.

Dunn, K. M., Baylie, R. L., & Nelson, M. T. (2012). TRPV4 channels tune astrocytic endfoot Ca2+ to optimize neurovascular coupling. *The FASEB Journal*, *26*(Suppl. 1), 685.610.

Dunn, K. M., Bonev, A. D., & Nelson, M. T. (2011). Functional evidence of TRPV4-mediated Ca2+ signals in cortical astrocytes. *The FASEB Journal*, *25* (1 Supplement), 1024.1023.

Dunn, K. M., Hill-Eubanks, D. C., Liedtke, W. B., & Nelson, M. T. (2013). TRPV4 channels stimulate Ca2+−induced Ca2+ release in astrocytic endfeet and amplify neurovascular coupling responses. *Proceedings of the National Academy of Sciences*, *110*(15), 6157–6162.

Filosa, J. A., Bonev, A. D., Straub, S. V., Meredith, A. L., Wilkerson, M. K., Aldrich, R. W., et al. (2006). Local potassium signaling couples neuronal activity to vasodilation in the brain. *Nature Neuroscience*, *9*(11), 1397–1403. https://doi.org/10.1038/nn1779.

Gerzanich, V., Woo, S. K., Vennekens, R., Tsymbalyuk, O., Ivanova, S., Ivanov, A., et al. (2009). De novo expression of Trpm4 initiates secondary hemorrhage in spinal cord injury. *Nature Medicine*, *15*(2), 185.

Girouard, H., Bonev, A. D., Hannah, R. M., Meredith, A., Aldrich, R. W., & Nelson, M. T. (2010). Astrocytic endfoot Ca2+ and BK channels determine both arteriolar dilation and constriction. *Proceedings of the National Academy of Sciences of the United States of America*, *107*(8), 3811–3816. https://doi.org/10.1073/pnas.0914722107.

Hochstetler, A. E., Smith, H. M., Preston, D. C., Reed, M. M., Territo, P. R., Shim, J. W., et al. (2020). TRPV4 antagonists ameliorate ventriculomegaly in a rat model of hydrocephalus. *JCI Insight*, *5*(18), e137646.

Iadecola, C. (2004). Neurovascular regulation in the normal brain and in Alzheimer's disease. *Nature Reviews Neuroscience*, *5*(5), 347.

Jang, Y., Jung, J., Kim, H., Oh, J., Jeon, J. H., Jung, S., et al. (2012). Axonal neuropathy-associated TRPV4 regulates neurotrophic factor-derived axonal growth. *The Journal of Biological Chemistry*, *287*(8), 6014–6024. https://doi.org/10.1074/jbc.M111.316315.

Jie, P., Hong, Z., Tian, Y., Li, Y., Lin, L., Zhou, L., et al. (2015). Activation of transient receptor potential vanilloid 4 induces apoptosis in hippocampus through downregulating PI3K/Akt and upregulating p38 MAPK signaling pathways. *Cell Death & Disease*, *6*. https://doi.org/10.1038/cddis.2015.146, e1775.

Koehler, R. C., Roman, R. J., & Harder, D. R. (2009). Astrocytes and the regulation of cerebral blood flow. *Trends in Neurosciences*, *32*(3), 160–169.

Koyanagi, I., Tator, C. H., & Lea, P. J. (1993). Three-dimensional analysis of the vascular system in the rat spinal cord with scanning electron microscopy of vascular corrosion casts. Part 2: Acute spinal cord injury. *Neurosurgery*, *33*(2), 285–292.

Kumar, H., Choi, H., Jo, M.-J., Joshi, H. P., Muttigi, M., Bonanomi, D., et al. (2018). Neutrophil elastase inhibition effectively rescued angiopoietin-1 decrease and inhibits glial scar after spinal cord injury. *Acta Neuropathologica Communications*, *6*(1), 73.

Kumar, H., Lee, S.-H., Kim, K.-T., Zeng, X., & Han, I. (2018). TRPV4: A sensor for homeostasis and pathological events in the CNS. *Molecular Neurobiology*, *55*(11), 8695–8708.

Kumar, H., Lim, C. S., Choi, H., Joshi, H. P., Kim, K.-T., Kim, Y. H., et al. (2020). Elevated TRPV4 levels contribute to endothelial damage and scarring in experimental spinal cord injury. *Journal of Neuroscience*, *40*(9), 1943–1955.

Kumar, H., Ropper, A. E., Lee, S.-H., & Han, I. (2017). Propitious therapeutic modulators to prevent blood-spinal cord barrier disruption in spinal cord injury. *Molecular Neurobiology*, *54*(5), 3578–3590.

Lanciotti, A., Brignone, M. S., Molinari, P., Visentin, S., De Nuccio, C., Macchia, G., et al. (2012). Megalencephalic leukoencephalopathy with subcortical cysts protein 1 functionally cooperates with the TRPV4 cation channel to activate the response of astrocytes to osmotic stress: Dysregulation by pathological mutations. *Human Molecular Genetics*, *21*(10), 2166–2180.

Lee, J. C., & Choe, S. Y. (2014). Age-related changes in the distribution of transient receptor potential vanilloid 4 channel (TRPV4) in the central nervous system of rats. *Journal of Molecular Histology*, *45*(5), 497–505. https://doi.org/10.1007/s10735-014-9578-z.

Lee, J. C., & Choe, S. Y. (2016). Region-specific changes in the distribution of transient receptor potential vanilloid 4 channel (TRPV4) in the central nervous system of Alzheimer's disease model mice. *Genes & Genomics*, *38*(7), 629–637.

Lee, J. C., Joo, K. M., Choe, S. Y., & Cha, C. I. (2012). Region-specific changes in the immunoreactivity of TRPV4 expression in the central nervous system of SOD1 G93A transgenic mice as an in vivo model of amyotrophic lateral sclerosis. *Journal of Molecular Histology*, *43*(6), 625–631.

Liedtke, W., Choe, Y., Marti-Renom, M. A., Bell, A. M., Denis, C. S., Sali, A., et al. (2000). Vanilloid receptor-related osmotically activated channel (VR-OAC), a candidate vertebrate osmoreceptor. *Cell*, *103*(3), 525–535.

Liedtke, W., & Friedman, J. M. (2003). Abnormal osmotic regulation in trpv4 −/− mice. *Proceedings of the National Academy of Sciences of the United States of America*, *100*(23), 13698–13703. https://doi.org/10.1073/pnas.1735416100.

Liu, M., Liu, X., Wang, L., Wang, Y., Dong, F., Wu, J., et al. (2018). TRPV4 inhibition improved myelination and reduced glia reactivity and inflammation in a cuprizone-induced mouse model of demyelination. *Frontiers in Cellular Neuroscience*, *12*, 392.

Lu, K. T., Huang, T. C., Tsai, Y. H., & Yang, Y. L. (2017). Transient receptor potential vanilloid type 4 channels mediate Na-K-Cl-co-transporter-induced brain edema after traumatic brain injury. *Journal of Neurochemistry*, *140*(5), 718–727.

Montell, C., Birnbaumer, L., & Flockerzi, V. (2002). The TRP channels, a remarkably functional family. *Cell*, *108*(5), 595–598.

Narita, K., Sasamoto, S., Koizumi, S., Okazaki, S., Nakamura, H., Inoue, T., et al. (2015). TRPV4 regulates the integrity of the blood-cerebrospinal fluid barrier and modulates transepithelial protein transport. *The FASEB Journal*, *29*(6), 2247–2259.

Nilius, B., Prenen, J., Wissenbach, U., Bödding, M., & Droogmans, G. (2001). Differential activation of the volume-sensitive cation channel TRP12 (OTRPC4) and volume-regulated anion currents in HEK-293 cells. *Pflügers Archiv*, *443*(2), 227–233.

Paulson, O. B., & Newman, E. A. (1987). Does the release of potassium from astrocyte endfeet regulate cerebral blood flow? *Science*, *237*(4817), 896–898.

Plant, T. D., & Strotmann, R. (2007). Trpv4. *Handbook of Experimental Pharmacology*, (179), 189–205. https://doi.org/10.1007/978-3-540-34891-7_11.

Rakers, C., Schmid, M., & Petzold, G. C. (2017). TRPV4 channels contribute to calcium transients in astrocytes and neurons during peri-infarct depolarizations in a stroke model. *Glia*, *65*(9), 1550–1561.

Senadheera, S., Bertrand, P. P., Grayson, T. H., Leader, L., Murphy, T. V., & Sandow, S. L. (2013). Pregnancy-induced remodelling and enhanced endothelium-derived hyperpolarization-type vasodilator activity in rat uterine radial artery: Transient receptor potential vanilloid type 4 channels, caveolae and myoendothelial gap junctions. *Journal of Anatomy*, *223*(6), 677–686. https://doi.org/10.1111/joa.12127.

Shibasaki, K., Ikenaka, K., Tamalu, F., Tominaga, M., & Ishizaki, Y. (2014). A novel subtype of astrocytes expressing TRPV4 (transient receptor potential vanilloid 4) regulates neuronal excitability via release of gliotransmitters. *The Journal of Biological Chemistry*, *289*(21), 14470–14480. https://doi.org/10.1074/jbc.M114.557132.

Shibasaki, K., & Tominaga, M. (2007). Implication that TRPV4 activation induces the excitation of astrocytes. In *Paper presented at the Proceedings of Annual Meeting of the Physiological Society of Japan Proceedings of Annual Meeting of the Physiological Society of Japan*.

Shibasaki, K., Yamada, K., Miwa, H., Yanagawa, Y., Suzuki, M., Tominaga, M., et al. (2020). Temperature elevation in epileptogenic foci exacerbates epileptic discharge through TRPV4 activation. *Laboratory Investigation*, *100*(2), 274–284.

Shimizu, N., Wada, N., Shimizu, T., Suzuki, T., Takaoka, E.-I., Kanai, A. J., et al. (2018). Effects of nerve growth factor neutralization on TRP channel expression in laser-captured bladder afferent neurons in mice with spinal cord injury. *Neuroscience Letters*, *683*, 100–103.

Simard, M., Arcuino, G., Takano, T., Liu, Q. S., & Nedergaard, M. (2003). Signaling at the gliovascular interface. *The Journal of Neuroscience*, *23*(27), 9254–9262.

Simard, M., & Nedergaard, M. (2004). The neurobiology of glia in the context of water and ion homeostasis. *Neuroscience*, *129*(4), 877–896. https://doi.org/10.1016/j.neuroscience.2004.09.053.

Straub, S. V., & Nelson, M. T. (2007). Astrocytic calcium signaling: The information currency coupling neuronal activity to the cerebral microcirculation. *Trends in Cardiovascular Medicine*, *17*(6), 183–190. https://doi.org/10.1016/j.tcm.2007.05.001.

Strotmann, R., Harteneck, C., Nunnenmacher, K., Schultz, G., & Plant, T. D. (2000). OTRPC4, a nonselective cation channel that confers sensitivity to extracellular osmolarity. *Nature Cell Biology*, *2*(10), 695–702. https://doi.org/10.1038/35036318.

Strotmann, R., Schultz, G., & Plant, T. D. (2003). Ca2+–dependent potentiation of the nonselective cation channel TRPV4 is mediated by a C-terminal calmodulin binding site. *The Journal of Biological Chemistry*, *278*(29), 26541–26549. https://doi.org/10.1074/jbc.M302590200.

Sullivan, M. N., Francis, M., Pitts, N. L., Taylor, M. S., & Earley, S. (2012). Optical recording reveals novel properties of GSK1016790A-induced vanilloid transient receptor potential channel TRPV4 activity in primary human endothelial cells. *Molecular Pharmacology*, *82*(3), 464–472. https://doi.org/10.1124/mol.112.078584.

Suzuki, M., Mizuno, A., Kodaira, K., & Imai, M. (2003). Impaired pressure sensation in mice lacking TRPV4. *The Journal of Biological Chemistry*, *278*(25), 22664–22668. https://doi.org/10.1074/jbc.M302561200.

Talavera, K., Nilius, B., & Voets, T. (2008). Neuronal TRP channels: Thermometers, pathfinders and life-savers. *Trends in Neurosciences*, *31*(6), 287–295. https://doi.org/10.1016/j.tins.2008.03.002.

Tanaka, K., Matsumoto, S., Yamada, T., Yamasaki, R., Suzuki, M., Kido, M. A., et al. (2020). Reduced post-ischemic brain injury in transient receptor potential vanilloid 4 knockout mice. *Frontiers in Neuroscience*, *14*, 453.

Tator, C. H., & Koyanagi, I. (1997). Vascular mechanisms in the pathophysiology of human spinal cord injury. *Journal of Neurosurgery*, *86*(3), 483–492. https://doi.org/10.3171/jns.1997.86.3.0483.

Tiruppathi, C., Freichel, M., Vogel, S. M., Paria, B. C., Mehta, D., Flockerzi, V., et al. (2002). Impairment of store-operated Ca2+ entry in TRPC4 –/– mice interferes with increase in lung microvascular permeability. *Circulation Research*, *91*(1), 70–76.

Vennekens, R., Owsianik, G., & Nilius, B. (2008). Vanilloid transient receptor potential cation channels: An overview. *Current Pharmaceutical Design*, *14*(1), 18–31.

Vergnolle, N., Cenac, N., Altier, C., Cellars, L., Chapman, K., Zamponi, G., et al. (2010). A role for transient receptor potential vanilloid 4 in tonicity-induced neurogenic inflammation. *British Journal of Pharmacology*, *159*(5), 1161–1173.

Vriens, J., Watanabe, H., Janssens, A., Droogmans, G., Voets, T., & Nilius, B. (2004). Cell swelling, heat, and chemical agonists use distinct pathways for the activation of the cation channel TRPV4. *Proceedings of the National Academy of Sciences of the United States of America*, *101*(1), 396–401. https://doi.org/10.1073/pnas.0303329101.

Walter, B. A., Purmessur, D., Moon, A., Occhiogrosso, J., Laudier, D. M., Hecht, A. C., et al. (2016). Reduced tissue osmolarity increases TRPV4 expression and pro-inflammatory cytokines in intervertebral disc cells. *European Cells & Materials*, *32*, 123–136.

Wang, Z., Zhou, L., An, D., Xu, W., Wu, C., Sha, S., et al. (2019). TRPV4-induced inflammatory response is involved in neuronal death in pilocarpine model of temporal lobe epilepsy in mice. *Cell Death & Disease*, *10*(6), 1–10.

Watanabe, H., Vriens, J., Janssens, A., Wondergem, R., Droogmans, G., & Nilius, B. (2003). Modulation of TRPV4 gating by intra- and extracellular Ca2+. *Cell Calcium*, *33*(5–6), 489–495.

Watanabe, H., Vriens, J., Suh, S. H., Benham, C. D., Droogmans, G., & Nilius, B. (2002). Heat-evoked activation of TRPV4 channels in a HEK293 cell expression system and in native mouse aorta endothelial cells. *The Journal of Biological Chemistry*, *277*(49), 47044–47051. https://doi.org/10.1074/jbc.M208277200.

Wegierski, T., Lewandrowski, U., Muller, B., Sickmann, A., & Walz, G. (2009). Tyrosine phosphorylation modulates the activity of TRPV4 in response to defined stimuli. *The Journal of Biological Chemistry*, *284*(5), 2923–2933. https://doi.org/10.1074/jbc.M805357200.

Wei, X., Edelmayer, R. M., Yan, J., & Dussor, G. (2011). Activation of TRPV4 on dural afferents produces headache-related behavior in a preclinical rat model. *Cephalalgia*, *31*(16), 1595–1600.

Woolums, B. M., McCray, B. A., Sung, H., Tabuchi, M., Sullivan, J. M., Ruppell, K. T., et al. (2020). TRPV4 disrupts mitochondrial transport and causes axonal degeneration via a CaMKII-dependent elevation of intracellular ca 2+. *Nature Communications*, *11*(1), 1–17.

Chapter 20

Autoantibodies in spinal cord injury

Daniel Garcia-Ovejero[a], Lukas Grassner[b,c,d,e,f], Eduardo Molina-Holgado[a], and Angel Arevalo-Martin[a]

[a]*Laboratory of Neuroinflammation, Hospital Nacional de Parapléjicos, SESCAM, Toledo, Spain,* [b]*Department of Neurosurgery, Medical University Innsbruck, Innsbruck, Austria,* [c]*Institute of Molecular Regenerative Medicine, Paracelsus Medical University, Salzburg, Austria,* [d]*Spinal Cord Injury and Tissue Regeneration Center Salzburg (SCI-TReCS), Paracelsus Medical University, Salzburg, Austria,* [e]*ParaMove, SCI Research Unit, BG Trauma Center Murnau, Murnau, Germany, and Paracelsus Medical University, Salzburg, Austria,* [f]*Spinal Cord Injury Center, Trauma Center Murnau, Murnau, Germany*

List of abbreviations

ALBU	albumin
ATPA	ATP synthase subunit alpha
CNS	central nervous system
CRMP2	collapsing response mediator protein-2
CSF	cerebrospinal fluid
ENOA	enolase alpha
GAPDH	glyceraldehyde 3-phosphate dehydrogenase
GFAP	glial fibrillary acidic protein
HBA	hemoglobin alpha
HBB	hemoglobin beta
IgG	immunoglobulin G
MBP	myelin basic protein
NFL	neurofilament light
NFM	neurofilament intermediate
NMDAR	*N*-methyl-D-aspartate receptors
PGK1	phosphoglycerate kinase 1
PPIA	peptidylprolyl isomerase A
SCI	spinal cord injury

Introduction

From a clinical perspective, immunodepression is the most obvious alteration of the immune system after spinal cord injury (SCI), exposed by the fact that infections are the leading cause of morbidity and mortality after SCI. However, the extent of alterations induced by SCI over the immune system is by far much more complex than just suffering from immunodepression. In fact, although it may seem paradoxical, patients with SCI also exhibit elevated levels of autoantibodies against both central nervous system and systemic antigens (Arevalo-Martin et al., 2018; Davies, Hayes, & Dekaban, 2007; Hayes et al., 2002; Hergenroeder et al., 2018; Hergenroeder, Moore, Schmitt, Redell, & Dash, 2016; Mizrachi et al., 1983; Palmers et al., 2016; Petrova, Ponomaryova, Alyoshkin, Eliseyev, & Yumashev, 1993; Taranova et al., 1992; Zajarías-Fainsod et al., 2012). While immunodepression is directly related to increased susceptibility to infections, the clinical consequences of elevated autoantibody levels are not that straightforward. The first study suggesting that autoantibodies could play a role in SCI pathology was that of Mizrachi and collaborators (Mizrachi et al., 1983), who described that patients with SCI exhibited higher blood serum levels of anti-MBP (myelin basic protein) and/or anti-GM1 ganglioside. In spite that the authors did not provide direct proof on the contribution of autoantibodies to pathology, they proposed that autoantibodies could be a factor impeding neural repair after SCI since GM1 ganglioside is involved in neuron maturation and serum from patients, enriched in anti-GM1 antibodies, inhibited neurite outgrowth in vitro. Later on, it was demonstrated that SCI increases the levels of IgM and IgG autoantibodies also in mice, and the injection of these IgG antibodies with unknown specificities into the spinal cord of intact mice induced necrotic inflammatory lesions and ipsilateral paralysis of hind limbs

Cellular, Molecular, Physiological, and Behavioral Aspects of Spinal Cord Injury. https://doi.org/10.1016/B978-0-12-822427-4.00020-4

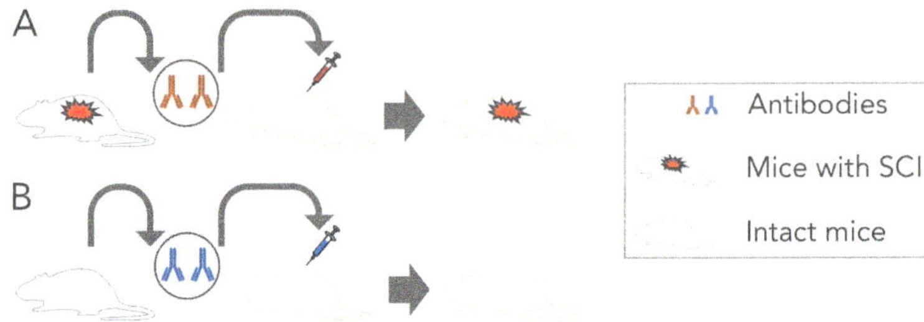

FIG. 1 Spinal cord injury induces pathogenic autoantibodies in mice. (A) Antibodies purified from mice with chronic SCI are pathogenic when injected into the spinal cord of intact mice, while (B) antibodies from intact mice do not elicit any pathogenic effect.

(Fig. 1) (Ankeny, Guan, & Popovich, 2009; Ankeny, Lucin, Sanders, McGaughy, & Popovich, 2006). But since the targets of these pathogenic antibodies in mice have not been identified, it remains still unknown which specific autoantibody or profile of autoantibodies was responsible for the pathological effects. Actually, it could happen that among the autoantibodies that increase in chronic SCI in mice some are pathogenic and some others are not. Accordingly, in a number of disorders, autoantibodies are increased but are not directly related to pathology. In these situations, in spite that autoantibodies are not potential therapeutic targets, they may be still useful for the clinical practice as diagnostic or prognostic biomarkers. Also, autoantibodies may provide further insights into the pathophysiological mechanisms that are taking place since they specifically point to the antigens against which the immune system is reacting. Whether autoantibodies are directly involved in the pathophysiology of human SCI or are just part of a secondary response to spinal cord damage is a question that remains open and constitutes a hot research topic in the field. Here we review the current knowledge about the origin, identities, and the known roles of autoantibodies that increase after human SCI.

Targets and possible origin of autoantibodies increased after SCI in humans

Currently, 24 different targets of autoantibodies that increase in peripheral blood after human SCI have been described (Table 1). Interestingly, at least 17 of these immunoglobulin G (IgG) autoantibodies are already elevated in the first hours/days after injury, before primary IgG responses to novel antigens may develop (Table 1) (Arevalo-Martin et al., 2018; Palmers et al., 2016). Such a rapid increase suggests that these autoantibodies are a part of pre-existing humoral immune responses, as occurs when the immune system is re-exposed to an antigen after an effective vaccination. Indeed, this is the case: the increased autoantibodies in the early stages after SCI are naturally occurring antibodies, i.e., antibodies naturally found in healthy subjects. The existence of naturally occurring autoantibodies (or natural autoantibodies) was reported almost 80 years ago (Kidd & Friedewald, 1942) and it has been a subject of research that challenges the classical view of autoantibodies as just effectors of an autoimmune disease (Cohen, 2007; Cohen & Cooke, 1986; Nataf, 2017). A hallmark of natural autoantibodies that is also observable in the autoantibodies increased in the early phases of SCI (Arevalo-Martin et al., 2018) is that they target altered or degraded proteins that arise under conditions of cellular stress or death (Fig. 2). Actually, many of the targets of the autoantibodies increased after injury are part of the SCI degradome, the group of proteins that are actively degraded by proteolytic enzymes activated after injury (Abou-El-Hassan et al., 2020). For this reason, natural autoantibodies have been suggested to participate in physiological responses to tissue damage, by pointing to cellular debris for their cleaning and, thus, promoting tissue healing (Fiskesund et al., 2010; Lutz, 2007; Schwartz-Albiez, Monteiro, Rodriguez, Binder, & Shoenfeld, 2009). Besides potentiating tissue repair, natural autoantibodies may be protective by counteracting pathological processes. For example, in patients who suffered from stroke, a neurological pathology shares some pathological mechanisms with SCI, the patients with natural anti-NMDAR (*N*-methyl-D-aspartate receptors) autoantibodies exhibit less expansion of the lesion area than the patients without these antibodies (provided that the blood-brain barrier was preserved before the insult) (Zerche et al., 2015). Suggesting that natural autoantibodies could also play a protective or repairing role after SCI, a commercial human intravenous IgG preparation—which is a pool of IgGs from thousands of healthy donors containing natural autoantibodies against antigens expressed by neurons, astrocytes, oligodendrocytes, and microglia—is therapeutic when administered immediately after SCI in mice (Brennan et al., 2016). Also, rats vaccinated with modified MBP after SCI present less neurological damage and a better functional recovery (Hauben et al., 2001). However, natural autoantibodies are proposed to be also the origin of pathogenic autoantibodies when the inflammatory response persists in time (Cohen, 2007; Cohen & Cooke, 1986; Nataf, 2017), as occurs in SCI. In this regard, as mentioned above, IgG antibodies from mice with chronic SCI (42 days

TABLE 1 Autoantibodies increased after human SCI.

Target	SCI stage[a]	References
AEBP1	Subacute	Palmers et al. (2016)
ALBU	Acute and subacute	Arevalo-Martin et al. (2018)
ATPA	Acute and subacute	Arevalo-Martin et al. (2018)
CAH2	Acute and subacute	Arevalo-Martin et al. (2018)
CRMP2	Subacute	Hergenroeder et al. (2018)
DHE3	Acute and subacute	Arevalo-Martin et al. (2018)
ENOA	Acute and subacute	Arevalo-Martin et al. (2018)
GalC	Chronic	Taranova et al. (1992)
GAPDH	Acute and subacute	Palmers et al. (2016), Arevalo-Martin et al. (2018)
GFAP	From acute to chronic	Hergenroeder et al. (2016), Arevalo-Martin et al. (2018), Hergenroeder et al. (2018)
GM1 ganglioside	From subacute to chronic	Mizrachi et al. (1983), Wang et al. (1995), Taranova et al. (1992), Hayes et al. (2002), Davies et al. (2007)
HBA	Acute and subacute	Arevalo-Martin et al. (2018)
HBB	Acute and subacute	Arevalo-Martin et al. (2018)
HSP7C	Acute and subacute	Arevalo-Martin et al. (2018)
IgG Fc	From subchronic to chronic	Petrova et al. (1993)
MBP	From acute to chronic	Mizrachi et al. (1983), Zajarías-Fainsod et al. (2012), Arevalo-Martin et al. (2018)
MYEOV2	Subacute	Palmers et al. (2016)
NFL	Acute and subacute	Arevalo-Martin et al. (2018)
NFM	Acute and subacute	Arevalo-Martin et al. (2018)
PEBP1	Acute and subacute	Arevalo-Martin et al. (2018)
PGK1	Acute and subacute	Arevalo-Martin et al. (2018)
PPIA	Acute and subacute	Arevalo-Martin et al. (2018)
PSMD4	Subacute	Palmers et al. (2016)
S100ß	Acute and subacute	Palmers et al. (2016)

AEBP1, adipocyte enhancer-binding protein 1; *ALBU*, albumin; *ATPA*, ATP synthase subunit alpha; *CAH2*, carbonic anhydrase 2; *CRMP2*, collapsing response mediator protein-2; *DHE3*, glutamate dehydrogenase 1; *ENOA*, enolase alpha; *GalC*, galactosylceramidase; *GAPDH*, glyceraldehyde 3-phosphate dehydrogenase; *GFAP*, glial fibrillary acidic protein; *GM1 ganglioside*, monosialotetrahexosylganglioside; *HBA*, hemoglobin subunit alpha; *HBB*, hemoglobin subunit beta; *HSP7C*, heat shock cognate 71 KDa protein; *IgG Fc*, constant fraction of immunoglobulin G; *MBP*, myelin basic protein; *MYEOV2*, myeloma-overexpressed gene 2; *NFL*, neurofilament light; *NFM*, neurofilament intermediate; *PEBP1*, phosphatidylethanolamine-binding protein 1; *PGK1*, phosphoglycerate kinase 1; *PPIA*, peptidylprolyl isomerase A; *PSMD4*, 26S proteasome non-ATPase regulatory subunit 4; *S100-ß*, protein S100-beta.
[a]*Stages: acute, from injury to 7 days after injury. Subacute, 2–6 weeks after injury. Subchronic, 3–6 months after injury. Chronic, more than 1 year after injury.*

after injury) are pathogenic when injected into the spinal cord of intact mice (Ankeny et al., 2009). Also, IgM natural autoantibodies against annexin IV and phospholipids have been shown to be produced by innate-like B1 cells in the peritoneum after SCI in mice and to be pathogenic (Narang et al., 2017).

The levels of autoantibodies after SCI are independent of lesion level and severity

Immunodepression following after SCI is more pronounced when the lesions occur at cervical or high thoracic spinal segments (above the 5th thoracic segment, T5) both in experimental models and in humans (Brommer et al., 2016; Lucin, Sanders, Jones, Malarkey, & Popovich, 2007; Lucin, Sanders, & Popovich, 2009). A detailed explanation about the pathogenic mechanisms underlying this can be found in the chapter "Infections and spinal cord injury: COVID-19 and beyond"

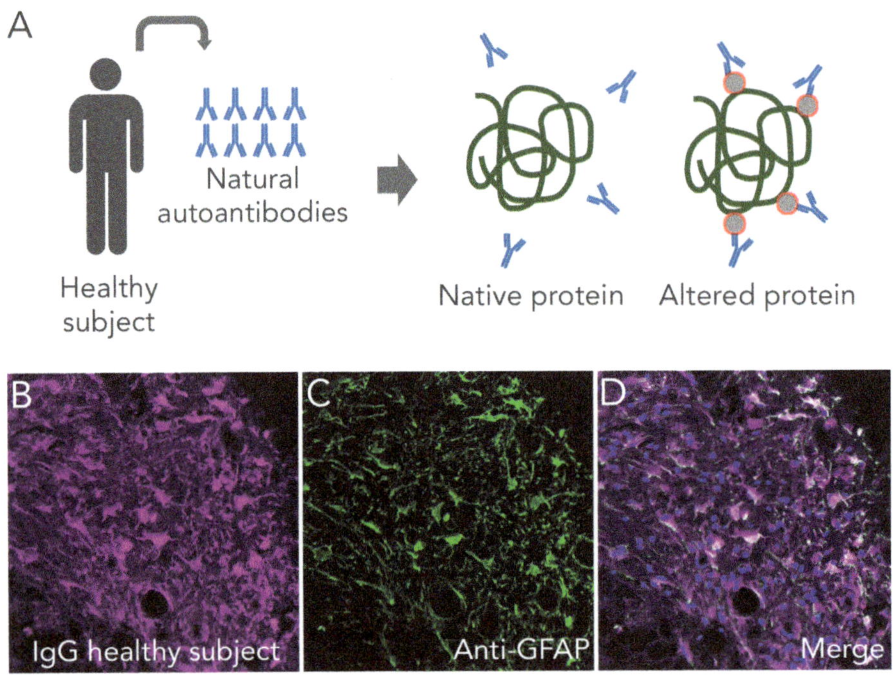

FIG. 2 Natural autoantibodies are preferentially directed against altered proteins. (A) Autoantibodies are normally present in healthy individuals. These naturally occurring autoantibodies (or natural autoantibodies) do not bind to native proteins but to proteins altered under stress conditions such as oxidative damage, inflammation, or cell death. (B) IgG antibodies from healthy human subjects bind to the damaged spinal cord of rats with SCI. (C, D) Among the targets of these natural autoantibodies are antigens expressed by astrocytes (GFAP-positive cells) surrounding the lesion site.

in this same book. Briefly, the spinal sympathetic preganglionic neurons located below the lesion level lose their inhibitory supraspinal descendent input, which results in their overactivation. As a consequence, there is an excessive release of norepinephrine in lymphoid organs, which binds to beta-2-adrenergic receptor on immune cells and drives immunosuppressive responses (Lucin et al., 2007). Since most of the sympathetic innervation of lymphoid organs arises from preganglionic neurons located below T5, lesions above this level result in a greater norepinephrine overflow of lymphoid organs than lesions that partially or totally preserve the normal sympathetic outflow to lymphoid organs. Therefore, it could be reasoned that autoantibody titers might be also affected by the lesion level and, thus, expected that titers should be lower in patients with cervical or high thoracic lesions. However, the autoantibody levels induced by SCI have not been found to be dependent on lesion level by five different studies (Table 2), either comparing patients with tetraplegia and those with paraplegia or comparing lesions above T5 and below T5 (Arevalo-Martin et al., 2018; Davies et al., 2007;

TABLE 2 Association of autoantibody levels with clinical and demographical variables.

Study	Severity (AIS grade)	Lesion level[a]	Gender	Age
Mizrachi et al. (1983)	–		–	–
Hayes et al. (2002)	–*	–		
Davies et al. (2007)		–		
Zajarías-Fainsod et al. (2012)	–*			
Palmers et al. (2016)		–		
Hergenroeder et al. (2018)	–	–		
Arevalo-Martin et al. (2018)	–	–	–	–

AIS, American Spinal Injury Association Impairment Scale.
–, no statistically significant association; +, statistically significant association; –*, differences between AIS grades, but not statistically significant.
[a]*Comparison between tetraplegia and paraplegia in all cases, and additionally above versus below 5th thoracic segment in the study by Arevalo-Martin et al. (2018).*

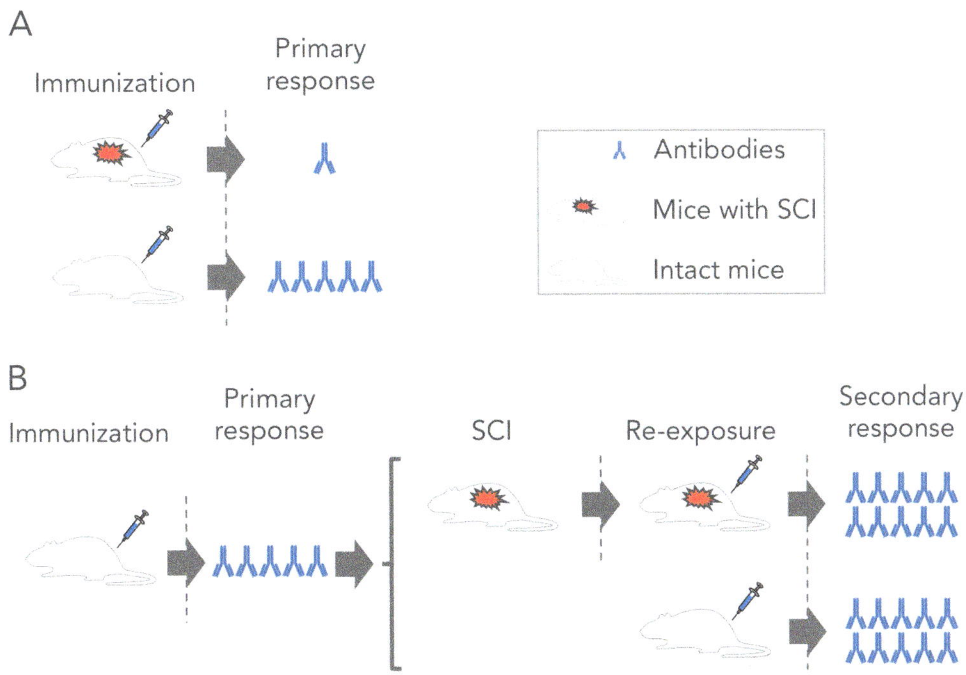

FIG. 3 Secondary antibody responses are preserved after SCI. (A) Primary antibody responses triggered by immunization against new antigens are impaired in mice with SCI. (B) However, the enhanced secondary antibody responses that are triggered after re-exposure to a previously encountered antigen are not different between mice with SCI and intact mice. Antibody levels are indicated by the number of them represented in the figure.

Hayes et al., 2002; Hergenroeder et al., 2018; Palmers et al., 2016). Therefore, the production of autoantibodies after SCI seems to be unsensitive to the main pathological mechanism driving immunodepression after SCI. Among the possible explanations, the B cell subpopulations that produce the autoantibodies elevated in SCI could lack beta-2-adrenergic receptors or these cells may not be reached by sympathetic terminals. Regardless of the molecular mechanism why autoantibodies increase in these otherwise immunodepressed patients, the fact that these antibodies have been identified as pre-existing natural autoantibodies agrees with previous reports demonstrating that SCI in mice preserves secondary humoral responses (increase of pre-existing antibodies) and only impairs primary humoral responses (Fig. 3) (Oropallo et al., 2012).

Also, five studies have reported that autoantibody levels are not associated with lesion severity measured by AIS grade (Table 2) (Arevalo-Martin et al., 2018; Hayes et al., 2002; Hergenroeder et al., 2018; Mizrachi et al., 1983; Zajarías-Fainsod et al., 2012). Among these, two reports have observed that the levels of autoantibodies are different depending on lesion severity, but the differences failed to reach statistical significance (Hayes et al., 2002; Zajarías-Fainsod et al., 2012).

Autoimmune diseases are more frequent in women than men. Although this could lead us to think that sex could make a difference in the autoantibody levels after SCI, these have not been found to be different when comparing by sex (Table 2) (Arevalo-Martin et al., 2018; Mizrachi et al., 1983).

Age is another factor affecting the development of autoimmune diseases, but the levels of autoantibodies after SCI do not vary depending on it (Table 2) (Arevalo-Martin et al., 2018; Mizrachi et al., 1983).

Therefore, data indicate that autoantibody levels rise up after SCI independently of lesion level, severity, gender, and age.

Autoantibodies increased after SCI target both CNS and peripheral antigens

Early studies on the specificities of autoantibodies increased after SCI was carried out by hypothesis-driven methodologies, specifically searching for autoantibodies directed against CNS antigens. With this approach, neurons (targeted by anti-GM1 ganglioside antibodies) and oligodendrocytes (targeted by anti-MBP and anti-GalC antibodies) were proposed as the CNS cellular objectives of a potential SCI-induced autoimmune response (Hayes et al., 2002; Mizrachi et al., 1983; Taranova et al., 1992). But the more recent use of hypothesis-free methodological approaches has expanded the list of autoantibody targets in the CNS to astrocytes (anti-GFAP antibodies) and to proteins that are expressed systemically or that are not considered to be expressed in the CNS (Arevalo-Martin et al., 2018; Hergenroeder et al., 2016; Palmers et al., 2016). Interestingly, this last group of autoantibodies is directed to proteins that seem to be related to secondary disorders occurring after SCI, like anemia, metabolic alterations, or immune depression. For example, at the acute phase patients suffer from anemia and concomitantly exhibit elevated levels of autoantibodies against hemoglobin alpha (HBA) and beta (HBB) (Arevalo-Martin et al., 2018). Also, the metabolic rate is altered after SCI and patients show increased levels of antibodies against the metabolic enzymes glyceraldehyde 3-phosphate dehydrogenase (GAPDH), phosphoglycerate kinase 1 (PGK1), and enolase alpha (ENOA), as well as against the ATP synthase subunit alpha (ATPA) (Arevalo-Martin et al., 2018; Palmers et al., 2016). And the same applies to the antibody against anti-peptidylprolyl isomerase A (PPIA, also known as cyclophilin A), which targets the receptor for the immunosuppressor cyclosporine (PPIA) in patients who concomitantly suffer from immunodepression (Arevalo-Martin et al., 2018). Even if these antibodies do not play a pathogenic role, their increase could indicate that alterations in their target proteins are involved in the development of the systemic alterations after SCI, providing new research perspectives for these disorders.

Antibody levels elevate after their target antigens get exposed to the immune system or after normally exposed antigens are modified or degraded. Accordingly, among the autoantibody targets reported so far (Table 1) it has been observed that at least ENOA, neurofilament light (NFL), neurofilament intermediate (NFM), GFAP, and MBP elevate in the cerebrospinal fluid (CSF) and blood of patients with SCI, most probably as a consequence of cell damage (Ahadi et al., 2015; Hayakawa et al., 2012; Kuhle et al., 2015; Pouw et al., 2014; Wolf et al., 2014). However, it is no so evident why autoantibodies increase against targets that are not thought to be expressed by CNS cells, like HBA and HBB. A potential explanation is that anti-HBA and anti-HBB antibodies may increase after their antigens are released from erythrocytes accumulated in spinal cord hemorrhagic areas. But it should not be discarded that HBA and HBB have been described to be expressed also by neurons and to play a role in response to injury (Richter, Meurers, Zhu, Medvedeva, & Chesselet, 2009). Regardless of whether hemoglobins come from spinal cord bleeding or by endogenous production, their levels rise up in the CNS after SCI since HBB is detected in the CSF of patients with SCI as early as 1 day after injury (Sengupta et al., 2014). Similarly, antibodies against albumin (ALBU, Table 1), a protein mainly synthesized by the liver, increase after injury. Albumin is normally exposed to the immune system as a freely circulating protein in blood and as a normal constituent of CSF. Therefore, it seems that more likely anti-albumin antibodies could rise due to the modification or degradation of albumin rather than to changes in their blood levels. Accordingly, two of the three albumin isoforms targeted by anti-albumin antibodies exhibit a lower molecular weight than expected (Arevalo-Martin et al., 2018).

Summarizing, in spite that it seems evident to think that autoantibodies that increase after SCI must be triggered by damage to the spinal cord and must drive autoimmune responses against the CNS, it should not be discarded that autoantibody levels could be triggered by damage outside the CNS and could drive autoimmune responses outside the CNS.

Local versus systemic production of autoantibodies

One important question to solve for understanding how autoantibody responses take place after SCI is whether autoantibodies are produced locally, in the damaged spinal cord, or systemically. Supporting that intraspinal production may take place, B cells have been shown to accumulate in the spinal cord of mice with SCI, forming structures similar to lymphoid follicles (Ankeny et al., 2006, 2009). But in humans, a detailed histopathological analysis from the acute to chronic SCI failed to detect B cells infiltrating the spinal cord after SCI neither in injured areas nor in their surroundings (Fleming et al., 2006). In a more recent study, a pronounced B cell infiltration was found only in two cases from the 22 patients evaluated (Zrzavy et al., 2021). Nevertheless, even though if autoantibodies are produced majorly outside the CNS in human SCI, as data suggest, blood-brain barrier is damaged after SCI, so antibodies could rapidly enter into the injured spinal cord. Accordingly, antibodies binding to cell profiles reminiscent of glial cells and neurons are detected in the spinal cord of rats as soon as 1 day after injury (Arevalo-Martin et al., 2018).

Association of autoantibodies with neuropathic pain development

One of the most common secondary complications affecting patients with SCI is the development of neuropathic pain, caused among other factors by the inflammatory response in the peripheral and central nervous system. In fact, abrogating the inflammatory response in experimental models ameliorates neuropathic pain. Thus, since autoantibodies are specific biomarkers of the immune responses taking place, it is reasonable to think that some of the increased autoantibodies could serve as biomarkers of neuropathic pain. Accordingly, anti-GM1 ganglioside antibodies have been related to the development of neuropathic pain (Davies et al., 2007), since patients suffering from it exhibit higher levels of anti-GM1 antibodies than patients without it. However, this relationship is not statistically significant after adjusting for severity (measured as AIS grade) and paraplegia versus tetraplegia in multivariate linear regression models.

Also, the levels of anti-GFAP antibodies in the acute/subacute phase (at 16 ± 7 days after injury) have been shown to be significantly associated with the development of neuropathic pain (Hergenroeder et al., 2018). In fact, if instead of considering the raw anti-GFAP autoantibody levels, a titer of 1:128 is established as a threshold to classify patients among anti-GFAP-positive or -negative, the odds of developing neuropathic pain increase by almost eight times in anti-GFAP-positive patients (Hergenroeder et al., 2018). In addition, although anti-collapsing response mediator protein-2 (CRMP2) antibodies are not useful to predict neuropathic pain development on their own, they improve the predictive ability of anti-GFAP antibodies. The odds of developing neuropathic pain increases by 9.5 times in patients positive for anti-GFAP and/or anti-CRMP2 antibodies, remaining significant after adjusting for clinical and demographical variables. Whether anti-GFAP antibodies are involved in neuropathic pain pathophysiology or are just biomarkers is a question that remains unsolved.

Roles of the autoantibodies increased after SCI in other pathologies

In spite that there is a lack of a direct experimental demonstration about the pathogenic role of autoantibodies in human SCI, there is evidence for a pathogenic role of some of them in other neurological disorders. As commented earlier, anti-GM1 ganglioside antibodies were proposed to underlie the inhibition of neuritogenesis exerted by sera of patients with SCI on neuron cell cultures, but a direct evidence of anti-GM1 antibodies underlying this effect has not been provided (Mizrachi et al., 1983). However, anti-GM1 antibodies are known blockers of nerve currents and are associated with paralysis in Guillen-Barre syndrome, post-polio syndrome, and other neurological disorders, including amyotrophic lateral sclerosis (Illa et al., 1995; Takigawa et al., 1995; Yu & Usuki, 2014).

Recently, in 2016, a new meningoencephalomyelitis called Autoimmune Glial Fibrillary Acidic Protein Astrocytopathy was described (Fang et al., 2016). This pathology is characterized by high levels of pathogenic anti-GFAP autoantibodies and astrocyte damage (Kunchok, Zekeridou, & McKeon, 2019; Yang et al., 2019; Zhou, Yu, & Hong, 2021). Noteworthy, this pathological condition has been reported that may exhibit clinical characteristics of Brown-Sequard syndrome (Zhou et al., 2021).

After SCI, a progressive degradation of myelin is observed in the spinal cord of patients (Buss et al., 2004). Interestingly, anti-MBP antibodies from multiple sclerosis (MS) patients were shown to cleave MBP peptides and cause myelin damage (Ponomarenko et al., 2006a). Moreover, the hydrolytic activity of anti-MBP antibodies correlates with the Expanded Disability Status Scale (EDSS) used to quantify disability in MS patients (Ponomarenko et al., 2006b).

Also, anti-GalC antibodies open Ca^{2+} channels in oligodendrocytes that lead to cytoskeleton reorganization and contraction of myelin sheets (Dyer & Benjamins, 1990). In vivo, the implantation of a hybridoma secreting anti-GalC antibodies in the spinal cord dorsal column of early postnatal rats impairs myelination in the surroundings of the implant (Rosenbluth, Liu, Guo, & Schiff, 1994).

Overall, evidence support the notion that autoantibodies increasing after SCI may have a physiopathological role and constitute new therapeutic targets.

Applications to other areas of neuroscience

In this chapter, we have reviewed the current knowledge about the origin, specificities, and the potential physiopathological role of the autoantibodies that increases after spinal cord injury (SCI). Regarding their origin, we have discussed the data providing evidence that the increased autoantibodies are pre-existing natural antibodies. In other neurological diseases that share physiopathological mechanisms with SCI, like stroke, it has been previously reported that autoantibodies that are triggered after the insult are natural antibodies that depending on the previous state of the blood-brain barrier may act as a counteractive response or as a pathogenic factor (Zerche et al., 2015). Despite its clinical relevance, the idea of pre-existing antibodies in healthy subjects is still a controversial question (Ehrenreich, 2018). The observations performed in SCI may help to further support that the elevation of natural autoantibodies in neurological diseases is not anecdotic and should be taken into consideration for clinical studies.

Also, immunodepression and elevated levels of autoantibodies have been described to occur after stroke and in traumatic brain injuries (Javidi & Magnus, 2019). Both phenomena are classically considered to be mutually exclusive. However, the observations that after SCI pre-existing humoral immune responses are preserved and that increased autoantibodies are pre-existing natural antibodies may help to understand why immunodepression and elevated levels of autoantibodies also co-occur after stroke and traumatic head injuries.

Mini-dictionary of terms

AIS grade: One of the five categories included in the American Spinal Injury Association (ASIA) of severity of spinal cord injury based on the assessments following the International Standards for the Neurological Classification of Spinal Cord Injury (ISNCSCI). It includes four categories for patients with sensorimotor impairments, AIS A to AIS D, and an additional category, AIS E, for patients recovering sensorimotor function evaluated by ISNCSCI.

Autoantibodies: Antibodies recognizing self-antigens. May drive destructive immune responses against self-constituents.

Beta-2-adrenergic receptor: G protein-coupled receptor for norepinephrine. Expressed in multiple cell types, including leukocytes.

Galactosylceramidase (GalC): Enzyme that hydrolyzes the galactose ester bond from galactosylceramidase and other lipids in the cellular membrane. Expressed by oligodendrocytes in the central nervous system.

Glial fibrillary acidic protein (GFAP): Type III intermediate filament, constituent of astrocytes cytoskeleton.

GM1 ganglioside: Ganglioside involved in the normal development of central nervous system, in neural plasticity and neural repair. Its therapeutic potential has been tested in several neurological disorders, including spinal cord injury.

Immunodepression: Strong diminution of the immune responses. Clinically manifested as suffering from higher susceptibility to infections.

Myelin basic protein (MBP): A major protein component of myelin sheaths in the central and peripheral nervous system. Located in the myelin major dense lines, necessary for the correct packaging of myelin.

Neuropathic pain: Pain occurring after damage to the central or peripheral nervous systems that is unrelated to the existence of a nociceptive stimulus. Frequent among spinal cord injury patients and usually described as burning sensation and hypersensitive to touch.

Norepinephrine: Also called noradrenaline, is a neurotransmitter of the catecholamine family secreted by post-ganglionic sympathetic neurons. It may also act as a hormone.

Primary humoral responses: Antibody-mediated immune response occurring against an antigen when it is first encountered. After several steps and a period of time of 14–21 days, IgG-specific antibodies are generated.

Secondary humoral responses: Antibody-mediated immune response against an antigen previously recognized. These responses are much faster than primary responses since specific IgGs were already previously generated in the course of the first encounter with the antigen.

Key facts of autoantibodies

- Autoantibodies are antibodies that instead of being directed against external components (such as microorganisms) are directed to normal components of the body.
- Healthy individuals present autoantibodies called natural autoantibodies (shortening of naturally occurring autoantibodies).
- Natural autoantibodies play a role in the maintenance and repair of body tissues as well as in the control of immune reactions.
- Intravenous immunoglobulins are pools of antibodies extracted from thousands of healthy donors that are enriched in natural autoantibodies and are used as therapy for both autoimmune diseases and immunodepression.
- After persistent exposure to their molecular target, natural autoantibodies may give rise to pathogenic autoantibodies, i.e., autoantibodies that drive immune reactions against self-structures.
- Not all autoantibodies originate from natural autoantibodies. In this regard, a commonly proposed origin of autoantibodies are antibodies directed against microorganisms that cross-react with self-structures.
- Autoantibodies are active players in numerous autoimmune diseases, like multiple sclerosis, systemic lupus erythosus, or rheumatoid arthritis.

Summary points

- The levels of autoantibodies targeting 24 different antigens are increased in patients with spinal cord injury.
- The increased autoantibodies target both central nervous system specific antigens as well as systemic antigens.
- Autoantibody levels are independent of lesion level, severity, age, and gender.
- The increased autoantibodies have been shown to be, at least in the early phases, natural antibodies.

- IgG autoantibodies extracted from mice with chronic spinal cord injury (42 days after injury) elicit a necrotic inflammatory lesion when injected into the spinal cord of intact mice.
- Currently, there is not a direct demonstration of the physiopathological role of the autoantibodies that increase after human SCI. However, the levels of anti-GFAP antibodies are related (with statistical significance after adjusting for clinical and demographical variables) to the development of neuropathic pain in patients with spinal cord injury.
- Some of the autoantibodies elevated after spinal cord injury are known to be pathogenic in other neurological diseases.

References

Abou-El-Hassan, H., Bsat, S., Sukhon, F., Assaf, E. J., Mondello, S., Kobeissy, F., et al. (2020). Protein degradome of spinal cord injury: Biomarkers and potential therapeutic targets. *Molecular Neurobiology*, *57*, 2702–2726.

Ahadi, R., Khodagholi, F., Daneshi, A., Vafaei, A., Mafi, A. A., & Jorjani, M. (2015). Diagnostic value of serum levels of GFAP, pNF-H, and NSE compared with clinical findings in severity assessment of human traumatic spinal cord injury. *Spine*, *40*, E823–E830.

Ankeny, D. P., Guan, Z., & Popovich, P. G. (2009). B cells produce pathogenic antibodies and impair recovery after spinal cord injury in mice. *Journal of Clinical Investigation*, *119*, 2990–2999.

Ankeny, D. P., Lucin, K. M., Sanders, V. M., McGaughy, V. M., & Popovich, P. G. (2006). Spinal cord injury triggers systemic autoimmunity: Evidence for chronic B lymphocyte activation and lupus-like autoantibody synthesis. *Journal of Neurochemistry*, *99*, 1073–1087.

Arevalo-Martin, A., Grassner, L., Garcia-Ovejero, D., Paniagua-Torija, B., Barroso-Garcia, G., Arandilla, A. G., et al. (2018). Elevated autoantibodies in subacute human spinal cord injury are naturally occurring antibodies. *Frontiers in Immunology*, *9*, 2365.

Brennan, F. H., Kurniawan, N. D., Vukovic, J., Bartlett, P. F., Käsermann, F., Arumugam, T. V., et al. (2016). IVIg attenuates complement and improves spinal cord injury outcomes in mice. *Annals of Clinical and Translational Neurology*, *3*, 495–511.

Brommer, B., Engel, O., Kopp, M. A., Watzlawick, R., Müller, S., Prüss, H., et al. (2016). Spinal cord injury-induced immune deficiency syndrome enhances infection susceptibility dependent on lesion level. *Brain*, *139*, 692–707.

Buss, A., Brook, G. A., Kakulas, B., Martin, D., Franzen, R., Schoenen, J., et al. (2004). Gradual loss of myelin and formation of an astrocytic scar during Wallerian degeneration in the human spinal cord. *Brain*, *127*, 34–44.

Cohen, I. R. (2007). Biomarkers, self-antigens and the immunological homunculus. *Journal of Autoimmunity*, *29*, 246–249.

Cohen, I. R., & Cooke, A. (1986). Natural autoantibodies might prevent autoimmune disease. *Immunology Today*, *7*, 363–364.

Davies, A. L., Hayes, K. C., & Dekaban, G. A. (2007). Clinical correlates of elevated serum concentrations of cytokines and autoantibodies in patients with spinal cord injury. *Archives of Physical Medicine and Rehabilitation*, *88*, 1384–1393.

Dyer, C. A., & Benjamins, J. A. (1990). Glycolipids and transmembrane signaling: Antibodies to galactocerebroside cause an influx of calcium in oligodendrocytes. *Journal of Cell Biology*, *111*, 625–633.

Ehrenreich, H. (2018). Autoantibodies against N-methyl-d-aspartate receptor 1 in health and disease. *Current Opinion in Neurology*, *31*, 306–312.

Fang, B., McKeon, A., Hinson, S. R., Kryzer, T. J., Pittock, S. J., Aksamit, A. J., et al. (2016). Autoimmune glial fibrillary acidic protein astrocytopathy: A novel meningoencephalomyelitis. *JAMA Neurology*, *73*, 1297–1307.

Fiskesund, R., Stegmayr, B., Hallmans, G., Vikstrom, M., Weinehall, L., de Faire, U., et al. (2010). Low levels of antibodies against phosphorylcholine predict development of stroke in a population-based study from Northern Sweden. *Stroke*, *41*, 607–612.

Fleming, J. C., Norenberg, M. D., Ramsay, D. A., Dekaban, G. A., Marcillo, A. E., Saenz, A. D., et al. (2006). The cellular inflammatory response in human spinal cords after injury. *Brain*, *129*, 3249–3269.

Hauben, E., Agranov, E., Gothilf, A., Nero, U., Cohen, A., Smirnov, I., et al. (2001). Posttraumatic therapeutic vaccination with modified myelin self-antigen prevents complete paralysis while avoiding autoimmune disease. *Journal of Clinical Investigation*, *108*, 591–599.

Hayakawa, K., Okazaki, R., Ishii, K., Ueno, T., Izawa, N., Tanaka, Y., et al. (2012). Phosphorylated neurofilament subunit NF-H as a biomarker for evaluating the severity of spinal cord injury patients, a pilot study. *Spinal Cord*, *50*, 493–496.

Hayes, K. C., Hull, T. C. L., Delaney, G. A., Potter, P. J., Sequeira, K. A. J., Campbell, K., et al. (2002). Elevated serum titers of proinflammatory cytokines and CNS autoantibodies in patients with chronic spinal cord injury. *Journal of Neurotrauma*, *19*, 753–761.

Hergenroeder, G. W., Moore, A. N., Schmitt, K. M., Redell, J. B., & Dash, P. K. (2016). Identification of autoantibodies to glial fibrillary acidic protein in spinal cord injury patients. *Neuroreport*, *27*, 90–93.

Hergenroeder, G. W., Redell, J. B., Choi, H. A., Schmitt, L. H., Donovan, W., Francisco, G. E., et al. (2018). Increased levels of circulating GFAP and CRMP2 autoantibodies in the acute stage of spinal cord injury predict the subsequent development of neuropathic pain. *Journal of Neurotrauma*, *35*, 2530–2539.

Illa, I., Leon-Monzon, M., Agboatwalla, M., Ilyas, A., Latov, N., & Dalakas, M. C. (1995). Antiganglioside antibodies in patients with acute polio and post-polio syndrome. *Annals of the New York Academy of Sciences*, *753*, 374–377.

Javidi, E., & Magnus, T. (2019). Autoimmunity after ischemic stroke and brain injury. *Frontiers in Immunology*, *10*, 686.

Kidd, J. G., & Friedewald, W. F. (1942). A natural antibody that reacts in vitro with a sedimentable constituent of normal tissue cells : II. Specificity of the phenomenon: General discussion. *Journal of Experimental Medicine*, *76*, 557–578.

Kuhle, J., Gaiottino, J., Leppert, D., Petzold, A., Bestwick, J. P., Malaspina, A., et al. (2015). Serum neurofilament light chain is a biomarker of human spinal cord injury severity and outcome. *Journal of Neurology, Neurosurgery and Psychiatry*, *86*, 273–279.

Kunchok, A., Zekeridou, A., & McKeon, A. (2019). Autoimmune glial fibrillary acidic protein astrocytopathy. *Current Opinion in Neurology*, *32*, 452–458.

Lucin, K., Sanders, V., Jones, T., Malarkey, W., & Popovich, P. (2007). Impaired antibody synthesis after spinal cord injury is level dependent and is due to sympathetic nervous system dysregulation. *Experimental Neurology*, *207*, 75–84.

Lucin, K. M., Sanders, V. M., & Popovich, P. G. (2009). Stress hormones collaborate to induce lymphocyte apoptosis after high level spinal cord injury. *Journal of Neurochemistry*, *110*, 1409–1421.

Lutz, H. U. (2007). Homeostatic roles of naturally occurring antibodies: An overview. *Journal of Autoimmunity*, *29*(4), 287–294. https://doi.org/10.1016/j.jaut.2007.07.007.

Mizrachi, B. Y., Ohry, A., Aviel, A., Rozin, R., Brooks, M. E., & Schwartz, M. (1983). Systemic humoral factors participating in the course of spinal cord injury. *Paraplegia*, *21*, 287–293.

Narang, A., Qiao, F., Atkinson, C., Zhu, H., Yang, X., Kulik, L., et al. (2017). Natural IgM antibodies that bind neoepitopes exposed as a result of spinal cord injury, drive secondary injury by activating complement. *Journal of Neuroinflammation*, *14*, 120.

Nataf, S. (2017). Autoimmunity as a driving force of cognitive evolution. *Frontiers in Neuroscience*, *11*, 582.

Oropallo, M. A., Held, K. S., Goenka, R., Ahmad, S. A., O'Neill, P. J., Steward, O., et al. (2012). Chronic spinal cord injury impairs primary antibody responses but spares existing humoral immunity in mice. *Journal of Immunology*, *188*, 5257–5266.

Palmers, I., Ydens, E., Put, E., Depreitere, B., Bongers-Janssen, H., Pickkers, P., et al. (2016). Antibody profiling identifies novel antigenic targets in spinal cord injury patients. *Journal of Neuroinflammation*, *13*, 1–11.

Petrova, N. V., Ponomaryova, A. M., Alyoshkin, V. A., Eliseyev, A. T., & Yumashev, G. S. (1993). Serum rheumatoid factors in spinal cord injury patients. *Paraplegia*, *31*, 265–268.

Ponomarenko, N. A., Durova, O. M., Vorobiev, I. I., Belogurov, A. A., Kurkova, I. N., Petrenko, A. G., et al. (2006a). Autoantibodies to myelin basic protein catalyze site-specific degradation of their antigen. *Proceedings of the National Academy of Sciences of the United States of America*, *103*, 281–286.

Ponomarenko, N. A., Durova, O. M., Vorobiev, I. I., Belogurov, A. A., Telegin, G. B., Suchkov, S. V., et al. (2006b). Catalytic activity of autoantibodies toward myelin basic protein correlates with the scores on the multiple sclerosis expanded disability status scale. *Immunology Letters*, *103*, 45–50.

Pouw, M. H., Kwon, B. K., Verbeek, M. M., Vos, P. E., Van Kampen, A., Fisher, C. G., et al. (2014). Structural biomarkers in the cerebrospinal fluid within 24 h after a traumatic spinal cord injury: A descriptive analysis of 16 subjects. *Spinal Cord*, *52*, 428–433.

Richter, F., Meurers, B. H., Zhu, C., Medvedeva, V. P., & Chesselet, M. F. (2009). Neurons express hemoglobin α- and β-chains in rat and human brains. *Journal of Comparative Neurology*, *515*, 538–547.

Rosenbluth, J., Liu, Z., Guo, D., & Schiff, R. (1994). Inhibition of CNS myelin development in vivo by implantation of anti-GalC hybridoma cells. *Journal of Neurocytology*, *23*, 699–707.

Schwartz-Albiez, R., Monteiro, R. C., Rodriguez, M., Binder, C. J., & Shoenfeld, Y. (2009). Natural antibodies, intravenous immunoglobulin and their role in autoimmunity, cancer and inflammation. *Clinical and Experimental Immunology*, *158*, 43–50.

Sengupta, M. B., Basu, M., Iswarari, S., Mukhopadhyay, K. K., Sardar, K. P., Acharyya, B., et al. (2014). CSF proteomics of secondary phase spinal cord injury in human subjects: Perturbed molecular pathways post injury. *PLoS One*, *9*, e110885.

Takigawa, T., Yasuda, H., Kikkawa, R., Shigeta, Y., Saida, T., & Kitasato, H. (1995). Antibodies against GM1 ganglioside affect K+ and NA+ currents in isolated rat myelinated nerve fibers. *Annals of Neurology*, *37*, 436–442.

Taranova, N. P., Makarov, A. I., Amelina, O. A., Luchakova, O. S., Loboda, E. B., & Leikin, I. B. (1992). The production of autoantibodies to nerve tissue glycolipid antigens in patients with traumatic spinal cord injuries. *Zhurnal Voprosy Neĭrokhirurgii Imeni N. N. Burdenko*, *4-5*, 21–24.

Wang, R., Chen, J., Zhou, S., Li, C., Yuang, G., Xu, W., et al. (1995). Enzyme-linked immunoadsorbent assays for myelin basic protein and antibodies to myelin basic protein in serum and CSF of patients with diseases of the nervous system. *Hua Xi Yi Ke Da Xue Xue Bao*, *26*, 131–134.

Wolf, H., Krall, C., Pajenda, G., Leitgeb, J., Bukaty, A. J., Hajdu, S., et al. (2014). Alterations of the biomarker S-100B and NSE in patients with acute vertebral spine fractures. *Spine Journal*, *14*, 2918–2922.

Yang, X., Huang, Q., Yang, H., Liu, S., Chen, B., Liu, T., et al. (2019). Astrocytic damage in glial fibrillary acidic protein astrocytopathy during initial attack. *Multiple Sclerosis and Related Disorders*, *29*, 94–99.

Yu, R. K., & Usuki, S. (2014). Human anti-glycosphingolipids antibodies in Guillain-Barre syndrome. In Y. Shoenfeld, P. L. Meroni, & M. E. Gershwin (Eds.), *Autoantibodies* (3rd ed., pp. 581–594). Elsevier.

Zajarías-Fainsod, D., Carrillo-Ruiz, J., Mestre, H., Grijalva, I., Madrazo, I., & Ibarra, A. (2012). Autoreactivity against myelin basic protein in patients with chronic paraplegia. *European Spine Journal*, *21*, 964–970.

Zerche, M., Weissenborn, K., Ott, C., Dere, E., Asif, A. R., Worthmann, H., et al. (2015). Preexisting serum autoantibodies against the NMDAR subunit NR1 modulate evolution of lesion size in acute ischemic stroke. *Stroke*, *46*, 1180–1186.

Zhou, Z. H., Yu, Q. Y., & Hong, M. F. (2021). Autoimmune glial fibrillary acidic protein astrocytopathy with the detection of anti-GM3 and anti-GM4 IgG antibodies, masquerading as brown-sequared syndrome. *JAMA Neurology*, *78*, 115–116. Online ahead of print.

Zrzavy, T., Schwaiger, C., Wimmer, I., Berger, T., Bauer, J., Butovsky, O., et al. (2021). Acute and non-resolving inflammation associate with oxidative injury after human spinal cord injury. *Brain*, *144*, 144–161. awaa360.

Chapter 21

Calpain role in the pathophysiology of spasticity after spinal cord injury

Nejada Dingu, Hélène Bras, and Frédéric Brocard
Faculté de Médecine de la Timone, Institut de Neurosciences de la Timone, UMR7298 CNRS and AMU, Marseille, France

List of abbreviations

$[Cl^-]_i$	intracellular chloride concentration
AIS	axonal initial segment
Ca^{2+} PIC or I_{CaP}	calcium persistent inward current
$Ca_v1.3$	voltage-gated calcium channel 1.3
CPG	central pattern generator
E_{Cl^-}	chloride equilibrium potential
$GABA_A$	γ-aminobutyric acid type A receptor
KCC2	potassium-chloride cotransporter 2
Na^+PIC or I_{NaP}	sodium persistent inward current
$Na_v1.6$	voltage-gated sodium channel 1.6
PEST	proline, glutamic acid, serine, threonine
PICs	persistent inward currents
ROS	reactive oxygen species
SCI	spinal cord injury

Introduction

Spinal cord injury (SCI) is a major cause of permanent disability with physical and psychological secondary complications for patients (Table 1). Neurological damage after SCI occurs in two phases, referred to as primary and secondary injury, characterized by peculiar cellular and molecular modifications of the spinal microenvironment. The primary SCI is the direct result of the initial mechanical trauma that induces cell death within the lesion epicenter. It additionally causes anatomical disconnection with axon tracts along with the lesion undergoing *Wallerian degeneration* (Hill, Beattie, & Bresnahan, 2001). Vascularization is also disrupted and the reintroduction of oxygen to ischemic areas after surgical decompression of the spinal cord instigates *reperfusion injury* leading to oxidative damage and the release of reactive oxygen species (ROS) that further contribute to cell death and inflammation (Kalogeris, Baines, Krenz, & Korthuis, 2012).

The secondary SCI lasts until the *glial scar* appears and it dramatically expands the primary injury from the lesion epicenter into the adjacent spinal segments. It occurs due to a variety of concomitant factors, such as hemorrhage, ischemia, excitotoxicity, inflammation, that exacerbate neurological deficits and outcomes. The inflammatory response is supported by activated microglia and infiltrating immune cells that easily cross the disrupted blood-spinal cord barrier and massively release pro-inflammatory cytokines and chemo-attractant molecules. Inflammatory processes also contribute to ROS formation, which causes DNA oxidative stress, protein oxidation, and lipid peroxidation, further promoting cellular death (Hall, Wang, Bosken, & Singh, 2016; Liu et al., 2020). In turn, damaged cells release excitatory amino acids, such as glutamate and aspartate, which overexcite neighboring neurons and trigger a highly disruptive process named *excitotoxicity*. In addition, a loss of ionic homeostasis is established, with intracellular calcium concentration increasing and persisting high for long periods (Fan et al., 2018). Lastly, hypercalcemia triggers a series of destructive events, including mitochondrial dysfunction, cell death, ROS production, acidosis, activation of calcium-dependent enzymes like calpains.

Cellular, Molecular, Physiological, and Behavioral Aspects of Spinal Cord Injury. https://doi.org/10.1016/B978-0-12-822427-4.00021-6

TABLE 1 Secondary conditions after spinal cord injury (SCI).

Physical conditions after SCI	Nonphysical conditions after SCI
Spinal shock	Sleep and circadian rhythm disorders
Spasticity	Depression
Autonomic dysreflexia	Anxiety
Orthostatic hypotension	Posttraumatic stress disorder
Bradycardia	
Thermal dysregulation	
Respiratory problems	
Bladder dysfunction	
Urinary tract infections	
Bowel dysfunction	
Sexual dysfunction	
Pregnancy complications	
Osteoporosis and osteoporotic limb fractures	
Joint and muscle pain	
Neuropathic pain (at and below level)	
Pressure ulcers	

Secondary conditions impact health and psychosocial well-being in individuals with SCI.

Hereinafter, we will focus on spasticity, one of the main pathophysiological consequences of SCI. We will present an overview of principal animal models for the study of this highly impairing condition. We will then extensively describe spinal changes taking place after SCI and leading to spasticity. This will be finally followed by a discussion of our recent studies on calpains, a family of calcium-activated proteases, as new therapeutic targets against spasticity.

Spasticity after spinal cord injury (SCI)

Spasticity is a pathological condition related to several neurological disorders inducing a primary and secondary injury of the motor system. The classical definition of spasticity states that it is a sensory-motor disorder characterized by heightened muscle tone (hypertonia) due to a velocity-dependent increase in tonic stretch reflexes (Lance, 1980). In other words, the faster a muscle stretch is performed, the higher the muscle resistance will be, showing the characteristic velocity-dependence of spasticity that results from abnormal spinal processing of proprioceptive input.

After SCI there is a 60–80% risk of developing spasticity one-year postinjury (Adams & Hicks, 2005), hence spasticity directly correlates with a significant economic burden. The time course of the disease is characterized by an initial period of flaccid muscle paralysis and a failure or depression of spinal reflexes below the level of the injury called *spinal shock* (Atkinson & Atkinson, 1996). Then, maladaptive mechanisms promoting generalized spinal hyperexcitability take place and typical clinical signs of spasticity appear, including pathological reflexes (or hyperreflexia), clonus (involuntary and rhythmic muscular contractions), spasms (involuntary contraction of a muscle or group of muscles), and hypertonia, that lead to muscle alterations and joint deformations.

There are several diagnostic tools for the measurement of spasticity, such as the Ashworth scale. However, clinical diagnosis is still operator-sensitive, lacks reliability, and does not provide detailed validation studies of such tools in the literature. This leads to erroneous management of spastic patients who likely undergo misclassification or delayed-treatment administration. The most adopted therapeutic strategy concerns the use of oral antispastic medications, which show limited effectiveness and various systemic side effects, especially in cases of severe and drug-resistant spasticity. Thus, systematic and in-depth investigations are required, by recurring to laboratory animal models.

Animal models of spasticity after SCI

Preclinical animal models of spasticity provide an essential tool to uncover pathological processes and to identify new and effective therapeutic strategies. Rodents are the most commonly used species in preliminary studies aimed at relating signs of sensory-motor dysfunction to network, cellular and molecular mechanisms (Sharif-Alhoseini et al., 2017). Rat models are relatively inexpensive, readily available having high reproductive rates, easy to handle and care, with a well-understood anatomy and pathology progress that mimics the human condition. Mice have the additional advantage of allowing for genetic manipulations and advanced optical techniques (e.g., optogenetics and calcium imaging) of identified neuron sub-populations, whereas the generation of transgenic models in rats is very limited.

Several rodent models have been developed in adult animals for the study of spasticity. Complete transection of the spinal cord is the most extensively used type of injury to induce spasticity, since it allows evoking long-lasting clinical symptoms closely resembling the human pathology, it is relatively easy to perform and very well reproducible. Our team has been investigating pathophysiological mechanisms of spasticity in a complete SCI model at a low thoracic level (T8-T9) in adult rats (Boulenguez et al., 2010; Brocard et al., 2016; Sadlaoud et al., 2020). This kind of injury causes complete paralysis of the hindlimbs, with animals using their forelimbs to move. As a result of the severe sensory-motor impairment, animals show co-contractions of antagonist muscles of the hindlimb that faithfully mimic clinical symptoms (Fig. 1A). Autonomic dysfunctions are also present and attentive caregiving is required, such as daily assisted bladder emptying, treatment of urinary infections and sores, etc. To preserve normal locomotor function and avoid autonomic signs, other groups opt for a sacral transection (S2 level) in adult rats, which evokes a spastic syndrome restricted to the tail muscles (Bennett et al., 1999). Importantly, the reduced diameter of the sacral spinal cord makes it viable for in vitro recordings of intracellular activities (Bennett, Li, & Siu, 2001), whereas big limitations are encountered in maintaining alive the lumbar spinal cord of adult animals ex vivo. To combine electrophysiology with genetic tools, the model of sacral SCI has recently been translated to adult mice (Bellardita, Marcantoni, Löw, & Kiehn, 2018), allowing to clarify mechanisms underlying muscle spasms after chronic SCI (Bellardita et al., 2017) and to identify potential therapeutic targets to prevent spasticity (Marcantoni et al., 2020).

Neonatal rodent models are also available for the study of spasticity. We have previously described a highly reproducible model where a complete transection is performed at T8-T9 spinal level in neonate rats at birth (Boulenguez et al., 2010; Brocard et al., 2016; Plantier et al., 2019). Interestingly, in neonates signs of spasticity emerge 4–5 days after SCI (Fig. 1B), with spinal hyperexcitability appearing already 48 h postinjury in the sublesional spinal cord (Plantier et al., 2019). On the contrary, adult animals undergo spinal shock during acute SCI and several weeks are necessary for spasticity to be firmly established (Bellardita et al., 2017; Bennett et al., 1999; Corleto et al., 2015). Therefore, the early appearance of

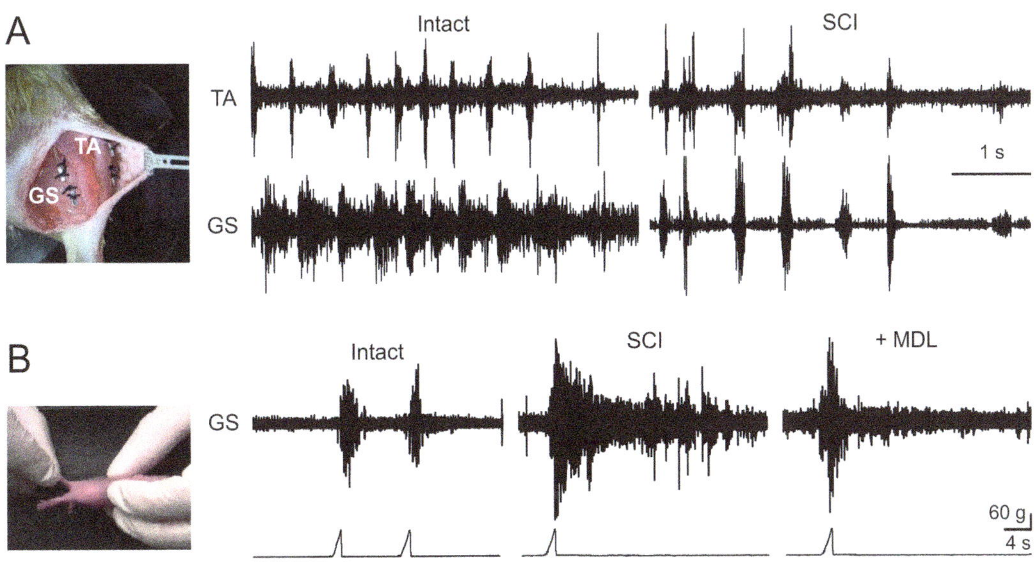

FIG. 1 Muscle spasms in animal models of spasticity after SCI. Electromyographic activity of hindlimb muscles recorded in adult rats (A), showing alternated contractions between antagonist muscles of the hindlimb in intact animals and co-contractions in paraplegic rats after SCI. In neonates (B), muscle activity was evoked by pressing the tail on a calibrated platform, and calpain inhibition with MDL28170 reduced huge muscle responses induced by the lesion. (*TA*, tibialis anterior; *GS*, gastrocnemius). *Panel (B): modified from Plantier, V., Sanchez-Brualla, I., Dingu, N., Brocard, C., Liabeuf, S., Gackière, F., et al. (2019) Calpain fosters the hyperexcitability of motoneurons after spinal cord injury and leads to spasticity.* eLife, *8, e51404.*

spasticity after SCI makes neonatal models very appealing, since they allow for collecting data in a quite shorter time. The greater viability for in vitro recordings and the almost unneeded postsurgical care should also be taken into account.

Network alterations contributing to spasticity after SCI

Animal models of spasticity allow shedding some light on the maladaptive plasticity of spinal networks. In in vitro spinal cords isolated at postnatal day 4/5 from SCI neonates, we observe a dramatic increase in spinal excitability, both in terms of spontaneous activity and evoked reflexes, strongly linked to spasticity (Fig. 2A–C; Plantier et al., 2019).

Spontaneous activities generally correspond to big bursts of uncorrelated tonic activity recorded as a network output from spinal ventral roots, collecting efferent fibers of α-motoneurons (Fig. 2B). Their origin is unclear since the spinal circuits responsible for their emergence are unknown. Sometimes spontaneous bursts appear as oscillatory rhythmic activities in lumbar ventral roots (Fig. 2B'). Given that the lumbar spinal nerves are responsible for muscle innervation of the hindlimbs, we assume that this oscillatory activity closely resembling *locomotor-like episodes* might likely arise from spinal networks belonging to the *central pattern generator* (CPG) for locomotion. This observation confirms a generalized hyperexcitability of the spinal cord, where circuits usually inactive like the CPG, that require pharmacological or electrical stimuli to be triggered, become spontaneously activated as a result of network plastic changes induced by the lesion. Furthermore, we can speculate that the clinical correlate of these big oscillatory bursts might be the rhythmic muscle contractions observed during clonus, whereas nonoscillatory bursts might correspond to involuntary muscle spasms. Indeed, the massive and prolonged activation of motoneuron pools is likely to induce a large release in acetylcholine at the neuromuscular junction promoting muscle fiber contraction compatible with spasticity symptoms.

Pathological mechanisms of exaggerated spinal reflexes are better understood and they relate to abnormal processing of proprioceptive input after SCI. Hyperexcitable reflex activities are characteristic in spasticity since several inhibitory pathways lose their regulatory role. Spastic reflexes typically arise as long responses lasting several seconds after stimulation and showing a huge polysynaptic component, referred to as *hyperreflexia* (Fig. 2C). Notably, long-lasting reflexes cannot be recorded during spinal shock in adult spinalized animals, neither as muscle spasms (Bennett, Sanelli, Cooke, Harvey, & Gorassini, 2004) nor as ventral spinal activities (Li, Harvey, Li, & Bennett, 2004), and at least a 2 -week period after SCI is required for their emergence. As previously remarked, neonates do not undergo areflexia, thus they present

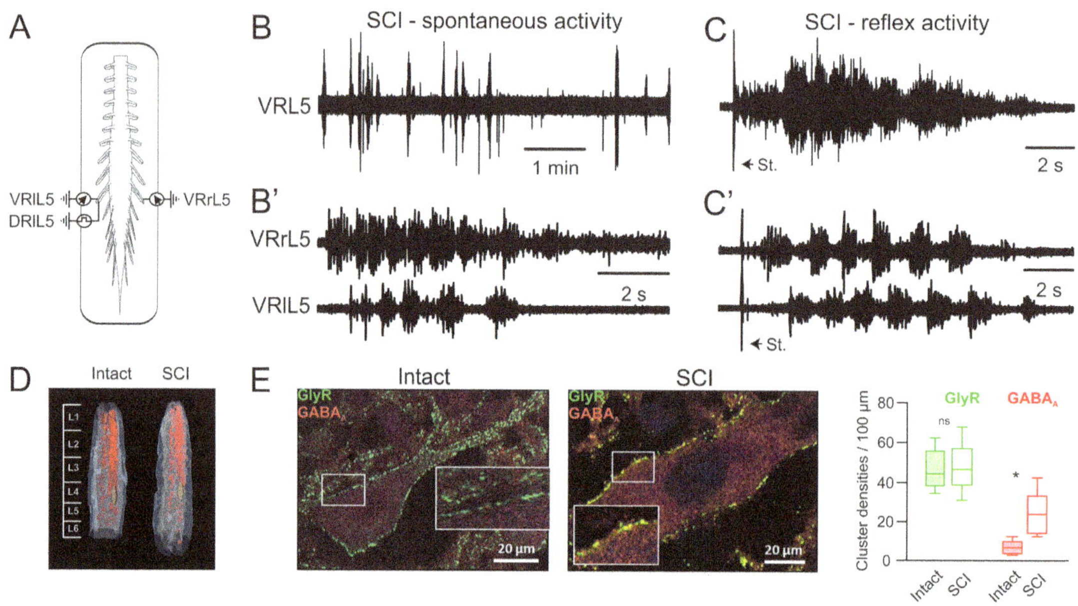

FIG. 2 Spinal network modifications related to spasticity after SCI. Ventral root recordings from isolated spinal cords in vitro (A), depicting exaggerated spontaneous (B) and evoked (C) spinal activities after SCI recorded at postnatal day 4/5. Sometimes activities arise as locomotor-like episodes (B'–C') suggesting activation of interneuronal networks of the locomotor CPG (VR/DR L5: ventral/dorsal root lumbar 5; r: right; l: left; St.: stimulation). Lumbar premotor interneurons show a trend of reduction after SCI (D). The expression of inhibitory receptors on motoneuron membranes is altered, with $GABA_A$ receptors being significantly enhanced postlesion (E). *Panel (D): modified from Khalki, L., Sadlaoud, K., Lerond, J., Coq, J.O., Brezun, J. M., Laurent, L., et al. (2018) Changes in innervation of lumbar motoneurons and organization of premotor network following training of transected adult rats.* Experimental Neurology, 299, *1–14.; Panel (E): modified from Bras, H, & Liabeuf, S. (2020) Differential effects of spinal cord transection on glycinergic and GABAergic synaptic signaling in sublesional lumbar motoneurons.* Journal of Chemical Neuroanatomy, *9, 101847.*

exaggerated reflexes as soon as 48 h postinjury (Plantier et al., 2019). Strikingly, a single electrical pulse is sufficient to trigger an episode of locomotor-like activity in spastic neonates after SCI (Fig. 2C'; Plantier et al., 2019), whereas trains of electrical stimuli at 1–10 Hz are usually required in control (Marchetti, Marco Beato, & Nistri, 2001). This further confirms the occurrence of maladaptive network changes in the sublesional spinal cord, involving not only motoneuron pools but also interneuron circuits of the locomotor CPG.

Cellular and molecular alterations contributing to spasticity after SCI

Spasticity is a condition related to increased central excitability where spinal reflexes are enhanced. The electrophysiological properties of motoneurons are closely involved in the emergence of hyperreflexia. Different pathophysiological mechanisms take place after SCI, responsible for an excitatory/inhibitory imbalance of spinal motoneurons. As a result, motoneuron intrinsic properties and synaptic transmission undergo maladaptive changes promoting an increased excitation (hyperexcitability) and a reduced inhibition (disinhibition) that allow for synaptic inputs to evoke exaggerated activity.

Long-lasting reflexes and muscle spasms are related to prolonged motoneuron firing in response to transient sensory transmission. Self-sustained firing is associated with the activation of persistent inward currents (PICs), which are slowly activating/inactivating voltage-sensitive conductances (Heckmann, Gorassini, & Bennett, 2005). PICs mediate *plateau potentials* in spinal motoneurons, consisting of a prolonged depolarization with self-sustained spiking triggered by a brief stimulation (Hounsgaard, Hultborn, Jespersen, & Kiehn, 1984). Therefore, plateau potentials are responsible for regulating spinal excitability by amplifying synaptic inputs and providing a sustained excitatory drive that allows motoneurons to fire even after stimulus termination, ultimately generating exaggerated and sustained reflex responses typical of spasticity.

Sublesional motoneurons exhibit large plateau potentials supported by an increase in PICs amplitude (Fig. 3C). PICs underlying the plateau are mediated by calcium (Ca^{2+}) currents through nifedipine-sensitive L-type Ca^{2+} channels ($Ca_v1.3$; Li & Bennett, 2003; Marcantoni et al., 2020) and sodium (Na^+) conductances through voltage-gated sodium channels $Na_v1.6$, densely expressed in the axonal initial segment (AIS) of lumbar motoneurons (Brocard et al., 2016). Ca^{2+}- and Na^+-dependent PICs are markedly depressed during acute SCI (spinal shock), but they spontaneously recover in motoneurons of chronic spinal rats some weeks after lesion, enabling huge plateaus, hyperreflexia, and muscle spasms (Bennett, Li, Harvey, & Gorassini, 2001; Bennett, Li, & Siu, 2001; Li & Bennett, 2003). The characteristic recovery of the chronic phase is concomitant with an increased expression of constitutively active serotonin receptors, which restore huge PICs in motoneurons by making them supersensitive to serotonin (Li, Murray, Harvey, Ballou, & Bennett, 2007; Murray et al., 2010). Interestingly, Ca^{2+}-PICs (or I_{CaP}) appear negligible in neonates, where Na^+-PICs (or I_{NaP}) seem to be the main component supporting exaggerated reflexes (Plantier et al., 2019). On the one hand, this is in line with delayed maturation of L-type Ca^{2+} channels peaking at the third postnatal week (Jiang et al., 1999). On the other hand, it fits with our previous work demonstrating that I_{NaP} is critical for driving plateau potentials in lumbar motoneurons during the perinatal period (Bouhadfane, Tazerart, Moqrich, Vinay, & Brocard, 2013).

Besides the aforementioned mechanisms of motoneuron hyperexcitability, processes underlying spinal disinhibition are of great relevance in the pathophysiology of spasticity (refer to schema in Fig. 4F). In mature neurons, a low intracellular chloride concentration ($[Cl^-]_i$) is maintained by the potassium-chloride cotransporter KCC2, which extrudes chloride anions hyperpolarizing the Cl^- *equilibrium potential* (E_{Cl}^-). When chloride-permeable γ-aminobutyric acid type A ($GABA_A$) and glycine receptors open, there is a net influx of Cl^- that induces neuron inhibition. The expression of KCC2 in motoneuron membranes of adult animals decreases in pathological conditions, including spasticity after SCI (Fig. 4A and B; Boulenguez et al., 2010; Bos et al., 2013). In neonates, GABA and glycine are not fully inhibitory, thus the physiological up-regulation of KCC2 contributes to the excitatory/inhibitory switch. On the contrary, after SCI the membrane expression of KCC2 in neonate motoneurons remains weak and promotes the early onset of spasticity typically observed in the perinatal period (Jean-Xavier, Pflieger, Liabeuf, & Vinay, 2006; Plantier et al., 2019). As a consequence, in both adult and neonate SCI models, the $[Cl^-]_i$ increases causing an efflux of Cl^- which depolarizes the E_{Cl}^-, thereby promoting spinal disinhibition (Fig. 4C–F). Besides KCC2 altered expression on motoneurons, inhibitory receptors are also affected in SCI adults, with $GABA_A$being long-term over-expressed on the plasma membrane (Fig. 2E; Sadlaoud et al., 2020). In lesioned neonates, the physiological up-regulation of glycine receptors is preserved in sublesional motoneurons, whereas the developmental down-regulation of $GABA_A$ does not take place (Sadlaoud et al., 2010). Inhibitory neurotransmission is also differentially influenced by the lesion, since presynaptic glycinergic inputs onto motoneurons show a similar density as in intact animals, whereas GABAergic boutons are increased (Khalki et al., 2018). This suggests an impairment of the GABAergic transmission and relative preservation of the glycinergic system in sublesional motoneurons (Bras & Liabeuf, 2020).

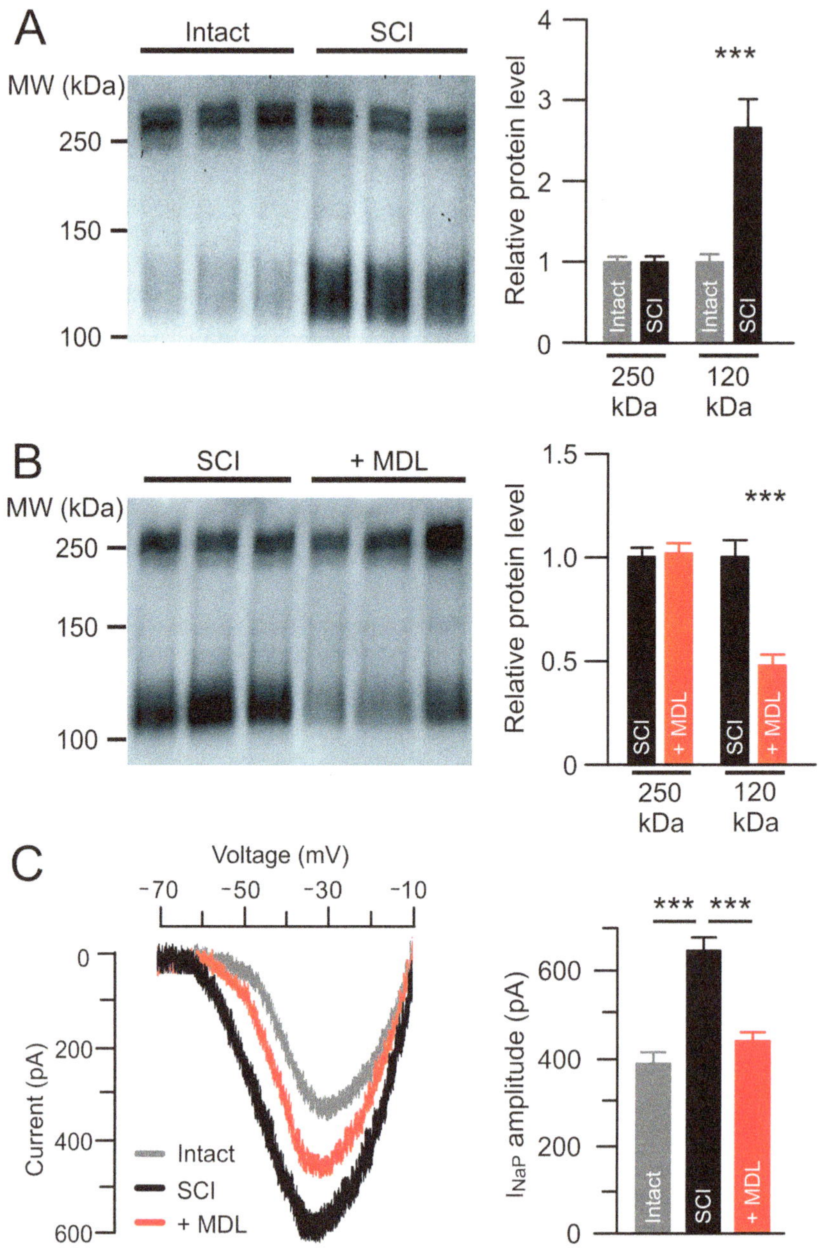

FIG. 3 $Na_v1.6$ channels as calpain targets after SCI. $Na_v1.6$ channels are cleaved after SCI, as confirmed by the appearance of a ~120 kDa band in Western blots (A). Calpains are responsible for channel cleavage, since MDL28170 reduces the amount of breakdown fragments (B). The amplitude of I_{NaP} is concomitantly increased after injury and recovers after MDL28170 application (C). *Modified from Brocard, C., Plantier, V., Boulenguez, P., Liabeuf, S., Bouhadfane, M., Viallat-Lieutaud, A., et al. (2016) Cleavage of Na(+) channels by calpain increases persistent Na(+) current and promotes spasticity after spinal cord injury.* Nature Medicine, *22, 404–411.*

Calpains and their role in spasticity after SCI

Calpains (*cal*cium-dependent proteases with pa*pain*-like activity) belong to a superfamily of cytoplasmic cysteine proteases activated by calcium. To date, 14 calpain members have been identified with both ubiquitous and tissue-specific localization (Sorimachi, Hata, & Ono, 2011). The best-characterized members are the μ- and m- calpains (or calpain 1 and 2), also referred to as "ubiquitous" or "conventional" as they are present in all mammalian tissues and cell types. Ubiquitous calpains differ in their in vitro calcium sensitivity, in the μM range for μ-calpain and in the mM range for m-calpain (Goll, Thompson, Li, Wei, & Cong, 2003). Calpains are mostly inactive in the cytosol and translocate to the membrane in response to increased intracellular calcium concentrations, where calcium-induced structural changes allow for their activation (Suzuki, Hata, Kawabata, & Sorimachi, 2004). Since calpains are involved in several pathophysiological processes, their activation is highly regulated by the endogenous peptidic inhibitor, calpastatin (Wendt, Thompson, & Goll, 2004). Calpains have a myriad of substrates and they need PEST (proline, glutamic acid, serine, and threonine) or PEST-like

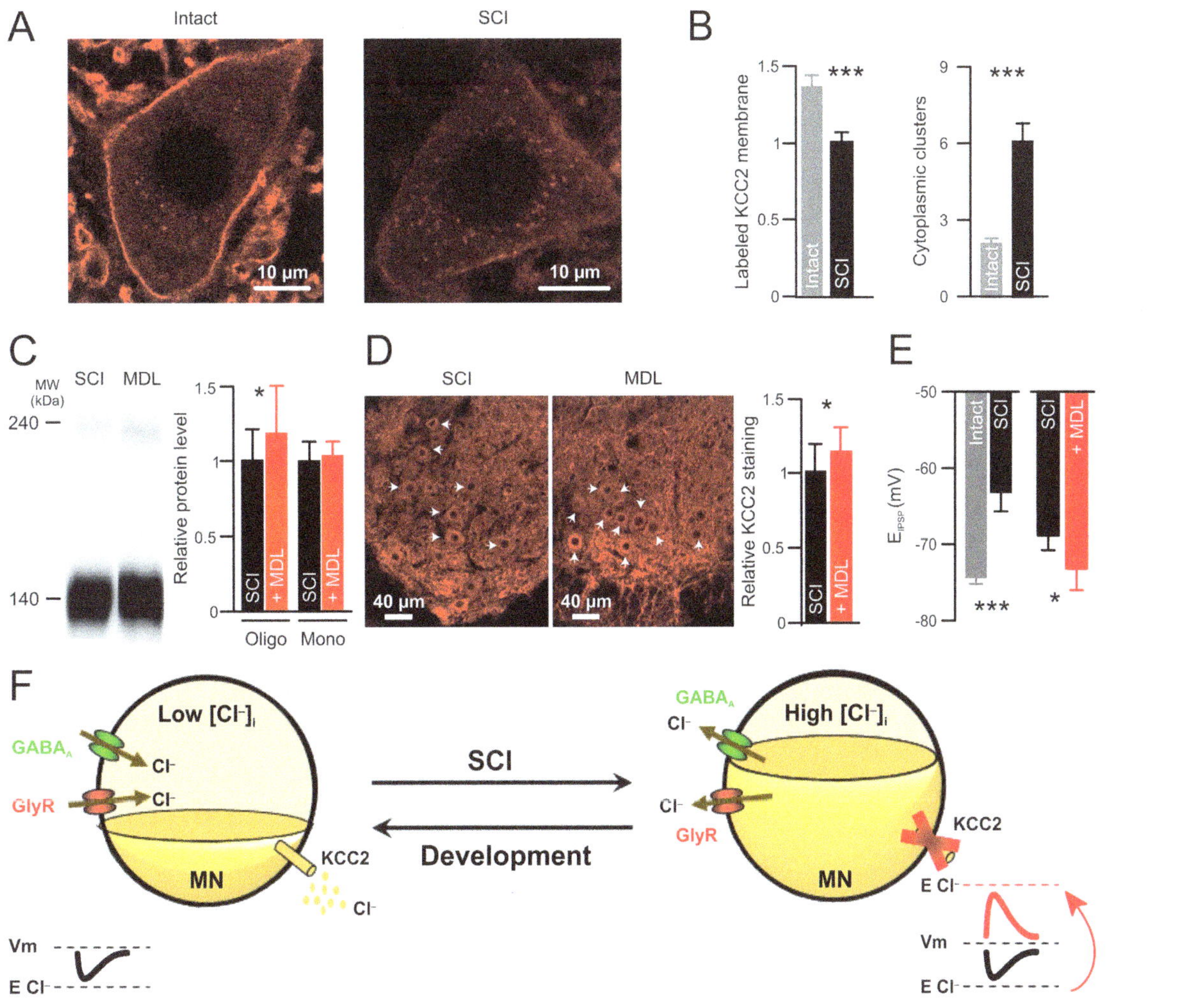

FIG. 4 KCC2 as calpain target after SCI. KCC2 is down-regulated at motoneuron membranes and internalized in the cytoplasm after SCI (A–B). MDL28170 recovers the functionally active oligomeric form of KCC2 on the plasma membrane (C–D). Note that the chloride equilibrium potential (E_{Cl^-}) affects the strength of synaptic inhibition, therefore we refer to it as E_{IPSP} (equilibrium potential of inhibitory postsynaptic potentials). E_{IPSP} undergoes a depolarizing shift after SCI and returns back to more hyperpolarized values when MDL28170 is applied (E). Summary schema showing that KCC2 down-regulation after SCI increases the intracellular Cl^- concentration and leads to spinal disinhibition (F). As pinpointed by the backward arrow in (F), the opposite condition takes place during development. *Panel (A–B): modified from Boulenguez, P., Liabeuf, S., Bos, R., Bras, H., Jean-Xavier, C., Brocard, C., et al. (2010) Down-regulation of the potassium-chloride cotransporter KCC2 contributes to spasticity after spinal cord injury.* Nature Medicine, *16, 302–307.; Panel (C–E): modified from Plantier, V., Sanchez-Brualla, I., Dingu, N., Brocard, C., Liabeuf, S., Gackière, F., et al. (2019) Calpain fosters the hyperexcitability of motoneurons after spinal cord injury and leads to spasticity.* eLife, *8, e51404.*

sequences for their recognition, although the existence of other restriction regions cannot be excluded (Tompa et al., 2004). Here we will address calpain expression and activity after SCI, followed by a discussion on identified and putative calpain targets related to spasticity, and final considerations on the role of calpain inhibition as an innovative antispastic treatment.

Calpain expression and activity after SCI

Excitotoxicity and intracellular hypercalcemia promote calpain activation and up-regulation during secondary SCI. The expression of ubiquitous calpains is increased at the lesion site and in the adjacent area during acute SCI, although it is still unclear whether this is due to either a selective rise in calpain 1/2 or both of them. We showed that calpain 1 (μ-calpain) is selectively enhanced in neonate rats (Plantier et al., 2019), whereas both calpains are augmented in adult rats acutely spinalized (Li, Hogan, & Banik, 1996). The pathophysiological meaning of this finding is not fully understood, although we speculate that higher intracellular calcium concentrations are required in the adult for calpain activation. Indeed, in adult animals calpains are posttranscriptionally regulated by many mechanisms, including calcium requirement, autoproteolytic cleavage, and the endogenous inhibitor calpastatin (Molinari & Carafoli, 1997). On the contrary, little is known about calpain expression in chronic SCI, albeit a progressive decrease in calpastatin and a parallel increase in calpain levels have been reported (Wienecke, Westerdahl, Hultborn, Kiehn, & Ryge, 2010). However, which of the calpain isoforms is over-expressed during the chronic phase remains unclear and further investigation is required.

Calpain activity is generally determined by quantifying breakdown products of different protein targets. There are several substrates for calpains that are degraded in lesioned spinal cords. We showed that calpains cleave $Na_v1.6$ channels inducing the appearance of a lower molecular weight band of ~120 kDa in immunoblots from lumbar segments, in addition to the full-length band of ~250 kDa consistent with the native α-subunit of $Na_v1.6$ channels, both in adult (Brocard et al., 2016) and neonate (Plantier et al., 2019) spinal rats (Fig. 3A and B).

Identified and putative calpain targets involved in the pathophysiology of spasticity

Calpains contribute to spasticity by mediating the excitatory/inhibitory imbalance typically occurring in lumbar motoneurons after SCI. Calpain cleavage of several targets promotes hyperexcitability and disinhibition in lesioned spinal cords. Here, we will discuss the role of identified calpain substrates in spasticity and speculate on new potential targets that might impact on this pathological condition.

Several mechanisms of spinal hyperexcitability are supported by calpain-mediated proteolysis of ion channels and receptors. Our team demonstrated that calpains cleave $Na_v1.6$ channels, highly expressed in the AIS of lumbar motoneurons, by producing a breakdown product of ~120 kDa (Fig. 3A and B; Brocard et al., 2016; Plantier et al., 2019). We also pinpointed a direct relationship between the calpain-mediated cleavage of $Na_v1.6$ and the up-regulation of I_{NaP} (Fig. 3C). Although the exact mechanism remains to be defined, we hypothesize that calpain cleavage of the α-subunit of $Na_v1.6$ channels interferes with their inactivation process, thus increasing the amplitude of I_{NaP} and supporting motoneuron hyperexcitability and muscle spasms. Interestingly, the amount of full-length $Na_v1.6$ channels (~250 kDa) is unchanged at the plasma membrane after SCI, likely due to regulatory mechanisms for the expression of the native form supported by calpain-mediated proteolytic fragments (Onwuli et al., 2017).

Hyperexcitability can also be sustained by calcium-dependent cellular processes since calpains are typically activated by high intracellular calcium. It is well-established that glutamate receptors induce intracellular calcium overload leading to calpain activation (Doshi & Lynch, 2009). In turn, calpains directly target all of the major glutamate receptors, disrupting their function and leading to excitotoxicity. However, direct evidence linking calpain-mediated dysregulation of glutamatergic transmission to spasticity after SCI is still missing, although glutamate receptors seem to be good candidates. Similarly, L-type calcium channels also allow for calcium influx in neurons, contributing to calpain activation (Schön, Paquet-Durand, & Michalakis, 2016). They are catabolized by calpains at the C-terminus in the heart, skeletal muscle, and in some regions of the brain (Abele & Yang, 2012), albeit no data are currently available in the spinal cord.

A reduced or impaired inhibition also contributes to spasticity and we recently reported a role for calpains in motoneuron disinhibition. We demonstrated that the membrane down-regulation of KCC2 after SCI, previously reported by the team (Fig. 4A and B; Boulenguez et al., 2010; Bos et al., 2013), is actually due to calpain activation (Fig. 4C–E; Plantier et al., 2019). KCC2 is present on motoneuron membranes both in oligomeric and monomeric forms, with the oligomeric form having the highest sensitivity to calpains. We argue that KCC2 is cleaved at the C-terminus due to the presence of two predicted PEST sequences (Acton et al., 2012; Mercado, Broumand, Zandi-Nejad, Enck, & Mount, 2006). Given that phosphorylation regulates substrate vulnerability to calpains (Yamashita, Teramoto, & Kwak, 2016), calpain sensitivity of the oligomeric form might be due to an SCI-mediated dephosphorylation of serine 960 near the PEST motifs (Modol, Mancuso, Alé, Francos-Quijorna, & Navarro, 2014). Moreover, since KCC2 requires a functional C-domain to extrude Cl^- (Acton et al., 2012; Mercado et al., 2006), the depolarized E_{Cl^-} might be a direct consequence of calpain proteolysis.

Physiological and pharmacological inhibition of calpains and effects on spasticity

Calpains have a plethora of targets relevant to the pathophysiology of spasticity. Physiologically, calpains coexist with their endogenous inhibitor calpastatin in the cytosol and membrane (Kawasaki & Kawashima, 1996). Calpastatin has four inhibitory domains, each inhibiting one molecule of calpain with an inhibitory rate of 1:4 (Wendt et al., 2004). Interestingly, calpastatin protein and mRNA levels gradually increase in the central nervous system during development, showing a rapid rise at postnatal days 5–10 (Li, Bondada, Joshi, & Geddes, 2009). Therefore, a lower calpastatin expression might account for the earlier emergence of spasticity in neonates with respect to adult animals. Although calpastatin up-regulation has been promoted in many pathological conditions to counteract calpain activity, its therapeutic role in SCI must still be elucidated. On the one side, there are encouraging data showing some calpastatin-induced neuroprotection after SCI

(Yu et al., 2020). On the other side, there is clear evidence that the ratio calpain/calpastatin is increased after SCI, as a result of calpain-mediated degradation of calpastatin which disrupts its regulatory role and makes it unsuitable as a therapeutic agent for calpain inhibition (Ray, Hogan, & Banik, 2003).

A huge research effort has been focused on identifying effective therapeutic strategies to inhibit calpain activity in SCI. To date, several classes of pharmacological inhibitors are available. Between them, MDL28170 is a small aldehyde compound capable of penetrating the cell membrane by passive diffusion. We have demonstrated that the pharmacological use of MDL28170 is efficient in preventing both the cleavage of $Na_v1.6$ channels (Fig. 3B) and the down-regulation of KCC2 on the plasma membrane (Fig. 4C and D), thus restoring the excitatory/inhibitory equilibrium of motoneurons and protecting against spasticity (Fig. 3C, Fig. 4E; Brocard et al., 2016; Plantier et al., 2019).

The major issue with calpain inhibitors is their lack of specificity since they act also on other proteases. Moreover, they target both calpain 1 and 2 and no selective drug is currently available for either isoform. Given that calpains have important physiological roles that might be affected by systemic inhibition, local therapeutic strategies for calpain silencing are advisable. Next-generation treatments should allow for cell-specific manipulations, such as therapies selectively targeting motoneurons innervating spastic muscles, therefore avoiding systemic adverse effects related to oral drug administration. Lastly, choosing calpain as a possible hit would allow targeting the main upstream molecule accounting for multiple spinal mechanisms leading to spasticity, instead of singularly acting on downstream players (Fig. 5).

Applications to other areas of neuroscience

Increased calpain activation has been implicated not only in spinal cord injury (SCI) but also in a number of other neurological disorders including traumatic brain injury (Ray, Dixon, & Banik, 2002), cerebral ischemia (Sun, Zhao, & Xu, 2008), neurodegenerative diseases like Alzheimer's (Saito, 1993), Parkinson's (Mouatt-Prigent, 1996), Huntington's (Gafni & Ellerby, 2002) diseases and amyotrophic lateral sclerosis (Stifanese et al., 2014). Although the etiology underlying each disorder varies, the main pathologic mechanism converges on calcium homeostasis dysregulation leading to calpain activation and overexpression. Increased intracellular calcium is generally due to excitotoxicity, promoted by uncontrolled glutamate release and impaired re-uptake, ultimately resulting in augmented calcium influx through the hyperactivation of glutamate receptors. This cascade causes the pathologic activation of calpains that catalyze a myriad of substrates promoting neuronal death. Indeed, calpains induce apoptosis by activating caspases and cause necrosis by disrupting lysosomal membranes that release proteases, such as cathepsins. The production of calpain-mediated neurotoxic protein fragments is also a common feature in the pathophysiology of these neurological disorders. Interestingly, the deregulation of calpain activity represents a key cytotoxic event also in epilepsy, where calpain over-activation has been shown to down-regulate the potassium-chloride cotransporter KCC2 on the plasma membrane (Li et al., 2018), in addition to neurotoxicity and constitutive activation of glutamate receptors. Therefore, the search for effective and highly selective strategies to inhibit calpains has obvious applicability not only in the treatment of spasticity after SCI but also in a wide and heterogeneous range of neurological disorders where calpains seem to play a critical role.

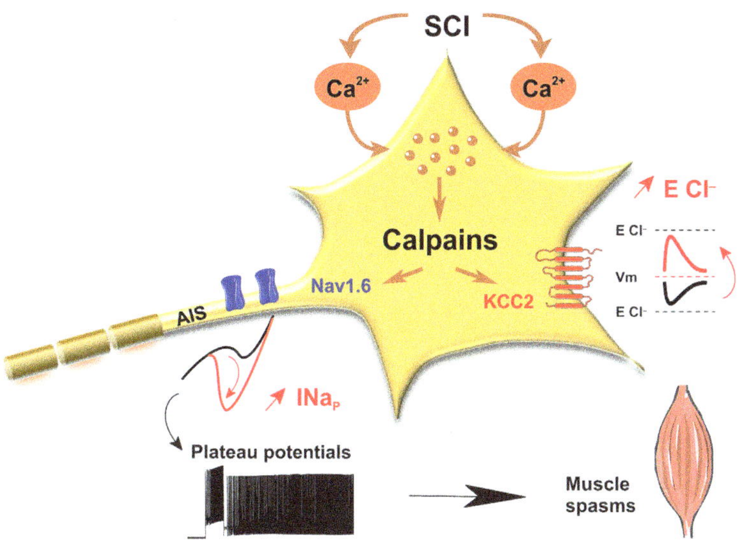

FIG. 5 Role of calpains in the pathophysiology of spasticity after SCI. The intracellular concentration of calcium increases after SCI inducing calpain activation. Calpains target several molecules, including $Na_v1.6$ channels (highly expressed at the axonal initial segment, AIS) and KCC2 that cause I_{NaP} up-regulation and a depolarizing shift of the E_{Cl}^{-}, respectively. As a result, an excitatory/inhibitory imbalance occurs in spinal motoneurons promoting plateau potentials and contributing to spasticity.

Mini-dictionary of terms

Wallerian degeneration: injury-induced program of axon degeneration consisting of deteriorating myelin ovoids and distal axon fragmentation, first described by Augustus Waller in 1850.

Reperfusion injury: cascade of events associated with the reestablishment of blood supply in ischemic tissues, which restores oxygenation, but also accelerates tissue necrosis through several mechanisms, such as the production of reactive oxygen species (ROS).

Glial scar: healing response supported by reactive astrogliosis and massive deposition of extracellular matrix at the injury site limiting functional repair after lesion.

Excitotoxicity: pathological process leading to neuronal death or damage arising from excessive or prolonged exposure to excitatory amino acids.

Spinal shock: temporary loss (areflexia) or depression (hyporeflexia) of spinal reflexes below the level of the injury occurring immediately after SCI.

Hyperreflexia: abnormal and exaggerated reflexes spontaneously appearing either at the end of the spinal shock phase in humans and in adult animals or few days postlesion in neonate animal models.

Central pattern generator (CPG): interneuron networks controlling rhythmicity and pattern formation in motor behaviors, such as locomotion, by providing the timing of motoneuron discharge and determining appropriate sequences of muscle contraction.

Locomotor-like episode: double alternation between flexor and extensor motor pools on the same side of the spinal cord and between left and right motor pools within the same spinal segment, which represents a hallmark of CPG activation in vitro.

Plateau potential: prolonged depolarization of the membrane potential supporting long-lasting discharges triggered by synaptic input or brief depolarizing pulse that activate persistent inward currents.

Equilibrium potential: membrane potential level where an ion is at the electrochemical equilibrium and there is no net flow of that ion across the membrane.

Key facts of "spinal hyperexcitability"

- It refers to molecular, cellular, and network modifications leading to an increased spinal excitation.
- We described a calpain-mediated mechanism involving the cleavage of voltage-gated sodium channels $Na_v1.6$ in SCI motoneurons.
- Calpain proteolysis likely affects the inactivation kinetics of $Na_v1.6$ channels.
- As a result, the sodium persistent inward current (I_{NaP}) is up-regulated and spinal motoneurons exhibit enhanced plateau potentials that contribute to exaggerated spinal activities and spasticity.
- The pharmacological inhibition of calpains prevents both the cleavage of $Na_v1.6$ channels and I_{NaP} up-regulation and protects from spasticity.

Key facts of "spinal disinhibition"

- It encloses spinal mechanisms causing a reduced or impaired inhibition.
- Calpains play a role by down-regulating the potassium-chloride cotransporter KCC2 on motoneuron membranes after SCI.
- The reduced expression of KCC2 causes an intracellular accumulation of chloride.
- When γ-aminobutyric acid type A ($GABA_A$) and glycine receptors open as a consequence of inhibitory neurotransmitter binding, chloride anions are extruded and there is a depolarizing shift of the chloride equilibrium potential (E_{Cl^-}).
- The reduced strength of the postsynaptic inhibition contributes to the onset of spasticity.
- The pharmacological blockage of calpains restores KCC2 on the plasma membrane recovering inhibitory transmission and preventing muscle spasms.

Summary points

- Spasticity is a pathological condition that develops in the majority of patients one year after SCI.
- Animal models for the study of spasticity after SCI represent an essential tool to uncover pathophysiological processes and to identify new therapeutic strategies.

- So far, several mechanisms responsible for an excitatory/inhibitory imbalance of spinal motoneurons have been related to spasticity after SCI.
- We reported that calpains have an upstream role in different spinal mechanisms leading to spasticity.
- On the one hand, calpains cleave voltage-gated sodium channels $Na_v1.6$ up-regulating the sodium persistent inward current (I_{NaP}); on the other hand, they down-regulate the potassium-chloride cotransporter KCC2 on motoneuron membranes depolarizing the chloride equilibrium potential (E_{Cl}^{-}).
- These mechanisms synergize in generating exaggerated activities in spinal motoneurons, lastly contributing to spasticity.
- Other molecules might be targeted by calpains and impact the development of spasticity, therefore supporting the concept that calpain inhibition might be an effective antispastic treatment.
- Strategies for calpain inhibition can be adopted not only to prevent spasticity after SCI but also to counteract several neurological disorders where calpains are critically involved.

Acknowledgments

Our research was financed by a grant from Agence National de la Recherche Scientifique (CalpaSCI, ANR-16-16CE-0004) and from the French Institut pour la Recherche sur la Moelle épinière et l'Encéphale (IRME).

References

Abele, K., & Yang, J. (2012). Regulation of voltage-gated calcium channels by proteolysis. *Sheng Li Xue Bao*, *64*, 504–514.

Acton, B. A., Mahadevan, V., Mercado, A., Uvarov, P., Ding, Y., Pressey, J., et al. (2012). Hyperpolarizing GABAergic transmission requires the KCC2 C-terminal ISO domain. *The Journal of Neuroscience*, *32*, 8746–8751.

Adams, M. M., & Hicks, A. L. (2005). Spasticity after spinal cord injury. *Spinal Cord*, *43*, 577–586.

Atkinson, P. P., & Atkinson, J. L. (1996). Spinal shock. *Mayo Clinic Proceedings*, *71*, 384–389.

Bellardita, C., Caggiano, V., Leiras, R., Caldeira, V., Fuchs, A., Bouvier, J., et al. (2017). Spatiotemporal correlation of spinal network dynamics underlying spasms in chronic spinalized mice. *eLife*, *6*, e23011.

Bellardita, C., Marcantoni, M., Löw, P., & Kiehn, O. (2018). Sacral spinal cord transection and isolated sacral cord preparation to study chronic spinal cord injury in adult mice. *Bio-Protocol*, *8*, e2784.

Bennett, D. J., Gorassini, M., Fouad, K., Sanelli, L., Han, Y., & Cheng, J. (1999). Spasticity in rats with sacral spinal cord injury. *Journal of Neurotrauma*, *16*, 69–84.

Bennett, D. J., Li, Y., Harvey, P. J., & Gorassini, M. (2001). Evidence for plateaupotentials in tail motoneurons of awake chronic spinal rats with spasticity. *Journal of Neurophysiology*, *86*, 1972–1982.

Bennett, D. J., Li, Y., & Siu, M. (2001). Plateau potentials in sacrocaudal motoneurons of chronic spinal rats, recorded in vitro. *Journal of Neurophysiology*, *86*, 1955–1971.

Bennett, D. J., Sanelli, L., Cooke, C. L., Harvey, P. J., & Gorassini, M. A. (2004). Spastic long-lasting reflexes in the awake rat after sacral spinal cord injury. *Journal of Neurophysiology*, *91*, 2247–2258.

Bos, R., Sadlaoud, K., Boulenguez, P., Buttigieg, D., Liabeuf, S., Brocard, C., et al. (2013). Activationof 5-HT2A receptors upregulates the function of the neuronal K-Cl cotransporter KCC2. *PNAS*, *110*, 348–353.

Bouhadfane, M., Tazerart, S., Moqrich, A., Vinay, L., & Brocard, F. (2013). Sodium-mediated plateau potentials in lumbar motoneurons of neonatal rats. *The Journal of Neuroscience*, *33*, 15626–15641.

Boulenguez, P., Liabeuf, S., Bos, R., Bras, H., Jean-Xavier, C., Brocard, C., et al. (2010). Down-regulation of the potassium-chloride cotransporter KCC2 contributes to spasticity after spinal cord injury. *Nature Medicine*, *16*, 302–307.

Bras, H., & Liabeuf, S. (2020). Differential effects of spinal cord transection on glycinergic and GABAergic synaptic signaling in sub-lesional lumbar motoneurons. *Journal of Chemical Neuroanatomy*, *9*, 101847.

Brocard, C., Plantier, V., Boulenguez, P., Liabeuf, S., Bouhadfane, M., Viallat-Lieutaud, A., et al. (2016). Cleavage of Na(+) channels by calpain increases persistent Na(+) current and promotes spasticity after spinal cord injury. *Nature Medicine*, *22*, 404–411.

Corleto, J. A., Bravo-Hernández, M., Kamizato, K., Kakinohana, O., Santucci, C., Navarro, M. R., et al. (2015). Thoracic 9 spinal transection-induced model of muscle spasticity in the rat: A systematic electrophysiological and histopathological characterization. *PLoS One*, *10*, e0144642.

Doshi, S., & Lynch, D. R. (2009). Calpain and the glutamatergic synapse. *Frontiers in Bioscience (Scholar Edition)*, *1*, 466–476.

Fan, B., Wei, Z., Yao, X., Shi, G., Cheng, X., Zhou, X., et al. (2018). Microenvironment imbalance of spinal cord injury. *Cell Transplantation*, *27*, 853–866.

Gafni, J., & Ellerby, L. M. (2002). Calpain activation in Huntington's disease. *The Journal of Neuroscience*, *22*, 4842–4849.

Goll, D. E., Thompson, V. F., Li, H., Wei, W., & Cong, J. (2003). The calpain system. *Physiological Reviews*, *83*, 731–801.

Hall, E. D., Wang, J. A., Bosken, J. M., & Singh, I. N. (2016). Lipid peroxidation in brain or spinal cord mitochondria after injury. *Journal of Bioenergetics and Biomembranes*, *48*, 169–174.

Heckmann, C. J., Gorassini, M. A., & Bennett, D. J. (2005). Persistent inward currents in motoneuron dendrites: Implications for motor output. *Muscle & Nerve*, *31*, 135–156.

Hill, C. E., Beattie, M. S., & Bresnahan, J. C. (2001). Degeneration and sprouting of identified descending supraspinal axons after contusive spinal cord injury in the rat. *Experimental Neurology*, *171*, 153–169.

Hounsgaard, J., Hultborn, H., Jespersen, B., & Kiehn, O. (1984). Intrinsic membrane properties causing a bistable behaviour of alpha-motoneurones. *Experimental Brain Research*, *55*, 391–394.

Jean-Xavier, C., Pflieger, J. F., Liabeuf, S., & Vinay, L. (2006). Inhibitory postsynaptic potentials in lumbar motoneurons remain depolarizing after neonatal spinal cord transection in the rat. *Journal of Neurophysiology*, *96*, 2274–2281.

Jiang, Z., Rempel, J., Li, J., Sawchuk, M. A., Carlin, K. P., & Brownstone, R. M. (1999). Development of L-type calcium channels and a nifedipine-sensitive motor activity in the postnatal mouse spinal cord. *The European Journal of Neuroscience*, *11*, 3481–3487.

Kalogeris, T., Baines, C. P., Krenz, M., & Korthuis, R. J. (2012). Cell biology of ischemia/reperfusion injury. *International Review of Cell and Molecular Biology*, *298*, 229–317.

Kawasaki, H., & Kawashima, S. (1996). Regulation of the calpain-calpastatin system by membranes. *Molecular Membrane Biology*, *13*, 217–224.

Khalki, L., Sadlaoud, K., Lerond, J., Coq, J. O., Brezun, J. M., Laurent, L., et al. (2018). Changes in innervation of lumbar motoneurons and organization of premotor network following training of transected adult rats. *Experimental Neurology*, *299*, 1–14.

Lance, J. W. (1980). Symposium synopsis. In R. G. Feldman, R. R. Young, & W. P. Koella (Eds.), *Spasticity: Disordered motor control* (pp. 485–494).

Li, Y., & Bennett, D. J. (2003). Persistent sodium and calcium currents cause plateau potentials in motoneurons of chronic spinal rats. *Journal of Neurophysiology*, *90*, 857–869.

Li, Y., Bondada, V., Joshi, A., & Geddes, J. W. (2009). Calpain 1 and Calpastatin expression is developmentally regulated in rat brain. *Experimental Neurology*, *220*, 316–319.

Li, Y., Harvey, P. J., Li, X., & Bennett, D. J. (2004). Spastic long-lasting reflexes of the chronic spinal rat studied in vitro. *Journal of Neurophysiology*, *91*, 2236–2246.

Li, Z., Hogan, E. L., & Banik, N. L. (1996). Role of calpain in spinal cord injury: Increased calpain immunoreactivity in rat spinal cord after impact trauma. *Neurochemical Research*, *21*, 441–448.

Li, X., Murray, K., Harvey, P. J., Ballou, E. W., & Bennett, D. J. (2007). Serotonin facilitates a persistent calcium current in motoneurons of rats with and without chronic spinal cord injury. *Journal of Neurophysiology*, *97*, 1236–1246.

Liu, Z., Yao, X., Jiang, W., Li, W., Zhu, S., Liao, C., et al. (2020). Advanced oxidation protein products induce microglia-mediated neuroinflammation via MAPKs-NF-κB signaling pathway and pyroptosis after secondary spinal cord injury. *Journal of Neuroinflammation*, *17*, 90.

Marcantoni, M., Fuchs, A., Löw, P., Bartsch, D., Kiehn, O., & Bellardita, C. (2020). Early delivery and prolonged treatment with nimodipine prevents the development of spasticity after spinal cord injury in mice. *Science Translational Medicine*, *12*, eaay0167.

Marchetti, C., Marco Beato, M., & Nistri, A. (2001). Alternating rhythmic activity induced by dorsal root stimulation in the neonatal rat spinal cord in vitro. *The Journal of Physiology*, *530*, 105–112.

Mercado, A., Broumand, V., Zandi-Nejad, K., Enck, A. H., & Mount, D. B. (2006). A C-terminal domain in KCC2 confersconstitutive K+-Cl- cotransport. *The Journal of Biological Chemistry*, *281*, 1016–1026.

Modol, L., Mancuso, R., Alé, A., Francos-Quijorna, I., & Navarro, X. (2014). Differential effects on KCC2 expression andspasticity of ALS and traumatic injuries to motoneurons. *Frontiers in Cellular Neuroscience*, *8*, 7.

Molinari, M., & Carafoli, E. (1997). Calpain: A cytosolic proteinase active at the membranes. *The Journal of Membrane Biology*, *156*, 1–8.

Mouatt-Prigent, A. (1996). Increased m-calpain expression in the mesencephalon of patients with Parkinson's disease but not in other neurodegenerative disorders involving the mesencephalon: A role in nerve cell death? *Neuroscience*, *73*, 979–987.

Murray, K., Nakae, A., Stephens, M. J., Rank, M., D'Amico, J., Harvey, P. J., et al. (2010). Recovery of motoneuron and locomotor function after spinal cord injury depends on constitutive activity in 5-HT2C receptors. *Nature Medicine*, *16*, 694–700.

Onwuli, D. O., Yanez-Bisbe, L., Pinsach-Abuin, M. L., Tarradas, A., Brugada, R., Greenman, J., et al. (2017). Do sodium channel proteolytic fragments regulate sodium channel expression? *Channels*, *11*, 476–481.

Plantier, V., Sanchez-Brualla, I., Dingu, N., Brocard, C., Liabeuf, S., Gackière, F., et al. (2019). Calpain fosters the hyperexcitability of motoneurons after spinal cord injury and leads to spasticity. *eLife*, *8*, e51404.

Ray, S. K., Dixon, C. E., & Banik, N. L. (2002). Molecular mechanisms in the pathogenesis of traumatic brain injury. *Histology and Histopathology*, *17*, 1137–1152.

Ray, S. K., Hogan, E. L., & Banik, N. L. (2003). Calpain in the pathophysiology of spinal cord injury: Neuroprotection with calpain inhibitors. *Brain Research. Brain Research Reviews*, *42*, 169–185.

Sadlaoud, K., Khalki, L., Brocard, F., Vinay, L., Boulenguez, P., & Bras, H. (2020). Alteration of glycinergic receptor expression in lumbar spinal motoneurons is involved in the mechanisms underlying spasticity after spinal cord injury. *Journal of Chemical Neuroanatomy*, *106*, 101787.

Sadlaoud, K., Tazerart, S., Brocard, C., Jean-Xavier, C., Portalier, P., Brocard, F., et al. (2010). Differential plasticity of the GABAergic and glycinergic synaptic transmission to rat lumbar motoneurons after spinal cord injury. *The Journal of Neuroscience*, *30*, 3358–3369.

Saito, K. (1993). Widespread activation of calcium-activated neutral proteinase (calpain) in the brain in Alzheimer disease: A potential molecular basis for neuronal degeneration. *Proceedings of the National Academy of Sciences of the United States of America*, *90*, 2628–2632.

Schön, C., Paquet-Durand, F., & Michalakis, S. (2016). Cav1.4 L-type calcium channels contribute to Calpain activation in degenerating photoreceptors of rd1 mice. *PLoS One*, *11*, e0156974.

Sharif-Alhoseini, M., Khormali, M., Rezaei, M., Safdarian, M., Hajighadery, A., Khalatbari, M. M., et al. (2017). Animal models of spinal cord injury: A systematic review. *Spinal Cord*, *55*, 714–721.

Sorimachi, H., Hata, S., & Ono, Y. (2011). Calpain chronicle—An enzyme family under multidisciplinary characterization. *Proceedings of the Japan Academy. Series B, Physical and Biological Sciences*, *87*, 287–327.

Stifanese, R., Averna, M., De Tullio, R., Pedrazzi, M., Milanese, M., Bonifacino, T., et al. (2014). Role of calpain-1 in the early phase of experimental ALS. *Archives of Biochemistry and Biophysics*, *562*, 1–8.

Sun, M., Zhao, Y., & Xu, C. (2008). Cross-talk between calpain and caspase-3 in penumbra and core during focal cerebral ischemia-reperfusion. *Cellular and Molecular Neurobiology*, *28*, 71–85.

Suzuki, K., Hata, S., Kawabata, Y., & Sorimachi, H. (2004). Structure, activation, and biology of calpain. *Diabetes*, *53*(Suppl), 12–18.

Tompa, P., Buzder-Lantos, P., Tantos, A., Farkas, A., Szilágyi, A., Bánóczi, Z., et al. (2004). On the sequential determinants of calpain cleavage. *The Journal of Biological Chemistry*, *279*, 20775–20785.

Wendt, A., Thompson, V. F., & Goll, D. E. (2004). Interaction of calpastatin with calpain: A review. *Biological Chemistry*, *385*, 465–472.

Wienecke, J., Westerdahl, A. C., Hultborn, H., Kiehn, O., & Ryge, J. (2010). Global gene expression analysis of rodent motor neurons following spinal cord injury associates molecular mechanisms with development of postinjury spasticity. *Journal of Neurophysiology*, *103*, 761–778.

Yamashita, T., Teramoto, S., & Kwak, S. (2016). Phosphorylated TDP-43 becomes resistant to cleavage by calpain: A regulatory role for phosphorylation in TDP-43 pathology of ALS/FTLD. *Neuroscience Research*, *107*, 63–69.

Yu, C. G., Bondada, V., Joshi, A., Reneer, D. V., Telling, G. C., Saatman, K. E., et al. (2020). Calpastatin overexpression protects against excitotoxic hippocampal injury and traumatic spinal cord injury. *Journal of Neurotrauma*, *37*, 2268–2276.

Chapter 22

Targeting mTOR signaling to promote autophagy for functional recovery after spinal cord injury

Nadia Al-Sammarraie and Swapan K. Ray
Department of Pathology, Microbiology, and Immunology, University of South Carolina School of Medicine, Columbia, SC, United States

List of abbreviations

Akt	Ak strain transforming or protein kinase B
AMPK	adenosine monophosphate-activated protein kinase
Bax	Bcl-2-associated protein X
GLP-1	glucagon-like peptide 1
LC3	microtubule-associated protein 1 light chain 3
mTOR	mechanistic target of rapamycin
p62/SQSTM1	sequestosome-1
p70S6K	the 70 kDa ribosomal protein S6 kinase
PI3KC3	phosphoinositide 3-kinase class 3 complex
PIKK	phosphatidylinositol 3-kinase-related kinase
SCI	spinal cord injury
VMNs	ventral motoeurons

Introduction

Spinal cord injury (SCI) is a devastating damage to the spinal cord causing significant impairments of the function of neurons distal to the site of injury (Drasites et al., 2020). SCI can lead to permanent debilitating neurological deficits, for example, paraplegia of the lower limbs, difficulties in breathing and respiratory insufficiency, altered sensation, and autonomous dysfunction of cardiovascular, gastrointestinal, respiratory, and urinary systems (Cheng, Liao, Liao, Chuang, & Shih, 2004; Galeiras Vázquez, Rascado Sedes, Mourelo Fariña, Montoto Marqués, & Ferreiro Velasco, 2013; Karlsson, 2006; Zompa, Cain, Everhart, Moyer, & Hulsebosch, 1997). SCI is a global problem with an annual incidence of 54 new cases per 1 million in the United States, with males affected more than females (Bracken, Freeman, & Hellenbrand, 1981; Jain et al., 2015). The etiology of SCI varies and it may occur due to internal and external causes that relate to patient's medical history, for example, primary or metastatic tumor of the spine or ischemic damage to the spinal cord manifesting a non-traumatic SCI, or most often, SCI occurs due to external causes due to severe physical trauma to the spine during traffic accidents on the road or in case of work-related injuries causing traumatic SCI (Ekong & Tator, 1985; Ge, Arul, & Mesfin, 2019; Lee, Cripps, Fitzharris, & Wing, 2014; Mauney et al., 1995). Traumatic SCI is the main focus of current research activities as it is more complex and more prevalent than non-traumatic SCI. From this point onward, we will use SCI only to refer to traumatic SCI. Current management of SCI varies based on the cause, location, duration, and severity of the trauma, and it is usually a combination of medical and surgical treatments that include immobilization, analgesia, and surgical intervention to decompress and stabilize the spine (Fehlings, Cadotte, & Fehlings, 2011).

The pathophysiology of SCI is complex and multifactorial, and it is characterized by immediate and late injuries to the neurons and surrounding tissue. Initial injury can lead to substantial neuronal loss, hemorrhage, and edema (Dumont et al., 2001; Nakamura & Okano, 2013). Following SCI initial damage, the neurons are exposed to further deleterious cascade of secondary damages, which cause further neuronal loss and demyelination leading to permanent neurological deficits

Cellular, Molecular, Physiological, and Behavioral Aspects of Spinal Cord Injury. https://doi.org/10.1016/B978-0-12-822427-4.00022-8

(Dumont et al., 2001; Nakamura & Okano, 2013). Hence, early intervention for minimizing the secondary damages is crucial to preserve both integrity and function of the neurons and to promote functional recovery after SCI.

Macroautophagy is a highly conserved recycling mechanism within the cell and it can be further aggravated during stressful conditions, for example, cell injury, hypoxia, or nutritional deprivations (Kulkarni, Chen, & Maday, 2018; Li et al., 2021; Stacchiotti & Corsetti, 2020). Macroautophagy is well-regulated sequential events that aim to clear the cell from old or damaged organelles and/or other macromolecules by lysosomal degradation, and to recycle and reuse the cellular building blocks (Feng, He, Yao, & Klionsky, 2014; Nakatogawa, 2020; Xie, Nair, & Klionsky, 2008). Autophagy is characterized by development and function of three main autophagy machinery structures: phagophore, autophagosome, and autolysosome. Phagophore is a cup-shaped lipid-rich membrane that surrounds and engulfs a portion of the cytoplasm containing the damaged cellular components (Feng et al., 2014; Nakatogawa, 2020; Xie et al., 2008). Phagophore becomes elongated and its two ends are then fused to form a double-membrane large circular vesicle called autophagosome (Feng et al., 2014; Nakatogawa, 2020; Xie et al., 2008). Autophagosome is then fused with the lysosome to form the autolysosome, which contains various hydrolase enzymes that degrade the sequestered cellular organelles and other macromolecules (Feng et al., 2014; Nakatogawa, 2020; Xie et al., 2008).

Autophagy flux is the measure of the autophagy degradation activity, which requires well development of autophagosome, its fusion with lysosome, and efficient lysosomal degradation to prevent buildup of toxic materials within the cells (du Toit, Hofmeyr, Gniadek, & Loos, 2018; Klionsky et al., 2012). Molecular regulation of autophagy involves interactions among several proteins, which are critical for promoting autophagy flux with the development of autophagy machinery components including Beclin-1, LC3 (microtubule-associated protein 1 light chain 3), p62/SQSTM1 (sequestosome-1), and others. Beclin-1 is a part of phosphoinositide 3-kinase class 3 complex (PI3KC3), which regulates formation of autophagosome, degradation of caspases, and attenuation of apoptosis (Xie, Kang, & Tang, 2016). LC3 I and II are critical proteins for the formation and maturation of autophagosome (Brier et al., 2019; Giatromanolaki et al., 2011; Park, Huang, Wu, Foster, & Sinicrope, 2013). During autophagy, the soluble cytosolic protein LC3 I with the aid of Beclin-1 is coupled to phosphatidylethanolamine phospholipids for its conversion into the membrane-bound form LC3 II, which is one of the major constituents of the autophagosome membrane (Brier et al., 2019; Giatromanolaki et al., 2011; Park et al., 2013). p62 is a polyubiquitin-binding protein that can bind to both LC3 II and the misfolded and/or ubiquitylated proteins (Pankiv et al., 2007; Park et al., 2013; Rusten & Stenmark, 2010). During autophagy, LC3 II binds to p62 to promote autophagy flux and degradation in which autophagic cargo as well as p62 is degraded by the autolysosome (Pankiv et al., 2007; Park et al., 2013; Rusten & Stenmark, 2010).

The role of autophagy in SCI is still debatable; however, growing evidence suggests that autophagy flux is impaired after SCI and it negatively impacts locomotor functions and neuronal survival (Li, Du, Lu, & Lin, 2019; Wang, Jiang, et al., 2018; Zhang et al., 2017). Studies in rodent models showed activation of autophagy shortly after compression or contusion injury to the spinal cord; however, autophagy flux was impaired in these models coupled with increases in neuronal death and impaired locomotor function (Li, Du, et al., 2019; Wang, Jiang, et al., 2018; Zhang et al., 2017). Molecular analysis has shown downregulation of Beclin-1 expression, reduction in LC3 II/LC3 I ratio, upregulation of p62, and increase in neuronal apoptosis as assessed by activation of caspase-3 and/or pro-apoptotic Bax (Bcl-2-associated X) proteins (Li, Du, et al., 2019; Wang, Jiang, et al., 2018; Zhang et al., 2017). These results suggested that promoting autophagy flux could be a potential therapeutic option to alleviate secondary damages to the neurons and to limit neuronal death following SCI.

The mechanistic target of rapamycin (mTOR) is a serine-threonine protein kinase that belongs to phosphatidylinositol 3-kinase-related kinase (PIKK) family, which regulates various cellular processes in developing as well as in mature neurons, for example, regulations of cell growth, cell proliferation, cell death, protein translation, and autophagy (Perluigi, Di Domenico, & Butterfield, 2015). mTOR is a key regulator of autophagy and it can be activated via the sensor of energy such as the adenosine monophosphate-activated protein kinase (AMPK)-dependent or AMPK-independent pathways (Perluigi et al., 2015). Within the scope of SCI, in vivo and in vitro studies suggest that mTOR and its downstream regulator of protein synthesis, p70S6K (the 70 kDa ribosomal protein S6 kinase) is upregulated following the injury and it negatively impacts autophagy flux in the injured neurons and enhances their apoptosis. So, inhibition of mTOR signaling is an attractive therapeutic strategy for molecular alterations for the promotion of autophagy flux and inhibition of apoptosis in SCI (Fig. 1). In subsequent parts, we will describe current advances in targeting mTOR signaling pathway with natural, genetic, and pharmacological mTOR inhibitors to potentiate autophagy flux, preserve neurons, and accelerate functional recovery in SCI. All these results strongly imply a neuroprotective role of the mTOR inhibitors via enhancing autophagy flux.

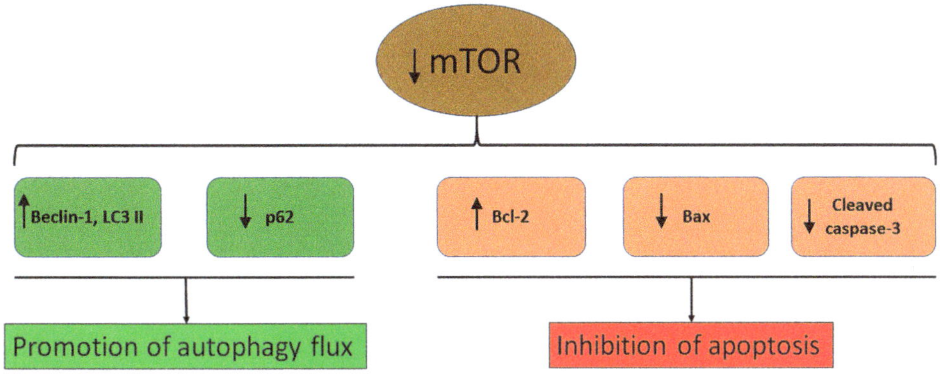

FIG. 1 Effect of mTOR inhibition on autophagy flux and apoptosis after SCI. Treatment with mTOR inhibitors promotes autophagy flux, which is characterized by increased conversion of LC3 I to LC3 II and decreasing the autophagy substrate p62. In contrast, mTOR inhibitors reduce apoptosis, which is marked by increasing anti-apoptotic protein Bcl-2, decreasing pro-apoptotic Bax protein, and preventing cleavage of pro-caspase-3 to caspase-3 in SCI.

Natural compounds for inhibition of mTOR signaling and promotion of autophagy flux and functional recovery after SCI

Studies show that various attempts have been made to inhibit mTOR signaling directly or indirectly in cell culture and animal models of SCI (Fig. 2). At first, we will describe the use of several natural compounds for inhibition of mTOR signaling to promote autophagy flux and enhance functional neuroprotection mostly in rodent models of SCI (Table 1).

Curcumin

It is a natural polyphenolic substance extracted from *Curcuma longa* plants and has been found to exert beneficial effects on neurological and non-neurological disorders via activation of autophagy (Fu et al., 2018; Kocaadam & Şanlier, 2017; Wang, Zhang, Teng, Zhang, & Li, 2014). Curcumin treatment has been shown to exert neuroprotective effects via activation of autophagy in rat SCI. Following contusion injury to spinal cord in rats, curcumin treatment potentiates autophagy, reduces neuronal apoptosis, and improves locomotor function via inhibition of protein kinase B or Akt /mTOR pathway (Li, Yao, et al., 2019).

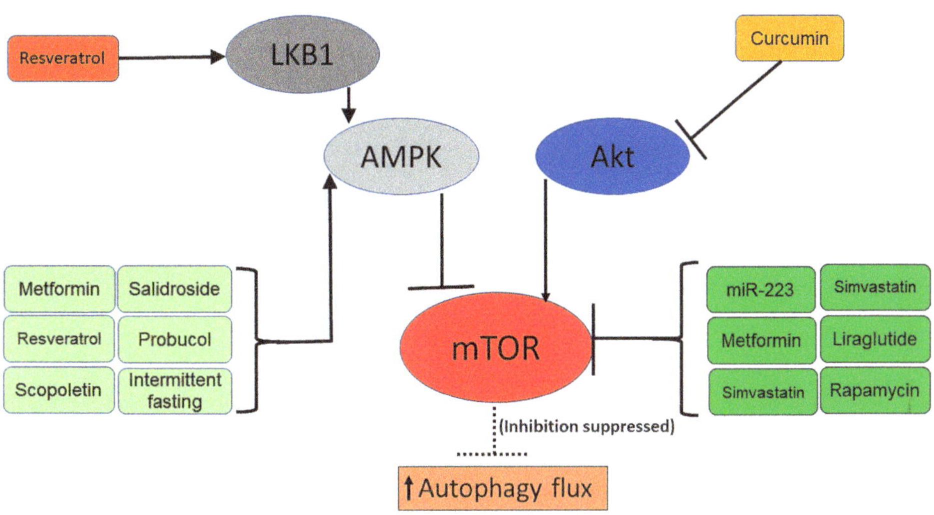

FIG. 2 Molecular mechanisms of mTOR inhibition using pharmacological and non-pharmacological approaches in SCI. Many mTOR inhibitors enhance autophagy flux via AMPK-dependent and AMPK-independent mechanisms in SCI.

TABLE 1 Effects of natural compounds on mTOR pathway and autophagy following SCI.

mTOR inhibitor	Class	SCI model	Effect on mTOR pathway	Effect on autophagy	Effect on SCI	References
Curcumin	Polyphenolic compound	Rat spinal cord contusion weight drop injury	Downregulates Akt/mTOR pathway	Enhances autophagy	Reduces neuronal apoptosis, improves functional recovery, preserves neurons, and enhances remyelination	Li, Yao, Li, Meng, and Sun (2019)
Resveratrol	Polyphenolic compound	Rat spinal cord contusion weight drop injury	Increases phosphorylation of AMPK and decreases phosphorylation of mTOR	Increases Beclin-1 and LC3 II/ LC3 I ratio	Improves functional recovery following SCI	Wang, Jiang, et al. (2018)
Resveratrol	Polyphenolic compound	Rat spinal cord contusion weight drop injury	Increases expression of LKB1 and phosphorylation of AMPK and reduces phosphorylation of mTOR and p70S6K	Increases LC3 II/LC3 I ratio and decreases p62	Reduces neuronal apoptosis via decreasing activation of caspase-3 and Bax/Bcl-2 ratio, preserving motor neurons, and improving functional recovery	Meng et al. (2018)
Resveratrol	Polyphenolic compound	H_2O_2-treated-PC12 cells (in vitro oxidative injury)	Increases LKB1 expression and phosphorylation of AMPK and reduces phosphorylation of mTOR and p70S6K	Increases LC3 II/LC3 I ratio and decreases p62	Not available	Wang, Jiang, et al. (2018)
Salidroside	Phenolic glycoside	Rat spinal cord crush injury with vascular clip	Increases phosphorylation of AMPK and reduces phosphorylation of mTOR and p70S6K	Increases LC3 II expression and decreases p62 expression	Reduces neuronal apoptosis and motor neuron loss and promotes locomotor function	Wang, Wang, et al. (2018)
Scopoletin	Phytoalexin compound	Rat spinal cord damage by impounder impact	Increases phosphorylation of AMPK and reduces phosphorylation of mTOR	Promotes autophagy	Reduces neuronal apoptosis and improves functional recovery after SCI	Zhou et al. (2020)

Resveratrol

It is a plant-derived polyphenolic compound (3,5,4′-trihydroxy-*trans*-stilbene), which is found in the skin of red grapes and blueberries and it provides various health benefits due to its antioxidant, antitumor, and anti-inflammatory effects (Salehi et al., 2018). Resveratrol has been shown to promote autophagy flux and manifest neuroprotective roles on SCI both in vivo and in vitro. In rat model of SCI, resveratrol treatment enhances autophagy, reduces neuronal apoptosis, preserves motoneurons, and improves locomotor function (Meng et al., 2018; Wang, Jiang, et al., 2018). Resveratrol-mediated autophagy activation occurs via activation of AMPK/mTOR/p70S6K pathway (Meng et al., 2018; Wang, Jiang, et al., 2018). Similarly, resveratrol treatment of the H_2O_2-treated neuronal PC12 cells potentiated autophagy flux via AMPK/mTOR/p70S6K pathway (Wang, Jiang, et al., 2018).

Salidroside

Salidroside is a phenolic glycoside, which is extracted from *Rhodiola sachalinensis* plant, and it has antioxidant, anti-inflammatory, and neuroprotective roles (Han et al., 2015; Li & Chen, 2001; Tang et al., 2016). Salidroside treatment of SCI rats reduces neuroinflammation, enhances locomotor function, decreases spinal cord lesion and apoptosis, and reduces loss of ventral motoneurons (VMNs) (Wang, Wang, et al., 2018). Molecular analysis shows that salidroside promotes autophagy flux via AMPK/mTOR/p70S6K pathway (Wang, Wang, et al., 2018).

Scopoletin

It is a phytoalexin, which is extracted from many plants such as *Crossostephium chinensis* and *Sinomonium acutum*, possessing anti-inflammatory and antioxidant properties (Chang et al., 2012; Shaw, Chen, Hsu, Chen, & Tsai, 2003). Scopoletin treatment of SCI rats showed activation of autophagy, reduction of neuronal apoptosis, and enhancement of locomotor function. Induction of autophagy for neuroprotection is mediated via AMPK/mTOR pathway (Zhou et al., 2020).

Pharmacological inhibition of mTOR signaling for enhancing autophagy flux and functional recovery following SCI

Some investigators have used various pharmacological agents for inhibiting mTOR signaling so as to enhance autophagy flux and promote functional recovery in animal models of SCI (Table 2). Here, we will describe five pharmacological agents (rapamycin, metformin, liraglutide, simvastatin, and probucol) that have been shown to inhibit mTOR signaling in treating SCI in animals to achieve an increase in autophagy flux and improvements in neurological functions.

Rapamycin

It is a macrocyclic lactone compound extracted from *Streptomyces hygroscopicus* and it has both antifungal and antibiotic effects (Pallet, Beaune, Legendre, & Anglicheau, 2006). Rapamycin is a well-known compound for mTOR inhibition. Mice model of SCI treated with rapamycin shows inhibition of the mTOR downstream target, p70S6K phosphorylation, increase in autophagy, decrease in neuronal cell death, and improvement in locomotor function (Sekiguchi et al., 2012). Similarly, post-SCI rapamycin treatment of rats enhances autophagy via Akt/mTOR pathway, upregulates Beclin-1, and reduces neuronal apoptosis and loss in SCI (Li, Du, et al., 2019).

TABLE 2 Effects of pharmacological inhibition of mTOR signaling on autophagy in SCI.

mTOR inhibitor	Class	SCI model	Effect on mTOR pathway	Effect on autophagy	Effect on SCI	References
Rapamycin	Specific mTOR inhibitor	Mice	Decreases phosphorylation of p70S6K	Increases LC3 and Beclin-1 expression	Reduces neuronal death and improves locomotor function	Sekiguchi, Kanno, Ozawa, Yamaya, and Itoi (2012)
Rapamycin	Specific mTOR inhibitor	Rat compressive SCI with aneurysm clip	Decreases phosphorylation of mTOR and increases phosphorylation of Akt	Increases Beclin-1	Reduces neuronal apoptosis via increase in Bcl-2, decreases in Bax, mitochondrial cytochrome *c* release, and active caspase-3	Li, Du, et al. (2019)

Continued

TABLE 2 Effects of pharmacological inhibition of mTOR signaling on autophagy in SCI—cont'd

mTOR inhibitor	Class	SCI model	Effect on mTOR pathway	Effect on autophagy	Effect on SCI	References
Metformin	Oral anti-diabetes medicine	Rat	Increases phosphorylation of AMPK and inhibits phosphorylation of mTOR	Increases formation of autophagosome and expression of Beclin-1 and LC3 II and Beclin-1 expression while reduction of p62 and ubiquitinated proteins	Reduces neuronal apoptosis and improves functional recovery	Zhang et al. (2017)
Metformin	Oral diabetes medicine	Rat	Decreases phosphorylation of mTOR and p70S6K	Increases Beclin-1 and LC3 II expression	Preserves motor neurons and enhances their functions	Wang et al. (2016)
Metformin	Oral anti-diabetic drug	Rat	Decreases mTOR and p70S6K	Increases autophagy	Reduces apoptosis, increases number of the survived neurons, and improves locomotor function	Guo et al. (2018)
Liraglutide	Glucagon-like peptide 1 (GLP-1) analog	Rat spinal cord compression injury with vascular clamp	Reduces phosphorylation of mTOR and p62 expression	Increases formation of autophagosome, Beclin-1, and LC3 II/LC3 I ratio	Reduces motor neuron loss and improves locomotor function	Chen et al. (2017)
Simvastatin	3-Hydroxy-3-methylglutaryl-coenzyme A reductase inhibitor	Rat spinal cord weight drop contusion injury	Reduces phosphorylation of mTOR and p70S6K	Increases Beclin-1 and LC3 IIB proteins	Inhibits apoptosis and enhances functional recovery	Gao et al. (2015)
Probucol	Anti-hyperlipidemic drug	Rat	Increases phosphorylation of AMPK and reduces phosphorylation of p70S6K	Increases Beclin-1 and LC3 IIB	Inhibits neural apoptosis and preserves motor neurons	Zhou et al. (2016)

Metformin

It is an oral biguanide commonly used to treat type 2 diabetes and it can potentiate autophagy via direct or indirect mTOR inhibition (Ling et al., 2017; Xiao et al., 2017). Metformin treatment of SCI rats potentiates autophagy, reduces neuronal death, decreases lesion size, and enhances functional recovery (Guo et al., 2018; Wang et al., 2016; Zhang et al., 2017). In these studies, metformin triggered autophagy via activation of AMPK/mTOR/p70S6K pathway or possibly via direct inhibition of mTOR. These results suggested that metformin could inhibit mTOR directly or indirectly via AMPK pathway.

Liraglutide

It is a glucagon-like peptide 1 (GLP-1) analog with anti-diabetic and anti-inflammatory properties (Hou et al., 2012). Liraglutide treatment of SCI rats enhances autophagy flux in neurons, reduces apoptosis, prevents neuronal loss, and enhances locomotor function. Liraglutide enhances autophagy flux via reduction of mTOR and p62 phosphorylation (Chen et al., 2017).

Simvastatin

It is a 3-hydroxy-3-methylglutaryl-coenzyme. As a reductase inhibitor, it has anti-hyperlipidemic properties and it is commonly used in the treatment of cardiovascular diseases (Li et al., 2020). Treatment of SCI in rats with simvastatin enhances autophagy, inhibits apoptosis, and improves functional recovery. Molecular analysis shows that the neuroprotective effects of simvastatin in SCI are regulated by activation of autophagy via inhibition of mTOR/p70S6K phosphorylation (Gao et al., 2015).

Probucol

It is an anti-hyperlipidemic drug with some anti-inflammatory and anti-diabetic effects (Mooranian et al., 2018). Treatment with probucol shows significant effects in increasing the number and function of motoneurons following SCI in rats. Treatment with probucol enhances autophagic activity and reduces induction of apoptosis via increasing phosphorylation of AMPK and decreasing phosphorylation of p70S6K (Zhou et al., 2016).

Genetic and non-genetic inhibitors of mTOR signaling to regulate autophagy in SCI

Some investigators have successfully used both genetic and non-genetic inhibitors for downregulating mTOR signaling for altering autophagy in the treatment of SCI in animals (Table 3). These approaches include miR-223 and intermittent fasting, which have beneficial effects in enhancing autophagy flux and locomotor function via inhibition of mTOR signaling pathways.

miR-223

It is a small non-coding single-stranded RNA that regulates gene expression in neurological and non-neuroglial diseases (Glasgow et al., 2013; Haneklaus, Gerlic, O'Neill, & Masters, 2013). Overexpression of miR-223 in lipopolysaccharide (LPS)-stimulated neuronal PC12 cells inhibits mTOR phosphorylation with downregulation of the histone demethylase protein RPH1, inhibits autophagy flux, blocks apoptosis, and improves cell viability. Results indicate that miR-223 reduces cell death and autophagy by targeting mTOR/RPH1 pathway in LPS-induced PC12 cells (Jia et al., 2017).

TABLE 3 Effects of genetic and non-genetic inhibition of mTOR on autophagy in SCI.

mTOR inhibitor	Class	SCI model	Effect on mTOR pathway	Effect on autophagy	Effect on SCI	References
miR-223 overexpression	Non-coding single-stranded RNA	LPS-stimulated PC12 cells	Inhibits phosphorylation of mTOR with downregulation of RPH1	Reduces Beclin-1 and LC3-II and increases p62	Inhibits neuronal apoptosis and enhances their survival	Jia, Niu, Li, and Zhang (2017)
Intermittent fasting	Dietary intervention	Rat	Increases phosphorylation of AMPK and decreases phosphorylation of mTOR	Increases Beclin-1 and LC3 II and decreases p62	Reduces neuronal apoptosis via decrease in Bax/Bcl-2 ratio and cleaved caspase-3	Yuan et al. (2021)

Intermittent fasting

It is a dietary intervention approach of short-term food restriction, which is known to promote autophagy in neurons (Alirezaei et al., 2010). Implementing intermittent fasting in rat model of SCI activates autophagy flux, reduces neuronal death, and increases survival and functional recovery following injury. This study found that intermittent fasting stimulated autophagy flux via inhibition of AMPK/mTOR pathway (Yuan et al., 2021).

Future directions

Impaired autophagy flux in SCI is a complex deregulated mechanism that is associated with increase in neuronal apoptosis and impaired functional recovery. Even though the role of autophagy in SCI could be protective or destructive (Ray, 2020), emerging evidence indicates that enhancing autophagy flux by inhibition of mTOR signaling has promising neuroprotective effects for functional recovery, particularly in severe contusion injury (Fig. 3). Although these studies have elucidated the role of AMPK-dependent or AMPK-independent mechanisms of mTOR inhibitors, further studies are required to understand more about the upstream molecular regulation of mTOR inhibition, the downstream target of mTOR that regulates autophagy flux, and how promotion of autophagy prevents apoptosis and neuronal loss in SCI. Further mechanistic studies are warranted to establish the precise role of autophagy in SCI.

Applications to other areas of neuroscience

So far recent studies strongly suggest that targeting mTOR signaling at an early phase of SCI can provide neuroprotective effects via promoting autophagy flux, reducing secondary damages, and improving functional recovery following SCI. In addition, mTOR signaling pathway is involved in deregulation of autophagy flux in other major neurodegenerative diseases including Alzheimer's disease and Parkinson's disease (Cai, Chen, He, Xiao, & Yan, 2015; Zhu et al., 2019). In Alzheimer's disease, deregulation of mTOR signaling and impairment of autophagy are found to be associated with increase in β-amyloid (Aβ) generation and deposition of amyloid precursor proteins (Cai et al., 2015). In Parkinson's disease, mTOR signaling is deregulated and it causes increase in accumulation of the aggregated harmful proteins such as α-synuclein, which is associated with autophagy impairment (Zhu et al., 2019). Similarly, combating mTOR signaling in these neurological diseases, with the use of pharmacological and non-pharmacological approaches, has shown to enhance autophagy flux via direct or indirect mTOR inhibition; however, it is still little known about the molecular regulation of upstream and downstream pathways to mTOR in these diseases. Although we have discussed the recent studies that highlight the role of mTOR inhibitors on locomotor recovery following SCI via natural, pharmacological, and genetic approaches, further studies will not only provide better understanding of the underlying mechanisms of the impaired autophagy flux in the pathogenesis of SCI but also will delineate the precise role of autophagy in other neurodegenerative diseases including Alzheimer's disease and Parkinson's disease, where deregulation of autophagy and mTOR signaling are well documented.

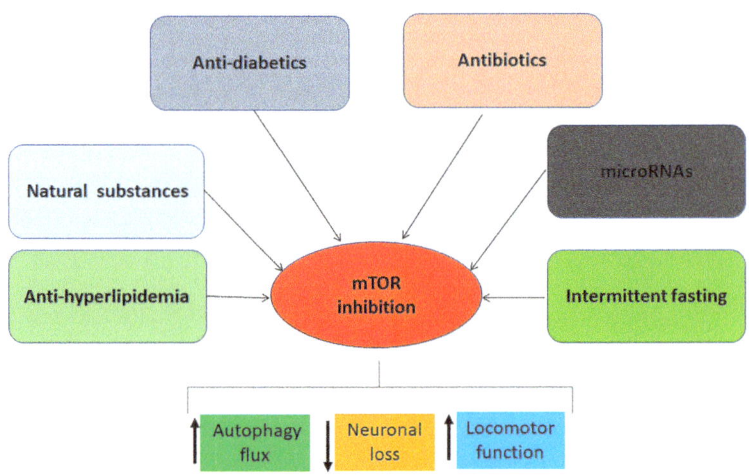

FIG. 3 Pharmacological and non-pharmacological inhibition of mTOR pathway in rodent model of SCI. Treatment with plant-derived natural substances, antibiotic drug, anti-diabetic drug, anti-hyperlipidemia drug, microRNAs, and intermittent fasting inhibits mTOR signaling, enhancing autophagy and functional recovery after SCI.

Mini-dictionary of terms

Traumatic spinal cord injury: it refers to direct physical injury to the spinal cord such as fracture of the spine during severe road traffic accident or falling from height.
Non-traumatic spinal cord injury: it refers to indirect injury to spinal cord that caused by progressive medical diseases, such as, in patients with vascular disease with poor circulation or abrupt impairment of blood supply to the spinal cord or in patients with chronic neurodegenerative diseases that involve the spinal cord.
Macroautophagy: it is a recycling process within the cell to clean the cytoplasm from the damaged or unused large molecules or organelles and reuse their degraded primary elements to maintain the function of the cell.

Key facts of targeting mTOR signaling in promotion of autophagy for functional recovery after SCI

- SCI is a serious neurodegenerative disorder, mostly traumatic in nature, and it frequently leads to long-life complications.
- Balance between autophagy and apoptosis is critical to minimize neuronal cell injury and improve functional recovery.
- Deregulation of macroautophagy, particularly autophagy flux, is evident shortly after SCI and it is associated with impairment of neuronal and neuroglial cell survival due to increase in apoptosis.
- Upregulation of mTOR has been reported in SCI and it is linked with the impaired autophagy flux.
- Therapeutic and natural inhibitions of mTOR signaling enhance autophagy flux and have shown promising beneficial effects on locomotor function following SCI.

Summary points

- Traumatic SCI is a deleterious global health problem with economic and medical burdens.
- Impairment of autophagy flux is one of the proposed secondary damaging mechanisms that is associated with poor functional recovery following SCI.
- Increase in mTOR signaling has been reported after SCI and it is thought to cause impairment in autophagy flux.
- Pharmacological and non-pharmacological inhibitions of mTOR signaling promote autophagy flux, increase neuronal cell survival, reduce apoptosis, and improve functional recovery following SCI.

Acknowledgments

The work was supported in part by the earlier R01 grants (CA-091460 and NS-057811) from the National Institutes of Health (Bethesda, MD, USA) and a current grant (SCIRF-2020-PD-01) from the State of South Carolina (Columbia, SC, USA).

References

Alirezaei, M., Kemball, C. C., Flynn, C. T., Wood, M. R., Whitton, J. L., & Kiosses, W. B. (2010). Short-term fasting induces profound neuronal autophagy. *Autophagy*, *6*(6), 702–710. https://doi.org/10.4161/auto.6.6.12376.

Bracken, M. B., Freeman, D. H., & Hellenbrand, K. (1981). Incidence of acute traumatic hospitalized spinal cord injury in the United States, 1970–1977. *American Journal of Epidemiology*, *113*(6), 615–622. https://doi.org/10.1093/oxfordjournals.aje.a113140.

Brier, L. W., Ge, L., Stjepanovic, G., Thelen, A. M., Hurley, J. H., & Schekman, R. (2019). Regulation of LC3 lipidation by the autophagy-specific class III phosphatidylinositol-3 kinase complex. *Molecular Biology of the Cell*, *30*(9), 1098–1107. https://doi.org/10.1091/mbc.E18-11-0743.

Cai, Z., Chen, G., He, W., Xiao, M., & Yan, L. J. (2015). Activation of mTOR: A culprit of Alzheimer's disease? *Neuropsychiatric Disease and Treatment*, *11*, 1015–1030. https://doi.org/10.2147/NDT.S75717.

Chang, T. N., Deng, J. S., Chang, Y. C., Lee, C. Y., Jung-Chun, L., Lee, M. M., ... Huang, G. J. (2012). Ameliorative effects of scopoletin from *Cross-ostephium chinensis* against Inflammation pain and Its mechanisms in mice. *Evidence-Based Complementary and Alternative Medicine: eCAM*, *2012*, 595603. https://doi.org/10.1155/2012/595603.

Chen, J., Wang, Z., Mao, Y., Zheng, Z., Chen, Y., Khor, S., ... Wang, X. (2017). Liraglutide activates autophagy via GLP-1R to improve functional recovery after spinal cord injury. *Oncotarget*, *8*(49), 85949–85968. https://doi.org/10.18632/oncotarget.20791.

Cheng, H., Liao, K. K., Liao, S. F., Chuang, T. Y., & Shih, Y. H. (2004). Spinal cord repair with acidic fibroblast growth factor as a treatment for a patient with chronic paraplegia. *Spine (Phila Pa 1976)*, *29*(14), E284–E288. https://doi.org/10.1097/01.brs.0000131217.61390.2c.

Drasites, K. P., Shams, R., Zaman, V., Matzelle, D., Shields, D. C., Garner, D. P., ... Banik, N. L. (2020). Pathophysiology, biomarkers, and therapeutic modalities associated with skeletal muscle loss following spinal cord injury. *Brain Sciences*, *10*(12), 933. https://doi.org/10.3390/brainsci10120933.

du Toit, A., Hofmeyr, J. S., Gniadek, T. J., & Loos, B. (2018). Measuring autophagosome flux. *Autophagy*, *14*(6), 1060–1071. https://doi.org/10.1080/15548627.2018.1469590.

Dumont, R. J., Okonkwo, D. O., Verma, S., Hurlbert, R. J., Boulos, P. T., Ellegala, D. B., & Dumont, A. S. (2001). Acute spinal cord injury, part I: Pathophysiologic mechanisms. *Clinical Neuropharmacology*, *24*(5), 254–264. https://doi.org/10.1097/00002826-200109000-00002.

Ekong, C. E., & Tator, C. H. (1985). Spinal cord injury in the work force. *Canadian Journal of Surgery*, *28*(2), 165–167.

Fehlings, M. G., Cadotte, D. W., & Fehlings, L. N. (2011). A series of systematic reviews on the treatment of acute spinal cord injury: A foundation for best medical practice. *Journal of Neurotrauma*, *28*(8), 1329–1333. https://doi.org/10.1089/neu.2011.1955.

Feng, Y., He, D., Yao, Z., & Klionsky, D. J. (2014). The machinery of macroautophagy. *Cell Research*, *24*(1), 24–41. https://doi.org/10.1038/cr.2013.168.

Fu, H., Wang, C., Yang, D., Wei, Z., Xu, J., Hu, Z., ... Cai, Q. (2018). Curcumin regulates proliferation, autophagy, and apoptosis in gastric cancer cells by affecting PI3K and P53 signaling. *Journal of Cellular Physiology*, *233*(6), 4634–4642. https://doi.org/10.1002/jcp.26190.

Galeiras Vázquez, R., Rascado Sedes, P., Mourelo Fariña, M., Montoto Marqués, A., & Ferreiro Velasco, M. E. (2013). Respiratory management in the patient with spinal cord injury. *BioMed Research International*, *2013*, 168757. https://doi.org/10.1155/2013/168757.

Gao, K., Wang, G., Wang, Y., Han, D., Bi, J., Yuan, Y., ... Mei, X. (2015). Neuroprotective effect of simvastatin via inducing the autophagy on spinal cord injury in the rat model. *BioMed Research International*, *2015*, 260161. https://doi.org/10.1155/2015/260161.

Ge, L., Arul, K., & Mesfin, A. (2019). Spinal cord injury from spinal tumors: Prevalence, management, and outcomes. *World Neurosurgery*, *122*, e1551–e1556. https://doi.org/10.1016/j.wneu.2018.11.099.

Giatromanolaki, A., Koukourakis, M. I., Koutsopoulos, A., Chloropoulou, P., Liberis, V., & Sivridis, E. (2011). High beclin 1 expression defines a poor prognosis in endometrial adenocarcinomas. *Gynecologic Oncology*, *123*(1), 147–151. https://doi.org/10.1016/j.ygyno.2011.06.023.

Glasgow, S. M., Laug, D., Brawley, V. S., Zhang, Z., Corder, A., Yin, Z., Wong, S.T., ... Deneen, B. (2013). The miR-223/nuclear factor I-A axis regulates glial precursor proliferation and tumorigenesis in the CNS. *The Journal of Neuroscience*, *33*(33), 13560–13568. https://doi.org/10.1523/JNEUROSCI.0321-13.2013.

Guo, Y., Wang, F., Li, H., Liang, H., Li, Y., Gao, Z., & He, X. (2018). Metformin protects against spinal cord injury by regulating autophagy via the mTOR signaling pathway. *Neurochemical Research*, *43*(5), 1111–1117. https://doi.org/10.1007/s11064-018-2525-8.

Han, J., Xiao, Q., Lin, Y. H., Zheng, Z. Z., He, Z. D., Hu, J., & Chen, L. D. (2015). Neuroprotective effects of salidroside on focal cerebral ischemia/reperfusion injury involve the nuclear erythroid 2-related factor 2 pathway. *Neural Regeneration Research*, *10*(12), 1989–1996. https://doi.org/10.4103/1673-5374.172317.

Haneklaus, M., Gerlic, M., O'Neill, L. A., & Masters, S. L. (2013). miR-223: Infection, inflammation and cancer. *Journal of Internal Medicine*, *274*(3), 215–226. https://doi.org/10.1111/joim.12099.

Hou, J., Manaenko, A., Hakon, J., Hansen-Schwartz, J., Tang, J., & Zhang, J. H. (2012). Liraglutide, a long-acting GLP-1 mimetic, and its metabolite attenuate inflammation after intracerebral hemorrhage. *Journal of Cerebral Blood Flow and Metabolism*, *32*(12), 2201–2210. https://doi.org/10.1038/jcbfm.2012.133.

Jain, N. B., Ayers, G. D., Peterson, E. N., Harris, M. B., Morse, L., O'Connor, K. C., & Garshick, E. (2015). Traumatic spinal cord injury in the United States, 1993–2012. *JAMA*, *313*(22), 2236–2243. https://doi.org/10.1001/jama.2015.6250.

Jia, D., Niu, Y., Li, D., & Zhang, Q. (2017). MicroRNA-223 alleviates lipopolysaccharide-induced PC-12 cells apoptosis and autophagy by targeting RPH1 in spinal cord injury. *International Journal of Clinical and Experimental Pathology*, *10*(9), 9223–9232.

Karlsson, A. K. (2006). Autonomic dysfunction in spinal cord injury: Clinical presentation of symptoms and signs. *Progress in Brain Research*, *152*, 1–8. https://doi.org/10.1016/S0079-6123(05)52034-X.

Klionsky, D. J., Abdalla, F. C., Abeliovich, H., Abraham, R. T., Acevedo-Arozena, A., Adeli, K., ... Zuckerbraun, B. (2012). Guidelines for the use and interpretation of assays for monitoring autophagy. *Autophagy*, *8*(4), 445–544. https://doi.org/10.4161/auto.19496.

Kocaadam, B., & Şanlier, N. (2017). Curcumin, an active component of turmeric (*Curcuma longa*), and its effects on health. *Critical Reviews in Food Science and Nutrition*, *57*(13), 2889–2895. https://doi.org/10.1080/10408398.2015.1077195.

Kulkarni, A., Chen, J., & Maday, S. (2018). Neuronal autophagy and intercellular regulation of homeostasis in the brain. *Current Opinion in Neurobiology*, *51*, 29–36. https://doi.org/10.1016/j.conb.2018.02.008.

Lee, B. B., Cripps, R. A., Fitzharris, M., & Wing, P. C. (2014). The global map for traumatic spinal cord injury epidemiology: Update 2011, global incidence rate. *Spinal Cord*, *52*(2), 110–116. https://doi.org/10.1038/sc.2012.158.

Li, H. B., & Chen, F. (2001). Preparative isolation and purification of salidroside from the Chinese medicinal plant Rhodiola sachalinensis by high-speed counter-current chromatography. *Journal of Chromatography A*, *932*(1–2), 91–95. https://doi.org/10.1016/s0021-9673(01)01232-8.

Li, X. G., Du, J. H., Lu, Y., & Lin, X. J. (2019). Neuroprotective effects of rapamycin on spinal cord injury in rats by increasing autophagy and Akt signaling. *Neural Regeneration Research*, *14*(4), 721–727. https://doi.org/10.4103/1673-5374.247476.

Li, W., He, P., Huang, Y., Li, Y. F., Lu, J., Li, M., ... Feng, D. (2021). Selective autophagy of intracellular organelles: recent research advances. *Theranostics*, *11*(1), 222–256. https://doi.org/10.7150/thno.49860.

Li, H., Rabearivony, A., Zhang, W., Chen, S., An, X., & Liu, C. (2020). Chronopharmacology of simvastatin on hyperlipidaemia in high-fat diet-fed obese mice. *Journal of Cellular and Molecular Medicine*, *24*(18), 11024–11029. https://doi.org/10.1111/jcmm.15709.

Li, W., Yao, S., Li, H., Meng, Z., & Sun, X. (2019). Curcumin promotes functional recovery and inhibits neuronal apoptosis after spinal cord injury through the modulation of autophagy. *Journal of Spinal Cord Medicine*, 1–9. https://doi.org/10.1080/10790268.2019.1616147.

Ling, S., Song, L., Fan, N., Feng, T., Liu, L., Yang, X., ... Li, Y. (2017). Combination of metformin and sorafenib suppresses proliferation and induces autophagy of hepatocellular carcinoma via targeting the mTOR pathway. *International Journal of Oncology*, *50*(1), 297–309. https://doi.org/10.3892/ijo.2016.3799.

Mauney, M. C., Blackbourne, L. H., Langenburg, S. E., Buchanan, S. A., Kron, I. L., & Tribble, C. G. (1995). Prevention of spinal cord injury after repair of the thoracic or thoracoabdominal aorta. *Annals of Thoracic Surgery*, *59*(1), 245–252. https://doi.org/10.1016/0003-4975(94)00815-O.

Meng, H. Y., Shao, D. C., Li, H., Huang, X. D., Yang, G., Xu, B., & Niu, H. Y. (2018). Resveratrol improves neurological outcome and neuroinflammation following spinal cord injury through enhancing autophagy involving the AMPK/mTOR pathway. *Molecular Medicine Reports*, *18*(2), 2237–2244. https://doi.org/10.3892/mmr.2018.9194.

Mooranian, A., Negrulj, R., Takechi, R., Mamo, J., Al-Sallami, H., & Al-Salami, H. (2018). The biological effects of the hypolipidaemic drug probucol microcapsules fed daily for 4 weeks, to an insulin-resistant mouse model: Potential hypoglycaemic and anti-inflammatory effects. *Drug Delivery and Translational Research*, *8*(3), 543–551. https://doi.org/10.1007/s13346-017-0473-5.

Nakamura, M., & Okano, H. (2013). Cell transplantation therapies for spinal cord injury focusing on induced pluripotent stem cells. *Cell Research*, *23*(1), 70–80. https://doi.org/10.1038/cr.2012.171.

Nakatogawa, H. (2020). Mechanisms governing autophagosome biogenesis. *Nature Reviews. Molecular Cell Biology*, *21*(8), 439–458. https://doi.org/10.1038/s41580-020-0241-0.

Pallet, N., Beaune, P., Legendre, C., & Anglicheau, D. (2006). Rapamycine and mTOR inhibitors: From bench to bedside. *Annales de Biologie Clinique (Paris)*, *64*(2), 107–115.

Pankiv, S., Clausen, T. H., Lamark, T., Brech, A., Bruun, J. A., Outzen, H., … Johansen, T. (2007). p62/SQSTM1 binds directly to Atg8/LC3 to facilitate degradation of ubiquitinated protein aggregates by autophagy. *Journal of Biological Chemistry*, *282*(33), 24131–24145. https://doi.org/10.1074/jbc.M702824200.

Park, J. M., Huang, S., Wu, T. T., Foster, N. R., & Sinicrope, F. A. (2013). Prognostic impact of beclin 1, p62/sequestosome 1 and LC3 protein expression in colon carcinomas from patients receiving 5-fluorouracil as adjuvant chemotherapy. *Cancer Biology & Therapy*, *14*(2), 100–107. https://doi.org/10.4161/cbt.22954.

Perluigi, M., Di Domenico, F., & Butterfield, D. A. (2015). mTOR signaling in aging and neurodegeneration: At the crossroad between metabolism dysfunction and impairment of autophagy. *Neurobiology of Disease*, *84*, 39–49. https://doi.org/10.1016/j.nbd.2015.03.014.

Ray, S. K. (2020). Modulation of autophagy for neuroprotection and functional recovery in traumatic spinal cord injury. *Neural Regeneration Research*, *15* (9), 1601–1612. https://doi.org/10.4103/1673-5374.276322.

Rusten, T. E., & Stenmark, H. (2010). p62, an autophagy hero or culprit? *Nature Cell Biology*, *12*(3), 207–209. https://doi.org/10.1038/ncb0310-207.

Salehi, B., Mishra, A. P., Nigam, M., Sener, B., Kilic, M., Sharifi-Rad, M., … Sharifi-Rad, J. (2018). Resveratrol: a double-edged sword in health benefits. *Biomedicines*, *6*(3), 91. https://doi.org/10.3390/biomedicines6030091.

Sekiguchi, A., Kanno, H., Ozawa, H., Yamaya, S., & Itoi, E. (2012). Rapamycin promotes autophagy and reduces neural tissue damage and locomotor impairment after spinal cord injury in mice. *Journal of Neurotrauma*, *29*(5), 946–956. https://doi.org/10.1089/neu.2011.1919.

Shaw, C. Y., Chen, C. H., Hsu, C. C., Chen, C. C., & Tsai, Y. C. (2003). Antioxidant properties of scopoletin isolated from Sinomonium acutum. *Phytotherapy Research*, *17*(7), 823–825. https://doi.org/10.1002/ptr.1170.

Stacchiotti, A., & Corsetti, G. (2020). Natural compounds and autophagy: Allies against neurodegeneration. *Frontiers in Cell and Development Biology*, *8*, 555409. https://doi.org/10.3389/fcell.2020.555409.

Tang, H., Gao, L., Mao, J., He, H., Liu, J., Cai, X., … Wu, T. (2016). Salidroside protects against bleomycin-induced pulmonary fibrosis: activation of Nrf2-antioxidant signaling, and inhibition of NF-κB and TGF-β1/Smad-2/-3 pathways. *Cell Stress & Chaperones*, *21*(2), 239–249. https://doi.org/10.1007/s12192-015-0654-4.

Wang, P., Jiang, L., Zhou, N., Zhou, H., Liu, H., Zhao, W., … Hu, Z. (2018). Resveratrol ameliorates autophagic flux to promote functional recovery in rats after spinal cord injury. *Oncotarget*, *9*(9), 8427–8440. https://doi.org/10.18632/oncotarget.23877.

Wang, C., Liu, C., Gao, K., Zhao, H., Zhou, Z., Shen, Z., … Mei, X. (2016). Metformin preconditioning provide neuroprotection through enhancement of autophagy and suppression of inflammation and apoptosis after spinal cord injury. *Biochemical and Biophysical Research Communications*, *477*(4), 534–540. https://doi.org/10.1016/j.bbrc.2016.05.148.

Wang, C., Wang, Q., Lou, Y., Xu, J., Feng, Z., Chen, Y., … Zhang, X. (2018). Salidroside attenuates neuroinflammation and improves functional recovery after spinal cord injury through microglia polarization regulation. *Journal of Cellular and Molecular Medicine*, *22*(2), 1148–1166. https://doi.org/10.1111/jcmm.13368.

Wang, C., Zhang, X., Teng, Z., Zhang, T., & Li, Y. (2014). Downregulation of PI3K/Akt/mTOR signaling pathway in curcumin-induced autophagy in APP/PS1 double transgenic mice. *European Journal of Pharmacology*, *740*, 312–320. https://doi.org/10.1016/j.ejphar.2014.06.051.

Xiao, Z., Gaertner, S., Morresi-Hauf, A., Genzel, R., Duell, T., Ullrich, A., & Knyazev, P. G. (2017). Metformin triggers autophagy to attenuate drug-induced apoptosis in NSCLC cells, with minor effects on tumors of diabetic patients. *Neoplasia*, *19*(5), 385–395. https://doi.org/10.1016/j.neo.2017.02.011.

Xie, Y., Kang, R., & Tang, D. (2016). Role of the beclin 1 network in the cross-regulation between autophagy and apoptosis. In M. A. Hayat (Ed.), *Vol. 8. Autophagy: Cancer, other pathologies, inflammation, immunity, infection, and aging* (pp. 75–88). Academic Press.

Xie, Z., Nair, U., & Klionsky, D. J. (2008). Atg8 controls phagophore expansion during autophagosome formation. *Molecular Biology of the Cell*, *19*(8), 3290–3298. https://doi.org/10.1091/mbc.e07-12-1292.

Yuan, W., He, X., Morin, D., Barrière, G., Liu, X., Li, J., & Zhu, Y. (2021). Autophagy induction contributes to the neuroprotective impact of intermittent fasting on the acutely injured spinal cord. *Journal of Neurotrauma*, *38*(3), 373–384. https://doi.org/10.1089/neu.2020.7166.

Zhang, D., Xuan, J., Zheng, B. B., Zhou, Y. L., Lin, Y., Wu, Y. S., … Zhang, X. L. (2017). Metformin improves functional recovery after spinal cord injury via autophagy flux stimulation. *Molecular Neurobiology*, *54*(5), 3327–3341. https://doi.org/10.1007/s12035-016-9895-1.

Zhou, Z., Chen, S., Zhao, H., Wang, C., Gao, K., Guo, Y., … Mei, X. (2016). Probucol inhibits neural cell apoptosis via inhibition of mTOR signaling pathway after spinal cord injury. *Neuroscience*, *329*, 193–200. https://doi.org/10.1016/j.neuroscience.2016.05.019.

Zhou, R., Kan, S., Cai, S., Sun, R., Yuan, H., & Yu, B. (2020). Scopoletin activates adenosine monophosphate-activated protein kinase/mammalian target of rapamycin signaling pathway and improves functional recovery after spinal cord injury in rats. *Pharmacology*, *105*(5–6), 349–359. https://doi.org/10.1159/000503866.

Zhu, Z., Yang, C., Iyaswamy, A., Krishnamoorthi, S., Sreenivasmurthy, S. G., Liu, J., … Li, M. (2019). Balancing mTOR signaling and autophagy in the treatment of Parkinson's disease. *International Journal of Molecular Sciences*, *20*(3), 728. https://doi.org/10.3390/ijms20030728.

Zompa, E. A., Cain, L. D., Everhart, A. W., Moyer, M. P., & Hulsebosch, C. E. (1997). Transplant therapy: Recovery of function after spinal cord injury. *Journal of Neurotrauma*, *14*(8), 479–506. https://doi.org/10.1089/neu.1997.14.479.

Chapter 23

Tertiary damage: Hippocampal and brain changes after spinal cord injury

Ignacio Jure and Florencia Labombarda
Laboratorio de Bioquímica Neuroendócrina, Instituto de Biología y Medicina Experimental, CONICET, Buenos Aires, Argentina

Abbreviations

BDNF	brain-derived neurotrophic factor
BrdU	bromodeoxiuridine
CCL2	chemokine (C-C motif) ligand 2
CCL2R	chemokine (C-C motif) ligand 2 receptor
CCL3	chemokine (C-C motif) ligand 3
CCL21	chemokine (C-C motif) ligand 21
DCX	doublecortin
GFAP	glial fibrillary acidic protein
GFP	green fluorescent protein
HPA	hypothalamic-pituitary-adrenal
IL-1β	interleukin 1 beta
IL-6	interleukin 6
IL-18	interleukin 18
NO	nitric oxide
ROS	reactive oxygen species
TNFα	tumor necrosis factor-alpha

Introduction

Given the inability of the CNS to regenerate lost neurons and their connections, spinal cord injury (SCI) leads to permanent motor, sensory, and autonomic dysfunction. Mechanic SCI occurs when the spinal cord is severely bruised, compressed, lacerated, or severed as a result of the traumatic impact. The mechanic impact leads to primary damage which involves cellular death, extracellular matrix changes, edema formation, and blood-brain barrier breakdown. This primary damage drives to the well-known secondary damage characterized by microglia and astrocytes reactivity, periphery immune cell invasion, inflammation, and oxidative stress. These events result in neuron and oligodendrocyte death, which generate a loss of function (Ahuja et al., 2017).

In this chapter, a new concept in SCI research will be developed. Compelling evidence suggests that SCI spreads to the brain affecting rostral distal areas and producing progressive neurodegeneration and neuroinflammation (Table 1). This damage will be considered *Tertiary Damage*.

Traditionally SCI investigation was focused on pathophysiological changes in the spinal cord with efforts being made to recover sensorimotor function and relief neuropathic pain. Long-distal areas, such as the brain, have been ignored for many years. Fortunately, this scenario has changed in the last decades and the brain of spinal cord injured patients has been explored by magnetic resonance imaging (MRI), diffusion tensor imaging (DTI), and transcranial magnetic stimulation. In addition, animal models have revealed abnormalities in distal brain regions related to cognition and mood, which result in behavior impairment.

Cellular, Molecular, Physiological, and Behavioral Aspects of Spinal Cord Injury. https://doi.org/10.1016/B978-0-12-822427-4.00023-X

TABLE 1 Effect of spinal cord injury on different brain regions.

	Neurons	Glial cells and neuroinflammation	References
Hippocampus	Neurogenesis reduction and NSCs inactivation Decrease of CA1 and granular neurons	Microglia and astrocyte activation CCL21, CCR2, CCL2 production	Felix et al. (2012) Jure, Pietranera, De Nicola, and Labombarda (2017) and Jure, De Nicola, Encinas, and Labombarda (2020) Wu, Stoica, et al. (2014) and Wu, Zhao, et al. (2014)
Thalamus	Neuronal reduction	Microglial activation CCL21, CCR2, CCL2 production	Li et al. (2020) Zhao, Waxman, and Hains (2007)
Sensorimotor cortex and pathways	Circuit reorganization, reduction of gray matter volume and corticoespinal tract atrophy	Myelin atrophy	Freund et al. (2013) Wrigley et al. (2009)
Other cortices	Neuronal reduction	Microglia activation	Wu, Stoica, et al. (2014) and Wu, Zhao, et al. (2014)
Whole brain studies		Production of pro-inflammatory cytokine mRNAs	Jure et al. (2020) Wu, Stoica, et al. (2014) and Wu, Zhao, et al. (2014)

Effects of spinal cord injury on neurons and glial cells in different brain regions during the chronic phase. *NSCs*, neural stem cells; *CA1*, pyramidal neurons.

Sensorimotor cortex and corticospinal tract alterations after SCI in humans

The sensorimotor cortex suffers deep plastic changes after SCI. An extensive circuit reorganization takes place as a consequence of cortical deafferentation. Neuroimaging and neurophysiological studies have shown changes in topographical maps of primary motor and sensory cortices which are related to the severity of the injury (Nardone et al., 2013). Although cortex circuit plasticity is involved in the sparse functional recovery, aberrant reorganization of the cerebral cortex also may have undesired consequences, such as phantom limb sensations or neuropathic pain. SCI drives not only to cortical reorganization but to the atrophy of the deafferentated sensorimotor cortex, corticospinal tract (CST), and to the spinal cord as well (Freund et al., 2013; Wrigley et al., 2009).

MRI studies using voxel-based morphometry have demonstrated that chronic lesion to the spinal cord reduces gray matter volume of the primary somatosensory and motor cortex as well as white matter volume of the CST. DTI measures the diffusion of water in tissues giving quantitative information on tissue microstructure. Alterations in DTI metrics reveal that the degeneration process includes microstructural changes related to myelin atrophy and axonal integrity of the CST and the subcortical area of the sensorimotor cortex (Freund et al., 2013; Wrigley et al., 2009). Anterograde and retrograde Wallerian degeneration spreads along white matter tracts producing axonal loss and demyelination. Notably, spinal cord atrophy, loss of CST integrity, and reduction of sensorimotor cortex volume are associated with poor clinical outcomes (Freund et al., 2013).

On the other hand, the survival of cortical axotomized motoneurons is a topic of debate and the information available comes from animal models. Few studies have described apoptotic neurons on Lamina V of the primary motor cortex. However, several reports have revealed that the number of upper-motoneurons persisted unchanged and no signs of apoptosis were observed after SCI (Brock et al., 2010; Lee et al., 2004). In spite of the discrepancy, both positions claim that large pyramidal motoneurons experience changes such as atrophy and shrinkage of the cell body. These changes are related to the atrophy of myelinated axons of the CST and may explain the reduction of cortical gray matter volume already mentioned.

Cognitive and emotional impairment after SCI

Humans

If little attention has been paid to regions of the brain related to motor and sensory functions, much less consideration has been given to those regions related to cognitive and emotional processes although cognitive impairment and emotional dysfunction have been observed in patients after SCI. In this regard, a number of reports have identified cognitive deficits in humans long-term after SCI. All these studies included comorbid traumatic brain injury (TBI) in the exclusion criteria as TBI is a well-known factor to impair cognition. Using standardized neuropsychological tests, several reports have described impairments in executive functioning, span memory, concentration ability, memory function, attention, learning, and processing speed (Craig et al., 2015; Lazzaro, Tran, Wijesuriya, & Craig, 2013; Murray et al., 2007; Roth et al., 1989).

These findings were underestimated by the scientific community because 60% of spinal cord injured patients suffer from depression, anxiety, and neuropathic pain (Li et al., 2020). Certainly, all these conditions contribute to decreasing quality of life, reducing social participation, and affecting cognition (Craig et al., 2015). To make matters even worse, patients often take multiple medications and may also consume alcohol and drugs of abuse. SCI produces permanent disabilities in patients changing completely their lifestyle; thus, it is very complex to discern what the causes of their cognitive deficits are. Therefore, it is crucial to develop animal models to study whether there are biological changes in the brain after SCI that lead to cognitive changes and behavioral alterations.

Animal models

Faden's work is very significant because it demonstrates for the first time that SCI causes emotional and cognitive deficits in animals. In this regard, his group has revealed an interesting relationship between rodent cognition and chronic SCI employing well-establish cognitive tests (Wu, Stoica, et al., 2014; Wu, Zhao, et al., 2014). In their work, mice received a moderate contusion injury in the thoracic segment of the spinal cord to let animals perform cognitive tests which implied locomotion. Spatial learning and retention memory were evaluated in the chronic phase using the Morris water maze (MWM) (Wu, Zhao, et al., 2014). Results indicated that lesioned mice employed more time to locate the hidden platform during the acquisition phase and showed worse retention memory as they spent less time looking for the platform in the target quadrant during the probe test than sham mice (Wu, Zhao, et al., 2014). Using the Y-Maze spontaneous alternation test, Faden's group also demonstrated that lesioned mice displayed a reduction of spontaneous alternation, an indicative of dysfunctional spatial working memory. The authors also evaluated recognition memory using the Novel object recognition (NOR) test, which measures recognition memory based on rodent innate preference to explore novel objects. Lesioned mice spent less time exploring the novel object in the recognition phase than sham ones demonstrating that injured mice could not discriminate between the novel and the familiar object, an indicative of impaired recognition memory (Wu, Zhao, et al., 2014).

Contextual and emotional memory was also evaluated by the authors using the step-down fear-avoidance test. During the conditioning phase, the mouse is placed on a platform and it receives a small electric shock when it steps down. During the testing phase, the animal is returned to the platform and the latency to descend is registered. Faden's work shows that lesioned mice decreased latency in the testing phase demonstrating impaired memory of the aversive experience. Finally, the authors have explored depression-like behavior using the tail suspension and the sucrose preference test, neither of them related to locomotion. The tail suspension is based on the observation that mice develop an immobile posture when they are hung by their tail and suffer an inescapable hemodynamic stress. On the other hand, the sucrose preference test is an indicator of anhedonia based on the mouse's interest in seeking a sweet rewarding drink instead of plain water. A decrease in the preference of sucrose indicates depression-like behavior. Injured mice were found to increase immobility time in the tail suspension tests and to reduce sweet water consumption during the sucrose preference, indicating depression-like behavior (Wu, Zhao, et al., 2014).

These findings are very significant because they demonstrate that cognitive and mood deficits are not induced by lifestyle change, drugs of abuse, and polimedication. They are related to changes in brain structures such as the hippocampus involved in cognition and emotion and not in sensorimotor functions.

The effect of SCI on the hippocampus

Hippocampal neurons

The hippocampus has an impressive capacity for structural reorganization and adaptive plasticity since neuronal circuits undergo constant modifications in dendritic complexity and the number of synapses (Zemla & Basu, 2017).

Since cognitive and emotional impairment described in humans and rodents after SCI is related to hippocampal-dependent functions, one of the aims of our laboratory is to study the impact of SCI on this structure. To achieve this goal we have used a compression model of the spinal lesion in rodents (Jure et al., 2017, 2020). We have recently demonstrated that SCI increased hippocampal vulnerability in mice. The neural density of both the dorsal granular cell layer (GCL) and pyramidal cell layer 1 (CA1) declined in the chronic phase following SCI (Jure et al., 2020). The number of mature neurons remained unchanged during the acute phase (Jure et al., 2020) even though BDNF production was down-regulated during the first-week post-injury (Fumagalli et al., 2009).

Faden's group has also shown that the number of total hippocampal neurons remained unchanged in the acute phase and decreased in the chronic phase (Wu, Stoica, et al., 2014; Wu, Zhao, et al., 2014). These results point out that hippocampal neurodegeneration after SCI is a long process that is verified in the chronic phase (at least 50 days after SCI in rodents). Moreover, they have shown that moderate and severe injury decreased the number of hippocampal neurons to the same extent, but mild injury preserved the neuronal population (Wu et al., 2016). These findings suggest that neuronal death is the result of an all-none response which is established after reaching a certain threshold.

Numerous reports describe that hippocampal dysfunction generates cognitive impairment in hippocampal-dependent learning such as retention and spatial memories (Ali, Badshah, Kim, & Kim, 2015; Belarbi et al., 2012; Li et al., 2016). In this regard, hippocampal failures are associated with low NOR and Y-Maze performances (Broadbent, Gaskin, Squire, & Clark, 2010; Cohen & Stackman Jr., 2015). In line with these findings, depressive-like behaviors are reported using the sucrose preference and the force swim test after hippocampal ischemia in mice (Luo et al., 2019). Therefore, cognitive and mood impairment described in patients and rodents could be explained through the neurodegeneration of hippocampal neurons.

It is important to clarify that although both cognition and emotional behavior involve the hippocampus, these functions also depend on other brain regions such as the amygdala, the hypothalamus as well as the prefrontal and perirhinal cortices (Spellman et al., 2015).

Neurogenesis

Neurogenesis during the acute phase

The subgranular zone (SGZ) of the dentate gyrus (DG) of the hippocampus is one of the brain regions where the generation of new neurons continues throughout life due to the persistence of a neural stem cell (NSC) population (Toda & Gage, 2018). Normally, NSCs are activated and produce highly proliferative transit-amplifying neural progenitors (ANPs). ANPs mostly die by apoptosis during the transition to immature neurons (Sierra et al., 2010). The surviving ANPs differentiate into DCX^+ immature neurons first and then become mature neurons that integrate into the hippocampal circuitry providing highly plastic properties and participating in the information processing (Toda & Gage, 2018) (Fig. 1).

The effect of SCI on neurogenesis during the acute phase is controversial. Our results, in accordance with Felix et al. (2012), have described a reduction in the number of DCX^+ cells during the first week after SCI (Fig. 2) (Felix et al., 2012; Jure et al., 2020). However, recent evidence has shown that neurogenesis increases after complete spinal cord transection 2 days post-injury but reaches control values after 3 months (Dehler et al., 2018). The use of different lesion models could explain the discrepancy in the results since transection lesions are usually characterized by a less rostral spread of secondary injury compared to the compression model used in our experiments (Beattie, Farooqui, & Bresnahan, 2000).

To further explore SCI effects on neurogenesis process, we have studied NSC and ANP activation in the neurogenic niche during the acute phase. Not only NSC and ANP differentiation but also their proliferation were unaltered in the acute

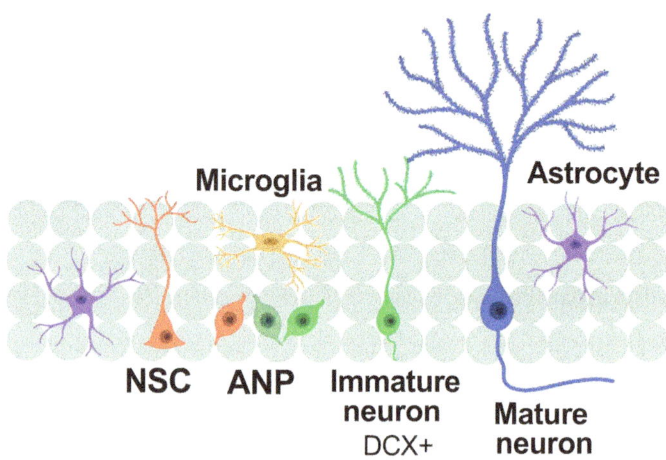

FIG. 1 Schematic summary of neurogenic process in the Subgranular zone. The scheme shows a neural stem cell (NSC) producing highly proliferative transit-amplifying neural progenitors (ANPs) which differentiate into immature DCX+ neurons. These cells become mature neurons that integrate into the hippocampal circuitry. *Image made with BioRender Create Illustration.*

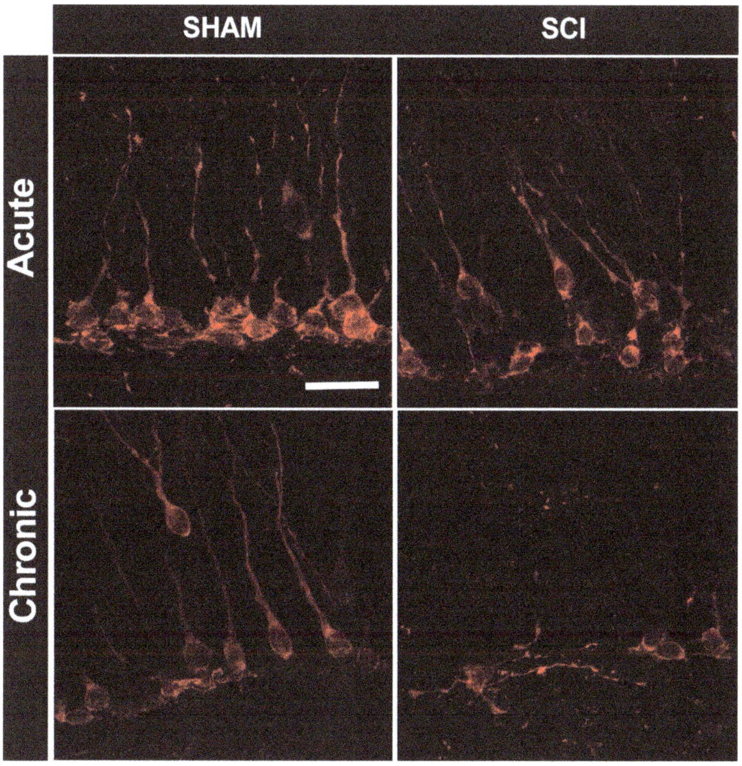

FIG. 2 Effect of spinal cord injury on neurogenesis during the acute and chronic phase. Immature neurons labeled with DCX at 7 and 50 days post-injury in spinal cord injured (SCI) and non-injured (sham) mice. The number of DCX+ cells decreases after injury in both phases. Scale bar: 30 μm. *Unpublished image.*

phase. Curiously, the number of total ANPs decreased while the number of NSCs remained unchanged after injury. Since NSC and ANP proliferation as well as the total number of NSCs were unaffected by SCI, the reduction of the ANP population could be due to ANP death. In accordance with this hypothesis the number of ANPs with abnormal nuclear morphology, an indicative of cellular death, increased in lesioned mice (Jure et al., 2020). These results were concurrent with the reduction of the number of DCX^+ cells aforementioned. Acute stress could explain the reduction of neurogenesis in the acute phase because glucocorticoids are well-known inhibitors of neurogenesis (Lucassen et al., 2015) and they are up-regulated during the first week after SCI (Popovich, Stuckman, Gienapp, & Whitacre, 2001).

Neurogenesis during the chronic phase

Long-term effects of SCI on neurogenesis are less controversial. In agreement with another group, we have described a reduction in the number of immature neurons (DCX^+ cells) in the SGZ after moderate and severe SCI during the chronic phase (Fig. 2) (Jure et al., 2017, 2020; Wu et al., 2016). Noteworthy, we have observed that in mildly injured rats neurogenesis remains unchanged, but in moderately and severely injured rats DCX^+ cells decreased in the same amount. Our data suggest that the down-regulation of neurogenesis could be an all-or-none response from a certain threshold (Jure et al., 2017). However, other authors have found that mild, moderate, and severe injuries decreased DCX^+ cells to the same extent and the reduction coincided with the expression of endoplasmic reticulum stress markers (Wu et al., 2016). In this regard, endoplasmic reticulum stress is related to newborn neurons´ death after traumatic brain injury (Hood et al., 2018).

We have also explored in detail the neurogenesis process long term after injury. In this regard, our results have shown that the number of total NSCs was unchanged, suggesting the survival of NSCs long term after SCI. However, dividing NSCs were down-regulated indicating their inactivation in the neurogenic niche (Jure et al., 2020). NSC quiescence is usually permanent and is related to neurodegenerative diseases, seizures, and aging (Martin-Suarez, Valero, Muro-Garcia, & Encinas, 2019; Sierra et al., 2015).

Regarding ANPs, we found a reduction in the number of both dividing and non-dividing precursor cells (Jure et al., 2020). This decrease could result from the inhibition of their proliferation and/or the reduction of NSC activation. Changes in ANP proliferation are frequently transient and reversible such as those observed in exercise, deep brain stimulation, fluoxetine treatment, and acute stress (Lucassen et al., 2015; Sierra et al., 2014).

Any condition that provokes impairment of hippocampal neurogenesis is likely to have an impact on cognitive performance and emotional behavior. In fact, the inhibition of neurogenesis by irradiation, pharmacological, and genetical tools impairs spatial-related memory and learning, pattern separation, and responses to fear and stress (Bergami et al., 2008; Oomen, Bekinschtein, Kent, Saksida, & Bussey, 2014). Studies in mice which ablate adult-generated neurons have demonstrated that these neurons are crucial to form and endure spatial and contextual fear memories evaluated by MWM and fear conditioning test (Arruda-Carvalho, Sakaguchi, Akers, Josselyn, & Frankland, 2011). On the other hand, several reports have described that neurogenesis up-regulation alleviates anxiety and reduces depressive-like behavior assessed by the elevated plus maze, sucrose preference, and force swim tests (O'Leary & Cryan, 2014).

Based on these findings the reduction of neurogenesis induced by SCI could explain in part the cognitive and emotional deficits described previously in rodents (Wu, Stoica, et al., 2014; Wu, Zhao, et al., 2014).

Hippocampal neuroinflammation

Glial reaction during the acute phase

Microglial cells alongside astrocytes manage the inflammatory response in the CNS. Microglia and astrocytes are plastic cells that respond to several kinds of molecules generating a different spectrum of functional phenotypes (Tang & Le, 2016). Homeostasis recovery after injury depends on the fine-tuning polarization of microglia and astrocytes from pro-inflammatory towards anti-inflammatory phenotypes. Microglia are maintained in a surveillant state of activation through several inhibitory signaling between microglia and other cells (Frank, Fonken, Annis, Watkins, & Maier, 2018).

There is compelling evidence demonstrating that SCI induces microglial changes in rodents (Jure et al., 2017, 2020; Wu, Zhao, et al., 2014). In this regard, our group has described that the number of microglial cells increased during the acute phase in the hilus, molecular layer, and GCL + SGZ of the DG in mice (Fig. 3). However, mRNA levels of pro-inflammatory cytokines remained unaltered in the acute phase (Jure et al., 2020). In agreement with these results, Faden's group has shown an increase in the number of microglial cells concurrent with the up-regulation of mRNA expression of cycle genes (cyclin A1, A2, D1) (Wu, Stoica, et al., 2014). Moreover, they described that microglial cells change their

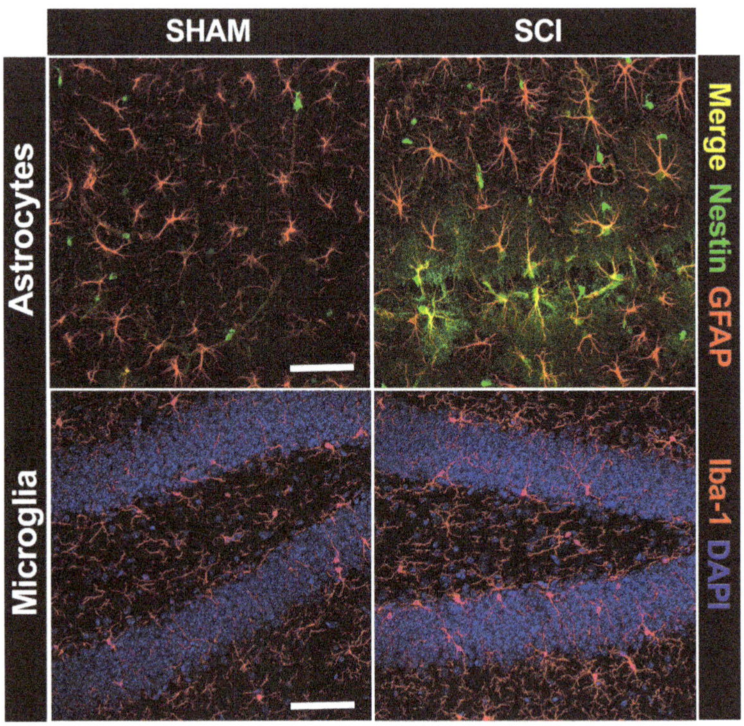

FIG. 3 Effect of spinal cord injury on glial cells during the acute phase. Immunohistochemistry for GFAP, Nestin, and Iba-1 in the dentate gyrus of spinal cord injured (SCI) and non-injured (sham) mice. Astrocytes are identified as GFAP+ cells while reactive astrocytes as GFAP+Nestin+ cells. Microglial cells are identified as iba-1+ cells. DAPI is a nuclear marker. The number of reactive astrocytes and microglial cells increases after SCI. Scale bar: 50 μm. *Unpublished image.*

morphology from ramified to hypertrophic and bushy, an indicative of microglia activation. However, coinciding with our results, mRNA levels of pro-inflammatory cytokines were experimented with no modifications 7 days post-injury (Felix et al., 2012; Wu, Stoica, et al., 2014).

Summarizing, during the acute phase microglial cells proliferate and acquire the morphology of an active cell although there is no production of pro-inflammatory cytokines. Thus, microglia remain in a primed state which has been characterized by a number of conditions such as infection, stress, and neurodegeneration (Frank et al., 2018). Furthermore, a hallmark of primed microglia is the fact that morphological changes are not followed by a pro-inflammatory profile.

Microglial priming occurs after acute stress and it is associated with the downregulation of CD200R (Frank et al., 2018). The increase of glucocorticoids after acute SCI could explain the prime of microglial cells. CD200:CD200R is a key signaling dyad that constrains microglial activation. CD200R is expressed only on myeloid cells while CD200 is widely distributed in neurons, oligodendrocytes, astrocytes, and endothelial cells. In this regard, we have demonstrated that the mRNA expression of CD200R was down-regulated after SCI, suggesting that the disruption of CD200:CD200R signaling could prime microglia to subsequent immune challenges (Jure et al., 2020). The disruption of CD200:CD200R signaling has also been involved in neuroinflammatory processes observed in aging, neuropathic pain, and Alzheimer's disease (Lyons et al., 2009). Regarding astrocytes, we have found that the number of activated astrocytes increased in the hilus, molecular layer, and GCL + SGZ during the acute phase (Fig. 3) (Jure et al., 2020).

Glial reaction during the chronic phase

The scenario definitely changes during the chronic phase. In agreement with Faden's results, we have shown that mRNA levels of pro-inflammatory cytokines such as IL-1β, TNFα, IL-18, and IL-6 were up-regulated. Furthermore, microglial cells presented an activated morphology displaying hypertrophic and bushy forms instead of the surveillant ramified phenotype. However, the number of total microglial cells decreased in the DG during the chronic phase (Jure et al., 2020; Wu, Stoica, et al., 2014). This can be explained because microglial cells die secondary to their activation in a process called activation induced cell death (Arroyo et al., 2013). Microglial activation might be due to neuron release of chemokines after SCI. In this regard, some studies have shown that CA1, CA3, and hilar neurons increased the expression of CCL2, CCR2, CCL3, and CCL21 in the chronic phase (Li et al., 2020). The effects of CCL2 and CCL21 on microglial activation and the key role of these chemokines on neuron-glia communication are well-known (Zhang, Jiang, & Gao, 2017). On the other hand, astrocytes also play a critical role in regulating neuroinflammation, and the number of activated cells increases in the hilus, molecular layer, and SGZ + GCL after chronic SCI (Jure et al., 2020).

Neurogenic niches are modified by surrounding microenvironment and pathological conditions. Indeed, neuroinflammation has been extensively reported to impair hippocampal neurogenesis (Sierra et al., 2014). In fact, neurogenesis decreases with aging, chronic stress, and neurodegenerative diseases, all conditions where neuroinflammation is a common landmark. Indeed, pro-inflammatory cytokines such as IL-1β, TNFα reduce the proliferation and survival of NSCs and ANPs (Kuzumaki et al., 2010). Astrocytes also can contribute to neurogenesis regulation as they secrete pro-neurogenic factors in normal conditions. However, reactive astrocytes which chronically release IL-6 reduce proliferation, survival, and differentiation of newborn cells (Sierra et al., 2014). The recruitment of new neurons to hippocampal circuits associated with encoding spatial information is also decreased during chronic inflammation leading to cognitive impairment in rodents (Sierra et al., 2014).

Chronic neuroinflammation also generates hippocampal dysfunction altering synaptic plasticity, decreasing cognitive performance such as retention and spatial memories, and inducing depressive-like behavior (Belarbi et al., 2012). In this regard, treatment with anti-inflammatory drugs such as minocycline, cannabinoids, or TNFα inhibitors improves recognition memory evaluated by NOR and spatial memory assessed by the Y-maze (Ali et al., 2015; Belarbi et al., 2012; Li et al., 2016). In addition, anti-inflammatory drugs improve fear memory, reduce anhedonia measured by the sucrose preference test and decrease immobility time in the force swimming test (Troubat et al., 2021).

Neuroinflammation in the chronic phase could also be responsible for the death of hippocampal neurons since it is common to various neurological disorders such as epilepsy, degenerative diseases, and multiple sclerosis where several neurons are usually lost. In fact, pro-inflammatory microglia releases TNFα, NO and ROS, which produce neuronal death (Kaindl et al., 2012; Lull & Block, 2010).

Based on these results, chronical neuroinflammation and the activation of hippocampal glial cells could cause neurogenesis reduction and neurodegeneration, which lead to cognitive and emotional impairment in hippocampal-dependent behaviors after SCI.

SCI and other brain regions

A report from Faden and colleagues further supports that neuroinflammation and neurodegeneration could be a widespread phenomenon long term after SCI. Indeed, they have described the reduction in the number of cortical and thalamic neurons as well as changes in microglial morphology which display activated forms in those regions (Wu, Stoica, et al., 2014; Wu, Zhao, et al., 2014). The neurodegenerative process is supported by the progressive enlargement of the ventricles and the increase of cerebrospinal fluid volume (Li et al., 2020). Regarding neuroinflammation, several reports have described microglial activation in thalamic neurons and periaqueductal gray matter (Li et al., 2020). These results could involve the development of neuropathic pain since the thalamus is involved in sensory and pain processing. In line with these findings, human MRI studies have demonstrated that SCI generates a progressive reduction in gray matter volume in areas related to the processing of emotional information and the modulation of attentional states such as the anterior cingulate, insular, medial prefrontal, and temporal cortices (Nicotra, Critchley, Mathias, & Dolan, 2006; Wrigley et al., 2009). Moreover, neuronal activity is reduced in the anterior cingulate cortex during emotional processing according to a functional MRI study (Nicotra et al., 2006).

On the other hand, the gene expression of the whole brain after acute and subacute SCI was recently elucidated using transcriptome analysis. Genes belonging to the oxidative phosphorylation pathway were increased during the acute phase (3 h after injury). However, the enriched brain pathways associated with the subacute phase (14 days post-injury) were both the inflammatory response (cytokines and chemokine signaling) and endoplasmic reticulum stress-related pathways (Baek, Cho, & Kim, 2017). These findings relate the pathophysiology of SCI with widespread neuroinflammation and cellular stress in the brain.

Possible mechanisms underlying the tertiary damage

The mechanisms underlying how SCI drives acute and chronic brain changes remain unknown. However, some speculative mechanisms can be delineated. On the one hand, the retrograde and anterograde Wallerian degeneration spreads along white matter tracts from the spinal cord to rostral areas of the brain, producing axonal loss, demyelination, neuronal shrinkage, and even neuronal death. Chemokines such as CCL21, CCL2, CCL3, and the chemokine receptor CCR2 are not expressed in the healthy brain. However, after axonal damage or glutamate exposure, the synthesis of CCL21 and CCL2 is stimulated. Damaged neurons transport and release those chemokines at distant regions resulting in remote microglial activation. In this regard, dorsal horn neurons after injury release CCL21 to the thalamus through the spinothalamic tract (Zhao et al., 2007). The activated thalamic microglia induce the expression of CCL21 in thalamic neurons, which project to other areas of the brain such as the hippocampus, propagating microglial activation. Furthermore, the microglial activator CCL2 and its receptor CCR2 are chronically expressed in neurons after severe spinal contusion in the thalamus, hippocampus, and periaqueductal gray matter (Knerlich-Lukoschus et al., 2011). This evidence suggests that chemokines can be transported through axons to distant regions from the epicenter of the lesion. On the other hand, the systemic immune and neuroendocrine functions are markedly altered after SCI. In this regard, a systemic inflammatory response is developed during the acute phase. Pro-inflammatory cytokines such as TNFα, IL-1β, and IL-6 increased in serum while leukocyte infiltration into tissues is promoted. In this context, the systemic inflammation produced immediately after SCI could stimulate microglial activation and/or induce microglial priming in the brain (Campbell et al., 2005). In addition to the inflammatory response, SCI activates the HPA axis releasing glucocorticoids during the acute phase. The high levels of corticosterone described during the first week after SCI could prime microglia and model the brain producing long-term alterations since the hippocampus and cerebral cortex are vulnerable to glucocorticoids (Lucassen et al., 2015; Popovich et al., 2001). Therefore, systemic inflammatory reaction and glucocorticoid action could prime microglia to subsequent immune challenges such as neural and axonal degeneration that takes place long term after SCI. Finally, factors released by the spinal tissue could reach and affect the brain through the cerebral spinal fluid.

Applications to other areas of neuroscience

In this chapter, we have reviewed the effects of SCI on the brain, especially on the hippocampus. These findings support the hypothesis that tertiary damage really occurs after SCI. Indeed, the hippocampus and the thalamus are deeply and chronically altered as a consequence of the damage to the spinal cord. Alterations in these structures might lead to the generation of neuropathic pain as well as cognitive and mood disorders seen in rodents and even humans. It is time to consider SCI as a brain neurodegenerative disease and not only as an event circumscribed to the spinal cord. This approach is opening a new area in neurobiology since therapeutic strategies are focused on preventing secondary injury and ignore the encephalopathy that develops at rostral sites. Future rehabilitation strategies should emphasize not only sensorimotor skills but also cognitive function and mood disorders. Understanding the mechanisms underlying hippocampal neurodegeneration further will help to develop new therapeutic strategies to treat spinal cord injured patients. For example, future therapies could include the inhibition of brain inflammatory response induced after SCI. On the other hand, cognitive and emotional impairment should be considered as a result of SCI and not as a consequence of lifestyle changes that occur after the disability. Brain neurodegeneration suggests that depressive and cognitive dysfunction might derive from a biological substrate. The study of brain pathology following SCI is a new exciting field of research whose development will provide valuable information for understanding the intimate interconnection among the spinal cord, the brain, and human behavior.

Mini-dictionary of terms

Primary damage: Damage to the spinal cord induced by mechanic impact.

Secondary damage: Cascade of cellular and molecular events that propagate the injury from the site of the mechanic impact to the rostrocaudal contiguous spinal segments.

Tertiary damage: Progressive neurodegeneration which affects distal parts of the brain after spinal cord injury.

Hippocampal neurogenesis: Production of newborn neurons from neural stem cells located in the subgranular zone of the dentate gyrus of the hippocampus.

Primed microglia: Activated microglia which does not release pro-inflammatory cytokines but produces an exaggerated response to future challenges.

Key facts of tertiary damage

- Tertiary Damage is a new concept in the field of Spinal cord injury (SCI).
- Tertiary Damage implies progressive neurodegeneration of distal parts of the brain after SCI.
- Axonal injury and microglial response are probably related to the development of Tertiary Damage.
- Tertiary Damage is associated with cognitive and emotional impairment described in animals and even humans after SCI.
- Tertiary Damage introduces new perspectives in rehabilitation and therapeutic strategies for spinal cord injured patients.

Summary points

- In this chapter, we propose the term tertiary damage to describe alterations suffered by distal parts of the brain after spinal cord injury (SCI).
- Traditionally, SCI investigation was focused on pathophysiological changes in the spinal cord with efforts being made to recover sensorimotor function and relieve neuropathic pain.
- Several studies have shown atrophy of the sensorimotor system (cortices and tracts).
- Although humans suffer from cognitive and emotional impairment, the study of brain areas related to these functions has been ignored for many years.
- Animal models have shown that SCI leads to cognitive deficits and depressive-like behavior which correlate with hippocampal neurodegeneration, neurogenesis reduction, and neuroinflammation.
- Neuroinflammation response and neurodegeneration after SCI reach other brain areas related to pain processing.
- The mechanisms underlying tertiary damage are unknown, but the release of chemokines at distance sites induced by axonal injury and the systemic response of the immune system and the stress axis could be involved.
- Finally, we discuss that rehabilitation and therapeutic strategies should include tertiary damage.

References

Ahuja, C. S., Wilson, J. R., Nori, S., Kotter, M. R. N., Druschel, C., Curt, A., et al. (2017). Traumatic spinal cord injury. *Nature Reviews. Disease Primers*, *3*, 17018.

Ali, T., Badshah, H., Kim, T. H., & Kim, M. O. (2015). Melatonin attenuates D-galactose-induced memory impairment, neuroinflammation and neurodegeneration via RAGE/NF-K B/JNK signaling pathway in aging mouse model. *Journal of Pineal Research*, *58*(1), 71–85.

Arroyo, D. S., Soria, J. A., Gaviglio, E. A., Garcia-Keller, C., Cancela, L. M., Rodriguez-Galan, M. C., et al. (2013). Toll-like receptor 2 ligands promote microglial cell death by inducing autophagy. *The FASEB Journal*, *27*(1), 299–312.

Arruda-Carvalho, M., Sakaguchi, M., Akers, K. G., Josselyn, S. A., & Frankland, P. W. (2011). Posttraining ablation of adult-generated neurons degrades previously acquired memories. *The Journal of Neuroscience*, *31*(42), 15113–15127.

Baek, A., Cho, S. R., & Kim, S. H. (2017). Elucidation of gene expression patterns in the brain after spinal cord injury. *Cell Transplantation*, *26*(7), 1286–1300.

Beattie, M. S., Farooqui, A. A., & Bresnahan, J. C. (2000). Review of current evidence for apoptosis after spinal cord injury. *Journal of Neurotrauma*, *17*(10), 915–925.

Belarbi, K., Jopson, T., Tweedie, D., Arellano, C., Luo, W., Greig, N. H., et al. (2012). TNF-alpha protein synthesis inhibitor restores neuronal function and reverses cognitive deficits induced by chronic neuroinflammation. *Journal of Neuroinflammation*, *9*, 23.

Bergami, M., Rimondini, R., Santi, S., Blum, R., Gotz, M., & Canossa, M. (2008). Deletion of TrkB in adult progenitors alters newborn neuron integration into hippocampal circuits and increases anxiety-like behavior. *Proceedings of the National Academy of Sciences of the United States of America*, *105*(40), 15570–15575.

Broadbent, N. J., Gaskin, S., Squire, L. R., & Clark, R. E. (2010). Object recognition memory and the rodent hippocampus. *Learning & Memory*, *17*(1), 5–11.

Brock, J. H., Rosenzweig, E. S., Blesch, A., Moseanko, R., Havton, L. A., Edgerton, V. R., et al. (2010). Local and remote growth factor effects after primate spinal cord injury. *The Journal of Neuroscience*, *30*(29), 9728–9737.

Campbell, S. J., Perry, V. H., Pitossi, F. J., Butchart, A. G., Chertoff, M., Waters, S., et al. (2005). Central nervous system injury triggers hepatic CC and CXC chemokine expression that is associated with leukocyte mobilization and recruitment to both the central nervous system and the liver. *The American Journal of Pathology*, *166*(5), 1487–1497.

Cohen, S. J., & Stackman, R. W., Jr. (2015). Assessing rodent hippocampal involvement in the novel object recognition task. A review. *Behavioural Brain Research*, *285*, 105–117.

Craig, A., Nicholson Perry, K., Guest, R., Tran, Y., Dezarnaulds, A., Hales, A., et al. (2015). Prospective study of the occurrence of psychological disorders and comorbidities after spinal cord injury. *Archives of Physical Medicine and Rehabilitation*, *96*(8), 1426–1434.

Dehler, S., Lou, W. P., Gao, L., Skabkin, M., Dallenbach, S., Neumann, A., et al. (2018). An immune-CNS axis activates remote hippocampal stem cells following spinal transection injury. *Frontiers in Molecular Neuroscience*, *11*, 443.

Felix, M. S., Popa, N., Djelloul, M., Boucraut, J., Gauthier, P., Bauer, S., et al. (2012). Alteration of forebrain neurogenesis after cervical spinal cord injury in the adult rat. *Frontiers in Neuroscience*, *6*, 45.

Frank, M. G., Fonken, L. K., Annis, J. L., Watkins, L. R., & Maier, S. F. (2018). Stress disinhibits microglia via down-regulation of CD200R: A mechanism of neuroinflammatory priming. *Brain, Behavior, and Immunity*, *69*, 62–73.

Freund, P., Weiskopf, N., Ashburner, J., Wolf, K., Sutter, R., Altmann, D. R., et al. (2013). MRI investigation of the sensorimotor cortex and the corticospinal tract after acute spinal cord injury: A prospective longitudinal study. *Lancet Neurology*, *12*(9), 873–881.

Fumagalli, F., Madaschi, L., Caffino, L., Marfia, G., Di Giulio, A. M., Racagni, G., et al. (2009). Acute spinal cord injury reduces brain derived neurotrohic factor expression in rat hippocampus. *Neuroscience*, *159*(3), 936–939.

Hood, K. N., Zhao, J., Redell, J. B., Hylin, M. J., Harris, B., Perez, A., et al. (2018). Endoplasmic reticulum stress contributes to the loss of newborn hippocampal neurons after traumatic brain injury. *The Journal of Neuroscience*, *38*(9), 2372–2384.

Jure, I., De Nicola, A. F., Encinas, J. M., & Labombarda, F. (2020). Spinal cord injury leads to hippocampal glial alterations and neural stem cell inactivation. *Cellular and Molecular Neurobiology*. https://doi.org/10.1007/s10571-020-00900-8. Online ahead of print.

Jure, I., Pietranera, L., De Nicola, A. F., & Labombarda, F. (2017). Spinal cord injury impairs neurogenesis and induces glial reactivity in the hippocampus. *Neurochemical Research*, *42*(8), 2178–2190.

Kaindl, A. M., Degos, V., Peineau, S., Gouadon, E., Chhor, V., Loron, G., et al. (2012). Activation of microglial N-methyl-D-aspartate receptors triggers inflammation and neuronal cell death in the developing and mature brain. *Annals of Neurology*, *72*(4), 536–549.

Knerlich-Lukoschus, F., Noack, M., von der Ropp-Brenner, B., Lucius, R., Mehdorn, H. M., & Held-Feindt, J. (2011). Spinal cord injuries induce changes in CB1 cannabinoid receptor and C-C chemokine expression in brain areas underlying circuitry of chronic pain conditions. *Journal of Neurotrauma*, *28*(4), 619–634.

Kuzumaki, N., Ikegami, D., Imai, S., Narita, M., Tamura, R., Yajima, M., et al. (2010). Enhanced IL-1beta production in response to the activation of hippocampal glial cells impairs neurogenesis in aged mice. *Synapse*, *64*(9), 721–728.

Lazzaro, I., Tran, Y., Wijesuriya, N., & Craig, A. (2013). Central correlates of impaired information processing in people with spinal cord injury. *Journal of Clinical Neurophysiology*, *30*(1), 59–65.

Lee, B. H., Lee, K. H., Kim, U. J., Yoon, D. H., Sohn, J. H., Choi, S. S., et al. (2004). Injury in the spinal cord may produce cell death in the brain. *Brain Research*, *1020*(1–2), 37–44.

Li, Y., Cao, T., Ritzel, R. M., He, J., Faden, A. I., & Wu, J. (2020). Dementia, depression, and associated brain inflammatory mechanisms after spinal cord injury. *Cell*, *9*(6).

Li, H. C., Xiao, J., Huang, Y. L., Li, L. J., Jiang, H., Huang, L. X., et al. (2016). Reduction in hypoxia-derived neuroinflammation and dysfunctional glutamate transporters by minocycline may restore hypoxia-injured cognition of neonatal rat. *Sheng Li Xue Bao*, *68*(2), 148–156.

Lucassen, P. J., Oomen, C. A., Naninck, E. F., Fitzsimons, C. P., van Dam, A. M., Czeh, B., et al. (2015). Regulation of adult neurogenesis and plasticity by (early) stress, glucocorticoids, and inflammation. *Cold Spring Harbor Perspectives in Biology*, *7*(9), a021303.

Lull, M. E., & Block, M. L. (2010). Microglial activation and chronic neurodegeneration. *Neurotherapeutics*, *7*(4), 354–365.

Luo, L., Li, C., Du, X., Shi, Q., Huang, Q., Xu, X., et al. (2019). Effect of aerobic exercise on BDNF/proBDNF expression in the ischemic hippocampus and depression recovery of rats after stroke. *Behavioural Brain Research*, *362*, 323–331.

Lyons, A., McQuillan, K., Deighan, B. F., O'Reilly, J. A., Downer, E. J., Murphy, A. C., et al. (2009). Decreased neuronal CD200 expression in IL-4-deficient mice results in increased neuroinflammation in response to lipopolysaccharide. *Brain, Behavior, and Immunity*, *23*(7), 1020–1027.

Martin-Suarez, S., Valero, J., Muro-Garcia, T., & Encinas, J. M. (2019). Phenotypical and functional heterogeneity of neural stem cells in the aged hippocampus. *Aging Cell*, *18*(4), e12958.

Murray, R. F., Asghari, A., Egorov, D. D., Rutkowski, S. B., Siddall, P. J., Soden, R. J., et al. (2007). Impact of spinal cord injury on self-perceived pre- and postmorbid cognitive, emotional and physical functioning. *Spinal Cord*, *45*(6), 429–436.

Nardone, R., Holler, Y., Brigo, F., Seidl, M., Christova, M., Bergmann, J., et al. (2013). Functional brain reorganization after spinal cord injury: Systematic review of animal and human studies. *Brain Research*, *1504*, 58–73.

Nicotra, A., Critchley, H. D., Mathias, C. J., & Dolan, R. J. (2006). Emotional and autonomic consequences of spinal cord injury explored using functional brain imaging. *Brain*, *129*(Pt. 3), 718–728.

O'Leary, O. F., & Cryan, J. F. (2014). A ventral view on antidepressant action: Roles for adult hippocampal neurogenesis along the dorsoventral axis. *Trends in Pharmacological Sciences*, *35*(12), 675–687.

Oomen, C. A., Bekinschtein, P., Kent, B. A., Saksida, L. M., & Bussey, T. J. (2014). Adult hippocampal neurogenesis and its role in cognition. *Wiley Interdisciplinary Reviews: Cognitive Science*, *5*(5), 573–587.

Popovich, P. G., Stuckman, S., Gienapp, I. E., & Whitacre, C. C. (2001). Alterations in immune cell phenotype and function after experimental spinal cord injury. *Journal of Neurotrauma*, *18*(9), 957–966.

Roth, E., Davidoff, G., Thomas, P., Doljanac, R., Dijkers, M., Berent, S., et al. (1989). A controlled study of neuropsychological deficits in acute spinal cord injury patients. *Paraplegia*, *27*(6), 480–489.

Sierra, A., Beccari, S., Diaz-Aparicio, I., Encinas, J. M., Comeau, S., & Tremblay, M. E. (2014). Surveillance, phagocytosis, and inflammation: How never-resting microglia influence adult hippocampal neurogenesis. *Neural Plasticity*, *2014*, 610343.

Sierra, A., Encinas, J. M., Deudero, J. J., Chancey, J. H., Enikolopov, G., Overstreet-Wadiche, L. S., et al. (2010). Microglia shape adult hippocampal neurogenesis through apoptosis-coupled phagocytosis. *Cell Stem Cell*, *7*(4), 483–495.

Sierra, A., Martin-Suarez, S., Valcarcel-Martin, R., Pascual-Brazo, J., Aelvoet, S. A., Abiega, O., et al. (2015). Neuronal hyperactivity accelerates depletion of neural stem cells and impairs hippocampal neurogenesis. *Cell Stem Cell*, *16*(5), 488–503.

Spellman, T., Rigotti, M., Ahmari, S. E., Fusi, S., Gogos, J. A., & Gordon, J. A. (2015). Hippocampal-prefrontal input supports spatial encoding in working memory. *Nature*, *522*(7556), 309–314.

Tang, Y., & Le, W. (2016). Differential roles of M1 and M2 microglia in neurodegenerative diseases. *Molecular Neurobiology*, *53*(2), 1181–1194.

Toda, T., & Gage, F. H. (2018). Review: Adult neurogenesis contributes to hippocampal plasticity. *Cell and Tissue Research*, *373*(3), 693–709.

Troubat, R., Barone, P., Leman, S., Desmidt, T., Cressant, A., Atanasova, B., et al. (2021). Neuroinflammation and depression: A review. *The European Journal of Neuroscience*, *53*(1), 151–171.

Wrigley, P. J., Gustin, S. M., Macey, P. M., Nash, P. G., Gandevia, S. C., Macefield, V. G., et al. (2009). Anatomical changes in human motor cortex and motor pathways following complete thoracic spinal cord injury. *Cerebral Cortex*, *19*(1), 224–232.

Wu, J., Stoica, B. A., Luo, T., Sabirzhanov, B., Zhao, Z., Guanciale, K., et al. (2014). Isolated spinal cord contusion in rats induces chronic brain neuroinflammation, neurodegeneration, and cognitive impairment. Involvement of cell cycle activation. *Cell Cycle*, *13*(15), 2446–2458.

Wu, J., Zhao, Z., Kumar, A., Lipinski, M. M., Loane, D. J., Stoica, B. A., et al. (2016). Endoplasmic reticulum stress and disrupted neurogenesis in the brain are associated with cognitive impairment and depressive-like behavior after spinal cord injury. *Journal of Neurotrauma*, *33*(21), 1919–1935.

Wu, J., Zhao, Z., Sabirzhanov, B., Stoica, B. A., Kumar, A., Luo, T., et al. (2014). Spinal cord injury causes brain inflammation associated with cognitive and affective changes: Role of cell cycle pathways. *The Journal of Neuroscience*, *34*(33), 10989–11006.

Zemla, R., & Basu, J. (2017). Hippocampal function in rodents. *Current Opinion in Neurobiology*, *43*, 187–197.

Zhang, Z. J., Jiang, B. C., & Gao, Y. J. (2017). Chemokines in neuron-glial cell interaction and pathogenesis of neuropathic pain. *Cellular and Molecular Life Sciences*, *74*(18), 3275–3291.

Zhao, P., Waxman, S. G., & Hains, B. C. (2007). Modulation of thalamic nociceptive processing after spinal cord injury through remote activation of thalamic microglia by cysteine cysteine chemokine ligand 21. *The Journal of Neuroscience*, *27*(33), 8893–8902.

Section C

Physiological and metabolic effects

Chapter 24

Hormonal events and spinal cord injury: A focus on vasopressin and natriuretic peptide

Jason H. Gumbel[a] and Charles H. Hubscher[a,b]

[a]*Department of Anatomical Sciences and Neurobiology, University of Louisville, Louisville, KY, United States,* [b]*Kentucky Spinal Cord Injury Research Center, University of Louisville, Louisville, KY, United States*

Abbreviations

ABRT	activity-based recovery training
AD	autonomic dysreflexia
ANP	atrial natriuretic peptide
AQP2	aquaporin 2 channels
AVP	vasopressin
BNP	brain-derived natriuretic peptide
cAMP	cyclic adenosine monophosphate
CNP	C-type natriuretic peptide
CORT	corticosterone
NP	natriuretic peptide
NPRA	natriuretic peptide receptor-A
PVN	paraventricular nucleus
RAAS	renin aldosterone angiotensin system
SCI	spinal cord injury
SCN	suprachiasmatic nucleus
SON	supraoptic nucleus
V1R	vasopressin receptor 1
V2R	vasopressin receptor 2
V3R	vasopressin receptor 3

Introduction

Dysfunctions of the upper and lower urinary tracts that arise after spinal cord injury (SCI) include detrusor-sphincter dyssynergia, incontinence, polyuria, urinary retention, and loss of sensation. Bladder emptying methods include intermittent catheterization (self or by an attendant), indwelling catheters (transurethral or suprapubic), reflex triggering, straining, manual compression (Crede), and sacral anterior root stimulation. A poorly understood complication that develops after SCI which impacts the daily frequency of conducting these management techniques is polyuria, the overproduction and/or passage of urine, which has been documented in both human SCI and pre-clinical contusion models (Goh et al., 2017; Hubscher et al., 2016; Szollar, Dunn, Brandt, & Fincher, 1997; Ward & Hubscher, 2012). This excessive volume of urine requires more frequent catheterizations, increases the risk of bladder and urinary tract infections, and prompts the need for nightly awakenings for bladder emptying (nocturia) which is disruptive to sleep and receiving enough rest. Oftentimes, SCI individuals will limit their fluid intake to avoid nocturia, which introduces further and potentially more serious confounding issues such as dehydration, which exasperates bowel management through constipation, and autonomic dysreflexia, a life-threatening sudden increase in arterial blood pressure that is elicited by noxious (e.g., pain) and/or innocuous stimuli (e.g., bladder filling) from below the level of injury (Dolinak & Balraj, 2007; Wan & Krassioukov, 2014).

Cellular, Molecular, Physiological, and Behavioral Aspects of Spinal Cord Injury. https://doi.org/10.1016/B978-0-12-822427-4.00024-1

The underlying mechanisms of polyuria/nocturia are unknown, but likely includes arginine vasopressin (AVP; also commonly referred to as anti-diuretic hormone), which regulates urine production and fluid homeostasis and has recently been shown to correlate with the incidence of SCI-induced polyuria (Denys et al., 2017; Montgomery & Hubscher, 2017). However, it remains undetermined what directly causes the decrease in AVP after SCI, as there are likely multiple factors contributing to polyuria (Fig. 1). These factors include the natriuretic peptides (NP), such as atrial natriuretic peptide (ANP), brain-derived natriuretic peptide (BNP), and C-type natriuretic peptide (CNP).

The focus of the current chapter review is upon SCI-induced changes in AVP and ANP as they play a major role in water/electrolyte balance. In addition, these neurohormones also impact the cardiovascular system. Substantial cardiovascular deficits present amongst the SCI population include orthostatic hypotension, bradycardia, daily fluctuations in blood pressure, and autonomic dysreflexia (Claydon, Steeves, & Krassioukov, 2006; Goh et al., 2017; Lindan, Joiner, Freehafer, & Hazel, 1980). Given these multiple functional roles of neurohormones, any imbalance is likely to impact the overall well-being of the SCI population. Additional hormones affected by SCI are listed in Table 1.

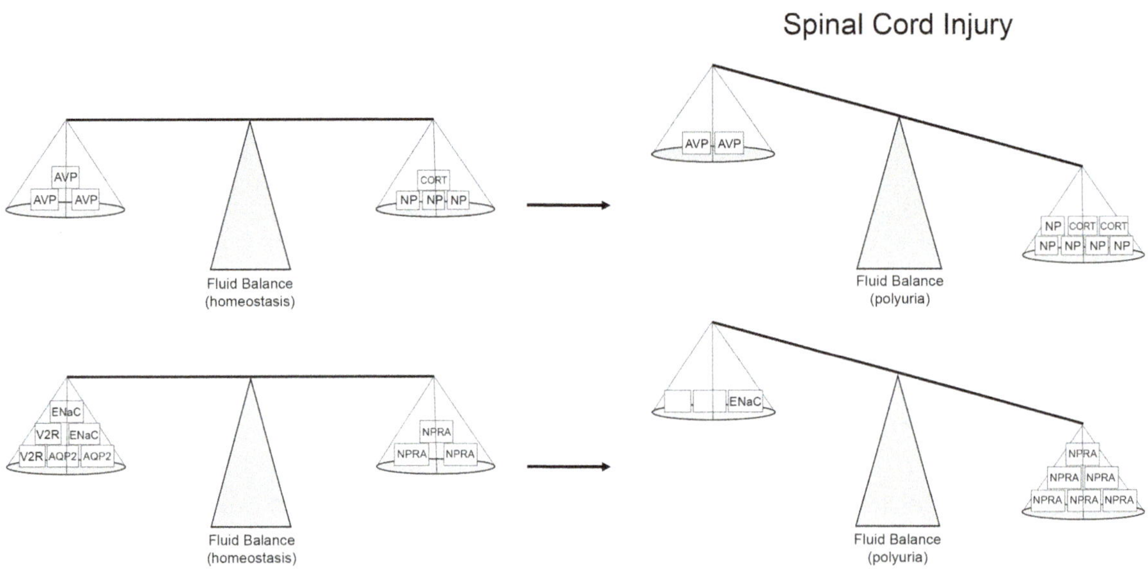

FIG. 1 Key hormone/receptors in fluid balance and how they are affected after SCI. Depiction of key hormones such as AVP, NPs, and CORT, along with paired receptors such as NPRA, V2R, AQP2, and ENaC, and how they contribute to fluid balance/homeostasis. In the non-injured state, these hormones and receptors work together to closely regulate fluid balance. After the injury, dysregulation occurs leading to an overproduction of water (polyuria/nocturia). Changes in the opposite direction would cause a decrease in urine production (oliguria).

TABLE 1 Hormone changes after SCI.

Major hormones affected by chronic spinal cord injury	
Adrenocorticotropic hormone (ACTH)	Increased
Atrial natriuretic peptide (ANP)	Increased
Cortisol	Increased
Human growth hormone	Decreased
Insulin-like growth factor (IGF-1)	Increased
Parathyroid hormone (PTH)	Decreased
Testosterone	Decreased
Vasopressin (AVP)	Decreased

This table outlines several important hormones that have been reported to be significantly altered after chronic spinal cord injury.
Note: Failure of SCI to alter the female menstrual cycle suggests no significant alteration in levels of 17 beta-estradiol and progesterone. Gonadal hormones have been examined therapeutically for their protective effects.

Function of vasopressin (AVP)

One of the bodies most important hormones required for the regulation of salt and water balance is AVP. The production of AVP takes place within neurons located in the supraoptic nucleus (SON), suprachiasmatic nucleus (SCN), and paraventricular nucleus (PVN) of the hypothalamus, regions of the brain limbic system having a central neuroendocrine function (Swanson & Eo Sawchenko, 1983; Van Leeuwen, Swaab, & De Raay, 1978). AVP is released from the posterior lobe of the pituitary in response to several different stimuli. First, under normal physiological circumstances, osmoreceptors located in the hypothalamus respond to an increase in blood osmolality (occurs with dehydration or a high sodium diet) (Cowley Jr, Skelton, Merrill, Quillen Jr, & Switzer, 1983; Kjeldsen et al., 1985) through physical shrinkage which triggers a signaling cascade to release AVP from the pituitary. Within the kidney, vasopressin 2 receptors (V2R) respond to elevated AVP circulating in the blood by increasing reabsorption of solute-free water back into the circulatory system through the opening of aquaporin 2 (AQP2) channels, which results in a decrease in urine volume. As many as 13 different types of aquaporins have been identified in mammals, of which at least seven are located within the kidney and function to transport water across membranes (Nielsen et al., 2002). However, AQP2 is the only kidney aquaporin known to be regulated by AVP.

There are three main AVP receptors; V1, V2, and V3. Each of these receptors has specific functions within different regions of the body. The V1 receptor (V1R) is mainly located in vascular smooth muscle and platelets, but can be found in brain, testis, superior cervical ganglion, liver, blood vessels, and kidney (Phillips et al., 1990). Functionally, the V1R is a G protein-coupled receptor and utilizes the activation of calcium influx, phospholipase A2, phospholipase C, and phospholipase D. The V1R functions mostly as a vasoconstrictor and in thrombosis, but has been shown to be associated with myocardial hypertrophy, glycogenolysis, and uterine contraction (Holmes, Landry, & Granton, 2003). V1Rs can also be found in the kidney where it also functions to promote vasoconstriction within the efferent arterioles, which reduces glomerular filtration rate.

The V2 receptor (V2R) is primarily located within the kidney and promotes a strong antidiuretic effect. As illustrated in Fig. 2, upon the binding of AVP to kidney V2 receptors, cAMP is activated which then triggers the fusion of AQP2 channels to the apical membrane of the collecting ducts (Harris, Zeidel, Jo, & Hammond, 1994). The increase in AQP2 channels promotes water reabsorption. For this reason, V2R is a target of pharmacological therapies to decrease urine production in cases like diabetes insipidus (Bichet et al., 1988) and SCI (Chancellor, Rivas, & Staas Jr, 1994; Zahariou, Karagiannis, Papaioannou, Stathi, & Michail, 2007). V2R is also located in vascular endothelium and smooth muscle cells and has vasodilation properties.

The V3 receptor (V3R) is also a G protein-coupled receptor but is far less distributed than either V1R or V2R. Less is known about V3R, but it is overexpressed in adrenocorticotropic hormone (ACTH)-secreting tumors. Additionally, its function is dependent on the concentration of AVP. One function of V3R is to promote ACTH release from the pituitary, but at other concentrations of AVP, V3R can increase DNA and cAMP synthesis, which is seen in tumor growth.

Function of natriuretic peptides (NP)

NPs include three structurally and functionally related hormone factors: ANP, BNP, and CNP, although there is evidence of a fourth, *Dendroaspis* natriuretic peptide (DNP, isolated from the green mamba snake). The primary function of NPs is to promote natriuresis (Potter, Yoder, Flora, Antos, & Dickey, 2009). Natriuretic peptide receptor A (NPRA) is the primary receptor for all three NPs. Although primarily found in the kidney as its function is to promote diuresis and natriuresis, NPRA is also located in lung, vasculature, heart, adrenal, adipose, and brain tissues (Goy et al., 2001; Lowe et al., 1989; Nagase, Katafuchi, Hirose, & Fujita, 1997). The ANP/NPRA interaction regulates blood pressure and therefore is essential for fluid balance homeostasis.

The release of both ANP and BNP is triggered by an increase in blood volume within the heart. While ANP is released primarily within the atria, BNP is released within the ventricles. Although ANP and BNP have a similar function, BNP has an estimated 10-fold lower affinity than ANP and therefore is considered a weaker diuretic. Clinically, serum ANP is used as a biomarker for cardiovascular disease, including myocardial infarction, stroke, coronary artery disease, and heart failure (Barbato et al., 2012; Sabatine et al., 2012; Wang et al., 2004). In clinical SCI research settings, ANP has been investigated as a therapeutic in ischemia/reperfusion injury (Nakayama et al., 2007), as well as for its role in exercise therapy (Yamamoto et al., 1999) and bladder distention (Yamamoto et al., 1999). Further work is necessary to elucidate the importance and extent to which the changes in these peptides after SCI affect urinary and cardiovascular health.

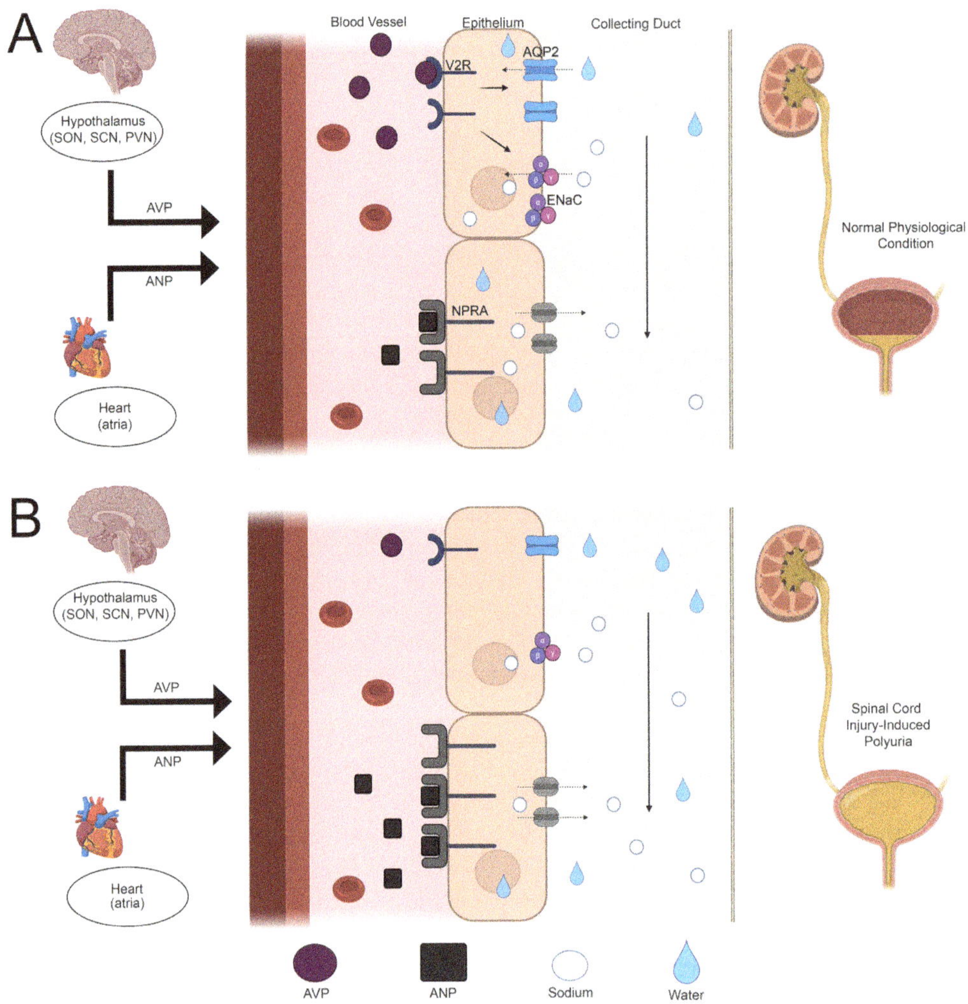

FIG. 2 Diagram of AVP/ANP and their receptors within the kidney under normal physiological conditions and after SCI. As part of normal physiological processes to maintain fluid homeostasis, AVP is released by the hypothalamus in response to conditions of hyperosmolality (A, upper portion). Once in the kidney, AVP binds to V2R, which triggers a response to open AQP2 channels to allow water flow from the kidney collecting duct back into the epithelium while ENaC allows for sodium to be reabsorbed. However, under conditions of elevated blood volume, ANP is released from the heart (A, lower portion) and upon binding to NPRA within the kidney, a cascade is triggered to permit salt and subsequently water excretions. As illustrated in B, there is a decrease in AVP, V2R, and ENaC as well as an increase in ANP and NPRA after SCI, which together contribute to the injury-induced occurrence of polyuria/nocturia. *Image created using Biorender.com.*

SCI-induced polyuria/nocturia

Nocturia is defined as the necessity to void one or more times at night, specifically while sleeping (Kerrebroeck et al., 2002). There are many potential causes of nocturia, including decreased bladder capacity, increase in fluid intake, and increased diuresis (Oelke et al., 2017). A common occurrence in the human SCI population is nocturnal polyuria or an excess of urine production/passage at night (Denys et al., 2017). This issue causes sleep disruptions, bladder overdistention, and an increased risk of acquiring lower urinary tract infections due to the necessity for additional intermittent catheterizations. Historically, SCI-induced polyuria was thought to be the effect of decreased vascular tone and pooling of fluid in the lower extremities, which upon fluid redistribution when supine at night-time causes "intravascular flooding" and subsequent diuresis (Krum, Louis, Brown, Jackman, & Howes, 1991; Williams et al., 1990). However, polyuria is present in animal SCI models (Gumbel, Montgomery, Yang, & Hubscher, 2020; Hubscher et al., 2016; Montgomery & Hubscher, 2017; Ward & Hubscher, 2012) which lack positional fluid redistribution, suggesting that other mechanisms are likely involved.

In chronic SCI patients, increased blood pressure at night versus day is associated with increased nighttime urine production (Goh et al., 2018). Further, nocturnal polyuria, SCI-induced or otherwise, can be caused by decreased AVP circulation, increased ANP, and cardiovascular insufficiency (Oelke, Adler, Marschall-Kehrel, Herrmann, & Berges,

2014). A study by Denys et al. revealed that SCI patients lacked circadian control of renal function (clearance of creatinine, free water, and solutes) (Denys et al., 2017). Together, these measured changes in circadian control of renal function, blood pressure, AVP, and ANP after SCI are all likely contributing to the mechanisms driving SCI-induced polyuria.

AVP after SCI

One potential factor that contributes to SCI-induced polyuria is a decrease in levels of AVP. Under normal conditions, humans exhibit an increase in circulating AVP at night which results in decreased sleep-time urine production and thus volume, thereby controlling the need for toileting. Clinically, children with nocturnal polyuria experience disruptions in the diurnal variation of AVP, specifically a decrease in circulating AVP, causing an abnormal increase in night-time urine production which may induce bed-wetting (Rittig, Knudsen, Norgaard, Pedersen, & Djurhuus, 1989). Similarly, the SCI population commonly experience disrupted diurnal variation of AVP (Kilinc, Akman, Levendoglu, & Özker, 1999; Szollar et al., 1997), leading to the need for bladder emptying one or more times a night. Additionally, individuals with SCI exhibit disrupted circadian control of renal creatinine clearance, water diuresis, and solute diuresis (Denys et al., 2017). This decrease in night-time AVP is likely a major contributor to SCI-induced polyuria/nocturia. Furthermore, persons with injuries above T6 tend to demonstrate an increase in nocturnal sodium excretion, which would suggest that the mechanisms behind SCI-induced polyuria include other factors such as ANP/BNP (Denys et al., 2017). However, polyuria has also been observed in individuals having varying severities and injury levels, trends consistent with metabolic cage results in the clinically relevant rat contusion model whereby polyuria was measured regardless of the extent of injury (Ward & Hubscher, 2012).

To further elucidate the potential mechanism of SCI-induced polyuria, pre-clinical experiments were conducted to investigate the changes in AVP and NPs in the rat SCI model. The reported data (Montgomery & Hubscher, 2017) are consistent with those described for the human SCI population. Beginning at 2-week post-SCI, male rats demonstrated a significant decrease in serum AVP (Montgomery & Hubscher, 2017), which continued at chronic time points as well (Gumbel et al., 2020). In addition to the sub-acute and chronic decrease in AVP, the key AVP receptor in the kidney, V2R, was significantly lower than uninjured surgical sham rats (Gumbel et al., 2020). This result is somewhat surprising as V2R is a G-coupled protein receptor and under normal physiological conditions, becomes desensitized to an abundance of its ligand (in this case, AVP), and inversely becomes sensitized under low levels of its ligand. Moreover, as illustrated in Fig. 2, protein levels of AQP2 channel and epithelial sodium channel (ENaC) within the kidney were also reduced along with V2R, further highlighting the importance of AVP in SCI-induced polyuria, as AQP2 channels are formed in response to the activation of V2R to promote water absorption, and ENaC functions to reabsorb sodium ions back into the bloodstream. Taken together, the changes in these receptor densities after SCI, as shown in Figs. 3 and 4, are in the direction expected to promote polyuria.

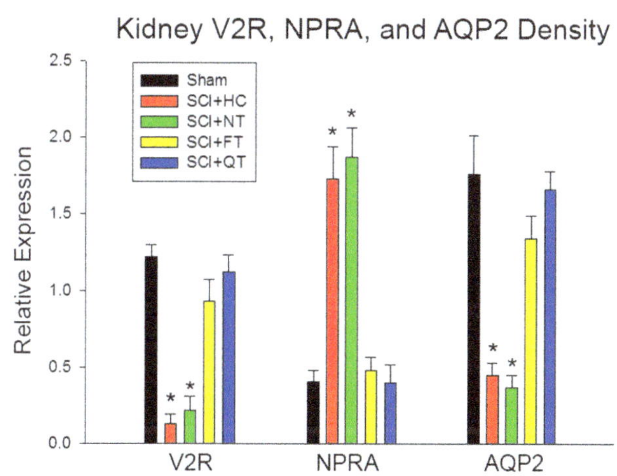

FIG. 3 Activity-based recovery training reverses several key fluid balance receptors in the kidney after SCI. After chronic SCI, V2R, NPRA, and AQP2 are significantly altered. Relative expression of proteins using Western blot reveals that V2R and AQP2 are significantly decreased while NPRA is significantly increased in groups of male Wistar rats that did not receive any therapeutic interventions (NT = non-trained; HC = home cage) after a moderate-severe contusion injury that yields approximately 11% white matter sparing at the T9 spinal level lesion epicenter. Animals that received ABRT involving quadrupedal or forelimb-only training (QT and FT, respectively) by stepping on a treadmill for 1 h a day, 7 days a week for 8 weeks beginning 2 weeks post-injury were found to have similar levels of these receptor/channel densities to surgical sham animals (spinal laminectomy but no contusion). Error bars: standard error of the mean.

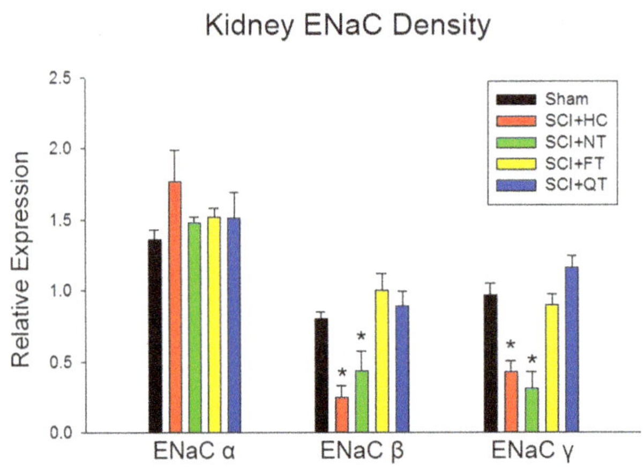

FIG. 4 Activity-based recovery training reverses ENaC β and γ sub-units deficit after SCI. After chronic SCI, both β and γ sub-units of ENaC are significantly decreased. However, after ABRT (SCI + FT and SCI + QT), the relative density of both β and γ sub-units was reversed to similar levels of surgical sham animals. Group abbreviations per Fig. 3. Error bars: standard error of the mean.

A significant increase in AVP has also been reported in female piglets 15 min after a cervical spinal transection, which is likely an acute response to cardiovascular changes to maintain homeostatic blood pressure (Zahra et al., 2010). This finding demonstrates that the hemodynamic changes seen after SCI are likely affecting the changes in AVP, as orthostatic hypotension is another common occurrence in the human SCI population (Claydon et al., 2006; Illman, Stiller, & Williams, 2000; Zerbe, Henry, & Robertson, 1983).

While desmopressin (a synthetic AVP analog) is used to suppress bedwetting in children (Glazener & Evans, 2002) and has yielded positive results in some cases of SCI-induced nocturnal polyuria (Chancellor et al., 1994; Zahariou, Karamouti, Karagiannis, & Papaioannou, 2008), its uses may be limited to those at lower risk of cardiovascular disease. Additionally, desmopressin only targets one of the several potential factors that lead to polyuria and leaves other pathways unchecked, such as ANP/NPRA.

ANP after SCI

Mechanisms driving the chronic decrease in AVP/V2R after SCI are unknown. Although blood osmolality is one driving force behind AVP release, non-osmotic stimuli such as NPs, glucocorticoids, and norepinephrine may control AVP levels and urine production (Dillingham & Anderson, 1986; Schrier, Berl, & Anderson, 1979). Of these stimuli, NPs (ANP/BNP) and corticosterone (CORT; primary glucocorticoid in rats) have been further investigated in the rat SCI model (see Fig. 1). While there is some evidence that SCI can induce changes in ANP in the clinical setting (Tajima et al., 1990), further preclinical and clinical experimental studies are necessary to further elucidate the impact of level and severity of the injury.

Changes in urinary ANP and CORT have been documented as early as two weeks after a moderate contusion injury in male rats, together with serum AVP (Montgomery & Hubscher, 2017). In a 2018 study of SCI rats with polyuria, both urinary ANP and CORT were significantly increased 2 weeks post- contusion, while surgical sham animals demonstrated no differences. Both ANP and CORT inhibit the function and release of AVP, therefore the significant changes in these hormones (increases in ANP and CORT and a decrease in AVP) are in the direction expected with respect to polyuria (Haack, Mohring, Mohring, Petri, & Hackenthal, 1977; Inoue, Nonoguchi, & Tomita, 2001). It is important to note that urinary BNP levels have not been shown to change after SCI. However, BNP is a weaker diuretic compared to ANP. It cannot be ruled out that plasma BNP changes occur after SCI, as its half-life is relatively short and potentially difficult to capture.

It has also been demonstrated that the significant changes in AVP and ANP continue chronically in SCI male rats (Gumbel et al., 2020). In a recent study, SCI animals having polyuria were randomized into one of two groups receiving different forms of activity-based recovery training (ABRT) on a treadmill or one of two groups having no therapeutic intervention (one group harnessed without any treadmill activity and one remaining in their home cage) (Gumbel et al., 2019). While all groups still demonstrated polyuria at the end of the study such that their 24-h urine volumes remained significantly increased from baseline pre-injury levels, the ABRT animals showed a significantly lower 24-h urine volume than the non-trained animals after the 8 weeks of daily 60-min training sessions. These findings were similar to those previously reported in animals (Hubscher et al., 2016; Ward et al., 2014) and human studies (Hubscher et al., 2018) demonstrating multi-system benefits of locomotor training. In addition to changes in AVP, ANP, and corticosterone, relative expression densities of NPRA were found to be significantly increased in the non-trained animals, but not the ABRT animal groups (see Fig. 3). The inverse was found for V2R and AQP2 as their expression were significantly decreased in only the non-trained animal groups, while the ABRT animals were not significantly different from surgical sham animals. These findings, illustrated in Fig. 3, suggest that one potential mechanism behind the improvement in polyuria after exercise therapy is the regulation of key receptors within the kidney. Note that although high-intensity exercise induces an acute increase in AVP in able-bodied subjects (Convertino, Keil, Bernauer, & Greenleaf, 1981; Inder, Hellemans, Swanney, Prickett, & Donald, 1998) as well as ANP (Ohba et al., 2001), these effects may be short-lived and may differ under conditions such as SCI. More studies are needed to further elucidate underlying mechanisms.

Potential mechanisms causing changes in AVP and/or ANP after SCI

Although many variables lead to the release of AVP, two important physical changes include hypertonicity and/or hypovolemia, which are important because orthostatic hypotension is a common occurrence amongst the SCI population (Claydon et al., 2006). Another issue for the SCI population, specifically with higher-level injuries, is autonomic dysreflexia (AD), a sudden onset of hypertension (Krassioukov & Weaver, 1995). Repeated and frequent occurrences of AD may have an impact on AVP release and subsequent concentration of V2R in the kidney. Such changes in alpha-adrenergic receptors (receptors for epinephrine and norepinephrine) have already been linked to AD (Arnold, Feng, Delaney, & Teasell, 1995). AD has also been characterized following severe injuries in the rat model of SCI (Maiorov, Fehlings, & Krassioukov, 1998).

One potential mechanism behind the observed increase in NPRA after SCI in rodents (Gumbel et al., 2020) is an increase in glucocorticoids. An increase in glucocorticoids can increase the glomerular filtration rate (Haack et al., 1977), decrease the level of plasma AVP, upregulate NPRA within the kidney (Liu et al., 2011), and increase ANP release (Dananberg & Grekin, 1992). Corticosterone, the main glucocorticoid in rodents (equivalent to cortisol in humans), is elevated in both humans and rodents after SCI, which could relate to stress (Lucin, Sanders, Jones, Malarkey, & Popovich, 2007; Popovich, Stuckman, Gienapp, & Whitacre, 2001). It is also documented that glucocorticoids have the ability to alter immune activity, whereby a significant increase may lead to immunosuppression (Coutinho & Chapman, 2011; DePasquale-Jardieu & Fraker, 1979; Wiegers, Stec, Klinkert, & Reul, 2000). After SCI, specifically high-level injuries, elevated glucocorticoids are associated with a significant decrease in immune responses, which can lead to an increased risk of infections, including that of the bladder and kidney (Lucin, Sanders, & Popovich, 2009).

It is noteworthy that the renin aldosterone angiotensin system (RAAS) also regulates cardiovascular homeostasis (blood pressure and volume) in addition to fluid and electrolyte balance (urine concentration and excretion). RAAS is an additional factor that can affect both AVP and ANP, and therefore needs further investigation in its potential role in SCI-induced polyuria. While there are studies in the clinical setting that indicate there are acute changes in renin and aldosterone after SCI (Mathias et al., 1975; Wall et al., 1994), the focus of these studies relates to cardiovascular function.

Redistribution of body fluids in the SCI population at bedtime when shifting from a sitting to supine position produces cardiovascular fluctuations, which in turn affects the delicate hormone balance that controls homeostatic fluid balance (Viaene et al., 2019). This fluid redistribution is one potential mechanism of SCI-induced polyuria. Support comes from a study where the use of compression stockings was shown to reduce leg edema during the day and subsequent urine production during the night (Viaene et al., 2019). However, animal models with SCI-induced polyuria that do not undergo fluid redistribution suggest other mechanisms are likely involved, such as hormones previously identified with respect to fluid balance (see Fig. 1). As further research progresses, additional key hormones/receptors will be identified along with AVP and ANP, and their roles elucidated with respect to SCI-induced polyuria as well as cardiovascular health.

Applications to other areas of neuroscience

Several associations have been made between altered AVP function and several neurological disorders besides SCI, including autism, depression, bipolar, and schizophrenia (Iovino et al., 2018). For example, altered binding of AVP and V1R in the CNS has been linked to autism disorders, (Wassink et al., 2004) and elevated CSF levels of neurophysin II (hypothalamic carrier protein for AVP) were found in patients with schizophrenia (Linkowski, Geenen, Kerkhofs, Mendlewicz, & Legros, 1984). Also, in multiple sclerosis, an autoimmune disorder that affects the myelin sheaths of nerves within the CNS, urinary tract dysfunctions include polyuria/nocturia (Panicker, Fowler, & Kessler, 2015). The potential mechanisms behind multiple sclerosis-associated polyuria overlap with SCI-induced polyuria, such as disrupted AVP secretion and nocturnal hypertension (Peyronnet et al., 2019).

Other neurological involvements related to NPs have been shown for studies involving the olfactory bulb (Wildey & Glembotski, 1986), Purkinje fibers of the cerebellum (McKenzie, Juan, Thomas, Berman, & Klein, 2001), the retina (Spes et al., 2020), and development (Cao & Yang, 2008). There have also been studies indicating a potential neuroprotective role for NPs, likely through increasing cGMP levels (Cao & Yang, 2008; Fiscus, Alex, & Chew, 2001). Together, it is clear that both NPs and AVP are two critical entities that should be considered in other settings in addition to SCI.

Mini-dictionary of terms

Polyuria: Overproduction and passage of urine; greater than 2.5–3 L/ 24 h in adults.
Nocturia: When a person is woken up at night to use the bathroom to urinate one or more times a night.
Vasopressin (AVP): Also known as an antidiuretic hormone; functions to help re-absorb water into the bloodstream as well as vasoconstriction.
Natriuretic peptides (NPs): Group of peptides that function to promote natriuresis and subsequent water excretion.
Corticosterone: Produced in the cortex of the adrenal glands; the main glucocorticoid in many animal species, including rodents (instead of cortisol which is the primary glucocorticoid in humans).
Aquaporins: A group of channel proteins that function to allow water transport between cells.
Autonomic dysreflexia: A condition seen more commonly with high-level SCI where the body reacts to a stimulus such as a full bladder or distended bowel and triggers the sympathetic nervous system to increase blood pressure. Due to injury, the body is unable to control the response which can lead to life-threatening hypertension.
Diuretic: Either a drug (usually taken to lower high blood pressure) or natural hormone/peptide that causes an increase in urine and/or salt excretion.
Orthostatic hypotension: A decrease in blood pressure (20 mmHg for systolic and 10 mmHg for diastolic) caused by a sudden position change, such as standing up from sitting.
Activity-based recovery training: Types of rehabilitation such as treadmill walking with the assistance that use activity to restore neurological function.

Key facts of vasopressin and natriuretic peptide changes after SCI

Key facts of vasopressin after SCI

- AVP is made by magnocellular neurons in the supraoptic, suprachiasmatic, and paraventricular nuclei of the hypothalamus.
- AVP is used for the treatments of various medical conditions.
- Both pre-clinical and clinical SCI studies demonstrate a chronic decrease and disrupted levels of AVP post-SCI.
- The dysregulation of AVP after SCI likely plays a major role in SCI-induced polyuria/nocturia.
- Intense exercise is known to increase serum AVP levels.

Key facts of natriuretic peptides

- Natriuretic peptides include ANP, BNP, and CNP.
- ANP is released by cardiomyocytes located in the atria of the heart.
- BNP is released by cardiomyocytes located in the ventricles of the heart and has less effectiveness as ANP.
- Natriuretic peptides all function to promote natriuresis.
- ANP levels increase chronically post-SCI.

Summary points

This chapter focuses on AVP and NP changes after SCI, particularly with respect to polyuria/nocturia.

- Both AVP and NP affect cardiovascular and upper urinary tract-related functions.
- Although AVP and NP changes after SCI are important, changes in the density of their subsequent receptors in the kidney contribute to mechanisms underlying SCI-induced polyuria.
- Hormonal changes after SCI, including altered AVP and NP levels and their receptors, influence multiple physiological systems, with resulting deficits that impact both daily activities and quality of life.
- Numerous associations have been made between altered AVP/NP functions and several other neurological disorders.

References

Arnold, J. M. O., Feng, Q.-P., Delaney, G. A., & Teasell, R. W. (1995). Autonomic dysreflexia in tetraplegic patients: Evidence for α-adrenoceptor hyperresponsiveness. *Clinical Autonomic Research*, *5*, 267–270.

Barbato, E., Bartunek, J., Marchitti, S., Mangiacapra, F., Stanzione, R., Delrue, L., et al. (2012). NT-proANP circulating level is a prognostic marker in stable ischemic heart disease. *International Journal of Cardiology*, *155*, 311–312.

Bichet, D. G., Razi, M., Lonergan, M., Arthus, M.-F., Papukna, V., Kortas, C., et al. (1988). Hemodynamic and coagulation responses to 1-desamino [8-D-arginine] vasopressin in patients with congenital nephrogenic diabetes insipidus. *New England Journal of Medicine*, *318*, 881–887.

Cao, L.-H., & Yang, X.-L. (2008). Natriuretic peptides and their receptors in the central nervous system. *Progress in Neurobiology*, *84*, 234–248.

Chancellor, M. B., Rivas, D. A., & Staas, W. E., Jr. (1994). DDAVP in the urological management of the difficult neurogenic bladder in spinal cord injury: Preliminary report. *The Journal of the American Paraplegia Society*, *17*, 165–167.

Claydon, V. E., Steeves, J. D., & Krassioukov, A. (2006). Orthostatic hypotension following spinal cord injury: Understanding clinical pathophysiology. *Spinal Cord*, *44*, 341–351.

Convertino, V. A., Keil, L. C., Bernauer, E. M., & Greenleaf, J. E. (1981). Plasma volume, osmolality, vasopressin, and renin activity during graded exercise in man. *Journal of Applied Physiology*, *50*, 123–128.

Coutinho, A. E., & Chapman, K. E. (2011). The anti-inflammatory and immunosuppressive effects of glucocorticoids, recent developments and mechanistic insights. *Molecular and Cellular Endocrinology*, *335*, 2–13.

Cowley, A. W., Jr., Skelton, M. M., Merrill, D. C., Quillen, E. W., Jr., & Switzer, S. J. (1983). Influence of daily sodium intake on vasopressin secretion and drinking in dogs. *American Journal of Physiology. Regulatory, Integrative and Comparative Physiology*, *245*. R860–R872.

Dananberg, J. A. M. I. E., & Grekin, R. J. (1992). Corticoid regulation of atrial natriuretic factor secretion and gene expression. *American Journal of Physiology. Heart and Circulatory Physiology*, *263*. H1377–H1381.

Denys, M.-A., Viaene, A., Goessaert, A.-S., Van Haverbeke, F., Hoebeke, P., Raes, A., et al. (2017). Circadian rhythms in water and solute handling in adults with a spinal cord injury. *The Journal of Urology*, *197*, 445–451.

DePasquale-Jardieu, P., & Fraker, P. J. (1979). The role of corticosterone in the loss in immune function in the zinc-deficient A/J mouse. *The Journal of Nutrition*, *109*, 1847–1855.

Dillingham, M. A., & Anderson, R. J. (1986). Inhibition of vasopressin action by atrial natriuretic factor. *Science*, *231*, 1572–1573.

Dolinak, D., & Balraj, E. (2007). Autonomic dysreflexia and sudden death in people with traumatic spinal cord injury. *The American Journal of Forensic Medicine and Pathology*, *28*, 95–98.

Fiscus, R. R., Alex, W. K. T., & Chew, S. B. C. (2001). Natriuretic peptides inhibit apoptosis and prolong the survival of serum-deprived PC12 cells. *Neuroreport*, *12*, 185–189.

Glazener, C. M. A., & Evans, J. H. C. (2002). Desmopressin for nocturnal enuresis in children. *Cochrane Database of Systematic Reviews*, (3), CD002112.

Goh, M. Y., Millard, M. S., Wong, E. C. K., Berlowitz, D. J., Graco, M., Schembri, R. M., et al. (2018). Comparison of diurnal blood pressure and urine production between people with and without chronic spinal cord injury. *Spinal Cord*, *56*, 847–855.

Goh, M. Y., Millard, M. S., Wong, E. C. K., Brown, D. J., Frauman, A. G., & O'Callaghan, C. J. (2017). Diurnal blood pressure and urine production in acute spinal cord injury compared with controls. *Spinal Cord*, *55*, 39–46.

Goy, M. F., Oliver, P. M., Purdy, K. E., Knowles, J. W., Fox, J. E., Mohler, P. J., et al. (2001). Evidence for a novel natriuretic peptide receptor that prefers brain natriuretic peptide over atrial natriuretic peptide. *Biochemical Journal*, *358*, 379–387.

Gumbel, J. H., Montgomery, L. R., Yang, C. B., & Hubscher, C. H. (2020). Activity-based training reverses spinal cord injury-induced changes in kidney receptor densities and membrane proteins. *Journal of Neurotrauma*, *37*, 555–563.

Gumbel, J. H., Steadman, C. J., Hoey, R. F., Armstrong, J. E., Fell, J. D., Yang, C. B., et al. (2019). Activity-based training on a treadmill with spinal cord injured Wistar rats. *Journal of Visualized Experiments*, e58983.

Haack, D., Mohring, J., Mohring, B., Petri, M., & Hackenthal, E. (1977). Comparative study on development of corticosterone and DOCA hypertension in rats. *American Journal of Physiology. Renal Physiology*, *233*. F403–F411.

Harris, H. W., Zeidel, M. L., Jo, I., & Hammond, T. G. (1994). Characterization of purified endosomes containing the antidiuretic hormone-sensitive water channel from rat renal papilla. *Journal of Biological Chemistry*, *269*, 11993–12000.

Holmes, C. L., Landry, D. W., & Granton, J. T. (2003). Science review: Vasopressin and the cardiovascular system part 1–receptor physiology. *Critical Care*, *7*, 427.

Hubscher, C. H., Herrity, A. N., Williams, C. S., Montgomery, L. R., Willhite, A. M., Angeli, C. A., et al. (2018). Improvements in bladder, bowel and sexual outcomes following task-specific locomotor training in human spinal cord injury. *PLoS One*, *13*, e0190998.

Hubscher, C. H., Montgomery, L. R., Fell, J. D., Armstrong, J. E., Poudyal, P., Herrity, A. N., et al. (2016). Effects of exercise training on urinary tract function after spinal cord injury. *American Journal of Physiology. Renal Physiology*, *310*. F1258–F1268.

Illman, A., Stiller, K., & Williams, M. (2000). The prevalence of orthostatic hypotension during physiotherapy treatment in patients with an acute spinal cord injury. *Spinal Cord*, *38*, 741–747.

Inder, W. J., Hellemans, J., Swanney, M. P., Prickett, T. C. R., & Donald, R. A. (1998). Prolonged exercise increases peripheral plasma ACTH, CRH, and AVP in male athletes. *Journal of Applied Physiology*, *85*, 835–841.

Inoue, T., Nonoguchi, H., & Tomita, K. (2001). Physiological effects of vasopressin and atrial natriuretic peptide in the collecting duct. *Cardiovascular Research*, *51*, 470–480.

Iovino, M., Messana, T., De Pergola, G., Iovino, E., Dicuonzo, F., Guastamacchia, E., et al. (2018). The role of neurohypophyseal hormones vasopressin and oxytocin in neuropsychiatric disorders. *Endocrine, Metabolic & Immune Disorders Drug Targets*, *18*, 341–347.

Kerrebroeck, V., Philip, P. A., Chaikin, D., Donovan, J., Fonda, D., Jackson, S., et al. (2002). The standardisation of terminology in nocturia: Report from the standardisation sub-committee of the international continence society. *Neurourology and Urodynamics*, *21*, 179–183.

Kilinc, S., Akman, M. N., Levendoglu, F., & Özker, R. (1999). Diurnal variation of antidiuretic hormone and urinary output in spinal cord injury. *Spinal Cord*, *37*.

Kjeldsen, S. E., Os, I., Forsberg, G., Aakesson, I., Skjøtø, J., Frederichsen, P., et al. (1985). Dietary sodium intake increases vasopressin secretion in man. *Journal of Clinical Hypertension*, *1*, 123–131.

Krassioukov, A. V., & Weaver, L. C. (1995). Episodic hypertension due to autonomic dysreflexia in acute and chronic spinal cord-injured rats. *American Journal of Physiology. Heart and Circulatory Physiology*, *268*. H2077–H2083.

Krum, H., Louis, W. J., Brown, D. J., Jackman, G. P., & Howes, L. G. (1991). Diurnal blood pressure variation in quadriplegic chronic spinal cord injury patients. *Clinical Science*, *80*, 271–276.

Lindan, R., Joiner, E., Freehafer, A. A., & Hazel, C. (1980). Incidence and clinical features of autonomic dysreflexia in patients with spinal cord injury. *Spinal Cord*, *18*, 285–292.

Linkowski, P., Geenen, V., Kerkhofs, M., Mendlewicz, J., & Legros, J. J. (1984). Cerebrospinal fluid neurophysins in affective illness and in schizophrenia. *European Archives of Psychiatry and Neurological Sciences*, *234*, 162–165.

Liu, C., Chen, Y., Kang, Y., Ni, Z., Xiu, H., Guan, J., et al. (2011). Glucocorticoids improve renal responsiveness to atrial natriuretic peptide by up-regulating natriuretic peptide receptor-A expression in the renal inner medullary collecting duct in decompensated heart failure. *Journal of Pharmacology and Experimental Therapeutics*, *339*, 203–209.

Lowe, D. G., Chang, M.-S., Hellmiss, R., Chen, E., Sujay, S., Garbers, D. L., et al. (1989). Human atrial natriuretic peptide receptor defines a new paradigm for second messenger signal transduction. *The EMBO Journal*, *8*, 1377–1384.

Lucin, K. M., Sanders, V. M., Jones, T. B., Malarkey, W. B., & Popovich, P. G. (2007). Impaired antibody synthesis after spinal cord injury is level dependent and is due to sympathetic nervous system dysregulation. *Experimental Neurology*, *207*, 75–84.

Lucin, K. M., Sanders, V. M., & Popovich, P. G. (2009). Stress hormones collaborate to induce lymphocyte apoptosis after high level spinal cord injury. *Journal of Neurochemistry*, *110*, 1409–1421.

Maiorov, D. N., Fehlings, M. G., & Krassioukov, A. V. (1998). Relationship between severity of spinal cord injury and abnormalities in neurogenic cardiovascular control in conscious rats. *Journal of Neurotrauma*, *15*, 365–374.

Mathias, C. J., Christensen, N. J., Corbett, J. L., Frankel, H. L., Goodwins, T. J., & Peart, W. S. (1975). Plasma catecholamines, plasma renin activity and plasma aldosterone in tetraplegic man, horizontal and tilted. *Clinical Science and Molecular Medicine*, *49*, 291–299.

McKenzie, J. C., Juan, Y.-W., Thomas, C. R., Berman, N. E. J., & Klein, R. M. (2001). Atrial natriuretic peptide-like immunoreactivity in neurons and astrocytes of human cerebellum and inferior olivary complex. *The Journal of Histochemistry and Cytochemistry*, *49*, 1453–1467.

Montgomery, L. R., & Hubscher, C. H. (2017). Altered vasopressin and natriuretic peptide levels in a rat model of spinal cord injury: Implications for the development of polyuria. *American Journal of Physiology. Renal Physiology*, *314*, F58–F66.

Nagase, M., Katafuchi, T., Hirose, S., & Fujita, T. (1997). Tissue distribution and localization of natriuretic peptide receptor subtypes in stroke-prone spontaneously hypertensive rats. *Journal of Hypertension*, *15*, 1235–1243.

Nakayama, T., Harada, N., Asano, M., Nomura, N., Saito, T., Mishima, A., et al. (2007). Atrial natriuretic peptide reduces ischemia/reperfusion-induced spinal cord injury in rats by enhancing sensory neuron activation. *Journal of Pharmacology and Experimental Therapeutics*, *322*, 582–590.

Nielsen, S., Frøkiær, J., Marples, D., Kwon, T.-H., Agre, P., & Knepper, M. A. (2002). Aquaporins in the kidney: From molecules to medicine. *Physiological Reviews*, *82*, 205–244.

Oelke, M., Adler, E., Marschall-Kehrel, D., Herrmann, T. R. W., & Berges, R. (2014). *Nocturia: State of the art and critical analysis of current assessment and treatment strategies*. Springer.

Oelke, M., De Wachter, S., Drake, M. J., Giannantoni, A., Kirby, M., Orme, S., et al. (2017). A practical approach to the management of nocturia. *International Journal of Clinical Practice*, *71*, e13027.

Ohba, H., Takada, H., Musha, H., Nagashima, J., Mori, N., Awaya, T., et al. (2001). Effects of prolonged strenuous exercise on plasma levels of atrial natriuretic peptide and brain natriuretic peptide in healthy men. *American Heart Journal*, *141*, 751–758.

Panicker, J. N., Fowler, C. J., & Kessler, T. M. (2015). Lower urinary tract dysfunction in the neurological patient: Clinical assessment and management. *The Lancet Neurology*, *14*, 720–732.

Peyronnet, B., Krupp, L. B., Reynolds, W. S., Gamé, X., Amarenco, G., Cornu, J.-N., et al. (2019). Nocturia in patients with multiple sclerosis. *Reviews in Urology*, *21*, 63.

Phillips, P. A., Abrahams, J. M., Kelly, J. M., Mooser, V., Trinder, D., & Johnston, C. I. (1990). Localization of vasopressin binding sites in rat tissues using specific VI and V2 selective ligands. *Endocrinology*, *126*, 1478–1484.

Popovich, P. G., Stuckman, S., Gienapp, I. E., & Whitacre, C. C. (2001). Alterations in immune cell phenotype and function after experimental spinal cord injury. *Journal of Neurotrauma*, *18*, 957–966.

Potter, L. R., Yoder, A. R., Flora, D. R., Antos, L. K., & Dickey, D. M. (2009). Natriuretic peptides: their structures, receptors, physiologic functions and therapeutic applications. In *cGMP: Generators, effectors and therapeutic implications* Springer.

Rittig, S., Knudsen, U. B., Norgaard, J. P., Pedersen, E. B., & Djurhuus, J. C. (1989). Abnormal diurnal rhythm of plasma vasopressin and urinary output in patients with enuresis. *American Journal of Physiology. Renal Physiology*, *256*. F664–F671.

Sabatine, M. S., Morrow, D. A., de Lemos, J. A., Omland, T., Sloan, S., Jarolim, P., et al. (2012). Evaluation of multiple biomarkers of cardiovascular stress for risk prediction and guiding medical therapy in patients with stable coronary disease. *Circulation*, *125*, 233–240.

Schrier, R. W., Berl, T., & Anderson, R. J. (1979). Osmotic and nonosmotic control of vasopressin release. *American Journal of Physiology. Renal Physiology*, *236*. F321–F332.

Spes, Š., Katarina, S. H., Werner, F., Koch, F., Völker, K., Krebes, L., et al. (2020). Natriuretic peptides attenuate retinal pathological neovascularization via cyclic guanosine monophosphate signaling in pericytes and astrocytes. *Arteriosclerosis, Thrombosis, and Vascular Biology*, *40*, 159–174.

Swanson, L. W., & Eo Sawchenko, P. (1983). Hypothalamic integration: Organization of the paraventricular and supraoptic nuclei. *Annual Review of Neuroscience*, *6*, 269–324.

Szollar, S. M., Dunn, K. L., Brandt, S., & Fincher, J. (1997). Nocturnal polyuria and antidiuretic hormone levels in spinal cord injury. *Archives of Physical Medicine and Rehabilitation*, *78*, 455–458.

Tajima, F., Sagawa, S., Iwamoto, J., Miki, K., Freund, B. J., Claybaugh, J. R., et al. (1990). Cardiovascular, renal, and endocrine responses in male quadriplegics during head-out water immersion. *American Journal of Physiology. Regulatory, Integrative and Comparative Physiology*, *258*. R1424–R1430.

Van Leeuwen, F. W., Swaab, D. F., & De Raay, C. (1978). Immunoelectronmicroscopic localization of vasopressin in the rat suprachiasmatic nucleus. *Cell and Tissue Research*, *193*, 1–10.

Viaene, A., Roggeman, S., Goessaert, A.-. S., Kessler, T. M., Mehnert, U., Van Besien, V., et al. (2019). Conservative treatment for leg oedema and the effect on nocturnal polyuria in patients with spinal cord injury. *BJU International*, *123*, E43–E50.

Wall, B. M., Runyan, K. R., Williams, H. H., Bobal, M. A., Crofton, J. T., Share, L., et al. (1994). Characteristics of vasopressin release during controlled reduction in arterial pressure. *The Journal of Laboratory and Clinical Medicine*, *124*, 554–563.

Wan, D., & Krassioukov, A. V. (2014). Life-threatening outcomes associated with autonomic dysreflexia: A clinical review. *The Journal of Spinal Cord Medicine*, *37*, 2–10.

Wang, T. J., Larson, M. G., Levy, D., Benjamin, E. J., Leip, E. P., Omland, T., et al. (2004). Plasma natriuretic peptide levels and the risk of cardiovascular events and death. *New England Journal of Medicine*, *350*, 655–663.

Ward, P. J., Herrity, A. N., Smith, R. R., Willhite, A., Harrison, B. J., Petruska, J. C., et al. (2014). Novel multi-system functional gains via task specific training in spinal cord injured male rats. *Journal of Neurotrauma*, *31*, 819–833.

Ward, P. J., & Hubscher, C. H. (2012). Persistent polyuria in a rat spinal contusion model. *Journal of Neurotrauma*, *29*, 2490–2498.

Wassink, T. H., Piven, J., Vieland, V. J., Pietila, J., Goedken, R. J., Folstein, S. E., et al. (2004). Examination of AVPR1a as an autism susceptibility gene. *Molecular Psychiatry*, *9*, 968–972.

Wiegers, G. J., Stec, I. E. M., Klinkert, W. E. F., & Reul, J. M. H. M. (2000). Glucocorticoids regulate TCR-induced elevation of CD4: Functional implications. *The Journal of Immunology*, *164*, 6213–6220.

Wildey, G. M., & Glembotski, C. C. (1986). Cross-linking of atrial natriuretic peptide to binding sites in rat olfactory bulb membranes. *Journal of Neuroscience*, *6*, 3767–3776.

Williams, H. H., Wall, B. M., Horan, J. M., Presley, D. N., Crofton, J. T., Share, L., et al. (1990). Nonosmotic stimuli alter osmoregulation in patients with spinal cord injury. *The Journal of Clinical Endocrinology & Metabolism*, *71*, 1536–1543.

Yamamoto, M., Tajima, F., Okawa, H., Mizushima, T., Umezu, Y., & Ogata, H. (1999). Static exercise-induced increase in blood pressure in individuals with cervical spinal cord injury. *Archives of Physical Medicine and Rehabilitation*, *80*, 288–293.

Zahariou, A., Karagiannis, G., Papaioannou, P., Stathi, K., & Michail, X. (2007). The use of desmopressin in the management of nocturnal enuresis in patients with spinal cord injury. *Europa Medicophysica*, *43*, 333–338.

Zahariou, A., Karamouti, M., Karagiannis, G., & Papaioannou, P. (2008). Maximal bladder capacity is a positive predictor of response to desmopressin treatment in patients with MS and nocturia. *International Urology and Nephrology*, *40*, 65–69.

Zahra, M., Samdani, A., Piggott, K., Gonzalez-Brito, M., Solano, J., Santo, R. D. L., et al. (2010). Acute changes in systemic hemodynamics and serum vasopressin after complete cervical spinal cord injury in piglets. *Neurocritical Care*, *13*, 132–140.

Zerbe, R. L., Henry, D. P., & Robertson, G. L. (1983). Vasopressin response to orthostatic hypotension: Etiologic and clinical implications. *The American Journal of Medicine*, *74*, 265–271.

Chapter 25

Linking sensorimotor plasticity, the motor cortex, and spinal cord injury

Raffaele Nardone[a,b], Stefan Golaszewski[b,*], and Eugen Trinka[b]
[a]*Department of Neurology, Hospital of Merano (SABES-ASDAA), Merano-Meran, Italy,* [b]*Department of Neurology, Christian Doppler Klinik, Salzburg, Austria*

Abbreviations

SCI	spinal cord injury
iSCI	incomplete spinal cord injury
TMS	transcranial magnetic stimulation
rTMS	repetitive transcranial magnetic stimulation
MRI	magnetic resonance imaging
fMRI	functional magnetic resonance imaging
M1	primary motor cortex
S1	primary somatosensory cortex
EEG	electroencephalography
CST	corticospinal tract
MEP	motor evoked potential
RMT	resting motor threshold
AMT	active motor threshold
CSP	cortical silent period
CMCT	central motor conduction time
SICI	short-interval intracortical inhibition
LICI	long-interval intracortical inhibition
SAI	short afferent inhibition
PAS	paired associative stimulation
CS	conditioning stimulus
TS	test stimulus
RC	recruitment curve
PSTHs	peristimulus time histograms
SVC	suppression of voluntary contraction
MVC	maximal voluntary contraction
VAC	volitional ankle control
MA	MEP-amplitudes with maximal stimulation
FE	facilitatory effects on MEP-amplitudes
APB	Abductor pollicis brevis
FCR	Flexor carpi radialis
BB	Biceps brachii
TA	Tibialis anterior

Sensorimotor plasticity after spinal cord injury

The treatment of spinal cord injury (SCI) is often frustrating and hopeless because of the remarkable morbidity and mortality, and restricted therapeutic options. SCI usually leads to loss of motor, sensory and autonomic functions below the

* In memoriam.

Cellular, Molecular, Physiological, and Behavioral Aspects of Spinal Cord Injury. https://doi.org/10.1016/B978-0-12-822427-4.00025-3

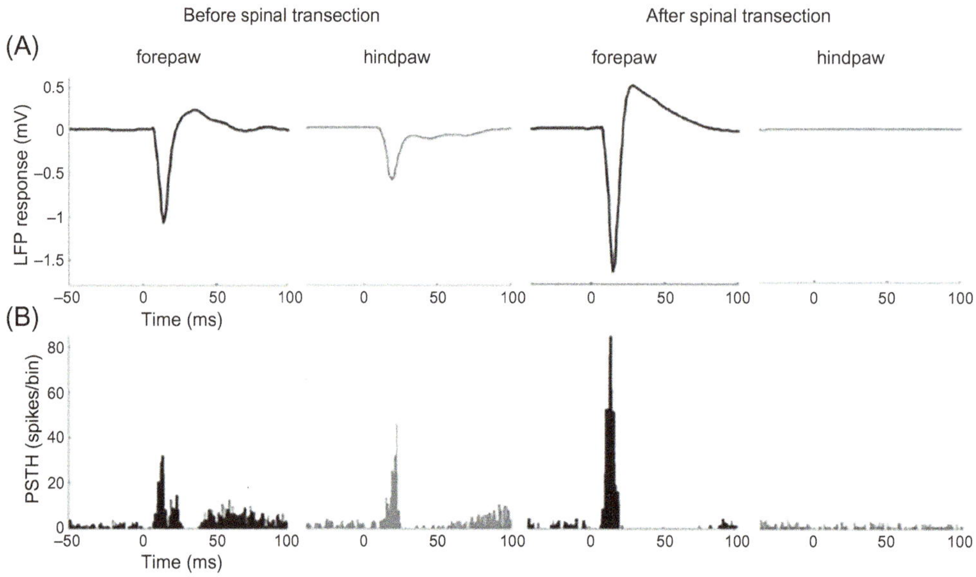

FIG. 1 (A) Responses in the forepaw cortex evoked by forepaw stimuli (*black*) and in the hindpaw cortex evoked by hindpaw stimuli (*gray*) before (*left*) and immediately after (*right*) complete thoracic spinal cord transection. (B) Peristimulus time histograms (PSTHs) represent multi-unit responses recorded from the same electrodes. Cortical responses and PSTHs were evaluated from 100 stimuli. Time 0 (x-axis) represents stimulus onset. The cortical responses are abolished by hindpaw stimuli, but enhanced by forepaw stimuli. *Reproduced with permission from Aguilar, J., Humanes-Valera, D., Alonso-Calvino, E., Yague, J.C., Moxon, K.A., Oliviero, A., et al. (2010). Spinal cord injury immediately changes the state of the brain.* Journal of Neuroscience, 30, *7528–7537.*

level of injury (Dietz & Curt, 2006), owing to the interruption of ascending and descending pathways (Freund et al., 2011; Jurkiewicz, Crawley, Verrier, Fehlings, & Mikulis, 2006; Wrigley et al., 2009).

Experimental electrophysiological studies have demonstrated extensive neuroplastic changes within the sensorimotor cortex after SCI. In anesthetized rats, an immediate functional reorganization has been observed in the primary somatosensory cortex (S1) following complete spinal transection (Aguilar et al., 2010) (Fig. 1). Cortical responses induced by somatosensory stimuli delivered below the level of the lesion are abolished, while those evoked by stimuli applied above the level of the lesion are enhanced. This increase could be explained by the unmasking of silent excitatory synapses, due to the elimination of inhibitory effects, similar to that typically reported after peripheral injuries in the thalamus (Faggin, Nguyen, & Nicolelis, 1997; Nicolelis, Lin, Woodward, & Chapin, 1993) and at the cortical level (Jacobs & Donoghue, 1991).

The augmented cortical responses following the deafferentation due to an SCI were found to be directly related to the slowdown of the cortical spontaneous activity, which becomes immediately overall slower and more silent, switching to a state of slow-wave activity. It has been hypothesized that the activity of brainstem structures and thalamic nuclei regulating cortical synchrony and arousal is directly affected by spinal transection, thus leading to these changes in cortical spontaneous activity (Aguilar & Castro-Alamancos, 2005). Accordingly, a slower cortical activity after SCI can also be detected in electroencephalography (EEG) recordings (Boord et al., 2008; Tran, Boord, Middleton, & Craig, 2004). This slow-wave activity probably plays an important role in shaping long-term cortical reorganization after deafferentation (Aton et al., 2009).

Functional imaging and electrophysiological studies have shown plasticity changes after SCI also in humans. It has been shown in EEG studies that changes of cortical motor activity to a posterior location occur after SCI (Green, Sora, Bialy, Ricamato, & Thatcher, 1998, 1999). This functional reorganization may result from a selective survival of axons arising in the somatosensory cortex and has also been found to have a significant relationship to prognosis in SCI patients with paraplegia.

Neuroimaging findings also suggest that SCI may lead to remarkable anatomical alterations in the human sensorimotor system. In particular, studies employing voxel-based morphometry of cortical volume have demonstrated a significant gray and white matter atrophy bilaterally in the S1. By contrast, some structural studies failed to find changes in gray matter volume in the primary motor cortex (M1) following SCI, whereas others have detected a significant decrease of gray matter volume also in M1 (Nardone et al., 2018). Cortical and subcortical mechanisms that underlie this functional reorganization of M1 and S1 are still incompletely understood.

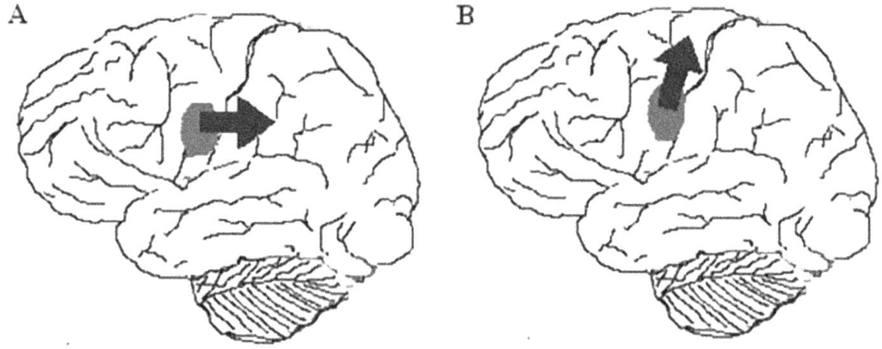

FIG. 2 (A) An upper limb motor task provokes a posterior shift of activation (represented by *shading*) toward the primary somatosensory cortex. (B) Activation of facial/upper limb movement is shifted in the direction of the deafferented upper/lower limb representation in para/tetraplegic subjects. *Reproduced with permission from Kokotilo, K.J., Eng, J.J., & Curt, A. (2009). Reorganization and preservation of motor control of the brain in spinal cord injury: A systematic review.* Journal of Neurotrauma, 26, *2113–2126.*

A functional MRI (fMRI) study showed that in SCI patients the representation of the non-impaired upper limb muscles is changed, but a topographical reorganization in M1 was not found (Jurkiewicz, Mikuli, McIlroy, Fehlings, & Verrier, 2007). The reduction of subcortical white matter volume in M1 reported in some studies could be indicative of atrophy due to retrograde degeneration (Beaud et al., 2008; Hains, Black, & Waxman, 2003), but might also depend on the decreased cortical connectivity related to remodeling in dendritic spines or a reduction in angiogenesis (Hutton, Draganski, Ashburner, & Weiskopf, 2009). Transneuronal degeneration leads to reduced cellular activity, which can induce neuronal atrophy in the S1 (Fields, 2008).

Cortical atrophy (Jurkiewicz et al., 2006; Wrigley et al., 2009) and reorganization in the sensorimotor cortex (Kokotilo, Eng, & Curt, 2009) (Fig. 2) have been reported in other fMRI studies. In particular, during recovery following traumatic SCI (Jurkiewicz et al., 2007), or in subjects who undergo surgical decompression of cervical spinal stenosis (Duggal et al., 2010), a significantly increased volume of activation in M1 has been observed. The increase in cerebral activation during lower limb movement correlated with spinal cord atrophy and functional impairment (Lundell et al., 2011).

Although several studies demonstrating cortical reorganization after human SCI (Kokotilo et al., 2009), the role of output-deprived corticospinal neurons has still not been clarified and previous studies yielded contradictory results. Different fMRI studies revealed an expansion of task-related brain activation (Jurkiewicz et al., 2007), which correlated with atrophy of the lateral part of the cervical cord (Lundell et al., 2011), but also a reduced brain activation of subjects with SCI with persistent paralysis (Jurkiewicz, Mikuli, McIlroy, Fehlings, & Verrier, 2010). The observed expansion of the cortical M1 hand area into the output-deprived M1 leg area is thought to be related to rewiring of damaged hind limb neurons (Ghosh et al., 2010), driven by compensatory use of a less affected part of the body, similar to that observed following rehabilitative training after stroke (Nudo, Milliken, Jenkins, & Milliken, 1996) or overuse (Elbert, Pantev, Wienbruch, Rockstroh, & Taub, 1995).

Cortical and subcortical mechanisms that underlie this reorganization of M1 and S1 remain still matter of debate. Interestingly, in patients with low cervical SCI, movement of the impaired hand was found to be associated with increased blood oxygenation level dependent (BOLD) responses, not only in the expected M1 hand area but also in the expected location of the motor output-deprived M1 leg area. However, the pathophysiological basis of these fMRI findings is still not well understood and whether this cortical reorganization translates into functional gain is still unknown. It should be considered, that the interpretation of functional fMRI findings is limited to the nature of the BOLD signal, which reflects indirect neuronal activation that is mediated by both motor and sensory inputs. Instead, neurophysiological techniques, in particular the transcranial magnetic stimulation (TMS), may offer a reliable tool to characterize important neurophysiological and pathophysiological aspects of motor cortex involvement after SCI.

Functional evaluation of motor cortex

TMS mapping studies

An increased excitability of motor pathways projecting to muscles rostral to the level of a spinal transection has been reported (Topka, Cohen, Cole, & Hallett, 1991), thus suggesting that a functional reorganization may occur either within cortical motor representation areas or at the spinal cord level.

In two quadriplegic adults, an important functional reorganization of the motor cortical projection system has been observed in those cortical areas usually eliciting digit movements instead activate proximal muscles of upper limbs (Levy, Amassian, & Traad, 1990).

Four patients with traumatic cervical (C5–6) SCI showed an expanded cortical map of the preserved contralateral biceps brachii (BB) muscle as early as 6 days (Streletz et al., 1995).

These mapping studies were performed with conventional scalp reference points, and indicate increased excitability of corticospinal projections to non-paralyzed muscles above the spinal cord lesion. However, scalp mapping with TMS is limited by the variation in brain anatomy with regard to scalp topography in individual subjects. Freund and colleagues first used the navigated TMS to investigate the physiological properties of the corticospinal system in the same SCI patients, and demonstrated abnormalities in the topography and excitability of the corticomotor projections to a forearm muscle (Freund, Rothwell, Craggs, Thompson, & Bestmann, 2011). Stereotaxic TMS revealed that the location of the center of gravity of the forearm muscles was shifted posteriorly toward the region of the anatomically defined hand knob in the central sulcus. Changes in the corticospinal projections and within cortico-cortical connections of partially deprived corticospinal neurons innervating the hand neurons may therefore represent an underlying mechanism that drives cortical reorganization and represent an additional neuronal substrate contributing to muscle activation of the impaired upper limb.

Recently, in a patient with SCI resulting in a left-sided hemiparesis for 4 weeks, which by chance had a pre-lesional navigated TMS motor mapping 2 years before, the same examination could be repeated during the acute (after 1 day), subacute (after 10 days) and chronic (after 2 years) phase to trace the cortical reorganization following SCI (Dias Leao, Wiesinger, Ziemann, Tatagiba, & Naros, 2020). While MRI showed no abnormal findings, navigated TMS mapping revealed a posterior shift of the abductor pollicis brevis (APB) muscle representation from the anatomical hand knob toward the S1 in the acute in comparison to the pre-lesional phase. Also, a slight (6%) increase of resting motor threshold (RMT) to TMS was found. An incomplete reversal of the posterior shift, in parallel with a clinical improvement of motor function, occurred within 10 days. Long-term follow-up revealed complete restitution of navigated TMS cortical mapping and motor function. This case report study clearly demonstrated a rapid cortical reorganization within a few days after a transient spinal shock. These findings provide further evidence that a posterior shift of motor cortical representation occurs following SCI.

Single-pulse TMS

The RMT and the suppression of voluntary contraction (SVC) induced by TMS were both found to be higher in patients with cervical incomplete SCI (iSCI, motor level C3–C8) than in healthy control subjects, while the latency of MEPs, the latency of SVC, and SVC-MEP latency difference were increased (Davey et al., 1998). The authors concluded that the longer latency difference between MEPs and SVC may be related to a weak or absent early component of cortical inhibition in SCI patients.

In a subsequent study, the modulation of single motor unit discharges to TMS using peristimulus time histograms (PSTHs) was assessed (Smith et al., 2000). The authors aimed at improving the resolution of the excitatory and inhibitory responses previously observed in surface EMG recordings. The mean threshold for the excitatory peak (excitation) or inhibitory trough (inhibition) in the PSTHs, as well as the mean latencies of excitation and inhibition, were increased in the patients when compared with the control subjects. The latency difference (inhibition-excitation) was also significantly longer in patients than in healthy controls. The authors postulated that increased thresholds and latencies of excitation and inhibition might reflect altered corticospinal transmission in the spinal cord, while the relatively greater increase in the latency of inhibition compared with excitation could reflect a reduction of corticospinal output in response to TMS.

A long pathway, which presumably comprises interneuronal circuits within the motor cortex, is thought to be involved in producing this inhibition.

In another study of the same research group, electrophysiological evaluation of corticospinal function was performed using TMS and EMG recordings from thenar muscles in 21 patients with cervical iSCI (Smith et al., 2000). Both procedures were applied at different time points, from 19 to 384 days to 124–1109 days post-SCI. The group data were pooled into time epochs of 50 or 100 days post-injury for analysis. On the first assessment, the mean latency for MEPs and inhibition of voluntary EMG significantly differ from those of control values. The authors concluded that the decreased inhibition observed following SCI occurs within the first few days after the SCI.

Furthermore, to explore whether crossed facilitatory interactions in the corticospinal pathway are impaired in subjects with SCI, MEPs have been examined in a resting hand, arm, or foot muscle when the contralateral side remained at rest or performed 70% of maximal voluntary contraction (MVC) into index finger abduction, elbow flexion, and ankle dorsiflexion, respectively (Bunday, Oudega, & Perez, 2013). MEPs remained unchanged in muscles at and within 5 segments below the injury during 70% of MVC compared to rest. Conversely, in muscles beyond 5 segments below the injury the size of MEPs increased similar to controls and in muscles distant (>15 segments) from the injury was even aberrantly high, twofold above controls.

A study that focused on plasticity processes of muscles cranial to the SCI level after complete SCI (Lotze, Laubis-Herrmann, & Topka, 2006) demonstrated different MEP recruitment curves (RC) of patients and control subjects. The APB muscle, which is supplied by motoneurons that are located more distinct from the cortical deafferented area than the BB muscle, showed decreased cortical excitability. Furthermore, other findings, such as the lack of relevant increase of MEP amplitudes with increasing stimulus intensities and prolonged cortical silent period (CSP) after SCI, point to reduced excitability of the cortical areas inferior-lateral to the deafferented area.

Since CSP reflect a long-lasting cortical inhibition mediated by $GABA_B$ most likely in the motor cortex (Ziemann et al., 2008), these results suggest an increase in excitability of cortical inhibitory circuits, which is indicative of reduced corticospinal excitability.

To further investigate the corticospinal excitability changes after SCI, the RC of the MEPs has been evaluated in five subjects with good recovery after traumatic cervical iSCI (Nardone et al., 2015). While RMT did not differ significantly between patients and controls, the slope of MEP RC was significantly increased in the patients. This abnormal finding suggests an adaptive response after SCI; the impaired ability of the motor cortex to generate proper voluntary movement may be compensated by increasing spinal excitability.

The easily performed measurement of MEP RC may thus provide a useful additional tool to improve the assessment and monitoring of motor cortical function in subjects with SCI.

Increasing our knowledge of the corticospinal excitability changes in the functional recovery after SCI may also support the development of effective therapeutic strategies.

Both MEP amplitude and latency were shown to have a relationship with the measures of volitional ankle control (VAC), such as tapping tasks, ankle strength, toe clearance during walking, and gait measures. In the studies that assessed VAC during tapping tasks, there was an association between maximal movement velocity and measures of CST transmission (Wirth, Van Hedel, & Curt, 2008a, 2008b, 2008c, 2008d). Two studies demonstrated an association between ankle strength with MEP latency (Wirth et al., 2008b) and MEP amplitude (Wirth et al., 2008c), while two others did not support a relationship between those measures (Labruyère, Zimmerli, & Van Hedel, 2013; Wirth et al., 2008d). The toe clearance during walking, which can be used to assess foot drop/toe drag, was measured in two MEP studies (Barthélemy et al., 2010; Barthélemy et al., 2013), which revealed that maximum toe elevation was associated with MEP amplitude and latency (Fig. 3).

A recent study aimed at establishing changes in corticospinal excitability with absent and partial body weight support; MEPs were reliably recorded at different sessions during stepping in healthy subjects (Knikou, Hajela, & Mummidisetty, 2013). The MEPs recorded from the tibialis anterior (TA) muscle were facilitated at heel contact and throughout the swing phase of the step cycle, while they progressively decrease during the stance phase. Conversely, the soleus MEPs were progressively increased at early-stance and depressed at the stance-to-swing transition as well as throughout the swing phase. These findings indicate that the strength of corticospinal drive will not be affected negatively during stepping under conditions of partial body loading.

Shimizu et al. first report three patients with cervical SCI who showed loss of the CSP after TMS. The absence of CSP was found in both the hand and foot muscles in two patients and in another one only in the foot muscle (Shimizu, Hino, Komori, & Hirai, 2000).

The active motor threshold (AMT) and the duration of CSP were increased in the SCI subjects (Freund, Rothwell, et al., 2011). There was also a negative association, as evaluated by multiple linear regression analysis between spinal cord atrophy and both, AMT and CSP duration. Increased AMT may reflect a reduced density of motoneurons and/or interneurons in low segments of the cervical cord that innervate the extensor digitorum communis muscle. Indeed, AMT differs from RMT in that provides a measure of corticospinal excitability with greater dependence on the spinal segmental level excitability (Hallett, 2000). Alternatively, the reduced density of surviving corticospinal neurons requires a higher intensity of TMS to recruit a given corticospinal output (Di Lazzaro et al., 2004).

The investigation of motor cortex excitability during fatiguing exercise may shed light on the role of exercise therapy in promoting brain reorganization and functional recovery in humans.

Therefore, the mechanisms of exercise-induced cortical plasticity have been evaluated in five patients with cervical iSCI (Nardone et al., 2013). The physiological lengthening of CSP during fatigue was not observed in the SCI patients. This reduced intracortical inhibition, probably secondary to decreased activity of the GABAergic inhibitory interneurons that modulate the corticomotoneuronal output, likely represents a "positive" neuroplastic response in an attempt to compensate for the loss of corticospinal axons.

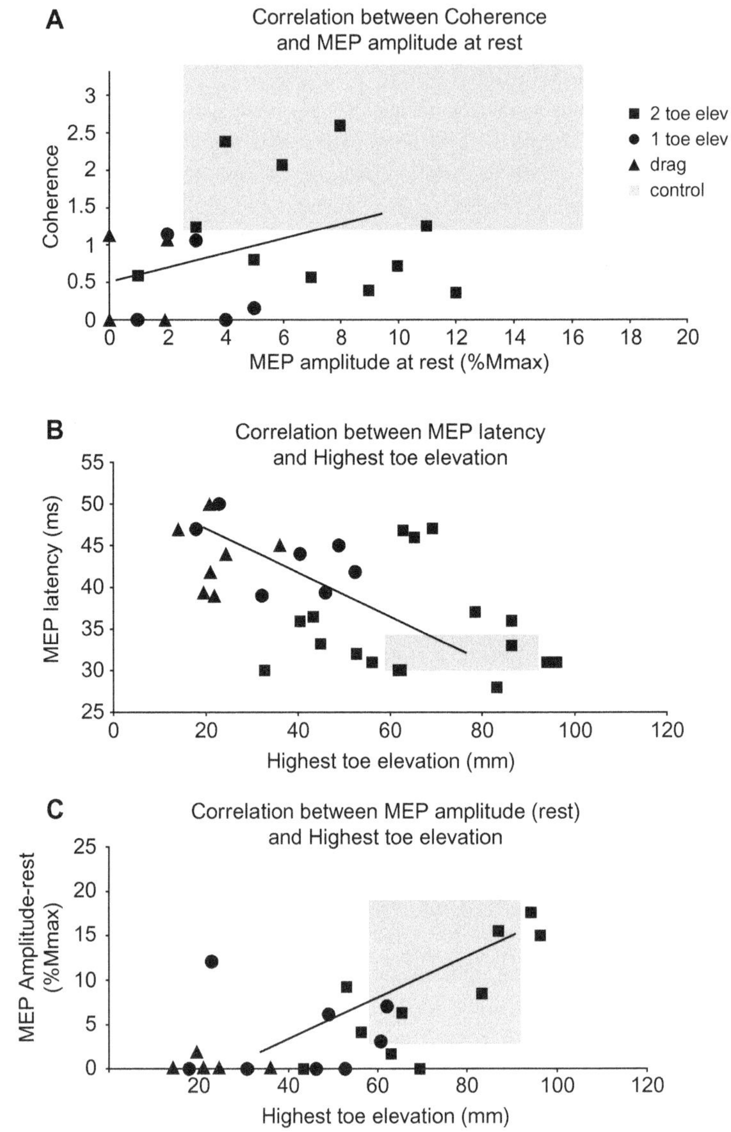

FIG. 3 Correlations between motor evoked potential (MEP), coherence, and kinematic measures. In (A) coherence is plotted against MEP amplitude assessed at rest. Each point represents a participant. In (B) MEP latency is correlated to the highest toe elevation and in (C) MEP amplitude at rest is correlated to the highest toe elevation (both legs of the participants were analyzed). *Reproduced with permission from Barthélemy, D., Willerslev-Olsen, M., Lundell, H., Conway, B.A., Knudsen, H., Biering-Sorensen, F., et al. (2010). Impaired transmission in the corticospinal tract and gait disability in spinal cord injured persons.* Journal of Neurophysiology, 104, *1167–1176.*

Paired-pulse TMS and paired associative stimulation

A paired-pulse TMS study revealed reduced excitability of inhibitory circuits as assessed in a single patient (Laubis-Herrmann, Dichgans, & Bilow, 2000). Two arm muscles distant to the SCI levels (T2–L3), the BB, and the APB muscles were studied in 13 patients. RMT, facilitatory effects on MEP-amplitudes (FE) with voluntary activation, MEP-amplitudes with maximal stimulation (MA), and MEP-RC were measured. Patients exhibited smaller MA from activated BB muscles, a tendency toward smaller FE and smaller RC-slopes. FE, MA, and RC-slopes tended to normalize.

Similarly, another TMS study in a single patient also showed reduced short-interval intracortical inhibition (SICI) (Saturno, Bonato, Miniussi, Di Lazzaro, & Callea, 2008). During posterior tibial nerve stimulation, the authors observed a contextual flexion of hand fingers contralateral to the stimulated lower limb, which suggests abnormal motor cortex excitability. TMS findings were consistent with reduced activity of motor cortex inhibitory circuits since SICI reflects mostly the $GABA_A$-mediated intracortical inhibitory interactions (Paulus et al., 2008). The authors postulated disinhibition of latent synapses within the motor cortex and the rewriting of a new motor cortical map.

Also in the study of Shimizu et al. (2000) paired-pulse TMS study in one SCI patient with pseudoathetotic hands showed reduced inhibition within the motor cortex.

Roy et al. compared SICI during a voluntary contraction in 16 patients with SCI and 14 control subjects, the latter group tested over a larger range of conditioning and test stimulus (CS and TS) intensities to best match the SCI patients data (Roy, Zewdie, & Gorassini, 2011). The main finding was that the average peak SICI in the TA muscle was typically 3–4 times lower in SCI subjects compared to controls. However, when matched for absolute TS intensity, in terms of maximum stimulator output, both U-shaped SICI RC were produced by similar CS intensities.

In another study, the effect of passive and active pedaling exercise on M1 leg area excitability in subjects with traumatic chronic cervical or thoracic SCI has been evaluated (Nardone et al., 2016). A significant effect of pedaling on SICI to paired TMS was found, thus a reduction of SICI, regardless of the experimental condition (active vs passive).

Mi and colleagues recorded long-interval intracortical inhibition (LICI), which reflects long-lasting cortical inhibitory mechanisms mediated by $GABA_B$ receptor transmission (Paulus et al., 2008), from the flexor carpi radialis (FCR) muscle during an isometric contraction equal to 15%–20% of MVC (Mi, Bailey, & Nelson, 2016). LICI was increased in the actively contracted FCR muscle in individuals with SCI compared with age-matched controls. These findings indicate that $GABA_B$ function mediating LICI is different in SCI vs controls. Increased LICI in SCI may be attributed to the medication baclofen or to changes in the neural mechanisms controlling contraction-related modulation of the LICI circuit.

More recently, the MEPs elicited by cortical and subcortical stimulation of corticospinal axons and SICI in the first dorsal interosseous muscle were examined in the preparatory phase of a reaction time task where individuals with chronic cervical iSCI and age-matched controls needed to suppress (NOGO) or initiate (GO) ballistic index finger isometric voluntary contractions (Federico & Perez, 2017) (Fig. 4).

MEPs remained unchanged compared to baseline in subjects with SCI but were suppressed in control subjects. During GO trials, MEPs amplitude increased to a similar extent in both groups but those elicited by subcortical stimulation

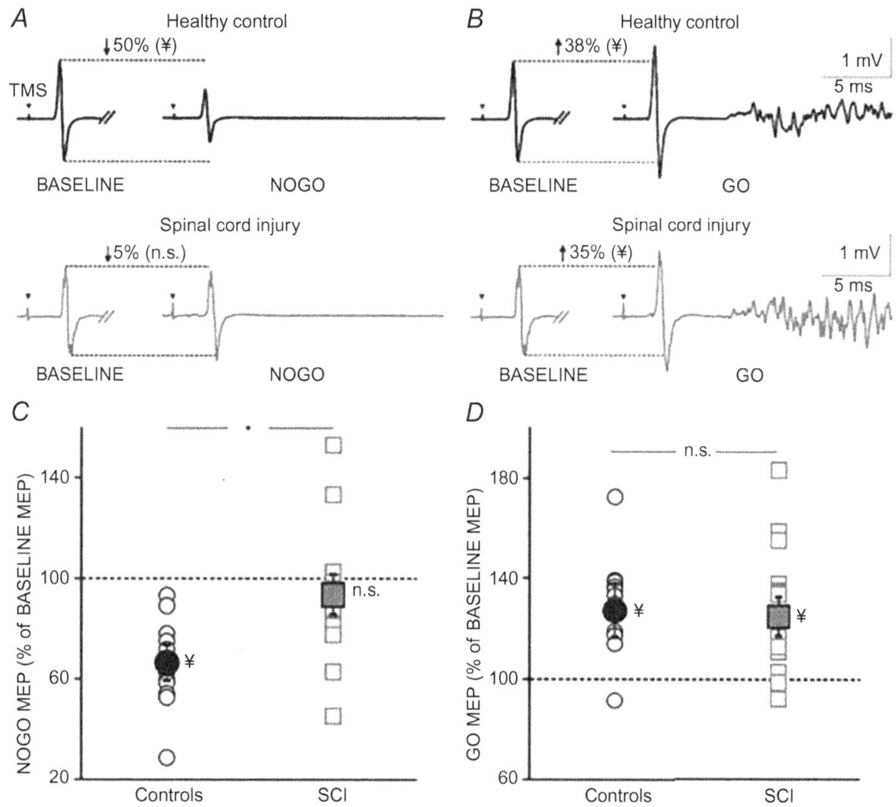

FIG. 4 (A and B) Raw traces of MEPs elicited by TMS in the first dorsal interosseous (FDI) muscle in representative control and SCI subject during NOGO (A) and GO (B) trials. Each waveform represents the average of 20 MEPs at baseline and during GO (control = *black*; SCI = *gray*) and NOGO (control = *black*; SCI = *gray*) trials. Amplitude scales are different for the SCI and control representative subjects to better show changes across the task. (C and D) Group data (controls, $n = 13$, *black circles*; SCI, $n = 18$, *gray squares*) show MEPs during NOGO (C) and GO (D). The *horizontal dotted line* shows the FDI MEP size at baseline. Data from individual subjects are shown in controls (*open circles*) and SCI subjects (*open squares*). Error bars indicate SEM. $^*P < 0.05$, comparison between groups; $^{¥}P < 0.05$, comparison between baseline and GO or NOGO trials. *Reproduced with permission from Federico, P., & Perez, M.A. (2017). Altered corticospinal function during movement preparation in humans with spinal cord injury.* Journal of Physiology, 595, *233–245.*

increased only in control subjects. The magnitude of SICI increased in control subjects but not in SCI patients during NOGO trials and decreased in both groups in GO trials.

Also, the investigation of the corticospinal I-waves, the so-called I-wave facilitation, revealed significant differences between SCI patients with normal and abnormal central motor conduction time (CMCT), and healthy controls (Nardone et al., 2015). SCI patients with normal CMCT showed increased I-wave facilitation, while those with prolonged CMCT showed lower I-wave facilitation compared to a control group.

Since the mechanisms responsible for the production of I-waves are mediated by GABA-related inhibition, these findings point to an increased motor cortical excitability in SCI patients with preserved corticospinal projections.

Another study examined short-latency afferent inhibition (SAI), which is thought to depend on the integrity of circuits linking sensory input and motor output (Sailer et al., 2003), in the FCR muscle in individuals SCI and uninjured controls (Bailey, Mi, & Nelson, 2015). The authors found that SAI was reduced in SCI in both the contracted and non-contracted muscles. MEP RC and RMT were decreased in SCI only in the active state and not the resting state. N20-P25 amplitude was similar between groups in both the resting and active states, even if sensory nerve action potentials were significantly reduced in SCI at rest. Reduced SAI in SCI can be attributed to neuroplasticity altering the intrinsic M1 circuitry mediating SAI and/or reduced afferent input traversing a direct thalamocortical route to M1.

The paired associative stimulation (PAS) protocol, which involves repeated pairs of electrical stimulation of a peripheral nerve (usually the median nerve) followed by TMS applied over the contralateral M1 hand area, can be considered a marker of motor cortical plasticity, and long-term plasticity-like mechanisms are thought to play a major role (Stefan, Kunesch, Benecke, Cohen, & Classen, 2002). PAS has been employed to determine whether the arrival of a corticospinal volley immediately prior to motoneuron discharge may enhance corticospinal transmission and voluntary motor control in patients with iSCI (Ellaway, Vásquez, & Craggs, 2014) (Fig. 5). A short period of paired-pulse stimulation, timed such that a corticospinal volley arrived 1–2 ms before the anti-dromic invasion of action potentials in motoneurons of the flexor digitorum longus muscle, resulted in an MEP facilitation lasting 30 min after PAS treatment.

Interestingly, the PAS protocol significantly enhanced corticospinal excitability in SCI patients with good motor recovery, while it was followed by a non-significant increase of MEP amplitude in the SCI patients with poor functional recovery (Versace et al., 2018).

Discussion

Neuroimaging and electrophysiological studies demonstrated that neuroplastic changes of large cortical networks occur after SCI and play a crucial role in the functional reorganization especially within the sensorimotor area. Besides the above reported TMS studies, increased EEG beta oscillations in the sensorimotor cortex in subjects with SCI may also be regarded as compensatory mechanisms activated to enhance neural plasticity (Simis, Uygur-Kucukseymen, Pacheco-Barrios, Mattistella, & Fregni, 2020).

SCI disrupts the pathways mediating efferent and afferent information flow between the spinal cord and the brain, resulting in dramatic cortical reorganization changes. Brain structures and functions react to missing or more often critically altered motor and sensory signals from peripheral nervous system. Indeed, in most subjects, some spared white matter still connects the caudal with the rostral spinal cord (Bunge, Puckett, Becerra, Marcillo, & Quencer, 1993), and part of surviving neurons remain receptive to synaptic input (Tseng & Prince, 1996; Hains et al., 2003).

A deeper understanding of the complex relationship between anatomical changes following SCI, sensorimotor reorganization, and upper limbs function is helpful for the identification of non-invasive biomarkers, which can quantify the impact of SCI upon the structural integrity, prognosis, and clinical trials for spinal cord repair. In turn, the investigation of the plastic changes which occur within the sensorimotor cortex after SCI could lead to a better understanding of the considerable complexity of the cortical reorganization, and may help to develop experimental and clinical models useful for managing that reorganization.

After SCI the functional recovery is often limited despite intensive efforts to improve the functional outcomes. Identification and better definition of the plasticity mechanisms in the adult human cortex after SCI is an important step toward the development of rationally founded treatment strategies in neurorehabilitation.

In particular, it would be desirable to find strategies for modulating the altered neural plasticity after SCI, not only in the spinal cord but also in higher-order structures, such as the sensorimotor cortex. The cortical reorganization induced by deafferentation may be functionally beneficial or maladaptive. Indeed, these plastic changes may contribute to the occurrence of secondary pathologic conditions, such as neuropathic pain, hyperalgesia, allodynia (Leemhuis, De Gennaro, & Pazzaglia, 2019). Since it is reasonable to expect that upregulation of SCI-induced cortical plasticity using non-invasive

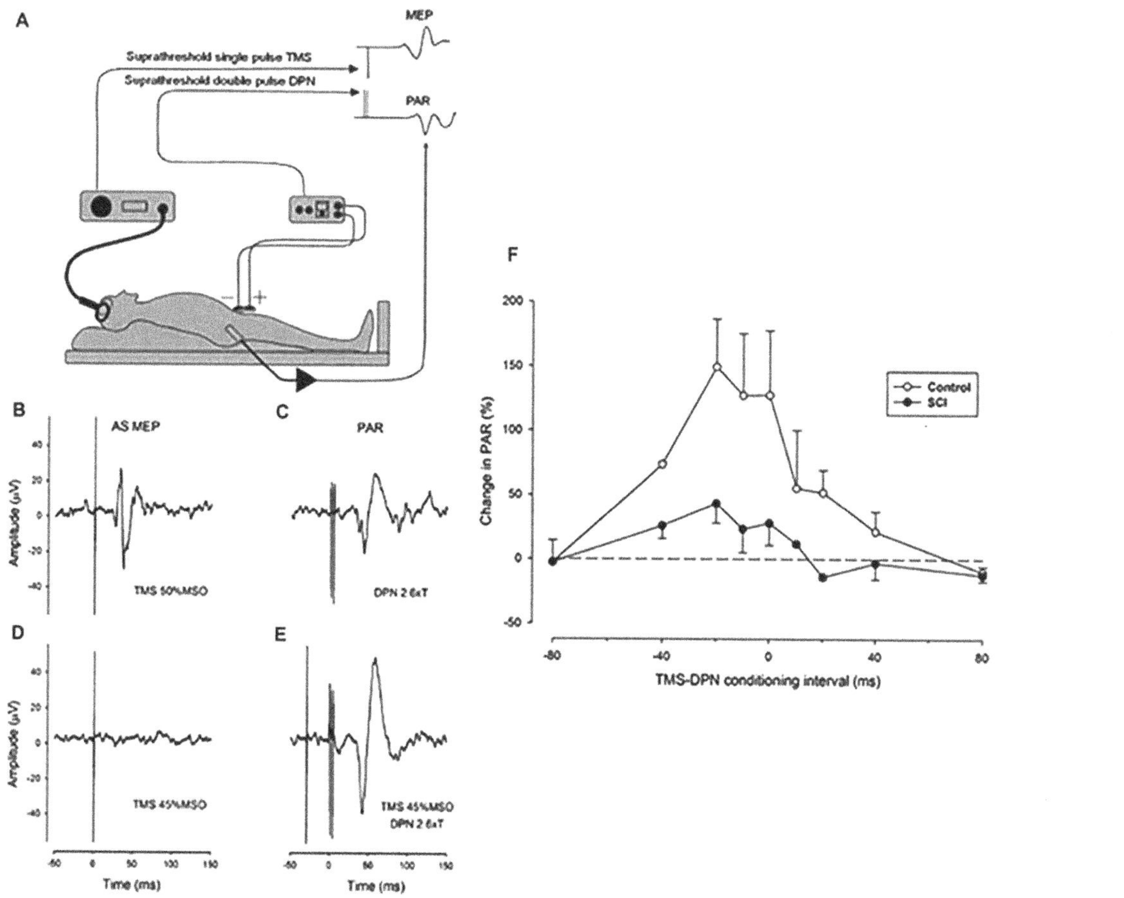

FIG. 5 Facilitation of the pudendal anal reflex (PAR) by single-pulse TMS. (A) Experimental set-up: dorsal penile nerve (DPN) stimulating electrodes, anal sphincter EMG recording electrode, and the TMS double cone coil. (*Insets*) Idealized PAR and anal sphincter MEP responses. (B–E) Anal sphincter EMG averaged ($n = 10$) evoked responses in a control subject. (B) MEP to cortical TMS at 50% of maximal stimulator output (MSO). (C) PAR response to stimulation of the DPN at 2.6 times sensory threshold (18 mA). (D) Lack of response to TMS at 45% MSO. (E) TMS at 45% MSO preceding DPN at 2.6 times sensory threshold by 30 ms. The DPN stimulus in E now evokes an increased PAR that is approximately double the peak-to-peak size of the unconditioned PAR in (C). (F) The degree of PAR facilitation of the PAR by TMS at different conditioning intervals. The graph shows the average percentage increase in the PAR for three control (*open symbols*) and three incomplete SCI (*closed symbols*) subjects. The *dashed horizontal line* indicates a level of zero facilitation. Facilitation is significantly greater for the control group (Wilcoxon signed-rank test $P = 0.008$). *Reproduced with permission from Ellaway, P.H, Vásquez, N., & Craggs, M. (2014). Induction of central nervous system plasticity by repetitive transcranial magnetic stimulation to promote sensorimotor recovery in incomplete spinal cord injury.* Frontiers in Integrative Neuroscience, 8, *42.*

brain stimulations could be an effective treatment method, it would be of interest to learn whether it is possible to purposely modulate the deafferentation-induced cerebral reorganization after SCI.

In fact, TMS can also be a useful tool for the functional assessment of the cortical reorganization following SCI, and rTMS could also play an important role as an intervention in SCI patients to enhance plasticity when used alone or in combination with other interventions to augment current pharmacotherapy and rehabilitation therapies for SCI rehabilitation.

Therefore, neural repair and neuroregeneration are critical goals and issues for rehabilitation in subjects with SCI, which also includes multi-modal neuromodulation techniques.

Magnetic/electrical modulation promotes neuroregeneration and neural repair by affecting signaling in the nervous system; namely, by exciting, inhibiting, or regulating neuronal and neural network activities to improve motor function and motor learning following SCI.

Interventions enhancing local plasticity such as rTMS or tDCS combined with robotic training also lead to an immediate increase in sensorimotor cortex activation, improvement in gait recovery, and a subsequent decrease in high-beta power on EEG.

If delivered repetitively, TMS can influence brain function and induce neuroplastic changes. Indeed, repetitive TMS (rTMS) can modulate cortical excitability, inducing lasting effects (Fitzgerald, Fountain, & Daskalakis, 2006; Lefaucheur et al., 2020). Therefore, rTMS has evolved into a powerful neuroscientific tool allowing to interfere transiently with specific brain functions.

RTMS techniques are known to be able to modulate neural plasticity in the human cerebral cortex. The therapeutic approaches in subjects SCI are usually based on techniques aimed at enhancing cortical excitability, in particular HF rTMS. However, experimental studies have demonstrated that the deafferentation-induced plastic changes following SCI can be up-regulated by motor cortical LF rTMS stimulation.

Overall, examination of different rTMS protocols will provide a better understanding on modulation of deafferentiation-induced brain plasticity after SCI and a useful method in the neurorehabilitation of these patients.

Applications to other areas of neuroscience

In this chapter we have reviewed neuroimaging and, in particular, neurophysiological studies demonstrating that neuroplastic changes of large cortical networks occur after SCI and play a critical role in the functional reorganization, especially within the sensorimotor cortex.

The interconnections between sensorimotor plasticity and functional recovery have also been explored in subjects with other neurological diseases, in particular stroke and traumatic brain injury. Therefore, neural repair and neuroregeneration are critical goals and issues for rehabilitation in patients with various neurological diseases and possibly include multimodal neuromodulation techniques. Furthermore, transcranial magnetic stimulation and other techniques of non-invasive brain stimulation can influence brain function and induce neuroplastic changes in different pathological conditions.

Key facts

- The treatment of spinal cord injury (SCI) is often frustrating and hopeless.
- Experimental studies have demonstrated neuroplastic changes within the sensorimotor cortex after SCI.
- An immediate functional reorganization following complete spinal transection has been observed in the somatosensory cortex of rats.
- A slower cortical activity after SCI can also be detected in electroencephalography recordings.
- Neuroimaging studies suggest that SCI may lead to significant anatomical alterations in the human sensorimotor system.
- Cortical and subcortical mechanisms that underlie functional reorganization of sensory-motor cortex are still incompletely understood.

Mini-dictionary of terms

Transcranial magnetic stimulation: Non-invasive form of brain stimulation using a changing magnetic field.

Repetitive transcranial magnetic stimulation: Treatment that involves magnetic pulses delivered repetitively to target specific areas of the brain.

Sensorimotor cortex: **B**rain area that covers the primary sensory and motor areas of the brain.

Neuroplasticity: The ability of the brain to undergo changes involving cortical remapping.

Spinal cord injury: Damage to any part of the spinal cord that causes temporary or permanent loss of its functions.

Summary points

- Deafferentation due to spinal cord injury (SCI) alters the state of large cortical networks,
- Changes within the sensorimotor cortex play a critical role in the functional reorganization after SCI.
- Transcranial magnetic stimulation (TMS) has shown to be a reliable tool to measure the plastic changes that occur in the human brain.
- Repetitively TMS (repetitive TMS), can be used for modulation of plasticity also after an SCI.
- Understanding the relationship between anatomical changes, sensorimotor reorganization, and upper limbs function is helpful for the identification of non-invasive biomarkers.
- Understanding the cortical reorganization mechanisms may help to develop interventions to maximize rehabilitation strategies.

References

Aguilar, J. R., & Castro-Alamancos, M. A. (2005). Spatiotemporal gating of sensory inputs in thalamus during quiescent and activated states. *Journal of Neuroscience*, *25*, 10990–11002.

Aguilar, J., Humanes-Valera, D., Alonso-Calvino, E., Yague, J. C., Moxon, K. A., Oliviero, A., et al. (2010). Spinal cord injury immediately changes the state of the brain. *Journal of Neuroscience*, *30*, 7528–7537.

Aton, S. J., Seibt, J., Dumoulin, M., Jha, S. K., Steinmetz, N., Coleman, T., et al. (2009). Mechanisms of sleep-dependent consolidation of cortical plasticity. *Neuron*, *61*, 454–466.

Bailey, A. Z., Mi, Y. P., & Nelson, A. J. (2015). Short-latency afferent inhibition in chronic spinal cord injury. *Translational Neuroscience*, *6*, 235–243.

Barthélemy, D., Knudsen, H., Willerslev-Olsen, M., Lundell, H., Nielsen, J. B., & Biering-Sørensen, F. (2013). Functional implications of corticospinal tract impairment on gait after spinal cord injury. *Spinal Cord*, *51*, 852–856.

Barthélemy, D., Willerslev-Olsen, M., Lundell, H., Conway, B. A., Knudsen, H., Biering-Sorensen, F., et al. (2010). Impaired transmission in the corticospinal tract and gait disability in spinal cord injured persons. *Journal of Neurophysiology*, *104*, 1167–1176.

Beaud, E., Schmidlin, T., Wannier, P., Freund, P., Bloch, J., & Mir, A. (2008). Nogo-A antibody treatment does not prevent cell body shrinkage in the motor cortex in adult monkeys subjected to unilateral cervical cord lesion. *BMC Neuroscience*, *9*, 5.

Boord, P., Siddall, P. J., Tran, Y., Herbert, D., Middleton, J., & Craig, A. (2008). Electroencephalographic slowing and reduced reactivity in neuropathic pain following spinal cord injury. *Spinal Cord*, *46*, 118–123.

Bunday, L., Oudega, M., & Perez, M. A. (2013). Aberrant crossed corticospinal facilitation in muscles distant from a spinal cord injury. *PLoS One*, *8*, E76747.

Bunge, R. P., Puckett, W. R., Becerra, J. L., Marcillo, A., & Quencer, R. M. (1993). Observations on the pathology of human spinal cord injury. A review and classification of 22 new cases with details from a case of chronic cord compression with extensive focal demyelination. *Advances in Neurology*, *9*, 75–89.

Davey, N. J., Smith, H. C., Wells, E., Maskill, D. W., Savic, G., Ellaway, P. H., et al. (1998). Responses of thenar muscles to transcranial magnetic stimulation of the motor cortex in patients with incomplete spinal cord injury. *Journal of Neurology, Neurosurgery and Psychiatry*, *65*, 80–87.

Di Lazzaro, V., Oliviero, A., Pilato, F., Saturno, E., Dileone, M., Mazzone, P., et al. (2004). The physiological basis of transcranial motor cortex stimulation in conscious humans. *Clinical Neurophysiology*, *115*, 255–266.

Dias Leao, M. T., Wiesinger, L., Ziemann, U., Tatagiba, M., & Naros, G. (2020). Rapid motor cortical reorganization following subacute spinal cord dysfunction. *Brain Stimulation*, *13*, 783–785.

Dietz, V., & Curt, A. (2006). Neurological aspects of spinal-cord repair: Promises and challenges. *Lancet Neurology*, *5*, 688–694.

Duggal, N., Rabin, D., Bartha, R., Barry, L., Gati, J. S., Kowalczyk, I., et al. (2010). Brain reorganisation in patients with spinal cord compression evaluated using fMRI. *Neurology*, *74*, 1048–1054.

Elbert, T., Pantev, C., Wienbruch, C., Rockstroh, B., & Taub, E. (1995). Increased cortical representation of the fingers of the left hand in string players. *Science*, *270*, 305–307.

Ellaway, P. H., Vásquez, N., & Craggs, M. (2014). Induction of central nervous system plasticity by repetitive transcranial magnetic stimulation to promote sensorimotor recovery in incomplete spinal cord injury. *Frontiers in Integrative Neuroscience*, *8*, 42.

Faggin, B. M., Nguyen, K. T., & Nicolelis, M. A. (1997). Immediate and simultaneous sensory reorganisation at cortical and subcortical levels of the somatosensory system. *Proceedings of the National Academy of Sciences of the United States of America*, *94*, 9428–9433.

Federico, P., & Perez, M. A. (2017). Altered corticospinal function during movement preparation in humans with spinal cord injury. *Journal of Physiology*, *595*, 233–245.

Fields, R. D. (2008). White matter in learning, cognition and psychiatric disorders. *Trends in Neurosciences*, *31*, 361–370.

Fitzgerald, P. B., Fountain, S., & Daskalakis, Z. J. (2006). A comprehensive review of the effects of rTMS on motor cortical excitability and inhibition. *Clinical Neurophysiology*, *117*, 2584–2596.

Freund, P., Rothwell, J., Craggs, M., Thompson, A. J., & Bestmann, S. (2011). Corticomotor representation to a human forearm muscle changes following cervical spinal cord injury. *European Journal of Neuroscience*, *34*, 1839–1846.

Freund, P., Weiskopf, N., Ward, N. S., Hutton, C., Gall, A., Ciccarelli, O., et al. (2011). Disability, atrophy and cortical reorganization following spinal cord injury. *Brain*, *134*, 1610–1622.

Ghosh, A. F., Haiss, F., Sydekum, E., Schneider, R., Gullo, M., Wyss, M. T., et al. (2010). Rewiring of hindlimb corticospinal neurons after spinal cord injury. *Nature Neuroscience*, *13*, 97–104.

Green, J. B., Sora, E., Bialy, Y., Ricamato, A., & Thatcher, R. W. (1998). Cortical sensorimotor reorganization after spinal cord injury: An electroencephalographic study. *Neurology*, *50*, 1115–1121.

Green, J. B., Sora, E., Bialy, Y., Ricamato, A., & Thatcher, R. W. (1999). Cortical motor reorganization after paraplegia: An EEG study. *Neurology*, *53*, 736–743.

Hains, B. C., Black, J. A., & Waxman, S. C. (2003). Primary cortical motor neurons undergo apoptosis after axotomizing spinal cord injury. *The Journal of Comparative Neurology*, *462*, 328–341.

Hallett, M. (2000). Transcranial magnetic stimulation and the human brain. *Nature*, *406*, 147–150.

Hutton, C., Draganski, B., Ashburner, J., & Weiskopf, N. (2009). A comparison between voxel-based cortical thickness and voxel-based morphometry in normal aging. *NeuroImage*, *48*, 371–380.

Jacobs, K. M., & Donoghue, J. P. (1991). Reshaping the cortical motor map by unmasking latent intracortical connections. *Science*, *251*, 944–947.

Jurkiewicz, M. T., Crawley, A. P., Verrier, M. C., Fehlings, M. G., & Mikulis, D. J. (2006). Somatosensory cortical atrophy after spinal cord injury: A voxel-based morphometry study. *Neurology*, *66*, 62–764.

Jurkiewicz, M. T., Mikuli, D. J., McIlroy, W. E., Fehlings, M. G., & Verrier, M. C. (2007). Sensorimotor cortical plasticity during recovery following spinal cord injury: A longitudinal fMRI study. *Neurorehabilitation and Neural Repair*, *21*, 527–538.

Jurkiewicz, M. T., Mikuli, D. J., McIlroy, W. E., Fehlings, M. G., & Verrier, M. C. (2010). Sensorimotor cortical activation in patients with cervical spinal cord injury with persisting paralysis. *Neurorehabilitation and Neural Repair*, *24*, 136–140.

Knikou, M., Hajela, N., & Mummidisetty, C. K. (2013). Corticospinal excitability during walking in humans with absent and partial body weight support. *Clinical Neurophysiology*, *124*, 2431–2438.

Kokotilo, K. J., Eng, J. J., & Curt, A. (2009). Reorganization and preservation of motor control of the brain in spinal cord injury: A systematic review. *Journal of Neurotrauma*, *26*, 2113–2126.

Labruyère, R., Zimmerli, M., & Van Hedel, H. J. (2013). Slowed down: Response time deficits in well-recovered subjects with incomplete spinal cord injury. *Archives of Physical Medicine and Rehabilitation*, *94*, 2020–2026.

Laubis-Herrmann, U., Dichgans, J., & Bilow, H. (2000). Motor reorganization after spinal cord injury: Evidence of adaptive changes in remote muscles. *Restorative Neurology and Neuroscience*, *17*, 75–181.

Leemhuis, E., De Gennaro, L., & Pazzaglia, A. M. (2019). Disconnected body representation: Neuroplasticity following spinal cord injury. *Journal of Clinical Medicine*, *8*, 2144.

Lefaucheur, J. P., Aleman, A., Baeken, C., Benninger, D. H., Brunelin, J., Di Lazzaro, V., et al. (2020). Evidence-based guidelines on the therapeutic use of repetitive transcranial magnetic stimulation (rTMS): An update (2014-2018). *Clinical Neurophysiology*, *131*, 474–528.

Levy, W. J., Amassian, V. E., & Traad, M. (1990). Focal magnetic coil stimulation reveals motor cortical system reorganized in humans after traumatic quadriplegia. *Brain Research*, *510*, 130–134.

Lotze, M., Laubis-Herrmann, U., & Topka, H. (2006). Combination of TMS and fMRI reveals a specific pattern of reorganization in M1 in patients after complete spinal cord injury. *Restorative Neurology and Neuroscience*, *24*, 97–107.

Lundell, H., Barthelemy, D., Skimminge, A., Dyrby, T. B., Biering-Sørensen, F., & Nielsen, J. B. (2011). Independent spinal cord atrophy measures correlate to motor and sensory deficits in individuals with spinal cord injury. *Spinal Cord*, *49*, 70–75.

Lundell, H., Christensen, M. S., Barthelemy, D., Willerslev-Olsen, M., Biering-Sørensen, F., & Nielsen, J. B. (2011). Cerebral activation is correlated to regional atrophy of the spinal cord and functional motor disability in spinal cord injured individuals. *NeuroImage*, *54*, 254–261.

Mi, Y. P., Bailey, A. Z., & Nelson, A. J. (2016). Short- and long-intracortical inhibition in incomplete spinal cord injury. *Canadian Journal of Neurological Sciences*, *43*, 183–191.

Nardone, R., Höller, Y., Bathke, A. C., Orioli, A., Schwenker, K., Frey, V., et al. (2015). Spinal cord injury affects I-wave facilitation in human motor cortex. *Brain Research Bulletin*, *116*, 93–97.

Nardone, R., Höller, Y., Brigo, F., Höller, P., Christova, M., Tezzon, F., et al. (2013). Fatigue-induced motor cortex excitability changes in subjects with spinal cord injury. *Brain Research Bulletin*, *99*, 9–12.

Nardone, R., Höller, Y., Sebastianelli, L., Versace, V., Saltuari, L., Brigo, F., et al. (2018). Cortical morphometric changes after spinal cord injury. *Brain Research Bulletin*, *137*, 107–119.

Nardone, R., Höller, Y., Thomschewski, A., Bathke, A. C., Ellis, A. R., Golaszewski, S. M., et al. (2015). Assessment of corticospinal excitability after traumatic spinal cord injury using MEP recruitment curves: A preliminary TMS study. *Spinal Cord*, *53*, 534–538.

Nardone, R., Langthaler, P. B., Bathke, A. C., Höller, Y., Brigo, F., Lochner, P., et al. (2016). Effects of passive pedaling exercise on the intracortical inhibition in subjects with spinal cord injury. *Brain Research Bulletin*, *124*, 144–149.

Nicolelis, M. A., Lin, R. C., Woodward, D. J., & Chapin, J. K. (1993). Induction of immediate spatiotemporal changes in thalamic networks by peripheral block of ascending cutaneous information. *Nature*, *361*, 533–536.

Nudo, R. J., Milliken, G. W., Jenkins, W. M., & Milliken, G. W. (1996). Use-dependent alterations of movement representations in primary motor cortex of adult squirrel monkeys. *Journal of Neuroscience*, *16*, 785–807.

Paulus, W., Classen, J., Cohen, L. G., Large, C. H., Di Lazzaro, V., Nitsche, M., et al. (2008). State of the art: Pharmacologic effects on cortical excitability measures tested by transcranial magnetic stimulation. *Brain Stimulation*, *1*, 151–163.

Roy, F. D., Zewdie, E. T., & Gorassini, M. A. (2011). Short-interval intracortical inhibition with incomplete spinal cord injury. *Clinical Neurophysiology*, *122*, 1387–1395.

Sailer, A., Molnar, G. F., Paradiso, G., Gunraj, C. A., Lang, A. E., & Chen, R. (2003). Short and long latency afferent inhibition in Parkinson's disease. *Brain*, *26*, 1883–1894.

Saturno, E., Bonato, C., Miniussi, C., Di Lazzaro, V., & Callea, L. (2008). Motor cortex changes in spinal cord injury: A TMS study. *Neurological Research*, *30*, 1084–1085.

Shimizu, T., Hino, T., Komori, T., & Hirai, S. (2000). Loss of the muscle silent period evoked by transcranial magnetic stimulation of the motor cortex in patients with cervical cord lesions. *Neuroscience Letters*, *286*, 199–202.

Simis, M., Uygur-Kucukseymen, E., Pacheco-Barrios, K., Mattistella, L. R., & Fregni, F. (2020). Beta-band oscillations as a biomarker of gait recovery in spinal cord injury patients: A quantitative electroencephalography analysis. *Clinical Neurophysiology*, *131*, 1806–1814.

Smith, H. C., Davey, N. J., Savic, G., Maskill, D. W., Ellaway, P. H., Jamous, M. A., et al. (2000). Modulation of single motor unit discharges using magnetic stimulation of the motor cortex in incomplete spinal cord injury. *Journal of Neurology, Neurosurgery, and Psychiatry*, *68*, 516–520.

Smith, H. C., Savic, G., Frankel, H. L., Ellaway, P. H., Maskill, D. W., Jamous, M. A., et al. (2000). Corticospinal function studied over time following incomplete spinal cord injury. *Spinal Cord*, *38*, 292–300.

Stefan, K., Kunesch, E., Benecke, R., Cohen, L. G., & Classen, J. (2002). Mechanisms of enhancement of human motor cortex excitability induced by interventional paired associative stimulation. *Journal of Physiology*, *543*, 699–708.

Streletz, L. J., Belevich, J. K., Jones, S. M., Bhushan, A., Shah, S. H., & Herbison, G. J. (1995). Transcranial magnetic stimulation: Cortical motor maps in acute spinal cord injury. *Brain Topography*, *7*, 245–250.

Topka, H., Cohen, L. G., Cole, R. A., & Hallett, M. (1991). Reorganization of corticospinal pathways following spinal cord injury. *Neurology*, *41*, 1276–1283.

Tran, Y., Boord, P., Middleton, J., & Craig, A. (2004). Levels of brain wave activity (8–13 Hz) in persons with spinal cord injury. *Spinal Cord*, *42*, 73–79.

Tseng, G. F., & Prince, D. A. (1996). Structural and functional alterations in rat corticospinal neurons after axotomy. *Journal of Neurophysiology*, *75*, 248–267.

Versace, V., Langthaler, P. B., Höller, Y., Frey, V. N., Brigo, F., Sebastianelli, L., et al. (2018). Abnormal cortical neuroplasticity induced by paired associative stimulation after traumatic spinal cord injury: A preliminary study. *Neuroscience Letters*, *664*, 167–171.

Wirth, B., Van Hedel, H. J. A., & Curt, A. (2008a). Ankle dexterity remains intact in patients with incomplete spinal cord injury in contrast to stroke patients. *Experimental Brain Research*, *191*, 353–361.

Wirth, B., Van Hedel, H. J. A., & Curt, A. (2008b). Ankle paresis in incomplete spinal cord injury: Relation to corticospinal conductivity and ambulatory capacity. *Journal of Clinical Neurophysiology*, *25*, 210–217.

Wirth, B., Van Hedel, H. J. A., & Curt, A. (2008c). Changes in corticospinal function and ankle motor control during recovery from incomplete spinal cord injury. *Journal of Neurotrauma*, *25*, 467–478.

Wirth, B., Van Hedel, H. J. A., & Curt, A. (2008d). Ankle dexterity is less impaired than muscle strength in incomplete spinal cord lesion. *Journal of Neurology*, *255*, 273–279.

Wrigley, P. J., Gustin, S. M., Macey, P. M., Nash, P. G., Gandevia, S. C., Macefiled, V. G., et al. (2009). Anatomical changes in human motor cortex and motor pathways following complete thoracic spinal cord injury. *Cerebral Cortex*, *19*, 224–232.

Ziemann, U., Paulus, W., Nitsche, M. A., Pascual-Leone, A., Byblow, W. D., Berardelli, A., et al. (2008). Consensus: Motor cortex plasticity protocols. *Brain Stimulation*, *1*, 164–182.

Chapter 26

Bone loss at the knee after spinal cord injury: Radiographic imaging, fracture risk, and treatment

Ifaz T. Haider[a], Narina Simonian[b,c], Thomas J. Schnitzer[b], and W. Brent Edwards[a]

[a]*Human Performance Lab and McCaig Institute for Bone and Joint Health, University of Calgary, Calgary, AB, Canada,* [b]*Department of Physical Medicine and Rehabilitation, Feinberg School of Medicine, Northwestern University Feinberg School of Medicine, Chicago, IL, United States,* [c]*Northwestern University Clinical and Translational Sciences Institute, Northwestern University Feinberg School of Medicine, Chicago, IL, United States*

Abbreviations

aBMD areal bone mineral density
BMD bone mineral density
BMC bone mineral content
DXA dual-energy X-ray absorptiometry
QCT quantitative computed tomography
pQCT peripheral quantitative computed tomography
SCI spinal cord injury

Introduction

Severe bone loss at sub-lesional locations is a recognized sequela of spinal cord injury (SCI). Changes are often attributed to the loss of mechanical stimuli (disuse) following injury; however, the rate of bone loss after SCI is typically greater than other disuse scenarios, such as spaceflight (Edwards & Schnitzer, 2015), suggesting that bone loss is likely exacerbated by the unique neurological and hormonal environment resulting from SCI (Dolbow et al., 2011). The clinical consequence of bone loss after SCI is an increased risk of fracture, with studies reporting prevalence in up to 34% of individuals with SCI (Lala et al., 2014; Lazo et al., 2001). Bone loss is particularly severe around skeletal regions of the knee, where fractures commonly occur during routine activities with little-to-no external trauma (Lala et al., 2014). While many individuals with SCI are already non-ambulatory prior to fracture, these injuries are still associated with high rates of morbidity and mortality (Carbone et al., 2014).

Medical imaging provides powerful, non-invasive tools to quantify and understand bone loss after SCI. Indeed, a large number of studies, across different modalities, have been published. Both 2D areal (Zehnder et al., 2004) and 3D volumetric (Eser et al., 2004) modalities have been used to characterize bone loss at the knee after SCI. These studies reported losses in bone minerals up to 88% (Haider, Lobos, Simonian, Schnitzer, & Edwards, 2018), depending on the region and site measured. Cross-sectional studies have also demonstrated a quantitative link between the amount of bone loss and fracture risk (Lala et al., 2014; Lazo et al., 2001); however, options for treatment are less clear and there is no standard of care for managing bone loss after SCI. Pharmacological (Oleson, Marino, Formal, Modlesky, & Leiby, 2020) and non-pharmacological (Biering-Sørensen, Hansen, & Lee, 2009) interventions have been explored in the recent past, but treatment efficacy at the knee is not well established.

The remainder of this chapter will further expand on these issues related to bone loss and fracture risk at the knee following SCI. We will first provide a more detailed overview of medical imaging studies characterizing bone loss after SCI. This will be followed by a brief discussion of studies that have helped quantify the link between measured bone loss and clinical fracture risk. Finally, we will conclude with an overview of interventions for the attenuation/recovery of bone loss following SCI.

Cellular, Molecular, Physiological, and Behavioral Aspects of Spinal Cord Injury. https://doi.org/10.1016/B978-0-12-822427-4.00026-5

Application to other areas of neuroscience

In this chapter, we review the phenomenon of bone loss after SCI and the clinical sequelae of skeletal fracture; however, the underlying mechanisms of bone loss after SCI are uncertain. Changes to the neurological environment, i.e., interruption of sensory and sympathetic nerve fibers in the lower limbs, have been implicated (Dolbow et al., 2011) but these pathways are not yet fully understood. This could be an important avenue for future research in neuroscience. Indeed, this work could help to better understand the multi-factorial nature of bone homeostasis and inform the development of new treatments for osteoporosis, perhaps even outside the context of SCI.

Other neurological conditions are also known to affect bone. For example, both stroke and cerebral palsy are associated with bone loss and elevated fracture risk, though there are important differences that distinguish bone loss from these conditions compared to SCI. The magnitude of bone loss is typically less severe than SCI, and there is greater involvement of sites proximal to the knee (hip, spine, ribs, and wrist); these suggest important differences in the underlying mechanisms of bone loss and fracture. However, the bone assessment techniques summarized here, i.e., different medical imaging modalities and subject-specific finite element (FE) modeling, are versatile and could be applied to sites most relevant for these diseases.

Bone loss at the knee after SCI

2D DXA imaging

DXA imaging allows for quantification of areal bone mineral density (aBMD) from 2D images of bone. As the name implies, two X-Ray energies are used, which allow this modality to isolate contributions from bone separate from soft tissue. DXA is the gold standard for the characterization of osteoporosis in non-disabled individuals, where scans are typically taken at the hip or spine; however, this modality can also be used to characterize bone loss at the knee. In the context of SCI, DXA measurements at the knee have been used in both cross-sectional and longitudinal studies (Table 1).

One of the earliest cross-sectional studies to examine bone loss at the knee in individuals with SCI reported that tibial BMC was, on average, only 50% of the value of age-matched controls (Biering-Sorensen et al., 1988), and this finding has been well supported by more recent studies (Garland et al., 1992, 2001; McPherson et al., 2014; Modlesky et al., 2004; Shields et al., 2005). Of course, cross-sectional studies comparing individuals with SCI to non-disabled controls at one time-point do not elucidate the time-course of bone loss; however, this phenomenon was explored by Zehnder, Risi, et al. (2004), who examined a large cross-section of 100 males with SCI 3–30 years post-injury, and identified a statistically significant relationship between time-since injury and decreasing aBMD z-score at the tibial diaphysis and epiphysis. Warden et al. (2002) longitudinally examined a group of 15 individuals over a 6-week period after SCI (30–180 days post-injury) and reported losses in tibial aBMD of −5.3% per week. A pair of studies over a longer duration (>1.5 years) have reported that bone loss is rapid soon after injury but eventually reaches a plateau, with steady-state measures that are 27–50% of the original baseline (Biering-Sørensen et al., 1990; Garland et al., 2004)—a finding that is consistent with cross-sectional literature.

Volumetric assessment of bone mineral and bone strength

DXA is a powerful clinical tool, but can only capture 2D areal projections of the bone, and potentially meaningful information may be lost. Bone is a complex three-dimensional structure, and cortical and trabecular compartments at different regions may have differing contributions to bone strength. As shown in Fig. 1, these regions can be assessed separately using quantitative computed tomography (QCT) or peripheral quantitative computed tomography (pQCT). In this section, we provide an overview of cross-sectional and longitudinal studies that used these volumetric imaging modalities in individuals with SCI (Table 2).

A number of large (≥60 individuals) cross-sectional studies used regional QCT and pQCT imaging of the knee in individuals with varying injury durations (2 months to 50 years post-SCI) (Edwards et al., 2015; Eser et al., 2004; Haider et al., 2018). From these cross-sectional data, the authors were able to infer the temporal and spatial patterns of bone loss after SCI. Consistent with the DXA studies outlined above, these studies demonstrated that bone loss was most rapid immediately after injury and reached a plateau 2.6–7.6 years after injury, depending on the measure. After plateau, losses were most significant at the epiphysis, up to 88% relative to recently injured SCI controls (Haider et al., 2018); however, the magnitude of bone loss decreased progressively at more diaphyseal locations. Finally, these studies demonstrated that cortical bone is lost primarily via endocortical resorption, with little change to the periosteal surface or to cortical density over time. Smaller investigations also reported results consistent with these findings (Coupaud et al., 2009; Dudley-Javoroski & Shields, 2010; McCarthy et al., 2012; Rittweger et al., 2010). Longitudinal studies further elucidated the

TABLE 1 Summary of radiographic studies in individuals with spinal cord injury that include dual-energy X-ray absorptiometry measures of the knee.

Study design	Author (year)	Study sample	Knee region(s) of interest	Key findings at knee
Cross-sectional	Biering-Sorensen, Bohr, and Schaadt (1988)	26 SCI (duration = 6–25 years) vs ND reference group from literature	PT	Tibia BMC after SCI approximately 50% of values from ND reference group
	Garland, Adkins, Stewart, Ashford, and Vigil (2001)	31 SCI (duration = 5.7 ± 2.3 years) vs 17 ND controls	Whole knee	38%–47% lower knee aBMD in SCI compared to controls
	Garland, Adkins, and Stewart (2005)	18 SCI (duration = 3–43 years) vs 10 ND controls	DF and PT	aBMD at knee in SCI < 50% of controls
	Lala et al. (2014)	19 SCI with fracture vs 51 SCI with no fracture history (duration >2 years)	DF and PT	Lower aBMD associated with an elevated risk of fracture
	McPherson et al. (2014)	12 acute SCI (duration = 2.1 ± 0.7 months) vs 34 chronic SCI (duration = 196.9 ± 111.4 months)	Epiphysis and metaphysis of DF; Epiphysis of PT	50% lower aBMD in all three regions, in chronic SCI vs acute SCI
	Modlesky, Majumdar, Narasimhan, and Dudley (2004)	10 SCI (duration >2 years) vs 8 ND controls	PT	Tibia aBMD 43% lower in SCI; included HR-MRI, which identified deterioration of trabecular microstructure in SCI vs controls
	Shields et al. (2005)	11 SCI (duration >2 years) vs 11 ND controls	DF and PT	aBMD at both sites approx. 25% lower in SCI compared to controls
	Zehnder, Risi, et al. (2004)	100 SCI (duration 0.25–30 years)	PT epiphysis and diaphysis	Significant association between decreasing aBMD (*Z*-score) and time
Longitudinal	Biering-Sørensen, Bohr, and Schaadt (1990)	8 SCI assessed 1–54 months post-injury	PT	BMC at steady-state was 40%–50% of baseline. Steady-state reached approx. 2 years after injury
	Garland et al. (2004)	6 SCI assessed up to 523 days post-injury	DF and PT	aBMD loss of 27% and 32% at DF and PT, respectively
	Warden et al. (2002)	15 SCI assessed 30–180 days post-injury	PT	Progressive losses in aBMD over duration; average loss of 5.3% after 6 weeks
Mixed	Garland et al. (1992)	25 recent SCI (duration = 114 ± 8.6 days, reassessed at 16 months); 20 chronic SCI (duration >5 years); 10 ND controls	DF and PT	38% and 36% lower aBMD in chronic SCI vs controls, at DF and PT respectively. No difference between recent SCI at 16 months vs chronic SCI group

SCI, spinal cord injury; ND, non-disabled; DF, distal femur; PT, proximal tibia; aBMD, areal bone mineral density; BMC, bone mineral content; HR-MRI, high-resolution magnetic resonance imaging.
This table summarizes the literature where dual-energy absorptiometry was used to assess the knee in individuals with SCI. An overview of the study sample (including time since injury), the imaged regions of interest, and the key findings are provided.

time-course of bone loss after SCI. Studies within the first year of SCI (Coupaud et al., 2015; Edwards et al., 2013a; Frey-Rindova et al., 2000) demonstrate losses in bone mineral at the knee of up to 20%, relative to baseline, while longer studies (de Bruin et al., 2000; Dudley-Javoroski & Shields, 2012; Frotzler, Berger, Knecht, & Eser, 2008) have further demonstrated that bone loss is most rapid immediately after injury and reaches a plateau after approximately 2–8 years post-SCI.

A few studies have reported the changes to bone microarchitecture concomitant with bone loss. Using high-resolution magnetic resonance imaging, Modlesky et al. (2004) reported that individuals with SCI (>2 years post-SCI) demonstrated

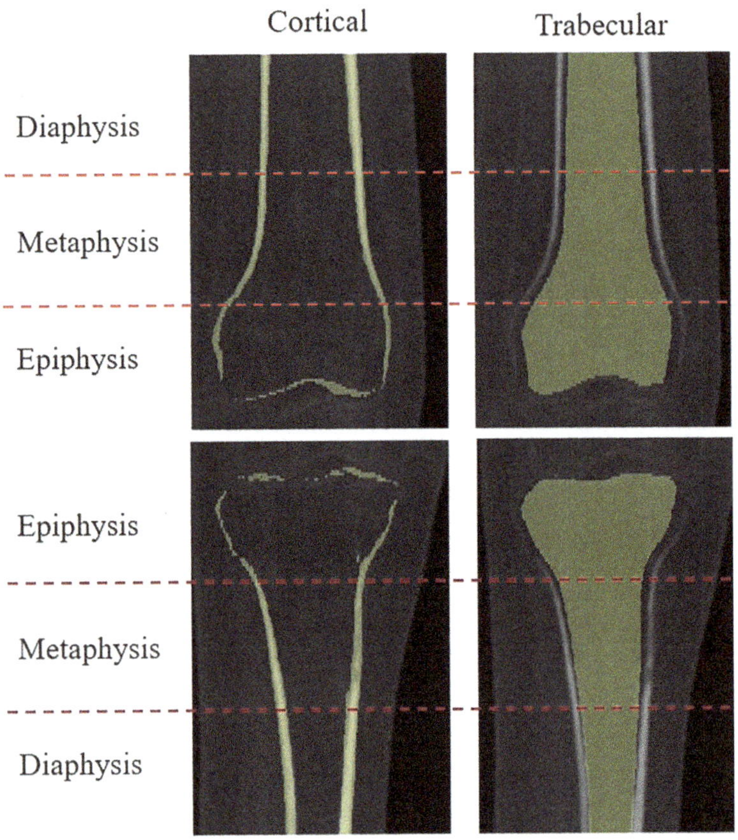

FIG. 1 Example of regional volumetric quantitative computed tomography analysis. This is an image of a frontal-plane slice from a three-dimensional computed tomography image of the knee. The distal femur (TOP) and proximal tibia (BOTTOM) were separated into epiphyseal, metaphysical, and diaphyseal regions. Cortical (LEFT) and trabecular (RIGHT) compartments were analyzed separately; this technique was used in a number of recent studies (Edwards, Schnitzer, & Troy, 2013a; Edwards, Simonian, Troy, & Schnitzer, 2015; Haider et al., 2018).

lower trabecular volume, lower trabecular number, and greater trabecular spacing (20%–44% reductions). A follow-up study by the same group (Slade et al., 2005) compared pre- and post-menopausal women with SCI against pre- and post-menopausal non-disabled women. Again, SCI was associated with 19%–26% lower trabecular number and 22%–33% reduced trabecular bone volume.

Changes in bone strength that occur as a result of bone loss after SCI have been estimated using finite element (FE) modeling, which are physics-based simulations derived from QCT imaging, and established relationships between bone density and local material properties. Tan et al. (2014) compared QCT and FE modeling data from individuals 0.12 to 37.5 years after SCI and confirmed a relationship between low bone mineral and low FE predicted stiffness. Another investigation (Edwards et al., 2013a, 2013b) looked at individuals <7 months post-SCI and used FE modeling to predict decreases in compressive stiffness of 1.3%–5.5%/month and 0.6%–5.9%/month across regions of the femur and tibia, respectively. Torsional stiffness also decreased by 1.3%–5.0%/month and 0.7%–5.2% at the femur and tibia, respectively. Moreover, losses in FE predicted stiffness and strength indices were approximately twice as large as losses in integral bone mineral, due to the highly non-linear relationship between bone mineral measures and the mechanical behavior of bone (Morgan, Bayraktar, & Keaveny, 2003). Furthermore, cross-sectional follow-up studies (Edwards et al., 2015; Haider et al., 2018) suggested that losses in stiffness accumulate to a total reduction of 40%–85% when bone loss plateaus. Torsional stiffness at the epiphysis of both the femur and tibia plateaued quickly, approximately 2 years post-SCI, suggesting that mechanically relevant changes occur rapidly. Again, losses in stiffness were greater than losses in bone mineral; at the tibial epiphysis, individuals with SCI had 33% of the integral BMC relative to the recently injured baseline, but only 16% of the torsional stiffness.

TABLE 2 Summary of radiographic studies that include volumetric assessment of the knee after spinal cord injury.

Study design	Author, year	Study sample	Modality	Knee region(s) of interest	Key findings at knee
Cross-sectional	Edwards et al. (2015)	60 SCI (duration = 0–50 years) vs 10 ND controls	QCT and FE	PT separated into three regions	Measures plateaued 2.1–2.7 years post-SCI, after which BMC, BMD, stiffness, and strength were 55%–69% lower than the reference group
	Coupaud, McLean, and Allan (2009)	47 SCI (duration = 0–42 years)	pQCT	DF and throughout length of tibia	Exponential decrease in BMD over time at both bones
	Eser et al. (2004)	89 SCI (duration = 0.2–50 years) vs 21 ND controls	pQCT	DF and PT; epiphysis and diaphysis	Total BMC lower by up to 60% at epiphysis and 35% at diaphysis; losses plateau after 3–8 years
	Haider et al. (2018)	101 SCI (duration = 0.2–50 years)	QCT and FE	DF and PT separated into three regions	Measures plateaued ≤3.6 years. After plateau, up to 88% lower stiffness and mineral measures compared to recently injured
	McCarthy, Bloomer, Gall, Keen, and Ferguson-Pell (2012)	8 acute SCI (duration <8 months); 9 chronic SCI (duration >48 months); 14 ND controls	pQCT	PT epiphysis and diaphysis + distal tibia epiphysis	Chronic SCI had significantly lower trabecular BMD at epiphyses and lower cross-section moment of inertia at PT diaphysis
	Rittweger, Goosey-Tolfrey, Cointry, and Ferretti (2010)	9 SCI (duration = 9–32 years) vs 9 ND controls	pQCT	Tibia, both ends + midshaft	47%–51% lower BMC at epiphysis. Differences more pronounced at epiphyses vs midshaft
	Tan et al. (2014)	27 SCI (duration = 0.1–38 years)	QCT and FE	DF	Reduced FE stiffness associated with BMD and years post-injury
	Modlesky et al. (2004)	10 SCI (duration = 2–20 years) vs 8 ND controls	HR-MRI	DF and PT	20%–27% lower bone volume and trabecular number, 33%–44% greater trabecular spacing
	Slade, Bickel, Modlesky, Majumdar, and Dudley (2005)	20 pre- and post-menopausal women with SCI vs 17 pre- and post-menopausal ND controls	HR-MRI	DF and PT	SCI had a greater effect than estrogen loss. SCI associated with 19%–26% fewer trabeculae and 22%–33% reduced trabecular volume
Longitudinal	de Bruin, Dietz, Dambacher, and Stüssi (2000)	12 SCI evaluated at 5 weeks and 2 years post-injury	pQCT	Tibia ankle and shaft	35% and 13% decrease in trabecular and cortical BMD
	Coupaud, McLean, Purcell, Fraser, and Allan (2015)	26 SCI (duration <5 weeks); baseline and 1-year follow-up	pQCT	DF and PT, shaft of both bones	BMC and BMD losses of 15%–20% at DF and PT, lower losses at more diaphyseal locations
	Edwards et al. (2013a); Edwards,	13 SCI (duration <4 months) follow-up after 3.5 months	QCT and FE	DF and PT, separated	Integral BMC losses of 0.4%–3.6%/month and BMD losses of 0.4%–3.4%/month at DF and PT.

Continued

TABLE 2 Summary of radiographic studies that include volumetric assessment of the knee after spinal cord injury—cont'd

Study design	Author, year	Study sample	Modality	Knee region(s) of interest	Key findings at knee
	Schnitzer, and Troy (2013b)			into three regions	FE predicted concomitant decreases in torsional and compressive stiffness
	Frey-Rindova, De Bruin, Stüssi, Dambacher, and Dietz (2000)	29 SCI evaluated at baseline, 6, 12, 24 months	pQCT	PT	Trabecular BMD reduced 5% and 15%, 6 and 12 months after injury; cortical BMD decreased 7% after 12 months after injury
	Frotzler et al. (2008)	39 SCI (duration = 0.9–34 years) follow-up at 15 and 30 months after baseline	pQCT	DF, PT, and midshaft of both bones	Differences over time were not significant; likely due to few individuals with injury duration <5 years
Mixed	Dudley-Javoroski and Shields (2012)	26 SCI scanned 0–5 years after SCI; 21 ND controls	pQCT	PT and tibia midshaft	BMD loss of 15%–35% in first year. Bone loss plateaued after 2–4 years, reaching BMD approx. 50% lower than controls
	Dudley-Javoroski and Shields (2010)	15 SCI received 1–3 scans 0.2–7 years post-injury; 10 ND controls	pQCT	DF metaphysis	Femur BMD losses 1.7%/month in first 2 years; bone loss appeared to plateau after 2 years. At end of study, BMD 40% lower than controls

SCI, spinal cord injury; ND, non-disabled; PT, proximal tibia; DF, distal femur; FE, finite element; (p)QCT, (peripheral) quantitative computed tomography; BMC, bone mineral content; BMD, bone mineral density; HR-MRI, high-resolution magnetic resonance imaging.
This table summarizes literature where volumetric imaging of the knee was used to assess bone minerals after spinal cord injury. The study sample, imaging modality, imaging region of interest, and key findings are reported.

Summary of bone loss after SCI—Temporal and spatial patterns

Radiographic studies of bone loss at the knee after SCI have demonstrated consistent results, and the temporal and spatial pattern of bone loss is clear. Losses occur most rapidly soon after injury, reaching a plateau 2–8 years after injury. Regionally, losses are most significant at the epiphysis of the distal femur and proximal tibia and become more attenuated at more diaphyseal sites; however, the relationship between bone mineral and bone mechanical properties is non-linear, and reductions in stiffness/strength may be twice as large as reductions in integral bone mineral measures.

Relationship between bone loss and fracture risk after SCI

In non-disabled individuals, a relationship between bone mineral measures and fracture risk is well established, and tools such as FRAX (Kanis et al., 2011) have been developed to estimate fracture risk based on BMD and other clinical risk factors. A similar tool does not exist for individuals with SCI; but a number of studies have quantified the relationship between bone loss and fracture risk in this context.

Lala et al. (2014), retrospectively assessed 70 individuals with SCI (19 fractured). After adjusting for motor completeness, logistic regression revealed that aBMD and pQCT measures at the knee were associated with a normalized odds ratio of 2.2–6.5 (the change in odds of fracture for a 1 standard deviation change in measure), with integral aBMD and trabecular BMD of the proximal tibia reported as the most sensitive measures. Another study (Lazo et al., 2001) identified a similar relationship between femoral neck aBMD and fracture risk, with a 2.2 and 2.8 increased odds of fracture per standard deviation decrease in *t*-score or *z*-score, respectively. Other radiographic studies have also reported statically lower aBMD in fractured vs non-fractured individuals (Biering-Sorensen et al., 1988; Zehnder et al., 2004), consistent with logistic regression analyses. Finally, a cross-sectional study by Garland et al. (2005), suggested a knee aBMD fracture threshold of 0.78 g/cm^2.

Importantly, however, radiographic measures may provide an incomplete representation of bone fragility. Cadaveric studies suggest that 25%–43% of the measured variation in bone strength is not accounted for by simple measures of bone mineral (Edwards et al., 2013b), and this variation can be better explained using FE models of bone. With this in mind, a recent study examined FE models from 50 individuals with SCI (14 fractures) (Haider, Simonian, Schnitzer, & Edwards, 2020) and demonstrated that FE predicted tibial strength was a more sensitive predictor of fracture risk compared to DXA (spine, hip, and knee) or CT measures (knee). It is important to note that additional risk factors for fracture after SCI include age, sex, severity of injury, excessive alcohol consumption, and duration of injury (Craven, Robertson, McGillivray, & Adachi, 2009); however, it is unclear if these measures provide information independent from bone density or strength.

Interventions for bone loss after SCI

There is currently no standard of care for bone loss after SCI, and intervention strategies remain an area of active research. Both pharmacological and non-pharmacological interventions have been explored, with varying success. There is a large breadth of research, spanning decades, and this work has been reviewed in the past (Cirnigliaro et al., 2017; Dolbow et al., 2011). In this section, we summarize this body of work and highlight key studies that are recent or most impactful.

Non-pharmaceutical intervention

Functional electrical stimulation (FES) is a technique that has been studied extensively in individuals with SCI. In this technique, the nervous system is stimulated via electrodes, and the resulting action potential causes muscle contraction and limb motion (Ho et al., 2014). Hypothetically, this loading stimulus could slow or reverse the adaptive remodeling processes that resorb bone during disuse, and there is some evidence of this in the literature. One recent study used FES rowing excise in 4 individuals with an SCI duration of less than 2 years (Lambach et al., 2020). After 60 sessions (over 6–8 months), 2 individuals gained 6% and 8% trabecular BMD at the knee, while another 2 individuals demonstrated a decrease in the rate of bone loss. Benefits have also been observed in individuals with more chronic SCI. For example, Frotzler, Coupaud, et al. (2008) studied FES cycling in individuals >3.6 post-injury and reported increased trabecular and total BMD at the distal femur (14% and 7% increase, respectively) after 12 months. While these results are encouraging, it should be noted that a number of studies report no therapeutic effect of FES training on the bone (Eser et al., 2003; Shields & Dudley-Javoroski, 2007). It is unclear why results differ among studies, but differences in training volume or load intensity may play a role (Biering-Sørensen et al., 2009; Morse et al., 2019). Furthermore, there is evidence that the benefits of FES training fade soon after training is ceased or reduced (Chen et al., 2005).

A few studies have demonstrated beneficial effects of weight-bearing or walking exercise (without electrical stimulation) soon after SCI, but with varying results. One relatively large study examined 54 individuals with recent (1 < year) SCI and found that those who were able to ambulate demonstrated higher BMD at the leg compared to those who could not (Alekna, Tamulaitiene, Sinevicius, & Juocevicius, 2008). There have also been controlled experiments like that of de Bruin et al. (1999), where harness-assisted weight-bearing exercise was performed at different volumes with three groups of individuals. This study found that individuals who received the early intervention had little or no bone loss in the tibia after 25 weeks, compared to a marked reduction in individuals who were immobile. These encouraging examples are balanced by the fact that there are a number of studies which demonstrate no effects of weight-bearing or exercise on bone in individuals with recent SCI (Frey-Rindova et al., 2000; Giangregorio et al., 2005), but the reason for differences in the literature is not clear. Finally, weight-bearing activity appears to have little effect in individuals with injury duration >1 year (Biering-Sørensen et al., 2009).

Pharmaceutical intervention

A number of studies have examined pharmaceutical intervention using anti-resorptive drugs to slow or reverse bone loss soon after SCI. The anti-resorptives etidonrate (Pearson, Nance, Leslie, & Ludwig, 1997) and pamidronate (Nance et al., 1999) have been shown to attenuate lower limb aBMD loss over 12 months when administered to individuals with recent SCI (<6 weeks), though effects were strongest in individuals who regained the ability to ambulate. Other anti-resorptives have been shown to attenuate bone loss at the hip, but efficacy at the knee remains unclear. Gilchrist et al. (2007) detected significant attenuation of aBMD loss at the hip after treatment with alendronate but did not examine the knee. Similarly, a number of studies have demonstrated the effects of zolendronate at the hip. In 2020, two separate studies (Goenka & Sethi, 2020; Oleson et al., 2020) reported that treatment was associated with attenuation of bone loss at the hip over 1 year, though benefits were not detected at the knee (Oleson et al., 2020) and older studies of this drug reported results consistent with these findings (Bubbear et al., 2011; Schnitzer et al., 2016; Shapiro et al., 2007). Finally, we note that recent investigations of the anti-resorptive denosumab have demonstrated promising early results. Gifre et al. (2016) treated 14 individuals (no control subjects) and found that denosumab was associated with statically significant increases to aBMD at the hip and spine (2%–8%), though the knee was not assessed. However, another recent study of denosumab (Cirnigliaro et al., 2020) demonstrated that treatment effectively maintained aBMD at the distal femur and attenuated losses at the tibia; however, pQCT measures did not demonstrate similarly robust significant effects, likely due to limited statistical power. Ultimately, these results are encouraging, but must be replicated before being put into practice. In particular, additional follow-up with a greater sample size and more robust volumetric imaging of the knee is needed.

A few studies have also examined treatment in individuals with a longer duration of SCI (>1 year). Alendronate has been used, but with mixed results. One study by Moran De Brito, Battistella, Saito, and Sakamoto (2005) reported no benefits to lower limb aBMD in individuals with SCI > 13 months. In contrast, Zehnder, Risi, et al. (2004) found that open-label alendronate effectively attenuated bone loss at the tibia; however, this study included individuals with a wide range of SCI durations (0.1–30 years prior to enrollment), and these results could be influenced by greater treatment effects in more recently injured individuals. A few other treatments have demonstrated statistically significant effects at the knee in individuals with longer SCI duration. For example, Bauman, Spungen, Morrison, Zhang, and Schwartz (2005) found that treatment with a vitamin D analog was associated with small, but statistically significant improvements to whole leg BMD after 24 months. Similarly, Morse et al. (2019) reported that FES exercise plus zolendronate was associated with improvement or mitigation of losses in bone geometric properties at the proximal tibia and distal femur, compared to FES exercise alone. Finally, Edwards et al. (2018) studied anabolic treatment with teriparatide. After 12 months, participants demonstrated treatment effect at the spine but not at the hip or the knee; however, after a 12-month open-label extension, individuals who received teriparatide for the entire 24 months demonstrated increased total hip aBMD (4%–7% compared to original baseline). Increases in QCT measures at the knee were demonstrated across all individuals receiving open-label drugs. Moreover, individuals who participated in a second extension with open-label alendronate (Haider et al., 2019) demonstrated that these benefits could be retained for at least 12 months after ceasing teriparatide. With that said, we note that all the studies discussed here demonstrated only modest improvements to the bone, and it is yet to be established whether differences are large enough to meaningfully influence fracture risk in this population.

Summary of interventions for bone loss after SCI

Non-pharmaceutical interventions to treat bone loss after SCI have been studied, with mixed results. There is some evidence that, at a sufficient training intensity and volume, FES exercise or assisted weight-bearing exercise (without FES) may help to attenuate bone loss at the knee in individuals with recent SCI (Biering-Sørensen et al., 2009), but the effects of treatment are likely to be lost after training is ended or reduced (Chen et al., 2005). A number of anti-resorptive drugs have been studied in individuals with recent SCI. In general, studies suggest that these drugs attenuate bone loss at the hip over the first 1–2 years after injury; however, efficacy at the knee in non-ambulatory individuals has not been established. Despite this limitation, these changes may still be clinically meaningful. Fractures at the hip can occur in this population, and these drugs may help reduce this risk. Moreover, a number of individuals will likely become ambulatory soon after SCI, and attenuation of bone loss at the spine and hip may help these individuals to rehabilitate quickly and safely. This may be particularly relevant in the future, as new or more effective treatments for SCI are introduced. Pharmaceutical intervention can also have statistically significant effects in individuals with a longer duration of SCI; however, we note that these studies typically report an attenuation of bone loss, with no large increases in bone mineral measures. It remains to be established if any treatments have a sufficiently large effect at the knee, to cause a meaningful reduction in clinical fracture risk.

There are a few important limitations in this body of work. We did not find any studies that evaluated recently injured individuals for a period greater than 2 years. In untreated individuals, fractures typically occur later than 2 years after injury (Gifre et al., 2014), highlighting the need to understand how treatment influences bone over these longer time periods. Furthermore, it is often unclear whether the magnitude of the treatment effect is sufficient to provide a clinically meaningful reduction in fracture risk. In the future, longitudinal follow-up with fracture surveillance may be warranted.

Summary and conclusions

There have been over two dozen radiographic studies of bone loss at the knee after SCI, across different modalities and study designs. Together, these studies unequivocally establish the fact that bone loss is rapid and severe. Furthermore, studies have also established a quantitative relationship between bone loss and fracture risk in this population, highlighting the potential benefit of preventing or reversing bone loss in these individuals; however, there is currently no standard of care for treatment of bone loss after SCI, and treatment options remain an area of active research. Anti-resorptive treatment can attenuate bone loss at the hip in individuals soon after SCI, but the efficacy of treatment and fracture risk reduction at the knee has not been well established in non-ambulatory individuals. Similarly, non-pharmaceutical intervention via FES exercise or assisted weight-bearing could plausibly help attenuate lower limb bone loss in individuals with recent SCI, but studies have yielded mixed results. In addition to exploring new types of treatment, we believe that more longitudinal investigation is needed to determine whether treatment effects can be retained over longer durations and provide clinically meaningful reductions in fracture risk.

Mini-dictionary terms

Cortical bone: Also called compact bone. This is low porosity (<15%) bone that is found on the outer shell of bones.
Diaphysis: The main body of long bones, located between the metaphysis (see definition below) of either end.
Dual-energy X-ray absorptiometry (DEXA): A 2D X-ray imaging technique, commonly used to obtain measurements of areal bone mineral density.
Finite element model: A computer representation of a system that simulates/predicts behavior by numerically solving the governing partial differential equations. In bone research, a finite element model can be used to predict the structural response (stiffness and strength) to the applied load.
Epiphysis: In long bones, this is the region above the growth plate, closest to the joint.
Quantitative computed tomography (QCT): An X-ray based imaging technique that combines information from multiple views (projections) around the object in order to build a 3D image. Commonly used to measure volumetric bone mineral density.
Metaphysis: In long bones, this region contains the growth plate and surrounding bone tissue that grows during development, to lengthen the bone.
Magnetic resonance imaging (MRI): A 3D imaging technique that identifies different materials/tissues by measuring the response to magnetic excitation.
Trabecular bone: Also called cancellous or spongy bone, a highly porous (>80%) bone that is found underneath the cortical layer, particularly at the ends of long bones. At the microstructure level, it is made of the rod (trabeculae) or plate-like structures.

Key facts

- The World Health Organization estimates that 250,000–500,000 individuals experience spinal cord injury each year. Thanks to modern medical practices, in-hospital mortality rates are <16% in Europe and America. Factors influencing long-term health, such as bone loss and fracture, are important concerns for the majority who survive.
- Clinical Dual-Energy Absorptiometry was first introduced in 1987, based on pioneering research by Dr. John Cameron in 1963 and Richard Mazess in 1981. Since then, it has become the gold standard for assessment of bone loss in non-disabled individuals and has seen substantial use in individuals with spinal cord injury.
- Bone strength is multi-factorial; the size, shape, and distribution of bone material can all influence bone fragility. However, these are not easily measured using dual-energy absorptiometry. Techniques like regional computed tomography analysis or finite element modeling are often used to better understand these factors.
- The modern finite element method was originally developed for the aerospace industry in the 1950s and 1960s. It was slowly adopted to other industries as digital computing became more accessible. Today, these techniques see widespread use in many fields, including bone research in individuals with spinal cord injury.
- Antiresorptive drugs are a mainstay of modern osteoporosis treatment in non-disabled individuals and are highly effective at reducing/reversing bone loss and reducing fracture risk in that population. However, spinal cord injury creates a unique mechanical and physiological environment that is challenging to bone health.

Summary points

- Bone loss is most rapid soon after spinal cord injury but plateaus 2–8 years after injury.
- After plateau, bone mineral is reduced by up to 88%. Losses are most severe at the epiphysis and become progressively less severe toward diaphyseal locations.
- Measures of bone mineral, and finite element predicted stiffness and strength, have been quantitatively linked to increased fracture risk.
- In recently injured (injury <1 year) individuals, pharmaceutical interventions have often demonstrated attenuation of bone loss at the hip, but efficacy at the knee in non-ambulatory individuals is not well established.
- Non-pharmaceutical intervention via functional electrical stimulation or assisted weight-bearing exercise may help attenuate bone loss after a recent spinal cord injury, but results have been mixed.

References

Alekna, V., Tamulaitiene, M., Sinevicius, T., & Juocevicius, A. (2008). Effect of weight-bearing activities on bone mineral density in spinal cord injured patients during the period of the first two years. *Spinal Cord*, *46*(11), 727–732. https://doi.org/10.1038/sc.2008.36.

Bauman, W. A., Spungen, A. M., Morrison, N., Zhang, R. L., & Schwartz, E. (2005). Effect of a vitamin D analog on leg bone mineral density in patients with chronic spinal cord injury. *Journal of Rehabilitation Research and Development*, *42*(5), 625–633. https://doi.org/10.1682/JRRD.2004.11.0145.

Biering-Sorensen, F., Bohr, H., & Schaadt, O. (1988). Bone mineral content of the lumbar spine and lower extremities years after spinal cord lesion. *Paraplegia*, *26*(5), 293–301. https://doi.org/10.1038/sc.1988.44.

Biering-Sørensen, F., Bohr, H. H., & Schaadt, O. P. (1990). Longitudinal study of bone mineral content in the lumbar spine, the forearm and the lower extremities after spinal cord injury. *European Journal of Clinical Investigation*, *20*(3), 330–335. https://doi.org/10.1111/j.1365-2362.1990.tb01865.x.

Biering-Sørensen, F., Hansen, B., & Lee, B. S. B. (2009). Non-pharmacological treatment and prevention of bone loss after spinal cord injury: A systematic review. *Spinal Cord*, *47*(7), 508–518. https://doi.org/10.1038/sc.2008.177.

Bubbear, J. S., Gall, A., Middleton, F. R. I., Ferguson-Pell, M., Swaminathan, R., & Keen, R. W. (2011). Early treatment with zoledronic acid prevents bone loss at the hip following acute spinal cord injury. *Osteoporosis International*, *22*(1), 271–279. https://doi.org/10.1007/s00198-010-1221-6.

Carbone, L. D., Chin, A. S., Burns, S. P., Svircev, J. N., Hoenig, H., Heggeness, M., et al. (2014). Mortality after lower extremity fractures in men with spinal cord injury. *Journal of Bone and Mineral Research*, *29*(2), 432–439. https://doi.org/10.1002/jbmr.2050.

Chen, S. C., Lai, C. H., Chan, W. P., Huang, M. H., Tsai, H. W., & Chen, J. J. J. (2005). Increases in bone mineral density after functional electrical stimulation cycling exercises in spinal cord injured patients. *Disability and Rehabilitation*, *27*(22), 1337–1341. https://doi.org/10.1080/09638280500164032.

Cirnigliaro, C. M., La Fountaine, M. F., Parrott, J. S., Kirshblum, S. C., McKenna, C., Sauer, S. J., et al. (2020). Administration of denosumab preserves bone mineral density at the knee in persons with subacute spinal cord injury: Findings from a randomized clinical trial. *JBMR Plus*, *4*(8). https://doi.org/10.1002/jbm4.10375.

Cirnigliaro, C. M., Myslinski, M. J., La Fountaine, M. F., Kirshblum, S. C., Forrest, G. F., & Bauman, W. A. (2017). Bone loss at the distal femur and proximal tibia in persons with spinal cord injury: Imaging approaches, risk of fracture, and potential treatment options. *Osteoporosis International*, *28*(3), 747–765. https://doi.org/10.1007/s00198-016-3798-x.

Coupaud, S., McLean, A. N., & Allan, D. B. (2009). Role of peripheral quantitative computed tomography in identifying disuse osteoporosis in paraplegia. *Skeletal Radiology*, *38*(10), 989–995. https://doi.org/10.1007/s00256-009-0674-1.

Coupaud, S., McLean, A. N., Purcell, M., Fraser, M. H., & Allan, D. B. (2015). Decreases in bone mineral density at cortical and trabecular sites in the tibia and femur during the first year of spinal cord injury. *Bone*, *74*, 69–75. https://doi.org/10.1016/j.bone.2015.01.005.

Craven, B. C., Robertson, L. A., McGillivray, C. F., & Adachi, J. D. (2009). Detection and treatment of sublesional osteoporosis among patients with chronic spinal cord injury: Proposed paradigms. *Topics in Spinal Cord Injury Rehabilitation*, *14*(4), 1–22. https://doi.org/10.1310/sci1404-1.

de Bruin, E. D., Dietz, V., Dambacher, M. A., & Stüssi, E. (2000). Longitudinal changes in bone in men with spinal cord injury. *Clinical Rehabilitation*, *14*(2), 145–152. https://doi.org/10.1191/026921500670532165.

de Bruin, E. D., Frey-Rindova, P., Herzog, R. E., Dietz, V., Dambacher, M. A., & Stüssi, E. (1999). Changes in tibia bending stiffness two years after spinal cord injury. *Archives of Physical Medicine and Rehabilitation*, *80*(February), 214–220.

Dolbow, D. R., Gorgey, A. S., Daniels, J. A., Adler, R. A., Moore, J. R., & Gater, D. R. (2011). The effects of spinal cord injury and exercise on bone mass: A literature review. *NeuroRehabilitation*, *29*(3), 261–269. https://doi.org/10.3233/NRE-2011-0702.

Dudley-Javoroski, S., & Shields, R. K. (2010). Longitudinal changes in femur bone mineral density after spinal cord injury: Effects of slice placement and peel method. *Osteoporosis International*, *21*(6), 985–995. https://doi.org/10.1007/s00198-009-1044-5.

Dudley-Javoroski, S., & Shields, R. K. (2012). Regional cortical and trabecular bone loss after spinal cord injury. *Journal of Rehabilitation Research and Development*, *49*(9), 1365–1376. https://doi.org/10.1682/jrrd.2011.12.0245.

Edwards, W. B., & Schnitzer, T. J. (2015). Bone imaging and fracture risk after spinal cord injury. *Current Osteoporosis Reports*, *13*(5), 310–317. https://doi.org/10.1007/s11914-015-0288-6.

Edwards, W. B., Schnitzer, T. J., & Troy, K. L. (2013a). Bone mineral and stiffness loss at the distal femur and proximal tibia in acute spinal cord injury. *Osteoporosis International*, *24*(9), 2461–2469. s00198-013-2557-5.

Edwards, W. B., Schnitzer, T. J., & Troy, K. L. (2013b). Torsional stiffness and strength of the proximal tibia are better predicted by finite element models than DXA or QCT. *Journal of Biomechanics*, *46*(10), 1655–1662. https://doi.org/10.1016/j.jbiomech.2013.04.016.

Edwards, W. B., Simonian, N., Haider, I. T., Anschel, A. S., Chen, D., Gordon, K. E., et al. (2018). Effects of teriparatide and vibration on bone mass and bone strength in people with bone loss and spinal cord injury: A randomized, controlled trial. *Journal of Bone and Mineral Research*, *33*(10), 1729–1740. Retrieved from http://doi.wiley.com/10.1002/jbmr.3525.

Edwards, W. B., Simonian, N., Troy, K. L., & Schnitzer, T. J. (2015). Reduction in torsional stiffness and strength at the proximal tibia as a function of time since spinal cord injury. *Journal of Bone and Mineral Research*, *30*(8), 1422–1430. https://doi.org/10.1002/jbmr.2474.

Eser, P., De Bruin, E. D., Telley, I., Lechner, H. E., Knecht, H., & Stüssi, E. (2003). Effect of electrical stimulation-induced cycling on bone mineral density in spinal cord-injured patients. *European Journal of Clinical Investigation*, *33*(5), 412–419. https://doi.org/10.1046/j.1365-2362.2003.01156.x.

Eser, P., Frotzler, A., Zehnder, Y., Wick, L., Knecht, H., Denoth, J., et al. (2004). Relationship between the duration of paralysis and bone structure: A pQCT study of spinal cord injured individuals. *Bone*, *34*(5), 869–880. https://doi.org/10.1016/j.bone.2004.01.001.

Frey-Rindova, P., De Bruin, E. D., Stüssi, E., Dambacher, M. A., & Dietz, V. (2000). Bone mineral density in upper and lower extremities during 12 months after spinal cord injury measured by peripheral quantitative computed tomography. *Spinal Cord*, *38*(1), 26–32. https://doi.org/10.1038/sj.sc.3100905.

Frotzler, A., Berger, M., Knecht, H., & Eser, P. (2008). Bone steady-state is established at reduced bone strength after spinal cord injury: A longitudinal study using peripheral quantitative computed tomography (pQCT). *Bone*, *43*(3), 549–555. https://doi.org/10.1016/j.bone.2008.05.006.

Frotzler, A., Coupaud, S., Perret, C., Kakebeeke, T. H., Hunt, K. J., de Donaldson, N. N., et al. (2008). High-volume FES-cycling partially reverses bone loss in people with chronic spinal cord injury. *Bone*, *43*(1), 169–176. https://doi.org/10.1016/j.bone.2008.03.004.

Garland, D. E., Adkins, R. H., Scott, M., Singh, H., Massih, M., & Stewart, C. (2004). Bone loss at the os calcis compared with bone loss at the knee in individuals with spinal cord injury. *Journal of Spinal Cord Medicine*, *27*(3), 207–211. https://doi.org/10.1080/10790268.2004.11753749.

Garland, D., Adkins, R., & Stewart, C. (2005). Fracture threshold and risk for osteoporosis and pathologic fractures in individuals with spinal cord injury. *Topics in Spinal Cord Injury Rehabilitation*, *11*(1), 61–69. https://doi.org/10.1310/G6TD-HPGC-XM3Q-7YJH.

Garland, D. E., Adkins, R. H., Stewart, C. A., Ashford, R., & Vigil, D. (2001). Regional osteoporosis in women who have a complete spinal cord injury. *Journal of Bone and Joint Surgery Series A*, *83*(8), 1195–1200. https://doi.org/10.2106/00004623-200108000-00009.

Garland, D. E., Stewart, C. A., Adkins, R. H., Hu, S. S., Rosen, C., Liotta, F. J., et al. (1992). Osteoporosis after spinal cord injury. *Journal of Orthopaedic Research*, *10*(3), 371–378. https://doi.org/10.1002/jor.1100100309.

Giangregorio, L. M., Hicks, A. L., Webber, C. E., Phillips, S. M., Craven, B. C., Bugaresti, J. M., et al. (2005). Body weight supported treadmill training in acute spinal cord injury: Impact on muscle and bone. *Spinal Cord*, *43*(11), 649–657. https://doi.org/10.1038/sj.sc.3101774.

Gifre, L., Vidal, J., Carrasco, J. L., Muxi, A., Portell, E., Monegal, A., et al. (2016). Denosumab increases sublesional bone mass in osteoporotic individuals with recent spinal cord injury. *Osteoporosis International*, *27*(1), 405–410. https://doi.org/10.1007/s00198-015-3333-5.

Gifre, L., Vidal, J., Carrasco, J., Portell, E., Puig, J., Monegal, A., et al. (2014). Incidence of skeletal fractures after traumatic spinal cord injury: A 10-year follow-up study. *Clinical Rehabilitation*, *28*(4), 361–369. https://doi.org/10.1177/0269215513501905.

Gilchrist, N. L., Frampton, C. M., Acland, R. H., Nicholls, M. G., March, R. L., Maguire, P., et al. (2007). Alendronate prevents bone loss in patients with acute spinal cord injury: A randomized, double-blind, placebo-controlled study. *Journal of Clinical Endocrinology and Metabolism*, *92*(4), 1385–1390. https://doi.org/10.1210/jc.2006-2013.

Goenka, S., & Sethi, S. (2020). Zoledronic acid attenuates early bone loss at forearm in patients with acute spinal cord injury. *The Indian Journal of Orthopaedics*, *54*(Suppl. 1), 127–133. https://doi.org/10.1007/s43465-020-00158-8.

Haider, I. T., Lobos, S. M., Simonian, N., Schnitzer, T. J., & Edwards, W. B. (2018). Bone fragility after spinal cord injury: reductions in stiffness and bone mineral at the distal femur and proximal tibia as a function of time. *Osteoporosis International*, *30*(8), 1422–1430. https://doi.org/10.1007/s00198-018-4733-0.

Haider, I. T., Simonian, N., Saini, A. S., Leung, F. M., Edwards, W. B., & Schnitzer, T. J. (2019). Open-label clinical trial of alendronate after teriparatide therapy in people with spinal cord injury and low bone mineral density. *Spinal Cord*. https://doi.org/10.1038/s41393-019-0303-3.

Haider, I. T., Simonian, N., Schnitzer, T. J., & Edwards, W. B. (2020). Stiffness and strength predictions from finite element models of the knee are associated with lower-limb fractures after spinal cord injury. *Annals of Biomedical Engineering*. https://doi.org/10.1007/s10439-020-02606-w.

Ho, C. H., Triolo, R. J., Elias, A. L., Kilgore, K. L., DiMarco, A. F., Bogie, K., et al. (2014). Functional electrical stimulation and spinal cord injury. *Physical Medicine and Rehabilitation Clinics of North America*, *25*(3), 631–654. ix https://doi.org/10.1016/j.pmr.2014.05.001.

Kanis, J. A., Hans, D., Cooper, C., Baim, S., Bilezikian, J. P., Binkley, N., et al. (2011). Interpretation and use of FRAX in clinical practice. *Osteoporosis International*, *22*(9), 2395–2411. https://doi.org/10.1007/s00198-011-1713-z.

Lala, D., Craven, B. C., Thabane, L., Papaioannou, A., Adachi, J. D., Popovic, M. R., et al. (2014). Exploring the determinants of fracture risk among individuals with spinal cord injury. *Osteoporosis International*, *25*(1), 177–185. https://doi.org/10.1007/s00198-013-2419-1.

Lambach, R. L., Stafford, N. E., Kolesar, J. A., Kiratli, B. J., Creasey, G. H., Gibbons, R. S., et al. (2020). Bone changes in the lower limbs from participation in an FES rowing exercise program implemented within two years after traumatic spinal cord injury. *Journal of Spinal Cord Medicine*, *43*(3), 306–314. https://doi.org/10.1080/10790268.2018.1544879.

Lazo, M. G., Shirazi, P., Sam, M., Giobbie-Hurder, A., Blacconiere, M. J., & Muppidi, M. (2001). Osteoporosis and risk of fracture in men with spinal cord injury. *Spinal Cord*, *39*(4), 208–214. https://doi.org/10.1038/sj.sc.3101139.

McCarthy, I. D., Bloomer, Z., Gall, A., Keen, R., & Ferguson-Pell, M. (2012). Changes in the structural and material properties of the tibia in patients with spinal cord injury. *Spinal Cord*, *50*(4), 333–337. https://doi.org/10.1038/sc.2011.143.

McPherson, J. G., Edwards, W. B., Prasad, A., Troy, K. L., Griffith, J. W., & Schnitzer, T. J. (2014). Dual energy X-ray absorptiometry of the knee in spinal cord injury: Methodology and correlation with quantitative computed tomography. *Spinal Cord*, *52*(11), 821–825. https://doi.org/10.1038/sc.2014.122.

Modlesky, C. M., Majumdar, S., Narasimhan, A., & Dudley, G. A. (2004). Trabecular bone microarchitecture is deteriorated in men with spinal cord injury. *Journal of Bone and Mineral Research*, *19*(1), 48–55. https://doi.org/10.1359/JBMR.0301208.

Moran De Brito, C. M., Battistella, L. R., Saito, E. T., & Sakamoto, H. (2005). Effect of alendronate on bone mineral density in spinal cord injury patients: A pilot study. *Spinal Cord*, *43*(6), 341–348. https://doi.org/10.1038/sj.sc.3101725.

Morgan, E. F., Bayraktar, H. H., & Keaveny, T. M. (2003). Trabecular bone modulus–density relationships depend on anatomic site. *Journal of Biomechanics*, *36*(7), 897–904. https://doi.org/10.1016/S0021-9290(03)00071-X.

Morse, L. R., Troy, K. L., Fang, Y., Nguyen, N., Battaglino, R., Goldstein, R. F., et al. (2019). Combination therapy with zoledronic acid and FES-row training mitigates bone loss in paralyzed legs: Results of a randomized comparative clinical trial. *JBMR Plus*, *3*(5). https://doi.org/10.1002/jbm4.10167, e10167.

Nance, P. W., Schryvers, O., Leslie, W., Ludwig, S., Krahn, J., & Uebelhart, D. (1999). Intravenous pamidronate attenuates bone density loss after acute spinal cord injury. *Archives of Physical Medicine and Rehabilitation*, *80*(3), 243–251. https://doi.org/10.1016/S0003-9993(99)90133-8.

Oleson, C. V., Marino, R. J., Formal, C. S., Modlesky, C. M., & Leiby, B. E. (2020). The effect of zoledronic acid on attenuation of bone loss at the hip and knee following acute traumatic spinal cord injury: A randomized-controlled study. *Spinal Cord*, *58*(8), 921–929. https://doi.org/10.1038/s41393-020-0431-9.

Pearson, E. G., Nance, P. W., Leslie, W. D., & Ludwig, S. (1997). Cyclical etidronate: Its effect on bone density in patients with acute spinal cord injury. *Archives of Physical Medicine and Rehabilitation*, *78*(3), 269–272. https://doi.org/10.1016/S0003-9993(97)90032-0.

Rittweger, J., Goosey-Tolfrey, V. L., Cointry, G., & Ferretti, J. L. (2010). Structural analysis of the human tibia in men with spinal cord injury by tomographic (pQCT) serial scans. *Bone*, *47*(3), 511–518. https://doi.org/10.1016/j.bone.2010.05.025.

Schnitzer, T. J., Kim, K., Marks, J., Yeasted, R., Simonian, N., & Chen, D. (2016). Zoledronic acid treatment after acute spinal cord injury: Results of a randomized, placebo-controlled pilot trial. *PM & R: The Journal of Injury, Function, and Rehabilitation*, *8*(9), 833–843. https://doi.org/10.1016/j.pmrj.2016.01.012.

Shapiro, J., Smith, B., Beck, T., Ballard, P., Dapthary, M., BrintzenhofeSzoc, K., et al. (2007). Treatment with zoledronic acid ameliorates negative geometric changes in the proximal femur following acute spinal cord injury. *Calcified Tissue International*, *80*(5), 316–322. https://doi.org/10.1007/s00223-007-9012-6.

Shields, R. K., & Dudley-Javoroski, S. (2007). Musculoskeletal adaptations in chronic spinal cord injury: Effects of long-term soleus electrical stimulation training. *Neurorehabilitation and Neural Repair*, *21*(2), 169–179. https://doi.org/10.1177/1545968306293447.

Shields, R. K., Schlechte, J., Dudley-Javoroski, S., Zwart, B. D., Clark, S. D., Grant, S. A., et al. (2005). Bone mineral density after spinal cord injury: A reliable method for knee measurement. *Archives of Physical Medicine and Rehabilitation*, *86*(10), 1969–1973. https://doi.org/10.1016/j.apmr.2005.06.001.

Slade, J. M., Bickel, C. S., Modlesky, C. M., Majumdar, S., & Dudley, G. A. (2005). Trabecular bone is more deteriorated in spinal cord injured versus estrogen-free postmenopausal women. *Osteoporosis International*, *16*(3), 263–272. https://doi.org/10.1007/s00198-004-1665-7.

Tan, C. O., Battaglino, R. A., Doherty, A. L., Gupta, R., Lazzari, A. A., Garshick, E., et al. (2014). Adiponectin is associated with bone strength and fracture history in paralyzed men with spinal cord injury. *Osteoporosis International*, *25*(11), 2599–2607. https://doi.org/10.1007/s00198-014-2786-2.

Warden, S. J., Bennell, K. L., Matthews, B., Brown, D. J., McMeeken, J. M., Wark, J. D., et al. (2002). Quantitative ultrasound assessment of acute bone loss following spinal cord injury: A longitudinal pilot study. *Osteoporosis International*, *13*(7), 586–592. https://doi.org/10.1007/s001980200077.

Zehnder, Y., Lüthi, M., Michel, D., Knecht, H., Perrelet, R., Neto, I., et al. (2004). Long-term changes in bone metabolism, bone mineral density, quantitative ultrasound parameters, and fracture incidence after spinal cord injury: A cross-sectional observational study in 100 paraplegic men. *Osteoporosis International*, *15*(3), 180–189. https://doi.org/10.1007/s00198-003-1529-6.

Zehnder, Y., Risi, S., Michel, D., Knecht, H., Perrelet, R., Kraenzlin, M., et al. (2004). Prevention of bone loss in paraplegics over 2 years with alendronate. *Journal of Bone and Mineral Research*, *19*(7), 1067–1074. https://doi.org/10.1359/JBMR.040313.

Chapter 27

Functional and morphological reorganization of the brain following spinal cord injury: Insights from MRI

Maria M. D'souza[a], Jeanne Maria Dsouza[b], Prabhjot Kaur[a], and Pawan Kumar[a]
[a]*Division of NMR, Institute of Nuclear Medicine and Allied Sciences, Delhi, India,* [b]*KMC Manipal Medical College, Madhav Nagar, Manipal, Karnataka, India*

Abbreviations

SCI	spinal cord injury
MRI	magnetic resonance imaging
DTI	diffusion tensor imaging
fMRI	functional MRI
MRS	magnetic resonance spectroscopy
VBM	voxel-based morphometry
SBA	surface-based analysis
FA	fractional anisotropy
MD	mean diffusivity
BOLD	blood oxygen level-dependent signal
rsfMRI	resting state fMRI

Introduction

The spinal cord is a vital conduit for the flow of information to and from the brain. Any disruption in the integrity of the spinal cord secondary to injury would lead to an impaired information flow. Deprived of their normal sources of activation, neurons in the brain undergo re-organization, a phenomenon that has been implicated both in the ensuing recovery of function as well as the occurrence of maladaptive changes (Kaas et al., 2008). Despite significant ongoing research, the ability to restore normal function following spinal cord injury (SCI) is limited. The functional and morphological changes following SCI could play a major role in hastening or impeding recovery. A better understanding of these changes would pave the way for more effective therapeutic and rehabilitative strategies, thus translating into a better clinical outcome.

The use of imaging techniques facilitates an effective understanding of the brain changes secondary to SCI. Magnetic resonance imaging (MRI) is regarded as the modality of choice for the evaluation of the nervous system, thanks to its excellent spatial resolution, multi-planar imaging capability, multi-parametric approach, and the use of non-ionizing radiation. Beyond the information it provides in the routine clinical work-up of the extent of local damage produced in SCI, MRI can unravel the underlying brain changes which ensure secondary to cortical plasticity, through the use of novel techniques such as diffusion tensor imaging (DTI), brain morphometry, functional MRI (fMRI), MR spectroscopy (MRS), etc. (Table 1). This chapter provides a glimpse of the information that these techniques can provide in the assessment of the structural and functional brain changes secondary to SCI.

Brain morphometry

Morphometry is the study of the size and shape of the brain and its various structures. MRI has the inherent capability to produce high-resolution images of the anatomical structure of the brain, which are useful to assess various metrics such as

Cellular, Molecular, Physiological, and Behavioral Aspects of Spinal Cord Injury. https://doi.org/10.1016/B978-0-12-822427-4.00027-7

TABLE 1 Advanced MR techniques for assessment of brain reorganization after SCI.

MR technique	Utility
Diffusion tensor imaging (DTI)	• High resolution imaging of fiber tracts in brain • Provides quantitative indices for assessment of microstructural integrity • Provides information on structural connectivity
Brain morphometry	• Quantification of anatomical features such as volume, surface area, curvature, cortical thickness, gyrification index • Provides information on subtle structural changes long before they are clinically apparent on conventional MR scanning
Functional MRI (fMRI)	• *Task-based fMRI*: Identification of brain regions which are functionally involved in a specific task performance • *Resting state fMRI*: Mapping of multiple functional neuronal networks in the absence of task performance • Provides information on the origin & organization of cognitive brain functioning in health and disease
MR spectroscopy (MRS)	• Quantitative measure of brain metabolites • Provides information on metabolic changes in the brain in various physiologic and pathologic states

the gray matter, white matter, and total brain volume, cortical surface area, cortical thickness, etc., through appropriate computational techniques. Two standard techniques in popular use at present in the study of intracranial changes post-SCI are voxel-based morphometry (VBM) and surface-based analysis (SBA). In VBM, one derives morphometric measures using a voxel-based volumetric approach, while SBA does so from geometric models of the cortical surface. These techniques have been used effectively in understanding several brain pathologies & processes such as aging, dementia, schizophrenia, multiple sclerosis, etc. (Mechelli, Price, Friston, & Ashburner, 2005) and show promise in the assessment of atrophic changes after SCI.

Diffusion tensor imaging (DTI)

DTI provides an excellent in vivo method of assessing the tissue microstructure, by measuring the diffusivity of water in multiple directions through brain tissue. Over and above the visualization of the white matter tracts in the brain, it provides useful quantitative indices to assess tissue integrity. These primarily include fractional anisotropy (FA) and mean diffusivity (MD). FA is a measure of the degree of directionality of water diffusion, while MD gives information on the overall diffusion. These indices are affected by cell size, shape, and integrity (Pierpaoli, Jezzard, Basser, Barnett, & Di Chiro, 1996) and get altered with cell degeneration as well as with disrupted tissue barriers, such as cell membranes and myelin sheaths (Basser, Pajevic, Pierpaoli, Duda, & Aldroubi, 2000). Other parameters such as axial diffusivity (AD) and radial diffusivity (RD), which report the extent of water diffusion along or across the axons, are thought to reflect the integrity of axons and myelin, respectively (Zhang et al., 2009).

Functional MRI (fMRI)

fMRI reflects blood oxygen level-dependent signal (BOLD) (Friston et al., 1995), which gets altered by hemodynamic changes secondary to neuronal activation and is thus an indirect marker of the same. fMRI is broadly of two types: task-based fMRI and resting state fMRI. Task-based fMRI is widely adopted to identify brain regions that are functionally involved in specific task performance (Logothetis, 2008). It is a commonly used tool both in clinical and cognitive neuroscience and is widely used in SCI research, but it relies heavily on patients' cognitive integrity and their cooperation in performing the task.

Resting state fMRI (rsfMRI) on the other hand, detects the spontaneous, temporally correlated changes in brain activity at rest, which occur in brain regions that are spatially distinct, but functionally related. As it entails performing the scan with the subject at rest, it is relatively free of compounding effects due to task difficulty, level of patient co-operation, practice and re-test, etc. The pattern of synchronous BOLD activity in different parts of the brain at rest, resembles the functional networks observed during the performance of cognitive tasks (Rigon, Duff, McAuley, Kramer, & Voss, 2016). rsfMRI has the ability to investigate several brain networks simultaneously such as the default mode network, frontoparietal network, central executive network, sensorimotor network, auditory network, visual network, etc. (Table 2).

TABLE 2 Brain neuronal networks commonly assessed on resting state functional MRI.

Resting state network	Main components	Salient function
Default mode network	Precuneus/posterior cingulate, lateral parietal, mesial prefrontal cortex	Active when individuals are focused on internal mental-state processes, e.g., interoception, autobiographical memory retrieval, etc.
Sensorimotor network	Precentral gyrus, post-central gyrus, supplementary motor area	Sensory and motor activity
Executive control network	Medial frontal gyrus, superior frontal gyrus, anterior cingulate cortex	Regulates the process of executive function, e.g., integration of sensory and motor information, working memory, etc.
Fronto-parietal network	Inferior frontal gyrus, medial frontal gyrus, precuneus, inferior parietal, angular gyrus	Co-ordinative control in a rapid, accurate, and flexible goal-driven manner
Visual network	*Medial visual*: Striate cortex and medial extra-striate eg. lingual gyrus *Lateral visual*: Occipital pole and occipito-temporal regions	Processing visual information
Auditory network	Superior temporal gyrus, Heschl's gyrus, insula, postcentral gyrus	Processing auditory information
Language network	Superior, middle, inferior frontal gyrus; superior and middle temporal gyrus, insula, angular gyrus, inferior parietal lobule	Language processing

MR spectroscopy

Magnetic resonance spectroscopy (MRS) provides a non-invasive assessment of various biochemical processes within the body. The key neurometabolites measured in the brain using MRS include *N*-acetylaspartate, choline, creatine, lactate, glutamate, glutamine, myoinositol, etc. (Bustillo, 2013). It is a widely used tool for the assessment of various brain pathologies such as tumors, infections, and degenerative processes. However, there are limited studies using MRS for assessing the brain changes secondary to SCI.

Structural reorganization—Insights from MRI

SCI leads to retrograde and anterograde degenerative processes within the spinal cord itself, leading to atrophy (Lundell et al., 2011). The disruption of tract integrity within the spinal cord secondary to injury (Fig. 1) has been elegantly demonstrated on DTI in previous studies (D'souza, Choudhary, Poonia, Kumar, & Khushu, 2017). These changes are however not limited to the spinal cord, but extend upwards into the brain.

In a cohort of 10 subjects with chronic SCI secondary to cervical injury, cord atrophy was seen to be accompanied by a decline in white matter volume in the brain using VBM, especially in the region of the cortico-spinal tract such as the pyramids and cerebral peduncles (Freund et al., 2011). These changes were observed to be accompanied by volumetric changes of gray matter in the denervated leg area of the primary motor (M1) and sensory cortex (S1). A correlation between the degree of cortical atrophy in the primary motor cortex and the extent of clinical impairment was noted in this study. Another study observed a similar correlation between the degree of volume loss in the somato-sensory cortex and the level of functional deficit (Jurkiewicz, Crawley, Verrier, Fehlings, & Mikulis, 2006).

A study performed at our center on a cohort of 30 cases of incomplete spinal cord injury analyzed surface-based measures (cortical thickness, cortical volume, and surface area) using the FreeSurfer 6.0 (Massachusetts General Hospital, Harvard Medical School) toolbox. The study revealed no significant volumetric changes at the early time point (namely within 15 days) post-injury, as compared to healthy, uninjured controls. However, on the follow-up study performed at approximately 6 months post-injury (range 5–8 months), there was a statistically significant decline in cortical surface area, thickness, and volume in the region of the bilateral pre- and post-central gyrus, paracingulate gyrus, the dorsolateral

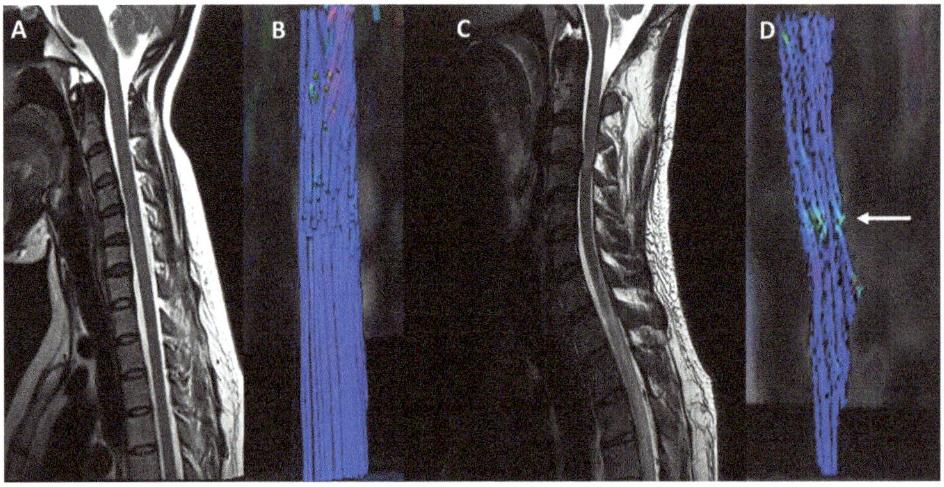

FIG. 1 Normal cervical cord on sagittal MRI (A) with normal tractography image on DTI (B). Sagittal MRI showing focal hyperintensity (*arrow*) in the cervical cord due to edema secondary to trauma (C) with corresponding tractography image (D) showing focal disruption (*arrow*).

pre-frontal and the orbitofrontal (also called the medial pre-frontal) cortex (Fig. 2). The pre- and post-central gyri represent the primary motor and sensory cortices, respectively. The dorsolateral pre-frontal cortex is involved in executive functions such as working memory, cognitive flexibility, planning, etc. The orbitofrontal cortex receives connections from the thalamus and limbic cortex and is thought to represent emotion and reward in decision making. Thus, the impairment of somatosensory input and motor output after SCI affects not just the sensory-motor cortices, but adjacent areas involved in higher-order cognitive function, emotion, etc. DTI studies performed using the tractography approach on the same cohort of incomplete SCI patients at the chronic time point, revealed a disruption in white matter integrity as evidenced by altered datametrics in the form of reduced FA and increased MD. The principal tracts to be involved were the bilateral corticospinal tracts (Fig. 3).

Over and above the observed changes within the sensorimotor systems, loss of gray matter volume has also been seen in the medial pre-frontal and anterior cingulate cortex in a cohort of patients with complete SCI, who sustained an injury to the thoracic cord (Wrigley, Gustin, et al., 2009). These areas play a crucial role in emotional processing. This study also revealed structural changes on whole-brain DTI analysis as evidenced by altered FA and MD in the above-mentioned areas and also in the superior cerebellar cortex. Additionally, structural abnormalities in the corticospinal and corticopontine tracts of the SCI subjects were observed on tractography. Most white matter pathways in the brain are oriented bilaterally. Although bilateral involvement is generally observed on DTI, it is usually asymmetrical (Freund et al., 2012). This

FIG. 2 Cortical regions showing decreased cortical thickness in SCI patients vs controls on surface-based morphometry (thresholded at $P < 0.05$). (A) and (B) Medial aspect of right and left cerebral hemispheres, respectively, with clusters of decreased cortical thickness in the pre- and post-central gyri and orbitofrontal cortex. (C) and (D) Lateral aspect of right and left cerebral hemispheres, respectively, with clusters of decreased cortical thickness in the pre- and post-central gyri and dorso-lateral pre-frontal cortex.

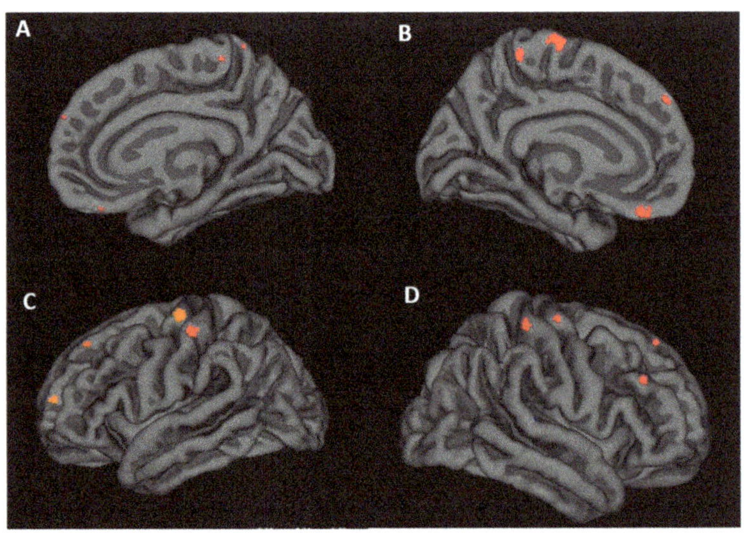

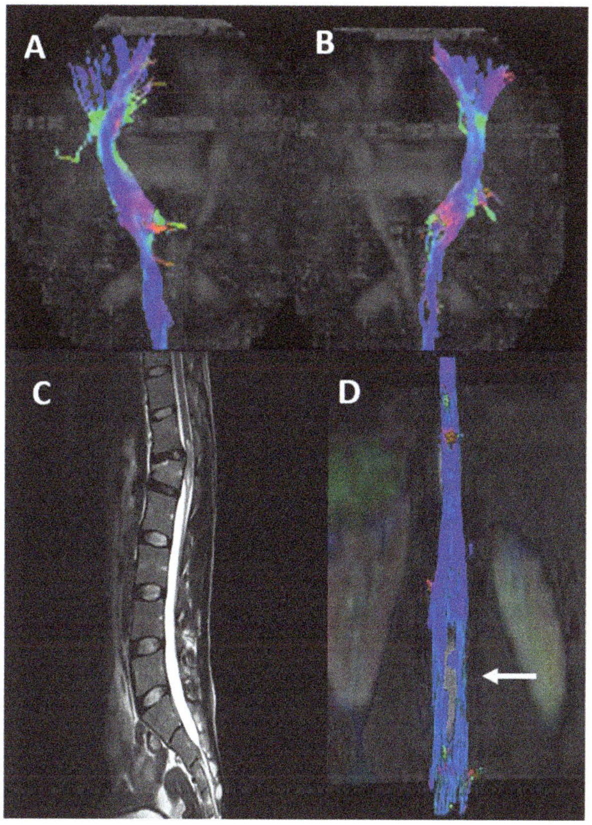

FIG. 3 Tractography images of right (A) and left (B) corticospinal tracts in a patient with dorsolumbar SCI secondary to a vertebral compression fracture as seen on sagittal MRI (C) with corresponding tractography image (D) showing the severed cord.

asymmetry could be partly owing to asymmetric use of bilateral limbs with compensatory overuse of the less-affected limb after SCI.

Changes in regional brain anatomy as determined on DTI, have also been linked to the phenomenon of persistent neuropathic pain after SCI. A study performed on 23 subjects with complete thoracic SCI (12 with neuropathic pain and 11 without) as well as 45 healthy controls, revealed an increase in MD on DTI in areas associated with nociceptive processing and the classic reward circuitry in patients experiencing neuropathic pain (Gustin, Wrigley, Siddall, & Henderson, 2010). These areas included the orbitofrontal, dorsolateral pre-frontal, and posterior parietal cortex as well as the nucleus accumbens. The degree of change in these areas was seen to correlate with the severity of pain.

Functional reorganization—Insights from MRI

Functional MRI (fMRI) studies

Evidence of functional reorganization of the brain was demonstrated using electrophysiological techniques several decades ago. Merzenich et al. used this technique to study the changes in the brain in monkeys post-amputation (Merzenich et al., 1983). They found that the S1 region which was originally innervated by inputs from the amputated digit was now invaded by neighboring S1 representations. Bruehlmeier et al. 18 observed similar changes in a cohort of complete thoracic SCI subjects, with a shift of the functional representation of upper limb digits into the deafferented cortex, originally receiving inputs from the lower limbs (Bruehlmeier et al., 1998). A task-based fMRI study on a cohort of complete thoracic SCI subjects, displayed functional re-organization in the M1 motor cortex, with a significant shift of the activated M1 maxima toward the deafferented cortex on performing tasks involving the spared upper limbs (Lotze, Laubis-Herrmann, Topka, Erb, & Grodd, 1999). A subsequent study observed that SCI patients displayed larger areas of M1 activation on performing tasks that involved muscle groups supplied proximal to the site of spinal injury (Cramer, Lastra, Lacourse, & Cohen, 2005). The degree of the shift was seen to increase with the time elapsed since the occurrence of injury, based on a study which made use of both fMRI and transcranial magnetic stimulation (Lotze, Laubis-Herrmann, & Topka, 2006).

Motor imagery (MI) is a popular technique, being employed in the rehabilitation of SCI patients with a view to the restoration of motor function. This cognitive approach involves imagining or mentally rehearsing movement representations. A study performed on 12 patients with chronic, complete SCI and 12 controls, utilized functional MRI during attempted, and during imagined, right foot movements to study the brain activation patterns (Cramer et al., 2005). It was observed that while many features of normal motor activation were preserved, there were several departures from normal as well. On attempted movement, the key areas to get activated in controls included the primary sensorimotor cortex, supplementary motor area, and cerebellum. Other areas which were activated included the left thalamus, right anterior cingulate, and left superior temporal gyrus. The authors observed a reduced volume of activation in several of these areas in patients with SCI. On imagined movement in controls, the activation pattern involved a cortical–subcortical motor network (involving the left mesial primary sensorimotor cortex, bilateral supplementary area, left basal ganglia, and right cerebellum), motor imitation areas (bilateral superior temporal gyrus and left ventral premotor cortex) as well as attention-related areas (bilateral inferior parietal lobule, left dorsolateral pre-frontal and left anterior cingulate cortex). Although SCI patients activated several of these areas, the volume of significant activation peaks was significantly smaller. Abnormal activation patterns were also observed in SCI subjects such as an increase in pallido–thalamocortical loop activity during attempted movement and abnormal processing in the primary sensorimotor cortex during imagined movement. Additionally, a change in task force level produced an altered modulation of the activation pattern in SCI patients as compared to controls. The authors concluded that while the fundamental aspects of motor system activation were largely intact in SCI patients, there were certain derangements in the activation patterns and their modulation with increasing task demands. This would need to be taken into consideration while planning rehabilitative strategies in SCI.

Another recent study undertaken in incomplete spinal cord injury patients to explore changes in the brain activation patterns during motor execution and motor imagery tasks found several nearly similar activation patterns in both patients and controls (Chen et al., 2016). The task here involved dorsi-plantar flexion of the right ankle. The motor imagery network was seen to be more preserved than the motor execution network in these patients, thus showing promise for the use of motor imagery in rehabilitative regimens. The neural mechanisms of motor imagery are quite similar to those that occur in the early stages of motor execution. Thus activating this network could help in rebuilding and strengthening neural connections. As compared to healthy controls, the incomplete spinal cord injury patients displayed greater activation than controls in the motor-related regions during the motor execution task and lesser activation in the parietal regions during both tasks. However, the cognitive-related regions (namely the left intraparietal lobule and supramarginal gyrus involved in motor attention and planning) showed decreased activation in SCI patients during both the motor execution and imagery tasks. Increased activation patterns in motor regions in SCI patients with complete cord injury as compared to healthy controls have also been observed in other studies (Alkadhi et al., 2005). The apparent differences in the level of activation between different studies could be due to variable study protocol including duration of injury, level of injury, and the specific activation task employed.

Resting-state fMRI studies

Task-based fMRI studies have an inherent limitation—the lack of standardized task protocols which can be uniformly applied to all subjects (Kokotilo, Eng, & Curt, 2009). Resting-state fMRI (rsfMRI) detects spontaneous low-frequency fluctuations in the blood oxygenation level-dependent (BOLD) contrast signal, while the patient is at rest. Hence it is a much more reproducible technique (Biswal, Yetkin, Haughton, & Hyde, 1995). As mentioned before, it has the added advantage of assessing multiple brain networks simultaneously. A rsfMRI study performed on 11 patients with complete cervical SCI in the chronic phase, revealed reorganization of the sensorimotor network with decreased functional connectivity in motor and sensory cortical regions when compared to controls (Oni-Orisan et al., 2016). Seed-based analysis was performed using motor and sensory regions of interest (ROI) in the bilateral pre- and post-central gyri, respectively. For both the motor ROIs, decreased connectivity was observed with bilateral primary motor and sensory regions as well as the interhemispheric region which serves as a connection between the primary motor and sensory areas. The sensory ROI in the left post-central gyrus revealed a similar pattern of decreased connectivity, while the ROI in the right post-central gyrus showed a significantly reduced connectivity only in the interhemispheric region. The authors conclude that a larger sample size would likely produce similar findings in the right-sided sensory ROI as well. Additionally, the left postcentral ROI demonstrated increased connectivity with the thalamus bilaterally. The increased thalamic connectivity may form the basis for the occurrence of neuropathic pain which is frequently observed in SCI subjects (Wrigley, Press, et al., 2009).

A study performed at our institution assessed the resting state networks on 30 incomplete SCI patients, at approximately 6 months post-injury using the independent component analysis and a dual regression approach. Decreased intra-network connectivity within the somatomotor network was observed in a cluster of regions principally involving the bilateral

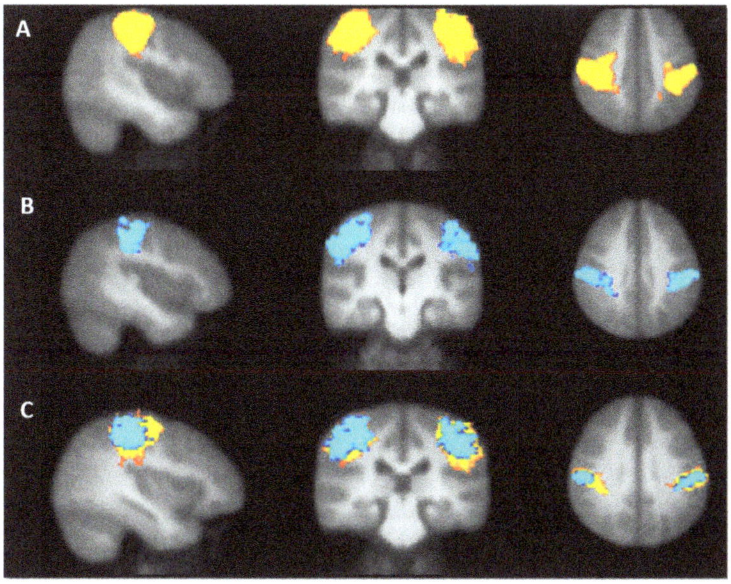

FIG. 4 Resting state fMRI map showing sagittal, coronal, and axial image of the somatomotor network in healthy control (A), SCI (B) and superimposition of healthy control on to SCI with evidence of decreased connectivity in SCI as compared to healthy controls.

pre- and post-central gyri (Fig. 4). There was also a significantly decreased temporal correlation in the default mode network in the region of the right superior frontal, bilateral anterior and posterior cingulate gyri, as compared to healthy controls. The default mode network, which is regarded as a key resting state network, supports internally directed mental activity. It also has an important influence on cognitive control through its interaction with several other brain networks. The diminished connectivity within the default mode network could have implications on cognitive task performance. A decline in motor output has been observed in acute SCI as well, in earlier studies (Hou et al., 2014). The study, performed on 25 patients with acute SCI and 25 healthy controls showed a decreased amplitude of low-frequency fluctuations in the bilateral primary sensorimotor cortex with an increase in the bilateral cerebellum and right orbitofrontal cortex. The study also revealed a decreased inter-hemispheric connectivity between the bilateral primary sensorimotor cortex, and increased intra-hemispheric connectivity within the motor network, including the primary sensorimotor cortex, premotor cortex, supplementary motor area, thalamus, and cerebellum.

Another group recently performed an assessment of resting state networks in a cohort of both complete and incomplete SCI patients (Hawasli et al., 2018). Seed-based analysis was performed on five networks: default-mode, dorsal-attention, salience, control, and somatomotor network. Decreased intra-network connectivity was observed within the somatomotor, default mode, and salience networks with a disrupted inter-network connectivity between the somato-motor and control networks. When the cohort of complete and incomplete SCI patients was analyzed separately with a group of age-matched controls, it was observed that the network connectivity changes were much more pronounced in the complete SCI group. It was also observed that connectivity changed with time: there was decreased connectivity between the primary motor and somatosensory cortex, decreased connectivity between the visual and the primary motor cortex, and increased connectivity between the visual and the sensory parietal cortex. The cerebellum is known to be anti-correlated with the primary motor and sensory cortices. However, this study demonstrated a disruption in this anti-correlation, with an increase in connectivity between the cerebellum and both primary motor and sensory cortices. This finding is corroborated by another recent study using a graph-theory of network efficiency, which found whole-brain connectivity changes, along with a disruption in the connectivity between the cerebellum and the cortex (Kaushal et al., 2017).

In another study on 19 patients with complete SCI (Karunakaran et al., 2020), a reorganization of thalamocortical connections has been observed with increased connectivity between the left pulvinar nucleus and the left inferior frontal gyrus and left inferior parietal lobule in the SCI group. Additionally, there was a decreased connectivity between the bilateral mediodorsal nucleus (of the thalamus) and right superior temporal gyrus and anterior cingulate cortex as well as the left ventrolateral thalamic nucleus and the left superior temporal gyrus. Disruptions in thalamic connectivity could have serious implications in sensory processing, as the thalamus is known to play an important role in multi-sensory integration and affective processing.

MR spectroscopy

This non-invasive technique allows brain chemistry to be studied by the in vivo measurements of specific metabolites such as *N*-acetylaspartate, choline. Creatine and phosphocreatine, etc., limited studies on SCI have been performed using MR spectroscopy. An MR spectroscopy of the thalamus was performed in SCI patients with paraplegia with and without neuropathic pain as compared to healthy controls (Puri et al., 1998). The study revealed that the concentration of *N*-acetyl aspartate was negatively correlated with pain intensity, thus pointing toward the fact that the presence of neuropathic pain after SCI is associated with biochemical changes in the thalamus. Another MRS study on patients with incomplete SCI observed an increased level of *N*-acetylaspartate, expressed relative to creatine in the motor cortex as compared to other cortical areas (Mirvis & Geisler, 1990). The authors interpreted the increase in *N*-acetylaspartate to be due to adaptive dendritic sprouting occurring in the reorganized brain secondary to injury.

Structural and functional reorganization—Insights from MRI

An integrated structural and functional assessment was performed by Freund et al. (2012) using a multi-modal MR protocol to study the relationship between spinal cord cross-sectional area, corticospinal tract integrity, and sensorimotor cortex reorganization in traumatic cervical SCI. The most significant finding of the study was that the reduction in corticospinal tract integrity (as assessed by DTI) was seen to be associated with changes in cortical motor function (as assessed by task-based fMRI). The tract integrity was assessed using an ROI approach at multiple levels—the pyramids, the internal capsule, the cerebral peduncle, and the hand area. The tract integrity in the region of the left pyramid was seen to predict increased task-related responses in the leg area of the primary motor cortex, while changes in the cerebral peduncle were predicted by reduced cord area. The increased task-related activation of the left primary motor cortex during right handgrip and its association with reduced FA in the left pyramid containing the corticospinal tract is suggestive of the fact that greater impairment engages additional neuronal resources to maximize motor output.

Another study which investigated reorganization of the primary somatosensory cortex in response to the loss of somatosensory drive from SCI, analyzed 20 subjects with complete thoracic SCI who underwent functional and structural MRI using fMRI and DTI, respectively (Henderson, Gustin, Macey, Wrigley, & Siddall, 2011). The cortical little finger representation (evoked by continuous brushing of the little finger at a pre-determined rate) was seen to move medially toward the deafferented lower body representation. A loss of gray matter volume was seen in the lower body representation, which however, was minimized as reorganization increased. Additionally, as the extent of reorganization increased, the fractional anisotropy decreased and the diffusion direction within the little finger representation was seen to be directed more medially (toward the region that would normally represent the leg) in SCI subjects. This phenomenon has been ascribed to the growth of new lateral connections as part of the re-organization process. The study elegantly demonstrates how the structural and functional changes go hand in hand and can be simultaneously demonstrated by optimizing the MR protocol.

Mechanisms underlying reorganization

Nearly 170 years ago, it was seen that subsequent to the injury of a peripheral nerve, axons lying distal to the injury undergo progressive degeneration, a phenomenon coined as Wallerian degeneration (Buss et al., 2004). Recent evidence suggests that retrograde and antegrade degeneration does indeed occur in the spinal cord of SCI subjects, with changes which extend further up into the brain. A decrease in both the size and numbers of corticospinal neurons as well as alterations in neuronal morphology and synaptic spine density have been reported (Ganchrow & Bernstein, 1985; Hains, Black, & Waxman, 2003; Tetzlaff et al., 1994). Although the use of immunohistochemical techniques provides an in-depth understanding of the morphological changes, it cannot be used in living humans. Electrophysiological studies have been used over the last few decades to glean information on cortical plasticity, post-SCI (Lacourse, Cohen, Lawrence, & Romero, 1999). Imaging studies which can non-invasively offer a high spatial resolution have shed further light on the structural and functional reorganization in the somatomotor cortex, extending into other parts of the brain (Mikulis et al., 2002).

What are the underlying mechanisms responsible for this well-established phenomenon of cortical re-organization that follows the loss of sensory drive and efferent output in SCI? The rapid unmasking of dormant synapses leading to facilitation or disinhibition of existing subthreshold inputs is believed to play a major role (Jacobs & Donoghue, 1991). In fact, investigators have reported that SI reorganization can occur in humans within hours of administering anesthesia, thus lending credence to the existence and unmasking of latent synapses (Björkman, Weibull, Rosén, Svensson, & Lundborg, 2009). With time, gradual mechanisms of subcortical rewiring (Ghosh et al., 2010) and lateral sprouting of dendrites (Henderson et al., 2011) enable the reinnervation of the cortex in SCI. Large cortical shifts that occur following massive sensory losses likely involve multiple mechanisms, which include the relatively rapid unmasking of already existing synapses, but more importantly, the slower growth of new dendrites. Subcortical structures also play a major role in reorganization through the processes mentioned above. For instance, a study on complete SCI in humans demonstrated that the ventral thalamic area that would normally respond to inputs from the upper limb responded to inputs from the neck and occiput (Lenz et al., 1987). This may also influence pain perception following SCI. Finally, it is yet unclear whether the phenomenon of neurogenesis, which is seen to have a definite impact on reorganization in animal studies, has a role to play in the human brain.

Challenges related to MRI evaluation of SCI data

Although substantial research has been undertaken in recent years on the MR study of brain plasticity after SCI, it is marked by sizeable heterogeneity. To begin with, the inclusion criteria in different studies vary from only incomplete SCI (of varying degrees), to complete SCI to a combination of both. Likewise, some investigators restrict themselves to cervical or thoracic cord injury, while others include both. The time since injury is also highly variable, ranging from acute to chronic, with a duration of injury that may span several years. Most studies have a limited sample size. There is also variability in other demographic characteristics such as age and sex of the patients, handedness, etc.

Another reason for the limited reproducibility of findings could be the confounding role played by a multitude of therapeutic and rehabilitative strategies which some patients adopt assiduously, while others do not. In fact, this could account for the differential extent and pattern of volume loss assessed by morphometric studies as the volumetric changes are dynamic, have distinct temporal patterns and not only represent degeneration but also activity-induced changes. An inherent limitation in DTI is its limited ability in detecting crossing and sprouting fibers, as well as unsuitability in the detection of changes in small bundles and nerves due to limited resolution and noise. Thus, miniscule changes induced by dendritic sprouting, etc. have the potential to go undetected. The heterogeneity in post-processing algorithms used in different MR studies could add to the variability of findings. Finally, the issue of assessing the reorganization at the spinal cord level itself remains unaddressed in most studies.

The institution of large-scale, longitudinal, multi-centric studies with uniform selection criteria, study protocols, and data processing algorithms would help to resolve the apparent discrepancy in findings between different studies. This would facilitate the use of MRI as a physiological marker for a comprehensive assessment of brain connectivity changes after SCI, prognostication of functional outcome, and therapeutic monitoring and optimization.

Applications to other areas of neuroscience

In this chapter, we describe the role of MRI in the evaluation of structural and functional reorganization in the brain after spinal cord injury. MRI is a versatile non-invasive modality, which serves as an excellent tool for neuroimaging due to its high spatial resolution, soft tissue contrast, and capability to assess both structure and function. It is thus a useful technique to study brain plasticity in response to a variety of neuropathological conditions arising outside the brain, such as spinal cord injury, limb amputation, brachial plexus injury, etc. It has also been used extensively for studying reorganization in response to intracranial pathologies such as stroke, ischemia, traumatic brain injury, multiple sclerosis, etc. Brain reorganization has also been documented on MRI in physiological states as diverse as normal aging to situations like long-distance space flights. The use of MRI has the potential to serve as a biomarker for prognostication of pathological conditions, as well as guiding and monitoring therapeutic and rehabilitative interventions.

Mini-dictionary of terms

Complete spinal cord injury: Complete loss of sensation and motor function below the level of spinal cord injury.
Incomplete spinal cord injury: Incomplete loss of sensation and motor function below the level of spinal cord injury.
Functional MRI: An MRI technique that measures neuronal activity by detecting changes in blood flow.
Diffusion tensor imaging: An MRI technique that assesses the location, orientation, and integrity of white matter tracts.
Brain plasticity: The ability of the brain to reorganize its structural and functional connections in response to a variety of situations.

Key facts

- An injury to the spinal cord could lead to an impaired information flow to and from the brain.
- The brain thus undergoes reorganization, in response to the loss of sensory input and motor output.

- The phenomenon of cortical plasticity underlies the ensuing recovery of function as well as mal-adaptive changes such as neuropathic pain phenomena.
- MRI is an effective, non-invasive method to assess the level and extent of reorganization.
- Several MR tools such as diffusion tensor imaging, brain morphometry, functional MRI, and MR spectroscopy provide an in-depth understanding of the structural and functional changes in the brain secondary to SCI.
- MRI thus has a tremendous potential to serve as an imaging biomarker for brain reorganization after SCI.

Summary points

- SCI produces morphological and functional reorganization in the brain which can be detected on advanced MRI techniques.
- Diffusion tensor imaging shows alteration in the microstructural integrity of white matter tracts, principally the corticospinal tracts.
- Brain morphometry using voxel-based or surface-based analysis reveals a decline in volume in the somatomotor cortex, as well as remote parts of the brain.
- Functional reorganization with shift of cortical representation toward the deafferented cortex from adjacent areas of the brain is noted on fMRI.
- An alteration in connectivity measures of the neuronal networks is observed on resting state fMRI.
- Perturbations in structure and function progress with time and are modified by rehabilitative interventions.
- Assessment of brain reorganization on MRI is a promising method to predict functional recovery and guide effective therapy in SCI patients.

References

Alkadhi, H., Brugger, P., Boendermaker, S. H., Crelier, G., Curt, A., Hepp-Reymond, M. C., et al. (2005). What disconnection tells about motor imagery: Evidence from paraplegic patients. *Cerebral Cortex*, *15*(2), 131–140. https://doi.org/10.1093/cercor/bhh116.

Basser, P. J., Pajevic, S., Pierpaoli, C., Duda, J., & Aldroubi, A. (2000). In vivo fiber tractography using DT-MRI data. *Magnetic Resonance in Medicine*, *44*(4), 625–632. https://doi.org/10.1002/1522-2594(200010)44:4<625::aid-mrm17>3.0.co;2-o.

Biswal, B., Yetkin, F. Z., Haughton, V. M., & Hyde, J. S. (1995). Functional connectivity in the motor cortex of resting human brain using echo-planar MRI. *Magnetic Resonance in Medicine*, *34*(4), 537–541. https://doi.org/10.1002/mrm.1910340409.

Björkman, A., Weibull, A., Rosén, B., Svensson, J., & Lundborg, G. (2009). Rapid cortical reorganisation and improved sensitivity of the hand following cutaneous anaesthesia of the forearm. *The European Journal of Neuroscience*, *29*(4), 837–844. https://doi.org/10.1111/j.1460-9568.2009.06629.x.

Bruehlmeier, M., Dietz, V., Leenders, K. L., Roelcke, U., Missimer, J., & Curt, A. (1998). How does the human brain deal with a spinal cord injury? *The European Journal of Neuroscience*, *10*(12), 3918–3922. https://doi.org/10.1046/j.1460-9568.1998.00454.x.

Buss, A., Brook, G. A., Kakulas, B., Martin, D., Franzen, R., Schoenen, J., et al. (2004). Gradual loss of myelin and formation of an astrocytic scar during Wallerian degeneration in the human spinal cord. *Brain: A Journal of Neurology*, *127*(Pt 1), 34–44. https://doi.org/10.1093/brain/awh001.

Bustillo, J. R. (2013). Use of proton magnetic resonance spectroscopy in the treatment of psychiatric disorders: A critical update. *Dialogues in Clinical Neuroscience*, *15*(3), 329–337. https://doi.org/10.31887/DCNS.2013.15.3/jbustillo.

Chen, X., Wan, L., Qin, W., Zheng, W., Qi, Z., Chen, N., et al. (2016). Functional preservation and reorganization of brain during motor imagery in patients with incomplete spinal cord injury: A pilot fMRI study. *Frontiers in Human Neuroscience*, *10*, 46. https://doi.org/10.3389/fnhum.2016.00046.

Cramer, S. C., Lastra, L., Lacourse, M. G., & Cohen, M. J. (2005). Brain motor system function after chronic, complete spinal cord injury. *Brain: A Journal of Neurology*, *128*(Pt 12), 2941–2950. https://doi.org/10.1093/brain/awh648.

D'souza, M. M., Choudhary, A., Poonia, M., Kumar, P., & Khushu, S. (2017). Diffusion tensor MR imaging in spinal cord injury. *Injury*, *48*(4), 880–884. https://doi.org/10.1016/j.injury.2017.02.016.

Freund, P., Weiskopf, N., Ward, N. S., Hutton, C., Gall, A., Ciccarelli, O., et al. (2011). Disability, atrophy and cortical reorganization following spinal cord injury. *Brain: A Journal of Neurology*, *134*(Pt 6), 1610–1622. https://doi.org/10.1093/brain/awr093.

Freund, P., Wheeler-Kingshott, C. A., Nagy, Z., Gorgoraptis, N., Weiskopf, N., Friston, K., et al. (2012). Axonal integrity predicts cortical reorganisation following cervical injury. *Journal of Neurology, Neurosurgery, and Psychiatry*, *83*(6), 629–637. https://doi.org/10.1136/jnnp-2011-301875.

Friston, K. J., Holmes, A. P., Worsley, K. J., Poline, J. B., Frith, C. D., & Frackowiak, R. S. J. (1995). Statistical parametric maps in functional imaging: A general linear approach. *Human Brain Mapping*, *2*, 189–210.

Ganchrow, D., & Bernstein, J. J. (1985). Thoracic dorsal funicular lesions affect the Bouton patterns on, and diameters of, layer VB pyramidal cell somata in rat hindlimb cortex. *Journal of Neuroscience Research*, *14*(1), 71–81. https://doi.org/10.1002/jnr.490140107.

Ghosh, A., Haiss, F., Sydekum, E., Schneider, R., Gullo, M., Wyss, M. T., et al. (2010). Rewiring of hindlimb corticospinal neurons after spinal cord injury. *Nature Neuroscience*, *13*(1), 97–104. https://doi.org/10.1038/nn.2448.

Gustin, S. M., Wrigley, P. J., Siddall, P. J., & Henderson, L. A. (2010). Brain anatomy changes associated with persistent neuropathic pain following spinal cord injury. *Cerebral Cortex*, *20*(6), 1409–1419. https://doi.org/10.1093/cercor/bhp205.

Hains, B. C., Black, J. A., & Waxman, S. G. (2003). Primary cortical motor neurons undergo apoptosis after axotomizing spinal cord injury. *The Journal of Comparative Neurology*, *462*(3), 328–341. https://doi.org/10.1002/cne.10733.

Hawasli, A. H., Rutlin, J., Roland, J. L., Murphy, R., Song, S. K., Leuthardt, E. C., et al. (2018). Spinal cord injury disrupts resting-state networks in the human brain. *Journal of Neurotrauma*, *35*(6), 864–873. https://doi.org/10.1089/neu.2017.5212.

Henderson, L. A., Gustin, S. M., Macey, P. M., Wrigley, P. J., & Siddall, P. J. (2011). Functional reorganization of the brain in humans following spinal cord injury: Evidence for underlying changes in cortical anatomy. *The Journal of Neuroscience: The Official Journal of the Society for Neuroscience*, *31*(7), 2630–2637. https://doi.org/10.1523/JNEUROSCI.2717-10.2011.

Hou, J. M., Sun, T. S., Xiang, Z. M., Zhang, J. Z., Zhang, Z. C., Zhao, M., et al. (2014). Alterations of resting-state regional and network-level neural function after acute spinal cord injury. *Neuroscience*, *277*, 446–454. https://doi.org/10.1016/j.neuroscience.2014.07.045.

Jacobs, K. M., & Donoghue, J. P. (1991). Reshaping the cortical motor map by unmasking latent intracortical connections. *Science (New York, N.Y.)*, *251* (4996), 944–947. https://doi.org/10.1126/science.2000496.

Jurkiewicz, M. T., Crawley, A. P., Verrier, M. C., Fehlings, M. G., & Mikulis, D. J. (2006). Somatosensory cortical atrophy after spinal cord injury: A voxel-based morphometry study. *Neurology*, *66*(5), 762–764. https://doi.org/10.1212/01.wnl.0000201276.28141.40.

Kaas, J. H., Qi, H. X., Burish, M. J., Gharbawie, O. A., Onifer, S. M., & Massey, J. M. (2008). Cortical and subcortical plasticity in the brains of humans, primates, and rats after damage to sensory afferents in the dorsal columns of the spinal cord. *Experimental Neurology*, *209*(2), 407–416. https://doi.org/10.1016/j.expneurol.2007.06.014.

Karunakaran, K. D., Yuan, R., He, J., Zhao, J., Cui, J. L., Zang, Y. F., et al. (2020). Resting-state functional connectivity of the thalamus in complete spinal cord injury. *Neurorehabilitation and Neural Repair*, *34*(2), 122–133.

Kaushal, M., Oni-Orisan, A., Chen, G., Li, W., Leschke, J., Ward, B. D., et al. (2017). Evaluation of whole-brain resting-state functional connectivity in spinal cord injury: A large-scale network analysis using network-based statistic. *Journal of Neurotrauma*, *34*, 1278–1282.

Kokotilo, K. J., Eng, J. J., & Curt, A. (2009). Reorganization and preservation of motor control of the brain in spinal cord injury: A systematic review. *Journal of Neurotrauma*, *26*(11), 2113–2126. https://doi.org/10.1089/neu.2008.0688.

Lacourse, M. G., Cohen, M. J., Lawrence, K. E., & Romero, D. H. (1999). Cortical potentials during imagined movements in individuals with chronic spinal cord injuries. *Behavioural Brain Research*, *104*(1–2), 73–88. https://doi.org/10.1016/s0166-4328(99)00052-2.

Lenz, F. A., Tasker, R. R., Dostrovsky, J. O., Kwan, H. C., Gorecki, J., Hirayama, T., et al. (1987). Abnormal single-unit activity recorded in the somatosensory thalamus of a quadriplegic patient with central pain. *Pain*, *31*(2), 225–236. https://doi.org/10.1016/0304-3959(87)90038-8.

Logothetis, N. K. (2008). What we can do and what we cannot do with fMRI. *Nature*, *453*(7197), 869–878. https://doi.org/10.1038/nature06976.

Lotze, M., Laubis-Herrmann, U., & Topka, H. (2006). Combination of TMS and fMRI reveals a specific pattern of reorganization in M1 in patients after complete spinal cord injury. *Restorative Neurology and Neuroscience*, *24*(2), 97–107.

Lotze, M., Laubis-Herrmann, U., Topka, H., Erb, M., & Grodd, W. (1999). Reorganization in the primary motor cortex after spinal cord injury—A functional magnetic resonance (fMRI) study. *Restorative Neurology and Neuroscience*, *14*(2–3), 183–187.

Lundell, H., Barthelemy, D., Skimminge, A., Dyrby, T. B., Biering-Sørensen, F., & Nielsen, J. B. (2011). Independent spinal cord atrophy measures correlate to motor and sensory deficits in individuals with spinal cord injury. *Spinal Cord*, *49*(1), 70–75. https://doi.org/10.1038/sc.2010.87.

Mechelli, A., Price, C., Friston, K., & Ashburner, J. (2005). Voxel-based morphometry of the human brain: Methods and applications. *Current Medical Imaging Reviews*, *1*.

Merzenich, M. M., Kaas, J. H., Wall, J., Nelson, R. J., Sur, M., & Felleman, D. (1983). Topographic reorganization of somatosensory cortical areas 3b and 1 in adult monkeys following restricted deafferentation. *Neuroscience*, *8*(1), 33–55. https://doi.org/10.1016/0306-4522(83)90024-6.

Mikulis, D. J., Jurkiewicz, M. T., McIlroy, W. E., Staines, W. R., Rickards, L., Kalsi-Ryan, S., et al. (2002). Adaptation in the motor cortex following cervical spinal cord injury. *Neurology*, *58*(5), 794–801. https://doi.org/10.1212/wnl.58.5.794.

Mirvis, S. E., & Geisler, F. H. (1990). Intraoperative sonography of cervical spinal cord injury: Results in 30 patients. *American Journal of Neuroradiology*, *11*(4), 755–761.

Oni-Orisan, A., Kaushal, M., Li, W., Leschke, J., Ward, B. D., Vedantam, A., et al. (2016). Alterations in cortical sensorimotor connectivity following complete cervical spinal cord injury: A prospective resting-state fMRI study. *PLoS One*, *11*(3). https://doi.org/10.1371/journal.pone.0150351, e0150351.

Pierpaoli, C., Jezzard, P., Basser, P. J., Barnett, A., & Di Chiro, G. (1996). Diffusion tensor MR imaging of the human brain. *Radiology*, *201*(3), 637–648. https://doi.org/10.1148/radiology.201.3.8939209.

Puri, B. K., Smith, H. C., Cox, I. J., Sargentoni, J., Savic, G., Maskill, D. W., et al. (1998). The human motor cortex after incomplete spinal cord injury: An investigation using proton magnetic resonance spectroscopy. *Journal of Neurology, Neurosurgery, and Psychiatry*, *65*(5), 748–754. https://doi.org/10.1136/jnnp.65.5.748.

Rigon, A., Duff, M. C., McAuley, E., Kramer, A. F., & Voss, M. W. (2016). Is traumatic brain injury associated with reduced inter-hemispheric functional connectivity? A study of large-scale resting state networks following traumatic brain injury. *Journal of Neurotrauma*, *33*(11), 977–989. https://doi.org/10.1089/neu.2014.3847.

Tetzlaff, W., Kobayashi, N. R., Giehl, K. M., Tsui, B. J., Cassar, S. L., & Bedard, A. M. (1994). Response of rubrospinal and corticospinal neurons to injury and neurotrophins. *Progress in Brain Research*, *103*, 271–286. https://doi.org/10.1016/s0079-6123(08)61142-5.

Wrigley, P. J., Gustin, S. M., Macey, P. M., Nash, P. G., Gandevia, S. C., Macefield, V. G., et al. (2009). Anatomical changes in human motor cortex and motor pathways following complete thoracic spinal cord injury. *Cerebral Cortex*, *19*(1), 224–232. https://doi.org/10.1093/cercor/bhn072.

Wrigley, P. J., Press, S. R., Gustin, S. M., Macefield, V. G., Gandevia, S. C., Cousins, M. J., et al. (2009). Neuropathic pain and primary somatosensory cortex reorganization following spinal cord injury. *Pain*, *141*(1–2), 52–59. https://doi.org/10.1016/j.pain.2008.10.007.

Zhang, J., Jones, M., DeBoy, C. A., Reich, D. S., Farrell, J. A., Hoffman, P. N., et al. (2009). Diffusion tensor magnetic resonance imaging of Wallerian degeneration in rat spinal cord after dorsal root axotomy. *The Journal of Neuroscience: The Official Journal of the Society for Neuroscience*, *29*(10), 3160–3171. https://doi.org/10.1523/JNEUROSCI.3941-08.2009.

Chapter 28

Cardiometabolic changes and upper exercise as an augmentative strategy in spinal cord injury

James Bresnahan (J.), Benjamin Scoblionko (R.), Patricia Orme, James Pendleton, and Roger Liu (H.)
Department of Rehabilitation Medicine, Thomas Jefferson University Hospital, Philadelphia, PA, United States

List of abbreviations

AB	able-bodied
ACE	arm crank ergometry
ACSM	American College of Sports Medicine
AIS	American Spinal Injury Association Impairment Scale
ANS	autonomic nervous system
BMR	basal metabolic rate
CMS	cardiometabolic syndrome
CRP	C-reactive protein
CVD	cardiovascular disease
FES	functional electrical stimulation
FFM	fat-free mass
FM	fat mass
GH	human growth hormone
HDL	high density
HOMA-IR	homeostatic model for assessing insulin resistance
IGF	insulin like growth factor
IMF	intramuscular fat
LDL	low-density lipoprotein
LOI	level of injury
OGTT	oral glucose tolerance testing
ROS	reactive oxygen species
SCI	spinal cord injury
VLDL	very low density lipoprotein
WHO	World Health Organization

Introduction

Spinal cord injury (SCI) is traditionally thought about in terms of obvious outward characteristics including paralysis, functional deficits, and loss of sensation. These deficits can grossly be categorized as somatic nervous system dysfunction. In addition to the aforementioned impairment and disabilities, individuals with SCI have less outwardly recognizable autonomic nervous system (ANS) dysfunction related to their injury (Krassioukov et al., 2012). ANS dysfunction leads to a myriad of initially sub-clinical states including neurogenic obesity (Farkas & Gater, 2018b) and neurogenic inflammation (Bloom, Herman, & Spungen, 2020) in addition to the more typically discussed neurogenic bowel, neurogenic bladder, and autonomic dysreflexia. To further complicate ANS dysfunction, level of injury (LOI) and completeness of injury (ISNSCI impairment scale) are correlated to the severity of ANS dysfunction (Bauman, Adkins, Spungen, & Waters, 1999; Bauman & Spungen, 1994; Schilero et al., 2014; Schmid et al., 2000).

Cellular, Molecular, Physiological, and Behavioral Aspects of Spinal Cord Injury. https://doi.org/10.1016/B978-0-12-822427-4.00028-9

Changes in neuropeptides after SCI lead to cardiovascular alterations including hypotension, decreased cardiac output, and decreased resting heart rate (Krassioukov & Claydon, 1990) as well as metabolic derangements including increased LDL, increased triglycerides, and decreased HDL (Gorgey et al., 2014). The broncho-pulmonary system is also significantly impacted by autonomic dysfunction resulting in obstructive lung disease caused by parasympathetic-induced bronchoconstriction and airway inflammation (Schilero et al., 2014; Schilero, Grimm, Bauman, Lenner, & Lesser, 2005) in addition to restrictive lung disease caused by chest wall paralysis (Schilero et al., 2014; Stone & Keltz, 1963).

Cardiovascular disease (CVD) is now the primary cause of death for individuals with SCI greater than 1 year after the initial injury (Bauman & Spungen, 2008; Garshick et al., 2005; Myers, Lee, & Kiratli, 2007). The combination of neurogenic obesity, neurogenic inflammation, cardiovascular alterations, broncho-pulmonary dysfunction, paralysis, barriers to exercise, and poor exercise tolerance due to sympathetic blunting ultimately promote a sedentary lifestyle that pleiotropically promotes cardiometabolic syndrome and increases morbidity and mortality (Farkas & Gater, 2018b; Gater & Farkas, 2017; Gorgey et al., 2014; Gorgey, Dolbow, Dolbow, Khalil, & Gater, 2015; Gorgey, Mather, Poarch, & Gater, 2011).

There is excellent evidence that exercise and increasing daily physical activity is effective in warding off comorbidities in the able-bodied (AB) population (Booth, Roberts, & Laye, 2012). There are also well-defined guidelines for exercise in the AB population (Piercy et al., 2018). Nearly three in four individuals with SCI are interested in being involved in an exercise program (Scelza, Kalpakjian, Zemper, & Tate, 2005) yet only one in four individuals participate in moderate physical activity on a regular basis (Mat Rosly et al., 2018). Unfortunately, individuals with SCI face unique barriers to exercise including functional, physiological, economic, and architectural considerations (Gorgey, 2014; Scelza et al., 2005). Recently, a multinational committee assessed SCI and physical activity with the goal of creating an exercise recommendation and ultimately disseminated a document to all major SCI and exercise organizations suggesting 30 min of moderate to vigorous activity 3 days per week (Martin Ginis, Jetha, MacK, & Hetz, 2010).

Application to other areas of neuroscience

The utility of exercise does not stop at reversing the detrimental effects cardiometabolic changes can have on the SCI patient. Exercise has a profound impact on neuroplasticity, the brain's ability to reprogram and create new neural connections in response to injury and disease. Specifically, in relation to SCI, neuroplasticity allows lost or impaired motor/sensory functions to be restored, via mechanisms of axonal sprouting and neural circuit reconstruction (Liu, Yang, Jiang, Wang, & Yang, 2012). Similarly, exercise in the post-stroke population has demonstrated positive results in promoting neuroplasticity for motor recovery (Mang, Campbell, Ross, & Boyd, 2013). Furthermore, activity and exercise have been linked in decreasing fatigue and depression in the SCI population. A meta-analysis examining physical activity and SCI patient self-report of well-being revealed a statistically significant positive relationship between physical activity and reported well-being (Martin Ginis et al., 2010). Exercise and physical activity has also been studied in regards to pain; in a study of SCI patients undergoing a 9-month exercise program, there was a reduction in pain and reported improvement in quality of life (Hicks et al., 2003). Exercise has also been linked to halting the development of neuropathic pain (Detloff, Smith, Quiros Molina, Ganzer, & Houlé, 2014), a common etiology of pain for the SCI population.

Main narrative text

Role of inflammation

The vast role of inflammation, local and systemic, acute and chronic has been widely explored within the general population. At its most fundamental level, inflammation is the biological response of our immune system to foreign materials, tissue damage, or toxic exposures (Chen et al., 2018). This response usually matches the degree of injury and result in healthy repair. Alternatively, this response may be disproportionate to the original injury, either in duration or intensity, at which point the inflammatory process can become harmful to the body. Chronically elevated levels of inflammation are associated with many comorbid conditions including metabolic syndrome and cardiovascular disease (Tracey, 2007). Inflammatory markers have been investigated more in recent years to better understand the mechanisms by which inflammation drives some of the physiologic and pathologic changes in patients following SCI.

The cytokine theory of disease established a direct connection between the immune system and the aforementioned inflammatory cascades (Tracey, 2007). Cytokines are small proteins locally produced at the site of tissue damage by immune cells (macrophages, T cells, B cells, natural killer cells). Interleukins (IL), tumor necrosis factors (TNF), colony-stimulating factors (CSF), and chemokines are all examples of inflammatory cytokines (Tracey, 2007). In addition, adipokines, a subset of cytokines, are pro-inflammatory mediators secreted by adipocytes (fat cells) (Farkas & Gater, 2018a). Adipokines provide a connection between body fat and inflammation.

Within minutes of a traumatic SCI, pro-inflammatory cytokines at the site of the injury promote activation of peripheral leukocytes and lymphocytes and migration (Burnside & Bradbury, 2014). Inflammation around the cord itself increases the permeability of the blood-spinal cord barrier for up to 8 weeks, allowing for toxins and larger molecules to pass (Schwab, Zhang, Kopp, Brommer, & Popovich, 2014). This places patients at high risk for infection and further damage to the cord.

Astrocytes are glial cells of the brain and spinal cord that support the endothelial cells of the blood-brain barrier. They also play a key role in the acute phase of SCI. Astrocytes release reactive oxygen species (ROS) and increase fibrin deposition and scarring at the site of the injury (Burnside & Bradbury, 2014). This limits axonal regeneration and neural plasticity, further worsening the secondary damage following SCI (Burnside & Bradbury, 2014).

Intraspinal macrophage concentration begins to rise within 1 h of SCI, peaks approximately 5–10 days after injury, and may stay elevated for several months (Fleming et al., 2006). Similarly, intraspinal concentrations of neutrophils rise within hours, and CD4 and CD8 T cells are elevated within weeks of injury (Fleming et al., 2006). This increase in immune cells is mirrored by an increase in inflammatory biomarkers (Kwon et al., 2010). Specifically, concentrations of IL-6, IL-8, IL-16, MCP-1, and IP-10 showed a proportionate increase to the severity of the injury (degree of completeness) and thus have the greatest elevation in American Spinal Injury Association Scale (AIS) A patients (Kwon et al., 2010).

This inflammatory response is the target of many therapeutic interventions in the acute phase following SCI. One of the most widely utilized and polarizing treatments for neural protection is glucocorticoids (Genovese et al., 2009). This has been extensively studied and large sample size has been unable to match the functional benefit to the theoretical, though it remains in favor with many surgeons (Wing, 2008).

In addition to the acute changes at the level of the injury, many chronic inflammatory changes are of particular concern in individuals with SCI as systemic inflammation directly impairs functional recovery (Arnold & Hagg, 2011). Systemic inflammation inversely correlates with mobility status, so higher levels of injury are associated with greater levels of inflammation (Schwab et al., 2014). This is in part because physical activity is a natural anti-inflammatory and individuals with lower levels of injury can be more active. In addition, individuals with a higher level of injury can activate fewer large muscle groups. Large muscle use promotes breakdown during activity and repair following. This energy expenditure reduces the amount of adipose tissue deposition, which is a potent producer of inflammatory mediators (Farkas & Gater, 2018a).

Individuals with SCI have increased adipose deposition compared to the able-bodied individuals with similar activity levels (Farkas & Gater, 2018a). This increases the systemic plasma concentration of adipokines which directly contributes to the rapid progression of metabolic disease and obesity (Farkas, Gorgey, Dolbow, Berg, & Gater, 2018). The level of injury does not change the distribution of new adipose tissue deposition but does impact the overall level of adipokines such that higher levels result in greater pro-inflammatory adipokine levels (Farkas et al., 2018). Following SCI, patients gain an average of 10 kg in the first 2 years following injury (Farkas & Gater, 2018a). In addition, 40%–60% of SCI patients will develop obesity (Shojaei, Alavinia, & Craven, 2017).

Catecholamines also play a key role in the modulation of the inflammatory process and help maintain homeostasis. Catecholamines (dopamine, epinephrine, and norepinephrine) are hormones produced and secreted by the adrenal medulla, which is innervated by T5–9 (Garstang & Miller-Smith, 2007). Injury at or above the T6 level results in decreased circulating plasma concentrations of catecholamines and IL-6 in response to strenuous activity suggesting that sympathetic tone plays a key role in the cytokine inflammatory response to exercise (Paulson, Goosey-Tolfrey, Lenton, Leicht, & Bishop, 2013), which is unfortunately blunted in patients with SCI at or above the T6 level (Bernstein, Damron, Schonberg, & Shapiro, 2009). Higher injuries cause a greater reduction in catecholamine production and sympathetic tone, which directly blunts the anti-inflammatory response (Paulson et al., 2013; Steinberg et al., 2000). In chronic stages, this may contribute to chronically elevated levels of inflammation.

Cardiometabolic changes

Cardiovascular disease

Individuals with spinal cord injuries are at uniquely high risk for the development of cardiovascular disease given the increased prevalence of risk factors and cardiometabolic changes following SCI including greater incidence of obesity, lipid disorders, cardiometabolic syndrome, and diabetes as compared to the AB population. These changes are due to the inflammatory response previously discussed, decrease energy expenditure due to paralysis, and hormonal changes after

SCI (Myers et al., 2007). A decrease in physical activity coupled with decreased anabolic hormone production leads to changes in body composition of muscles predisposing individuals with SCI to pre-mature aging and cardiovascular disease (CVD) (Gorgey et al., 2014). A study of 545 SCI subjects found that the risk of developing CVD was related to the level and extent of injury, finding a 44% increased rate of CVD in those with complete injuries (Groah, Weitzenkamp, Sett, Soni, & Savic, 2001).

Body composition

Following SCI, several structural changes to muscle tissue have been observed, including skeletal muscle atrophy, decreased lean mass, increase in intramuscular fat (IMF), and decrease in fat-free muscle (FFM) (Gorgey et al., 2014). In the first 6 months following SCI, cross-sectional area (CSA) of paralyzed muscles decreased 45%–80% compared to age and weight-matched AB controls (Castro, Apple Jr, Hillegass, & Dudley, 1999). Spungen et al. (2003) examined body composition in chronic SCI and found a significantly lower total body and regional lean mass and also higher fat mass, especially with advancing age Furthermore, in a study of six incomplete SCI patients matched with six controls, there was a threefold higher rate of IMF in the SCI group as well as a notable 26% increase in IMF after 3 months from the initial measurement in the SCI group (Gorgey & Dudley, 2007). A decrease in FFM in combination with decreased sympathetic output results in SCI individuals having a loss of metabolically active muscle mass and reduced basal metabolic rate (BMR) by as much as 14%–27% compared to AB controls (Buchholz & Pencharz, 2004).

Following SCI there are decreased levels of testosterone, human growth hormone (GH), and insulin growth factors (IGF) all of which lead to decreased cellular repair capacity and promote the loss of lean muscle mass and strength (Gorgey et al., 2014). In addition, a study found that individuals 45 years old or younger with SCI had significantly lower mean plasma IGF-I levels when compared to controls (Bauman et al., 1994). Individuals with SCI have decreased total and free testosterone and these values further decline with increasing duration of injury (Tsitouras, Zhong, Spungen, & Bauman, 1995). Consequently, with decreased anabolic hormones there is a resulting increased body FM and decreased FFM that also contributes to an increase in CVD risk for the SCI population (Gorgey et al., 2014). A summary of body composition changes can be found in Table 1.

Lipid disorders

The SCI population is predisposed to dyslipidemia (Schmid et al., 2000). A group of 100 tetraplegic and paraplegic SCI veterans were found to have 16% higher rates of diabetes, carbohydrate disorders at a younger age, decreased high-density lipoprotein (HDL) cholesterol levels, a direct correlation between peak plasma insulin and serum triglyceride (TG), and an inverse correlation between serum TG and HDL cholesterol (Bauman & Spungen, 1994). Lipid derangements post-SCI appear to be level dependent with tetraplegic individuals having worse dyslipidemia (Schmid et al., 2000), further contributing to this populations' increased risk of CVD (Myers et al., 2007).

Cardiometabolic syndrome

Those with SCI are also at high risk for the development of cardiometabolic syndrome (CMS). CMS occurs at a rate of 22.9% in the general US population whereas this increases to a rate of 50% in those with SCI (Beltrán-Sánchez, Harhay, Harhay, & McElligott, 2013; Dyson-Hudson & Nash, 2009). CMS is defined as having three or more of the following: abdominal (central) obesity, hypertension, insulin resistance, and dyslipidemia, all of which are at increased prevalence after SCI (Myers et al., 2007). The prevalence of obesity in the SCI population ranges from 40% to 66% across studies (Shojaei et al., 2017). The SCI population has a high occurrence of sarcopenic obesity (decrease muscle mass in combination with increased fat mass) as well as insulin resistance, further contributing to CMS (Pelletier, Miyatani, Giangregorio, & Craven, 2016). Insulin resistance is

TABLE 1 Body composition changes after SCI.

Increase	Decrease
Intramuscular fat	Fat-free muscle
Total body fat mass	Skeletal muscle
Percent fat per unit body mass index	Bone mass

Summary of changes after SCI as they relate to body composition.

TABLE 2 Hormonal and metabolic changes after SCI.

Increase	Decrease
Serum triglyceride	Total and free testosterone
Very low-density lipoprotein	Human growth hormone
Low-density lipoprotein	Insulin growth factors
Total cholesterol	High-density lipoprotein
Insulin resistance	Bone density

Summary of hormonal and metabolic changes after SCI.

significantly increased after SCI (Bauman et al., 1999; Gorgey et al., 2014) and can be further broken down into hepatic and peripheral insulin sensitivity. Hepatic insulin sensitivity is most commonly measured by fasting insulin concentration and the homeostatic model for assessing insulin resistance (HOMA-IR) while peripheral insulin sensitivity is most commonly measured by assessing insulin and glucose levels to create an "area under the curve" and oral glucose tolerance testing (OGTT) (Wallace, Levy, & Matthews, 2004). A summary of hormonal and metabolic changes can be found in Table 2.

Peripheral vascular regulation

After SCI there is known dysregulation of the autonomic nervous system, stemming from the loss of supraspinal control over the descending sympathetic pathways, leading to the decreased sympathetic output below the level of injury in conjunction with unopposed vagal parasympathetic outflow (Grigorean et al., 2009). The impaired sympathetic response results in low heart rate, low blood pressure, orthostatic hypotension, lack of adaptability, and loss of diurnal fluctuation of blood pressure (Grigorean et al., 2009). Due to the reduced sympathetic response, during exercise those with SCI lack the ability to maintain cardiac output have reduced heart rate variability, and decreased cardiac contractility further contributing to exercise intolerance and deconditioning (Myers et al., 2007; Teasell, Arnold, Krassioukov, & Delaney, 2000).

Role of exercise as a mitigating treatment

The World Health Organization (WHO) recommends at least 150 min per week of moderate-intensity aerobic activity or 75 min per week of vigorous-intensity aerobic activity), plus muscle-strengthening activities twice per week to reduced morbidity and mortality for the general population across a range of medical conditions (WHO, 2010). Physical activity has been shown to improve physical fitness and cardiometabolic health outcomes in people with chronic SCI (Hicks et al., 2011; Myers et al., 2007). Unfortunately, the WHO guidelines were not developed with specific attention to people with SCI, including exercise capacity and feasibility, upper-body over-use injuries (Requejo et al., 2008), skin breakdown (Groah, Schladen, Pineda, & Hsieh, 2015), autonomic dysreflexia (Eldahan & Rabchevsky, 2018), over-heating (Griggs, Price, & Goosey-Tolfrey, 2014), and bone mineral density (Dolbow et al., 2011).

A number of organizations, including the American College of Sports Medicine (ACSM) and SCI Action Canada, have provided recommendations for cardiorespiratory training in SCI (Evans et al., 2015; Ginis et al., 2011). There are differences in the prescribed exercise frequency, intensity, and duration, and resultant cardiometabolic health outcomes but there are clear areas of agreement: (1) aerobic exercise is necessary to maintain or improve cardiorespiratory fitness, (2) moderate to vigorous-intensity exercise is required to induce the metabolic demand necessary for a positive training effect, (3) the duration of aerobic exercise should be a minimum of 20–30 min per session at a frequency of at least 2–3 times per week. A wide variety of exercise intervention strategies have been applied to people with SCI. These interventions include arm-based exercises such as arm crank ergometry (ACE), functional electrical stimulation (FES)-augmented training (Figs. 1 and 2), and robotic-assisted activity training.

Arm-based exercise where the patient initiates and carries through movement has included ACE (Figs. 3 and 4), upper extremity resistance training, wheelchair propulsion, swimming, and circuit resistance training, among others (Nash, Jacobs, Mendez, & Goldberg, 2001; Verellen, Vanlandewijck, Andrews, & Wheeler, 2007). It is more difficult for people with higher-level SCI to participate in volitional exercise owing to functional capacity and increased requirements of staff and equipment. Still, a recent study found that 10 weeks of 30 min per day, 3 days per week ACE at 70% VO_{2Peak} in high

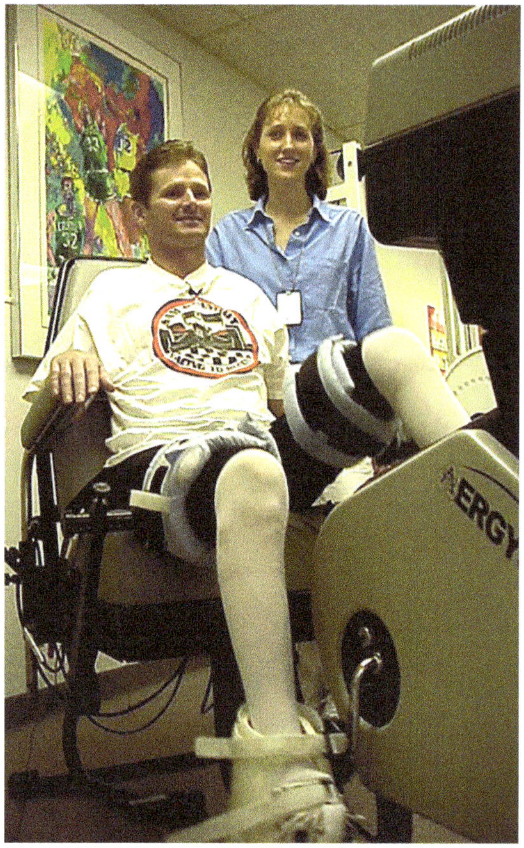

FIG. 1 FES cycling machine. *Permission to use granted from John W. McDonald, MD PhD, International Center for Spinal Cord Injury, Kennedy Krieger Institute. This is an example of an FES cycling machine set up with a patient using it.*

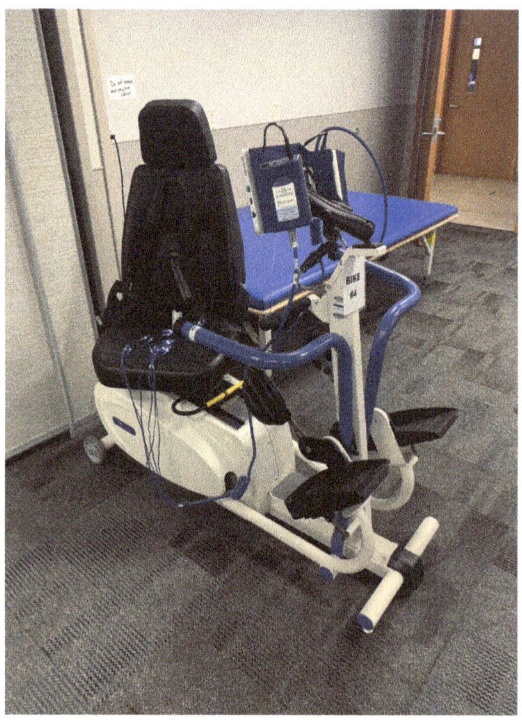

FIG. 2 FES cycling machine. Another example of an FES cycling machine without a patient.

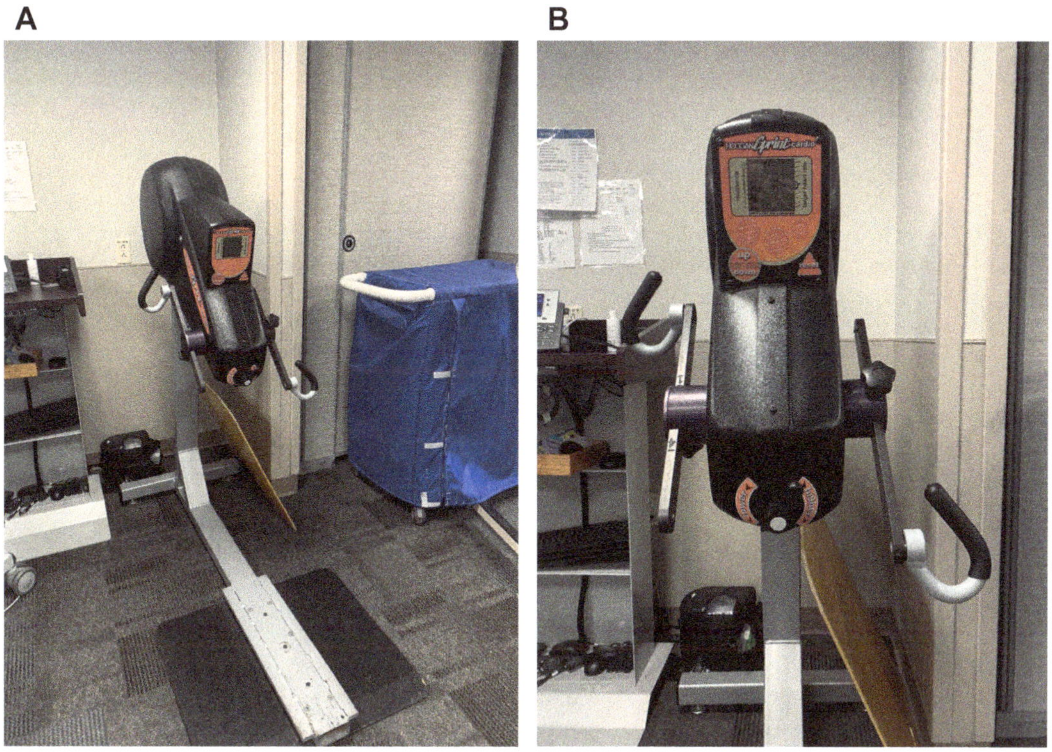

FIG. 3 ACE cycling machine, compact. ACE cycling machine at an appropriate height for wheelchair use.

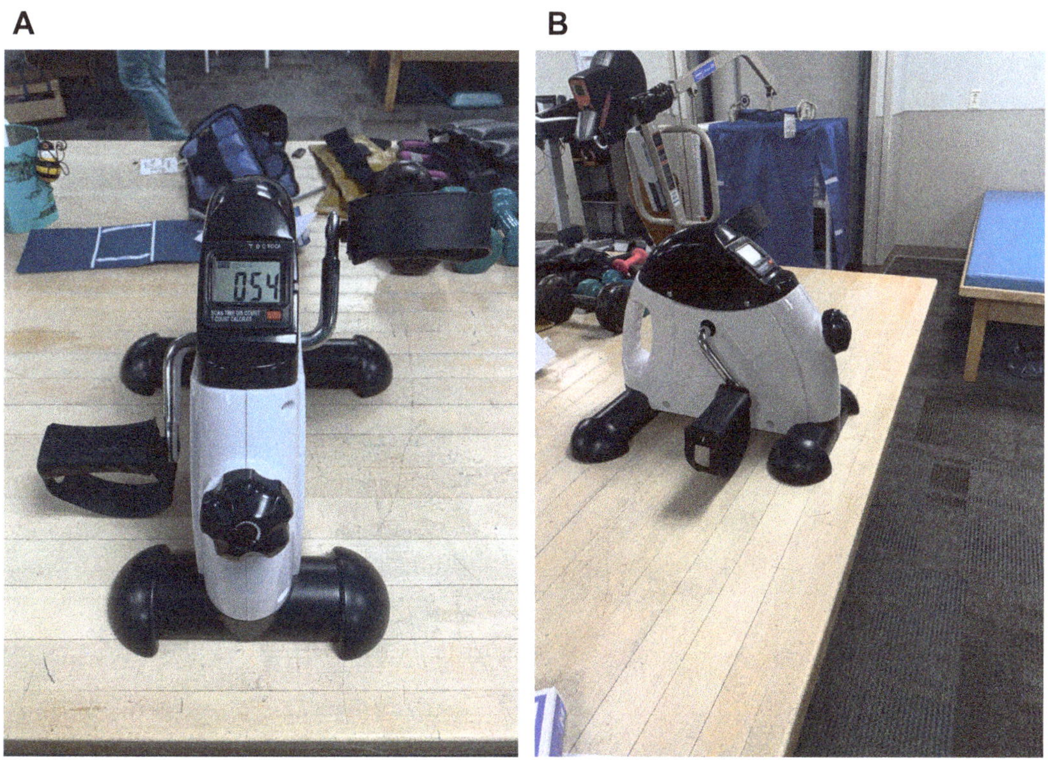

FIG. 4 ACE cycling machine. ACE cycling machine at table height that is portable.

motor complete SCI improves aerobic capacity, community mobility, and metabolic profiles independent of changes in body composition (Bresnahan, Farkas, Clasey, Yates, & Gater, 2019). A recent metaanalysis of 65 studies evaluating upper extremity aerobic exercise training at >75% maximum heart rate or equivalent VO_{2peak} found upper extremity training to be sufficient for improving endurance, waist circumference, and hepatic insulin sensitivity but insufficient for improving fasting glucose, lipid profiles, or resting blood pressure (Farrow et al., 2020). Upper extremity aerobic activity has been shown to improve hepatic insulin resistance but not peripheral insulin resistance (Bresnahan et al., 2019). Glucose intolerance after SCI is at least 70% due to the accumulation of IMF (Elder, Apple, Bickel, Meyer, & Dudley, 2004) and it is well established that upper extremity training alone also does not change body composition (Bresnahan et al., 2019; Farrow et al., 2020; Gorgey et al., 2014). The addition of FES to the lower extremities is beneficial for decreasing IMF (Gorgey, Mather, Cupp, & Gater, 2012) and there is an active trial ongoing to evaluate the potential benefit on insulin resistance as a function of changes in body composition (Gorgey et al., 2019).

Mini-dictionary of terms

Cardiovascular disease: A systemic group of disorders affecting the heart and blood vessels including coronary artery disease, cerebrovascular disease, peripheral arterial disease, rheumatic heart disease, congenital heart disease, deep vein thrombosis, and pulmonary embolism. Cardiovascular disease is the primary cause of death after 1-year post-injury in paraplegic individuals and probably the most common cause of death in tetraplegic individuals greater than 1-year post-injury.

Metabolic syndrome or cardiometabolic syndrome (CMS): A cluster of conditions that together increase the risk of heart attack and stroke. Metabolic syndrome is defined as having at least three of the following five metabolic risk factors: waistline of 35 in. or more for women or 40 or more inches for men, triglyceride level of 150 mg/dL or higher, HDL level of less than 50 mg/dL for women or less than 40 mg/dL for men, blood pressure of 130/85 or higher, and/or a fasting blood sugar of 100 mg/dL or higher.

Functional electrical stimulation (FES): The use of small electrical pulses applied over muscle groups in meaningful synchrony to cause muscle contractions and ultimately cause joints to move in a functionally meaningful manner for the sake of exercising paralyzed muscles and increasing aerobic capacity. FES can be coupled with cycling machines (Figs. 1 and 2) or rowing machines.

Arm-Crank ergometry (ACE): An arm crank or also known as arm cycle device that can be used as a reliable mode of upper extremity exercise to assess physiologic changes (Figs. 3 and 4).

VO_{2max}: The maximum rate of oxygen consumption during exercise, which can provide a quantitative surrogate for car diovascular endurance fitness. Measured in L/min.

Body mass index (BMI): Body mass divided by the square of the body height, expressed in kg/m^2. BMI is a standardized method for evaluating obesity.

Fat mass (FM): Total body mass composed of just fat.

Fat-free mass (FFM): Total body mass composed of internal organs, bone, muscle, water, and connective tissue.

Intramuscular fat (IMF): Fat stored inside muscle fibers in lipid droplets. Excess accumulation of IMF is associated with poor physical fitness, insulin resistance, and type 2 diabetes.

Basal metabolic rate (BMR) or resting energy expenditure (REE): Number of calories used by the body at rest throughout the day.

Total energy expenditure (TEE): Total number of calories used by the body throughout the day due to physical activity in addition to resting energy expenditure.

Key facts: Cardiometabolic changes and exercise in SCI

- Cardiometabolic syndrome occurs in more than 50% of those with SCI.
- Dyslipidemia occurs in a level-dependent fashion in individual with SCI with higher level lesions causing a more dyslipidemic state.
- After SCI there is a cascade of changes in body composition including a decrease in fat-free mass, significant skeletal muscle atrophy, increase in intramuscular fat accumulations and up to 25% decrease in basal metabolic rate.
- There is a level-dependent increase in adipokines and inflammatory markers (CRP, TNF-a, TLRs, IL-6, IL-10, and others) after SCI that contributes to a neurogenic pro-inflammatory state that likely influences the aforementioned cardiometabolic changes.

- Exercise, especially aerobic exercise activating lower extremity paralyzed muscle groups, can help reverse changes in body composition, improve aerobic fitness, improve cardiometabolic profiles, decrease neurogenic inflammation, and improve community mobility.
- Upper extremity exercise alone may not result in robust changes in body composition but can still improve fitness, alter cardiometabolic profiles, and improve community mobility.

Summary points

- Cardiometabolic syndrome is the number one long-term risk factor for death after SCI
- There are four distinct phases of neuroinflammatory response after SCI.
  1. Mechanical trauma leading to disruption of spinal vasculature and the blood-spinal cord barrier.
  2. Secondary injury increases inflammation, promotes cellular apoptosis.
  3. Glial scar formation by activated astrocytes.
  4. Spinal cord remodeling and structural tissue reorganization.
- Individuals with SCI are much less active than their able-bodied peers.
- Individuals with SCI express interest in using exercise as an augmentative strategy to combat cardiometabolic changes but have many physical barriers to accessible exercise programs.
- Exercise after SCI can help reverse changes in body composition, improve cardiometabolic profiles, decrease neurogenic inflammation, and improve community mobility.

References

Arnold, S. A., & Hagg, T. (2011). Anti-inflammatory treatments during the chronic phase of spinal cord injury improve locomotor function in adult mice. *Journal of Neurotrauma*, *28*(9), 1995–2002. https://doi.org/10.1089/neu.2011.1888.

Bauman, W. A., Adkins, R. H., Spungen, A. M., & Waters, R. L. (1999). The effect of residual neurological deficit on oral glucose tolerance in persons with chronic spinal cord injury. *Spinal Cord*, *37*(11), 765–771. https://doi.org/10.1038/sj.sc.3100893.

Bauman, W. A., & Spungen, A. M. (1994). Disorders of carbohydrate and lipid metabolism in veterans with paraplegia or quadriplegia: A model of premature aging. *Metabolism*, *43*(6), 749–756. https://doi.org/10.1016/0026-0495(94)90126-0.

Bauman, W. A., & Spungen, A. M. (2008). Coronary heart disease in individuals with spinal cord injury: Assessment of risk factors. *Spinal Cord*, *46*, 466–476. https://doi.org/10.1038/sj.sc.3102161.

Bauman, W., Spungen, A., Flanagan, S., Zhong, Y.-G., Alexander, L., & Tsitouras, P. (1994). Blunted growth hormone response to intravenous arginine in subjects with a spinal cord injury. *Hormone and Metabolic Research*, *26*(03), 152–156. https://doi.org/10.1055/s-2007-1000798.

Beltrán-Sánchez, H., Harhay, M. O., Harhay, M. M., & McElligott, S. (2013). Prevalence and trends of metabolic syndrome in the adult US population, 1999–2010. *Journal of the American College of Cardiology*, *62*(8), 697–703. https://doi.org/10.1016/j.jacc.2013.05.064.

Bernstein, I. M., Damron, D., Schonberg, A. L., & Shapiro, R. (2009). The relationship of plasma volume, sympathetic tone, and proinflammatory cytokines in young healthy nonpregnant women. *Reproductive Sciences*, *16*(10), 980–985. https://doi.org/10.1177/1933719109338876.

Bloom, O., Herman, P. E., & Spungen, A. M. (2020). Systemic inflammation in traumatic spinal cord injury. *Experimental Neurology*, *325*(2020), 1–11. https://doi.org/10.1016/j.expneurol.2019.113143.

Booth, F. W., Roberts, C. K., & Laye, M. J. (2012). Lack of exercise is a major cause of chronic diseases. *Comprehensive Physiology*, *2*(2), 1143–1211. https://doi.org/10.1002/cphy.c110025.

Bresnahan, J. J., Farkas, G. J., Clasey, J. L., Yates, J. W., & Gater, D. R. (2019). Arm crank ergometry improves cardiovascular disease risk factors and community mobility independent of body composition in high motor complete spinal cord injury. *Journal of Spinal Cord Medicine*, *42*(3), 272–280. https://doi.org/10.1080/10790268.2017.1412562.

Buchholz, A., & Pencharz, P. (2004). Energy expenditure in chronic spinal cord injury. *Current Opinion in Clinical Nutrition and Metabolic Care*, *7*(6), 635–639.

Burnside, E. R., & Bradbury, E. J. (2014). Review: Manipulating the extracellular matrix and its role in brain and spinal cord plasticity and repair. *Neuropathology and Applied Neurobiology*, *40*(1), 26–59. https://doi.org/10.1111/nan.12114.

Castro, M. J., Apple, D. F., Jr., Hillegass, E. A., & Dudley, G. A. (1999). Influence of complete spinal cord injury on skeletal muscle cross-sectional area within the first 6 months of injury. *European Journal of Applied Physiology and Occupational Physiology*, *80*(4), 373–378. https://doi.org/10.1007/s004210050606.

Chen, L., Deng, H., Cui, H., Fang, J., Zuo, Z., Deng, J., et al. (2018, January 23). Inflammatory responses and inflammation-associated diseases in organs. *Oncotarget*. https://doi.org/10.18632/oncotarget.23208.

Detloff, M. R., Smith, E. J., Quiros Molina, D., Ganzer, P. D., & Houlé, J. D. (2014). Acute exercise prevents the development of neuropathic pain and the sprouting of non-peptidergic (GDNF- and artemin-responsive) c-fibers after spinal cord injury. *Experimental Neurology*, *255*, 38–48. https://doi.org/10.1016/j.expneurol.2014.02.013.

Dolbow, D. R., Gorgey, A. S., Daniels, J. A., Adler, R. A., Moore, J. R., & Gater, D. R. (2011). The effects of spinal cord injury and exercise on bone mass: A literature review. *Neuropsychological Rehabilitation*. https://doi.org/10.3233/NRE-2011-0702. article, Netherlands.

Dyson-Hudson, T., & Nash, M. (2009). Guideline-driven assessment of cardiovascular disease and related risks after spinal cord injury. *Topics in Spinal Cord Injury Rehabilitation*, *14*(3), 32–45. https://doi.org/10.1310/sci1403-32.

Eldahan, K. C., & Rabchevsky, A. G. (2018). Autonomic dysreflexia after spinal cord injury: Systemic pathophysiology and methods of management. *Autonomic Neuroscience: Basic & Clinical*. https://doi.org/10.1016/j.autneu.2017.05.002.

Elder, C. P., Apple, D. F., Bickel, C. S., Meyer, R. A., & Dudley, G. A. (2004). Intramuscular fat and glucose tolerance after spinal cord injury—A cross-sectional study. *Spinal Cord*, *42*(12), 711–716. https://doi.org/10.1038/sj.sc.3101652.

Evans, N., Wingo, B., Sasso, E., Hicks, A., Gorgey, A. S., & Harness, E. (2015). Exercise recommendations and considerations for persons with spinal cord injury. *Archives of Physical Medicine and Rehabilitation*, *96*(9), 1749–1750. https://doi.org/10.1016/j.apmr.2015.02.005.

Farkas, G. J., & Gater, D. R. (2018a). Neurogenic obesity and systemic inflammation following spinal cord injury: A review. *The Journal of Spinal Cord Medicine*, *41*(4), 378–387. https://doi.org/10.1080/10790268.2017.1357104.

Farkas, G. J., & Gater, D. R. (2018, July 4). Neurogenic obesity and systemic inflammation following spinal cord injury: A review. *Journal of Spinal Cord Medicine*. https://doi.org/10.1080/10790268.2017.1357104. Taylor and Francis Ltd.

Farkas, G. J., Gorgey, A. S., Dolbow, D. R., Berg, A. S., & Gater, D. R. (2018). The influence of level of spinal cord injury on adipose tissue and its relationship to inflammatory adipokines and cardiometabolic profiles. *The Journal of Spinal Cord Medicine*, *41*(4), 407–415. https://doi.org/10.1080/10790268.2017.1357918.

Farrow, M., Nightingale, T. E., Maher, J., Mckay, C. D., Thompson, D., & Bilzon, J. L. J. (2020). Effect of exercise on Cardiometabolic risk factors in adults with chronic spinal cord injury: A systematic review. *Archives of Physical Medicine and Rehabilitation*, *S0003-9993*, 30283–30285. https://doi.org/10.1016/j.apmr.2020.04.020.

Fleming, J. C., Norenberg, M. D., Ramsay, D. A., Dekaban, G. A., Marcillo, A. E., Saenz, A. D., et al. (2006). The cellular inflammatory response in human spinal cords after injury. *Brain*, *129*(12), 3249–3269. https://doi.org/10.1093/brain/awl296.

Garshick, E., Kelley, A., Cohen, S. A., Garrison, A., Tun, C. G., Gagnon, D., et al. (2005). A prospective assessment of mortality in chronic spinal cord injury. *Spinal Cord*, *43*(7), 408–416. https://doi.org/10.1038/sj.sc.3101729.

Garstang, S. V., & Miller-Smith, S. A. (2007). Autonomic nervous system dysfunction after spinal cord injury. *Physical Medicine and Rehabilitation Clinics of North America*, *18*(2), 275–296. https://doi.org/10.1016/j.pmr.2007.02.003.

Gater, D. R., & Farkas, G. J. (2017). Alterations in body composition after SCI and the mitigating role of exercise. In J. A. Taylor, & L. Morse (Eds.), *The physiology of exercise in spinal cord injury* (pp. 175–199). New York, NY: Springer Healthcare LLC.

Genovese, T., Esposito, E., Mazzon, E., Crisafulli, C., Paterniti, I., Di Paola, R., et al. (2009). PPAR-α modulate the anti-inflammatory effect of glucocorticoids in the secondary damage in experimental spinal cord trauma. *Pharmacological Research*, *59*(5), 338–350. https://doi.org/10.1016/j.phrs.2009.01.018.

Ginis, K., Martin, A., Hicks, A. L., Latimer, A. E., Warburton, D. E. R., Bourne, C., et al. (2011). The development of evidence-informed physical activity guidelines for adults with spinal cord injury. *Spinal Cord*. https://doi.org/10.1038/sc.2011.63. article, England.

Gorgey, A. S. (2014). Exercise awareness and barriers after spinal cord injury. *World Journal of Orthopaedics*, *5*(3), 158–162. https://doi.org/10.5312/wjo.v5.i3.158.

Gorgey, A. S., Dolbow, D. R., Dolbow, J. D., Khalil, R. K., Castillo, C., & Gater, D. R. (2014). Effects of spinal cord injury on body composition and metabolic profile—Part I. *The Journal of Spinal Cord Medicine*, *37*(q), 693–702. https://doi.org/10.1179/2045772314Y.0000000245.

Gorgey, A. S., Dolbow, D. R., Dolbow, J. D., Khalil, R. K., & Gater, D. R. (2015). The effects of electrical stimulation on body composition and metabolic profile after spinal cord injury—Part II. *The Journal of Spinal Cord Medicine*, *38*(1), 23–37. https://doi.org/10.1179/2045772314Y.0000000244.

Gorgey, A. S., & Dudley, G. A. (2007). Skeletal muscle atrophy and increased intramuscular fat after incomplete spinal cord injury. *Spinal Cord*, *45*(4), 304–309. https://doi.org/10.1038/sj.sc.3101968.

Gorgey, A. S., Khalil, R. E., Davis, J. C., Carter, W., Gill, R., Rivers, J., et al. (2019). Skeletal muscle hypertrophy and attenuation of cardio-metabolic risk factors (SHARC) using functional electrical stimulation-lower extremity cycling in persons with spinal cord injury: Study protocol for a randomized clinical trial. *Trials*, *20*(1). https://doi.org/10.1186/s13063-019-3560-8.

Gorgey, A. S., Mather, K. J., Cupp, H. R., & Gater, D. R. (2012). Effects of resistance training on adiposity and metabolism after spinal cord injury. *Medicine and Science in Sports and Exercise*, *44*(1), 165–174. https://doi.org/10.1249/MSS.0b013e31822672aa.

Gorgey, A. S., Mather, K. J., Poarch, H. J., & Gater, D. R. (2011). Influence of motor complete spinal cord injury on visceral and subcutaneous adipose tissue measured by multi-axial magnetic resonance imaging. *The Journal of Spinal Cord Medicine*, *34*(1), 99–109. https://doi.org/10.1179/107902610X12911165975106.

Griggs, K. E., Price, M. J., & Goosey-Tolfrey, V. L. (2014). Cooling athletes with a spinal cord injury. *Sports Medicine*. https://doi.org/10.1007/s40279-014-0241-3.

Grigorean, V., Sandu, A., Popescu, M., Lacobini, M., Stoian, R., Neascu, C., et al. (2009). Cardiac dysfunctions following spinal cord injury. *Journal of Medicine and Life*, *2*(2), 133–145.

Groah, S. L., Schladen, M., Pineda, C. G., & Hsieh, C. H. J. (2015). Prevention of pressure ulcers among people with spinal cord injury: A systematic review. *PM & R: The Journal of Injury, Function, and Rehabilitation*, *7*(6), 613–636. https://doi.org/10.1016/j.pmrj.2014.11.014.

Groah, S. L., Weitzenkamp, D., Sett, P., Soni, B., & Savic, G. (2001). The relationship between neurological level of injury and symptomatic cardiovascular disease risk in the aging spinal injured. *Spinal Cord*, *39*. Retrieved from www.nature.com/sc.

Hicks, A. L., Martin, K. A., Ditor, D. S., Latimer, A. E., Craven, C., Bugaresti, J., et al. (2003). Long-term exercise training in persons with spinal cord injury: Effects on strength, arm ergometry performance and psychological well-being. *Spinal Cord*, *41*(1), 34–43. https://doi.org/10.1038/sj.sc.3101389.

Hicks, A. L., Martin Ginis, K. A., Pelletier, C. A., Ditor, D. S., Foulon, B., & Wolfe, D. L. (2011). The effects of exercise training on physical capacity, strength, body composition and functional performance among adults with spinal cord injury: A systematic review. *Spinal Cord*, *49*(11), 1103–1127. https://doi.org/10.1038/sc.2011.62.

Krassioukov, A., Biering-Sørensen, F., Donovan, W., Kennelly, M., Kirshblum, S., Krogh, K., et al. (2012). International standards to document remaining autonomic function after spinal cord injury. *Journal of Spinal Cord Medicine*, *35*(4), 201–210. https://doi.org/10.1179/1079026812Z.00000000053.

Krassioukov, A., & Claydon, V. E. (1990). The clinical problems in cardiovascular control following spinal cord injury: An overview. *Progress in Brain Research*, *152*. https://doi.org/10.1016/S0079-6123(05)52014-4.

Kwon, B. K., Stammers, A. M. T., Belanger, L. M., Bernardo, A., Chan, D., Bishop, C. M., et al. (2010). Cerebrospinal fluid inflammatory cytokines and biomarkers of injury severity in acute human spinal cord injury. *Journal of Neurotrauma*, *27*(4), 669–682. https://doi.org/10.1089/neu.2009.1080.

Liu, J., Yang, X., Jiang, L., Wang, C., & Yang, M. (2012). Neural plasticity after spinal cord injury. *Neural Regeneration Research*, *7*(5), 386–391. https://doi.org/10.3969/j.issn.1673-5374.2012.05.010.

Mang, C. S., Campbell, K. L., Ross, C. J. D., & Boyd, L. A. (2013). Perspective promoting neuroplasticity for motor rehabilitation after stroke: Brain-derived neurotrophic factor. *Physical Therapy*, *93*(12), 1707–1716.

Martin Ginis, K. A., Jetha, A., MacK, D. E., & Hetz, S. (2010). Physical activity and subjective well-being among people with spinal cord injury: A meta-analysis. *Spinal Cord*, *48*(1), 65–72. https://doi.org/10.1038/sc.2009.87.

Mat Rosly, M., Halaki, M., Hasnan, N., Mat Rosly, H., Davis, G. M., & Husain, R. (2018). Leisure time physical activity participation in individuals with spinal cord injury in Malaysia: Barriers to exercise. *Spinal Cord*, *56*(8), 806–818. https://doi.org/10.1038/s41393-018-0068-0.

Myers, J., Lee, M., & Kiratli, J. (2007). Cardiovascular disease in spinal cord injury: An overview of prevalence, risk, evaluation, and management. *American Journal of Physical Medicine & Rehabilitation*, *86*(2), 142–152. https://doi.org/10.1097/PHM.0b013e31802f0247.

Nash, M. S., Jacobs, P. L., Mendez, A. J., & Goldberg, R. B. (2001). Circuit resistance training improves the atherogenic lipid profiles of persons with chronic paraplegia. *Journal of Spinal Cord Medicine*, *24*(1), 2–9. https://doi.org/10.1080/10790268.2001.11753548.

Paulson, T. A., Goosey-Tolfrey, V. L., Lenton, J. P., Leicht, C. A., & Bishop, N. C. (2013). Spinal cord injury level and the circulating cytokine response to strenuous exercise. *Medicine & Science in Sports & Exercise*, *45*(9), 1649–1655. https://doi.org/10.1249/MSS.0b013e31828f9bbb.

Pelletier, C. A., Miyatani, M., Giangregorio, L., & Craven, B. C. (2016). Sarcopenic obesity in adults with spinal cord injury: A cross-sectional study. *Archives of Physical Medicine and Rehabilitation*, *97*(11), 1931–1937. https://doi.org/10.1016/j.apmr.2016.04.026.

Piercy, K. L., Troiano, R. P., Ballard, R. M., Carlson, S. A., Fulton, J. E., Galuska, D. A., et al. (2018). The physical activity guidelines for Americans. *Journal of the American Medical Association*, *320*(19), 2020–2028. https://doi.org/10.1001/jama.2018.14854.

Requejo, P. S., Mulroy, S. J., Haubert, L. L., Newsam, C. J., Gronley, J. A. K., & Perry, J. (2008). Evidence-based strategies to preserve shoulder function in manual wheelchair users with spinal cord injury. *Topics in Spinal Cord Injury Rehabilitation*. https://doi.org/10.1310/sci1304-86.

Scelza, W. M., Kalpakjian, C. Z., Zemper, E. D., & Tate, D. G. (2005). Perceived barriers to exercise in people with spinal cord injury. *American Journal of Physical Medicine and Rehabilitation*, *84*(8), 576–583. https://doi.org/10.1097/01.phm.0000171172.96290.67.

Schilero, G. J., Grimm, D. R., Bauman, W. A., Lenner, R., & Lesser, M. (2005). Assessment of airway caliber and bronchodilator responsiveness in subjects with spinal cord injury. *Chest*, *127*(1), 149–155. https://doi.org/10.1378/chest.127.1.149.

Schilero, G. J., Radulovic, M., Wecht, J. M., Spungen, A. M., Bauman, W. A., & Lesser, M. (2014). A center's experience: Pulmonary function in spinal cord injury. *Lung*, *192*, 339–346. https://doi.org/10.1007/s00408-014-9575-8.

Schmid, A., Halle, M., Stutzle, C., Konig, D., Baumstark, M. W., Storch, M. J., et al. (2000). Lipoproteins and free plasma catecholamines in spinal cord injured men with different injury levels. *Clinical Physiology*, *20*(4), 304–310. https://doi.org/10.1046/j.1365-2281.2000.00263.x.

Schwab, J. M., Zhang, Y., Kopp, M. A., Brommer, B., & Popovich, P. G. (2014). The paradox of chronic neuroinflammation, systemic immune suppression, autoimmunity after traumatic chronic spinal cord injury. *Experimental Neurology*, *258*, 121–129. https://doi.org/10.1016/j.expneurol.2014.04.023.

Shojaei, M. H., Alavinia, S. M., & Craven, B. C. (2017). Management of obesity after spinal cord injury: A systematic review. *The Journal of Spinal Cord Medicine*, *40*(6), 783–794. https://doi.org/10.1080/10790268.2017.1370207.

Spungen, A. M., Adkins, R. H., Stewart, C. A., Wang, J., Pierson, R. N., Waters, R. L., et al. (2003). Factors influencing body composition in persons with spinal cord injury: A cross-sectional study. *Journal of Applied Physiology*, *95*(6), 2398–2407. https://doi.org/10.1152/japplphysiol.00729.2002.

Steinberg, L. L., Lauro, F. A. A., Sposito, M. M. M., Tufik, S., Mello, M. T., Naffah-Mazzacoratti, M. G., et al. (2000). Catecholamine response to exercise in individuals with different levels of paraplegia. *Brazilian Journal of Medical and Biological Research*, *33*(8), 913–918. https://doi.org/10.1590/S0100-879X2000000800007.

Stone, D. J., & Keltz, H. (1963). The effect of respiratory muscle dysfunction on pulmonary function. Studies in patients with spinal cord injuries. *American Review of Respiratory Diseases*, *88*(5), 621–629.

Teasell, R. W., Arnold, J. M. O., Krassioukov, A., & Delaney, G. A. (2000). Cardiovascular consequences of loss of supraspinal control of the sympathetic nervous system after spinal cord injury. *Archives of Physical Medicine and Rehabilitation*, *81*(4), 506–516. https://doi.org/10.1053/mr.2000.3848.

Tracey, K. J. (2007). Physiology and immunology of the cholinergic antiinflammatory pathway. *Journal of Clinical Investigation*, *117*(2), 289–296. https://doi.org/10.1172/JCI30555.

Tsitouras, P., Zhong, Y.-G., Spungen, A., & Bauman, W. (1995). Serum testosterone and growth hormone insulin-like growth factor-I in adults with spinal cord injury. *Hormone and Metabolic Research*, *27*(06), 287–292. https://doi.org/10.1055/s-2007-979961.

Verellen, J., Vanlandewijck, Y., Andrews, B., & Wheeler, G. D. (2007). Cardiorespiratory responses during arm ergometry, functional electrical stimulation cycling, and two hybrid exercise conditions in spinal cord injured. *Disability and Rehabilitation. Assistive Technology*, *2*(2), 127–132. Article https://doi.org/10.1080/09638280600765712.

Wallace, T. M., Levy, J. C., & Matthews, D. R. (2004). Use and abuse of HOMA modeling. *Diabetes Care*. https://doi.org/10.2337/diacare.27.6.1487.

WHO. (2010). *Global recommendations on physical activity for health* (p. 60). Geneva: World Health Organization. Retrieved from http://medcontent.metapress.com/index/A65RM03P4874243N.pdf%5Cnhttp://scholar.google.com/scholar?hl=en&btnG=Search&q=intitle:Global+Recomendations+on+physical+activity+for+health#0.

Wing, P. C. (2008). Early acute management in adults with spinal cord injury: A clinical practice guideline for health-care providers. Who should read it? *The Journal of Spinal Cord Medicine*, *31*(4), 360. https://doi.org/10.1080/10790268.2008.11760737.

Chapter 29

Electrophysiological outcome measures in spinal cord injury: A new narrative

Radha Korupolu[a,b], Argyrios Stampas[a,b], Sudha S. Tallavajhula[b,c], and Lumy Adams Sawaki[d]

[a]*Physical Medicine and Rehabilitation, University of Texas Health Science Centre, Houston, TX, United States,* [b]*TIRR Memorial Hermann Hospital, Houston, TX, United States,* [c]*Department of Neurology, University of Texas Health Science Centre, Houston, TX, United States,* [d]*Department of Physical Medicine and Rehabilitation, University of Kentucky, Lexington, KY, United States*

List of abbreviations

ALS	amyotrophic lateral sclerosis
CHEPs	contact heat evoked potentials
CMAP	compound muscle action potential
EP	electrophysiological
EMG	electromyography
GABA	gamma-aminobutyric acid
HD	high density
H-reflex	Hoffman reflex
ITB	intrathecal baclofen
MEPs	motor evoked potentials
MUAP	motor unit action potential
MUNE	motor unit number estimation
MVC	maximum voluntary contraction
NCS	nerve conduction studies
PAD	post-activation depression
rMT	at resting motor thresholds
SCI	spinal cord injury
SP	silent period
SSEPs	somatosensory evoked potentials
SSR	sympathetic skin response
TMS	transcranial magnetic stimulation

Introduction

The assessment of residual sensory, motor, and autonomic function after spinal cord injury (SCI) is frequently measured by electrophysiological (EP) measures in clinical and research settings over the last 2 decades (Hubli et al., 2019; Korupolu, Stampas, Singh, Zhou, & Francisco, 2019). EP measures can detect subtle changes in the nervous system and improve prognostication and assessment of neuro-recovery after spinal cord injury (SCI) (Curt & Dietz, 1999; Curt, Rodic, Schurch, & Dietz, 1997; Hupp et al., 2018; Iseli, Cavigelli, Dietz, & Curt, 1999; Korupolu et al., 2019). EP measures can also evaluate the effects of an intervention on neurophysiology and provide valuable mechanistic data to guide future treatments. Electrophysiological assessments can complement clinical evaluations, such as the International Standards for Neurological Classification of SCI (ISNCSCI), and other standard clinical evaluations after SCI (Curt et al., 1997; Curt & Dietz, 1999; Hupp et al., 2018; Iseli et al., 1999; Korupolu et al., 2019; Xie & Boakye, 2008). The combination of EP measures used in conjunction with standard clinical evaluations can provide qualitative and quantitative information that can guide rehabilitation treatments and improve functional outcomes prediction. With the growing usage of EP measures in SCI research studies, there is a need to standardize the measurement and reporting methods of EP measures. Due to the lack of standard guidelines, there is significant variation in reporting of EP measures. Standardization of reporting of EP

Cellular, Molecular, Physiological, and Behavioral Aspects of Spinal Cord Injury. https://doi.org/10.1016/B978-0-12-822427-4.00029-0

measures in SCI clinical trials could provide better comparative data, which will benefit our field by improving our understanding of these measures in relation to neurologic recovery.

In addition, the sample size in SCI clinical trials in which EP measures were reported is small and lacks justification (Korupolu et al., 2019), limiting our ability to assess an intervention's efficacy. If studies consistently reported the EP measures in a standardized fashion, it would allow us to compare or perform a meta-analysis of studies, providing valuable information for the scientific community.

This chapter will discuss commonly reported EP outcome measures in SCI clinical trials, the clinical and research application of these measures (Table 1), current methods of measurement, and reporting. Electromyography (EMG), motor evoked potentials (MEPs), somatosensory evoked potentials (SSEPs), and H-reflex are typical EP outcome measures in SCI clinical trials. Other infrequently reported electrophysiological measures include reflex EMG activity, nerve conduction studies (NCS), silent period, contact heat evoked potentials (CHEPs), and sympathetic skin response (SSR) (Korupolu et al., 2019). EP outcomes were frequently used in interventions studying neuromodulation, stem cells, and gait training to assess neuro recovery (Korupolu et al., 2019). Infrequently, some studies utilized EP outcome measures to detect changes in response to medications, diet, and hypoxia.

Electromyography (EMG)

EMG is an electrodiagnostic tool to record muscles' electrical activity for evaluating neuromuscular disorders. Clinically, it can be measured with a needle or surface electrode to record spontaneous, insertional, and voluntary muscle activity. The needle electrodes measure spontaneous and insertional activity from target muscles, which is then analyzed to report the electrical activity duration, initial deflection of the waveform, amplitude, rate, and rhythm of the firing of electrical potentials. The motor unit action potential (MUAP) is an electrical potential produced by voluntary muscle activity. The MUAP

TABLE 1 Commonly used electrophysiological measures in people with spinal cord injury.

Type	Units of measure	Clinical and research application
Somatosensory evoked potentials	Latency, amplitude	To assess ascending sensory pathways.
Motor evoked potentials	Latency, amplitude	To assess descending motor pathways.
Contact heat evoked potentials	Latency, amplitude	To study small fiber peripheral neuropathy.
Cortical silent period	Duration and latency	To assess corticospinal excitability.
Cutaneous silent period	Duration and latency	To assess spinal motor neuron activity.
Electromyography (EMG)	EMG area, frequency, recruitment, peak amplitude, mean amplitude, motor unit number estimation	To study the pathophysiology of motor neurons and muscles. To assess motor recovery or improvement. To assess progression or improvement of neuromuscular disorders.
Nerve conduction studies	Latency, amplitude, conduction velocities	To assess peripheral nerve disorders.
H reflex	Latency, amplitude	To assess abnormalities of proximal segments of peripheral nerves. To evaluate malfunction of intrathecal baclofen delivery system. To evaluate spinal excitability.
Sympathetic skin response	Waveform, frequency of occurrence, latency, amplitude	To assess the sympathetic sudomotor function.

Describes commonly used electrophysiological measures, units of measures, and application of these measures in clinical and research.

waveform and motor unit firing pattern obtained help to differentiate between a peripheral nerve or muscle pathology. MUAP size in amplitude, number of phases, and frequency of firing are often analyzed clinically.

Characterization of abnormal insertional and spontaneous EMG activity helps differentiate between muscle fiber or nerve pathology. Long duration and large amplitude polyphasic potentials suggest reinnervation from collateral sprouting. Even after maximal muscle contraction, the firing of a few MUAPS suggests reduced recruitment and is commonly seen in nerve disorders. In contrast, the firing of many MUAPS, even with a mild muscle contraction, suggests a muscle disorder and known as early recruitment. EMG signals obtained during gait or upper extremity motion studies help analyze and evaluate abnormal muscle activation patterns in clinical settings. However, it is challenging to interpret EMG activity in people with chronic SCI due to atrophy of muscles and baseline abnormal activity below the spinal cord injury level, thus limiting its application to diagnose peripheral nerve and muscle disorders in people with SCI (Kirshblum, Lim, Garstang, & Millis, 2001).

Research application

In the majority of clinical trials, EMG signals/activity are acquired from surface electrodes placed directly on the skin (Chhabra et al., 2016; Gorassini, Norton, Nevett-Duchcherer, Roy, & Yang, 2009; Houldin, Luttin, & Lam, 2011; Kawashima, Nozaki, Abe, & Nakazawa, 2008; Lam, Wirz, Lunenburger, & Dietz, 2008; Lima et al., 2010; Mazzoleni et al., 2011; Trumbower, Hayes, Mitchell, Wolf, & Stahl, 2017). EMG activity obtained via surface electrodes is frequently reported as mean EMG amplitude, sometimes as peak EMG amplitude, and few times as the presence or absence of any EMG activity pre-post-intervention. Other infrequent methods of reporting EMG results include EMG area, median frequency, and recruitment.

In some studies, improvement in EMG activity was associated with improved motor strength and locomotion (Gorassini et al., 2009; Mazzoleni et al., 2011; Trumbower et al., 2017). In a systematic review, EMG activity was reported as the most commonly used EP measures in SCI clinical trials (Korupolu et al., 2019). More recently, newer EMG techniques such as high density (HD) surface EMG have emerged, which provides motor unit number estimation (MUNE) (Bromberg, 2004) and have been used to detect a change in motor innervation zones (Afsharipour, Sandhu, Rasool, Suresh, & Rymer, 2016).

Reflex EMG activity

Reflex EMG activity is applied in clinical trials as an outcome measure to study the effects of various interventions on reflex excitability. In these studies, researchers recorded reflex EMG activity to quantify spasticity before and after the application of an intervention (D'Amico, Li, Bennett, & Gorassini, 2013; Khan, Patrick, Roy, Gorassini, & Yang, 2016; Kumru et al., 2010). Reflex EMG activity is measured from target muscles using surface electrodes in response to cutaneous stimulation. Clinicians and researchers have used different terms to report this reflex activity. For example, in three studies (D'Amico et al., 2013; Khan et al., 2016; Zewdie, Roy, Yang, & Gorassini, 2015), reflex EMG activity was reported as a **cutaneous muscular reflex which** is evoked by electrical stimulation of the tibial nerve behind the medial malleolus or median arch of the foot (D'Amico et al., 2013; Khan et al., 2016; Zewdie et al., 2015). In another two studies, it was reported as **withdrawal reflex activity**, or flexion reflex, which is evoked by electrical stimulation of the medial arch of the foot (Kumru et al., 2010; Theiss, Hornby, Rymer, & Schmit, 2011), and sometimes by stimulation of the sural nerve (Hajela, Mummidisetty, Smith, & Knikou, 2013), to produce tibialis anterior EMG activity. The **T-reflex, or T-wave**, is recorded from the soleus muscle after tapping the Achilles tendon (Kumru et al., 2010; Murillo et al., 2011). Regardless of the terminology, reflex EMG activity is consistently reported as mean EMG amplitude. Reduction in reflex EMG activity was associated with decreased spasticity and improved walking (D'Amico et al., 2013; Khan et al., 2016; Zewdie et al., 2015).

Overall EMG activity provides an objective outcome measure in SCI clinical trials to study motor output to assess motor recovery and reflex excitability to assess spasticity, providing mechanistic data. It is essential to understand the mechanism to develop interventions with a specific target to maximize recovery. EMG can be measured non-invasively with minimal discomfort, and it can be localized to target a particular muscle, which makes it an excellent outcome tool. For example, We cannot quantify each muscle involved in elbow flexion (Biceps, brachialis, brachioradialis) separately via standard clinical exam, whereas we can study each muscle separately using EMG.

Evoked potentials

Evoked potentials are the evoked responses generated by the nervous system in response to various external stimuli. Evoked potentials are recorded as electrical signals from target areas to assess ascending sensory or descending motor pathways' integrity (Fig. 1). The most frequently reported evoked potentials in SCI clinical trials are motor evoked potentials (MEPs)

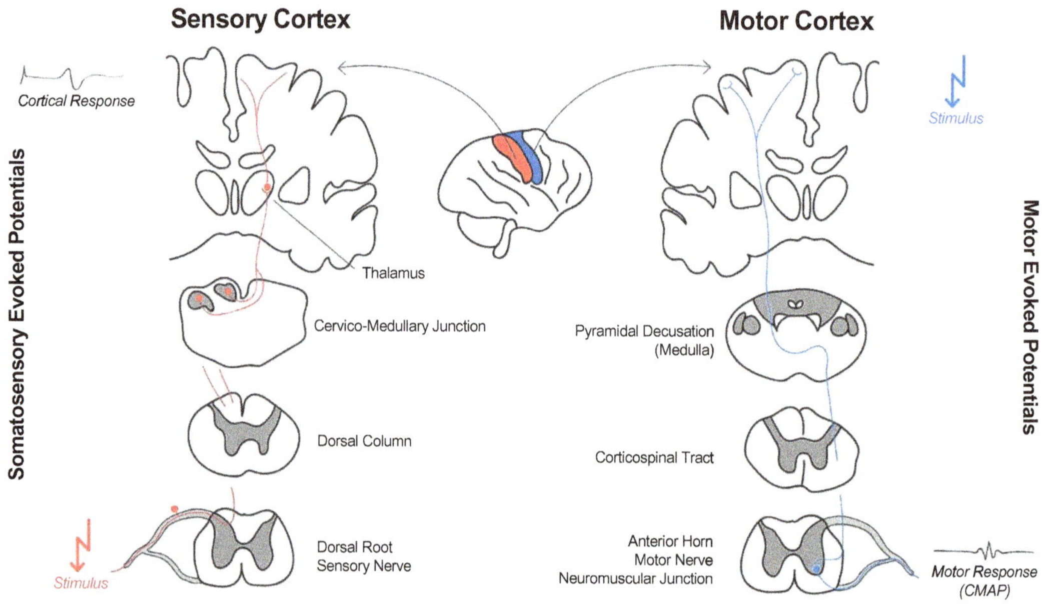

FIG. 1 Pathways of somatosensory and motor evoked potentials. *Courtesy of Maier, S., Goebel, U., Krause, S., Benk, C., Schick, M. A., Buerkle, H., et al. (2018). Principles of transcranial motor evoked potentials and somatosensory evoked potentials.* PLoS One. *Figure. https://doi.org/10.1371/journal.pone.0205410.g001. (An email was sent to PLOS to obtain permission. Permission was granted per their policy to use with acknowledgment.)*

followed by Somatosensory Evoked Potentials (SSEPs) (Korupolu et al., 2019). Contact heat evoked potentials (CHEPs) are infrequently used to study neuropathic pain (Kumru et al., 2012, 2013). We will discuss these evoked potentials in detail below.

Somatosensory evoked potentials (SSEPs)

SSEPs assess the ascending somatosensory conduction pathway integrity, from peripheral nerves to the spinal cord (posterior columns) to the contralateral sensory cortex (Caizhong et al., 2014). An SSEP is recorded from the scalp (over sensory cortex area), spine, and extremities following the electrical stimulation of mixed or sensory nerves with surface electrodes (Fig. 1). Commonly, SSEPs are recorded with stimulation of the median nerve at the wrist for upper extremities and the posterior tibial nerve at the ankle for lower extremities, clinically for diagnostic purposes and in research studies as outcome measures. An electrical stimulus to these nerves activates large myelinated fibers of mixed nerve, action potential travels up toward dorsal root ganglia to the ipsilateral dorsal columns of the spinal cord, then pass synapses in dorsal nuclei at the cervico-medullary junction and ascend toward sensory cortex via medial lemniscus. Clinically latencies of SSEPs are studied to assess abnormalities in ascending sensory pathways and for localization of lesions through ascending sensory pathways. SSEP latencies provide information on the speed of conduction of electrical potentials. For example, prolonged latency suggests abnormal conduction. SSEPs also help predict functional prognosis after SCI (Curt & Dietz, 1999; Iseli et al., 1999). In addition, the presence of tibial and pudendal SSEPs after acute SCI has been shown to predict bladder function recovery (Curt et al., 1997). Other routinely used function of SSEP is to monitor the integrity of the spinal cord during spine surgeries (Pajewski, Arlet, & Phillips, 2007).

Research application

In addition to obtaining SSEPs from the median nerve and tibial nerve; infrequently, SSEPs are obtained from the ulnar nerve and para-vertebral area in SCI clinical trials as an outcome measure. Para-vertebral SSEPs are obtained to report changes in sensory neurologic levels, pre-post intervention. Reporting of SSEPs in clinical trials is inconsistent. Researchers often reported the presence or absence of SSEPs pre-post intervention in most studies; some reported latencies, and others reported amplitudes and latencies (Adel, Hamdy, Afifi, & Mahmoud, 2009; Chhabra et al., 2016; Cristante et al., 2009; Frolov & Bryukhovetskiy, 2012; Grijalva et al., 2010; Lima et al., 2010; Mackay-Sim et al., 2008).

In one study, investigators suggested that recovery of SSEPs could indicate the formation of new synapses or remyelination following stem cell intervention (Cristante et al., 2009). In another stem cell study, an increase in SSEPs amplitudes was considered to indicate enlargement of the amount of the conducting axons. An improvement in latency was considered an indication of remyelination of upper sensory pathways (Frolov & Bryukhovetskiy, 2012).

It is quite challenging to obtain lower extremity SSEPs in people with SCI using surface electrodes in our experience. The presence of ankle edema and body habitus limits stimulation and recording over the spine in many cases. Furthermore, the lower extremity homunculus is located medially and not easily accessible to record SSEPs with surface electrodes. We experienced difficulty obtaining reproducible SSEPs above the injury site after stimulation of mixed nerves from the lower extremities below the injury site, even in people with incomplete SCI, because of these challenges. It is important to note that SSEPs provide information regarding the dorsal columns' connectivity with the sensory cortex.

Motor evoked potentials (MEPs)

Muscle action potentials elicited by stimulation of the motor cortex or spinal cord are known as motor evoked potentials (Fig. 1). MEPs are recorded over target muscles via surface electrodes and are commonly elicited by transcranial or trans-spinal stimulation, either electrical or magnetic. On stimulation of the motor cortex, an electrical signal travels from upper motor neurons located in the motor cortex to the corticospinal tract. It crosses to the contralateral side at the medulla oblongata level. It descends through alpha motor neurons to peripheral nerves resulting in muscle activation. Clinically, MEPs elicited by transcranial electrical stimulation (invasive and non-invasive) can monitor the integrity of the upper motor neuron pathways (motor cortex and corticospinal tract). Monitoring of MEPs is a reliable technique to assess spinal cord ischemia and injury during thoracoabdominal aortic aneurysm repair and spinal surgeries (Pajewski et al., 2007).

Clinical application of MEPs to assess upper motor neuron pathways is currently limited due to the need for expensive equipment, special skills, and training to measure MEPs.

Research application

Non-invasive methods of eliciting MEPs have become popular as an outcome measure in SCI clinical trials to assess neurologic motor recovery following an intervention (Beekhuizen & Field-Fote, 2008; Bolliger et al., 2018; Chhabra et al., 2009, 2016; El-Kheir et al., 2014; Frolov & Bryukhovetskiy, 2012; Gomes-Osman & Field-Fote, 2015a, 2015b; Hoffman & Field-Fote, 2010; Hur et al., 2016; Jette, Cote, Meziane, & Mercier, 2013; Korupolu et al., 2019; Kuppuswamy et al., 2011; Mackay-Sim et al., 2008; Murray et al., 2015; Oh et al., 2016; Shin et al., 2015; Tabakow et al., 2013; Vaquero et al., 2016, 2017; Zewdie et al., 2015; Zhao et al., 2017). MEPs can be non-invasively produced by electrical or magnetic stimulation of the motor cortex or spinal cord (Rossini et al., 2015). However, the use of high voltage electrical stimulation required to elicit MEPs is limited in clinical trials due to local discomfort at the stimulation site (Rossini et al., 2015). In the last two decades, the development of transcranial magnetic stimulation (TMS) to elicit MEPs has become popular as an outcome measure in SCI research studies (Beekhuizen & Field-Fote, 2008; Bolliger et al., 2018; Chhabra et al., 2009, 2016; El-Kheir et al., 2014; Frolov & Bryukhovetskiy, 2012; Gomes-Osman & Field-Fote, 2015a, 2015b; Hoffman & Field-Fote, 2010; Hur et al., 2016; Jette et al., 2013; Korupolu et al., 2019; Kuppuswamy et al., 2011; Mackay-Sim et al., 2008; Murray et al., 2015; Oh et al., 2016; Shin et al., 2015; Tabakow et al., 2013; Vaquero et al., 2017, 2016; Zewdie et al., 2015; Zhao et al., 2017). Magnetic coils or cortical stimulators can be of a different shape to target different areas of the motor cortex. After mapping the cortical region of interest, the devices are held over the target area to stimulate the motor cortex or spinal cord to record the MEPS over the target muscles (Fig. 2). Researchers map the motor cortex to identify the optimal location for obtaining MEPs by obtaining MEPs at resting motor thresholds (rMT). Cortical motor mapping indicates the location and strength of the cortical representation of a given muscle. The rMT is the lowest stimulator intensity required to produce MEPs with an amplitude of at least 50 μV. However, the amplitude size required to determine rMT vary. rMT is a measure of neuronal membrane excitability (Bunday & Perez, 2012; Chen, 2000; Ziemann, Chen, Cohen, & Hallett, 1998). It is reproducible across sessions and is thus extensively used in longitudinal studies (Mills & Nithi, 1997). MEPs can be elicited at various intensities above the resting motor threshold to obtain a recruitment curve. Various methods of collecting MEPs are reported in the current literature. Several researchers obtained MEPs in resting-state at various intensities ranging from 1.1 to 1.4 times resting motor threshold (Beekhuizen & Field-Fote, 2008; Gomes-Osman & Field-Fote, 2015a; Hoffman & Field-Fote, 2010; Jette et al., 2013; Kuppuswamy et al., 2011; Murray et al., 2015; Shin et al., 2015; Stetkarova & Kofler, 2013). Most commonly in SCI clinical trials, MEPs are obtained from the upper limbs, followed by lower limbs, likely due to the challenge in accessing the lower extremity homunculus. There are inconsistencies in methods and reporting of MEPs as an outcome measure in SCI clinic trials (Beekhuizen & Field-Fote, 2008; Chhabra et al., 2009, 2016; El-Kheir et al., 2014; Frolov & Bryukhovetskiy, 2012; Gomes-Osman & Field-Fote, 2015a, 2015b; Hoffman & Field-Fote, 2010; Hur et al., 2016; Korupolu et al., 2019; Kuppuswamy et al., 2011; Mackay-Sim et al., 2008; Murray et al., 2015; Oh et al., 2016; Shin et al., 2015;

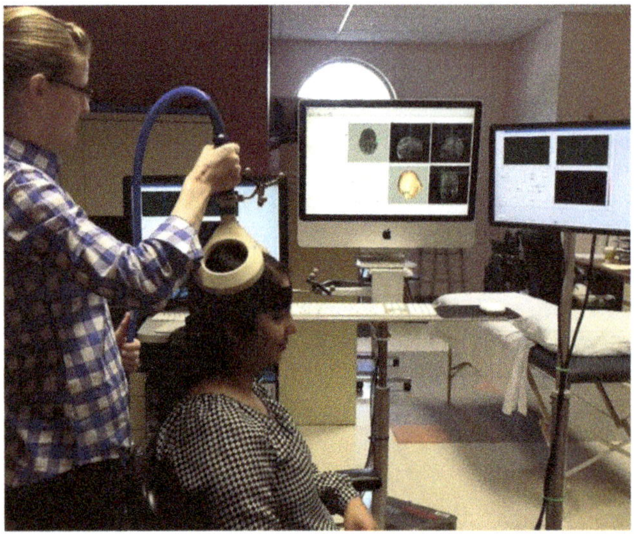

FIG. 2 Transmagenstic stimulation to elicit motor evoked potential. Mapping of the motor cortex to obtain motor evoked potentials using double cone trans magnetic coil for lower extremities.

Tabakow et al., 2013; Vaquero et al., 2016, 2017; Zewdie et al., 2015; Zhao et al., 2017). In a recently published systematic review, some studies reported MEP amplitudes obtained at resting state or during voluntary contraction of a muscle. Other MEP measures reported were MEP latencies, cortical motor map, and recruitment curve.

MEP latencies provide information on conduction in descending motor pathways. A change in latency suggests changes in the myelination, regeneration, and a change in connectivity of the corticospinal tract over time (Xie & Boakye, 2008). Investigators found an association between improvement in MEP amplitudes and improved muscle movement (Wirth, Van Hedel, & Curt, 2008). An increase in the cortical map's spatial volume has been associated with neuroplasticity (Hallett, 2001; Hoffman & Field-Fote, 2010; Sawaki et al., 2008; Wittenberg et al., 2003). The recruitment curve indicates the relationship between stimulus intensity and motor output. An increase in the recruitment curve slope is an indicator of increased neuroplasticity (Chen, 2000; Kaelin-Lang et al., 2002; Thomas & Gorassini, 2005). Investigators reported the presence and absence of MEPs pre-post intervention in few studies. Few researchers collected active MEPs obtained at various percentages of maximum voluntary contraction (MVC), which ranged from 10% to 70% of MVC (Gomes-Osman & Field-Fote, 2015a; Nardone et al., 2017). Definition of active MEPs and stimulator intensities to produce active MEPs also differed in each study (Beekhuizen & Field-Fote, 2008; Gomes-Osman & Field-Fote, 2015b; Mendonca et al., 2014; Nardone et al., 2017; Tabakow et al., 2013). Currently, there is lack of a standard method of collecting and reporting MEPs in SCI clinical trials which is essential to assess the validity of MEPs as an outcome tool to assess motor pathways (Caizhong et al., 2014; Korupolu et al., 2019).

Contact heat evoked potentials (CHEPs)

CHEPs have been used to study small fiber peripheral neuropathy (Kumru et al., 2012, 2013). CHEPs are obtained non-invasively by applying a heat stimulus of 32–51 °C over the skin, and the resulting evoked potentials are recorded over the sensory cortex. CHEPs amplitudes and latencies are utilized to detect small fiber pathologies in people with neuropathic pain. CHEPs are infrequently used to study the effects of an intervention on neuropathic pain in people with SCI (Kumru et al., 2013). CHEPs are not used routinely in the clinical diagnosis of peripheral neuropathy.

H-reflex

The Hoffman reflex (H-reflex) is named after Paul Hoffman, who originally described it in 1910 (Palmieri, Ingersoll, & Hoffman, 2004). It is an electrically induced reflex that bypasses the muscle spindle in contrast to mechanically induced stretch reflex (Knikou & Taglianetti, 2006; Palmieri et al., 2004). It is a compound muscle action potential elicited by low-threshold electrical stimulation of afferent fibers in the mixed nerve with subsequent monosynaptic excitation of alpha motoneurons (Knikou & Taglianetti, 2006; Palmieri et al., 2004). Subsequently, the stimulus is transmitted down the alpha motor neuron axon to the motor end plates and collected using surface electrodes (Burke, 2016; Knikou & Taglianetti, 2006;

Palmieri et al., 2004). H-reflex latencies help diagnose abnormalities in the peripheral nervous system's proximal segments, such as the plexus and nerve roots. Prolonged latencies suggest abnormal nerve conduction. It is commonly elicited by selectively stimulating the Ia fibers of the posterior tibial or median nerve and recorded over the soleus and flexor carpi radialis muscle, respectively (Fig. 3). Clinically, H-reflex latencies are used to study sensory radiculopathy. The H-reflex amplitude helps assess spasticity and study the malfunction of intrathecal baclofen pump delivery systems. A decrease in H-reflex amplitude was seen after intrathecal baclofen (ITB) therapy suggesting a reduction in spasticity (Stokic & Yablon, 2012; Stokic, Yablon, Hayes, Vesovic-Potic, & Olivier, 2006).

Research application

In SCI clinical trials, the H-reflex is used as an outcome measure to assess changes in spinal excitability in response to an intervention. It is necessary to normalize the H-reflex value to obtain between-subject comparisons. Different methods have been advocated to normalize the H-reflex between subjects for comparisons. The most common method of reporting the H-reflex is the ratio of maximum H-reflex to maximum compound muscle action potential (CMAP) amplitude, known as Hmax/Mmax. The M-wave is a CMAP produced by direct supra-maximal stimulation of motor axons. Changes in the ratio of the maximal H-reflex amplitude and maximal M-wave amplitude (Hmax/Mmax) provide a rough estimate of modulation in spinal excitability (D'Amico, Li, Bennett, & Gorassini, 2013; Khan, Patrick, Roy, Gorassini, & Yang, 2016). Higher Hmax/Mmax ratio suggests increased spinal excitability. Other alternatives used for normalization include soleus H-reflex conditioned by peroneal nerve stimulation and plantar stimulation,(Knikou & Mummidisetty, 2014; Piazza et al., 2018) H-reflex obtained at stimulation intensities adjusted with reference to Mmax (Chang et al., 2012, 2013; D'Amico et al., 2013; Knikou & Mummidisetty, 2014), and H-reflex collected at an intensity to evoke 50% of H-max (Piazza et al., 2018). Some researchers reported H-reflex post-activation depression (PAD) at various frequencies by comparing H-reflex amplitudes pre and post-intervention (Chang et al., 2012, 2013; Knikou & Mummidisetty, 2014; Piazza et al., 2018). The H-reflex can be depressed if the interstimulus gap is less than 10 s; this phenomenon is known as post-activation depression. PAD is decreased in people with upper motor neuron disorders, suggesting an increase in spinal excitability or hyperreflexia. A positive association has been found between decreased PAD and the severity of spasticity. Therefore, changes in PAD have been evaluated in clinical trials assessing interventions for spasticity (Adams & Hicks, 2011; Kumru et al., 2010; Murillo et al., 2011; Stetkarova & Kofler, 2013).

The H-reflex is a useful outcome measure to assess spasticity objectively following SCI. However, there are many complexities involved with its collection. In our experience, joint contractures and severe spasticity limit the proper positioning of a person with SCI, which is essentially required to record H-reflex reliably. Anti-spasticity medications alter H-reflex. It is essential to consider the concomitant usage (dosing and timing) of these medications during participation in a research study if H-reflex is recorded as an outcome tool.

Nerve conduction studies (NCS)

NCS provide information on the conduction properties of peripheral nerves and aid in diagnosing peripheral nerve disorders (Mallik & Weir, 2005). Sensory NCS are performed with electrical stimulation of sensory or mixed nerves at a proximal location, and sensory action potentials are recorded over sensory nerves distally. Latencies of sensory nerves of right and

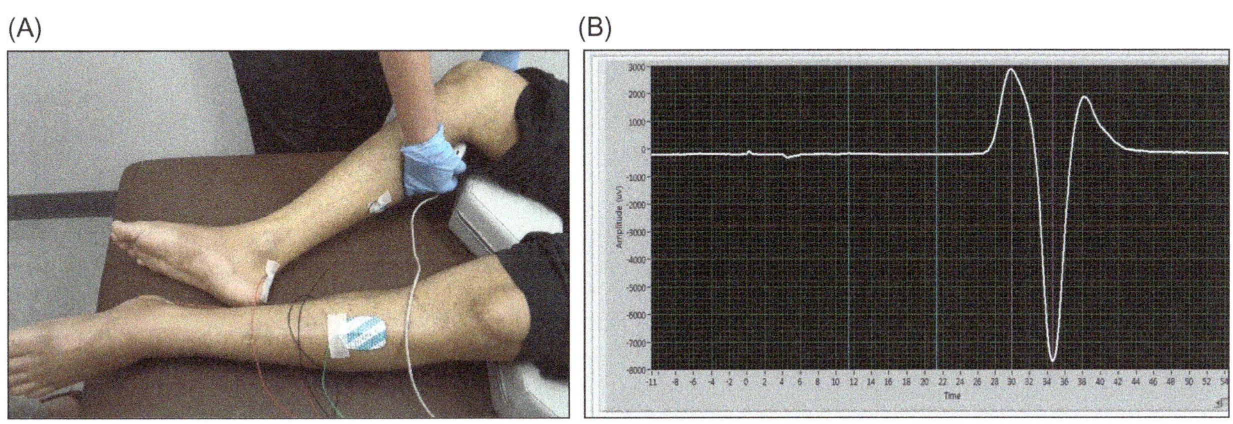

FIG. 3 (A) H reflex from a person with a spinal cord injury. (B) H wave.

left upper extremities are compared to study conduction abnormalities. Motor NCS are performed with electrical stimulation of motor or mixed nerve proximally, and CMAPs are collected via surface electrodes placed over a target muscle. CMAP amplitude and latencies are studied to study abnormalities in conduction. However, the usefulness of NCS in chronic SCI is limited due to muscle atrophy. There were no changes in sural, tibial and peroneal nerve latencies in a study of people with chronic complete cervical SCI, but amplitudes were diminished (Kirshblum et al., 2001). Due to changes in NCS after SCI, routine clinical application of NCS is limited in the diagnosis of peripheral nerve disorders in people with chronic SCI.

Research application

In SCI clinical trials, sensory and motor NCS were obtained from various peripheral nerves as an outcome measure. Conduction velocities were compared in some studies (Allison, Gabriel, Klentrou, Josse, & Ditor, 2017; Hur et al., 2016; Tabakow et al., 2013; Vaquero et al., 2017), and CMAP amplitudes were compared in a few other studies (Allison et al., 2017; Tabakow et al., 2013). CMAP amplitude is dependent on stimulation of the number of axons of a nerve band on the area of muscle fibers served by these axons. For these reasons, serial measurement of CMAP can be a useful tool to assess neuro recovery after acute SCI. Especially to study the recovery of muscles innervated by segments at and around the level of SCI.

F-wave is a low amplitude late response and is elicited by supramaximal electrical stimulation of a mixed or a motor nerve resulting in anti-dromic stimulation of alpha motor neurons (Mallik & Weir, 2005). F-waves may not appear after each stimulus and are inherently variable in shape, latency, and amplitude. F-waves are obtained for the diagnostic evaluation of proximal peripheral nerve lesions in the clinical setting. Infrequently, it is also used as an outcome measure in SCI clinical trials (Murray et al., 2015). F-waves were captured as an outcome measure in a study to investigate the effects of anodal transcranial direct current stimulation on spinal excitability (Murray et al., 2015).

Silent period (SP)

Cortical SP refers to the suppression of EMG activity following a suprathreshold transcranial magnetic stimulation during the contralateral target muscle's voluntary contraction (Poston, Kukke, Paine, Francis, & Hallett, 2012). Current evidence suggests that intracortical inhibition resulting in suppression of muscle activity is Gamma-aminobutyric acid (GABA) receptor-mediated (McDonnell, Orekhov, & Ziemann, 2006). As a result, cortical SPs are mainly used in research studies to assess GABA-mediated pathways. In SCI clinical trials, cortical SPs are studied to evaluate changes in corticospinal excitability in response to baclofen and repetitive trans magnetic stimulation (rTMS) (Kuppuswamy et al., 2011; Stetkarova & Kofler, 2013). In these studies, cortical SP was obtained at various TMS and maximal voluntary contraction intensities. In one study, there was no difference in duration and latency of cortical SP after five sessions of rTMS (Kuppuswamy et al., 2011). Cortical SP latency and duration increased progressively up to 3 h following the ITB dose administration in another study suggesting increased GABA mediated activity (Stetkarova & Kofler, 2013). This finding confirms the usefulness of this tool to study GABA-mediated interventions.

Cutaneous SP is a brief transient period with a loss of voluntary EMG activity following noxious cutaneous nerve stimulation. It is considered a spinal inhibitory reflex produced due to afferent sensory stimulation of small diameter, slow conducting A-delta fibers. Afferent stimulation of A-delta fibers results in suppression of spinal motor neuron activity (Kofler, Frohlich, & Saltuari, 2003; Kofler, Kronenberg, Brenneis, Felber, & Saltuari, 2003). Results of a study following intrathecal baclofen bolus dose suggest that GABA B receptor-mediated interventions do not modulate cutaneous SP reflex pathways (Stetkarova & Kofler, 2013). There was no change in cutaneous SP latency, end latency, and duration following the ITB bolus in this study. The mechanism of cutaneous SP is not known at this time.

Sympathetic skin response (SSR)

Sympathetic skin response (SSR) is a non-invasive measure to assess sympathetic sudomotor function. SSR is generated by efferent small unmyelinated fibers of sweat glands and elicited by stimulation of afferent large myelinated fibers (Vetrugno, Liguori, Cortelli, & Montagna, 2003). SSR is usually collected over the palms and soles after electrical stimulation of the contralateral median and tibial nerves, respectively (Vetrugno et al., 2003). However, SSR can be evoked by various stimuli, including deep breathing, intense emotion, cough, and loud noise (Vetrugno et al., 2003). Latency and amplitudes are compared to normative data to determine abnormality if present. Though SSR is a simple technique, it may be difficult to reproduce and obtain consistently. To our knowledge, the routine clinical use of SSR is not performed in people with SCI.

Research application

To our knowledge, SSR has been used in one SCI clinic trial to investigate the function of the sympathetic system (Kuppuswamy et al., 2011). In this study, SSR was obtained from surface electrodes from the hand's palmar and dorsal surface. SSR was elicited by applying magnetic stimulation to the back of the neck at 65% of maximum stimulator output. The waveform, frequency of occurrence, latency, and amplitude of SSR potentials were collected. However, researchers were able to collect SSR in less than 50% of the subjects. The inability to collect this measure consistently limits its usage as an outcome measure in SCI clinical trials at this time.

Discussion

EP outcome measures are valuable prognostic and diagnostic tools for the quantification of neuromuscular dysfunction after SCI. In SCI clinical trials, EP tools have been used to evaluate neuromodulation of sensory, motor, and autonomic function in response to an intervention. The EP outcome measures provide objective evaluation and eliminate biases of subjective evaluation. However, current evidence suggests a lack of consistency in collecting and reporting EP outcome measures. As a result of this heterogeneity, it is challenging to compare and replicate these studies. The development of standardized methods of collection and reporting of EP outcome measures is critically important to overcome these challenges. A limiting factor for routine incorporation of EP measures in clinical and research settings is the need for expensive equipment, lengthy testing, a standardized environment, and special training to perform these tests.

In summary, this chapter provides information on commonly used EP measures that can complement the clinical assessments and provide objective changes in neurophysiology after SCI. Clinical and functional outcomes cannot be applied during acute phases of injury due to altered mentation, sedation, and inability to follow commands, limiting their application to assess longitudinal recovery from acute to chronic SCI stages. Conversely, EP measures can be applied acutely, even in patients unable to follow commands, providing necessary objective measures. Thus, EP measures can be useful tools in assessing neuromuscular changes throughout the spectrum of recovery after SCI. However, before widespread application for clinical and research purposes, it is essential to validate methods and standardize reporting of these measures.

As our understanding of EP measures in the setting of SCI improves, the application of these measures in SCI research will continue to grow.

Applications to other areas of neuroscience

EP measures are widely used to diagnose and monitor disease progression and recovery of neuromuscular disorders such as neuropathy, amyotrophic lateral sclerosis (ALS), and myopathies. Abnormal EMG findings can assist in differentiating pathologies of lower motor neurons from a muscle disorder. Furthermore, EMG and nerve conduction studies can assist in localizing the site of the lesion. High-density surface EMG and motor unit number estimation are emerging non-invasive techniques for diagnosing and monitoring of amyotrophic lateral sclerosis (ALS) (Bromberg, 2004). MEPs and SEEPs are used extensively to diagnose multiple sclerosis and evaluate prognosis after stroke. MEPs and SSEPs are also recorded during spine and brain surgeries for intra-operative monitoring (Azad et al., 2018; Park, Park, Park, & Lee, 2017). CHEPs are obtained to detect changes in conduction in small fibers, which results in neuropathic pain in various neurological disorders (Madsen, Finnerup, & Baumgartner, 2014). H-reflex and F-waves help diagnose the peripheral nervous system's proximal segments, including plexus and nerve roots (Jerath & Kimura, 2019). Sympathetic skin response has been studied to assess autonomic dysfunction of neurological disorders including multiple sclerosis, Parkinson's disease, and ALS (Vetrugno et al., 2003).

Key facts of electrophysiological measures

- EP measures are useful tools to assess residual function after SCI for objective prognostication. In research studies, EP measures are widely used to assess changes in conduction, connectivity, and regeneration across the injury site.
- Commonly used EP measures in SCI clinical trials include EMG, SSEPs, MEPs, and H-reflex.
- There are no standard guidelines for the collection and reporting of EP measures in SCI clinical trials.

Summary points

- Electrophysiological measures help prognosticate, assess residual function after spinal cord injury, and evaluate pain and spasticity.
- Electrophysiological measures are widely used to evaluate response to intervention in SCI clinical trials.
- Electrophysiological measures are increasingly utilized due to their ability to provide objective measurements and detect subtle changes with quantitative data on neural function.
- Frequently reported electrophysiological outcomes in clinical trials of people with SCI include: electromyography activity, motor-evoked potentials, somatosensory evoked potentials, and H-reflex.
- There is heterogeneity in the methods of measurement and the reporting of EP outcome data that warrant the standardization of electrophysiological measures in SCI clinical trials.

References

Adams, M. M., & Hicks, A. L. (2011). Comparison of the effects of body-weight-supported treadmill training and tilt-table standing on spasticity in individuals with chronic spinal cord injury. *The Journal of Spinal Cord Medicine*, *34*(5), 488–494. https://doi.org/10.1179/2045772311Y.0000000028.

Adel, N. G. H., Hamdy, S., Afifi, L., & Mahmoud, H. (2009). Stem cell therapy in chronic spinal cord injuries. *Egyptian Journal of Neurology, Psychiatry and Neurosurgery*, *46*(2), 467–478.

Afsharipour, B., Sandhu, M. S., Rasool, G., Suresh, N. L., & Rymer, W. Z. (2016). Using surface electromyography to detect changes in innervation zones pattern after human cervical spinal cord injury. *Annual International Conference of the IEEE Engineering in Medicine and Biology Society*, *2016*, 3757–3760. https://doi.org/10.1109/EMBC.2016.7591545.

Allison, D. J., Gabriel, D. A., Klentrou, P., Josse, A. R., & Ditor, D. S. (2017). The influence of chronic inflammation on peripheral motor nerve conduction following spinal cord injury: A randomized clinical trial. *Topics in Spinal Cord Injury Rehabilitation*, *23*(4), 377–385. https://doi.org/10.1310/sci16-00045.

Azad, T. D., Pendharkar, A. V., Nguyen, V., Pan, J., Connolly, I. D., Veeravagu, A., … Grant, G. A. (2018). Diagnostic utility of intraoperative neurophysiological monitoring for intramedullary spinal cord tumors: Systematic review and meta-analysis. *Clinical Spine Surgery*, *31*(3), 112–119. https://doi.org/10.1097/BSD.0000000000000558.

Beekhuizen, K. S., & Field-Fote, E. C. (2008). Sensory stimulation augments the effects of massed practice training in persons with tetraplegia. *Archives of Physical Medicine and Rehabilitation*, *89*(4), 602–608. https://doi.org/10.1016/j.apmr.2007.11.021.

Bolliger, M., Blight, A. R., Field-Fote, E. C., Musselman, K., Rossignol, S., Barthelemy, D., … Steeves, J. (2018). Lower extremity outcome measures: Considerations for clinical trials in spinal cord injury. *Spinal Cord*, *56*(7), 628–642. https://doi.org/10.1038/s41393-018-0097-8.

Bromberg, M. B. (2004). Motor unit number estimation: New techniques and new uses. *Supplements to Clinical Neurophysiology*, *57*, 120–136. https://doi.org/10.1016/s1567-424x(09)70350-2.

Bunday, K. L., & Perez, M. A. (2012). Impaired crossed facilitation of the corticospinal pathway after cervical spinal cord injury. *Journal of Neurophysiology*, *107*(10), 2901–2911. https://doi.org/10.1152/jn.00850.2011.

Burke, D. (2016). Clinical uses of H reflexes of upper and lower limb muscles. *Clinical Neurophysiology Practice*, *1*, 9–17. https://doi.org/10.1016/j.cnp.2016.02.003.

Caizhong, X., Chunlei, S., Beibei, L., Zhiqing, D., Qinneng, D., & Tong, W. (2014). The application of somatosensory evoked potentials in spinal cord injury rehabilitation. *NeuroRehabilitation*, *35*(4), 835–840. https://doi.org/10.3233/NRE-141158.

Chang, Y. J., Liang, J. N., Hsu, M. J., Lien, H. Y., Fang, C. Y., & Lin, C. H. (2013). Effects of continuous passive motion on reversing the adapted spinal circuit in humans with chronic spinal cord injury. *Archives of Physical Medicine and Rehabilitation*, *94*(5), 822–828. https://doi.org/10.1016/j.apmr.2012.11.035.

Chang, S. H., Tseng, S. C., McHenry, C. L., Littmann, A. E., Suneja, M., & Shields, R. K. (2012). Limb segment vibration modulates spinal reflex excitability and muscle mRNA expression after spinal cord injury. *Clinical Neurophysiology*, *123*(3), 558–568. https://doi.org/10.1016/j.clinph.2011.08.001.

Chen, R. (2000). Studies of human motor physiology with transcranial magnetic stimulation. *Muscle & Nerve. Supplement*, *9*, S26–S32.

Chhabra, H. S., Lima, C., Sachdeva, S., Mittal, A., Nigam, V., Chaturvedi, D., … Khan, T. A. H. (2009). Autologous mucosal transplant in chronic spinal cord injury: An Indian pilot study. *Spinal Cord*, *47*(12), 887–895. https://doi.org/10.1038/sc.2009.54.

Chhabra, H. S., Sarda, K., Arora, M., Sharawat, R., Singh, V., Nanda, A., … Tandon, V. (2016). Autologous bone marrow cell transplantation in acute spinal cord injury—An Indian pilot study. *Spinal Cord*, *54*(1), 57–64. https://doi.org/10.1038/sc.2015.134.

Cristante, A. F., Barros-Filho, T. E., Tatsui, N., Mendrone, A., Caldas, J. G., Camargo, A., … Marcon, R. M. (2009). Stem cells in the treatment of chronic spinal cord injury: Evaluation of somatosensitive evoked potentials in 39 patients. *Spinal Cord*, *47*(10), 733–738. https://doi.org/10.1038/sc.2009.24.

Curt, A., & Dietz, V. (1999). Electrophysiological recordings in patients with spinal cord injury: Significance for predicting outcome. *Spinal Cord*, *37*(3), 157–165. https://doi.org/10.1038/sj.sc.3100809.

Curt, A., Rodic, B., Schurch, B., & Dietz, V. (1997). Recovery of bladder function in patients with acute spinal cord injury: Significance of ASIA scores and somatosensory evoked potentials. *Spinal Cord*, *35*(6), 368–373. https://doi.org/10.1038/sj.sc.3100423.

D'Amico, J. M., Li, Y., Bennett, D. J., & Gorassini, M. A. (2013). Reduction of spinal sensory transmission by facilitation of 5-HT1B/D receptors in noninjured and spinal cord-injured humans. *Journal of Neurophysiology*, *109*(6), 1485–1493. https://doi.org/10.1152/jn.00822.2012.

El-Kheir, W. A., Gabr, H., Awad, M. R., Ghannam, O., Barakat, Y., Farghali, H. A., … Sabaawy, H. E. (2014). Autologous bone marrow-derived cell therapy combined with physical therapy induces functional improvement in chronic spinal cord injury patients. *Cell Transplantation*, *23*(6), 729–745. https://doi.org/10.3727/096368913X664540.

Frolov, A. A., & Bryukhovetskiy, A. S. (2012). Effects of hematopoietic autologous stem cell transplantation to the chronically injured human spinal cord evaluated by motor and somatosensory evoked potentials methods. *Cell Transplantation*, *21*(Suppl. 1), S49–S55. https://doi.org/10.3727/096368912X633761.

Gomes-Osman, J., & Field-Fote, E. C. (2015a). Cortical vs. afferent stimulation as an adjunct to functional task practice training: A randomized, comparative pilot study in people with cervical spinal cord injury. *Clinical Rehabilitation*, *29*(8), 771–782. https://doi.org/10.1177/0269215514556087.

Gomes-Osman, J., & Field-Fote, E. C. (2015b). Improvements in hand function in adults with chronic tetraplegia following a multiday 10-Hz repetitive transcranial magnetic stimulation intervention combined with repetitive task practice. *Journal of Neurologic Physical Therapy*, *39*(1), 23–30. https://doi.org/10.1097/NPT.0000000000000062.

Gorassini, M. A., Norton, J. A., Nevett-Duchcherer, J., Roy, F. D., & Yang, J. F. (2009). Changes in locomotor muscle activity after treadmill training in subjects with incomplete spinal cord injury. *Journal of Neurophysiology*, *101*(2), 969–979. https://doi.org/10.1152/jn.91131.2008.

Grijalva, I., Garcia-Perez, A., Diaz, J., Aguilar, S., Mino, D., Santiago-Rodriguez, E., … Madrazo, I. (2010). High doses of 4-aminopyridine improve functionality in chronic complete spinal cord injury patients with MRI evidence of cord continuity. *Archives of Medical Research*, *41*(7), 567–575. https://doi.org/10.1016/j.arcmed.2010.10.001.

Hajela, N., Mummidisetty, C. K., Smith, A. C., & Knikou, M. (2013). Corticospinal reorganization after locomotor training in a person with motor incomplete paraplegia. *BioMed Research International*, *2013*, 516427. https://doi.org/10.1155/2013/516427.

Hallett, M. (2001). Functional reorganization after lesions of the human brain: Studies with transcranial magnetic stimulation. *Revue Neurologique (Paris)*, *157*(8–9 Pt 1), 822–826. 829&population=N&View=abstract.

Hoffman, L. R., & Field-Fote, E. C. (2010). Functional and corticomotor changes in individuals with tetraplegia following unimanual or bimanual massed practice training with somatosensory stimulation: A pilot study. *Journal of Neurologic Physical Therapy*, *34*(4), 193–201. https://doi.org/10.1097/NPT.0b013e3181fbe692.

Houldin, A., Luttin, K., & Lam, T. (2011). Locomotor adaptations and aftereffects to resistance during walking in individuals with spinal cord injury. *Journal of Neurophysiology*, *106*(1), 247–258. https://doi.org/10.1152/jn.00753.2010.

Hubli, M., Kramer, J. L. K., Jutzeler, C. R., Rosner, J., Furlan, J. C., Tansey, K. E., & Schubert, M. (2019). Application of electrophysiological measures in spinal cord injury clinical trials: A narrative review. *Spinal Cord*, *57*(11), 909–923. https://doi.org/10.1038/s41393-019-0331-z.

Hupp, M., Pavese, C., Bachmann, L. M., Koller, R., Schubert, M., & EMSCI Study Group. (2018). Electrophysiological multimodal assessments improve outcome prediction in traumatic cervical spinal cord injury. *Journal of Neurotrauma*, *35*(24), 2916–2923. https://doi.org/10.1089/neu.2017.5576.

Hur, J. W., Cho, T. H., Park, D. H., Lee, J. B., Park, J. Y., & Chung, Y. G. (2016). Intrathecal transplantation of autologous adipose-derived mesenchymal stem cells for treating spinal cord injury: A human trial. *The Journal of Spinal Cord Medicine*, *39*(6), 655–664. https://doi.org/10.1179/2045772315Y.0000000048.

Iseli, E., Cavigelli, A., Dietz, V., & Curt, A. (1999). Prognosis and recovery in ischaemic and traumatic spinal cord injury: Clinical and electrophysiological evaluation. *Journal of Neurology, Neurosurgery, and Psychiatry*, *67*(5), 567–571. https://doi.org/10.1136/jnnp.67.5.567.

Jerath, N., & Kimura, J. (2019). F wave, a wave, H reflex, and blink reflex. *Handbook of Clinical Neurology*, *160*, 225–239. https://doi.org/10.1016/B978-0-444-64032-1.00015-1.

Jette, F., Cote, I., Meziane, H. B., & Mercier, C. (2013). Effect of single-session repetitive transcranial magnetic stimulation applied over the hand versus leg motor area on pain after spinal cord injury. *Neurorehabilitation and Neural Repair*, *27*(7), 636–643. https://doi.org/10.1177/1545968313484810.

Kaelin-Lang, A., Luft, A. R., Sawaki, L., Burstein, A. H., Sohn, Y. H., & Cohen, L. G. (2002). Modulation of human corticomotor excitability by somatosensory input. *The Journal of Physiology*, *540*(Pt 2), 623–633.

Kawashima, N., Nozaki, D., Abe, M. O., & Nakazawa, K. (2008). Shaping appropriate locomotive motor output through interlimb neural pathway within spinal cord in humans. *Journal of Neurophysiology*, *99*(6), 2946–2955. https://doi.org/10.1152/jn.00020.2008.

Khan, A. S., Patrick, S. K., Roy, F. D., Gorassini, M. A., & Yang, J. F. (2016). Training-specific neural plasticity in spinal reflexes after incomplete spinal cord injury. *Neural Plasticity*, *2016*, 6718763. https://doi.org/10.1155/2016/6718763.

Kirshblum, S., Lim, S., Garstang, S., & Millis, S. (2001). Electrodiagnostic changes of the lower limbs in subjects with chronic complete cervical spinal cord injury. *Archives of Physical Medicine and Rehabilitation*, *82*(5), 604–607. https://doi.org/10.1053/apmr.2001.22348.

Knikou, M., & Mummidisetty, C. K. (2014). Locomotor training improves premotoneuronal control after chronic spinal cord injury. *Journal of Neurophysiology*, *111*(11), 2264–2275. https://doi.org/10.1152/jn.00871.2013.

Knikou, M., & Taglianetti, C. (2006). On the methods employed to record and measure the human soleus H-reflex. *Somatosensory & Motor Research*, *23*(1–2), 55–62. https://doi.org/10.1080/08990220600702715.

Kofler, M., Frohlich, K., & Saltuari, L. (2003). Preserved cutaneous silent periods in severe entrapment neuropathies. *Muscle & Nerve*, *28*(6), 711–714. https://doi.org/10.1002/mus.10595.

Kofler, M., Kronenberg, M. F., Brenneis, C., Felber, A., & Saltuari, L. (2003). Cutaneous silent periods in intramedullary spinal cord lesions. *Journal of the Neurological Sciences*, *216*(1), 67–79. https://doi.org/10.1016/s0022-510x(03)00211-9.

Korupolu, R., Stampas, A., Singh, M., Zhou, P., & Francisco, G. (2019). Electrophysiological outcome measures in spinal cord injury clinical trials: A systematic review. *Topics in Spinal Cord Injury Rehabilitation*, *25*(4), 340–354. https://doi.org/10.1310/sci2504-340.

Kumru, H., Murillo, N., Samso, J. V., Valls-Sole, J., Edwards, D., Pelayo, R., ... Pascual-Leone, A. (2010). Reduction of spasticity with repetitive transcranial magnetic stimulation in patients with spinal cord injury. *Neurorehabilitation and Neural Repair*, *24*(5), 435–441. https://doi.org/10.1177/1545968309356095.

Kumru, H., Soler, D., Vidal, J., Navarro, X., Tormos, J. M., Pascual-Leone, A., & Valls-Sole, J. (2013). The effects of transcranial direct current stimulation with visual illusion in neuropathic pain due to spinal cord injury: An evoked potentials and quantitative thermal testing study. *European Journal of Pain*, *17*(1), 55–66. https://doi.org/10.1002/j.1532-2149.2012.00167.x.

Kumru, H., Soler, D., Vidal, J., Tormos, J. M., Pascual-Leone, A., & Valls-Sole, J. (2012). Evoked potentials and quantitative thermal testing in spinal cord injury patients with chronic neuropathic pain. *Clinical Neurophysiology*, *123*(3), 598–604. https://doi.org/10.1016/j.clinph.2011.07.038.

Kuppuswamy, A., Balasubramaniam, A. V., Maksimovic, R., Mathias, C. J., Gall, A., Craggs, M. D., & Ellaway, P. H. (2011). Action of 5 Hz repetitive transcranial magnetic stimulation on sensory, motor and autonomic function in human spinal cord injury. *Clinical Neurophysiology*, *122*(12), 2452–2461. https://doi.org/10.1016/j.clinph.2011.04.022.

Lam, T., Wirz, M., Lunenburger, L., & Dietz, V. (2008). Swing phase resistance enhances flexor muscle activity during treadmill locomotion in incomplete spinal cord injury. *Neurorehabilitation and Neural Repair*, *22*(5), 438–446. https://doi.org/10.1177/1545968308315595.

Lima, C., Escada, P., Pratas-Vital, J., Branco, C., Arcangeli, C. A., Lazzeri, G., ... Peduzzi, J. D. (2010). Olfactory mucosal autografts and rehabilitation for chronic traumatic spinal cord injury. *Neurorehabilitation and Neural Repair*, *24*(1), 10–22. https://doi.org/10.1177/1545968309347685.

Mackay-Sim, A., Feron, F., Cochrane, J., Bassingthwaighte, L., Bayliss, C., Davies, W., ... Geraghty, T. (2008). Autologous olfactory ensheathing cell transplantation in human paraplegia: A 3-year clinical trial. *Brain*, *131*(Pt 9), 2376–2386. https://doi.org/10.1093/brain/awn173.

Madsen, C. S., Finnerup, N. B., & Baumgartner, U. (2014). Assessment of small fibers using evoked potentials. *Scandinavian Journal of Pain*, *5*(2), 111–118. https://doi.org/10.1016/j.sjpain.2013.11.007.

Mallik, A., & Weir, A. I. (2005). Nerve conduction studies: Essentials and pitfalls in practice. *Journal of Neurology, Neurosurgery, and Psychiatry*, *76* (Suppl. 2), ii23–31. https://doi.org/10.1136/jnnp.2005.069138.

Mazzoleni, S., Boldrini, E., Laschi, C., Carrozza, M. C., Stampacchia, G., & Rossi, B. (2011). Changes on EMG activation in healthy subjects and incomplete SCI patients following a robot-assisted locomotor training. *IEEE International Conference on Rehabilitation Robotics*, *2011*, 5975467. https://doi.org/10.1109/ICORR.2011.5975467.

McDonnell, M. N., Orekhov, Y., & Ziemann, U. (2006). The role of GABA(B) receptors in intracortical inhibition in the human motor cortex. *Experimental Brain Research*, *173*(1), 86–93. https://doi.org/10.1007/s00221-006-0365-2.

Mendonca, M. V., Larocca, T. F., de Freitas Souza, B. S., Villarreal, C. F., Silva, L. F., Matos, A. C., ... dos Santos, R. R. (2014). Safety and neurological assessments after autologous transplantation of bone marrow mesenchymal stem cells in subjects with chronic spinal cord injury. *Stem Cell Research & Therapy*, *5*(6), 126. https://doi.org/10.1186/scrt516.

Mills, K. R., & Nithi, K. A. (1997). Corticomotor threshold to magnetic stimulation: Normal values and repeatability. *Muscle & Nerve*, *20*(5), 570–576. https://doi.org/10.1002/(SICI)1097-4598(199705)20:5<570::AID-MUS5>3.0.CO;2-6 (pii).

Murillo, N., Kumru, H., Vidal-Samso, J., Benito, J., Medina, J., Navarro, X., & Valls-Sole, J. (2011). Decrease of spasticity with muscle vibration in patients with spinal cord injury. *Clinical Neurophysiology*, *122*(6), 1183–1189. https://doi.org/10.1016/j.clinph.2010.11.012.

Murray, L. M., Edwards, D. J., Ruffini, G., Labar, D., Stampas, A., Pascual-Leone, A., & Cortes, M. (2015). Intensity dependent effects of transcranial direct current stimulation on corticospinal excitability in chronic spinal cord injury. *Archives of Physical Medicine and Rehabilitation*, *96*(Suppl. 4), S114–S121. https://doi.org/10.1016/j.apmr.2014.11.004.

Nardone, R., Langthaler, P. B., Orioli, A., Holler, P., Holler, Y., Frey, V. N., ... Trinka, E. (2017). Effects of intermittent theta burst stimulation on spasticity after spinal cord injury. *Restorative Neurology and Neuroscience*, *35*(3), 287–294. https://doi.org/10.3233/RNN-160701.

Oh, S. K., Choi, K. H., Yoo, J. Y., Kim, D. Y., Kim, S. J., & Jeon, S. R. (2016). A phase III clinical trial showing limited efficacy of autologous mesenchymal stem cell therapy for spinal cord injury. *Neurosurgery*, *78*(3), 436–447. discussion 447 https://doi.org/10.1227/NEU.0000000000001056.

Pajewski, T. N., Arlet, V., & Phillips, L. H. (2007). Current approach on spinal cord monitoring: The point of view of the neurologist, the anesthesiologist and the spine surgeon. *European Spine Journal*, *16*(Suppl. 2), S115–S129. https://doi.org/10.1007/s00586-007-0419-6.

Palmieri, R. M., Ingersoll, C. D., & Hoffman, M. A. (2004). The Hoffmann reflex: Methodologic considerations and applications for use in sports medicine and athletic training research. *Journal of Athletic Training*, *39*(3), 268–277.

Park, T., Park, J., Park, Y. G., & Lee, J. (2017). Intraoperative neurophysiological monitoring for spinal cord tumor surgery: Comparison of motor and somatosensory evoked potentials according to tumor types. *Annals of Rehabilitation Medicine*, *41*(4), 610–620. https://doi.org/10.5535/arm.2017.41.4.610.

Piazza, S., Torricelli, D., Gomez-Soriano, J., Serrano-Munoz, D., Avila-Martin, G., Galan-Arriero, I., ... Taylor, J. (2018). Assessing sensorimotor excitability after spinal cord injury: A reflex testing method based on cycling with afferent stimulation. *Medical & Biological Engineering & Computing*, *56*(8), 1425–1434. https://doi.org/10.1007/s11517-018-1787-2.

Poston, B., Kukke, S. N., Paine, R. W., Francis, S., & Hallett, M. (2012). Cortical silent period duration and its implications for surround inhibition of a hand muscle. *The European Journal of Neuroscience*, *36*(7), 2964–2971. https://doi.org/10.1111/j.1460-9568.2012.08212.x.

Rossini, P. M., Burke, D., Chen, R., Cohen, L. G., Daskalakis, Z., Di Iorio, R., ... Ziemann, U. (2015). Non-invasive electrical and magnetic stimulation of the brain, spinal cord, roots and peripheral nerves: Basic principles and procedures for routine clinical and research application. An updated report from an I.F.C.N. committee. *Clinical Neurophysiology*, *126*(6), 1071–1107. https://doi.org/10.1016/j.clinph.2015.02.001.

Sawaki, L., Butler, A. J., Leng, X., Wassenaar, P. A., Mohammad, Y. M., Blanton, S., ... Wittenberg, G. F. (2008). Constraint-induced movement therapy results in increased motor map area in subjects 3 to 9 months after stroke. *Neurorehabilitation and Neural Repair*, *22*(5), 505–513. Doi:22/5/505 [pii] https://doi.org/10.1177/1545968308317531.

Shin, J. C., Kim, K. N., Yoo, J., Kim, I. S., Yun, S., Lee, H., … Park, K. I. (2015). Clinical trial of human fetal brain-derived neural stem/progenitor Cell transplantation in patients with traumatic cervical spinal cord injury. *Neural Plasticity*, *2015*, 630932. https://doi.org/10.1155/2015/630932.

Stetkarova, I., & Kofler, M. (2013). Differential effect of baclofen on cortical and spinal inhibitory circuits. *Clinical Neurophysiology*, *124*(2), 339–345. https://doi.org/10.1016/j.clinph.2012.07.005.

Stokic, D. S., & Yablon, S. A. (2012). Effect of concentration and mode of intrathecal baclofen administration on soleus H-reflex in patients with muscle hypertonia. *Clinical Neurophysiology*, *123*(11), 2200–2204. https://doi.org/10.1016/j.clinph.2012.04.007.

Stokic, D. S., Yablon, S. A., Hayes, A., Vesovic-Potic, V., & Olivier, J. (2006). Dose-response relationship between the H-reflex and continuous intrathecal baclofen administration for management of spasticity. *Clinical Neurophysiology*, *117*(6), 1283–1289. https://doi.org/10.1016/j.clinph.2006.02.018.

Tabakow, P., Jarmundowicz, W., Czapiga, B., Fortuna, W., Miedzybrodzki, R., Czyz, M., … Raisman, G. (2013). Transplantation of autologous olfactory ensheathing cells in complete human spinal cord injury. *Cell Transplantation*, *22*(9), 1591–1612. https://doi.org/10.3727/096368912X663532.

Theiss, R. D., Hornby, T. G., Rymer, W. Z., & Schmit, B. D. (2011). Riluzole decreases flexion withdrawal reflex but not voluntary ankle torque in human chronic spinal cord injury. *Journal of Neurophysiology*, *105*(6), 2781–2790. https://doi.org/10.1152/jn.00570.2010.

Thomas, S. L., & Gorassini, M. A. (2005). Increases in corticospinal tract function by treadmill training after incomplete spinal cord injury. *Journal of Neurophysiology*, *94*(4), 2844–2855. doi:00532.2005 [pii] https://doi.org/10.1152/jn.00532.2005.

Trumbower, R. D., Hayes, H. B., Mitchell, G. S., Wolf, S. L., & Stahl, V. A. (2017). Effects of acute intermittent hypoxia on hand use after spinal cord trauma: A preliminary study. *Neurology*, *89*(18), 1904–1907. https://doi.org/10.1212/WNL.0000000000004596.

Vaquero, J., Zurita, M., Rico, M. A., Bonilla, C., Aguayo, C., Fernandez, C., … Neurological Cell Therapy Group. (2017). Repeated subarachnoid administrations of autologous mesenchymal stromal cells supported in autologous plasma improve quality of life in patients suffering incomplete spinal cord injury. *Cytotherapy*, *19*(3), 349–359. https://doi.org/10.1016/j.jcyt.2016.12.002.

Vaquero, J., Zurita, M., Rico, M. A., Bonilla, C., Aguayo, C., Montilla, J., … Neurological Cell Therapy Group. (2016). An approach to personalized cell therapy in chronic complete paraplegia: The Puerta de Hierro phase I/II clinical trial. *Cytotherapy*, *18*(8), 1025–1036. https://doi.org/10.1016/j.jcyt.2016.05.003.

Vetrugno, R., Liguori, R., Cortelli, P., & Montagna, P. (2003). Sympathetic skin response: Basic mechanisms and clinical applications. *Clinical Autonomic Research*, *13*(4), 256–270. https://doi.org/10.1007/s10286-003-0107-5.

Wirth, B., Van Hedel, H. J., & Curt, A. (2008). Changes in corticospinal function and ankle motor control during recovery from incomplete spinal cord injury. *Journal of Neurotrauma*, *25*(5), 467–478. https://doi.org/10.1089/neu.2007.0472.

Wittenberg, G. F., Chen, R., Ishii, K., Bushara, K. O., Eckloff, S., Croarkin, E., … Cohen, L. G. (2003). Constraint-induced therapy in stroke: Magnetic-stimulation motor maps and cerebral activation. *Neurorehabilitation and Neural Repair*, *17*(1), 48–57.

Xie, J., & Boakye, M. (2008). Electrophysiological outcomes after spinal cord injury. *Neurosurgical Focus*, *25*(5). https://doi.org/10.3171/FOC.2008.25.11.E11, E11.

Zewdie, E. T., Roy, F. D., Yang, J. F., & Gorassini, M. A. (2015). Facilitation of descending excitatory and spinal inhibitory networks from training of endurance and precision walking in participants with incomplete spinal cord injury. *Progress in Brain Research*, *218*, 127–155. https://doi.org/10.1016/bs.pbr.2014.12.005.

Zhao, Y., Tang, F., Xiao, Z., Han, G., Wang, N., Yin, N., … Dai, J. (2017). Clinical study of NeuroRegen scaffold combined with human mesenchymal stem cells for the repair of chronic complete spinal cord injury. *Cell Transplantation*, *26*(5), 891–900. https://doi.org/10.3727/096368917X695038.

Ziemann, U., Chen, R., Cohen, L. G., & Hallett, M. (1998). Dextromethorphan decreases the excitability of the human motor cortex. *Neurology*, *51*, 1320–1324.

Chapter 30

Features and physiology of spinal stretch reflexes in people with chronic spinal cord injury

Aiko K. Thompson[a] and Thomas Sinkjær[b]

[a]*Department of Health Sciences and Research, College of Health Professions, Medical University of South Carolina, Charleston, SC, United States,* [b]*Department of Health Science and Technology, Aalborg University, Aalborg, Denmark*

List of abbreviations

A-P GRF	anterior-posterior ground reaction force
CP	cerebral palsy
CNS	central nervous system
EMG	electromyography
LLR	(spinal) long latency reflex
MLR	(spinal) medium latency reflex
SCI	spinal cord injury
SE	standard error (of mean)
SLR	(spinal) short latency reflex
TA	tibialis anterior

Introduction

The activity of spinal reflex pathways is task- and phase-dependently modulated in normal motor control (Duysens, Bastiaanse, Smits-Engelsman, & Dietz, 2004; Stein & Capaday, 1988; Zehr & Stein, 1999). During locomotion, different sensory afferents and associated spinal pathways contribute to the generation of locomotor muscle activity (Grey, Nielsen, Mazzaro, & Sinkjaer, 2007; Mazzaro, Grey, do Nascimento, & Sinkjaer, 2006; Mazzaro, Grey, & Sinkjaer, 2005; Sinkjaer, Andersen, & Larsen, 1996). After spinal cord injury (SCI) disrupts the activity of supraspinal and propriospinal pathways, spinal reflexes are often altered (Crone, Johnsen, Biering-Sorensen, & Nielsen, 2003; Hiersemenzel, Curt, & Dietz, 2000; Stein, Yang, Belanger, & Pearson, 1993; Thompson, Estabrooks, Chong, & Stein, 2009; Yang et al., 1991). Abnormal neuronal behaviors and reflex behaviors during static motor tasks or isolated joint motion in individuals with SCI (Gorassini, Knash, Harvey, Bennett, & Yang, 2004; Hornby, Kahn, Wu, & Schmit, 2006; Hultborn, 2003; Li, Gorassini, & Bennett, 2004) have led to our fields' general assumption that spinal reflex abnormalities contribute to movement disorders (Dietz & Sinkjaer, 2007; Nielsen, Crone, & Hultborn, 2007). A typical example is spasticity, which is commonly characterized by a velocity-dependent increase in tonic stretch reflexes with exaggerated tendon jerks (Lance, 1980). Spasticity, which is reported to affect 65%–78% of people after SCI (Adams & Hicks, 2005; Maynard, Karunas, & Waring, 1990; Skold, Levi, & Seiger, 1999) is negatively associated with quality of life after SCI (Westerkam, Saunders, & Krause, 2011). While different symptoms are likely reported as spasticity in clinical practice, including the hyperexcitability of tonic (e.g., hypertonia) and phasic (e.g., clonus) stretch reflexes (Adams & Hicks, 2005), spastic movement disorders are often viewed as problems of hyperactivity in spinal stretch reflex pathways. However, an increasing body of findings in stroke, multiple sclerosis, and cerebral palsy (CP) questions this view (see opinion paper by Nielsen, Christensen, Farmer, & Lorentzen, 2020).

This chapter reviews and discusses functions and contributions of spinal stretch reflex pathways in normal locomotion, characteristics of stretch reflexes in different task conditions (e.g., in resting and actively contracting muscles) in people after SCI, and their potential implications in spastic movement disorders post-SCI.

Cellular, Molecular, Physiological, and Behavioral Aspects of Spinal Cord Injury. https://doi.org/10.1016/B978-0-12-822427-4.00030-7

Spinal stretch reflexes

An abrupt stretch of a muscle elicits stretch reflexes in that muscle. In an actively contracting lower extremity muscle, three components of stretch reflexes can be observed (Fig. 1): spinal short-latency reflex (SLR, also known as M1, mainly Ia muscle spindle afferent origin), spinal medium-latency reflex (MLR, also known as M2, presumably mainly spindle afferent II mediated), and long-latency reflex (LLR, also known as M3, suggested to be transcortical or subcortical) (Af Klint, Mazzaro, Nielsen, Sinkjaer, & Grey, 2010; Corna, Grasso, Nardone, & Schieppati, 1995; Schieppati & Nardone, 1997; Sinkjaer, Andersen, Nielsen, & Hansen, 1999). Normally, SLR is the most dominant and persistent among the three components. In order to investigate Ia excitation of homonymous motoneurons, the Hoffmann (or H-) reflex is frequently used as a convenient substitute to eliciting SLR, as these responses share the same pathways (Henneman & Mendell, 1980; Mendell & Henneman, 1971). However, SLR and H-reflex do not necessarily respond to modulatory input in the same way (Sinkjaer et al., 1996). The H-reflex is an electrically-elicited spinal reflex that arises mainly from synchronous axonal excitation of large Ia (and some Ib/II) afferents, whereas the stretch reflexes elicited by rapid joint rotation arise from less synchronous activation of Ia and II afferents and reflect γ-motoneuron mediated fusimotor control (Henneman & Mendell, 1980; Magladery, Porter, Park, & Teasdall, 1951). Ib afferents arising from Golgi tendon organ also fire in response to muscle stretch, and depending on the motor task, also excite motoneurons via interneurons (Donelan & Pearson, 2004; Duysens, Clarac, & Cruse, 2000; Gossard, Brownstone, Barajon, & Hultborn, 1994; Jankowska & Edgley, 2010; Jankowska & McCrea, 1983; Nichols, 2018; Schafer, Dadfar, Hartel, Haupts, & Fischer, 1999; Vincent et al., 2017). Thus, investigation of stretch reflexes with mechanical perturbation provides important information for understanding normal and impaired motor control; it is not readily substituted by the H-reflex studies.

Function of the soleus stretch reflex pathways in intact human locomotion

During locomotion, stretch and load receptor afferents (i.e., Ia, II, and Ib) fire in a phasic manner, in accord with joint rotations that change the muscle length and loading/unloading (i.e., stance/swing), and contribute to the excitation of plantarflexor motoneurons through reflex pathways (Donelan & Pearson, 2004; Gossard et al., 1994; Mayer et al., 2018). To understand how different sensory afferent pathways contribute to soleus activity during the stance phase of locomotion, Sinkjaer and his colleagues applied fast and slow dorsiflexion or plantarflexion perturbations while participants walked on a treadmill (Grey et al., 2007; Mazzaro et al., 2005, 2006; Sinkjaer, Andersen, Ladouceur, Christensen, & Nielsen, 2000). During the stance phase, when the naturally occurring dorsiflexion is blocked or plantarflexion is imposed, the ankle plantarflexors become unloaded and the afferent firing would decrease. This unloading manipulation results in a significant reduction of soleus activity (Sinkjaer et al., 2000). Since this unloading response is unaffected by an ischemic block of Ia afferent or block of the common peroneal nerve that causes reciprocal inhibition of the plantarflexors, feedback from spindle afferent II and/or load receptor afferent Ib from the plantarflexors would be the major contributors to naturally occurring (i.e., undisturbed) soleus activity in the mid-stance phase (Sinkjaer et al., 2000). When additional dorsiflexion is imposed during the same (mid-stance) phase, it elicits spinal stretch reflexes (Mazzaro et al., 2005). A picture that emerged through those studies was that Ia, II, and Ib contribute differently to the stance phase soleus activity; that is,

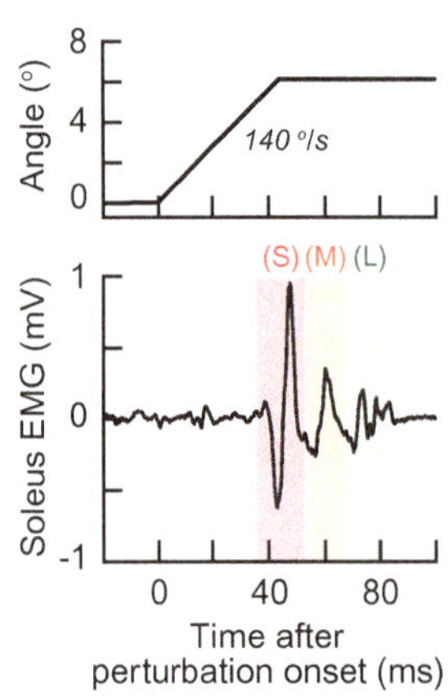

FIG. 1 Soleus stretch reflexes. An example of the soleus short (SLR), medium (MLR), and long (LLR) latency stretch reflexes elicited by 6° of ankle dorsiflexion at 140°/s during standing in an individual with no known neurological injuries.

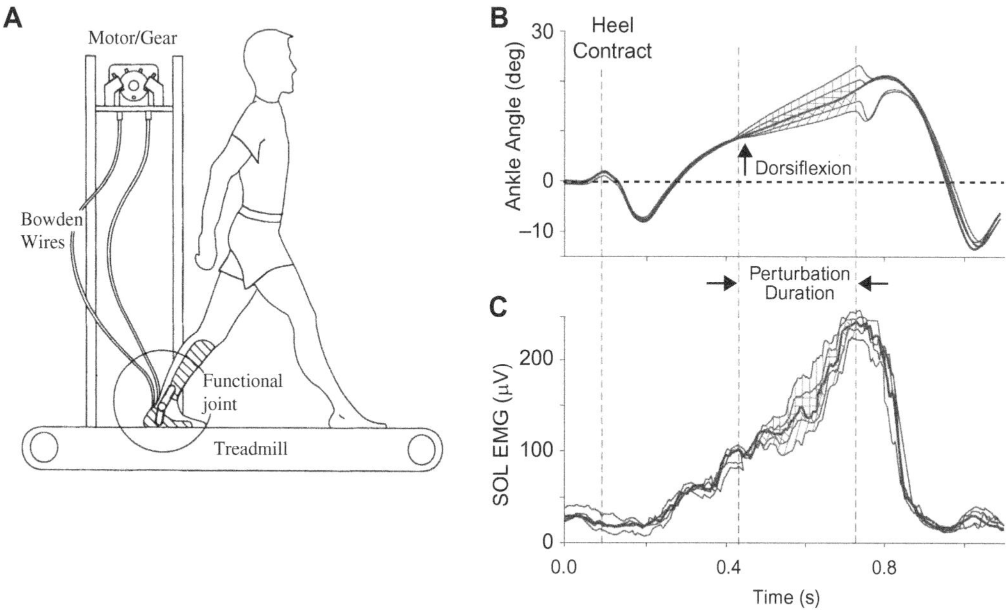

FIG. 2 Proprioceptive afferent input contributes to the soleus EMG activity during walking. Slow and small ankle joint rotation perturbations (enhancement or reduction of dorsiflexion) that manipulate the activity of stretch and load receptor afferents change the soleus EMG activity during the stance phase of walking (Mazzaro et al., 2005). (A) Experimental setup. The participant wears a motorized mechanical joint, which can impose precisely timed and controlled (both in amount and speed) ankle joint rotation at any part of the gait cycle (Sinkjaer et al., 1996). (B) A thick line indicates ankle joint angle during non-perturbed steps and thin lines indicate dorsiflexion enhancement and reduction perturbations (±2° at ±6°/s and ± 5° at ±16°/s). During the stance phase of non-perturbed steps, ≅10° of dorsiflexion occurred at 33°/s. (C) Rectified and filtered soleus EMG during perturbed and non-perturbed steps that correspond to (B). *(Modified from Mazzaro, N., Grey, M. J., & Sinkjaer, T. (2005). Contribution of afferent feedback to the soleus muscle activity during human locomotion.* Journal of Neurophysiology, *93(1), 167–177. https://doi.org/10.1152/jn.00283.2004).*

SLR originated from Ia excitation is present when rapid perturbation triggers a corrective reaction, whereas feedback from Ib and/or II afferents likely contribute to the soleus activation in unperturbed steps (Sinkjaer et al., 2000). In line with locomotor-related Ib excitation found in cats (Donelan & Pearson, 2004; Gossard et al., 1994), the contribution of force-sensitive afferent pathways to the ongoing soleus activation in unperturbed steps has been repeatedly confirmed in humans during more natural-like walking conditions (Af Klint et al., 2010; Grey et al., 2007; Mazzaro et al., 2005, 2006) (Fig. 2). These studies show that Ia-mediated stretch reflex pathways are accessible on a moment-to-moment basis when there is a need to increase the muscle activity in response to small or large perturbation, but the primary pathways that generate the normal, undisturbed stance phase soleus activation during locomotion are likely not Ia but load sensitive afferents mediated pathways (Grey et al., 2007).

Features of spinal stretch reflexes in people with SCI

Spinal reflexes change after SCI (Crone et al., 2003; Hiersemenzel et al., 2000; Stein et al., 1993; Thompson et al., 2009; Yang et al., 1991). In the soleus muscle of people with chronic incomplete SCI, task-dependent modulation of the H-reflex becomes absent or diminished (Boorman, Lee, Becker, & Windhorst, 1996; Thompson et al., 2009); Ib inhibition becomes absent (Morita et al., 2006); reciprocal inhibition between the soleus and tibialis anterior becomes abnormal (Ashby & Wiens, 1989; Boorman et al., 1996; Boorman, Hulliger, Lee, Tako, & Tanaka, 1991; Crone et al., 2003; Thompson et al., 2009; Xia & Rymer, 2005); and recurrent inhibition may increase (Shefner, Berman, Sarkarati, & Young, 1992). Changes of neuronal (i.e., motoneurons and interneurons) behaviors in the spinal cord (Gorassini et al., 2004; Hultborn, 2003; Li et al., 2004) would also likely be part of these reflex abnormalities. If and how these abnormal reflex characteristics, which are often defined in non-dynamic postures and tasks, affect control and execution of dynamic motion, such as locomotion, are yet to be fully understood.

When SCI results in affecting the behaviors of different spinal pathways, what happens to the spinal stretch reflexes? In spastic individuals with chronic SCI, the soleus stretch reflex excitability is elevated during sitting or lying in supine (Mirbagheri, Barbeau, Ladouceur, & Kearney, 2001; Nakazawa, Kawashima, & Akai, 2006); and repeated ankle joint rotations facilitate the soleus reflexes and reflex-induced plantarflexion torque velocity-dependently (Hornby et al., 2006).

Unlike the observations in people with spastic hemiparesis due to stroke, in whom the reflexes are hyperexcitable at rest or in passive conditions but near normal during active muscle contractions (e.g., Berger, Horstmann, & Dietz, 1988; Sinkjaer & Magnussen, 1994), in spastic individuals with SCI, the stretch reflex gain and reflex stiffness do not decrease with muscle activation (rather, they remain high) (Mirbagheri et al., 2001). Similarly, abnormal modulation and large soleus reflexes during voluntary muscle contraction have also been found with the H-reflex in people with chronic SCI (Kim, Corcos, & Hornby, 2015; Taylor, Ashby, & Verrier, 1984; Thompson et al., 2009). Do these observations made in static postures and tasks account for stretch reflex behaviors that impair locomotion in people with SCI?

Soleus stretch reflexes during locomotion in people with chronic incomplete SCI

In people with chronic incomplete SCI, clonus and abnormal modulation of the locomotor soleus H-reflex activity are frequently present (Barbeau, Ladouceur, Mirbagheri, & Kearney, 2002; Corcos, Gottlieb, Penn, Myklebust, & Agarwal, 1986; Fung & Barbeau, 1989, 1994; Hidler & Rymer, 1999; Knikou, Angeli, Ferreira, & Harkema, 2009; Stein et al., 1993; Yang et al., 1991). A recent study showed that the SLR, MLR, and H-reflex are abnormally modulated during walking in individuals with chronic incomplete SCI (Thompson, Mrachacz-Kersting, Sinkjaer, & Andersen, 2019). During the swing phase, the SLR and H-reflex were suppressed in individuals without injuries but not in an individual with SCI (Fig. 3), suggesting reduced inhibition of Ia excitatory pathways in this phase in individuals with SCI. MLR was modulated similarly to SLR; in individuals with SCI, MLR amplitudes were very large in the mid-late swing phase (Fig. 3). Such exaggerated MLR could be due to reduced inhibition of group II afferents, increased oligo or polysynaptic Ia excitation, and/or increased excitation of other afferents (e.g., actions of excitatory and inhibitory interneurons that receive input from Ib afferents could also affect MLR modulation (Jankowska & Edgley, 2010; Schafer et al., 1999). Unique firing behaviors of Ia, II, and Ib afferents (Donelan & Pearson, 2004; Henneman & Mendell, 1980), together with altered modulation of their interneuron excitability, may also partly explain abnormal soleus EMG burst or large MLR activity in the late-swing phase of locomotion in people with SCI.

It is noteworthy that, in addition to abnormal modulation of locomotor stretch reflexes, a normal positive correlation between locomotor EMG and reflex excitability (Stein & Capaday, 1988) was rare in individuals with SCI (Thompson et al., 2019). This suggests that, while useful in characterizing the pathways altered by SCI, the reflex abnormalities measured in selected passive states, during isolated joint movements, and/or during static motor tasks may not explain how they impact locomotion after SCI. Hence, the question arises; does the hyperexcitability of stretch reflex pathways impair locomotion in people with incomplete SCI?

Stretch reflexes in spastic gait after SCI

Normally, the soleus stretch reflexes are phase-dependently modulated during walking, in accord with soleus motoneuron excitability level (expressed indirectly through the EMG activity), which is high in the mid-stance phase and little-to-none in the swing phase (Sinkjaer et al., 1996, 1999). In individuals with SCI, soleus EMG is often unsuppressed during the swing phase but reduced in the mid-stance phase (e.g., Fig. 3); at the same time, the stretch reflexes are unsuppressed or often larger in the mid-late swing phase. These observations support a possibility that inadequately hyperactive soleus elevating the reflex excitability and/or unsuppressed reflex pathways contributing to the reduced soleus suppression in the swing phase. Interestingly, and in consistently with observations in hemiplegic stroke and CP (Dietz & Sinkjaer, 2007, 2012), the soleus stretch reflexes are not hyperactive in mid-stance, during which group II and Ib afferents can positively contribute to the generation of locomotor EMG and Ia afferent input can further increase the EMG burst (see above). That is, the stretch reflex pathways are hyperactive and powerful when their activation is meaningless (e.g., during the swing), whereas they are "near" normal when the muscle is functionally active (i.e., during stance).

A clear example of not-suppressed stretch reflex pathways negatively affecting movements is clonus, which is triggered or produced by hyperexcitable stretch reflexes (Corcos et al., 1986; Latash, Penn, Corcos, & Gottlieb, 1989; Rossi, Mazzocchio, & Scarpini, 1990; however, see also Beres-Jones, Johnson, & Harkema, 2003). On the contrary to a common conception, elicitation of stretch reflexes requires only a degree or two of joint rotation (Stein & Kearney, 1995). In spastic individuals, small joint displacement such as repositioning a foot in the wheelchair (Wallace, Ross, & Thomas, 2012) or a foot's ground contact can induce clonus (Fig. 4). Clonus triggered by ground contact, which is a small joint motion perturbation, often continues into the stance phase, exaggerates breaking force, and reduces forward propulsion (Turns, Neptune, & Kautz, 2007). Furthermore, clonic bursts are often followed by more clonic bursts and/or a smaller fused push-off burst in the mid-late stance phase (Fig. 4), which would result in the ineffective generation of propulsive force

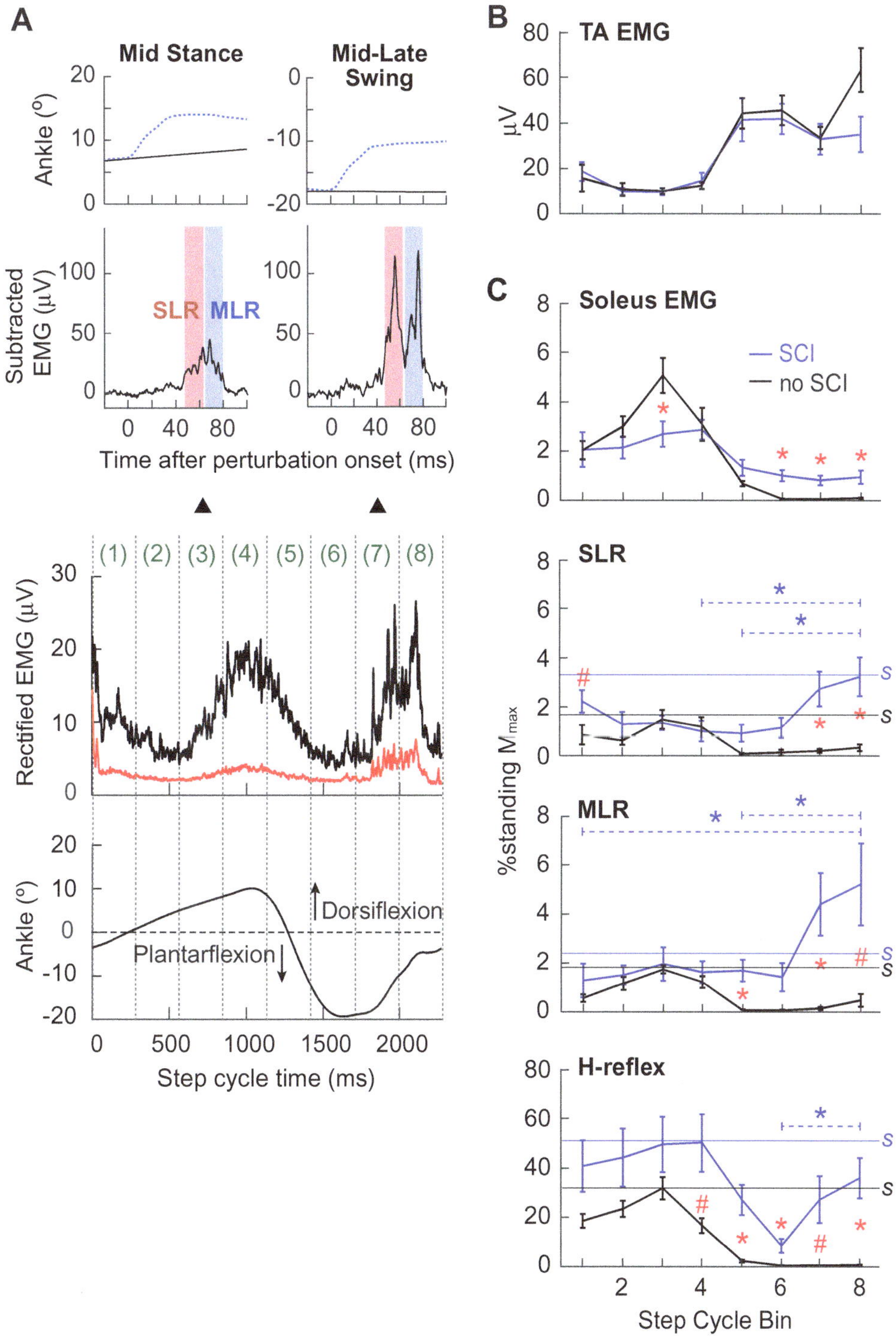

FIG. 3 Soleus stretch reflexes during walking in people with chronic incomplete SCI. (A) Soleus stretch reflexes elicited by 6° dorsiflexion perturbation at 250°/s in mid-stance and mid-late swing, soleus, and tibialis anterior (TA) EMG, and ankle joint motion in individuals with chronic incomplete SCI. The top panel shows the average ankle joint motion around the time of perturbation in unperturbed steps (solid line) and perturbed steps (dotted line). The second panel shows the averaged [perturbed – unperturbed] EMG response. 10–15 responses were averaged together. The third panel shows locomotor EMG activity in the soleus (black) and TA (red). The fourth panel shows the average ankle joint motion of unperturbed steps. For the third and fourth panels, >50 unperturbed steps were averaged together. (B) Locomotor TA EMG (mean ± SE) in individuals with chronic incomplete SCI ($N = 9$, blue) and individuals without injuries ($N = 9$, black). In general, bins 1–4 correspond to the stance phase, bin 5 to the stance-swing transition, and bins 6–8 to the swing phase. (C) Locomotor soleus EMG and reflexes in the same individuals with B. All values are normalized to the standing maximum M-wave (M_{max}) value. For reflexes, ≥10 responses were averaged together for each individual's each bin. Horizontal lines labeled with "S" indicate the average reflex sizes during standing. Statistical differences between the groups and between the bins (SCI group only) by Student *t*-test with Bonferroni correction are indicated with * ($\alpha = 0.05$) and # ($\alpha = 0.1$). *(Modified from Thompson, A. K., Estabrooks, K. L., Chong, S., & Stein, R. B. (2009). Spinal reflexes in ankle flexor and extensor muscles after chronic central nervous system lesions and functional electrical stimulation.* Neurorehabilitation and Neural Repair, 23*(2), 133–142. https://doi.org/10.1177/1545968308321067).*

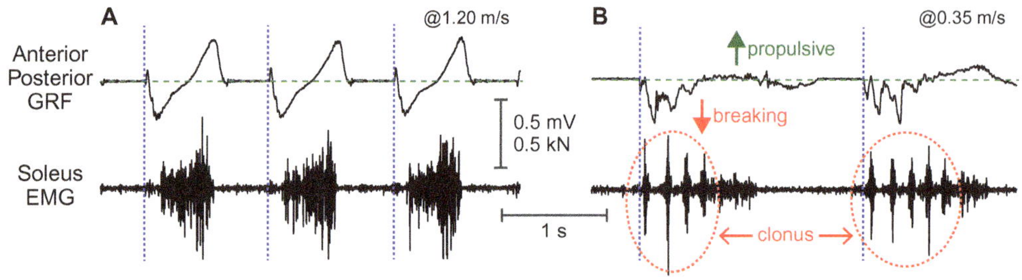

FIG. 4 Anterior-posterior (A-P) ground reaction force (GRF) and soleus EMG during walking. Positive A-P GRF represents propulsion and the positive area under the curve is the propulsive impulse; inversely, negative A-P GRF represents breaking and the negative area under the curve is the breaking impulse (Bowden et al., 2006). (A) In an individual with no injuries, after ground contact (indicated by vertical dotted lines), soleus EMG activity increases gradually through mid-late stance while the A-P GRF transitions smoothly from breaking to propulsion. (B) In a spastic individual, clonus triggered by ground contact and ineffective push-off burst are accompanied by exaggerating breaking force and little-to-no propulsive force.

(Turns et al., 2007) and reduction of gait speed (Bowden, Balasubramanian, Neptune, & Kautz, 2006). Thus, clonus impairs locomotion in people with incomplete SCI.

Do hyperexcitable stretch reflexes contribute to spastic gait in people with SCI? Among several spinal and supraspinal pathways possibly involved in spastic movement disorders (Burke, Wissel, & Donnan, 2013; Dietz & Sinkjaer, 2007; Hultborn, 2003; Nielsen et al., 2007), current opinions on post-stroke spastic gait are dismissive of hyperexcitable stretch reflexes (Dietz & Sinkjaer, 2015; Nielsen et al., 2020). In people with spastic hemiparesis due to stroke, stretch reflexes are hyperexcitable at rest or in passive conditions but near normal or even suppressed during active muscle contractions (Berger et al., 1988; Sinkjaer & Magnussen, 1994), and anti-spastic medication aiming to reduce hyperreflexia and hypertonus is not necessarily helpful (Dietz & Sinkjaer, 2012). Thus, the role of hyperexcitable stretch reflexes in impaired locomotion post-stroke is not immediately clear. Since the stretch reflex gain and stiffness in active versus passive conditions (Mirbagheri et al., 2001) and the effects of anti-spastic medicine such as baclofen and tizanidine (Hornby et al., 2006; Mirbagheri, Chen, & Rymer, 2010) seem to differ between spastic SCI and hemiparetic stroke, the origin of reflex hyperexcitability may be different, and further investigation would be essential in evaluating the role of hyperexcitable stretch reflexes in post-SCI spastic gait.

Consideration of CNS plasticity in addressing the reflex hyperexcitability in chronic SCI

Abnormal reflex behaviors observed in chronic, impaired movements do not necessarily mean that such reflex behaviors are the primary causes of impairments; they may be the results of (non-functional) CNS plasticity, adaptation to or compensation for other problems caused by injuries or disease. Thus, while impaired use of sensory afferent input or abnormally functioning spinal pathways may pose or exacerbate movement disorders, simply removing or disabling them may not solve the problem. This might be the case in treating post-stroke hemiparetic gait (Dietz & Sinkjaer, 2007, 2015; Nielsen et al., 2020) with pharmacological interventions that aim to tame stretch reflex hyperexcitability.

In people with incomplete SCI, as summarized above, behaviors of multiple spinal pathways including stretch reflex pathways become abnormal. When the spinal cord is in such states, feedback from certain afferents (e.g., muscle spindle afferents) and their associated reflexes (e.g., stretch reflexes) would be, not only useless but also detrimental in the execution of the motor function (e.g., Figs. 3 and 4). A rather simple CNS strategy to deal with this problem would be to circumvent it by avoiding activation of problem-causing afferents and their pathways. As a result, the muscles and limb with hyperreflexia would be engaged less, leading to a cascade of other complications, such as disuse atrophy, step and postural asymmetry, and other muscular mechanical changes (Dietz & Sinkjaer, 2007, 2012, 2015). To prevent such maladaptive problems, one might opt for surgical (Smyth & Peacock, 2000) or pharmacological approaches (Dario & Tomei, 2004; Sheean, 2006; Ward, 2008). However, suppressing the transmission of spinal pathways or eliminating afferent feedback does not normalize the excitability of a specific pathway in a targeted manner, and necessitates compensation and adaptation to pharmacologically or surgically induced changes (some of which are irreversible). In fact, in spastic stroke, the problem may well be that the afferent integration is too suppressed during walking (Mazzaro, Nielsen, Grey, & Sinkjaer, 2007). Functioning abnormally or being unusable does not mean that stretch reflex pathways should be uniformly suppressed or eliminated for good. Instead, making these pathways more usable by restoring their excitability modulation may help to attain better functional recovery. Indeed, in an individual with spastic hyperreflexia due to chronic incomplete SCI, when the soleus H-reflex excitability was reduced through operant down-conditioning, their locomotor EMG activity improved in the conditioned soleus and other lower and upper leg muscles, and walking speed increased significantly (Thompson, Pomerantz, & Wolpaw, 2013; Thompson & Wolpaw, 2019). Interestingly but not surprisingly, decreasing the H-reflex excitability did not decrease but increased soleus activity in the mid-late stance phase (Fig. 5 in Thompson et al., 2013), likely contributed to increased walking speed in those individuals with SCI. These observations are in line with the above argument. An emerging picture is this: when the Ia excitatory pathway is not usable (because its excitability is too high and cannot be readily modulated) and its inappropriate activation hinders movement execution, the associated spinal circuitry (and thereby the associated muscles) would become less used as if it is to avoid activating the "annoying" Ia pathway. When the Ia excitatory pathway becomes modulable and thus usable (e.g., through operant conditioning; Thompson & Wolpaw, 2015; Wolpaw, 2018), it becomes unnecessary to avoid engaging that leg/muscle; this would allow the associated circuitry, including other pathways (e.g., group II and Ib pathways) to become functionally viable, enabling them to contribute to locomotor muscle activation. Regardless of whether or not hyperexcitable stretch or Ia reflex pathways are the causes of spastic gait, they are often part of it in people with SCI (Thompson et al., 2019), and their excitability reduction is accompanied by gait function improvement (Thompson et al., 2013; Thompson & Wolpaw, 2019). Stretch reflex stiffness and intrinsic muscle stiffness also decrease along with gait improvements resulting from long-term functional electrical stimulation assisted walking (Mirbagheri, Ladouceur, Barbeau, & Kearney, 2002) and robotic-assisted step training (Mirbagheri, Kindig, & Niu, 2015) in spastic individuals with chronic incomplete SCI.

Conclusion

Recently, the validity of hyperexcitable stretch reflexes as a cause of spastic hemiparetic gait disorders has been questioned (Dietz & Sinkjaer, 2007; Nielsen et al., 2020). So, we considered the following questions in spastic individuals with SCI. Do the soleus stretch reflexes become abnormal after SCI? Do the stretch reflexes behave abnormally during locomotion, and if so, do the abnormal reflexes contribute to their gait disorders?

The currently available data from spastic individuals with chronic SCI indicate that the stretch reflex gain and reflex stiffness are high (Mirbagheri et al., 2001); the soleus H-reflex is also large during voluntary muscle contraction (Kim et al., 2015; Taylor et al., 1984; Thompson et al., 2009). During walking, the soleus stretch reflexes (SLR, MLR) are near normal in the mid-late stance phase but are facilitated in the swing phase (Thompson et al., 2019). Unsuppressed stretch reflexes in mid-late swing through early stance trigger clonus, which very likely negatively affect locomotion in chronic SCI.

It would be important to remind ourselves that abnormal reflex behaviors observed during chronically impaired locomotion do not necessarily mean that they are the causes of impairments. Hyperexcitable stretch or Ia reflex pathways may be the partial causes of spastic gait or the products of compensatory or adaptive plasticity to chronic SCI; regardless of this, they are often part of post-SCI spastic gait (Thompson et al., 2019). When one considers these issues, simply removing or disabling stretch reflex pathways seems far from an optimal solution. Instead, making these pathways more usable by restoring their excitability modulation through neurobehavioral training (e.g., locomotor training and reflex operant conditioning) could be a better solution for enhancing functional recovery after SCI.

Applications to other areas of neuroscience

After SCI, many spinal reflexes become abnormal. Supported by abnormal reflex behaviors observed in muscles at rest or during static motor tasks, there has been a general assumption that abnormal spinal reflexes, especially hyperactive spinal stretch reflexes, contribute to spastic movement disorders in individuals with SCI. However, the cumulative evidence in other CNS populations such as stroke, multiple sclerosis, and cerebral palsy questions this view.

The findings in SCI suggest a negative impact of stretch reflex hyperexcitability on post-SCI spastic gait in the swing and swing-stance transition phases, whereas in the mid-late stance phase (where the plantarflexors are active) the behavior is near normal, as seen in stroke and CP individuals. This suggests that some parts of underlying neural mechanisms of hyperreflexia and spasticity are alike, whereas some other parts may differ between SCI and other CNS impaired populations. These differences and how they can help to guide the restoration of walking need to be carefully investigated in the future.

Regardless of the etiology of spastic movement disorders, when examining abnormal reflex behaviors, one must be careful not to presume a causal relationship between a malfunctioning reflex and motor impairments. In the case of SCI, the hyperexcitability of stretch reflex pathways may be a cause or a result of (i.e., adaptation to) chronically impaired movement. In another CNS population, another kind of reflex abnormality may be more prominent. Neurobehavioral training approach, such as reflex operant conditioning, may therefore become a useful tool for improving the excitability and behavior of a specific reflex pathway that is a part of an individual's motor deficits, so as to enhance motor function recovery in each and every individual.

Mini-dictionary of terms

- Stretch reflex. A fast spinal reflex originated from multiple sensory afferents. Stretch of the nerve endings in the muscle spindle excites group Ia and II afferents, which in turn excites α-motoneurons in the spinal cord. The gain of this reflex is affected by γ-motoneuron mediated fusimotor control. Muscle stretch (muscle tension change) also excites group Ib Golgi tendon organ afferents.
- Hoffmann (or H-) reflex. An artificially elicited spinal reflex, commonly viewed as an electrical analog of the spinal short-latency stretch reflex. With electrical stimulation of a mixed nerve (e.g., tibial nerve), group Ia (and some II) afferent axons and group Ib Golgi tendon organ afferent axons are excited electrically.
- Clonus. An involuntary, reflexive, rhythmic, cyclic muscle activation. While its mechanisms are not completely understood, it is generally accompanied by hyperreflexia due to CNS lesions.
- Spasticity. A condition that is commonly characterized by a velocity-dependent increase in tonic stretch reflexes with exaggerated tendon jerks.
- Dorsiflexion. Upward flexion of the foot. Upward rotation of the ankle joint.
- Plantarflexion. Downward flexion of the foot. Downward rotation of the ankle joint.
- Soleus. A muscle in the lower calf, situated below the gastrocnemius muscles. Its action is plantarflexion. It is innervated by the tibial nerve.
- Motoneuron (or α-motoneuron). Part of a pool of large neurons located in the anterior/ventral portion of the spinal cord. Motoneurons innervate extrafusal muscle fibers of skeletal muscle, and their excitation produces muscle contraction. α-motoneurons are distinct from γ motoneurons, which innervate intrafusal muscle fibers of muscle spindles.
- Interneuron. A neuron that transmits signals between other neurons.
- Hemiparesis. A unilateral paresis (a weakness of one side of the body). An inability to move one side of the body. Hemiparesis is often caused by stroke or other brain conditions and injuries. Its most severe form is called hemiplegia, which is complete paralysis of half of the body.

Key facts of spinal reflexes

- Although they are involuntary responses to stimuli applied to the periphery, spinal reflexes are posture and task-dependently modulated.
- Excitability of spinal reflexes is phase-dependently modulated during dynamic motion such as locomotion.
- In humans, specific patterns of reflex behaviors are acquired use- and activity-dependently in early childhood and maintained throughout life.
- Aging affects the excitability and behaviors of spinal reflexes.
- Spinal reflexes are affected by injuries and diseases. Some of the reflex changes caused by injuries and/or diseases can be reversed through various treatments and interventions.

Summary points

- In spastic individuals with chronic SCI, the stretch reflex gain and reflex stiffness are high at rest and do not decrease with muscle activation.
- In the mid-late stance phase of walking, the stretch reflex behavior is near normal, as seen in post-stroke and CP individuals.
- Hyperexcitable stretch reflexes in mid-late swing through swing-stance transition trigger clonus and negatively affect locomotion in people with chronic SCI.
- This suggests that some of the underlying neural mechanisms of hyperreflexia and spasticity are alike whereas some other mechanisms may differ between SCI and other CNS impaired populations.
- Different from pharmacologically or surgically disabling a reflex pathway, changing a reflex through operant conditioning can change the reflex behavior in a desirable direction and induce CNS multi-site plasticity in the brain and the spinal cord.
- Changing an abnormal reflex behavior through neurobehavioral training (e.g., locomotor training and reflex conditioning) could provide a new means to enhance motor function recovery in people with SCI.

Funding acknowledgments

The presented work was supported in part by the National Institutes of Health [NS069551 and NS114279 to AKT, U44 NS114420 (Clements, Thompson, Wolpaw), P41-EB018783 (Wolpaw), P2C HD086844 (Kautz), U54-GM104941 (Binder-MacLeod)]; the South Carolina Spinal Cord Injury Research Fund (AKT); the New York State Spinal Cord Injury Research Trust Fund (AKT and Wolpaw); and Juhl Memorial Scholarship Denmark (TS).

References

Adams, M. M., & Hicks, A. L. (2005). Spasticity after spinal cord injury. *Spinal Cord*, *43*(10), 577–586. https://doi.org/10.1038/sj.sc.3101757.

Af Klint, R., Mazzaro, N., Nielsen, J. B., Sinkjaer, T., & Grey, M. J. (2010). Load rather than length sensitive feedback contributes to soleus muscle activity during human treadmill walking. *Journal of Neurophysiology*, *103*(5), 2747–2756. https://doi.org/10.1152/jn.00547.2009.

Ashby, P., & Wiens, M. (1989). Reciprocal inhibition following lesions of the spinal cord in man. *Journal of Physiology*, *414*, 145–157.

Barbeau, H., Ladouceur, M., Mirbagheri, M. M., & Kearney, R. E. (2002). The effect of locomotor training combined with functional electrical stimulation in chronic spinal cord injured subjects: Walking and reflex studies. *Brain Research. Brain Research Reviews*, *40*(1–3), 274–291.

Beres-Jones, J. A., Johnson, T. D., & Harkema, S. J. (2003). Clonus after human spinal cord injury cannot be attributed solely to recurrent muscle-tendon stretch. *Experimental Brain Research*, *149*(2), 222–236. https://doi.org/10.1007/s00221-002-1349-5.

Berger, W., Horstmann, G. A., & Dietz, V. (1988). Spastic paresis: Impaired spinal reflexes and intact motor programs. *Journal of Neurology, Neurosurgery and Psychiatry*, *51*(4), 568–571.

Boorman, G., Hulliger, M., Lee, R. G., Tako, K., & Tanaka, R. (1991). Reciprocal Ia inhibition in patients with spinal spasticity. *Neuroscience Letters*, *127*(1), 57–60.

Boorman, G. I., Lee, R. G., Becker, W. J., & Windhorst, U. R. (1996). Impaired "natural reciprocal inhibition" in patients with spasticity due to incomplete spinal cord injury. *Electroencephalography and Clinical Neurophysiology*, *101*(2), 84–92.

Bowden, M. G., Balasubramanian, C. K., Neptune, R. R., & Kautz, S. A. (2006). Anterior-posterior ground reaction forces as a measure of paretic leg contribution in hemiparetic walking. *Stroke*, *37*(3), 872–876. https://doi.org/10.1161/01.STR.0000204063.75779.8d.

Burke, D., Wissel, J., & Donnan, G. A. (2013). Pathophysiology of spasticity in stroke. *Neurology*, *80*(3 Suppl. 2), S20–S26. https://doi.org/10.1212/WNL.0b013e31827624a7.

Corcos, D. M., Gottlieb, G. L., Penn, R. D., Myklebust, B., & Agarwal, G. C. (1986). Movement deficits caused by hyperexcitable stretch reflexes in spastic humans. *Brain*, *109*(Pt. 5), 1043–1058.

Corna, S., Grasso, M., Nardone, A., & Schieppati, M. (1995). Selective depression of medium-latency leg and foot muscle responses to stretch by an alpha 2-agonist in humans. *The Journal of Physiology*, *484*(Pt. 3), 803–809.

Crone, C., Johnsen, L. L., Biering-Sorensen, F., & Nielsen, J. B. (2003). Appearance of reciprocal facilitation of ankle extensors from ankle flexors in patients with stroke or spinal cord injury. *Brain*, *126*(Pt. 2), 495–507.

Dario, A., & Tomei, G. (2004). A benefit-risk assessment of baclofen in severe spinal spasticity. *Drug Safety*, *27*(11), 799–818.

Dietz, V., & Sinkjaer, T. (2007). Spastic movement disorder: Impaired reflex function and altered muscle mechanics. *Lancet Neurology*, *6*(8), 725–733. https://doi.org/10.1016/S1474-4422(07)70193-X.

Dietz, V., & Sinkjaer, T. (2012). Spasticity. *Handbook of Clinical Neurology*, *109*, 197–211. https://doi.org/10.1016/B978-0-444-52137-8.00012-7.

Dietz, V., & Sinkjaer, T. (2015). Secondary changes after damage of the central nervous system: Significance of spastic muscle tone in rehabilitation. In V. Dietz, & N. Ward (Eds.), *Oxford textbook of neurorehabilitation* Oxford University Press.

Donelan, J. M., & Pearson, K. G. (2004). Contribution of force feedback to ankle extensor activity in decerebrate walking cats. *Journal of Neurophysiology*, *92*(4), 2093–2104.

Duysens, J., Bastiaanse, C. M., Smits-Engelsman, B. C., & Dietz, V. (2004). Gait acts as a gate for reflexes from the foot. *Canadian Journal of Physiology and Pharmacology*, *82*(8–9), 715–722.

Duysens, J., Clarac, F., & Cruse, H. (2000). Load-regulating mechanisms in gait and posture: Comparative aspects. *Physiological Reviews*, *80*(1), 83–133.

Fung, J., & Barbeau, H. (1989). A dynamic EMG profile index to quantify muscular activation disorder in spastic paretic gait. *Electroencephalography and Clinical Neurophysiology*, *73*(3), 233–244.

Fung, J., & Barbeau, H. (1994). Effects of conditioning cutaneomuscular stimulation on the soleus h-reflex in normal and spastic paretic subjects during walking and standing. *Journal of Neurophysiology*, *72*(5), 2090–2104.

Gorassini, M. A., Knash, M. E., Harvey, P. J., Bennett, D. J., & Yang, J. F. (2004). Role of motoneurons in the generation of muscle spasms after spinal cord injury. *Brain*, *127*(Pt. 10), 2247–2258. https://doi.org/10.1093/brain/awh243.

Gossard, J. P., Brownstone, R. M.; Barajon, I., & Hultborn, H. (1994). Transmission in a locomotor-related group ib pathway from hindlimb extensor muscles in the cat. *Experimental Brain Research*, *98*(2), 213–228. https://doi.org/10.1007/BF00228410.

Grey, M. J., Nielsen, J. B., Mazzaro, N., & Sinkjaer, T. (2007). Positive force feedback in human walking. *The Journal of Physiology*, *581*(Pt. 1), 99–105. https://doi.org/10.1113/jphysiol.2007.130088.

Henneman, E., & Mendell, L. M. (1980). Functional organization of motoneuron pool and its inputs. In *Handbook of physiology: Section 1: The nervous system volume ii, part 1 &2: Motor control* (pp. 423–507). American Physiological Society. https://doi.org/10.1002/cphy.cp010211.

Hidler, J. M., & Rymer, W. Z. (1999). A simulation study of reflex instability in spasticity: Origins of clonus. *IEEE Transactions on Rehabilitation Engineering*, *7*(3), 327–340.

Hiersemenzel, L. P., Curt, A., & Dietz, V. (2000). From spinal shock to spasticity: Neuronal adaptations to a spinal cord injury. *Neurology*, *54*(8), 1574–1582.

Hornby, T. G., Kahn, J. H., Wu, M., & Schmit, B. D. (2006). Temporal facilitation of spastic stretch reflexes following human spinal cord injury. *The Journal of Physiology*, *571*(Pt 3), 593–604. https://doi.org/10.1113/jphysiol.2005.102046.

Hultborn, H. (2003). Changes in neuronal properties and spinal reflexes during development of spasticity following spinal cord lesions and stroke: Studies in animal models and patients. *Journal of Rehabilitation Medicine*, (41 Suppl), 46–55.

Jankowska, E., & Edgley, S. A. (2010). Functional subdivision of feline spinal interneurons in reflex pathways from group ib and ii muscle afferents; an update. *The European Journal of Neuroscience*, *32*(6), 881–893. https://doi.org/10.1111/j.1460-9568.2010.07354.x.

Jankowska, E., & McCrea, D. A. (1983). Shared reflex pathways from ib tendon organ afferents and ia muscle spindle afferents in the cat. *The Journal of Physiology*, *338*, 99–111. https://doi.org/10.1113/jphysiol.1983.sp014663.

Kim, H. E., Corcos, D. M., & Hornby, T. G. (2015). Increased spinal reflex excitability is associated with enhanced central activation during voluntary lengthening contractions in human spinal cord injury. *Journal of Neurophysiology*, *114*(1), 427–439. https://doi.org/10.1152/jn.01074.2014.

Knikou, M., Angeli, C. A., Ferreira, C. K., & Harkema, S. J. (2009). Soleus h-reflex modulation during body weight support treadmill walking in spinal cord intact and injured subjects. *Experimental Brain Research*, *193*(3), 397–407. https://doi.org/10.1007/s00221-008-1636-x.

Lance, J. W. (1980). The control of muscle tone, reflexes, and movement: Robert wartenberg lecture. *Neurology*, *30*(12), 1303–1313.

Latash, M. L., Penn, R. D., Corcos, D. M., & Gottlieb, G. L. (1989). Short-term effects of intrathecal baclofen in spasticity. *Experimental Neurology*, *103*(2), 165–172.

Li, Y., Gorassini, M. A., & Bennett, D. J. (2004). Role of persistent sodium and calcium currents in motoneuron firing and spasticity in chronic spinal rats. *Journal of Neurophysiology*, *91*(2), 767–783.

Magladery, J. W., Porter, W. E., Park, A. M., & Teasdall, R. D. (1951). Electrophysiological studies of nerve and reflex activity in normal man. Iv. The two-neurone reflex and identification of certain action potentials from spinal roots and cord. *Bulletin of the Johns Hopkins Hospital*, *88*(6), 499–519.

Mayer, W. P., Murray, A. J., Brenner-Morton, S., Jessell, T. M., Tourtellotte, W. G., & Akay, T. (2018). Role of muscle spindle feedback in regulating muscle activity strength during walking at different speed in mice. *Journal of Neurophysiology*, *120*(5), 2484–2497. https://doi.org/10.1152/jn.00250.2018.

Maynard, F. M., Karunas, R. S., & Waring, W. P., 3rd. (1990). Epidemiology of spasticity following traumatic spinal cord injury. *Archives of Physical Medicine and Rehabilitation*, *71*(8), 566–569.

Mazzaro, N., Grey, M. J., do Nascimento, O. F., & Sinkjaer, T. (2006). Afferent-mediated modulation of the soleus muscle activity during the stance phase of human walking. *Experimental Brain Research*, *173*(4), 713–723. https://doi.org/10.1007/s00221-006-0451-5.

Mazzaro, N., Grey, M. J., & Sinkjaer, T. (2005). Contribution of afferent feedback to the soleus muscle activity during human locomotion. *Journal of Neurophysiology*, *93*(1), 167–177. https://doi.org/10.1152/jn.00283.2004.

Mazzaro, N., Nielsen, J. F., Grey, M. J., & Sinkjaer, T. (2007). Decreased contribution from afferent feedback to the soleus muscle during walking in patients with spastic stroke. *Journal of Stroke and Cerebrovascular Diseases*, *16*(4), 135–144. https://doi.org/10.1016/j.jstrokecerebrovasdis.2007.01.003.

Mendell, L. M., & Henneman, E. (1971). Terminals of single ia fibers: Location, density, and distribution within a pool of 300 homonymous motoneurons. *Journal of Neurophysiology*, *34*(1), 171–187.

Mirbagheri, M. M., Barbeau, H., Ladouceur, M., & Kearney, R. E. (2001). Intrinsic and reflex stiffness in normal and spastic, spinal cord injured subjects. *Experimental Brain Research*, *141*(4), 446–459.

Mirbagheri, M. M., Chen, D., & Rymer, W. Z. (2010). Quantification of the effects of an alpha-2 adrenergic agonist on reflex properties in spinal cord injury using a system identification technique. *Journal of Neuroengineering and Rehabilitation*, *7*, 29. https://doi.org/10.1186/1743-0003-7-29.

Mirbagheri, M. M., Kindig, M. W., & Niu, X. (2015). Effects of robotic-locomotor training on stretch reflex function and muscular properties in individuals with spinal cord injury. *Clinical Neurophysiology*, *126*(5), 997–1006. https://doi.org/10.1016/j.clinph.2014.09.010.

Mirbagheri, M. M., Ladouceur, M., Barbeau, H., & Kearney, R. E. (2002). The effects of long-term fes-assisted walking on intrinsic and reflex dynamic stiffness in spastic spinal-cord-injured subjects. *IEEE Transactions on Neural Systems and Rehabilitation Engineering*, *10*(4), 280–289.

Morita, H., Shindo, M., Momoi, H., Yanagawa, S., Ikeda, S., & Yanagisawa, N. (2006). Lack of modulation of ib inhibition during antagonist contraction in spasticity. *Neurology*, *67*(1), 52–56. https://doi.org/10.1212/01.wnl.0000223399.59212.f4.

Nakazawa, K., Kawashima, N., & Akai, M. (2006). Enhanced stretch reflex excitability of the soleus muscle in persons with incomplete rather than complete chronic spinal cord injury. *Archives of Physical Medicine and Rehabilitation*, *87*(1), 71–75. https://doi.org/10.1016/j.apmr.2005.08.122.

Nichols, T. R. (2018). Distributed force feedback in the spinal cord and the regulation of limb mechanics. *Journal of Neurophysiology*, *119*(3), 1186–1200. https://doi.org/10.1152/jn.00216.2017.

Nielsen, J. B., Christensen, M. S., Farmer, S. F., & Lorentzen, J. (2020). Spastic movement disorder: Should we forget hyperexcitable stretch reflexes and start talking about inappropriate prediction of sensory consequences of movement? *Experimental Brain Research*, *238*(7–8), 1627–1636. https://doi.org/10.1007/s00221-020-05792-0.

Nielsen, J. B., Crone, C., & Hultborn, H. (2007). The spinal pathophysiology of spasticity—From a basic science point of view. *Acta Physiologica (Oxford, England)*, *189*(2), 171–180.

Rossi, A., Mazzocchio, R., & Scarpini, C. (1990). Clonus in man: A rhythmic oscillation maintained by a reflex mechanism. *Electroencephalography and Clinical Neurophysiology*, *75*(2), 56–63.

Schafer, S. S., Dadfar, F., Hartel, J., Haupts, S., & Fischer, M. (1999). The period of latency before a muscle receptor generates an action potential as a response to a muscle stretch. *Brain Research*, *843*(1–2), 36–47.

Schieppati, M., & Nardone, A. (1997). Medium-latency stretch reflexes of foot and leg muscles analysed by cooling the lower limb in standing humans. *The Journal of Physiology*, *503*(Pt. 3), 691–698.

Sheean, G. (2006). Botulinum toxin treatment of adult spasticity: A benefit-risk assessment. *Drug Safety*, *29*(1), 31–48.

Shefner, J. M., Berman, S. A., Sarkarati, M., & Young, R. R. (1992). Recurrent inhibition is increased in patients with spinal cord injury. *Neurology*, *42*(11), 2162–2168.

Sinkjaer, T., Andersen, J. B., Ladouceur, M., Christensen, L. O., & Nielsen, J. B. (2000). Major role for sensory feedback in soleus EMG activity in the stance phase of walking in man. *Journal of Physiology*, *523*(Pt. 3), 817–827.

Sinkjaer, T., Andersen, J. B., & Larsen, B. (1996). Soleus stretch reflex modulation during gait in humans. *Journal of Neurophysiology*, *76*(2), 1112–1120.

Sinkjaer, T., Andersen, J. B., Nielsen, J. F., & Hansen, H. J. (1999). Soleus long-latency stretch reflexes during walking in healthy and spastic humans. *Clinical Neurophysiology*, *110*(5), 951–959.

Sinkjaer, T., & Magnussen, I. (1994). Passive, intrinsic and reflex-mediated stiffness in the ankle extensors of hemiparetic patients. *Brain*, *117*(Pt. 2), 355–363. https://doi.org/10.1093/brain/117.2.355.

Skold, C., Levi, R., & Seiger, A. (1999). Spasticity after traumatic spinal cord injury: Nature, severity, and location. *Archives of Physical Medicine and Rehabilitation*, *80*(12), 1548–1557.

Smyth, M. D., & Peacock, W. J. (2000). The surgical treatment of spasticity. *Muscle and Nerve*, *23*(2), 153–163.

Stein, R. B., & Capaday, C. (1988). The modulation of human reflexes during functional motor tasks. *Trends in Neurosciences*, *11*(7), 328–332.

Stein, R. B., & Kearney, R. E. (1995). Nonlinear behavior of muscle reflexes at the human ankle joint. *Journal of Neurophysiology*, *73*(1), 65–72.

Stein, R. B., Yang, J. F., Belanger, M., & Pearson, K. G. (1993). Modification of reflexes in normal and abnormal movements. *Progress in Brain Research*, *97*, 189–196.

Taylor, S., Ashby, P., & Verrier, M. (1984). Neurophysiological changes following traumatic spinal lesions in man. *Journal of Neurology, Neurosurgery, and Psychiatry*, *47*(10), 1102–1108.

Thompson, A. K., Estabrooks, K. L., Chong, S., & Stein, R. B. (2009). Spinal reflexes in ankle flexor and extensor muscles after chronic central nervous system lesions and functional electrical stimulation. *Neurorehabilitation and Neural Repair*, *23*(2), 133–142. https://doi.org/10.1177/1545968308321067.

Thompson, A. K., Mrachacz-Kersting, N., Sinkjaer, T., & Andersen, J. B. (2019). Modulation of soleus stretch reflexes during walking in people with chronic incomplete spinal cord injury. *Experimental Brain Research*, *237*(10), 2461–2479. https://doi.org/10.1007/s00221-019-05603-1.

Thompson, A. K., Pomerantz, F. R., & Wolpaw, J. R. (2013). Operant conditioning of a spinal reflex can improve locomotion after spinal cord injury in humans. *The Journal of Neuroscience*, *33*(6), 2365–2375. https://doi.org/10.1523/JNEUROSCI.3968-12.2013.

Thompson, A. K., & Wolpaw, J. R. (2015). Restoring walking after spinal cord injury: Operant conditioning of spinal reflexes can help. *The Neuroscientist*, *21*(2), 203–215. https://doi.org/10.1177/1073858414527541.

Thompson, A. K., & Wolpaw, J. R. (2019). H-reflex conditioning during locomotion in people with spinal cord injury. *The Journal of Physiology*. https://doi.org/10.1113/JP278173.

Turns, L. J., Neptune, R. R., & Kautz, S. A. (2007). Relationships between muscle activity and anteroposterior ground reaction forces in hemiparetic walking. *Archives of Physical Medicine and Rehabilitation*, *88*(9), 1127–1135. https://doi.org/10.1016/j.apmr.2007.05.027.

Vincent, J. A., Gabriel, H. M., Deardorff, A. S., Nardelli, P., Fyffe, R. E. W., Burkholder, T., et al. (2017). Muscle proprioceptors in adult rat: Mechanosensory signaling and synapse distribution in spinal cord. *Journal of Neurophysiology*, *118*(5), 2687–2701. https://doi.org/10.1152/jn.00497.2017.

Wallace, D. M., Ross, B. H., & Thomas, C. K. (2012). Characteristics of lower extremity clonus after human cervical spinal cord injury. *Journal of Neurotrauma*, *29*(5), 915–924. https://doi.org/10.1089/neu.2010.1549.

Ward, A. B. (2008). Spasticity treatment with botulinum toxins. *Journal of Neural Transmission*, *115*(4), 607–616. https://doi.org/10.1007/s00702-007-0833-2.

Westerkam, D., Saunders, L. L., & Krause, J. S. (2011). Association of spasticity and life satisfaction after spinal cord injury. *Spinal Cord*, *49*(9), 990–994. https://doi.org/10.1038/sc.2011.49.

Wolpaw, J. R. (2018). The negotiated equilibrium model of spinal cord function. *The Journal of Physiology*, *596*(16), 3469–3491. https://doi.org/10.1113/JP275532.

Xia, R., & Rymer, W. Z. (2005). Reflex reciprocal facilitation of antagonist muscles in spinal cord injury. *Spinal Cord*, *43*(1), 14–21. https://doi.org/10.1038/sj.sc.3101656.

Yang, J. F., Fung, J., Edamura, M., Blunt, R., Stein, R. B., & Barbeau, H. (1991). H-reflex modulation during walking in spastic paretic subjects. *Canadian Journal of Neurological Sciences*, *18*(4), 443–452.

Zehr, E. P., & Stein, R. B. (1999). What functions do reflexes serve during human locomotion? *Progress in Neurobiology*, *58*(2), 185–205.

Chapter 31

Metabolic syndrome in spinal cord injury: Impact on health

Arcangelo Barbonetti, Chiara Castellini, Sandro Francavilla, Felice Francavilla, and Settimio D'Andrea
Andrology Unit, Department of Life, Health and Environmental Sciences, University of L'Aquila, L'Aquila, Italy

List of abbreviations

ATP	adenosine triphosphate
BMI	body mass index
CRP	C-reactive protein
CVD	cardiovascular disease
eNOS	endothelial nitric oxide synthase
ET-1	endothelin-1
FFM	fat-free mass
FPG	fasting plasma glucose
GH	growth hormone
GLUT	glucose transporter
HDL	high-density lipoprotein
IDF	International Diabetes Federation
IFG	impaired fasting glucose
IGT	impaired glucose tolerance
IGF-1	insulin-like growth factor-1
IL-6	interleukin-6
IMF	intramuscular fat
IR	insulin resistance
IRS	insulin receptor substrate
LDL	low-density lipoprotein
LH	luteinizing hormone
LTPA	leisure time physical activity
MAPK	mitogen activated protein kinase
MetS	metabolic syndrome
NAFLD	non-alcoholic fatty liver disease
NCEP:ATPIII	National Cholesterol Education Program/third Adult Treatment Panel
NEFA	non-esterified fatty acid
NO	nitric oxide
OCN	osteocalcin
PI3K	phosphatidylinositol 3-kinase
SAT	subcutaneous adipose tissue
SCI	spinal cord injury
SERCA	sarco(endo)plasmic reticulum calcium-ATP-ase
SHBG	sex hormone-binding globulin
T2DM	type 2 diabetes mellitus
TNF-α	tumor-necrosis factor-α
VAT	visceral adipose tissue
VLDL	very low-density lipoprotein
WC	waist circumference
WHO	World Health Organization
WHR	waist-to-hip ratio

Cellular, Molecular, Physiological, and Behavioral Aspects of Spinal Cord Injury. https://doi.org/10.1016/B978-0-12-822427-4.00031-9

Introduction

In the last decades, advances in healthcare and clinical management of the acute post-injury phase resulted in an obvious improvement in the life expectancy of people with spinal cord injury (SCI). In chronic SCI (lasting more than 1 year), clinical concerns closely related to neurological damage largely overlap with those from different fields of medicine with a substantial impact on quality of life, morbidity, and mortality. This chapter outlines the pathophysiology and diagnostic challenges of metabolic syndrome (MetS) in people with SCI, where the cluster of clinical and metabolic factors underlying the MetS definition, together with the systemic inflammation associated with visceral obesity, results in cardiovascular health deterioration. Emerging correlates of MetS, including androgen deficiency and hypovitaminosis D, will be also described, as they are highly prevalent in chronic SCI and establish complex multi-directional relationships with visceral obesity.

Changes in body composition after spinal cord injury

In people with SCI, the lack of neurotrophic influences of motor nerve projections results in a dramatic and early muscle wasting below the level of the lesion up to sarcopenia (Gorgey et al., 2014). Skeletal muscle adaptive molecular responses to denervation would occur within just a few days after SCI with upregulation of genes involved in the proteolytic ubiquitin-proteasome pathway, leading to degradation of contractile proteins (Urso et al., 2007). In keeping with findings from animal models, where muscle atrophy can be demonstrated within the first few months following spinal cord transection (Roy, Baldwin, & Edgerton, 1991), Castro, Apple Jr, Hillegass, and Dudley (1999) reported that in patients with complete cervical/thoracic SCI, the average cross-sectional area in lower extremity muscles, as assessed by magnetic resonance, was up to 80% smaller than that of age- and weight-matched able-bodied controls 24 weeks post-injury. Muscle atrophy does not spare individuals with incomplete SCI, exhibiting an approximately 30% smaller average cross-sectional area in affected muscles of lower extremities with respect to controls (Shah et al., 2006).

Biopsy data showed qualitative other than quantitative muscle changes in the early time period after SCI with an increased ratio between fast- (type II) and slow- (type I) twitch fibers and a decrease in the proportion of fibers with the slow isoform (type 2) of sarco(endo)plasmic reticulum calcium- adenosine triphosphate (ATP)-ase (SERCA2), which transports calcium from the cytosol into the sarcoplasmic reticulum (Talmadge, Castro, Apple Jr, & Dudley, 2002). These shifts make the skeletal muscle more fatigable. In fact, while slow-twitch muscle fibers are aerobic fatigue-resistant fibers supporting endurance physical activity, the anaerobic fast-twitch fibers provide more powerful forces, but for a shorter time and fatigue quickly. Moreover, the fast SERCA isoform (SERCA1), which is prevalent in muscle from people with SCI, is more susceptible to repeated contraction-relaxation cycles, resulting in lower calcium uptake by the sarcoplasmic reticulum between contractions. Poorer intracellular calcium stores and, hence, lower calcium release from sarcoplasmic reticulum during contraction would result in decreased myofilament activation and force output.

The loss of muscle trophism and performance ability underlies a substantial decrease in overall energy expenditure, by as much as 54% (Gorgey et al., 2014). Energy expenditure is the combination of basal metabolic rate, thermic effect of food digestion, and thermic effect of physical activity, with the former accounting for 60%–70% of the total (Gater Jr., 2007). The major contribution to the basal metabolic rate is provided by the fat-free mass (FFM), 85% of which is accounted for by the skeletal muscle (Gater Jr., 2007). Therefore, a loss in skeletal muscle mass significantly affects both physical activity-related energy expenditure and basal metabolic rate, with obvious reflections on total daily energy expenditure. On this basis, in people with SCI, body energy balance is overwhelmed since energy intake easily exceeds energy expenditure thus leading to fat mass increase. When compared to waist circumference (WC)-matched able-bodied individuals, people with SCI exhibit greater volumes of visceral adipose tissue (VAT). Indeed, in this population, for every centimeter increase in WC and for every unit increase in body mass index (BMI), VAT increases by 30% and 20%, respectively (Cirnigliaro et al., 2015), resulting in a high ratio of VAT to subcutaneous adipose tissue (SAT), which represents a well-known risk factor for insulin resistance (IR) and endocrine-metabolic disorders.

Insulin resistance in spinal cord injury: Pathophysiology and clinical features

As visceral body fat is highly sensitive to lipolytic stimuli, its accumulation is accompanied by an increased generation of non-esterified fatty acids (NEFA), which are released in the blood circulation and deposited into muscle and liver cells. This could explain the very high prevalence of non-alcoholic fatty liver disease (NAFLD) (Barbonetti, Caterina Vassallo, et al., 2016) as well as the accumulation of intramuscular fat (IMF) (Elder, Apple, Bickel, Meyer, & Dudley, 2004), peculiar to people with SCI. The presence of NEFA in such ectopic sites triggers serine phosphorylations of the insulin receptors, thus negatively affecting the downstream signaling cascade of insulin receptor substrates (IRSs) and phosphatidylinositol 3-kinase (PI3K) (Gallagher, LeRoith, & Karnieli, 2008).

In skeletal muscle, impaired insulin signaling results in inhibited translocation of glucose transporters-4 (GLUT-4) to the cell membrane with decreased glucose entrance into the cell and, hence, post-prandial hyperglycemia (Gallagher et al., 2008). In individuals with SCI, these events can be amplified by physical inactivity which contributes to decreasing insulin sensitivity in skeletal muscle. Accordingly, a high prevalence of diabetes has been reported in SCI (Cragg, Noonan, Krassioukov, & Borisoff, 2013), where IMF accumulation and muscle atrophy would account for 70% of glucose intolerance (Elder et al., 2004). In the liver, NEFA accumulation and IR negatively affect both glucose and lipid metabolism: higher rates of gluconeogenesis result in fasting hyperglycemia; meanwhile, an increased generation of very low-density (VLDL) and low-density (LDL) lipoproteins is accompanied by a slowed synthesis of apolipoprotein A, with a consequent decrease in high-density lipoproteins (HDL) (Kolovou, Anagnostopoulou, & Cokkinos, 2005). These pathogenic processes underpin the unfavorable lipid profile peculiar to people with SCI exhibiting low HDL, high triglycerides, and high LDL (Gater Jr., 2007).

An additional mechanism by which adipose tissue contributes to IR involves the adipocyte release of proinflammatory cytokines, such as interleukin-6 (IL-6) and tumor-necrosis factor-α (TNF-α), which interfere with insulin signaling (Farkas & Gater, 2018). IL-6 and TNF-α also account for low-grade vascular inflammation by promoting the hepatic synthesis of acute-phase reactants, including C-reactive protein (CRP) (Farkas & Gater, 2018). Accordingly, when compared to matched able-bodied controls, individuals with SCI exhibit higher circulating CRP levels (Liang, Mojtahedi, Chen, & Braunschweig, 2008), which are positively associated with both WC and percentage of body fat (Gibson et al., 2008) and negatively associated with HDL (Liang et al., 2008).

Complex pathogenetic mechanisms also underlie the relationship of visceral obesity with hypertension. Fat-derived proinflammatory cytokines non only induce a systemic vascular inflammation but also promote the hepatic synthesis of angiotensinogen, thus activating the renin-angiotensin-aldosterone system (Engeli, Negrel, & Sharma, 2000). Moreover, VAT secretes leptin, a hormone which reaches high circulating concentrations in individuals with SCI (Wang et al., 2005) and affects blood pressure control by stimulating the sympathetic nervous system activity (Wofford & Hall, 2004). An additional mechanism reflects an imbalance in the biosynthesis of vasodilator and vasoconstrictor molecules. Indeed, at the cellular level, the vascular response to insulin reflects a balance between PI3K-dependent insulin signaling pathways that regulate the endothelial generation of the vasodilator nitric oxide (NO) by the endothelial NO synthase (eNOS), and mitogen activated protein kinase (MAPK)-dependent pathways regulating the secretion of the vasoconstrictor endothelin-1 (ET-1). Endothelial IR is typically characterized by a defective PI3K-NO pathway accompanied by an intact or enhanced MAPK-ET-1 pathway: the net effect is a vasoconstriction that increases peripheral vascular resistance and, hence, diastolic blood pressure (Muniyappa & Sowers, 2013).

Diagnostic challenges of metabolic syndrome in people with spinal cord injury

The constellation of central obesity and its correlates, including IR, glycemic dysregulation, hypertension, and dyslipidemia has been referred to as "metabolic syndrome" (MetS), a cluster of clinical and metabolic factors that increase the risk of stroke and cardiovascular disease (CVD) (Zafar, Khaliq, Ahmad, Manzoor, & Lone, 2018).

Over the past decade, different diagnostic criteria for MetS have been proposed (Table 1). The first definition by the World Health Organization (WHO), focusing on insulin resistance as the key element for the diagnosis (Alberti & Zimmet, 1998), was followed by the recommendations from the National Cholesterol Education Program/third Adult Treatment Panel (NCEP:ATP III), where no single factor was required for establishing the diagnosis, but instead, central obesity, dyslipidemia, hypertension, and impaired fasting glucose (IFG)/T2DM were placed on the same level (NCEP:ATP III, 2002). The subsequent definition by the International Diabetes Federation (IDF) emphasized abdominal obesity as a prerequisite for the MetS diagnosis, while maintaining unchanged the remainder of the NCEP:ATP III criteria (Ford, 2005).

Unfortunately, all these definitions that are widely used in the able-bodied general population result to be inappropriate in people with SCI, where the diagnosis of obesity is challenged by the deep changes in body composition. According to the WHO criteria, overweight and obesity are defined as BMI values $\geq$25 kg/m^2 and $\geq$ 30 kg/m^2, respectively (WHO, 1995). As lean tissue, fat tissue, and bone mass together contribute to the body weight, for the same BMI value, people with SCI exhibit a higher percentage of body fat than able-bodied individuals, due to the muscle wasting below the level of the lesion: this means that, in this population, the use of traditional BMI cut-off points grossly underestimates the diagnosis of overweight and obesity (Jones, Legge, & Goulding, 2003). Indeed, bioelectrical impedance analyses showed that, in adults with chronic SCI, a BMI of 22–25 kg/m^2 translates to >30% body fat (Nash et al., 2019), which is well above the accepted threshold for obesity of 22% body fat in the in able-bodied population. Accordingly, the recent guidelines for identification and management of cardiometabolic risk after SCI (Nash et al., 2019), recommended a BMI $\geq$22 kg/m^2 as the best cutoff point when used as a surrogate marker for obesity in this population. It has been also suggested that WC could represent a better measure of obesity than is BMI in this population (Ravensbergen, Lear, & Claydon, 2014). However, no specific WC

TABLE 1 Main definitions of metabolic syndrome.

WHO, 1998	NCEP:ATP III, 2001	IDF, 2006
IGT or IFG or T2DM and/or IR[a] plus any 2 or more of the following:	Any three or more of the following:	WC ≥94 cm for men and ≥ 80 cm for women[b] plus any 2 or more of the following
• BMI >30 Kg/m^2 or WHR >0.90 for men and > 0.85 for women	• WC ≥102 cm for men and ≥ 88 cm for women	
• Blood pressure ≥ 140/90 mmHg	• Blood pressure ≥ 130/85 mmHg	• Blood pressure ≥ 130/85 mmHg or on treatment
• Triglycerides ≥150 mg/dL or on treatment • HDL <35 mg/dL for men and < 39 mg/dL for women • Urinary albumin excretion rate ≥ 20 μg/min or urine albumin/creatine ratio ≥ 30 mg/g	• Triglycerides ≥150 mg/dL or on treatment • HDL <40 mg/dL for men and < 50 mg/dL for women • FPG ≥100 mg/dL or on treatment for T2DM	• Triglycerides ≥150 mg/dL or on treatment • HDL <40 mg/dL for men and < 50 mg/dL for women • FPG ≥100 mg/dL or on treatment for T2DM

Abbreviations: BMI, body mass index; FPG, fasting plasma glucose; HDL, high-density lipoprotein; IDF, International Diabetes Federation; IFG, impaired fasting glucose; IGT, impaired glucose tolerance; IR, insulin resistance; NCEP:ATPIII, National Cholesterol Education Program/third Adult Treatment Panel; T2DM, type 2 diabetes mellitus; WC, waist circumference; WHO, World Health Organization; WHR, waist-to-hip ratio.
[a]*As assessed by the euglycemic hyperinsulinemic clamp method.*
[b]*The WC criterion can be assumed if BMI ≥30 kg/m^2.*

cut-off points have been validated so far in large series of individuals with SCI, where WC may be expanded due to the abdominal muscle paralysis, leading to a substantial overestimate of central obesity.

An equally thorny issue is the diagnosis of hypertension which can be confounded by the SCI-related neurogenic derangement of blood pressure control. On this basis, traditional models for the assessment of cardiovascular risk appear to be largely unsuitable in people with SCI (Barton et al., 2020).

In this scenario, inconsistent results produced by studies investigating the prevalence of MetS in people with SCI not only reflect differences in clinical characteristics of the study populations, but also the use of different criteria that have to deal with the complexity of MetS diagnosis in SCI. In clinically different small series from different countries, the prevalence of MetS largely varied from 3.6% to 19.0% according to WHO definition, from 12.3% to 43.0% when using NCEP: ATP III definition and from 19.3% to 31.0% when the IDF criteria were adopted (Yahiro, Wingo, Kunwor, Parton, & Ellis, 2020). Recently, among 473 veterans with SCI, aged 56.0 ± 13.1 years, where IDF criteria were modified by the inclusion of a BMI ≥22 Kg/m^2 as an SCI-appropriate surrogate marker of obesity, the prevalence of MetS reached 57%, with high prevalence rates of single components of MetS: 76.7% with BMI ≥22 kg/m^2, 55.1% with hypertension, 49.7% with T2DM and 69.7% with dyslipidemia (Gater Jr., Farkas, Berg, & Castillo, 2019). This highly prevalent clustering of vascular/metabolic risk factors, together with physical inactivity, proatherogenic systemic inflammation, and impaired autonomic cardiovascular control, largely accounts for the unfavorable cardiovascular and cardiometabolic profile of people with chronic SCI.

Cardiovascular reflections of metabolic syndrome in spinal cord injury

While most of the research dealing with the relationship between SCI and CVD has documented higher prevalence rates of cardiovascular risk factors, only a few studies have specifically assessed CVD prevalence and risk estimates. In a British multi-centric study on medical records from 545 individuals with long-lasting SCI (Groah, Weitzenkamp, Sett, Soni, & Savic, 2001), the age-adjusted rates of all CVD ranged from 21.2 to 35.2 per 1000 SCI person-years, according to the level of the lesion. The risk increased with both advancing age and the severity of SCI: tetraplegic state conferred an excess 16% risk of all CVD and a fivefold risk of cerebrovascular when compared with paraplegia; meanwhile more complete SCI resulted in an excess 44% all CVD risk. More recently, in a large 4-year follow-up cohort study on 2806 SCI individuals from Taiwan's National Health Insurance Research Database (Wu et al., 2012), the incidence rate of hospitalized stroke (5.96 per 1000 person-years) was approximately 3 times higher than that in 28,060 able-bodied controls (2.04 per 1000 person-years). Finally, a cross-sectional analysis of a representative sample of more than 60,000 individuals from the Canadian Community Health Survey (Cragg, Noonan, Dvorak, et al., 2013) revealed a significant and independent association of SCI with both self-reported heart disease (OR: 2.72, 95% CI: 1.94–3.82) and stroke (OR: 3.72, 95% CI:

2.22–6.23), after adjustment for age and gender. These findings are in line with mortality data. In recent years, major advances in life support treatments and medical care resulted in a substantial improvement in life expectancy in this population. While a larger number of patients survive now the acute post-injury stage, for long-term SCI, morbidity, and mortality from cardiovascular diseases have become increasingly meaningful, also as a result of better clinical management of renal and pulmonary comorbidities, the primary causes of SCI-related mortality in the previous decades. In a 70-year retrospective analysis on more than 5000 individuals who were injured between 1943 and 2010 and survived first-year post-injury (Savic et al., 2017), mortality rates showed improvement over time for most cases, and, interestingly, among the leading causes of mortality, cardiovascular and cerebrovascular diseases (26.7% of all certified causes) were second only to respiratory disorders (29.3%) and followed by cancer (13.9%), urogenital complications (11.5%), digestive causes (5.3%) and suicide (4.5%).

As a further threat to metabolic and cardiovascular health, a decline in anabolic hormones has been also reported following SCI. Indeed, chronic SCI has been postulated to accelerate age-related hormonal changes that can contribute to maintaining the deterioration in body composition (Gorgey et al., 2014). These abnormalities not only include a possible decrease in circulating levels of growth hormone (GH)/insulin-like growth factor-1 (IGF-1), likely due to the drastic drop in daily physical activity (Bauman et al., 1994), but also androgen deficiency, which represents an emerging concern in SCI male population.

Pathophysiology of androgen deficiency in men with chronic spinal cord injury

A decline in circulating testosterone levels in men following SCI has been repeatedly reported in the last decades (Barbonetti et al., 2014, 2019; Barbonetti, Sperandio, et al., 2016; Barbonetti, Caterina Vassallo, et al., 2016; Bauman, Fountaine, & Spungen, 2014; Clark et al., 2008; D'Andrea, Castellini, et al., 2020; Kostovski, Iversen, Birkeland, Torjesen, & Hjeltnes, 2008; Schopp et al., 2006). The very high proportion of men with biochemical androgen deficiency during the acute post-injury stage, reaching 83% (Schopp et al., 2006), can reflect the impact of severe physical distress and systemic illness on testosterone secretion. Nevertheless, when compared to age-matched able-bodied controls, also men with chronic SCI tend to exhibit higher prevalence rates of low testosterone (Bauman et al., 2014; Kostovski et al., 2008).

The pathophysiology of androgen deficiency in this population, albeit not yet fully elucidated, seems to be highly complex and multi-factorial. The aforementioned low-grade systemic inflammation related to obesity is exacerbated by recurrent urinary and respiratory infections, as well as by the presence of severe pressure sores in many patients. The resultant increase in the levels of proinflammatory cytokines can lower testosterone levels by suppressing the pituitary secretion of luteinizing hormone (LH). Moreover, adipocytes express aromatase activity accounting for higher levels of estrogens which exert a well-known inhibitory effect on LH secretion in males (Giagulli, Kaufman, & Vermeulen, 1994), thus promoting a "non-hypergonadotropic" form of androgenic deficiency peculiar to men with chronic SCI (Barbonetti et al., 2014, 2019; Barbonetti, Caterina Vassallo, et al., 2016; Bauman et al., 2014; Kostovski et al., 2008) (Fig. 1). Estrogens also increase the circulating concentrations of sex hormone-binding globulin (SHBG), resulting in lower free testosterone levels (Fig. 1). Androgen deficiency, in turn, would make obesity worse, as testosterone drives the pluripotent stem cell commitment into myogenic rather than adipogenic lineage (Bhasin et al., 2003), thus establishing a vicious cycle (Fig. 2).

A possible additional physiopathological factor is osteocalcin (OCN), the key mediator of the so-called "bone–testis axis." OCN is a non-collagenous protein synthesized by osteoblasts that, after carboxylation of glutamic residues, is deposited on the extracellular matrix of the bone (Neve, Corrado, & Cantatore, 2013). As bone resorption is accompanied by OCN released from the bone matrix into the blood circulation, serum levels of OCN represent a sensitive marker of both bone formation and turnover (Neve et al., 2013). Interestingly, experimental data from the mouse model indicate that the circulating undercarboxylated fraction of OCN can exert extra-skeletal endocrine effects, leading to an increase in insulin secretion and insulin sensitivity, a decrease of visceral fat (Lee et al., 2007), and, notably, an increase of testosterone biosynthesis via the Gprc6a receptor of Leydig cells (Oury et al., 2011) (Fig. 3). However, in large population-based studies, the positive association between OCN and testosterone levels remains controversial, likely due to the preeminent role of a well-functioning hypothalamus–pituitary axis in stimulating testosterone biosynthesis. Following SCI, although circulating levels of bone remodeling markers, including OCN, are usually high during the acute post-injury stage, they undergo a substantial decrease during the chronic phase (Sabour et al., 2014). Interestingly, we recently demonstrated a significant positive association of OCN with both total and calculated free testosterone in a series of men with chronic SCI. This suggests that, in this population, the association would be unmasked by the blunted pituitary LH response (Barbonetti et al., 2019).

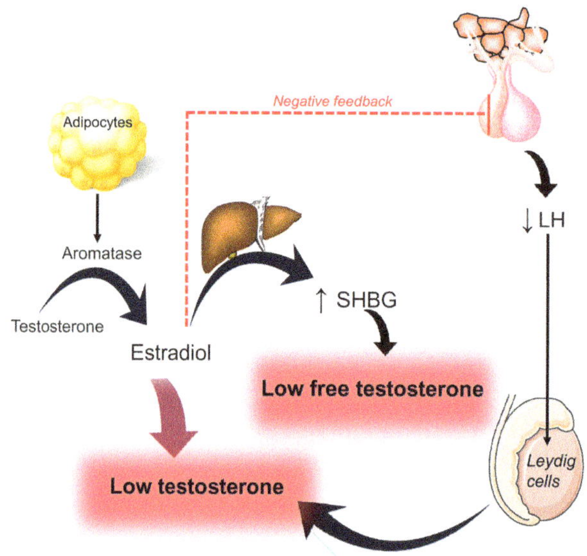

FIG. 1 Aromatase activity of adipocytes converts testosterone into estrogens: the levels of estrogens increases, while testosterone decreases. High levels of estrogens exert an inhibitory effect on luteinizing hormone (LH) secretion, leading to a "non-hypergonadotropic" form of androgenic deficiency. Estrogens also increase the circulating concentrations of sex hormone-binding globulin (SHBG), which binds to testosterone, resulting in lower free testosterone levels.

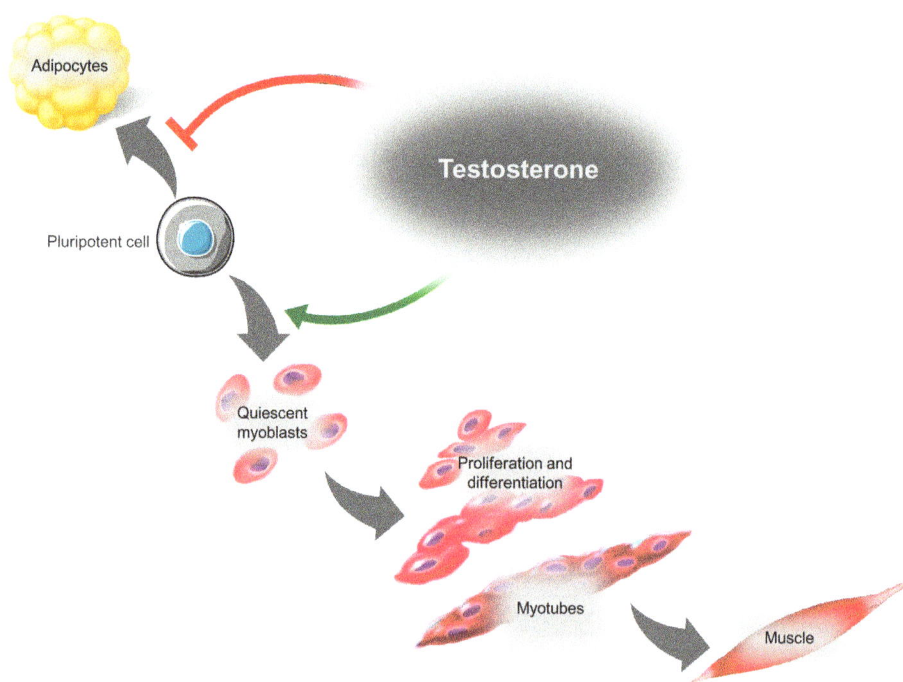

FIG. 2 Testosterone drives the pluripotent stem cell commitment into myogenic rather than adipogenic lineage.

Correlates of androgen deficiency in chronic spinal cord injury

Whether the SCI-related biochemical androgen deficiency can translate into clinical hypogonadism still remains an unsolved matter. The diagnosis of "androgen deficiency" as a biochemical and clinical syndrome requires the demonstration of unequivocally low testosterone levels as well as characteristic symptoms and signs (Bhasin et al., 2010), with sexual symptoms (poor morning erection, decreased libido, and erectile dysfunction) displaying a significant syndromic association with low testosterone in aging men (Wu et al., 2010) (Fig. 4). Unfortunately, in men with SCI, not only sexual

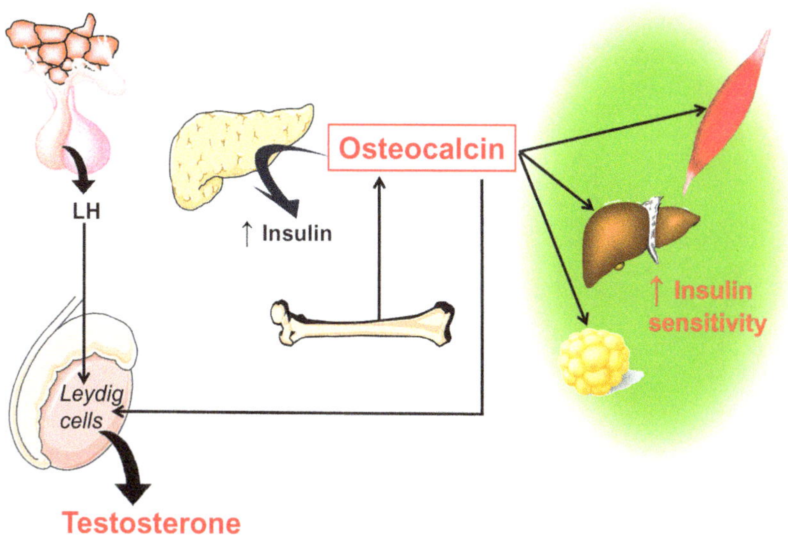

FIG. 3 Circulating undercarboxylated fraction of osteocalcin, released during bone resorption, exerts extra-skeletal endocrine effects on the pancreas, skeletal muscle, liver, and adipocytes, resulting in increased insulin secretion and sensitivity. Osteocalcin also promotes testosterone biosynthesis by acting on Leydig cells.

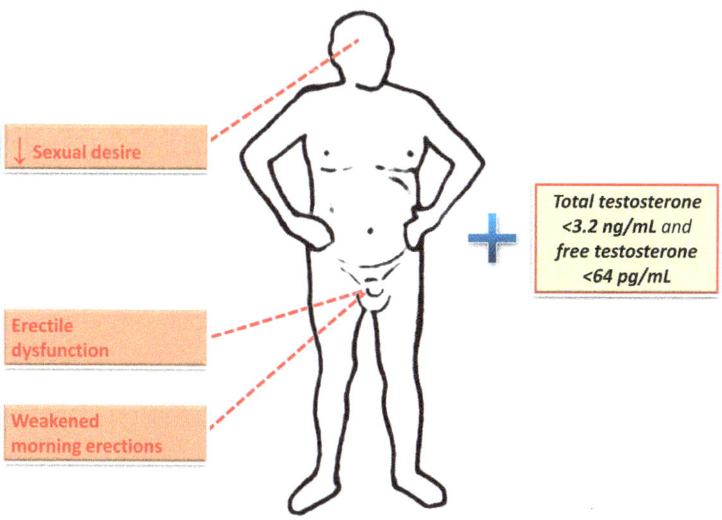

FIG. 4 Hypogonadism is a biochemical and clinical syndrome, where unequivocally low testosterone levels are associated with characteristic symptoms and signs. In aging men, sexual symptoms (poor morning erection, decreased libido, and erectile dysfunction) exhibit a significant syndromic association with low testosterone levels.

symptoms but also many other clinical features which can suggest a diagnosis of hypogonadism (e.g., body composition changes, anemia, osteoporosis, and depression) could represent direct or indirect reflections of neurological injury and disability (Fig. 5). In a series of men with chronic SCI, modifiable lifestyle-related risk factors, such as high BMI and poor leisure-time physical activity (LTPA), as well as a decreased sexual desire were found to be significant and independent predictors of low testosterone (Barbonetti et al., 2014). Notably, due to the very high prevalence of androgen deficiency, chronic SCI could represent an attractive clinical model to explore emerging correlates of low testosterone; among these, hypovitaminosis D deserves to be mentioned.

Leydig cells express the CYP2R1 gene, which encodes 25-hydroxylase, the enzyme involved in the first step of the cholecalciferol activation process (Blomberg Jensen et al., 2010). Therefore, it has been hypothesized that Leydig cell dysfunctions could result in both low testosterone and 25(OH)D deficiency (Foresta et al., 2011). Actually, in a recent meta-analysis of 18 case-control studies, hypovitaminosis D was associated with a slight decrease in serum testosterone levels, although the pooled estimate was burdened with a large heterogeneity between the included studies (D'Andrea et al., 2021).

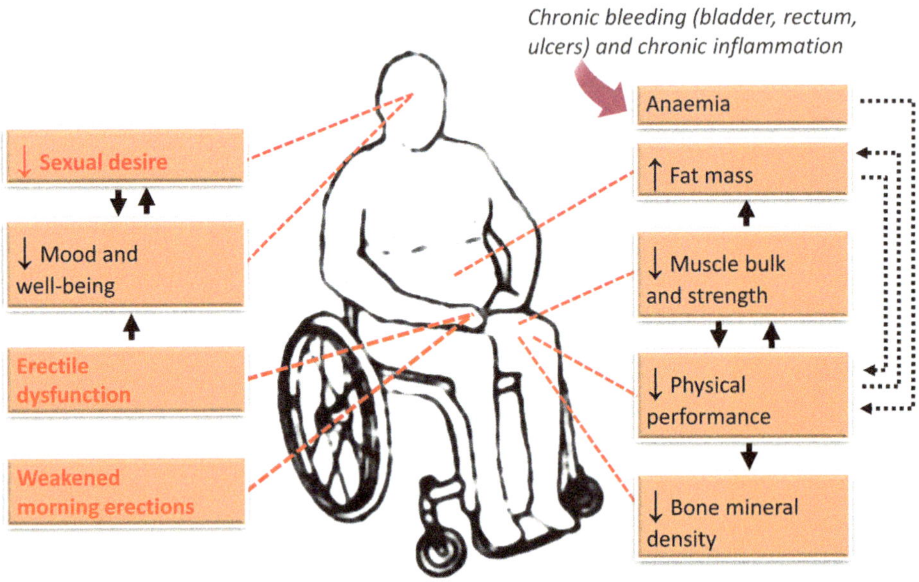

FIG. 5 In spinal cord-injured men, not only sexual symptoms (highlighted in red) but also many other clinical features suggesting a diagnosis of hypogonadism (e.g., body composition changes, anemia, osteoporosis, and depression) could represent direct or indirect reflections of neurological injury and disability.

Notably, at a sub-group analysis according to health-related characteristics of the study populations, a significant positive association between testosterone and vitamin D only arose from studies enrolling patients with clinical frailty (D'Andrea et al., 2021). Indeed, a combination of obesity, inadequate diet, and poor sunlight exposure could contribute to vitamin D deficiency in frail individuals. In particular, obesity, a recognized risk factor for androgen deficiency, is also associated with hypovitaminosis D, as the sequestration in fat mass decreases the bioavailability of the lipophilic 25(OH)D in blood circulation. Therefore, both androgen deficiency and hypovitaminosis D might represent markers of poor health conditions, sharing common pathogenetic mechanisms. In this light, it is not surprising that the strongest association was demonstrated in men with chronic SCI, who exhibit a very high prevalence of comorbidity-related risk factors predisposing to both low testosterone and vitamin D deficiency (Barbonetti, Sperandio, et al., 2016).

Hypovitaminosis D plays a key role in the weakening of bone mineral density in this population; however, vitamin D is now regarded as a pleiotropic hormone and the ubiquitous distribution of its receptor provides biological plausibility to a wide spectrum of systemic effects on a variety of target tissues, including nervous system (Eyles, Smith, Kinobe, Hewison, & McGrath, 2005) and skeletal muscle (Bischoff et al., 2001). Intriguingly, in people with chronic SCI, we demonstrated that lower vitamin D exhibits an independent association with both depression (Barbonetti et al., 2017) and poorer physical function (Barbonetti, Vassallo, Felzani, Francavilla, & Francavilla, 2016) and predicts the physical function worsening over time (Barbonetti et al., 2018). Obviously, the demonstration of direct causal links could be produced only by intervention studies assessing the effects of vitamin D supplementation.

Applications to other areas of neuroscience

In this chapter, we have reviewed the pathogenetic mechanisms underlying metabolic syndrome (MetS), the diagnostic challenges, and its reflections on cardiovascular and endocrine/metabolic health in people with spinal cord injury (SCI).

Although chronic SCI represents a unique clinical model to study the effects of an increased fat mass to lean mass ratio, to a lesser extent, the loss of muscle mass and physical inactivity represent hallmarks for many other neurological disorders. Indeed, increasing evidence indicates that in neurodegenerative diseases the pathophysiological links with MetS could be bidirectional: on one hand, muscle hypotrophy/weakening and physical disability lower energy expenditure, thus promoting adiposity accumulation; on the other hand, MetS is associated with systemic inflammation (Farkas & Gater, 2018) which is thought to induce a breakdown of the blood–brain barrier (Van Dyken & Lacoste, 2018). A higher permeability of the blood–brain-barrier could predispose to infiltration of immune cells and neuroinflammation (Van Dyken & Lacoste, 2018) which is involved in the pathogenesis of several neurological disorders. Intriguingly, the relationship of obesity and diabetes with Alzheimer's disease has been well documented (Nguyen, Killcross, & Jenkins, 2014;

Pugazhenthi, Qin, & Reddy, 2017) and it has been reported that the presence of obesity during adolescence is associated with a higher odd for developing multiple sclerosis later on (Hedström, Olsson, & Alfredsson, 2012, 2016). It has been suggested that the infiltration of immune cells, accompanied by the generation of inflammatory cytokines and reactive oxygen species, would decrease the threshold for autoimmune activation and stimulation of myelin-reactive cells (Van Dyken & Lacoste, 2018). Further studies are warranted to assess whether these pathophysiological events could also occur in people with chronic SCI.

Mini-dictionary of terms

Androgen deficiency: biochemical and clinical syndrome characterized by unequivocally low testosterone levels as well as symptoms and signs of hypogonadism.

Energy expenditure: the sum of basal metabolic rate, thermic effect of food digestion, and thermic effect of physical activity.

Erectile dysfunction: sexual dysfunction characterized by recurrent inability to achieve or maintain an adequate penis erection.

Fat-free mass (FFM): the total amount of lean (nonfat) components of the body, consisting of bone, muscle, and organs.

Impaired fasting glucose (IFG): fasting glucose levels between 100 and 125 mg per dL (5.6 to 6.9 mmol per L).

Insulin resistance: metabolic disorder in which the biological effect produced by a known concentration of insulin is lower than expected due to poor responsiveness of cells to the hormone action.

Leisure-time physical activity (LTPA): physical activities that individuals choose to do in free time independently of rehabilitation (walking/wheeling, playing sport).

Non-alcoholic fatty liver disease (NAFLD): liver phenotype of metabolic syndrome in which an excess of fat builds up in the liver with no obvious causes such as alcohol abuse.

Sarcopenia: extensive loss of muscle mass, strength, and function, strongly correlated with a physical disability.

Visceral obesity: excess of fat deposition in the abdominal viscera and omentum tissue.

Key facts of metabolic syndrome in spinal cord injury: Impact on health

- Metabolic syndrome (MetS) is a cluster of clinical and metabolic factors that increases the risk of developing cardiovascular and metabolic diseases, highly prevalent in people with chronic spinal cord injury (SCI).
- MetS features include visceral obesity, insulin resistance, hyperglycemia, hypertension, and dyslipidemia.
- In people with SCI, the use of the traditional cut-off point of body mass index (BMI) for obesity ($\geq$30 kg/m^2) grossly underestimates the diagnosis due to the muscle wasting below the level of the lesion: the same value of BMI reflects higher percentages of body fat than able-bodied individuals.
- In people with SCI the measure of waist circumference could overestimate visceral obesity due to abdominal muscle paralysis.
- Multi-factorial pathogenetic mechanisms, largely related to visceral obesity and MetS, could account for an accelerated age-dependent decline in testosterone levels peculiar to men with chronic SCI.
- Androgen deficiency is a biochemical and clinical syndrome where unequivocally low testosterone levels are associated with symptoms and signs of hypogonadism.
- In men with SCI, clinical features suggestive of hypogonadism could represent a reflection of neurological damage and disability, thus challenging the diagnosis of clinical hypogonadism.

Summary points

- Following spinal cord injury (SCI), a dramatic and early muscle wasting occurs below the level of the lesion
- This underlies a substantial decrease in overall energy expenditure which is easily overwhelmed by energy intake, thus leading to a fat mass increase
- Accumulation of visceral fat mass promotes insulin resistance, the key element of metabolic syndrome (MetS) which increases the risk of developing cardiometabolic diseases
- All definitions, that are widely accepted in the general able-bodied population, result to be inappropriate in people with SCI, where the diagnosis of obesity is challenged by the changes in body composition.
- The deranged metabolic profile of chronic SCI also includes a decline in testosterone levels that can further contribute to the deterioration in body composition.

- Due to the very high prevalence of androgen deficiency, chronic SCI could represent an attractive clinical model to study hypogonadism and the emerging correlates of low testosterone.

References

Alberti, K. G., & Zimmet, P. Z. (1998). Definition, diagnosis and classification of diabetes mellitus and its complications. Part 1: Diagnosis and classification of diabetes mellitus provisional report of a WHO consultation. *Diabetic Medicine*, *15*, 539–553.

Barbonetti, A., Caterina Vassallo, M. R., Cotugno, M., Felzani, G., Francavilla, S., & Francavilla, F. (2016). Low testosterone and non-alcoholic fatty liver disease: Evidence for their independent association in men with chronic spinal cord injury. *The Journal of Spinal Cord Medicine*, *39*, 443–449.

Barbonetti, A., Cavallo, F., D'Andrea, S., Muselli, M., Felzani, G., Francavilla, S., et al. (2017). Lower vitamin D levels are associated with depression in people with chronic spinal cord injury. *Archives of Physical Medicine and Rehabilitation*, *98*, 940–946.

Barbonetti, A., D'Andrea, S., Martorella, A., Felzani, G., Francavilla, S., & Francavilla, F. (2018). Low vitamin D levels are independent predictors of 1-year worsening in physical function in people with chronic spinal cord injury: A longitudinal study. *Spinal Cord*, *56*, 494–501.

Barbonetti, A., D'Andrea, S., Samavat, J., Martorella, A., Felzani, G., Francavilla, S., et al. (2019). Can the positive association of osteocalcin with testosterone be unmasked when the preeminent hypothalamic-pituitary regulation of testosterone production is impaired? The model of spinal cord injury. *Journal of Endocrinological Investigation*, *42*, 167–173.

Barbonetti, A., Sperandio, A., Micillo, A., D'Andrea, S., Pacca, F., Felzani, G., et al. (2016). Independent Association of Vitamin D with Physical Function in people with chronic spinal cord injury. *Archives of Physical Medicine and Rehabilitation*, *97*, 726–732.

Barbonetti, A., Vassallo, M. R., Felzani, G., Francavilla, S., & Francavilla, F. (2016). Association between 25(OH)-vitamin D and testosterone levels: Evidence from men with chronic spinal cord injury. *The Journal of Spinal Cord Medicine*, *39*, 246–252.

Barbonetti, A., Vassallo, M. R., Pacca, F., Cavallo, F., Costanzo, M., Felzani, G., et al. (2014). Correlates of low testosterone in men with chronic spinal cord injury. *Andrology*, *2*, 721–728.

Barton, T. J., Low, D. A., Bakker, E. A., Janssen, T., de Groot, S., van der Woude, L., et al. (2020). Traditional cardiovascular risk factors strongly underestimate the 5-year occurrence of cardiovascular morbidity and mortality in spinal cord injured individuals. *Archives of Physical Medicine and Rehabilitation*, *S0003-9993*(20), 30529–3.

Bauman, W. A., Fountaine, M. F., & Spungen, A. M. (2014). Age-related prevalence of low testosterone in men with spinal cord injury. *The Journal of Spinal Cord Medicine*, *37*, 32–39.

Bauman, W. A., Spungen, A. M., Flanagan, S., Zhong, Y. G., Alexander, L. R., & Tsitouras, P. D. (1994). Blunted growth hormone response to intravenous arginine in subjects with a spinal cord injury. *Hormone and Metabolic Research*, *26*, 152–156.

Bhasin, S., Cunningham, G. R., Hayes, F. J., Matsumoto, A. M., Snyder, P. J., Swerdloff, R. S., et al. (2010). Testosterone therapy in men with androgen deficiency syndromes: An Endocrine Society clinical practice guideline. *The Journal of Clinical Endocrinology and Metabolism*, *95*, 2536–2559.

Bhasin, S., Taylor, W. E., Singh, R., Artaza, J., Sinha-Hikim, I., Jasuja, R., et al. (2003). The mechanisms of androgen effects on body composition: Mesenchymal pluripotent cell as the target of androgen action. *The Journals of Gerontology. Series A, Biological Sciences and Medical Sciences*, *58*, M1103–M1110.

Bischoff, H. A., Borchers, M., Gudat, F., Duermueller, U., Theiler, R., Stähelin, H. B., et al. (2001). In situ detection of 1,25-dihydroxyvitamin D3 receptor in human skeletal muscle tissue. *The Histochemical Journal*, *33*, 19–24.

Blomberg Jensen, M., Nielsen, J. E., Jørgensen, A., Rajpert-De Meyts, E., Kristensen, D. M., Jørgensen, N., et al. (2010). Vitamin D receptor and vitamin D metabolizing enzymes are expressed in the human male reproductive tract. *Human Reproduction*, *25*, 1303–1311.

Castro, M. J., Apple, D. F., Jr., Hillegass, E. A., & Dudley, G. A. (1999). Influence of complete spinal cord injury on skeletal muscle cross-sectional area within the first 6 months of injury. *European Journal of Applied Physiology and Occupational Physiology*, *80*, 373–378.

Cirnigliaro, C. M., LaFountaine, M. F., Dengel, D. R., Bosch, T. A., Emmons, R. R., Kirshblum, S. C., et al. (2015). Visceral adiposity in persons with chronic spinal cord injury determined by dual energy X-ray absorptiometry. *Obesity (Silver Spring)*, *23*, 1811–1817.

Clark, M. J., Schopp, L. H., Mazurek, M. O., Zaniletti, I., Lammy, A. B., Martin, T. A., et al. (2008). Testosterone levels among men with spinal cord injury. *American Journal of Physical Medicine & Rehabilitation*, *87*, 758–767.

Cragg, J. J., Noonan, V. K., Dvorak, M., Krassioukov, A., Mancini, G. B., & Borisoff, J. F. (2013). Spinal cord injury and type 2 diabetes: Results from a population health survey. *Neurology*, *81*, 1864–1868.

Cragg, J. J., Noonan, V. K., Krassioukov, A., & Borisoff, J. (2013). Cardiovascular disease and spinal cord injury: Results from a national population health survey. *Neurology*, *81*, 723–728.

D'Andrea, S., Castellini, C., Minaldi, E., Totaro, M., Felzani, G., Francavilla, S., et al. (2020). Testosterone, level of the lesion and age are independently associated with prostate volume in men with chronic spinal cord injury. *Journal of Endocrinological Investigation*, *43*, 1599–1606.

D'Andrea, S., Martorella, A., Coccia, F., Castellini, C., Minaldi, E., Totaro, M., et al. (2021). Relationship of vitamin D status with testosterone levels: A systematic review and meta-analysis. *Endocrine*, *72*, 49–61.

Elder, C. P., Apple, D. F., Bickel, C. S., Meyer, R. A., & Dudley, G. A. (2004). Intramuscular fat and glucose tolerance after spinal cord injury—A cross-sectional study. *Spinal Cord*, *42*, 711–716.

Engeli, S., Negrel, R., & Sharma, A. M. (2000). Physiology and pathophysiology of the adipose tissue renin-angiotensin system. *Hypertension*, *35*, 1270–1277.

Eyles, D. W., Smith, S., Kinobe, R., Hewison, M., & McGrath, J. J. (2005). Distribution of the vitamin D receptor and 1 alpha-hydroxylase in human brain. *Journal of Chemical Neuroanatomy*, *29*, 21–30.

Farkas, G. J., & Gater, D. R. (2018). Neurogenic obesity and systemic inflammation following spinal cord injury: A review. *The Journal of Spinal Cord Medicine*, *41*, 378–387.

Ford, E. S. (2005). Prevalence of the metabolic syndrome defined by the international diabetes federation among adults in the U.S. *Diabetes Care*, *28*, 2745–2749.

Foresta, C., Strapazzon, G., De Toni, L., Perilli, L., Di Mambro, A., Muciaccia, B., et al. (2011). Bone mineral density and testicular failure: Evidence for a role of vitamin D 25-hydroxylase in human testis. *The Journal of Clinical Endocrinology and Metabolism*, *96*, E646–E652.

Gallagher, E. J., LeRoith, D., & Karnieli, E. (2008). The metabolic syndrome—From insulin resistance to obesity and diabetes. *Endocrinology and Metabolism Clinics of North America*, *37*(559–79), vii.

Gater, D. R., Jr. (2007). Obesity after spinal cord injury. *Physical Medicine and Rehabilitation Clinics of North America*, *18*, 333–vii.

Gater, D. R., Jr., Farkas, G. J., Berg, A. S., & Castillo, C. (2019). Prevalence of metabolic syndrome in veterans with spinal cord injury. *The Journal of Spinal Cord Medicine*, *42*, 86–93.

Giagulli, V. A., Kaufman, J. M., & Vermeulen, A. (1994). Pathogenesis of the decreased androgen levels in obese men. *The Journal of Clinical Endocrinology and Metabolism*, *79*, 997–1000.

Gibson, A. E., Buchholz, A. C., Martin Ginis, K. A., & SHAPE-SCI Research Group. (2008). C-reactive protein in adults with chronic spinal cord injury: Increased chronic inflammation in tetraplegia vs paraplegia. *Spinal Cord*, *46*, 616–621.

Gorgey, A. S., Dolbow, D. R., Dolbow, J. D., Khalil, R. K., Castillo, C., & Gater, D. R. (2014). Effects of spinal cord injury on body composition and metabolic profile—Part I. *The Journal of Spinal Cord Medicine*, *37*, 693–702.

Groah, S. L., Weitzenkamp, D., Sett, P., Soni, B., & Savic, G. (2001). The relationship between neurological level of injury and symptomatic cardiovascular disease risk in the aging spinal injured. *Spinal Cord*, *39*, 310–317.

Hedström, A. K., Olsson, T., & Alfredsson, L. (2012). High body mass index before age 20 is associated with increased risk for multiple sclerosis in both men and women. *Multiple Sclerosis Journal*, *18*, 1334–1336.

Hedström, A., Olsson, T., & Alfredsson, L. (2016). Body mass index during adolescence, rather than childhood, is critical in determining MS risk. *Multiple Sclerosis Journal*, *22*, 878–883.

Jones, L. M., Legge, M., & Goulding, A. (2003). Healthy body mass index values often underestimate body fat in men with spinal cord injury. *Archives of Physical Medicine and Rehabilitation*, *84*, 1068–1071.

Kolovou, G. D., Anagnostopoulou, K. K., & Cokkinos, D. V. (2005). Pathophysiology of dyslipidaemia in the metabolic syndrome. *Postgraduate Medical Journal*, *81*, 358–366.

Kostovski, E., Iversen, P. O., Birkeland, K., Torjesen, P. A., & Hjeltnes, N. (2008). Decreased levels of testosterone and gonadotrophins in men with long-standing tetraplegia. *Spinal Cord*, *46*, 559–564.

Lee, N. K., Sowa, H., Hinoi, E., Ferron, M., Ahn, J. D., Confavreux, C., et al. (2007). Endocrine regulation of energy metabolism by the skeleton. *Cell*, *130*, 456–469.

Liang, H., Mojtahedi, M. C., Chen, D., & Braunschweig, C. L. (2008). Elevated C-reactive protein associated with decreased high-density lipoprotein cholesterol in men with spinal cord injury. *Archives of Physical Medicine and Rehabilitation*, *89*, 36–41.

Muniyappa, R., & Sowers, J. R. (2013). Role of insulin resistance in endothelial dysfunction. *Reviews in Endocrine & Metabolic Disorders*, *14*, 5–12.

Nash, M. S., Groah, S. L., Gater, D. R., Dyson-Hudson, T. A., Lieberman, J. A., Myers, J., et al. (2019). Identification and Management of Cardiometabolic Risk after spinal cord injury. *The Journal of Spinal Cord Medicine*, *42*, 643–677.

National Cholesterol Education Program (NCEP) Expert Panel on Detection, Evaluation, and Treatment of High Blood Cholesterol in Adults (Adult Treatment Panel III). (2002). Third report of the National Cholesterol Education Program (NCEP) expert panel on detection, evaluation, and treatment of high blood cholesterol in adults (adult treatment panel III) final report. *Circulation*, *106*, 3143–3421.

Neve, A., Corrado, A., & Cantatore, F. P. (2013). Osteocalcin: Skeletal and extra-skeletal effects. *Journal of Cellular Physiology*, *228*, 1149–1153.

Nguyen, J. C. D., Killcross, A. S., & Jenkins, T. A. (2014). Obesity and cognitive decline: Role of inflammation and vascular changes. *Frontiers in Neuroscience*, *8*, 375.

Oury, F., Sumara, G., Sumara, O., Ferron, M., Chang, H., Smith, C. E., et al. (2011). Endocrine regulation of male fertility by the skeleton. *Cell*, *144*, 796–809.

Pugazhenthi, S., Qin, L., & Reddy, P. H. (2017). Common neurodegenerative pathways in obesity, diabetes, and Alzheimer's disease. *Biochimica et Biophysica Acta*, *1863*, 1037–1045.

Ravensbergen, H., Lear, S., & Claydon, V. (2014). Waist circumference is the best index for obesity-related cardiovascular disease risk in individuals with spinal cord injury. *Journal of Neurotrauma*, *31*, 292–300.

Roy, R. R., Baldwin, K. M., & Edgerton, V. R. (1991). The plasticity of skeletal muscle: Effects of neuromuscular activity. *Exercise and Sport Sciences Reviews*, *19*, 269–312.

Sabour, H., Norouzi Javidan, A., Latifi, S., Larijani, B., Shidfar, F., Vafa, M. R., et al. (2014). Bone biomarkers in patients with chronic traumatic spinal cord injury. *The Spine Journal*, *14*, 1132–1138.

Savic, G., DeVivo, M. J., Frankel, H. L., Jamous, M. A., Soni, B. M., & Charlifue, S. (2017). Causes of death after traumatic spinal cord injury-a 70-year British study. *Spinal Cord*, *55*, 891–897.

Schopp, L. H., Clark, M., Mazurek, M. O., Hagglund, K. J., Acuff, M. E., Sherman, A. K., et al. (2006). Testosterone levels among men with spinal cord injury admitted to inpatient rehabilitation. *American Journal of Physical Medicine & Rehabilitation*, *85*, 678–684.

Shah, P. K., Stevens, J. E., Gregory, C. M., Pathare, N. C., Jayaraman, A., Bickel, S. C., et al. (2006). Lower-extremity muscle cross-sectional area after incomplete spinal cord injury. *Archives of Physical Medicine and Rehabilitation*, *87*, 772–778.

Talmadge, R. J., Castro, M. J., Apple, D. F., Jr., & Dudley, G. A. (2002). Phenotypic adaptations in human muscle fibers 6 and 24 wk after spinal cord injury. *Journal of Applied Physiology*, *92*, 147–154.

Urso, M. L., Chen, Y. W., Scrimgeour, A. G., Lee, P. C., Lee, K. F., & Clarkson, P. M. (2007). Alterations in mRNA expression and protein products following spinal cord injury in humans. *The Journal of Physiology*, *579*, 877–892.

Van Dyken, P., & Lacoste, B. (2018). Impact of metabolic syndrome on neuroinflammation and the blood-brain barrier. *Frontiers in Neuroscience*, *12*, 930.

Wang, Y. H., Huang, T. S., Liang, H. W., Su, T. C., Chen, S. Y., & Wang, T. D. (2005). Fasting serum levels of adiponectin, ghrelin, and leptin in men with spinal cord injury. *Archives of Physical Medicine and Rehabilitation*, *86*, 1964–1968.

Wofford, M. R., & Hall, J. E. (2004). Pathophysiology and treatment of obesity hypertension. *Current Pharmaceutical Design*, *10*, 3621–3637.

World Health Organization. (1995). Physical status: the use and interpretation of anthropometry. In *Report of a WHO Expert Committee. WHO technical report series 854*. Geneva, Switzerland: WHO Press.

Wu, J. C., Chen, Y. C., Liu, L., Chen, T. J., Huang, W. C., Cheng, H., et al. (2012). Increased risk of stroke after spinal cord injury: A nationwide 4-year follow-up cohort study. *Neurology*, *78*, 1051–1057.

Wu, F. C., Tajar, A., Beynon, J. M., Pye, S. R., Silman, A. J., Finn, J. D., et al. (2010). Identification of late-onset hypogonadism in middle-aged and elderly men. *The New England Journal of Medicine*, *363*, 123–135.

Yahiro, A. M., Wingo, B. C., Kunwor, S., Parton, J., & Ellis, A. C. (2020). Classification of obesity, cardiometabolic risk, and metabolic syndrome in adults with spinal cord injury. *The Journal of Spinal Cord Medicine*, *43*, 485–496.

Zafar, U., Khaliq, S., Ahmad, H. U., Manzoor, S., & Lone, K. P. (2018). Metabolic syndrome: An update on diagnostic criteria, pathogenesis, and genetic links. *Hormones (Athens, Greece)*, *17*, 299–313.

Chapter 32

Body composition and spinal cord injury

Katherine J. Desneves[a,c], Nicole Kiss[a], Robin M. Daly[a], and Leigh C. Ward[b]

[a]*Institute for Physical Activity and Nutrition (IPAN), School of Exercise and Nutrition Sciences, Faculty of Health, Deakin University, Burwood, VIC, Australia,* [b]*School of Chemistry and Molecular Biosciences, The University of Queensland, St Lucia, QLD, Australia,* [c]*Nutrition and Dietetics Department, Division of Allied Health, Austin Health, Heidelberg, VIC, Australia*

List of abbreviations

2-C two-compartment model
3-C three-compartment model
4-C four-compartment model
AIS ASIA impairment scale
ALS amyotrophic lateral sclerosis
BIA bioelectrical impedance analysis
BIS bioimpedance spectroscopy
BMI body mass index
CSA cross-sectional area
DXA dual X-ray absorptiometry
EXF extracellular fluid
FFM fat-free mass
FM fat mass
ICF intracellular fluid
IMF intramuscular fat
LTM lean tissue mass
MRI magnetic resonance imaging
MS multiple sclerosis
PD Parkinson's disease
SCI spinal cord injury
WHO World Health Organization

Introduction

Adverse body composition changes, including gains in fat mass and losses in lean muscle mass, occur following spinal cord injury (SCI) due to multiple factors including the metabolic response to trauma, secondary complications such as urinary tract infections and pressure injuries, and their associated inflammatory cascades. Inactivation of skeletal muscle caused by paralysis, immobility, and the subsequent reduced loading and decrease in habitual physical activity also contribute to these changes (Gorgey et al., 2014). There is also evidence that decreased production of sex hormones due to acute stress from trauma, hypothalamic-pituitary axis dysfunction, and certain medications and related side effects can all impact body composition (Bryce, 2010). Loss of muscle mass in the early period following traumatic SCI has equivalent consequences to those described in the general population including higher acute care mortality, adverse events, and postoperative complications (Banaszek et al., 2019). In the longer term, ongoing muscle wasting results in decreased muscular strength, power, and endurance, increased fatigue, and reduced levels of physical function. Furthermore, muscle wasting leads to a reduced resting energy expenditure that predisposes individuals with SCI to weight gain, other co-morbid complications and chronic diseases such as diabetes and cardiovascular disease (Gorgey et al., 2014), and an increased risk for falls and fragility fractures (Galea, 2012).

In this section, we review the effects of traumatic SCI on body composition (muscle and fat). We also provide an overview of the available methods for assessing body composition and the considerations for use in individuals with SCI.

Cellular, Molecular, Physiological, and Behavioral Aspects of Spinal Cord Injury. https://doi.org/10.1016/B978-0-12-822427-4.00032-0

Body composition

Assessment of body composition is important in clinical practice as it provides information on nutritional status and health risk at a single time point and when measured longitudinally, is a surrogate measure of the adequacy of nutrition and/or exercise interventions (Teigen, Kuchnia, Mourtzakis, & Earthman, 2017). Body composition is commonly described in terms of the number of physiologically functional compartments assessed using atomic, molecular, cellular, or tissue models (Siervo & Jebb, 2010), which are largely determined by the technology available to measure these compartments rather than based on anatomical measurements as these are not readily practicable. Fig. 1 presents commonly used models representing physiologically functional compartments along with the percentage contribution of each compartment to total body mass. Models variously subdivide body mass into two, three, four, or more compartments. In the simplest two-compartment model (2-C), body mass is divided into fat-free mass (FFM) and fat mass (FM), in the three-compartment (3-C) model lean tissue mass, fat, and bone and in the four-compartment (4-C) model water, fat, protein and bone.

Body composition changes following traumatic spinal cord injury

Body composition changes substantially during the first year following spinal cord injury (SCI) and persists in individuals with long-standing SCI. The most rapid muscle wasting occurs in the first 3 months following injury but gradual muscle loss continues for up to 12 months post-injury (Castro, Apple Jr, Hillegass, & Dudley, 1999; Gorgey & Dudley, 2007; Singh, Rohilla, Saini, & Kaur, 2014). The extent of muscle wasting varies across the different regions of the body and according to the level and severity of injury (Singh et al., 2014). For instance, a 12-month longitudinal study following acute SCI (within 3 days post-injury) using dual-energy X-ray absorptiometry (DXA) observed a 20% loss of leg lean tissue mass (LTM) in individuals with tetraplegia and paraplegia (Singh et al., 2014). As shown in Fig. 2, differences were observed in the magnitude of loss in the arms and trunk, where individuals with tetraplegia lost 12% more LTM and gained 6% more fat in the arms than those with paraplegia (Singh et al., 2014). Similarly, people with motor complete injuries were reported to lose 4% to 6% more LTM in the trunk and legs and gain 8% more fat mass in the legs compared to those with incomplete injuries who had an 8% decrease in leg fat mass (Singh et al., 2014). These alterations in body composition are persistent, with cross-sectional studies using DXA reporting that individuals with long-standing (>1 year) SCI have 3–11 kg lower total body LTM and 9%–12% higher total body fat and trunk fat mass than controls (Beck, Lamb, Atkinson, Wuermser, & Amin, 2014; Buchholz, McGillivray, & Pencharz, 2003; Maggioni et al., 2003; Spungen et al., 2003). In addition, there are marked variations in segmental body composition changes between tetraplegia and paraplegia and complete and incomplete categories of SCI (Spungen et al., 2003). The temporal nature of body composition, time since injury, the severity of the injury

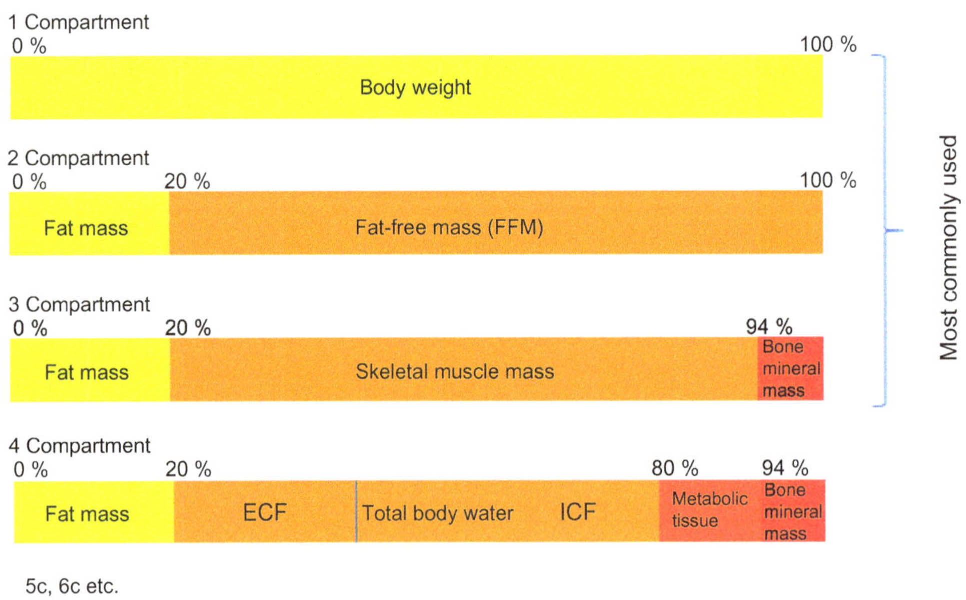

FIG. 1 Models of body composition. 1-C = one-compartment model; 2-C = two-compartment model, e.g., as measured by bioelectrical impedance analysis (BIA); 3-C = three-compartment model, e.g., as measured by dual-energy X-ray absorptiometry (DXA); 4-C = four-compartment model, e.g., as measured by isotope dilution (TBW) combined with DXA (BMC) and air-displacement plethysmography (body volume and density); *ECF*, extracellular fluid; *ICF*, intracellular fluid, measured by tracer dilution or bioimpedance spectroscopy.

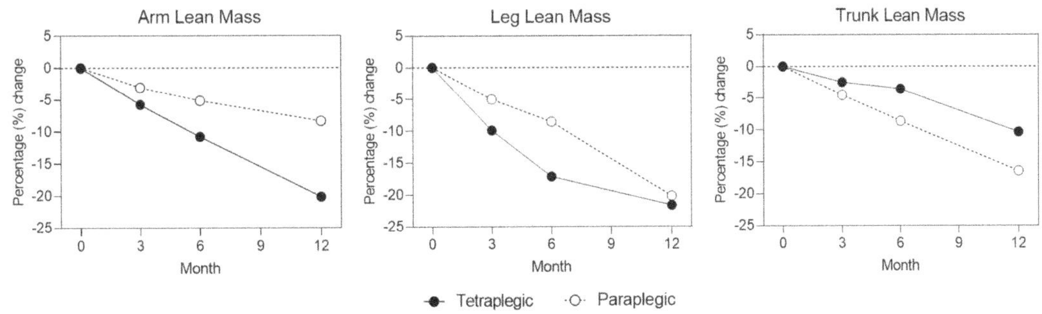

FIG. 2 Longitudinal changes in regional lean body mass in tetraplegia and paraplegia in 12 months following acute SCI. Body composition changes during the 12 months following spinal cord injury are show for arms, legs and trunk of people with tetraplegia and paraplegia. *Arm-T*, mean both arms tetraplegia (g); *arm-P*, mean both arms paraplegia (g); *trunk-T*, mean trunk tetraplegia (g); *trunk -P*, mean trunk paraplegia (g); *leg-T*, mean both legs tetraplegia (g); *leg-P*, mean both legs paraplegia (g). *(Modified from Singh, R., Rohilla, R. K., Saini, G., & Kaur, K. (2014). Longitudinal study of body composition in spinal cord injury patients.* Indian Journal of Orthopaedics, *48(2), 168.)*

and segmental changes, along with normal age-related changes, all need to be considered when assessing the nutritional status of individuals with SCI.

Studies using more advanced imaging techniques such as quantitative computed tomography (QCT) and magnetic resonance imaging (MRI) to quantify changes in muscle cross-sectional area (CSA) have likewise reported marked and heterogeneous muscle losses according to the severity of injury (Castro et al., 1999; Gorgey & Dudley, 2007; Modlesky et al., 2004). For instance, a study in six people with incomplete SCI using MRI found that mid-thigh skeletal muscle CSA was 33% smaller and intramuscular fat CSA 126% greater than able-bodied controls at 6 weeks after injury (Fig. 3). In a 3-month follow-up, thigh skeletal muscle CSA did not change significantly whereas intramuscular fat increased by a further 26% (Gorgey & Dudley, 2007). A study using thigh skeletal muscle biopsy described a dramatic 22% decrease in average muscle fiber CSA from 6 to 11 weeks post-injury, with a more gradual decline (10%) from 11 to 24 weeks. At week 24, average muscle fiber CSA in individuals with complete SCI was a third the size of age and weight-matched able-bodied controls (Castro et al., 1999). These changes in skeletal muscle morphology in SCI patients are also persistent. A study using MRI and DXA in eight individuals followed $\geq$2 years following SCI found a 17% lower mid-thigh skeletal muscle mass and two times higher mid-thigh percentage fat (34%) in individuals with complete SCI than controls (Modlesky et al., 2004). Another study using peripheral quantitative computed tomography (pQCT) found tibial midshaft muscle CSA was 33% smaller in men with paraplegia of $\sim$6 years duration compared with age-matched able-bodied men (Dionyssiotis et al., 2019). Clinically these losses in muscle are important as low muscle mass has been associated with significantly more adverse events (including post-operatively) and in-hospital mortality in acute SCI patients (Banaszek et al., 2019); however, the impact of low muscle mass on morbidity and mortality in people with long-term SCI has not been investigated.

In summary, there is a dramatic loss of muscle below the level of the lesion in the first 6–12 weeks following acute spinal cord injury that differs regionally according to the level and severity of the injury and persists long term. Generally, higher level and complete injuries result in greater muscle wasting than lower level and incomplete injuries. This loss in muscle mass may have implications for morbidity and mortality and has a profound effect on the energy needs of individuals with SCI that are described in another chapter.

Sarcopenia and SCI

Sarcopenia is a term historically used to describe low skeletal muscle mass, but in more recent years it has evolved to include a combination of low muscle mass, low muscle strength, and/or impaired physical function (Cruz-Jentoft et al., 2019). While there is currently no international consensus on the diagnostic criteria for sarcopenia, there is considerable evidence that sarcopenia is associated with an increased risk for many chronic diseases, including cardiac and respiratory diseases, insulin resistance and type 2 diabetes, impaired cognition, and dementia, falls and fractures and premature mortality (Pacifico et al., 2020). Sarcopenia is also associated with a reduced ability to perform activities of daily living, impaired mobility, loss of independence, higher levels of disability and increases the risk of hospitalization, healthcare costs, and mortality (Cruz-Jentoft et al., 2019). Individuals with SCI are at increased risk of secondary (disease-related) sarcopenia due to neurological impairment, immobility, and disability (Cruz-Jentoft et al., 2019); however, there is an absence of research on the assessment and impacts of sarcopenia in SCI, and whether the same sarcopenia criteria used in able body individuals are appropriate for people with SCI. Two studies in individuals with chronic SCI report a prevalence of sarcopenia defined as low muscle mass only of 57%–97% (Dionyssiotis et al., 2019; Pelletier, Miyatani,

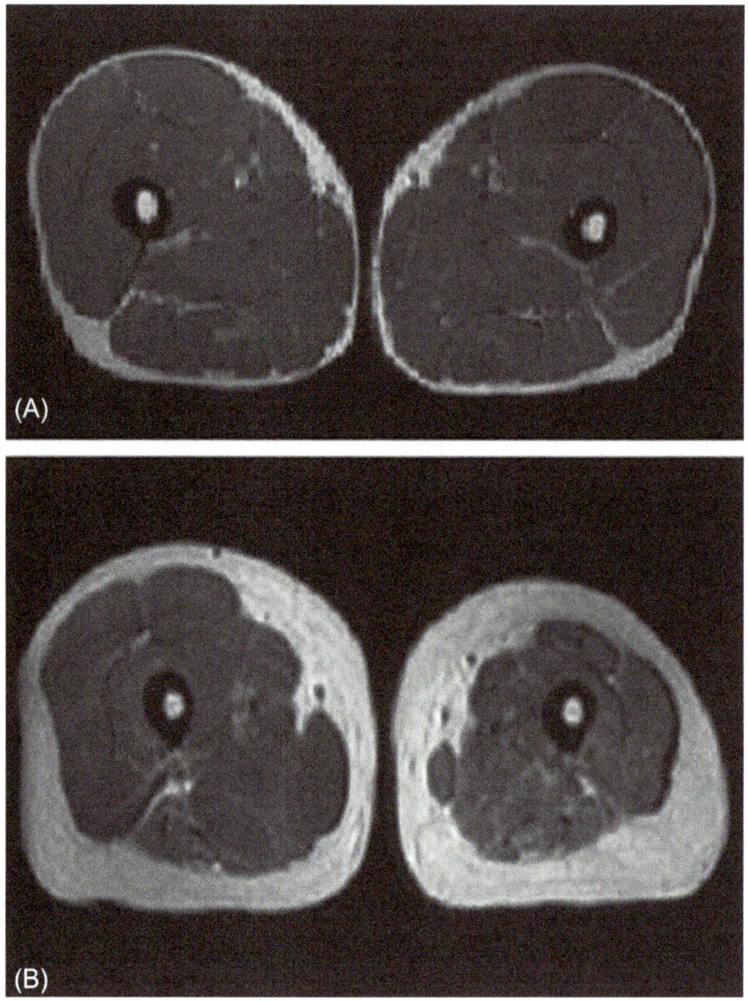

FIG. 3 Comparison of MRI muscle cross-sectional area, subcutaneous and intramuscular fat at the mid-thigh in able-bodied control versus an incomplete SCI patient 6 weeks following injury. The mid-thigh of an able-bodied individual (A) and a person with incomplete SCI 6 weeks following injury (B) are shown. Note that the mid-thigh muscle of the person with incomplete SCI is a third smaller than the able-bodied person with a marked increase in subcutaneous and intramuscular fat *(white regions). (From Gorgey, A., & Dudley, G. (2007). Skeletal muscle atrophy and increased intramuscular fat after incomplete spinal cord injury.* Spinal Cord, *45(4), 304, with permission obtained from the Publishers.)*

Giangregorio, & Craven, 2016). In these studies, sarcopenia was diagnosed using a DXA-derived appendicular lean mass index (ALMI), calculated from appendicular lean tissue mass (kg) divided by height $(m)^2$ using cut-points of $\leq$7.26 kg/m^2 for men and $\leq$ 5.50 kg/m^2 for women, which were based on the criteria (cut-points) defined for able-bodied populations (Cruz-Jentoft et al., 2019). Another study that used total psoas muscle CSA derived from QCT to diagnose sarcopenia reported significantly more adverse events, post-operative adverse events, and in-hospital mortality in sarcopenic SCI patients (Banaszek et al., 2019). Collectively, these findings highlight that low muscle mass is highly prevalent following SCI, but sarcopenia is not well studied nor defined in this population. Despite this, there is strong evidence to support the routine assessment of muscle mass in people with SCI, and whether the inclusion of muscle strength and/or function provides further clinically important information requires further investigation.

Another condition that might be of relevance to SCI is sarcopenic obesity, which is characterized by the coexistence of sarcopenia and excess fat mass (Pelletier et al., 2016). There is some evidence that sarcopenic obesity is more strongly linked to metabolic-related diseases such as insulin resistance, metabolic syndrome, dyslipidemia, inflammation, hypertension, and diabetes than either obesity or sarcopenia alone (Roh & Choi, 2020). Although not well studied in SCI, one study in 136 people with chronic SCI (>15 years post-injury) reported a prevalence of sarcopenic obesity of 42% using percentage body fat $\geq$25% for men and $\geq$ 35% for women with the ALMI cut-offs mentioned previously (Pelletier et al., 2016). No reports have described the impact of sarcopenic obesity on metabolic risk or mortality in SCI.

In summary, preserving skeletal muscle mass and reducing fat mass are important goals to avoid secondary comorbidities following SCI. Monitoring changes in body composition is critical to inform and monitor outcomes of diet and/or exercise interventions. The following section summarizes the technical and practical considerations when choosing a method for assessing body composition in individuals with SCI.

Body composition assessment

Bedside and field methods

A variety of field and bedside methods are available to measure body composition, but there are a number of practical considerations specific to using and interpreting these methods in people with SCI, as outlined in Table 1.

TABLE 1 Bedside and field body composition assessment methods, strengths, limitations, and considerations for use in individuals with SCI.

Method	Strengths	Limitations	Considerations for use in individuals with SCI
Anthropometry	Fast Non-invasive Minimal equipment High mobility		
Weight		Does not distinguish between changes in muscle and fat	Choice of chair, hoist, platform depends on mobility and balance Aim for a consistent method Ensure legs are not touching any surfaces Subtract the weight of wheelchair and any accessories or sling attachment 5%–10% weight loss ≤6 months or 10%–20% weight loss ≥6 months not appropriate for malnutrition diagnosis in acute SCI Use SCI adjusted healthy weight range for chronic SCI
Height		Requires ability to stand and balance or accessibility to a bed	Measure supine length for people unable to stand
BMI		Does not distinguish between components of weight or their distribution	Unclear which cut-off to use for acute SCI Proposed SCI-specific BMI cut-off >22 kg/m^2 for chronic SCI
Waist circumference		Proposed methods for SCI require accessibility to a bed	Proposed SCI-specific waist circumference cut-offs 1. ≥ 94 cm measured supine at the narrowest part of the waist after normal expiration 2. ≥ 96 cm measured supine at the midpoint between the top of the iliac crest and the lower margin of the last palpable rib in the mid-axillary line after several consecutive natural breaths OR measurement at the lowest rib after normal expiration Use consistent method
Skinfolds and circumferences		Require meticulous site location and standardized measurement techniques Not sensitive enough to detect small changes	Underestimate percentage body fat in SCI Poor mobility and sitting balance limits assessment of suprailiac and subscapular sites Excess weight and obesity limit assessment of triceps and subscapular skinfolds Lack of SCI population-specific equations and normative values Use of skinfold measurements discouraged in SCI Thigh and calf circumferences and sagittal and transverse abdominal diameters may be used in SCI population-specific energy requirement prediction equation

Continued

TABLE 1 Bedside and field body composition assessment methods, strengths, limitations, and considerations for use in individuals with SCI—cont'd

Method	Strengths	Limitations	Considerations for use in individuals with SCI
Other methods			
Bioimpedance	Portable Affordable Precision 2.7%–4.0% Performed easily at bedside Minimal training required Promptly repeatable Estimates muscle mass ICF and ECF can be differentiated with BIS and MF-BIA	Validity affected by assumptions including body geometry, constant hydration, body resistivity and is influenced by sex, age, ethnicity Precision and accuracy influenced by fever, fluid, and electrolyte disturbances Differences between devices and software SF-BIA and MF-BIA rely on proprietorial, undisclosed prediction equations to estimate whole-body and segmental body composition	BIS validated in acute and chronic SCI SCI population-specific FFM prediction equation exists BIS can be used in chronic SCI to predict FFM and LTM of the body regions BIS may have utility to assess and monitor the presence of edema from extracellular fluid to intracellular fluid volume ratios
Ultrasound	Portable Affordable Performed easily at bedside Promptly repeatable Ability to differentiate between muscle and subcutaneous adipose tissue Can measure muscle quality and support diagnosis of sarcopenia	Lack of consensus on an appropriate protocol Longitudinal measures influenced by inconsistent technique and protocol Single site measurement assumed to represent whole-body composition	Not widely used in SCI Moderate correlation between DXA and ultrasound for android region, percent fat in SCI Research needed to validate prediction equations for use in SCI Lack of SCI-specific protocol and cut-off points

Portable body composition assessment methods that can be used at the bedside or in the community and factors to consider for use in people with spinal cord injury are described. *SCI*, spinal cord injury; *BMI*, body mass index; *MF-BIA*, multi-frequency bioimpedance; *SF-BIA*, single-frequency bioimpedance; *ICF*, intracellular fluid; *ECF*, extracellular fluid; *BIS*, bioimpedance spectroscopy; *FFM*, fat free mass; *LTM*, lean tissue mass; *DXA*, dual-energy X-ray absorptiometry.
Modified from for SCI Affuso, O., Pradhan, L., Zhang, C., Gao, S., Wiener, H. W., Gower, B., ... Allison, D. B. (2018). A method for measuring human body composition using digital images. *PLoS One, 13*(11), e0206430. doi:10.1371/journal.pone.0206430; Borga, M., West, J., Bell, J. D., Harvey, N. C., Romu, T., Heymsfield, S. B., & Leinhard, O. D. (2018). Advanced body composition assessment: From body mass index to body composition profiling. *Journal of Investigative Medicine: The Official Publication of the American Federation for Clinical Research, 66*, 887–895; Cirnigliaro, C. M., La Fountaine, M. F., Emmons, R., Kirshblum, S. C., Asselin, P., Spungen, A. M., & Bauman, W. A. (2013). Prediction of limb lean tissue mass from bioimpedance spectroscopy in persons with chronic spinal cord injury. *Journal of Spinal Cord Medicine, 36*(5), 443–453. doi:10.1179/2045772313Y.0000000108; Clasey, J. L., & Gater Jr, D. R. (2005). A comparison of hydrostatic weighing and air displacement plethysmography in adults with spinal cord injury. *Archives of Physical Medicine and Rehabilitation, 86*(11), 2106–2113. doi:10.1016/j.apmr.2005.06.013; Desneves, K. J., Panisset, M. G., Galea, M. P., Kiss, N., Daly, R. M., & Ward, L. C. (2020). Comparison of segmental lean tissue mass in individuals with spinal cord injury measured by dual energy X-ray absorptiometry and predicted by bioimpedance spectroscopy. *Spinal Cord*, 1–8. doi:10.1038/s41393-020-00568-3; Academy of Nutrition and Dietetics (2009). Spinal Cord Injury (SCI) evidence-based nutrition practice guideline; Evidence analysis library. Retrieved October 2018, https://www.andeal.org/topic.cfm?menu=5292; Dionyssiotis, Y., Skarantavos, G., Petropoulou, K., Galanos, A., Rapidi, C.-A., & Lyritis, G. P. (2019). Application of current sarcopenia definitions in spinal cord injury. *Journal of Musculoskeletal & Neuronal Interactions, 19*(1), 21; Edwards, L. A., Bugaresti, J. M., & Buchholz, A. C. (2008). Visceral adipose tissue and the ratio of visceral to subcutaneous adipose tissue are greater in adults with than in those without spinal cord injury, despite matching waist circumferences. *American Journal of Clinical Nutrition, 87*(3), 600–607; Fosbol, M. O., & Zerahn, B. (2015). Contemporary methods of body composition measurement. *Clinical Physiology and Functional Imaging, 2*, 81. doi:10.1111/cpf.12152; Fraser, C., Iruthayarajah, J., McIntyre, A., Serrato, J., Benton, B., & Teasell, R. W. (2016). Nutrition issues following spinal cord injury. Version 6.0.; Garshick, E., Ashba, J., Tun, G. C., Lieberman, L. S., & Brown, R. (1997). Assessment of stature in spinal cord injury. *The Journal of Spinal Cord Medicine, 20*(1), 36–42; Kocina, P., & Heyward, V. (1997). Validation of a bioimpedance equation for estimating fat-free mass of spinal cord injured adults 319. *Medicine & Science in Sports & Exercise, 29*(5), 55; Kyle, U. G., Bosaeus, I., De Lorenzo, A. D., Deurenberg, P., Elia, M., Gómez, J. M., ... Pirlich, M. (2004). Bioelectrical impedance analysis—Part I: Review of principles and methods. *Clinical Nutrition, 23*(5), 1226–1243; Lee, S. Y., & Gallagher, D. (2008). Assessment methods in human body composition. *Current Opinion in Clinical Nutrition and Metabolic Care, 11*(5), 566; Maggioni, M., Bertoli, S., Testolin, G., Merati, G., Margonato, V., & Veicsteinas, A. (2003). Body composition assessment in spinal cord injury subjects. *Acta Diabetologica, 40*(Suppl. 1), S183–S186. doi:10.1007/s00592-003-0061-7; Modlesky, C. M., Bickel, C. S., Slade, J. M., Cureton, K. J., Dudley, G. A., & Meyer, R. A. (2004). Assessment of skeletal muscle mass in men with spinal cord injury using dual-energy X-ray absorptiometry and magnetic resonance imaging. *Journal of Applied Physiology, 96*(2), 561–565. doi:10.1152/japplphysiol.00207.2003; Panisset, M. G., Desneves, K., Ward, L. C., Rafferty, J., Rodi, H., Roff, G., ... Galea, M. P. (2018). Bedside quantification of fat-free mass in acute spinal cord injury using bioelectrical impedance analysis: A psychometric study. *Spinal Cord, 56*(4), 355; Siervo, M., & Jebb, S. A. (2010). Body composition assessment: Theory into practice: Introduction of multicompartment models. *IEEE Engineering in Medicine and Biology Magazine, 29*(1), 48–59; Silveira, S., Ledoux, T., Robinson-Whelen, S., Stough, R., & Nosek, M. (2017). Methods for classifying obesity in spinal cord injury: A review. *Spinal Cord, 55*(9), 812; Teigen, L. M., Kuchnia, A. J., Mourtzakis, M., & Earthman, C. P. (2017). The use of technology for estimating body composition: Strengths and weaknesses of common modalities in a clinical setting. *Nutrition in Clinical Practice, 32*(1), 20–29.

Anthropometry

Anthropometry, the external measurement of body proportions and size (e.g., weight, height, waist circumference, skinfold thickness, and mid-arm muscle and calf circumferences), provides body composition information based on the one or two-compartment models and/or an assessment of health risk (Gibson, 2005). A brief overview of these methods specific to SCI is provided below.

Weight and height

Percentage weight loss is frequently used as an indicator of malnutrition risk (Cederholm et al., 2019). However, it is common to see a weight loss of 5–10 kg following traumatic SCI (Cox et al., 1985) and it is difficult to distinguish between unavoidable weight loss due to denervation and avoidable weight loss due to inadequate energy intake. Consequently, the usual phenotypic criteria of 5%–10% weight loss within 6 months or 10%–20% loss of weight over greater than 6 months (Cederholm et al., 2019) are not appropriate for the diagnosis of malnutrition in acute SCI. However, weight should be routinely monitored following SCI.

Height may be assessed by measuring supine length, heel to crown between metal bookends, using a rigid tape measure (Garshick, Ashba, Tun, Lieberman, & Brown, 1997). However, standards for healthy body weight for height have not been established or validated in SCI. Guidelines recommend adjusting the Metropolitan Life Insurance tables for individuals of an equivalent height and weight by 10%–15% or 7–9 kg and 5%–10% or 5–7 kg lower than the table weight for tetraplegia and paraplegia respectively (Academy of Nutrition and Dietetics, 2009). However, this recommendation is based on consensus (Academy of Nutrition and Dietetics, 2009) and does not account for relative changes in lean muscle and fat mass.

Body mass index

Body mass index (BMI) is a weight for height indicator that does not distinguish between the individual components of weight or body composition. BMI ranges associated with the risk of developing chronic diseases provided by WHO are based on data for able-bodied individuals but individuals with SCI within the normal BMI range (<25) often have lower LTM, higher total fat mass, and percentage body fat than equivalent abled-bodied individuals (Silveira, Ledoux, Robinson-Whelen, Stough, & Nosek, 2017). Hence, a lower SCI-specific BMI cut-off (>22 kg/m^2) to identify obesity and estimate cardiometabolic health risk has been proposed and validated in individuals with chronic SCI (Silveira et al., 2017). However, it is unclear whether WHO criteria or chronic SCI-specific BMI cut-offs should be used in acute SCI.

Waist circumference

Waist circumference is a predictor of intra-abdominal fat and has been associated with an increased risk for a range of cardiometabolic-related diseases (Gandy, 2014). A 26% higher mean total abdominal tissue, 58% greater mean visceral adipose tissue and a 48% greater ratio of visceral adipose tissue to subcutaneous adipose tissue has been reported in individuals with SCI compared with controls (Edwards, Bugaresti, & Buchholz, 2008). Consequently, lower SCI-specific waist circumference cut-offs (≥94 cm and ≥ 96 cm) have been proposed to identify health risks compared to non-disabled individuals. However, various studies have used different waist circumference measurement sites and proposed a range of alternative cut-offs (Edwards et al., 2008; Ravensbergen, Lear, & Claydon, 2014). In summary, there is no consensus with regard to the specific assessment or cut-points to define a waist circumstance threshold that is associated with increased health risks in SCI populations (Silveira et al., 2017). One of the challenges for this cohort is that it is not always practical within the clinical setting to obtain a supine measurement. However, if measuring waist circumference to define disease risk or monitor changes over time, it is important to use an SCI-specific cut-off value and a consistent method for assessment.

Skinfolds and circumferences

Changes in body fat provide an indirect estimate of alterations in energy balance. Total body fat or percentage body fat can be estimated using one or more skinfold thickness measurements at indicator sites for subcutaneous adipose tissue, with or without limb circumference measurements in empirical regression equations (Gibson, 2005). The combination of skinfold thickness at the triceps, biceps, mid-thigh and calf with circumference of the same limb enables calculation of limb adipose area which is more accurate than a single skinfold (Gibson, 2005). Generally, the measurement of skinfolds at multiple sites such as triceps, biceps, subscapular, and suprailiac provides a more accurate estimate of total body fat content than a single site (Gibson, 2005). Likewise, estimates of total body muscle mass and subsequently protein nutritional status can be predicted from mid-upper arm muscle circumference and area as both are correlated with measures of total muscle mass. However, these methods have a number of limitations including requiring highly standardized protocols and measurement techniques due to substantial intra- and interobserver variability, they are not sufficiently sensitive to detect small changes, have low accuracy and precision in obese people with measurements influenced by hydration (Fosbol & Zerahn, 2015). It is also important to be aware of the population in which predictive algorithms were developed and that data are compared to appropriate reference norms (Fosbol & Zerahn, 2015), which are not available for the SCI population.

For individuals with SCI, practical difficulties such as poor mobility and sitting balance may limit assessment of some skinfold measures (e.g., suprailiac and subscapular) and excess weight and obesity limit assessment at other sites (e.g., triceps, subscapular, and mid-thigh skinfolds) (Desport et al., 2000). Additionally, mid-upper arm muscle circumference may be disproportionate to total muscle mass in manual wheelchair users. Studies have reported that anthropometric methods using multiple skinfolds with or without circumferences underestimate the percentage body fat in people with SCI (Maggioni et al., 2003; Spungen, Bauman, Wang, & Pierson, 1995). There is also a lack of SCI population-specific equations for estimation of percentage body fat and SCI-specific normative values (Maggioni et al., 2003); consequently, guidelines discourage the use of skinfold measurements to assess body composition in persons with SCI due to the various limitations and inaccuracy (Academy of Nutrition and Dietetics, 2009).

Bioelectrical impedance

Bioelectrical impedance analysis (bioimpedance, BIA) is a non-invasive, portable, inexpensive bedside method that provides measures of total body water, FFM, FM, and body fat percentage (Kyle et al., 2004) using a 2-C model. There are three types of bioimpedance devices; single frequency (SFBIA), multi-frequency (MFBIA), and bioimpedance spectroscopy (BIS). SFBIA and MFBIA devices are generally limited by their reliance on proprietary undisclosed prediction equations to estimate whole-body and segmental body composition, which may not be suitable to use in individuals with SCI. In contrast, BIS data predicts body composition (ECF, ICF, FFM, LTM, and FM) from impedance measures using fundamental model-based equations/algorithms developed against the reference method deuterium dilution (Teigen et al., 2017). The overall reproducibility/precision of BIS is 2.7%–4.0% (Kyle et al., 2004). BIS is a promising tool that has been validated for the assessment of body composition in individuals with acute (Panisset et al., 2018) and chronic SCI when a population-specific prediction equation is used (Kocina & Heyward, 1997). BIS has also been used in chronic SCI (Cirnigliaro et al., 2013) and more recently acute SCI to predict FFM and LTM of the body regions and may have utility to assess and monitor the presence of edema from the extracellular fluid to intracellular fluid volume ratios to monitor fluid shifts and distribution (Desneves et al., 2020). To minimize the variability due to hydration it is important to perform impedance measurements under standardized euhydration conditions. The Academy of Nutrition and Dietetics Spinal Cord Injury Evidence-Based Nutrition Practice Guidelines recommend using bioimpedance to assess body composition (Academy of Nutrition and Dietetics, 2009) in SCI; however, SCI-specific population normative values for interpretation of whole and regional body composition measures are lacking. Despite the lack of population-specific reference data, it is appropriate to use bioimpedance to measure body composition changes over time in people with SCI; however, precision must be taken into consideration when interpreting results.

Ultrasound

Ultrasound can also assess regional distribution of body fat, quantify abdominal subcutaneous fat and intra-abdominal diameter and be used in severely obese subjects (Rashmi & Snekhalatha, 2019). Ultrasound has the advantage of being non-invasive, portable and can be used in real time at the bedside (Teigen et al., 2017). However, it currently has limited value in monitoring longitudinal muscle changes due to the lack of consensus on standardized measurement protocols for assessing muscle mass, primarily site selection and the amount of force or muscle compression to apply (Teigen et al., 2017). It is also assumed that measurements obtained at a single site are representative of whole-body composition (Teigen et al., 2017). To date, ultrasound has not been widely used to assess body composition in SCI, though one study in individuals with chronic SCI found a moderate correlation between DXA and ultrasound for the android region, percent fat (Emmons et al., 2011). Further research is needed to determine the utility of ultrasound in people with SCI.

Laboratory and research methods

More advanced body composition assessment tools, which are often less readily available as they are limited to a hospital or research setting, including emerging technologies used to assess body composition based on the two, three, or four-compartment models, and the strengths, limitations, and considerations for use in individuals with SCI, are summarized in Table 2.

TABLE 2 Laboratory and research body composition assessment methods, strengths, limitations and considerations for use in individuals with SCI.

Method	Strengths	Limitations	Considerations for use in individuals with SCI
Isotope dilution	Gold standard for determination of TBW Reference measure for validation of other methods Excellent precision, 1%–2% for TBW and ECW Medium-high safety Medium-high mobility	Limited to research Medium-high cost Medium-highly complex methodology Assumes fat-free mass hydration is stable	Feasible to collect urine samples with bladder management
Air-displacement plethysmography	Based on 2C-model Non-invasive Safe Fast No radiation exposure No need for sedation Precision for FM ~ 1.7%–4.5%	Limited to research and clinical Affected by variations in bone mineral content and hydration	Measurement of thoracic gas volume problematic in SCI above T3 due to paralysis of intercostal and abdominal muscles Population-specific predicted thoracic gas volume equation is lacking Inbuilt scale device inaccessible for wheelchair users
Dual-energy X-ray absorptiometry	Gold standard measure of bone density Based on 3-C model Used to validate other body composition estimating devices and equations High precision 0.8%–2.7% whole-body FM, 0.4%–1.3% whole-body LTM Useful for sarcopenia diagnosis Provides whole-body and regional data Widely available	Medium cost Low dose radiation Low mobility Medium-high methodology Scanners have weight and size limitations Doesn't differentiate between organ versus skeletal muscle lean tissue	Physical height of DXA scanning beds limit accessibility for SCI Challenges positioning individuals with SCI, especially those with spasticity Presence of metal implants is a source of error Evidence-based SCI guidelines suggest assessing and monitoring body composition using DXA Validity in SCI may be compromised due to altered hydration
MRI	Gold standard for body composition measurement at tissue level High accuracy and reproducibility Excellent precision, ~1.2%–2% error Measures whole-body and regional adipose tissue and skeletal muscle Used for sarcopenia diagnosis	High cost Time-consuming image analysis Patient positioning important to minimize error Limited availability Claustrophobic persons cannot be scanned Scanners cannot accommodate very large people (BMI > 40 kg/m^2) Only available when performed as part of routine care	Accessibility of scanning bed for individuals with SCI Challenges positioning individuals with SCI, especially those with spasticity

Continued

TABLE 2 Laboratory and research body composition assessment methods, strengths, limitations and considerations for use in individuals with SCI—cont'd

Method	Strengths	Limitations	Considerations for use in individuals with SCI
CT	Gold standard for body composition measurement at tissue level Provide reference measures of body composition Excellent precision ~2% Able to differentiate skeletal muscle, adipose and organ tissue Prognostic tool to establish muscle quantity and outcomes Mid-thigh muscle area is a good predictor of whole-body skeletal muscle and sensitive to change Can be used when diagnosing sarcopenia	High cost Only available when requested for medical treatment purposes High dose of radiation Assumes single slice measure is representative of whole-body composition Repeatability limited by patient positioning Density measures influenced by physiological variability e.g., edema. Differences in devices/software Not advised for serial measurements in longitudinal studies	Accessibility of scanning bed for individuals with SCI Challenges positioning individuals with SCI, especially those with spasticity Total psoas muscle CSA on lumbar CT used to predict early post-injury mortality and adverse events
Peripheral computed tomography	Minimum radiation exposure Precision for limb muscle area < 5%, low for intramuscular adipose tissue (3%–42%)	Medium cost Limited to appendicular sites (arms and legs) Scans easily influenced by movement artifacts Single slice assessment at a given site	Contractures may limit knee and ankle range of motion Spasms may create movement artifacts
Computerized digital image analysis	Novel approach. Simple Portable Relatively inexpensive Reduced bias as requires little human input during photography and processing	Early stages of development. Unclear if generalizable to very lean or very obese individuals, all race-ethnic groups	Requires subject to stand motionless with arms and legs abducted Applicability for people unable to stand for photographs is unknown Further research is required to create norms for mobility status

Sophisticated body composition assessment methods that are usually used in research laboratories and factors to consider for use in people with spinal cord injury are described. *TBW*, total body water; *ECW*, extracellular water; *FM*, fat mass; *LTM*, lean tissue mass; *2C-model*, 2 compartment model; *3-C model*, 3 compartment model; *SCI*, spinal cord injury; *DXA*, dual-energy X-ray absorptiometry; *MRI*, magnetic resonance imaging; *CT*, computed tomography; *CSA*, cross-sectional area.

Modified from for SCI Affuso, O., Pradhan, L., Zhang, C., Gao, S., Wiener, H. W., Gower, B., … Allison, D. B. (2018). A method for measuring human body composition using digital images. *PLoS One, 13*(11), e0206430. doi:10.1371/journal.pone.0206430; Borga, M., West, J., Bell, J. D., Harvey, N. C., Romu, T., Heymsfield, S. B., & Leinhard, O. D. (2018). Advanced body composition assessment: From body mass index to body composition profiling. *Journal of Investigative Medicine: The Official Publication of the American Federation for Clinical Research 66*, 887–895; Cirnigliaro, C. M., La Fountaine, M. F., Emmons, R., Kirshblum, S. C., Asselin, P., Spungen, A. M., & Bauman, W. A. (2013). Prediction of limb lean tissue mass from bioimpedance spectroscopy in persons with chronic spinal cord injury. *Journal of Spinal Cord Medicine, 36*(5), 443–453. doi:10.1179/2045772313Y.0000000108; Clasey, J. L., & Gater Jr, D. R. (2005). A comparison of hydrostatic weighing and air displacement plethysmography in adults with spinal cord injury. *Archives of Physical Medicine and Rehabilitation, 86*(11), 2106–2113. doi:10.1016/j.apmr.2005.06.013; Cruz-Jentoft, A. J., Bahat, G., Bauer, J., Boirie, Y., Bruyère, O., Cederholm, T., … Writing Group for the European Working Group on Sarcopenia in Older People 2 (EWGSOP2), and the Extended Group for EWGSOP2. (2019). Sarcopenia: Revised European consensus on definition and diagnosis. *Age and Ageing, 48*, 16–31. doi:10.1093/ageing/afy169; Academy of Nutrition and Dietetics, (2009). Spinal Cord Injury (SCI) evidence-based nutrition practice guideline; Evidence analysis library. Retrieved October 2018 https://www.andeal.org/topic.cfm?menu=5292; Dionyssiotis, Y., Skarantavos, G., Petropoulou, K., Galanos, A., Rapidi, C.-A., & Lyritis, G. P. (2019). Application of current sarcopenia definitions in spinal cord injury. *Journal of Musculoskeletal & Neuronal Interactions, 19*(1), 21; Erlandson, M., Lorbergs, A., Mathur, S., & Cheung, A. (2016). Muscle analysis using pQCT, DXA and MRI. *European Journal of Radiology, 85*(8), 1505–1511.; Fosbol, M. O., & Zerahn, B. (2015). Contemporary methods of body composition measurement. *Clinical Physiology and Functional Imaging, 2*, 81. doi:10.1111/cpf.12152; Fraser, C., Iruthayarajah, J., McIntyre, A., Serrato, J., Benton, B., & Teasell, R. W. (2016). Nutrition issues following spinal cord injury. Version 6.0.; Lee, S. Y., & Gallagher, D. (2008). Assessment methods in human body composition. *Current Opinion in Clinical Nutrition and Metabolic Care, 11*(5), 566; Mitsiopoulos, N., Baumgartner, R., Heymsfield, S., Lyons, W., Gallagher, D., & Ross, R. (1998). Cadaver validation of skeletal muscle measurement by magnetic resonance imaging and computerized tomography. *Journal of Applied Physiology, 85* (1), 115–122; Modlesky, C. M., Bickel, C. S., Slade, J. M., Cureton, K. J., Dudley, G. A., & Meyer, R. A. (2004). Assessment of skeletal muscle mass in men with spinal cord injury using dual-energy X-ray absorptiometry and magnetic resonance imaging. *Journal of Applied Physiology, 96*(2), 561–565. doi:10.1152/japplphysiol.00207.2003; Siervo, M., & Jebb, S. A. (2010). Body composition assessment: Theory into practice: Introduction of multicompartment models. *IEEE Engineering in Medicine and Biology Magazine, 29*(1), 48–59; Teigen, L. M., Kuchnia, A. J., Mourtzakis, M., & Earthman, C. P. (2017). The use of technology for estimating body composition: Strengths and weaknesses of common modalities in a clinical setting. *Nutrition in Clinical Practice, 32*(1), 20–29.

Tracer dilution

Tracer dilution is a common method to assess body composition based on the principle that all the body water is present in the FFM (body fat contains no water), and the assumption that 73.2% of the FFM is composed of water (Fosbol & Zerahn, 2015). To assess TBW a dose of tracer-labeled water (either deuterium and/or oxygen-18) is ingested after collection of a baseline urine sample to determine background isotope enrichment of body water (Siervo & Jebb, 2010). Urine samples are collected again after an equilibration period of 2–3 h following the given dose of tracer-labeled water (Siervo & Jebb, 2010). Isotope dilution methods are based on the assumption that the tracer only distributes in the body pool which consists of plasma, interstitial and intracellular fluids, is equally distributed in this pool, that equilibration occurs rapidly and the tracer is not metabolized during the equilibration time. Since water is found exclusively within FFM, the amount of FFM can be determined from TBW using an assumed hydration constant for FFM of 0.732, which represents 73.2% water in adults (Siervo & Jebb, 2010). FM is then derived by subtracting FFM from total body weight. This technique is considered the reference method for TBW and has a precision of 1%–2% for TBW (Siervo & Jebb, 2010) but is expensive due to the high cost of isotopes and analytical procedures but is a feasible approach in SCI as tracer administration and urine sampling can be undertaken in the home.

Air-displacement plethysmograph

Air-displacement plethysmograph (ADP) is based on the two-compartment model and measures the volume of air displaced by the body inside an enclosed chamber to provide an estimate of total body volume (Siervo & Jebb, 2010). Body density is then calculated by the ratio between volume and mass (weight). Using known densities of FFM and FM estimated from chemical analysis of cadavers and derived formulas, percentage body fat is calculated (Siervo & Jebb, 2010). Precision for FM is 1.7%–4.5% (Siervo & Jebb, 2010). Given that the density of FFM is known to vary between different groups (e.g., children, older adults, obese, blacks), a limitation of this technique is that FFM is assumed to have a constant density (Fields, Goran, & McCrory, 2002). In addition, the use of ADP is limited to research settings and there are practical difficulties associated with measuring lung volumes which must be accounted for when estimating total body volume in individuals with SCI above T3 due to paralysis of intercostal and abdominal muscles (Clasey & Gater Jr, 2005). For SCI populations, measuring weight and access into the measurement chamber are also limitations that may hinder its use clinically (Clasey & Gater Jr, 2005).

Imaging techniques

Dual-energy X-ray absorptiometry

Dual-energy X-ray absorptiometry (DXA) is a two-dimensional X-ray imaging procedure based on the 3-C model that differentiates the body into bone mineral, fat mass, and non-fat (lean) mass (Erlandson, Lorbergs, Mathur, & Cheung, 2016). This method is based on the attenuation characteristics of different tissues in the body exposed to two X-rays of high and low energy, which allow bone attenuation to be separated from soft tissue attenuation (Erlandson et al., 2016). As a result, DXA can provide estimates of total body and regional (arms, legs, trunk) fat mass, percent body fat, lean tissue mass, and bone mineral mass [bone mineral content (BMC, g) and bone mineral density (BMD, g/cm^2)] (Teigen et al., 2017). DXA is considered the gold standard for the diagnosis of osteoporosis (Lee & Gallagher, 2008) and has been used for defining low total and ALM which is part of the sarcopenia criteria (Teigen et al., 2017). The precision of DXA estimates of whole-body FM and lean tissue mass in sequential measurements range from 0.8%–2.7% and 0.4%–1.3% respectively and the precision of regional, particularly trunk fat mass are poorer than whole-body measures (Fosbol & Zerahn, 2015). While the majority of body composition studies in the SCI population have utilized DXA and its' use is recommended in The Academy of Nutrition and Dietetics Spinal Cord Injury Evidence-Based Nutrition Practice Guidelines (Academy of Nutrition and Dietetics, 2009) physical access to the scanning bed and difficulty positioning people with SCI limit its' routine use.

Magnetic resonance imaging

Magnetic resonance imaging (MRI) is highly precise, ~1.2%–2% (Mitsiopoulos et al., 1998), well suited for serial measurements (Erlandson et al., 2016), and able to provide detailed three-dimensional images of the body, and thus can be used to determine the quantity and distribution of adipose tissue (visceral, subcutaneous and intermuscular) and skeletal muscle (muscle cross-sectional area) at any site in the body (Siervo & Jebb, 2010). However, there are a number of limitations. It is time consuming and costly, requires highly trained personnel, is not appropriate for people with implants (e.g., pacemakers), is claustrophobic for some people (Fosbol & Zerahn, 2015), is difficult to access and maintain a prolonged supine position for people with impaired mobility such as SCI (Erlandson et al., 2016) and movement artifacts may arise from spasms and contractures.

Computed tomography

Computed tomography (CT) estimates total and regional adipose tissue (visceral, subcutaneous, and intermuscular), skeletal muscle, and bone (Borga et al., 2018) and is also highly precise ~2% (Mitsiopoulos et al., 1998). CT is considered a reference measure of body composition but due to high cost and radiation when used for whole-body composition, usage is limited to diagnostic purposes (Fosbol & Zerahn, 2015). CT measures at specific sites such as L3 and mid-thigh regions obtained are correlated with DXA whole-body FFM and skeletal muscle mass and have been suggested as alternatives to whole-body measurements for detecting low muscle mass (Cruz-Jentoft et al., 2019); however, this has not been investigated in people with SCI. The psoas muscle CSA measured using CT has also been proposed as an alternative tool to predict muscle quantity (Cruz-Jentoft et al., 2019) and has been used in one study as a measure of sarcopenia in individuals with acute traumatic SCI (Banaszek et al., 2019). While newer CT machines have lower radiation doses, accessibility for individuals with disabilities such as SCI remains problematic and spasms may create movement artifacts. Furthermore, the use of CT measures at specific sites to predict whole-body muscle mass warrants further investigation in people with SCI.

Peripheral quantitative computer tomography

Peripheral quantitative computed tomography (pQCT) is a newer imaging tool that uses similar technology to CT and produces a three-dimensional cross-sectional image of the tissue structure (BMD, BMC, muscle, and fat CSA) of a limb (Erlandson et al., 2016). pQCT has a lower radiation exposure, a shorter scan time, and the equipment is cheaper than a whole-body CT scanner, with precision for limb muscle area of <5% (Erlandson et al., 2016). The main limitations of pQCT are that it is largely used as a research tool, precision is low for intramuscular adipose tissue (3%–42%), accuracy is lower than MRI, scanning is limited to a single slice at appendicular sites (arms and legs), and a gold standard protocol for image analysis is lacking (Erlandson et al., 2016). Furthermore, movement artifacts, e.g., spasms and contractures may limit application in people with SCI.

Novel technique

Computerized digital image analysis

This novel approach is under development but entails deriving body shape and volume from back and side photographic images of the body while standing motionless with arms and legs abducted (Affuso et al., 2018). Percentage body fat predicted from computer algorithm using support vector regression models correlates moderately well with DXA (Affuso et al., 2018). This method has not been used in people with impaired mobility, such as SCI.

Summary of body composition assessment methods

In summary, it is currently recommended that weight, BMI, waist circumference, and body composition are routinely monitored following SCI. Choice of a body composition method will be dependent on a range of factors as outlined in the following section.

Choosing a body composition method for use in SCI

When selecting a method(s) to quantify body composition in individuals with SCI, a number of factors should be taken into consideration, including the:

- compositional variable to be quantified
- validity and precision of the method
- ease of use
- cost
- safety and convenience for the individual undergoing measurement

- measurement frequency, i.e., single point in time or longitudinal
- existence of population-specific prediction equations and reference measures or cut-off values developed and cross-validated for use in SCI
- accessibility to equipment for people with impaired mobility

When selecting a body composition technique, it is critical to use population-specific prediction equations and reference measures or cut-off values developed and cross-validated for use in SCI when available or to be aware of the limitations of using those developed in non-disabled people.

Applications to other areas of neuroscience

In this chapter, we reviewed the effects of traumatic spinal cord injury on body composition. We also provided an overview of the methods for assessing body composition and the considerations for use in individuals with spinal cord injury. Heterogeneous alterations in body composition are also seen in other neurological disorders and there are numerous etiological factors that are disease genotype and phenotype specific. These include whether the disease is neurodegenerative, stage of the disease, inflammatory abnormalities, endocrine, metabolic and motor disorders, and decreased physical activity (Çekici & Acar Tek, 2020).

Neurological diseases show a variety of impacts on body composition. A high prevalence of sarcopenia and segmental body composition changes similar to spinal cord injury have been described following stroke and in men with Multiple Sclerosis (Wingo, Young, & Motl, 2018). Whereas, asymmetrical disease effects, i.e., greater muscle wasting on the affected side of the body are seen in amyotrophic lateral sclerosis (Desport et al., 2003) and stroke, with a greater decline in lean mass of limbs on the hemiplegic side (Chang, Wu, Huang, & Han, 2020). Analogous to spinal cord injury, level of impairment impacts body composition in persons with relapsing-remitting multiple sclerosis; whole body, regional fat mass, and percentage body fat are higher in those with moderate compared with mild disability (Pilutti & Motl, 2019). In contrast, progressive weight loss composed of decreases in fat mass, percentage body fat, trunk, and peripheral fat mass with the maintenance of fat-free mass has been described in Parkinson's disease (Tan et al., 2018; Yong et al., 2020). However, the reason for the maintenance of fat-free mass is unclear (Tan et al., 2018; Yong et al., 2020). Differences in body composition between neurological disease subtypes have also been reported in Parkinson's disease (Femat-Roldán et al., 2020) and subgroups of Spinocerebellar ataxias (Leite et al., 2020). The diverse changes in the whole body and segmental body composition seen in spinal cord injury are reflected in other neurological diseases.

Considerations for body composition assessment in spinal cord injury described in Tables 1 and 2 are also applicable to other neurological conditions. For example, use of skinfold and bioimpedance prediction equations generated in normal-able bodied adults are inaccurate compared with dual-energy X-ray absorptiometry and oxygen-18 for prediction of percentage body fat in adults with cerebral palsy (Hildreth, Johnson, Goran, & Contompasis, 1997) and not valid in patients with amyotrophic lateral sclerosis (Desport et al., 2003). Difficulties accessing suprailiac, abdomen, mid-thigh, and subscapular skinfolds in wheelchair-bound subjects have also been reported (Hildreth et al., 1997). A population-specific bioimpedance equation has been validated (Desport et al., 2003) and recommended in guidelines for amyotrophic lateral sclerosis (Burgos et al., 2018). Analogous to spinal cord injury, the conventional BMI cut-off for classifying obesity underestimates the true adiposity in multiple sclerosis compared with dual-energy X-ray absorptiometry (Pilutti & Motl, 2016). In summary, when selecting a body composition technique it is critical to use population-specific prediction equations and reference measures or cut-off values developed and cross-validated for use in the specific neurological condition when available or to be aware of the limitations of using those developed in able-bodied people.

Key facts

Muscle wasting and increases in body fat, particularly around the waist occur following SCI.
Muscle wasting is more dramatic in severe SCI, such as tetraplegia and complete injuries.
Muscle wasting and body fat increases are greater in tetraplegia than paraplegia.
People with tetraplegia lose more muscle and gain more fat in the arms than those with paraplegia.
Increased body fat around the waist increases the risk of health problems such as heart disease and diabetes in people with SCI.
It is unclear whether low muscle mass increases health risk and early death in people with SCI.
Standards that define a healthy body weight normalised for height/ supine length have not been established.

Recommended waist circumference standards that define an increased health risk are lower than that for the healthy population, however the optimal cut point is not established.
Physical access to anthropometric and sophisticated body composition techniques may be limited by impaired mobility and sitting balance.
Bioelectrical impedance analysis shows promise as an accurate and practical tool to measure body composition.
Equations that estimate percentage body fat using skinfold thickness and fat and muscle using bioimpedance developed in healthy populations are inaccurate in SCI.
Population-specific bioimpedance equations developed in spinal cord-injured people estimate body composition more accurately.
SCI-specific reference ranges and cut-offs for muscle and fat have not been defined.

Mini-dictionary of terms

Lean tissue mass: Sum of all soft tissue minus bone mineral content and excluding fat mass, e.g., measured using dual-energy X-ray absorptiometry.
Fat-free mass: Lean tissue mass plus bone mineral content, e.g., measured using bioimpedance or deuterium dilution.
Fat mass: Bodyweight minus fat-free mass.
Extracellular fluid: Fluid contained outside of body cells, including interstitial fluid, lymph, vascular fluid.
Intracellular fluid: Fluid contained within the body cells.
Total body water: Water content of the body, composed of extracellular fluid plus intra-cellular fluid.
Sarcopenia: Low muscle mass, combined with reduced strength and functional impairment.

Summary points

This chapter focuses on the alterations that occur to body composition and body composition assessment methods suitable for use in clinical practice following spinal cord injury.

- Muscle wasting occurs below the level of the injury and body fat increases.
- Alterations in body composition are more dramatic in people with more severe injuries such as tetraplegia and complete injuries.
- Muscle wasting and gains in body fat are associated with poor health outcomes and increased risk of cardiovascular disease and diabetes.
- Anthropometric methods used routinely to measure and monitor body composition in non-disabled people are not appropriate or convenient to use in people with SCI.
- Reference "norms" to classify health risks in non-disabled people are not appropriate to use in people with SCI.
- Anthropometry may be suitable to monitor changes over time in an individual with SCI if a consistent method is used.
- Bioimpedance is a valid and convenient tool to measure body composition in people with SCI when population-specific equations are used.
- Imaging methods such as MRI, CT, and DXA are not easily accessible for people with SCI but are important for diagnostic purposes.

References

Academy of Nutrition and Dietetics. (2009). *Spinal cord injury (SCI) evidence-based nutrition practice guideline. Evidence analysis library*. Retrieved October 2018 https://www.andeal.org/topic.cfm?menu=5292.

Affuso, O., Pradhan, L., Zhang, C., Gao, S., Wiener, H. W., Gower, B., … Allison, D. B. (2018). A method for measuring human body composition using digital images. *PLoS One*, *13*(11), e0206430. https://doi.org/10.1371/journal.pone.0206430.

Banaszek, D., Inglis, T., Ailon, T., Charest-Morin, R., Dea, N., Fisher, C. G., … Street, J. (2019). P153. Sarcopenia, independent of age, predicts mortality and acute care adverse events in patients with traumatic spinal cord injury. *The Spine Journal*, *19*(9), S229.

Beck, L. A., Lamb, J. L., Atkinson, E. J., Wuermser, L. A., & Amin, S. (2014). Body composition of women and men with complete motor paraplegia. *Journal of Spinal Cord Medicine*, *37*(4), 359–365. https://doi.org/10.1179/2045772313Y.0000000151.

Borga, M., West, J., Bell, J. D., Harvey, N. C., Romu, T., Heymsfield, S. B., & Leinhard, O. D. (2018). Advanced body composition assessment: From body mass index to body composition profiling. *Journal of Investigative Medicine: The Official Publication of the American Federation for Clinical Research*, *66*, 887–895.

Bryce, T. N. (2010). *Spinal cord injury*. New York: Demos Medical.

Buchholz, A. C., McGillivray, C. F., & Pencharz, P. B. (2003). Differences in resting metabolic rate between paraplegic and able-bodied subjects are explained by differences in body composition. *American Journal of Clinical Nutrition*, *77*(2), 371–378.

Burgos, R., Bretón, I., Cereda, E., Desport, J. C., Dziewas, R., Genton, L., … Bischoff, S. C. (2018). ESPEN guideline clinical nutrition in neurology. *Clinical Nutrition*, *37*(1), 354–396. https://doi.org/10.1016/j.clnu.2017.09.003.

Castro, M. J., Apple, D. F., Jr., Hillegass, E. A., & Dudley, G. A. (1999). Influence of complete spinal cord injury on skeletal muscle cross-sectional area within the first 6 months of injury. *European Journal of Applied Physiology and Occupational Physiology*, *80*(4), 373–378.

Cederholm, T., Jensen, G., Correia, M., Gonzalez, M., Fukushima, R., Higashiguchi, T., … Coats, A. (2019). GLIM criteria for the diagnosis of malnutrition—A consensus report from the global clinical nutrition community. *Journal of Cachexia, Sarcopenia and Muscle*, *10*(1), 207–217.

Çekici, H., & Acar Tek, N. (2020). Determining energy requirement and evaluating energy expenditure in neurological diseases. *Nutritional Neuroscience*, *23*(7), 543–553.

Chang, K., Wu, W., Huang, K., & Han, D. (2020). Segmental body composition transitions in stroke patients: Trunks are different from extremities and strokes are as important as hemiparesis. *Clinical Nutrition*, *39*(6), 1968–1973.

Cirnigliaro, C. M., La Fountaine, M. F., Emmons, R., Kirshblum, S. C., Asselin, P., Spungen, A. M., & Bauman, W. A. (2013). Prediction of limb lean tissue mass from bioimpedance spectroscopy in persons with chronic spinal cord injury. *Journal of Spinal Cord Medicine*, *36*(5), 443–453. https://doi.org/10.1179/2045772313Y.0000000108.

Clasey, J. L., & Gater, D. R., Jr. (2005). A comparison of hydrostatic weighing and air displacement plethysmography in adults with spinal cord injury. *Archives of Physical Medicine and Rehabilitation*, *86*(11), 2106–2113. https://doi.org/10.1016/j.apmr.2005.06.013.

Cox, S., Weiss, S. M., Posuniak, E. A., Worthington, P., Prioleau, M., & Heffley, G. (1985). Energy expenditure after spinal cord injury: An evaluation of stable rehabilitating patients. *The Journal of Trauma*, *25*(5), 419–423.

Cruz-Jentoft, A. J., Bahat, G., Bauer, J., Boirie, Y., Bruyère, O., Cederholm, T., … Writing Group for the European Working Group on Sarcopenia in Older People 2 (EWGSOP2), and the Extended Group for EWGSOP2. (2019). Sarcopenia: Revised European consensus on definition and diagnosis. *Age and Ageing*, *48*, 16–31. https://doi.org/10.1093/ageing/afy169.

Desneves, K. J., Panisset, M. G., Galea, M. P., Kiss, N., Daly, R. M., & Ward, L. C. (2020). Comparison of segmental lean tissue mass in individuals with spinal cord injury measured by dual energy X-ray absorptiometry and predicted by bioimpedance spectroscopy. *Spinal Cord*, 1–8. https://doi.org/10.1038/s41393-020-00568-3.

Desport, J. C., Preux, P. M., Bouteloup-Demange, C., Clavelou, P., Beaufrere, B., Bonnet, C., & Couratier, P. P. (2003). Validation of bioelectrical impedance analysis in patients with amyotrophic lateral sclerosis. *The American Journal of Clinical Nutrition*, *77*(5), 1179.

Desport, J. C., Preux, P. M., Guinvarc'h, S., Salle, J. Y., Daviet, J. C., Dudognon, P., … Ritz, P. (2000). Total body water and percentage fat mass measurements using bioelectrical impedance analysis and anthropometry in spinal cord-injured patients. *Clinical Nutrition*, *19*(3), 185–190. https://doi.org/10.1054/clnu.1999.0122.

Dionyssiotis, Y., Skarantavos, G., Petropoulou, K., Galanos, A., Rapidi, C.-A., & Lyritis, G. P. (2019). Application of current sarcopenia definitions in spinal cord injury. *Journal of Musculoskeletal & Neuronal Interactions*, *19*(1), 21.

Edwards, L. A., Bugaresti, J. M., & Buchholz, A. C. (2008). Visceral adipose tissue and the ratio of visceral to subcutaneous adipose tissue are greater in adults with than in those without spinal cord injury, despite matching waist circumferences. *American Journal of Clinical Nutrition*, *87*(3), 600–607.

Emmons, R. R., Garber, C. E., Cirnigliaro, C. M., Kirshblum, S. C., Spungen, A. M., & Bauman, W. A. (2011). Assessment of measures for abdominal adiposity in persons with spinal cord injury. *Ultrasound in Medicine and Biology*, *37*(5), 734–741.

Erlandson, M., Lorbergs, A., Mathur, S., & Cheung, A. (2016). Muscle analysis using pQCT, DXA and MRI. *European Journal of Radiology*, *85*(8), 1505–1511.

Femat-Roldán, G., Gaitán Palau, M. A., Castilla-Cortázar, I., Elizondo Ochoa, G., Moreno, N. G., Martín-Estal, I., & Jiménez Yarza, M. (2020). Altered body composition and increased resting metabolic rate associated with the postural instability/gait difficulty Parkinson's disease subtype. *Parkinson's Disease*, 1–9. https://doi.org/10.1155/2020/8060259. 20420080.

Fields, D. A., Goran, M. I., & McCrory, M. A. (2002). Body-composition assessment via air-displacement plethysmography in adults and children: A review. *American Journal of Clinical Nutrition*, *75*(3), 453–467. https://doi.org/10.1093/ajcn/75.3.453.

Fosbol, M. O., & Zerahn, B. (2015). Contemporary methods of body composition measurement. *Clinical Physiology and Functional Imaging*, *2*, 81. https://doi.org/10.1111/cpf.12152.

Galea, M. P. (2012). Spinal cord injury and physical activity: Preservation of the body. *Spinal Cord*, *50*(5), 344–351. https://doi.org/10.1038/sc.2011.149.

Gandy, J. (2014). *Manual of dietetic practice*. Hoboken, United Kingdom: John Wiley & Sons, Incorporated.

Garshick, E., Ashba, J., Tun, G. C., Lieberman, L. S., & Brown, R. (1997). Assessment of stature in spinal cord injury. *The Journal of Spinal Cord Medicine*, *20*(1), 36–42.

Gibson, R. S. (2005). *Principles of nutritional assessment*. USA: Oxford University Press.

Gorgey, A. S., Dolbow, J. D., Khalil, R. K., Dolbow, D. R., Castillo, C., & Gater, D. R. (2014). Effects of spinal cord injury on body composition and metabolic profile—Part I. *Journal of Spinal Cord Medicine*, *37*(6), 693–702.

Gorgey, A., & Dudley, G. (2007). Skeletal muscle atrophy and increased intramuscular fat after incomplete spinal cord injury. *Spinal Cord*, *45*(4), 304.

Hildreth, H. G., Johnson, R. K., Goran, M. I., & Contompasis, S. H. (1997). Body composition in adults with cerebral palsy by dual-energy X-ray absorptiometry, bioelectrical impedance analysis, and skinfold anthropometry compared with the 18O isotope-dilution technique. *The American Journal of Clinical Nutrition*, *66*(6), 1436–1442.

Kocina, P., & Heyward, V. (1997). Validation of a bioimpedance equation for estimating fat-free mass of spinal cord injured adults 319. *Medicine & Science in Sports & Exercise*, *29*(5), 55.

Kyle, U. G., Bosaeus, I., De Lorenzo, A. D., Deurenberg, P., Elia, M., Gómez, J. M., … Pirlich, M. (2004). Bioelectrical impedance analysis—Part I: Review of principles and methods. *Clinical Nutrition*, *23*(5), 1226–1243.

Lee, S. Y., & Gallagher, D. (2008). Assessment methods in human body composition. *Current Opinion in Clinical Nutrition and Metabolic Care*, *11* (5), 566.

Leite, C. D. M. B. A., Schieferdecker, M. E. M., Frehner, C., Munhoz, R. P., Ashizawa, T., & Teive, H. A. G. (2020). Body composition in spinocerebellar ataxia type 3 and 10 patients: Comparative study with control group. *Nutritional Neuroscience*, *23*(1), 49–54.

Maggioni, M., Bertoli, S., Testolin, G., Merati, G., Margonato, V., & Veicsteinas, A. (2003). Body composition assessment in spinal cord injury subjects. *Acta Diabetologica*, *40*(Suppl. 1), S183–S186. https://doi.org/10.1007/s00592-003-0061-7.

Mitsiopoulos, N., Baumgartner, R., Heymsfield, S., Lyons, W., Gallagher, D., & Ross, R. (1998). Cadaver validation of skeletal muscle measurement by magnetic resonance imaging and computerized tomography. *Journal of Applied Physiology*, *85*(1), 115–122.

Modlesky, C. M., Bickel, C. S., Slade, J. M., Cureton, K. J., Dudley, G. A., & Meyer, R. A. (2004). Assessment of skeletal muscle mass in men with spinal cord injury using dual-energy X-ray absorptiometry and magnetic resonance imaging. *Journal of Applied Physiology*, *96*(2), 561–565. https://doi.org/10.1152/japplphysiol.00207.2003.

Pacifico, J., Geerlings, M. A. J., Reijnierse, E. M., Phassouliotis, C., Lim, W. K., & Maier, A. B. (2020). Prevalence of sarcopenia as a comorbid disease: A systematic review and meta-analysis. *Experimental Gerontology*, *131*. https://doi.org/10.1016/j.exger.2019.110801.

Panisset, M. G., Desneves, K., Ward, L. C., Rafferty, J., Rodi, H., Roff, G., … Galea, M. P. (2018). Bedside quantification of fat-free mass in acute spinal cord injury using bioelectrical impedance analysis: A psychometric study. *Spinal Cord*, *56*(4), 355.

Pelletier, C. A., Miyatani, M., Giangregorio, L., & Craven, B. C. (2016). Original research: Sarcopenic obesity in adults with spinal cord injury: A cross-sectional study. *Archives of Physical Medicine and Rehabilitation*, *97*, 1931–1937. https://doi.org/10.1016/j.apmr.2016.04.026.

Pilutti, L. A., & Motl, R. W. (2016). Body mass index underestimates adiposity in persons with multiple sclerosis. *Archives of Physical Medicine and Rehabilitation*, *97*(3), 405–412.

Pilutti, L. A., & Motl, R. W. (2019). Body composition and disability in people with multiple sclerosis: A dual-energy X-ray absorptiometry study. *Multiple Sclerosis and Related Disorders*, *29*, 41–47. https://doi.org/10.1016/j.msard.2019.01.009.

Rashmi, R., & Snekhalatha, U. (2019). Evaluation of body composition parameters using various diagnostic methods: A meta analysis study. *Obesity Medicine*, *16*. https://doi.org/10.1016/j.obmed.2019.100150.

Ravensbergen, H. J. C., Lear, S. A., & Claydon, V. E. (2014). Waist circumference is the best index for obesity-related cardiovascular disease risk in individuals with spinal cord injury. *Journal of Neurotrauma*, *31*(3), 292–300. https://doi.org/10.1089/neu.2013.3042.

Roh, E., & Choi, K. M. (2020). Health consequences of sarcopenic obesity: A narrative review. *Frontiers in Endocrinology*, *11*, 332.

Siervo, M., & Jebb, S. A. (2010). Body composition assessment: Theory into practice: Introduction of multicompartment models. *IEEE Engineering in Medicine and Biology Magazine*, *29*(1), 48–59.

Silveira, S., Ledoux, T., Robinson-Whelen, S., Stough, R., & Nosek, M. (2017). Methods for classifying obesity in spinal cord injury: A review. *Spinal Cord*, *55*(9), 812.

Singh, R., Rohilla, R. K., Saini, G., & Kaur, K. (2014). Longitudinal study of body composition in spinal cord injury patients. *Indian Journal of Orthopaedics*, *48*(2), 168.

Spungen, A. M., Adkins, R. H., Stewart, C. A., Wang, J., Pierson, R. N., Jr., Waters, R. L., & Bauman, W. A. (2003). Factors influencing body composition in persons with spinal cord injury: A cross-sectional study. *Journal of Applied Physiology*, *95*(6), 2398.

Spungen, A. M., Bauman, W. A., Wang, J., & Pierson, R. N. (1995). Measurement of body fat in individuals with tetraplegia: A comparison of eight clinical methods. *Spinal Cord*, *33*(7), 402.

Tan, A. H., Hew, Y. C., Lim, S.-Y., Ramli, N. M., Kamaruzzaman, S. B., Tan, M. P., … Lang, A. E. (2018). Altered body composition, sarcopenia, frailty, and their clinico-biological correlates, in Parkinson's disease. *Parkinsonism & Related Disorders*, *56*, 58–64. https://doi.org/10.1016/j.parkreldis.2018.06.020.

Teigen, L. M., Kuchnia, A. J., Mourtzakis, M., & Earthman, C. P. (2017). The use of technology for estimating body composition: Strengths and weaknesses of common modalities in a clinical setting. *Nutrition in Clinical Practice*, *32*(1), 20–29.

Wingo, B. C., Young, H.-J., & Motl, R. W. (2018). Body composition differences between adults with multiple sclerosis and BMI-matched controls without MS. *Disability and Health Journal*, *11*(2), 243–248. https://doi.org/10.1016/j.dhjo.2017.10.003.

Yong, V. W., Tan, Y. J., Ng, Y.-D., Choo, X. Y., Sugumaran, K., Chinna, K., … Tan, A. H. (2020). Progressive and accelerated weight and body fat loss in Parkinson's disease: A three-year prospective longitudinal study. *Parkinsonism & Related Disorders*, *77*, 28–35. https://doi.org/10.1016/j.parkreldis.2020.06.015.

Chapter 33

Energy requirements and spinal cord injury

Katherine J. Desneves[a,c], Nicole Kiss[a], Robin M. Daly[a], and Leigh C. Ward[b]

[a]*Institute for Physical Activity and Nutrition (IPAN), School of Exercise and Nutrition Sciences, Faculty of Health, Deakin University, Burwood, VIC, Australia,* [b]*School of Chemistry and Molecular Biosciences, The University of Queensland, St Lucia, QLD, Australia,* [c]*Nutrition and Dietetics Department, Division of Allied Health, Austin Health, Heidelberg, VIC, Australia*

List of abbreviations

BMR	basal metabolic rate
DLW	doubly labeled water
FFM	fat free mass
IC	indirect calorimetry
PAEE	physical activity energy expenditure
REE	resting energy expenditure
RMR	resting metabolic rate
SCI	spinal cord injury
TEE	total energy expenditure
TEF	thermic effect of food

Introduction

Muscle wasting following spinal cord injury (SCI) leads to a decreased resting energy expenditure, which combined with reduced levels of physical activity pre-disposes individuals with SCI to weight gain, other co-morbid complications, and chronic diseases such as diabetes and cardiovascular disease (Gorgey et al., 2014). It is important to assess and monitor the energy needs of people with SCI to prescribe diet and exercise interventions that preserve muscle mass and prevent positive energy balance, weight gain, and adiposity. In this section, we review the effects of traumatic SCI on energy expenditure. We also provide an overview of the available methods for assessing energy requirements and considerations for use in individuals with SCI.

Energy expenditure

Total energy expenditure (TEE) is the energy used by an individual over a 24-h period and comprises basal metabolic rate (BMR), the thermic effect of food (TEF), and physical activity-related energy expenditure (PAEE) (Gandy, 2014). Each of these terms is defined in the mini-dictionary. BMR is often substituted in the scientific literature with resting metabolic rate (RMR) and resting energy expenditure (REE). BMR is the largest element of TEE, comprising 60%–75% of TEE in a healthy person (Nevin, Steenson, Vivanti, & Hickman, 2016) and is influenced by age, gender, body weight, and body composition. Fat free mass (FFM) is the greatest predictor of RMR in non-disabled and SCI people and explains 70% of the variation in RMR (Buchholz & Pencharz, 2004). The second largest and most variable component is PAEE, which is usually only 25%–30% of TEE (Buchholz & Pencharz, 2004), whereas the third component, the TEF generally accounts for 5%–10% TEE (Buchholz & Pencharz, 2004).

Energy expenditure following SCI

It has been suggested that the energy needs of individuals with SCI vary over time (Nevin et al., 2016). During the acute phase of injury energy requirements are thought to be raised due to neurogenic shock (Nevin et al., 2016), surgical

Cellular, Molecular, Physiological, and Behavioral Aspects of Spinal Cord Injury. https://doi.org/10.1016/B978-0-12-822427-4.00033-2

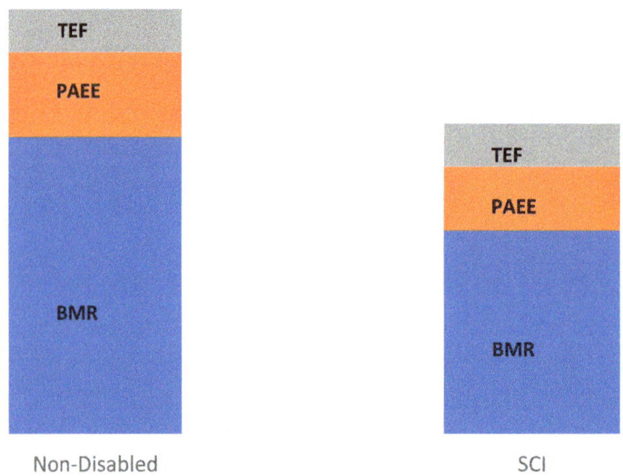

FIG. 1 Total energy expenditure in non-disabled and SCI. Comparison of components of total energy expenditure in non-disabled people and people with long-term spinal cord injury; *BMR*, basal metabolic rate; *TEF*, thermic effect of food; *PAEE*, physical activity energy expenditure. Note that total energy expenditure is lower in SCI due to a reduction in BMR and PAEE (Buchholz & Pencharz, 2004; Gorgey et al., 2014; Nevin et al., 2016).

intervention, and medical complications (Long, Schaffel, Geiger, Schiller, & Blakemore, 1979). However, a systematic review of studies investigating measured and predicted energy needs found no evidence of this theorized hypermetabolism in the acute phase (<1-month post-injury) (Nevin et al., 2016). It is plausible that reductions in PAEE due to paralysis and immobility may neutralize any stress-induced increases in BMR in the early stages following injury (Gandy, 2014). In contrast, it is well documented that individuals with chronic SCI have a reduced RMR, indicative of hypometabolism, with cross-sectional studies in SCI patients $\geq$1.5 years post-injury reporting an RMR 14%–27% less than non-injured controls (Buchholz & Pencharz, 2004). This reduced RMR is attributed to lower FFM and decreased sympathetic nervous system activity (Buchholz & Pencharz, 2004). Injury severity also affects energy expenditure in chronic SCI, with a ~19% lower BMR observed in individuals with tetraplegia compared with paraplegia (Farkas, Gorgey, Dolbow, Berg, & Gater, 2019). Physical activity levels have also been reported to be 39% lower in tetraplegia than paraplegia (Kyriakides et al., 2019) and 18% lower in persons with complete compared with incomplete SCI (Buchholz & Pencharz, 2004). However, no changes to the TEF have been reported in people with SCI (Gorgey et al., 2014). In summary, it is difficult to draw any conclusions about the energy needs in the acute phase post-injury due to a lack of studies. However, it is well documented that individuals with chronic SCI have a reduced RMR that is directly related to loss of FFM and injury severity. This reduced RMR combined with declines in PAEE contributes to positive energy balance which likely explains fat mass gains and the increased risk for multiple co-morbidities in people with SCI. These differences in RMR and PAEE in non-disabled and long-term SCI are illustrated in Fig. 1.

Methods for determining energy requirements

It is important to accurately assess the energy needs of people with SCI for two reasons:

(1) To preserve lean (muscle) mass.
(2) To prevent positive energy balance that contributes to weight and fat gain.

The following section details methods available and considerations when choosing a method for predicting energy expenditure in people with SCI.

Doubly labeled water

The reference method for the determination of total energy expenditure (TEE) is isotope dilution using two stable isotopes, deuterium (^2H_2O) and ^{18}O-labeled water ($^2H_2^{18}O$, or oxygen-18) (DeLany, 1997). After ingestion, doubly-labeled water (DLW) rapidly reaches equilibrium with total body water (TBW) and enters the metabolic pool. Deuterium is eliminated as water, and oxygen-18 is eliminated both as water and carbon dioxide. Isotope ratio mass spectroscopy analysis is used to determine the elimination rates of the isotopes in urine. The difference between the rates of disappearance of the two isotopes permits calculation of the CO_2 produced and using respiratory equations, the total energy expended. This method is

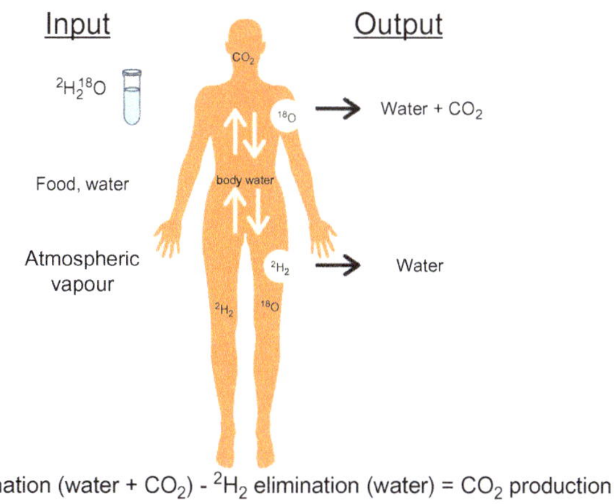

FIG. 2 Doubly labeled water method. After ingestion, doubly-labeled water (DLW-$^2H_2^{18}O$) rapidly reaches equilibrium with total body water (TBW) and enters the metabolic pool. Deuterium ($^2H_2$) is eliminated as water, and oxygen-18 (^{18}O) is eliminated both as water (H_2O) and carbon dioxide (CO_2). The difference between the rates of disappearance of the two isotopes permits calculation of the CO_2 produced and using respiratory equations, the total energy expended (DeLany, 1997).

illustrated in Fig. 2. An additional advantage of the DLW technique is that it concomitantly allows calculation of body composition from dilution of either tracer (usually deuterium).

Indirect calorimetry

Indirect calorimetry (IC) is considered the gold standard for assessing BMR and REE/RMR (Gandy, 2014; Nevin et al., 2016) and is recommended for use in critically ill adults (McClave et al., 2016) and people with SCI (Academy of Nutrition and Dietetics, 2009). IC involves the measurement of oxygen consumption and CO_2 production. Energy expenditure is then calculated using a respiratory equation, the modified Weir equation (Haugen, Chan, & Li, 2007) shown in Table 1. Both doubly-labeled water and indirect calorimetry methods are typically limited to a research setting due to cost, practical considerations, and availability (Gandy, 2014; Nevin et al., 2016).

TABLE 1 Common energy prediction equations developed in non-disabled populations.

Source	Age range	Equation
Schofield (1985) (MJ/d)	18–30	Males BMR = (0.063 × W) + 2.896
		Females BMR = (0.062 × W) + 2.036
	30–60	Males BMR = (0.048 × W) + 3.653
		Females BMR = (0.034 × W) +3.538
	>60	Males BMR = (0.049 × W) + 2.459
		Females BMR = (0.038 × W) + 2.755
Henry (2005) (MJ/d)	18–30	Males BMR = (0.0669 × W) + 2.28
		Females BMR = (0.0546 × W) + 2.33
	30–60	Males BMR = (0.0592 × W) + 2.48
		Females BMR = (0.0407 × W) + 2.90
	60+	Males BMR = (0.0563 × W) + 2.15
		Females BMR = (0.0424 × W) + 2.38
Harris and Benedict (1918) (kJ/d)		Males BMR = 278 + (57.5 × W) + (20.9 × H)–(28.3 × A)
		Females BMR = 2741 + (40 × W) + (7.7 × H)–(19.6 × A)
Weir (kcal/d) (Haugen et al., 2007)		[(VO_2 (L/min)×3.941) + (VCO_2 (L/min)×1.11)]×1440

Common equations used to predict energy needs in non-disabled populations and the modified Weir equation used to predict resting energy expenditure (REE) from oxygen consumption and carbon dioxide production. *W*, weight in kg; *A*, age in years; *H*, height in cm (Harris & Benedict, 1918; Haugen et al., 2007; Henry, 2005; Schofield, 1985).

Predictive equations

In clinical practice, published predictive equations incorporating a combination of age, weight, or sex are commonly used to estimate REE when IC is unavailable. The most commonly used predictive equations are shown in Table 1. The Harris and Benedict (1918), Schofield (1985), and Henry (2005) equations were derived from able-bodied populations. These are adjusted by multiplying REE by activity and/or stress factors to predict the total energy expenditure of an individual, their use is subjective and has received much criticism due to their inaccuracy (Reeves & Capra, 2003). Another more convenient, commonly used method is to calculate energy requirements based simply on energy values per kilogram of body weight, e.g., 25 kcal/kg, with adjustments for certain conditions, such as hypermetabolism or malabsorption that are not based on scientific evidence (Reeves & Capra, 2003). This method also doesn't take into consideration age, gender, metabolic state, or body composition, all factors known to affect energy expenditure (Gandy, 2014).

A less utilized method in clinical practice is to calculate energy requirements based on published prediction equations that incorporate body composition measurements (Cunningham, 1991). Recently SCI-specific equations that include body composition measures (FFM, FM, and/or circumferences) have emerged for use in chronic SCI (Chun, Shin, & Kim, 2017; Kocina & Heyward, 1997; Nightingale & Gorgey, 2018). These are summarized in Table 2. The accuracy and validity of predictive equations for use in the various phases following SCI is discussed in the next section.

Predictive equations validated in SCI

Energy prediction equations and activity factors developed in non-disabled people are inaccurate when used in SCI (Nevin et al., 2016). Guidelines suggest using Harris-Benedict with an injury factor of 1.2 and an activity factor of 1.1 if the SCI patient is in the acute phase, and that individuals with long term paraplegia and tetraplegia have their energy needs to be estimated using 22.7 and 27.9 kcal/kg, respectively (Academy of Nutrition and Dietetics, 2009). However, a systematic review found that energy prediction equations including age, weight, or sex developed in non-disabled populations are inaccurate in SCI, overestimating energy requirements by up to 90% (Nevin et al., 2016). The simple kcal/kg method also shows the poor predictive value (Desneves et al., 2019; Ramirez et al., 2018; Sedlock & Laventure, 1990), as do prediction equations developed in non-disabled people that include FFM (Chun et al., 2017; Nightingale & Gorgey, 2018; Sedlock & Laventure, 1990). Compounding this, the use of the able-bodied physical activity correction factor of 1.2 when applied to Harris and Benedict (1918), Schofield (1985), and Henry (2005) equations has been found to inaccurately estimate TEE in people with acute SCI (Desneves et al., 2019) and an SCI-specific correction factor of 1.15 has been suggested (Farkas et al., 2019). However, the reliability and validity of this recommendation have not been tested (Farkas et al., 2019). The use of equations and activity factors developed in non-disabled people and the kcal/kg method are inaccurate and not recommended for predicting energy needs in people with SCI.

Contemporary research has focused on the development and validation of SCI-specific regression equations utilizing FFM with or without anthropometric measurements to predict BMR in chronic SCI (Andersen, Sweet, Reid, Sydney, & Plourde, 2018; Buchholz et al., 2003; Chun et al., 2017; Nightingale & Gorgey, 2018). The Buchholz et al. (2003) RMR prediction equation developed in men and women with paraplegia that includes FFM has been cross-validated in people with acute and chronic SCI. A study in people with acute SCI (<2 months post-injury) found that the Buchholz et al., RMR equation multiplied by a combined stress and activity factor of 1.3 was the most accurate estimate of TDEE compared with

TABLE 2 SCI-specific energy prediction equations.

Source	Equation
Buchholz, McGillivray, and Pencharz (2003) (kJ/d)	RMR =10,682–1238(ln age)–521(sex)–24(H) + 87(FFM)
Chun et al. (2017) (kcal/d)	BMR = 24.5 × FFM + 244.4
Nightingale and Gorgey (2018) (kcal/day)	BMR = 23.469 × FFM + 294.330 BMR = 23.995 × FFM + 6.189 × SAD +6.384 × TAD–6.948 × TC + 275.211 BMR = 19.789 × FFM + 5.156 × W + 8.090 × H – 15.301 × C BMR = 13.202 × H + 11.329 × W – 16.729 × TAD – 1185.445

Equations used to predict energy needs developed specifically for use in people with spinal cord injury are listed; *W*, weight in kg; *H*, height in cm; *A*, age in years; FFM in kg; *SAD*, sagittal abdominal diameter in cm; *TAD*, transverse abdominal diameter in cm; *TC*, thigh circumference in cm; *C*, calf circumference in cm (Buchholz et al., 2003; Chun et al., 2017; Nightingale & Gorgey, 2018).

DLW, whereas a study in individuals with chronic SCI found that predicted RMR was not significantly different to measured RMR using IC. Others have reported a non-significant difference (bias minus 84 kJ/day) between predicted and measured BMR using the Chun et al. (2017), a population-specific equation that also includes FFM. A newer SCI-specific BMR equation incorporating FFM, weight, height, and calf circumference as predictor variables has not been cross-validated with a larger, independent group of males and females with SCI (Nightingale & Gorgey, 2018). These equations need the input of FFM which requires access to equipment such as bioimpedance. While FFM predicted from bioimpedance spectroscopy using the SCI-population specific Kocina and Heyward FFM equation (Kocina & Heyward, 1997) is comparable to FFM calculated from DLW (Desneves et al., 2019) it is unclear whether it is practical and feasible to use bioimpedance spectroscopy to obtain FFM for use in energy prediction equations in clinical practice.

In summary, prediction equations and activity factors derived from non-disabled populations have not been successfully validated in SCI patients (Nevin et al., 2016) but there is mounting evidence that population-specific prediction equations based on body composition measures and SCI-specific activity factors are more accurate when predicting energy requirements in both acute and chronic SCI.

Choosing an energy requirement method for use in SCI

In the absence of indirect calorimetry, it is preferable to use an SCI population-specific energy prediction equation and an activity factor. However, clinical judgment should be used when choosing any energy prediction method and when interpreting requirement calculations. Calculations should only be used as a starting point and SCI patients should be reviewed regularly and requirements recalculated in alignment with substantial alterations in clinical condition, physical activity levels, and treatment goals.

Applications to other areas of neuroscience

In this chapter, we have reviewed the effects of traumatic spinal cord injury on energy requirements. We also provided an overview of the available methods for assessing energy requirements and the considerations for use in individuals with spinal cord injury. Heterogeneous alterations in energy expenditure are also seen in other neurological disorders and there are numerous etiological factors that are disease genotype and phenotype-specific. These include whether the disease is neurodegenerative, stage of the disease, inflammatory abnormalities, endocrine, metabolic, motor disorders, and decreased physical activity (Çekici & Acar Tek, 2020).

The effect of neurological disease on energy expenditure is diverse. Hypometabolism has been reported in some neurological diseases, whereas increased total energy expenditure due to increased physical activity resulting from motor disorders including tremor, rigidity, dystonia, dyskinesia, and chorea has been reported in others. Further details can be found in the review: Determining energy requirement and evaluating energy expenditure in neurological disease (Çekici & Acar Tek, 2020).

Indirect calorimetry is the ideal method for determining energy requirements in healthy individuals, spinal cord injury, and some other neurological diseases. In addition to the usual considerations such as availability, cost, and need for trained personnel, it is difficult to use indirect calorimetry in patients with excessive movement such as Huntington's disease due to severe chorea movements and in children with Cerebral palsy because it is hard to determine oxygen levels as a result of breathing difficulties (Çekici & Acar Tek, 2020). Since each neurological disease has different impacts on energy expenditure due to varying etiologies, it is recommended that resting energy expenditure and total energy expenditure are measured on an individual basis if possible or that neurological disease-specific equations and activity adjustment factors are used to predict energy requirements (Çekici & Acar Tek, 2020).

Mini-dictionary of terms

Basal metabolic rate: The energy an individual needs to maintain body functions at rest. Basal metabolic rate is measured using indirect calorimetry and following a strict protocol; including mental and physical rest, following a 12 h fast, at 27–29°C, with no vigorous physical activity or stimulants within the previous 24 h, calibration of gases and establishment of steady-state.

Resting energy expenditure or resting metabolic rate: A measurement is taken using indirect calorimetry when any criteria for basal metabolic rate are not met.

Thermic effect of food: The energy used digesting, absorbing, and transporting nutrients within the body.

Physical activity energy expenditure: Energy used for incidental or planned movement above rest.

Total energy expenditure: The energy used by an individual consisting of basal metabolic rate, thermic effect of food, and physical activity.

Hypermetabolism: A basal metabolic rate higher than predicted for healthy individuals of the same age and weight due to injury or illness.

Hypometabolism: A basal metabolic rate below that predicted for healthy, non-injured individuals.

Key facts of energy metabolism

- It is unclear if energy needs are elevated immediately following spinal cord injury.
- People with long-term spinal cord injury have lower energy needs than non-disabled people due to muscle loss, reduced metabolism, and lower physical activity levels.
- Energy needs are lower in people with more severe injuries.
- It is recommended that indirect calorimetry is used to measure energy needs when available because energy prediction equations developed in healthy people are inaccurate when used in people with spinal cord injury.
- Energy prediction equations developed in spinal cord injury that include body composition measures are more accurate to use in people with spinal cord injury.
- A spinal cord injury-specific activity correction factor of 1.15 has been suggested but the reliability and validity have not been tested.

Summary points

This chapter focuses on the alterations that occur to energy requirements following spinal cord injury:

- The total energy expenditure of people with long-term spinal cord injury is lower than non-disabled people due to a reduced basal metabolic rate and lower levels of physical activity. These predispose people with spinal cord injury to positive energy balance and weight gain.
- It is important to assess energy needs accurately to preserve muscle and prevent weight and fat gain.
- Clinical guidelines recommend using indirect calorimetry to assess energy needs in people with spinal cord injury, however, this method is not usually available in clinical practice.
- Energy prediction equations are often used in clinical practice but equations developed in nondisabled people are inaccurate when used in people with spinal cord injury.
- Spinal cord injury-population-specific equations to predict energy needs incorporating body composition measures have been developed and validated for use in people with spinal cord injury.
- Bioimpedance can be used to predict fat-free mass for use in these equations.
- It is preferable to use a spinal cord injury-specific activity factor. Although the reliability and validity have not been tested.
- Clinical judgment should be used when choosing any energy prediction method and when interpreting requirement calculations.
- Calculations should be used as a starting point and spinal cord injury patients should be reviewed regularly and requirements recalculated in alignment with substantial alterations in clinical condition, physical activity levels, and treatment goals.

References

Academy of Nutrition and Dietetics. (2009). *Spinal cord injury (SCI) evidence-based nutrition practice guideline. Evidence analysis library*. Retrieved October 2018 https://www.andeal.org/topic.cfm?menu=5292.

Andersen, R. E., Sweet, S. N., Reid, R. E. R., Sydney, F., & Plourde, H. (2018). Accuracy of two generic prediction equations and one population-specific equation for resting energy expenditure in individuals with spinal cord injury. *Canadian Journal of Dietetic Practice and Research: A Publication of Dietitians of Canada*, *79*, 1–6. https://doi.org/10.3148/cjdpr-2018-021.

Buchholz, A. C., McGillivray, C. F., & Pencharz, P. B. (2003). Differences in resting metabolic rate between paraplegic and able-bodied subjects are explained by differences in body composition. *American Journal of Clinical Nutrition*, *77*(2), 371–378.

Buchholz, A. C., & Pencharz, P. B. (2004). Energy expenditure in chronic spinal cord injury. *Current Opinion in Clinical Nutrition and Metabolic Care*, *7* (6), 635–639.

Çekici, H., & Acar Tek, N. (2020). Determining energy requirement and evaluating energy expenditure in neurological diseases. *Nutritional Neuroscience*, *23*(7), 543–553.

Chun, S. M., Shin, H. I., & Kim, H. R. (2017). Estimating the Basal metabolic rate from fat free mass in individuals with motor complete spinal cord injury. *Spinal Cord*, *55*(9), 844–847.

Cunningham, J. J. (1991). Body composition as a determinant of energy expenditure: A synthetic review and a proposed general prediction equation. *The American Journal of Clinical Nutrition*, *54*(6), 963–969.

DeLany, J. P. (1997). *Doubly labeled water for energy expenditure in emerging technologies for nutrition research: Potential for Assessing Military Performance Capability Committee on Military Nutrition Research, Institute of Medicine (Chapter 12, pp. 281)*. National Academies Press, ISBN:0-309-55338-5.

Desneves, K. J., Panisset, M. G., Rafferty, J., Rodi, H., Ward, L. C., Nunn, A., & Galea, M. P. (2019). Comparison of estimated energy requirements using predictive equations with total energy expenditure measured by the doubly labelled water method in acute spinal cord injury. *Spinal Cord*, *57*(7), 562–570.

Farkas, G. J., Gorgey, A. S., Dolbow, D. R., Berg, A. S., & Gater, D. R. (2019). Caloric intake relative to total daily energy expenditure using a spinal cord injury—Specific correction factor. *American Journal of Physical Medicine & Rehabilitation*, *98*(11), 947–952.

Gandy, J. (2014). *Manual of dietetic practice*. Hoboken, United Kingdom: John Wiley & Sons.

Gorgey, A. S., Dolbow, J. D., Khalil, R. K., Dolbow, D. R., Castillo, C., & Gater, D. R. (2014). Effects of spinal cord injury on body composition and metabolic profile—Part I. *Journal of Spinal Cord Medicine*, *37*(6), 693–702.

Harris, J. A., & Benedict, F. G. (1918). A biometric study of human basal metabolism. *Proceedings of the National Academy of Sciences*, *4*(12), 370–373.

Haugen, H. A., Chan, L. N., & Li, F. (2007). Indirect calorimetry: A practical guide for clinicians. *Nutrition in Clinical Practice*, *22*(4), 377–388.

Henry, C. (2005). Basal metabolic rate studies in humans: Measurement and development of new equations. *Public Health Nutrition*, *8*(7a), 1133–1152.

Kocina, P., & Heyward, V. (1997). Validation of a bioimpedance equation for estimating fat-free mass of spinal cord injured adults 319. *Medicine & Science in Sports & Exercise*, *29*(5), 55.

Kyriakides, A., Poulikakos, D., Galata, A., Konstantinou, D., Panagiotopoulos, E., & Chroni, E. (2019). The effect of level of injury and physical activity on heart rate variability following spinal cord injury. *Journal of Spinal Cord Medicine*, *42*(2), 212–219. https://doi.org/10.1080/10790268.2017.1383709.

Long, C. L., Schaffel, N., Geiger, J. W., Schiller, W. R., & Blakemore, W. S. (1979). Metabolic response to injury and illness: Estimation of energy and protein needs from indirect calorimetry and nitrogen balance. *Journal of Parenteral and Enteral Nutrition*, *3*(6), 452–456.

McClave, S. A., Taylor, B. E., Martindale, R. G., Warren, M. M., Johnson, D. R., Braunschweig, C., … Cresci, G. A. (2016). Guidelines for the provision and assessment of nutrition support therapy in the adult critically ill patient: Society of Critical Care Medicine (SCCM) and American Society for Parenteral and Enteral Nutrition (ASPEN). *Journal of Parenteral and Enteral Nutrition*, *40*(2), 159–211.

Nevin, A. N., Steenson, J., Vivanti, A., & Hickman, I. J. (2016). Investigation of measured and predicted resting energy needs in adults after spinal cord injury: A systematic review. *Spinal Cord*, *54*(4), 248.

Nightingale, T. E., & Gorgey, A. S. (2018). Predicting Basal metabolic rate in men with motor complete spinal cord injury. *Medicine and Science in Sports and Exercise*, *50*(6), 1305–1312.

Ramirez, C. L., Pelekhaty, S., Massetti, J. M., Galvagno, S., Harmon, L., Botwinick, I., … Stein, D. M. (2018). Validation of predictive equations to assess energy expenditure in acute spinal cord injury. *Journal of Trauma and Acute Care Surgery*, *85*(5), 984–991.

Reeves, M. M., & Capra, S. (2003). Predicting energy requirements in the clinical setting: Are current methods evidence based? *Nutrition Reviews*, *61*(4), 143–151.

Schofield, W. (1985). Predicting basal metabolic rate, new standards and review of previous work. *Human Nutrition. Clinical Nutrition*, *39*, 5–41.

Sedlock, D. A., & Laventure, S. J. (1990). Body composition and resting energy expenditure in long term spinal cord injury. *Spinal Cord*, *28*(7), 448.

Chapter 34

Virtual walking and spinal cord injury neuropathic pain

Elizabeth J. Richardson[a] and Zina Trost[b]

[a]Department of Behavioral & Social Sciences, University of Montevallo, Montevallo, AL, United States, [b]Department of Physical Medicine and Rehabilitation, Virginia Commonwealth University, Richmond, VA, United States

List of abbreviations

2D	two-dimensional
3D	three-dimensional
fMRI	functional magnetic resonance imaging
HMD	head-mount display
SCI	spinal cord injury
tDCS	transcranial direct current stimulation
VR	virtual reality

Introduction

An unfortunate reality for many with spinal cord injury (SCI) is the number of secondary complications that can arise beyond the immediate physical impairments from their injury such as sensorimotor disruptions (Kirshblum & Lin, 2019; Richardson, Richards, & Boyer, 2008; see Table 1). Persistent pain is a common secondary complication following spinal cord injury (SCI), affecting over two-thirds of individuals with SCI (Bonica, 1991). Chronic pain following SCI is not a unitary phenomenon, as there are multiple subtypes of SCI-related pain with various underlying pathophysiological mechanisms. Individuals who are many years post-SCI may experience chronic musculoskeletal pain from prolonged manual wheelchair use (Siddall & Loeser, 2001), while neuropathic forms of pain may develop soon after the SCI (Siddall, McClelland, Rutkowski, & Cousins, 2003; Siddall, Taylor, McClelland, Rutkowski, & Cousins, 1999). Unlike musculoskeletal forms of post-SCI pain, SCI-related neuropathic pain is often described as burning or "electric" and is typically experienced at or below the level of injury in areas with sensory disturbance (Bryce et al., 2007, 2012).

SCI-related neuropathic pain affects more than half of those with SCI (Burke, Fullen, Stokes, & Lennon, 2017) and for many, it persists in its course (Finnerup, 2013; Siddall, 2009) or even worsens with time (Jensen, Kuehn, Amtmann, & Cardenas, 2007). Thus, it is not surprising that SCI-related neuropathic pain is associated with disruptions in several life activities and an overall decline in quality of life (Burke, Lennon, & Fullen, 2018). Finding more effective forms of treatment for SCI-related neuropathic pain has become increasingly important given these negative psychosocial effects as well as the shortcomings of currently available pharmacological agents (Cardenas & Jensen, 2006; Teasell et al., 2010). Recent evidence suggests virtual walking as a potentially viable non-pharmacological alternative or at least an effective adjunctive treatment for SCI-related neuropathic pain. The mechanisms underlying this treatment approach draw upon conceptualizing SCI-related neuropathic pain as a result of central nervous system changes that arise from the SCI.

This chapter will begin by discussing SCI-related forms of neuropathic pain and the theoretical concepts of its development and maintenance that may underlie the efficacy of virtual reality (VR) walking therapies. We will define VR and identify the components of VR with therapeutic utility. Lastly, this chapter will describe the evidence to date on the use of VR walking or other VR protocols to treat SCI-related pain as well as future directions for further validating the clinical utility for this form of therapy for SCI-related neuropathic pain.

Cellular, Molecular, Physiological, and Behavioral Aspects of Spinal Cord Injury. https://doi.org/10.1016/B978-0-12-822427-4.00034-4

TABLE 1 Common secondary complications that can arise following spinal cord injury (SCI).

Autonomic dysreflexia
Bowel and bladder dysfunction (including urinary tract infection)
Cardiovascular dysfunction
Muscle spasticity below the level of injury
Pain
Pressure sores
Pulmonary/respiratory complications
Sexual dysfunction

Based on Richardson, E. J., Richards, J. S., & Boyer, B. A. (2008). Spinal cord injury. In B. A. Boyer & M. I. Paharia (Eds.), *Comprehensive handbook of clinical health psychology*. Hoboken, NJ: Wiley; Kirshblum, S., & Lin, V. W. (Eds). (2019). *Spinal cord medicine*. New York: Demos Medical Publishing.

SCI-related neuropathic pain as a deafferentation pain

Adding to the clinical complexity of SCI-related neuropathic pain is that it can present in different forms that may be associated with different symptoms and outcomes (Finnerup, Johannesen, Fuglsang-Frederiksen, et al., 2003; Jordan & Richardson, 2016; Richardson & Redden, 2014). *At-level* neuropathic pain is often experienced as a "band-like" pain within three dermatomes of the level of injury (Bryce et al., 2012) and likely involves peripheral mechanisms to some degree, such as nerve root damage (Siddall & Loeser, 2001). In contrast, changes within the central nervous system are believed to primarily mediate *below-level* neuropathic pain, which occurs in regions well below the level of injury (Siddall, Taylor, & Cousins, 1997). Such centrally mediated pain is associated with an altered spinothalamic function that can be measured by evoked sensitivity thresholds (Bowsher, 1996). The severity of below-level SCI-related neuropathic pain among those with SCI is related to the degree of *at-level* neuronal hyperexcitability, and likely hyperexcitability at supraspinal levels as well (Finnerup et al., 2003). In rodent models of SCI (not specific to SCI-related *pain*), neuronal hyperexcitability was found specifically within the somatosensory cortex (Yague, Foffani, & Aguilar, 2011). In humans, functional magnetic resonance imaging (fMRI) has revealed functional reorganization in the sensorimotor cortices following SCI, with the most notable changes occurring among those with below-level SCI-related neuropathic pain (Wrigley, Press, Gustin, et al., 2009; see Fig. 1). Anatomically, there are also differences in several brain regions among those with below-level SCI-related neuropathic pain, including the posterior parietal cortices, the nucleus accumbens, and the prefrontal cortex, as well as regions known to be associated with pain processing, such as the thalamus (Gustin, Wrigley, Siddall, & Henderson, 2010). Differences in resting blood flow within regions of the cerebellum and prefrontal cortex in those with SCI-related neuropathic pain have also been observed, potentially relating to the affective-motivational and cognitive-evaluative aspects of the pain experience (Richardson, Deutsch, Deshpande, & Richards, 2020).

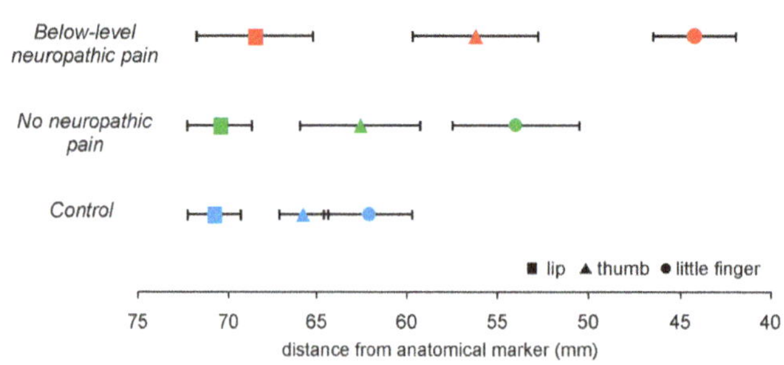

FIG. 1 Reorganization of the primary somatosensory cortex was found by Wrigley et al. (2009). Individuals with SCI, with and without pain, received tactile stimulation to the lip, thumb, and little finger. These researchers measured the distance between the point of maximum activation and a common point: the intersection of the longitudinal and central sulcus. Compared to able-bodied subjects, those with SCI showed encroachment of somatosensory cortical activation formerly representing the legs, with those who had SCI-related neuropathic pain showing the greatest shift. *(Data from Wrigley, P. J., Press, S. R., Gustin, S. M., et al. (2009). Neuropathic pain and primary somatosensory cortex reorganization following spinal cord injury.* Pain, 141, *52–59, with permission.)*

These findings, particularly the functional reorganization that occurs in somatosensory cortices among those with SCI-related neuropathic pain (Wrigley et al., 2009) parallel other deafferentation pain syndromes such as phantom limb pain. Phantom limb pain can also persist many years following amputation (Ephraim, Wegener, MacKenzie, Dillingham, & Pezzin, 2005) and has been conceptualized as a product of neuroplastic changes that occur as a result of deafferentation. Specifically, somatosensory neurons that formerly processed sensory information from the deafferented limb may subsequently become responsive to sensory input from nearby regions of the body represented on the homunculus (Ramachandran & Hirstein, 1998).

While such models often focus on changes specific to the somatosensory cortex, it is important to note that multiple forms of sensory input (vision, proprioception, and touch, for example) together with motor output are involved in creating an overall body percept (Tsakiris, 2017). Deafferentation and the loss of sensory input may disrupt this system (e.g., descending motor messages without the congruent sensory or proprioceptive feedback) and create the sensation of pain (Flor, Nikolajsen, & Jensen, 2006; Harris, 1999).

Mirror therapy: A precursor to virtual walking

Disrupted equilibrium between sensory/afferent and motor/efferent feedback (understood to underpin central pain) is highlighted by reductions in phantom pain observed when a person is given the visual illusion that the amputated limb has returned and is moving (Chan et al., 2007; Ramachandran & Altschuler, 2009). Mirror box therapy, or simply *mirror therapy*, has also shown benefit with other centrally mediated pain conditions, such as complex regional pain syndrome (Sayegh et al., 2013)—a pain syndrome that is also characterized by changes in bodily percepts (Moseley, 2005). Interestingly, mirror therapy may recalibrate the dysfunctional cortical reorganization accompanying phantom limb pain. Foell, Bekrater-Bodmann, Diers, and Flor (2014) found reorganization in the somatosensory cortex of the hemisphere that previously processed the amputated limb, but following mirror therapy, the functional organization in the somatosensory cortices of both hemispheres became more similar, suggesting a reversal of the cortical changes associated with amputation. However, it is important to consider that maladaptive cortical reorganization may not be a cause of pain, but rather a consequence or a correlate. Behavioral and psychological factors, including catastrophic appraisals of pain, also appear to contribute to the development of phantom pain following amputation (Andoh, Milde, Tsao, & Flor, 2018; Richardson, Glenn, Horgan, & Nurmikko, 2007).

What is VR?

Visual stimulation has now extended from the use of mirrors to virtual reality (VR), which offers a broader array of sensory stimulation. There has not been a singular definition of VR, as the technology with which to deliver VR has rapidly evolved in recent years and continues to do so (Trost, France, Anam, & Shum, 2020). Nonetheless, the goal of VR has remained the same: to simulate an altered experience by some degree of sensory stimulation (Chi, Chau, Yeo, & Ta, 2019; Mallari, Spaeth, Goh, & Boyd, 2019). VR therapies can be deconstructed into several components, including *VR Configuration Factors*, *User Experiential Factors*, and *Pain Targets* (Trost et al., 2020). VR Configuration Factors mostly apply to the technology itself and how real-time data, such as the user's location in space and their movement is incorporated and transposed into a virtual environment that is experienced by the user (Trost et al., 2020). Configuration of the VR, in turn, influences key User Experiential Factors of *interactivity* (the degree to which the user can manipulate the virtual environment), *immersion* and *presence* (the degree to which the user feels present and absorbed in the virtual environment), and *embodiment* (the degree to which the user feels a virtual body to be their own) (Burdea & Coifett, 2003; Matamala-Gomez et al., 2019; Trost et al., 2020). It is suggested that user experiential factors be considered as a continuum of degree rather than a dichotomy (e.g., a VR experience can vary in degree of immersion) as this will depend on the VR equipment used as well as individual factors, such as one's susceptibility to being "absorbed" into the virtual environment (Baños et al., 1999; Richardson et al., 2020). Nonetheless, it is these experiential factors that ultimately mediate changes in one's pain experience, whether that is via the cognitive, emotional, behavioral, or sensory aspects of pain (Trost et al., 2020).

There are also two primary pain targets for using VR to treat pain: (1) distraction, or shifting one's attention from pain, and (2) promoting central nervous system plasticity (Austin & Siddall, 2019; Gupta, Scott, & Dukewich, 2018). Generally, VR therapies for acute and procedural pain have focused on distraction as a primary target, whereas neuroplasticity is a primary target for chronic forms of pain (Trost et al., 2020).

Considering the above components, VR therapy targeting SCI-related neuropathic pain has focused on configuration factors that aid in creating the simulated experience of lower leg movement, such as walking, to reduce the intensity of below and at-level pain (Donati, Shokur, Morya, et al., 2016; Jordan and Richardson, 2016; Moseley, 2007; Özkul,

Kılınç, Yıldırım, Topçuoğlu, & Akyüz, 2015; Richardson et al., 2019; Soler et al., 2010; Soler, Moriña, Kumru, Vidal, & Navarro, 2021; Sumitani et al., 2008; Villiger et al., 2013). The targeted mechanism by which pain is reduced by VR walking or leg movement is functional central nervous system reorganization, by potentially reversing the maladaptive plasticity that is believed to underpin the pain (Austin & Siddall, 2019; Eick & Richardson, 2015; Wrigley et al., 2009). There is some evidence that these forms of therapy activate functionally reorganized cortical regions. In one small pilot study by Eick and Richardson (2015), individuals with SCI (though not necessarily with neuropathic pain) were provided a non-immersive illusion of walking as well as wheeling while undergoing functional magnetic resonance imaging (fMRI). Greater activation was observed in the somatosensory cortex, whereas abled-bodied individuals showed little to no activation in this region (Eick & Richardson, 2015; see Fig. 2).

Non- or partially-immersive virtual walking in SCI

Much of the research on virtual walking to treat SCI-related neuropathic pain has been done using non- or partially immersive modalities, such as having the individual seated in front of 2D or 3D screens to view walking imagery (Austin & Siddall, 2019). This likely reflects the time in which the studies were conducted, as study design and implementation of published results may not have kept pace with the rapidly evolving VR technology, such as cost-effective commercially available VR setups, smaller and more compact equipment, and ease of use with various computer interfaces.

A pilot trial by Sumitani et al. (2008) using mirror therapy in a small mixed sample of individuals with incomplete SCI, peripheral nerve injury, or amputation, resulted in a reduction in pain intensity. Using this paradigm, participants viewed the reflection of an unaffected limb, which resulted in a significant reduction of pain in the affected limb or phantom. However, Moseley (2007) was one of the first to create a full analog to mirror therapy to treat SCI-related neuropathic pain. In this original method, a mirror was used to reflect the participant's upper torso; this mirror was placed above a screen that projected a video of an actor's legs (waist down) walking (see Fig. 3). Thus, individuals observed their upper body coordinated with the video of walking legs. Participants experienced a significant (65%) reduction in pain lasting approximately 30 min following treatment. Soler et al. (2010) replicated Moseley's (2007) method in participants with SCI-related neuropathic pain but complemented this paradigm with transcranial direct current stimulation (tDCS) over the primary motor cortex. Illusory walking alone significantly reduced levels of neuropathic pain, but the pain-reducing benefits were enhanced when the motor cortex was simultaneously stimulated. When considering subtypes of SCI-related neuropathic pain, these studies in combination suggest that those with at-level (Moseley, 2007) and below-level (Soler et al., 2010) experienced pain reduction from virtual walking.

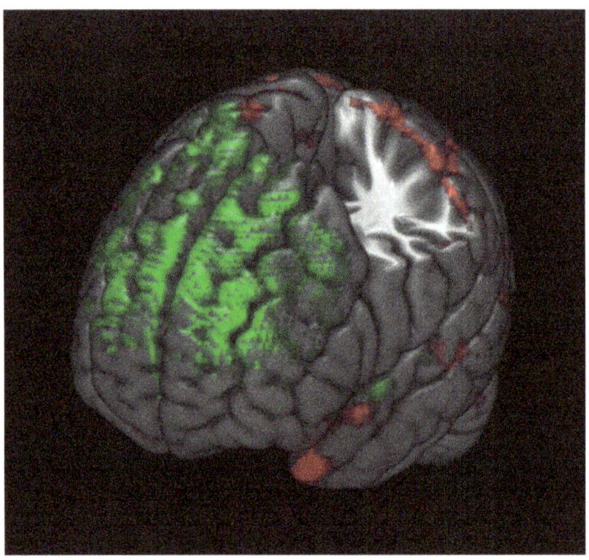

FIG. 2 fMRI cortical activation during virtual walking (non-immersive) among individuals with and without SCI from Eick and Richardson (2015). fMRI brain activity during illusory walking when controlling for brain activation during illusory wheeling. Significant activation was observed in the bilateral somatosensory cortex, right paracentral lobule, and to a lesser degree, medial motor areas among those with SCI *(red)*. Individuals without SCI *(green)* had greater activation in frontal and premotor regions. *(From Eick, J., & Richardson, E. J. (2015). Cortical activation during visual illusory walking in persons with spinal cord injury: A pilot study.* Archives of Physical Medicine and Rehabilitation, 96, *750–753, with permission.)*

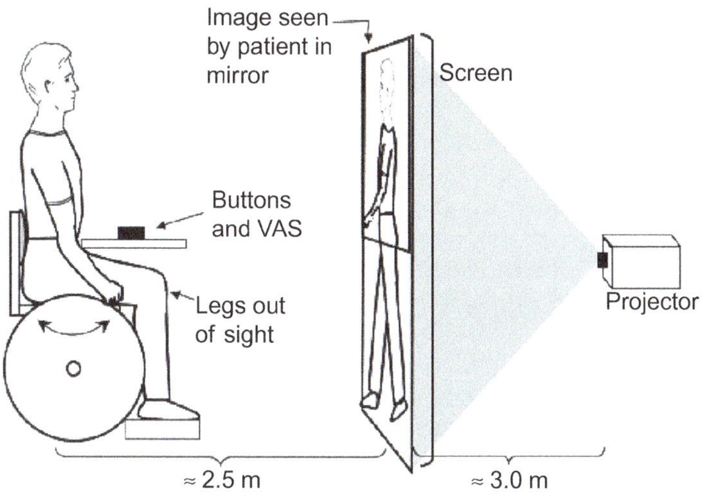

FIG. 3 Virtual walking setup from Moseley (2007). A mirror was used to reflect the participant's upper body and a screen, projecting legs of an actor walking, was placed below the mirror to provide the illusion that the individual's body was walking. Participants' upper body movements were synchronized with the actor's gait. Participants rated pain using a visual analog scale (VAS) by pressing right and left buttons. *(From Moseley, G. L. (2007). Using visual illusion to reduce at-level neuropathic pain in paraplegia.* Pain, 130, *294–298. https://doi.org/10.1016/j.pain.2007.01.007, with permission.)*

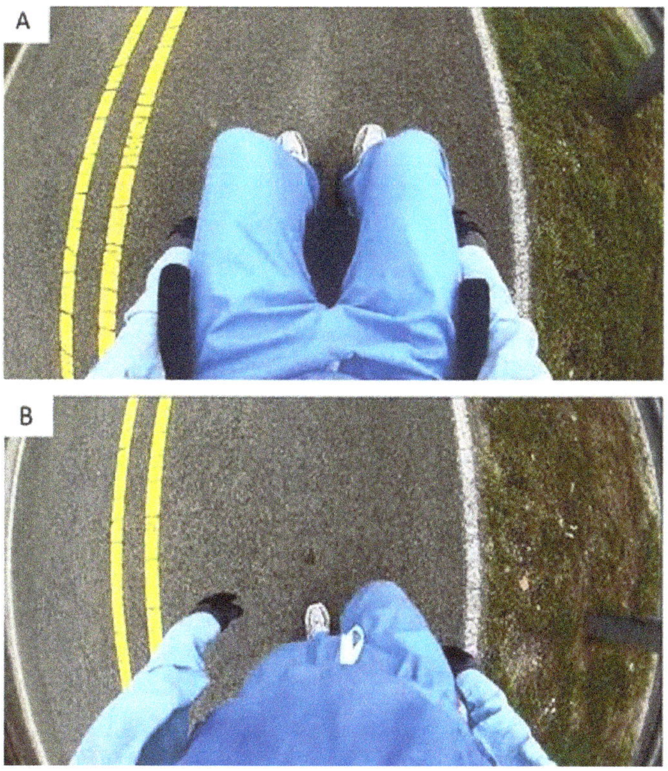

FIG. 4 First-person view in a partially immersive virtual walking protocol by Richardson et al. (2019). (A) is the first-person view of the virtual wheeling condition, and (B) is the virtual walking condition. Video was of an actor walking and was recorded using a stereoscopic (3D) camera, hence the "fish-eye" effect of the 2D image above. *(From Richardson, E. J., McKinley, E. C., Fazlur Rahman, A. K. M., Klebine, P., Redden, D. T., & Richards, J. S. (2019). Effects of virtual walking on spinal cord injury-related neuropathic pain: A randomized, controlled trial.* Rehabilitation Psychology, 64*(1), 13–24. https://doi.org/10.1037/rep0000246, with permission.)*

More recently, we examined the efficacy of a partially immersive virtual walking paradigm to treat SCI-related neuropathic pain in a randomized controlled trial (Richardson et al., 2019). Unlike the studies above, this trial used virtual reality technology as it is understood today. Expanding on the illusory walking paradigm used by Moseley (2007) and Soler et al. (2010), participants with SCI-related neuropathic pain were randomized to watch a 20-min stereoscopic 3D video of a first-person view of an actor's legs walking (walking condition) or seated in a manual wheelchair while propelling down the same path (wheeling condition). Video stills of both conditions are shown in Fig. 4. After one session, participants' levels of pain declined in both the walking and wheeling conditions; however, there was a significant pre- to post-treatment reduction in SCI-pain irrespective of subtype of pain (musculoskeletal, neuropathic, or mixed forms of pain) in the walking condition, but not the wheeling condition. When considering SCI-related neuropathic pain specifically, the walking condition resulted in a significant reduction in the pain unpleasantness rating of their neuropathic pain, but not the intensity of it. Further, SCI-related neuropathic pain that was experienced as "cold," "deep" or associated with increased skin sensitivity was significantly reduced following virtual walking. Quantitative sensory testing to examine at-level neuronal hyperexcitability (evoked sensitivity within 3 dermatomes of their neurologic level of injury) was performed in a subset of participants in this study (Jordan and Richardson, 2016). Declines in SCI-related neuropathic pain were greatest in the virtual walking conditioning, regardless of whether the pain was at or below the level of injury. While it was not statistically significant, there was a trend for the relationship between increased at-level evoked sensitivity and less favorable below-level SCI-related neuropathic pain outcomes after virtual walking treatment (Jordan and Richardson, 2016). Soler et al. (2021) have since replicated their virtual walking and tDCS paradigm, and while they again found a reduction in SCI-related neuropathic pain with the combined treatment, there were no differences with respect to different sensory profiles of SCI-related neuropathic pain. Taken together, these results suggest that the mechanisms potentially underlying subtypes of SCI-related neuropathic pain are not entirely clear. At- and below-level forms of neuropathic pain may have different initial precipitating mechanisms, but both may eventually be maintained by supraspinal changes (Meacham, Shepherd, Mohapatra, & Haroutounian, 2017).

Though the existing studies are sparse and comprised of small sample sizes, non-immersive virtual walking therapies appear to offer some benefit for both at and below-level SCI-related neuropathic pain (Austin & Siddall, 2019; Chi et al., 2019). The larger controlled trial (Richardson et al., 2019), which was not included in Austin and Siddall's (2019) or Chi et al.'s (2019) review, suggest that non-immersive virtual walking paradigms may be effective with only certain aspects of SCI-related neuropathic pain, such as specific sensory-discriminative aspects or the affective-motivational component of this form of pain (Richardson et al., 2019). However, this trial only examined one session of virtual walking, and it is, therefore, unclear if there is a dose-dependent effect on neuropathic pain and whether pain intensity would be significantly reduced with additional sessions. One small, randomized trial of 24 patients found that a single exposure to non-immersive virtual walking that did not use a first-person view significantly reduced pain but the effect did not last after 2 weeks (Özkul et al., 2015). Studies that implemented repeated virtual walking or limb movement sessions across time showed the largest reductions in SCI-related neuropathic pain (Austin & Siddall, 2019).

Immersion, embodiment, and interactivity

As noted, immersion, interactivity, presence, and embodiment are key experiential indices of virtual reality experience (Trost et al., 2020). The illusory or virtual walking interventions described above to a large degree include the passive (non-interactive) observation that incorporates elements of the participant's body via reflection (Moseley, 2007; Soler et al., 2010) or a first-person view (Richardson et al., 2019) to create the illusion of walking outside a fully immersive context. Emerging applications of modern virtual technologies to illusory walking allow us to examine the potential utility of additional experiential factors. For instance, Villiger et al. (2013) were among the first to integrate *interactivity*, albeit in a non-immersive context. Specifically, Villiger et al. (2013) used a virtual reality-augmented paradigm in a sample of 14 participants with incomplete SCI who also had some motor sparing. In this paradigm, participant's real legs and feet were mapped in real-time to 3D virtual lower limb displayed, in the first-person perspective, on a 2D monitor (see Fig. 5). Participants were required to engage in various tasks that incorporated elements of gaming, such as mastering various levels of difficulty or earning point values for completing specific tasks. VR sessions were 45-min in duration, and sessions were repeated 16–20 times within 4 weeks. Of the 9 participants who experienced SCI-related pain (at- or below-level), 6 experienced clinically meaningful reductions in pain, which persisted 12–16 weeks later (Villiger et al., 2013). This suggests that including goal-directed, gaming-like tasks may increase user interactivity, which in turn improves effectiveness. Volitional interactive use similarly serves as a key component in exergaming to improve physical functioning among those with a physical disability (Mat Rosly, Mat Rosly, Davis Oam, Husain, & Hasnan, 2017).

Although not strictly a virtual reality intervention, an investigation by Pozeg et al. (2017) focused on the effects of illusory embodiment. These researchers examined the effects of inducing a virtual leg illusion or full-body illusion using an HMD in a sample of 20 individuals (11 of whom had at or below-level neuropathic pain). Rather than creating illusory movement, using the HMD, these researchers allowed participants to view a stationary dummy to which tactile stimulation was provided. Using an embodiment paradigm traditionally applied in rubber hand/body research (Christ & Reiner, 2014), participants received visuotactile stimulation to their actual back as well as dummy legs, which were observed from a first-person perspective on the HMD. To create a full-body illusion, Pozeg et al. (2017) had the participants view, in real-time, video of their own back receiving tactile stimulation through the HMD as they were experiencing it. For both conditions,

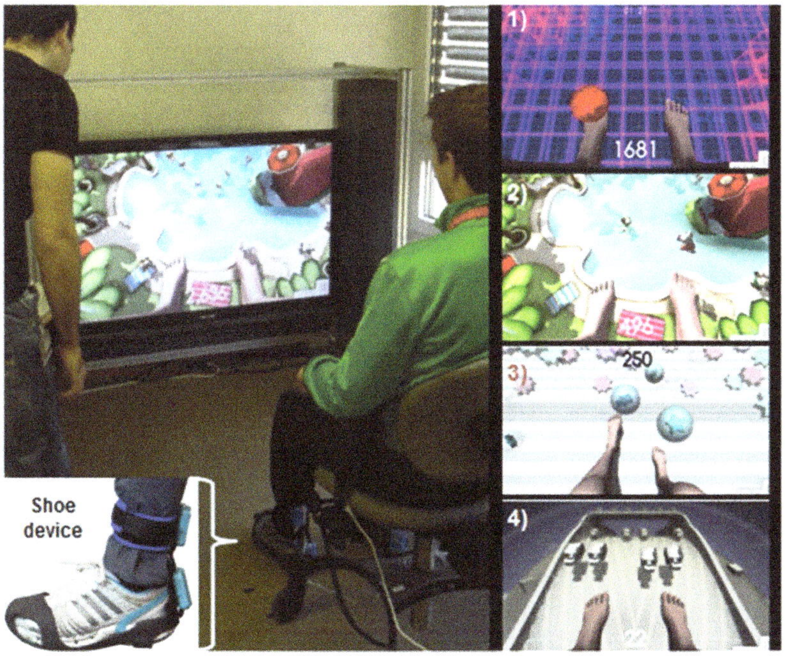

FIG. 5 First-person view VR lower limb movement protocol by Villiger et al. (2013). Incorporation of participant movement in a virtual environment among individuals with incomplete SCI using movement sensors fitted to participants' feet. Goal-directed and motivated interaction with the virtual environment was also used. *(From Villiger, M., Bohli, D., Kiper, D., Pyk, P., Spillmann, J., Meilick, B., et al. (2013). Virtual reality-augmented neurorehabilitation improves motor function and reduces neuropathic pain in patients with incomplete spinal cord injury.* Neurorehabilitation and Neural Repair, *27(8), 675–683. https://doi.org/10.1177/1545968313490999, with permission.)*

tactile and visual stimulation occurred synchronously or asynchronously. Synchronous visuotactile stimulation in either illusory condition created a sense of embodiment, and even without movement, this appeared to offer at least modest relief with SCI-neuropathic pain following a single exposure (Pozeg et al., 2017).

To date, participants in non-immersive VR environments have not been fully disconnected from stimuli within the real world, such as surrounding vision or sound (Chi et al., 2019). Conversely, more immersive virtual reality methods may reduce or attempt to eliminate stimuli from the real world by integrating full visual and tactile stimulation or by allowing the participant to have motor control of the virtual environment (Trost et al., 2020). Greater immersion increases the potency of virtual walking to where single interventions may become more clinically effective (Austin & Siddall, 2019). Perhaps one of the more striking studies to date that addressed immersion and multiple experiential factors is that of Donati et al. (2016). These researchers investigated the potential for locomotion recovery in 8 individuals with paraplegia using brain-machine interface-based gait neurorehabilitation over 12 months. Their intervention made use of a fully immersive virtual environment using an HMD, brain-controlled 3D avatar, and robotic gain training, as well as the integration of visual and tactile feedback. The participants in the study described pain experienced below the level of their injury, and while the pain was not a primary outcome of this study, participants demonstrated improved ability to localize their pain and reported a decrease in day-to-day pain intensity outside of sessions (Donati et al., 2016). Functional changes (assessed by electroencephalography) in leg representation areas of the primary sensorimotor cortices following the intervention were also observed (Donati et al., 2016).

A recent study by Trost and colleagues (Anam et al., 2019) is of particular note, as it is the first study to leverage the interactive, immersive, and embodiment characteristics of a portable virtual reality interface to provide individuals with SCI-related neuropathic pain volitional control over their virtual legs. Participants' virtual ambulation within a 360-degree virtual environment was self-guided by hand-held controllers; participants moved their arms using gestures analogous to natural gait which translated into leg movement in the virtual world (see Fig. 6). Participants matched their avatar to personal characteristics and viewed their virtual arms and legs from a first-person perspective through an HMD. The "VRWalk" intervention incorporated exploratory gaming elements to promote interactivity and motivation. The study included 18 individuals with complete paraplegia, who completed twice daily 5-min VR sessions in their own homes for a total of 20 sessions. Participants reported strong enjoyment and embodiment effects ("I felt like the legs were my own") following the intervention. The investigators observed statistically and clinically significant decrement in neuropathic pain intensity ratings from prior to following individual sessions, as well as the full intervention. Similar improvements in mood and affect were also observed. Improved pain ratings persisted at 2 weeks and 1 month following intervention. The results of this study tentatively support the value of strong immersion/presence, embodiment, and

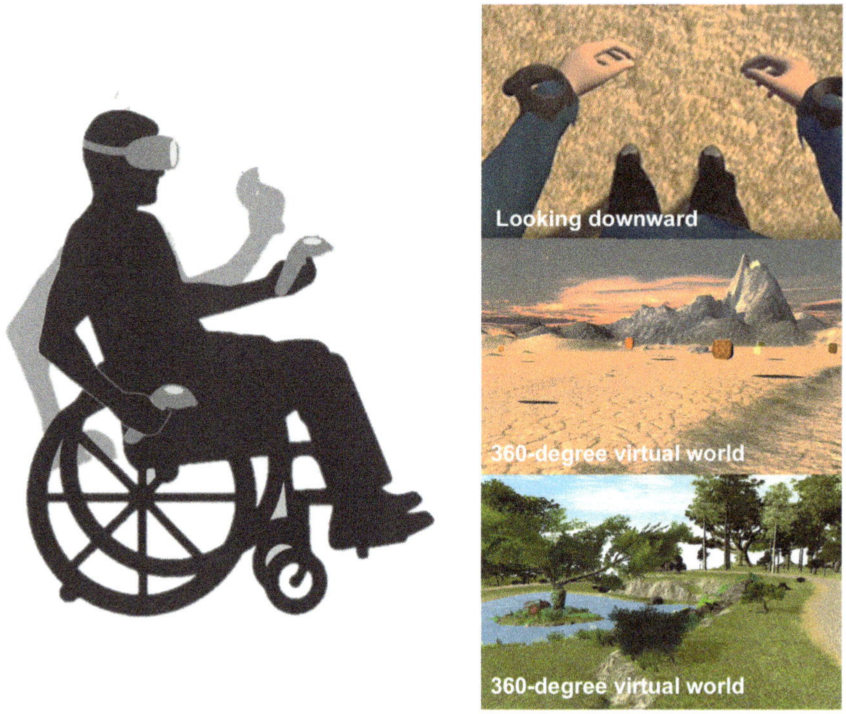

FIG. 6 The VRWalk intervention described in Anam et al. (2019). Participants were immersed in a virtual scene via HMD with a 360-degree, first-person view. Arm movements to simulate walking and motion sensing controllers generated a virtual gait for the avatar. *(Unpublished image by the author (Trost).)*

motivation in SCI neuropathic pain intervention, and perhaps most significantly, suggest the acceptability and generalizability of a self-guided VR intervention to home use.

Applications to other areas of neuroscience

Just over a third of existing studies examining the application of VR for pain have been done on those with chronic pain, and within that category only approximately 8% have been conducted on individuals experiencing SCI-related chronic pain (Trost et al., 2020). Clearly, there is a need for more rigorously controlled and comparative studies, and there remains much to be understood regarding the most effective VR configurations, their dose, and which user experiential factors are most potent in producing an analgesic effect. However, there is reason to believe that VR applications that incorporate interactive, goal-directed walking rather than passive visual input have a greater potential of targeting neuroplastic changes that may underlie deafferented neuropathic pain (Foell et al., 2014). For example, individuals with upper limb amputation showed significantly increased activation in the primary motor and somatosensory cortices performing a movement task with the intact hand that displayed the phantom moving within a virtual space (Andoh et al., 2020). Tasks with functional value or volitional interactive use versus simply imagining a movement are also associated with increased activity in the primary somatosensory cortex (Raffin, Mattout, Reilly, & Giraux, 2012). As suggested by more recent studies, the use of multisensory integration (e.g., sound, haptic, and visual stimuli) can increase the user's sense of embodiment of the virtual legs or body, thus increasing the pain-alleviating effects of VR (Matamala-Gomez et al., 2019).

It is also important to note that the experience of pain, regardless of its underlying mechanism, is best understood as having physiological, cognitive, social, and emotional components (Melzack, 1999). While much of the work focuses on the central nervous system changes associated with SCI-related neuropathic pain, individuals with SCI may also experience pain-related fear-conditioned from interoceptive, proprioceptive, or external stimuli (Vlaeyen & Linton, 2012). Thus, VR configurations may combine immersion, interactivity, and gaming in a virtual environment that exposes that individual to feared movements, positions, or contexts or distracts from negative pain-related cognitions (Freitas & Spadoni, 2019). Such an application may work through enhancing activity within the brain's reward circuit, which tends to override pain processing (Becker, Gandhi, Pomares, Wager, & Schweinhardt, 2017). Specifically, Becker et al. (2017) observed that PFC was involved in inhibiting other cortical regions that were involved in establishing the relative threat value of pain.

Thus, experiencing a reward, or even the expectation of reward, may diminish the drive to avoid pain to continue pursuit of the desired reward.

In a related vein, individuals with SCI may face several environmental barriers that interfere with social participation (Tsai et al., 2017), yet greater positive social interaction is associated with better functioning among individuals with physical disabilities (Jensen, Moore, Bockow, Ehde, & Engel, 2011) and lower risk for developing phantom pain following amputation (Hanley et al., 2004). Interpersonal interactions may reduce pain similarly through the brain's reward pathway, via dopamine systems (Younger, Aron, Parke, Chatterjee, & Mackey, 2010). Thus, VR configurations that incorporate gaming environments with social elements may best target the social and affective components of the pain experience for those with SCI.

With the increasing availability of commercially available VR computer interfaces and hardware in recent years, it is expected that relevant lines of research will broaden to encompass identifying the VR configuration and user experiential factors that are the strongest mediators of cortical reorganization in SCI-related neuropathic pain. Moreover, user experiential factors associated with VR walking modalities can be altered to address cognitive, emotional, and social factors in order to comprehensively target a wider range of pain outcomes in those with SCI-related pain.

Mini-dictionary of terms

Autonomic dysreflexia: Bouts of abrupt increases in blood pressure due to disruptions in autonomic pathways.
Avatar: A representation of one's body, in part or whole, in a virtual environment.
Deafferentation: The state in which afferent neural fibers no longer transmit information to receiving regions, typically discussed as disrupted peripheral input to the brain.
Exergaming: The use of gaming or other motivating tasks in exercise.
Functional reorganization: Changes in neural activity in certain regions of the brain due to changes in input, environmental stimulation, or behavior of the organism.
Interoceptive: A type of stimulus that comes from within the body.
Neuropathic pain: Pain arising from peripheral or central nervous system damage.
Mirror therapy: A therapeutic modality that provides the visual illusion, through the use of mirrors, that an area of the body is intact and functioning normally.
Supraspinal: Regions of the central nervous system above the spinal cord.
Virtual reality: A general modality that alters one's perception of the environment, themselves, or both in various degrees.

Key facts of virtual walking and spinal cord injury neuropathic pain

Key facts of SCI neuropathic pain

- Neuropathic pain following SCI is persistent and often refractory to currently available treatments.
- SCI-related neuropathic pain is associated with interference in life activities and a decline in quality of life.
- SCI-related neuropathic pain is associated with functional cortical reorganization, particularly in somatosensory regions.
- These neurological changes have led to conceptualizing SCI-related neuropathic pain as a deafferentation pain, similar to phantom limb phenomena.
- Several lines of research have examined ways to adapt visual stimulation forms of therapy used for a phantom limb to spinal cord injury.

Key facts of VR use for SCI neuropathic pain

- Virtual reality (VR) is a modality that alters an individual's sensory experience as it pertains to the sense of self or one's environment.
- VR can vary in degrees of immersiveness, presence, interactivity, and embodiment.
- VR that simulates walking has shown promise with reducing levels of neuropathic pain in SCI.
- VR that creates embodiment or simulates other types of lower limb movement is also associated with reduced pain.
- VR modalities that incorporate multisensory input and goal-directed activity, such as gaming, may increase treatment effect.
- VR may also target other biopsychosocial aspects of SCI-related pain.

Summary points

- SCI-related neuropathic pain is difficult to treat with traditional pharmacological therapies.
- It is experienced at or below the neurological level of injury.
- SCI-related neuropathic pain is associated with maladaptive neuroplastic changes in supraspinal regions of the central nervous system.
- It can be described as a deafferentation pain, similar to post-amputation phantom limb pain.
- Visual illusory modalities incorporating movement of the lower extremities to treat phantom limb pain have been extended to SCI-related neuropathic pain.
- Less immersive virtual walking treatments have shown some benefits in reducing pain.
- Using currently available technology to create more immersive, multisensory virtual reality (VR) walking modalities will likely offer greater benefit.

References

Anam, M., Sizemore, K., Mansour, H., Seward, J., Mitchell, L., Shum, C., et al. (2019). Virtual reality walking for neuropathic pain in spinal cord injury: Preliminary efficacy findings. *The Journal of Pain*, *20*(4, Suppl), S51.

Andoh, J., Milde, C., Diers, M., Bekrater-Bodmann, R., Trojan, J., Fuchs, X., et al. (2020). Assessment of cortical reorganization and preserved function in phantom limb pain: A methodological perspective. *Scientific Reports*, *10*, 11504.

Andoh, J., Milde, C., Tsao, J. W., & Flor, H. (2018). Cortical plasticity as a basis of phantom limb pain: Fact or fiction? *Neuroscience*, *387*, 85–91. https://doi.org/10.1016/j.neuroscience.2017.11.015.

Austin, P. D., & Siddall, P. J. (2019). Virtual reality for the treatment of neuropathic pain in people with spinal cord injuries: A scoping review. *The Journal of Spinal Cord Medicine*. https://doi.org/10.1080/10790268.2019.1575554.

Baños, R., Botella, C., García-Palacios, A., Villa, H., Perpiñá, C., & Gallardo, M. (1999). Psychological variables and reality judgment in virtual environments: The roles of absorption and dissociation. *Cyberpsychology & Behavior*, *2*, 143–148. https://doi.org/10.1089/cpb.1999.2.143.

Becker, S., Gandhi, W., Pomares, F., Wager, T. D., & Schweinhardt, P. (2017). Orbitofrontal cortex mediates pain inhibition by monetary reward. *Social Cognitive and Affective Neuroscience*, *12*, 651–661. https://doi.org/10.1093/scan/nsw173.

Bonica, J. J. (1991). Introduction: Semantic, epidemiologic, and educational issues. In K. L. Casey (Ed.), *Pain and central nervous system disease: The central pain syndromes* (pp. 13–29). New York, NY: Raven Press.

Bowsher, D. (1996). Central pain: Clinical and physiological characteristics. *Journal of Neurology, Neurosurgery, and Psychiatry*, *61*, 62Y9.

Bryce, T. N., Biering-Sørensen, F., Finnerup, N. B., Cardenas, D. D., Defrin, R., Lundeberg, T., et al. (2012). International spinal cord injury pain classification: Part I. Background and description. *Spinal Cord*, *50*, 413–417.

Bryce, T. N., Budh, C. N., Cardenas, D. D., Dijkers, M., Felix, E. R., Finnerup, N. B., et al. (2007). Pain after spinal cord injury: An evidence-based review for clinical practice and research. *Journal of Spinal Cord Medicine*, *30*, 421–440.

Burdea, G. C., & Coifett, P. (2003). *Virtual reality technology* (2nd ed.). Hoboken, NJ: John Wiley & Sons. Available from https://www.wiley.com/en-us/Virtual+Reality+Technology%2C+2nd+Edition-p-9780471360896 (Accessed 21 January 2020).

Burke, D., Fullen, B. M., Stokes, D., & Lennon, O. (2017). Neuropathic pain prevalence following spinal cord injury: A systematic review and meta-analysis. *European Journal of Pain*, *21*(1), 29–44. https://doi.org/10.1002/ejp.905.

Burke, D., Lennon, O., & Fullen, B. M. (2018). Quality of life after spinal cord injury: The impact of pain. *European Journal of Pain*, *22*(9), 1662–1672. https://doi.org/10.1002/ejp.1248.

Cardenas, D. D., & Jensen, M. P. (2006). Treatments for chronic pain in persons with spinal cord injury: A survey study. *Journal of Spinal Cord Medicine*, *29*, 109–117.

Chan, B. L., Witt, R., Charrow, A. P., Magee, A., Howard, R., Pasquina, P. F., et al. (2007). Mirror therapy for phantom limb pain. *New England Journal of Medicine*, *357*, 2206–2207.

Chi, B., Chau, B., Yeo, E., & Ta, P. (2019). Virtual reality for spinal cord injury-associated neuropathic pain: Systematic review. *Annals of Physical Medicine & Rehabilitation*, *62*, 49–57. https://doi.org/10.1016/j.rehab.2018.09.006.

Christ, O., & Reiner, M. (2014). Perspectives and possible applications of the rubber hand and virtual hand illusion in non-invasive rehabilitation: Technological improvements and their consequences. *Neuroscience and Biobehavioral Reviews*, *44*, 33–44.

Donati, A., Shokur, S., Morya, E., et al. (2016). Long-term training with a brain-machine interface-based gait protocol induces partial neurological recovery in paraplegic patients. *Scientific Reports*, *6*, 30383. https://doi.org/10.1038/srep30383.

Eick, J., & Richardson, E. J. (2015). Cortical activation during visual illusory walking in persons with spinal cord injury: A pilot study. *Archives of Physical Medicine and Rehabilitation*, *96*, 750–753.

Ephraim, P. L., Wegener, S. T., MacKenzie, E. J., Dillingham, T. R., & Pezzin, L. E. (2005). Phantom pain, residual limb pain, and back pain in amputees: Results of a national survey. *Archives of Physical Medicine and Rehabilitation*, *86*, 1910–1919. https://doi.org/10.1016/j.apmr.2005.03.031.

Finnerup, N. B. (2013). Pain in patients with spinal cord injury. *Pain*, *154*(Suppl. 1), S71–S76.

Finnerup, N. B., Johannesen, I. L., Fuglsang-Frederiksen, A., et al. (2003). Sensory function in spinal cord injury patients with and without central pain. *Brain*, *126*, 57Y70.

Flor, H., Nikolajsen, L., & Jensen, T. S. (2006). Phantom limb pain: A case of maladaptive CNS plasticity? *Nature Reviews Neuroscience*, *7*, 873–881. https://doi.org/10.1038/nrn1991.

Foell, J., Bekrater-Bodmann, R., Diers, M., & Flor, H. (2014). Mirror therapy for phantom limb pain: Brain changes and the role of body representation. *European Journal of Pain*, *18*, 729–739. https://doi.org/10.1002/j.1532-2149.2013.00433.x.

Freitas, D. M., & Spadoni, V. S. (2019). Is virtual reality useful for pain management in patients who undergo medical procedures? *Einstein (Sao Paulo)*, *17* (2), eMD4837. https://doi.org/10.31744/einstein_journal/2019MD4837.

Gupta, A., Scott, K., & Dukewich, M. (2018). Innovative technology using virtual reality in the treatment of pain: Does it reduce pain via distraction, or is there more to it? *Pain Medicine*, *19*(1), 151–159.

Gustin, S. M., Wrigley, P. J., Siddall, P. J., & Henderson, L. A. (2010). Brain anatomy changes associated with persistent neuropathic pain following spinal cord injury. *Cerebral Cortex*, *20*, 1409–1419.

Hanley, M. A., Jensen, M. P., Ehde, D. M., Hoffman, A. J., Patterson, D. R., & Robinson, L. R. (2004). Psychosocial predictors of long-term adjustment to lower-limb amputation and phantom limb pain. *Disability and Rehabilitation*, *26*, 882–893.

Harris, A. J. (1999). Cortical origin of pathological pain. *Lancet*, *354*, 1464–1466.

Jensen, M. P., Kuehn, C. M., Amtmann, D., & Cardenas, D. D. (2007). Symptom burden in persons with spinal cord injury. *Archives of Physical Medicine and Rehabilitation*, *88*, 638–645.

Jensen, M. P., Moore, M. R., Bockow, T. B., Ehde, D. M., & Engel, J. M. (2011). Psychosocial factors and adjustment to chronic pain in persons with physical disabilities: A systematic review. *Archives of Physical Medicine and Rehabilitation*, *92*, 146–160.

Jordan, M., & Richardson, E. J. (2016). Effects of virtual walking treatment on spinal cord injury-related neuropathic pain: Pilot results and trends related to location of pain and at-level neuronal hypersensitivity. *American Journal of Physical Medicine & Rehabilitation*, *95*(5), 390–396. https://doi.org/10.1097/PHM.0000000000000417.

Kirshblum, S., & Lin, V. W. (Eds.). (2019). *Spinal cord medicine*. New York: Demos Medical Publishing.

Mallari, B., Spaeth, E. K., Goh, H., & Boyd, B. S. (2019). Virtual reality as an analgesic for acute and chronic pain in adults: A systematic review and meta-analysis. *Journal of Pain Research*, *12*, 2053–2085.

Mat Rosly, M., Mat Rosly, H., Davis Oam, G. M., Husain, R., & Hasnan, N. (2017). Exergaming for individuals with neurological disability: A systematic review. *Disability and Rehabilitation*, *39*(8), 727–735. https://doi.org/10.3109/09638288.2016.1161086.

Matamala-Gomez, M., Donegan, T., Bottiroli, S., Sandrini, G., Sanchez-Vives, M. V., & Tassorelli, C. (2019). Immersive virtual reality and virtual embodiment for pain relief. *Frontiers in Human Neuroscience*, *13*, 279. https://doi.org/10.3389/fnhum.2019.00279.

Meacham, K., Shepherd, A., Mohapatra, D. P., & Haroutounian, S. (2017). Neuropathic pain: Central vs. peripheral mechanisms. *Current Pain and Headache Reports*, *21*(6), 28. https://doi.org/10.1007/s11916-017-0629-5.

Melzack, R. (1999). From the gate to the neuromatrix. *Pain, Suppl. 6*, S121–S126.

Moseley, G. L. (2005). Distorted body image in complex regional pain syndrome. *Neurology*, *65*, 773. https://doi.org/10.1212/01.wnl.0000174515.07205.11.

Moseley, G. L. (2007). Using visual illusion to reduce at-level neuropathic pain in paraplegia. *Pain*, *130*, 294–298. https://doi.org/10.1016/j.pain.2007.01.007.

Özkul, Ç., Kılınç, M., Yıldırım, S. A., Topçuoğlu, E. Y., & Akyüz, M. (2015). Effects of visual illusion and transcutaneous electrical nerve stimulation on neuropathic pain in patients with spinal cord injury: A randomised controlled crossover trial. *Journal of Back and Musculoskeletal Rehabilitation*, *28*, 709–719. https://doi.org/10.3233/BMR-140573.

Pozeg, P., Palluel, E., Ronchi, R., Solcà, M., Al-Khodairy, A. W., Jordan, X., et al. (2017). Virtual reality improves embodiment and neuropathic pain caused by spinal cord injury. *Neurology*, *89*, 1894–1903. https://doi.org/10.1212/WNL.0000000000004585.

Raffin, E., Mattout, J., Reilly, K. T., & Giraux, P. (2012). Disentangling motor execution from motor imagery with the phantom limb. *Brain*, *135*, 582–595. https://doi.org/10.1093/brain/awr337.

Ramachandran, V. S., & Altschuler, E. L. (2009). The use of visual feedback, in particular mirror visual feedback, in restoring brain function. *Brain*, *132*, 1693–1710.

Ramachandran, V. S., & Hirstein, W. (1998). The perception of phantom limbs. *Brain*, *121*, 1603–1630.

Richardson, E. J., Deutsch, G., Deshpande, H. D., & Richards, J. S. (2020). Differences in resting cerebellar and prefrontal cortical blood flow in spinal cord injury-related neuropathic pain: A brief report. *The Journal of Spinal Cord Medicine*. https://doi.org/10.1080/10790268.2020.1786321.

Richardson, C., Glenn, S., Horgan, M., & Nurmikko, T. (2007). A prospective study of factors associated with the presence of phantom limb pain six months after major lower limb amputation in patients with peripheral vascular disease. *The Journal of Pain*, *8*, 793–801.

Richardson, E. J., McKinley, E. C., Fazlur Rahman, A. K. M., Klebine, P., Redden, D. T., & Richards, J. S. (2019). Effects of virtual walking on spinal cord injury-related neuropathic pain: A randomized, controlled trial. *Rehabilitation Psychology*, *64*(1), 13–24. https://doi.org/10.1037/rep0000246.

Richardson, E. J., & Redden, D. T. (2014). Moving towards multiple site outcomes in spinal cord injury pain clinical trials: An issue of clustered observations in trial design and analysis. *Journal of Spinal Cord Medicine*, *37*(3), 278–287. https://doi.org/10.1179/2045772313Y.0000000165.

Richardson, E. J., Richards, J. S., & Boyer, B. A. (2008). Spinal cord injury. In B. A. Boyer, & M. I. Paharia (Eds.), *Comprehensive handbook of clinical health psychology*. Hoboken, NJ: Wiley.

Sayegh, S. A., Filén, T., Johansson, M., Sandström, S., Stiewe, G., & Butler, S. (2013). Mirror therapy for complex regional pain syndrome (CRPS): A literature review and an illustrative case report. *Scandinavian Journal of Pain*, *4*(4), 200–207. https://doi.org/10.1016/j.sjpain.2013.06.002.

Siddall, P. J. (2009). Management of neuropathic pain following spinal cord injury: Now and in the future. *Spinal Cord*, *47*, 352–359.

Siddall, P. J., & Loeser, J. D. (2001). Pain following spinal cord injury. *Spinal Cord*, *39*(2), 63–73. https://doi.org/10.1038/sj.sc.3101116.

Siddall, P. J., McClelland, J. M., Rutkowski, S. B., & Cousins, M. J. (2003). A longitudinal study of the prevalence and characteristics of pain in the first 5 years following spinal cord injury. *Pain, 103*, 249–257.

Siddall, P. J., Taylor, D. A., & Cousins, M. J. (1997). Classification of pain following spinal cord injury. *Spinal Cord, 35*, 69Y75.

Siddall, P. J., Taylor, D. A., McClelland, J. M., Rutkowski, S. B., & Cousins, M. J. (1999). Pain report and the relationship of pain to physical factors in the first 6 months following spinal cord injury. *Pain, 81*, 187–197.

Soler, M. D., Kumru, H., Pelayo, R., Vidal, J., Tormos, J. M., Fregni, F., et al. (2010). Effectiveness of transcranial direct current stimulation and visual illusion on neuropathic pain in spinal cord injury. *Brain, 133*, 2565–2577.

Soler, D., Moriña, D., Kumru, H., Vidal, J., & Navarro, X. (2021). Transcranial direct current stimulation and visual illusion effect according to sensory phenotypes in patients with spinal cord injury and neuropathic pain. *The Journal of Pain, 22*, 86–96. https://doi.org/10.1016/j.jpain.2020.06.004.

Sumitani, M., Miyauchi, S., McCabe, C. S., Shibata, M., Maeda, L., Saitoh, Y., et al. (2008). Mirror visual feedback alleviates deafferentation pain, depending on qualitative aspects of the pain: A preliminary report. *Rheumatology, 47*, 1038–1043. https://doi.org/10.1093/rheumatology/ken170.

Teasell, R. W., Mehta, S., Aubut, J. L., Foulon, B., Wolfe, D. L., Hsieh, J. T. C., et al. (2010). A systematic review of pharmacological treatments of pain following spinal cord injury. *Archives of Physical Medicine and Rehabilitation, 91*, 816–831. https://doi.org/10.1016/j.apmr.2010.01.022.

Trost, Z., France, C. R., Anam, M., & Shum, C. (2020). Virtual reality approaches to pain: Toward a state of the science. *Pain*. https://doi.org/10.1097/j.pain.0000000000002060.

Tsai, I. H., Graves, D. E., Chan, W., Darkoh, C., Lee, M. S., & Pompeii, L. A. (2017). Environmental barriers and social participation in individuals with spinal cord injury. *Rehabilitation Psychology, 62*, 36–44. https://doi.org/10.1037/rep0000117.

Tsakiris, M. (2017). The multisensory basis of the self: From body to identity to others. *Quarterly Journal of Experimental Psychology, 70*, 597–609. https://doi.org/10.1080/17470218.2016.1181768.

Villiger, M., Bohli, D., Kiper, D., Pyk, P., Spillmann, J., Meilick, B., et al. (2013). Virtual reality-augmented neurorehabilitation improves motor function and reduces neuropathic pain in patients with incomplete spinal cord injury. *Neurorehabilitation and Neural Repair, 27*(8), 675–683. https://doi.org/10.1177/1545968313490999.

Vlaeyen, J. W., & Linton, S. J. (2012). Fear-avoidance model of chronic musculoskeletal pain: 12 years on. *Pain, 153*, 1144–1147. https://doi.org/10.1016/j.pain.2011.12.009.

Wrigley, P. J., Press, S. R., Gustin, S. M., et al. (2009). Neuropathic pain and primary somatosensory cortex reorganization following spinal cord injury. *Pain, 141*, 52–59.

Yague, J. G., Foffani, G., & Aguilar, J. (2011). Cortical hyperexcitability in response to preserved spinothalamic inputs immediately after spinal cord hemisection. *Experimental Neurology, 227*, 252–263.

Younger, J., Aron, A., Parke, S., Chatterjee, N., & Mackey, S. (2010). Viewing pictures of a romantic partner reduces experimental pain: Involvement of neural reward systems. *PLoS One, 5*(10), e13309. https://doi.org/10.1371/journal.pone.0013309.

Chapter 35

Cervical spinal cord injury and thermoregulatory processes: A new narrative

Charbel Moussalem[a], Louna Ftouni[b], Farah Mneimneh[c], Shadi Bsat[a], Mohamad Nabih El Houshiemy[a], Sarah Kawtharani[a], Adham Halaoui[a], Safwan Al Omari[a], Rana Sarieddine[a], Firas Kobeissy[d], and Ibrahim Omeis[a,e]

[a]*Department of Neurosurgery, American University of Beirut Medical Center, Beirut, Lebanon,* [b]*Faculty of Medicine, Beirut Arab University, Beirut, Lebanon,* [c]*Department of Engineering and Architecture, American University of Beirut, Beirut, Lebanon,* [d]*Department of Biochemistry and Molecular Genetics, Faculty of Medicine, American University of Beirut, Beirut, Lebanon,* [e]*Department of Neurosurgery, Baylor College of Medicine, Houston, TX, United States*

List of abbreviations

AEE	active energy expenditure
ANS	autonomic nervous system
ASIA	American Spinal Injury Association
BSCB	blood-spinal cord barrier
cSCI	cervical spinal cord injury
MnPO	median preoptic nucleus
NSCISC	National Spinal Cord Injury Statistical Center
REE	resting energy expenditure
SCI	spinal cord injury
SNS	sympathetic nervous system
VO2	oxygen consumption

Introduction

Cervical spinal cord injury (cSCI) is an insult to the cervical region of the spinal cord that can terminate in tetraplegia and even death, depending on the site of injury. The higher the level of injury in the cervical spine, the worse the outcome is. Although uncommon, complete neurological recovery may occur after such insults. According to the National Spinal Cord Injury Statistical Center (NSCISC), there are 17,810 new cases of SCI yearly, with a total approaching 300,000 cases in the United States. Around 80% of the cases are males and the most affected race/ethnicity is the non-Hispanic white population. All age groups are susceptible to the event; however, people in their 40s have the highest tendency to suffer from SCI (National Spinal Cord Injury Statistical Center, 2020). The causes of SCI are many, but motor vehicle accidents have been in the lead for the past 40 years, followed by falls, crimes and violence, sports injuries, medical and surgical complications, among others (Chen, Tang, Vogel, & Devivo, 2013; National Spinal Cord Injury Statistical Center, 2020).

Regardless of the spinal cord injury level, SCI can be classified as either complete or incomplete. According to the American Spinal Injury Association (ASIA), a complete injury refers to the total loss of the sensory and motor functions below the injury level. Whereas, in an incomplete injury, there is the maintenance of certain sensory and/or motor functions below the injury level. SCI can be categorized into 5 grades, according to its severity, as follows:

ASIA A: Complete loss of both motor and sensory functions below the injury level.
ASIA B: Preservation of certain sensory functions with loss of all motor functions below the injury level.
ASIA C: Preservation of certain motor functions below the injury level; more than half of the key muscles have a power grade less than 3.

Cellular, Molecular, Physiological, and Behavioral Aspects of Spinal Cord Injury. https://doi.org/10.1016/B978-0-12-822427-4.00035-6

ASIA D: Preservation of motor functions below the injury level; at least half of the key muscles have a power grade greater than or equal to 3.

ASIA E: All sensory and motor functions are intact (American Spinal Cord Injury Association, 2019; Martinez-Perez, Cepeda, Paredes, Alen, & Lagares, 2017).

Furthermore, cSCI does not impact motor and sensory functions only; there can be loss of bladder and bowel control, loss of reflexes, sexual dysfunction, respiratory complications, thromboembolic events, spasticity, pain, and physiological and psychological disorders (depression and suicide), depending on the injury level and type (Abrams & Wakasa, 2019). Physiological disturbances are numerous, the most important of which are the disorders caused by the dysfunction of the autonomic nervous system (ANS). The loss of the sympathetic nervous system control and the compensatory attempt by the parasympathetic one may result in bradycardia, hypotension, thermoregulatory disturbances, skin blood flow and microcirculatory system dysfunction, and a fatal complication known as autonomic dysreflexia, among others (Karlsson, 2006; Wecht et al., 2015).

This chapter aims to discuss the thermoregulatory disturbances that develop in individuals with cSCI. The principle of homeostasis and the process of thermoregulation in able-bodied individuals are first introduced for a better understanding of the thermoregulatory changes that occur in cSCI (Figs. 1–4).

Homeostasis

Homeostasis is the ability of an organism to maintain a well-balanced internal environment under varying stress levels. There are a plethora of internal processes finely tuned by homeostasis, among which are the control of the arterial blood pressure and the ionic concentration, the regulation of oxygen and carbon dioxide exchange, the management of the waste products, and the adjustment of the core body temperature, among others (Chovatiya & Medzhitov, 2014; Pear, 2012).

Thermoregulatory process in able-bodied individuals

One of the fundamental homeostatic mechanisms in the body is the regulation of the core body temperature, i.e., the temperature of the internal organs and deeper tissues. Under normal conditions, the core body temperature is 37°C ± 0.5°C, despite any fluctuations in the ambient temperature. A slight deviation from this range may damage the function of the vital organs and disrupt homeostasis (Hall, 2015). On the other hand, the skin temperature is more flexible and adapts to a wider range of ambient temperatures. It is an important factor in aiding the body to acquire or eliminate excess heat (Tansey & Johnson, 2015). The nervous system, principally the hypothalamus, controls the core body temperature. The median preoptic nucleus (MnPO) situated in the anterior hypothalamus is central in the thermoregulatory processes. Inside the MnPO nucleus, heat and cold-sensitive neurons act as sensors for the body temperature. To note that, there exist peripheral thermal receptors that are sensitive to either cold or heat (Mrowka & Reuter, 2016). When the MnPO senses excessive heat, it sends different signals: a signal to terminate thermogenesis, and another one to the cutaneous blood vessels to dilate and induce perspiration, all of which contribute to the body's coolness and restoration of its balance. On the contrary, when the cold sensors detect a drop in temperature, the MnPO sends a stimulus to activate thermogenesis and another one to the skin to constrict its blood vessels and prevent any loss of heat, thereby bringing the temperature back to its set point (Hall, 2015). The diameter of cutaneous blood vessels and consequently the blood flow to the skin are dependent on the autonomic nervous system, specifically the sympathetic nervous system (SNS). Likewise, the sweat glands are also innervated by the sympathetic fibers. Activation of the SNS leads to the vasoconstriction of the cutaneous blood vessels and stimulation of thermogenesis, while its inhibition entails the opposite changes (Tansey & Johnson, 2015). Apart from the SNS, hormonal and behavioral tunings play a role in the regulation of the core body temperature. Hormones like thyroxine, leptin, and neuropeptide Y are among the factors that affect thermoregulation. Furthermore, behavioral adjustments are key components in the adaptation to the weather while supporting the internal mechanisms in maintaining thermoregulatory homeostasis (Hall, 2015).

Differences between healthy individuals and cSCI patients

Alterations in the body composition and energy expenditure after cSCI

There is a decreased level of physical activity and an increased prevalence of corpulence in patients with SCI compared to the general population (Holmlund, Ekblom-Bak, Franzen, Hultling, & Wahman, 2018). In addition to a lower resting energy expenditure (REE) and activity energy expenditure (AEE) in such patients (Jaeger & Blight, 1997), it was also

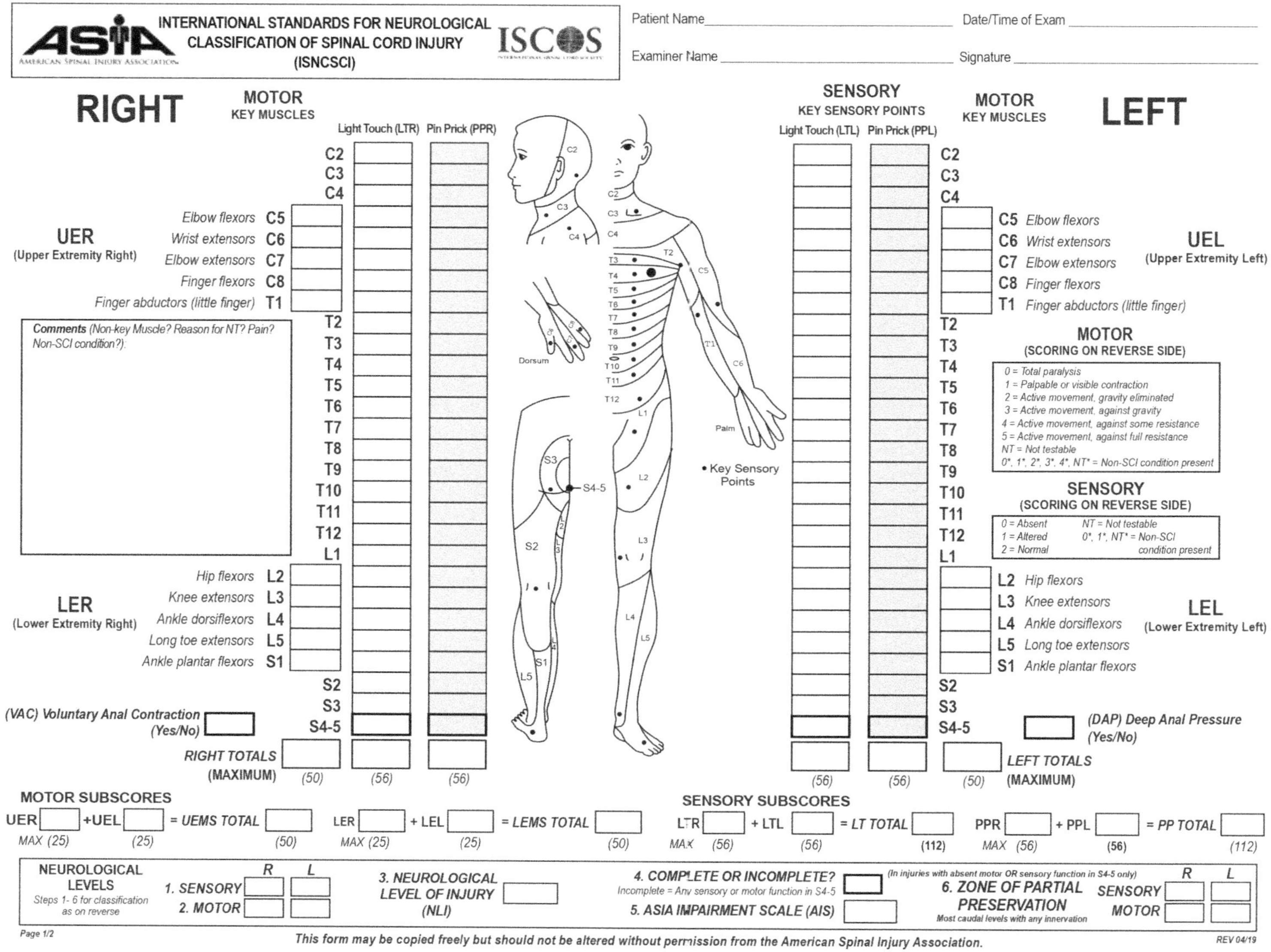

INTERNATIONAL STANDARDS FOR NEUROLOGICAL CLASSIFICATION OF SPINAL CORD INJURY (ISNCSCI)

ASIA — AMERICAN SPINAL INJURY ASSOCIATION; ISCOS

Patient Name ______ Date/Time of Exam ______

Examiner Name ______ Signature ______

RIGHT — MOTOR KEY MUSCLES; SENSORY KEY SENSORY POINTS: Light Touch (LTR), Pin Prick (PPR)

LEFT — MOTOR KEY MUSCLES; SENSORY KEY SENSORY POINTS: Light Touch (LTL), Pin Prick (PPL)

C2, C3, C4

UER (Upper Extremity Right) / UEL (Upper Extremity Left):
C5 Elbow flexors
C6 Wrist extensors
C7 Elbow extensors
C8 Finger flexors
T1 Finger abductors (little finger)

T2, T3, T4, T5, T6, T7, T8, T9, T10, T11, T12, L1

Comments (Non-key Muscle? Reason for NT? Pain? Non-SCI condition?):

LER (Lower Extremity Right) / LEL (Lower Extremity Left):
L2 Hip flexors
L3 Knee extensors
L4 Ankle dorsiflexors
L5 Long toe extensors
S1 Ankle plantar flexors

S2, S3, S4-5

(VAC) Voluntary Anal Contraction (Yes/No)

(DAP) Deep Anal Pressure (Yes/No)

RIGHT TOTALS (MAXIMUM) (50) (56) (56)

LEFT TOTALS (MAXIMUM) (56) (56) (50)

MOTOR (SCORING ON REVERSE SIDE)
0 = Total paralysis
1 = Palpable or visible contraction
2 = Active movement, gravity eliminated
3 = Active movement, against gravity
4 = Active movement, against some resistance
5 = Active movement, against full resistance
NT = Not testable
0*, 1*, 2*, 3*, 4*, NT* = Non-SCI condition present

SENSORY (SCORING ON REVERSE SIDE)
0 = Absent
1 = Altered
2 = Normal
NT = Not testable
0*, 1*, NT* = Non-SCI condition present

MOTOR SUBSCORES
UER ___ + UEL ___ = UEMS TOTAL ___ MAX (25) (25) (50)
LER ___ + LEL ___ = LEMS TOTAL ___ MAX (25) (25) (50)

SENSORY SUBSCORES
LTR ___ + LTL ___ = LT TOTAL ___ MAX (56) (56) (112)
PPR ___ + PPL ___ = PP TOTAL ___ MAX (56) (56) (112)

NEUROLOGICAL LEVELS (Steps 1- 6 for classification as on reverse): 1. SENSORY R L; 2. MOTOR R L

3. NEUROLOGICAL LEVEL OF INJURY (NLI)

4. COMPLETE OR INCOMPLETE? Incomplete = Any sensory or motor function in S4-5

5. ASIA IMPAIRMENT SCALE (AIS)

6. ZONE OF PARTIAL PRESERVATION (In injuries with absent motor OR sensory function in S4-5 only) Most caudal levels with any innervation: SENSORY R L; MOTOR R L

Page 1/2

This form may be copied freely but should not be altered without permission from the American Spinal Injury Association.

REV 04/19

FIG. 1 ASIA grading system. *ANS*, autonomic nervous system; *SNS*, sympathetic nervous system; *BSCB*, blood-spinal cord barrier. *(© 2021 American Spinal Injury Association. Reprinted with permission.)*

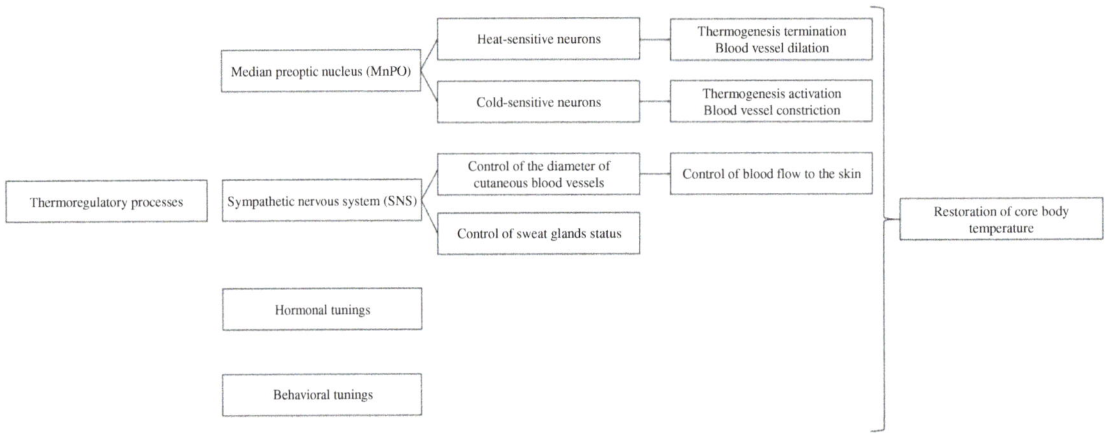

FIG. 2 Thermoregulatory process in able-bodied individuals.

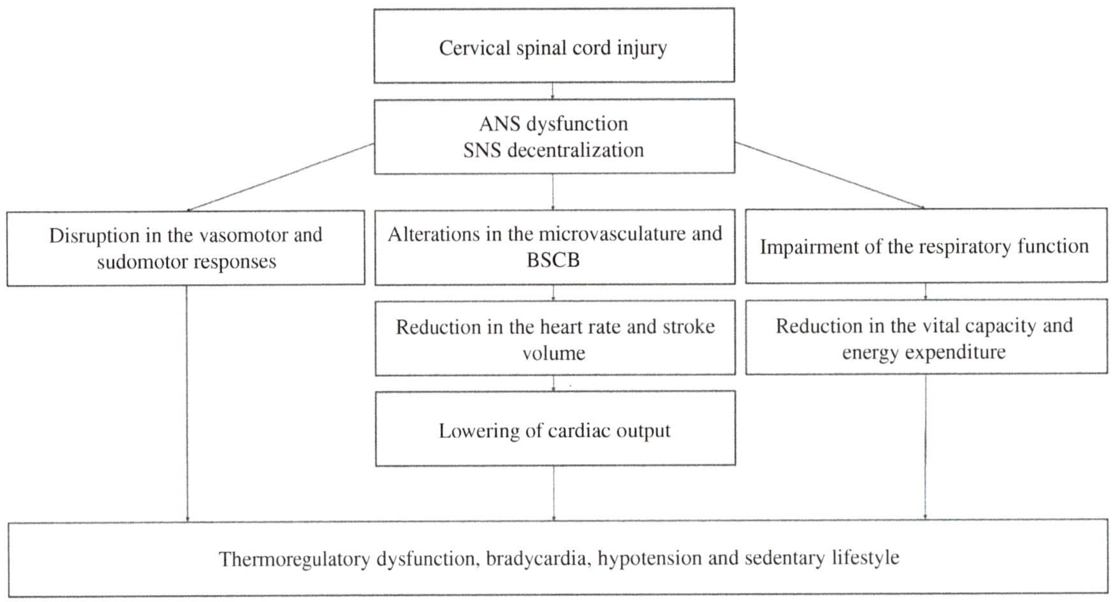

FIG. 3 The negative effects of the cervical spinal cord injury on the human body.

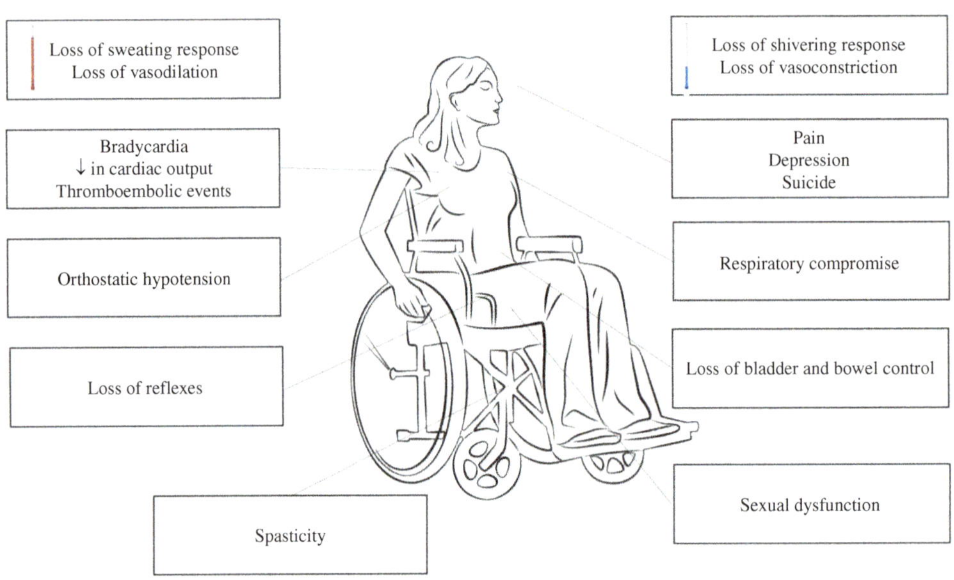

FIG. 4 Detailed negative effects of the cervical spinal cord injury on the human body parts.

shown that it will take a longer time to consume the same amount of energy while performing the same kind of exercise in tetraplegic patients compared to paraplegic patients (Collins et al., 2010). This is because complete motor injuries above thoracic level 6 affect the autonomic nervous system and more muscle tissues (Collins et al., 2010). This will lead to a decrease in the communication between the parasympathetic and the sympathetic nervous systems resulting in an impaired respiratory function with lower vital capacity and maximal heart rate (Currie, West, Hubli, Gee, & Krassioukov, 2015). All of which leads to an energy supply through anaerobic processes with the buildup of lactate, thus affecting the exercise performance and duration (Currie et al., 2015). Holmlund et al. performed a cross-sectional study to compare the resting energy expenditure and VO2 levels for sedentary non-exercise activities and exercise activities in patients with complete motor tetraplegia. They also compared the same parameters between tetraplegic and paraplegic patients. They found out that tetraplegic patients can increase their energy expenditure and VO2 levels three to four times when performing non-exercise physical activity compared to sedentary activities (Holmlund et al., 2018).

In addition, it showed no major differences in resting energy expenditure (REE) between tetraplegic and paraplegic groups. There was an increase in AEE up to 16 times for people with tetraplegia and 14 times for people with paraplegia during exercise activities. This can be attributed to an impairment in the ANS and a greater portion of reduced muscle mass incomplete tetraplegia (Holmlund et al., 2018).

Alterations in the cardiovascular system after cSCI

Traumatic injury to the spinal cord has a profound effect on the microvasculature and the blood-spinal cord barrier (BSCB). It renders the microvasculature permeable to the circulating proteins along the entire axis of the cord not just at the site of injury (Noble, Mautes, & Hall, 1996). Noble et al. demonstrated that in healthy individuals the luminal plasma membranes of the spinal cord show carbohydrate moieties and anionic sites that are like the ones present in the cerebral vessels. However, SCI causes alteration in the composition of the plasma membrane, leading to the loss of the anionic sites with the breakdown of the barrier and an increase in the expression of B-D-galactosamine residues (Noble et al., 1996). Additionally, SCI leads to a reduction in the diameter of blood vessels below the injury level secondary to SNS dysfunction and sedentary lifestyle. This phenomenon will prompt a limitation in the dissipation and preservation of heat resulting in hyperthermia and hypothermia, respectively (Mneimneh, Moussalem, Ghaddar, Aboughali, & Omeis, 2019).

Similar to blood vessel diameter, the cardiac output is also reduced which restricts the individual's tolerance to exercise and culminates in a sedentary lifestyle (Mneimneh et al., 2019).

As previously mentioned, SCI damages the BSCB but in as little as 5 min post-injury (Maikos & Shreiber, 2007). There is a breakdown of this barrier which leads to additional harm to the injured area as well as to the healthy unaffected segments of the spinal cord (Bartanusz, Jezova, Alajajian, & Digicaylioglu, 2011). Several factors are involved in the disruption of this fundamental barrier including the release of toxic free radicals, the disruption of the endothelial tight junctions, and the ischemic condition resulting in the expression of endothelin-1, among many others (Bartanusz et al., 2011).

Disruption in the vasomotor, sudomotor, and shivering responses after cSCI

One of the consequences of cSCI is autonomic dysreflexia (AD), which as the name implies, is the inability of the spinal cord to synchronize the human body sympathetic activity resulting in the disruption of the vasomotor (blood vessel vasodilatation/vasoconstriction) and sudomotor (shivering/sweating) responses. AD, which is potentially fatal, is characterized by bouts of acute hypertension reaching extremely high levels and compensatory bradycardia mediated by the baroreceptors (Eldahan & Rabchevsky, 2018).

Hence, people with tetraplegia are unable to maintain a balanced body temperature due to the loss of the sweating and vasodilatation responses and the vasoconstriction and shivering responses which normally occur when the body temperature is above or below its thermoneutral state, respectively (Gunduz & Binak, 2012). The pathophysiology of such disruption is due to the impairment of the stellate ganglion, a star-shaped sympathetic collection of nerves located in the cervical region, which regulates the involuntary sympathetic nervous system functions such as vaso-regulation, sweating, and pain (Kirshblum et al., 2011).

The sudomotor response is usually activated when the human body temperature is above its thermoneutral state resulting in sweating to cool down the body temperature. It is estimated that 100% of the sudomotor responses are lost in case of complete cSCI compared to 50%–75% loss in case of incomplete cSCI, leading to a disrupted body temperature (Yaggie, Niemi, & Buono, 2002). Hence, damage to the ANS post cSCI hinders the human body's ability from regulating its temperature at extreme thermal stress conditions. Due to the absence of shivering and sweating responses, people with complete cSCI are at an increased risk of acute and chronic health conditions, such as hyperthermia and hypothermia leading to the disruption of the normal function of the proteins and enzymes leading to cell death, on the molecular level, (Cheung, Lee, & Oksa, 2016) and to multi-organ failure. Hyperthermia, in particular, even in its mildest forms, can lead to cognitive problems including memory and attention deficits, and neurological deficits that can become persistent and even permanent (Walter & Carraretto, 2016). In contrast, hypothermia is more likely to affect the myocardium causing ischemia and infarctions, and the blood causing bleeding diathesis (Sessler, 2001).

Alterations in the blood pressure after cSCI

Due to the loss of the central sympathetic tone (supraspinal neuronal centers), patients with cSCI develop acute hypotension that might require vasopressive treatment to maintain vital blood pressure, although the myogenic (smooth muscle) peripheral tone is preserved in such patients (Piepmeier, Lehmann, & Lane, 1985). Particularly, the development of hypotension occurs in the peripheral blood vessels that are below the lesion. Thus, patients with the highest cervical injuries, develop hypotension the most (West, Mills, & Krassioukov, 2012). Besides resting hypotension, patients with cSCI develop orthostatic hypotension which occurs when they are transferred from supine to an upright position (Mathias & Frankel, 1983). In a healthy individual, orthostatic hypotension is avoided by the sympathetic outflow leading to the baroreflex-mediated peripheral vasoconstriction to preserve blood pressure and cerebral perfusion (Bush, Wight, Brown, & Hainsworth, 2000). It has been suggested that decreased baroreceptor sensitivity and sympathetic hypoactivity are the causes of orthostatic hypotension post cSCI. Besides, disturbed salt/water balance and cardiovascular deconditioning have been also proposed as causes of orthostatic hypotension in such patients (Munakata, Kameyama, Nunokawa, Ito, & Yoshinaga, 2001). To note that problems in orthostasis can last for many years and have the potential to become permanent in tetraplegic individuals (Krassioukov, Eng, Warburton, & Teasell, 2009).

Applications to other areas of neuroscience

In this chapter, we have reviewed the thermoregulatory dysfunctions that occur in patients with cSCI. These changes can be incapacitating to the individuals who might already be suffering on numerous scales including financially, physically, psychologically, and emotionally. Despite this serious unparalleled injury being neglected for years, there are currently hundreds of ongoing research projects worldwide attempting to help individuals live a better quality of life using advanced devices and there are efforts made towards curing this condition using, for example, stem cell technology.

In this section, we will be briefly reviewing some of the advances in the field of treatment of cSCI, and hence in the prevention of thermoregulatory dysfunctions. In addition to the early surgical intervention for decompression which has been proven to improve the outcome (Fehlings et al., 2012; La Rosa, Conti, Cardali, Cacciola, & Tomasello, 2004), there are several clinical trials on pharmacological and non-pharmacological treatment options.

Among the drugs being tested as neuroprotective agents in SCI, we cite glibenclamide (an antidiabetic drug) (Minnema et al., 2019), minocycline (an antibiotic) (Casha et al., 2012), riluzole (a drug for amyotrophic lateral sclerosis) (Fehlings et al., 2016), and certain growth factors such as hepatocyte growth factor, granulocyte colony-stimulating factor, and basic fibroblast growth factor (Badhiwala, Ahuja, & Fehlings, 2018).

The use of cell therapy and stem cell therapy are growing due to their scaffolding, regenerative, and to a lesser extent neuroprotective properties, providing an adequate environment for the spinal cord to heal and regenerate (Badner, Siddiqui, & Fehlings, 2017).

As for the non-pharmacological lines of treatment, there are reports on functional electrical stimulation and cooling of the spinal cord as being effective means to reduce the deleterious effects of cSCI. Functional electrical stimulation was shown to be a safe way to increase the degree of independence and ambulation in SCI, yet larger trials are being conducted to prove its effectiveness (Kapadia et al., 2014; Thrasher, Flett, & Popovic, 2006). On the other hand, inducing hypothermia in SCI using various devices and approaches (e.g. skin surface cooling using jackets or blankets, cooling catheters, chilled saline, and so on) has been assessed in numerous experimental and clinical studies to enhance the outcomes and prevent, or at least halt the progression on the injury (Ahmad, Wang, & Levi, 2014; Zhu, 2018), yet larger studies and randomized control trials are needed to support the adoption of systemic hypothermia as a treatment line in cSCI.

Until the results of large multi-center trials confirm or refute the efficacy of the treatment measures mentioned above, the current guidelines for the treatment of cSCI remain vague.

Mini-dictionary of terms

Tetraplegia: A paralysis due to a spinal cord injury in the cervical region.
Paraplegia: A paralysis due to a spinal cord injury in the thoracic, lumbar, or sacral region.
Thromboembolic events: The movement of a blood clot (thrombus) via blood from one site to another to plug the vessels.
Autonomic nervous system: It is the involuntary nervous system that is responsible for bodily functions, including the heart and respiratory rate, digestion, and urination, among others. It is divided into the sympathetic and parasympathetic nervous systems.
Sympathetic nervous system: It is responsible for the fight and flight responses, including the sudomotor and vasomotor systems.
Parasympathetic nervous system: It is responsible for the rest and digest responses.
Resting energy expenditure: It refers to the amount of energy a body requires during a non-active 24-h period.
Activity energy expenditure: It refers to the energy consumed during planned and unplanned physical activity.
Vital capacity: It refers to the maximal amount of air expelled from the lungs after a full deep breath.

Key facts of spinal cord injury

- SCI is an insult to the spinal cord that can occur at any level.
- The severity of SCI depends on the injury level and type.
- There are profound economic, psychological, and social burdens resulting from SCI.
- Numerous causes have been described but motor vehicle accidents and falls are the most frequently encountered.
- Most of the physiological disturbances are due to the involvement of the autonomic nervous system.
- SCI leads to a variety of dysregulations, most notably, the homeostatic imbalance and thermoregulatory dysfunction.
- People with SCI develop resting and orthostatic hypotension.

Key facts of American Spinal Injury Association

- The American Spinal Injury Association (ASIA) was launched in 1973.
- It is concerned with the care of patients living with spinal cord injury (SCI) and promotes knowledge about SCI both for the public and healthcare personnel.
- It is actively engaged in and encourages research in all fields related to SCI starting with its prevention and not ending with ways to enhance the quality of life of affected patients and finding new therapies.
- It relies on a multidisciplinary approach to SCI.
- It publishes the International Standards for Neurological Classification of Spinal Cord Injury (ISNCSCI), a widely used assessment tool to determine the level and severity of SCI.

Summary points

- Motor vehicle accidents are the leading cause of traumatic SCI which is increasing every year.
- SCI can be classified into 5 grades depending on the preservation percentage of the sensory and motor functions below the injury level.
- In people with SCI, the sympathetic nervous system loses its function and the parasympathetic one attempts to take control instead.
- People with SCI cannot maintain their core body temperature stable under ambient conditions due to thermoregulatory disturbances.
- Thermoregulatory disruptions are due to the loss of the sudomotor (shivering/sweating) and vasomotor (vasoconstriction/dilation) responses below the injury level.
- People with SCI have a more sedentary lifestyle and thus lower REE and AEE.

References

Abrams, G. M., & Wakasa, M. (2019). In M. J. Aminoff, & M. J. Silveira (Eds.), *Chronic complications of spinal cord injury and disease* UptoDate.

Ahmad, F. U., Wang, M. Y., & Levi, A. D. (2014). Hypothermia for acute spinal cord injury—A review. *World Neurosurgery*, *82*(1–2), 207–214.

American Spinal Cord Injury Association. (2019). *International standards for neurological classification of spinal cord injury*. Retrieved 26/09/2020, from https://asia-spinalinjury.org/international-standards-neurological-classification-sci-isncsci-worksheet/.

Badhiwala, J. H., Ahuja, C. S., & Fehlings, M. G. (2018). Time is spine: A review of translational advances in spinal cord injury. *Journal of Neurosurgery. Spine*, *30*(1), 1–18.

Badner, A., Siddiqui, A. M., & Fehlings, M. G. (2017). Spinal cord injuries: How could cell therapy help? *Expert Opinion on Biological Therapy*, *17*(5), 529–541.

Bartanusz, V., Jezova, D., Alajajian, B., & Digicaylioglu, M. (2011). The blood-spinal cord barrier: Morphology and clinical implications. *Annals of Neurology*, *70*(2), 194–206.

Bush, V. E., Wight, V. L., Brown, C. M., & Hainsworth, R. (2000). Vascular responses to orthostatic stress in patients with postural tachycardia syndrome (POTS), in patients with low orthostatic tolerance, and in asymptomatic controls. *Clinical Autonomic Research*, *10*(5), 279–284.

Casha, S., Zygun, D., McGowan, M. D., Bains, I., Yong, V. W., & Hurlbert, R. J. (2012). Results of a phase II placebo-controlled randomized trial of minocycline in acute spinal cord injury. *Brain*, *135*(Pt. 4), 1224–1236.

Chen, Y., Tang, Y., Vogel, L. C., & Devivo, M. J. (2013). Causes of spinal cord injury. *Topics in Spinal Cord Injury Rehabilitation*, *19*(1), 1–8.

Cheung, S. S., Lee, J. K., & Oksa, J. (2016). Thermal stress, human performance, and physical employment standards. *Applied Physiology, Nutrition, and Metabolism*, *41*(6 Suppl. 2), S148–S164.

Chovatiya, R., & Medzhitov, R. (2014). Stress, inflammation, and defense of homeostasis. *Molecular Cell*, *54*(2), 281–288.

Collins, E. G., Gater, D., Kiratli, J., Butler, J., Hanson, K., & Langbein, W. E. (2010). Energy cost of physical activities in persons with spinal cord injury. *Medicine and Science in Sports and Exercise*, *42*(4), 691–700.

Currie, K. D., West, C. R., Hubli, M., Gee, C. M., & Krassioukov, A. V. (2015). Peak heart rates and sympathetic function in tetraplegic nonathletes and athletes. *Medicine and Science in Sports and Exercise*, *47*(6), 1259–1264.

Eldahan, K. C., & Rabchevsky, A. G. (2018). Autonomic dysreflexia after spinal cord injury: Systemic pathophysiology and methods of management. *Autonomic Neuroscience*, *209*, 59–70.

Fehlings, M. G., Nakashima, H., Nagoshi, N., Chow, D. S., Grossman, R. G., & Kopjar, B. (2016). Rationale, design and critical end points for the Riluzole in Acute Spinal Cord Injury Study (RISCIS): A randomized, double-blinded, placebo-controlled parallel multi-center trial. *Spinal Cord*, *54*(1), 8–15.

Fehlings, M. G., Vaccaro, A., Wilson, J. R., Singh, A. W., Cadotte, D., Harrop, J. S., et al. (2012). Early versus delayed decompression for traumatic cervical spinal cord injury: Results of the Surgical Timing in Acute Spinal Cord Injury Study (STASCIS). *PLoS One*, *7*(2), e32037.

Gunduz, H., & Binak, D. F. (2012). Autonomic dysreflexia: An important cardiovascular complication in spinal cord injury patients. *Cardiology Journal*, *19*(2), 215–219.

Hall, J. E. (2015). Body temperature regulation and fever. In *Guyton and Hall textbook of medical physiology* (p. 1168). Philadelphia, PA: Elsevier.

Holmlund, T., Ekblom-Bak, E., Franzen, E., Hultling, C., & Wahman, K. (2018). Energy expenditure after spinal cord injury in people with motor-complete tetraplegia or motor-complete paraplegia. *Spinal Cord*, *56*(3), 274–283.

Jaeger, C. B., & Blight, A. R. (1997). Spinal cord compression injury in guinea pigs: Structural changes of endothelium and its perivascular cell associations after blood-brain barrier breakdown and repair. *Experimental Neurology*, *144*(2), 381–399.

Kapadia, N., Masani, K., Catharine Craven, B., Giangregorio, L. M., Hitzig, S. L., Richards, K., et al. (2014). A randomized trial of functional electrical stimulation for walking in incomplete spinal cord injury: Effects on walking competency. *The Journal of Spinal Cord Medicine*, *37*(5), 511–524.

Karlsson, A. K. (2006). Autonomic dysfunction in spinal cord injury: Clinical presentation of symptoms and signs. *Progress in Brain Research*, *152*, 1–8.

Kirshblum, S. C., Burns, S. P., Biering-Sorensen, F., Donovan, W., Graves, D. E., Jha, A., et al. (2011). International standards for neurological classification of spinal cord injury (revised 2011). *The Journal of Spinal Cord Medicine*, *34*(6), 535–546.

Krassioukov, A., Eng, J. J., Warburton, D. E., & Teasell, R. (2009). A systematic review of the management of orthostatic hypotension after spinal cord injury. *Archives of Physical Medicine and Rehabilitation*, *90*(5), 876–885.

La Rosa, G., Conti, A., Cardali, S., Cacciola, F., & Tomasello, F. (2004). Does early decompression improve neurological outcome of spinal cord injured patients? Appraisal of the literature using a meta-analytical approach. *Spinal Cord*, *42*(9), 503–512.

Maikos, J. T., & Shreiber, D. I. (2007). Immediate damage to the blood-spinal cord barrier due to mechanical trauma. *Journal of Neurotrauma*, *24*(3), 492–507.

Martinez-Perez, R., Cepeda, S., Paredes, I., Alen, J., & Lagares, A. (2017). MRI prognostication factors in the setting of cervical spinal cord injury secondary to trauma. *World Neurosurgery*, *101*, 623–632.

Mathias, C. J., & Frankel, H. L. (1983). Clinical manifestations of malfunctioning sympathetic mechanisms in tetraplegia. *Journal of the Autonomic Nervous System*, *7*(3–4), 303–312.

Minnema, A. J., Mehta, A., Boling, W. W., Schwab, J., Simard, J. M., & Farhadi, H. F. (2019). SCING-spinal cord injury neuroprotection with glyburide: A pilot, open-label, multicentre, prospective evaluation of oral glyburide in patients with acute traumatic spinal cord injury in the USA. *BMJ Open*, *9*(10), e031329.

Mneimneh, F., Moussalem, C., Ghaddar, N., Aboughali, K., & Omeis, I. (2019). Influence of cervical spinal cord injury on thermoregulatory and cardiovascular responses in the human body: Literature review. *Journal of Clinical Neuroscience*, *69*, 7–14.

Mrowka, R., & Reuter, S. (2016). Thermoregulation. *Acta Physiologica*, *217*(1), 3–5.

Munakata, M., Kameyama, J., Nunokawa, T., Ito, N., & Yoshinaga, K. (2001). Altered Mayer wave and baroreflex profiles in high spinal cord injury. *American Journal of Hypertension*, *14*(2), 141–148.

National Spinal Cord Injury Statistical Center. (2020). *Facts and figures at a glance*. Retrieved 22/09/2020, from https://www.nscisc.uab.edu/Public/Facts%20and%20Figures%202020.pdf.

Noble, L. J., Mautes, A. E., & Hall, J. J. (1996). Characterization of the microvascular glycocalyx in normal and injured spinal cord in the rat. *The Journal of Comparative Neurology*, *376*(4), 542–556.

Pear, J. J. (2012). Physiological homeostasis and learning. In N. M. Seel (Ed.), *Encyclopedia of the sciences of learning* (pp. 2628–2631). Boston, MA: Springer US.

Piepmeier, J. M., Lehmann, K. B., & Lane, J. G. (1985). Cardiovascular instability following acute cervical spinal cord trauma. *Central Nervous System Trauma*, *2*(3), 153–160.

Sessler, D. I. (2001). Complications and treatment of mild hypothermia. *Anesthesiology*, *95*(2), 531–543.

Tansey, E. A., & Johnson, C. D. (2015). Recent advances in thermoregulation. *Advances in Physiology Education*, *39*(3), 139–148.

Thrasher, T. A., Flett, H. M., & Popovic, M. R. (2006). Gait training regimen for incomplete spinal cord injury using functional electrical stimulation. *Spinal Cord*, *44*(6), 357–361.

Walter, E. J., & Carraretto, M. (2016). The neurological and cognitive consequences of hyperthermia. *Critical Care*, *20*(1), 199.

Wecht, J., Lafountaine, M., Handrakis, J., West, C., Phillips, A., Ditor, D., et al. (2015). Autonomic nervous system dysfunction following spinal cord injury: Cardiovascular, cerebrovascular, and thermoregulatory effects. *Current Physical Medicine and Rehabilitation Reports*, *3*, 197–205.

West, C. R., Mills, P., & Krassioukov, A. V. (2012). Influence of the neurological level of spinal cord injury on cardiovascular outcomes in humans: A meta-analysis. *Spinal Cord*, *50*(7), 484–492.

Yaggie, J. A., Niemi, T. J., & Buono, M. J. (2002). Adaptive sweat gland response after spinal cord injury. *Archives of Physical Medicine and Rehabilitation*, *83*(6), 802–805.

Zhu, L. (2018). Hypothermia used in medical applications for brain and spinal cord injury patients. *Advances in Experimental Medicine and Biology*, *1097*, 295–319.

Chapter 36

Spinal cord injury and the gut microbiota

Kristina A. Kigerl[a,b,c] and Phillip G. Popovich[a,b,c]

[a]Center for Brain and Spinal Cord Repair, The Ohio State University College of Medicine, Columbus, OH, United States, [b]Belford Center for Spinal Cord Injury, The Ohio State University College of Medicine, Columbus, OH, United States, [c]Department of Neuroscience, The Ohio State University College of Medicine, Columbus, OH, United States

List of abbreviations

ANS	autonomic nervous system
APOA1	apolipoprotein A-I
CFS	chronic fatigue syndrome
CNS	central nervous system
GALT	gut-associated lymph tissue
GI	gastrointestinal
HPA	hippocampus pituitary axis
HDL	high density lipoprotein
KEGG	Kyoto Encyclopedia of Genes and Genomes
LPA	lipoprotein A
MAG	metagenome assembled genome
NAFLD	non-alcoholic fatty liver disease
NE	norepinephrine
SCFA	short-chain fatty acid
SCI	spinal cord injury
SPN	sympathetic pre-ganglionic neuron

Introduction

The gut microbiota comprises all of the bacteria, viruses, fungi, and other microbes that live in the gastrointestinal tract, specifically within the small and large intestines. The density of this community of microbes is at least as numerous as mammalian cells (Sender, Fuchs, & Milo, 2016) and may outnumber them ~10:1 (Gill et al., 2006; Hollister, Gao, & Versalovic, 2014). This diverse ecosystem, and associated genome (i.e., the microbiome), is critical for regulating host metabolism, digestion, nutrient absorption, and immune system development and function (Hooper, Littman, & Macpherson, 2012; Nicholson et al., 2012; Round & Mazmanian, 2009). The types and diversity of microbiota vary as a function of location in the gut. The small intestine has a more limited microbial community structure enriched with Lactobacillaceae and Enterobacteriaceae (Donaldson, Lee, & Mazmanian, 2016; Gu et al., 2013) while a more diverse and abundant microbiome exists in the colon. Because there is lower oxygen tension in the colon, it is enriched with anaerobic microbiota (Donaldson et al., 2016; Gu et al., 2013). In addition to regulating inflammation and GI function, the intestinal microbiota can influence disease development throughout the host, including in the CNS (Collins, Surette, & Bercik, 2012; Wang & Kasper, 2014).

The gut and the CNS communicate with one another via several routes. Gut microbes, through activation of immune cells in the gut-associated lymphoid tissues (GALT) and subsequent release of cytokines, affect CNS function (Bercik et al., 2010). Many (~80%) of the body's immune cells are found in the GALT, and this neuro-immune axis is likely an important route of communication between the gut and brain.

Gut microbes play a role in regulating the body's response to stress, including cortisol release (Crumeyrolle-Arias et al., 2014; Sudo et al., 2004). Germ-free mice develop an exaggerated hypothalamic-pituitary-adrenal (HPA) axis response to stress, which can be completely or partially reversed with a fecal transplant from conventional mice (Sudo et al., 2004). Similar exaggerated stress response has been seen in germ-free rats (Crumeyrolle-Arias et al., 2014), suggesting that the ability of gut microbes to regulate stress is conserved across species.

Cellular, Molecular, Physiological, and Behavioral Aspects of Spinal Cord Injury. https://doi.org/10.1016/B978-0-12-822427-4.00036-8

Finally, gut microbiota produces neuroactive metabolites (i.e., short-chain fatty acids, choline) and neurotransmitters (GABA, noradrenaline, serotonin, dopamine, acetylcholine). These gut metabolites can act locally on neurons and glia that make up the enteric nervous system (Clarke et al., 2014; Tillisch, 2014) but they can also enter the bloodstream. After crossing the blood-brain barrier or by modulating signaling of the afferent vagus nerve in the periphery, circulating metabolites born in the gut can directly influence CNS structure/function (Clarke et al., 2014; Forsythe, Bienenstock, & Kunze, 2014; Wikoff et al., 2009).

In the healthy gut, a symbiosis exists between the gut microbiota, intestinal epithelia, and immune system. Altering the composition of the gut microbiota creates a state of "dysbiosis" where the balance between beneficial and pathogenic or inflammatory bacteria is skewed, usually favoring the latter. Common causes of gut dysbiosis include antibiotic use, stress, and gut dysfunction (Bailey et al., 2011, 2010; El Aidy, van den Bogert, & Kleerebezem, 2015; Hawrelak & Myers, 2004; Hill & Artis, 2010; Hooper et al., 2012; Round & Mazmanian, 2009). Many of these triggers of gut dysbiosis also compromise the mucosal barrier, leading to bacterial translocation, a process whereby gut bacteria migrate from the intestinal lumen into mesenteric lymph nodes and throughout the body (Balmer et al., 2014; Balzan, de Almeida Quadros, de Cleva, Zilberstein, & Cecconello, 2007; Macpherson & Smith, 2006). Gut dysbiosis may also influence the progression of or predispose individuals to develop autoimmune diseases (e.g., multiple sclerosis, type I diabetes, rheumatoid arthritis), allergy or metabolic disorders (Berer et al., 2011; Cao, Feehley, & Nagler, 2014; Kriegel et al., 2011; Lee, Menezes, Umesaki, & Mazmanian, 2011; Tilg & Kaser, 2011). Similarly, dysbiosis has been implicated in the onset or progression of neurological diseases including autism, pain, depression, anxiety, and stroke (Benakis et al., 2016; de Theije et al., 2014; Foster & McVey Neufeld, 2013; Hsiao et al., 2013; Rousseaux et al., 2007; Winek et al., 2016).

Spinal cord injury-induced dysautonomia

Traumatic spinal cord injury (SCI) causes loss of motor and sensory function below the level of the injury, resulting in paralysis and permanent sensory dysfunction. SCI also permanently damages the sympathetic branch of the autonomic nervous system, leading to persistent dysautonomia. SCI-induced dysautonomia contributes to chronic multi-organ pathology and dysfunction including loss of bladder and bowel function, immune suppression, depression and anxiety, fatigue, and cardiovascular dysfunction and disease.

Sympathetic pre-ganglionic neurons (SPNs) reside in the thoracic and upper lumbar spinal cord. SPNs are cholinergic neurons that innervate adrenergic neurons located outside the spinal cord within pre-/post-vertebral ganglia. Post-ganglionic adrenergic sympathetic neurons release norepinephrine (NE) in target organs including (but not limited to) the immune system, liver, cardiovascular system, and the gastrointestinal (GI) tract (Felten, Ackerman, Wiegand, & Felten, 1987; Felten & Olschowka, 1987). SPN neurons can be directly injured during SCI. However, even if SPNs are spared, their function becomes disrupted due to loss or damage to pre-sympathetic axons (from the brain/brainstem) or the intraspinal neuronal networks that regulate autonomic reflexes below the level of injury.

In the gut, sympathetic post-ganglionic innervation controls motility, mucosal secretions, vascular tone, and immune function (Cervi, Lukewich, & Lomax, 2014; Elenkov, Wilder, Chrousos, & Vizi, 2000). SPNs controlling the small and large intestines are located in the intermediolateral cell column in thoracic spinal segments T5–10 and T10-L4, respectively (Fig. 1) (Browning & Travagli, 2014; Levatte, Mabon, Weaver, & Dekaban, 1998; Mabon, LeVatte, Dekaban, & Weaver, 1997). Adrenergic post-ganglionic nerve fibers from the celiac-mesenteric ganglia innervate the stomach, small intestine, and, to some extent, the proximal large intestine. Neurons in the inferior mesenteric ganglia receive input from SPNs located in the T11 and upper lumbar spinal cord then send post-ganglionic projections to innervate the large intestine (Furness & Costa, 1974). Therefore, an injury at any spinal level will adversely affect the GI tract, although these effects are exacerbated when SCI occurs at high spinal levels and progressively more pre-sympathetic control over spinal sympathetic networks that control the gut are lost. Common GI complications after SCI include constipation, fecal incontinence, and decreased colonic transit time (Hou & Rabchevsky, 2014; Qualls-Creekmore, Tong, & Holmes, 2010; Tate et al., 2016). After SCI, impaired autonomic control of gastrointestinal function likely contributes to the development of dysbiosis, which is then maintained or exacerbated by stress and the need for frequent antibiotic use – both are common in people living with SCI (Schönenberg et al., 2014; Wong et al., 2017) (Fig. 1).

Gut dysbiosis after SCI: Pre-clinical studies

SCI-induced gut dysbiosis has been documented in multiple independent pre-clinical studies and, in most cases, 16 s rRNA sequencing was used to analyze the composition and relative abundance of bacteria in fecal samples (Jing et al., 2019; Kigerl et al., 2016; Myers et al., 2019; O'Connor et al., 2018; Schmidt et al., 2020). In SCI mice, robust and lasting

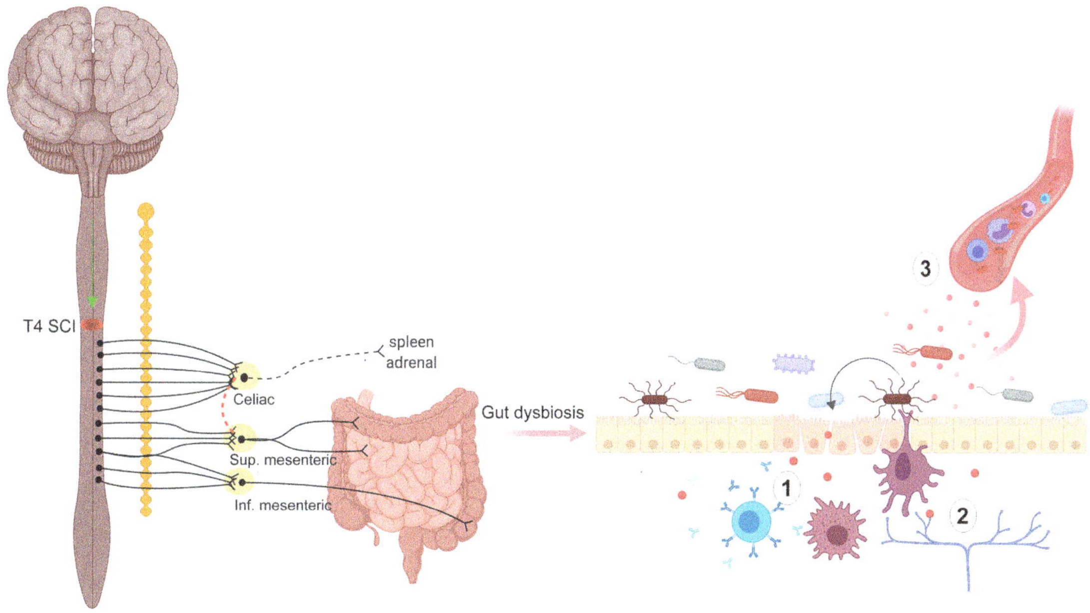

FIG. 1 Spinal cord injury disrupts autonomic control over the gastrointestinal tract, leading to the development of gut dysbiosis. Gut dysbiosis can influence (1) immune system function, (2) vagal signaling, and (3) circulating metabolites that can contribute to the development of common SCI-associated disorders such as immune dysfunction, metabolic and cardiovascular disease, pain, and mood disorders. Created with BioRender.com.

gut dysbiosis develops and is characterized by inverse changes in *Bacteroidales* and *Clostridiales*, the two most prevalent bacterial taxa in mouse gut: *Bacteroidales* decreases while *Clostridiales* increases after SCI (Jing et al., 2019; Kigerl et al., 2016). Minor taxa (*Anaeroplasmatales*, *Turicibacterales* and *Lactobacillales)* also are affected by SCI (Jing et al., 2019; Kigerl et al., 2016). The relative abundance of bacteria from the phylum *Proteobacteria* also increases after SCI in mice (Myers et al., 2019). Gut dysbiosis has similarly been described in SCI rats (O'Connor et al., 2018; Schmidt et al., 2020), in some cases lasting at least 8 weeks post-injury (O'Connor et al., 2018).

In addition to changing the composition and relative abundance of bacteria, SCI increases gut permeability and alters the expression of tight junction proteins in the gut (Jing et al., 2019; Kigerl et al., 2016). Increased gut permeability likely contributes to bacterial translocation (Bai et al., 2011; Kigerl et al., 2016; Liu et al., 2004) and increased levels of circulating endotoxin after SCI (Myers et al., 2019). Combined with gut dysbiosis, bacterial translocation and circulating endotoxin may cause systemic inflammation and contribute to the development of comorbidities associated with SCI, including increased incidence of cardiovascular disease, depression/anxiety, NAFLD, and metabolic syndrome (Foster & McVey Neufeld, 2013; Gérard, 2016; Safari & Gérard, 2019). Schmidt et al. showed that gut dysbiosis contributes to anxiety in SCI rats and this can be reversed by treating SCI rats with a fecal transplant derived from healthy donor rats (Schmidt et al., 2020). Fecal transplantation has been shown to normalize gut dysbiosis and confer therapeutic benefits in recurrent *Clostridium difficile* infection and inflammatory bowel disease. Fecal transplantation is also under investigation for Multiple Sclerosis, Parkinson's Disease, autism, and metabolic syndrome (Vendrik et al., 2020).

The onset of gut dysbiosis after SCI is also associated with intestinal inflammation. Cytokine expression increases in the colon and the GALT after SCI (Jing et al., 2019; Kigerl et al., 2016; O'Connor et al., 2018), and these changes correlate with changes in bacterial taxa (Jing et al., 2019; O'Connor et al., 2018). Increased intestinal permeability and CCL2 expression positively correlate with the abundance of *Clostridiales* and negatively correlate with *Lactobacillales* (Jing et al., 2019). Expression of IL-1β, IL-12, and CXCL2 also correlate with changes in microbiota after SCI (O'Connor et al., 2018). Increased intestinal inflammation, along with the impaired function of the gut microbiome, may exacerbate the gastrointestinal motility deficits caused by SCI. Treatments that reverse gut dysbiosis have been shown to increase gastrointestinal transit time in SCI mice (Jing et al., 2019).

Recently, it has become possible to predict the functional implications of changes in microbial community structure after SCI. Indeed, computational tools now exist that can predict microbial function from 16 s rRNA data (i.e., PICRUSt) or more directly from whole-genome metagenomic sequencing (Langille et al., 2013; Niu et al., 2017). Using these tools, consistent changes were noted in both SCI rats and mice. Notably, SCI appears to affect the gut microbiota's ability to carry out normal carbohydrate metabolism, bile acid biosynthesis, metabolism of cofactors and vitamins, and lipid biosynthesis

(Schmidt et al., 2020; Du et al., 2021). Using metagenome-assembled genomes (MAGs), Du et al. found that a subset of these functions can be attributed to discrete bacterial taxa. Thus, using data from MAGs, one can identify then target specific bacterial taxa to improve or block microbial functions that influence host physiology (Frioux, Singh, Korcsmaros, & Hildebrand, 2020).

So far, only a limited number of pre-clinical studies have tried to manipulate the gut microbiota for therapeutic gain, although those studies have yielded promising results. Post-injury feeding of a medical-grade probiotic formula improved locomotor recovery, reduced lesion pathology and modulated the immune response in the GALT (Kigerl et al., 2016). Fecal suspensions procured from healthy rat donors then transplanted into SCI rats ameliorated SCI-induced gut dysbiosis and reduced anxiety-like behaviors caused by SCI (Schmidt et al., 2020). Melatonin treatment reduced gut dysbiosis, improved gastrointestinal transit time, and modulated intestinal inflammation (Jing et al., 2019). Additional research is needed to further advance our understanding of how the gut microbiota influences neurological recovery and systemic comorbidities after SCI.

Gut dysbiosis after SCI: Clinical studies

SCI also causes gut dysbiosis in humans (Gungor, Adiguzel, Gursel, Yilmaz, & Gursel, 2016; Li et al., 2020; Lin et al., 2020; Zhang et al., 2019, 2018). SCI-induced alterations in the human gut microbiota are long-lasting; the abundance of butyrate-producing gut bacteria was found to be reduced for at least 1-year post-SCI (Gungor et al., 2016). Butyrate and other short-chain fatty acids (SCFAs) have potent immune-modulatory properties and are critical for the maintenance of intestinal epithelium homeostasis (Tan et al., 2014). Other serum metabolites also change after SCI and these changes correlate with altered microbial community structure. For example, Zhang et al. showed that serum glucose, HDL, APOA1, and LPA levels correlate with changes in gut microbiota (based on 16S rRNA sequencing data). Then, using predictive algorithms, they showed that KEGG pathways related to lipid metabolism and fatty acid metabolism decrease in the microbiota of SCI individuals (Zhang et al., 2019). Li et al. reported that bacterial families associated with impaired glucose and lipid metabolism and intestinal inflammation increase in the gut in individuals with SCI (compared to able-bodied controls) (Li et al., 2020). Collectively, these studies suggest that various serum metabolites may serve as biomarkers of SCI-induced changes in gut microbiota and that gut dysbiosis may contribute to metabolic disorders that occur after SCI. Indeed, metabolic function is significantly impaired in paraplegics and tetraplegics; both have higher body fat content compared to able-bodied individuals (Gater, 2007; Gorgey et al., 2014; Gorgey & Gater, 2011). Future studies should determine if gut dysbiosis is causal in the high incidence of metabolic disease and adiposity in SCI individuals (Gorgey and Gater, 2007; Gorgey et al., 2014; Gorgey & Gater, 2011; Manns, McCubbin, & Williams, 2005; Maruyama et al., 2008; Nelson et al., 2007).

Demographic factors (injury level, injury completeness, age, sex)

The magnitude of change and relative composition of the gut microbiota may vary as a function of spinal injury level. Indeed, injuries occurring at high spinal levels will cause more significant or more complete deficits in autonomic tone than an SCI that occurs at lower spinal levels (Holmes & Blanke, 2019). That is because SPNs controlling the small and large intestines are located primarily in thoracic spinal segments T5–10 and T10-L4, respectively (Fig. 1) (Browning & Travagli, 2014; Levatte et al., 1998; Mabon et al., 1997). Consequently, an SCI at or above the most rostrally located SPNs will remove all supraspinal control over spinal sympathetic reflexes, including those that regulate the GI tract. Indeed, using metagenomics to sequence fecal samples obtained from mice with a T4 or T10 SCI, we recently discovered that SCI-induced changes in gut microbial community structure are spinal level-dependent (Du et al., 2021). Although the composition of the gut microbiota in SCI mice was distinct from sham-operated mice, regardless of injury level, gut dysbiosis was exacerbated in T4 SCI mice. These level-dependent differences were evident when evaluating bacterial, functional, and viral components of the microbiome (Du et al., 2021). In clinical studies, gut microbiota composition was also shown to vary as a function of injury level; SCI at cervical or upper thoracic levels produced a gut microbiota that was distinct from that found in individuals with lower thoracic or lumbar SCI (Gungor et al., 2016; Zhang et al., 2018).

Age and sex also influence gut microbial composition (Jašarević, Morrison, & Bale, 2016; Markle et al., 2013; Sheng et al., 2017; Stilling et al., 2015; Thevaranjan et al., 2017). SCI still occurs most often in young adults, but the average age at the time of injury has steadily increased from 29 in the 1970s to 43 in 2015 (National Spinal Cord Injury Statistical Center, 2020). Advanced age at the time of injury is a risk factor for the poorer outcome and increased complications (Devivo, Kartus, Rutt, Stover, & Fine, 1990; Wilson, Cadotte, & Fehlings, 2012). To date, there has been no attempt to evaluate how age at the time of SCI influences the development of gut dysbiosis nor have there been studies that determine how the gut microbiota changes as people age with SCI. Given the effects of SCI-induced dysautonomia on metabolism, immune function, and gastrointestinal function, the progression of gut dysbiosis in SCI individuals as they age is expected to be unique when compared with age-matched, able-bodied individuals. Likewise, no studies have evaluated whether gut dysbiosis is similarly affected by SCI in males and females. Since published data indicate that the gut microbiota affects serotonin synthesis, metabolism, and neurotransmission differently in males and females, there is reason to predict sex-dependent differences in gut microbial homeostasis after SCI (Bolnick et al., 2014; Clarke et al., 2013; O'Mahony, Clarke, Borre, Dinan, & Cryan, 2015).

Gut dysbiosis and health/disease after SCI

Because gut microbes normally affect the physiological function of organ systems throughout the body, the onset of gut dysbiosis after SCI is likely to cause or contribute to various comorbidities that affect this patient population. Indeed, gut dysbiosis affects able-bodied individuals and has been linked to major depressive disorders and gastrointestinal and metabolic diseases (e.g., obesity, diabetes, Chron's disease, irritable bowel syndrome, etc.) (Carding, Verbeke, Vipond, Corfe, & Owen, 2015; Hsiao et al., 2013; O'Mahony et al., 2015). Dysbiosis and "leaky gut" also have been implicated in the onset and progression of chronic fatigue syndrome (CFS). Chronic fatigue syndrome (CFS) is a multi-system disease characterized by persistent fatigue, cognitive impairment, mood changes, and GI disturbances. Many of these same symptoms also plague people with SCI (Altindag, Karagullu, & Gur, 2014; Jensen, Hirsh, Molton, & Bamer, 2009; Wulff, Gatti, Wettstein, & Foster, 2010). Fatigue affects most SCI individuals (Jensen, Kuehn, Amtmann, & Cardenas, 2007). Fatigue after SCI is contributed to by changes in physiological and behavioral factors including innate neuromuscular ability, motivation, resilience, depressive-like mood swings and is influenced by one's overall sense of the quality of life (Craig, Tran, Wijesuriya, & Middleton, 2012; Wijesuriya, Tran, Middleton, & Craig, 2012). As such, fatigue is closely linked to social and mental health after SCI and gut dysbiosis can significantly affect mental health (Foster & McVey Neufeld, 2013; Jensen et al., 2007).

SCI also enhances the risk or frequency of developing heart disease or metabolic dysfunction (Bauman & Spungen, 2000; Inskip et al., 2010). Although the increased incidence of obesity, diabetes, and liver dysfunction after SCI is often attributed to inactivity, changes in muscle mass and adiposity as well as the onset of insulin resistance and hyperinsulinemia are delayed consequences of SCI. Similarly, cardiovascular disease and the effects of chronic low-grade systemic inflammation do not develop soon after SCI. SCI individuals also suffer from dysfunctional immune responses that exacerbate the above conditions and impair host defense, rendering SCI individuals more susceptible to infection (Brommer et al., 2016; DeVivo, Black, & Stover, 1993; Riegger et al., 2009). This multi-organ "failure", is a slow and insidious disease process that develops in parallel with GI dysfunction, dysbiosis and leaky gut. Although a causal relationship remains unproven, existing therapies could be used to reveal the effects of gut dysbiosis on these disease states. For example, oral probiotics can restore homeostasis to the gut microbiota and in turn, have been shown to positively affect nutrient absorption, mental health, whole-body metabolism, and systemic immune function (Frei, Akdis, & O'mahony, 2015; Judkins, Archer, Kramer, & Solch, 2020; Martin et al., 2008; Mörkl, Butler, Holl, Cryan, & Dinan, 2020). A similar approach might be used to treat people with SCI. We have just begun to understand how changes in the gut microbiome may impact the health of the CNS. Applying newer metagenomic and "multi-omic" tools, it should be possible to reveal how novel, rapidly evolving ecological niches that form in the gut under conditions of dysbiosis, regulate the function and phenotype of cells in the brain or spinal cord. For example, the gut microbiome is already known to shape the transcriptome of adult microglia (Thion et al., 2018), and gut microbiota can alter microglial structure and maturity (Erny et al., 2015). In germ-free mice, adding discrete microbial-derived metabolites (e.g., short-chain fatty acids), even in the absence of the bacteria themselves, can reverse deficits in microglia function (Erny et al., 2015). Under physiological conditions, the gut microbiome regulates the metabolism of several key neurotransmitters. As such, the gut microbiome is intimately associated with normal CNS structure/function. Indeed, most ($\sim$90%) of the body's serotonin is produced in the gut (Gershon & Tack, 2007), and its synthesis is regulated by gut microbiota (Wikoff et al., 2009; Yano et al., 2015). Microbe-dependent metabolism of tryptophan also produces small molecules that can directly influence astrocyte function and alter the glial response to neuroinflammation (Rothhammer et al., 2018, 2016). As research continues to identify small molecules produced by gut bacteria and the genes responsible for their production, it will likely become feasible to manipulate these bacteria or their "drug-like" payloads to treat a range of human neurological disorders. Indeed, gut dysbiosis occurs in people with multiple sclerosis, Parkinson's disease, Alzheimer's disease, epilepsy, autism spectrum disorders, and after stroke and brain injury (Cryan, O'Riordan, Sandhu, Peterson, & Dinan, 2020). Interventional trials using probiotics or diet interventions to treat gut dysbiosis have already shown beneficial effects on various clinical aspects of multiple sclerosis, Parkinson's disease, Alzheimer's disease, autism spectrum disorders, and epilepsy (Cryan et al., 2020) demonstrating the breadth of gut microbe effects on brain function.

Mini-dictionary of terms

Gut microbiota: the bacteria, viruses, fungi, and archaea that live in the gastrointestinal tract.
Gut microbiome: the collective genome of the gut microbiota.
Gut dysbiosis: an imbalance in the normal community structure of gut microbes.
Bacterial translocation: the process of gut microbiota moving from the intestinal lumen across epithelial cells into lymphatics, blood, or other tissues in the body.
GALT: The lymphoid tissues, primarily the mesenteric lymph nodes and Peyer's patches, that reside in the gastrointestinal tract.
Dysautonomia: A composite of medical conditions caused by disruption of normal autonomic nervous system function.

Key facts of the gut microbiome

- The gastrointestinal tract is inhabited by trillions of microbes called the gut microbiota. Most of these microbes are located in the colon.
- The collective genomes of these microbes in the gut microbiome, and these genes outnumber genes in the human genome 150:1
- Gut microbiota harvest energy from food, improve gut motiligy, reinforce the gut barrier, and produce metabolites, vitamins and hormones that are important for health.
- Disruption of the composition of gut microbiota is called *gut dysbiosis*. Gut dysbiosis has been linked to the development or worsening of many health conditionings, including those that affect the CNS.

Summary points

- The gut microbiota affects the metabolic function, mental health, and immune function of the host.
- The gut microbiota communicates to the CNS through immune function, the vagus nerve, and microbial-derived metabolites in the circulation.
- Gut dysbiosis develops after SCI in pre-clinical and clinical studies
- Gut dysbiosis correlates with increased intestinal permeability and inflammation after SCI
- Gut dysbiosis is observed in many CNS diseases, and therapies directed at treating gut dysbiosis are beneficial in multiple sclerosis, Parkinson's disease, epilepsy, and autism.

References

Altindag, O., Karagullu, H., & Gur, A. (2014). Sleep disturbances in patients with spinal cord injury. *Orthopedic & Muscular System: Current Research, 3*, 164.

Bai, C., An, H., Wang, S., Jiang, D., Fan, W., & Nie, H. (2011). Treatment and prevention of bacterial translocation and endotoxemia with stimulation of the sacral nerve root in a rabbit model of spinal cord injury. *Spine (Phila Pa 1976), 36*, 363–371.

Bailey, M. T., Dowd, S. E., Galley, J. D., Hufnagle, A. R., Allen, R. G., & Lyte, M. (2011). Exposure to a social stressor alters the structure of the intestinal microbiota: Implications for stressor-induced immunomodulation. *Brain, Behavior, and Immunity, 25*, 397–407.

Bailey, M. T., Dowd, S. E., Parry, N. M., Galley, J. D., Schauer, D. B., & Lyte, M. (2010). Stressor exposure disrupts commensal microbial populations in the intestines and leads to increased colonization by *Citrobacter rodentium. Infection and Immunity, 78*, 1509–1519.

Balmer, M. L., Slack, E., de Gottardi, A., Lawson, M. A., Hapfelmeier, S., Miele, L., et al. (2014). The liver may act as a firewall mediating mutualism between the host and its gut commensal microbiota. *Science Translational Medicine, 6*, 237.

Balzan, S., de Almeida Quadros, C., de Cleva, R., Zilberstein, B., & Cecconello, I. (2007). Bacterial translocation: Overview of mechanisms and clinical impact. *Journal of Gastroenterology and Hepatology, 22*, 464–471.

Bauman, W. A., & Spungen, A. M. (2000). Metabolic changes in persons after spinal cord injury. *Physical Medicine and Rehabilitation Clinics of North America, 11*, 109–140.

Benakis, C., Brea, D., Caballero, S., Faraco, G., Moore, J., Murphy, M., et al. (2016). Commensal microbiota affects ischemic stroke outcome by regulating intestinal γδ T cells. *Nature Medicine, 22*, 516–523.

Bercik, P., Verdu, E. F., Foster, J. A., Macri, J., Potter, M., Huang, X., et al. (2010). Chronic gastrointestinal inflammation induces anxiety-like behavior and alters central nervous system biochemistry in mice. *Gastroenterology, 139*, 2102–2112.

Berer, K., Mues, M., Koutrolos, M., Rasbi, Z. A., Boziki, M., Johner, C., et al. (2011). Commensal microbiota and myelin autoantigen cooperate to trigger autoimmune demyelination. *Nature, 479*, 538–541.

Bolnick, D. I., Snowberg, L. K., Hirsch, P. E., Lauber, C. L., Org, E., Parks, B., et al. (2014). Individual diet has sex-dependent effects on vertebrate gut microbiota. *Nature Communications, 5*, 4500.

Brommer, B., Engel, O., Kopp, M. A., Watzlawick, R., Müller, S., Prüss, H., et al. (2016). Spinal cord injury-induced immune deficiency syndrome enhances infection susceptibility dependent on lesion level. *Brain, 139*, 692–707.

Browning, K. N., & Travagli, R. A. (2014). Central nervous system control of gastrointestinal motility and secretion and modulation of gastrointestinal functions. *Comprehensive Physiology, 4*, 1339–1368.

Cao, S., Feehley, T. J., & Nagler, C. R. (2014). The role of commensal bacteria in the regulation of sensitization to food allergens. *FEBS Letters, 588*(22), 4258–4266.

Carding, S., Verbeke, K., Vipond, D. T., Corfe, B. M., & Owen, L. J. (2015). Dysbiosis of the gut microbiota in disease. *Microbial Ecology in Health and Disease, 26*, 26191.

Cervi, A. L., Lukewich, M. K., & Lomax, A. E. (2014). Neural regulation of gastrointestinal inflammation: Role of the sympathetic nervous system. *Autonomic Neuroscience, 182*, 83–88.

Clarke, G., Grenham, S., Scully, P., Fitzgerald, P., Moloney, R. D., Shanahan, F., et al. (2013). The microbiome-gut-brain axis during early life regulates the hippocampal serotonergic system in a sex-dependent manner. *Molecular Psychiatry, 18*, 666–673.

Clarke, G., Stilling, R. M., Kennedy, P. J., Stanton, C., Cryan, J. F., & Dinan, T. G. (2014). Gut microbiota: The neglected endocrine organ. *Molecular Endocrinology, 28*(8), 1221–1238.

Collins, S. M., Surette, M., & Bercik, P. (2012). The interplay between the intestinal microbiota and the brain. *Nature Reviews. Microbiology, 10*, 735–742.

Craig, A., Tran, Y., Wijesuriya, N., & Middleton, J. (2012). Fatigue and tiredness in people with spinal cord injury. *Journal of Psychosomatic Research, 73*, 205–210.

Crumeyrolle-Arias, M., Jaglin, M., Bruneau, A., Vancassel, S., Cardona, A., Daugé, V., et al. (2014). Absence of the gut microbiota enhances anxiety-like behavior and neuroendocrine response to acute stress in rats. *Psychoneuroendocrinology, 42*, 207–217.

Cryan, J. F., O'Riordan, K. J., Sandhu, K., Peterson, V., & Dinan, T. G. (2020). The gut microbiome in neurological disorders. *Lancet Neurology, 9*(2), 179–194.

de Theije, C. G., Wopereis, H., Ramadan, M., van Eijndthoven, T., Lambert, J., Knol, J., et al. (2014). Altered gut microbiota and activity in a murine model of autism spectrum disorders. *Brain, Behavior, and Immunity, 37*, 197–206.

DeVivo, M. J., Black, K. J., & Stover, S. L. (1993). Causes of death during the first 12 years after spinal cord injury. *Archives of Physical Medicine and Rehabilitation, 74*, 248–254.

Devivo, M. J., Kartus, P. L., Rutt, R. D., Stover, S. L., & Fine, P. R. (1990). The influence of age at time of spinal cord injury on rehabilitation outcome. *Archives of Neurology, 47*, 687–691.

Donaldson, G. P., Lee, S. M., & Mazmanian, S. K. (2016). Gut biogeography of the bacterial microbiota. *Nature Reviews. Microbiology, 14*, 20–32.

Du, J., Zayed, A. A., Kigerl, K. A., Zane, K., Sullivan, M. B., & Popovich, P. G. (2021). Spinal cord injury changes the structure and functional potential of gut bacterial and viral communities in a spinal-level dependent manner. *mSystems, 6*(3). https://doi.org/10.1128/mSystems.01356-20, e01356-20.

El Aidy, S., van den Bogert, B., & Kleerebezem, M. (2015). The small intestine microbiota, nutritional modulation and relevance for health. *Current Opinion in Biotechnology, 32*, 14–20.

Elenkov, I. J., Wilder, R. L., Chrousos, G. P., & Vizi, E. S. (2000). The sympathetic nerve—An integrative interface between two supersystems: The brain and the immune system. *Pharmacological Reviews, 52*, 595–638.

Erny, D., de Angelis, A. L., Mahlakoiv, T., Buch, T., Chun, E., Schwierzeck, V., et al. (2015). Host microbiota constantly control maturation and function of microglia in the CNS. *Nature Neuroscience, 18*, 965–977.

Felten, D. L., Ackerman, K. D., Wiegand, S. J., & Felten, S. Y. (1987). Noradrenergic sympathetic innervation of the spleen: I. Nerve fibers associate with lymphocytes and macrophages in specific compartments of the splenic white pulp. *Journal of Neuroscience Research, 18*(28–36), 118–121.

Felten, S. Y., & Olschowka, J. (1987). Noradrenergic sympathetic innervation of the spleen: II. Tyrosine hydroxylase (TH)-positive nerve terminals form synapticlike contacts on lymphocytes in the splenic white pulp. *Journal of Neuroscience Research, 18*, 37–48.

Forsythe, P., Bienenstock, J., & Kunze, W. A. (2014). Vagal pathways for microbiome-brain-gut axis communication. *Advances in Experimental Medicine and Biology, 817*, 115–133.

Foster, J. A., & McVey Neufeld, K. A. (2013). Gut-brain axis: How the microbiome influences anxiety and depression. *Trends in Neurosciences, 36*, 305–312.

Frei, R., Akdis, M., & O'mahony, L. (2015). Prebiotics, probiotics, synbiotics, and the immune system: Experimental data and clinical evidence. *Current Opinion in Gastroenterology, 31*, 153–158. 1.

Frioux, C., Singh, D., Korcsmaros, T., & Hildebrand, F. (2020). From bag-of-genes to bag-of-genomes: Metabolic modelling of communities in the era of metagenome-assembled genomes. *Computational and Structural Biotechnology Journal, 18*, 1722–1734.

Furness, J. B., & Costa, M. (1974). The adrenergic innervation of the gastrointestinal tract. *Ergebnisse der Physiologie, biologischen Chemie und experimentellen Pharmakologie, 69*(0), 2–51.

Gater, D. R. (2007). Obesity after spinal cord injury. *Physical Medicine and Rehabilitation Clinics of North America, 18*(333–51), vii.

Gérard, P. (2016). Gut microbiota and obesity. *Cellular and Molecular Life Sciences, 73*(1), 147–162.

Gershon, M. D., & Tack, J. (2007). The serotonin signaling system: From basic understanding to drug development for functional GI disorders. *Gastroenterology, 132*, 397–414.

Gill, S. R., Pop, M., Deboy, R. T., Eckburg, P. B., Turnbaugh, P. J., Samuel, B. S., et al. (2006). Metagenomic analysis of the human distal gut microbiome. *Science, 312*(5778), 1355–1359.

Gorgey, A. S., Dolbow, D. R., Dolbow, J. D., Khalil, R. K., Castillo, C., & Gater, D. R. (2014). Effects of spinal cord injury on body composition and metabolic profile—Part I. *The Journal of Spinal Cord Medicine, 37*(6), 693–702.

Gorgey, A. S., & Gater, D. R. (2007). Prevalence of obesity after spinal cord injury. *Topics in Spinal Cord Injury Rehabilitation, 12*, 1–7.

Gorgey, A. S., & Gater, D. R. (2011). Regional and relative adiposity patterns in relation to carbohydrate and lipid metabolism in men with spinal cord injury. *Applied Physiology, Nutrition, and Metabolism, 36*, 107–114.

Gu, S., Chen, D., Zhang, J. N., Lv, X., Wang, K., Duan, L. P., et al. (2013). Bacterial community mapping of the mouse gastrointestinal tract. *PLoS One, 8*, e74957.

Gungor, B., Adiguzel, E., Gursel, I., Yilmaz, B., & Gursel, M. (2016). Intestinal microbiota in patients with spinal cord injury. *PLoS One, 11*, e0145878.

Hawrelak, J. A., & Myers, S. P. (2004). The causes of intestinal dysbiosis: A review. *Alternative Medicine Review, 9*, 180–197.

Hill, D. A., & Artis, D. (2010). Intestinal bacteria and the regulation of immune cell homeostasis. *Annual Review of Immunology, 28*, 623–667.

Hollister, E. B., Gao, C., & Versalovic, J. (2014). Compositional and functional features of the gastrointestinal microbiome and their effects on human health. *Gastroenterology, 146*, 1449–1458.

Holmes, G. M., & Blanke, E. N. (2019). Gastrointestinal dysfunction after spinal cord injury. *Experimental Neurology, 320*, 113009.

Hooper, L. V., Littman, D. R., & Macpherson, A. J. (2012). Interactions between the microbiota and the immune system. *Science, 336*, 1268–1273.

Hou, S., & Rabchevsky, A. G. (2014). Autonomic consequences of spinal cord injury. *Comprehensive Physiology, 4*, 1419–1453.

Hsiao, E. Y., McBride, S. W., Hsien, S., Sharon, G., Hyde, E. R., McCue, T., et al. (2013). Microbiota modulate behavioral and physiological abnormalities associated with neurodevelopmental disorders. *Cell, 155*, 1451–1463.

Inskip, J., Plunet, W., Ramer, L., Ramsey, J. B., Yung, A., Kozlowski, P., et al. (2010). Cardiometabolic risk factors in experimental spinal cord injury. *Journal of Neurotrauma, 27*, 275–285.

Jašarević, E., Morrison, K. E., & Bale, T. L. (2016). Sex differences in the gut microbiome—Brain axis across the lifespan. *Philosophical Transactions of the Royal Society of London. Series B, Biological Sciences, 371*(1688), 20150122.

Jensen, M. P., Hirsh, A. T., Molton, I. R., & Bamer, A. M. (2009). Sleep problems in individuals with spinal cord injury: Frequency and age effects. *Rehabilitation Psychology, 54*, 323–331.

Jensen, M. P., Kuehn, C. M., Amtmann, D., & Cardenas, D. D. (2007). Symptom burden in persons with spinal cord injury. *Archives of Physical Medicine and Rehabilitation, 88*, 638–645.

Jing, Y., Yang, D., Bai, F., Zhang, C., Qin, C., Li, D., et al. (2019). Melatonin treatment alleviates spinal cord injury-induced gut dysbiosis in mice. *Journal of Neurotrauma, 36*, 2646–2664.

Judkins, T. C., Archer, D. L., Kramer, D. C., & Solch, R. J. (2020). Probiotics, nutrition, and the small intestine. *Current Gastroenterology Reports, 22*(1), 2.

Kigerl, K. A., Hall, J. C. E., Wang, L., Mo, X., Yu, Z., & Popovich, P. G. (2016). Gut dysbiosis impairs recovery after spinal cord injury. *The Journal of Experimental Medicine, 213*, 2603–2620.

Kriegel, M. A., Sefik, E., Hill, J. A., Wu, H. J., Benoist, C., & Mathis, D. (2011). Naturally transmitted segmented filamentous bacteria segregate with diabetes protection in nonobese diabetic mice. *Proceedings of the National Academy of Sciences of the United States of America, 108*, 11548–11553.

Langille, M. G. I., Zaneveld, J., Caporaso, J. G., McDonald, D., Knights, D., Reyes, J. A., et al. (2013). Predictive functional profiling of microbial communities using 16S rRNA marker gene sequences. *Nature Biotechnology, 31*, 814–821.

Lee, Y. K., Menezes, J. S., Umesaki, Y., & Mazmanian, S. K. (2011). Proinflammatory T-cell responses to gut microbiota promote experimental autoimmune encephalomyelitis. *Proceedings of the National Academy of Sciences of the United States of America, 108*(Suppl), 4615–4622.

Levatte, M. A., Mabon, P. J., Weaver, L. C., & Dekaban, G. A. (1998). Simultaneous identification of two populations of sympathetic preganglionic neurons using recombinant herpes simplex virus type 1 expressing different reporter genes. *Neuroscience, 82*, 1253–1267.

Li, J., Van Der Pol, W., Eraslan, M., McLain, A., Cetin, H., Cetin, B., et al. (2020). Comparison of the gut microbiome composition among individuals with acute or long-standing spinal cord injury vs. able-bodied controls. *The Journal of Spinal Cord Medicine, 4*, 1–9.

Lin, R., Xu, J., Ma, Q., Chen, M., Wang, L., Wen, S., et al. (2020). Alterations in the fecal microbiota of patients with spinal cord injury. *PLoS One, 15*(8), e0236470.

Liu, J., An, H., Jiang, D., Huang, W., Zou, H., Meng, C., et al. (2004). Study of bacterial translocation from gut after paraplegia caused by spinal cord injury in rats. *Spine (Phila Pa 1976), 29*, 164–169.

Mabon, P. J., LeVatte, M. A., Dekaban, G. A., & Weaver, L. C. (1997). Identification of sympathetic preganglionic neurons controlling the small intestine in hamsters using a recombinant herpes simplex virus type-1. *Brain Research, 753*, 245–250.

Macpherson, A. J., & Smith, K. (2006). Mesenteric lymph nodes at the center of immune anatomy. *The Journal of Experimental Medicine, 203*, 497–500.

Manns, P. J., McCubbin, J. A., & Williams, D. P. (2005). Fitness, inflammation, and the metabolic syndrome in men with paraplegia. *Archives of Physical Medicine and Rehabilitation, 86*, 1176–1181.

Markle, J. G. M., Frank, D. N., Mortin-Toth, S., Robertson, C. E., Feazel, L. M., Rolle-Kampczyk, U., et al. (2013). Sex differences in the gut microbiome drive hormone-dependent regulation of autoimmunity. *Science, 339*, 1084–1088.

Martin, F. P. J., Wang, Y., Sprenger, N., Yap, I. K. S., Lundstedt, T., Lek, P., et al. (2008). Probiotic modulation of symbiotic gut microbial-host metabolic interactions in a humanized microbiome mouse model. *Molecular Systems Biology, 4*, 157.

Maruyama, Y., Mizuguchi, M., Yaginuma, T., Kusaka, M., Yoshida, H., Yokoyama, K., et al. (2008). Serum leptin, abdominal obesity and the metabolic syndrome in individuals with chronic spinal cord injury. *Spinal Cord, 46*, 494–499.

Mörkl, S., Butler, M. I., Holl, A., Cryan, J. F., & Dinan, T. G. (2020). Probiotics and the microbiota-gut-brain axis: Focus on psychiatry. *Current Nutrition Reports, 9*, 171–182.

Myers, S. A., Gobejishvili, L., Saraswat Ohri, S., Garrett Wilson, C., Andres, K. R., Riegler, A. S., et al. (2019). Following spinal cord injury, PDE4B drives an acute, local inflammatory response and a chronic, systemic response exacerbated by gut dysbiosis and endotoxemia. *Neurobiology of Disease, 124*, 353–363.

National Spinal Cord Injury Statistical Center. (2020). *Spinal cord injury facts and figures at a glance 2020 SCI data sheet.*

Nelson, M. D., Widman, L. M., Abresch, R. T., Stanhope, K., Havel, P. J., Styne, D. M., et al. (2007). Metabolic syndrome in adolescents with spinal cord dysfunction. *The Journal of Spinal Cord Medicine*, *30*(Suppl. 1), S127–S139.

Nicholson, J. K., Holmes, E., Kinross, J., Burcelin, R., Gibson, G., Jia, W., et al. (2012). Host-gut microbiota metabolic interactions. *Science*, *336*, 1262–1267.

Niu, S. Y., Yang, J., McDermaid, A., Zhao, J., Kang, Y., & Ma, Q. (2017). Bioinformatics tools for quantitative and functional metagenome and meta-transcriptome data analysis in microbes. *Briefings in Bioinformatics*, *19*, 1415–1429.

O'Connor, G., Jeffrey, E., Madorma, D., Marcillo, A., Abreu, M. T., Deo, S. K., et al. (2018). Investigation of microbiota alterations and intestinal inflammation post-spinal cord injury in rat model. *Journal of Neurotrauma*, *35*, 2159–2166.

O'Mahony, S. M., Clarke, G., Borre, Y. E., Dinan, T. G., & Cryan, J. F. (2015). Serotonin, tryptophan metabolism and the brain-gut-microbiome axis. *Behavioural Brain Research*, *277*, 32–48.

Qualls-Creekmore, E., Tong, M., & Holmes, G. M. (2010). Time-course of recovery of gastric emptying and motility in rats with experimental spinal cord injury. *Neurogastroenterology and Motility*, *22*(62–9), e27–e28.

Riegger, T., Conrad, S., Schluesener, H. J., Kaps, H. P., Badke, A., Baron, C., et al. (2009). Immune depression syndrome following human spinal cord injury (SCI): A pilot study. *Neuroscience*, *158*, 1194–1199.

Rothhammer, V., Borucki, D. M., Tjon, E. C., Takenaka, M. C., Chao, C. C., Ardura-Fabregat, A., et al. (2018). Microglial control of astrocytes in response to microbial metabolites. *Nature*, *557*, 724–728.

Rothhammer, V., Mascanfroni, I. D., Bunse, L., Takenaka, M. C., Kenison, J. E., Mayo, L., et al. (2016). Type I interferons and microbial metabolites of tryptophan modulate astrocyte activity and central nervous system inflammation via the aryl hydrocarbon receptor. *Nature Medicine*, *22*, 586–597.

Round, J. L., & Mazmanian, S. K. (2009). The gut microbiota shapes intestinal immune responses during health and disease. *Nature Reviews. Immunology*, *9*, 313–323.

Rousseaux, C., Thuru, X., Gelot, A., Barnich, N., Neut, C., Dubuquoy, L., et al. (2007). *Lactobacillus acidophilus* modulates intestinal pain and induces opioid and cannabinoid receptors. *Nature Medicine*, *13*, 35–37.

Safari, Z., & Gérard, P. (2019). The links between the gut microbiome and non-alcoholic fatty liver disease (NAFLD). *Cellular and Molecular Life Sciences*, *76*(8), 1541–1558.

Schmidt, E. K. A., Torres-Espin, A., Raposo, P. J. F., Madsen, K. L., Kigerl, K. A., Popovich, P. G., et al. (2020). Fecal transplant prevents gut dysbiosis and anxiety-like behaviour after spinal cord injury in rats. *PLoS One*, *15*(1), e0226128.

Schönenberg, M., Reimitz, M., Jusyte, A., Maier, D., Badke, A., & Hautzinger, M. (2014). Depression, posttraumatic stress, and risk factors following spinal cord injury. *International Journal of Behavioral Medicine*, *21*(1), 169–176.

Sender, R., Fuchs, S., & Milo, R. (2016). Revised estimates for the number of human and bacteria cells in the body. *PLoS Biology*, *14*(8), e1002533.

Sheng, L., Jena, P. K., Liu, H. X., Kalanetra, K. M., Gonzalez, F. J., French, S. W., et al. (2017). Gender differences in bile acids and microbiota in relationship with gender dissimilarity in steatosis induced by diet and FXR inactivation. *Scientific Reports*, *7*, 1–12.

Stilling, R. M., Ryan, F. J., Hoban, A. E., Shanahan, F., Clarke, G., Claesson, M. J., et al. (2015). Microbes & neurodevelopment—Absence of microbiota during early life increases activity-related transcriptional pathways in the amygdala. *Brain, Behavior, and Immunity*, *50*, 209–220.

Sudo, N., Chida, Y., Aiba, Y., Sonoda, J., Oyama, N., Yu, X. N., et al. (2004). Postnatal microbial colonization programs the hypothalamic-pituitary-adrenal system for stress response in mice. *The Journal of Physiology*, *558*, 263–275.

Tan, J., McKenzie, C., Potamitis, M., Thorburn, A. N., Mackay, C. R., & Macia, L. (2014). The role of short-chain fatty acids in health and disease. *Advances in Immunology*, *121*, 91–119.

Tate, D. G., Forchheimer, M., Rodriguez, G., Chiodo, A., Cameron, A. P., Meade, M., et al. (2016). Risk factors associated with neurogenic bowel complications and dysfunction in spinal cord injury. *Archives of Physical Medicine and Rehabilitation*, *97*, 1679–1686.

Thevaranjan, N., Puchta, A., Schulz, C., Naidoo, A., Szamosi, J. C., Verschoor, C. P., et al. (2017). Age-associated microbial dysbiosis promotes intestinal permeability, systemic inflammation, and macrophage dysfunction. *Cell Host & Microbe*, *21*, 455–466.

Thion, M. S., Low, D., Silvin, A., Chen, J., Grisel, P., Schulte-Schrepping, J., et al. (2018). Microbiome influences prenatal and adult microglia in a sex-specific manner. *Cell*, *172*, 500–516.

Tilg, H., & Kaser, A. (2011). Gut microbiome, obesity, and metabolic dysfunction. *The Journal of Clinical Investigation*, *121*, 2126–2132.

Tillisch, K. (2014). The effects of gut microbiota on CNS function in humans. *Gut Microbes*, *5*(3), 404–410.

Vendrik, K. E. W., Ooijevaar, R. E., de Jong, P. R. C., Laman, J. D., van Oosten, B. W., van Hilten, J. J., et al. (2020). Fecal microbiota transplantation in neurological disorders. *Frontiers in Cellular and Infection Microbiology*, *10*, 98.

Wang, Y., & Kasper, L. H. (2014). The role of microbiome in central nervous system disorders. *Brain, Behavior, and Immunity*, *38*, 1–12.

Wijesuriya, N., Tran, Y., Middleton, J., & Craig, A. (2012). Impact of fatigue on the health-related quality of life in persons with spinal cord injury. *Archives of Physical Medicine and Rehabilitation*, *93*, 319–324.

Wikoff, W. R., Anfora, A. T., Liu, J., Schultz, P. G., Lesley, S. A., Peters, E. C., et al. (2009). Metabolomics analysis reveals large effects of gut microflora on mammalian blood metabolites. *Proceedings of the National Academy of Sciences of the United States of America*, *106*, 3698–3703.

Wilson, J. R., Cadotte, D. W., & Fehlings, M. G. (2012). Clinical predictors of neurological outcome, functional status, and survival after traumatic spinal cord injury: A systematic review. *Journal of Neurosurgery. Spine*, *17*, 11–26.

Winek, K., Engel, O., Koduah, P., Heimesaat, M. M., Fischer, A., Bereswill, S., et al. (2016). Depletion of cultivatable gut microbiota by broad-spectrum antibiotic pretreatment worsens outcome after murine stroke. *Stroke*, *47*, 1354–1363.

Wong, S., Santullo, P., Hirani, S. P., Kumar, N., Chowdhury, J. R., García-Forcada, A., et al. (2017). Use of antibiotics and the prevalence of antibiotic-associated diarrhoea in patients with spinal cord injuries: An international, multi-centre study. *Journal of Hospital Infection*, *97*(2), 146–152.

Wulff, K., Gatti, S., Wettstein, J. G., & Foster, R. G. (2010). Sleep and circadian rhythm disruption in psychiatric and neurodegenerative disease. *Nature Reviews. Neuroscience*, *11*, 589–599.

Yano, J. M., Yu, K., Donaldson, G. P., Shastri, G. G., Ann, P., Ma, L., et al. (2015). Indigenous bacteria from the gut microbiota regulate host serotonin biosynthesis. *Cell*, *161*, 264–276.

Zhang, C., Jing, Y., Zhang, W., Zhang, J., Yang, M., Du, L., et al. (2019). Dysbiosis of gut microbiota is associated with serum lipid profiles in male patients with chronic traumatic cervical spinal cord injury. *American Journal of Translational Research*, *11*, 4817–4834.

Zhang, C., Zhang, W., Zhang, J., Jing, Y., Yang, M., Du, L., et al. (2018). Gut microbiota dysbiosis in male patients with chronic traumatic complete spinal cord injury. *Journal of Translational Medicine*, *16*, 353.

Section D

Behavioral and psychological effects

Chapter 37

Risk factors and predictors of depression after spinal cord injury: Emphasis on the inflammatory process

Caroline Cunha do Espírito Santo[a], Fernando da Silva Fiorin[b], and Luiz Fernando Freire Royes[c]

[a]Neuromotor System Laboratory, Spinal Cord Injury Rehabilitation Research Group, NuLeMe, Santa Catarina State University, Florianópolis, SC, Brazil, [b]Graduate Program in Neuroengineering, Edmond and Lily Safra International Institute of Neuroscience, Santos Dumont Institute, Macaíba, RN, Brazil, [c]Exercise Biochemistry Laboratory, Center of Physical Education and Sports, Federal University of Santa Maria, Santa Maria, RS, Brazil

Abbreviations

5-HT	5-hydroxtryptamine or serotonin
ACTH	adrenocorticotropic hormone
AIS	American spinal injury association impairment scale
ASIA	American spinal injury association
BDI	beck depression inventory
CEDS	center for epidemiological depression scale
CRH	corticotropin-releasing hormone
CSS	clinically significant symptomatology
GC	glucocorticoids
GM-CSF	granulocyte macrophage-colony stimulating factor
GR	glucorticoids receptor
HADS	hospital anxiety and depression scale
HPA	hypothalamus-pituitary-adrenal axis
IDO	indoleamine 2,3-dioxygenase
IL	interleukins
IL-2R	interleukin-2 receptor
INF-(γ and α)	interferon-gamma and -alpha
ISNCSCI	international standards for neurological classification of spinal cord injury
MDD	mood depressive disorders
OAHMQ	older adult health and mood questionnaire
PHQ-9	patient health questionnaire–9
PMD	probable major depression
SCI	spinal cord injury
SCI-QOL	spinal cord injury–quality of life
TNF-α	tumor necrosis factor-alpha
YLD	years lived with disability

Introduction

Spinal cord injury (SCI) is a significant public health concern that imposes many challenges on survivors. Individuals face physical limitations, including pain, loss of functional independence, and reduced self-care capacity, with a significant number of patients suffering from mental disorders, such as depression (Bombardier, Richards, Krause, Tulsky, & Tate, 2004; Hoffman, Bombardier, Graves, Kalpakjian, & Krause, 2011; Oh, Shin, Paik, Yoo, & Ku, 2006). Table 1 provides more detail on the clinical aspects of individuals with SCI.

Cellular, Molecular, Physiological, and Behavioral Aspects of Spinal Cord Injury. https://doi.org/10.1016/B978-0-12-822427-4.00037-X

TABLE 1 Summary of clinical aspects encountered in individuals with spinal cord injury.

Motor outcomes	Muscle weakness	Sensory outcomes	Loss of light touch sensation	Autonomic outcomes	Bowel dysfunction	Mental and cognitive outcomes	Depression
	Muscle atrophy		Loss of proprioception		Bladder dysfunction		Anxiety
	Muscle contracture		Paresthesia		Autonomic dysreflexia		Deficit in information processing speed
	Spasticity		Chronic pain		Orthostatic hypotension		Deficit in learning
	Fatigue		Pressure ulcer		Sexual dysfunction		Decline in verbal fluency

Source: Authors based in Cieza, A., Kirchberger, I., Biering-Sørensen, F., Baumberger, M., Charlifue, S., Post, M. W., et al. (2010). ICF Core sets for individuals with spinal cord injury in the long-term context. *Spinal Cord, 48*(4), 305–312 and Chiaravalloti, N. D., Weber, E., Wylie, G., Dyson-Hudson, T., & Wecht, J. M. (2020). Patterns of cognitive deficits in persons with spinal cord injury as compared with both age-matched and older individuals without spinal cord injury. *The Journal of Spinal Cord Medicine, 43*(1), 88–97.

Research has demonstrated that post-SCI depression negatively affects rehabilitation (Dryden et al., 2005) and the individual's reintegration into society (Craig, Nicholson Perry, Guest, Tran, & Middleton, 2015). This concerning scenario underlines the importance of adequate clinical management of depression, including diagnosis and treatment, in preventing the worsening of adverse motor outcomes in SCI patients.

Although the diagnosis of depression has improved, its etiological factors are still poorly known. Significant efforts have been made to establish a relationship between the physical impairment (injury level and severity) and post-SCI depression (Arango-Lasprilla, Ketchum, Starkweather, Nicholls, & Wilk, 2011; Bombardier et al., 2012; Shin, Goo, Yu, Kim, & Yoon, 2012) but conclusions are yet to be consistent. In contrast, biological factors such as inflammation, which accompanies SCI, have a well-known role in depression (Capuron & Miller, 2011; Dantzer, O'Connor, Freund, Johnson, & Kelley, 2008).

In humans with SCI, increased circulating pro-inflammatory mediators, such as interleukin (IL)-2, IL-6, IL-1α, and TNF-α (tumor necrosis factor-alpha) (Davies, Hayes, & Dekaban, 2007; Gibson, Buchholz, & Ginis, 2008; Hayes et al., 2002), have been previously evidenced and associated with depressive symptoms after SCI (Allison & Ditor, 2015). Similarly, SCI animal models have revealed a correlation between inflammation and depression-like behavior (Do Espírito Santo et al., 2019; Maldonado-Bouchard et al., 2016), and these findings led to the inflammatory process being recently recognized as an etiological agent of depression post-SCI.

This chapter will address the role of the systemic inflammatory process as a robust biological predictor and summarize the demographic, social, environmental, and behavioral risk factors for the development of depression after SCI.

Pathophysiological overview: From local inflammation to systemic inflammation after SCI

Spinal cord injury is a clinical condition with no cure and is usually associated with damage that culminates in numerous symptoms responsible for chronic disability in survivors. As a result, extensive research has been proposed and focused on restorative treatments, such as cellular and molecular therapies and rehabilitation strategies. According to the National Spinal Cord Injury Statistical Center, the incidence of SCI is approximately 54 cases/million population/year in the United States of America, with a prevalence of 285 thousand individuals injured every year (NSCISC, 2016). Furthermore, SCI contributes to increased morbidity and mortality rates in both men and women (aged 32.4 years), although more men are affected than women (ratio 4:1) (Rahimi-Movaghar et al., 2013).

This neurological condition is caused by disruption of ascending and descending white matter tracts in the spinal cord after SCI and is classified using the International Standards for Neurological Classification of Spinal Cord Injury (ISNCSCI) (ASIA and ISCoS, 2019). This standardized method, in association with the American Spinal Injury Association (ASIA) Impairment Scale (AIS), defines the neurological level of injury (tetraplegia or paraplegia) and severity (complete or incomplete) of SCI (Kirshblum, Snider, Rupp, & Read, 2020).

Traumatic SCI is a predominant form of SCI (Halvorsen et al., 2019) whose pathophysiology involves primary mechanical injury (e.g., contusion, compression, and/or laceration) to the spinal cord parenchyma (Kakulas, 2019) and secondary endogenous lesion, which leads to severe tissue damage mainly due to the inflammatory process. Regardless of the type of primary mechanical injury, the initial impact on the spinal cord causes edema, hemorrhage, and necrosis (Kakulas & Kaelan, 2015). Moreover, mRNA transcript levels of pro-inflammatory cytokines increase in the immediate stage of SCI (Bartholdi & Schwab, 1997; Salewski, Emrani, & Fehlings, 2013) (Fig. 1A). The robust inflammation in

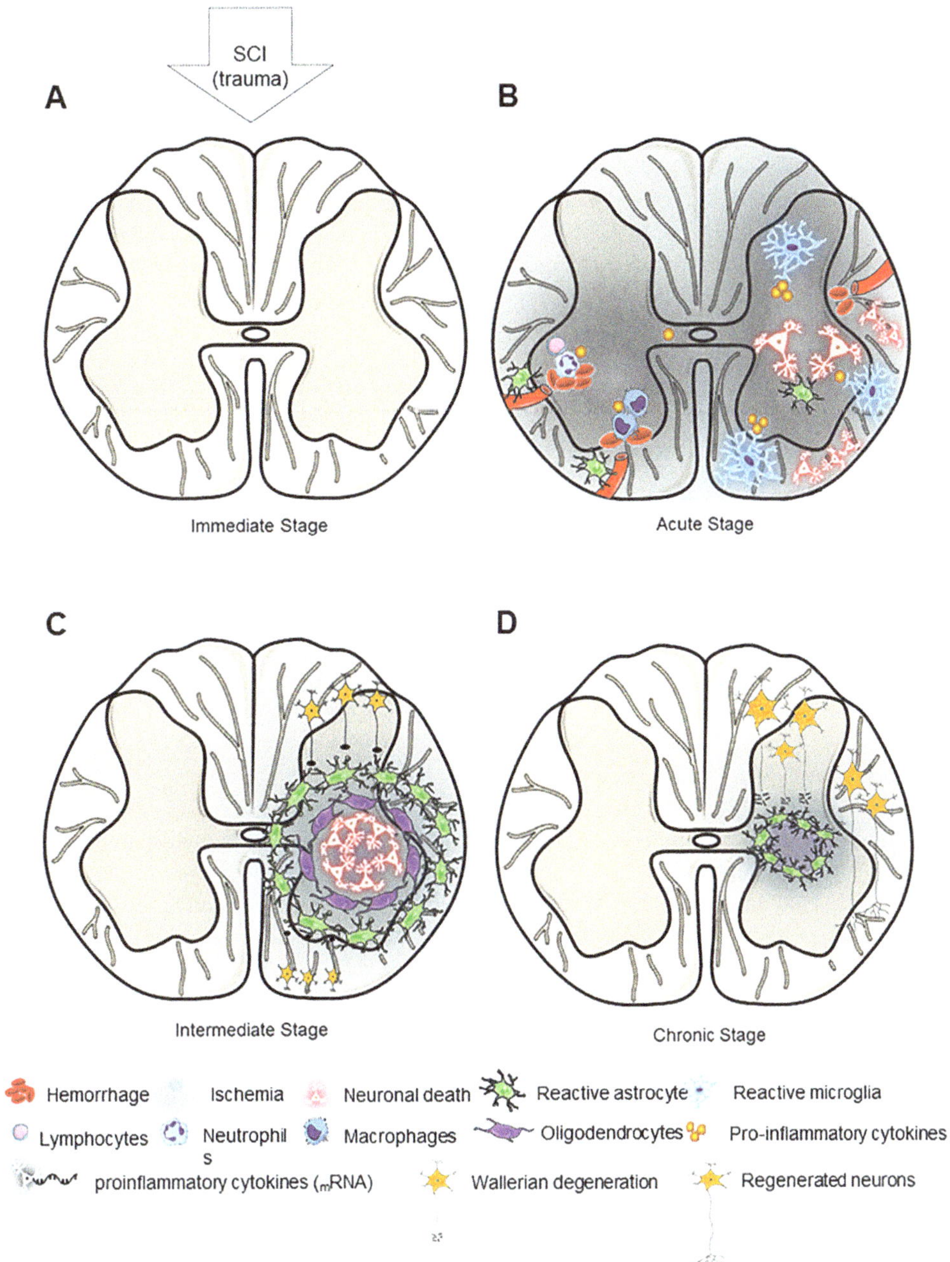

FIG. 1 Pathophysiology of spinal cord injury. Primary and secondary mechanisms of spinal cord trauma. (A) Immediate stage; (B) acute stage; (C) intermediate stage; (D) chronic stage. *Source: Authors.*

response to the immediate events results in secondary endogenous lesion, which is subdivided into acute, intermediate, and chronic phases (Salewski et al., 2013) (Fig. 1B–D).

After SCI, immune system activation is involved in the progression and maintenance of the inflammatory state. The release of TNF-α and IL-1β by resident microglia attracts innate immune cells, including neutrophils and monocytes, to the injury site. The monocyte-derived macrophage increases inflammation by releasing more pro-inflammatory cytokines (David & Kroner, 2011). In addition, T-lymphocyte infiltration increases interferon-gamma (INF-γ) levels (Sun et al., 2018), corroborating macrophage polarization from M1 to M2 phenotype (Kigerl et al., 2009) (Fig. 1B). The hostile microenvironment provokes glial fibrillary acidic protein overexpression, astrocyte proliferation, and astroglial scar formation in the damaged area (Kakulas, 2019; Salewski et al., 2013) (Fig. 1C). In the last stage, injured axons undergo demyelination and fragmentation of the distal segment in a phenomenon described as Wallerian degeneration (Kakulas & Kaelan, 2015; Salewski et al., 2013) (Fig. 1D).

Inflammation is an important part of SCI pathophysiology with evident peripheral immune system involvement in response to spinal cord damage. In addition to local inflammation, there is the occurrence of systemic inflammation, whose causes and consequences are still debatable. Serum concentrations of IL-2, IL-2R, IL-6, TNF-α, IL-1β, and INF-γ are elevated in the acute, subchronic, and chronic stages in animal models (Do Espírito Santo et al., 2019) and humans with SCI (Allison, Josse, Gabriel, Klentrou, & Ditor, 2017; Davies et al., 2007; Hayes et al., 2002).

After SCI, the systemic inflammatory response is counterbalanced by a compensatory anti-inflammatory response responsible for releasing molecules such as IL-10 and glucocorticoids (GC). However, the disruption in central sympathetic control triggers an unregulated anti-inflammatory response (Meisel, Schwab, Prass, Meisel, & Dirnagl, 2005; Schwab, Zhang, Kopp, Brommer, & Popovich, 2014), leading to chronic immune dysfunction, which may alter the functioning of multiple organs, including the brain (Schwab et al., 2014; Sun et al., 2016). Cognitive and psychological dysfunctions observed after SCI point to impairments in distinct brain areas and have been associated with neuroinflammation in the cerebral cortex and hippocampus (Maldonado-Bouchard et al., 2016; Wu et al., 2014), which are both anatomic structures related to emotion (Liu et al., 2017). Together, these findings help illustrate the crosstalk between local inflammation, systemic inflammation, and depressive disorders after SCI.

Depression and spinal cord injury

Epidemiology, assessment, and risk factors of depression after SCI

According to the World Health Organization (WHO), depression is a prevalent mental disorder in the world population. On a global scale, roughly 4.4% of individuals are diagnosed with depression. People of all ages, sexes and social classes are susceptible to developing mental disorders of this nature, although the risk of becoming depressed is more significant in adult women living in underdeveloped and developing countries (WHO, 2017). The concern of these psychiatric disorders is consistent with the high rates of disability in the world (WHO, 2017). The years lived with disability (YLD) represented 7.5% of the years of productive life lost due to depression in 2015. In 2020, depression was predicted to occupy the global position as the second-largest cause of YLD (Ferrari et al., 2013).

Depressive disorders are divided into subcategories, with major depressive disorder (MDD) being the category that best represents the classic scenario of depression. According to the American Psychiatric Association, the presence of at least four symptoms (out of 9 established by the Diagnostic and Statistical Manual of Mental Disorders - DSM–5), besides the core symptoms of depressed mood and anhedonia, serve as a basis for diagnosing MDD. These symptoms occur most of the day, almost every day, for at least 2 weeks (American Psychiatric Association, 2013).

Post-SCI depression has been recognized by specialists for over 40 years, with initial studies dating back to the 1970s. Since then, physical problems and behavioral changes such as reduced social activities, family conflicts, and fewer interactions with friends have been reported in SCI survivors (North, 1999). More current data have shown that the prevalence of MDD after SCI ranges from 18.7% to 26.3% (Williams & Murray, 2015), is estimated as at least three times more frequent than in the world population.

Depressive symptoms can be present in the acute and chronic stages after SCI (Table 2). It is estimated that approximately 10% of patients present depression during rehabilitation in the acute stage, 8.6% in discharge, and 14.1% 6 months after discharge (Craig et al., 2015). After 1 year, depressive symptoms range from 11.9% (Arango-Lasprilla et al., 2011) to 20.6% (Hoffman et al., 2011), while this spectrum varies from 9.7% (Arango-Lasprilla et al., 2011) to 18.1% (Hoffman et al., 2011) at 5 years. Regarding the long-term prevalence of depression, 8.7% of individuals with depression after 1 year remain in similar clinical conditions 5 years after SCI (Hoffman et al., 2011). Additionally, clinically significant symptomatology (CSS) and probable major depression (PDM) 2 years after SCI have higher chances of persisting over time,

TABLE 2 Contribution of neurological condition for developing depressive disorders after spinal cord injury.

Authors	Participants/% of participants with depression	Neurological conditions	Neurological conditions vs depressive symptoms	Risk factors to depression
Krause, Kemp, and Coker (2000)	*n*: 171 participants Clinically significant symptoms[a]: 24% Probable major depression[a]: 24%	+T/P I/+C Chronic SCI	Tetraplegia vs Paraplegia: Ø Complete vs incomplete: Ø	Aged, sex or ethnicity, education and income.
Bombardier et al. (2004)	*n*: 849 participants Mild, moderate and severe[b]: 20%, 11.9% and 4.4%	+T/P +I/C Chronic SCI	Tetraplegia vs Paraplegia: Ø Complete vs incomplete: Ø	–
Oh et al. (2006)	*n*: 102 participants Mild, moderate and severe[c]: 3.9%, 23.5% and 69.9%	T/+P - Chronic SCI	Tetraplegia vs Paraplegia: Ø	Sex and dependence for perform catheterization
Bombardier et al. (2012)	*n*: 244 participants Mild[b]: 39%	+T/P +I/C Acute and chronic SCI	- Complete vs incomplete: ↑ incomplete	Lower availability of reinforcement in the environment (Limited positive reinforced related to health)
Fann et al. (2011)	*n*: 947 participants Moderate[b]: 23% Moderately severe and severe depression[b]: 9%	+T/P - Acute and chronic SCI	Tetraplegia vs Paraplegia: Ø	–
Hassanpour et al. (2012)	*n*: 130 participants Mild[c]: 30% Moderate to severe[c]: < 5%	T/+P +I/C Acute and chronic SCI	Tetraplegia vs Paraplegia: Ø Complete vs incomplete: Ø	
Arango-Lasprilla et al. (2011)	*n: 2.830 participants* At year 1[b]: 11.9% At year 5[b]: 9.7%	+T/P +I/C Chronic SCI	Tetraplegia vs Paraplegia: ↑ tetraplegia Complete vs incomplete: ↑ complete	Aged (35–55 years old), unemployed, having an indwelling catheter, voiding, no bladder management at discharge, and higher scores on ASIA motor index
Hoffman et al. (2011)	*n*: 1035 participants At year 1[b]: 20.6% At year 5[b]: 18.1%	+T/P +I/C Chronic SCI	Tetraplegia vs Paraplegia: Ø Complete vs incomplete: Ø	Pain, declining health status, and decrease in unsafe alcohol use
Shin et al. (2012)	*n*: 36 participants Mild[c]: 63,9%	+T/P I/C	Tetraplegia vs Paraplegia: Ø Complete vs incomplete: ↑ complete	–
Craig et al. (2015)	*n*: 88 participants Inpatient rehabilitation[d]: 10.3% Discharge[d]: 8.6% 6mo postdischarge[d]: 14.1%	T/+P +I/C Acute SCI	Tetraplegia vs Paraplegia: Ø Complete vs incomplete: Ø	–
Kim, Lee, Park, and Jeon (2020)	*n*: 103 participants -	+T/P I/+C Chronic SCI	Level of the injury was identified as a predictor of depression Complete vs incomplete: Ø	Lower physical activity levels

"–" information not reported; "+" predominant cases; "↑" more depressive symptoms.; "Ø" no clear association between depressive symptoms and the level and/or severity of the injury; *mo*: months; *n*: sample; *P*: paraplegic; *SCI*: spinal cord injury; *T*: tetraplegic.
[a]*OAHMQ: Older Adult Health and Mood Questionnaire.*
[b]*PHQ-9: Patient Health Questionnaire–9.*
[c]*BDI: Beck Depression Inventory.*
[d]*HADS: Hospital Anxiety and Depression Scale.*

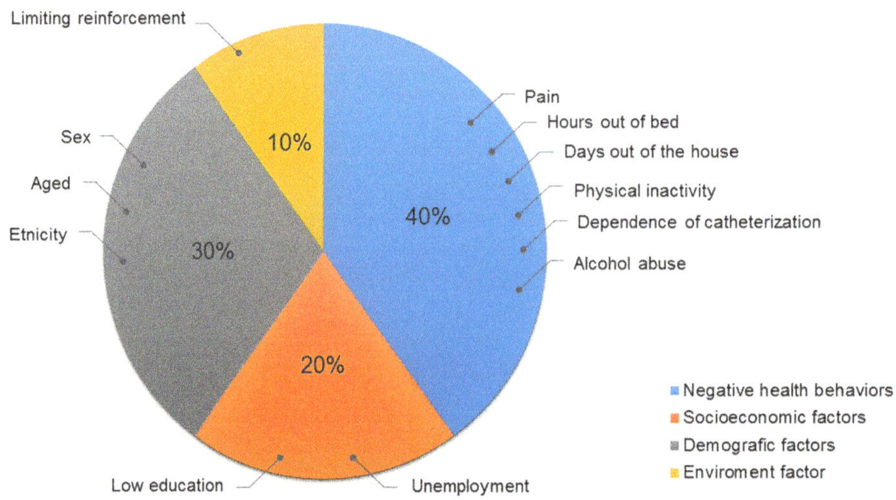

FIG. 2 Risk factors for depression after spinal cord injury. Number of studies (represented in percentage) that reported the risk factors for the development of depression after spinal cord injury (SCI). *Source: Authors based in reference citation Table 2.*

while the likelihood of developing these complications decreases to 20.5% and 8.2%, respectively, in non-depressed patients (Saunders, Krause, and Focht, 2012).

In other clinical conditions such as stroke, higher depressive symptoms are associated with increased illness severity (i.e., predisposition to depression is higher in patients most affected by stroke) (Alghwiri, 2016). Nevertheless, there is still no clear evidence that the severity and/or level of injury are underlying factors of the symptomology of MDD after SCI (Table 2). In contrast, among demographic factors, it is evident that white and black women in the productive age (35–59 years) have higher risks of depression (Krause et al., 2000; Arango-Lasprilla et al., 2011; Saunders, Krause, and Focht, 2012) (Table 2). In addition, research has suggested that the development of MDD post-SCI is associated with adverse health behaviors (Arango-Lasprilla et al., 2011; Hoffman et al., 2011; Kim et al., 2020; Oh et al., 2006; Saunders, Krause, and Focht, 2012), and socioeconomic (Arango-Lasprilla et al., 2011; Krause et al., 2000) and environmental factors (Bombardier et al., 2012) (Table 2; Fig. 2).

It is important to note that the implications of depression can be drastic for individuals with SCI, including elevated stress levels, lower motor scores, and reduced quality of life (Shin et al., 2012). Furthermore, there is a positive correlation between depression and anxiety, limited daily life activity, and general health (Hassanpour et al., 2012). Knowledge concerning the effects of depression on patients comes with the growing number of diagnostic tools to measure depressive symptoms after SCI. Currently, some instruments have been widely used in the literature due to their psychometric properties, including the Older Adult Health and Mood Questionnaire (OAHMQ), Patient Health Questionnaire–9 (PHQ-9), Beck Depression Inventory (BDI), Hospital Anxiety and Depression Scale (HADS), Center for Epidemiologic Studies Depression (CESD) scale, and Spinal Cord Injury–Quality of Life (SCI-QOL) (for more details, see https://www.sralab.org/).

Etiological factors for depression: The role of inflammation

The inflammatory hypothesis proposed by Smith in the 1990s (Smith, 1991) seems to be closely associated with depression. Data compiled from the literature shows that IL-1, INF-α, and TNF-α produced by active macrophages are important mediators of depression since volunteers who received systemic administration of these cytokines showed symptoms of MDD (Smith, 1991). However, the mechanisms of which peripheral cytokines were associated with depression had previously been hypothesized. In 1989, Dantzer and Kelley reported that the systemic administration of pro-inflammatory cytokines induced a non-specific adaptive response known as sickness behavior. Sickness behavior comprises feeling sick, loss of energy or fatigue, loss of interest in usual activities, poor appetite, significant weight loss, sleep changes, and fever, all of which are induced by infectious agents in the periphery (Dantzer & Kelley, 1989). Nonetheless, physiological and behavioral symptoms of sickness are understood by the action of peripherally produced cytokines in the brain (Dantzer et al., 2008). There is a similarity (behaviorally predominant) between sickness behavior and depression. The prominent hypothesis is that depression in humans (or depressive-like behavior in rodents) results from a maladaptive version of cytokine-induced sickness. This maladaptive response is provoked by uncontrolled intensity and/or duration of the innate immune response, leading to the exacerbation of sickness and consequent depression (Dantzer et al., 2008).

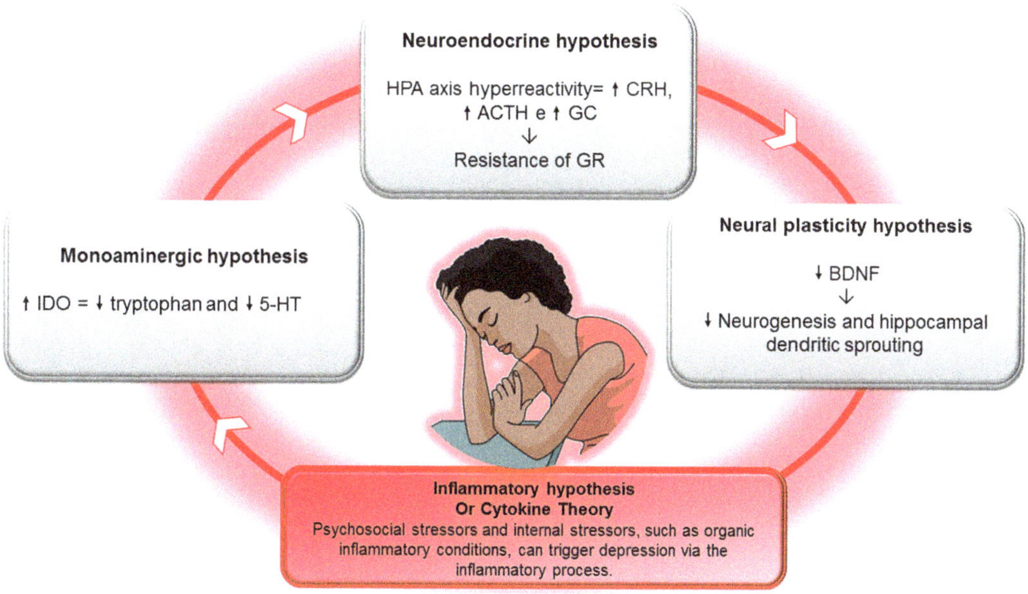

FIG. 3 Flowchart of the inflammatory hypothesis of depression interacting with classic theories postulated for the development of depression (read from left to right). Regarding the monoaminergic hypothesis of depression (*first left arrow*), peripheral cytokines stimulate the enzyme indoleamine 2,3-dioxygenase (IDO), leading to tryptophan depletion. Consequently, serotonin (5-hydroxytryptamine [5-HT]) and tryptophan levels decreased to the detriment of high kynurenine levels. In the neuroendocrine hypothesis, stress-mediated pro-inflammatory signaling directly affects cytokine activation in the hypothalamic and pituitary structures. This provokes the release of the hormones from the corticotropin-releasing hormone (CRH) and adrenocorticotropic hormone (ACTH), which stimulates the release of glucocorticoids (GC) by the adrenal glands. The hyper-reactivity hypothalamus-pituitary-adrenal (HPA) axis is attributed to glucocorticoid receptor (GR) resistance in the hypothalamus. Lastly, inflammation damages synaptic plasticity, which is mainly due to the loss of neuronal integrity in the hippocampus. It is recognized that kynurenines, high GC levels, and reduced trophic factors, such as brain-derived neurotrophic factor (BDNF), may decrease neurogenesis and dendritic sprouting in the dentate gyrus of the hippocampus. ↑: increase; ↓: reduction. *Source: Authors based in Dantzer, R., O'Connor, J. C., Freund, G. G., Johnson, R. W., & Kelley, K. W. (2008). From inflammation to sickness and depression: When the immune system subjugates the brain.* Nature Reviews Neuroscience, *9(1), 46–56.; Maes, M., Leonard, B. E., Myint, A. M., Kubera, M., & Verkerk, R. (2011). The new '5-HT' hypothesis of depression: Cell-mediated immune activation induces indoleamine 2, 3-dioxygenase, which leads to lower plasma tryptophan and an increased synthesis of detrimental tryptophan catabolites (TRYCATs), both of which contribute to the onset of depression.* Progress in Neuro-Psychopharmacology and Biological Psychiatry, *35(3), 702–721.; Capuron, L., & Miller, A. H. (2011). Immune system to brain signaling: Neuropsychopharmacological implications.* Pharmacology & Therapeutics, *130(2), 226–238.*.

Reports elucidate that abundant peripheral pro-inflammatory cytokines, such as IL-1β, IL-6, and TNF-α (Dowlati et al., 2010; Russo & Nestler, 2013), are found in individuals with depression and have their inflammatory signals transmitted and amplified in different regions of the brain. In response to these signals, important changes in the physiological domains of depression occur, including the metabolism of neurotransmitters (especially serotonin or 5-hydroxytryptamine [5-HT]), neuroendocrine function, and neural plasticity (Capuron & Miller, 2011) (Fig. 3).

Evidence shows that serotonergic transmission in cortico-limbic circuits is susceptible to disruption in the presence of systemic inflammation. In rodents, the elevated TNF-α and IL-1β systemic levels were associated with the stimulation of 5-HT transporter activity in affective processing areas, leading to behavior-like depression, including hopelessness and anhedonia (Zhu et al., 2010). Similarly, a link between depression and stress-induced inflammation is generally accepted. Indeed, tricyclic antidepressant drugs reduce GC and IL-6 plasma levels and social withdrawal provoked by stress, further reinforcing that neuroendocrine dysregulation and inflammation are components of depression-related illness (Ramirez & Sheridan, 2016). Other studies have reported that synaptic plasticity and neurotrophic factor synthesis are effects of prolonged neuroinflammation. Tang, Lin, Pan, Guan, and Li (2016) observed that IL-6, TNF-α, and IL-1β levels in rodent brains were associated with impaired hippocampal neurogenesis during anhedonia, with both conditions being reverted by BDNF administration (Briones & Woods, 2013), suggesting a possible protecting role of BDNF on neuroplasticity hippocampal in depressive-like behavior.

In this context, it is recognized that depression is a complex psychological condition that involves numerous pathophysiological mechanisms, including inflammation. After SCI, the role of the inflammatory process in depression has been expanded to explore a "gap" that still exists regarding the underlying mechanisms of this heterogeneous disease that afflicts part of individuals injured.

Inflammatory aspects of depression after SCI

Research has emphasized that inflammation plays a pivotal role in the development of depression after SCI. Inflammation is a critical component of secondary endogenous injury and potentially leads to the circulation of soluble inflammatory mediators post-trauma. Elevated serum IL-2, IL-2R, IL-6, TNF-α, IL-1β, and INF-γ levels were observed in individuals with subchronic and chronic SCI (Allison et al., 2017; Davies et al., 2007; Hayes et al., 2002). In humans, the chronic systemic inflammatory response has been observed to predispose to depression through a negative correlation between improved mood and tryptophan bioavailability (amino acid precursors of serotonin) and reduced IFN-γ and IL-1β serum levels in patients with chronic SCI submitted to a 12-week anti-inflammatory diet (Allison & Ditor, 2015). Additional data support that women are more vulnerable to psychomotor agitation, anhedonia, and/or mood depression after SCI (Bombardier et al., 2012; Krause et al., 2000; Oh et al., 2006). Healthy women and those with chronic diseases present more mood depression and feelings of social disconnection symptoms in systemic IL-6 and TNF-α responses than men (Moieni et al., 2015). Thus, predisposition to depression may be influenced by sex and inflammatory conditions after SCI.

Animal models of SCI have provided further insight into the development of inflammation-associated depression. Maldonado-Bouchard et al. (2016) showed that immune system reactivity increased in parallel with depressive-like behavior in injured mice. The authors associated hopelessness, social isolation, and psychomotor retardation with higher chemokine and cytokine levels, including granulocyte macrophage-colony stimulating factor (GM-CSF), IL-1β, and TNF-α serum levels, IL-18 and TNF-α in the spinal cord, and IL-1α and TNF-α in the hippocampus in rodents after contusion SCI (Maldonado-Bouchard et al., 2016).

The link between systemic inflammation, neuroinflammation, and depression-like behavior after contusion SCI is also corroborated by Wu et al. (2014, 2016)). The authors reported that reactive microglia co-localized with TNF-α and IL-6 mRNA expressions in the hippocampus increased in rodents with anhedonia and hopelessness (Wu et al., 2014). Besides microglial activation, the impaired hippocampal neurogenesis and increased degeneration in the hippocampus and cerebral cortex were observed in paraplegic mice (Wu et al., 2014, 2016), providing evidence that chronic SCI leads to neuroimmune dysregulation.

Additionally, there is data that neuroinflammation may reduce BDNF levels in the hippocampus and increase corticosterone serum levels in rodents with depressive-like behavior after SCI (Li et al., 2017). Considering that BDNF is involved in synaptic plasticity and is downregulated by corticosterone in response to stress-induced depression (Masi & Brovedani, 2011), Li et al. (2017) showed that treatment with ZBD-2 (a synthetic protein of indirect anti-inflammatory activity) has antidepressant effects able to restore serum corticosterone and hippocampal BDNF levels, reducing microglial reactivity in the spinal cord in rodents with SCI.

In general, the systemic levels and type of pro-inflammatory cytokines tend to vary between SCI stages (acute, subchronic, and chronic) and the animals' psychological well-being (social isolation and anhedonia). However, the early systemic inflammatory peak (48 h after compression SCI) appears to be predictive of developing depressive-like behavior (but not anxiety). In the acute and chronic stages (7 and 35 days post-SCI, respectively), the inflammatory peaks of IL-1β, TNF-α, INF-γ, and IL-6 were negatively correlated with reduced social exploration time. In the subchronic to chronic stages (22 days post-SCI), a similar profile was observed for anhedonia, given that a significant correlation took place between early systemic inflammation and loss of interest/pleasure in rodents (Do Espírito Santo et al., 2019).

Notably, anhedonic behavior has been reported to have lasting effects, enduring for over 2.8 months after secondary SCI-induced neuroinflammation (Wu et al., 2014, 2016), while social isolation manifests for at least 1 month (Do Espírito Santo et al., 2019). Both behavioral conditions have been associated with early and persistent local and systemic inflammatory responses observed in female rats (Do Espírito Santo et al., 2019). These data deserve attention for two reasons: depressive symptoms generally have long-term effects on individuals with SCI, and human studies have not yet been performed on the female sex, although evidence points to this direction.

Lastly, all studies and review articles have a significant scientific value since depression is treatable as long as symptoms and causes are recognized early. Thus, knowledge of biological predictors, including inflammation, may provide an easily accessible prognostic marker for SCI individuals that are more susceptible to inflammation-mediated depression.

Conclusion

For a considerable part of the population of individuals with SCI, physical and psychological problems coexist. In this chapter, the role of inflammation in the etiology of depression was described and the following considered: (1) the risk factors inherent to SCI (age, sex, social status and, less likely, the level and severity of SCI), and (2) the possible biological process in which inflammation overlaps with random depression conditions (monoaminergic system, neuroendocrine function, and neural plasticity). Therefore, there is evidence to support those systemic pro-inflammatory cytokines propagated at the damage site in the spinal cord have a multifaceted role in SCI and are associated with several behavioral changes, including anhedonia and mood depression (in humans) and anhedonia, social isolation, and hopelessness (in rodents). In this context, it is concluded that chronic inflammation can predispose to depression after SCI.

Applications to other areas of neuroscience

In this chapter, we reviewed the risk factors and predictors of depression after spinal cord injury (SCI) while emphasizing the role of inflammation in the development of depressive symptoms. Depression is a complex, multifactorial clinical condition with the absence of a clear understanding of how it arises and should be treated. Thus, it is speculated that the inaccurate understanding of neurobiological mechanisms may lead to incomplete information to support the etiology of depression (Perugi, Frare, Toni, Ruffolo, & Torti, 2002). This is even more concerning for neurological disorders such as SCI because several maladaptive responses are triggered by spinal cord trauma, including inflammation. Therefore, establishing the relationship between inflammation and depression and exploring the (1) systemic pro-inflammatory cytokines involved, (2) brain areas subject to neuroinflammation caused by traumatic injury, and better understanding (3) inflammation by delving into classic theories regarding the development of depression (all discussed in the present chapter) are important factors that may provide a possible easy-access and low-cost prognostic marker of post-SCI depression.

Furthermore, this knowledge is highly relevant, given that depression is also associated with systemic inflammation in stroke (Chen et al., 2018) and traumatic brain injury (Bodnar, Morganti, & Bachstetter, 2018). Hence, regardless of neurological injury (i.e., trauma or vascular insult to the central nervous system), depression appears to share the same etiologic factor, making inflammation a new target for intervention. Besides inflammation, we also interlink the risk factors inherent to SCI in the development of depression. Indeed, the manifestation of depression is typically masked by behavioral abnormalities (distress, difficulty in dealing with daily life situations, reduced social activities, and family conflicts), consequently delaying diagnosis and treatment. Knowing the demographic factors (age and sex), socioeconomic conditions (education and employment), and social environment is vital to identifying people at risk of developing depression after SCI.

Mini-dictionary of terms

American Spinal Injury Association Impairment Scale: used to classify the severity of spinal cord injury from A to E. A being the most severe, D being the least severe, and E being normal.

Anhedonia: loss of interest or pleasure.

Clinically significant symptoms of depression: presence of depressive symptoms but do not meet the criteria for major depression.

Complete spinal cord injury: the absence of sensory or motor function in the sacral segments (S4-S5).

Cytokines: glycoprotein regulators of the host's defense in response to infection, inflammation, or disease and classified as interleukins, interferons, tumor necrosis factor-alpha, and tumor growth factors.

Hopelessness: similar to behavioral despair in people with depression.

Incomplete spinal cord injury: preservation of any sensory and/or motor function in most caudal sacral segments (S4–S5).

Inflammation: a tissue's response to injury, which is increased by pro-inflammatory cytokines.

Mood depression: a low mood state (i.e., mood disturbance) related to depression.

Paraplegia: paralysis of the lower limbs due to lower cervical cord injury.

Probable major depression: psychological condition attributed by the Patient Health Questionnaire (PHQ-9) and Older Adult Health and Mood Questionnaire that does not meet the criteria for major depression.

Sickness behavior: an adaptive response to pathogenic infection. When the pathogen is cleared, the adaptive response is completely reversed.

Social isolation: reduced social exploration due to loss of motivation.

Tetraplegia: paralysis of all four limbs and torso due to cervical cord injury that may manifest with or without autonomic dysfunctions.

Key facts of depression

- It is estimated to be the second-largest cause of years lived with disability.
- It is prevalent comorbidity in several neurological disorders and negatively affects rehabilitation.

- The biological mechanisms of depression are not entirely understood.
- Systemic inflammation is considered a biological predictor of the onset and/or maintenance of depression.
- Sex, age, ethnicity, income, education, and employment are risk factors for depression.

Summary points

- This chapter focuses on the role of inflammation in the development of depression after SCI.
- Secondary endogenous damage stimulates the activity of the immune system during the pathological stages of SCI.
- Spinal cord injury leads to increased systemic pro-inflammatory cytokines that exert detrimental effects on mental health.
- Cytokines play an important role in the neuroinflammatory responses in depression.
- Women are more susceptible to becoming depressed than men, although other risk factors are associated with depression.

References

Alghwiri, A. A. (2016). The correlation between depression, balance, an physical functioning post stroke. *Journal of Stroke and Cerebrovascular Diseases*, *25*(2), 475–479.

Allison, D. J., & Ditor, D. S. (2015). Targeting inflammation to influence mood following spinal cord injury: A randomized clinical trial. *Journal of Neuroinflammation*, *12*(204), 1–10.

Allison, D. J., Josse, A. R., Gabriel, D. A., Klentrou, P., & Ditor, D. S. (2017). Targeting inflammation to influence cognitive function following spinal cord injury: A randomized clinical trial. *Spinal Cord*, *55*(1), 26–32.

American Psychiatric Association. (2013). *Diagnostic and statistical manual of mental disorders (DSM-5)* (5th ed.). American Psychiatric Association.

Arango-Lasprilla, J. C., Ketchum, J. M., Starkweather, A., Nicholls, E., & Wilk, A. R. (2011). Factors predicting depression among persons with spinal cord injury 1 to 5 years post injury. *NeuroRehabilitation*, *29*(1), 9–21.

ASIA and ISCoS International Standards Committee. (2019). The 2019 revision of the international standards for neurological classification of spinal cord injury (ISNCSCI)-what's new? *Spinal Cord*, *57*(10), 815–817.

Bartholdi, D., & Schwab, M. E. (1997). Expression of pro-inflammatory cytokine and chemokine mRNA upon experimental spinal cord injury in mouse: An in situ hybridization study. *European Journal of Neuroscience*, *9*(7), 1422–1438.

Bodnar, C. N., Morganti, J. M., & Bachstetter, A. D. (2018). Depression following a traumatic brain injury: Uncovering cytokine dysregulation as a pathogenic mechanism. *Neural Regeneration Research*, *13*(10), 1693–1704.

Bombardier, C. H., Fann, J. R., Tate, D. G., Richards, J. S., Wilson, C. S., Warren, A. M., et al. (2012). An exploration of modifiable risk factors for depression after spinal cord injury: Which factors should we target? *Archives of Physical Medicine and Rehabilitation*, *93*(5), 775–781.

Bombardier, C. H., Richards, J. S., Krause, J. S., Tulsky, D., & Tate, D. G. (2004). Symptoms of major depression in people with spinal cord injury: Implications for screening. *Archives of Physical Medicine and Rehabilitation*, *85*(11), 1749–1756.

Briones, T. L., & Woods, J. (2013). Chronic binge-like alcohol consumption in adolescence causes depression-like symptoms possibly mediated by the effects of BDNF on neurogenesis. *Neuroscience*, *254*, 324–334.

Capuron, L., & Miller, A. H. (2011). Immune system to brain signaling: Neuropsychopharmacological implications. *Pharmacology & Therapeutics*, *130*(2), 226–238.

Chen, H., Luan, X., Zhao, K., Qiu, H., Liu, Y., Tu, X., et al. (2018). The association between neutrophil-to-lymphocyte ratio and post-stroke depression. *Clinica Chimica Acta*, *486*, 298–302.

Craig, A., Nicholson Perry, K., Guest, R., Tran, Y., & Middleton, J. (2015). Adjustment following chronic spinal cord injury: Determining factors that contribute to social participation. *British Journal of Health Psychology*, *20*(4), 807–823.

Dantzer, R., & Kelley, K. W. (1989). Stress and immunity: An integrated view of relationships between the brain and the immune system. *Life Sciences*, *44*(26), 1995–2008.

Dantzer, R., O'Connor, J. C., Freund, G. G., Johnson, R. W., & Kelley, K. W. (2008). From inflammation to sickness and depression: When the immune system subjugates the brain. *Nature Reviews Neuroscience*, *9*(1), 46–56.

David, S., & Kroner, A. (2011). Repertoire of microglial and macrophage responses after spinal cord injury. *Nature Reviews Neuroscience*, *12*(7), 388–399.

Davies, A. L., Hayes, K. C., & Dekaban, G. A. (2007). Clinical correlates of elevated serum concentrations of cytokines and autoantibodies in patients with spinal cord injury. *Archives of Physical Medicine and Rehabilitation*, *88*(11), 1384–1393.

Do Espírito Santo, C. C., da Silva Fiorin, F., Ilha, J., Duarte, M. M. M. F., Duarte, T., & Santos, A. R. S. (2019). Spinal cord injury by clip-compression induces anxiety and depression-like behaviours in female rats: The role of the inflammatory response. *Brain, Behavior, and Immunity*, *78*, 91–104.

Dowlati, Y., Herrmann, N., Swardfager, W., Liu, H., Sham, L., Reim, E. K., et al. (2010). A meta-analysis of cytokines in major depression. *Biological Psychiatry*, *67*(5), 446–457.

Dryden, D. M., Saunders, L. D., Rowe, B. H., May, L. A., Yiannakoulias, N., Svenson, L. W., et al. (2005). Depression following traumatic spinal cord injury. *Neuroepidemiology*, *25*(2), 55–61.

Fann, J. R., Bombardier, C. H., Richards, J. S., Tate, D. G., Wilson, C. S., Temkin, N., et al. (2011). Depression after spinal cord injury: Comorbidities, mental health service use, and adequacy of treatment. *Archives of Physical Medicine and Rehabilitation*, *92*(3), 352–360.

Ferrari, A. J., Charlson, F. J., Norman, R. E., Patten, S. B., Freedman, G., Murray, C. J., et al. (2013). Burden of depressive disorders by country, sex, age, and year: Findings from the global burden of disease study 2010. *PLoS Medicine*, *10*(11), e1001547.

Gibson, A. E., Buchholz, A. C., & Ginis, K. M. (2008). C-reactive protein in adults with chronic spinal cord injury: Increased chronic inflammation in tetraplegia vs paraplegia. *Spinal Cord*, *46*(9), 616–621.

Halvorsen, A., Pettersen, A. L., Nilsen, S. M., Halle, K. K., Schaanning, E. E., & Rekand, T. (2019). Non-traumatic spinal cord injury in Norway 2012–2016: Analysis from a national registry and comparison with traumatic spinal cord injury. *Spinal Cord*, *57*(4), 324–330.

Hassanpour, K., Hotz-Boendermaker, S., Dokladal, P., Curt, A., & European Multicenter Study for Human Spinal Cord Injury Study group. (2012). Low depressive symptoms in acute spinal cord injury compared to other neurological disorders. *Journal of Neurology*, *259*(6), 1142–1150.

Hayes, K. C., Hull, T. C. L., Delaney, G. A., Potter, P. J., Sequeira, K. A. J., Campbell, K., et al. (2002). Elevated serum titers of proinflammatory cytokines and CNS autoantibodies in patients with chronic spinal cord injury. *Journal of Neurotrauma*, *19*(6), 753–761.

Hoffman, J. M., Bombardier, C. H., Graves, D. E., Kalpakjian, C. Z., & Krause, J. S. (2011). A longitudinal study of depression from 1 to 5 years after spinal cord injury. *Archives of Physical Medicine and Rehabilitation*, *92*(3), 411–418.

Kakulas, B. A. (2019). Spinal cord injuries the facts of neuropathology: Opportunities and limitations. *Current Neuropsychiatry and Clinical Neuroscience Reports*, *1*(1), 1–5.

Kakulas, B. A., & Kaelan, C. (2015). The neuropathological foundations for the restorative neurology of spinal cord injury. *Clinical Neurology and Neurosurgery*, *129*, S1–S7.

Kigerl, K. A., Gensel, J. C., Ankeny, D. P., Alexander, J. K., Donnelly, D. J., & Popovich, P. G. (2009). Identification of two distinct macrophage subsets with divergent effects causing either neurotoxicity or regeneration in the injured mouse spinal cord. *Journal of Neuroscience*, *29*(43), 13435–13444.

Kim, D. I., Lee, J., Park, H., & Jeon, J. Y. (2020). The relationship between physical activity levels and mental health in individuals with spinal cord injury in South Korea. *International Journal of Environmental Research and Public Health*, *17*(12), 4423–4431.

Kirshblum, S., Snider, B., Rupp, R., & Read, M. S. (2020). Updates of the international standards for neurologic classification of spinal cord injury: 2015 and 2019. *Physical Medicine and Rehabilitation Clinics*, *31*(3), 319–330.

Krause, J. S., Kemp, B., & Coker, J. (2000). Depression after spinal cord injury: Relation to gender, ethnicity, aging, and socioeconomic indicators. *Archives of Physical Medicine and Rehabilitation*, *81*(8), 1099–1109.

Li, X. M., Meng, J., Tao Li, L., Guo, T., Yang, L. K., Shi, Q. X., et al. (2017). Effect of ZBD-2 on chronic pain, depressive-like behaviors, and recovery of motor function following spinal cord injury in mice. *Behavioural Brain Research*, *322*, 92–99.

Liu, W., Ge, T., Leng, Y., Pan, Z., Fan, J., Yang, W., et al. (2017). The role of neural plasticity in depression: From hippocampus to prefrontal cortex. *Neural Plasticity*, 1–11.

Maldonado-Bouchard, S., Peters, K., Woller, S. A., Madahian, B., Faghihi, U., Patel, S., et al. (2016). Inflammation is increased with anxiety-and depression-like signs in a rat model of spinal cord injury. *Brain, Behavior, and Immunity*, *51*, 176–195.

Masi, G., & Brovedani, P. (2011). The hippocampus, neurotrophic factors and depression. *CNS Drugs*, *25*(11), 913–931.

Meisel, C., Schwab, J. M., Prass, K., Meisel, A., & Dirnagl, U. (2005). Central nervous system injury-induced immune deficiency syndrome. *Nature Reviews Neuroscience*, *6*(10), 775–786.

Moieni, M., Irwin, M. R., Jevtic, I., Olmstead, R., Breen, E. C., & Eisenberger, N. I. (2015). Sex differences in depressive and socioemotional responses to an inflammatory challenge: Implications for sex differences in depression. *Neuropsychopharmacology*, *40*(7), 1709–1716.

North, N. T. (1999). The psychological effects of spinal cord injury: A review. *Spinal Cord*, *37*(10), 671–679.

NSCISC. (2016). *National Spinal Cord Injury Statistical Center (facts and figures at a glance)*. AL: Department of Physical Medicine and Rehabilitation, University of Alabama at Birmingham.

Oh, S. J., Shin, H. I., Paik, N. J., Yoo, T., & Ku, J. H. (2006). Depressive symptoms of patients using clean intermittent catheterization for neurogenic bladder secondary to spinal cord injury. *Spinal Cord*, *44*(12), 757–762.

Perugi, G., Frare, F., Toni, C., Ruffolo, G., & Torti, C. (2002). Open-label evaluation of venlafaxine sustained release in outpatients with generalized anxiety disorder with comorbid major depression or dysthymia: Effectiveness, tolerability and predictors of response. *Neuropsychobiology*, *46*(3), 145–149.

Rahimi-Movaghar, V., Sayyah, M. K., Akbari, H., Khorramirouz, R., Rasouli, M. R., Moradi-Lakeh, M., et al. (2013). Epidemiology of traumatic spinal cord injury in developing countries: A systematic review. *Neuroepidemiology*, *41*(2), 65–85.

Ramirez, K., & Sheridan, J. F. (2016). Antidepressant imipramine diminishes stress-induced inflammation in the periphery and central nervous system and related anxiety-and depressive-like behaviors. *Brain, Behavior, and Immunity*, *57*, 293–303.

Russo, S. J., & Nestler, E. J. (2013). The brain reward circuitry in mood disorders. *Nature Reviews Neuroscience*, *14*(9), 609–625.

Salewski, R., Emrani, H., & Fehlings, M. G. (2013). Neural stem/progenitor cells for spinal cord regeneration. In S. Wislet-Gendebien (Ed.), *Trends in cell signaling pathways in neuronal fate decision* (pp. 271–304). Crotia: In Tech.

Saunders, L. L., Krause, J. S., & Focht, K. L. (2012). A longitudinal study of depression in survivors of spinal cord injury. *Spinal Cord*, *50*(1), 72–77.

Schwab, J. M., Zhang, Y., Kopp, M. A., Brommer, B., & Popovich, P. G. (2014). The paradox of chronic neuroinflammation, systemic immune suppression, autoimmunity after traumatic chronic spinal cord injury. *Experimental Neurology*, *258*, 121–129.

Shin, J. C., Goo, H. R., Yu, S. J., Kim, D. H., & Yoon, S. Y. (2012). Depression and quality of life in patients within the first 6 months after the spinal cord injury. *Annals of Rehabilitation Medicine*, *36*(1), 119–125.

Smith, R. S. (1991). The macrophage theory of depression. *Medical Hypotheses*, *35*(4), 298–306.

Sun, X., Jones, Z. B., Chen, X. M., Zhou, L., So, K. F., & Ren, Y. (2016). Multiple organ dysfunction and systemic inflammation after spinal cord injury: A complex relationship. *Journal of Neuroinflammation*, *13*(1), 260–271.

Sun, G., Yang, S., Cao, G., Wang, Q., Hao, J., Wen, Q., et al. (2018). γδ T cells provide the early source of IFN-γ to aggravate lesions in spinal cord injury. *Journal of Experimental Medicine*, *215*(2), 521–535.

Tang, M. M., Lin, W. J., Pan, Y. Q., Guan, X. T., & Li, Y. C. (2016). Hippocampal neurogenesis dysfunction linked to depressive-like behaviors in a neuroinflammation induced model of depression. *Physiology & Behavior*, *161*, 166–173.

Williams, R., & Murray, A. (2015). Prevalence of depression after spinal cord injury: A meta-analysis. *Archives of Physical Medicine and Rehabilitation*, *96*(1), 133–140.

World Health Organization. (2017). *Depression and other common mental disorders: Global health estimates*. World Health Organization.

Wu, J., Zhao, Z., Kumar, A., Lipinski, M. M., Loane, D. J., Stoica, B. A., et al. (2016). Endoplasmic reticulum stress and disrupted neurogenesis in the brain are associated with cognitive impairment and depressive-like behavior after spinal cord injury. *Journal of Neurotrauma*, *33*(21), 1919–1935.

Wu, J., Zhao, Z., Sabirzhanov, B., Stoica, B. A., Kumar, A., Luo, T., et al. (2014). Spinal cord injury causes brain inflammation associated with cognitive and affective changes: Role of cell cycle pathways. *Journal of Neuroscience*, *34*(33), 10989–11006.

Zhu, C. B., Lindler, K. M., Owens, A. W., Daws, L. C., Blakely, R. D., & Hewlett, W. A. (2010). Interleukin-1 receptor activation by systemic lipopolysaccharide induces behavioral despair linked to MAPK regulation of CNS serotonin transporters. *Neuropsychopharmacology*, *35*(13), 2510–2520.

Chapter 38

Spirituality, hope, and resilience in the recovery and adaptation process following spinal cord injury

Kate Jones[a], Julie Pryor[b], Malcolm Anderson[c], Candice Care-Unger[b], and Grahame Simpson[d]

[a]*Institute for Ethics and Society, University of Notre Dame Australia, Sydney, NSW, Australia,* [b]*Royal Rehab, Ryde, NSW, Australia,* [c]*School of Nursing and Health, Avondale University, Cooranbong, NSW, Australia,* [d]*John Walsh Centre for Rehabilitation Research, Sydney School of Medicine, University of Sydney, NSW, Australia*

List of abbreviations

SCI spinal cord injury

Introduction

A spinal cord injury (SCI) is a catastrophic event that abruptly interrupts the lives of an individual and their family, changing everything in a moment (Palmer, Kriegsman, & Palmer, 2008). Described as one of the most devastating conditions to occur in both developed and developing countries, estimates of the global incidence of traumatic SCI range from 3.6 to 195.4 per million across the world (Jazayeri, Beygi, Shokranch, Hagen, & Rahimi-Movaghar, 2015). According to the World Health Organisation (2013), between 250,000 and 500,000 people sustain a SCI every year and most are due to preventable causes. Furthermore, those with SCI are more likely to die prematurely and experience lower rates of economic participation (World Health Organisation, 2013). Damage to the spinal cord may affect an individual's physical functioning in a number of ways, impacting upon mobility, continence, respiration, sexual and reproductive health (Kirshblum & Benevento, 2009; Palmer et al., 2008), as well as triggering neuropathic pain (Hagen & Rekand, 2015).

The substantial psychological impacts of SCI that accompany the bodily impairments and functional alterations have been extensively documented. The prevalence of depression after SCI, as found in a meta-analysis based on 19 studies, is estimated at 22.2%, compared with 16% in the general U.S. population (Williams & Murray, 2015). Furthermore, the risk for suicide is up to five times greater than for the general population (Simpson & Brenner, 2019). The course of psychological distress such as anxiety and depression peaks in the acute stages of treatment just before discharge from hospital (Kennedy & Rogers, 2000), but can last up to 5 years and longer post-SCI (Hoffman, Bombardier, Graves, Kalpakjian, & Krause, 2011). Risk factors for psychological distress include increased pain, worsening health status, and unsafe alcohol use and have all been associated with elevated depression 5 years post-injury (Hoffman et al., 2011). Life satisfaction has also been reported to be affected by SCI, and is significantly determined by the level of injury, pain, and secondary impairments (Van Koppenhagen et al., 2008).

Research examining the impact of an SCI within family members and broader social connections has also emphasized the challenges (Baker, Barker, Sampson, & Martin, 2017). A significant number of studies have reported elevated levels of psychological distress and burden experienced by family caregivers post-SCI (Baker et al., 2017). Furthermore, these impacts may worsen over time, with a rare longitudinal study finding that spouses in the UK who played a caregiving role for their partner with SCI experienced greater depressive affect, physical and emotional stress, burnout, fatigue, and anger and resentment than spouses not in this role (Weitzenkamp, Gerhart, Charlifue, Whiteneck, & Savic, 1997) (see Fig. 1).

Although this body of research is of unquestionable value, the focus on the investigation of psychopathology, embodying deficits and vulnerability-oriented approach, has left other dimensions of the psychological adjustment process post-SCI under-researched and poorly understood (White, Driver, & Warren, 2008). Over the last two decades, alternative paradigms from positive psychology and strengths-based approaches (Saleebey, 2006; Seligman & Csikszentmihalyi,

Cellular, Molecular, Physiological, and Behavioral Aspects of Spinal Cord Injury. https://doi.org/10.1016/B978-0-12-822427-4.00038-1

FIG. 1 Challenges for family caregivers of relatives with SCI (Weitzenkamp et al., 1997).

2000) have been utilized in broadening our understanding of individual and family adjustment to SCI (White et al., 2008). These involve a shift from a sole emphasis on pathology and deficits to health promotion and the nurturing of strengths (Connor & Davidson, 2003; Richardson, 2002). Research has demonstrated that some individuals and family members have been able to identify positive gains, benefits, and areas of growth after SCI (Byra, 2015; Chun & Lee, 2013; Kalpakjian et al., 2014; Middleton et al., 2014). However, key constructs which may be of importance within such paradigms and their application within the context of SCI are still not well understood.

This chapter, therefore, provides an overview of three central constructs, namely spirituality, hope, and resilience. These constructs all play a mediating role between explanatory factors such as demographic, injury, and psychosocial variables (e.g., sex, age, injury severity, functional status, relationship status, pre-injury psychiatric status) and outcomes such as psychological distress and carer burden post-SCI. Also, the theoretical literature and initial empirical research suggest that these constructs are interrelated.

The first section of this chapter introduces the *quest* narrative, a dynamic context within which to understand the role these constructs play in response to SCI. The following sections will then define each of these constructs and provide a brief outline of the nascent quantitative and qualitative research, both about the person with SCI, and family members. Following this, a model derived from qualitative interviews that integrates spirituality, hope and resilience is outlined, followed by quantitative studies that also demonstrate the inter-relationship among these three constructs. The final section of the chapter highlights approaches to integrating spirituality, hope, and resilience into the clinical practice of SCI rehabilitation.

Responses to trauma: The quest narrative and the role of spirituality, hope, and resilience

Investigations examining the process of adjustment to traumatic events have often characterized this process in terms of a journey (Campbell, 2008). In *The Wounded Storyteller*, Frank (1995) introduces the *quest* narrative as one of three possible narratives for those with a long-standing illness or injury, alongside *restitution* and *chaos* narratives (see Fig. 2). A *restitution* narrative goes "Yesterday I was healthy, today I'm sick, but tomorrow I'll be healthy again" (p. 77). Such a narrative views the body as something needing to be "fixed," like a television set or other piece of equipment. Smith and Sparkes (2005) have suggested that for those with an SCI the restitution narrative is about recovery and "concrete hope," about returning to being able-bodied. The *chaos* narrative is the opposite of the *restitution* narrative. The plot of the *chaos* narrative is that life will never get better and the story is one of "vulnerability, futility, and impotence" (p. 97). Hope is lost and replaced with despair.

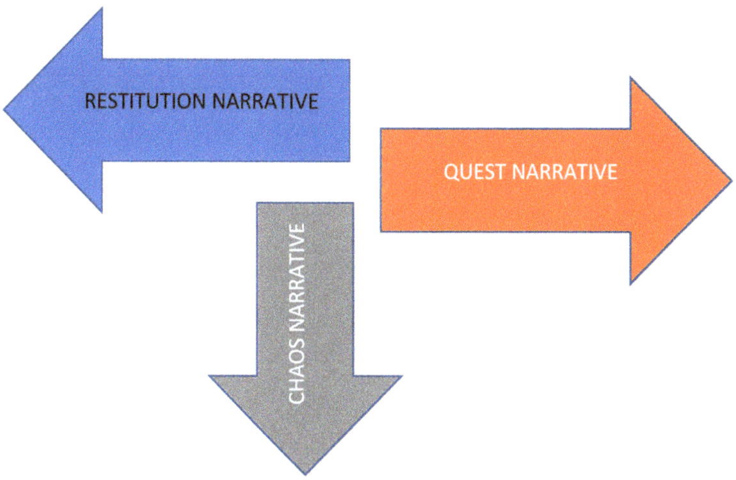

FIG. 2 The three narratives (Frank, 1995).

The *quest* narrative is different from both the *restitution* narrative and the *chaos* narrative. It reframes the interruption to an individual's life as a challenge. Illness or injury becomes a journey, and the teller has a story to tell. According to Frank quest stories meet

> *...suffering head on; they accept illness (or injury) and seek to use it. Illness (or injury) is the occasion of a journey that becomes a quest. What is quested for may never be wholly clear, but the quest is defined by the ill person's belief that something is to be gained through the experience (Frank, 1995, p. 115).*

Smith and Sparkes (2005) suggest that the *quest* narrative fosters a deeper "transcendent hope," one that embraces uncertainty, and moves towards reconstructing identity and meaning. This process of adaptation in response to challenge is also at the heart of resilience, as will be demonstrated through this chapter.

Spirituality after SCI

Spirituality is recognized as playing an important role in enabling individuals with SCI and their family members to thrive and move forward amid adversity (Jones, Dorsett, Simpson, & Briggs, 2018; White, Driver, & Warren, 2010). Spirituality has been defined as "the aspect of humanity that refers to the way individuals seek and express meaning and purpose, and the way they experience their connectedness to the moment, to self, to others, to nature and to the significant or sacred" (Puchalski et al., 2009). This definition implies that spirituality is a broad construct, encompassing a range of different sources of meaning and connection. Religious faith, which is sometimes considered to be interchangeable with spirituality, may be one source of spiritual strength (Davis et al., 2015), however, spirituality may also encompass connection with the natural world, other people, and oneself (Canda & Furman, 2009; Davis et al., 2015) (see Fig. 3). Within the broader healthcare field, there is an increasing awareness and acknowledgment of the importance of spirituality for individuals at times of illness or disability (Cobb, Puchalski, & Rumbold, 2012; Jones, Pryor, Care-Unger, & Simpson, 2018; Puchalski et al., 2009).

Most studies of spirituality after SCI have focused upon the individual with SCI (Jones, Simpson, Briggs, & Dorsett, 2016). A scoping review of the literature (Jones et al., 2016) found among quantitative studies, spirituality was positively associated with other indicators of adjustment after SCI, including life satisfaction, quality of life, and perceived health (see Fig. 4). These findings have been confirmed and extended in recent research (Jones, Simpson, Briggs, Dorsett, & Anderson, 2019; Siddall, McIndoe, Austin, & Wrigley, 2016; Wilson, Forchheimer, Heinemann, Warren, & McCullumsmith, 2017), with spirituality also being associated with lower levels of depression and anxiety. Furthermore, one study (Rahnama et al., 2015) has observed that lower levels of spirituality and negative spirituality may contribute towards greater levels of anxiety and depression.

Two qualitative studies (McColl et al., 2000a; Mundle, 2015) identified in the scoping review (Jones et al., 2016) investigated the concept of spirituality after SCI. The *quest* narrative featured in one of these. Mundle (2015) identified the *restitution, chaos,* and *quest* narratives, during a series of semi-structured interviews with a woman identifying with the

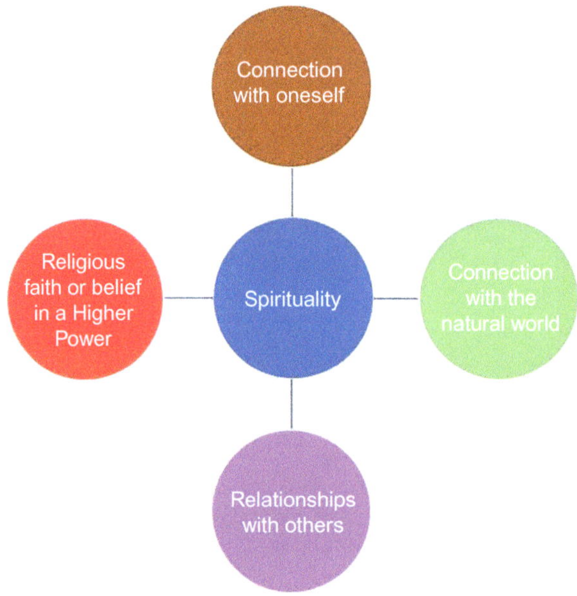

FIG. 3 Possible sources of spiritual strength (Canda & Furman, 2009; Davis et al., 2015).

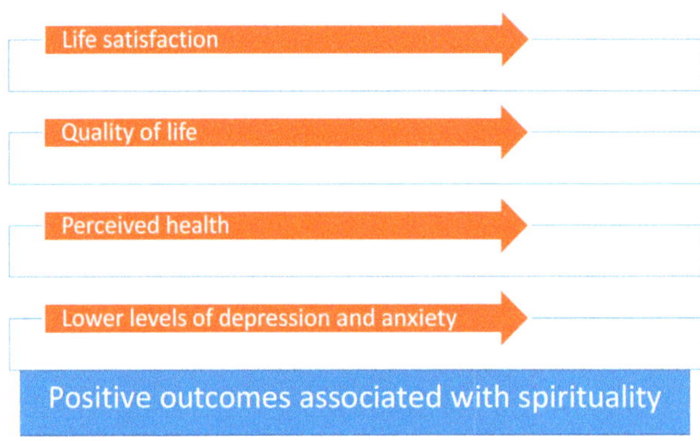

FIG. 4 Positive health outcomes associated with spirituality (Jones et al., 2016; Wilson et al., 2017).

Catholic faith. Mundle suggested that the woman moved from one narrative type to another, eventually arriving at a narrative of "testimony," where others observed her be an "actor in her own healing drama who, in turn, could offer healing and spiritual guidance to others" (p. 41).

The second qualitative study identified to consider spirituality after SCI (McColl et al., 2000a) stands alone as the only one in the scoping review to specifically consider the concept of spirituality as a relational construct. From their interviews with individuals with brain injury or SCI, McColl et al. (2000a) identified five themes which they argued impact upon all aspects of spirituality: (i) awareness, relating to a greater appreciation of life, others, and the world; (ii) closeness, encompassing increased intimacy, particularly within the family; (iii) trust, involving allowing others to provide help due to new dependency and reliance; (iv) vulnerability, as participants "recognised that they were not invincible or immortal"; and (v) purpose, due to the purpose of life which had changed for many participants, with some specifically referring to God or a Creator having a plan for them (p. 821). These findings led to the development of a broad framework for consideration of spiritual issues associated with disability, which encompassed intra-personal, inter-personal, and trans-personal elements (McColl et al., 2000b).

Hope after SCI

Hope has been recognized to be an important element in recovery from SCI (Dorsett, 2010). Hope is "a multi-dimensional dynamic life force characterised by a confident yet uncertain expectation of achieving a future good which….is realistically possible and personally significant" (Dufault & Martocchio, 1985, p. 380). Dorsett (2010, pp. 85, 86) has suggested that hope "provides a reason to go on living, helps maintain motivation, positive expectations, and may mediate the effect of depression."

Within the field of SCI, hope has played an important role in supporting positive psychological adjustment after SCI. Higher hope agency, as measured by the State Hope Scale (Snyder et al., 1996), has been associated with higher levels of acceptance after SCI (Kennedy, Evans, & Sandhu, 2009). Kortte, Gilbert, Gorman, and Wegener (2010) observed that hope and positive effect demonstrated a significant positive relationship with life satisfaction during the initial period of acute rehabilitation after SCI, and contributed to the prediction of life satisfaction at a three-month follow-up.

A range of qualitative studies have also considered the process of hope for individuals after SCI (Babamohamadi, Negarandeh, & Dehghan-Nayeri, 2011; Dorsett, 2010; Lohne, 2009; Lohne & Severinsson, 2005; Parashar, 2015; Smith & Sparkes, 2005). Hope has been observed to be often associated with a desire for complete recovery (Dorsett, 2010; Parashar, 2015). However, over time, hope has also been expressed for a satisfying quality of life, even in the absence of recovery (Dorsett, 2010; Lohne, 2009; Parashar, 2015). Hope has in some circumstances been closely associated with having religious faith (Babamohamadi et al., 2011).

Smith and Sparkes (2005) observed three kinds of hope among men who sustained SCI through sport: concrete hope, transcendent hope, and despair (loss of hope). Drawing upon the work of Frank (1995) the authors suggested that each kind of hope was shaped by a particular narrative type. Those reflecting the "*restitution* narrative" focused upon walking again, being cured, and returning to their pre-injury lifestyles. The "*quest* narrative," was associated with embracing "uncertainty and finitude, celebrating surprise, play, novelty, mystery, and openness to change" (Smith & Sparkes, 2005, p. 1099). In contrast, the "*chaos* narrative" is characterized as individuals who have experienced a loss of hope, perceiving life to be over as a consequence of SCI.

The presence of hope is common across different spiritual traditions. For example, Babamohamadi et al. (2011) found that hope was one of the most common coping strategies among Muslims in Iran. This included hope for successful surgery, hope in God for divine healing or miracles, hope for medical progress, and hope in the future.

Hope has also been observed independent of religion or spirituality, and something which may change over time. In a series of qualitative studies, Lohne and colleagues (Lohne, 2009; Lohne & Severinsson, 2005) considered hope in people with SCI at different stages post-injury. A year after the SCI, Lohne and Severinsson (2005) observed that individuals with SCI associated the process of hope with will power and struggle. Three to four years after SCI Lohne (2009) identified three main themes from interviews with the same participants; life-related hopes, body-related hopes, and creative and expanding hopes, where hope became about enjoying life within the limits experienced. In another longitudinal study Dorsett (2010) identified three foci of hope: (i) hope for a complete recovery, (ii) hope for a cure, and (iii) hope for a satisfying quality of life. For many participants in this study hope was an important factor in coping with their SCI, and a strong motivator to survive and "get on with life" (p. 95).

Finally, in a longitudinal study conducted over 2 years with individuals with SCI in India, Parashar (2015) considered the continuum of hope and influencers of hope. They found that early in rehabilitation, participants were hopeful and optimistic, which Parashar labeled "inevitable optimism" (p. 566). During this period religion and spirituality were tools several participants relied upon. As time since SCI increased and participants realized more fully the extent of their injury, hope moved to a more "tempered optimism" where the focus shifted from "recovery to rehabilitation" (p. 566). Finally, according to Parashar, the continuum of hope brought participants to a stage of "inevitable realism (p. 567)" where participants "aligned their expectations to the challenges, as well as the possibilities of the present" (p. 567). In the words of one participant, hope "builds onto something else" (p. 567). These longitudinal studies demonstrate how hope can change over time, and although initial hope for recovery may fade, hope does not disappear altogether. Rather it is invested in one's quality of life or returning to activities once previously enjoyed.

Resilience after SCI

Resilience has also been applied to the recovery process after SCI. Historically, resilience was perceived as a personality trait only displayed by a subset of the population. However, research over the past three decades has clarified that resilience is a set of skills that anyone can acquire (Newman, 2005). It is "a dynamic process encompassing positive adaptation within the context of significant adversity" (Luthar, Cicchetti, & Becker, 2000, p. 543), comprising a "a process of coping with disruptive, stressful, or challenging life events in a way that provides the individual with additional protective and coping skills than prior to the disruption" (Richardson, Neiger, Jensen, & Kumpfer, 1990, p. 34). Resilience is relevant to individuals, couples, families, and communities. It entails more than simply bouncing back, but growth in the face of the challenges arising from disruption to an individual's usual life (Richardson, 2002, p. 313).

Resilience after SCI has been positively associated with higher levels of self-efficacy (Driver et al., 2016; Kilic, Dorstyn, & Guiver, 2013), spirituality or religion (Driver et al., 2016; Duggan, Wilson, DiPonio, Trumpower, & Meade, 2016; White et al., 2010), satisfaction with life (Jones et al., 2019; White et al., 2010), positive affect (Jones et al., 2019), hope (Simpson, Anderson, Jones, Genders, & Gopinath, 2020) and social support (Duggan et al., 2016; Monden et al., 2014) as well as lower levels of depression (Driver et al., 2016; Jones et al., 2019; Kilic et al., 2013; Monden et al., 2014; White et al., 2010). The smaller number of studies focused on the experience of family members identified that resilience is closely associated with a positive outlook, increased levels of social support, higher levels of positive affect and lower levels of carer burden and negative affect (Elliot, Berry, Richards, & Shewchuk, 2014; Simpson & Jones, 2013).

The relationship between spirituality, hope, and resilience after SCI

The relationship between spirituality, hope, and resilience has been explored within theoretical frameworks of individual and family adaptation, identified in models generated from the lived experience of people with SCI, and reported by empirical studies. Starting with theoretical frameworks, Walsh (2003) has proposed a model of family resilience that "involves key processes over time that foster the ability to 'struggle well,' surmount obstacles, and go on to live and love fully" (p. 1). Walsh suggests that making meaning out of crises, facilitating a hopeful, positive outlook, and transcendent or spiritual moorings are three key processes that contribute towards family resilience. In the field of traumatic brain injury Nalder, Hartman, Hunt, and King (2019) have suggested that hope and spirituality may be two of several personal characteristics which contribute towards enabling beliefs and resiliency-related outcomes; reconstructing identity, adjusting participation patterns or preferences, and re-engaging in activities.

The interrelationship between spirituality, hope, and resilience has been identified in studies of the lived experience of people with SCI and their family members. The model presented in Fig. 5 brings together concepts of spirituality, hope, and resilience within the dynamic journey of adjustment, drawing upon the lived experience of both the person with SCI and family caregiver (Jones, Dorsett, Simpson, & Briggs, 2018). This model was developed from a study conducted with 10 family dyads at two time points, at a spinal injury unit in Sydney, Australia. The model highlights how people with SCI and their family members can draw upon a range of sources of spiritual strength. These sources of spiritual strength included the natural world, faith in God, oneself, and meaningful connectedness with others. In many cases these sources of spiritual strength can be tested after an SCI, leading to a meaning-making process. For people with SCI meaning-making responses included the perception that they had been given a second chance at life, that they could pray for healing, thinking positively about the situation, and comparing themselves with others who were less fortunate. Three key outcomes associated with these meaning-making responses were gratitude, hope, and deepening relationships with others, each of which may

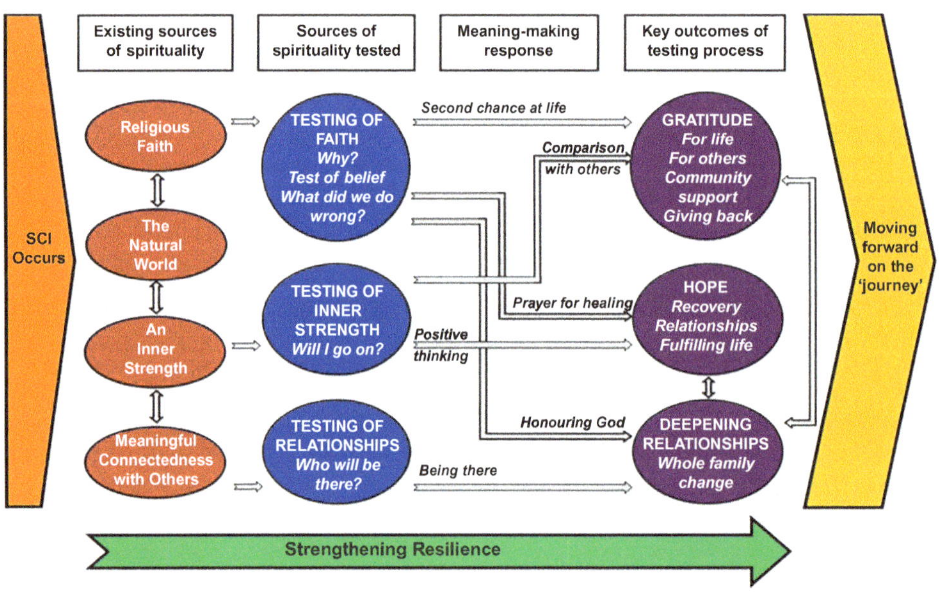

FIG. 5 Family resilience after SCI. *(Adapted from Jones, K. F., Dorsett, P., Simpson, G., & Briggs, L. (2018). Moving forward on the journey: Spirituality and family resilience after spinal cord injury.* Rehabilitation Psychology, 63*(4), 521–531. https://doi.org/10.1037/rep0000229, Figure 1 (p. 6).)*

contribute to the ongoing process of strengthening resilience. Hope included hope for physical recovery, hope for deeper and stronger relationships, and hope for a meaningful life together. Each outcome was observed to contribute to the ongoing process of resilience-building. Like the *quest* narrative, this model reveals that adaptation after SCI can be a journey, with people with SCI and their family members moving forward, stronger together.

The interrelationship between spirituality, hope, and resilience found in this model is echoed in two other qualitative studies. Monden et al. (2014) identified six factors that may contribute to resilience after SCI: spirituality or faith, alongside psychological strength, social support, perspective, adaptive coping, and serving as a role model or inspiring others. In a qualitative study with 475 people with SCI, Duggan et al. (2016), found resilience to be one of eight factors associated with happiness; the others included religion or faith in a higher power, general outlook on life, social support, and social relationships, physical health and functioning, mood, social comparisons, and resources.

Structural equation modeling has found an interrelationship among spirituality, hope, and resilience in two independent samples (Jones et al., 2019; Simpson et al., 2020). In both models, spirituality contributed to resilience, rather than vice versa, with the regression coefficients (ranging from 0.39 to 0.71) indicating a strong relationship. In the second of these studies, hope was also introduced, finding that in addition to the direct relationship between spirituality and resilience, spirituality also had an indirect relationship with resilience mediated by hope (Simpson et al., 2020).

Higher levels of spirituality, hope, and resilience were strongly associated with positive outcomes post-SCI. These included positive affect and satisfaction with life (Jones et al., 2019; Simpson et al., 2020). Higher levels of these three constructs were also associated with lower levels of negative outcomes from SCI, such as depression and negative affect (Jones et al., 2019; Simpson et al., 2020).

Implications for SCI rehabilitation: Person-centered care that embraces spirituality, hope, and resilience

The associations between higher levels of spirituality, hope, and resilience and better outcomes for individuals with SCI and their families need further testing and investigation. If a growing body of evidence supports the associations among these constructs and positive outcomes after SCI, then it raises a number of important implications for clinical practice.

Rehabilitation staff who have been surveyed, generally hold positive attitudes towards spirituality, even those who do not hold personal spiritual beliefs (Jones, Dorsett, Briggs, & Simpson, 2018; Jones, Pryor, Care-Unger, & Simpson, 2020a, 2020b, 2020c). Furthermore, many of them valued rehabilitation approaches that incorporated spirituality as part of a person-centered approach (Jones et al., 2020b). Staff in an SCI unit who participated in two focus groups perceived spirituality as potentially playing an important role in the adjustment of many individuals and their families after SCI (Jones, Dorsett, Briggs, & Simpson, 2018).

Despite the putative importance attributed to the role of spirituality in facilitating the adjustment process following SCI, focus groups and survey data found that few felt equipped to address the spiritual needs of rehabilitation clients and deemed that more training was necessary (Jones et al., 2020b). A brief (two and a half hours) spiritual care training program developed for rehabilitation staff was trialed at Royal Rehab in Sydney Australia. The training comprised a one-hour online self-study component followed by a one and a half hour face-to-face workshop that employed lectures, videoed interviews, role-plays and individual exercises to foster knowledge and skills (Jones et al., 2020c).

The quantitative evaluation of the program revealed that the competency, confidence and comfort levels in delivering spiritual care for those in the intervention group were significantly higher at post and follow-up time points than those in a matched control group (Jones et al., 2020a). An analysis of interviews with 16 participants 6–8 weeks after the training indicated that the overarching take-home message for staff was that "spirituality is everybody's business" (Jones et al., 2020c). Other themes included increased awareness of the nature of spirituality, realization of the importance of spirituality to clients, a desire to keep spirituality on the radar, identifying barriers to providing spiritual care, incorporating spirituality into practice, and recognizing spirituality as personally meaningful.

These findings suggest that incorporating spiritual care into healthcare practice brings many positives. However, some authors have advised caution. For instance, McSherry and Ross (2002) have highlighted some of the dilemmas of conducting spiritual assessments. These include questions regarding how spirituality is defined, staff motives for undertaking spiritual assessment, the timing and comprehensiveness of assessments, direct questioning versus observation, and which staff should undertake a spiritual assessment.

In addition to spirituality, it is also possible to focus on resilience as a modifiable factor that could be targeted for intervention. For example, the Strength 2 Strength program was developed to build resilience among family members supporting relatives with an SCI or a traumatic brain injury (Simpson et al., 2011). A program that focuses on resilience could provide options for individuals with SCI or family members who do not identify as having a spiritual dimension to their life. Future research could examine whether increased levels of hope could be a secondary outcome from such programs.

As promising as the current research has been, there are still many questions to be answered. Can programs be developed that support individuals with SCI to draw upon their spirituality to facilitate adjustment to SCI. Will these programs lead to increased levels of hope and resilience? Does the training of rehabilitation staff in spiritual care lead to measurable changes in their practice with clients and can this be captured within the research process? If programs to build resilience demonstrate efficacy among family caregivers supporting individuals with SCI, could such a program have similar benefits for individuals with SCI themselves?

Research into the role of spirituality in facilitating the process of adaptation to SCI is only just beginning. It has a strong potential to expand our understanding of the process of adjustment to SCI. The common references to spirituality in the lived experience accounts of individuals with SCI and their family members provide a strong rationale for continuing to investigate this domain. Observational studies have found strong associations between spirituality and other key clinical constructs such as hope and resilience. In combination, spirituality, hope, and resilience seem to play a significant role in enhancing the quality of life and well-being after SCI and provide a buffer to psychopathology and burden. More sophisticated longitudinal observational studies and experimental trials will be a key to moving this promising area of research forward into the future.

Applications to other areas of neuroscience

This section draws heavily upon Newberg, A. B. (2014). The neuroscientific study of spiritual practices. *Frontiers in Psychology*, 5, 215.

The chapter has focused on research documenting the subjective experience of spirituality within the context of adjustment to SCI. The neurobiological correlates of religious and spiritual phenomena have been investigated, generally among healthy populations, with meditation practices the most frequent area examined. Objective measures have evaluated the association between religious/spiritual practices and/or experiences and physiological parameters such as blood pressure and heart rate, as well as hormone and immune function. Neurobiological parameters such as electroencephalographic activity, cerebral blood flow, cerebral metabolism, and changes in neurotransmitter levels can be measured by techniques including electroencephalography and functional neuroimaging (positron emission tomography, functional magnetic resonance imaging fMRI). These techniques have been found to have both advantages and limitations. For example, fMRI scans require people to lie in a confined space with a significant underlying level of noise. This presents logistic challenges for subjects who are asked to undertake a series of spiritual practices such as prayer while being scanned. Future research could be expanded to include clinical populations, such as individuals with SCI, to help increase understanding of the association between spiritual states, experiences, and the brain.

Mini-dictionary of terms

Spirituality: Spirituality is a broad and dynamic construct that encompasses a range of sources of ultimate meaning-making and connection. Although religious faith may be one source of spirituality, it is not the only one. Other sources of spirituality may include the natural world, music, art, others, and oneself.

Hope: Hope is a dynamic life force focused on achieving something positive or good and is future based. Hope provides motivation and a reason for living.

Resilience: Resilience is a dynamic process that helps a person to overcome adversity and to move forward with growth and adaptation.

Key facts of "Spirituality, hope, and resilience in the recovery and adaptation process following spinal cord injury"

- In 1984 at the 37th World Health Assembly, the spiritual dimension was integrated into the World Health Organisation Member States' strategies for health and understood to play "a great role in motivating people's achievement in all aspects of life."
- In the most comprehensive review of the literature to date, Koenig (2012) identified that in many studies religion/spirituality was significantly associated with better mental health, greater social support, and lower levels of depression, suicide rates and anxiety.

- Hope has been described as understandable, measurable and essential as a coping strategy (Snyder, 1995).
- According to the American Psychology Association (Newman, 2005), resilience is a skill that can be learned by almost everyone.
- A holistic approach to healthcare, which acknowledges the biological, psychological, social and spiritual aspects of a person's well-being, leads to better patient outcomes.

Summary points

- The majority of research associated with coping and adaptation after SCI has focused on the physical, psychological and social challenges confronting the injured person and their family members.
- Recent research has highlighted the potential role that positive attributes such as spirituality, hope, and resilience may play in the process of recovery and adaption from SCI. The quest narrative described by Frank (1995) may be a helpful depiction of this process.
- Spirituality has been associated with a range of positive outcomes after SCI, including greater life satisfaction, quality of life, and lower levels of depression and anxiety.
- Hope has been identified to be an important coping mechanism for people after SCI, and something which may change over time.
- Resilience has been defined as a dynamic growth process, to which spirituality and hope may contribute.
- A study of individuals with SCI and their family members has generated a model of resilience, which depicts the whole family drawing upon different sources of spiritual strength to move forward together on a journey of adaptation and coping.
- A recent spiritual care training program conducted with health professionals has been successful in increasing awareness of the importance of spirituality after traumatic injury, and increasing staff confidence, comfort, and competency levels in spiritual care delivery.

References

Babamohamadi, H., Negarandeh, R., & Dehghan-Nayeri, N. (2011). Coping strategies used by people with spinal cord injury: A qualitative study. *Spinal Cord*, *49*, 832–837.

Baker, A., Barker, S., Sampson, A., & Martin, C. (2017). Caregiver outcomes and interventions: A systematic scoping review of the traumatic brain injury and spinal cord injury literature. *Clinical Rehabilitation*, *31*(1), 45–60.

Byra, S. (2015). Posttraumatic growth in people with traumatic long-term spinal cord injury: Predictive role of basic hope and coping. *Spinal Cord*, *54*, 478–482.

Campbell, J. (2008). *The hero with a thousand faces* (3rd ed.). Novato, CA: New World Library.

Canda, E. R., & Furman, L. D. (2009). *Spiritual diversity in social work practice: The heart of helping*. Cary, NC: Oxford University Press.

Chun, S., & Lee, Y. (2013). "I am just thankful": The experience of gratitude following traumatic spinal cord injury. *Disability and Rehabilitation*, *35*(1), 11–19.

Cobb, M., Puchalski, C. M., & Rumbold, B. (Eds.). (2012). *Oxford textbook of spirituality in healthcare*. Oxford: Oxford University Press.

Connor, K. M., & Davidson, J. R. (2003). Development of a new resilience scale: The Connor-Davidson resilience scale (CD-RISC). *Depression and Anxiety*, *18*, 76–82.

Davis, D. E., Rice, K., Hook, J. N., Van Tongeren, D. R., DeBlaere, C., Choe, E., et al. (2015). Development of the sources of spirituality scale. *Journal of Counselling Psychology*, *62*(3), 503–513.

Dorsett, P. (2010). The importance of hope in coping with severe acquired disability. *Australian Social Work*, *63*(1), 83–102.

Driver, S., Warren, A. M., Reynolds, M., Agtarap, S., Hamilton, R., Trost, Z., et al. (2016). Identifying predictors of resilience at inpatient and 3-month post-spinal cord injury. *The Journal of Spinal Cord Medicine*, *39*(1), 77–84. https://doi.org/10.1179/2045772314Y.0000000270.

Dufault, K., & Martocchio, B. C. (1985). Hope: Its spheres and dimensions. Symposium on compassionate care and the dying experience. *Nursing Clinics of North America*, *20*(2), 379–391.

Duggan, C., Wilson, C., DiPonio, L., Trumpower, B., & Meade, M. A. (2016). Resilience and happiness after spinal cord injury: A qualitative study. *Topics in Spinal Cord Injury Rehabilitation*, *22*(2), 99–110. https://doi.org/10.1310/sci2202-99.

Elliot, T. R., Berry, J. W., Richards, J. S., & Shewchuk, R. (2014). Resilience in the initial year of caregiving for a family member with a traumatic spinal cord injury. *Journal of Consulting and Clinical Psychology*, *82*(6), 1072–1086.

Frank, A. W. (1995). *The wounded storyteller: Body, illness and ethics*. Chicago: University of Chicago Press.

Hagen, E. M., & Rekand, T. (2015). Management of neuropathic pain associated with spinal cord injury. *Pain and therapy*, *4*(1), 51–65. https://doi.org/10.1007/s40122-015-0033-y.

Hoffman, J., Bombardier, C., Graves, D., Kalpakjian, C., & Krause, J. (2011). A longitudinal study of depression 1 to 5 years after spinal cord injury. *Archives of Physical Medicine and Rehabilitation*, *92*(3), 411–418.

Jazayeri, S. B., Beygi, S., Shokranch, F., Hagen, E. M., & Rahimi-Movaghar, V. (2015). Incidence of traumatic spinal cord injury worldwide: A systematic review. *European Spine Journal*, *24*, 905–918.

Jones, K. F., Dorsett, P., Briggs, L., & Simpson, G. (2018). The role of spirituality in spinal cord injury (SCI) rehabilitation: Exploring health professional perspectives. *Spinal Cord Series And Cases*, *4*(1), 54. https://doi.org/10.1038/s41394-018-0078-3.

Jones, K. F., Dorsett, P., Simpson, G., & Briggs, L. (2018). Moving forward on the journey: Spirituality and family resilience after spinal cord injury. *Rehabilitation Psychology*, *63*(4), 521–531. https://doi.org/10.1037/rep0000229.

Jones, K. F., Pryor, J., Care-Unger, C., & Simpson, G. (2018). Spirituality and its relationship with positive adjustment following traumatic brain injury: A scoping review. *Brain Injury*, 1–11. https://doi.org/10.1080/02699052.2018.1511066.

Jones, K. F., Pryor, J., Care-Unger, C., & Simpson, G. (2020a). Effectiveness of spiritual care training for rehabilitation professionals: An exploratory controlled trial. *NeuroRehabilitation*. https://doi.org/10.3233/NRE-203221. Advance online publication.

Jones, K. F., Pryor, J., Care-Unger, C., & Simpson, G. (2020b). Rehabilitation health professionals' perceptions of spirituality and spiritual care: The results of an online survey. *NeuroRehabilitation*, *46*(1), 17–30. https://doi.org/10.3233/NRE-192857.

Jones, K. F., Pryor, J., Care-Unger, C., & Simpson, G. (2020c). "Spirituality is everybody's business": An exploration of the impact of spiritual care training upon the perceptions and practice of rehabilitation professionals. *Disability and Rehabilitation*, 1–10. https://doi.org/10.1080/09638288.2020.1820586. Advance online publication.

Jones, K. F., Simpson, G., Briggs, L., & Dorsett, P. (2016). Does spirituality facilitate adjustment and resilience among individuals and their families after SCI? *Disability and Rehabilitation*, *38*(10), 921–935.

Jones, K. F., Simpson, G., Briggs, L., Dorsett, P., & Anderson, M. (2019). A study of whether individual and dyadic relations between spirituality and resilience contribute to psychological adjustment among individuals with spinal cord injuries and their family members. *Clinical Rehabilitation*, *33*(9), 1503–1514. https://doi.org/10.1177/0269215519845034.

Kalpakjian, C. Z., McCullumsmith, C. B., Fann, J. R., Richards, J. S., Stoelb, B. L., Heinemann, A. W., et al. (2014). Post-traumatic growth following spinal cord injury. *The Journal of Spinal Cord Medicine*, *37*(2), 218–225. https://doi.org/10.1179/2045772313Y.0000000169.

Kennedy, P., Evans, M., & Sandhu, N. (2009). Psychological adjustment to spinal cord injury: The contribution of coping, hope and cognitive appraisals. *Psychology, Health & Medicine*, *14*(1), 17–33.

Kennedy, P., & Rogers, B. A. (2000). Anxiety and depression after spinal cord injury: A longitudinal analysis. *Archives of Physical Medicine and Rehabilitation*, *81*(7), 932–937.

Kilic, S. A., Dorstyn, D., & Guiver, N. G. (2013). Examining factors that contribute to the process of resilience following spinal cord injury. *Spinal Cord*, *51*, 553–557.

Kirshblum, S., & Benevento, B. (2009). Understanding spinal cord injury and advances in recovery. In E. Druin, & M. M. Sliwinski (Eds.), *Spinal Cord Injuries: Management and Rehabilitation* (pp. 1–17). Saint Louis: Mosby.

Koenig, H. (2012). Religion, spirituality, and health: The research and clinical implications. *ISRN Psychiatry*, *8*. https://doi.org/10.5402/2012/278730.

Kortte, K. B., Gilbert, M., Gorman, P., & Wegener, S. T. (2010). Positive psychological variables in the prediction of life satisfaction after spinal cord injury. *Rehabilitation Psychology*, *55*(1), 40–47.

Lohne, V. (2009). Back to life again—Patients' experiences of hope three to four years after a spinal cord injury—A longitudinal study. *Canadian Journal of Neuroscience Nursing*, *31*(2), 20–25.

Lohne, V., & Severinsson, E. (2005). Patients' experiences of hope and suffering during the first year following acute spinal cord injury. *Journal of Clinical Nursing*, *14*, 285–293.

Luthar, S. S., Cicchetti, D., & Becker, B. (2000). The construct of resilience: A critical evaluation and guidelines for future work. *Child Development*, *71* (3), 543–562.

McColl, M. A., Bickenbach, J., Johnston, J., Nishihama, S., Schumaker, M., Smith, K., et al. (2000a). Changes in spiritual beliefs after traumatic disability. *Archives of Physical Medicine and Rehabilitation*, *81*(6), 817–823.

McColl, M. A., Bickenbach, J., Johnston, J., Nishihama, S., Schumaker, M., Smith, K., et al. (2000b). Spiritual issues associated with traumatic-onset disability. *Disability and Rehabilitation*, *22*(12), 555–564.

McSherry, W., & Ross, L. (2002). Dilemmas of spiritual assessment: Considerations for nursing practice. *Journal of Advanced Nursing*, *38*(5), 479–488.

Middleton, J., Simpson, G., De Wolf, A., Quirk, R., Descallar, J., & Cameron, I. (2014). Health status, quality of life and strain in carers during community integration after SCI. *Archives of Physical Medicine and Rehabilitation*, *95*, 1312–1319. https://doi.org/10.1016/j.apmr.2014.03.017.

Monden, K. R., Trost, Z., Catalano, D., Garner, A. N., Symcox, J., Driver, S., et al. (2014). Resilience following spinal cord injury: A phenomenological view. *Spinal Cord*, *52*, 197–201.

Mundle, R. (2015). "I should be closer to god because of this"-a case study of embodied narratives and spiritual reconstruction in spinal cord injury and stroke rehabilitation. *Journal of Disability and Religion*, *19*(1), 30–49.

Nalder, E., Hartman, L., Hunt, A., & King, G. (2019). Traumatic brain injury resiliency model: A conceptual model to guide rehabilitation research and practice. *Disability and Rehabilitation*, *41*(22), 2708–2717.

Newman, R. (2005). APA's resilience initiative. *Professional Psychology: Research and Practice*, *36*, 227–229.

Palmer, S., Kriegsman, K. H., & Palmer, J. B. (2008). *Spinal cord injury: A guide for living*. Baltimore: John Hopkins University Press.

Parashar, D. (2015). The trajectory of hope: Pathways to find meaning and reconstructing the self after a spinal cord injury. *Spinal Cord*, *53*, 565–568.

Puchalski, C., Ferrell, B., Virani, R., Otis-Green, S., Baird, P., Bull, J., et al. (2009). Improving the quality of spiritual care as a dimension of palliative care: The report of the consensus conference. *Journal of Palliative Medicine*, *12*, 885–904.

Rahnama, P., Javidan, A. N., Saberi, H., Montazeri, A., Tavakkoli, S., Pakpour, A. H., et al. (2015). Does religious coping and spirituality have a moderating role on depression and anxiety in patients with spinal cord injury? A study from Iran. *Spinal Cord*, *53*, 870–874.

Richardson, G. E. (2002). The metatheory of resilience and resiliency. *Journal of Clinical Psychology*, *58*(3), 307–321.

Richardson, G. E., Neiger, B. L., Jensen, S., & Kumpfer, K. L. (1990). The resiliency model. *Health Education*, *21*(6), 33–39.

Saleebey, D. (2006). The strengths perspective: Possibilities and problems. In D. Saleebey (Ed.), *The strengths perspective in social work practice* (4th ed., pp. 279–303). Boston: Pearson.

Seligman, M. E., & Csikszentmihalyi, M. (2000). Positive psychology: An introduction. *American Psychologist*, *55*(1), 5–14.

Siddall, P. J., McIndoe, L., Austin, P., & Wrigley, P. J. (2016). The impact of pain on spiritual well-being in people with a spinal cord injury. *Spinal Cord*, 1–7.

Simpson, G., Anderson, M., Jones, K. F., Genders, M., & Gopinath, B. (2020). Do spirituality, resilience and hope mediate outcomes among family caregivers after traumatic brain injury or spinal cord injury? A structural equation modelling approach. *NeuroRehabilitation*, *46*(1), 3–15.

Simpson, G., & Brenner, L. A. (2019). *Suicide prevention after neurodisability: An evidence-informed approach.* New York: Oxford University Press.

Simpson, G., & Jones, K. (2013). How important is resilience among family members supporting relatives with traumatic brain injury or spinal cord injury? *Clinical Rehabilitation*, *27*(4), 367–377.

Simpson, G., Jones, K., Pfeiffer, D., Unger, C., Oosthuizen, H., Francis, J., et al. (2011). *Strength2Strength: Building resilience among families supporting relatives with traumatic injury.* Sydney, Australia: Lifetime Care and Support Authority.

Smith, B., & Sparkes, A. (2005). Men, sport, spinal cord injury, and narratives of hope. *Social Science & Medicine*, *61*, 1095–1105.

Snyder, C. R. (1995). Conceptualizing, measuring, and nurturing hope. *Journal of Counseling and Development*, *73*(3), 355–360. https://doi.org/10.1002/j.1556-6676.1995.tb01764.x.

Snyder, C. R., Sympson, S. C., Ybasco, F. C., Borders, T. F., Babyak, M. A., & Higgins, R. L. (1996). Development and validation of the State Hope Scale. *Journal of Personality and Social Psychology*, *70*(2), 321–335.

Van Koppenhagen, C. F., Post, M. W., Van der Woude, L. H., De Witte, L. P., Van Ashbeck, F. W., De Groot, S., et al. (2008). Changes and determinants of life satisfaction after spinal cord injury: A cohort study in the Netherlands. *Archives of Physical Medicine and Rehabilitation*, *89*, 1733–1740.

Walsh, F. (2003). Family resilience: A framework for clinical practice. *Family Process*, *42*(1), 1–18.

Weitzenkamp, D. A., Gerhart, K. A., Charlifue, S., Whiteneck, G., & Savic, G. (1997). Spouse of spinal cord injury survivors: The added impact of caregiving. *Archives of Physical Medicine and Rehabilitation*, *78*, 822–827.

White, B., Driver, S., & Warren, A. (2008). Considering resilience in the rehabilitation of people with traumatic disabilities. *Rehabilitation Psychology*, *53* (1), 9–17.

White, B., Driver, S., & Warren, A. (2010). Resilience and indicators of adjustment during rehabilitation from a spinal cord injury. *Rehabilitation Psychology*, *55*(1), 23–32.

Williams, R., & Murray, A. (2015). Prevalence of depression after spinal cord injury: A meta-analysis. *Archives of Physical Medicine and Rehabilitation*, *96*(1), 133–140.

Wilson, C. S., Forchheimer, M., Heinemann, A. W., Warren, A. M., & McCullumsmith, C. (2017). Assessment of the relationship of spiritual well-being to depression and quality of life for persons with spinal cord injury. *Disability and Rehabilitation*, *39*(5), 491–496.

World Health Organisation. (2013). *Spinal cord injury*. Retrieved from http://www.who.int/mediacentre/factsheets/fs384/en/.

Chapter 39

Wellness intervention for persons with spinal cord injury

Stephanie L. Silveira[a], Emma V. Richardson[b], and Robert W. Motl[a]

[a]Department of Physical Therapy, School of Health Professions, University of Alabama at Birmingham, Birmingham, AL, United States, [b]School of Sport and Exercise Science, University of Worcester, Worcester, United Kingdom

List of abbreviations

BCT behavior change technique
HAPA health action process model
MS multiple sclerosis
QOL quality of life
SCI spinal cord injury
SCT social cognitive theory
TBI traumatic brain injury
TPB theory of planned behavior
TTM transtheoretical model

Introduction

Spinal cord injury (SCI), defined by damage to the spinal cord that causes a temporary or permanent change in function, is often a devastating injury that impacts approximately 17,700 people in the United States (US) per year (NSCISC, 2020). From the onset, SCI requires immediate and long-term intervention to address changes in function such as paralysis, loss of sensation, bowel and bladder dysfunction, reduced muscle strength, and pain (NSCISC, 2020). Acquiring an SCI can have a devastating impact on the psychological well-being of a person, with depression, anxiety, a loss of self and identity, and hope for the future (Sparkes & Smith, 2002) being elements of SCI that the majority of people, at some time, will experience. Socially, an SCI can negatively affect an individual's independence, social relationships, and overall quality of life (QOL) (Kennedy, Lude, Elfström, & Smithson, 2011).

Persons with an SCI can, however, live a positive and fulfilling life filled with great meaning and happiness (Chun & Lee, 2020). The process of experiencing long-term wellness after SCI is long, complex, and unique to each individual. The unique nature of SCI regarding the level of injury, degree of completeness, the individual person and immediate social support and surrounding areas underscores the importance of innovative, multifaceted approaches that promote and maintain independence, wellness, and QOL.

This chapter provides readers with tools for designing and assessing wellness interventions that may improve the lives of persons with SCI. We first define wellness and describe unique attributes that distinguish wellness interventions from other health behavior change interventions. Following this, we discuss common measures to assess wellness outcomes of interest. We next present modifiable health-promoting and health-risk behaviors that may be targeted for improving wellness outcomes. We then outline previous literature examining behavior change theories utilized for promoting health behavior change among persons with SCI. We conclude with a brief review of previous wellness interventions and provide considerations that guide the development of future wellness interventions for persons with SCI.

What is wellness?

The concept of wellness has evolved from the mid-20th century to now and differs between health disciplines. In 1948, the World Health Organization defined wellness as a state of complete physical, mental, and social well-being and not merely

Cellular, Molecular, Physiological, and Behavioral Aspects of Spinal Cord Injury. https://doi.org/10.1016/B978-0-12-822427-4.00039-3

FIG. 1 The six wellness dimensions. The six primary wellness dimensions outlined by Hettler 1980.

the absence of disease or infirmity. With greater political and socio-cultural movements from disability studies, and evolution within psychological health sciences regarding the concepts of wellness and illness (Goodwin & Jamison, 2007), wellness has become a much more multidimensional construct that can mean different things to people from different disciplines. For example, Lebensohn, Dodds, Benn, Brooks, and Birch (2013) defined wellness as "an active process of making choices in multiple dimensions of body, mind, and spirit that lead toward healthier ways of living." Alternatively, from a psychological standpoint, wellness is understood as optimal psychological function and experience (Ryan & Deci, 2001). A hedonic tradition understands wellness to be the pursuit of happiness and seeking to obtain a "good life" (Keyes, Shmotkin, & Ryff, 2002), while a eudemonic tradition defines wellness through realizing one's potential and flourishing (Ryff & Keyes, 1995). Further, wellness is often operationalized into six dimensions: occupational, social, intellectual, physical, emotional, and spiritual (Fig. 1). For this chapter, we define wellness as the existence of positive health in an individual with a focus on functioning, health, and QOL (Corbin & Pangrazi, 2001), but draw upon the aforementioned concepts to highlight the ways interventions can target and enhance the various aspects of wellness among persons with SCI.

Wellness interventions aim to modify behaviors for improving health, functioning, and QOL. In the field of rehabilitation, wellness interventions are specifically characterized by approaches that focus on improving aspects of well-being in the context of a health condition as opposed to concentration on medical management (Stuifbergen, Morris, Jung, Pierini, & Morgan, 2010). Promotion of health behaviors via wellness interventions in persons with SCI are therefore critical given reduced overall QOL reported in this population, both immediately post-injury and long-term. Indeed, persons with SCI report significantly lower overall QOL than the general population as well as higher rates of secondary health conditions such as depression, pain, and cardiovascular disease (Decker & Schulz, 1985; Post, de Witte, van Asbeck, van Dijk, & Schrijvers, 1998). The high incidence of poor QOL, and dramatic changes in life circumstances following SCI, support the interest in wellness interventions targeting health behavior change.

There is an immediate, immense impact of SCI on well-being such that most individuals must learn a completely new way of living. Over the short term during inpatient and acute treatment, healthcare practitioners focus on restoring activities of daily living and minimizing functional impairment through pharmaceutical treatments as well as physical and occupational therapies. Moreover, during the years following acquiring an SCI, a majority of individuals experience clinically significant depressive symptoms and increases in adiposity based on changes in energy expenditure (Gater Jr, 2007; Krause, Kemp, & Coker, 2000). Traditionally, interventions exclude persons with SCI in this sensitive period the first year following injury, yet this period may be a critical gateway for promoting long-term wellness that requires tailored approaches. Further research is needed regarding an appropriate timeline for emphasizing health behaviors in the rehabilitation period following SCI.

Wellness interventions across the lifespan in SCI must consider the changing demographics in the SCI population. Indeed, the average age at injury is now 41 years old compared to 29 years old in the 1970s (NSCISC, 2020). Further, years post-injury, age, gender, marital status, education, employment status, and income are associated with differing levels of QOL among persons with SCI (Dijkers, 1999; Putzke, Elliott, & Richards, 2001; Putzke, Richards, Hicken, & DeVivo, 2002). Such factors may be important considerations when focusing on wellness in diverse groups with SCI. Age may be a focal factor considering life expectancy following SCI has significantly increased based on advances in medicine and assistive device technology. One study of older adults with SCI reported that average activity limitations measured using the Spinal Cord Independence Measure were moderate (mean total score 65.2, range 8–100) and were associated with severity and level of injury (Jörgensen, Iwarsson, & Lexell, 2017). Satisfaction with life among older adults was associated with marital status, vocational status, and bladder functioning (Jörgensen et al., 2017). Collectively, promotion of wellness is important in the SCI population immediately following injury, longitudinally, and among older adults, and innovation strategies may be warranted that address needs and preferences across the lifespan.

Measurement of wellness in SCI

The study of wellness interventions requires appropriate measurement tools, both quantitative and qualitative. From a quantitative perspective, primary outcomes among wellness interventions often focus on QOL, satisfaction with life, or other measures of subjective well-being. This focus is based on the need for global measures that capture an individual's perception of their overall life rather than solely focus on physical functioning. Overall QOL or satisfaction with life is often measured using Diener's 5-item Satisfaction with Life Scale (SWLS) or the 9-item Life Situation Questionnaire (LSQ) (Diener, Emmons, Larsen, & Griffin, 1985; Fugl-Meyer, Bränholm, & Fugl-Meyer, 1991). Health-related QOL (HRQOL) among persons with SCI is widely measured using the SF-12 of SF-36, and SCI-specific measures of QOL have been developed such as the Ferrans and Powers Quality of Life Index (QLI-SCI) (May & Warren, 2002). The focus on HRQOL is based on a need to assess perceptions regarding physical and mental functioning, and overall wellness.

Qualitative methods provide another assessment approach to guide the creation and refinement of wellness interventions. Indeed, exploratory qualitative studies are a crucial first step when designing interventions as they can provide an empirical foundation regarding a person's current wellness state, and identifying preferences for future interventions. Qualitative methods can be incorporated as part of feasibility assessments of interventions, and guide further refinement of interventions for larger clinical trials. Qualitative studies have much to contribute to wellness interventions in SCI as they afford rich detail regarding experience, perceptions, satisfaction, and recommendations for improving an intervention design, delivery, and assessment, mainly through employing interview methods. For example, semi-structured interviews were conducted with four males with SCI following a center-based wellness program and highlighted that the program positively influenced the physical, mental, and social well-being of participants (Ekelman et al., 2017). Findings from this qualitative study contributed to knowledge regarding the utility of the program as well as the value of social and occupational components.

Health behaviors: Targets for wellness interventions

An important consideration of wellness intervention design is the deliberate inclusion of multiple health behaviors to reflect the multifaced nature of wellness. For example, interventions typically target modifiable health-promoting behaviors such as physical activity, nutrition, and stress management as well as health-risk behaviors such as tobacco use. Behaviors of interest may be targeted synchronously or asynchronously. The health behaviors outlined below are but a few samples of possible targets for wellness interventions among persons with SCI.

Physical activity is the most widely studied health-promoting behavior among persons with SCI. Physical activity is broadly defined as any bodily movement produced by skeletal muscles that require energy expenditure – including activities undertaken while working, playing, carrying out household chores, traveling, and engaging in recreational pursuits (WHO, 2018). Exercise is a subset of physical activity that is planned, structured, repetitive, and targets improvements or maintenance of overall fitness (WHO, 2018). Physical activity interventions for persons with SCI may include the promotion of lifestyle physical activity or structured exercise training programs. Exercise training is a key part of rehabilitation immediately following injury and often focuses on functional recovery. Longitudinally, exercise training is essential for persons living with an SCI to increase strength and fitness to maintain and improve independence, participation, and overall health. Physical activity is often targeted as an intervention not only for its purported physical benefits, but also its associations improving subjective well-being in persons with SCI (Ginis, Jetha, Mack, & Hetz, 2010; Warms, Belza, Whitney, Mitchell, & Stiens, 2004). This highlights the potential impact of health-enhancing behaviors on multiple facets of wellness.

Nutrition interventions focus on the adjustment of personal practices and habits to improve nutritional status (WHO, 2020). Such interventions are critical for the persons with SCI to maintain overall health, reduce chances of osteoporosis, and manage bowel and bladder dysfunction. For persons with SCI, these types of interventions may be initiated at various phases of rehabilitation from immediate management of protein needs during the acute phase to the long-term promotion of a balanced diet during the chronic phase. Recent evidence suggests that low carbohydrate or ketogenic diets may improve neurological recovery during the acute phase, immediately following injury (Yarar-Fisher et al., 2018). However, wellness interventions generally focus on healthy eating to promote weight management and overall health for persons living with SCI. One recent systematic review and meta-analysis examining persons with SCI reported imbalances in fiber intake and micronutrients that may be targeted in wellness interventions (Farkas, Pitot, Berg, & Gater, 2019).

Stress management is a critical component of wellness interventions for persons with SCI. Adaptive stress management strategies are key to healthy living following SCI given the traumatic nature of many injuries and the need to adapt to a new life. Individual stress management interventions aim to help individuals identify stressors and develop strategies to cope aiming to reduce stress and increase well-being (Holman, Johnson, & O'Connor, 2018). Group stress management interventions often take the form as support groups for persons with SCI (Holman et al., 2018). One study pilot tested a coping effectiveness training program for persons with SCI that demonstrated reductions in symptoms of anxiety and depression (King & Kennedy, 1999). Stress management is primarily included as part of an intervention targeting other health-promoting behaviors and relevant components for wellness interventions for persons with SCI.

Tobacco use is the primary health-risk behavior of interest in the SCI literature given the prevalence of smoking among persons with SCI is significantly higher than the general population (Saunders, Krause, Saladin, & Carpenter, 2015). Tobacco use is detrimental to rehabilitation following SCI given the nature of CNS dysfunction following injury and preservation of respiratory system function is vital. Smoking behavior may exacerbate secondary health conditions such as pain and impede participation in other health behaviors such as physical activity (Richardson, Richards, Stewart, & Ness, 2012). Smoking cessation has not been a focal target in wellness interventions among persons with SCI. Therefore, tobacco use has been identified as an often overlooked, but vital topic for future wellness interventions.

Behavior change theory for supporting behavior change interventions in SCI

Wellness interventions target change in health behaviors for improving psychosocial outcomes in SCI, and success depends on the actual change and maintenance of behavior change. To that end, such interventions should be informed and supported based on behavior change models or theories. Theory is critical for identifying targets of behavior change and then guiding the selection of behavior change techniques that serve as the active ingredients of the intervention. Common behavior change theories for health promotion among persons with SCI include Social Cognitive Theory, The Transtheoretical Model, Health Action Process Approach Model, and The Theory of Planned Behavior. Researchers should be aware of the underlying assumptions of a theory, its causal structure and variables for behavior change strategies and then align the intervention target, goals, and population with the theory. Theories outlined in this chapter represent the most commonly applied in SCI; however, this is not an exclusive list as there are dozens of theories of human behavior.

Social cognitive theory (SCT) is one of the most widely applied behavior change theories for health promotion among persons with SCI, including wellness interventions (Zemper et al., 2003). The underlying premise of SCT is that behavior is socially constructed through observation (Bandura, 1998). Specifically, behavior is constructed and modified through dynamic interactions among the person, their environment, and behavior (Fig. 2). Self-efficacy is a primary tenet of SCT in which individual confidence in their ability to perform a behavior (i.e., self-efficacy) that guides behavior change and maintenance (Bandura, 1998). Indeed, sources of self-efficacy include mastery experiences, vicarious experiences, verbal persuasion, and physiological or emotional responses (Bandura, Freeman, & Lightsey, 1999). Other central tenets of SCT include outcome expectations, goal setting, barriers, and impediments. Applications of SCT in the assessment of behavior in persons with SCI have largely focused on physical activity behavior (Martin Ginis et al., 2011), however applications to other health behaviors are plausible based on extensive application of SCT in diverse populations.

The Health Action Process Model (HAPA) is a more recent psychological theory that blends components of SCT and the Transtheoretical Model (Fig. 3). The HAPA model includes at least two phases, motivation and volition, and further splits volition phase into planning, action, and maintenance phases (Schwarzer, Lippke, & Luszczynska, 2011). The HAPA model outlines the process of behavior change across various phases (Schwarzer et al., 2011). For example, self-efficacy is highlighted as a crucial component through all phases, whereas risk perceptions play a primary role during the contemplation phase. Schwarzer et al. (2011) provide an overview of how HAPA can be applied in a rehabilitation setting for persons with chronic diseases, such as SCI, however most of the current evidence draws on research in physical activity.

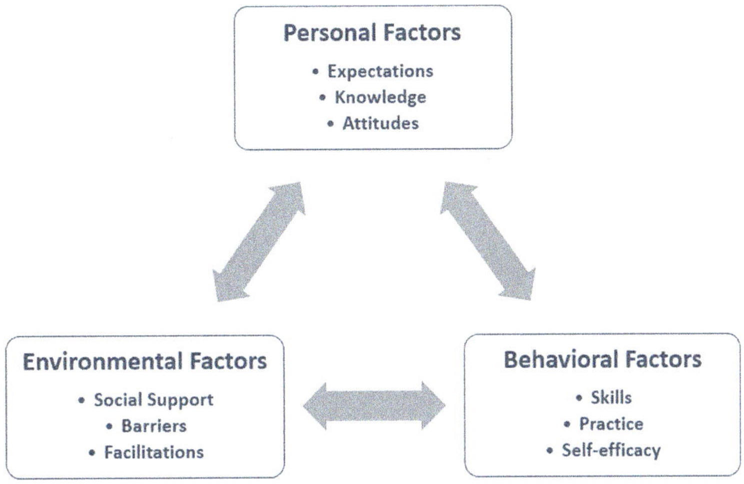

FIG. 2 Social cognitive theory model. Bandura's concept of triadic reciprocal determinism outlines the dynamic interaction between the individual, their environment, and behaviors.

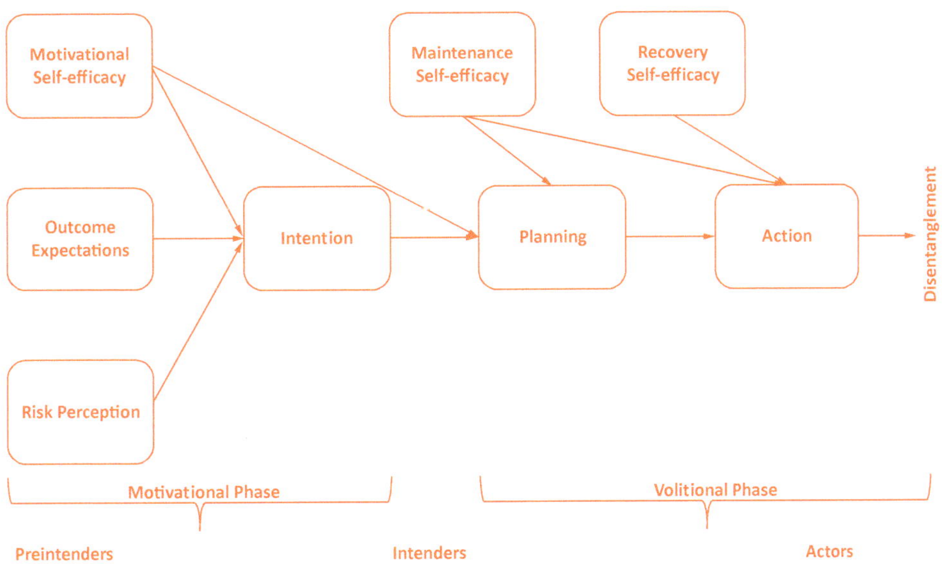

FIG. 3 The health action's process model. The health action's process model outlines various stages and the primary influences that interact within phases that results in actions.

The Transtheoretical Model (TTM) is also known as the Stage of Change Model focuses on the staged and gradual process of health behavior change (Prochaska & Velicer, 1997). TTM was originally developed as a model for smoking behavior change in which individuals change behavior and maintain change based on readiness for change. The stages of change depicted in Fig. 4 include Precontemplation, Contemplation, Preparation, Action, and Maintenance (Prochaska & Velicer, 1997). The stages are predicted by self-efficacy and decisional balance. TTM is unique as its primary utility is to provide a model for the application of behavior change theory along the stages of change. The TTM has been applied in nutrition assessment among adults with SCI as well as physical activity assessment and intervention development (Tsunoda, Inayama, Hata, & Oka, 2015).

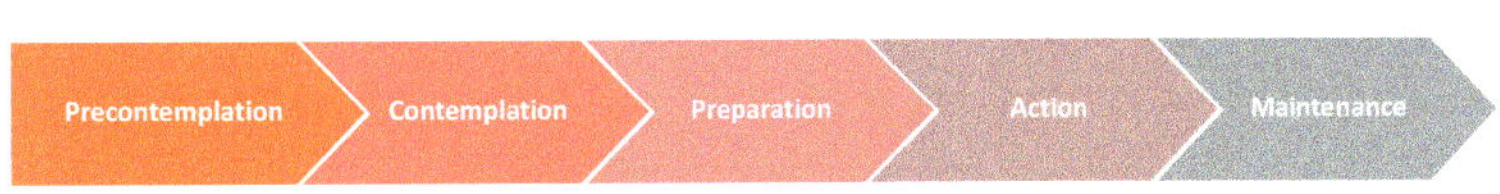

FIG. 4 The transtheoretical model. Five discrete stages are outlined in the transtheoretical model, however individuals may progress and regress between stages during various phases of behavior change.

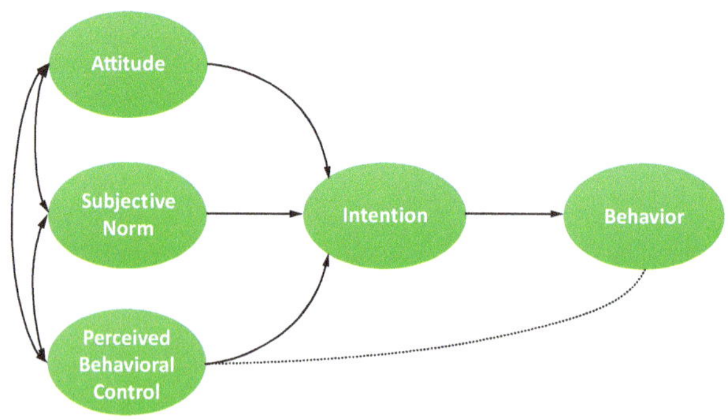

FIG. 5 Theory of planned behavior. The theory of planned behavior model focuses on the influences and factors that guide intentions and subsequent behavior.

The Theory of Planned Behavior (TPB) focuses on behavioral intentions or an individual's plans to engage in a behavior at a specific time and place (Ajzen, 1991). TPB has been widely applied in behavior change interventions for smoking, substance use, health services utilization, and breastfeeding. The two foundational components of TPB are motivation and ability that are guided by perceptions. Underlying concepts associated with TPB are attitudes, intentions, norms (subjective and social), and perceptions of power and behavioral control (Ajzen, 1991) (Fig. 5). TPB has been examined as a determinant of secondary condition management (i.e., skincare) and physical activity in persons with SCI (Latimer & Martin Ginis, 2005; Sheppard, Kennedy, & Mackey, 2006).

Exemplar wellness interventions for persons with SCI: Examples, settings, and results

A thorough background regarding previous wellness interventions for persons with SCI is critical for designing future interventions. The foundational wellness interventions for persons with SCI were delivered via workshops (i.e., short-term educational courses). For example, one randomized controlled trial examined the efficacy of six half-day wellness workshops for adults with SCI over 3 months. Each workshop included an expert-led portion on lifestyle management, nutrition, exercise, and secondary conditions (Zemper et al., 2003). Preliminary results indicated that there were no between-group differences. However, secondary analyses indicated that treatment response among the intervention group participants was related to marital status and age, specifically, single/divorced participants were more likely to not respond to the intervention and further participants who were single/divorced and ≤ 50 years old were most likely to not respond to the intervention (Silveira et al., 2020). Overall, short-term workshop interventions are a feasible option for providing wellness information for persons with SCI; however, demographics such as marital status and age may impact response and longitudinal positive impact is not established.

Wellness programs for persons with SCI have been delivered in community fitness settings. Some centers have unique resources such as one-on-one personal trainers who have a background in working with individuals with SCI. A recent qualitative study examined perceptions of four males with SCI following a center-based community wellness program reported positive physical, social, and mental outcomes. Participants reported that the center provided them with "a place for hope" and "a supportive community" (Ekelman et al., 2017). One primary barrier to the implementation of center-based programs is pervasive transportation issues among persons with SCI and the lack of adaptive community centers that are equipped to deliver quality interventions.

Several hospitals are creating wellness programs as part of SCI rehabilitation practices that may be delivered by peers or health care professionals. One recent wellness program created for adults with SCI and spina bifida was delivered through an integrated delivery system and focused on prevention and treatment of pressure sores, depression, neurogenic bladder and bowel, diet, and physical activity (Dicianno et al., 2016). Participants were provided with a wellness practitioner who provided patient education, case management services including goal setting, and incentives for making progress toward goals. Following the 2 year intervention, participants reported better QOL, overall health, lower depressive symptoms, and health care costs were attenuated (Dicianno et al., 2016). The provision of robust integrated delivery system programs is challenging due to time and cost considerations; however, there is some evidence of longitudinal benefits.

Hospital-based wellness programs utilizing a peer mentor approach may help improve self-efficacy to manage life following SCI (Gassaway et al., 2017; Ljungberg, Kroll, Libin, & Gordon, 2011). A recent randomized controlled trial tested the efficacy of an inpatient-based peer mentoring program on self-efficacy and hospital readmissions (Gassaway et al., 2017). Participants in the intervention group received an initial consultation with a peer liaison followed by weekly meetings with a peer mentor in-person during inpatient care and weekly phone, e-mail, or in-person contact for 90 days following discharge. Positive outcomes were reported with higher rates of self-efficacy in the intervention group compared to the standard-care control group and fewer unplanned hospital stays (Gassaway et al., 2017). Therefore, peer mentoring programs during the initial phases of SCI rehabilitation demonstrate efficacy for improving self-efficacy for managing medical care and life following SCI.

Telehealth approaches to promoting wellness among persons with SCI have received increasing attention. Many telehealth interventions for persons with SCI focus on providing knowledge and resources to manage health promoting behaviors in addition to increased contact with health care providers. For example, CareCall system was created for persons with various spinal cord dysfunctions combining patient education, cognitive-behavioral interventions, screening and referrals, and alerts to a nurse telerehabilitation coordinator for direct non-emergent phone follow-up (Houlihan et al., 2013). Pilot testing of the CareCall system demonstrated efficacy for reducing pressure sores, depression, and patient reports of health care availability, but not health care utilization (Houlihan et al., 2013). Given the efficacy of wellness interventions for persons with SCI, there is a pressing need for further applications of telehealth interventions for promoting health behavior change (McIntyre et al., 2020).

Design of future wellness interventions for persons with SCI

Future wellness interventions for persons with SCI should be founded on behavioral theory and identify behavior change techniques (BCTs) that align with theory. As mentioned, physical activity is currently the most widely researched health-promoting behavior for persons with SCI. Physical activity researchers have provided a roadmap for researchers in other areas to follow and build upon. For example, a review of BCTs in physical activity interventions for persons with SCI coded BCTs based on the BCT Taxonomy V1 and reported the most commonly used were: 4.1. instruction on how to perform the behavior, 1.1. goal setting (behavior), 3.2. social support (practical), 9.1. credible source, 1.2. problem solving, and 1.4. action planning (Tomasone et al., 2018). Compilation of such information is key to identifying the most widely used and the most effective BCTs for future interventions. Separate health behaviors likely require distinctive strategies and techniques for initiating and maintaining health behavior change that aligns with appropriate theory—this is the major challenge for multiple health behavior change interventions.

Multiple health behavior change interventions may further consider aligning with an established model of wellness or QOL. Stuifbergen's Model of QOL in individuals with chronic disabling conditions has established the factors preceding QOL in persons with various chronic disabling conditions, such as SCI (Stuifbergen, Seraphine, & Roberts, 2000) (Fig. 6). This model considers the various interactions among severity of illness/condition, barriers, resources, acceptance, self-efficacy, and health-promoting behaviors. Though more research is needed to determine the validity of the direct versus indirect paths in persons with SCI, the factors of interest are well-established as strong correlates of QOL in persons with SCI. This model is provided as an example in which behavior change theories may be applied for creating theory-based wellness interventions for persons with SCI in various settings.

One primary barrier for creating and disseminating wellness interventions is the timeliness of scientific research to reach community settings. One branch of the Neurorecovery Network aims to provide rapid translation of evidence-based techniques and technologies for promoting health among persons with SCI and other neurological diseases (Tolle et al., 2018).

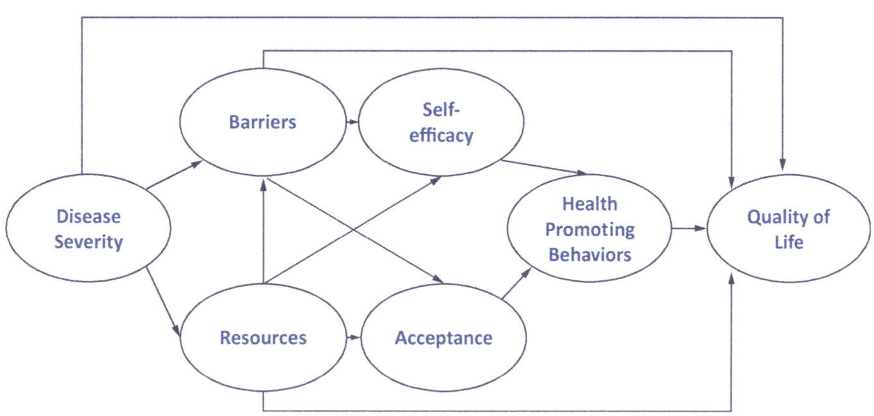

FIG. 6 Model of health promotion and quality of life in chronic disabling conditions. This explanatory model of health promotion and quality of life is a foundation for developing wellness interventions.

The Community fitness and wellness facilities network was created to provide opportunities for promoting wellness across the care continuum (i.e., inpatient through community settings) (Tolle et al., 2018). The network partners utilize standardized intervention strategies and outcome measurements to assess interventions in community settings, bringing together researchers and community partners to bridge the gap between research and practice. The Community fitness and wellness facilities network is an in-person approach wherein individuals work with trained staff and adaptive equipment and therefore is likely not an all-encompassing approach as many individuals with SCI live in rural areas or face significant barriers to transportation; however, this model may be adapted using technology to provide essential wellness resources and support.

Advances in technology provide unique opportunities for telehealth interventions that overcome barriers for persons with SCI such as transportation and accessibility. Researchers and practitioners are examining the efficacy of wellness interventions delivered through various online mediums such as websites and mobile apps. For example, the iMHere system is a website portal and app that includes modules on medication management, skincare, bowel and bladder programs, mood tracking, and messaging that aims to connect people with SCI with their healthcare providers and improve wellness outcomes (Kryger et al., 2019). Preliminary results from the iMHere study indicate reductions in urinary tract infections, but not psychosocial outcomes (Kryger et al., 2019). Such technologies may be a promising foundation for wellness interventions for persons with SCI such that future iterations may include other modifiable behaviors such as nutrition and exercise.

Overall, the design of wellness interventions for persons with SCI is a considerable task that requires inquiry, partnership, and trust among participants, researchers, and the health care community. Mixed-methods approaches are pivotal to the design of wellness interventions that meet the unique needs and preferences of individuals with SCI who differ by demographics, clinical characteristics, and lived experiences. Exploratory Sequential Design is one mixed-method approach that has been recommended for developing self-management interventions for persons with SCI (Munce, Guetterman, & Jaglal, 2020). Exploratory sequential approaches juxtapose qualitative and quantitative methods in the creation of intervention, specifically, researchers inquire about preferences using qualitative methods such as interviews in conjunction with quantitative methods to identify appropriate outcomes such as surveys. This approach allows flexibility to incorporate peers and caregivers to create tailored interventions and has demonstrated efficacy for developing interventions for persons with SCI.

Applications to other areas of neuroscience

Wellness interventions are becoming a pillar of rehabilitation given them potential as second-line approaches for improving symptoms, overall health, and QOL. Wellness interventions for persons with SCI may draw from, or inform, other neurological diseases such as multiple sclerosis (MS), traumatic brain injury (TBI), stroke, and Parkinson's disease. Barriers to engaging in health-promoting behaviors among persons with various neurological diseases overlap such as accessibility, lack of knowledge, transportation, and debilitating symptoms such as pain. Further, as depicted in Stuifbergen's model of QOL, the factors related to QOL overlap in meaningful ways that guided successful wellness interventions for persons with various chronic disabling conditions.

Many wellness interventions have been created for persons with MS that may be adapted for persons with SCI. Symptoms impacting persons with SCI and persons with MS overlap, especially among persons with MS who have lesions in their spinal cord. The National Multiple Sclerosis Society recently convened a Wellness Research Working Group that provides focal reviews of literature regarding physical activity, diet, and emotional well-being (Motl et al., 2017). The Wellness Research Working Group further provided a clear research agenda to guide the development of wellness interventions for persons with MS. The priority questions posited by the Wellness Research Working Group such as "To what extent does resilience affect emotional health and/or the course of the disease" are pressing topics in SCI rehabilitation (Motl et al., 2017).

We underscore the importance to draw on other populations with neurological dysfunction, however note the unique trauma associated with SCI. Persons with SCI often start their rehabilitation journey inpatient with vastly different resources. The rehabilitation journey of each person with SCI may influence the appropriate components of wellness promotion. For example, a person with SCI who experienced a traumatic car accident may require more emotional health resources whereas someone who experiences a fall may require resources focused on mobility and function. In this regard, drawing upon wellness interventions for persons with TBI or stroke may be more appropriate. The period immediately following injury/trauma serves as a sensitive period in which individuals are learning to live a whole new way that warrants focal examination.

Mini-dictionary of terms

Wellness: A multidimensional construct that is characterized by the existence of positive health in an individual with a focus on functioning, health, quality of life.

Well-being: Sense of overall positive health focused on global perceptions of an individual's current status. Well-being has various sub-categories: physical, material, social, productive, emotional, and civic.

Quality of life: Individual evaluation regarding their overall satisfaction with life and health status.

Behavioral intervention: Coordinated sets of activities designed to change specified behavior patterns by providing instruction and application of skills, strategies, and resources for behavior change.

Behavior change techniques: Observable and replicable components are used as strategies in behavioral interventions to initiate or maintain health behavior change. Behavior change techniques are often referred to as the "active ingredients" in behavioral interventions and are frequently used in combination.

Physical activity: Any bodily movement produced by skeletal muscles that require energy expenditure including movements undertaken while working, playing, carrying out household chores, traveling, and engaging in recreational pursuits.

Exercise: A subset of physical activity that is planned, structured, repetitive, and targets improvements or maintenance of overall fitness.

Qualitative research: Research designs that focus on collection and analysis of non-numerical data to understand concepts, develop theories, or inquire about opinions and experiences.

Quantitative research: Research designs that focus on the collection and analysis of numerical data can identify patterns, test causal inferences, and assess generalizability to populations.

Self-efficacy: Individual confidence in ability to follow through on a task or behavior. Self-efficacy is not static and may vary greatly depending on the specific task or behavior.

Key facts of wellness interventions

- Wellness interventions are a second-line approach of interest among persons with spinal cord injury (SCI) and their health care providers for improving outcomes following SCI.
- Wellness interventions focus on behavior change to maximize health within the context of a disease or condition.
- Wellness interventions are not focused on medical management of a specific disease.
- Wellness interventions can be initiated at any time to improve overall health, but may be tailored for different subgroups depending on age, sex, or other demographic and clinical characteristics.
- Wellness interventions may incorporate various health behaviors including behaviors that promote health such as physical activity and nutrition as well as risk behaviors that should be limited such as tobacco use.

Summary points

- Wellness is a major priority for spinal cord injury (SCI) rehabilitation given significant changes in overall health and quality of life.
- Wellness interventions target health behaviors such as physical activity, nutrition, stress management, and tobacco use to improve quality of life over time.
- Rigorous wellness interventions are founded on behavioral theory such as Social Cognitive Theory, the Health Action Process Model, Transtheoretical Model, or Theory of Planned Behavior.
- Previous wellness interventions for persons with SCI have widely been in-person at community centers or hospitals that do not address significant barriers among persons with SCI.
- Future wellness interventions for persons with SCI require qualitative and quantitative approaches for identifying the needs and priorities among persons with SCI and options for optimal delivery within the rehabilitation team.

References

Ajzen, I. (1991). The theory of planned behavior. *Organizational Behavior and Human Decision Processes*, *50*(2), 179–211.

Bandura, A. (1998). Health promotion from the perspective of social cognitive theory. *Psychology and Health*, *13*(4), 623–649.

Bandura, A., Freeman, W., & Lightsey, R. (1999). *Self-efficacy: The exercise of control*. Springer.

Chun, S., & Lee, Y. (2020). "Racing made me feel strong and more positive": Experiencing personal meaning in life through leisure following traumatic spinal cord injury. *Leisure Sciences*, 1–18.

Corbin, C. B., & Pangrazi, R. P. (2001). *Toward a uniform definition of wellness: A commentary*. President's council on physical fitness and sports research digest.

Decker, S. D., & Schulz, R. (1985). Correlates of life satisfaction and depression in middle-aged and elderly spinal cord-injured persons. *American Journal of Occupational Therapy*, *39*(11), 740–745.

Dicianno, B. E., Lovelace, J., Peele, P., Fassinger, C., Houck, P., Bursic, A., et al. (2016). Effectiveness of a wellness program for individuals with spina bifida and spinal cord injury within an integrated delivery system. *Archives of Physical Medicine and Rehabilitation*, *97*(11), 1969–1978.

Diener, E., Emmons, R. A., Larsen, R. J., & Griffin, S. (1985). The satisfaction with life scale. *Journal of Personality Assessment*, *49*(1), 71–75.

Dijkers, M. P. (1999). Correlates of life satisfaction among persons with spinal cord injury. *Archives of Physical Medicine and Rehabilitation*, *80*(8), 867–876.

Ekelman, B. A., Allison, D. L., Duvnjak, D., DiMarino, D. R., Jodzio, J., & Iannarelli, P. V. (2017). A wellness program for men with spinal cord injury: Participation and meaning. *OTJR: Occupation, Participation and Health*, *37*(1), 30–39.

Farkas, G. J., Pitot, M. A., Berg, A. S., & Gater, D. R. (2019). Nutritional status in chronic spinal cord injury: A systematic review and meta-analysis. *Spinal Cord*, *57*(1), 3–17.

Fugl-Meyer, A. R., Bränholm, I.-B., & Fugl-Meyer, K. S. (1991). Happiness and domain-specific life satisfaction in adult northern swedes. *Clinical Rehabilitation*, *5*(1), 25–33.

Gassaway, J., Jones, M. L., Sweatman, W. M., Hong, M., Anziano, P., & DeVault, K. (2017). Effects of peer mentoring on self-efficacy and hospital readmission after inpatient rehabilitation of individuals with spinal cord injury: A randomized controlled trial. *Archives of Physical Medicine and Rehabilitation*, *98*(8), 1526–1534. e1522.

Gater, D. R., Jr. (2007). Obesity after spinal cord injury. *Physical Medicine and Rehabilitation Clinics of North America*, *18*(2), 333–351.

Ginis, K. M., Jetha, A., Mack, D., & Hetz, S. (2010). Physical activity and subjective well-being among people with spinal cord injury: A meta-analysis. *Spinal Cord*, *48*(1), 65–72.

Goodwin, F. K., & Jamison, K. R. (2007). *Manic-depressive illness: Bipolar disorders and recurrent depression. Vol. 2*. Oxford University Press.

Holman, D., Johnson, S., & O'Connor, E. (2018). Stress management interventions: Improving subjective psychological well-being in the workplace. In *Handbook of well-being*. Salt Lake City, UT: DEF Publishers.

Houlihan, B. V., Jette, A., Friedman, R. H., Paasche-Orlow, M., Ni, P., Wierbicky, J., et al. (2013). A pilot study of a telehealth intervention for persons with spinal cord dysfunction. *Spinal Cord*, *51*(9), 715–720.

Jörgensen, S., Iwarsson, S., & Lexell, J. (2017). Secondary health conditions, activity limitations, and life satisfaction in older adults with long-term spinal cord injury. *PM & R : The Journal of Injury, Function, and Rehabilitation*, *9*(4), 356–366.

Kennedy, P., Lude, P., Elfström, M. L., & Smithson, E. F. (2011). Psychological contributions to functional independence: A longitudinal investigation of spinal cord injury rehabilitation. *Archives of Physical Medicine and Rehabilitation*, *92*(4), 597–602.

Keyes, C. L., Shmotkin, D., & Ryff, C. D. (2002). Optimizing well-being: The empirical encounter of two traditions. *Journal of Personality and Social Psychology*, *82*(6), 1007.

King, C., & Kennedy, P. (1999). Coping effectiveness training for people with spinal cord injury: Preliminary results of a controlled trial. *British Journal of Clinical Psychology*, *38*(1), 5–14.

Krause, J. S., Kemp, B., & Coker, J. (2000). Depression after spinal cord injury: Relation to gender, ethnicity, aging, and socioeconomic indicators. *Archives of Physical Medicine and Rehabilitation*, *81*(8), 1099–1109.

Kryger, M. A., Crytzer, T. M., Fairman, A., Quinby, E. J., Karavolis, M., Pramana, G., et al. (2019). The effect of the interactive mobile health and rehabilitation system on health and psychosocial outcomes in spinal cord injury: Randomized controlled trial. *Journal of Medical Internet Research*, *21*(8), e14305.

Latimer, A. E., & Martin Ginis, K. A. (2005). The theory of planned behavior in prediction of leisure time physical activity among individuals with spinal cord injury. *Rehabilitation Psychology*, *50*(4), 389.

Lebensohn, P., Dodds, S., Benn, R., Brooks, A. J., & Birch, M. (2013). Resident wellness behaviors. *Family Medicine*, *45*(8), 541–549.

Ljungberg, I., Kroll, T., Libin, A., & Gordon, S. (2011). Using peer mentoring for people with spinal cord injury to enhance self-efficacy beliefs and prevent medical complications. *Journal of Clinical Nursing*, *20*(3–4), 351–358.

Martin Ginis, K. A., Latimer, A. E., Arbour-Nicitopoulos, K. P., Bassett, R. L., Wolfe, D. L., & Hanna, S. E. (2011). Determinants of physical activity among people with spinal cord injury: A test of social cognitive theory. *Annals of Behavioral Medicine*, *42*(1), 127–133.

May, L. A., & Warren, S. (2002). Measuring quality of life of persons with spinal cord injury: External and structural validity. *Spinal Cord*, *40*(7), 341–350.

McIntyre, A., Marrocco, S. L., McRae, S. A., Sleeth, L., Hitzig, S., Jaglal, S., et al. (2020). A scoping review of self-management interventions following spinal cord injury. *Topics in Spinal Cord Injury Rehabilitation*, *26*(1), 36–63.

Motl, R. W., Mowry, E. M., Ehde, D. M., LaRocca, N. G., Smith, K. E., Costello, K., et al. (2017). Wellness and multiple sclerosis: The National MS Society establishes a wellness research working group and research priorities. *Multiple Sclerosis Journal*, *24*(3), 262–267.

Munce, S. E. P., Guetterman, T. C., & Jaglal, S. B. (2020). Using the exploratory sequential design for complex intervention development: Example of the development of a self-management program for spinal cord injury. *Journal of Mixed Methods Research*, 1558689820901936.

National Spinal Cord Injury Statistical Center (NCSISC). (2020). *Spinal cord injury facts and figures at a glance*. https://www.nscisc.uab.edu/Public/Facts%20and%20Figures%202020.pdf.

Post, M. W., de Witte, L. P., van Asbeck, F. W., van Dijk, A. J., & Schrijvers, A. J. (1998). Predictors of health status and life satisfaction in spinal cord injury. *Archives of Physical Medicine and Rehabilitation*, *79*(4), 395–401.

Prochaska, J. O., & Velicer, W. F. (1997). The transtheoretical model of health behavior change. *American Journal of Health Promotion*, *12*(1), 38–48.

Putzke, J. D., Elliott, T. R., & Richards, J. S. (2001). Marital status and adjustment 1 year post-spinal-cord-injury. *Journal of Clinical Psychology in Medical Settings*, *8*(2), 101–107.

Putzke, J. D., Richards, J. S., Hicken, B. L., & DeVivo, M. J. (2002). Predictors of life satisfaction: A spinal cord injury cohort study. *Archives of Physical Medicine and Rehabilitation*, *83*(4), 555–561.

Richardson, E. J., Richards, J. S., Stewart, C. C., & Ness, T. J. (2012). Effects of nicotine on spinal cord injury pain: A randomized, double-blind, placebo controlled crossover trial. *Topics in Spinal Cord Injury Rehabilitation*, *18*(2), 101–105.

Ryan, R. M., & Deci, E. L. (2001). On happiness and human potentials: A review of research on hedonic and eudaimonic well-being. *Annual Review of Psychology*, *52*(1), 141–166.

Ryff, C. D., & Keyes, C. L. M. (1995). The structure of psychological well-being revisited. *Journal of Personality and Social Psychology*, *69*(4), 719.

Saunders, L. L., Krause, J. S., Saladin, M., & Carpenter, M. J. (2015). Prevalence of cigarette smoking and attempts to quit in a population-based cohort with spinal cord injury. *Spinal Cord*, *53*(8), 641–645.

Schwarzer, R., Lippke, S., & Luszczynska, A. (2011). Mechanisms of health behavior change in persons with chronic illness or disability: The Health action process approach (HAPA). *Rehabilitation Psychology*, *56*(3), 161.

Sheppard, R., Kennedy, P., & Mackey, C. A. (2006). Theory of planned behaviour, skin care and pressure sores following spinal cord injury. *Journal of Clinical Psychology in Medical Settings*, *13*(4), 358–366.

Silveira, S. L., Ledoux, T. A., Johnston, C. A., Kalpakjian, C., O'Connor, D. P., Cottingham, M., et al. (2020). Differences among participants' response to a wellness intervention for individuals with spinal cord injuries. *Disability and Health Journal*, *13*(1), 100842.

Sparkes, A. C., & Smith, B. (2002). Sport, spinal cord injury, embodied masculinities, and the dilemmas of narrative identity. *Men and Masculinities*, *4*(3), 258–285.

Stuifbergen, A. K., Morris, M., Jung, J. H., Pierini, D., & Morgan, S. (2010). Benefits of wellness interventions for persons with chronic and disabling conditions: A review of the evidence. *Disability and Health Journal*, *3*(3), 133–145.

Stuifbergen, A. K., Seraphine, A., & Roberts, G. (2000). An explanatory model of health promotion and quality of life in chronic disabling conditions. *Nursing Research*, *49*(3), 122–129.

Tolle, H., Rapacz, A., Weintraub, B., Shogren, C., Harkema, S. J., & Gibson, J. L. (2018). Establishing the neurorecovery network community fitness and wellness facilities: Multi-site fitness facilities provide activity-based interventions and assessments for evidence-based functional gains in neurologic disorders. *Disability and Rehabilitation*, *40*(25), 3086–3093.

Tomasone, J. R., Flood, S. M., Ma, J. K., Scime, N. V., Burke, S. M., Sleeth, L., et al. (2018). Physical activity self-management interventions for adults with spinal cord injury: Part 1–A systematic review of the use and effectiveness of behavior change techniques. *Psychology of Sport and Exercise*, *37*, 274–285.

Tsunoda, N., Inayama, T., Hata, K., & Oka, J. (2015). Vegetable dishes, dairy products and fruits are key items mediating adequate dietary intake for Japanese adults with spinal cord injury. *Spinal Cord*, *53*(11), 786–790.

Warms, C. A., Belza, B. L., Whitney, J. D., Mitchell, P. H., & Stiens, S. A. (2004). Lifestyle physical activity for individuals with spinal cord injury: A pilot study. *American Journal of Health Promotion*, *18*(4), 288–291.

World Health Organizaion (WHO). (2018, February 23). *Physical activity*. https://www.who.int/news-room/fact-sheets/detail/physical-activity.

World Health Organization (WHO). (2020). *Nutrition interventions*. https://www.who.int/elena/intervention/en/.

Yarar-Fisher, C., Kulkarni, A., Li, J., Farley, P., Renfro, C., Aslam, H., et al. (2018). Evaluation of a ketogenic diet for improvement of neurological recovery in individuals with acute spinal cord injury: A pilot, randomized safety and feasibility trial. *Spinal Cord Series And Cases*, *4*(1), 88.

Zemper, E. D., Tate, D. G., Roller, S., Forchheimer, M., Chiodo, A., Nelson, V. S., et al. (2003). Assessment of a holistic wellness program for persons with spinal cord injury. *American Journal of Physical Medicine & Rehabilitation*, *82*(12), 957–968.

Chapter 40

Sexual life in individuals with spinal cord injury and management

Seçil Taylan[a] and İlknur Özkan[b]

[a]Surgical Nursing Department, Kumluca Faculty of Health Sciences, Akdeniz University, Kumluca-Antalya, Turkey, [b]İnternal Medicine Nursing Department, Kumluca Faculty of Health Sciences, Akdeniz University, Kumluca-Antalya, Turkey

List of abbreviations

AD	autonomic dysreflexia
ED	erectile dysfunction
EEJ	electro-ejaculation
ICI	intracavernous injections
IIEF	International Index of Erectile Function
IPP	inflatable penile implants
L	lumbar
PDE5-Is	phosphodiesterase 5-inhibitors
PM	prostate massage
PPI	penile prosthesis implantation
PVS	penile vibrator stimulation
S	sacral
SCI	spinal cord injury
T	thoracal

Introduction

Significant physiological changes related to sensory, voluntary motor, and autonomic functions developing due to spinal cord injury (SCI) affect sexual interest and satisfaction negatively by causing problems in sexual arousal, orgasm potential, ejaculation in men, and sexual position for both genders (Stoffel, Van der Aa, Wittmann, Yande, & Elliott, 2018). SCIs usually occur in separate regions of the neuraxis, and their effects on sexual function can be predicted depending on the spinal motor and autonomic pathways that are affected. Three spinal segments are particularly important for sexual function: T11-L2, sympathetic; S2-S4, parasympathetic; and somatic centers. Depending on the level of injury and severity of these pathways, some changes in genital sexual arousal (erection problems in men and decreased lubrication in women) occur (Stoffel et al., 2018). Besides, bladder and bowel incontinence and related problems affect sexual self-esteem, sexual expression, and sexual satisfaction as secondary health consequences of SCI including autonomic dysreflexia, nociceptive and neuropathic pain, depression, and pharmacological management of such conditions (Adriaansen et al., 2016; Courtois, Alexander, & McLain, 2017).

Several studies reported that sexual dysfunctions experienced in SCI are associated with psychiatric outcomes, such as poor quality of life, anxiety, depression, and changes in personality structure (Choi, Kang, & Shin, 2015; Cramp, Courtois, & Ditor, 2015; Le & Dorstyn, 2016). Therefore, sexual satisfaction is an important aspect of the life of people with SCI, and it is important to diagnose and manage sexual dysfunctions with a multidisciplinary approach to improve the quality of life of these individuals (Elliott, Hocaloski, & Carlson, 2017). In this section, the problems experienced in sexual life after SCI and their management are examined separately for men and women as a guide for health professionals (Fig. 1).

Cellular, Molecular, Physiological, and Behavioral Aspects of Spinal Cord Injury. https://doi.org/10.1016/B978-0-12-822427-4.00040-X

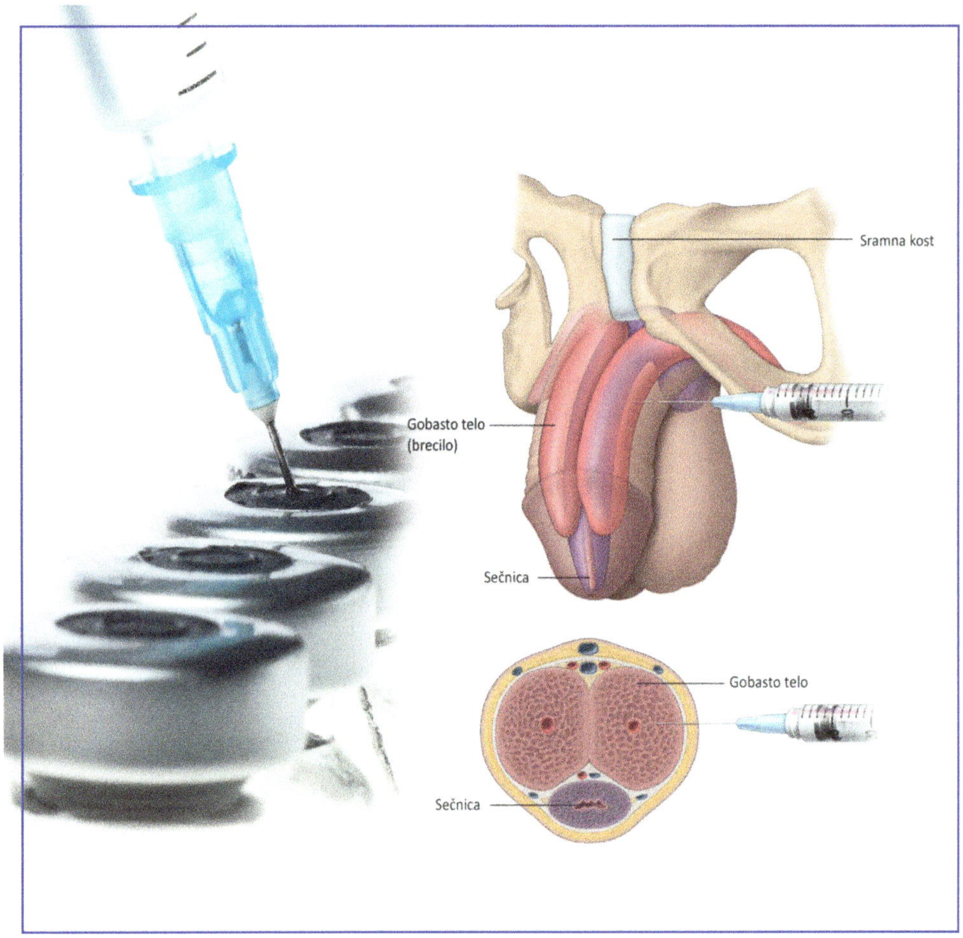

FIG. 1 Intracavernous injections.

Sexual life in men with SCI

It is estimated that 35%–80% of men become inactive after SCI. Following (Choi et al., 2015; Dahlberg, Alaranta, Kautiainen, & Kotila, 2007; Gomes et al., 2017) SCI, most men experience severe impairment in their sexual health and reproductive function due to erectile and/or ejaculatory dysfunction and semen abnormalities (Aikman, Oliffe, Kelly, & McCuaig, 2018). Male sexual dysfunction includes erectile dysfunction, decreased libido, and abnormal ejaculation. The normal male sexual function requires interactions between vascular, neurological, hormonal, and psychological systems (Courtois & Charvier, 2015). The first requirement for male sexual activity is the achievement and maintenance of penile erection (Afferi et al., 2020). In this section, first, the physiology of normal sexual function is explained so that sexual dysfunction in men with SCI can be understood (Fig. 2).

Physiology of normal sexual function

Parasympathetic innervation

The preganglionic neurons of parasympathetic efferents come out of the intermediolateral region at the level of sacral 2–4 (S2–4) segments of the spinal cord known as the sacral parasympathetic nucleus, reach the pelvic plexus, and feed the lower urinary system and genital system. Parasympathetic efferent stimulation loosens up corpus cavernosum and vascular (cavernosal artery) smooth muscles in the penis, which results in blood filling (erection) into the penis (Elliott, 2018; Sabharwal, 2013).

FIG. 2 Vacuum device.

Sympathetic innervation

Sympathetic efferents emerge from thoracal 10 and lumbar 2 (T10-L2) spinal segments and feed the bladder floor, internal sphincter, proximal urethra, prostate, penis, and smooth muscles of seminal tract in men. Sympathetic efferent stimulation contracts bladder outlet, prostate, and seminal tract smooth muscles, cavernosal tissue and vascular smooth muscles in the penis, epididymis, vas deferens, and seminal vesicle smooth muscles through alpha 1 adrenergic receptors (Alwaal, Breyer, & Lue, 2015; Elliott, 2018). Thus, while sympathetic innervation suppresses erection, it plays an important role in the formation of the emission phase of urine storage and ejaculation (Krassioukov & Elliott, 2017; Sabharwal, 2013).

Neurophysiology of erection

Erections can be psychogenic or reflexogenic. Normally, these two mechanisms act in synergy. Sensory stimulation of the genitals leads to reflexogenic erection in men. This reflex occurs at the S2–S4 level of the spinal cord. Afferents are the pudendal nerve that carries sensory transmitters from the genitals, and the efferent pathway passes through the pelvic nerve. Psychogenic erection, on the other hand, involves more complex pathways (Alwaal et al., 2015; Elliott, 2018). In psychogenic erection, various audio, visual, tactile, and/or creative afferent stimuli are processed centrally. After passing through different paths, these fibers move through the hypogastric nerve, reaching the pelvic plexus where the action is integrated with the parasympathetic function (Krassioukov, 2009) (Fig. 3).

Ejaculation

Ejaculation is a neurologically more complex phenomenon and relies on the coordination of the sympathetic (T11-L2) and parasympathetic nervous systems (S2–4) in addition to the somatic nervous system using the pudendal nerve (S2–5). Semen ejection occurs by the rhythmic contraction of the urethral smooth muscle (via sympathetic innervation) and the ischiocavernosus and bulbocavernosus muscles (somatic innervation). The autonomic component of ejaculation is emission, and its somatic component is expulsion. The emission phase of ejaculation describes the accumulation of epididymis, seminal vesicles, and prostate secretions in the posterior urethra, while the expulsion phase describes the excretion

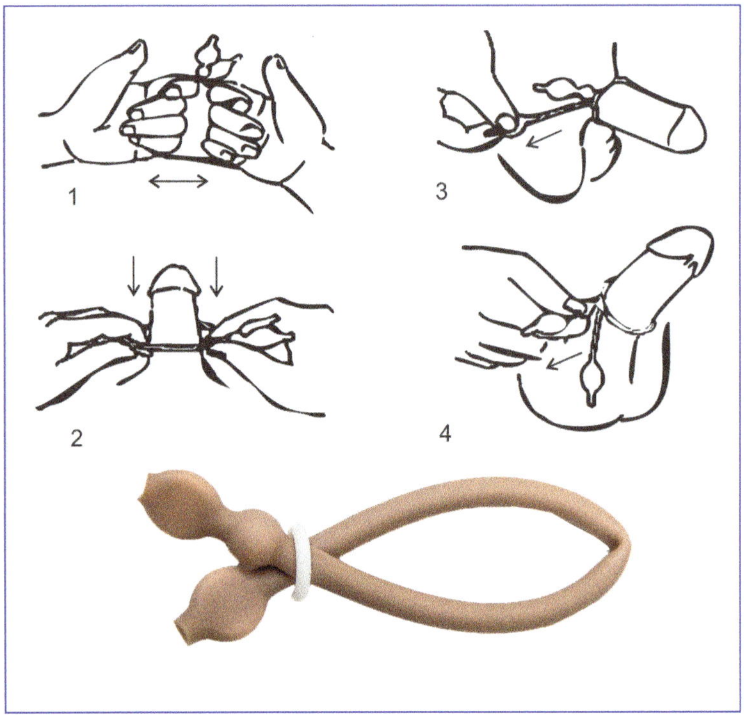

FIG. 3 Penile rings.

of the accumulated seminal secretion from the urethra (Alwaal et al., 2015). Both parasympathetic and sympathetic tonus that provide ejaculation are achieved by integrating genital sensory and/or cerebral erotic stimuli at the spinal cord level (Sabharwal, 2013).

Problems affecting sexual life in men with SCI

Erectile dysfunction in men with SCI

Erectile dysfunction is defined as constant or recursive inadequacy that leads to a failure in achieving or maintaining an erection with full rigidity and duration for sexual intercourse (Burnett et al., 2018). Depending on the innervation of the male genitals, the erectile potential following SCI may be affected by the level and completeness of the lesion. As a general rule, while higher lesions retain better reflex activity and greater sexual potential (especially through multiseptic reflexes), lower lesions impair reflex activity and only retain psychogenic potential. Higher lesions in the cervical or thoracic spinal segments preserve all reflexes beneath the lesion, including multisegmental reflexes. Reflex erections are therefore possible. In contrast, psychogenic erection usually disappears (Aikman et al., 2018; Courtois & Charvier, 2015; Everaert et al., 2010).

While reflex erections are seen in spinal cord lesions between the thoracolumbar region and sacral centers, psychogenic erections do not usually occur (Courtois & Charvier, 2015; Elliott, 2018; Ibrahim, Lynne, & Brackett, 2016). While psychogenic erections persist in spinal cord lesions of the sacral center and below this center, reflex erections disappear (Courtois & Charvier, 2015; Elliott, 2018; Ibrahim et al., 2016).

Ejaculation disorders in men with SCI

The potential for discharge following SCI is similarly affected by the level and completeness of the lesion (Aikman et al., 2018). Accordingly, while autonomic ejaculation is possible in the thoracolumbar region or in a lesion above, nocturnal ejaculation does not occur. Autonomic, somatic, nocturnal, and vibro ejaculations disappear in spinal cord lesions under the thoracolumbar region, but electro-ejaculations can be seen because postsynaptic fibrils remain intact (Alexander, Aisen, Alexander, & Aisen, 2017; Courtois & Charvier, 2015).

Orgasm in men with SCI

Orgasm is the sexual cycle phase accompanied by spinal cord reflexes showing up with subjective pleasure, delight, and relaxation that occurs with the activation of the amygdala and hippocampus in the brain at the end of ejaculation (Alwaal et al., 2015). Studies conducted on men with SCI show that they can experience orgasm even with complete lesions in the spinal cord (Alexander et al., 2017; Courtois & Charvier, 2015; Krassioukov & Elliott, 2017). The responses in individuals with SCI (especially those with lesions above T6) are similar to symptoms of autonomic hyperreflexia during orgasm in healthy individuals (Alexander et al., 2017; Courtois & Charvier, 2015; Krassioukov & Elliott, 2017).

Semen abnormalities and infertility

Although sperm concentration is less affected, impaired sperm motility is common after SCI. The causes are probably multifactorial and have not been fully understood. Immotility of semen associated with seminal vesicle dysfunction has been documented. High levels of cytokines and leukocytes have been reported in semen, suggesting an inflammatory component (Aikman et al., 2018; Ibrahim et al., 2016; Sabharwal, 2013; Stoffel et al., 2018). Only a small percentage of men with SCI can become a father without medical intervention (Sabharwal, 2013).

Management of sexual problems in men with SCI

The management of sexual problems in individuals with SCI should be handled with a holistic approach, and this is possible with a multidisciplinary or interdisciplinary teamwork. Although an interdisciplinary approach in patient-centered care is recommended for a more holistic approach, some patients may prefer limited, individual multidisciplinary appointments to team evaluations due to the sensitivity of the issue of sexuality. The members of a multidisciplinary team usually consist of doctors, nurses, occupational therapists, physiotherapists, psychologists, sexual therapists, social service experts, and peer support workers (Elliott et al., 2017).

Evaluation of neurological factors and medical conditions that may affect sexual function is an important aspect of the management of sexual problems. A detailed neurological, musculoskeletal, and functional evaluation should be made. Examination that is carried out using the International Standards for the Neurological Classification of SCI should include special attention to sensory protection coming from T10-L2 and S2–5, as well as determining the presence of voluntary anal contraction and reflexes to assess sexual function (Sabharwal, 2013).

The most widely employed method for evaluating sexual function is the International Index of Erectile Function (IIEF) in men. IIEF is a self-administered scale consisting of 15 questions and five sub-dimensions, including erectile dysfunction, orgasm function, sexual desire, sexual intercourse satisfaction, and general satisfaction. IIEF is recommended for use in male individuals with SCI to evaluate sexual dysfunction. Another assessment scale is the four-point Erection Hardness Grading Scale, developed for use in various drug studies (Alexander et al., 2009; Steadman & Hubscher, 2016). There is no scale to evaluate ejaculation capacity in men with SCI. Clinically, reflex assessments allow men to predict ejaculatory capacity. The bulbocavernosis reflex can be visualized by stimulating the glans penis and recording the responses of the bulbospongiosus muscle or anal sphincter via electromyography. The presence of the bulbocavernosis reflex indicates that the S2-S4 level is intact. The hip flexor response is a pathogenic reflex initiated by stimulation of the sole that leads to hip flexion typically seen in individuals with SCI. This reflex suggests lesions at the S2-S4 level, if present. The presence of both bulbocavernosis reflex and hip flexor response indicates that the individual can ejaculate with the help of PVS (Penile Vibratory Stimulation). These spinal reflex assessments are therefore clinically useful as indicators of sexual prognosis and as a tool to assist in the selection of future treatment plans (Courtois & Charvier, 2015; Steadman & Hubscher, 2016).

Treatment of erectile dysfunction

Treatment options for Erectile Dysfunction (ED) range from Phosphodiesterase 5-inhibitors to Intracavernous Injections (ICI), vacuum devices, penile rings, and penile prostheses. The choice of treatment depends on individual factors and patient compliance; yet, usually, the least invasive is the first option (Afferi et al., 2020).

Phosphodiesterase 5-inhibitors (PDE5-Is)

PDE5-Is (Sildenafil, Tadalafil, and Vardenafil) are recommended as first-level therapy in men with neurogenic ED (Afferi et al., 2020). The effectiveness of this drug in managing ED depends on the extent and location of the neurological lesion.

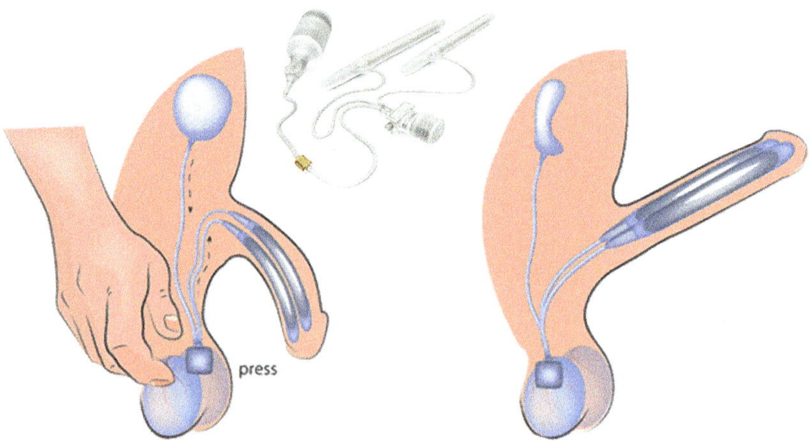

FIG. 4 Penile prostheses.

PDE5I is highly effective in the population with SCI, except for people with cauda equina and conus medullaris injuries (Courtois & Charvier, 2015; Sabharwal, 2013; Stoffel et al., 2018) (Fig. 4).

Intracavernous injections (ICI)

Intracavernosal vasoactive agent injections can be used in patients who do not respond to oral treatments, use nitrates, or cannot achieve reflexogenic erections. Particularly, patients with SCI who have lower lesion levels are more likely to respond to ICI compared to oral pharmacotherapy (Groen et al., 2016). The first product tested in ICI was phentolamine, followed by papaverine and prostaglandins. All of these trigger an erection without the need for sexual stimulation. Self-injections begin with minimal dosages (usually fail), then continue in small increments (0.1 mL increments) until a satisfactory dose is reached (Afferi et al., 2020; Pawar et al., 2019). Patients who start injection therapy should be warned especially about the risk of hematoma and priapism. (Afferi et al., 2020; Courtois & Charvier, 2015; Sabharwal, 2013).

Intraurethral medication

Alprostadil can be used intraureterally for the treatment of erectile dysfunction. Intraurethral administration of alprostadil is less invasive than intracavernosal injection. (Afferi et al., 2020; Courtois & Charvier, 2015).

Vacuum device and penile rings

With the help of a pump, a vacuum is created that promotes an increase in cavernous blood flow, thus causing penile erection. Placing a constriction ring at the base of the penis prevents venous outflow and thus maintains erection. The constriction ring should never be left in place for more than 30 min. The device requires hand skill or partner assistance to some degree. The use of anticoagulant medication is a contraindication (Afferi et al., 2020; Ibrahim et al., 2016; Sabharwal, 2013).

Penile prostheses

The last level of treatment in ED treatment is Penile Prosthesis Implantation (PPI). Penile prostheses are divided into two types: inflatable and malleable. Malleable prostheses consist of semi-rigid rods that can bend upwards during sexual intercourse. The disadvantages of malleable prostheses are the constant rigidness, which can irritate the user and cause him to become asocial. Inflatable Penile Implants (IPP) consist of two cylinders that are placed in the scrotum and provide fluid transfer from the balloon to the cylinder chambers when erection is desired, a reservoir balloon, and a pump. This implant has the advantage of providing a rigid erection on demand and better flaccidity when deflated (Afferi et al., 2020; Courtois & Charvier, 2015; Ibrahim et al., 2016).

Ejaculation dysfunction treatment

The main methods used for sperm retrieval in patients with SCI include penile vibrator stimulation (PVS), electro-ejaculation (EEJ), prostate massage (PM), and surgical sperm retrieval technique (Fenstermaker, Dupree, Hadj-Moussa, & Ohl, 2018).

Penile vibrator stimulation

Penile Vibrator Stimulation (PVS) is recommended as the first choice for anejaculation in patients with SCI due to its reliability, provision of the highest rate of sperm retrieval as antegrade ejaculation, and cost-effectiveness. PVS can be applied in supine or sitting position. The vibrator is placed in the dorsum, frenulum, or both sides of the penis. If there is no ejaculation within 2 min, the procedure is stopped and continued after a 1–2 min interval. In this way, a total of 10 min of stimulation can be maintained. Meanwhile, the penile skin is checked for edema and abrasions (Courtois & Charvier, 2015; Fenstermaker et al., 2018; Ibrahim et al., 2016).

Electroejaculation

Electroejaculation can be preferred in patients in whom a penile vibrator has not been successful. In this case, semen is retrieved by giving an electrical stimulation through an electrical probe placed in the rectal mucosa in the prostate and seminal vesicle localization from the rectal route. The procedure may be uncomfortable for patients; in this case, spinal or general anesthesia may be needed (Alwaal et al., 2015; Ibrahim et al., 2016).

Prostate massage (PM)

Prostate massage is a simple method of semen extraction applied by a doctor who presses the prostate and seminal vesicles with a finger through the rectum in men with spinal cord injuries. In cases where PVS fails and EEJ is unavailable or requires anesthesia, this method can be recommended as another option before surgical sperm retrieval (Ibrahim et al., 2016).

Surgical sperm retrieval

If PVS, EEJ, and PM are unsuccessful or contraindicated, surgical sperm retrieval methods can be used as the last option for sperm retrieval. Surgical sperm retrieval is more costly and more invasive than PVS and EEJ methods (Ibrahim et al., 2016).

Psychological support

Beyond the well-known physiological changes that cause sexual dysfunction in men with traumatic spinal cord injuries, other factors have been reported to have a potential effect on sexual responses and function. These factors include individual characteristics, such as post-injury self-esteem and body image problems, relationship status, previous sexual attitudes and experiences, and openness to the sexual experience (Cehic et al., 2016; Eisenberg, Andreski, & Mona, 2015). For example, studies have shown that men who adhere to masculine norms and are overly self-confident increase their negative perceptions of their physical limitations after SCI, inhibit their desire to engage in emotional and social support, and end up with higher rates of depression (Burns, Hough, Boyd, & Hill, 2010). To address these concerns, interventions, including sexual training, counseling (both individual and couple counseling), cognitive behavioral therapy, sex therapy, and peer support, are often helpful (Aikman et al., 2018; Eisenberg et al., 2015).

Sex life in women with SCI

Women with SCI need information and counseling about sexual life as well as all their life experiences. Women experience some problems in their sexual life after SCI. These problems include physiological changes, such as difficulty in genital arousal, reduced sensation or loss of sensation, vaginal dryness, difficulties in reaching orgasm, bladder and bowel incontinence, movement and self-positioning difficulty, as well as multidimensional psychological changes, such as unattractiveness or feeling less attractive, lower self-esteem, and maintaining relationships with the partner or finding a partner (Ferreiro-Velasco et al., 2005). Women also need more support and follow-up than other women's processes regarding their fertility status, contraception methods, and gestational periods (Bertschy et al., 2020; Bertschy, Bostan, Meyer, & Pannek, 2016) (Fig. 5).

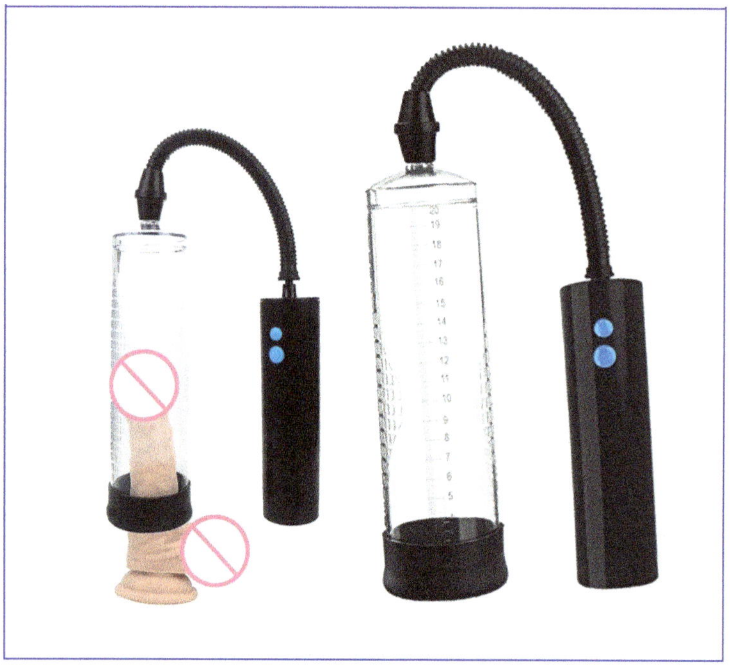

FIG. 5 Penile vibrator stimulator.

Problems affecting sexual life in women with SCI and their management

Compared to the options available for the treatment of sexual dysfunction in men, the treatment options available to women are rather limited. To improve sexual life and sexual motivation in women, it is very important to first manage the conditions that affect sexual life. Managing biological or medical factors (e.g., low testosterone or estrogen replacement), treating depression (for low libido and poor genital arousal), handling bladder or bowel incontinence, and managing fatigue or pain positively affect sex life and sexual motivation to a great extent (Stoffel et al., 2018). Problems that occur in the lives of women with SCI and their management are discussed in the following sections.

Difficulty in genital arousal

Women with SCI experience physical and psychological genital arousal difficulties (Otero-Villaverde et al., 2015). Tingling sensations (35.3%) and spasms (35%) are the most frequently reported physical sensory problems in women with SCI (Stoffel et al., 2018). In coping with this problem, it may be recommended to try positions and activities different from pre-injury sexual intercourse positions. The multidisciplinary team can support women with SCI and their partners to talk about their needs and aspirations for these new activities and positions. PDE5 inhibitors are used to treat erectile dysfunction in men. These oral medications have been tested in women to increase sexual responses such as arousal. The use of these drugs in women has been reported to increase blood flow in the perineum and vaginal stimulation, which affects sexual satisfaction positively (M. S. Alexander, Biering-Sorensen, Elliott, Kreuter, & Sonksen, 2011; Otero-Villaverde et al., 2015).

Orgasm

One of the important sexual life problems experienced by women after SCI is the difficulty in reaching orgasm. Orgasm is associated with mental excitement. It is reported that in approximately 40–50% of women with spinal cord injuries, the stimulation required to reach orgasm is prolonged compared to pre-injury experiences. Women with SCI can achieve self-defined or laboratory-recorded orgasmic release despite this prolonged orgasm process (Alexander & Rosen, 2008). In women with SCI, conditions such as complete or partial injury and having genital sensation affect orgasm (Stoffel et al., 2018). The use of vibrators may be recommended by a multidisciplinary team for injuries below the T6 level. Counseling can be provided as to whether the drugs used by women can be adjusted to minimize the effect on sexual responses (Courtois et al., 2017).

Another condition that affects orgasm is Autonomic Dysreflexia (AD). Sympathetic activity, which is activated by a stimulus such as genital stimulation, cannot be controlled by the upper centers, causing a continuous and strong sympathetic release (Krassioukov, 2009). Although hypertension is life-threatening, it may not always progress with the same clinical picture and the same severity. Systolic blood pressure in these patients can reach up to 300 mmHg and is important for the diagnosis of AD. Bradycardia, tachycardia, flushing, headache, sweating, tremors, nasal congestion, weakness, blurred vision, spasms are other symptoms of AD (Khastgir, Drake, & Abrams, 2007). AD can make sex unappealing or lead to active abstinence from having sex in some patients with SCI. It is important to give couples time to find the best position for sexual stimulation and orgasm. Longer foreplay and stimulation of body senses (breast, ear lobes, inner thigh) may be recommended for sexual pleasure (Hess & Hough, 2012; Taylan, Gozuyesil, Manav, & Isik, 2019).

Position problems

Positioning problems during foreplay and sexual intercourse are among the most important changes affecting the sexual life of women with SCI. It is important to give couples time to find the best position for sexual stimulation and orgasm. Studies have recommended putting a pillow under the hips to relieve the pressure experienced regarding the position and to provide support during sexual intercourse. Women must be encouraged to try different positions (Courtois et al., 2017; Hess & Hough, 2012).

Psychosocial problems

Apart from the level of injury, some determinants related to the psychosocial life of the person also affect sexual life. In patients with SCI, having a supportive partner (Valtonen, Karlsson, Siosteen, Dahlof, & Viikari-Juntura, 2006), partner satisfaction, and the quality of the relationship (Hess & Hough, 2012) positively affect sexual life. It is emphasized that especially patients with partner support have a better adaptation to their sexual life over time (Valtonen et al., 2006). It is important that women speak openly and honestly to their partners about their sexual needs and desires, and that the partner is understanding and sensitive to these needs. Women may take on the role of giving pleasure to their partner instead of getting pleasure (Taylan et al., 2019). Therefore, the greatest sexual concern of women after injury may include feelings, such as the inability to achieve sexual satisfaction, abandonment by their spouse, emptiness, and exclusion (Maasoumi, Zarei, Merghati-Khoei, Lawson, & Emami-Razavi, 2018).

It is very important to support socialization in coping with psychological difficulties, which draws attention to the fact that SCI affects sexual life in women. Regarding sexual rehabilitation, supporting women in terms of depression and stress, attractiveness, self-esteem, and the ability to overcome the difficulties of managing their relationship with their partner, or finding a partner is important in raising their attention to their preserved sensations during sexual activity (Taylan et al., 2019). For this reason, the multidisciplinary team should provide partner cooperation, including the partner in training and consultancy services, and provide consultancy for appropriate communication by supporting communication between spouses (Stoffel et al., 2018).

Bowel and bladder incontinence

Bowel and bladder continence problems are among the most important problems affecting the sexual life of women with SCI. Urinary incontinence is more common especially in women with complete tetraplegia (Lombardi, Del Popolo, Macchiarella, Mencarini, & Celso, 2010). Bladder management is a source of concern for most women with SCI. It is important to provide counseling to reduce the possibility of urinary accidents during sexual activities. Fluid intake before sexual intercourse can be limited (Hubscher et al., 2018). Women using intermittent catheterization for bladder treatment can empty their bladder before engaging in sexual activity. Women who use suprapubic or Foley catheters may have concerns about the presence of the catheter. Women with SCI can be informed that the Foley catheter is attached to the urethra, the urethra is a separate region from the vagina, and that it will not interfere with sexual intercourse (Lombardi et al., 2010). During sexual intercourse, women may be recommended to fix (using some adhesive tape) the Foley catheter to the thigh or abdomen, taking care not to bend it, and to attach an empty urine bag to the tip of the Foley. If they have the opportunity to reattach the Foley catheter after sexual intercourse, they may also have the option of pulling the Foley catheter before sexual intercourse (Hubscher et al., 2018).

Bowel management is another problem that affects the sex life of women with SCI. The best way to prevent incontinence during sex is to establish a consistent bowel management program. Once a routine has been established, an accident is much less likely to occur. For greater confidence, it is important to empty the bowel before sexual intercourse and to avoid eating before sexual intercourse (Hubscher et al., 2018).

Fertility

SCI also affects fertility as well as sexual life in both men and women. Neither the level of injury nor the completeness of the lesion appears to have an effect on the menstrual cycle over time (Stoffel et al., 2018). However, most women with SCI experience a brief pause in their menstrual cycle after SCI. This pause is normal and can last up to 6 months after injury (Bertschy et al., 2020). Women with SCI may experience menstrual symptoms, such as increased bladder and muscle spasms, before and during menstruation. Injury can rarely cause menopause in premenopausal women. Except for an increase in mood disorders, menopausal symptoms are similar to those of other women (Stoffel et al., 2018).

Fertility treatment strategies in individuals with SCI are similar to those of the general population. However, it is important to understand the options, consequences, and potential complications that may be specific to SCI, and to inform women about these possible changes (Stoffel et al., 2018).

Birth control

After a period of transient amenorrhea, most women with SCI return to their fertile state before the injury. Although hormonal birth control can be administered to women with SCI, the increased thromboembolic risk should be taken into account. Hormonal contraception should be avoided in women with SCI who continue to smoke and have a history of cardiovascular circulation problems (Stoffel et al., 2018).

Pregnancy

SCI does not necessarily reduce a woman's desire or capacity to give birth. However, several factors, including urological complications, thromboembolism, and autonomic dysfunction, should be considered during pregnancy and delivery (Bertschy et al., 2020). Pregnancy complications with a relatively higher risk of occurrence compared to other pregnant women are urinary tract infections, pyelonephritis, venous thrombosis, pulmonary embolism, premature rupture of membrane, and preterm delivery (Bertschy et al., 2016). Women with SCI often take more than one medication because of the associated symptoms. Therefore, they should have a planned pregnancy. Early prenatal anesthetic consultation is recommended. Anesthetic difficulties may arise, especially in the case of AD. The absence of contractions in women with injuries above T 10 can result in unplanned home births and unattended hospital births. Pregnant women with SCI should be trained about uterine palpation techniques to detect contractions at home (Bertschy et al., 2020).

Maternity care for women with SCI is unavailable in the structural and procedural processes of many high-income countries. Different care philosophies in gynecology/obstetrics and rehabilitation medicine can cause miscommunication, tension, and even antagonism in the continuity of care. These issues must be resolved to benefit women with SCI. In this period, gynecology/obstetrics should be included in the multidisciplinary team (Bertschy et al., 2020).

Conclusion

Problems that may occur in sexual life after injury in individuals with SCI are among the primary issues that should be addressed in both sexes. Treatment options, problems with sexual function and fertility, and possible complications should be addressed with a multidisciplinary approach. It is important that the multidisciplinary team should be aware of the feelings and problems experienced by patients and that patients should be provided with counseling about appropriate coping methods. Informing individuals about possible changes in sexuality and methods for handling them after SCI, and supporting individuals in managing this process will alleviate their pain and anxiety and enable them to be more motivated in their sexual lives (Table 1).

Applications to other areas of neuroscience

As in spinal cord injuries, sexual life problems are common in neurological disorders. Interruption of the long spinal cord pathways between the cortex and sacral cord or pelvic autonomic nerves affects genital blood circulation, erections, ejaculation, and orgasm. Neurological disorders; Intimacy during sexual activity can make it difficult to maintain sexual arousal, sexual intercourse and urinary and bowel continence.

TABLE 1 Problems affecting their sexual life in individuals with spinal cord injuries.

	Physical problems	Psychological problems
Male	Erectile dysfunction Decreased libido Abnormal ejaculation	Anxiety Low self-esteem Depression Decreased sexual satisfaction Changes in personality structure
Female	Genital arousal reduced sensation or loss of sensation Vaginal dryness Difficulties in reaching orgasm Bladder and bowel incontinence Movement and self-positioning difficulty	

Sexual problems experienced in individuals with neurological disorders such as stroke, severe brain trauma, epilepsy, Parkinson's, multiple sclerosis, cauda equina syndrome, perpheric neuropathy are similar to those with spinal cord injuries. Problems such as difficulty in genital arousal, orgasm problems, position problems, bowel and bladder incontinence are common in women. Problems such as erectile dysfunction, ejaculation disorders, orgasm problems, semen abnormalities and infertility are experienced in men. All these problems in both sexes affect the sexual life satisfaction of individuals.

As in SCI individuals, it is important to be aware of the sexual problems of individuals with neurological disorders and to be managed with the same approach. Management of sexual problems should be handled with a holistic approach. Because reduced body image, fatigue, loss of independence, depression, and changes in interpersonal relationships can damage sexual function as much or more than impaired neural circuits. In addition to medical treatment options, patients should be provided with psychological support, talk about their sexual problems, and provide counseling in solving these problems. Although an interdisciplinary approach in patient-centered care is recommended for a more holistic approach, some patients may prefer limited, individual multidisciplinary appointments for team evaluations due to the sensitivity of sexuality.

Mini-dictionary of terms

Autonomic dysreflexia: It is a clinical picture that occurs as a result of uncontrollable sympathetic activity, usually triggered by a sensory stimulus, in spinal cord injuries with injury level T6 and above.
Bladder incontinence: The involuntary leakage of urine.
Bowel Incontinence: The loss of bowel control and causes unexpected stool excretion.
Bulbocavernosus reflex: It is an involuntary reaction that contracts the breech muscles due to stimulation of the clitoris or glans penis.
Ejeculation dysfunction: The inability to efficiently ejaculate sperm from the penis at the time of sexual orgasm.
Erectile dysfunction: It is that the duration and strength of erection in the penis is not sufficient for sexual intercourse.
Neuropathic pain: Type of pain caused by damage or disease affecting the somatosensory nervous system.
Nociceptive pain: The type of pain caused by structural dysfunction is.
Orgasm: It is the sexual cycle phase accompanied by subjective pleasure, pleasure, and relaxation that occurs with the activation of the amygdala and hippocampus in the brain.
Phosphodiesterase 5-inhibitors (PDE5-Is): The first-line drugs used in the treatment of erectile dysfunction (ED).
Priapism: It is a prolonged erection of the penis that is not caused by sexual arousal.
Psychogenic erection: It is an erection that results from audiovisual stimuli or fantasies that generate an impulse from the brain to the spinal erection centers.
Reflexogenic erection: It is an erection caused by direct stimulation of the penis.
Sexual arousal: It is the stimulation of sexual desire during or anticipation of sexual activity.
Sexual dysfunction: It is the difficulty experienced by an individual or a couple at any stage of normal sexual activity, including physical pleasure, desire, preference, arousal, or orgasm.

Key facts of sexual life in individuals with SCI and management

- Spinal cord injuries greatly affect the sexual lives of both men and women.
- The degree of individual sexual dysfunction depends on the level and severity of the SCI.

- It is estimated that 35%–80% of men become inactive after SCI Following.
- Tingling sensations (35.3%) and spasms (35%) are the most frequently reported in women with SCI.
- Male sexual dysfunction includes erectile dysfunction, decreased libido, and abnormal ejaculation.
- Female sexual dysfunction includes difficulty in genital arousal, vaginal dryness, difficulty reaching orgasm.
- Interventions for patients' sexual life, including sexual education, counseling cognitive behavioral therapy, are often helpful

Summary points

- In this section, the effects of SCI on sexual life are discussed.
- It includes male sexual dysfunction, erectile dysfunction, decreased libido, and abnormal ejaculation.
- Female sexual disorders are physiological problems, feeling less attractive, and lack of self-confidence.
- Sexual dysfunctions experienced are associated with psychiatric consequences such as low quality of life, anxiety.
- It is important to diagnose and manage sexual dysfunctions with a multidisciplinary approach.

References

Adriaansen, J. J., Ruijs, L. E., van Koppenhagen, C. F., van Asbeck, F. W., Snoek, G. J., van Kuppevelt, D., et al. (2016). Secondary health conditions and quality of life in persons living with spinal cord injury for at least ten years. *Journal of Rehabilitation Medicine*, *48*(10), 853–860.

Afferi, L., Pannek, J., Burnett, A. L., Razaname, C., Tzanoulinou, S., Bobela, W., et al. (2020). Performance and safety of treatment options for erectile dysfunction in patients with spinal cord injury: A review of the literature. *Andrology*, 1–14. https://doi.org/10.1111/andr.12878.

Aikman, K., Oliffe, J. L., Kelly, M. T., & McCuaig, F. (2018). Sexual health in men with traumatic spinal cord injuries: A review and recommendations for primary health-care providers. *American Journal of Men's Health*, *12*(6), 2044–2054.

Alexander, M. S., Aisen, C. M., Alexander, S. M., & Aisen, M. L. (2017). Sexual concerns after spinal cord injury: An update on management. *NeuroRehabilitation*, *41*(2), 343–357. https://doi.org/10.3233/NRE-172202.

Alexander, M. S., Biering-Sorensen, F., Elliott, S., Kreuter, M., & Sonksen, J. (2011). International spinal cord injury female sexual and reproductive function basic data set. *Spinal Cord*, *49*(7), 787–790. https://doi.org/10.1038/sc.2011.7.

Alexander, M. S., Brackett, N. L., Bodner, D., Elliott, S., Jackson, A., Sonksen, J., et al. (2009). Measurement of sexual functioning after spinal cord injury: Preferred instruments. *The Journal of Spinal Cord Medicine*, *32*(3), 226–236. https://doi.org/10.1080/10790268.2009.11760777.

Alexander, M., & Rosen, R. C. (2008). Spinal cord injuries and orgasm: A review. *Journal of Sex & Marital Therapy*, *34*(4), 308–324. https://doi.org/10.1080/00926230802096341.

Alwaal, A., Breyer, B. N., & Lue, T. F. (2015). Normal male sexual function: Emphasis on orgasm and ejaculation. *Fertility and Sterility*, *104*(5), 1051–1060.

Bertschy, S., Bostan, C., Meyer, T., & Pannek, J. (2016). Medical complications during pregnancy and childbirth in women with SCI in Switzerland. *Spinal Cord*, *54*(3), 183–187. https://doi.org/10.1038/sc.2015.205.

Bertschy, S., Schmidt, M., Fiebag, K., Lange, U., Kues, S., & Kurze, I. (2020). Guideline for the management of pre-, intra-, and postpartum care of women with a spinal cord injury. *Spinal Cord*, *58*(4), 449–458. https://doi.org/10.1038/s41393-019-0389-7.

Burnett, A. L., Nehra, A., Breau, R. H., Culkin, D. J., Faraday, M. M., Hakim, L. S., et al. (2018). Erectile dysfunction: AUA guideline. *The Journal of Urology*, *200*(3), 633–641.

Burns, S. M., Hough, S., Boyd, B. L., & Hill, J. (2010). Men's adjustment to spinal cord injury: The unique contributions of conformity to masculine gender norms. *American Journal of Men's Health*, *4*(2), 157–166.

Cehic, E., Kasum, M., Simunic, V., Oreskovic, S., Vujic, G., & Grgic, F. (2016). Fertility in men with spinal cord injury. *Gynecological Endocrinology*, *32*(12), 937–941. https://doi.org/10.1080/09513590.2016.1202231.

Choi, Y., Kang, J., & Shin, H.-I. (2015). Sexual activity and sexual satisfaction in Korean men with spinal cord injury. *Spinal Cord*, *53*(9), 697–700.

Courtois, F., Alexander, M., & McLain, A. B. J. (2017). Women's sexual health and reproductive function after SCI. *Topics in Spinal Cord Injury Rehabilitation*, *23*(1), 20–30. https://doi.org/10.1310/sci2301-20.

Courtois, F., & Charvier, K. (2015). Sexual dysfunction in patients with spinal cord lesions. *Handbook of Clinical Neurology*, *130*, 225–245.

Cramp, J. D., Courtois, F. J., & Ditor, D. S. (2015). Sexuality for women with spinal cord injury. *Journal of Sex & Marital Therapy*, *41*(3), 238–253.

Dahlberg, A., Alaranta, H., Kautiainen, H., & Kotila, M. (2007). Sexual activity and satisfaction in men with traumatic spinal cord lesion. *Journal of Rehabilitation Medicine*, *39*(2), 152–155.

Eisenberg, N. W., Andreski, S.-R., & Mona, L. R. (2015). Sexuality and physical disability: A disability-affirmative approach to assessment and intervention within health care. *Current Sexual Health Reports*, *7*(1), 19–29.

Elliott, S. (2018). *Sexual dysfunction and infertility in individuals with spinal cord disorders. Spinal cord medicine* (pp. 411–435). Springer Publishing Company.

Elliott, S., Hocaloski, S., & Carlson, M. (2017). A multidisciplinary approach to sexual and fertility rehabilitation: The sexual rehabilitation framework. *Topics in Spinal Cord Injury Rehabilitation*, *23*(1), 49–56.

Everaert, K., De Waard, W., Van Hoof, T., Kiekens, C., Mulliez, T., & D'herde, C. (2010). Neuroanatomy and neurophysiology related to sexual dysfunction in male neurogenic patients with lesions to the spinal cord or peripheral nerves. *Spinal Cord*, *48*(3), 182–191.

Fenstermaker, M., Dupree, J. M., Hadj-Moussa, M., & Ohl, D. A. (2018). Management of Erectile Dysfunction and Infertility in the male spinal cord injury patient. *Current Urology Reports*, *19*(7), 47. https://doi.org/10.1007/s11934-018-0806-z.

Ferreiro-Velasco, M. E., Barca-Buyo, A., de la Barrera, S. S., Montoto-Marques, A., Vazquez, X. M., & Rodriguez-Sotillo, A. (2005). Sexual issues in a sample of women with spinal cord injury. *Spinal Cord*, *43*(1), 51–55. https://doi.org/10.1038/sj.sc.3101657.

Gomes, C. M., Miranda, E. P., de Bessa Jr, J., Bellucci, C. H. S., Battistella, L., Abdo, C. H. N., et al. (2017). Erectile function predicts sexual satisfaction in men with spinal cord injury. *Sexual Medicine*, *5*(3), e148–e155.

Groen, J., Pannek, J., Castro Diaz, D., Del Popolo, G., Gross, T., Hamid, R., et al. (2016). Summary of European Association of Urology (EAU) guidelines on neuro-urology. *European Urology*, *69*(2), 324–333. https://doi.org/10.1016/j.eururo.2015.07.071.

Hess, M. J., & Hough, S. (2012). Impact of spinal cord injury on sexuality: Broad-based clinical practice intervention and practical application. *The Journal of Spinal Cord Medicine*, *35*(4), 211–218. https://doi.org/10.1179/2045772312Y.0000000025.

Hubscher, C. H., Herrity, A. N., Williams, C. S., Montgomery, L. R., Willhite, A. M., Angeli, C. A., et al. (2018). Improvements in bladder, bowel and sexual outcomes following task-specific locomotor training in human spinal cord injury. *PLoS One*, *13*(1). https://doi.org/10.1371/journal.pone.0190998, e0190998.

Ibrahim, E., Lynne, C. M., & Brackett, N. (2016). Male fertility following spinal cord injury: An update. *Andrology*, *4*(1), 13–26.

Khastgir, J., Drake, M. J., & Abrams, P. (2007). Recognition and effective management of autonomic dysreflexia in spinal cord injuries. *Expert Opinion on Pharmacotherapy*, *8*(7), 945–956. https://doi.org/10.1517/14656566.8.7.945.

Krassioukov, A. (2009). Autonomic function following cervical spinal cord injury. *Respiratory Physiology & Neurobiology*, *169*(2), 157–164. https://doi.org/10.1016/j.resp.2009.08.003.

Krassioukov, A., & Elliott, S. (2017). Neural control and physiology of sexual function: Effect of spinal cord injury. *Topics in Spinal Cord Injury Rehabilitation*, *23*(1), 1–10.

Le, J., & Dorstyn, D. (2016). Anxiety prevalence following spinal cord injury: A meta-analysis. *Spinal Cord*, *54*(8), 570–578.

Lombardi, G., Del Popolo, G., Macchiarella, A., Mencarini, M., & Celso, M. (2010). Sexual rehabilitation in women with spinal cord injury: A critical review of the literature. *Spinal Cord*, *48*(12), 842–849. https://doi.org/10.1038/sc.2010.36.

Maasoumi, R., Zarei, F., Merghati-Khoei, E., Lawson, T., & Emami-Razavi, S. H. (2018). Development of a sexual needs Rehabilitation framework in women Post-spinal cord injury: A study from Iran. *Archives of Physical Medicine and Rehabilitation*, *99*(3), 548–554. https://doi.org/10.1016/j.apmr.2017.08.477.

Otero-Villaverde, S., Ferreiro-Velasco, M. E., Montoto-Marques, A., Salvador de la Barrera, S., Arias-Pardo, A. I., & Rodriguez-Sotillo, A. (2015). Sexual satisfaction in women with spinal cord injuries. *Spinal Cord*, *53*(7), 557–560. https://doi.org/10.1038/sc.2015.53.

Pawar, P., Patil, S., Tamhankar, A., Kulkarni, V., Savalia, A., Shah, N., et al. (2019). An audit of andrological needs in traumatic paraplegic male patients: Five year assessment in a single paraplegic Centre. *Journal of Clinical and Diagnostic Research*, *13*(12).

Sabharwal, S. (2013). *Essentials of spinal cord Medicine*. New York: Demosmedical.

Steadman, C. J., & Hubscher, C. H. (2016). Sexual function after spinal cord injury: Innervation, assessment, and treatment. *Current Sexual Health Reports*, *8*(2), 106–115.

Stoffel, J. T., Van der Aa, F., Wittmann, D., Yande, S., & Elliott, S. (2018). Fertility and sexuality in the spinal cord injury patient. *World Journal of Urology*, *36*(10), 1577–1585. https://doi.org/10.1007/s00345-018-2347-y.

Taylan, S., Gozuyesil, E., Manav, A. I., & Isik, E. I. (2019). An evaluation of the factors that affect the sexual satisfaction of people with spinal cord injuries. *The Journal of Spinal Cord Medicine*, 1–8. https://doi.org/10.1080/10790268.2019.1672955.

Valtonen, K., Karlsson, A. K., Siosteen, A., Dahlof, L. G., & Viikari-Juntura, E. (2006). Satisfaction with sexual life among persons with traumatic spinal cord injury and meningomyelocele. *Disability and Rehabilitation*, *28*(16), 965–976. https://doi.org/10.1080/09638280500404362.

Chapter 41

Depressive symptoms in rehabilitation post-spinal cord injury

Swati Mehta[a,b,c], Steve Orenczuk[b], Eldon Loh[a,c], and Robert Teasell[a,c]

[a]*Lawson Health Research Institute, Parkwood Institute Main Bldg., Department of Physical Medicine and Rehabilitation, London, ON, Canada,* [b]*St. Joseph's Health Care London, Parkwood Institute Main Bldg., Department of Psychology, London, ON, Canada,* [c]*St. Joseph's Health Care London, Department of Physical Medicine and Rehabilitation, London, ON, Canada*

Abbreviations

ABI	acquired brain injury
CBT	cognitive behavioral therapy
CES-D	Center for Epidemiologic Studies Depression Scale
DSM5	diagnostic and statistical manual of mental disorders
HPA	hypothalamic-pituitary-adrenal
MDD	major depressive disorder
MINI	Mini-International Neuropsychiatric Interview
NIH PROMIS	National Institutes of Health Patient-Reported Outcomes Measurement Information System
PHQ-2, PHQ-9	Patient Health Questionnaire-2 and -9
SCI	Spinal Cord Injury
SCID-5	Structured Clinical Interview for DSM5
SNRI	serotonin-norepinephrine reuptake inhibitors
SSRI	serotonin reuptake inhibitors
TBI	traumatic brain injury

Introduction

Depression is a common concern among persons with spinal cord injury (SCI). Given the losses and innumerable adjustments required, individuals with an SCI will likely encounter repeated strains upon their available coping resources. As many as 40% of people suffering an SCI experience depression during rehabilitation and around 1 in 5 people experience depression a year after the injury (Williams & Murray, 2015). The experience of depression interferes with rehabilitation efforts, and compromises capacity building, thereby impeding the potential to prevent complications. Depression after SCI is associated with several negative outcomes post-injury including longer hospitalization, decreased longevity, higher levels of pain, increased rates of suicide, reduced health, daily functioning, and limited community participation; it is likely best viewed as a secondary complication or sequelae rather than part of an adaptive process facilitating overall emotional adjustment (Consortium for Spinal Cord Medicine, & Paralyzed Veterans of America, 1998). Depression post-SCI can interfere with recovery and rehabilitation and is related to longer hospital stays, higher levels of pain, and lower quality of life after injury. Management of depression post-SCI is often overlooked due to failure to recognize symptoms and/or access to appropriate services (Paralyzed Veterans of America, 2020). SCI rehabilitation should focus not only on acute care but also on longer-term psychological care (Craig, Tran, & Middleton, 2009). Failure to adopt a multi-modal approach to SCI rehabilitation, one that also targets psychological distress may result in added pressures to health care resources (Craig et al., 2009.; Tate & Pledger, 2003). This chapter summarizes the state of depression among persons with SCI and reviews evidence for its management.

Cellular, Molecular, Physiological, and Behavioral Aspects of Spinal Cord Injury. https://doi.org/10.1016/B978-0-12-822427-4.00041-1

Screening and assessment of depression post-SCI

While depression was once considered a universal aspect of grieving and adjustment following catastrophic SCI illness or injury (Hohmann, 1975), the adverse impact of clinical depression upon the course of recovery has been more recently well established. Depression is associated with fewer functional improvements and longer inpatient rehabilitation lengths of stay (Dobrez, Heinemann, Deutsch, Durkin, & Almagor, 2010). When compared with non-depressed peers, persons living with SCI in the community encounter greater health-related complications (e.g., pressure ulcers and urinary tract infections) and associated medical costs (Craig et al., 2009). A recent (2015) meta-analysis estimates that the prevalence rates for depression in persons with spinal cord injury to be between 19% and 26% (Williams & Murray, 2015). Depression is not a necessary consequence of SCI. Longitudinal profiles of those with SCI have found that there is a gradual increase in depressive symptoms between 24 and 48 weeks post-injury; followed by a significant decrease post-discharge (Kennedy & Rogers, 2000; Fig. 1). Unfortunately, evidence indicates persons with SCI and significant depression often go unrecognized, or when identified, may not receive sufficient pharmacologic or psychological intervention. In one study, only 29% of persons with probable major depressive disorder (MDD) received anti-depressants and 11% received psychotherapy within the past 3 months (Fann et al., 2011).

From a practical perspective, how best to identify individuals with SCI who are experiencing significant depressive symptoms has long been a focus of clinical concern in rehabilitation. In 2020, the Consortium for Spinal Cord Medicine published a set of clinical practice guidelines for health care providers specific to the management of mental health disorders, substance use disorders, and suicide in adults with spinal cord injury (Karnes et al., 2020). It was recommended that individuals with SCI be screened for major depression early during the initial inpatient hospital or rehabilitation stay, with repeated screening to monitor symptoms at the initial follow-up post-discharge as well as annually (or more frequently) thereafter. Factors including premorbid mental health issues, positive depression screenings following the onset of SCI, and significant chronic pain (Craig et al., 2013) should be considered in determining an individualized screening strategy. The process of regular symptom monitoring can offer an opportunity for earlier intervention to potentially reduce morbidity and mortality in persons "at risk" for developing depressive symptoms following SCI. In a recent study, suicide mortality among those with SCI decreased over three injuries, but it still remained three times higher than that of the general population (Cao, Massaro, Krause, Chen, & Devivo, 2014).

Structured diagnostic interviews (e.g., Structured Clinical Interview for DSM5 (SCID-5) or the Mini-International Neuropsychiatric Interview (MINI)) are often considered the "gold standard" for establishing mental health diagnoses (First, Williams, Karg, & Spitzer, 2016; Sheehan et al., 1998) (Table 1); however, the specialized staff training and burden to clients required to complete these assessments can be impractical in high volume settings such as acute care, rehabilitation, or outpatient clinic settings. Accordingly, a simpler screening strategy to identify depressive symptoms in individuals with a possible diagnosis and facilitate ongoing monitoring becomes of obvious importance in providing ongoing care for

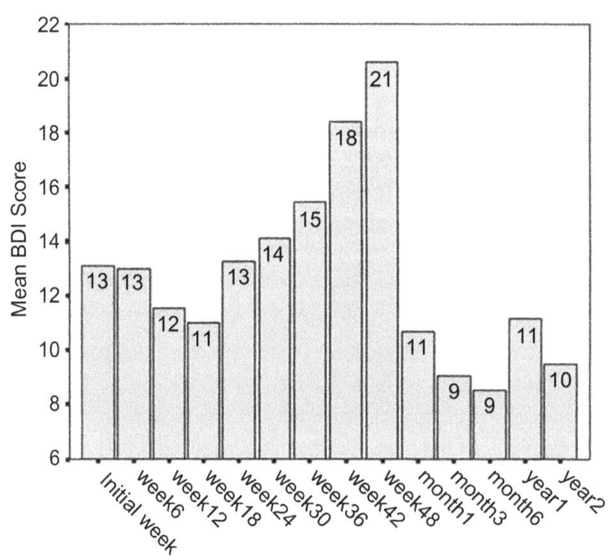

FIG. 1 Mean scores on Beck Depression Inventory over 2 years among those with SCI; Clinical cut-off at >14. *Reprint from: Kennedy, P., & Rogers, B. A. (2000). Anxiety and depression after spinal cord injury: a longitudinal analysis.* Archives of Physical Medicine and Rehabilitation, 81*(7), 932–937.*

TABLE 1 DSM-5 major depressive disorder diagnostic criteria.

A. Five (or more) of the following symptoms have been present during the same 2-week period and represent a change from previous functioning; at least one of the symptoms is either (1) depressed mood or (2) loss of interest or pleasure. Note: Do not include symptoms that are clearly attributable to another medical condition.

1. Depressed mood most of the day, nearly every day, as indicated by either subjective report (e.g., feels sad, empty, hopeless) or observation made by others (e.g., appears tearful). (Note: In children and adolescents, can be irritable mood.)
2. Markedly diminished interest or pleasure in all, or almost all, activities most of the day, nearly every day (as indicated by either subjective account or observation).
3. Significant weight loss when not dieting or weight gain (e.g., a change of more than 5% of body weight in a month) or decrease or increase in appetite nearly every day. (Note: In children, consider failure to make expected weight gain.)
4. Insomnia or hypersomnia nearly every day.
5. Psychomotor agitation or retardation nearly every day (observable by others, not merely subjective feelings of restlessness or being slowed down).
6. Fatigue or loss of energy nearly every day.
7. Feelings of worthlessness or excessive or inappropriate guilt (which may be delusional) nearly every day (not merely self-reproach or guilt about being sick).
8. Diminished ability to think or concentrate, or indecisiveness, nearly every day (either by subjective account or as observed by others).
9. Recurrent thoughts of death (not just fear of dying), recurrent suicidal ideation without a specific plan, or a suicide attempt or a specific plan for committing suicide.

B. The symptoms cause clinically significant distress or impairment in social, occupational, or other important areas of functioning.

C. The episode is not attributable to the physiological effects of a substance or another medical condition.

Note: Criteria A–C represent a major depressive episode.

Note: Responses to a significant loss (e.g., bereavement, financial ruin, losses from a natural disaster, a serious medical illness or disability) may include the feelings of intense sadness, rumination about the loss, insomnia, poor appetite, and weight loss noted in Criterion A, which may resemble a depressive episode. Although such symptoms may be understandable or considered appropriate to the loss, the presence of a major depressive episode in addition to the normal response to a significant loss should also be carefully considered. This decision inevitably requires the exercise of clinical judgment based on the individual's history and the cultural norms for the expression of distress in the context of loss.
In distinguishing grief from a major depressive episode (MDE), it is useful to consider that in grief the predominant affect is feelings of emptiness and loss, while in an MDE it is persistent depressed mood and the inability to anticipate happiness or pleasure. The dysphoria in grief is likely to decrease in intensity over days to weeks and occurs in waves, the so-called pangs of grief. These waves tend to be associated with thoughts or reminders of the deceased. The depressed mood of an MDE is more persistent and not tied to specific thoughts or preoccupations. The pain of grief may be accompanied by positive emotions and humor that are uncharacteristic of the pervasive unhappiness and misery characteristic of an MDE. The thought content associated with grief generally features a preoccupation with thoughts and memories of the deceased, rather than the self-critical or pessimistic ruminations seen in an MDE. In grief, self-esteem is generally preserved, whereas in an MDE feelings of worthlessness and self-loathing are common. If self-derogatory ideation is present in grief, it typically involves perceived failings vis-à-vis the deceased (e.g., not visiting frequently enough, not telling the deceased how much he or she was loved). If a bereaved individual thinks about death and dying, such thoughts are generally focused on the deceased and possibly about "joining" the deceased, whereas in an MDE such thoughts are focused on ending one's own life because of feeling worthless, undeserving of life, or unable to cope with the pain of depression.

D. The occurrence of the major depressive episode is not better explained by schizoaffective disorder, schizophrenia, schizophreniform disorder, delusional disorder, or other specified and unspecified schizophrenia spectrum and other psychotic disorders.

E. There has never been a manic episode or a hypomanic episode.

This table is recreated from the American Psychiatric Association (2013). Diagnostic and statistical manual of mental disorders (5th ed.). https://doi.org/10.1176/appi.books.9780890425596

persons with SCI. In a practical two-stage strategy, individuals who test positive on short screening tools can then be more carefully (re)evaluated to confirm the persistence of symptoms, understand factors underlying them, arrive at a diagnosis (e.g., MDD, adjustment disorder, other) and then determine appropriate interventions.

The need for a brief, valid measure of depression with established sensitivity and specificity has long been sought. A wide variety of self-report depression measures have been employed in studies with persons following spinal cord injury

TABLE 2 Validity of depression screeners compared with major depression diagnosis.

Study	Screening measure	Cut point	Criterion measure	Sensitivity %	Specificity %	PPV	NPV	Prevalence %	N
Radnitz et al. (1996)	BDI	≥18	SCID DSM-III-R	83.3	90.8	50.0	98.1	9.7	124
Tate, Forchheimer, Maynard, Davidoff, and Dijkers (1993)	Zung SDS	≥55	DSM-III-R	86.0	67.0	42.9	93.8	23	30
Bombardier et al. (2012)	PHQ-9	≥10	SCID DSM-IV	100	80	36.0	100	10	142
Bombardier et al. (2012)	PHQ-9	≥11	SCID DSM-IV	100	84	40.0	100	10	142
Krause et al. (2009)	QAHMQ	≥11	PHQ-9	89.7	88.8	48.4	98.6	10.7	727

Abbreviations: *BDI*, Beck Depression Inventory; *NPV*, negative predictive value; *OAHMQ*, Older Adult Health and Mood Questionnaire; *PHQ-9*, 9-item Patient Health Questionnaire-9; *PPV*, positive predictive value; *SCID DSM-III-R*, Structured Clinical Interview for the Diagnostic and Statistical Manual of Mental Disorders, 3rd edition, revised; *SCID DSM-IV*, Structured Clinical Interview for the Diagnostic and Statistical Manual of Mental Disorders, 4th edition; *Zung SDS*, Zung Self-Rating Depression Scale.
This table is recreated from Paralyzed Veterans of America. (2020). Management of mental health disorders, substance use disorders, and suicide in adults with spinal cord injury. Retrieved from https://www.iscos.org.uk/uploads/sitefiles/Newsletters/PVA_CPG_Mental_Health_and_substa.pdf

(Table 2). A longstanding concern is an impact of including somatic symptoms of depression (e.g., low energy/fatigue, sleep, and appetite disturbances) in screening measures—and the apparent risk that such "transdiagnostic" or overlapping symptoms may serve to inflate depression scores when used with persons with spinal cord injury. A 2017 study of retrospective responses to the CES-D, PHQ-9, and the NIH PROMIS depression scales supported an "inclusive" view toward somatic symptomatology (regardless of etiology) and did not find these three depression scales to have performed differently in persons with SCI when compared to general and community samples. Accordingly, the authors offered that these measures can be applied with greater confidence in helping to identify persons with MDD providing an endorsement of at least one of the two cardinal symptoms (anhedonia; low mood) (Cook et al., 2017).

Other concerns regarding the use of "case finding" screening measures of depression in general, and in persons following SCI include the range of length of the surveys, issues involving manual dexterity when completing pencil/paper assessments, social desirability, accurate item comprehension, or reading challenges. Further, many of the screening measures do not map easily onto established diagnostic classification (DSM5) systems (symptoms and/or time frames) and intervention planning. The PHQ-9 has the advantages of broad clinical application, brevity, no cost, and the 9-item version corresponds directly to the symptoms of MDD in the DSM5. Enhancing its clinical utility, the measure is appropriate for adults and adolescents (as young as 12 years of age), is available in multiple languages and has been widely used across a variety of rehabilitation settings (including acquired brain injury (ABI) and stroke populations). It has been well investigated in persons with SCI (both in inpatient rehabilitation and community samples) and demonstrates strong psychometric properties in identifying clinical cases—though evidence for PHQ-9 as a severity measure and aid to monitoring treatment outcomes is less clear (Pettersson, Boström, Gustavsson, & Ekselius, 2015). Importantly, it includes an item specific to identifying self-destructive ideation. Scores can range from 0 (none)—27 (severe) with 5-point increments in severity. A process that employs an initial screen using the 2-item Patient Health Questionnaire (PHQ-2) can offer further efficiency, such that if neither anhedonia nor depressed mood is endorsed, screening can be discontinued, while if present, the remaining PHQ-9 items can be administered.

In general use, a score of 10 on the PHQ-9 is considered of "moderate" severity prompting consideration of a treatment plan (psychological and/or pharmacological), and a score of 5–9 warrants a discussion of symptoms with their provider and periodic monitoring of their ongoing adjustment. For persons with SCI, it has been suggested that an optimal cut off for detecting major depression with the PHQ-9 during inpatient rehabilitation is a score of 11 (or higher) acknowledging the potential impact of non-specific factors during inpatient rehabilitation and resulting in 100% sensitivity and 84% specificity with persons endorsing at least 1 cardinal symptom of depression (Bombardier et al., 2012). Further investigation into the implementation and outcomes of routine depression screening in persons with SCI is viewed as critical toward supporting emotional well-being and preventing the negative consequences of untreated depression on health and community participation (Titman, Liang, & Craven, 2019).

Theoretical correlates of depression post-SCI

As mentioned previously, depression is not inevitable following SCI, thus it is noteworthy that depression is related to modifiable factors that play a role in its development and maintenance. Several factors, including demographics, injury characteristics, pre-injury behaviors and psychopathology, personality factors, social/environmental factors, and pathophysiology, have been implicated as correlates of depression post-SCI.

Personal factors such as pre-injury psychological distress and psychiatric disorders as well as post-injury difficulties adjusting to SCI have often been shown to be correlated with the presence of depressive symptoms (Arango-Lasprilla, Ketchum, Starkweather, Nicholls, & Wilk, 2011; Krueger, Noonan, Williams, Trenaman, & Rivers, 2013; Williams, Smith, & Papathomas, 2014). Secondary physical complications such as the presence of pain, spasticity, bowel/bladder dysfunction have also been associated with greater levels of depressive symptoms (Hartoonian et al., 2014). However, the self-management model demonstrates that individuals with greater levels of self-efficacy in managing the physical complications of SCI are less likely to experience depression and/or depressive symptoms (Hartoonian et al., 2014). Studies have demonstrated that those that have strong problem-solving skills and self-confidence report fewer symptoms of depression (Elliott, Godshall, Herrick, Witty, & Spruell, 1991).

A higher proportion of depressive symptoms have been correlated with race, particularly Latinos reporting greater level of depressive symptoms compared to African American or Caucasians (Kemp, Krause, & Adkins, 1999). Socioeconomic status, specifically having lower income, has been shown to predict high scores on self-reported depression after controlling for demographic and injury characteristics (Khazaeipour, Taheri-Otaghsara, & Naghdi, 2015). There is conflicting evidence for the role of gender in the prevalence of depression. While some studies indicate women experience greater rates of depression (Khazaeipour et al., 2015), others indicate that there's no significant difference in prevalence rates based on gender (Kalpakjian & Albright, 2006). Environmental factors such as lack of social support or having difficulty having care needs met have been associated with depressive symptoms (Khazaeipour et al., 2017; Zürcher et al., 2019).

There is limited clinical research on pathophysiological models of depression post-SCI. Depression is characterized by changes in neurochemical and neuroendocrine systems, including monoaminergic systems and the hypothalamic-pituitary-adrenal (HPA) axis (Brown, Steinberg, & Van Praag, 1994). These factors are reported in clinical and animal studies during various conditions that result in immunity system activation. In the non-SCI clinical population, studies have found an association between elevated levels of pro-inflammatory cytokines and depression (Dowlati, Herrmann, Swardfager, Reim, & Lanctot, 2010; Liu, Ho, & Mak, 2012). Animal models indicate that among those with a depression-like profile there are higher levels of pro-inflammatory cytokines peripherally. These changes in inflammation were not associated with severity of injury indicating that the correlation between inflammation and depression-like behaviors is independent of impairment (Maldonado-Bouchard et al., 2016). In addition, chronic microglia activation resulting from sustaining an SCI has been shown to increase cognitive dysfunction and depression-like symptoms in mouse models (Wu et al., 2014). Thus there may be a role for the immune system in the development of depression-like symptoms. It is important to keep in mind that these pathophysiological factors alone do not result in decreased emotional well-being. Rather, they may interact with existing maladaptive environmental and psychosocial factors that lead to increased distress (Maldonado-Bouchard et al., 2016).

Management of depression post-SCI

Based on the biopsychosocial model, appropriate management of depression and related secondary complications should be through a multi-disciplinary approach that is best able to deal with a complex number of diverse variables beyond just the actual SCI. Multi-disciplinary SCI rehabilitation programs based on a biopsychosocial model are effective (Budh, Kowalski, & Lundeberg, 2006; Perry, Nicholas, & Middleton, 2010). However, resource limitations can restrict the ability of service providers to deliver these integrative biopsychosocial approaches in the community.

Pharmacological

As there is a lack of studies specific to the SCI population for pharmacologic management of depression, treatment decisions usually follow the basic principles that guide the management of depression in the general population. Psychotherapy and pharmacotherapy in combination are more effective than either alone; they should be offered together if possible (Cuijpers, Dekker, Hollon, & Andersson, 2009; Paralyzed Veterans of America, 2020; Richelson, 2001). Specific studies within the SCI population demonstrate improvements in mood following combined treatment with pharmacotherapy and

CBT/supportive psychotherapy (Judd, Stone, Webber, Brown, & Burrows, 1989; Kahan, Mitchell, Kemp, & Adkins, 2006; Kemp, Kahan, Krause, Adkins, & Nava, 2004).

In general, the addition of pharmacotherapy is recommended in those with at least moderate depression based on symptom severity scales and/or functional impairment (Kennedy et al., 2016). Anti-depressant management in those with mild depression may be considered in certain cases, including patient preference, past response to anti-depressants, and failure of non-pharmacologic management (Kennedy et al., 2016).

Pharmacologic targets for depression management have primarily focused on the neurotransmitters dopamine, serotonin, and dopamine (Richelson, 2001). Anti-depressants may block different neurotransmitter receptors, block the reuptake of different neurotransmitters, or inhibit monoamine oxidase (a mitochondrial enzyme) (Richelson, 2001). This alters the magnitude of neurotransmitter response at the synapse (Richelson, 2001).

In general, first-line pharmacologic treatments include selective serotonin reuptake inhibitors (SSRIs), serotonin-norepinephrine reuptake inhibitors (SNRIs), agomelatine, bupropion, mirtazapine, and vortioxetine (Kennedy et al., 2016). When choosing an anti-depressant, factors such as patient preference, prior medication response, relative efficacy and effectiveness, safety, tolerability, anticipated side effects, co-occurring psychiatric and medical conditions, drug interactions, ease of use, half-life, and cost should be considered (Gelenberg et al., 2010; Kennedy et al., 2016). The selection of a first-line treatment for depression is individualized, as differences between these medications are small (Kennedy et al., 2016).

Venlafaxine, an SNRI, is the only anti-depressant that has demonstrated improvement in depressive symptoms when assessing pharmacology-only depression management after SCI (Fann et al., 2015). In those with nociceptive pain, venlafaxine has also been shown to improve pain interference of mood (Richards et al., 2015). Amitriptyline, gabapentin, and carbamazepine have not demonstrated an effect on depressive symptoms after SCI, but the evidence is limited (Rintala et al., 2007; Salinas, Lugo, & García, 2012). Bupropion, mirtazapine, and vortioxetine have not been systematically evaluated within the SCI population (Paralyzed Veterans of America, 2020) nor has agomelatine.

Within the SCI population, caution should be exercised with the SSRI fluoxetine (Paralyzed Veterans of America, 2020), as increased spasticity risk has been described in a case report (Stolp-Smith & Wainberg, 1999). Tricyclic anti-depressants, which are considered second-line for depression in the general population (Kennedy et al., 2016), should also be used with caution in those with SCI given their anticholinergic side effects. An increased risk of certain anti-depressant side effects after SCI may also result from polypharmacy, including prolonged QTc interval and serotonin syndrome (Paralyzed Veterans of America, 2020).

Given the potential for increased risk of side effects from anti-depressants after SCI, frequent monitoring after initiation of treatment is recommended (Paralyzed Veterans of America, 2020). Initiation of anti-depressants at half their usual dose with titration at half the usual rate is also recommended (Paralyzed Veterans of America, 2020).

Non-pharmacological

Evidence for non-pharmacological management of depression post-SCI is limited. However, recent clinical practice guidelines recommend cognitive behavior therapy (CBT) (Paralyzed Veterans of America, 2020). The application of CBT-based approaches has been widely used on a range of psychosocial issues associated with SCI. CBT has been shown to improve an individual's psychological health by modifying maladaptive emotional, behavioral, and cognitive responses (Tolin, 2010). CBT aims to teach specific coping strategies to manage stressful stimuli and develop strategies to help modify an individual's response to the stressor (Kennedy, 2008). CBT also focuses on problem-solving, cognitive restructuring, increasing access and willingness to engage in rewarding activities, relaxation training, as well as self-efficacy and coping (Swett & Richardson, 2004). Several systematic reviews among the SCI population demonstrate that CBT facilitates adjustment and has been found effective in improving psychological outcomes such as depression, anxiety, coping, and self-efficacy for up to 6 months in follow-up (Dorstyn, Mathias, & Denson, 2010; Mehta et al., 2011). More recent studies have found that online psychological interventions are effective for reducing depressive symptoms among adults with SCI who struggle with mild to moderate depressive symptoms (Burke et al., 2019; Dear et al., 2018; Mehta et al., 2020; Migliorini, Sinclair, Brown, Tonge, & New, 2016). With online interventions, persons with SCI may be able to access therapy that otherwise would have been inaccessible in person.

Other management non-pharmacological management strategies include exercise programs which included stretching, aerobic arm ergometry, and resistance exercises among those with SCI. The studies found a significant reduction in depressive symptoms post-SCI after exercise treatment (Crane, Hoffman, & Reyes, 2017; Hicks et al., 2003; Latimer, Ginis, Hicks, & McCartney, 2004; Ginis et al., 2003). There is limited evidence for external brain stimulation interventions including repetitive transcortical magnetic stimulation or transcranial direct current stimulation in decreasing depressive symptoms with comorbid pain (Defrin, Grunhaus, Zamir, & Zeilig, 2007; Fregni, Boggio, Nitsche, Rigonatti, & Pascual-Leone, 2006; Tan, Hartung, Sharp, & Temel, 2011). However, follow-up times for these treatments have been relatively short (6 weeks); long-term effectiveness is yet to be determined. In addition, access to these forms of treatments may be limited.

Applications to other areas of neuroscience

Other neurological conditions such as traumatic brain injury (TBI), stroke, multiple sclerosis, epilepsy, Alzheimer's diseases also experience a higher prevalence of depressive symptoms compared to the general population (Bulloch et al., 2015) with impaired physical health and quality of life across these neurological conditions (Prisnie et al., 2018). The dispositional factors and pathophysiological pathways are seen in the development of depressive symptoms among those with SCI have also been implicated among other neurological conditions. Consistent findings have linked the presence of depressive symptoms with the hyperactivity of the HPA axis and activation of specific immune responses across neurological disorders (Benedetti, Bernasconi, & Pontiggia, 2006).

Several management strategies reviewed in the current chapter are also translatable among other neurological conditions. Similar to the management of depression among those with SCI, recommendations include the use of SSRIs and CBT among other neurological conditions (Benedetti et al., 2006; Fernie, Kollmann, & Brown, 2015; Hellmann-Regen et al., 2013). Though these management strategies are effective across neurological conditions, several differences exist. Pharmacological treatments like SSRIs (e.g., fluoxetine) may be more tolerable among certain neurological conditions such as with TBI without the risk of spasticity seen among those with SCI (Fann, Hart, & Schomer, 2009). In addition, management strategies involving a strong cognitive component such as CBT may not be appropriate for individuals with severe cognitive or linguistic impairments such as those seen in Alzheimer's disease, TBI, and stroke. Given these differences, it's important to examine the distinct profiles of each neurological condition when developing personalized management plans.

Mini-dictionary of terms

- **Adjustment**: adjustment in a psychological setting, is the process by which an individual attempt to adjust to the challenges from their environment.
- **Anhedonia**: a condition in which an individual is unable to experience pleasure from pleasurable experiences.
- **Cytokines**: a variety of substances that are secreted by cells of the immune system and affect other cells.
- **Depression**: a common medical condition that negatively affects how the individual feels, thinks, and acts.
- **Dispositional Factors**: internal factors or individual characteristics that affect the individual's behavior and actions.
- **Half-life**: the length of time it takes for the concentration of a drug to decrease to half of its starting dose.
- **Hypothalamic pituitary adrenal (HPA) axis**: comprises the hypothalamus, the pituitary gland, and the adrenal glands. It plays an important role in homeostasis and the body's response to stress.
- **Monoaminergic systems**: are a network of neurons that use monoamine neurotransmitters and are important mediators of emotion, arousal, and certain types of memory.
- **Neurotransmitters**: are chemical agents released by neurons to stimulate neighboring neurons or muscle cells, resulting in the transmission of impulses from one cell to the next.
- **QT interval**: is a measurement of the total time from ventricular depolarization to complete repolarization.
- **Sensitivity**: is the proportion of people with the condition who test positive on a specific test.
- **Specificity**: is the proportion of people without the condition who test negative on a specific test

Key facts of screening and diagnosis

- Depression is a common secondary condition post-SCI.
- It can be a complicated process that significantly impedes adjustment and activities of daily living.
- Depression is not universal and though many experience grief after SCI, not everyone experiences persistence of these symptoms into depression once in the community.
- Identifying symptoms of depression can be difficult since comorbid physical concerns may endorse similar somatic symptoms (e.g., fatigue, insomnia, changes in appetite).
- Self-report measures are important screening tools to help clinicians identify any psychosocial distress. These should be followed up with an appropriate referral to management.

Key facts of management

- Combined psychotherapy and pharmacotherapy are effective at managing depression post-SCI than either alone.
- Pharmacological management of depression post-SCI is primarily through SSRIs and SNRIs.
- Due to the increased risk of side effects, it is important to monitor treatment.
- The current recommendation for non-pharmacological management include CBT to improve depression post-SCI
- There is limited evidence for other non-pharmacological strategies including exercise and physical stimulation interventions such as trans magnetic stimulation and transcranial direct current stimulation.

Summary points

- This chapter focuses on depression which is a common secondary condition post-SCI with a prevalence of up to 29%.
- Depression impedes recovery post-SCI and results in several negative outcomes including increased hospitalization, increased morbidity, greater levels of pain, reduced quality of life, and decreased functional capacity.
- Clinical practice guidelines recommend regular screening and monitoring of depressive symptoms among those with SCI.
- Screening should be followed by appropriate diagnosis using the SCID-5 or MINI which is considered the gold standard for establishing diagnoses.
- Due to the existence of transdiagnostic somatic symptoms of depression among persons with SCI, there is a potential for inflation of depression scores while using some outcome measures; however, measures such as the CES-D, PHQ-9, and NIH PROMIS depression scales were found to be psychometrically sound in persons with SCI regardless of the presence of somatic symptoms.
- Several factors have been established as correlates of depression post-SCI including pre-injury psychological distress, race, secondary physical complications, socioeconomic status, gender, and neurotransmitters.
- Optimal management should include both psychotherapy and pharmacotherapy. In terms of pharmacotherapy, current CPGs recommend the use of SSRIs and SNRIs as first-line treatments. CBT is the primary non-pharmacological strategy recommended by the CPGs.
- The depression profile of those with SCI is similar to other neurological condition populations. However, it is important to examine the differences among each condition when developing a personalized management plan.

References

American Psychiatric Association. (2013). *Diagnostic and statistical manual of mental disorders* (5th ed.). American Psychiatric Publishing, Inc.

Arango-Lasprilla, J. C., Ketchum, J. M., Starkweather, A., Nicholls, E., & Wilk, A. R. (2011). Factors predicting depression among persons with spinal cord injury 1 to 5 years post injury. *NeuroRehabilitation*, *29*(1), 9–21.

Benedetti, F., Bernasconi, A., & Pontiggia, A. (2006). Depression and neurological disorders. *Current Opinion in Psychiatry*, *19*(1), 14–18.

Bombardier, C. H., Kalpakjian, C. Z., Graves, D. E., Dyer, J. R., Tate, D. G., & Fann, J. R. (2012). Validity of the patient health questionnaire-9 in assessing major depressive disorder during inpatient spinal cord injury rehabilitation. *Archives of Physical Medicine and Rehabilitation*, *93*(10), 1838–1845.

Brown, S. L., Steinberg, R. L., & Van Praag, H. M. (1994). The pathogenesis of depression: Reconsideration of neurotransmitter data. In *Handbook of depression and anxiety: A biological approach* (pp. 317–347). New York, NY: Marcel Dekker, Inc.

Budh, C. N., Kowalski, J., & Lundeberg, T. (2006). A comprehensive pain management programme comprising educational, cognitive and behavioural interventions for neuropathic pain following spinal cord injury. *Journal of Rehabilitation Medicine*, *38*(3), 172–180.

Bulloch, A. G., Fiest, K. M., Williams, J. V., Lavorato, D. H., Berzins, S. A., Jetté, N., … Patten, S. B. (2015). Depression—A common disorder across a broad spectrum of neurological conditions: A cross-sectional nationally representative survey. *General Hospital Psychiatry*, *37*(6), 507–512.

Burke, D., Lennon, O., Blake, C., Nolan, M., Barry, S., Smith, E., … Fullen, B. M. (2019). An internet-delivered cognitive behavioural therapy pain management programme for spinal cord injury pain: A randomized controlled trial. *European Journal of Pain*, *23*(7), 1264–1282.

Cao, Y., Massaro, J. F., Krause, J. S., Chen, Y., & Devivo, M. J. (2014). Suicide mortality after spinal cord injury in the United States: Injury cohorts analysis. *Archives of Physical Medicine and Rehabilitation*, *95*(2), 230–235.

Consortium for Spinal Cord Medicine, & Paralyzed Veterans of America. (1998). *Depression following spinal cord injury: A clinical practice guideline for primary care physicians*. Consortium for Spinal Cord Medicine.

Cook, K. F., Kallen, M. A., Bombardier, C., Bamer, A. M., Choi, S. W., Kim, J., … Amtmann, D. (2017). Do measures of depressive symptoms function differently in people with spinal cord injury versus primary care patients: The CES-D, PHQ-9, and PROMIS®-D. *Quality of Life Research*, *26*(1), 139–148.

Craig, A., Tran, Y., & Middleton, J. (2009). Psychological morbidity and spinal cord injury: A systematic review. *Spinal Cord*, *47*(2), 108–114.

Craig, A., Tran, Y., Siddall, P., Wijesuriya, N., Lovas, J., Bartrop, R., & Middleton, J. (2013). Developing a model of associations between chronic pain, depressive mood, chronic fatigue, and self-efficacy in people with spinal cord injury. *The Journal of Pain*, *14*(9), 911–920.

Crane, D. A., Hoffman, J. M., & Reyes, M. R. (2017). Benefits of an exercise wellness program after spinal cord injury. *The Journal of Spinal Cord Medicine*, *40*(2), 154–158.

Cuijpers, P., Dekker, J., Hollon, S. D., & Andersson, G. (2009). Adding psychotherapy to pharmacotherapy in the treatment of depressive disorders in adults: A meta-analysis. In *Database of abstracts of reviews of effects (DARE): Quality-assessed reviews [internet]* Centre for Reviews and Dissemination (UK).

Dear, B. F., Fogliati, V. J., Fogliati, R., Gandy, M., McDonald, S., Talley, N., … Jones, M. (2018). Transdiagnostic internet-delivered cognitive-behaviour therapy (CBT) for adults with functional gastrointestinal disorders (FGID): A feasibility open trial. *Journal of Psychosomatic Research*, *108*, 61–69.

Defrin, R., Grunhaus, L., Zamir, D., & Zeilig, G. (2007). The effect of a series of repetitive transcranial magnetic stimulations of the motor cortex on central pain after spinal cord injury. *Archives of Physical Medicine and Rehabilitation*, *88*(12), 1574–1580.

Dobrez, D., Heinemann, A. W., Deutsch, A., Durkin, E. M., & Almagor, O. (2010). Impact of mental disorders on cost and reimbursement for patients in inpatient rehabilitation facilities. *Archives of Physical Medicine and Rehabilitation*, *91*(2), 184–188.

Dorstyn, D. S., Mathias, J. L., & Denson, L. A. (2010). Psychological intervention during spinal rehabilitation: A preliminary study. *Spinal Cord*, *48*(10), 756–761.

Dowlati, Y., Herrmann, N., Swardfager, W. L., Reim, E. K., & Lanctot, K. L. (2010). Efficacy and tolerability of antidepressants for treatment of depression in coronary artery disease: A meta-analysis. *The Canadian Journal of Psychiatry*, *55*(2), 91–99.

Elliott, T. R., Godshall, F. J., Herrick, S. M., Witty, T. E., & Spruell, M. (1991). Problem-solving appraisal and psychological adjustment following spinal cord injury. *Cognitive Therapy and Research*, *15*(5), 387–398.

Fann, J. R., Bombardier, C. H., Richards, J. S., Tate, D. G., Wilson, C. S., Temkin, N., & PRISMS Investigators. (2011). Depression after spinal cord injury: Comorbidities, mental health service use, and adequacy of treatment. *Archives of Physical Medicine and Rehabilitation*, *92*(3), 352–360.

Fann, J. R., Bombardier, C. H., Richards, J. S., Wilson, C. S., Heinemann, A. W., Warren, A. M., … Tate, D. G. (2015). Venlafaxine extended-release for depression following spinal cord injury: A randomized clinical trial. *JAMA Psychiatry*, *72*(3), 247–258.

Fann, J. R., Hart, T., & Schomer, K. G. (2009). Treatment for depression after traumatic brain injury: A systematic review. *Journal of Neurotrauma*, *26*(12), 2383–2402.

Fernie, B. A., Kollmann, J., & Brown, R. G. (2015). Cognitive behavioural interventions for depression in chronic neurological conditions: A systematic review. *Journal of Psychosomatic Research*, *78*(5), 411–419.

First, M. B., Williams, J. B., Karg, R. S., & Spitzer, R. L. (2016). *SCID-5-CV: Structured clinical interview for DSM-5 disorders, clinician version*. American Psychiatric Association Publishing.

Fregni, F., Boggio, P. S., Nitsche, M. A., Rigonatti, S. P., & Pascual-Leone, A. (2006). Cognitive effects of repeated sessions of transcranial direct current stimulation in patients with depression. *Depression and Anxiety*, *23*(8), 482–484.

Gelenberg, A. J., Freeman, M. P., Markowitz, J. C., Rosenbaum, J. F., Thase, M. E., Trivedi, M. H., … Silbersweig, D. A. (2010). Practice guideline for the treatment of patients with major depressive disorder third edition. *The American Journal of Psychiatry*, *167*(10), 1.

Ginis, K. A. M., Latimer, A. E., McKechnie, K., Ditor, D. S., McCartney, N., Hicks, A. L., … Craven, B. C. (2003). Using exercise to enhance subjective well-being among people with spinal cord injury: The mediating influences of stress and pain. *Rehabilitation Psychology*, *48*(3), 157.

Hartoonian, N., Hoffman, J. M., Kalpakjian, C. Z., Taylor, H. B., Krause, J. K., & Bombardier, C. H. (2014). Evaluating a spinal cord injury–specific model of depression and quality of life. *Archives of Physical Medicine and Rehabilitation*, *95*(3), 455–465.

Hellmann-Regen, J., Piber, D., Hinkelmann, K., Gold, S. M., Heesen, C., Spitzer, C., … Otte, C. (2013). Depressive syndromes in neurological disorders. *European Archives of Psychiatry and Clinical Neuroscience*, *263*(2), 123–136.

Hicks, A. L., Martin, K. A., Ditor, D. S., Latimer, A. E., Craven, C., Bugaresti, J., & McCartney, N. (2003). Long-term exercise training in persons with spinal cord injury: Effects on strength, arm ergometry performance and psychological well-being. *Spinal Cord*, *41*(1), 34–43.

Hohmann, G. W. (1975). Psychological aspects of treatment and rehabilitation of the spinal cord injured person. *Clinical Orthopaedics and Related Research*, *112*, 81–88.

Judd, F. K., Stone, J., Webber, J. E., Brown, D. J., & Burrows, G. D. (1989). Depression following spinal cord injury: A prospective in-patient study. *The British Journal of Psychiatry*, *154*(5), 668–671.

Kahan, J. S., Mitchell, J. M., Kemp, B. J., & Adkins, R. H. (2006). The results of a 6-month treatment for depression on symptoms, life satisfaction, and community activities among individuals aging with a disability. *Rehabilitation Psychology*, *51*(1), 13.

Kalpakjian, C. Z., & Albright, K. J. (2006). An examination of depression through the lens of spinal cord injury: Comparative prevalence rates and severity in women and men. *Women's Health Issues*, *16*(6), 380–388.

Karnes, J. H., Rettie, A. E., Somogyi, A. A., Huddart, R., Fohner, A. E., Formea, C. M., … Callaghan, J. T. (2020). Clinical pharmacogenetics implementation consortium (CPIC) guideline for CYP2C9 and HLA-B genotypes and phenytoin dosing: 2020 update. *Clinical Pharmacology and Therapeutics*.

Kemp, B. J., Kahan, J. S., Krause, J. S., Adkins, R. H., & Nava, G. N. (2004). Treatment of major depression in individuals with spinal cord injury. *The Journal of Spinal Cord Medicine*, *27*(1), 22–28.

Kemp, B., Krause, J. S., & Adkins, R. (1999). Depression among African Americans, Latinos, and Caucasians with spinal cord injury: A exploratory study. *Rehabilitation Psychology*, *44*(3), 235.

Kennedy, S. H., Lam, R. W., McIntyre, R. S., Tourjman, S. V., Bhat, V., Blier, P., … CANMAT Depression Work Group. (2016). Canadian network for mood and anxiety treatments (CANMAT) 2016 clinical guidelines for the management of adults with major depressive disorder: Section 3. Pharmacological treatments. *The Canadian Journal of Psychiatry*, *61*(9), 540–560.

Kennedy, F. (2008). The use of formulation in inpatient settings. In *Cognitive behaviour therapy for acute inpatient mental health units* (pp. 57–80). Routledge.

Kennedy, P., & Rogers, B. A. (2000). Anxiety and depression after spinal cord injury: A longitudinal analysis. *Archives of Physical Medicine and Rehabilitation*, *81*(7), 932–937.

Khazaeipour, Z., Ahmadipour, E., Rahimi-Movaghar, V., Ahmadipour, F., Vaccaro, A. R., & Babakhani, B. (2017). Association of pain, social support and socioeconomic indicators in patients with spinal cord injury in Iran. *Spinal Cord*, *55*(2), 180–186.

Khazaeipour, Z., Taheri-Otaghsara, S. M., & Naghdi, M. (2015). Depression following spinal cord njury: Its relationship to demographic and socioeconomic indicators. *Topics in Spinal Cord Injury Rehabilitation*, *21*(2), 149–155.

Krause, J. S., Saunders, L. L., Reed, K. S., Coker, J., Zhai, Y., & Johnson, E. (2009). Comparison of the Patient Health Questionnaire and the Older Adult Health and Mood Questionnaire for self-reported depressive symptoms after spinal cord injury. *Rehabilitation Psychology*, *54*(4), 440–448. https://doi.org/10.1037/a0017402.

Krueger, H., Noonan, V. K., Williams, D., Trenaman, L. M., & Rivers, C. S. (2013). The influence of depression on physical complications in spinal cord injury: Behavioral mechanisms and health-care implications. *Spinal Cord*, *51*(4), 260–266.

Latimer, A. E., Ginis, K. A. M., Hicks, A. L., & McCartney, N. (2004). An examination of the mechanisms of exercise-induced change in psychological well-being among people with spinal cord injury. *Journal of Rehabilitation Research and Development*, *41*(5).

Liu, Y., Ho, R. C. M., & Mak, A. (2012). Interleukin (IL)-6, tumour necrosis factor alpha (TNF-α) and soluble interleukin-2 receptors (sIL-2R) are elevated in patients with major depressive disorder: A meta-analysis and meta-regression. *Journal of Affective Disorders*, *139*(3), 230–239.

Maldonado-Bouchard, S., Peters, K., Woller, S. A., Madahian, B., Faghihi, U., Patel, S., … Hook, M. A. (2016). Inflammation is increased with anxiety- and depression-like signs in a rat model of spinal cord injury. *Brain, Behavior, and Immunity*, *51*, 176–195.

Mehta, S., Hadjistavropoulos, H., Nugent, M., Karin, E., Titov, N., & Dear, B. F. (2020). Guided internet-delivered cognitive-behaviour therapy for persons with spinal cord injury: A feasibility trial. *Spinal Cord*, *58*(5), 544–552.

Mehta, S., Orenczuk, S., Hansen, K. T., Aubut, J. A. L., Hitzig, S. L., Legassic, M., & Teasell, R. W. (2011). An evidence-based review of the effectiveness of cognitive behavioral therapy for psychosocial issues post-spinal cord injury. *Rehabilitation Psychology*, *56*(1), 15.

Migliorini, C., Sinclair, A., Brown, D., Tonge, B., & New, P. (2016). A randomised control trial of an internet-based cognitive behaviour treatment for mood disorder in adults with chronic spinal cord injury. *Spinal Cord*, *54*(9), 695–701.

Paralyzed Veterans of America. (2020). *Management of mental health disorders, substance use disorders, and suicide in adults with spinal cord injury*. Retrieved from https://www.iscos.org.uk/uploads/sitefiles/Newsletters/PVA_CPG_Mental_Health_and_substa.pdf.

Perry, K. N., Nicholas, M. K., & Middleton, J. W. (2010). Comparison of a pain management program with usual care in a pain management center for people with spinal cord injury-related chronic pain. *The Clinical Journal of Pain*, *26*(3), 206–216.

Pettersson, A., Boström, K. B., Gustavsson, P., & Ekselius, L. (2015). Which instruments to support diagnosis of depression have sufficient accuracy? A systematic review. *Nordic Journal of Psychiatry*, *69*(7), 497–508.

Prisnie, J. C., Sajobi, T. T., Wang, M., Patten, S. B., Fiest, K. M., Bulloch, A. G., … Jette, N. (2018). Effects of depression and anxiety on quality of life in five common neurological disorders. *General Hospital Psychiatry*, *52*, 58–63.

Radnitz, C. L., Broderick, C. P., Perez-Strumolo, L., Tirch, D. D., Festa, J., Schlein, I. S., et al. (1996). The prevalence of psychiatric disorders in veterans with spinal cord injury: A controlled comparison. *Journal of Nervous and Mental Disease*, *184*(7), 431–433. https://doi.org/10.1097/00005053-199607000-00006.

Richards, J. S., Bombardier, C. H., Wilson, C. S., Chiodo, A. E., Brooks, L., Tate, D. G., … Fann, J. R. (2015). Efficacy of venlafaxine XR for the treatment of pain in patients with spinal cord injury and major depression: A randomized, controlled trial. *Archives of Physical Medicine and Rehabilitation*, *96*(4), 680–689.

Richelson, E. (2001). Pharmacology of antidepressants. In *Vol. 76*. *Mayo Clinic Proceedings* (pp. 511–527). Elsevier. No. 5.

Rintala, D. H., Holmes, S. A., Courtade, D., Fiess, R. N., Tastard, L. V., & Loubser, P. G. (2007). Comparison of the effectiveness of amitriptyline and gabapentin on chronic neuropathic pain in persons with spinal cord injury. *Archives of Physical Medicine and Rehabilitation*, *88*(12), 1547–1560.

Salinas, F. A., Lugo, L. H., & García, H. I. (2012). Efficacy of early treatment with carbamazepine in prevention of neuropathic pain in patients with spinal cord injury. *American Journal of Physical Medicine & Rehabilitation*, *91*(12), 1020–1027.

Sheehan, D. V., Lecrubier, Y., Sheehan, K. H., Amorim, P., Janavs, J., Weiller, E., … Dunbar, G. C. (1998). The Mini-international neuropsychiatric interview (MINI): The development and validation of a structured diagnostic psychiatric interview for DSM-IV and ICD-10. *The Journal of Clinical Psychiatry*, *59*(Suppl 20), 22–33.

Stolp-Smith, K. A., & Wainberg, M. C. (1999). Antidepressant exacerbation of spasticity. *Archives of Physical Medicine and Rehabilitation*, *80*(3), 339–342.

Swett, S. E., & Richardson, W. S. (2004). Cognitive-behavioral therapy. In F. Chan, N. L. Berven, & K. R. Thomas (Eds.), *Counseling theories and techniques for rehabilitation health professionals* (pp. 159–176). NewYork, NY: Springer Publishing Co.

Tan, S. K., Hartung, H., Sharp, T., & Temel, Y. (2011). Serotonin-dependent depression in Parkinson's disease: A role for the subthalamic nucleus? *Neuropharmacology*, *61*(3), 387–399.

Tate, D. G., Forchheimer, M., Maynard, F., Davidoff, G., & Dijkers, M. (1993). Comparing two measures of depression in spinal cord injury. *Rehabilitation Psychology*, *38*(1), 53–61. https://doi.org/10.1037/h0080287.

Tate, D. G., & Pledger, C. (2003). An integrative conceptual framework of disability: New directions for research. *American Psychologist*, *58*(4), 289.

Titman, R., Liang, J., & Craven, B. C. (2019). Diagnostic accuracy and feasibility of depression screening in spinal cord injury: A systematic review. *The Journal of Spinal Cord Medicine*, *42*(sup1), 99–107.

Tolin, D. F. (2010). Is cognitive–behavioral therapy more effective than other therapies?: A meta-analytic review. *Clinical Psychology Review*, *30*(6), 710–720.

Williams, R., & Murray, A. (2015). Prevalence of depression after spinal cord injury: A meta-analysis. *Archives of Physical Medicine and Rehabilitation*, *96*(1), 133–140.

Williams, T. L., Smith, B., & Papathomas, A. (2014). The barriers, benefits and facilitators of leisure time physical activity among people with spinal cord injury: A meta-synthesis of qualitative findings. *Health Psychology Review*, *8*(4), 404–425.

Wu, J., Zhao, Z., Sabirzhanov, B., Stoica, B. A., Kumar, A., Luo, T., … Faden, A. I. (2014). Spinal cord injury causes brain inflammation associated with cognitive and affective changes: Role of cell cycle pathways. *Journal of Neuroscience*, *34*(33), 10989–11006.

Zürcher, C., Tough, H., Fekete, C., & SwiSCI Study Group. (2019). Mental health in individuals with spinal cord injury: The role of socioeconomic conditions and social relationships. *PLoS One*, *14*(2), e0206069.

Chapter 42

Self-harm behaviors in patients with spinal cord injuries: From non-adherence to suicide

Maggi A. Budd[a,b,e], Herb Ames[c,d], and John C. Bradley[b,f,g]

[a]*Department of Spinal Cord Injury/Disorders, Boston VA Healthcare System, Boston, MA, United States,* [b]*Department of Psychiatry, Harvard Medical School, Boston, MA, United States,* [c]*Department of Spinal Cord Injury, Michael E. DeBakey VA Medical Center, Houston, TX, United States,* [d]*Department of Physical Medicine and Rehabilitation, Baylor College of Medicine, Houston, TX, United States,* [e]*Department of Bioethics, Harvard Medical School, Boston, United States,* [f]*Department of Mental Health, VA Boston Healthcare System, Brockton, MA, United States,* [g]*Department of Psychiatry, Boston University, Boston, MA, United States*

List of abbreviations

SCI spinal cord injury

Introduction

Harming oneself seems unnatural. Yet, many people knowingly engage in behaviors that are not in their best interest and, if prolonged, some behaviors can result in serious harm to oneself, hence "self-harm." In medical settings, self-harm can emerge in an assortment of behaviors and range from mismanaging medications, failing to attend medical or mental health appointments, sleeping too much or little—to frank self-neglect, complete social isolation, overuse of pain medication, deliberate alcohol and illicit drug abuse, candid "giving up", decisions to medically hasten death, or suicide attempts.

Spinal cord injury (SCI) forces people to suddenly learn how to do everything in their lives in new ways, an abundantly stressful event for most people. Many factors affect a person's response after SCI—physical health, economic and social factors, comorbid psychological and substance use disorders, and more (Craig, Tran, & Middleton, 2017; Kishi, Robinson, & Kosier, 2001; Martz, Livneh, Priebe, Wuermser, & Ottomanelli, 2005; Pollard & Kennedy, 2007). Olkin (1993) states these influences on adaptation persist with a "...response curve, while steeper at first, does not ever level off at some mythical stage of adjustment and acceptance" (p. 15). Rather, "adaptation" (preferred term over adjustment) to SCI is a continuous process in a non-linear fashion (Craig et al., 2017; Pollard & Kennedy, 2007) and is intensely personal.

Physicians, nurses, psychologists, social workers, and other health care providers (referred to as "clinicians" going forward) have an ethical obligation to benefit the patient (beneficence), avoid or minimize harm, and respect patients' values and preferences (Beauchamp & Childress, 2009; Torke & Sachs, 2008). Under most contexts, patients are presumed to want help to improve their health and well-being. However, sometimes patients decline or ignore vital medical interventions and act in ways that can or do cause themselves harm, interfering with the clinician's obligation for beneficence. This can be a source of pointed frustration, realistic concern, and/or moral distress for clinicians (Torke & Sachs, 2008).

Self-harm behaviors and suicidal behaviors

"Self-harm" is usually automatically associated with suicide and an intent to die. However, there are additional forms of self-harm that have been less well described but potentially as lethal (Bostwick, 2015; Bostwick & Cohen, 2009). Some self-harm behaviors are conscious choices or expressions of autonomous self-determination. Some behaviors are to seek attention or help, to escape a challenging situation, or to express "just how bad" things are for them. Whatever the underlying source, self-harm behaviors can be symptomatic of treatable conflicts and problems such a depression, misinformation, hampered mastery and autonomy, or social or financial concerns.

Cellular, Molecular, Physiological, and Behavioral Aspects of Spinal Cord Injury. https://doi.org/10.1016/B978-0-12-822427-4.00042-3

A category of patients theoretically mislabeled as suicidal can be non-adherent patients. World Health Organization defines adherence as "the extent to which a person's behavior—taking medication, following a diet, and/or executing lifestyle changes corresponds with agreed recommendations from a health care provider" (World Health Organization, 2003). Medications, rehabilitation therapies, and self-care behaviors will not work if patients do not follow prescribed plans. In addition to clinicians' frustrations when patients act directly against their best interest, persistent self-harm behaviors pose ethical and practical challenges for clinicians because not intervening would be violating our duties to avoid or mitigate harm when possible (Torke & Sachs, 2008).

When chronic or severe enough, the fears of clinicians and loved ones are realistic: a patient's behaviors can produce serious self-harm or even death. Bostwick and Cohen (2009) catalogued two accounts that can help distinguish true suicide from other forms of self-harm by exploring the presence of (1) a death wish or intent to die, and, (2) complicity or collaboration with clinicians and/or family members. Particularly applicable to medical settings, these two criteria help disentangle authentic suicide from non-adherence, physician-assisted suicide, and palliative/hospice care (Bostwick, 2015). A true suicide has both (1) an intent to die, and, (2) is not collaborative with clinicians or loved ones. Many self-harm behaviors observed in medical settings, such as non-adherence, are not directed at death, and clinicians and loved ones are not complicit with patient's poor choices. Other "self-harm" behaviors such as declining life-sustaining interventions may be motivated toward death yet do not have the secrecy and exclusion of clinicians as in suicide.

Chapter clarifications

This chapter speaks to clinical experiences about a selection of self-harm behaviors observed in SCI populations. Building on the work of Macleod (Macleod, 1988), behaviors are conceptualized as a sign of core conflicts signifying clinical needs that SCI clinicians can detect and with which they can empathize, and consequently treat. As such, self-harm behaviors can be symptomatic of treatable circumstances such as problems with adapting to a disability, depression, existential drives, medical mistrust, or authentic patient preferences. The goal is to expand a clinician's understanding and repertoire of responses when patients with SCI make choices that may result in self-harm, including suicidal behaviors.

The term "self-harm" is used broadly, with an assumed range of severity and types of behavior, and without suggestion of causation nor attention to idiosyncratic situations or individual attributes. Individuals with cognitive problems, delirium, or psychopathologies may require different approaches. There is no one-size-fits-all clinical response, and numerous factors influence a patient's self-care decisions (e.g., medical status, state of well-being, time since injury, pain, coping abilities, financial security, social support) that are too vast to consider here. In addition, suggested responses for each self-harm behavior are not exhaustive and will overlap (e.g., strength-focused, patient-centered reactions apply to all). Lastly, the ideas are offered with the utmost respect and acknowledgment that most people adapt well to SCI, without psychopathology, and with amazing stability over time (Pollard & Kennedy, 2007). Nevertheless, adopting "different" ways to conceptualize self-harm behaviors observed in SCI clinics can enhance appreciation and stimulate an array of approaches to hopefully halt further serious or devasting harm.

Regarding language in this chapter, the term "adherence" is preferred over "compliance" because adherence retains focus on the patient as an active collaborator rather than "obeying" the clinician (Merbitz, 2016). The term "adapting" is preferred over "adjustment" based on the literature from people with disabilities (Olkin, 1993).

Self-harm behaviors as presenting symptoms

Mark Twain said, "Actions speak louder than words." Self-harm *behavior* is a multi-dimensional phenomenon symptomatic of countless contributors. Five select examples underlying self-harm behaviors to enhance understanding and inform clinician responses are described: 1. Problems related to adapting to disability; 2. Major depression; 3. An existential wish to not live; 4. Medical mistrust; and, 5. A conscientious lifestyle choice. General information from the literature relating each issue to SCI will be presented along with various core tensions, behavioral signs, and ideas for clinician responses. Comments applying and contrasting Bostwick's (2015) criteria for true suicide (intention to die and caregiver corroboration) versus other self-harm behaviors are incorporated. Table 1 summarizes the five self-harm behaviors, possible etiologies, and clinical responses.

Adapting to disability

Numerous factors can disturb adaptation to SCI that emanate from self-harm behaviors. Frequently cited obstacles to adaption are premorbid mental health problems, and substance abuse (Craig et al., 2017; Kolakowsky-Hayner et al.,

TABLE 1 Self-harm Behaviors: Underlying causes and conflicts, presenting symptoms and interventions.

Possible etiology	Core conflicts/tensions	Behaviors	Clinician responses
Adapting to disability	Loss of autonomy and individuation; external locus of control; limited coping resources for the demands	Anger, indifference, non-adherence, or inconsistent compliance	Promote mastery of new and existing life-valued activities; bolster coping skills; instruct how to direct one's care; supportive therapy; peer support; meaning-making
Depression	Feelings of loss; Feeling hopeless; Inappropriate guilt; Suicidal ideation	Agitation; negative withdrawal; "giving up"; suicidal talk	Anti-depressants; evidence-based psychotherapies; supportive therapy; process grief; increase or expand life valued activities; practice gratitude; identify future goals of value
Existential/ palliative or hospice	Informed choice for personal quality of life; preference of quality over quantity of life; self-determination; Skewed perception of living with disability	Decline rehabilitation services; decline medical interventions; act personally resolved with their decision; appear righteous	Confirm informed consent; comprehensively address physical, emotional, and social needs; consult ethics committees; family meetings; peer support; therapeutic trial period
Medical mistrust	Feel compelled to be guarded/overly cautious; defensive; fear someone will harm or impede their autonomy	Act in anger; withdrawal; dismiss recommendations; negative expectations; miss appointments; question motives	Validate mistrust; validate prior experiences; appreciate cultural factors in context; educate with facts; offer options; bolster therapeutic alliance
Patient preferences	Not wanting to change; Angry with the interference of SCI on quality of life; Latent fears or misperceived expectations about living with SCI	Same pre-injury behavior; non-standard or inconsistent healthcare adherence; resign themselves to a victim/sick role.	Evaluate capacity; continue to offer interventions while respecting autonomous choices; persist with healthy recommendations and excellent care; peer support if they agree; collaborative care plans; accentuate values for life and living

Examples of core tensions, behavioral signs, and ideas for clinician responses.

2002; Martz et al., 2005). Poor self-care happens when SCI challenges the availability of coping resources for people with prior mental health issues. Excessive stress impacts mental health symptoms, in general, and especially for those predisposed (Russell, Turner, & Joiner, 2009).

Substance use is repeatedly associated with SCI. It has been reported that at the time of acquiring their SCI, 31%–50% of individuals were abusing alcohol, 16%–33% were using illegal drugs, and 26% were using a combination of each (Tétrault & Courtois, 2014). A considerable number of people with SCI continue to use alcohol or drugs after SCI (Kolakowsky-Hayner et al., 2002), placing themselves at risk for secondary problems with physical and mental health, managing pain, and social and other life challenges (Skidmore & Budd, 2016). Cumulatively, untreated mental health and substance abuse issues complicate self-management, and mental and physical health consequences can be severe if mismanaged (Furlanetto & Stefanello, 2011; Kishi et al., 2001). Also impacting adaptation to SCI are reduced social participation and autonomy, fatigue, loss of employment, secondary infections, pain (Craig et al., 2017; Martz et al., 2005) self-efficacy, and social support (Bhattarai, Jin, Smedema, Cadel, & Baniya, 2021; Kishi et al., 2001; Russell et al., 2009).

Underlying poor adaptation to SCI can be conflicts with loss of autonomy and individuation leaving some feeling a "helpless dependence." Patients may withdraw, acting in anger or indifference. Nonadherence behaviors give people an element of control and autonomy over their situations (MacLeod, 1998). Self-harm behaviors stemming from problems with adapting to SCI may be a form of suicide if the intention is to "give up" on living while not being forthcoming with clinicians and loved ones about their intention (e.g., could constitute a passive form of suicide).

Treatments to facilitate adaptation to SCI include promoting a sense of mastery by encouraging personal responsibility, cultivating self-efficacy (Bhattarai et al., 2021), and collaboration with planning and making decisions for the future

(Russell et al., 2009). Specifically, individuals with SCI desire more focus on well-being interventions, financial and other social worries, assistance with bowel and bladder concerns, increased physical activity, occupation guidance, and peer and family support (Simpson, Villeneuve, & Clifton, 2020). Attending to these while validating the scope and intensity of the challenges, and teaching how acceptance of disability changes over time, can proffer empathy and hope (Craig et al., 2017; Kripke, 2017; Ubel, Loewenstein, Schwarz, & Smith, 2005).

Depression

Underlying depression can manifest a range of behaviors for people with SCI. Pollard and Kennedy's (2007) 10-year longitudinal study determined about 30% of people with SCI are depressed after SCI, and these rates of depression are consistent over time in this population (Craig et al., 2017). Coping at 12-weeks post-injury was most predictive of lasting depression, indicating a need for early detection and intervention (Pollard & Kennedy, 2007).

Depression has been associated with self-harm behaviors ranging from poor self-care to severe self-neglect, and suicidal ideation to suicidal behaviors (Bostwick, 2015; Kishi et al., 2001). A history of depression increases the risk of depression after SCI. People with depression exhibit pervasively low mood and energy, disturbed sleep and appetite, suicidal ideation, and expressed negativism, all of which consume personal resources. Caring for oneself with SCI requires a certain amount of energy and diligence. Even minor inattention can worsen situations. For example, ignoring pressure relief or improper cushion inflation for 1 day can result in pressure ulcers affecting mobility, impacting participation [that can have further ramifications] and future ulcers (Sprigle, McNair, & Sonenblum, 2020). In general, prolonged inattention to self-care contributes to feelings of hopelessness. Self-harm behaviors driven by clinical depression demands immediate attention and comprehensive assessment to rule out suicidal ideation, intention, and plans. Unfortunately, an expectation of depression following SCI may delay appropriate diagnosis and needed treatment (Craig et al., 2017).

Accurate depression diagnosis and treatment can be critical (Kishi et al., 2001; Lewis, Anderson, & Feuchtinger, 2014). Fortunately, combined anti-depressant and psychotherapy interventions frequently resolve depression and the patient resumes more healthful behaviors (Kishi et al., 2001). See the chapter on Depression in this book for a comprehensive review of depression and SCI.

Existential (palliative/hospice)

Individuals with an existential self-harm/neglect are less common (Macleod, 1988). Self-harm behaviors with an underlying existential element may encompass requests like ventilation removal, or voluntary cessation of eating and drinking. These individuals express a forthright desire for death despite the absence of clinical depression. If their situation were different, these individuals would not be contemplating life-ending activities. Applying Bostwick and Cohen's (2009) criteria, this is not suicide. These patient's meaning to hasten death is based on intense circumstances and correlated life values. Unlike the devastation following a death by suicide, the patient's survivors usually proactively collaborate with these thoughts and accept the arrangement. These decisions carry ethical and moral conflicts. A thorough assessment is warranted before deciding to assist or oppose death-hastening behaviors. Notably, existential choices are not "passive suicide" but a non-punitive selection of autonomy (Macleod, 1988) for how they want to live their lives, and perhaps die.

Ethically and legally, choices to decline treatment or stop an existing treatment are permissible if the person has decision-making capacity demonstrating an appreciation, understanding, and rational reasoning of the risks and benefits of their choices, even if it causes harm, or result in death (Beauchamp & Childress, 2009).

Motivations to medically hasten death are often when the patient wants to exercise self-determination and elect death over a life that they perceive as more burdens and less quality than they want to experience. Supporters are usually proponents of physician-assisted suicide and "rational suicide." Commonly, like depression, these plans are often transient and ameliorate over time and many people ultimately elect to live (Kishi et al., 2001; Pollard & Kennedy, 2007).

Suggested treatments for these patients involve addressing physical needs (e.g., pain control), maximizing social and clinical support, and discussing the impact of decisions with loved ones. Consult hospital ethics committees and conduct team-family meetings for thoroughness. Repeat offers of recommended care with non-judgmental acceptance of their choice (i.e., giving the patient the right to say "no" may sway them to say "yes"). Encourage a trial period of living with disability and educate that disability is not a disease (Kripke, 2017), and describe the occurrence of the "Disability Paradox"—i.e., most people with SCI experience a good quality of life; until they were disabled, they too believed it impossible (Ubel et al., 2005).

Medical mistrust

Patient trust is central to therapeutic relationships and treatment adherence to materialize best outcomes (Suchman & Matthews, 1988); behaviors in response to medical mistrust can spawn unintended self-harm. Previous medical mistakes or bad experiences may have disenfranchised the patient for years before SCI. Now with SCI, medical needs are inescapable, and there are no options but reliance on medical experts and dependence upon others for assistance. This may be especially applicable for patients in marginalized populations or certain cultural groups (Gonzalez et al., 2018). Disbelieving their SCI and prognosis (Craig et al., 2017), reading misinformation on the internet, and clinician attitudes (VanPuymbrouck, Friedman, & Feldner, 2020) also can generate mistrust, affect the therapeutic relationship, and influence clinical decision-making (Baron & Berinsky, 2019; Suchman & Matthews, 1988). Self-harm behaviors emitted from medical mistrust are not suicidal because the person prefers to live, but they are guarded and reluctant to fully trust medical recommendations. To them, they are not the ones precipitating the self-harm.

In any group, prior unpleasant encounters or medical mistakes may make one delay care or affect medical decision making and selecting care (Eipper-Mains, Kao, Gitlin, & Peteet, 2019). Medical distrust is a protective defense with pathways to mend (Baron & Berinsky, 2019).

If medical mistrust is suspected, validate the apprehensions and attempt to understand the patient's decision-making process in the context of the specific medical need. Maximize control whenever possible while furthering patient understanding and appreciation of medical facts. Explain medical details about interaction effects with concerning behaviors. Practice shared-decision making and offer all alternative solutions, even if a compromise is less than ideal to both parties (Suchman & Matthews, 1988).

Patient preferences

Some behavioral choices can deviate far from medical recommendations and have the potential to cause serious self-harm. Some patients choose to mix alcohol with certain medications or refuse a procedure or treatment that would help them, for example. Patient-centered care is grounded on a balance of deep respect for the uniqueness of the person and empirically scientific generalizations, balanced in the context of that person and that situation (Epstein & Street, 2011). Ethically and legally, patients with decision-making capacity have the right to accept or refuse offered treatment interventions, including in circumstances in which the decision to forgo treatment may be dangerous or even result in death (Beauchamp & Childress, 2009). Decisions to decline care are supported ethically under the strongly held principle of patient autonomy. Patients have the right to make unhealthy choices.

However, honoring patient preferences occurs at the same time clinicians honor the integrity of their profession. Patients do not always know what is best, and our duty is to offer specialized services along with benefits and risks for choosing or declining. For example, a drug may be what a patient wants but not what they need. We welcome their participation but can only offer treatments fitting within professional standards of care, which they can decline or accept. The goal is a partnership, solidarity, and collaboration with both parties having autonomous roles.

Electing harmful behaviors over healthier options may not have suicidal intentions, but rather reflect an unwillingness to change a pre-injury lifestyle. For them, quality of life has personal benefits (pleasures congruent with lifelong values) that outweigh prospects for consequential self-harm. For example, a young man who designed his life as a professional gambler traveling across the country becomes paraplegic. Doctors warn against this lifestyle due to a host of pre-existing problems combined with SCI. The man continues life-valued activities despite risks and the likelihood of additive harm. This can be distinguished from suicide because he has no intention to die, and he likely shares his preferences with clinicians and loved ones.

Clinical responses to scenarios of behaviors precipitated by patient preferences are shared with approaches when existential motivations underly self-harm (e.g., confirm decision-making capacity; respect autonomy; persist with healthy recommendations) along with peer support and concentration on the hierarchy of the person's values to aid in decision-making.

Suicide behaviors and SCI

While the majority adapt well, individuals with SCI are at increased risk for suicide when compared with the general population. Suicide is the 7th leading cause of death in traumatic SCI, befalling 4%–11% of people with SCI (Kennedy & Garmon-Jones, 2017). Prominently, 91% of suicides happen within the first 10 years after injury then recedes the years following (Savic et al., 2018). The risk for suicide is more pronounced when SCI was caused by a suicide attempt

(Kennedy & Garmon-Jones, 2017) and when there were prior attempts (Furlanetto & Stefanello, 2011). The means of committing suicide depends upon accessibility and the person's physical abilities. For example, firearms are rather accessible in the US and used more frequently than in Europe, where drug poisoning/overdose (Savic et al., 2018) was a common method for suicide for people with SCI. Associations of the level of injury and suicide are too mixed to be conclusive (Kennedy & Garmon-Jones, 2017).

Risk mitigation

Warning signs for suicide

Consider as suicidal all self-harm behaviors that (a) have an intention for dying, and (b) are not corroborative with caregivers/clinicians (Bostwick, 2015). Since secretive in nature, we must look for other signs of suicidal self-harm. Changes in behavior, thoughts, or emotions can be warning signs. Look for medication non-adherence, increased substance use, disengagement from social contacts, increased verbal or physical aggression, mood swings, and preparatory behaviors (e.g., selling belongings, researching ways to die by suicide, purchasing firearms, hoarding pills). The types of thoughts and the way the person experiences thoughts can also be warning signs. For example, concerning thoughts include passive suicidal thoughts about preferring to be dead, desiring peace or a sense of control, experiencing many thoughts rapidly, feeling overwhelmed, or of their "mind going blank." Emotional changes to watch for include shame, hopelessness, guilt, anger, sadness, anxiety, irritability. See Table 2.

What to assess

Psychological, social, and medical risk factors should be assessed along with protective factors (see Fig. 1). Psychological risk factors include current suicidal ideation, prior suicide attempts, mood or substance abuse disorders, personality disorders, hopelessness, and prior psychiatric hospitalization (Fuller-Thomson, Tulipano, & Song, 2012; Kennedy & Garmon-Jones,

TABLE 2 Suicidal self-harm behaviors: underlying causes and conflicts, presenting symptoms and interventions.

Possible etiology	Core conflict	Behaviors	Interventions
SCI is perceived as a fate worse than death; Financial anxieties; Residential options	Existential conflict of life not being worth living; perceived burdensomeness; catastrophizing pain experience; desiring peace or sense of control; shame; hopelessness; guilt; anger; sadness; anxiety; financial and social concerns	Changes in behavior—disengagement from others, extreme irritability, or aggression; Hoarding guns or medications; selling or giving away belongings; mood swings; increased substance use; researching ways to die; suicidal ideation, statements, or attempt	Mental health evaluation or hospitalization; treat any mental health conditions; Safety planning; reduce all lethal means—lock or give away firearms; restrict medication access; rally support system; all other self-harm interventions can apply; focus on reasons for living

Examples of core tensions, behavioral signs, and ideas for clinician responses.

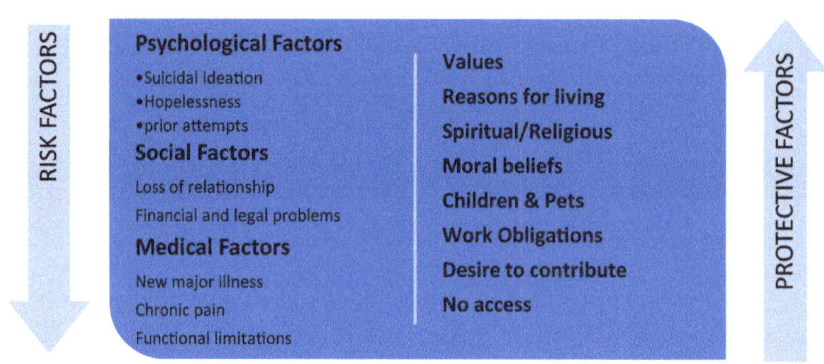

FIG. 1 Risk and protective factors for suicide. As protective factors increase, risk factors decrease. As Protective Factors increase, Risk Factors decrease.

2017). Social risk factors include stressful life events (e.g., loss of a relationship; illness of self or family member), financial problems, legal problems, lack of social support). A new diagnosis of a major illness is a suicide risk factor, along with chronic pain, worsening medical illness, or functional limitation.

Assessment of biopsychosocial risk factors for suicide is beneficial for risk mitigation. For example, attention to intrapsychic factors such as mental health symptomatology can reduce emotional distress while at the same time enhance a sense of connectedness with a helping professional, building a channel for reaching out in times of crisis. Reflecting on core values advances relatedness to life- as opposed to death-orientation. Simple authentic interest in listening about someone's psychic pain can decrease the experience of loneliness and foster hope.

Social isolation

Social support is key to providing patients with SCI with appropriate emotional, educational, and instrumental (e.g., caregiver) resources to meet physical, functional, and emotional needs. One area often associated with elevated suicide risk is feeling like a burden (Joiner, 2005). Whether perceived burdensomeness reflects a primary intrapsychic struggle or an actual interpersonal conflict, demands exploration. Reconciling being dependent on others as a previously independent adult is fundamental to SCI adaptation, and its intensity and persistence vary among individuals based on many factors. The "usefulness" or "value" that someone places on themself, if unemployed, warrants consideration in the context of their culture. Research has increasingly demonstrated the association of work and psychosocial health in SCI (Goetz, Ottomanelli, Barnett, Sutton, & Njoh, 2018).

Reduced social isolation's protection for suicidal risks may be evidenced indirectly in other studies. For example, employment's negative association with suicide is perhaps because of the increased socialization while working. Studies that suggested tetraplegics have reduced risk of suicide is because their reliance on other people for activities of daily living requires much human interaction (Kennedy & Garmon-Jones, 2017).

Addressing environmental and social factors cannot be overlooked. There is great significance for emotional functioning and psychosocial adjustment associated with the stability, or lack thereof, of accessible housing and transportation, caregiver arrangements, and access to valued leisure and recreational activities.

Reduce access to lethal means

Conversations about and direct action regarding lethal means are vital. As stated, the most common methods of suicide are gunshots and drugs. Firearm access is an independent suicide risk factor independent of age, gender, and mental health diagnosis (Studdert et al., 2020). Importantly, any action that increases the time from suicidal ideation to availability of means reduces the risk for completion. Examples include storing firearms in a locked safe, separating ammunition and firearm, removing firing pins, and giving the firearm safe key to a trusted person. Reduce medication as a lethal choice by limiting supplies, using pill bubbles and less toxic medicines, or holding them safely. The main point is to add time for reconsideration of suicide, and hopefully focus on better solutions.

Applications to other areas of neuroscience

Patients with other neurological disorders share similar psychological responses, and risks for self-harm as people with SCI – pre-injury personality, chronic pain, level of independence, use of alcohol and other substances, cognitive appraisals, sleep quality, perceived social supports, social conditions, and financial considerations—are all pertinent both at the time of diagnosis and throughout their lives. All underlying symptoms and ideas to manage self-harm behaviors discussed for SCI apply to other neurological populations. Many physical, mental, and social stressors are similar and also contextual.

The initial months following a neurological diagnosis pose the highest risk for suicide (Kishi et al., 2001), particularly for patients with multiple hospitalizations (Roy-Byrne, 2020), unmarried (Russell et al., 2009), Caucasian, and younger age (Fuller-Thomson et al., 2012; Furlanetto & Stefanello, 2011). Roy-Byrne (2020) recently determined suicidal risks were largest for amyotrophic lateral sclerosis and Huntington disease followed by multiple sclerosis, head injury, epilepsy, and stroke. Common risk factors for suicidal ideation in neurological patients are hopelessness, depression, social isolation (Kishi et al., 2001; Lewis et al., 2014), and functional limitations (Fuller-Thomson et al., 2012). However, like SCI, over time most people with a new onset of a neurological disorder do adapt after a period of shock and suicide risks decrease (Kripke, 2017). Depression, if present, is considered the most important factor in preventing self-harm and suicidal ideation and behaviors (Kishi et al., 2001). It is important to not assume clinical depression is a normal reaction and that depression does not need to be assessed and addressed.

Three of four top correlates for suicidal ideation in medical patients are modifiable—anxiety, depression, and poor social support (Furlanetto & Stefanello, 2011). The fourth factor, prior suicide attempts, is unmodifiable but with further exploration with the patient, discussion can yield valuable information for a personalized approach. Concentrated treatments can perhaps halt further or future suicidal ideation.

Identifying depression is critical. As stated, depression is a known and treatable, risk factor for suicide for people with neurological problems. Kishi et al. (2001) found 25% of people with an acute medical illness with depression had suicidal ideation, of these 7.3% had suicidal plans. Depressed people perceive events more negatively, and thus may perceive their medical condition as a stressor beyond their capacity to manage (Furlanetto & Stefanello, 2011; Russell et al., 2009). Studies show treatment of depression ameliorates related suicidal ideation in medical populations (Kishi et al., 2001).

Suicide mitigation safety planning for all populations

When a patient is at risk for suicide or self-harm behaviors but is not considered a level of severity for involuntary hospitalization, a Safety Plan can be lifesaving. Safety plans help people identify risks and help individuals develop personalized plans that describe warning signs, internal coping strategies, external factors to provide healthy distractions, people to contact (friends, family, professionals), and ways to make the environment safe. See Fig. 2 for content inclusion and examples.

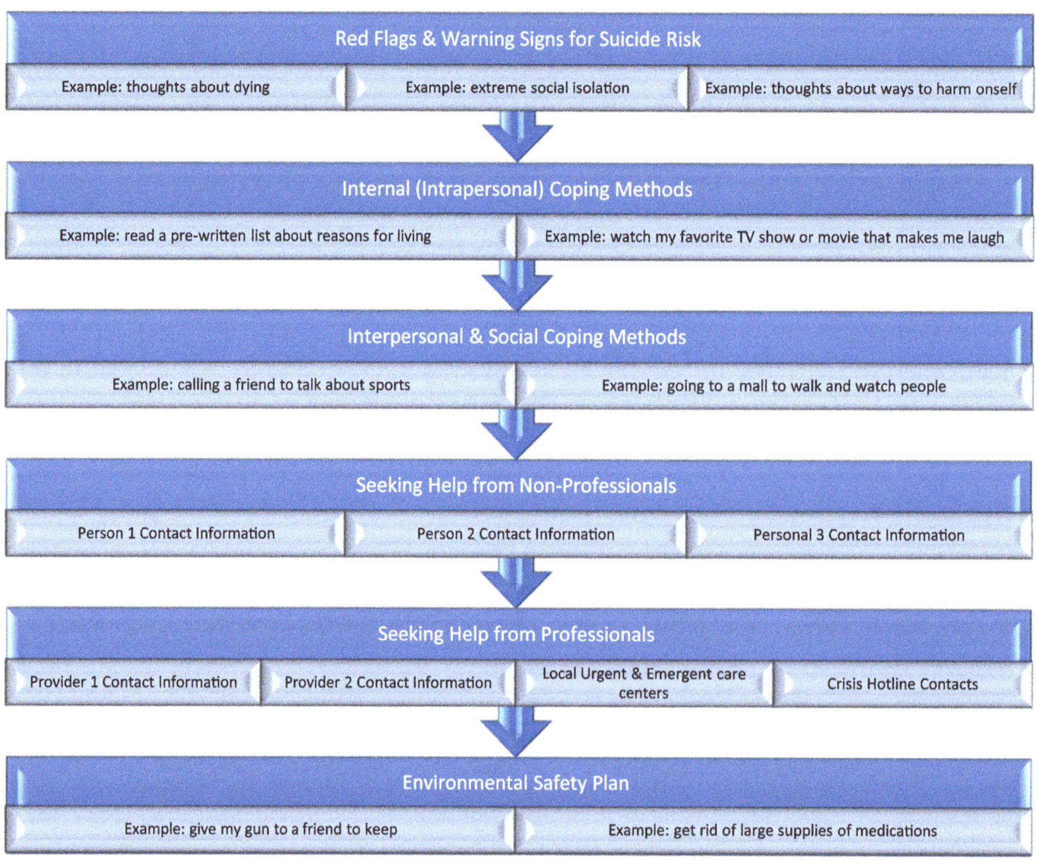

Safety Plan content includes: Warning signs for suicide risk, plus ways to mitigate risk by making a plan highlighting Internal Coping Methods, Social Coping Methods, Self-help Seeking from non-professionals and professionals, and keeping the environment safe.

FIG. 2 Suicide mitigation strategies.

Conclusion

Suicide is "transdiagnostic"—suicidal ideation and behaviors cross medical, mental, and behavioral health diagnoses and 54% of suicides were not associated with mental health conditions. Rather, contributors were relationship issues, substance use, health, and financial problems (Stone et al., 2018), not depression (Kishi et al., 2001). This suggests that suicide prevention strategies do not limit assessment to individuals with known mental health diagnoses. It has been documented that suicidal patients in medical settings have different profiles from suicidal patients in psychiatric hospitals (Furlanetto & Stefanello, 2011; Russell et al., 2009).

Behaviors can be informative and add precision in our care. Identifying sources motivating self-harm behaviors can promote a more compassionate understanding of a patient's core tensions and problems and generate interventions relevant to each situation. A key element is a therapeutic relationship. Empathic validation and genuine shared-decision making often dissipate distress and increases openness toward healthier behaviors. Self-harm behaviors fitting under the rubric of classic suicide are behaviors with a goal of death and are not corroborative with caregivers/loved ones (Bostwick, 2015). Self-harm and suicide are often extreme acts to get relief. Multiple biological, psychological, and social factors affect one's response to serious neurological disorders (Martz et al., 2005), and this chapter was directed to increase understanding about some reasons some people may indulge in the gamut of self-harm behaviors. Rather than ask why the suicide or self-harm, ask, what are the struggles or challenges driving it?

Mini-dictionary of terms

Adherence: "the extent to which a person's behavior—taking medication, following a diet, and/or executing lifestyle changes corresponds with agreed recommendations from a health care provider" (World Health Organization).

Beneficence: the state of producing or doing good; state of being beneficent.

Patient-centered care: the patient's specific health needs and their desired outcomes are the mainspring for health care decisions and assessment. Patient preferences are respected but within an acceptable scope of the clinician's practice, balanced in the context of that person and that situation.

Self-harm: any type of self-injurious behavior along a continuum of severity ranging from medical non-adherence to passive or active suicide.

Suicidal ideation: thoughts about or considering suicide.

Suicide: the act of taking one's life.

Key facts of self-harm

- It is a myth that asking someone about suicide makes things worse; discussing suicide is the most helpful way to get the person's help.
- There are often no notes before a suicide attempt.
- Clinicians must be aware of bias in assessing self-harm in people with spinal cord injury because either under-or overestimating risk can harm the therapeutic relationship (Budd, Haque, & Stein, 2020).
- CDC and Joint Commission (2019) recommend universal suicide risk screening for all patients in medical settings, across the lifespan and regardless of the time since disability.
- All patients who screen positive should have a safety plan, resources (National Suicide Lifeline, 1–800-273-TALK (8255), and Crisis Text Line, Text HOME to 741–741), and means restriction education.
- Focusing on reasons for living is a strength-based approach to facilitate motivation toward healthful behaviors.
- ASQ toolkit www.nimh.gov/asq includes scripts, flyers, and resources in several languages for youth and adults in three settings: Emergency Department; Inpatient Medical/Surgical Unit; Outpatient Primary Care/Specialty Clinics.

Summary points

- Human behaviors may be symptomatic of an underlying treatable problem.
- Self-harm phenomena in SCI medical settings can take a variety of forms.
- Identification and addressing underlying problems can ameliorate self-harm behaviors.
- People with SCI experience a significantly higher quality of life than clinicians expect, which can affect the options presented and decision-making.
- A person can be suicidal without a diagnosed mental health condition.

- A new diagnosis of a major illness is a suicide risk factor, along with chronic pain, worsening medical illness, or functional limitation.
- Good therapeutic relationships yield the best outcomes.
- Self-harm behaviors that *are* suicidal have (a) an intent to die and (b) no complicity or corroboration with caregivers.
- Depression and social circumstances are key risk factors for suicide in neurological populations.
- People who survive a suicide attempt are more likely to do it again, more if their neurological injury was caused by a suicide attempt.

References

Baron, R. J., & Berinsky, A. J. (2019). Mistrust in science—A threat to the patient–physician relationship. *The New England Journal of Medicine*, *381*(2), 182–185. https://doi.org/10.1056/NEJMms1813043.

Beauchamp, T. L., & Childress, J. F. (2009). *Principles of biomedical ethics*. New York: Oxford University Press.

Bhattarai, M., Jin, Y., Smedema, S. M., Cadel, K. R., & Baniya, M. (2021). The relationships among self-efficacy, social support, resilience, and subjective well-being in persons with spinal cord injuries. *Journal of Advanced Nursing*, *77*(1), 221–230. https://doi.org/10.1111/jan.14573.

Bostwick, J. M. (2015). When suicide is not suicide: Self-induced morbidity and mortality in the general hospital. *Rambam Maimonides Medical Journal*, *6* (2), 1–6. https://doi.org/10.5041/RMMJ.10197.

Bostwick, J. M., & Cohen, L. M. (2009). Differentiating suicide from life-ending acts and end-of-life decisions: A model based on chronic kidney disease and dialysis. *Psychosomatics*, *50*(1), 1–7.

Budd, M. A., Haque, O. S., & Stein, M. A. (2020). Biases in the evaluation of self-harm in patients with disability due to spinal cord injury. *Spinal Cord Series And Cases*, *6*(1), 43. https://doi.org/10.1038/s41394-020-0293-6.

Craig, A., Tran, Y., & Middleton, J. (2017). Theory of adjustment following severe neurological injury: Evidence supporting the spinal cord injury adjustment model. In A. Costa, & E. Villalba (Eds.), *Horizons in neuroscience research* (pp. 117–139). Nova Science Publishers.

Eipper-Mains, J. E., Kao, L. E., Gitlin, D. F., & Peteet, J. R. (2019). Life-threatening mistrust: Assessing and enhancing capacity to trust. *Psychosomatics*, *60*(6), 606–611. https://doi.org/10.1016/j.psym.2019.02.001.

Epstein, R. M. M. D., & Street, R. L. P. (2011). The values and value of patient-centered care. *Annals of Family Medicine*, *9*(2), 100–103. https://doi.org/10.1370/afm.1239.

Fuller-Thomson, E., Tulipano, M. J., & Song, M. (2012). The association between depression, suicidal ideation, and stroke in a population-based sample. *International Journal of Stroke*, *7*(3), 188–194. https://doi.org/10.1111/j.1747-4949.2011.00702.x.

Furlanetto, L., & Stefanello, B. (2011). Suicidal ideation in medical inpatients: Psychosocial and clinical correlates. *General Hospital Psychiatry*, *33*(6), 572–578. https://doi.org/10.1016/j.genhosppsych.2011.08.002.

Goetz, L. L., Ottomanelli, L., Barnett, S. D., Sutton, B., & Njoh, E. (2018). Relationship between comorbidities and employment among veterans with spinal cord injury. *Topics in Spinal Cord Injury Rehabilitation*, *24*(1), 44–53. https://doi.org/10.1310/sci16-00047.

Gonzalez, C. M., Deno, M. L., Kintzer, E., Marantz, P. R., Lypson, M. L., & McKee, M. D. (2018). Patient perspectives on racial and ethnic implicit bias in clinical encounters: Implications for curriculum development. *Patient Education and Counseling*, *101*(9), 1669–1675. https://doi.org/10.1016/j.pec.2018.05.016.

Joiner, T. (2005). *Why people die by suicide*. Harvard University Press.

Kennedy, P., & Garmon-Jones, L. (2017). Self-harm and suicide before and after spinal cord injury: A systematic review. *Spinal Cord*, *55*(1), 2–7. https://doi.org/10.1038/sc.2016.135.

Kishi, Y., Robinson, R. G., & Kosier, J. T. (2001). Suicidal ideation among patients with acute life-threatenng physical illness: Patients with stroke, traumatic brain injury, myocardial infarction, and spinal cord injury. *Psychosomatics*, *42*(5), 382–390. https://doi.org/10.1176/appi.psy.42.5.382.

Kolakowsky-Hayner, S. A., Iii, E. V. G., Kreutzer, J. S., Marwitz, J. H., Meade, M. A., & Cifu, D. X. (2002). Post-injury substance abuse among persons with brain injury and persons with spinal cord injury. *Brain Injury*, *16*(7), 583–592. https://doi.org/10.1080/02699050110119475.

Kripke, C. (2017). Patients with disabilities: Avoiding unconscious bias when discussing goals of care. *American Family Physician*, *96*(3), 192–195. https://www.aafp.org/afp/2017/0801/p192.html.

Lewis, D. S., Anderson, K. H., & Feuchtinger, J. (2014). Suicide prevention in neurology patients: Evidence to guide practice. *The Journal of Neuroscience Nursing*, *46*(4), 241–248. https://doi.org/10.1097/JNN.0000000000000062.

Macleod, A. D. (1988). Self-neglect of spinal injured patients. *Spinal Cord*, *26*(5), 340–349. https://doi.org/10.1038/sc.1988.48.

Martz, E., Livneh, H., Priebe, M., Wuermser, L. A., & Ottomanelli, L. (2005). Predictors of psychosocial adaptation among people with spinal cord injury or disorder. *Archives of Physical Medicine and Rehabilitation*, *86*(6), 1182–1192. https://doi.org/10.1016/j.apmr.2004.11.036.

Merbitz, N. (2016). Treatment adherence. In M. A. Budd, S. Hough, S. T. Wegener, & W. Stiers (Eds.), *Practical psychology in medical rehabilitation* Springer. https://doi.org/10.1007/978-3-319-34034-0_28.

Olkin, R. (1993). Crips, gimps and epileptics explain it all to you. *Readings*, *8*, 13–17.

Pollard, C., & Kennedy, P. (2007). A longitudinal analysis of emotional impact, coping strategies and post-traumatic psychological growth following spinal cord injury: A 10-year review. *British Journal of Health Psychology*, *12*(3), 347–362. https://doi.org/10.1348/135910707X197046.

Roy-Byrne, P. (2020). Heightened risks for suicide in patients with neurological disorders. *NEJM Journal Watch. Psychiatry*. https://www.jwatch.org/na50835/2020/02/12/heightened-risks-suicide-patients-with-neurological.

Russell, D., Turner, R. J., & Joiner, T. E. (2009). Physical disability and suicidal ideation: A community-based study of risk/protective factors for suicidal thoughts. *Suicide & Life-Threatening Behavior*, *39*(4), 440–451. https://doi.org/10.1521/suli.2009.39.4.440.

Savic, G., DeVivo, M. J., Frankel, H. L., Jamous, M. A., Soni, B. M., & Charlifue, S. (2018). Suicide and traumatic spinal cord injury-a cohort study. *Spinal Cord*, *56*(1), 2–6. https://doi.org/10.1038/sc.2017.98.

Simpson, B., Villeneuve, M., & Clifton, S. (2020). The experience and perspective of people with spinal cord injury about well-being interventions: A systematic review of qualitative studies. *Disability and Rehabilitation*, 1–15. https://doi.org/10.1080/09638288.2020.1864668. Advance online publication.

Skidmore, W. C., & Budd, M. A. (2016). Alcohol and substance use disorders in medical rehabilitation. In M. A. Budd, S. Hough, S. T. Wegener, & W. Stiers (Eds.), *Practical psychology in medical rehabilitation* Springer. https://doi.org/10.1007/978-3-319-34034-0_28.

Sprigle, S., McNair, D., & Sonenblum, S. (2020). Pressure ulcer risk factors in persons with mobility-related disabilities. *Advances in Skin & Wound Care*, *33*(3), 146–154. https://doi.org/10.1097/01.ASW.0000653152.36482.7d.

Stone, D. M., Simon, T. R., Fowler, K. A., Kegler, S. R., Yuan, K., Holland, K. M., et al. (2018). Vital signs: Trends in state suicide rates—United States, 1999-2016 and circumstances contributing to suicide—27 states, 2015. *MMWR. Morbidity and Mortality Weekly Report*, *67*(22), 617–624. https://doi.org/10.15585/mmwr.mm6722a1.

Studdert, D. M., Zhang, Y., Swanson, S. A., Prince, L., Rodden, J. A., Holsinger, E. E., et al. (2020). Handgun ownership and suicide in California. *New England Journal of Medicine*, *382*(23), 2220–2229. https://doi.org/10.1056/NEJMsa1916744.

Suchman, A. L., & Matthews, D. A. (1988). What makes the patient-doctor relationship therapeutic? Exploring the connexional dimension of medical care. *Annals of Internal Medicine*, *108*(1), 125–130. https://doi.org/10.7326/0003-4819-108-1-125.

Tétrault, M., & Courtois, F. (2014). Use of psychoactive substances in persons with spinal cord injury: A literature review. *Annals of Physical and Rehabilitation Medicine*, *57*(9), 684–695. https://doi.org/10.1016/j.rehab.2014.10.002.

The Joint Commission. (2019). *National Patient Safety Goals (Report No. 15.01.01)*. https://www.jointcommission.org/resources/patient-safety-topics/suicide-prevention/.

Torke, A. M., & Sachs, G. A. (2008). Self-neglect and resistance to intervention: Ethical challenges for clinicians. *Journal of General Internal Medicine*, *23* (11), 1926–1927. https://doi.org/10.1007/s11606-008-0807-6.

Ubel, P. A., Loewenstein, G., Schwarz, N., & Smith, D. (2005). Misimagining the unimaginable: The disability paradox and health care decision making. *Health Psychology*, *24*(4S), S57–S62. https://doi.org/10.1037/0278-6133.24.4.S57.

VanPuymbrouck, L., Friedman, C., & Feldner, H. (2020). Explicit and implicit disability attitudes of healthcare providers. *Rehabilitation Psychology*, *65* (2), 101–112. https://doi.org/10.1037/rep0000317.

World Health Organization. (2003). *Adherence to long-term therapies: Evidence for action*. World Health Organization. https://www.who.int/chp/knowledge/publications/adherence_report/en/.

Index

Note: Page numbers followed by *f* indicate figures and *t* indicate tables.

B

C

D

E

Q

R

S